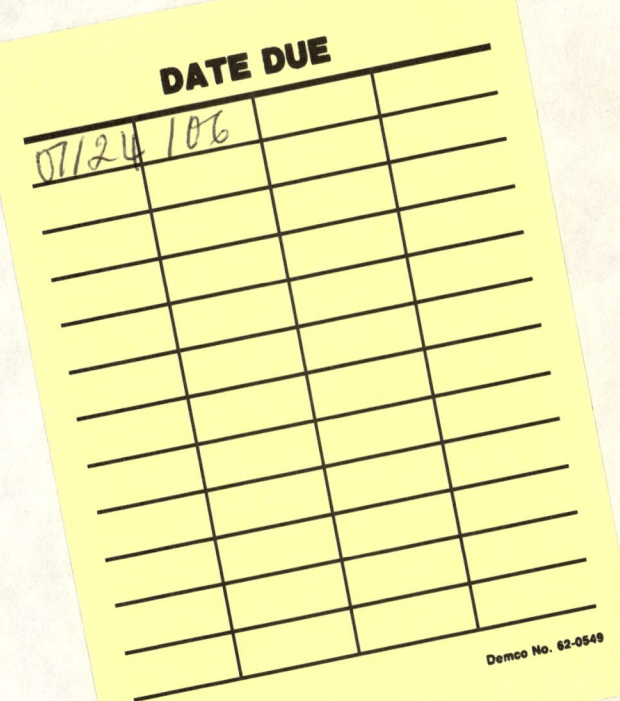

PHYSICS DEMONSTRATION EXPERIMENTS

IN TWO VOLUMES

VOLUME I
MECHANICS AND WAVE MOTION

VOLUME II
HEAT, ELECTRICITY AND MAGNETISM, OPTICS, ATOMIC AND NUCLEAR PHYSICS

Editor

HARRY F. MEINERS is Professor of Physics at Rensselaer Polytechnic Institute. He was for many years a member of the American Association of Physics Teachers Committee on Apparatus for Educational Institutions, and served as its Acting Chairman. Professor Meiners has been consultant on physics apparatus to the State University of New York and film consultant to the National Film Board of Canada. He is co-author of *Analytical Laboratory Physics* and *Laboratory Physics*. In 1967 the American Association of Physics Teachers awarded him its Distinguished Service Citation for his contributions in the form of new laboratory experiments, new lecture demonstrations, and new uses for films. During 1967, 1968, and 1969 he was a Staff Scientist on the National Science Foundation Liaison Staff, New Delhi, India where he received a special citation from the National Physical Laboratory for his work on apparatus at the laboratory, and a commendation from the Indian National Council on Science Education for his contributions in the area of visual aids.

Project Directors

HARRY F. MEINERS (co-chairman), *Rensselaer Polytechnic Institute*

ROBERT RESNICK (co-chairman), *Rensselaer Polytechnic Institute*

Advisory Committee

GEORGE D. FREIER, *University of Minnesota*

JOHN G. KING, *Massachusetts Institute of Technology*

MELBA PHILLIPS, *University of Chicago*

HOWARD A. ROBINSON, *Adelphi University*

ROLF M. STEFFEN, *Purdue University*

KARL S. WOODCOCK, *Bates College*

PHYSICS DEMONSTRATION EXPERIMENTS

Edited by HARRY F. MEINERS

Volume II

HEAT, ELECTRICITY AND MAGNETISM, OPTICS, ATOMIC AND NUCLEAR PHYSICS

SPONSORED BY THE AMERICAN ASSOCIATION OF PHYSICS TEACHERS

ROBERT E. KRIEGER PUBLISHING COMPANY
MALABAR, FLORIDA

Original Edition 1970
Reprint Edition 1985

Printed and Published by
ROBERT E. KRIEGER PUBLISHING COMPANY, INC.
KRIEGER DRIVE
MALABAR, FLORIDA 32950

Printed in the United States of America.

Library of Congress Cataloging in Publication Data

Main entry under title:

Physics demonstration experiments.

"Sponsored by the American Association of Physics Teachers."
Reprint. Originally published: New York: Ronald Press, c1970.
Includes index.
1. Physics—Experiments. 2. Physics—Study and teaching. I. Meiners, Harry F. II. American Association of Physics Teachers.
QC33.P45 1985 530'.07'8 84-23409
ISBN O-89874-821-6

CONTENTS

VOLUME II

LECTURE TECHNIQUES

20 *Closed-Circuit TV in Physics Demonstrations* 655
Rosalie Hoyt, Bryn Mawr College

21 *Films as a Lecture Aid* 670
Alfred Leitner, Rensselaer Polytechnic Institute

22 *Corridor Demonstrations* 698
Everett M. Hafner, University of Rochester

23 *The Overhead Projector in the Physics Lecture* 719
Walter Eppenstein, Rensselaer Polytechnic Institute

24 *Stroboscopic Effects* 731
Harry F. Meiners, Rensselaer Polytechnic Institute

HEAT

25 *Temperature* 749

26 *Heat and Thermodynamics* 755

27 *Gases and Kinetic Theory* 783

28 *Low-Temperature Phenomena* 826

ELECTRICITY AND MAGNETISM

29 *Electrostatics* 835

30 *Electrical Properties of Matter* 880

31 *Magnetism* 908

32 *Magnetic Properties of Matter* 953

33 *Electromagnetic Oscillations* 990

OPTICS

34 *Geometrical Optics* 1035

35 *Physical Optics* 1064

ATOMIC AND NUCLEAR PHYSICS

36 *Gas Laser Demonstrations* 1127
Haven Whiteside, Oberlin College

37 *Holograms* 1147
Tung H. Jeong, Lake Forest College

38 *Special Relativity and Quantum Effects* 1164

39 *Atomic Physics* 1202

40 *Solid State Physics* 1220

41 *Nuclear Physics* 1240

Appendices to Demonstrations: Construction Details and Materials Lists 1285–1395

List of Other Reference Sources for Demonstrations

List of Contributors

List of Manufacturers

Index

LECTURE TECHNIQUES

20

CLOSED-CIRCUIT TV IN PHYSICS DEMONSTRATIONS

By Rosalie Hoyt
Bryn Mawr College

20–1 Introduction

This article will be concerned with the use of closed-circuit television (CCTV) as a demonstration aid, such as in magnifying small objects that are difficult or impossible to project. No very detailed discussion will be given here of the use of CCTV for multiple classroom teaching, where the complexity of the system and of its operation may approach that of the professional studio.[1] The discussion of the equipment will in general be limited to small, moderately priced systems (video and black and white only) that can be operated in the class by the instructor alone or with the help of a single assistant. Such equipment is normally of the type developed for educational, commercial, and industrial monitoring and for telemetering applications. The quality of the resulting picture can be as high as, and for some models considerably higher than, that of more elaborate and costly broadcast systems. A few of the typical accessories required for multiple classroom use will also be mentioned.

The first part of the article is devoted to a brief résumé of the basic equipment, general operating characteristics, and critical specifications that should be considered when the relative merits of different makes and models are being judged prior to purchase. In addition, brief mention is made of useful auxiliary equipment. In the second part, lecture room planning, such as the positioning and number of monitors required, is very briefly touched upon. In the last part, the use of CCTV for demonstrations is discussed, and a few typical demonstrations are listed.[2]

[1] H. E. White, *Am. J. Phys.*, **28**, 368 (1960).

[2] The author acknowledges with pleasure the cooperation of the representatives of the manufacturers of CCTV equipment listed in Section 20–2. The author especially wishes to thank Henry Groves of Pierce Phelps Co. for his time and for his patience in answering questions, and J. Nathaniel Marshall and Walter C. Michels for their helpful and constructive criticisms. Any errors of statement are those of the author and not of the persons mentioned. Sincere thanks are also given to William L. Millard of Rensselaer Polytechnic Institute for reviewing this article and for his helpful suggestions which were incorporated into several sections.

20–2 Equipment

The TV equipment now available and suited for classroom demonstration use varies widely in cost and quality. The line of models put out by any one manufacturer is continually changing as older models are discontinued and newer ones are developed. New manufacturers of CCTV equipment are making their appearance at the same time that others are leaving the field. Company names change as they are taken over by different parent organizations, and local distributors and technical sales representatives are continually changing the "horses in their stable." Prospective purchasers of CCTV equipment must therefore go to the yellow pages of the nearest large city directory for the names of manufacturers and their local representatives; additional sources of information are the Audio-Visual Equipment Directory, published by the National Audio Visual Association, Inc., Fairfax, Virginia, and the Broadcast Equipment Buyers Guide, published by the Mactier Publishing Corporation, New York, New York.

The list below represents just a few of the many manufacturers of CCTV equipment and is far from complete. The list merely consists of those manufacturers from whom the author received demonstrations and/or technical literature in preparing this article.

Cohu Electronics Inc., San Diego, California
Dage-Bell Corp., Michigan City, Indiana
Diamond Electronics, Lancaster, Ohio
General Precision Inc., GPL Division, Aerospace Group, Pleasantville, New York
RCA, Closed-Circuit Television Dept., Camden, New Jersey
Sylvania Electric Products, Inc., Commercial Electronics Division, Bedford, Massachusetts

Since the quality and cost of the basic equipment available vary widely, arrangements should be made for live demonstrations of the equipment of several different manufacturers for the demonstrations to be performed in the physics lecture room on subjects of the professor's choosing, the difficult ones (e.g., diffusion cloud chamber) as well as the easy ones (e.g., a 2-in. panel meter). Only in this way can one make the difficult decision between high quality, and accompanying greater usefulness at a higher cost, versus lower quality, and thus more limited usefulness at lower cost. A side-by-side demonstration, if it can be arranged, may be worthwhile.

20–2.1 Basic Equipment

The basic units of a small (portable) CCTV system are a camera, an electronic control unit, and a monitor (Fig. 20–1). Figure 20–2 shows a CCTV studio in operation. The camera consists of a camera tube, a lens system for focusing light onto the sensitive surface of the camera tube, electronic circuitry for forming the scanning beam and producing the image-modulated video signal, and a preamplifier to produce a low noise signal to feed into the control unit. The control unit consists of the various power supplies needed for operation of the camera tube, video amplifier, and synchronizing signal generator for superimposing synchronizing pulses to produce the composite video signal to feed into the coaxial cable leading to the monitor. The monitor consists of the kinescope tube plus associated power supplies, amplifiers, and scanning and synchronizing circuits such that the composite

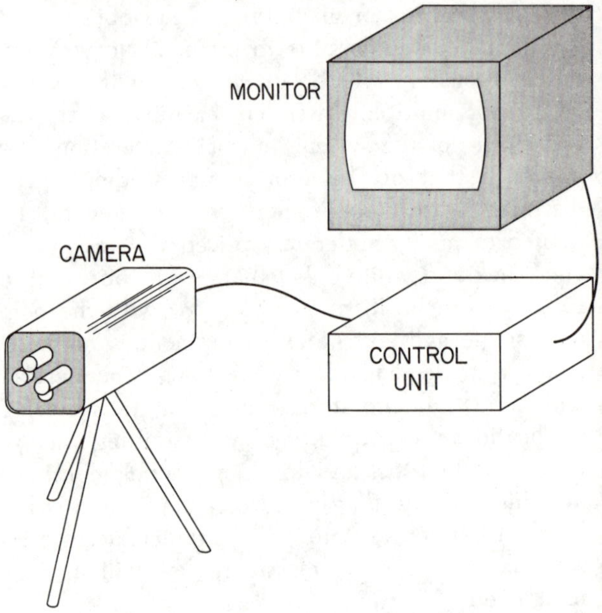

Fig. 20–1 A Small CCTV System.

Fig. 20–2 A Typical ETV Studio at Thornton Township High School, Harvey, Illinois. (Photo courtesy of Diamond Electronics.)

video signal causes modulated scanning across the screen in synchronism with the camera scanning.

The Camera Tube

Three types of tubes are in common use, the vidicon, the plumbicon, and the image orthicon. The vidicon is used in most monitoring and telemetering service, while the plumbicon or image orthicon is used in most, though not all, studio and broadcast systems. Up until recently the vidicon suffered from quite poor light sensitivity compared to that of the image orthicon. However, the newer types of vidicons have excellent sensitivity. The cost of a vidicon is much less than that of an image orthicon, and at the same time the cost and complexity of the associated equipment is much less for vidicon cameras than for image orthicons. Finally, the lifetime of a vidicon is considerably longer than that of an image orthicon.

The recently introduced plumbicon tube combines advantages of the vidicon in terms of low cost and long life and of the orthicon in terms of sensitivity and elimination of image "lag." Associated electronic circuitry is similar to that employed in the vidicon camera and is consequently of lower cost than that of the orthicon. Plumbicon camera equipment is not presently available in the variety of formats in which vidicon types are, but this situation should improve with time.

Orthicon camera tubes have been standard in the broadcast industry for years and are used in non-broadcast applications where extreme sensitivity and the picture reproduction characteristics of the orthicon tube are required. During recent years lower cost nonviewfinder orthicon cameras have become more readily available for scientific and educational applications.

This discussion will be restricted to vidicon types of equipment since the apparatus has evolved to a more sophisticated state in nonbroadcast areas. Similar types of telecasting activities could, of course, be handled by either plumbicon or orthicon equipment.

Lenses

Some cameras are supplied with 16-mm movie camera lenses, but the 16-mm frame size is not optimum for the ½ x ⅜-in. sensitive area of the 1-in. vidicon tube. Lens manufacturers now make a line of vidicon lenses, but still use the standard 16-mm C-mount, and CCTV manufacturers have also agreed on this standard mount for their cameras. A large line of interchangeable lenses is thus available, ranging from slow, fixed-focus, nonadjustable lenses, all the way to wide angle, telephoto and zoom lenses. Some cameras are equipped with a four-lens turret for quick interchange of lenses. An assortment of two or more lenses is probably a necessary requirement to allow flexibility in the type of demonstration televised. If the budget will stand it, a zoom lens can be very useful. For example, an overall view of some demonstration setup can be shown first and the various component parts pointed out. The lens can then be zoomed in to show some one part in closeup magnified detail. The pictures of Fig. 20–3, (a) through (d), were taken with a zoom lens on the TV camera.

Control Circuitry

No attempt is made here to discuss the many varied aspects of the circuitry that goes into producing the composite video output. Rather, a few of the aspects which should be considered when weighing the relative merits of different systems will be mentioned.

1. First, of course, are the requirements of stability, regulation of all power supplies, adequate margin of safety in specified ratings of components, fail-safe provision to protect the vidicon tube, etc.

2. Some systems have only manual control of the target voltage (light sensitivity control), while others have an automatic target control. The latter has many advantages over the former. The target voltage is automatically adjusted to maintain a constant integrated video signal under extreme variations of overall light level. Panning the camera from a very bright scene to a quite dim one, or vice versa, requires no adjustment on the part of the operator. Still other systems have coupled automatic target voltage, video gain, and clamped dark current, maintaining optimum signal-to-noise ratio and optical contrast at all times.

3. Another consideration is the type of interlace system used. It has become standard practice to divide the scanning of a single field into two parts (frames), alternate lines being scanned in each frame. Thus in a 600-line system (that is, 600 scanning lines per field), lines 2, 4, 6, . . . , 600 are scanned first, and then lines 1, 3, 5, . . . , 599 are scanned. Such a system, however, requires rigidly synchronized interlacing of the two frames (so-called 2:1 interlacing); otherwise the vertical resolution is reduced. This interlacing allows a relatively low scanning frequency (30 fields/sec, 60 frames/sec) without undue flicker. In the cheaper systems so-called random interlacing is used, leading to bunching of the scanning lines. In the author's judgment, fixed 2:1 interlacing is well worth the extra cost.

4. In most models the number of lines per field, or raster, is the same as that used in the broadcast industry, i.e., 525 lines, and this immediately sets an upper limit on the vertical resolution. CCTV manufacturers have provided some models for closer scanning (e.g., 600, 800, or even higher), with resulting greater vertical resolution.[3] Note that the number of scanning lines per field and the vertical resolution are not synonymous. Consider, for example, a system using 525 scanning lines per field with 2:1 interlacing. First, only about 500 of the scanning lines will appear on the kinescope screen of the monitor. Second, two horizontal lines must be separated vertically by more than the distance between two scanning lines in order to be resolved. A convention commonly employed for defining the vertical resolution consists of multiplying the number of useful lines by the factor 0.7. Thus a 525-line scanning system is said to have a vertical resolution of $0.7 \times 500 = 350$ lines.

(3) As will be discussed in more detail in the next section, scanning at other than 525 lines prevents the use of commercial home receivers as monitors.

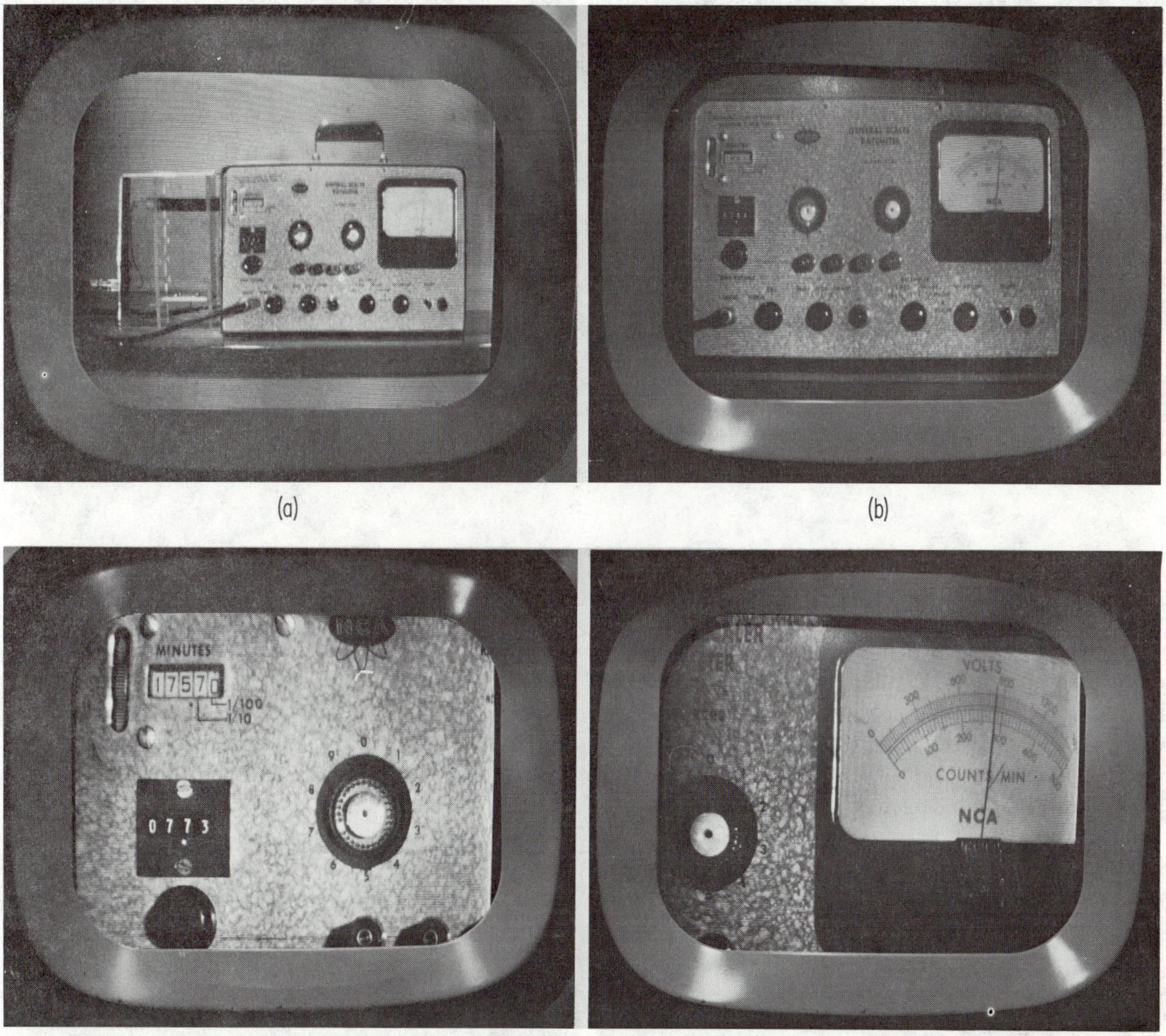

(a) (b)

(c) (d)

Fig. 20–3 Radioactivity Demonstration. (a) Scaler-ratemeter, counter, and stand for samples and absorbers. (b) Closer view of the scaler-ratemeter. (c) Close-up of the clock timer, totalizing counter, and one of the decade counter tubes. (d) Close-up view of the other decade counting tube and the voltmeter.

5. The horizontal resolution, again expressed in terms of lines, depends on the band width of the video amplifiers used. This can range from 400 lines or under in the cheaper models, to 600 lines, 800 lines, and even up to 1000 lines in high resolution systems. The pictures shown in Figs. 20–4, (a) and (b), and 20–5, (a) and (b), indicate the difference between low and high resolution systems. Both the monitor and the camera system affect the resolution, and a single statement of resolution by a manufacturer does not tell the whole story. The resolution at the corners may be very much less than that at the center unless the circuits have been very carefully designed. A high quality, 1000-line system approaches the upper limit set by the 1-in. vidicon tube itself. One and one-half inch and 2-in. vidicon tubes have been developed, but the 2-in. tubes are not generally available.

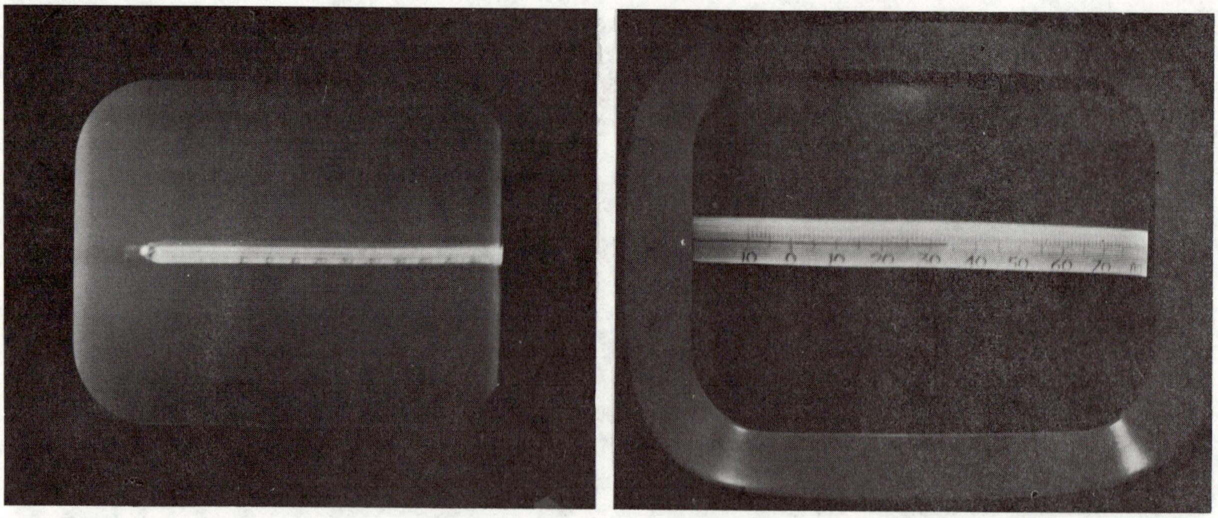

(a) (b)

Fig. 20–4 Mercury in Glass Thermometer viewed at (a) low resolution and (b) high resolution.

The equipment available from different manufacturers, and in the different models available from any one manufacturer, varies considerably in many respects. If the department budget can meet it, a camera plus control unit and monitor with good stability, high sensitivity, and high resolution (at least 600 lines), both vertical and horizontal, is well worth the extra cost. It will enable a much wider range of classroom demonstrations to be satisfactorily televised.

The Monitor

Some CCTV systems require special video monitors, while others have an rf generator included in the control unit so that the output can be fed directly to the antenna lead-in of commercial home TV receivers. The one advantage of the latter system is that the same receivers can be used either as CCTV monitors or for off-the-air reception of commercial and educational broadcast channels.

(a) (b)

Fig. 20–5 A Two-Inch Ammeter viewed at (a) low resolution and (b) high resolution.

However, a serious disadvantage for CCTV work lies in the reduced band width that accompanies conversion to rf. A video system capable of 800 lines or more horizontal resolution is immediately cut to 300 line resolution or less when displayed. A further disadvantage lies in the necessity of using 525 line scanning rather than higher scanning, with loss in vertical resolution. There is no question that the CCTV systems that remain at video bandwidths throughout have much better resolution capabilities and also eliminate the necessity of tuning the receivers to the rf signal.[4] The monitor should be chosen to have bandwidths that can utilize the full resolution of the camera unit feeding it.

An alternative solution if commercial reception is desired is to add a television audio-video tuner to the system. The separate audio and video outputs, demodulated from the rf carrier, can then be fed in separately to the video monitors and to a lecture room sound system.

Modifications of the Basic Design

Some manufacturers have models on the market in which the control unit is miniaturized and included in the camera unit (Fig. 20–6). The self-contained units can be used with conventional camera tripods and require only a 110-V power cord plus a coaxial line leading to the monitor(s). The bulky and expensive cable leading from the camera to the control unit is eliminated. However, since 10–20 ft of cable from the camera to the control unit is adequate for most purposes, its elimination may not be an important advantage. In general, the self-contained units on the market today can achieve a level of performance approaching that of the highest quality two-unit system.

A second modification consists of the inclusion of a small monitor in the back of the camera. Such viewfinder cameras (Fig. 20–7(a) and (b)) allow the operator to see the picture appearing on the monitor(s). The studio type of camera is almost invariably a viewfinder. In a lecture hall the monitors are mounted for student viewing and may be invisible to the operator located at the lecture desk. In such cases a viewfinder camera may be of real advantage. It should be noted, however, that the extra cost of a viewfinder camera may be much greater than that of a separate small monitor which may be clamped onto the camera head or placed on a rotating stand that can serve the operator and possibly some of the front row students as well. However, if two or more cameras are contemplated, as in a multiple classroom setup where one camera is to be focused on the lecturer and blackboard and another on a demonstration, one of the two might well be a viewfinder type.

A third type of modification lies in the design of special cameras for use in abnormal environments, e.g., high temperature, high background radiation, etc. These should only be considered if such additional departmental use, over and above the classroom use, is contemplated.

[4] The rf generator is tuned to one of the channels unused in the neighborhood.

20–2.2 Accessories

Tripod

The first type of accessory to be mentioned is probably an essential one, namely, some sort of tripod for mounting the camera unit. A conventional camera tripod, with manual height, pan and tilt adjustments will serve for most purposes. A dolly on which the tripod can be mounted is also useful when the camera is to be moved from distance to close-up shots.

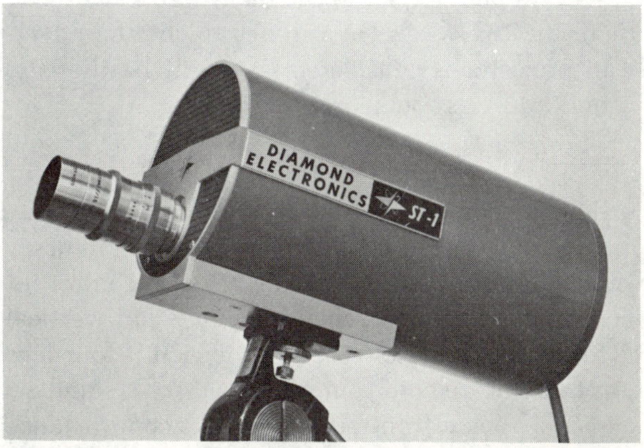

Fig. 20–6 Model ST-1 Vidicon Camera, Diamond Electronics, provides clear, sharp, virtually noise-free pictures with 650-line horizontal resolution and excellent geometry. An rf output is available, so a TV receiver can be used in place of, or in addition to, a closed-circuit TV monitor. If a receiver is used, its limitations will determine the resolution of the system. (Photo courtesy of Diamond Electronics.)

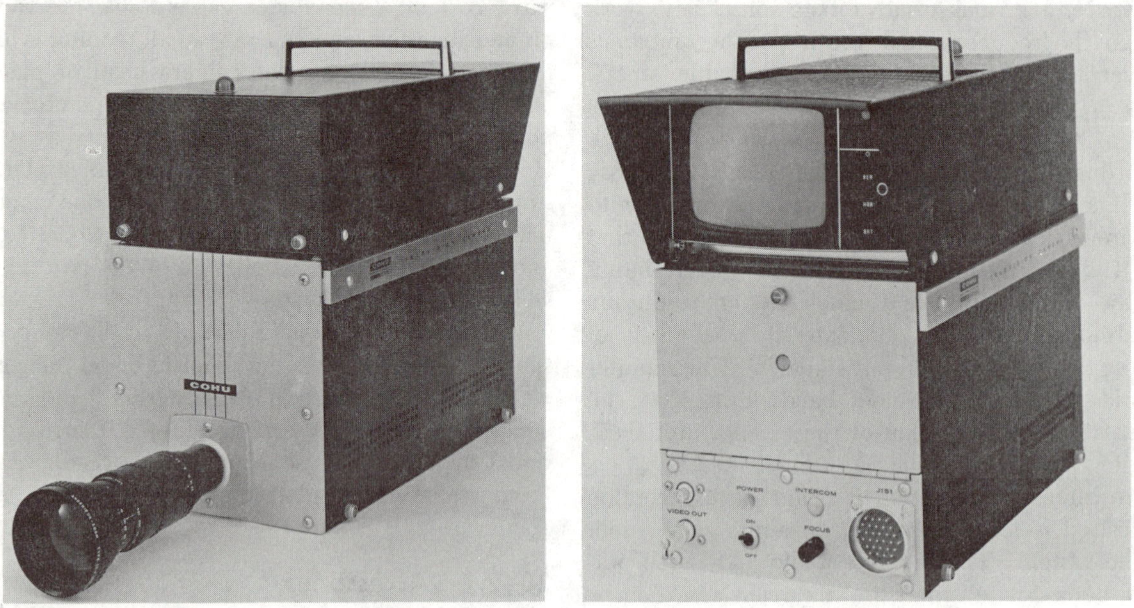

(a) (b)

Fig. 20–7 3200 Series Viewfinder Camera, Cohu Electronics: (a) front view, (b) rear view. (Photo courtesy of Cohu Electronics.)

Vertical Mount

If use of the TV camera in conjunction with a microscope is contemplated, a sturdy vertical mount is essential. (A microscope adapter will also be required.) Such a vertical mount is also essential if use of the CCTV system as replacement of an overhead projector is considered, as might be the case in multiple classroom use.

Remote Control Apparatus

Remote control of camera position and of the optical adjustments are accessories offered by many manufacturers. These include remote control of the tripod, both horizontal panning and vertical tilt; remote control of lens selection on a four-lens turret mount; remote control of focus of a standard lens; and remote control of iris and zoom distance on a zoom lens. Remote control panels for the electronic camera control unit are also available in some systems. The need for remote controls, both optical and electronic, must be judged in each case individually. For lecture room use they are probably not essential, but they may be of use in other physics department applications such as the monitoring of accelerators, hot labs, etc.

Wide Screen

The inclusion of a wide screen projection kinescope, either as an accessory or as an alternate to a number of monitors or receivers, should be considered. Such a projection kinescope is shown in Fig. 20–8. The light output and associated optics of the better models are sufficient to provide good illumination filling a 12 x 16-ft projection screen, with viewing by 500 to 1000 persons or more.

Multiplexer

Another accessory that may be useful consists of a film chain multiplexer that allows alternate or overlaid camera pickup from one or more 16 mm movie projectors, 2 x 2-in. automatic slide projectors and film strips. This is of particular interest in multiple classroom use. The 16 mm movie projectors are specially designed so that the 24 frames/sec of the film can be used with the 30 frames/sec (60 fields/sec) of the TV scanning.

Video Switcher

Still another accessory that is especially useful in multiple classroom operation is a video switcher (or switcher and fader) to allow switching from one camera output to another. For example, one camera may be trained on the lecturer, a second on the apparatus on the lecture desk, and a third on a film chain multiplexing system.

Recording Equipment

Should the department budget allow, the inclusion of either 16 mm kinescope or of video tape recording equipment will allow the canning of difficult-to-perform demonstrations, or of entire lecture presentations, for future playback over the years. The relative merits of the two systems will not be discussed here, but must be judged individually. Inexpensive videotape machines compare favorably in cost to that of the small television system primarily considered here. Physics departments in institutions with audio-video centers might well investigate selective use of such equipment. However, a high-resolution camera system when viewed directly on a monitor may have its resolution seriously reduced when recorded and played back.

Synchronizing Generator

If any of the kinescope or video tape recordings are ever to be rebroadcast over a commercial or educational station, it is essential that they be recorded with Electronics Industry Association (EIA) synchronization. An external EIA synchronizing generator is then an essential accessory.

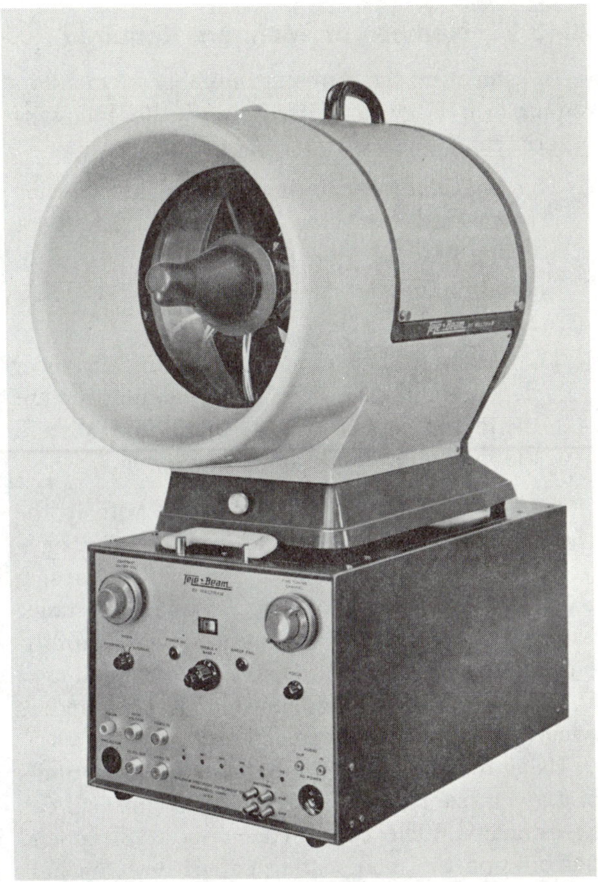

Fig. 20–8 Projection Kinescope, General Precision, GPL Division. (Photo courtesy of General Precision.)

20–3 Systems Planning

20–3.1 Sources of Background Information

Studies have been made and are continuing on the optimum design and layout of class and lecture rooms for the use of television. Reference should be made here to the brochure "Design for ETV," published by the Educational Facilities Laboratories, Inc., 477 Madison Ave., New York, N.Y. 10022, which contains much useful information for anyone contemplating the use of TV, whether as a demonstration aid or for multiple classroom teaching. The EFL have also given financial backing to the "New Spaces for Learning" project of the School of Architecture of Rensselaer Polytechnic Institute, and to the AIP-AAPT project that resulted in the book *Modern Physics Buildings*, by R. Ronald Palmer and William Maxwell Rice, published by Reinhold Publishing Corp. Both a project report by the Rensselaer School of Architecture and the Palmer and Rice book contain information of interest. Additional information can be obtained from the *Educational Television Guidebook*, by Philip L. Lewis, McGraw-Hill Book Co., and the article by Harvey E. White in the *American Journal of Physics*, 28, 368 (1960). No one who saw any of the programs of Dr. White's physics course on "Continental Classroom" can doubt that TV can be useful in bringing students into more direct contact with what is happening on the lecture desk.

The reader is referred to the sources mentioned above for detailed discussions. Only a few of the major points of interest will be summarized here.

20–3.2 Number of Monitors Required

In estimating the minimum number of monitors required in a given lecture room, the following criteria can be used:

The maximum horizontal angle of viewing (measured from the centerline) should be about 40 deg.

The maximum vertical viewing angle should be about 20 deg.

The maximum and minimum viewing distances depend on the width W of the monitor screen: the maximum should be 11 or 12 times W, and the minimum should be 3 or 4 times W.[5]

An additional set of criteria, useful both in the choice of monitor size and the optical and electronic magnifications required for a given presentation, is that which connects viewing distance with alpha-numeric symbols indicating readability. Figures from Eastman Kodak Co. show that $S = D/32$, where S is the minimum symbol size in inches when viewed at a distance of D feet.

Using the above criteria, the number of monitors of a given size required for a given lecture hall can be estimated. This estimated number will depend not only on the total seating capacity of the hall, but also upon its shape and dimensions. For many lecture halls the number comes out to be, for 20-in. monitors, about one monitor for each twenty-five students.

20–3.3 Room and Camera Lighting

Good monitors give a picture of sufficient brilliance that a darkened room is not necessary. Normal, or slightly subdued, ambient room lighting can be used, enabling the students to take lecture notes as well as to observe the blackboard and lecture desk. The positioning and angling of the monitors should be such as to avoid glare from the house lights as well as from the spot and flood lights used to illuminate the lecture desk demonstrations.

TV cameras using a sensitive vidicon tube can give adequate pictures at light levels as low as 1 to 10 foot-candles (fc) when used with an f/1.5 lens opened wide. However, levels of 100 fc or more will allow stopping down of the lens and will give pictures of lower noise level, better definition, and greater depth of field. Spot and flood lighting may therefore be necessary, and several different varieties (e.g., Fresnel lens spotlights, diffusing lamps in a "scoop" mount, klieg lights, etc.) should be permanently mounted above the lecture desk. Although sharp shadows may occasionally be useful, for most purposes diffuse lighting with a minimum of shadow is best. This is especially true in the case of metal instruments such as steel scales. The lighting for the vernier caliper of Fig. 20–9 was obtained by aiming a GE spot lamp on the ceiling. In this way diffuse lighting of sufficient intensity was obtained.

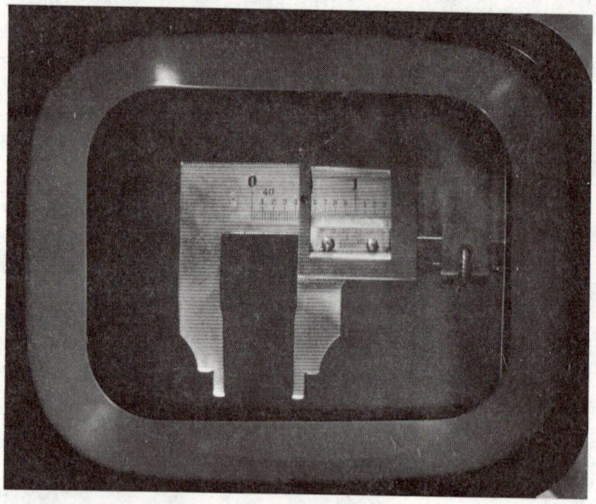

Fig. 20–9 Venier Caliper viewed with a medium resolution system.

20–3.4 Distribution System

The monitors are fed by coaxial cable from the control unit. Most control units have a video output impedance of 75 Ω and are designed to feed into a RG-59/U or RG-11/U coaxial cable.[6] The video monitors are of high input impedance (e.g., 470 kΩ in parallel with 15 $\mu\mu$F) so that more than one monitor can be bridged across the line. The monitors also normally include a 75-Ω termination that can be switched in on the last monitor fed by the line.

[5] With systems of high resolution (800 or more lines), this minimum viewing distance may be reduced.

[6] Balanced output feeding into a balanced 125-Ω shielded video pair cable is also possible and is advantageous when long distances, such as several miles, are involved.

With long cable lengths (greater than 1000 ft for the RG-59/U or 2000 ft for the RG-11/U) and many monitors, attenuation may become serious, especially in its effect on the frequency response. In such cases an equalizing amplifier, and/or a distribution amplifier for feeding into two or more separate lines, must be added. This will greatly improve performance.

20–4 Use of CCTV in Physics Demonstrations

The following discussion is illustrative rather than complete. The number of uses of CCTV as a demonstration aid are almost endless. Only a few typical uses will be mentioned; the reader can add many more.

No substitute exists for the trial and error type of practice with the TV equipment by the instructor and his assistant. Most of the problems involved in TV camera work are the same as those involved in any form of photography, except that in addition to the optical considerations and adjustments such as choice of lens, magnification, angle, lighting and iris control, there are now added electronic adjustments such as brightness, contrast, vertical and horizontal holds, and rf tuning if an rf stage is used. However, most CCTV systems are quite stable and will hold their adjustments for long periods of time.

With a little care, glare can be avoided from glass-faced instruments, and closeups of 2-in. panel meters, stopwatches, and thermometers can be made so perfectly visible (with a medium-to-good resolution system) that the class can take the readings rather than the instructor. (Figs. 20–3, 20–4, 20–5, and 20–10). The same can be done with slide rules, steel scales, vernier calipers, etc. (Figs. 20–9 and 20–11). As mentioned earlier, diffuse rather than spot lighting is essential here.

By focusing the TV camera on the screen of a cathode ray oscilloscope, the pattern on a 5-in. oscilloscope can be blown up to a 21-in. or larger size (Fig. 20–12 (a) and (b)). Transients as well as stationary patterns can be viewed.

A variety of techniques can be used to show diffraction and interference effects. One difficulty, though, lies in the very great intensity ratio between the zero and higher orders. For example, when a line source of monochromatic light is viewed with a Ronchi ruling of 200 lines/in. held close to the eye, at least 15 orders on either side of the central one are clearly visible. Figure 20–13 (a) shows the pattern obtained from a low-sensitivity and low-resolution CCTV system. A system with medium resolution and high sensitivity gave the picture in Fig. 20–13 (b). In these examples a slit was placed immediately in front of a sodium vapor lamp

Fig. 20–10 **Stopwatch** viewed with a medium resolution system.

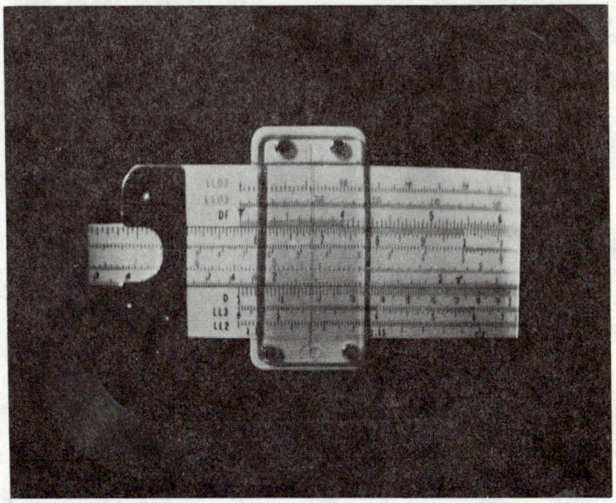

Fig. 20–11 **Sliderule** viewed with a medium resolution system.

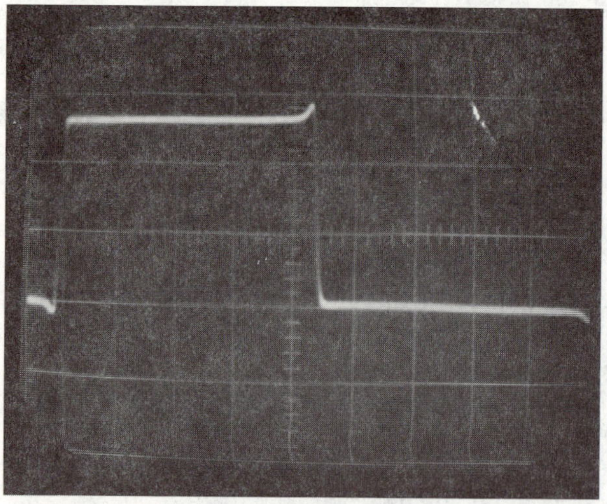

(a)

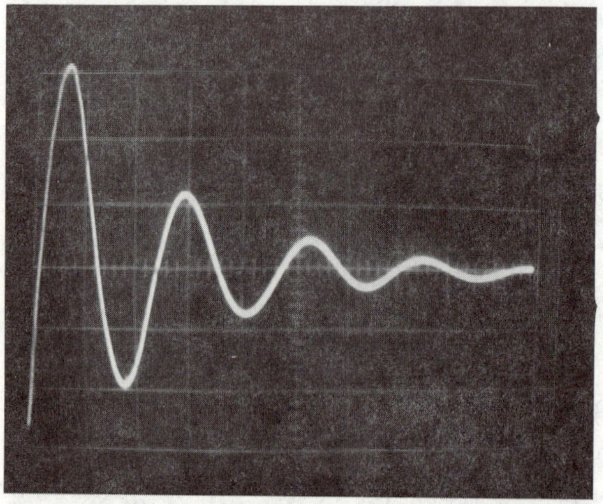

(b)

Fig. 20–12 Cathode Ray Oscilloscope Presentation: (a) 1 msec square wave; (b) LRC transient of about 1000 cps.

and the TV camera was focused onto this slit. The Ronchi ruling was then fastened to the front of the TV camera lens with tape. For comparison, Fig. 20–13 (c) was obtained without the TV system. Instead, the Ronchi ruling was taped directly onto the camera lens. These examples show that a low or medium quality TV system cannot compete with direct vision of interference fringes by the human eye. Although not shown here, a high quality system comes close to doing so. The cost of Cornell slits, the Kodak grating material, etc. is so small that handing out individual slits and grating to the students is a much more satisfactory method of showing these effects, especially since the black and white TV can not show the beautiful color patterns with nonmonochromatic light. However, the TV system can be useful here in allowing the instructor to point out to the students what it is that they are to look for when they hold the slits and rulings up to their eyes.

The technique shown in Fig. 20–3, and discussed earlier in connection with zoom lenses, is often useful, namely, for first showing an overall (front seat) view of a demonstration, and then switching to a closeup view for the taking of readings, etc. An additional technique used success-

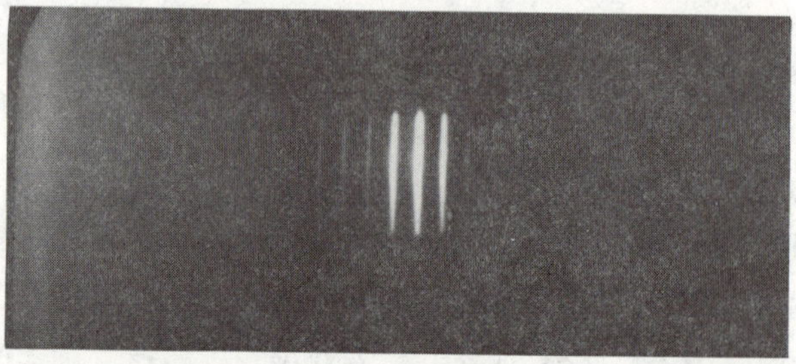

(a)

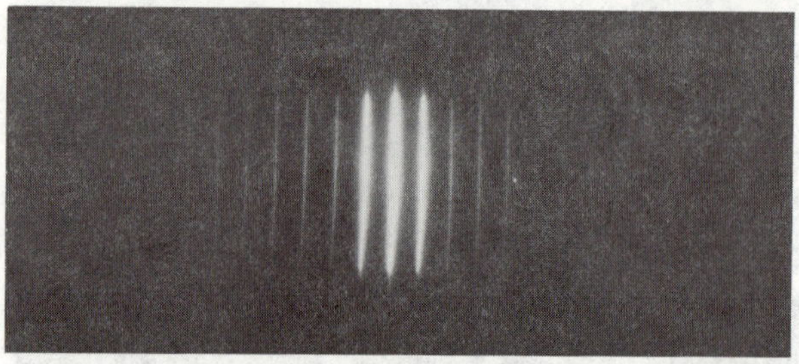

(b)

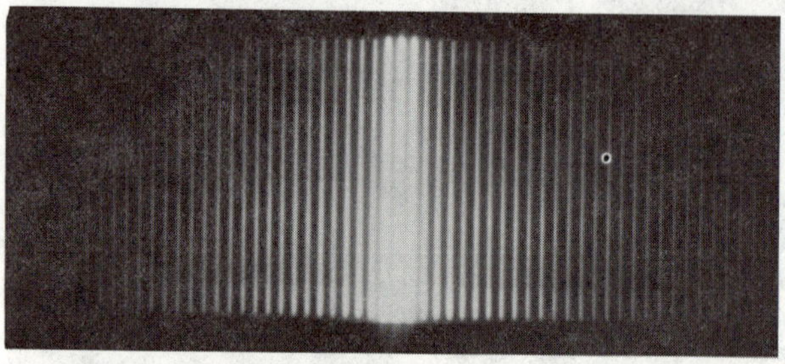

(c)

Fig. 20–13 Interference Patterns from a coarse Ronchi grating of 200 lines/in.: (a) low resolution and low light sensitivity system; (b) medium resolution and high light sensitivity system; (c) Ronchi grating placed directly onto the lens of the camera.

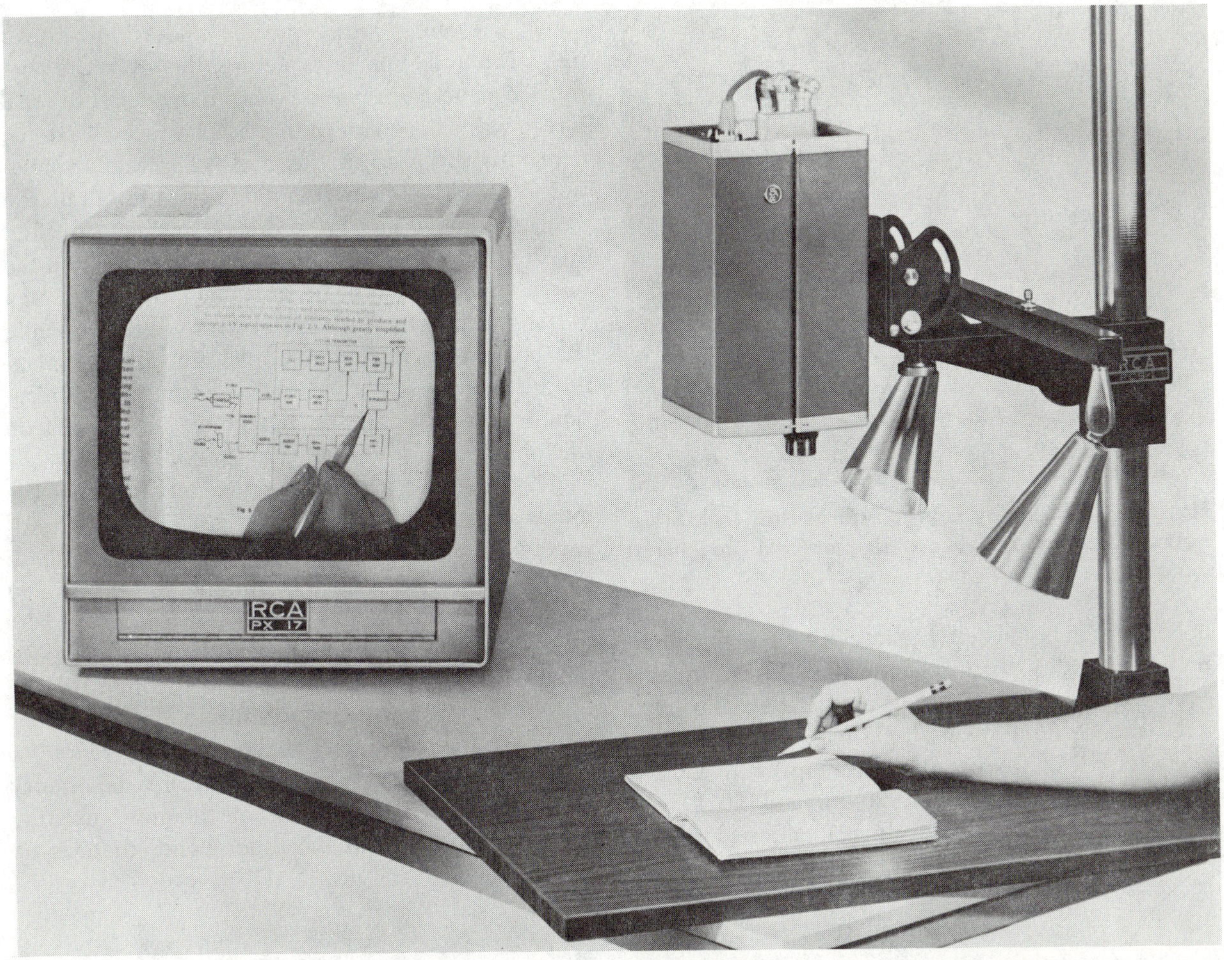

Fig. 20–14 CCTV System used for opaque projection. (Photo courtesy of RCA.)

fully by the author consists of using the TV as a supplement to shadow projection. For example, with an electroscope leaf system shadow projected onto a translucent screen, the TV camera can give the students a behind-the-scenes view of the manipulations performed by the instructor, bringing charged wands up to the electroscope knob, grounding it, etc.

With medium or high resolution, opaque projection of graphs, diagrams from books and journals, etc. can be shown as easily as with an overhead projector (Figs. 20–14 and 20–15). As mentioned earlier, a sturdy vertical camera support with adjustable height and permanent lighting is essential here. Using a ground glass screen illuminated from below, shadow projection of such things as iron filings surrounding magnets as well as black and white slides can be shown (Fig. 20–16). These

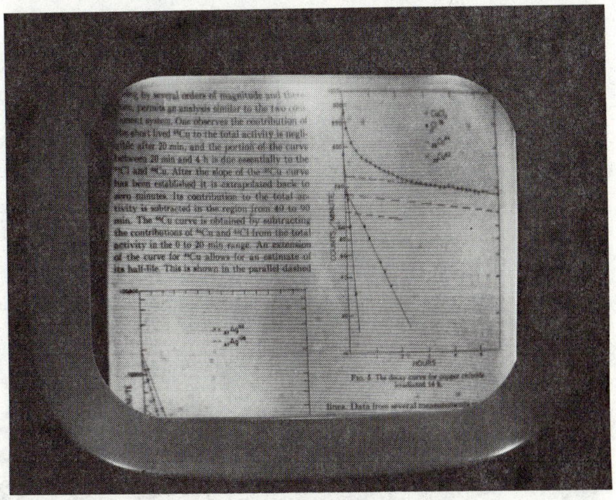

Fig. 20–15 Display with TV Camera focused on Fig. 5 of article by J. H. Taylor and A. P. West, *Am. J. Phys.*, **33,** 1063 (1965).

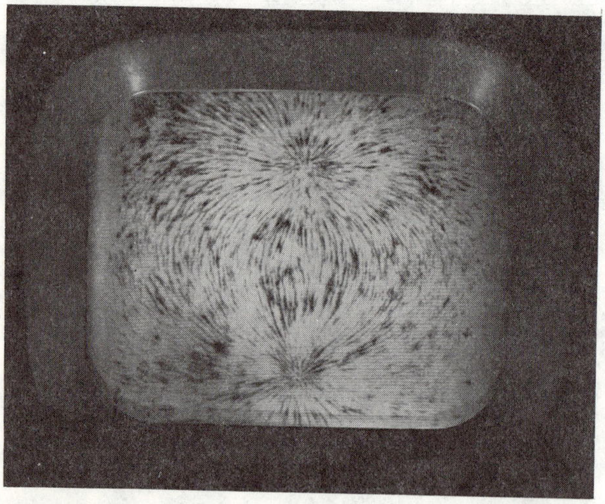

Fig. 20–16 Display from a Pair of Half-Inch Magnets covered by a ground glass plate with iron filings scattered on top.

items are included not because they can be done better by CCTV than by conventional techniques—some of them can't—but because they illustrate what can be done for multiple classroom teaching.

Examples of the use of microscopy with CCTV are shown in Figs. 20–17, 20–18, and 20–19. In Fig. 20–18, the Ronchi ruling mentioned above (200 lines/in.) was viewed under low-power magnification. For this picture the eyepiece and the TV camera lens were both removed and the real image was focused on to the vidicon face. Extreme care must be used in this and other similar situations to avoid allowing too much light to fall on the vidicon, damaging the sensitive surface. In Fig. 20–19 a slide of a muscle section was placed on the microscope stage and the TV camera mounted on a vertical support. An adapter was used to couple the camera to the microscope. The optical magnification was 40×, and the electronic magnification (ratio of monitor diagonal, 14 in., to vidicon diagonal, ⅝ in.) was 22×.

The examples given above are only a few of the many situations where CCTV can be used to advantage in lecture room demonstrations. In these and other uses, careful thought should be given in each case as to whether or not CCTV is the best way to do the job. On the one hand, a poor TV picture of α-tracks with a poor to medium quality system is no substitute for a diffusion cloud chamber setup in the laboratory where small groups of students can take five minutes out of a laboratory period to view directly one of the most exciting sights in physics. On the other hand, there is no

Fig. 20–17 Arrangement for Showing Mouthparts of Mosquito. (Photo courtesy of RCA.)

question that CCTV can be of great use as a routine demonstration aid, as many of the examples have shown. Again it should be emphasized that the quality of the system determines the extent of its usefulness. For demonstration of α-tracks or of Brownian motion under a microscope, a system of high quality is required. For routine magnification of instruments, a medium quality system will do.

Fig. 20–18 Microscopic Picture of 200 lines/in. Ronchi ruling under low power.

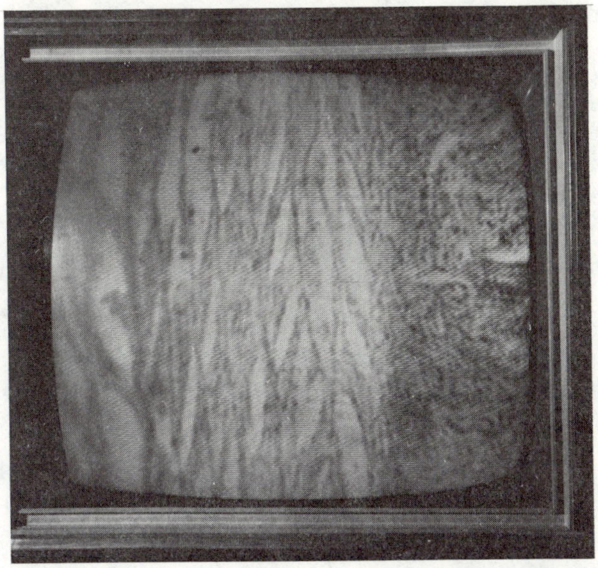

Fig. 20–19 Muscle Section under 40× optical magnification.

21

FILMS AS A LECTURE AID

By Alfred Leitner

Rensselaer Polytechnic Institute

21–1 Introduction[1]

These two volumes include descriptions of existing equipment designed for demonstrations in physics. Is it proper to talk here of films? It is a sign of the times that we must.

There is a tradition for the "live" demonstration lecture. Demonstration lecturing is a valuable pedagogic device in college physics teaching, especially at the introductory level. I am an enthusiastic practitioner of the "art." Yet, I submit, it is difficult to do a good job, and not enough people are demonstrating in their introductory lectures. Preparations are time-consuming. The activity receives insufficient intradepartmental reward in many places. In other places, where the balance between research and teaching may be fair to both, one sometimes finds that funds and facilities are lacking. A partial solution is the use of physics films.

Films demonstrating physics are being made in large numbers. They may threaten the time-honored traditions. Yet, they provide a way of bringing experimental demonstrations into more classrooms.

A list of films available (or in production) as of 1968 appears in Sec. 21–6.

In discussing and listing films for demonstrating physical phenomena, I have chosen a liberal interpretation of the word "demonstration." If you wish to interest a student in the study of physics, a single filmed demonstration experiment may be less valuable than a film that demonstrates what a certain field of physics is like, or one that shows what physicists are like, especially what great physicists are like, or what they accomplished.

(1) The author wishes to thank Harvard Project Physics for encouraging and supporting the indexing of films relating to physics at various educational levels. Thanks are also due M. Jacques Parent of the National Film Board of Canada, Professors Walter Eppenstein and Harry F. Meiners of Rensselaer Polytechnic Institute, and Dr. James Strickland of the Educational Development Center. Miss L. Parmenter and Mrs. W. Timmons have been invaluable in maintaining and expanding the index.

21–2 Films Made by Physicists

During the past decade entirely new concepts have been introduced into the production of educational films about science, and, in particular, about physics. The following are without doubt the two most important innovations.

1. Films are now made whose production is controlled, completely, by highly qualified physicists and physics teachers. This makes the end result eminently useful in teaching physics. Films made by physicists, moreover, paint a truer picture of what physics really is. Such films give a deeper exposition of how physicists think. At the same time, let us be frank and confess that such films can be cinematographically inferior.

2. There has appeared on the scene the silent film loop in a cartridge. The film loop, coupled with the projector available for showing it, may still require many improvements. We shall discuss this in Sec. 21–5. But the system has become exciting to many people. Everybody seems to be making film loops!

These developments are a part of the major educational revolution in which the world finds itself. This revolution is still in process. Its results, whether it be good or bad, cannot yet be fully assessed.

In physics, the upheaval was begun in the United States by the high school course of the Physical Sciences Study Committee. Its novel approach, especially to film-making (innovation 1 above) has had wide influence, not only in physics, and not just at the high school level.

Some of the PSSC films are, in fact, useful in teaching college physics as well, and many of them are listed at the end of this article. The film studio at Educational Development Center (EDC), originally organized for PSSC, has long since gone into making college level films in physics and in other areas.[2] Moreover, physicists independent of EDC have begun to make films. The first group of loop films for use in physics was made by F. Miller, Jr.

21–3 Films Not Made by Physicists

Professional film producers may criticize some of these developments, and with justice. The films referred to in Sec. 21–2 are made by people who are *scientists* first. Cinematic qualities are secondary to them, or to some of them, understandably.

The professional film makers may be right. Before shooting a film, the physicist may not take sufficient pains (a) to construct the story lucidly, (b) to write the narration—if it is a sound film—with extreme care and clarity, (c) to interweave the narration with the images with equal care, (d) to use only the clearest demonstration apparatus, and (e), last but not least, to check all he has prepared with an expert physicist and teacher besides himself, *and* with his camera man.

There are many fine commercial and noncommercial film units around the world. Naturally, the control of subjects filmed by them is in the hands of highly qualified film makers. But they are *film makers* first. Scientific qualities are secondary to them, or to some of them, understandably.

It is now the scientist's turn to criticize! Understanding of science, and information about science, is somewhat limited outside the scientific profession. It often suffers from common misconceptions, especially concerning pure science. There are exceptions, of course.

The science films originating from some professional film studios tend to be beautiful to look at, polished, and well edited. But, at the same time, the exposition of the scientific content may be superficial, or merely descriptive. The films tend to be hardware-oriented. They may be fine training films, but they are not films presenting science. They end before they could be useful in teaching.

(2) Notably in fluid dynamics, a field which nowadays belongs more to engineering than physics. Some of these films have such a fundamental quality of approach to the subject that they are included in the list at the end of this article. For example, if you wish to teach about the meaning of the curl of a vector (a concept which originated in fluid dynamics), you will find most convincing experimental demonstrations of a curl meter in *Vorticity* (a 16-mm film; also excerpted as 8-mm film-loop cartridges, *Vorticity Meter*, Parts I and II, and *The Bathtub Vortex*).

Other science films tend to be arty and high-flown. Scientific ideas are introduced in the form of extremely beautiful, allegorical pictures which, however, detract from the ideas and overpower them. This type of film is too self-conscious, cinematically, for teaching purposes.

Fortunately, the gap between the two groups of film makers is narrowing. Certainly, each group (the physicists and the nonphysicists) making physics films can learn from each other. It is clear that the first group has made its point with the second (with help from government and foundations). Professional film studios are entering this field of film production, consulting qualified physicists and physics teachers. There is now a market for such films. The end is not in sight.

21–4 16-mm Films with Sound

The 16-mm sound films tend to be long compared with the length of class lectures. The narration interferes with the lecturer's domain over the classroom. In spite of this, films in this category can be very useful.

Excellent films are now available in this format. Many of them cannot, or should not, be cut up into silent excerpts. Further production of good 16-mm physics films, with sound, should be encouraged.

A number of universities and colleges schedule film showings separately from formal classes. The students themselves can administer these showings. The showings can be on an interdisciplinary basis, with science and engineering departments participating, because fine teaching films are becoming available in many academic areas. Or, students may be provided projectors in a common room of their dormitories, along with access to a film library.

There are several categories of 16-mm sound film on physics. See also at the end of this chapter (Sec. 21–7) an excerpt from the 1961 M.I.T. Science Teaching Center Proposal to the National Science Foundation on the "Use of Moving Picture Film," by F. L. Friedman.[3]

1. *The demonstration lecture on film.* If its subject is commonly taught in our courses, and if the demonstrations in it are brief, then this may be equally useful as a series of excerpts, short and silent, preferably in the form of loops in cartridges. A large number of excerpts of existing 16-mm sound films is becoming available at present (unfortunately in 8-mm, which limits their usefulness in large classes).

However, if the experiments involved are long, excerpting is impossible. Moreover, a long experiment may as well be narrated by the people who made the film. (Examples: *Ultimate Speed* and *Time Dilation.* See Sec. 21–6.1.)

Filmed demonstration lectures in fields not commonly discussed in detail by our sources are useful, especially if they bring modern physics and the frontiers of physics nearer to the student.

2. *Well-known physicists on film.* Such films may be less valuable than a visit in person, but more students can get to see the physicist in this way. If the well-known physicist is, in addition, a good teacher, a film of his lecture can be a source of inspiration. (Example: Feynman's *1964 Messenger Lectures.*) Films of this type also have historical value of their own.

3. *Documentaries on physics and physicists.* There are not enough of these. Our students could benefit from an objective view of what physics is, or what physicists do.

Documentaries are hard to make, especially if physicists make them. More than in any other type of film, documentaries require an experienced professional film maker.

I know of only one documentary, available at present, which paints a true and full picture of physicists at work: *An Experiment in Physics.* It was made in the laboratory of the late D. K. C. MacDonald at the National Research Council of Canada and is narrated by him. The film is credible, unposed, witty, and interesting.

Harvard Project Physics sponsored production of two documentaries made by well-known film directors. One follows a group of high-energy physicists and their students through an experiment and into their lives. Another is *The World of Fermi,* a reconstruction, with interviews, existing film footage, and photographs, of his life and work. Both documentaries are now available.

[3] Courtesy J. G. King, Massachusetts Institute of Technology.

Reconstructing acts of great discovery on film can be a source of inspiration to the student, if it is done properly. The best science film that I have seen of this type is not about physics: *William Harvey and the Circulation of Blood,* a film made in England in 1957 by the Wellcome Film Unit under the sponsorship of the Royal College of Physicians and Surgeons.

21–5 The Film Loop in a Cartridge

1. *Single-concept films.* The short and silent single-concept film has great potential for use in physics classes. Demonstration experiments on film can save faculty the time needed to set up and adjust apparatus for live demonstrations. It is important that the film be silent. The teacher can include it in his own lecture and make use of it as he sees fit.

The short and silent single-concept film also has potential in the physics laboratory. A filmed record of an actual event can become a laboratory project. Consider the loop films listed below under *Collisions* as an example. These films record collisions, in one and in two dimensions, of two or more balls, using slow motion. The student can project this film on to a sufficiently large sheet of paper and study this event in great quantitative detail, with satisfactory precision.

2. *Film loops.* If the short film is, in addition, available as a loop, it can—*in principle, but not necessarily in practice*—be run through the projector as often as is needed. The demonstration lecturer can repeat the film right away in order to clinch points in his discussion which he may wish to emphasize. And in the laboratory uses mentioned above, the student can take data as often as is deemed advisable.

3. *Still-picture button.* If it is possible, as the film is being shown, to stop the film and project any frame as a still picture—*without damaging the film*—the versatility of this format is advanced further. Now the instructor can also make special points anywhere along the film (and the student can measure positions at leisure, with precision).

4. *Cartridge loading.* Finally, if a film loop is supplied in a convenient cartridge, and if loading the cartridge into the projector requires no threading of film by the user, the system would seem to be ideal.

The reader is doubtless aware that this "ideal system" and certain presently "available systems" share many properties. Moreover, physics teachers and film producers have gone into the production of films fitting one or more of the "available systems," particularly those which use 8-mm film. The market is being flooded with such films, on many extremely useful teaching subjects, and the long list at the end of this article reflects this fact. However, *the available systems for the projection of easily loaded loops in cartridges fall very short of the ideal.* These systems were originally designed for the home movie market, even if the manufacturers claim otherwise.

The film loop wears out rapidly in the projectors now available. But the repeatability requirement mentioned under item 2 above is absolutely essential in the teaching uses of the film loop. Requirement 3 above is almost as essential. Projectors with this "still-picture" feature are available. A filter is interposed mechanically between the light bulb and the film at the time when a button is pushed, which also stops the film. This device also wears out the film rapidly.

The presently available film-loop systems which, in principle, satisfy requirements 2, 3, and 4 are 8-mm systems (the Technicolor cartridge and projector, the Fairchild cartridge and projector, and the Kodak Super-8 system).

The 8-mm format is adequate for viewing films in small groups only. It is adequate for use in small classes or by individual students. It is also fine for the laboratory uses referred to in the second paragraph of Sec. 21–5, item 1.

The principal limitation in the 8-mm format is upon the maximum light intensity. This is due entirely to the fact that the individual frames are small. And Super-8, although it has larger frame size (and therefore better picture quality), is still too small to project intense images to large groups in a classroom. Experience showed long ago that the 16-mm film has a frame size sufficiently large to project images that can be seen clearly from most seats of common lecture rooms in use nowadays.

It is unfortunate that film loops are being made mostly for 8-mm systems. These are inadequate in lecture demonstrations. I have heard of teachers who move the projector to different parts of the room and run the film several times so that all in the audience can see. This should not be necessary.

Film loops for use in demonstration lectures should be on 16-mm film. Projectors should (and eventually no doubt will) be developed which can be loaded, without threading, with 16-mm film loop cartridges, and which have a still-picture device as well.

21–6 Film List

Large though it is, the list below is selective. The films in it are useful in teaching physics. The emphasis in the selection was toward the "intensive" rather than the "descriptive" type of science film. Ignorance about some films suited to this list may have caused omissions for which the author wishes to apologize.

Five film lists in this area are due to R. L. Weber.[(4)] Reviews of some of the films listed can be found in the *American Journal of Physics.* W. R. Riley[(5)] is continuing R. I. Hulsizer's project to publish an annotated critical bibliography of physics films in the *American Journal of Physics.*

All films in Sec. 26–6.1, 16-mm films, are sound films unless marked silent. All films in Sec. 26–6.2, film loops, are silent 8-mm films in a cartridge except for *Wave Motion* (Sets 1 and 2), which are 16-mm loops without cartridges. These can be threaded into standard projectors.

The film loops vary in presentation time to a maximum of five minutes, with the average between three and four minutes. Times are not given in this list for film loops, but they are given for 16-mm films when known. Some distributors have recognized that 8-mm film is inadequate for use in large classes and have made the film loops available as 16-mm film on standard reels.

Topics in which a given film is likely to be useful are listed in a column to the left of the titles, according to a key of abbreviations given just below. The names of these topics are also listed in the alphabetical film lists of Sec. 21–6.1 (16-mm films) and 21–6.2 (8-mm film loops). Under their headings appear film titles, as cross-references.

Key to Topics

AP	Atomic physics
D	Documentary film
EM	Electricity and magnetism
EP	Elementary particles
FM	Fluid mechanics
GP	General physics
HT	Heat and thermodynamics
KT	Kinetic theory
Lab	A film containing either instruction on the uses of instruments, or scenes inviting direct data-taking off the projected image, as a laboratory exercise.
LTP	Low-temperature physics
M	Mechanics
MP	Modern physics
NP	Nuclear physics
O	Optics
Q	A film that presents quantitative data, in tabular or other form, arising out of the experiment being shown, as a problem for the viewer.
QM	Quantum mechanics
R	Relativity
SM	Statistical mechanics
SSP	Solid state physics
V	Vectors and vector analysis
W	Waves and wave phenomena

Other Abbreviations

A,a	Author
AAPT	American Association of Physics Teachers
BBC	British Broadcasting Corporation
BTL	Bell Telephone Laboratories
B & W	Black and white
C	Color
Chem Study	Chemical Education Materials Study
d	Distributor (Number after d refers to a specific distributor. See Sec. 21–6.3 for list of distributors.)
EDC	Educational Development Center (formerly ESI; see d,8)
ESI	Educational Services, Inc. (see d,8)
HPP	Harvard Project Physics, Cambridge, Mass.
IWF	Institut für Wissenschaftlichen Film (see d,10)

(4) R. L. Weber, *Am. J. Phys.*, **17**, 408 (1949); **22**, 54 (1953); **29**, 222 (1961); **30**, 321 (1962); **36**, 302 (1968).

(5) W. R. Riley, *Am. J. Phys.*, **36**, 475 (1968).

NCFMF National Committee for Fluid Mechanics Films
NFB National Film Board of Canada (see d,17)
OSU The Ohio State University (see d,19)
p Producer
PSSC Physical Science Study Committee
reel–16 16-mm film on reel
reg regular 8-mm film lop in cartridge
reg–reel regular 8-mm film on reel
SEEC Semiconductor Electronics Education Committee
SUNY State University of New York
sup super–8-mm film loop in cartridge
sup–reel super–8-mm film on reel

21–6.1 Sixteen-Millimeter Films

AAPT Physics Series. 12 films (mostly animation), 8 listed here; B & W; p, d, 14. See *Carnot Cycle (Kelvin Temperature Scale)*, 8 min; *Circular Motion,* 8 min; *Doppler Effect,* 8 min; *Progressive Waves, Transverse and Longitudinal,* 19 min; *Simple Harmonic Motion,* 10 min; *Stationary Longitudinal Waves,* 9 min; *Stationary Transverse Waves,* 9 min; *U–238 Radioactive Series,* 9 min.

Advanced Science Series. Fifteen films, 1 listed here. B & W; p, Educational Foundation for Visual Arts, Great Britain; d, 14. See *Discharge Through Gases.*

M,V *Angular Momentum, a Vector Quantity.* B & W, 27 min; a, A. Lemonick; p, ESI; d, 16, 26. Spinning wheels, torque and precession.

EM,O *Angular Momentum of Circularly Polarized Radiation.* B & W, 19 min; a, J. Meyer and J. Ladish; p, ESI; d, 16, 26. Torque on a torsion pendulum due to circularly polarized light.

AP Atomic Physics. See *Bohr, The Small World of Niels; Electrical Discharge in Gases; Franck-Hertz Experiment; Mass of Atoms; Size of Atoms from an Atomic Beam Experiment; Stern-Gerlach Experiment; Velocity Distribution of Atoms in a Beam.*

AP *Atoms, Size of, from an Atomic Beam Experiment, The.* B & W, 28 min; a, J. King; p, ESI; d, 16, 26. Beam of K atoms scattered by A atoms at different argon pressures. Atom size inferred from each run found independent of pressure.

W *Barrier Penetration.* Waves over a deep channel of variable width between two shallow regions. See *Ripple Tank Wave Phenomena Series.*

AP,D *Bohr, The Small World of Niels.* B & W, 58 min. Lecture by E. Teller; p, SUNY, Stony Brook; d, 24.

MP,SSP, W *Bragg Reflection.* Diffraction of water waves from two-dimensional arrays of pegs. See *Ripple Tank Wave Phenomena Series.*

D,SSP *Brattain on Semiconductor Physics.* B & W, 30 min; p, BTL; d, 2. Booklet on request.

KT *Brownian Motion.* B & W, 6 min; a, W. Lochte-Holtgreven; p, IWF; d, 1, 10. Model and actual. Fat in milk, colloid suspension, living cells.

HT *Carnot Cycle (Kelvin Temperature Scale).* Animation. See *AAPT Physics Series.*

M *Cavendish Balance, a Measurement of the Universal Gravitational Constant (Purdue Series)*; C, 27 min; a, D. J. Tendam; p, d, 20.

M *Center of Mass, Moving with the* (PSSC). B & W, 27 min; a, H. Branson; p, ESI; d, 16. Conservation of energy and momentum in a system of several magnetic Dry Ice pucks.

FM *Channel Flow of a Compressible Fluid.* B & W, 29 min; a, D. Coles; p, NCMF–ESI; d, 9.

D,GP, MP *Character of Physical Law, The.* Seven lectures by R. Feynman to general undergraduate audiences. See *Messenger Lectures.*

Chem Study Series. 27 titles, 2 listed here. B & W, or C; a, various; p, various, sponsored by Chemical Education Material Study; d, 16. See *Crystals and Their Structures, Gases and How They Combine.*

M,V *Circular Motion, Uniform.* See *AAPT Physics Series.*

Computer-generated Films. See *Computer Studies of Fluid Dynamics; Force, Mass and Motion; Harmonic Phasors; Quantum Mechanical Harmonic Oscillator; Quantum Mechanical Wave Packets from Potential Well and Barrier, Scattering of; Vector Kinematics* (parts only).

FM,V *Computer Studies of Fluid Dynamics.* B & W, silent, 8 min; p, d, 12. Computer-generated flow fields.

EM *Coulomb Force Constant* (PSSC). B & W, 34 min; a, E. Rogers; p, ESI; d, 16.

EM *Coulomb's Law* (PSSC). B & W, 30 min; a, E. Rogers; p, ESI; d, 16.

EM *Counting Electrical Charges in Motion* (PSSC). B & W, 22 min; a, J. Strickland; p, ESI; d, 16.

SSP *Crystallization, Experiments with Models on.* B & W, 4.5 min; a, G. Bekow, G. Wasserman; d, 10.

SSP *Crystals* (PSSC). B & W, 25 min; a, A. Holden; p, ESI; d, 16. Formation, growth, micro-motion-photography.

SSP *Crystals, an Introduction.* C, 25 min; a, E. Wood; p, BTL; d, 2. Symmetries, cleavage, x-ray diffraction.

D,SSP *Crystals, Imperfections in.* B & W, 69 min; p, d, 6. Lecture by Frederick Seitz.

SSP *Crystals and Their Structures* (*Chem Study Series*). B & W, 22 min; a, J. A. Campbell; p, ESI; d, 16. Formation, growth, cleavage, models, x-ray diffraction with ripple tank analogy experiments on Bragg reflections.

M,V *Deflecting Forces* (PSSC). B & W, 30 min; a, N. Frank; p, ESI; d, 16. Dynamics of curvilinear motion.

FM,V *Deformation of Continuous Media.* B & W, 38 min; a, J. Lumley; p, NCMF–ESI; d, 9. Kinematics of continuum deformation.

D Dirac, P. A. M. See *Physical Ideas, Evolution of.*

MP *Discharge Through Gases.* B & W, 12 min; p, Educational Foundation for Visual Aids, Great Britain; d, 14 (*Advanced Science Series*). Discharge tube pressure progressively reduced.

D DOCUMENTARY FILMS ON PHYSICS. See *Bohr, The Small World of Niels* (by E. Teller); *Brattain on Semiconductor Physics; Crystals, Imperfections in* (by F. Seitz); *Einstein, The Large World of Albert* (by E. Teller); *Experiment in Physics* (by D. K. C. MacDonald); *Fabric of the Atom* (by P. Morrison); *Fermi, The World of Enrico. Flows, Low Reynolds Number* (by Sir G. I. Taylor); *Franck-Hertz Experiment; Inertia,* and *Inertial Mass* (by E. M. Purcell); *Messenger Lectures* (by R. Feynman); *Minority Carriers in Semiconductors* (by R. Haynes and W. Shockley); *People and Particles; Stellar Evolution* (by G. Gamow); *Strangeness Minus Three* (Feynman, Gell-Mann, Ne'eman); *Synchrotron.*

EM,SSP *Domains and Hysteresis in Ferromagnetic Materials.* C, 39 min; a, H. J. Williams, J. R. Friedman; p, BTL; d, 2. Advanced; can be shown to students after course introduction or after 45-min 35-mm film strip and LP record entitled, *The Formation of Ferromagnetic Domains* (d, 2). See also elementary version of *Ferromagnetic Domains.*

W *Doppler Effect. Animation.* See *AAPT Physics Series.*

W *Doppler Effect and Shock Waves.* See *Ripple Tank Wave Phenomena Series.*

D,R *Einstein, The Large World of Albert.* B & W, 56 min. A lecture by Edward Teller; p, SUNY at Stony Brook; d, 24.

M *Elastic Collisions and Stored Energy* (PSSC). B & W, 27 min; a, J. Strickland; p, ESI; d, 16. Measurements of kinetic energy of two magnetized Dry-Ice pucks lead to potential energy.

EM *Electric Fields* (PSSC). B & W, 25 min; a, F. Bitter, J. Waymouth; p, ESI; d, 16.

EM,V *Electric Lines of Force* (PSSC). B & W, 7 min; a, A. Joseph; p, ESI; d, 16. Grass seed demonstrating patterns.

MP *Electrical Discharge in Gases.* 9.5 min; a, E. Flegler; d, 10.

EM ELECTRICITY AND MAGNETISM. See *Angular Momentum of Circularly Polarized Radiation; Coulomb Force Constant; Coulomb's Law; Counting Electrical Charges in Motion; Domains and Hysteresis in Ferromagnetic Materials; Electric Fields; Electric Lines of Force; Electromagnetic Waves; Electrons in a Uniform Magnetic Field; Elementary Charges and Transfer of Kinetic Energy; Ferromagnetic Domains; Harmonic Phasors; Magnet Laboratory; Magnetohydrodynamics; Millikan Experiment; Momentum of Electrons; Pressure of Light; Speed of Light; Synchrotron; Ultimate speed.*

EM,O, W *Electromagnetic Waves* (PSSC). B & W, 33 min; a, G. Wolga; p, ESI; d, 16. Experiments showing radiation by accelerated charges. Young's double slit by x-ray, visible light, micro- and radiowave.

MP *Electron Emission Series.* 3 short films; a, A. P. French; p, ESI; d, 16, 26. *Field Emission of Electrons,* B & W, 4 min. *Photo-Emission of Electrons,* B & W, 4 min. *Thermionic Emission of Electrons,* C, 6 min. (Also listed separately).

EM *Electrons in a Uniform Magnetic Field* (PSSC). B & W, 11 min; a, D. D. Montgomery; p, ESI; d, 16. Uses Leybold's e/m tube.

EM *Elementary Charges and Transfer of Kinetic Energy* (PSSC). B & W, 34 min; a, F. L. Friedman; p, ESI; d, 16. Uses diode and calorimetry.

EP ELEMENTARY PARTICLES. See *People and Particles; Positron–Electron Annihilation; Strangeness Minus Three; Synchrotron; Time Dilation, an Experiment with μ-Mesons; Ultimate Speed, an Exploration with High-Energy Electrons.*

M *Elliptic Orbits* (PSSC). B & W, 18.5 min; a, A. V. Baez; p, ESI; d, 16. Geometric methods used graphically, and otherwise, to prove Kepler's 1st and 2nd laws for inverse square central force.

D,LTP, SSP *Experiment in Physics, An.* B & W, 15 min; a, D. K. C. MacDonald; p, d, 18. Excellent documentary on research atmosphere. Study of thermopower of gold at low temperature, using adiabatic demagnetization.

D,MP *Fabric of the Atom.* 10 films of a lecture series by Philip Morrison. B & W, 30 min each; p, BBC-TV; d, 8 (loan only), 28 (purchase only). Titles: (1) *On the Stability of Matter.* (2) *The Graininess of Energy.* (3) *The Quantum of Action.* (4) *The Fusion of Cause and Chance.* (5) *Matter Waves.* (6) *The Principle of Uncertainty.* (7) *The Physics of Identity.* (8) *The Quantum Fabric of Matter.* (9) *Creation and Destruction.* (10) *Elementary Particles, the Elusive Ultimate.*

D,MP *Fermi, The World of Enrico* (tentative title); a, HPP and J. Kemeny; p, HPP; d, 32 (release 1969).

EM,SSP *Ferromagnetic Domains, Part I: Magnetism and Domains,* C, 11 min; *Part II: How Domains are Formed,* C, 11 min; a, E. A. Nesbitt; p, BTL; d, 2. Elementary version, suitable for introductory college physics. See also advanced version, *Domains and Hysteresis in Ferromagnetic Materials.*

D Feynman, Richard. See *Messenger Lectures; Strangeness Minus Three.*

MP *Field Emission of Electrons.* Emission from metal by strong electric fields. See *Electron Emission Series.*

FM,V *Flow Visualization.* B & W, 31 min; a, S. J. Kline; p, NCFMF–ESI; d, 9. Many types of laminar and turbulent flow made visible by hydrogen bubble and dye injection techniques.

D,FM *Flows, Low Reynolds Number.* C, 33 min; a, Sir G. I. Taylor; p, NCFMF–ESI; d, 9.

FM *Fluid Dynamics of Drag.* B & W. Part I, 21 min, Part II, 32 min, Part III, 37 min, Part IV, 29 min; a, A. H. Shapiro; p, NCFMF–ESI; d, 9. Viscosity, Reynolds number, laminar and turbulent flows, viscous and pressure drag, streamlining.

FM FLUID MECHANICS. See *Computer Studies of Fluid Dynamics; Fluid Mechanics Series; Liquid Helium II, the Superfluid; Schlieren.*

FM *Fluid Mechanics Series.* 18 titles as of summer 1968, more to be released, 11 listed here; a, various; p, NCFMF–ESI. See *Channel Flow of a Compressible Fluid; Deformation of Continuous Media; Flow Visualization; Flows, Low Reynolds Number; Fluid Dynamics of Drag; Magnetohydrodynamics; Pressure Fields and Fluid Acceleration; Rheological Behavior of Fluids; Surface Tension in Fluid Mechanics; Vorticity; Waves in Fluids.* (See also film loops in Sec. 21–6.2.)

M,V *Force, Mass and Motion.* B & W, 10 min; a, F. W. Sinden; p, BTL; d, 8 (loan only), 11 (sale). Computer-generated; motions under central forces. (See also *Mechanics Series* (*Sinden*) of 8-mm film loops in Sec. 21–6.2.)

M *Forces* (PSSC). B & W, 23 min; a, J. R. Zacharias; p, ESI; d, 16. Introductory.

Qualitative Cavendish experiment. Gravitational and electrical forces compared.

M — *Forces* (PSSC), *An Excerpt.* B & W, 8 min; a, J. R. Zacharias; p, ESI; d, 8. Cavendish experiment. Gravitational and electrical forces compared.

GP,M,R — *Frames of Reference* (PSSC). B & W, 28 min; a, P. Hume and D. Ivey; p, ESI; d, 16. Inertial, accelerated and rotating frames of reference. Outstanding.

M,R — *Frames of Reference* (PSSC), *Excerpt I.* B & W, 7 min; a, P. Hume and D. Ivey; p, ESI; d, 8. Falling body in different frames of reference.

M,R — *Frames of Reference* (PSSC), *Excerpt II.* B & W, 6 min; a, P. Hume and D. Ivey; p, ESI; d, 8. Dry Ice puck in a rotating frame.

AP,D, MP — *Franck-Hertz Experiment* (PSSC). B & W, 30 min; a, B. Youtz; p, ESI; d, 16. With epilogue by James Franck.

M — *Free Fall and Projectile Motion* (PSSC). B & W, 27 min; a, N. Frank; p, ESI; d, 16.

M,R — *Galilean Relativity Demonstrated by Projectile Motion* (*Purdue Series*). B & W, 4 min; a, D. J. Tendam; p, d, 20. Slow-motion and multiple-flash photography of ball projected upward on a moving cart.

D — Gamow, George. See *Stellar Evolution.*

SSP — *Gap Energy and Recombination Light in Germanium.* B & W, 37 min; a, J. I. Pankove, R. B. Adler; p, SEEC–ESI; d, 8.

HT,KT — *Gases, Behavior of* (PSSC). B & W, 15 min; a, L. Grodzins; p, ESI; d, 16. Brownian motion, analogues, and Boyle's law.

KT — *Gases and How They Combine* (*Chem Study Series*). C, 22 min; a, G. Pimentel; p, Davidson Films; d, 16. Demonstrates law of simple and multiple proportions, leads to Avogadro's first hypothesis.

D — Gell-Mann, Murray. See *Strangeness Minus Three.*

GP — GENERAL PHYSICS. See *Fabric of the Atom; Frames of Reference; Messenger Lectures; Physical Ideas, the Evolution of; Symmetry.*

W — *Group and Phase Velocity.* B & W, silent, 10 min; a, E. David and G. Bekow; p, 10; d, 1 (loan only), 10; Clear animation, and demonstrations with water waves.

EM,M, V — *Harmonic Phasors.* B & W, silent, 7 min; a, W. Huggins; p, BTL; d, 8. Linear oscillations and waves described by rotating vectors. Computer-generated.

HT — HEAT AND THERMODYNAMICS. See *Carnot Cycle; Gases, Behavior of; Gases and How They Combine; Liquid Helium II, the Superfluid.*

HPP (Harvard Project Physics). See *Fermi, the World of Enrico; People and Particles; Synchrotron.* (See also *HPP Series* of 8-mm film loops in Sec. 21–6.2.)

D,M — *Inertia* (PSSC). B & W, 26 min; a, E. M. Purcell; p, ESI; d, 16. $F \propto a$ at constant mass. Uses Dry Ice pucks.

D,M — *Inertial Mass* (PSSC). B & W, 19 min; a, E. M. Purcell; p, ESI; d, 16. Continuation of *Inertia* (PSSC).

W — *Interference and Diffraction.* See *Ripple Tank Wave Phenomena Series.*

O,MP, QM — *Interference of Photons* (PSSC). B & W, 13 min; a, J. King; p, ESI; d, 16. Dim light source, double slit, photomultiplier. Less than one photon in apparatus at a time, on average. Statistics and uncertainty.

KT — KINETIC THEORY. See *Brownian Motion; Gases, Behavior of; Gases and How they Combine.*

Lab — *Laboratory High-Vacuum Technique.* B & W, 30 min; a, J. Orsula; p, ESI; d, 16, 26. Instructions for laboratory practice. (Note: this film replaces *Solder Glass Technique.*)

O — LASER. See *Optical Maser, Principles of the.*

FM,HT, LTP — *Liquid Helium II, the Superfluid.* B & W, 38 min; a, A. Leitner; p, d, 15. λ-point transition. Viscosity paradox: superleak vs. bulk flow. Fountain effect. Rollin Film. 2nd sound. 2-fluid model.

SSP — *Low-Energy Electron Diffraction Pattern from Germanium (100) and (111) Surfaces.* B & W, silent, 15 min; a, J. J. Lander and J. Morrison; p, d, 2.

LTP — LOW-TEMPERATURE PHYSICS. See *Experiment in Physics; Liquid Helium II, the Superfluid; Superconductivity, an Introduction to.*

D MacDonald, D. K. C. See *Experiment in Physics.*

EM *Magnet Laboratory, A* (PSSC). B & W, 21 min; a, F. Bitter and J. Waymouth; p, ESI; d, 16. Large conventional magnets. Strong currents and fields.

EM,FM *Magnetohydrodynamics.* B & W, 27 min; a, J. A. Shercliff; p, NCFMF–ESI; d, 9. Experiments on conducting liquids.

AP,MP, NP *Mass of Atoms, The.* B & W, 47 min (2 reels); a, R. Hertz and C. Brewer; p, ESI; d, 16, 26. Part I (20 min): sample of Po is weighed, sealed, and decays. Part II (27 min): 3 weeks later Po activity and He mass measured. Data also determines He, Po atomic masses.

M MECHANICS. See the following: *Angular Momentum, A Vector Quantity; Cavendish Balance, A Measurement of the Universal Gravitational Constant; Center of Mass, Moving with the; Circular Motion, Uniform; Deflecting Forces; Elastic Collisions and Stored Energy; Elliptic Orbits; Force, Mass and Motion; Forces; Forces, an Excerpt; Frames of Reference; Frames of Reference, Excerpts I and II; Free Fall and Projectile Motion; Galilean Relativity Demonstrated by Projectile Motion; Harmonic Phasors; Inertia; Inertial Mass; Million to One, A; Newton's Equal Areas; Periodic Motion; Satellite Orbits; Simple Harmonic Motion; Time Dilation, an Experiment with μ-Mesons; Ultimate Speed, an Exploration with High-Energy Electrons; Vector Kinematics.*

D,GP, MP *Messenger Lectures at Cornell University, The 1964.* 7 excellent films of R. Feynman lecturing on the Character of Physical Law to a general undergraduate audience. B & W; p, BBC-TV; d, 8. Accompanying book, entitled *The Character of Physical Law* (BBC-TV, London, 1965). Film titles: (1) *The Law of Gravitation, an Example of Physical Law* (55 min); (2) *The Relation of Mathematics to Physics* (57 min); (3) *The Great Conservation Principles* (56 min); (4) *Symmetry in Physical Law* (58 min); (5) *The Distinction of Past and Future* (47 min); (6) *Probability and Uncertainty—The Quantum Mechanical View of Nature* (59 min); (7) *Seeking New Laws* (59 min.)

EM,MP *Millikan Experiment* (PSSC). B & W, 30 min; a, F. L. Friedman and A. Redfield; p, ESI; d, 16.

M *Million to One, A* (PSSC). B & W, with music, 5 min; a, p, ESI; d, 16 (purchase only). Dry ice puck explained, trip to flea circus, flea accelerates a massive puck. Entertaining, instructive spoof.

D,SSP *Minority Carriers in Semiconductors.* B & W, 26 min; a, R. Haynes and W. Shockley; p, SEEC–ESI; d, 8.

MP MODERN PHYSICS. See *Bragg Reflection; Discharge Through Gases; Experiment in Physics; Fabric of the Atom; Field Emission of Electrons; Franck-Hertz Experiment; Interference of Photons; Mass of Atoms; Messenger Lectures; Momentum of Electrons; Optical Maser, Principles of the; People and Particles; Photoelectric Effect; Photoelectric Effect, an Excerpt; Photo-Emission of Electrons; Pressure of Light; Rutherford Atom; Stern–Gerlach Experiment; Thermionic Emission; Time Dilation, an Experiment with μ-Mesons; Ultimate Speed, an Exploration with High-Energy Electrons.*

EM,MP *Momentum of Electrons.* C, 10 min; a, J. King; p, ESI; d, 16, 26. Measured with a vane on a torsion pendulum inside a solder-glass sealed vacuum tube.

D Morrison, Philip. See *Fabric of the Atom.*

D Ne'eman, Y. See *Strangeness Minus Three.*

M *Newton's Equal Areas.* C, 8 min; a, A. Bork, and B. and K. Cornwell; p, B. and K. Cornwell; d, 29. Newton's derivation, using brief impact forces, valid for any central force law, of Kepler's second law. Animation.

NP NUCLEAR PHYSICS. See *Mass of Atoms; Positron–Electron Annihilation; Random Events; Rutherford Atom; U-238 Radioactive Series.*

O,MP *Optical Maser, Principles of the.* C, 30 min; a, C. G. B. Garrett; p, BTL; d, 2. Gas laser operation, and elements of theory.

O OPTICS. See *Angular Momentum of Circularly Polarized Radiation; Electromagnetic Waves; Interference of Photons; Photons; Optics; Introduction to; Optical Maser, Principles of the; Schlieren;*

Speed of Light. See also under *Ripple Tank Wave Phenomena Series,* and WAVES.

O *Optics, Introduction to* (PSSC). C, 23 min; a, E. P. Little; p, ESI; d, 16.

D,EP *People and Particles* (HPP). B & W, 28 min; a, M. Butler, P. Ferguson; p, R. W. Gardner, G. Holton; d, 32. Story of an experiment, and of the group involved, in positron–electron pair production at high energy with the Cambridge (Mass.) electron accelerator, using a wide-gap spark chamber.

M *Periodic Motion* (PSSC). B & W, 33 min; a, P. Hume and D. Ivey; p, ESI; d, 16. Elements of simple harmonic and uniform circular motion, using Dry Ice puck. See also *Simple Harmonic Motion.*

MP *Photoelectric Effect* (PSSC). C, 28 min; a, J. Strong; p, ESI; d, 16. Quantitative experimentation. See also *Photoelectric Effect, an Excerpt; Photo-Emission of Electrons.*

MP *Photoelectric Effect* (PSSC), *an Excerpt.* C, 9 min; a, J. Strong; p, ESI; d, 8. Qualitative experimentation.

MP *Photo-Emission of Electrons.* See *Electron Emission Series.*

MP,O *Photons* (PSSC). B & W, 19 min; a, J. King; p, ESI; d, 16. Demonstrates particle nature of light carefully, using photomultiplier, oscilloscope, etc.

D,GP *Physical Ideas, The Evolution of.* B & W, 29 min; a lecture by P. A. M. Dirac; p, SUNY at Stony Brook; d, 24.

EP,NP, R *Positron–Electron Annihilation.* B & W, 28 min; a, S. Berko; p, ESI; d, 16, 26. Detects γ-rays moving in opposite directions, measuring energy, proves annihilation.

FM,V *Pressure Fields and Fluid Acceleration.* B & W, 30 min; a, A. H. Shapiro; p, NCFMF–ESI; d, 9. Intuitive relations of pressure gradient to velocity fields in steady flows, developed by demonstration experiments.

EM,MP, O *Pressure of Light* (PSSC). B & W, 23 min; a, J. R. Zacharias; p, ESI; d, 16. Light incident on a thin foil on torsion fibre in vacuo causes oscillation in foil. Crookes radiometer unmasked.

W *Progressive Waves, Transverse and Longitudinal.* Mainly animated, clear exposition, useful. See under *AAPT Physics Series.*

PSSC Series. 56 titles. 34 titles and 4 excerpts listed here, as well as separately with details. B & W or C; a, various; p, ESI; d, 16. *Center of Mass, Moving with the; Coulomb Force Constant; Coulomb's Law; Counting Electrical Charges in Motion; Crystals; Deflecting Forces; Elastic Collisions and Stored Energy; Electric Fields; Electric Lines of Force; Electromagnetic Waves; Electrons in a Uniform Magnetic Field; Elementary Charges and Transfer of Kinetic Energy; Elliptic Orbits; Forces; Forces, an Excerpt; Frames of Reference; Frames of Reference, Excerpts I* and *II; Franck-Hertz Experiment; Free Fall and Projectile Motion; Gases, Behavior of; Inertia; Inertial Mass; Interference of Photons; Magnet Laboratory; Millikan Experiment; Million to One; Optics, Introduction to; Periodic Motion; Photoelectric Effect; Photoelectric Effect, an Excerpt; Pressure of Light; Random Events; Rutherford Atom; Sound Waves in Air; Speed of Light; Vector Kinematics.*

Purdue Series. 22 titles; 2 listed here, and separately. B & W; a, D. J. Tendam; p, d, 20. See *Cavendish Balance, a Measurement of the Universal Gravitational Constant; Galilean Relativity Demonstrated by Projectile Motion.*

D Purcell, Edward M. See *Inertia; Inertial Mass.*

QM *Quantum Mechanical Harmonic Oscillator.* B & W, silent, 4 min; a, A. Bork; p, ESI; d, 8. Computer-generated. Solutions of time-independent Schroedinger equation displayed for uniformly increasing energy values. (Also available as 8-mm film loop, see Sec. 21–6.2.)

QM *Quantum Mechanical Wave Packets from Potential Well and Barrier, Scattering of.* C, 10 min; a, A. Goldberg, H. M. Schey, J. L. Schwartz; p, Lawrence Rad. Lab; d, 25. Six events. Computer-generated.

QM QUANTUM MECHANICS. See *Fabric of the Atom; Interference of Photons; Liquid Helium II, the Superfluid; Messenger Lectures; Photons; Quantum Mechanical Harmonic Oscillator; Quan-*

tum Mechanical Wave Packets from Potential Well and Barrier, Scattering of; Superconductivity, An Introduction to.

NP *Random Events* (PSSC). B & W, 31 min; a, P. Hume and D. Ivey; p, ESI; d, 16. Statistics of random events; relation to radioactive decay.

W *Reflection and Refraction.* See *Ripple Tank Wave Phenomena Series.*

R RELATIVITY. See *Einstein, The Large World of Albert; Frames of Reference; Frames of Reference, Excerpts I* and *II. Galilean Relativity Demonstrated by Projectile Motion; Positron–Electron Annihilation; Time Dilation, an Experiment with μ-Mesons; Ultimate Speed, an Exploration with High-Energy Electrons.*

FM *Rheological Behavior of Fluids.* B & W, 22 min; a, H. Markovitz; p, NCFMF–ESI; d, 9.

W,MP, SSP *Ripple Tank Wave Phenomena Series.* 5 titles. B & W; a, J. Strickland; p, ESI; d, 16, 26. See *Barrier Penetration* (8 min); *Bragg Reflection* (10 min); *Doppler Effect and Shock Waves* (8 min); *Interference and Diffraction* (19 min); *Reflection and Refraction* (17 min). (Also available as a series of 8-mm loop films, see Sec. 26–6.2.) (Note also ripple tank experiments on Bragg reflections in *Crystals and Their Structure.*)

MP,NP *Rutherford Atom* (PSSC). B & W, 40 min; a, R. I. Hulsizer; p, ESI; d, 16. Alpha particle scattering and model experiments.

M *Satellite Orbits.* C, 17 min; a, H. F. Meiners; p, d, 21. Six orbital elements. Kepler's three laws. Regression of an orbit. Scale orbits of Vanguard I, Sputnik III, Explorer IV, and Discoverer V, moving simultaneously around earth.

FM,O, W *Schlieren.* C, 20 min; a, National Physical Laboratory, England; p, Shell Film Unit; d, 23. Technique explained, used on waves, flows, air foils, nozzles, flames, convection.

D Seitz, Frederick. See *Crystals, Imperfections in.*

D Shockley, William. See *Minority Carriers in Semiconductors.*

W *Similarities in Wave Behavior.* B & W, 26.5 min; a, J. N. Shive; p, BTL; d, 2. Uses Shive-Bell wave demonstrator and emphasizes common properties of waves in all fields. See also *Simple Waves.*

M *Simple Harmonic Motion.* See *AAPT Physics Series.* Exposition mainly by animation. Clear. See also *Periodic Motion.*

W *Simple Waves* (PSSC). B & W, 27 min; a, J. N. Shive; p, ESI; d, 16. See also *Similarities in Wave Behavior.*

Lab *Solder Glass Technique.* This has been replaced by *Laboratory High-Vacuum Technique.*

SSP SOLID STATE PHYSICS. See *Bragg Reflection; Brattain on Semiconductor Physics; Crystallization, Experiments with Models on; Crystals; Crystals, Imperfections in* (F. Seitz); *Crystals, an Introduction; Crystals and Their Structures; Domains and Hysteresis in Ferromagnetic Materials; Electron Emission Series; Experiment in Physics; Ferromagnetic Domains, Parts I* and *II; Gap Energy and Recombination Light in Germanium; Low-Energy Electron Diffraction from Germanium (100) and (111) Surfaces; Minority Carriers in Semiconductors* (R. Haynes and W. Shockley); *Superconductivity, An Introduction to; Surface Diffusion with the Aid of Optical Microscopy (Potassium on Quartz), Investigations into; Symmetry; Transistor Structure and Technology; Wave Number Space, Parts I* and *II.*

W *Sound Waves in Air* (PSSC). B & W, 35 min; a, R. H. Bolt; p, ESI; d, 16.

EM,O *Speed of Light* (PSSC). B & W, 21 min; a, W. Siebert; p, ESI; d, 16. Measured in air using spark, fixed mirror, photocell, and oscilloscope. Compared, air to water, with rotating mirror.

W *Stationary Longitudinal Waves.* See *AAPT Physics Series.* Clear exposition, using animation. Pulse, its reflection; then, standing wave. See also *Progressive Waves, Transverse and Longitudinal; Stationary Transverse Waves.*

W *Stationary Transverse Waves.* See *AAPT Physics Series.* Clear exposition, using animation. Pulse, its reflection; then, standing wave. See also *Progressive Waves, Transverse and Longitudinal; Stationary Longitudinal Waves.*

D *Stellar Evolution.* B & W, 60 min. A lecture by George Gamow; d, 6.

AP,MP *Stern–Gerlach Experiment.* B & W, 26 min; a, J. R. Zacharias; p, ESI; d, 16, 26. Beam of Cs atoms split in two on passing through a non-uniform magnetic field.

D,EP *Strangeness Minus Three.* B & W, 45 min; a, R. Feynman, M. Gell-Mann, Y. Ne'eman, N. Samios; p, BBC-TV; d, 28. Elementary particle symmetries and the omega-minus particle as of 1966.

LTP, SSP *Superconductivity, An Introduction to.* B & W, 48 min (one reel); a, A. Leitner; p, d, 15. Part I, Resistance: attempts to measure fail; transition temperature. Part II, Magnetic Effects: on transition temperature; Meissner effect; shielding currents; persistent current; floating magnet. Part III, Energy Gap: electron tunneling junction experiment.

SSP *Surface Diffusion with the Aid of Optical Microscopy (Potassium on Quartz), Investigations into.* B & W, silent, 8.5 min; a, H. Wegener; p, d, 10.

FM *Surface Tension in Fluid Mechanics.* C, 29 min; a, L. M. Trefethen; p, NCFMF–ESI; d, 9. Liquid surfaces, interfaces, films. Formation and dynamics of bubbles.

GP,SSP *Symmetry.* C, 10 min; a, J. Bregman, R. Davisson, G. Forell, A. Holden, P. Stapp; p, Sturgis Grant Productions, Inc.; d, 5. All problems of two-dimensional symmetry presented in beautiful artistic animation. Booklet accompanying film is essential.

D,EM, EP *Synchrotron.* C, 12 min; a, W. Shurcliff; p, HPP; d, 32. Simple tour of the 2 GeV Cambridge (Mass.) Electron Accelerator.

D Taylor, Sir Geoffrey I. See *Flows, Low Reynolds Number.*

D Teller, Edward. See *Bohr, The Small World of Niels; Einstein, The Large World of Albert.*

THERMODYNAMICS. See HEAT AND . . .

MP *Thermionic Emission of Electrons.* See *Electron Emission Series.*

EP,R *Time Dilation, An Experiment with μ-Mesons.* B & W, 36 min; a, D. H. Frisch, and J. H. Smith; p, ESI; d, 16, 26. Cosmic ray muons moving between 0.995c and 0.997c are stopped in scintillator on top of a mountain and at sea level. Decay time statistics obtained and tantamount to a clock. Time dilation of 9. See *Am. J. Phys.*, **31**, 342 (1963). Outstanding.

SSP *Transistor Structure and Technology.* C, 38 min; p, SEEC–ESI; d, 8.

NP *U-238 Radioactive Series.* Traces stage of decay, U-238 to lead. See *AAPT Physics Series.*

EM,EP, R *Ultimate Speed, an Exploration with High-Energy Electrons.* B & W, 38 min; a, W. Bertozzi; p, ESI; d, 16, 26. Linac used for 0.5 to 15 MeV electrons. Speed, measured by time of flight, has limit *c*. Energy determined with a charge counter and calorimetry. See *Am. J. Phys.*, **32**, 551 (1964). Outstanding.

M,V *Vector Kinematics* (PSSC). B & W, 16 min; F. L. Friedman; p, ESI; d, 16. Velocity and acceleration traced in 2-dimensional motions. Some footage computer-generated.

V VECTORS AND VECTOR ANALYSIS. See *Circular Motion, Uniform; Computer Studies of Fluid Dynamics; Deflecting Forces; Deformation of Continuous Media; Electric Lines of Force; Flow Visualization; Force, Mass and Motion; Harmonic Phasors; Pressure Fields and Fluid Acceleration; Vector Kinematics; Vorticity.*

AP *Velocity Distribution of Atoms in a Beam.* B & W, 16 min; a, J. King; p, ESI; d, 16, 26. Beam of K atoms chopped by a rotating drum. Distribution of flight times measured.

FM,V *Vorticity.* B & W, 44 min (one reel); a, A. H. Shapiro; p, ESI; d, 9. One of the most instructive aspects of this film is the curl (or vorticity) meter used, which shows the curl of a velocity field. It also shows the irrotational vortex of efflux of water from a tub due to earth rotation. Film presents basic theorems on circulation with many demonstrations of flows. Outstanding.

SSP *Wave Number Space, Part I* and *Part II.* 2 films; C, silent, 5 min each; a, G. L. Salinger and H. B. Huntington; p, d, 21. *Part I. Construction of the Wave Vector,* traveling waves pass over 2-dimensional lattice, vector is defined, constructed.

Part II. Construction of Brillouin Zones, extends *Part I* to shorter wavelength; periodicity; zone construction.

W — WAVES. See *Barrier Penetration; Bragg Reflection; Crystals and Their Structures; Doppler Effect; Doppler Effect and Shock Waves; Electromagnetic Waves; Group and Phase Velocity; Interference and Diffraction; Magnetohydrodynamics; Quantum Mechanical Wave Packets from Potential Well and Barrier, Scattering of; Reflection and Refraction; Ripple Tank Wave Phenomena Series; Schlieren; Similarities in Wave Behavior; Wave Number Space, I* and *II; Waves in Fluids.*

FM,W — *Waves in Fluids.* B & W, 33 min; a, A. E. Bryson; p, NCFMF–ESI; d, 9. Gravity waves in water demonstrated and thoroughly described, including shocks, breakers, and turbulence.

21–6.2 Eight-Millimeter Film Loops

Some of the films listed in this section are also available on reels, rather than as cartridged loops. Where this fact is known, it is recorded according to a key of abbreviations.

One set of film loops (with the title Wave Motion) listed here are available as 16-mm loops without cartridge.

AP,MP,O — *Absorption Spectra.* Na vapor, hemoglobulin, and didymium glass. See *Miller Series.*

Lab,M — *Acceleration Due to Gravity* (2 loops), *Method I* and *Method II. Method I:* slow-motion sequence permits student measurement of average speed of bowling ball at two different parts of a free fall. High-speed camera frame rate given; student times projector's frame delivery rate and converts to real time. *Method II:* slow-motion sequence allows determination of average speed of freely falling bowling ball at several points of the motion. Distances given; real time conversion as in *Method I.* See *HPP Series.*

Acceleration Vector. See *Vector Kinematics Series.*

D,FM — *Airfoil, Generation of Circulation and Lift from an.* B & W; a, L. Prandtl. Vortex creation at start, counter-vortex creation at stop of motion, conserving circulation (angular momentum). Classical research footage; approx. 1930. See *Fluid Mechanics Series.*

FM — *Airfoil, Velocities Near an.* B & W; a, D. Hazen and S. J. Kline. Flow patterns (by smoke in wind tunnel, by hydrogen-bubble technique in water) at zero and positive lift, and at stall. See *Fluid Mechanics Series.*

M,Q — AIR TABLE. See *Kinetic Theory on an Air Table Series; Mechanics on an Air Table Series.*

Lab,M — *Air Track Works, How an.* Instruction on operation. See *Mechanics on an Air Track Series, Newtonian.*

AMATEUR LOOPS: May be many more than are listed here. J. M. Fowler, Univ. of Maryland: *Lenz's Law.* R. E. Garrett, Univ. of Florida: *Heat Transfer—Dewar Flasks.* T. Norton, Schenectady, N.Y., 3 loops: *Chain Reaction; Regelation; Current in an Electrolyte in a Magnetic Field.* G. Rees and G. Griswold, Washtenaw Community College, Ypsilanti, Michigan: *Series Circuits.* G. Schwarz, Florida State Univ.: *Reference Circles.* T. Spickman, Ohio College of Applied Science.: *Does Impulse Equal Momentum Change?* F. W. Zurheide, Southern Illinois Univ., and J. T. Brooks, Washington Univ., St. Louis: *Wave Motion.*

AP,MP,NP — Aston, F. W. See *Mass Spectrograph, Aston's.*

AP — ATOMIC PHYSICS. See *Absorption Spectra; Mass Spectrograph, Aston's; Rutherford Scattering; Thomson Model of the Atom; Thomson's Positive Ray Parabola.*

W — BALLISTIC PENDULUM. See *Finding the Speed of a Rifle Bullet* (2 loops).

W — *Barrier Penetration by Waves.* Demonstration, using water waves, of matching problem at boundary between media with different phase velocities. See *Ripple Tank Wave Phenomena Series.*

FM,V — *Bathtub Vortex, The.* B & W; a, A. H. Shapiro. Efflux of water initially at rest from a small central hole in bottom of large cistern situated on northern hemisphere of earth. Curl(vorticity)-meter floats in water. Counterclockwise irrotational vortex forms. Fascinating; instructive. See *Fluid Mechanics Series.* (See also: *Sink Vortex.*)

FM *Boundary Layer Formation.* B & W; a, A. H. Shapiro. See *Fluid Mechanics Series.*

FM,W *Bow Waves in Hypersonic Flow.* C. Schlierens of Mach 7 and 10 flows. See *Fluid Mechanics Series.* (See also *Shock Waves, Formation of.*)

D,SSP Bragg, Sir Lawrence. See *Bubble Model of a Crystal.*

SSP,W *Bragg Reflection of Waves.* See *Ripple Tank Wave Phenomena Series.*

KT,SM BROWNIAN MOTION. See *Random Walk and Brownian Motion.*

D,SSP *Bubble Model of a Crystal* (2 loops), (1) *Structure and Boundaries,* (2) *Deformation and Dislocations.* a, Sir L. Bragg; p, Royal Institution, London; d, 7 (reg, sup, sup–reel). Analogy experiments on two-dimensional crystal; geometrical and dynamical properties.

FM *Bubbles, Formation of.* C; a, E. M. Trefethen. Nucleation at vapor pockets in cracks, etc. See *Fluid Mechanics Series.*

EM *Capacitors and Dielectrics.* Two metal plates, electroscope. Effects of area, separation, and dielectric. See *Electrostatics Series.*

M *Cavendish Experiment–A Measurement of "G."* Time-lapse photography of the oscillations. Uses the Leybold apparatus. See *Miller Series.*

M *Center of Mass.* Motion of variously shaped bodies. The center of mass is shown to obey particle dynamics. See *Mechanics on an Air Table Series.*

M *Center-of-Mass Pendulum.* Physical pendulum hangs from moving glider on a linear air track. Oscillation depends on glider to pendulum mass ratio; c.m. moves uniformly. See *Mechanics on an Air Track Series, Newtonian.*

Lab,M *Central Forces–Iterated Blows.* Computer-generated. Particle in uniform motion given blows at equal time intervals. The blows are central ("in" or "out"). Tracing projected image, student can verify Kepler's law of areas. See *HPP Series.*

M CENTRAL FORCES. See *Central Forces–Iterated Blows; Jupiter Satellite Orbits; Kepler's Laws; Mechanics Series* (Sinden); *Orbits, Unusual; Program Orbit* (2 loops).

MP,NP *Chain Reaction.* a, T. Norton. See AMATEUR LOOPS.

EM *Charge Distribution* (2 loops), (1) *Concentration and Point Discharge,* (2) *Faraday Ice Pail Experiment.* See *Electrostatics Series.*

M COLLISIONS. See *Collisions, One-Dimensional; Collisions, Two-Dimensional; Collisions in Two Dimensions; Collisions with an Unknown Object; Conservation of Linear and Angular Momentum; Conservation of Momentum* (2 loops); *Elastic Collisions, Inelastic Collisions; Finding the Speed of a Rifle Bullet* (2 loops); *Relative Motion, A Matter of; Rutherford Scattering; Scattering of a Cluster of Objects.*

Lab,M *Collisions, One Dimensional* (3 loops), *Part I, Part II,* and *Inelastic.* Each loop contains 2 "events" in slow motion, permitting determination of velocities in relative units, to verify momentum conservation and to discuss energy-elasticity questions quantitatively. *Part I* shows setup; balls, hanging from 35-ft piano wires, released simultaneously. Stroboscopic still photographs were also made. See *HPP Series.* (See also *Collisions with an Unknown Object.*)

Lab,M, V *Collisions, Two-Dimensional* (3 loops), *Part I, Part II,* and *Inelastic.* (Description as for *Collisions, One-Dimensional.*) See *HPP Series.* (See also *Explosion of a Cluster of Objects; Scattering of a Cluster of Objects.*)

M,Q *Collisions in Two Dimensions.* Momentum and energy in two-puck elastic and inelastic collisions. Collisions with identical pucks demonstrate scattering angle. Angular momentum about center of mass produces rotation effects. See *Mechanics on an Air Table Series.*

Lab,M, NP *Collisions with an Unknown Object.* Chadwick's problem with neutron, by analogy. Steel balls hung from 35-ft piano wires in one-dimensional collisions. "Unknown" ball (neutron) strikes like ball (proton) and heavy ball (N nucleus) of given masses. Student can measure latters' speeds, infer "neutron" mass and speed. See *HPP Series.*

COMPUTER-GENERATED FILM LOOPS. See *Diffusion; Program Orbit* (2 loops); *Central Forces—Iterated Blows; Image Methods in Electrostatics; Irreversibility and Fluctuations* (2 loops); *Kepler's Laws; Mechanics Series* (Sinden) (5 loops); *Orbits, Unusual; Quantum Mechanical Harmonic Oscillator; Quantum Physics Series* (6 loops); *Rutherford Sattering; Vector Kinematics Series.*

M,Q *Conservation of Energy—Potential to Kinetic.* (a) constant force device, (b) inclined air track, and (c) linear spring, acting on a glider, provide data for student problems. See *Mechanics on an Air Track Series, Newtonian.*

M,Q *Conservation of Linear and Angular Momentum.* Two identical ice pucks connected by rod. A ball launched by a spring whose momentum can be deduced by student next strikes the system. Data allow checking the momentum conservations, with reference to the center of mass. See *Meiners Series.*

M,Q *Conservation of Momentum* (2 loops), (1) *Elastic Collisions,* (2) *Inelastic Collisions.* Each loop shows collisions between gliders and presents data of experiments performed. See *Mechanics on an Air Track Series, Newtonian.*

M,Q *Constant Velocity and Uniform Acceleration.* Gliders on air track at 3 different inclinations, and level. Motion timed by photo-cells. See *Mechanics on an Air Track Series, Newtonian.*

M *Coupled Oscillators* (ESI) (3 loops), (1) *Energy Transfer,* (2) *Normal Modes,* (3) *Other Oscillators.* C; a, A. Holden; p, ESI; d, 4 (reg, sup), 7 (sup, sup–reel), 13 (reg, reg–reel, reel–16), 14 (reg, in series sets only), 16 (reg, reg–reel, reel–16), 22 (reg, reg–reel), 26 (reg), 30 (reg, sup), 31 (reg, sup). Two coupled pendulums. Energy transfer shown by trace of sand pendulums. (See also *Coupled Oscillators*—Miller; *Sand Pendulum Series; Wilberforce Pendulum.*)

M *Coupled Oscillators* (Miller) (2 loops), (1) *Equal Masses,* (2) *Unequal Masses.* Three identical springs on a line to hold carts of (un)equal masses. Normal modes, combinatory motions, and other effects. See *Miller Series.* (See also *Coupled Oscillators* (ESI); *Sand Pendulum Series; Wilberforce Pendulum.*)

HT *Critical Temperature.* Shows entire transition for ether, in slow motion. See *Miller Series.*

D,SSP CRYSTALS. See *Bubble Model of a Crystal* (2 loops, by Sir L. Bragg).

FM,V CURL-METER. A small waterwheel floating with the flow, demonstrates the curl of a velocity field. See *Bathtub Vortex; Vorticity Meter, Visualization of Vorticity with.*

EM *Current in Electrolyte in a Magnetic Field.* a, T. Norton. See *Amateur Loops.*

O,SSP, W DIFFRACTION. See also *Bragg Reflection of Waves.*

O,W *Diffraction, Double Slit.* Optical pattern; in color; either of (a) slit separation and (b) wavelength varied continuously as other is held constant. See *Miller Series.*

O,W *Diffraction, Single Slit.* Optical pattern in color. Either of (a) slit width and (b) wavelength varied continuously as the other is constant. See *Miller Series.*

O,W *Diffraction and Scattering Around Obstacles.* See *Ripple Tank Wave Phenomena Series.*

O,W *Diffraction at an Aperture.* C; p, in England; d, 9 (reg, sup). "Plane" waves, by ripple tank, through apertures of different widths. (See also *Ripple Tank Wave Phenomena Series.*)

O,W *Diffraction at Obstacles of Plane and Spherical Waves.* C; p, in England; d, 9 (reg, sup). By ripple tank. (See also *Ripple Tank Wave Phenomena Series.*)

O,W *Diffraction of Waves* (2 loops), (1) *Multiple Slit,* (2) *Single Slit.* See *Ripple Tank Wave Phenomena Series.*

KT,SM *Diffusion.* A population of pucks on an air table, in "temperature" equilibrium, invaded by other pucks. Effects of mass on times of diffusion studied and histograms plotted of the data. See *Kinetic Theory on an Air Table Series.*

KT,SM *Diffusion* (2 loops), *Part I* and *Part II.* B & W; p, BTL; d, 2. Computer-generated. Changes in distribution curve plotted simultaneously with transport processes shown.

D — DOCUMENTARY FOOTAGE IN FILM LOOPS. See Bragg, Sir L.; Prandtl, L; *Tacoma Narrows Bridge Collapse; Wave Fronts, Non-recurrent.*

W — *Doppler Effect.* See *Ripple Tank Wave Phenomena Series.*

FM,GP — *Drops, Breakup of Liquid into.* C; a, E. M. Trefethen. Excellent slow-motion photography. See *Fluid Mechanics Series.*

M,Q — *Dynamics of Pendulums.* The bob is a puck on a tilted air table. Effect of tilt angle is studied quantitatively. See *Mechanics on an Air Table Series.*

M,Q — *Dynamics of Springs.* Puck on an air table between two springs. Mass, force constant varied. Data given. Accelerometer on puck. See *Mechanics on an Air Table Series.*

EM — ELECTRICITY AND MAGNETISM. See *Current in Electrolyte in a Magnetic Field; Electrostatics Series* (10 loops). (See also *Image Methods in Electrostatics; Lenz's Law;* (MAGNETOHYDRODYNAMICS:) *Current Induced Instabilities of a Mercury Jet; Paramagnetism of Liquid Oxygen; Series Circuits; Standing Electromagnetic Waves.*)

EM — *Electroscope, The.* See *Electrostatics Series.*

EM — *Electrostatic Induction.* Charging by contact, by induction. See *Electrostatics Series.*

EM — *Electrostatics, Introduction to.* Charging by rubbing; electrostatic force. See *Electrostatics Series.*

EM — *Electrostatics, Problems in.* Electrophorus, Leyden jar, Van de Graaff generator. See *Electrostatics Series.*

EM — *Electrostatics Series* (10 loops, all separately listed). C; a, A. E. Walters; p, d, 7 (sup, sup–reel). *Capacitors and Dielectrics; Charge Distribution* (2 loops), (1) *Concentration and Point Discharge,* (2) *Faraday Ice Pail Experiment; Electroscope; Electrostatic Induction; Electrostatics, Introduction to; Electrostatics, Problems in; Insulators and Conductors; Photoelectric Effect; Van de Graaff Generator.*

SM,Q — *Equipartition of Energy.* By trailing pucks on an air table, the film compares velocity distributions and rms speed of each species in a population of different pucks of masses M and 2M. Repeated for different species. See *Kinetic Theory on an Air Table Series.*

Lab,M — *Explosion of a Cluster of Objects.* Five steel balls of given masses, hung from 35-ft piano wires, in a cluster at rest around a powder charge. Student by measurement can calculate momentum and energy of system. See *HPP Series.*

SSP — *Ferromagnetic Domain Wall Motion.* See *Miller Series.*

Lab,M — *Finding the Speed of a Rifle Bullet* (2 loops), *Method I* and *Method II.* Wood block of given mass and dimension, hangs from 35-ft wires; bullet mass given. Student measures: (in I) speed of wood block in slow-motion sequence, with conversion procedure to real time indicated; in II, the vertical rise of block's excursion. See *HPP Series.*

D,FM — *Flow Separation and Vortex Shedding.* B & W; a, L. Prandtl. Filmed in 1930 (approx). See *Fluid Mechanics Series.*

FM — *Fluid Mechanics Series* (108 loops, as of May 1968, 25 titles listed here and separately). B & W or C; a, various; p, NCFMF–ESI; d, 9 (reg, sup). *Airfoil, Generation of Circulation and Lift for an* (B & W); *Airfoil, Velocity Near an* (B & W); *Bathtub Vortex* (B & W); *Boundary Layer Formation* (B & W); *Bow Waves in Hypersonic Flow* (C); *Bubbles, Formation of* (C); *Drops, Breakup of Liquids into* (C); *Flow Separation and Vortex Shedding* (B & W); (MAGNETOHYDRODYNAMICS:) *Current Induced Instabilities of a Mercury Jet* (C); *Magnus Effect* (B & W); *Potential Flows, Hele–Shaw Analog* (C) (2 loops), (1) *Sources and Sinks,* (2) *Sources and Sinks in Uniform Flow; Shear Deformation of Viscous Fluids* (B & W); *Sink Vortex* (B & W); *Surface Tension* (C) (3 loops), (1) *and Contact Angles,* (2) *and Curved Surfaces,* (3) *Examples of; Surfaces, Motions Caused Along Liquid* (C) (3 loops), (1) *by Composition Gradients,* (2) *by Electrical and Chemical Effects,* (3) *by Temperature Gradients; Torna-*

does in Nature and in the Laboratory (B & W); *Turbulence, The Occurrence of* (B & W); *Venturi Passage* (B & W); *Vorticity Meter, Visualization of Vorticity with* (2 loops) (B & W).

M *Frames, Rotating Reference.* Cart in uniform motion sprays dots on fixed and rotating backgrounds. Effects of varying cart speed and background rate of rotation are shown. See *Meiners Series.*

R GALILEAN. See under RELATIVITY.

KT,Q *Gas, Properties of.* Two-dimensional gas model of pucks on an air table. Effect on pressure of temperature, density, and volume. Pressure is measured by counting pucks hitting wall. See *Kinetic Theory on Air Table Series.*

GP GENERAL PHYSICS. See *Surface Tension* (3 loops); *Surfaces, Motions Caused Along Liquid* (3 loops); *Tornadoes in Nature and in the Laboratory.*

KT *Gravitational Distribution.* Study of a two-dimensional gas of pucks on a tilted air table. Histograms of density vs elevation (barometer formula). Effect on mixtures studied. See *Kinetic Theory on an Air Table Series.*

HT *Heat Transfer—Dewar Flasks.* a, R. E. Garrett. See *Amateur Loops.*

HT HEAT AND THERMODYNAMICS. See *Critical Temperature; Heat Transfer—Dewar Flask; Phase Change; Regelation; Temperature Waves.*

HPP (Harvard Project Physics) *Series* (48 loops as of fall 1968, 31 listed here). C; a, various; p, NFB–HPP; d, 7 (sup, sup–reel), 32 (sup, sup–reel). See *Acceleration Due to Gravity* (2 loops); *Central Forces—Iterated Blows; Collisions, One-Dimensional* (3 loops); *Collisions, Two-Dimensional* (3 loops); *Collisions with an Unknown Object; Explosion of a Cluster of Objects; Finding the Speed of a Rifle Bullet* (2 loops); *Jupiter Satellite Orbits; Kepler's Laws; Orbits, Unusual; Program Orbit* (2 loops); *Relative Motion, A Matter of; Relativity, Galilean* (3 loops); *Rutherford Scattering; Scattering of a Cluster of Objects; Standing Electromagnetic Waves; Superposition; Thomson Model of the Atom; Vibrations* (4 loops), *of a Drum, of a Metal Plate, of a Rubber Hose, of a Wire.*

EM *Image Methods in Electrostatics.* B & W; a, R. Blum; p, d, 33 (reg, reel–16). Computer-generated. Point charge and conducting sphere.

M *Impulse Equal Momentum Change?, Does.* a, T. Spickman. See *Amateur Loops.*

M *Inertial Forces* (2 loops), (1) *Translational Acceleration,* elevator problems, (2) *Centripetal Acceleration,* amusement park "roto-ride." See *Miller Series.*

EM *Insulators and Conductors.* The classical experiments with charged rods, electroscopes, proof-planes on strips of paper, foil, etc. See *Electrostatics Series.*

O,W *Interference.* C; p, in England; d, 9 (reg, sup). Ripple tank interference field of two sources. (See also *Ripple Tank Wave Phenomena Series.*)

O,W *Interference of Waves.* See *Ripple Tank Wave Phenomena Series.*

SM *Irreversibility and Fluctuations* (2 loops), (1) *Approach to Equilibrium and Irreversible Behavior,* (2) *Ordinary and Rare Fluctuations.* B & W; a, F. Reif, B. J. Alder; d, 14 (reg, reel–16). Computer-generated. Two-dimensional model gas in a box.

M,Q *Jupiter Satellite Orbits.* Actual time-lapse photography of the motion of Io about Jupiter, at 1-min intervals. Student can calculate the mass of Jupiter. See *HPP Series.*

Lab,M *Kepler's Laws.* Computer-generated. Two planetary orbits; positions displayed at equal time intervals. All 3 laws can be verified. See *HPP Series.*

KT,SM *Kinetic Theory on an Air Table Series* (6 loops). C; a, H. Daw; p, d, 7 (sup, sup–reel). Two-dimensional model gas. See *Diffusion; Equipartition of Energy; Gas, Properties of; Gravitational Distribution; Maxwellian Speed Distribution; Random Walk and Brownian Motion.*

Lab LABORATORY FILM LOOPS. These contain either instruction on the use of instruments, or scenes inviting data-taking off the projected image as a laboratory exercise. See *Acceleration Due to Gravity* (2 loops); *Air Track Works, How an; Collisions, One-Dimensional* (3 loops); *Collisions, Two-Dimensional* (3 loops);

Collisions with an Unknown Object; Explosion of a Cluster of Objects; Finding the Speed of a Rifle Bullet (2 loops); *Kepler's Laws; Program Orbit* (2 loops); *Scattering of a Cluster of Objects; Solder Glass Vacuum Technique Series* (7 loops).

EM *Lenz's Law.* a, J. M. Fowler. See *Amateur Loops.*

M LINEAR AIR TRACK (J. Stull). See *Mechanics on an Air Track Series, Newtonian.*

EM,FM (MAGNETOHYDRODYNAMICS:) *Current Induced Instabilities of a Mercury Jet.* C; a, A. Dattner. Vertical jet carries electric current; axial magnetic field applied. See *Fluid Mechanics Series.*

D,FM *Magnus Effect, The.* B & W; a, L. Prandtl (1930 approx). The phenomena of starting vortices and of stopping vortices. See *Fluid Mechanics Series.*

AP,MP, NP *Mass Spectrograph, Aston's.* C; p, in England; d, 9 (reg, sup). Mostly animation to explain operation. Shows photographs of Aston's original, and other spectrographs. (See also *Thomson's Positive Ray Parabolas.*)

KT,SM *Maxwellian Speed Distribution.* Two dimensional model gas of pucks on air table, studied by "white trail" animation, to plot histograms of speeds vs number of particles. See *Kinetic Theory on an Air Table Series.*

M MECHANICS. See *Acceleration Due to Gravity* (2 loops); *Cavendish Experiment—A Measurement of "G"; Central Forces—Iterated Blows; Collisions, One-Dimensional* (3 loops); *Collisions, Two-Dimensional* (3 loops); *Collisions with an Unknown Object; Coupled Oscillators* (ESI) (3 loops); *Coupled Oscillators* (Miller) (2 loops); *Explosion of a Cluster of Objects; Finding the Speed of a Rifle Bullet* (2 loops); *Impulse Equal Momentum Change?, Does; Inertial Forces* (2 loops); *Jupiter Satellite Orbits; Kepler's Laws; Mechanics on an Air Table Series* (5 loops); *Mechanics on an Air Track Series, Newtonian* (9 loops); *Mechanics Series* (Sinden) (5 loops); *Meiners Series* (3 loops); NORMAL MODES; *Oscillations, Forced Damped Harmonic; Orbits, Unusual;* PENDULUMS; *Pendulum, Sand* (4 loops); *Pendulums, Ring of Coupled* (2 loops); *Program Orbit* (2 loops); *Relative Motion, A Matter of; Relativity, Galilean* (3 loops); *Sand Pendulum Series* (7 loops); *Scattering of a Cluster of Objects; Superposition; Tacoma Narrows Bridge Collapse; Vector Kinematics Series* (6 loops); *Wilberforce Pendulum.*

M *Mechanics on an Air Table Series* (11 loops, 5 listed here). C; a, D. Kutliroff; p, d, 7 (sup, sup-reel). See *Center of Mass; Collisions in Two Dimensions; Dynamics of Pendulums; Dynamics of Springs; Simple Harmonic Motion.*

M *Mechanics on an Air Track Series, Newtonian* (9 loops). C; a, J. L. Stull; p, d, 7 (sup, sup-reel). *Air Track Works, How an; Center-of-Mass Pendulum; Conservation of Energy—Potential to Kinetic; Conservation of Momentum* (2 loops); *Constant Velocity and Uniform Acceleration; Newton's First and Second Laws; Newton's Third Law; Simple Harmonic Motion—The Stringless Pendulum.*

M *Mechanics Series* (Sinden) (5 loops); B & W; a, F. Sinden; p, EDC; d, release fall 1968, contact 8. Computer-generated. *Newton's Laws of Motion; Orbiting Bodies* (4 loops); (1) *Fixed System of;* (2) *in Various Force Fields, Part I—Negative Power Laws;* (3) *in Various Force Fields, Part II—Positive Power Laws;* (4) *Moving System of.* (See also 16-mm *Force, Mass and Motion* film listed in Sec. 21–6.1.

M *Meiners Series* (3 loops). C; a, H. F. Meiners; p, RPI; d, 7 (sup, sup-reel). *Conservation of Linear and Angular Momentum; Frames, Rotating Reference; Motion, One-Dimensional.*

O *Michelson Interferometer.* See *Miller Series.*

Miller Series (19 loops). C (except one loop); a, F. Miller, Jr.; p, OSU; d, 3 (reel-16), 4 (reg, reg-reel, reel-16), 7 (sup, sup-reel), 13 (reg, reg-reel, reel-16), 19 (reel-16), 22 (reel-16), 27 (reel-16). See *Absorption Spectra; Cavendish Experiment—A Measurement of "G"; Coupled Oscillators* (2 loops), (1) *Equal Masses,* (2) *Unequal Masses; Critical Temperature; Diffraction* (2 loops) (1) *Double Slit,* (2) *Single Slit; Ferromagnetic Domain Wall Motion; Inertial Forces* (2 loops), (1) *Centripetal Acceleration,* (2) *Translational Acceleration; Michelson Interferometer; Paramagnetism of Liquid Oxygen; Radioactive Decay; Resolving Power; Scintillation Spectrometry; Tacoma Narrows Bridge Collapse; Temperature Waves; Wavefronts, Non-recurrent* (B & W); *Wilberforce Pendulum.*

MP MODERN PHYSICS. See *Absorption Spectra; Chain Reaction; Mass Spectrograph, Aston's; Paramagnetism of Liquid Oxygen; Photoelectric Effect* (2 loops, 1 in *Electrostatics Series,* 1 in *UNESCO Physics Series*); *Radioactive Decay; Rutherford–Royds' Identification of α-particles; Rutherford Scattering; Scintillation Spectrometry; Thomson Model of the Atom; Thomson's Positive Ray Parabolas.*

M *Motion, One-Dimensional.* Cart on track in different motions; x–y recorder plots position, velocity and acceleration simultaneously. See *Meiners Series.*

M,Q *Newton's First and Second Laws.* 10-meter air track; free glider; glider under constant force. Data shown. See *Mechanics on an Air Track Series, Newtonian.*

M *Newton's Laws of Motion.* Computer-generated. See *Mechanics Series* (Sinden).

M,Q *Newton's Third Law.* Two gliders, at rest, forced apart by spring; varying mass, i.e., acceleration ratios. Oscillating spring-joined glider pairs. Data shown. See *Mechanics on an Air Track Series.*

M NORMAL MODES. See *Coupled Oscillators* (ESI); *Coupled Oscillators* (Miller); *Pendulums, Ring of Coupled* (2 loops), (1) *Mixture of Normal Modes,* (2) *Normal Modes.*

NP NUCLEAR PHYSICS. See *Chain Reaction; Collisions with an Unknown Object; Mass Spectrograph, Aston's; Radioactive Decay; Rutherford-Royds' Identification of α-particles; Rutherford Scattering; Scintillation Spectrometry.*

O OPTICS. See *Absorption Spectra; Diffraction at an Aperture; Diffraction at Obstacles of Plane and Spherical Waves. Diffraction* (2 loops), (1) *Double Slit,* (2) *Single Slit; Interference; Michelson Interferometer; Pinhole Camera; Reflected Light—Glass in Liquids; Resolving Power; Ripple Tank Wave Phenomena Series.*

M *Orbiting Bodies* (4 loops); (1) *Fixed System of;* (2) *in Various Force Fields, Part I—Negative Power Laws;* (3) *in Various Force Fields, Part II—Positive Power Laws;* (4) *Moving System of.* Computer-generated. See *Mechanics Series* (Sinden).

M *Orbits, Unusual.* Computer-generated. Nearly inverse-square field, precessing or spiraling orbits. See *HPP Series.*

Oscillations, Forced Damped Harmonic. See *Sand Pendulum Series.*

EM,MP *Paramagnetism of Liquid Oxygen.* Strong attraction into intense part of inhomogeneous magnetic field. See *Miller Series.*

QM *Particle in a Box.* Computed-generated. See *Quantum Physics Series.*

M PENDULUMS. See *Center-of-Mass Pendulum. Coupled Oscillators* (ESI); *Dynamics of Pendulums; Finding the Speed of a Rifle Bullet* (2 loops); *Oscillations, Forced Damped Harmonic; Pendulum, Sand* (4 loops); *Pendulums, Ring of Coupled* (2 loops); *Simple Harmonic Motion—the Stringless Pendulum; Wilberforce Pendulum.*

Pendulum, Sand (4 loops), (1) *Multiple Saddle Suspension,* (2) *Point Suspension,* (3) *Saddle Suspension,* (4) *Short Saddle Suspension.* See *Sand Pendulum Series.*

Pendulums, Ring of Coupled (2 loops), (1) *Mixture of Normal Modes,* (2) *Normal Modes.*

HT *Phase Change.* C; p, d, 9 (reg, sup). Pressure dependence of transition temperatures in H_2O.

O,W *Phase Difference Between Two Sources, The Effect of.* See *Ripple Tank Wave Phenomena Series.*

MP *Photoelectric Effect.* Electroscope, Zn surface, charge, light. Clear qualitative demonstration. Two different loops of this title. See *Electrostatics Series* (C); *UNESCO Physics Series* (B & W).

O *Pinhole Camera.* Still photographs taken of same object by 6 different pinhole apertures from 0.07 to 2 mm. Optimum aperture 0.35 mm. Diffraction vs rectilinear propagation. Excellent qualitative loop. See *UNESCO Physics Series.*

FM,V *Potential Flows, Hele–Shaw Analog* (2 loops). (1) *Sources and Sinks,* (2) *Sources and Sinks in Uniform Flow.* C. Streamline patterns, by dye injection, of velocity fields with interesting sidelights on divergence of vectors. See *Fluid Mechanics Series.*

D Prandtl, L. Fluid flow film footage taken 1927–33. See *Airfoil, Generation of Circulation and Lift from an; Flow Separation and Vortex Shedding; Magnus Effect; Sink Vortex.*

Lab,M *Program Orbit* (2 loops), *Part I* and *Part II.* Computer-generated. Orbit calculated iteratively first as done by student in laboratory project, then with increasing accuracy as number of iterations increased. Laboratory related. See *HPP Series.*

W *Pulses, Superposition of.* See *Ripple Tank Wave Phenomena Series.*

Q QUANTITATIVE-PROBLEM LOOPS. See *Collision in Two Dimensions; Conservation of Energy—Potential to Kinetic; Conservation of Linear and Angular Momentum; Conservation of Momentum* (2 loops), (1) *Elastic Collisions,* (2) *Inelastic Collisions; Equipartition of Energy; Gas, Properties of; Jupiter Satellite Orbits; Simple Harmonic Motion; Simple Harmonic Motion—The Stringless Pendulum.*

QM *Quantum Mechanical Harmonic Oscillator.* B & W; a, A. Bork; p, ESI; d, 8 (reg, sup, reel–16). Computer-generated solutions of time-independent Schrödinger equation displayed for uniformly increasing energy values, including eigenvalues.

QM *Quantum Physics Series* (6 loops). B & W; a, J. L. Schwartz; p, EDC; d, 4 (reg, sup), 7 (sup, sup–reel), 13 (reg, reg–reel, reel–16), 22 (reg, reg–reel), 26 (reg, sup), 30 (reg, sup), 31 (reg, sup). Computer-generated. See *Particle in a Box; Scattering in One Dimension* (4 loops), (1) *Barriers,* (2) *Edge Effects,* (3) *Momentum Space,* (4) *Wells; Wave Packets, Free.*

NP *Radioactive Decay.* Time-lapse photography of read-out of 400-channel analyzer; several decays. See *Miller Series.*

KT,SM *Random Walk and Brownian Motion.* One puck, of a two-dimensional model gas of pucks on an air table, is followed by "white trail" animation over many free paths. Histogram of path lengths plotted. Large puck in a gas of small ones shown. Also, Brownian motion of smoke particles. See *Kinetic Theory on an Air Table Series.*

Reference Circles. a, G. Schwarz. See *Amateur Loops.*

O *Reflected Light—Glass in Liquids.* Four liquids. Glass fragments vanish if index is right. See *UNESCO Physics Series.*

O,W *Reflection at Plane and Curved Surfaces.* C; p, in England; d, 9 (reg, sup). By ripple tank.

O,W *Reflection of Circular Waves from Various Barriers.* See *Ripple Tank Wave Phenomena Series.*

O,W *Reflection of Straight Waves from Straight Barriers.* See *Ripple Tank Wave Phenomena Series.*

O,W *Reflection of Waves from Concave Barriers.* See *Ripple Tank Wave Phenomena Series.*

O,W *Refraction.* C; p, in England; d, 9 (reg, sup). Ripple tank refraction of straight waves by a straight interface, by a prism, and by a lens.

O,W *Refraction of Waves.* See *Ripple Tank Wave Phenomena Series.*

HT *Regelation.* a, T. Norton. See *Amateur Loops.*

M,R *Relative Motion, A Matter of.* Elastic head-on collision between identical objects, one initially at rest. Filmed by camera at rest, at speed of incoming object, at ½ speed. Newtonian relativity. See *HPP Series.*

R RELATIVITY. See *Relative Motion, a Matter of; Relativity, Galilean* (3 loops).

M,R *Relativity, Galilean* (3 loops), (1) *Ball Dropped from Mast of Ship,* (2) *Object Dropped from Aircraft,* (3) *Projectile Fired Vertically.* The latter fired from a Ski-doo moving over snow. See *HPP Series.*

O *Resolving Power.* Pinhole, two-pinhole sources viewed in telescope of variable aperture. See *Miller Series.*

O,W *Ripple Tank Wave Phenomena Series* (14 loops). B & W; a, J. Strickland; p, ESI; d, 4 (reg, sup), 7 (sup, sup–reel), 13 (reg, reg–reel, reel–16), 14 (reg; series sets only), 16 (reg, reg–reel, reel–16), 22 (reg, reg–reel), 26 (reg, sup), 30 (reg, sup), 31 (reg, sup). *Barrier Penetration by Waves; Bragg Reflection of Waves; Diffraction and Scattering Around Obstacles; Diffraction of Waves* (2 loops), (1) *Multiple Slit,* (2) *Single Slit; Doppler Effect; Interference of Waves; Phase Difference Between Two Sources, Effect of; Pulses, Superposition of; Reflection of Circular Waves from Various Barriers; Reflection of Straight Waves from Straight Barriers; Reflection of Waves from Concave Barriers; Refraction of Waves; Shock Waves, Formation of.*

AP,MP, NP *Rutherford Scattering.* Computer-generated. Fixed inverse-square repulsion center affecting incoming particles. See *HPP Series.*

MP,NP *Rutherford–Royds' Identification of Alpha Particles.* C; p, in England; d, 9 (reg, sup). Shows original apparatus by which α-particles were shown to be He nuclei; then explains by animation.

M *Sand Pendulum Series* (7 loops). a, J. Bregman, R. Davisson, A. Holden; p, SUNY–EDC; d, release fall 1968, contact 8. *Oscillations, Forced Damped Harmonic; Pendulum, Sand* (4 loops), (1) *Multiple Saddle Suspension,* (2) *Point Suspension,* (3) *Saddle Suspension,* (4) *Short Saddle Suspension; Pendulums, Ring of Coupled* (2 loops), (1) *Mixture of Normal Modes,* (2) *Normal Modes.*

QM *Scattering in One Dimension* (4 loops), (1) *Barriers,* (2) *Edge Effects,* (3) *Momentum Space,* (4) *Wells.* Computer-generated. See *Quantum Physics Series.*

Lab,M, V *Scattering of a Cluster of Objects.* Six steel balls of given masses, hung from 35-ft piano wire and at rest in a cluster, are struck by a seventh. Slow-motion sequence allows finding of all the velocities. See *HPP Series.*

NP *Scintillation Spectrometry.* Mn^{56} γ-rays, scintillator, photomultiplier, oscilloscope. Pulse height, Compton edge, back scatter studied. See *Miller Series.*

EM *Series Circuits.* a, G. Griswold, G. Rees. See *Amateur Loops.*

FM,V *Shear Deformation of Viscous Fluids.* B & W; a, A. H. Shapiro. See *Fluid Mechanics Series.*

FM,W *Shock Waves, Formation of.* See *Ripple Tank Wave Phenomena Series.*

M,Q *Simple Harmonic Motion.* (See also next entry.) Uniform circular and simple harmonic motion on an air table, and a simple pendulum, all of equal periods, studied in detail. See *Mechanics on an Air Table Series.*

M SIMPLE HARMONIC MOTION. (See also neighboring entries.) Also *Dynamics of Pendulums; Dynamics of Springs; Velocity and Acceleration in Simple Harmonic Motion; Velocity in Circular and Simple Harmonic Motion.*

M,Q *Simple Harmonic Motion—The Stringless Pendulum.* Curved air track, data given to find radius of curvature, glider period. Find g. Accelerometer sequences. See *Mechanics on an Air Track Series, Newtonian.*

D,FM *Sink Vortex.* B & W; a, L. Prandtl (1930 approx). Water enters tangentially into a circular tank with drain at center. See *Fluid Mechanics Series.* (See also *Bathtub Vortex.*)

SIXTEEN-MILLIMETER FILM LOOPS. See *Wave Motion, Sets 1* and *2.*

M *Soap Film Oscillations.* B & W; a, A. Hudson; p, ESI; d, 4 (reg, sup), 7 (sup, sup–reel), 13 (reg, reg–reel, reel–16), 14 (reg), 16 (reg, reg–reel, reel–16), 22 (reg, reg–reel), 26 (reg, sup), 30 (reg, sup), 31 (reg, sup).

Lab *Solder Glass Vacuum Technique Series* (7 loops), (1) *Preparing the Getter Header,* (2) *Preparing and Baking a Solder Glass Seal,* (3) *Making the Anode Structure,* (4) *Preparing the Diode Header,* (5) *Sealing off the Small Header Tube,* (6) *Testing for Leaks and Outgassing the Elements,* (7) *Firing the Getters.* B & W; a, J. Orsula; p, EDC; d, 4 (reg, sup), 7 (sup, sup–reel), 13 (reg, reg–reel, reel–16), 14 (reg; series sets only), 16 (reg, reg–reel, reel–16), 22 (reg, reg–reel), 26 (reg, sup), 30 (reg, sup), 31 (reg, sup). See also 16-mm film, *Laboratory High-Vacuum Technique* in Sec. 21–6.1.

SSP — SOLID STATE PHYSICS. See *Bragg Reflection of Waves; Bubble Model of a Crystal* (2 loops); *Ferromagnetic Domain Wall Motion.*

EM,W — *Standing Electromagnetic Waves.* 435 M/sec waves in a cavity. See *HPP Series.*

SM — STATISTICAL MECHANICS. See *Irreversibility and Fluctuations* (2 loops), and *Kinetic Theory on an Air Table Series* for 6 related loops.

M,W — *Superposition.* Cathode ray oscilloscope face displays three traces: third is superposition of first two. Both fixed and progressive sinusoids, including standing waves, are shown. See *HPP Series.*

FM,GP — *Surface Tension* (3 loops), (1) *and Contact Angles,* (2) *and Curved Surfaces,* (3) *Examples of.* C; a, E. M. Trefethen. Showing, (1) soap films, intersecting films, formation of drops in water, mercury; (2) that pressure on concave side is the greater, the smaller the radius of curvature (smoke ejected from bursting soap bubbles, drop of water formed at faucet, capillary vise, etc.); (3) many examples proving the existence of surface tension. Excellent. See *Fluid Mechanics Series.* (See also *Drops, Break up of Liquids into.*)

FM,GP — *Surfaces, Motions Caused Along Liquid* (3 loops), (1) *by Composition Gradients,* (2) *by Electrical and Chemical Effects,* (3) *by Temperature Gradients.* C; a, E. M. Trefethen. In (1) many demonstrations including effect of soap on water surface, and camphor boat; (2) shows oscillations on Hg surface touched by nail, and Hg drops chasing potassium dichromate particles; (3) shows effects with air and vapor bubbles. See *Fluid Mechanics Series.*

D,M — *Tacoma Narrows Bridge Collapse.* A famous example of resonance filmed in color. See *Miller Series.*

HT — *Temperature Waves.* Thermometers stationed along a brass rod display effect of time-dependent temperature gradient. See *Miller Series.*

AP,MP — *Thomson Model of the Atom.* An experimental analogy of the plum pudding model. Magnets on floats in water in radial magnetic field are added, one by one, and find symmetric positions on the water surface. See *HPP Series.*

AP,MP — *Thomson's Positive Ray Parabolas.* C; p, in England; d, 9 (reg, sup). Shows J. J. Thomson's 1911 spectograph, and plates, measuring. q/m of positive ions, and explains experiment by animation.

FM,GP — *Tornadoes in Nature and in the Laboratory.* B & W; a, A. H. Shapiro, E. S. Taylor. See *Fluid Mechanics Series.*

FM — *Turbulence, The Occurrence of.* B & W; a, S. Corrsin. See *Fluid Mechanics Series.*

UNESCO Physics Series (11 loops, 3 listed here). B & W; p, Univ. Sao Paulo, Brazil; d, unknown. See *Photoelectric Effect; Pinhole Camera; Reflected Light —Glass in Liquids.*

Lab — VACUUM TECHNIQUE. See *Solder Glass Vacuum Technique Series* (7 loops).

EM — *Van de Graaff Generator, The.* Construction, operation and properties of a demonstration model. See *Electrostatics Series.*

M,V — *Vector Kinematics Series* (6 loops). B & W; a, F. L. Friedman; p, ESI; d, 4 (reg, sup), 7 (sup, sup–reel), 13 (reg, reg–reel, reel–16), 14 (reg; series sets only), 16 (reg, reg–reel, reel–16), 22 (reg, reg–reel), 26 (reg, sup), 30 (reg, sup). Computer-generated. See *Acceleration Vector. Velocity and Acceleration* (3 loops), (1) *in Circular Motion,* (2) *in Free Fall,* (3) *in Simple Harmonic Motion. Velocity in Circular and Simple Harmonic Motion. Velocity Vector.*

V — VECTORS AND VECTOR ANALYSIS. See *Bathtub Vortex,* (demonstrating an irrotational field); *Collisions, Two-Dimensional* (lab loops requiring vector addition); CURL-METER; *Potential Flows, Hele–Shaw Analog* (2 loops demonstrating fields with sources and sinks); *Scattering of a Cluster of Objects* (lab loop requiring addition of 7 vectors); *Shear Deformation of Viscous Fluids; Vector Kinematics Series* (6 loops); *Vorticity Meter, Visualization of Vorticity with* (2 loops demonstrating vector fields with curls).

M,V *Velocity and Acceleration* (3 loops), (1) *in Circular Motion,* (2) *in Free Fall,* (3) *in Simple Harmonic Motion.* Computer-generated. See *Vector Kinematics Series.*

M,V *Velocity in Circular and Simple Harmonic Motion.* Computer-generated. See *Vector Kinematics Series.*

M,V *Velocity Vector, The.* Computer-generated. See *Vector Kinematics Series.*

F,M *Venturi Passage.* B & W; a, A. H. Shapiro. Pressure distribution in flows; jet (suction) pump; cavitation by constriction. See *Fluid Mechanics Series.*

W *Vibrations* (4 loops), (1) *of a Drum,* (2) *of a Metal Plate,* (3) *of a Rubber Hose,* (4) *of a Wire.* Slow-motion footage stroboscopically "slows" modes, displaying shapes in (1) and (4), displaying the fate of sand strewn over (Chladni) plate in (2). Each loop shows many modes, and (3) is so written that infinity of modes is suggested after showing 15 modes. Drum, plate are loud-speaker driven. Drum is a circular rubber membrane. In (4) straight and circular wires are driven electromagnetically in transverse modes. See *HPP Series.* (See also *Standing Electromagnetic Waves.*)

FM,V *Vorticity Meter, Visualization of Vorticity with* (2 loops), *Part I* and *Part II.* B & W; a, A. H. Shapiro. The curl of velocity fields in many flows actually shown by small floating water wheel (curl-meter). Instructive for vector analysis. See *Fluid Mechanic Series.* (See also *Bathtub Vortex.*)

D,W *Wave Fronts, Non-recurrent.* Breaking water waves, seiche in river Seine, nuclear bomb shocks, row-of-people shocks, reflection of light pulse in interstellar gas. See *Miller Series.*

W *Wave Motion.* a, J. T. Brooks, F. W. Zurheide. See *Amateur Loops.*

W *Wave Motion, Set 1* (5 loops) and *Set 2* (8 loops). B & W; 16-mm film loops only, silent, 20–30 sec each; a, J. J. Heileman; p, d, 27.

QM,W *Wave Packets, Free.* Computer-generated. See *Quantum Physics Series.*

W Waves and Wave Phenomena. See *Bow Waves in Hypersonic Flow; Bragg Reflection of Waves; Diffraction at an Aperture; Diffraction at Obstacles of Plane and Spherical Waves; Diffraction, Double Slit; Diffraction, Single Slit; Interference; Michelson Interferometer; Pinhole Camera; Quantum Mechanical Harmonic Oscillator; Quantum Physics Series* (6 loops); *Reflected Light–Glass in Liquids; Reflection at Plane and Curved Surfaces; Refraction; Resolving Power; Ripple Tank Wave Phenomena Series* (14 loops); *Soap Film Oscillations; Standing Electromagnetic Waves; Superposition; Tacoma Narrows Bridge Collapse; Vibrations* (4 loops), (1) *of a Drum,* (2) *of a Metal Plate,* (3) *of a Rubber Hose,* (4) *of a Wire; Wave Fronts, Non-recurrent; Wave Motion; Wave Motion, Sets 1* and *2.*

M *Wilberforce Pendulum.* Resonance between torsional and linear modes of a system of helical springs and mass. See *Miller Series.*

21–6.3 Distributors

1. American Institute of Physics, New York. Loan only.
2. Bell Telephone System, local business offices or, Film Library, Bell Telephone Laboratories, Murray Hill, N.J. Free loan only.
3. Cambosco Scientific Co., Boston, Mass.
4. Canadian Laboratory Supplies, Ltd., Toronto, Canada.
5. Contemporary Films, Inc., New York.
6. Convair Motion Picture Section, San Diego, Calif. Loan only.
7. Ealing Corporation, Cambridge, Mass.
8. Film Librarian, Educational Development Center (formerly Educational Services, Inc.), Newton, Mass.
9. Encyclopaedia Britannica Educational Corporation (formerly Encyclopaedia Britannica Films), Chicago, Ill. Regional offices in Anchorage, Atlanta, Baraboo (Wis.), Boston, Bryn Mawr, Dallas, Dayton, Detroit, Edina (Minn.), Hollywood (Calif.), Honolulu, Kansas City (Mo.), Mexico City, Nashville, New York, Oakland (Calif.), Portland (Ore.), Ra-

leigh, Rome (Italy), Salt Lake City, San Leandro (Calif.), Skokie (Ill.), Sydney (Australia), Tokyo, Toronto, and Washington (D.C.).

10. Institut für den Wissenschaftlichen Film, Nonnenstieg 72, Göttingen, West Germany.
11. Lab–T.V., New York.
12. Report Librarian, Los Alamos Scientific Laboratory, Los Alamos, N.M. Loan only.
13. Macalaster Scientific Corporation, Nashua, New Hampshire.
14. Text-Film Division, McGraw-Hill Book Company, New York. Does not rent films.
15. Instructional Media Center, Michigan State University, East Lansing, Mich.
16. Modern Learning Aids, New York. Regional offices in Atlanta, Boston, Buffalo, Cedar Rapids, Chicago, Cincinnati, Cleveland, Dallas, Denver, Detroit, Honolulu, Kansas City (Mo.), Los Angeles, Milwaukee, Minneapolis, Omaha, Philadelphia, San Francisco, Seattle, St. Louis, Summit (N.J.), Toronto, and Washington (D.C.).
17. National Film Board of Canada, Montreal, Quebec, Canada; also in New York.
18. National Research Council of Canada, Ottawa, Canada.
19. Motion Picture Division, The Ohio State University, Columbus, Ohio.
20. Audio-Visual Center, Purdue University, Lafayette, Ind.
21. Office of Institutional Research, Rensselaer Polytechnic Institute, Troy, N.Y.
22. Science Electronics, Nashua, N.H.
23. Shell Films Library, Shell Oil Company, Flushing, N.Y.; also in Indianapolis (Ind.), San Mateo (Calif.).
24. Educational Television Office, State University of New York, New York.
25. Regional Film Libraries, Division of Public Information, United States Atomic Energy Commission, at Aiken (S.C.), Albuquerque (N.M.), Argonne (Ill.), Berkeley (Calif.), Grand Junction (Colo.), Idaho Falls, New York, Oak Ridge (Tenn.), Richland (Wash.), and Washington (D.C.).
26. Universal Education & Visual Arts, a division of Universal City Studios, Inc. (formerly United World Films, Inc.), New York.
27. Sargent-Welch Scientific Company, Skokie, Ill.
28. P. M. Robeck Co., Inc., New York.
29. International Film Bureau, Inc., Chicago, Ill.
30. National Instructional Films, Nanuet, N.Y.
31. Audio Visual Division, Popular Science Publishing Co., Inc., New York.
32. Holt, Rinehart and Winston, Inc., New York.
33. Film Center, Argonne National Laboratories, Argonne, Ill.

21–7 Use of Moving Picture Film[5]

Among the various teaching aids which have not been adequately exploited probably the most important is the moving picture. We have examined the use of film in the PSSC high school physics course, and also considered other educational films. This examination reveals several distinct uses of film which should be considered in college science teaching. Before specifying these uses and discussing the kind of film that should be made in order to explore them, we should emphasize our belief that very little is known about the use of films. We still need to determine the effectiveness of film in college teaching rather than assuming that we know what it will be, but there are strong indications that film will be useful, and there are several types of films that we are sure deserve a thorough exploration. We wish to consider moving picture film use as a set of experiments.

Looking at film which is already made, we can sort out the following types:

1. *Introductory films which are made to stimulate interest and to indicate the scope of a field which will be studied over a period of a month or more.* Examples of this type can be found among the PSSC films. Characteristically they show a large set of experiments which would be very costly to assemble for a showing of only twenty minutes. By running through these experiments a series of significant questions can be raised, and the questions can be sufficiently defined that the students already know the kind of problem on which they are about to work as well as the general nature of the techniques they will use. In an introductory

[5] This section, written by F. L. Friedman of the Massachusetts Institute of Technology, is an excerpt from the 1961 Science Teaching Center Proposal to the National Science Foundation.

film the pace is far too rapid to pretend that details will be learned. The object is to raise enough interest to encourage a student to go more willingly through the material at the subsequent slower pace of standard pedagogy and to go through it with better understanding because he has a general knowledge of where he is going. In the PSSC course the "Introduction to Optics" and the film "Forces" are good examples.

The need for films of this type in college is probably less severe than in high school. Some colleges can afford the large outlay to produce the apparatus and the expense of setting it up even for a short demonstration. Nevertheless, there are times when film will have two advantages. It may be more economical to run a film than it is to build up and repeat expensive demonstrations. Second, and more important, because of the possibilities of close-up, high-speed, and time-lapse photography the filmed demonstrations in some instances will be clearer than live demonstrations.

2. *Films to extend laboratory experience of students.* During the course of standard pedagogy, demonstrations in lecture are often used to supplement the student's laboratory experience. These demonstrations are usually qualitative and rapid. Films may be used to extend a student's laboratory experience in a different direction. When an experiment is too expensive in terms of time, or in terms of the student's technique, it is possible to make a careful film study of the experimental apparatus and of the actual performance of an experiment including taking the relevant data. Beyond this point the data may be interpreted, or we may leave it to the student to make the interpretation just as if he, himself, had performed the experiment. In the PSSC high school, examples of this kind are the film on the "Millikan Experiment," the film on the "Pressure of Light," and the film showing the interference pattern of two slits being painted by individual photons. Each of these films should be appropriate for beginning or intermediate college physics, and at the college level many more such films appear desirable. Eventually, these films should probably differ from any high school version. For example, we would like to experiment with the existing "Millikan Experiment" film, so modifying it that the interpretation of the data is largely eliminated and left to the student rather than carried out on the film as has been done for high school students. Undoubtedly, a completely different "Pressure of Light" experiment should be shown for college students, going beyond a demonstration of the existence of light pressure and allowing the students to establish a quantitative relation between energy flow and momentum flow in the light. As an example of a new film that falls in this category and which should be made for college use, we may suggest filming the set of interrelated experiments in modern particle physics which were developed for use in the elementary course at the University of Illinois. The techniques of counting and the time required for collecting data have proved to be excessive, whereas the ideas involved in detecting and identifying particles and examining their behavior as they pass through matter seem to be well explored. A film in which the experiment is carried out efficiently would avoid the problems of technique and the expense in student's time, while the interpretation of the data would give the students the same experience with the ideas.

3. *Short demonstrations.* Short lecture demonstrations are often used to avoid long experiments. Many of these demonstrations are hard to see and their brevity depends on the investment of a large amount of time in the preparation of apparatus. Such apparatus, once built, can often be profitably photographed and the short demonstration seen more clearly on film. Other demonstrations which normally take a longer time can be shortened by the use of time-lapse photography. The Cavendish experiment determining the constant of universal gravitation can become such a demonstration, although it normally is difficult and time consuming. The motions of electron beams in a magnetic field have already been made far more visible by long-time exposures and stop-frame techniques in the film "Electrons in a Uniform Magnetic Field" by Dr. D. Montgomery, one of the PSSC films. This film uses a standard demonstration apparatus, but the demonstration is greatly improved by the use of film, and the cost of repeating it is greatly reduced.

The short demonstrations should be carefully distinguished from film that give an abstract schematic in place of a real experiment. The demonstrations are to show the actual apparatus and the actual phenomena. It seems likely that they can be done with a minimum of interpretation so that they can be used by individual teachers at any time they wish and presented from the point of view of the individual instructor. There are a large number of demonstrations which will be necessary to bring modern physics into the immediate experience of a beginning student, and there is a fairly general agreement as to quite a number of these demonstrations. Consequently, these films and the others mentioned above should have use in a large number of colleges, universities, and engineering schools rather than being restricted to the local needs of one institution.

When new pedagogic experiments and demonstrations must be developed, as we believe is necessary for quantum physics and for relativity, the investment in their development is apt to be rather large, and even reproducing individual copies or modifications of the resulting apparatus will often prove to be expensive. We believe that there will be times when it will be cheaper to film an experiment performed with such newly developed apparatus and distribute the experiment on film rather than making many copies of the equipment. The additional cost of making film of the demonstration is likely to run in the region of four to ten thousand dollars in any reasonable number of copies, and when used over several years this cost is almost sure to be less than the cost of building separate pieces of apparatus for the individual colleges. The possible economy of filmed demonstration is of course not the only consideration and we do not look forward to replacing all live demonstration by film.

4. In addition to the categories mentioned above, and far short of putting a complete course of standard lectures on film, there may be a need for some films which deal with certain problems which are frequently met in teaching beginning science courses. Two examples of different kinds will illustrate what we have in mind. First is the discussion of the handling of errors in the interpretation of experimental data. On the level of beginning courses the procedure for handling the theory of errors is not likely to be one in which the teacher is carrying on research and to which he might bring special examples and illumination. It seems likely that a film showing actual laboratory examples and discussing the handling of errors, as well as their origin, would be of use in many places and at several times during the study of the elements of science. Such a film would relieve teachers of the burden of frequently repeating what is usually to them a dull but necessary discussion.

The second example probably has a wider significance which may be partially shared by the first one. The logical development and understanding of some of the abstract ideas in the theoretical structure of science are quite complex, and for many students the mathematical development of these ideas is not very revealing. It is common to use diagrams to illustrate the nature of mathematical statements, the solutions of mathematical equations, and the proofs of mathematical theorems. It seems clear that when these visual representations are complex, they should be built up as a sequence of steps. For this purpose the diagrams that can be put on a blackboard or commonly published in a textbook fall far short of the possibilities of schematic representation that can be done on film. Take a very concrete example: tracing the geometric and temporal relations that enter in discussing Huygens' principle by the method given in Rossi's textbook on optics. The logical argument is extremely powerful and extremely complicated. It can surely be made clearer by the use of film. A somewhat simpler specific instance is the solution of a differential equation and its extension to the set of solutions of differential equations of the same type involving different coefficients. How the solution changes as the coefficients are changed is something that can be calculated by a computing machine and displayed visually. A filmed record of this display would be far cheaper than supplying computing machines every time we wanted to discuss the characteristics of a family of solutions.

5. There are still a number of different possibilities for the use of film. These include (a) complete "lectures" which give alternative presentations so that a student can use them much as he should use alternative textbooks; (b) small sets of films on subjects which are not part of the essential core of a basic course and which can be used by students to learn the elements of a later specialty; (c) complete standard "lectures" which might relieve teachers of the preparation and delivery of the initial presentation of a subject and thus free

them to spend their valuable time in later discussion; and (d) training films to cut down the time and personnel involved in transmitting techniques as distinct from the content of science. Sooner or later all these film usages should be tried out experimentally.

In the next few years at M.I.T. we would like to start to experiment with some of the possibilities for the use of film. In particular, we would like to use films to investigate the possibility of extending laboratory experience. For this purpose we believe that a good deal of material can be assembled by selecting appropriate PSSC films and modifying them. Also, we believe that one or two films designed specifically for the needs of the college course and benefiting by being conceived and produced solely for this purpose should be tried out. Secondly, we believe that short demonstrations should be presented on film. Here again we think that some testing can be done by making excerpts and modifications of existing films, while a fairly large number of films should be made in the next few years specifically for the purpose. Third, to the extent that we can find films available which do any of the other jobs discussed above, we should like to try them out carefully with experimental groups of students. If necessary, we would like to make one or two films that run for an entire class period in order to eliminate lectures and evaluate the efficacy of film followed by discussion groups as compared with lecture and discussion. Lastly, we have in mind the possibility of making some short films illustrating the carrying out of a few specific procedures or techniques. The purpose of this kind of film is to diminish the time and the emphasis placed on the acquisition of technique as compared with the acquisition of ideas and general experience in the beginning science courses. A trial of a short film demonstrating how to make vacuum systems by a relatively simple modern technique indicates that we can save a large amount of time both of students and of instructors by teaching such techniques on film rather than teaching them almost individually to every student.

In the immediate future we do not intend to make films to train college teachers rather than college students. Also we hope that other groups will take up the problems of representing abstract ideas, clarifying the sequential steps in mathematical theorems, and visualizing the solutions of equations. These are important problems and in the longer run we may be impelled to work on them. However, most of the initial group would prefer to center our first efforts around the physical phenomena and their interpretation rather than the abstract logic of the purely mathematical problems.

We wish again to emphasize the fact that we view all film use at this stage as experimental. To do good experiments, however, we need not only students, but also films which have been designed specifically for the kinds of use that we think may prove valuable.

22

CORRIDOR DEMONSTRATIONS

By Everett M. Hafner
University of Rochester

22–1 Introduction

At least 30 departments of physics in the colleges and universities of the United States have created sequences of exhibits in the corridors of their buildings. Many others—especially those departments that are expanding into new and spacious quarters—plan to use corridors as exhibition areas. There is a growing awareness among teachers of physics that this technique can play a strong role in the development of a student's ideas, for the display can be more than an attractive chart, a passive model, a collection of apparatus, a film loop, or even a working demonstration of a single interesting phenomenon. All such things are useful when constructed with care and are conventional material in many excellent displays. But they do not exhaust the possibilities of the method. The demonstration can be a fairly complete experiment, an essential and well-integrated extension of a course of lectures, supplementing or even replacing an exercise in a conventional laboratory. In the presence of rapidly growing populations of students, increasing costs of equipment and its maintenance, and the perennial problem of staffing instructional laboratories, a semi-automatic corridor experiment can bring a student into close interaction with a piece of physics that might otherwise escape his interest.

This chapter is a review, mainly by example, of the many ways in which corridor demonstrations are currently being used for physics instruction. It incorporates only such material as has come to the author's attention through tours and correspondence; it therefore does not constitute a complete guide to the field, but it attempts to cover typical motives, techniques, and difficulties in the broad spectrum of present practice. It also provides information concerning institutions where much of the current activity is to be found. Finally, it suggests a set of ideas which may be worthy of pursuit and which, to the author's knowledge, have not yet been successfully developed.

22–2 Science Museums

Some of the most thoughtfully designed physics exhibits of our time are on display in the great science museums of the world. Much of this work exemplifies the best that can be done in the effort to bring the principles and applications of physics to an extremely varied audience. What one finds most frequently is a working demonstration of an elementary phenomenon, with apparatus that the spectator can set into operation. A brief explanation of the display, sometimes with references to more thorough treatments, is usually located nearby. Figure 22–1, showing a display at the Museum of Science and Industry in Chicago, illustrates the conventional technique. The spectator varies the angle of an inclined plane, reads the tangential component of force, and discovers a law with the help of simple graphs. Hundreds of such exhibits, many of which were designed by Professor Harvey Lemon of the University of Chicago, make the physics balcony of the museum a complete laboratory for investigation of elementary laws.

Another beautifully organized museum is the Cranbrook Institute of Science in Bloomfield Hills, Michigan. Since 1958, under the leadership of its energetic director, Robert T. Hatt, the Institute has been constructing an ambitious set of physics exhibits. In describing the origins of the project,[1] Dr. Hatt says:

"By reason of our basic objectives—the promotion of scientific literacy and the diversion of an occasional individual into a research career—we faced several choices of technique. First, we decided that we should deal with *principles* rather than *products,* injecting matters of technology only when they best illustrated a concept of basic physics or, by introduction of the familiar, gave easier transition to the abstract. . . .

"We faced a major question: How may one best introduce physics through museum displays? For economy, one might aim for completely static displays—portraits, historic writings, equipment from the period of emerging science, or replicas of equipment which led to the discovery of physical principles. . . .

"At the other extreme—and very expensive because it would involve endless hours in development and in maintenance—would be the completely automated displays, those either constantly in motion or activated by pushbuttons, levers, or concealed electric systems. . . .

"A third plan considered, and that which was in essence adopted, was the open laboratory, in which

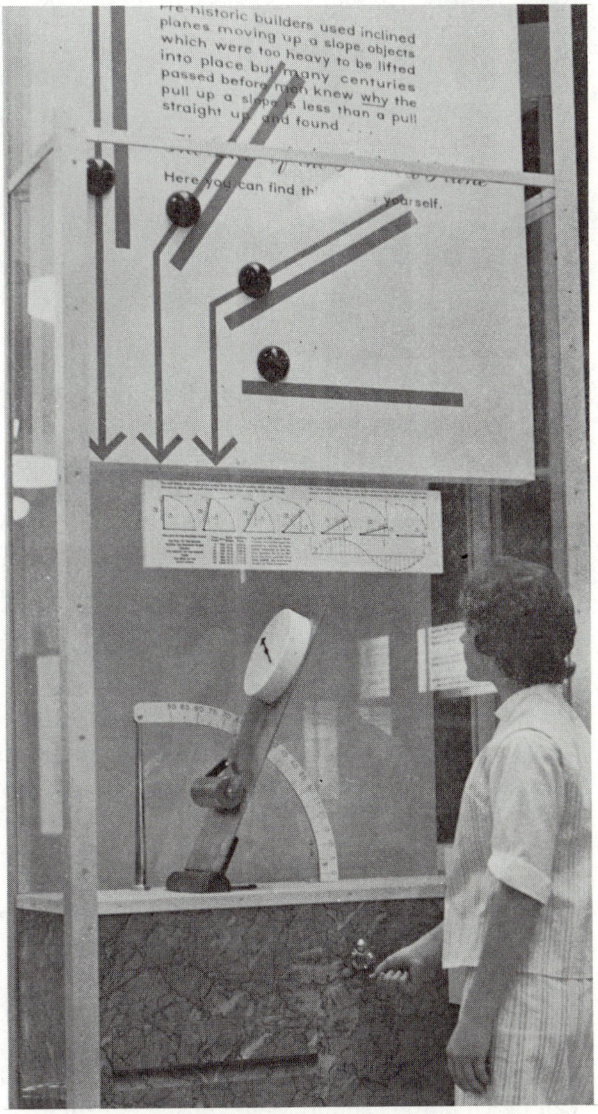

Fig. 22–1 A Corridor Demonstration of Forces on an Inclined Plane. (Photograph courtesy of the Museum of Science and Industry, Chicago.)

(1) R. T. Hatt, "Giving a New Look to Physics," *Museum News,* published by the American Association of Museums, May 1965.

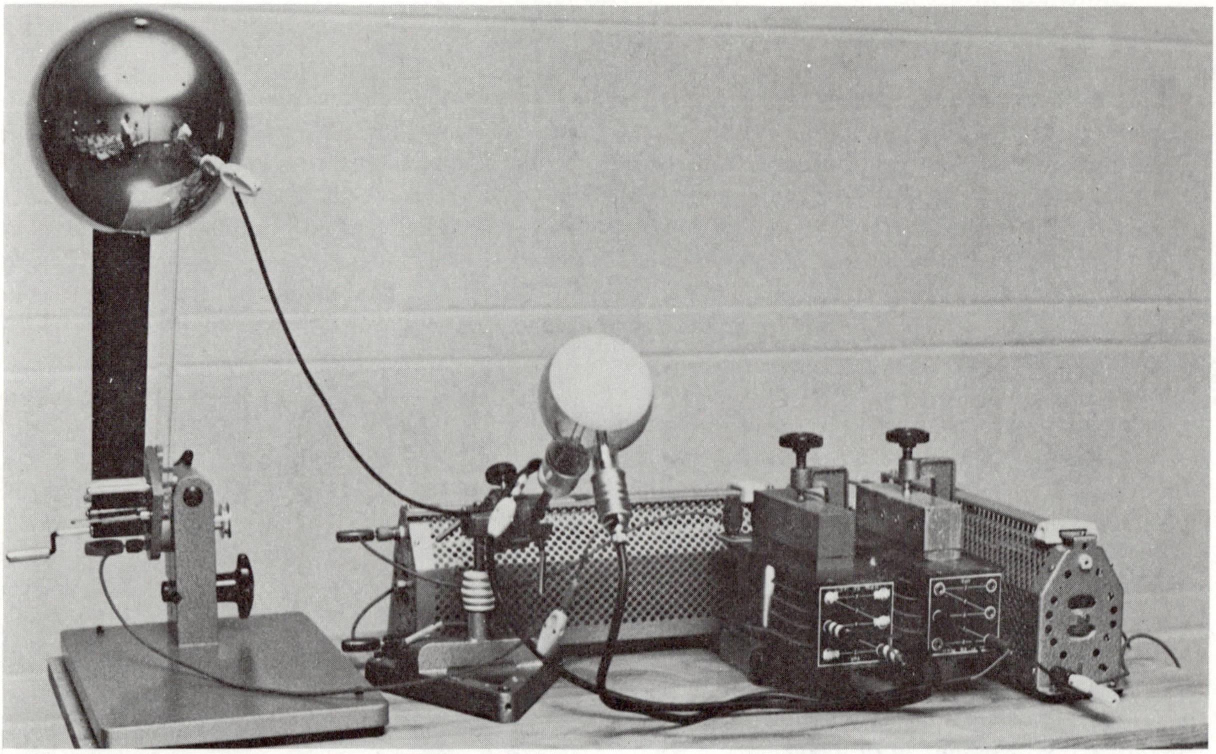

Fig. 22–2 Working Display of Field Emission Microscope. (Photograph courtesy of Cranbrook Institute of Science.)

one or more demonstrators would be always at hand to discuss an exhibit. . . . In the main experimental area a demonstrator's desk could become a focus of attention. Nearby, a reader's alcove and supply of books would assist the leisured guest. . . ."

Actually, as Dr. Hatt and other museum directors have found, all three modes of exhibition have their appropriate places. Historic displays, modeled after the great Arts-et-Métiers exhibit in Paris, are conventionally static. But many of them (Galileo's apparatus, the Foucault pendulum, Oersted's experiment, and a host of others) are also easily made into working demonstrations for which very little maintenance is necessary. Even some recent and important research developments in physics can now be seen as automatic exhibits. For example, the Cranbrook Institute contains a Müller field emission microscope[(2)] (Fig. 22–2) excited by a small Van de Graaff generator. The apparatus exhibits a pattern of atoms in a crystal; Fig. 22–3 is one of Dr. Müller's own photographs obtained from platinum.

A well-informed demonstrator, handling and describing apparatus that is too complex or dangerous for spectators to touch, plays an essential role in the third mode of exhibition mentioned by Dr. Hatt. The technique has evolved successfully in several museums, notably at the Palais de la Découverte in Paris. Figure 22–4 shows one of their demonstrators at work with a high-voltage electrostatic balance, part of an extensive gallery of exhibits that are shown daily. The spectator sometimes takes part; he can, for example, enter a Faraday cage (Fig. 22–5) on which the demonstrator sprays charge.

There is a fourth mode of corridor experiment which has evolved at several universities and museums. The idea is to present the spectator with equipment, to suggest a question, and to challenge him to provide the answer. At the Palais de la Découverte these experiments constitute what is called the "Salle des Tables d'Expérience." A visitor reads the following introduction:

(2) For a review of the technique: E. W. Müller, *Science* **149**, 591 (1965).

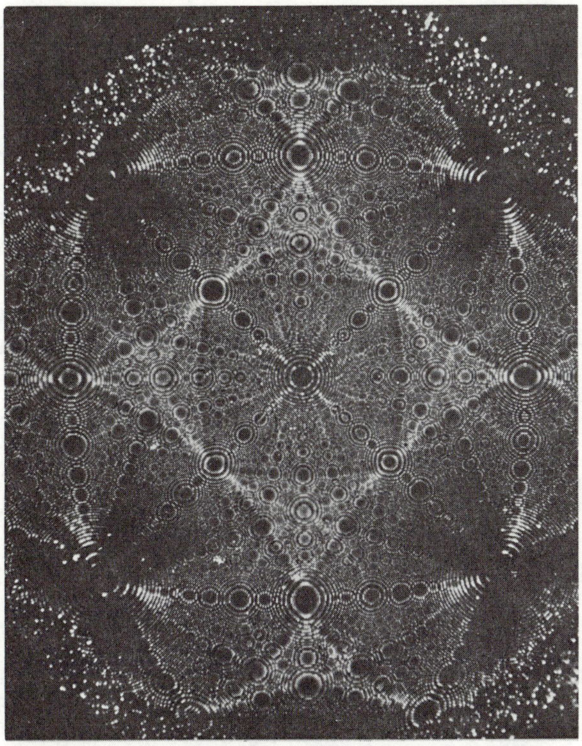

Fig. 22–3 Field Emission Microphotograph of platinum crystals. (Photograph courtesy of E. W. Müller.)

"This demonstration room is reserved for those who have completed one full year of physics. Problems marked with green spots are the easiest; the yellow are intermediate; the red are difficult. Please take good care of the demonstration equipment."

Figure 22–6 shows one of the elementary prob-

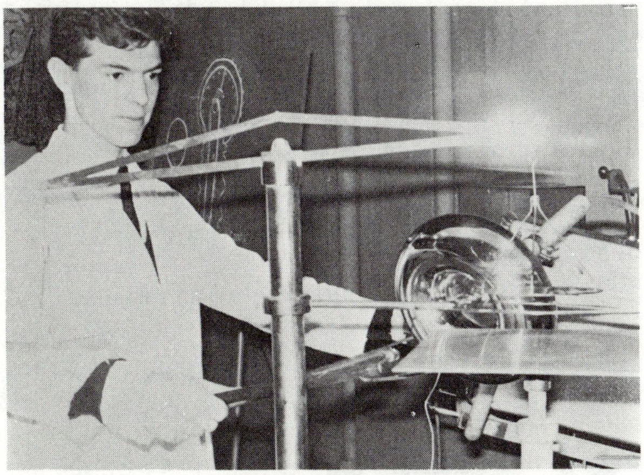

Fig. 22–4 Demonstration of Electrostatic Balance. (Photograph courtesy of Palais de la Découverte, Paris.)

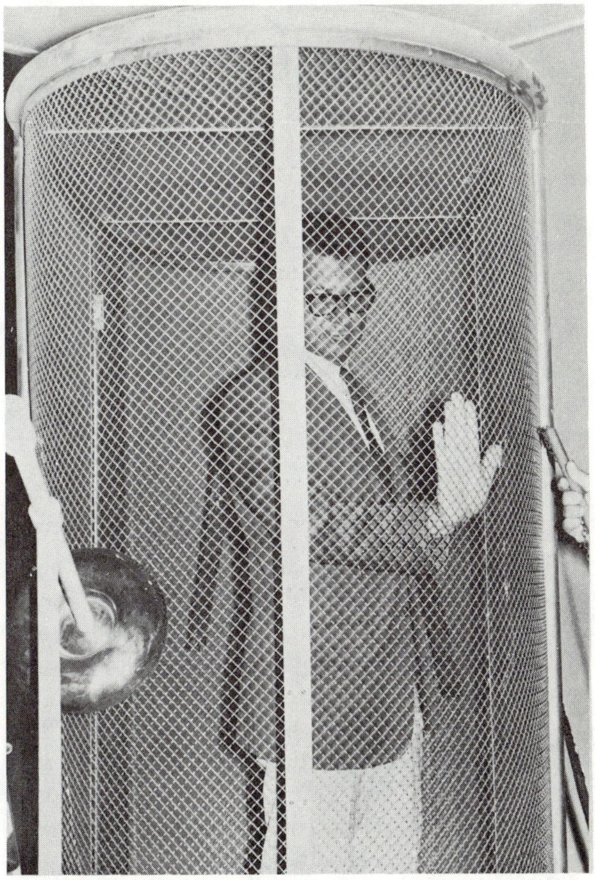

Fig. 22–5 Faraday Cage, electrostatic gallery. (Photograph courtesy of Palais de la Découverte, Paris.)

lems. It consists of a spinning disk, pierced with regular holes, between a pulsed neon lamp and a photocell. The cell output drives a loudspeaker. A fluorescent lamp illuminates the whole disk. With both lamp frequencies given, the problem is to choose a speed of rotation for which a frequency of sound can be calculated. An essential point (which the spectator is not told) is that a solution is possible only when one uses stroboscopic effects on the pattern at the center of the disk.

Perhaps the most ambitious attempt to build this mode into a complete laboratory course is the work[3] at the Lawrence Hall of Science in Berkeley. The "self-instruction demonstration exhibits" created by this group during the last few years include experiments on diffraction of sound, detection of gamma rays, refraction of light, and many

(3) H. E. White *et al.*, *Am. J. Phys.*, 34, 660 (1966).

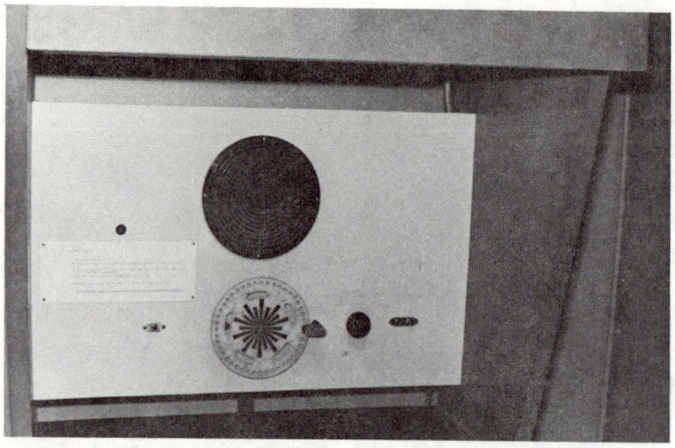

Fig. 22–6 Stroboscopic Problem, Salle des Tables d'Expérience. (Photograph courtesy of Palais de la Découverte, Paris.)

other traditional exercises of the elementary physics course. The experiments are quantitative and completely automatic; one pushes buttons, turns knobs, and records data without ever touching the inner components. After some experience with a pilot program in which 500 students worked on 11 of these exhibits, the staff has concluded that "this kind of laboratory operation, properly set up, will cost less than 20 percent of the generally accepted college or high school laboratory operations, and 40 experiments per year per student can readily be performed instead of 20. These laboratory practices are more effective than lecture demonstrations and are infinitely better than no laboratory at all."

There is little doubt that an exhibit, no matter how completely automated, is better than no laboratory experience. But the notion that it is somehow "more effective" than a lecture demonstration may be more difficult to accept. In what sense is effectiveness defined, and how is it assessed? The authors of the Berkeley report tell us that, in the typical lecture demonstration, "the instructor tabulates the data, makes calculations, derives the results, and draws conclusions on the spot. This activity demonstrates the experimental method, performed by an expert, and for that reason has considerable merit. It does *not,* however, allow the student to perform an experiment on his own. . . . He should discover for himself that the same law of nature can be formulated by his data, as well as by the different numerical data recorded by another."

The measure of effectiveness, then, appears to be the degree to which the student can call the data his own. But, even in this respect, the style of the Berkeley exhibits leads us to some difficult questions. Is it true that the student has "the opportunity of deciding for himself what measurements to make" when, in the acoustic interference exhibit, frequency and distance are fixed and the only variable is the angle of the microphone? Can the student expect to obtain "different numerical data" when the positions of maxima and minima are the same for all observers, and when built-in scales provide "direct measures of the two distances, thereby eliminating the necessity for angle measurements and the use of trigonometry"? And the hardest question: can an observer lay claim to data from apparatus which he did not help to design or assemble? According to the Berkeley report:

"In all graduate work in which a Ph.D. candidate does research for his dissertation, the laboratory setup time may take one or two years, whereas the recording of data with the properly working apparatus requires but a short period of time and provides the essential information for discovering a natural law. Here then is one direct clue as to how large numbers of students might be permitted to perform individual laboratory experiments at relatively low cost. Eliminate the students' preliminary setup time."

But the fact is that many of us, convinced that the design of apparatus is essential training, would be reluctant to pass a Ph.D. candidate who had eliminated the setup time.

The Berkeley experience is an interesting and important step toward a new laboratory technique. Its weaknesses can perhaps be remedied by the creation of exercises which contain elements of preparation, design, and surprise within the limitations of a corridor display. Such things as the little stroboscopic puzzle in the Paris museum may suggest, to those who are interested in the laboratory question, a path worth following.

Returning to a tour of the great museums, one finds a sequence of extremely ingenious corridor demonstrations at the Science Museum in South Kensington, London. Mr. V. K. Chew of the Physics Department is especially interested in the demonstration of oscillations and waves. His exhibits are small masterpieces of the art. Figure 22–7 shows his apparatus for demonstrating forced vibration

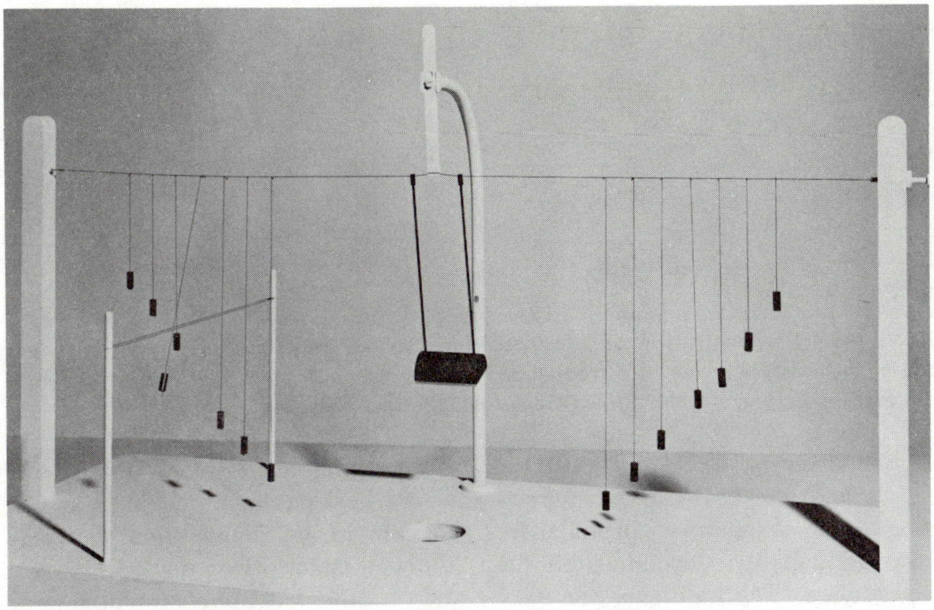

Fig. 22–7 Demonstration of Forced Vibration and Resonance, Science Museum, Kensington. The length is 32 in. Driving is maintained by a flow of air from an opening in the base. Damping of the left group of pendula is relatively small; the resonance is correspondingly sharp. (British Crown Copyright. Science Museum, London.)

and resonance, about which he writes[4] as follows:

"Most of the phenomena associated with forced vibration and resonance can be shown by means of an apparatus devised[5] in 1918. In our version of this classic demonstration, the forcing vibration can be maintained at constant amplitude for as long as is required.

"The apparatus consists of a thread, supported horizontally, from which are suspended two groups of simple pendula. Their lengths are so adjusted that all the bobs of a group lie on a straight line passing through an end of the suspending thread. In one group (on the left in the photograph) the bobs are of brass; when these are in motion, the effect of air damping is relatively small. The bobs in the other group are of balsa wood; damping is relatively large. I shall call the two groups 'undamped' and 'damped,' respectively.

"The supporting thread is attached at its midpoint to the end of a vertical blade, symmetrically disposed to which are two thin aluminum rods suspended from the thread and supporting the massive bob of a driving pendulum: a semicylinder of light wood with a flat piece of lead attached to its base. A closely spaced pair of horizontal threads is arranged to guide the motion of the fourth pendulum of the undamped group, since this motion may become elliptical if the apparatus is disturbed by air currents.

"When the driving pendulum is set in motion,

[4] V. K. Chew, private communication to the author.

[5] Barton and Browning, *Phil. Mag.*, **36**, 169 (1918).

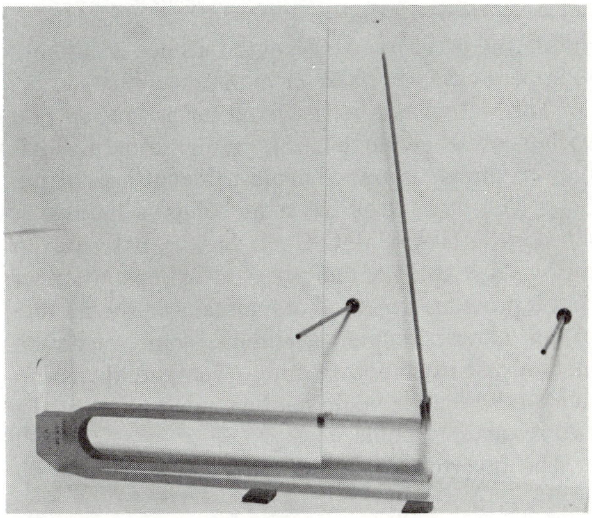

Fig. 22–8 Fork-Driven Inverted Pendulum, Science Museum, Kensington. (British Crown Copyright. Science Museum, London.)

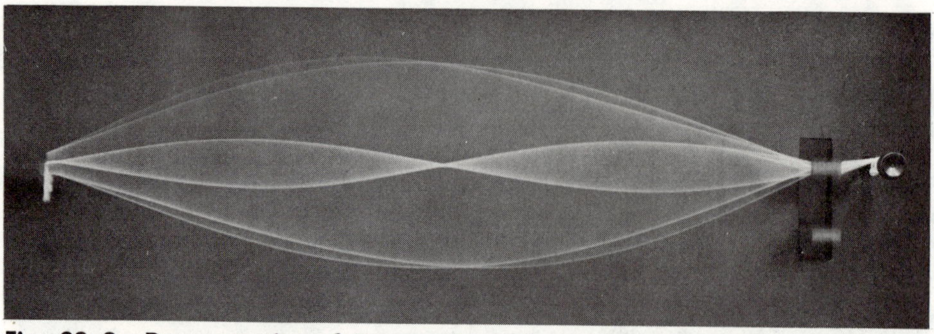

Fig. 22–9 Demonstration of Superposition, Science Museum, Kensington. The left end of the string is driven at a frequency f, the fundamental; the right end at $2f$. This is a triple exposure. (British Crown Copyright. Science Museum, London.)

we see transient phenomena associated with the coexistence of free and forced vibrations. Directing our attention to one of the driven pendula, we see beats of varying rapidity, depending on the length of the pendulum. The transients die away rapidly in the damped group and very slowly in the undamped group. Finally only the forced vibrations remain, and the resonant undamped pendulum is observed to have the largest amplitude. The diminution of amplitude from one pendulum to the next is much more pronounced in the undamped group. Also, in this group, the three shorter pendula are in phase with the driving pendulum; the longer ones are in opposite phase. The phase change is more gradual through the damped group and the resonant pendulum is 90 deg out of phase with the drive. In the undamped group the resonance is so sharp that it is difficult to give a pendulum the exact resonant length. Hence it is likely to appear either in phase or in opposite phase.

"The system can be sustained for as long as one wishes by a gentle current of air from a small blower, directed upward against the flat face of the bob. This effect itself has some points of interest[(6)] (it does not work if the bob has its flat side on top). Its virtue in the present demonstration is that it provides a method of maintaining the motion of an almost simple pendulum; some electrical methods of excitation require a compound pendulum whose length would not be equal to that of the two resonant pendula."

The inverted pendulum, interesting to us both as a pleasing curiosity and a good exercise[(7,8,9,10)] in analytical mechanics, appears at the Kensington Museum in the simple form shown in Fig. 22–8. Although motor-driven devices[(11)] are more rugged and versatile, Mr. Chew's tuning-fork drive is easier to improvise. He gives us the following hints about this:

"If a tuning fork is available which can be cut down until its frequency is slightly greater than double that of the ac supply, it can be driven by an electromagnet acting on the lower prong. A rod of balsa wood, with a metal bearing at the driven end, is mounted on the upper prong. The rod rests against a stop until the fork is excited. A tuning slide is mounted on one prong of the fork."

An excellent corridor demonstration of the principle of superposition, as applied to stationary waves on a string, also appears in the Kensington Museum. The method is to drive one end at the frequency of the ac supply and the other end at twice the frequency. In the version shown (Fig. 22–9), the left end is driven at the lower frequency by three short pieces of hacksaw blade clamped together at one end, excited by a solenoid and a biasing permanent magnet. The right end passes through one prong of a tuning fork driven at the double frequency and is attached to a screw for adjusting tension. One can drive each end sepa-

(6) Jacob P. Den Hartog, *Mechanical Vibrations* (McGraw-Hill Book Company, Inc., New York and London, 1956), 4th ed., p. 301.

(7) Inglis, *J. Inst. Civil Eng.*, **22**, 322, 350 (1943).

(8) H. C. Corben and P. Stehle, *Classical Mechanics* (John Wiley & Sons, Inc., New York, 1960), 2nd ed., p. 67.

(9) F. M. Phelps and J. H. Hunter, *Am. J. Phys.*, **33**, 285 (1965).

(10) L. Blitzer, *Am. J. Phys.*, **33**, 1076 (1965).

(11) F. M. Phelps and J. H. Hunter, *Am. J. Phys.*, **34**, 533 (1966).

Fig. 22–10 Reproduction of Archimedes' Test of a Crown. When a button is pressed, the beakers rise and the balance is destroyed. (Photograph courtesy of Deutsches Museum, Munich.)

rately and then observe the superposition when both ends are driven. The photograph is a triple exposure of the three cases. A slight change in the tension of the string will alter the phase difference between the two vibrations and hence the shape of the superposed wave. Surprisingly beautiful patterns arise in this demonstration, especially with relatively large amplitudes of the faster vibration.

The Deutsches Museum in Münich is one of the great repositories of original scientific artifacts, re-

Fig. 22–11 Demonstration of Ohm's Law. The switches are linked to external controls. (Photograph courtesy of Deutsches Museum, Munich.)

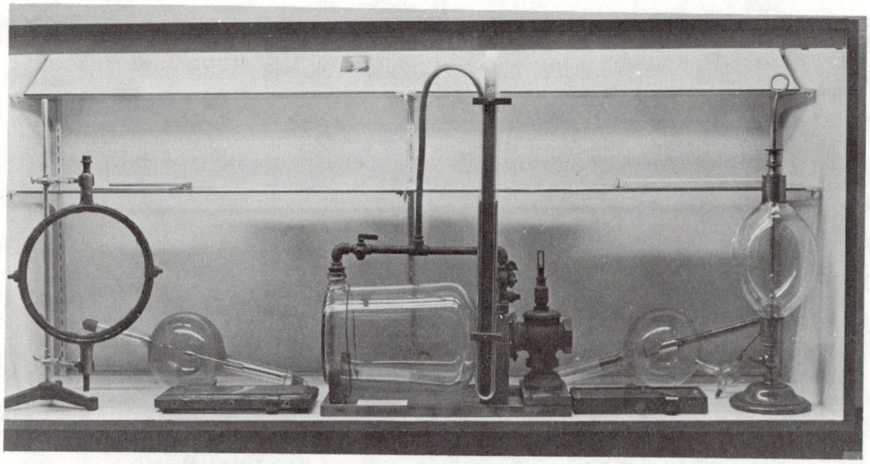

Fig. 22–12 Early X-ray Tubes and Cloud-Forming Devices constructed by Professor Carl Barus. (Photograph courtesy of Brown University.)

productions of crucial episodes in the history of physics and working demonstrations of a didactic nature. Archimedes' notorious assay is replayed in a pushbutton demonstration (Fig. 22–10); the original Magdeburg hemispheres are on display, together with a working model of the famous demonstration; Georg Simon Ohm, 1787–1854, is of course commemorated in an elegant display (Fig. 22–11); the Hahn-Strassman apparatus is preserved. The museum is extraordinarily rich with elaborate exhibits on a grand scale: a complete alchemist's laboratory; an enormous, working diffusion cloud chamber; an operating 300-kV plant; an extensive cryogenic laboratory. But there are also many simple and ingenious demonstrations: ripple tanks, microwave detectors, experiments in fluid dynamics, bell-jar demonstrations, and so on. Magnetic effects of current are exhibited in many ways, one

Fig. 22–13 Galley of Exhibits on Transmission of Radiation in Glass. (Photograph courtesy of the Corning Glass Center—Hall of Science and Industry.)

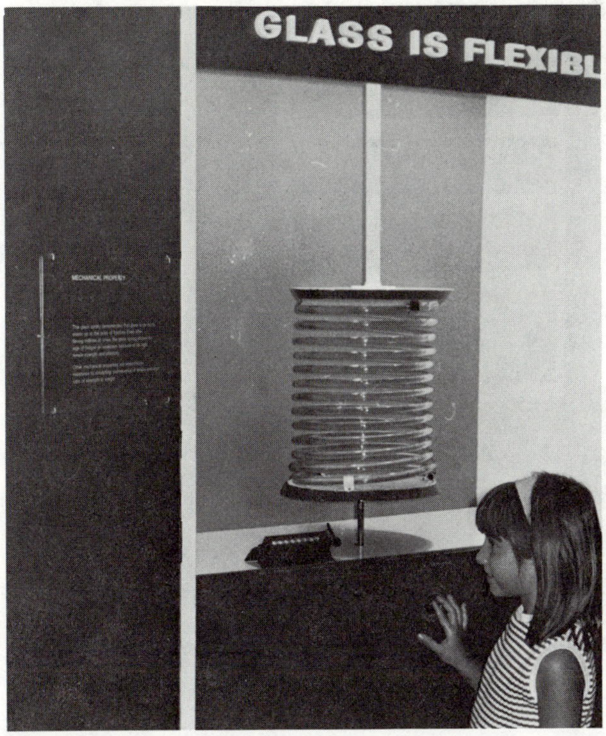

Fig. 22–14 Pushbutton Demonstration of Glass Flexibility. (Photograph courtesy of Corning Glass Center—Hall of Science and Industry.)

of which is especially appealing. A flexible conducting ribbon hangs loosely in the neighborhood of a bar magnet. When current is passed through the ribbon, it snaps into a helix around the magnet.

Interest in historical exhibits of physics is not confined to museums; it is also found in our academic departments. At Brown University one sees displays of apparatus from as far back as 1783 that have been used for research or instruction. Figure 22–12 is a photograph of one of their showcases, about which Professor Robert T. Beyer writes the following:

"The X-ray tubes in this case date back to 1914. Included in the display are pertinent letters and a photograph of Carl Barus, Professor of Physics at Brown from 1895 to 1926. Professor Barus was one of the first presidents of the American Physical Society. In the early 1900s he did some pioneering research on cloud formation, and the equipment in the lower center and left of the case were items used in his research. One of the letters in the case above is from Lord Rutherford in 1904, commending Barus on the excellence of his cloud-forming devices. They are some of the earliest of cloud chambers, which Wilson first began to study in 1897."

At the University of North Carolina, the Department maintains an historical exhibit featuring apparatus used for the teaching of physics in the nineteenth century. The Department of Physics at the University of Pennsylvania prepares displays built around the lives of eminent physicists and devotes space to a variety of rare topics—a recent example was a description of 17th century Ph.D. requirements in France. The showcases at Pomona College contain items connected with the life and work of Millikan. Southwestern College at Memphis is collecting old and interesting pieces of optical physics equipment to be displayed in the lobby of their new science center. It has generally been found that exhibits of this kind, especially those with local historical meaning, arouse considerable interest among students and visitors.

Ideas also come to use from small museums with special interests, some of which are operated by scientific industries. The Hall of Science and Industry of the Corning Glass Center, for example, is steadily developing its exhibits on physical properties of glass. Figure 22–13 shows a group of five working demonstrations on absorption of radiation; Fig. 22–14 shows an exhibit on elasticity in which the glass spring is expanded and compressed with large amplitude. Dr. P. N. Perrot, Director of the

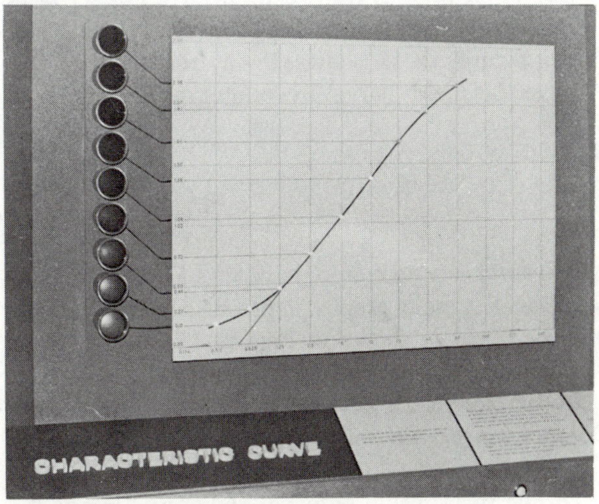

Fig. 22–15 Point-by-Point Demonstration of the Hurter-Driffield Curve of Photographic Response. (Photograph courtesy of George Eastman House.)

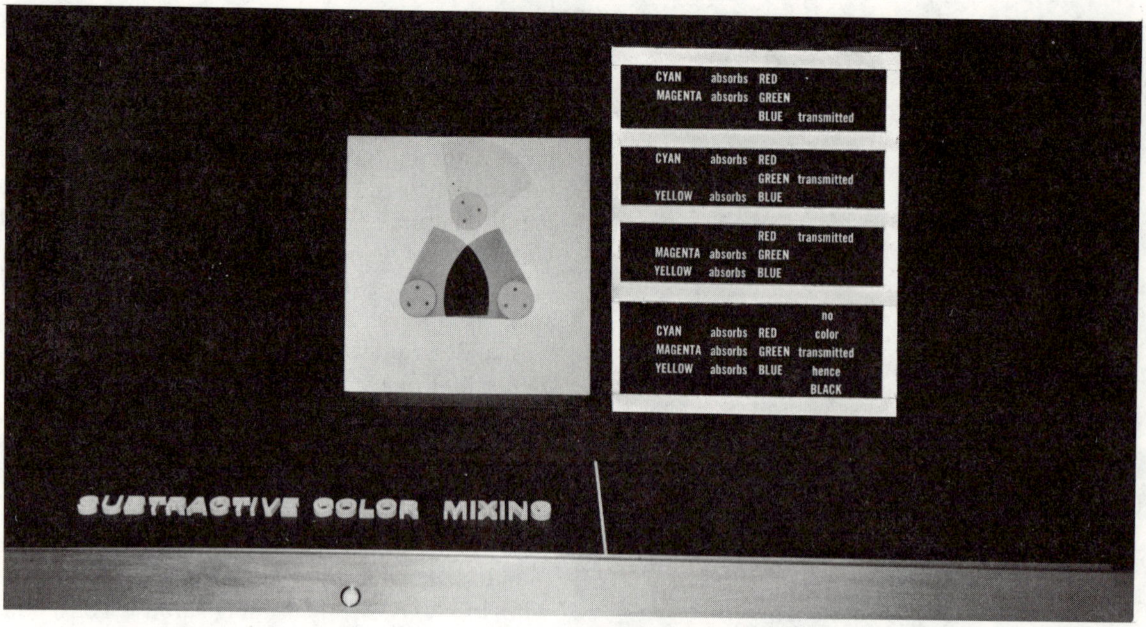

Fig. 22–16 Demonstration of Subtractive Color Mixing. (Photograph courtesy of George Eastman House.)

Corning Museum of Glass of the Corning Glass Center says: "We expect in years to come to alter the presentation greatly and to include far more didactic material—particularly to describe the physical properties of glass with diagrams and push-button exhibits."

The George Eastman House in Rochester, a museum devoted to the history, technology, and esthetics of photography, contains a set of scientific exhibits that are famous for their clarity and artistry. Figure 22–15 shows their treatment of the Hurter-Driffield curve of photographic response. Points on the curve glow in sequence; corresponding film densities appear simultaneously on the set of disks to the left. Visitors to the Eastman House also learn about geometrical optics, illumination measurement, aberration, color processes, spectroscopy, xerography, and other aspects of photographic science in a series of working exhibits. Figure 22–16 shows the demonstration of subtractive color mixing in which yellow, cyan, and magenta filters form blue, green, and red light when transmitting in pairs, and complete extinction when all three are in the beam.

22–3 Educational Institutions

Several demonstrators have had the idea of generating luminous graphs, more or less in the style of the example shown in Fig. 22–15. One of them is suggested by Marvin J. Pryor of the State University of New York at Albany. It is a vivid way of demonstrating the thin lens formula. He describes it as follows:

"The relationship between object and image distances for a lens of 20-cm focal length is plotted at full scale on a board that is about one meter square. A 10-W lamp with clear glass and a plane filament is the object, at the origin of the graph. Six prisms, equally spaced along the p-axis, reflect the light through identical lenses onto screens located on the hyperbola, producing a set of focused images of diminishing size. A slowly moving shutter at the source produces an attractive sequential variation in the images."

Professor Byron E. Cohn of the University of Denver has supplied details of his version of the field emission microscope (Fig. 22–2), which operates as a corridor demonstration. Figure 22–17 is a

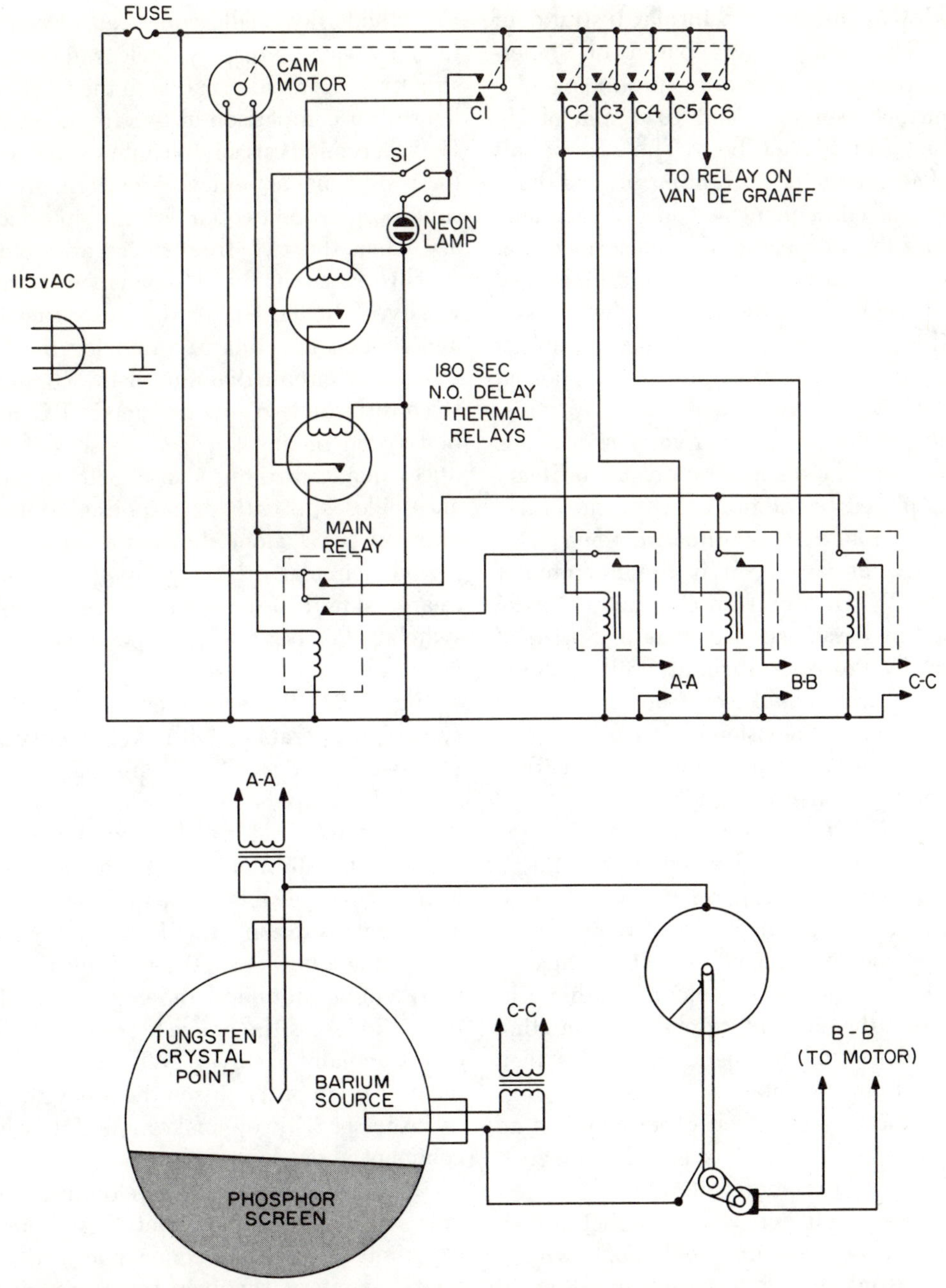

Fig. 22–17 Diagram for Self-Operating Field Ion Microscope, as exhibited at the University of Denver.

schematic diagram of the system. The spectator closes *S1*, starting the cam motor. After 8 sec, *C1* switches, the neon lamp goes out, and the spectator releases *S1*. The subsequent cam switches then operate the demonstration in a 6-min cycle: a heater cleans the surface of the tungsten point for 48 sec; the high voltage is then applied and the pure tungsten pattern is visible for about a minute; with the voltage still on, the barium source is then heated and migration of barium atoms to tungsten is visible for about 3 min; the high voltage is removed, the tungsten is cleaned, and the system turns off.

A demonstration of light pressure can be seen as

a corridor display at the California Institute of Technology. The design[12] is based on solder-glass and ion-pumping techniques, producing ultimate vacuum between 10^{-8} and 10^{-7} mm of Hg in a sealed-off soft glass tube. (The use of soft glass has the advantage of preventing inward diffusion of helium and consequent slow loss of vacuum.) A fused quartz torsion pendulum, whose mechanical parameters are shown, carries a half-blackened aluminum vane and a quartz mirror. Typical systems built by Professor Henry V. Neher deflect through one radian under pressure of radiation at power levels between 10^{-3} and 10^{-2} W/cm^2—levels that are easily available from a microscope lamp. In the corridor demonstration, an ordinary radiometer is placed beside the light-pressure tube. The two systems rotate in opposite directions.

Also on view at Cal Tech is the micromotor constructed by W. H. McLellan of Electro-Optical Systems, Inc., in Pasadena. He won the prize of $1000 offered by Professor Richard Phillips Feynman to anyone who could construct an electric motor with maximum dimension of $\frac{1}{64}$ in. Spectators operate the motor by pressing a button; they observe its rotation under a magnifier.

Professor J. H. Howey reports from Georgia Institute of Technology that his staff is building a group of "semiqualitative" experiments which students are expected to observe as part of their laboratory training. One experiment, now operating, is an electron beam e/m measurement. With push-button controls, the student applies accelerating voltages and deflecting magnetic fields; he then computes e/m from his data.

Another of the Georgia Tech experiments is an oscilloscope measurement of flux in a bar magnet. Professor Howey's description follows:

"We use a $\frac{5}{8}$-in. cylinder Alnico magnet, a 600-turn pickup coil, and a 5-in. oscilloscope with a sensitivity of 50 mV/cm at a sweep speed of 10 cm/sec. The magnet is moved by hand, in synchronism with the sweep. One then computes the flux from a visual estimate of area under the curve. With reasonable care, it is possible to measure the flux to 15 percent. The method is certainly adequate for an order of magnitude measurement, and it gives a direct feeling for a basic physical phenomenon."

(12) H. V. Neher, *Am. J. Phys.*, **29**, 666 (1961).

Cathode ray oscilloscopes can, of course, be used in many ways as display devices in corridor demonstrations. As an alternative to the familiar Lissajous figures for comparison of two frequencies, Professor Howey reminds us of the interesting technique[13] for producing an epicycle pattern. It is done by superposition of two circles; the epicycle is stationary when the two frequencies are commensurate.

This idea has been developed[14] at the University of Rochester for demonstrating Copernican and Ptolemaic orbits in an analog display. Figure 22–18 is a circuit diagram of the complete system. It consists of two low-frequency RC oscillators, a buffer and mixing stage, a phase shifter that produces quadrature outputs at both frequencies, and an oscilloscope with dc response. When the spectator opens *S1* alone, he sees a small, rapid (1-sec period) circular orbit. Opening *S2* alone, he sees a large slow (10-sec period) circle. Opening both switches, he sees the corresponding Ptolemaic epicycle.

Figure 22–19 is a time exposure of the complete system in operation, with both oscillators running (*S1* and *S2* open). The amplitude ratio of the two sinusoidal outputs in this example is 5.2; the frequency ratio is 11.8. These parameters represent the sun-earth-Jupiter system. Precession of the pattern is easily visible on a high-persistence screen. The demonstration stimulates thought on certain interesting questions. What would be the pattern representing motion of the earth in the Jupiter rest frame? How is the epicycle related to what astronomers actually observe? Where would an observer have to be in order to see the epicycle?

Professor Howey makes the following general comment about his program:

"In our situation (physics required of all students), it is my opinion that these semiqualitative experiments, used in connection with our scheduled laboratory sections, are more effective than casual corridor demonstrations. Therefore, in planning our new physics building, we are providing space for such experiments in small rooms adjacent to instructional laboratories. One of these rooms will be accessible both from the corridor and from the laboratory."

(13) J. G. Brainerd, ed., *UHF Techniques* (D. Van Nostrand Co., Inc., Princeton, N.J., 1942), p. 217.

(14) E. M. Hafner and O. M. Bilaniuk, *Am. J. Phys.*, **30**, 615 (1962).

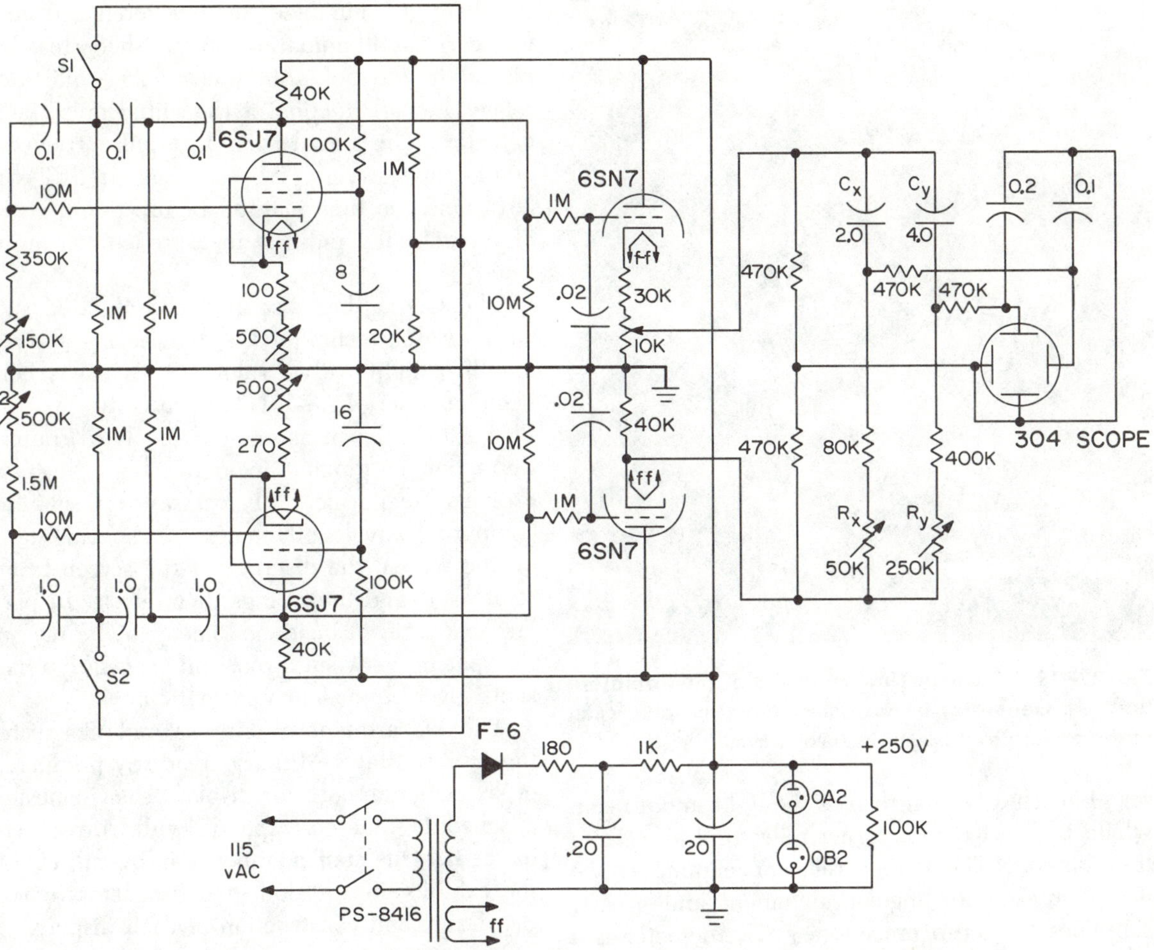

Fig. 22–18 Circuit for Epicycle Generator as demonstrated at the University of Rochester. Capacitances are in microfarads. The upper oscillator has a period of about 1 sec; the lower, about 10 sec.

Plans for the new physics building at the University of Southern California also include a corridor demonstration facility closely related to an instructional laboratory. Professor G. L. Weissler, in charge of this program, feels that the better students should be offered the opportunity of putting together their own experiments. His idea is to have them build their apparatus on movable tables which can then be placed at 12-ft windows for demonstration where they can be seen and operated from the adjacent hallway. An experiment can be explained to visitors by the student who built it. The same windows can, of course, be used for other purposes, including the showing of films.

Emphasis on simple quantitative corridor experiments is found once again in the work of the Science Teaching Center at M.I.T., where a large number of displays have been developed and tried. A particularly successful example is a measurement of the velocity of light, the technique for which is described by J. L. Burkhardt, Laboratory Director of the Center:

"The experiment consists of a pulsed light beam, a photomultiplier detector, and an oscilloscope with a fast triggered sweep. The velocity of light is found by observing the time required for the pulse to travel to a distant mirror and return.

"The light source is a spark generator operated at a 60-cycle repetition rate. The pulsing circuit (Fig. 22–20) was designed by Professor Harold Edgerton. An 11-in., second-surface parabolic mir-

Fig. 22–19 A 15-sec Time Exposure of the Complete Epicycle Generator in operation. Note the precession of the pattern as it begins its second cycle.

ror (Bausch & Lomb) focuses the light into a beam which is directed to a corner reflector 150 m from the source. (The corner reflector eliminates the need for accurate angular alignment and greatly simplifies the setup procedure.) The detector is a 931-A photomultiplier with an opaque hood and collimator to minimize the interference from normal corridor illumination. A polished steel ball is placed in the field of view of the collimator to reflect a small fraction of the initial pulse into the detector. The signal from this pulse triggers the oscilloscope sweep (0.1 μsec/cm on a Tektronix 581), and the time elapsed before the appearance of the reflected pulse is measured directly on the trace."

The Center has also had considerable success with a laser technique,[15] "measuring the wavelength of light with a ruler." They use a helium-neon gas laser (LAS-101, Electro Optics Associates, Palo Alto); the measurement involves Fraunhofer diffraction at grazing incidence on rulings of an ordinary steel scale. All necessary information for computing wavelength comes from spacings of spots on the screen, the distance of the screen from the steel scale, and the rulings themselves. In particular, the angle of incidence need not be measured; the spacing between direct and specularly reflected spots gives equivalent information.

The Department of Physics and Chemistry at the United States Military Academy produces five or six elaborate corridor displays each semester, on a schedule that corresponds with current course material. One staff member is in overall charge of the exhibits; in addition, two instructors work out the design and construction of each display. They

[15] A. L. Schawlow, *Am. J. Phys.*, **33**, 922 (1965).

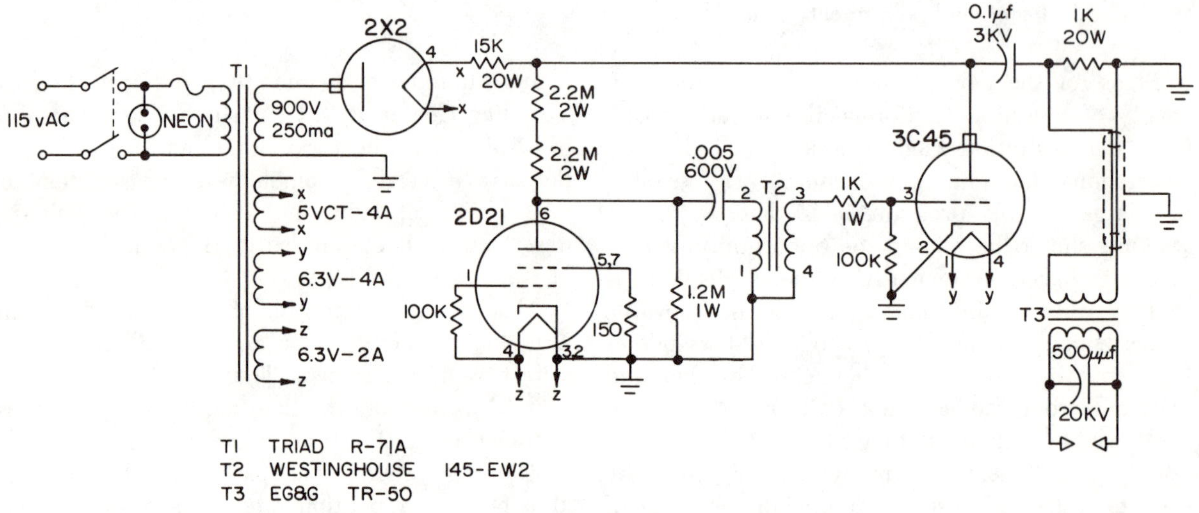

Fig. 22–20 The 50-nsec Spark Generator, Edgerton design, used for the velocity-of-light corridor demonstration at the Massachusetts Institute of Technology.

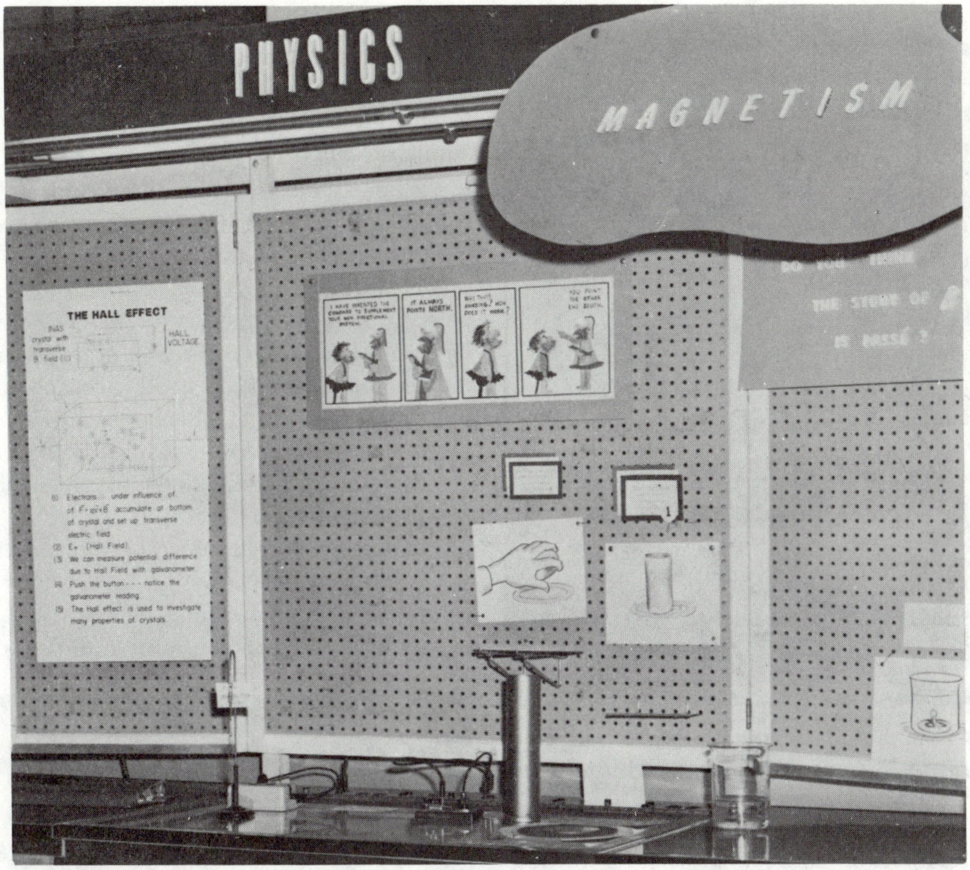

Fig. 22–21 Corridor Demonstrations of Magnetic Phenomena, U.S. Military Academy. (U.S. Army photograph.)

draw on a large stock of material stored from previous exhibits, and they are assisted by the imaginative staff of people in the graphic arts. Figure 22–21 shows a demonstration of magnetism in which the observer detects fields, measures the Hall effect in indium arsenide, and sets up for himself a sequence of simple experiments in heating effects, repulsion, tone generation, and transformer action.

For the last 13 years, stimulated by the interest and enthusiasm of his students, Professor Wallace A. Hilton of William Jewell College in Liberty, Missouri, has been constructing and exhibiting corridor displays of several kinds. His version[16] of the Fabry-Perot interferometer is shown in Fig. 22–22. His other contributions, familiar to readers of pedagogical journals, include satellite detection equipment, air sampling of short-lived isotopes,

[16] W. A. Hilton, *Am. J. Phys.*, **30**, 724 (1962).

Fig. 22–22 Continuous Demonstration of Fabry-Perot Interferometer, one of the many corridor experiments at William Jewell College. (Photo by Wallace A. Hilton.)

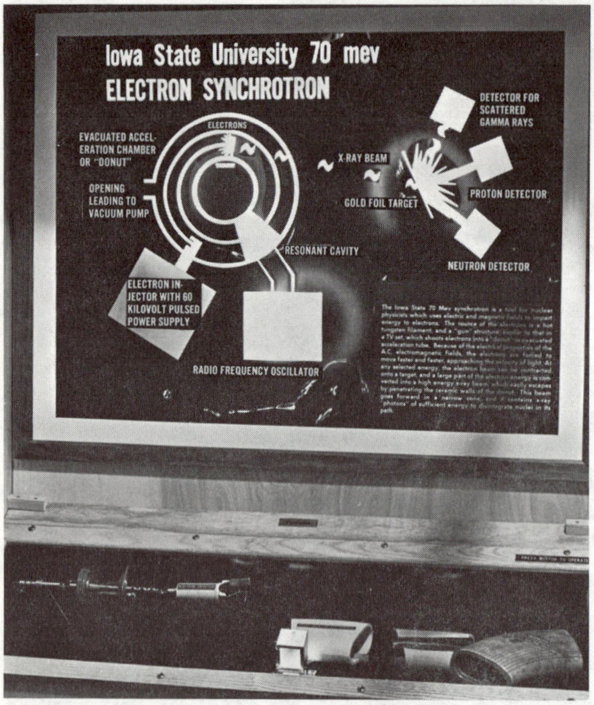

Fig. 22–23 Animated Demonstration of an Electron Synchrotron.* (Photograph courtesy of Iowa State University in Ames.)

models of retardation plates, and switching devices for corridor demonstrations.

The Department of Physics at Iowa State University in Ames maintains an excellent group of corridor displays showing a variety of modes of demonstration. Ideas for the displays usually come from the research activities of the department, which are often sophisticated. One display, for example, shows samples of the nonstoichiometric compound Na_xWO_3, sodium tungsten bronze, whose color depends on the sodium content. Another display is an animated diagram (Fig. 22–23) of the University's 70-MeV electron synchrotron. At the bottom of the cabinet, one sees actual parts of the accelerator: an electron injector, a piece of the "donut," and a resonant cavity.

A simple and dramatic demonstration at Iowa State is shown in Fig. 22–24. It exhibits the Peltier effect in semiconductors by indicating the direction of current flow and permitting the observer to detect the temperature change by placing his hand on the element at the bottom of the display. The legend attached to the display is an excellent example of concise exposition:

"When an electric current flows through two dissimilar solids joined end to end, a change in temperature occurs at the junction. If the current is reversed, the change in temperature is also reversed. This thermoelectric effect (Peltier effect) has been known for over a hundred years, but for metals the amount of heating or cooling at the junction is very small. Recently, new semiconducting materials have given many times more thermoelectric cooling (or heating) than can be obtained from metals.

"In this display, cooling of the junction occurs when the current flows from the *N*-type to the *P*-type semiconductor, and heating of the junction occurs when the current flows in the opposite direc-

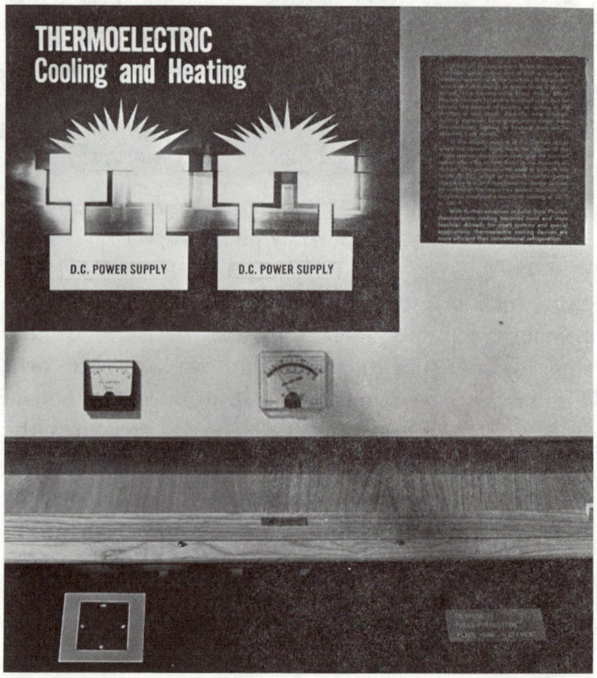

Fig. 22-24 Working Demonstration of the Peltier Effect in Semiconductors. (Photograph courtesy of Iowa State University in Ames.)

*The Iowa State 70 Mev synchrotron is a tool for nuclear physicists which uses electric and magnetic fields to impart energy to electrons. The source of the electrons is a hot tungsten filament, and a "gun" structure, similar to that in a TV set, which shoots electrons into a "donut" or evacuated acceleration tube. Because of the electrical properties of the A.C. electromagnetic fields, the electrons are forced to move faster and faster, approaching the velocity of light. At any selected energy, the electron beam can be contracted onto a target, and a large part of the electron energy is converted into a high energy x-ray beam, which easily escapes by penetrating the ceramic walls of the donut. This beam goes forward in a narrow cone, and it contains x-ray "photons" of sufficient energy to disintegrate nuclei in its path.

tion. The semiconductor used is bismuth telluride, which is made *N*-type or *P*-type by control of the impurities added. Bismuth telluride has produced a maximum cooling of 70°C.

"With further advances in solid state physics, thermoelectric cooling becomes more and more feasible. Already for small systems and special applications thermoelectric cooling devices are more efficient than conventional refrigeration."

A quantitative corridor demonstration at Iowa State provides a measurement of thermal diffusivity. A sinusoidally varying temperature is applied to the end of a long metal rod, producing heat pulses which propagate down the rod with decreasing amplitude. Two thermocouples measure the amplitude decrement and propagation velocity. The relation for diffusivity is a solution of the appropriate boundary-value problem.

An interesting approach to the demonstration of a complicated technique, also developed by the group at Ames, uses models and animation to provide what they call a "dynamic outline" of steps in the measurement. For example, they demonstrate a technique for determining the internal resistivity of diamond. The problem is to give the observer a feeling for the actual technique, which has four features:

1. An external *E* field is applied to the diamond sample.
2. Gamma rays ionize the sample, and ions separate.
3. Some ions are trapped by impurities.
4. The *E* field is removed, the internal field is slowly neutralized by free carriers, and the rate of decrease gives a measurement of resistivity.

These steps are simulated in the display through the use of a "model" diamond in which one can observe the trapping and neutralization of carriers.

As a final example from the superb corridor demonstrations at this university, Fig. 22–25 shows their animated display of nuclear magnetic resonance. Exploiting the analogy between a top precessing in a *g*-field and a nucleus precessing in a *B*-field, the demonstration exhibits the NMR technique and displays (in animation on a model CRO) two absorption peaks for Na^{23}. It makes a point of the fact that the resonant frequency for a given nucleus depends on its atomic environment. Thus, it shows the peak occurring at 11.27 Mc in sodium chloride and at 11.28 Mc in sodium metal. Graphs at the top of the display show resonances for other nuclei.

In addition to the academic institutions mentioned so far, others have been active in the development of corridor demonstrations. Among them are the California State Polytechnic College in Pomona, the United States Coast Guard Academy in New London, the University of Connecticut in Storrs, the University of Florida in Gainesville, the University of Illinois in Urbana, Kansas State College in Pittsburgh, Lake Forest College in Illinois, the University of Louisville, Michigan State University in East Lansing, Michigan Technological University in Sault Ste. Marie, the United States Naval Academy in Annapolis, the University of Nevada in Reno, New York University, Occidental College in Los Angeles, the University of Oklahoma in Norman, Pennsylvania State University in State College, Southern Methodist University in Dallas, and Washington and Lee University in Lexington, Virginia. There are undoubtedly many that have not come to the author's attention.

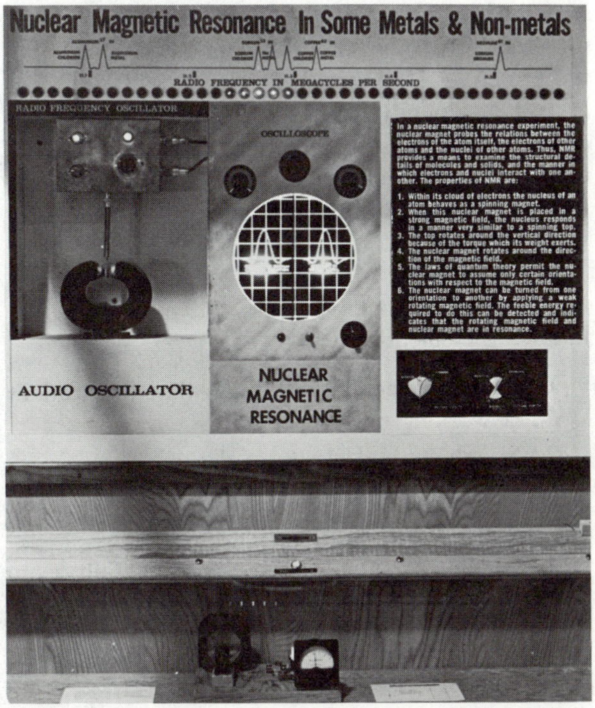

Fig. 22–25 Schematic Demonstration of Nuclear Magnetic Resonance, showing the analogy with a spinning top. (Photograph courtesy of Iowa State University in Ames.)

22–4 Requirements for Effective Corridor Demonstrations

Our survey of the current state of physics exhibits leads us to suspect that, although an enormous number of ingenious ideas have been developed by talented people, the art of corridor demonstration has not yet reached maturity. Of the four modes of presentation that we have identified (static display, self-contained pushbutton demonstration, experiment with demonstrator on hand, and complete laboratory exercise), only the first two have found extensive expression, most of which appears in the museums of science. The purposes of academic instruction are probably not well served by these two modes alone. However fascinated the spectator may be by the apparatus, descriptive legend, or curious phenomenon set before him, he is not likely to become deeply involved with the fundamental ideas at stake. Even when he gets his hands on the equipment he may not learn very much from it. At the University of Illinois, an air-track apparatus has been placed in the corridor where students can operate it themselves. Professor Almy reports that the results are disappointing: "My impression is that although the students and others play with it a good deal, they rarely relate it specifically to the principles of physics." Dr. Burkhardt says that, at M.I.T., "purely passive demonstrations (spark chamber, double-slit diffraction, Brownian motion, electron beams, field emission, etc.) are examined but not studied, and we are not sure that they are worth the effort."

If an exhibit is to play an effective part in teaching physics to people, I believe it must have the following elements:

1. Close connection with ideas that the student is working on,
2. Simplicity and clarity of design,
3. Capacity for catching the eye and engaging the mind,
4. Challenge to one's ability and ingenuity,
5. Variability under control of the spectator,
6. Temptation toward follow-up, and
7. Reward for solving a problem.

One or more of these elements can, of course, be found in many displays of the kind described in this chapter. The excellent demonstration of forced vibration (Fig. 22–7) deals with an idea that every student copes with and is also simple and transparent in design. An inverted pendulum (Fig. 22–8) is an engaging curiosity; the stroboscope problem (Fig. 22–9) is a challenge. But the last three features seem almost always to be missing. Students have little or no control over the experiment (we have discussed this point in connection with the Berkeley demonstrations, suggesting that they may be far less flexible than their authors claim). And, since there are few demands on the spectator, his reward is correspondingly meager.

Can effective experiments in fact be created? Let us look at this question in constructive terms. Suppose, for example, that the aim of one of our corridor exhibits is to draw attention to the ratio $\gamma = c_p/c_v$ of specific heats in a gas. A familiar technique for a determination of γ is the Rüchhardt method, in which a sphere oscillates in the neck of a closed container. One can show that the ratio is given by

$$\gamma = \frac{64mV}{d^4pt^2},$$

where m and d are the mass and diameter of the sphere, p and V are the equilibrium pressure and volume of the gas, and t is the period of oscillation. In its simplest form, the experiment requires one to drop a ball into a closely fitting tube and to time as many of its oscillations as one can observe.

A major improvement, which has occurred to several workers and which makes the method far more suitable for demonstration, is achieved by flowing the gas through the container (Fig. 22–26) and using a tube that has a very slight taper with its wider end at the top. A slow flow of gas transfers enough energy to the ball to maintain its oscillation indefinitely. The precision of the measurement is thus enormously improved. Typical data obtained at the University of Rochester, for γ at 15°C., are:

argon: $\gamma = 1.6635 \pm 0.0015$,
nitrogen: $\gamma = 1.4055 \pm 0.0012$,
carbon dioxide: $\gamma = 1.3134 \pm 0.0011$,

with errors that reflect uncertainties in all of the five measured parameters.

It is easy to imagine this experiment in the form of a corridor exhibit with sufficient equipment to enable the spectator to vary several parameters and so to investigate the form of the law. He can also be asked to derive the formula and to consider some questions that suggest themselves. Why is it that, although the result for nitrogen is in excellent agreement with data from other techniques, the argon γ is slightly low and the carbon dioxide γ is slightly high? Is the simple theory of the experiment exact when the gas is flowing? If not, can a correction be arrived at by calculation or by experiment? Can the accuracy and precision of the method be further improved?

The results that a student obtains from this experiment are, of course, no more "his own" than the positions of maxima and minima in the Berkeley interference demonstration. He plays no part in designing the apparatus and, if his data are carefully taken, they are the same as for other observers. Our list of criteria for effectiveness did not include the idea of uniqueness, which seems impossible to achieve with apparatus whose design is frozen.

Another failing of our hypothetical experiment is that, although the equipment is self-operating, it cannot *automatically* present the challenge of provocative questions at appropriate points in the procedure. Listed in the instruction sheet (like the "questions to be answered in the report" in our conventional laboratories), they are far less effective than when they arise in the course of the experiment itself. Furthermore, a question should be raised for the student only when he fails to raise it himself. The ideal situation is one in which a patient and perceptive instructor stands nearby offering suggestions, raising questions, remaining silent when things go smoothly, and rewarding good work with a "Well done."

It is possible to hope that contingent monitoring of this kind will soon be accomplished by computer methods. Physics programs in linguistic modes have already been devised[17] and tested; they include examples of laboratory exercises in which the student reports his data to a computer and receives guidance, criticism, and a grade.

If a computer facility were added to the corri-

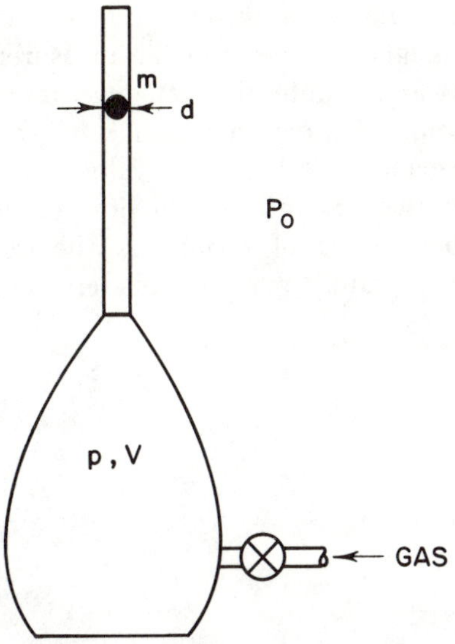

Fig. 22–26 Rüchhardt Method for Determining γ.

dor demonstration of our γ-measurement, a hypothetical dialogue might proceed as follows:

"What parameters of the system, in addition to the period of oscillation, should one measure to determine γ?"

"*m, p, V.*"

"Yes, these are essential, but you need at least one more. Try a dimensional analysis."

"We need a length. Maybe the radius of the ball."

"Right. On what power of the radius will γ depend?"

"The fourth."

"Well, it's the *inverse* fourth power. Now a certain gas is flowing in the system. Take the necessary data and report your value of γ."

"1.65."

"Good. That's close to my value. What is the most likely molecular structure of this gas?"

"Diatomic."

"No. It's monatomic. Let's see why. . . ."

—and so on.

In its ultimate form, the demonstration can contain a variety of readout devices (slide projector, tape, film) under control of a computer, with a style of exhibit to be chosen by the spectator. When he selects DISPLAY, the equipment operates and he sees or hears a brief description. Selecting DEMONSTRATE, he obtains a longer display, exhibit-

(17) For a report of the status in 1965, see "The Computer in Physics Instruction," published by the Commission on College Physics, Ann Arbor, Michigan.

ing the full range of the experiment. If he is a student, assigned to perform the measurement, he selects LAB and initiates the dialogue with an appropriate computer program which keeps records of his performance as a basis for evaluation.

Finally, there is an excellent prospect for corridor demonstrations of computers themselves, inviting the spectator to vary parameters in an analog computation, to program from a keyboard, or simply to observe an automatic solution of a problem in physics. On-line systems for research are developing rapidly[18] and their applications to instruction are only beginning to be explored.

[18] See, for example, W. J. Karplus, ed., *On-Line Computing in Time-Shared Man-Computer Systems* (McGraw-Hill Book Co., Inc., New York, 1965).

23

THE OVERHEAD PROJECTOR IN THE PHYSICS LECTURE

By Walter Eppenstein
Rensselaer Polytechnic Institute

23–1 Introduction

The publication of this demonstration reference with its many contributions indicates the continued interest of physicists in the presentation of demonstrations before small and large groups of students. As pointed out in several articles in these two volumes, demonstration lectures in physics have been given for a long time and have always played a significant role in the teaching of physics. Many of the so-called visual aids that have found their way into the classrooms and lecture halls in recent years in almost every subject have been employed by the physicist for many years. It is no coincidence that so many of the recent innovations were first used in physics classes.

23–2 Visual Aids in the Physics Lecture

The reason for the importance of all kinds of audio-visual aids in physics lectures is, of course, the simple fact that the most beautiful demonstration is of no use at all if it cannot be seen or heard by all students present. Probably the biggest mistake any lecturer can make is the performance of an experiment that cannot be followed by all students, even the ones in the last row of the room. Unfortunately, many of our experiments cannot be enlarged easily; in this case they should either be dropped or used with an appropriate device for magnification. In most physics courses, of course, students also have a laboratory in which they have the chance of actually handling equipment. Many of our demonstration experiments properly belong in the laboratory.

Let us assume that we have picked a demonstration we do want to show and explain in lecture. We must then determine the best possible way to make the equipment visible to everybody. There is no unique way to accomplish this; the type of device or visual aid employed depends on many factors. We may want to make use of live closed-circuit television in order to magnify the setup, or we may decide that a video tape recorded ahead of time would be better in our particular case (Chapter 20). Possibly a slide, 2 x 2 in. or 3¼ x 4 in., or a transparency on the overhead projector can help us explain the equipment. A short film may be the best solution for some special demonstration experiments (Chapter 21).

Of course the demonstration equipment may be made so large that everybody can see it, and we do not have to worry about projection techniques. This, however, is not practical in many cases. Some experiments just cannot be performed if the equipment is too big. Large pieces of apparatus also add to the difficulties of setting up the demonstration as well as to the storage problem. There is also the concern for not obscuring the view of other experiments, the blackboard (if used) or the screen.

One possibility to make apparatus visible to large groups is shadow projection, a method that has been used for a long time very effectively. With the introduction of the overhead projector, however, we have a highly efficient optical system all ready for shadow projection which avoids the time-consuming adjustments required very often if the projection system is home-made. It is therefore suggested that the overhead projector be used as one of the means at our disposal to help students view lecture demonstrations.

23–3 The Overhead Projector

Originally, the overhead projector was introduced to replace the blackboard, and this is still its primary function. The blackboard, one of the oldest visual aids used in the classroom, has distinct disadvantages in the physics lecture hall. Very often the lecturer does not realize how large he has to write in order to have the student in the back of the room be able to read every single word or symbol. Drawings on the board become more difficult to make for most physicists as the complexity increases. Although colored chalk is used frequently, good chalks are not easy to obtain.

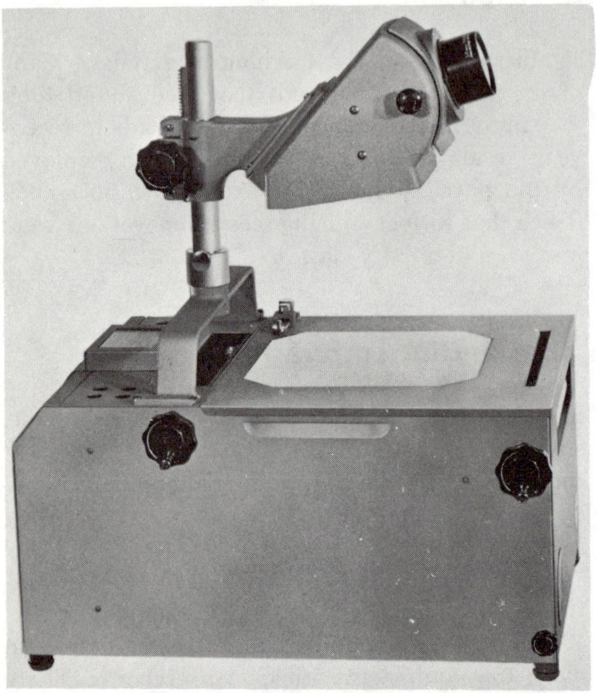

Fig. 23–1 A Typical Overhead Projector for use in lecture halls. (Courtesy Projection Optics Co. Inc.)

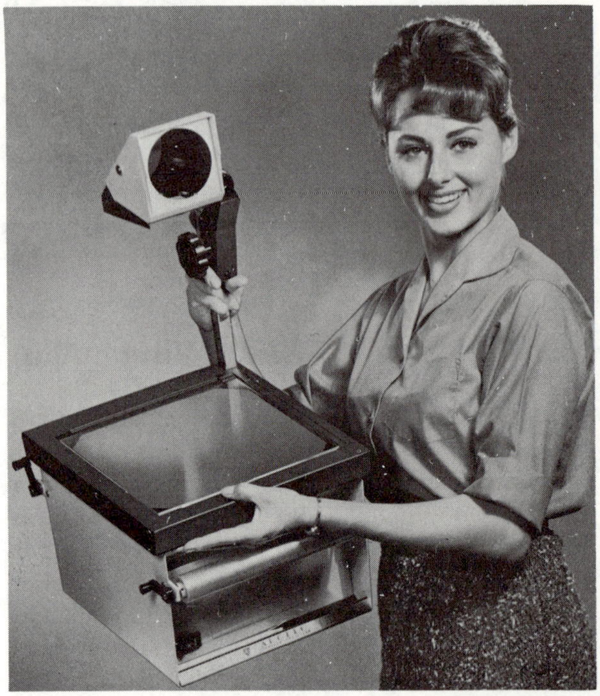

Fig. 23–2 A Portable Overhead Projector weighing 15 lb. (Courtesy of the American Optical Company.)

Erasing during the lecture is a nuisance and usually ends up in a rather dirty blackboard as well as a lecturer covered with a layer of dust. Equipment on the lecture table, traditionally located between the student and the board, very often obscures the view of the blackboard, at least for some of the students. Blackboards must be well illuminated, whereas the showing of slides, films, television and shadow projection requires a somewhat darkened room.

One advantage in lecturing with the overhead projector is the fact that the lecturer faces his class. He is able to watch the students and thereby judge their reactions to his presentation, and the students have the benefit of being able to see their instructor at all times. The lecturer may stay in one place while using the projector and he does not have to erase part of his lecture. If he uses a commercially available roll of acetate on the projector, he can bring back any part of his lecture at any time.

A large number of overhead projector models are presently available from many different manufacturers; some of these are listed at the end of this chapter. Figure 23–1 shows a typical projector as used in lecture halls, and Fig. 23–2 pictures a portable projector of lightweight construction. Figure 23–3 shows an overhead projector mounted for the projection of materials on a vertical stage instead of a horizontal one. This system is of particular interest in chemistry demonstrations[1] but may also have application in physics.

23–4 Transparencies

23–4.1 The Use of Transparencies

The use of the overhead projector has the added advantage of enabling the lecturer to have materials prepared ahead of time. Transparencies for the overhead projector may be produced just prior to the lecture by using a grease pencil on acetate sheets. They can also be produced by an office copying machine such as Thermofax or Xerox or by an artist in a professional studio using a diazo or printing process, or they may be purchased when commercially available. Once a transparency has been prepared by any method, it can be stored for future use and is always available. When a file of overlays has been built up, the task of preparing a lecture will be very much simplified.

Problems, problem solutions, and quizzes can, of course, also be put on transparencies and projected at the appropriate time. How an individual transparency is used depends entirely on the lecturer. It must be realized that most of the visual aids do not teach; they are simply an aid to the lecturer to simplify and improve his teaching. A lecture should not be built around available teaching devices; such devices, e.g. transparencies, should be used only when they really add to the presentation.

Single transparencies replace the regular slide as well as the blackboard. The lecturer's ability to add to or complete such a transparency makes it much more useful than the common small slide. A bubble chamber photograph, for instance, may be projected first. The lecturer can then show the tracks he is interested in and write out the nuclear reactions in appropriate places. Any marks made on the transparency with a grease pencil or a colored marker can easily be removed.

While the single transparency is a substitute for slides or blackboards, the colored overlay uses techniques not possible by any other means. A complicated picture or diagram may be built up in front of the students by using up to six or eight visuals of different colors. With a basic transparency mounted in a frame, colored overlays can be flipped on or taken off from all four sides. In Fig. 23–4, a crystal model is shown as an example. By means of overlays, one atomic plane after the other is projected. Up to four different sets of planes can be shown, flipped over from different sides. The lecturer can, of course, write on the trans-

[1] This projector was designed for use with TOPS (Tested Overhead Projector Series) experiments described in the *Journal of Chemical Education. Editor's Note:* "TOPS" is a monthly series by Hubert N. Alyea, Princeton University, appearing in the *J. Chem. Ed.* since 1962. A reprint of all TOPS articles from 1962 to July 1967 is available from the author; title *TOPS in General Chemistry*, 3rd ed., 1967. Also available are *Tested Demonstrations in General Chemistry*, by Hubert N. Alyea and Frederick B. Dutton, 6th ed., 1967; and *Microchemistry Projected—TOPS*, 1st ed., 1968.

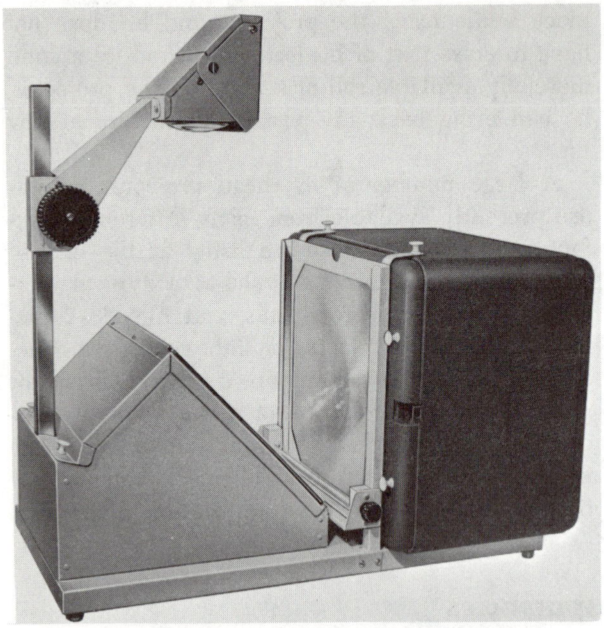

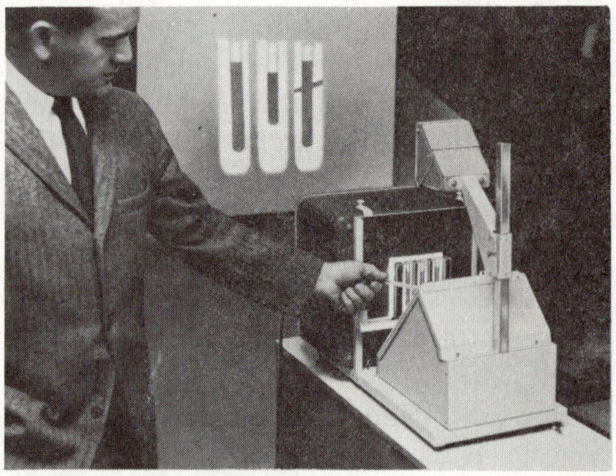

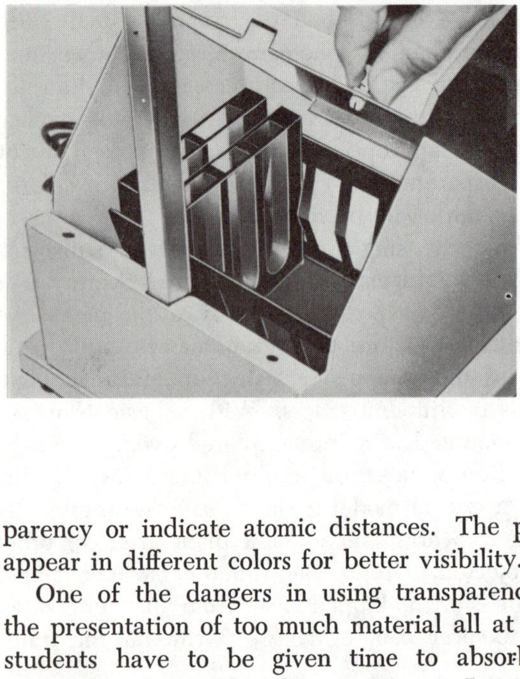

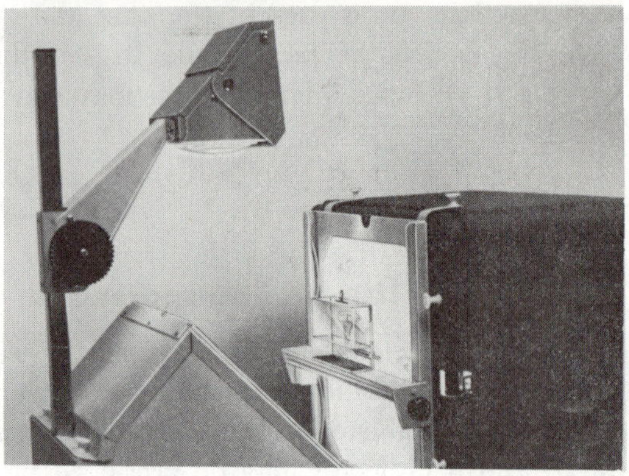

Fig. 23–3 The Beseler Porta-Lab Science Experiment Projector. (Courtesy of the Charles Beseler Company.)

parency or indicate atomic distances. The planes appear in different colors for better visibility.

One of the dangers in using transparencies is the presentation of too much material all at once; students have to be given time to absorb the material and possibly take some notes. It is never desirable to have a lengthy derivation or a complicated problem suddenly appear on the screen. One could use a number of overlays and thereby slowly build up the final picture, or one could uncover the transparency as the ideas are being discussed. An unfinished transparency has proved to be very effective; the lecturer has the basic outline but can complete it as he talks about it.

In some cases overlays may be used to illustrate physical principles difficult to show by other means. An example is the interference pattern resulting from two coherent sources, such as in Young's experiment, which can be illustrated easily by means of movable overlays as shown in Fig. 23–5. By slowly moving one overlay with respect to the other one, the relation between the angles subtended by successive maxima and source separation can be established. Three different situations are shown respectively in (a), (b), and (c).

It should be made clear in this case that we are dealing with a visual aid and not a physical demonstration of interference. It is suggested that the

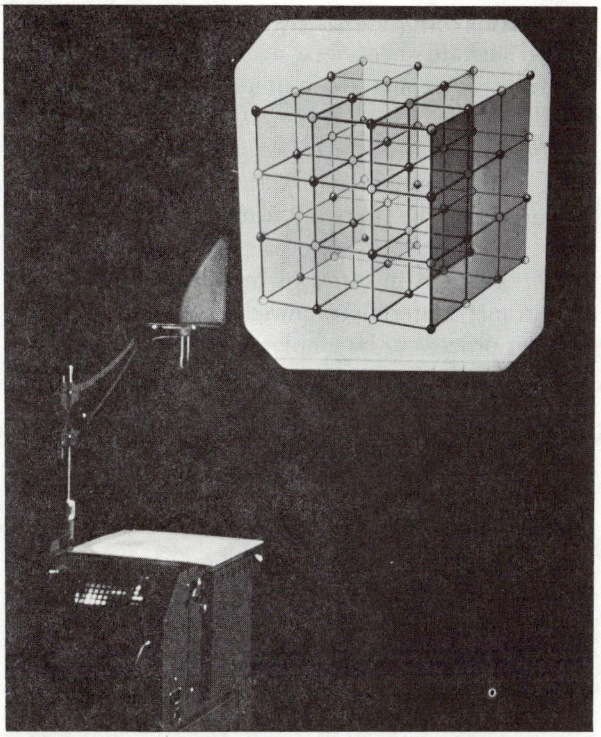

Fig. 23–4 A Transparency of a Crystal Model is projected. The various atomic planes are shown in different colors.

transparency model be followed by an actual interference demonstration, such as in a ripple tank. If the ripple tank itself is not shown, the film loop on interference of two sources in a ripple tank may be shown.(2) This will illustrate the similarity between the resultant interference pattern due to actual waves (water waves in case of the ripple tank) and the movable transparencies.

The relation between the physical demonstration and a model discussed above is just one example; there are many others. For instance, the use of Technamated transparencies with a Polaroid spinner gives the impression of motion across the screen; this is a visual aid. When we discuss polarization, however, this visual aid now becomes a physical demonstration. The process of Technamation uses the principle of rotary polarization. Special materials are available that consist of this birefringent, polarized plastics arranged and permanently mounted on acetate sheets.(3) The apparent motion is activated by placing a rotating

(2) Interference and Diffraction, listed in Section 21–6.2.

(3) Materials produced by Technical Animations, Inc., and distributed by various companies such as the American Optical Co. and the Tecnifax Corp.

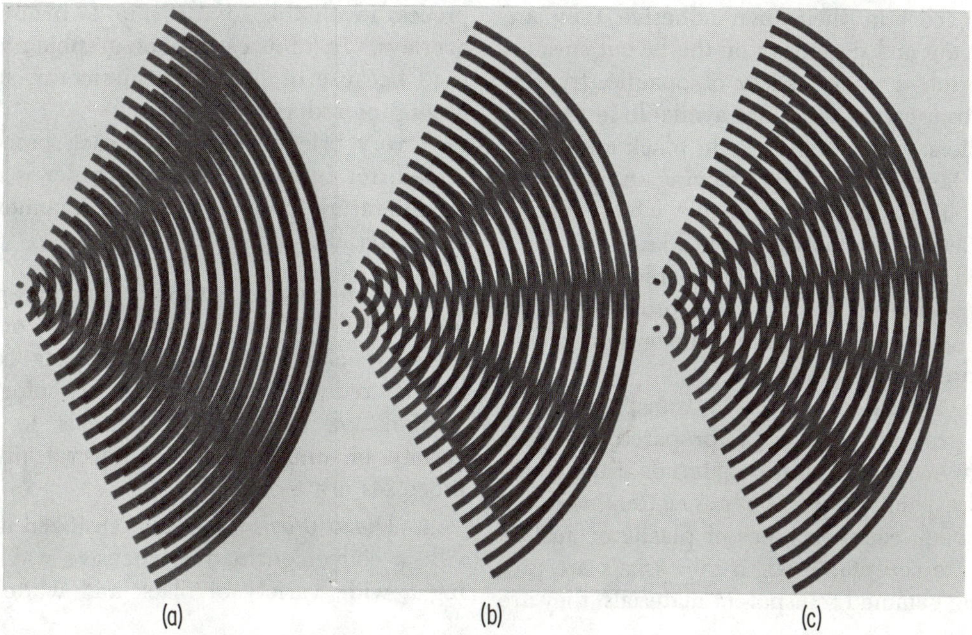

Fig. 23–5 Two Identical Transparencies with Concentric Circles are moved with respect to each other to illustrate interference effects.

polarizing disk[4] (polarized spinner) between the light passing through the transparency and the screen.

23–4.2 Design of Transparencies

If a permanent professional-looking transparency is desired, some attention has to be paid to the design of the visuals. Here the physicist may want to consult the artist with some experience in the design of transparencies. Such principles as repetition, balance, emphasis, unity and contrast must be considered. It is important to choose proper line width, color, size, texture, etc. to make the transparency as effective as possible. This can be accomplished best by an understanding between the content authority and the artist.

The finished artwork is accomplished through the skillful application of art materials to fulfill the elements of design contained in the layout. The artist of today has available, in addition to his own skill in creating and producing visual materials, a wide assortment of prepared art materials. These include sheets of letters, numbers, designs, symbols, textures, patterns, figures, and shadings. These materials can be obtained in a wide variety of sizes and colors—either opaque, translucent, or transparent. Since they contain their own adhesive they are simply cut out and pasted up on the layout surface.

In addition, a wide variety of opaque, translucent, and transparent tapes are available in varying widths, colors, and designs. Plain black and white or colored tapes are used for drawing straight lines or circles. Tapes are also available which contain numbers, letters, symbols and the like.

Finished artwork may be prepared either for direct "as is" use on the overhead projector or for reproduction purposes in making one or more copies on transparent materials.

Artwork for direct use as a transparency for overhead projection must be prepared on clear transparent acetate or plastic materials using transparent inks, paints, tapes, screens, letters, and the like. Although common types of plastic or acetate materials are acceptable when the visuals are prepared using gummed transparent materials, they are not recommended for inking or printing. Specially treated acetate sheeting is available where acetate type inks or paints are required in preparing the visual.

Finished artwork to be used as a master visual from which reproduction transparency copies are produced may be prepared using a variety of materials. The process used in preparing the artwork is usually selected in terms of the efficiency of preparation, art materials required, and especially the printing process to be employed.

23–4.3 Reproduction of Transparencies

The four main types of reproduction processes available for the preparation of overhead transparencies are (1) photographic, (2) diazo, (3) thermal, and (4) electrostatographic.

The diazo process proves to be the cheapest as well as the simplest to use, especially in connection with color overlays in which carefully prepared artwork is desirable. For the case in which color is not important, the thermal process is certainly the most convenient one. The lecturer can produce a transparency on an office copying machine, such as Thermo-Fax, from any printed or written material only minutes before his class, making this type of process invaluable for last minute black and white overlays. In a few cases photographic processes are used because of special requirements, such as enlarging or reducing originals.

A very brief discussion of each process is presented; for further details the reader is referred to the literature published by the manufacturers of the materials.

1. *Photographic process.* Photography offers great flexibility in the preparation of overhead transparencies. Artwork and objects can be enlarged, reduced, printed and rephotographed in combination with other materials to achieve a variety of unusual effects. Several photographic processes are available.

2. *Diazo process.* Diazo-sensitized films[5] provide a convenient and inexpensive way of preparing a wide variety of black and white and color

[4] Polarizing spinners for overhead projectors are available from the Charles Beseler Co., the American Optical Co., the Tecnifax Corp., and others.

[5] Diazo materials are available from the Tecnifax Corp., the General Aniline and Film Corp., the Keuffel & Esser Co., and others.

transparencies by contact printing methods. Diazo films are available in at least ten colors, and secondary colors can be produced by superposing one color over another. This makes it possible to prepare multicolor transparencies for overhead projection, usually at a lower cost per visual than by full color photography.

Diazo printing is a direct reproduction process by which a positive original produces a positive copy and a negative original produces a negative copy.

A print is made by exposing the diazo-sensitized material to ultraviolet light through the original artwork. The exposed diazo film is produced in an aqua-ammonia vapor where the diazo combines with a coupler to form a dye-image. This diazo printing process may be carried out in any normally lighted room if the film is not subjected to direct sunlight or prolonged exposure to fluorescent lights.

Printing and developing equipment may range from the very simple to the rather expensive apparatus. For the production of transparencies in small quantities, an ordinary sunlamp has been used as the ultraviolet light source, with a photographic printing frame to hold the film and artwork, and a jar of ammonia with a sponge as the developer. A variety of portable printers are available for faster production of transparencies.[6] Use can also be made of the more expensive ammonia developing diazo-type machines found in printing establishments and graphic arts shops.

3. *Thermal process.* Projection transparencies from a variety of original materials can be prepared in seconds using office-type thermal process machines.[7] Both positive and negative copies can be prepared from the same original. Information not wanted on the transparency can be masked out before printing. By double printing, separate visual materials may be combined in a single transparency.

Originals can be opaque or translucent and printed on one or both sides, but they must contain a carbon base or the equivalent if they are to be reproduced by the thermal process. Xerox copies may be used as originals for Thermofax transparencies.

(6) Most of the companies selling diazo-sensitized film also have printers available.

(7) The Thermofax copying machine, manufactured by the Minnesota Mining & Manufacturing Co., can be used to obtain transparencies in four seconds.

Office copying machines generally produce 8½ x 11-in. transparencies, while the overhead projector stage is usually 10 x 10-in. Although color film material is available, it requires a liquid processing solution and is not yet practical.

4. *Electrostatographic process.* The electrostatographic process, such as xerography, produces copies from any type of visual material. Original matter, printed text and artwork, drawings, and the like are reproduced on office-type copying machines, usually exactly to size. Transparent copying sheets for Xerox machines are now commercially available.

23–4.4 Mounting Transparencies

The proper mounting of all transparencies is important for ease of handling as well as for safe storage. Most manufacturers of printing materials also sell frames and other mounting accessories. When several visuals are used it is most important to properly register the slides so that the resulting transparency becomes meaningful. Numerous devices are commercially available to facilitate this operation.

A spiral binding can be used to mount a set of transparencies together with teacher notes;[8] a hinge system for mounting is also used.[9]

23–4.5 Availability of Transparencies

In many cases the lecturer will want to prepare his own transparencies to fit his particular need, and in the preceding sections it was assumed that this is the case. Sometimes, however, a commercially prepared transparency may save considerable time. These may be available either as masters or as finished transparencies. Several companies sell series of transparency masters[10] from which one can prepare transparencies by any of the processes discussed.

Finished transparencies in many fields are sold

(8) One such mounting system has been developed by Visual Systems, Inc.

(9) One such mounting system has been developed by the Tecnifax Corp.

(10) Keuffel & Esser produces a number of series for use in high schools, including a physics series: Physics Diazo Transparency Masters, available through all K & E distributors. A series of finished transparencies for high school physics is also available from Modern Learning Aids.

by companies dealing with overhead projectors and diazo materials as well as others; relatively few transparency series are available in physics. Recently several publishers have started to distribute transparency series,[11] some designed for use with specific texts.[12]

23–5 Demonstrations on the Overhead Projector

The use of a projector for lecture demonstrations is not new. Various still and movable models, projection meters, etc. for 3¼ x 4-in. slide projectors have been in use for decades. The projection of actual demonstrations using a slide projector was described in the literature many years ago.[13] At that time, however, slide projectors with a small stage had to be used. With the development of the 10 x 10-in. stage and an efficient optical system on modern overhead projectors, experiments can be carried out with greater ease and better visibility.

Any physicist using an overhead projector will soon find it a useful projection technique in a variety of situations. Old demonstration experiments, formerly made visible by means of carbon arcs or other shadow projection apparatus, can now be projected without any special setups or adjustments by making use of the stage of an overhead projector.

The physics lecturer should differentiate between the use of an overhead projector as a visual aid and its use for the demonstration of physical phenomena or actual experiments. Using the projector to shadow-project the components of an electric circuit is a visual aid; when the switch is closed, however, and a real measurement is taken, the visual aid has turned into an actual experiment of a qualitative or quantitative nature. The overhead projector, although originally designed solely as a visual aid, can easily be used to make real physical demonstrations and experiments visible to large groups of students.

It is certainly not proposed that all lecture demonstrations be performed on the 10 x 10-in. stage of an overhead projector. Students should watch the lecturer actually perform numerous experiments using equipment sufficiently large and loud to be seen or heard by everyone in the room. Other types of visual aids such as closed-circuit TV, shadow projection, films, slides, stroboscopic illumination and large models have their place in any physics lecture. It was found, however, that the visibility of many old demonstrations could be improved considerably by the use of overhead projection.

One definite advantage of working with the overhead projector is the ease of storing it and its rather small accessories, but any equipment used for projection purposes must be stored in a dust-free place, preferably covered with a plastic bag or put into a closed cabinet.

The overhead projector serves as both the light source and the projection system for making visible a large number of conventional demonstrations. Many of these are discussed throughout this book, and every lecturer using overhead projectors will soon find new applications. The following are just a few examples to illustrate the types of demonstration possible.

23–5.1 Motorizing the Projector

Most demonstrations do not require any modification of the projector. In some cases, however, it is convenient to attach special motors or other devices; some of the possibilities are shown in Fig. 23–6. The variable-speed motor D is used to drive the rotating stage shown in Fig. 23–7. See also Chapter 15, Exp. 15–1.11, for an application of the acetate roll drive.

23–5.2 Polarization Demonstrations

With a sheet of Polaroid measuring 10 x 10 in. on the stage of the projector and a Polaroid spinner or a second sheet of Polaroid, many of the demonstrations in polarization can be performed

(11) A set of 38 transparencies in physics is available from McGraw-Hill Film Text Department.

(12) A transparency series for PSSC Physics is available from Macalaster Scientific Corporation, and a similar series for Harvard Project Physics is distributed by Holt, Rinehart and Winston, Inc. A college level series for use with Halliday and Resnick and other texts of this level is published by John Wiley & Sons, Inc.

(13) *Am. J. Chem. Ed.*, **16**, 314 (1939), and **17**, 210 (1940).

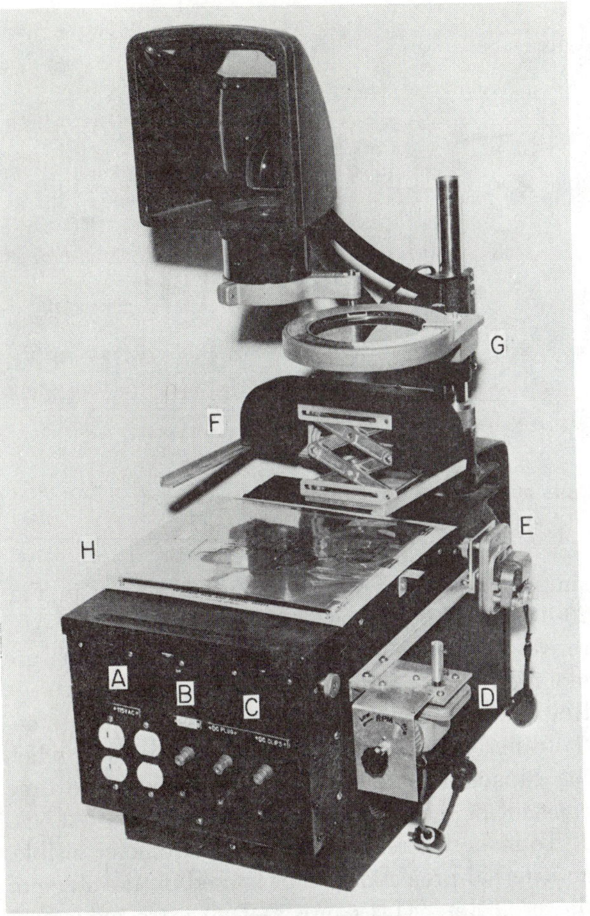

Fig. 23–6 A Vu-Graph Overhead Projector with the following accessories: A, supplementary 115-V ac outlets. B, 6-V ac outlet. C, 6-V or 110-V dc outlets. D, variable-speed motor to drive rotating stage or wave demonstrations. E, constant-speed motor driving acetate roll. F, frame for mounting breadboards for electrical connections. G, Polaroid spinner for use with technamated slides. H, mounted colored overlay.

with the greatest ease. These may range from the very simple to the complex. A crumbled piece of cellophane from a pack of cigarettes, for instance, will cause beautiful colors changing with the rotation of one of the Polaroid sheets. This is very effective in a darkened room and should, of course, be explained by the students.

The principles of photoelasticity can be demonstrated to large groups by projecting strained, two-dimensional photoelastic models. A special device for these demonstrations has been constructed.(14)

(14) "A Demonstration Polaroscope Used with an Overhead Projector," by Byron J. Pelan, *J. of Eng. Educ.*, 460, 57 (Feb. 1967).

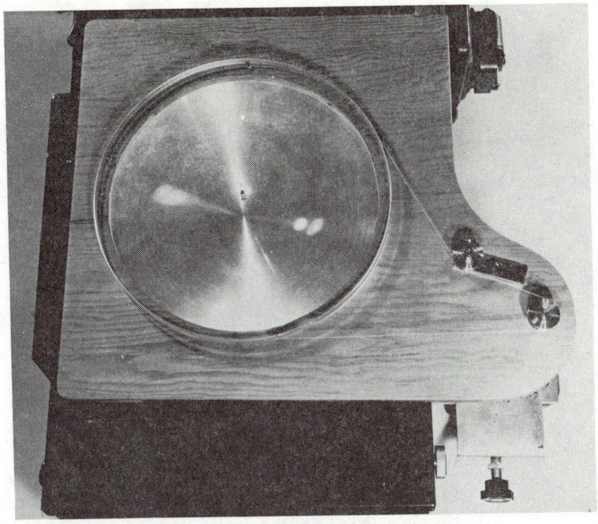

Fig. 23–7 Rotating Stage for overhead projector.

23–5.3 Magnetic Fields and Induced Electromotive Forces

A projection galvanometer is placed on the stage of the projector. When connected across a straight wire, the induced emf can be observed as the wire is passed across the pole pieces of a magnet also on the stage of the projector. The speed and the angles of the wire may be varied, of course.

The galvanometer may also be connected to various coils placed on the projector and induced emf's due to moving magnets are observed or measured. In Fig. 23–8 the primary coil is connected across a battery, and a galvanometer across the secondary shows once again the induced emf.

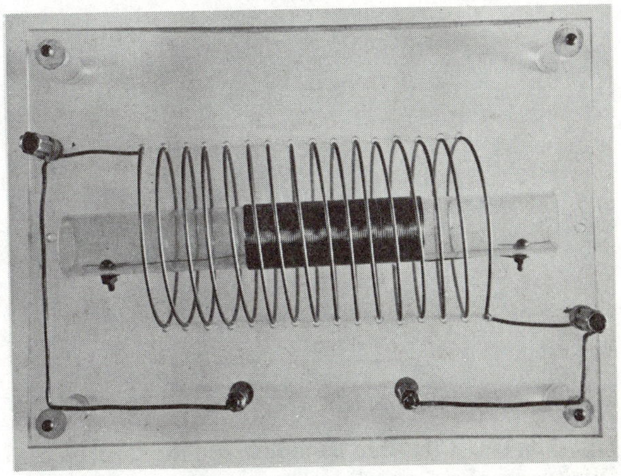

Fig. 23–8 Primary and Secondary Coils for use on overhead projectors.

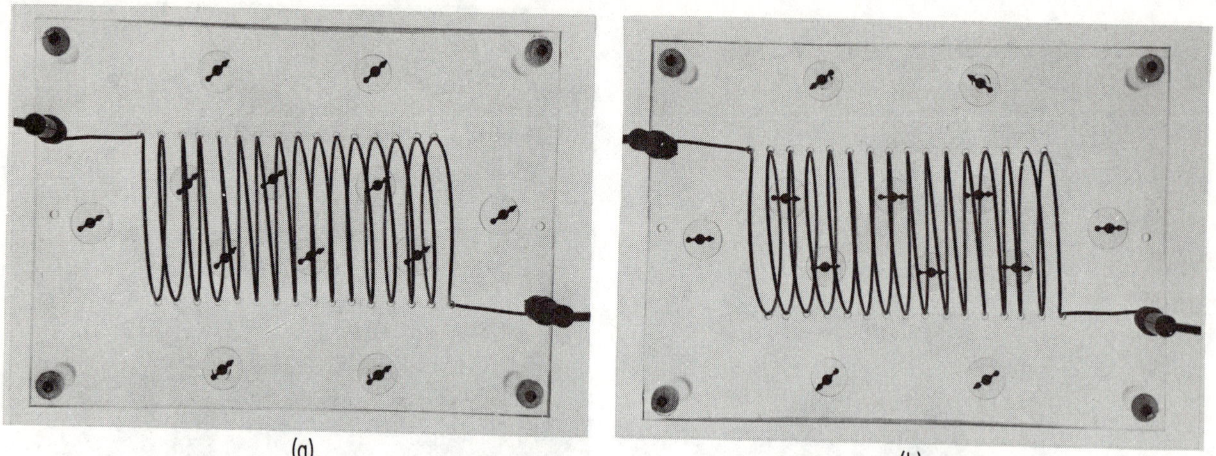

(a) (b)

Fig. 23–9 Magnetic Field as shown by compass needles: (a) no current and (b) current flowing through the coil.

A steel rod can be pushed through the inner coil when permeabilities are discussed.

The primary coil can also be used to illustrate magnetic fields. In Fig. 23–9(a) no current passes through the coil, while in Fig. 23–9(b) there is a current flowing, as shown by the direction of the compass needles. When performing this experiment, a projection ammeter is connected in series with the coil and placed on the stage of the projector.

23–5.4 Electric Circuits

Working electric circuits can be shown in shadow projection. An example is the demonstration of the Hall effect described in Chapter 40. Another simple example of such a circuit is shown in Fig. 23–10. A 30-V battery, a 1-μF capacitor, a 7-MΩ resistor and a single-pole double-throw switch are mounted on a plastic breadboard as shown. A Keithley electrometer is used to measure the potential difference across the capacitor or resistor when the capacitor is charged or discharged. The time constant of the circuit can easily be measured.

To make the reading on the electrometer visible, an external projection meter is used on the stage of the projector. This consists of an extra movement for the electrometer mounted in a clear plastic case with a transparent photograph of the original scale, as shown in Fig. 23–11. A switch on the side of

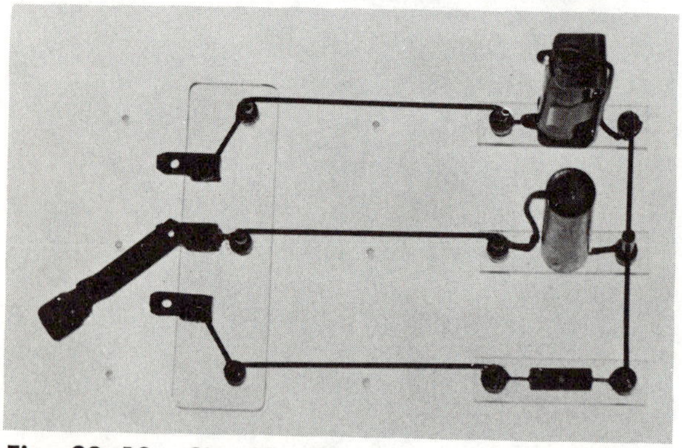

Fig. 23–10 Circuit to Demonstrate the Charge and Discharge of a Capacitor.

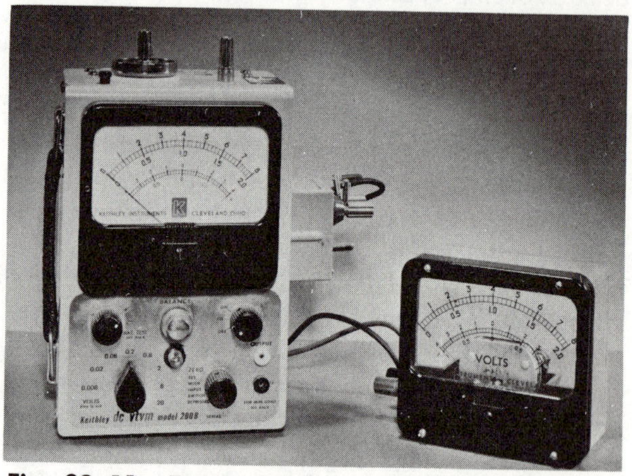

Fig. 23–11 External Projection Unit Connected to a Keithley Electrometer.

the electrometer enables us to use this instrument with or without the external projection unit.

When projecting electric circuits mounted on plastic breadboards, one has to ensure proper focusing of the projector by having all components, the face of the projection meters, etc. in the same horizontal plane. Letters and symbols can be added. Sometimes a transparency of the circuit may be shown first; after its discussion the actual working circuit is then superimposed.

23–5.5 Deflection of an Electron Beam by Magnetic Fields

There are situations where the light source of the projector can be replaced by a different source, using just the lens system for projection purposes. One example is the use of a cathode ray tube commercially available to demonstrate deflections of charged particles due to magnetic fields.[15] The tube is first shadow-projected, making use of the overhead projector light source. When the spark coil supplying the high voltage across the tube is turned on, the projector light is turned off, and the greenish looking electron beam is focused onto the screen. A magnet brought close to the tube will deflect the beam. The magnet can also be shadow-projected first. For the electron beam to be clearly visible, however, the room has to be in complete darkness.

(15) Central Scientific Co., No. 71555, or W. M. Welch Scientific Co., No. 2145.

23–6 Models on the Overhead Projector

A large number of models and analog devices can be constructed for use on the stage of the overhead projector; many of these are described in either volume of this book, such as the two-dimensional kinetic theory model, the superposition model, and the Rutherford scattering model. These models may be very simple but still effective. The following two examples illustrate a few possibilities and are not meant to be a complete list; once again each lecturer will have his own ideas depending on his individual needs.

23–6.1 Hydrogen Molecule Model

When discussing equipartition of energy, specific heats, and degrees of freedom, a model of a dumbbell-shaped molecule is useful. It is shown easily that the rotational energy should consist of two terms. Although the two spheres of the model first seem to be connected by a rigid rod, as shown in Fig. 23–12(a), the connector is actually a spring, and this new model gives rise to vibrational energy, as shown in Fig. 23–12(b).

23–6.2 Rotating Vector (Phasor) Model for Interference Effects

In an introductory course multiple slit interference patterns, diffraction gratings, as well as single slit diffraction are usually explained by the use of rotating vectors, called phasors. Phasor models

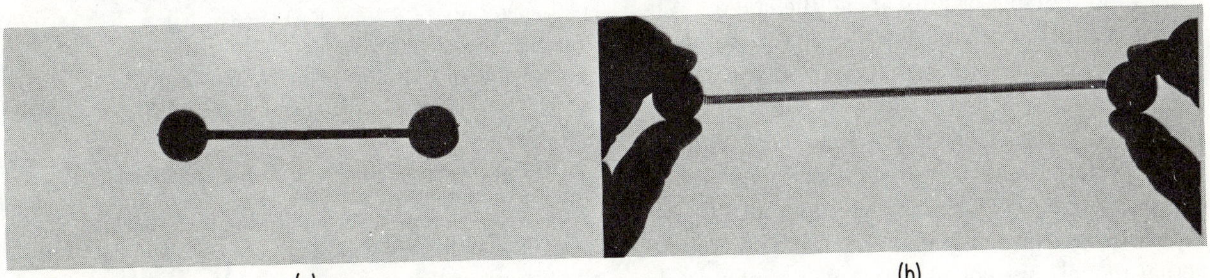

(a) (b)

Fig. 23–12 Dumbbell Model for the Hydrogen Molecule. (a) Two spheres of model appear to be connected by a rigid rod. (b) The connector is a spring.

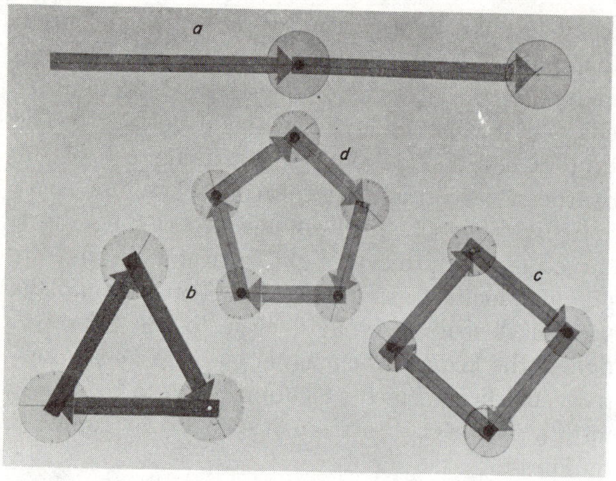

Fig. 23–13 Phasor Models for Explaining Multiple-Slit Interference Patterns.

have been constructed for two, three, four, and five slits and are used on the stage of the overhead projector to explain the approximate position of maxima and minima and the relative intensity of all peaks in the pattern.[16] In Fig. 23–13, a two-slit model *a* is shown for maximum intensity (constructive interference), while models for three, four, and five slits, *b*, *c*, and *d*, respectively, show zero intensity (destructive interference).

In single slit diffraction, we are dealing with a large number of small phasors; these can be represented by a chain, a flexible wire or an O-ring cut open. It is easy to show the approximate position of minima as well as the relative intensity of the maxima.[17]

23–7 Sources of Equipment and Materials

Fundamentally there seems to be no great difference between the various models of overhead projectors now on the market. Most 10 x 10-in. models have 1000-W lamps, type T-20, medium prefocus (MPF). The most common lens is 14 in. *f*/3.5 (4 in. diam). Most, but not all manufacturers use Fresnel lenses between the light source and the stage. Most models include 110-V outlets to plug in various accessories.

The reflection of light into the eyes of the lecturer as well as the noise of the cooling fan may prove annoying in some cases. There has been a trend to lower projection heads in order to avoid obstructions. These projectors are designed, of course, for transparencies and overlays and not necessarily for models and demonstration equipment. For this reason some types will not focus on planes more than 1 or 2 in. above the stage. When the projector has a very low head, it may also be difficult to perform an experiment on the stage of the projector.

Overhead projectors range from very light and easily portable units to rather heavy ones intended for permanent installations. Because of the very rapid changes presently made by all manufacturers, actual models are not listed. Anyone interested in overhead projectors and accessories as well as supplies for the design and construction of transparencies should consult the latest issue of equipment directories.[18]

Most scientific equipment distributors, such as the Central Scientific Co. and the Sargent-Welch Scientific Co., carry overhead projectors and accessories. Among the many companies producing or distributing projectors or accessories and willing to supply free literature are the following:

American Optical Co., Instrument Division, Buffalo, N.Y.
Applied Science, Inc., Cleveland, Ohio
Buhl Optical Co., Pittsburgh, Pa.
Charles Beseler Co., East Orange, N.J.
Charles Bruning Co., Mount Prospect, Ill.
Eastman Kodak Co., Rochester, N.Y.
Keuffel & Esser Co., Audiovisual Division, Hoboken, N.J.
Minnesota Mining and Manufacturing Co., 3M Visual Products, St. Paul, Minn.
General Aniline and Film Corp., New York, N.Y.
Projection Optics Co., Inc., East Orange, N.J.
Tecnifax Corp., Holyoke, Mass.
Technical Animations Inc., Port Washington, N.Y.
United Transparencies, Inc., Binghamton, N.Y.
Visual Systems, Inc., Milwaukee, Wis.

(16) See David Halliday and Robert Resnick, *Physics*, Part II (John Wiley & Sons, Inc., New York, London, Sydney, 1966) Fig. 45–5, for a five-phasor diagram.

(17) See Halliday and Resnick, *Physics*, Fig. 44–7.

(18) Such as the *Audio-Visual Equipment Directory*, published by the National Audio-Visual Association, Inc., Fairfax, Va., and *A Materials Handbook*, Indiana University, Audio-Visual Center, Bloomington, Indiana.

24

STROBOSCOPIC EFFECTS

By *Harry F. Meiners*
Rensselaer Polytechnic Institute

24–1 Introduction

High-speed photography was introduced by Mach over a century ago. He used exposure light from high-voltage spark gaps to photograph bullets and shock waves on glass plates. Until recently, because of its complex technique, high-speed photography was used only in research laboratories.

Recent contributions to the development of electronic-flash photography by H. E. Edgerton of M.I.T. helped greatly to simplify procedure. Today, high school and college students use the stroboscope as a convenient research tool. Edgerton's artistic and scientific photographs are also a constant source of inspiration to everyone.

Combination of the stroboscope with the Polaroid camera enables the investigator to obtain nearly instant single and multiple-flash photographs. If a photograph of an event is not satisfactory, another can be taken almost immediately and the results checked quickly. Hence, the stroboscope can be used with great effectiveness by students in the laboratory and by the lecturer.

Table 24–1 contains a list of additional stroboscopic experiments, not discussed in this chapter, and should be consulted for experiments and techniques useful in lecture.

Table 24–1

Volume I	Volume II
Chapter 6, Measurement, Simple Machines, and Vectors, 6–2.7	Chapter 27, Gases and Kinetic Theory, 27–7.3
Chapter 7, Kinematics; 7–1.4, 7–1.14, 7–1.15, 7–2.1, 7–2.4, 7–2.5	Chapter 31, Magnetism; 31–1.30
Chapter 8, Particle Dynamics; 8–1.6	
Chapter 9, Energy and Momentum; 9–1.9, 9–5.14	
Chapter 10, Experiments with Low Friction I; 10–2.2, 10–3.1, 10–3.2, 10–3.3	
Chapter 15, Oscillations; 15–1.3, 15–7.2	
Chapter 18, Waves in Elastic Media; 18–6.2, 18–7.1	
Chapter 19, Sound Waves; 19–4.12	

24–2 Properties of a Stroboscope

Before discussing applications of the stroboscope, some of its unique properties will be considered. Since most of the experiments in this chapter employ the General Radio 1531 Strobotac® electronic stroboscope, an outline of its characteristics will be helpful. Material in this section (24–2.1 to 24–2.3) is based on information in the *Handbook of High Speed Photography,* which can be obtained from the General Radio Company, West Concord, Massachusetts. The reader should refer to this excellent handbook for more details.

24–2.1 Light Characteristics

The Strobotac® provides high-intensity light flashes of extremely short duration. At the highest flash-rate setting, flashes of less than 1 microsecond are obtained. The longest flashes are only 3 μsec (measured between one-third-peak intensity points). With these unusually short light flashes, high photographic resolution of rapidly moving objects can be easily obtained. In Table 24–2, approximate total light-output values are given for the operating limits of each range-switch setting of the Strobotac. Table 24–2 should not be used for exposure settings, because of the difficulty in converting light-intensity data into appropriate guide numbers; see Table 24–5 for guide numbers.

When the reflector is attached to the flash lamp, the light output is concentrated into a 10° beam, as shown in Fig. 24–1 and Table 24–3. Outside the 10° cone, the light intensity falls off sharply. The *diameter* of the beam cone is a good approximation to the useful illumination area. To improve the effective beam diameter, *remove the reflector* and use only the flash lamp. Although a larger surface area can be illuminated in this manner, the incident light density decreases. This fall-off in density can be minimized by attaching smooth household

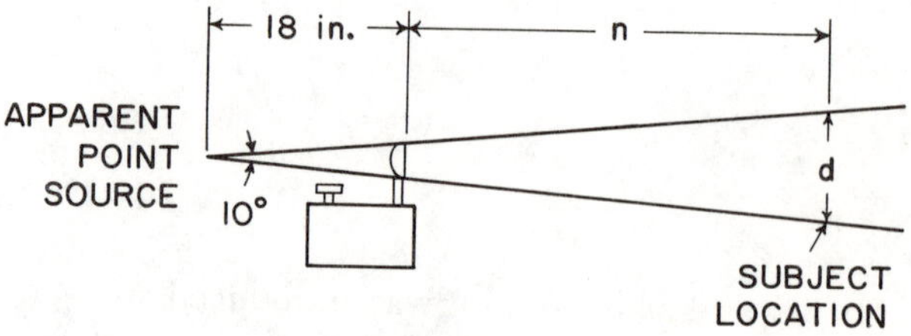

Fig. 24–1

Table 24–2

Light Output Data

Type 1531-A Strobotac® Electronic Stroboscope

(Courtesy the General Radio Company)

Speed-Range Switch Setting	Flashes per Minute	Bare Lamp Peak Candlepower *	Peak Beam Intensity (million beam candlepower)	Flash Duration Measured at ⅓ Peak Intensity (μsec)	Total Light Output (BCPS) †
RPM	25,000	4,200	0.21	0.8	0.17
(high)	4,000	11,000	0.55		0.44
RPM	4,200	24,000	1.2	1.2	1.4
(medium)	600	44,000	2.2		2.6
RPM	700	85,000	4.2		13
(low)	100	140,000	7.0	3.0	21
60∼ Line	3,600	30,000	1.5	1.6	2.4
EXT INPUT					
Low Intensity	Single	11,000	0.55	0.8	0.44
Medium Intensity	Single	44,000	2.2	1.2	2.6
High Intensity	Single	140,000	7.0	3.0	21

* Measured 2 ft from bare lamp with arc perpendicular to phototube–lamp line.
† Beam-candlepower–seconds.

Table 24–3

(Courtesy the General Radio Company)

Lamp-to-Subject Distance (*n*) (in feet)	Diameter of Illuminated Spot at Half-Peak Intensity Points (in inches)
1	5
2	7
3	9
4	12
5	14
6	16
7	18
8	20
9	22
10	24

aluminum foil, brightest side inward, over half of the flash lamp between the two spark electrodes.

Because the flash lamp *without reflector* can be considered a point light source, the beam can be used for the study of high-speed phenomena. Some examples discussed below are: photographing a bullet in flight; and investigating the shockwaves produced by a bullet in flight.

Figure 24–2 is a remarkable photograph of a playing card being split by a .22-calibre rifle bullet traveling at approximately 1150 feet per second. Note how the slug started to tumble from contact

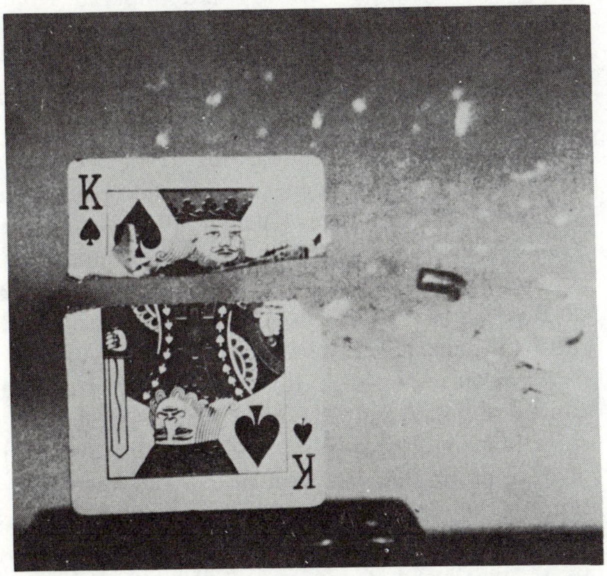

Fig. 24–2 A Playing Card Being Split by a .22-Calibre Bullet. (Photograph courtesy of the General Radio Company, West Concord, Mass.)

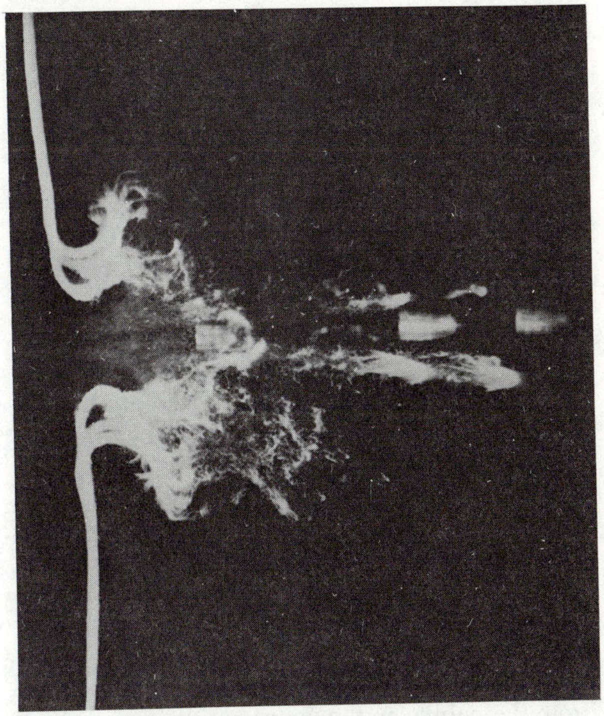

Fig. 24–3 Strobe Photograph of a Bullet Cutting a String. (From H. E. Edgerton, Massachusetts Institute of Technology; courtesy of the General Radio Company, West Concord, Mass.)

with the card. To trigger the single high-intensity flash, a microphone was placed forward of the rifle muzzle and connected to the Strobotac through a small voltage amplifier. The flash lamp was located about 14 in. from the playing card. A camera with a 135-mm Schneider lens and a Kodak Portra + 3 close-up attachment was positioned about 10 in. from the card, and the lens set at f/8. Polaroid Type 47 film was used at a 3-μsec exposure. See also Fig. 24–3, which shows a bullet cutting a string.

Eldridge, Skinner, and Tsepas[(1)] obtained the unusual shadowgraphs in Figs. 24–4 and 24–5 of a bullet passing through a soap bubble filled with Freon–12. Again, only the flash lamp without reflector was used. The Strobotac range selector was set at "ext input, 25,000 rpm max" to provide a short light pulse of 0.7–4 μsec duration. The bias (sensitivity) for external triggering was adjusted by rotating the RPM dial completely counterclockwise and then back about 20 degrees.

Instead of a camera, a Polaroid 4 x 5-in. film

(1) D. C. Eldridge, E. M. Skinner, and J. Tsepas, *Am. J. Phys.*, 30, 921 (1962).

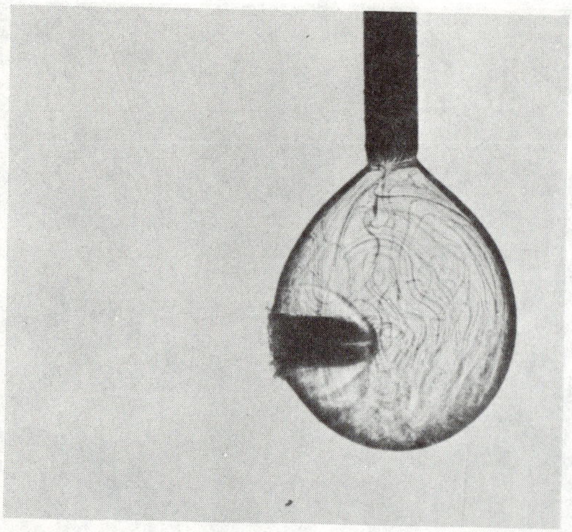

Fig. 24–4 Shadowgraph of a Bullet Entering a Soap Bubble Filled with Freon-12. Note the intensification of the shock wave. (Photograph courtesy of D. C. Eldridge, E. M. Skinner, J. Tsepas, and the General Radio Company, West Concord, Mass.)

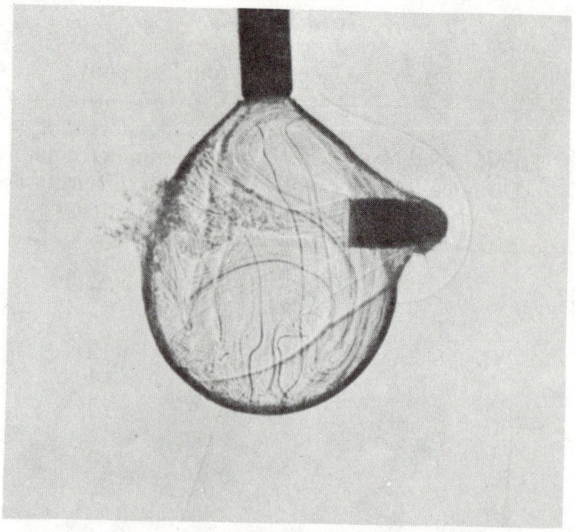

Fig. 24–5 Shadowgraph of a Bullet Leaving a Soap Bubble Filled with Freon-12. Note how strongly the turbulence stands out inside the bubble. (Photograph courtesy of D. C. Eldridge, E. M. Skinner, J. Tsepas, and the General Radio Company, West Concord, Mass.)

pack with Polaroid 3000 film was placed 3½ in. from the bubble on the opposite side from the light source. The flash lamp was in line with the direction of the subject and about 9 ft distant. A completely darkened room was used in order that only the flash triggered by the bullet would expose the film.

A microphone placed directly below the film plate was used to detect the shock waves from the bullet. The pulse detected by the microphone was amplified and sent to the input jack of the strobe. The amplifier (part of a Bell & Howell movie projector) was set on maximum bass, and the volume turned to the minimum position at which firing would trigger a flash.

H. E. Edgerton[2] evolved an extremely useful technique for studying intense shock waves and other fluid disturbances of a similar nature. In this method, a Scotchlite screen is used as a reflective backdrop for the shadowgraphs. To obtain good definition, the flash lamp without reflector must be placed next to and slightly behind the camera lens, as illustrated in Fig. 24–6, so that light is reflected from the screen back through the subject plane and onto the film. The camera lens is *focused sharply on the Scotchlite screen,* but not on the subject. There should be at least six inches between the screen and the subject.

(2) H. E. Edgerton, *Rev. Sci. Instr.,* **20,** 121 (1958).

Scotchlite, manufactured by the Minnesota Mining & Manufacturing Company, has the peculiar characteristic of reflecting the light striking its surface back in the direction from which it came with an efficiency nearly 200 times that of an ordinary white surface. It is packaged with a pressure-sensitive and adhesive backing in rolls from 5 to 7 yd long and widths up to 36 in. Although many colors are available, silver (#3270) and imperial white (#3280) are highly reflective and should be used for photographic purposes.[3]

Figure 24–7 is a shadowgraph of a .30-calibre bullet being fired from a Springfield rifle. The spherical sound wave emanating from the gun's muzzle can be easily seen. A 4-ft reflective screen of silver Scotchlite was located 10 ft from the camera. The rifle was 5 ft from the camera. A 135-mm lens was used at f/4.7 on Polaroid Type 52 positive film (ASA 200). A microphone triggered the high-intensity Strobotac flash.

Light output up to 15 times that of the Strobotac can be obtained with the type 1532-D Strobolume shown in Fig. 24–8. Table 24–4 gives data for both of its operating modes. Notice that the flash duration for the high-intensity operation is much longer

(3) For further information on Scotchlite, see William G. Hyzer, *Engineering and High Speed Photography* (The Macmillan Company, New York, 1962), pp. 324–25. See also pp. 424–29 for more details on shadow photography and Edgerton's technique.

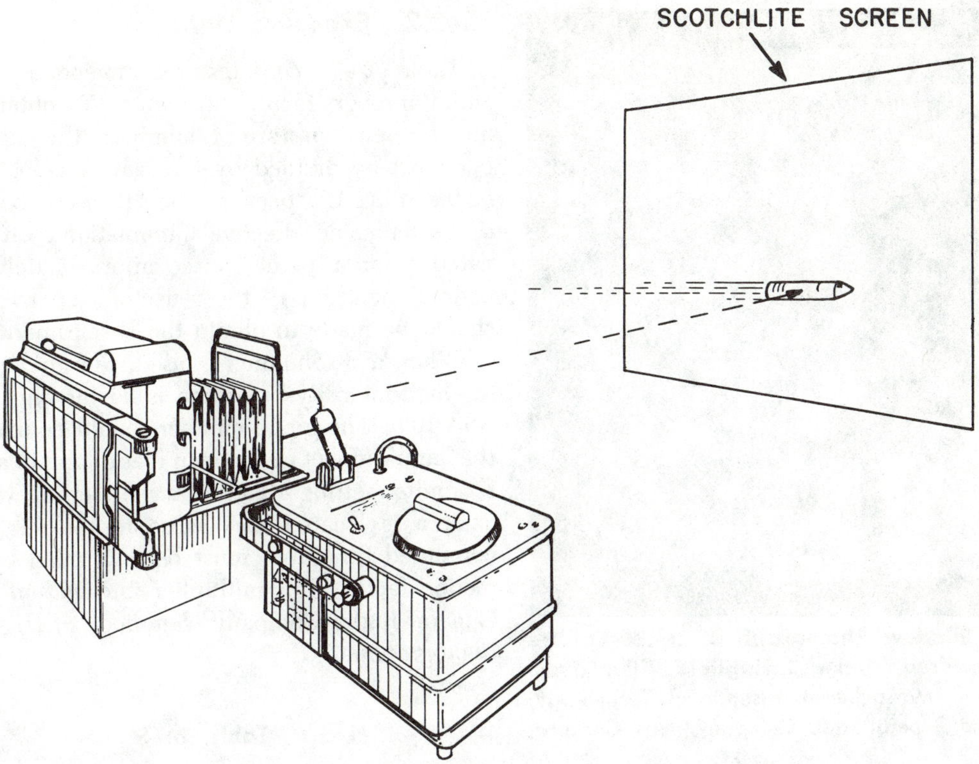

Fig. 24–6 Strobotac® Electronic Stroboscope Used as a Point Light Source. The spark gap must be oriented parallel to the camera subject axis and as close beside the camera lens as possible. (Courtesy of the General Radio Company, West Concord, Mass.)

than the shortest flash duration of the Strobotac. Due to the long flash time, the photographic blur of moving objects will increase, and care must be taken with film exposure.

When coupled to the Strobotac by a trigger cable, the Strobolume can be operated as a slave unit. Since it contains no internal oscillator for repetitive flashing, the flash rate must always be controlled externally by contacts or by an electrical input.

Table 24–4

Light Output Data

Type 1532-D Strobolume

(Courtesy the General Radio Company)

	Maximum Flashes per Minute		Peak Beam Intensity (beam candle-power, continuous operation)	Approximate Flash Duration (μsec)	Total Light Output (BCPS)
	Continuous	Intermittent			
Low intensity	3000	3000	140,000	10	1.4
High intensity	60	1200	10,000,000	30	300

Very good results can be secured in a large darkened lecture hall with a Strobolume if black "flock" paper or black velvet is used as a nonreflective backdrop. The brilliant white light and wide beam angle of the Strobolume are extremely useful when a large area is to be illuminated, or when the phenomena to be studied must be seen by a large group of students.

Experiment 7–1.15 describes apparatus used in a large auditorium to study quantitatively the motion of falling bodies. The Strobolume is flashed by a microswitch that rides in the notches of a rotating toothed wheel mounted on the top of a 25-ft tower. At regular intervals, the wheel drops ½-in.-diam steel ball bearings supplied from a helical copper-tubing reservoir. Successive balls are visually stopped by the strobe light at successive

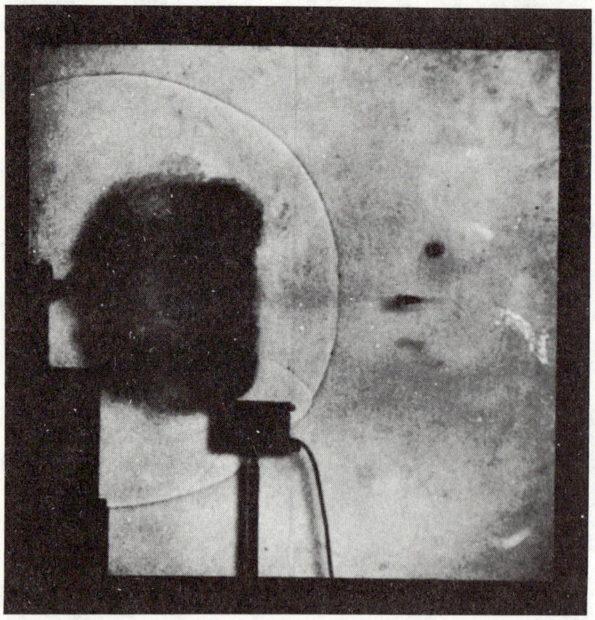

Fig. 24–7 Shadow Photograph of a .30-Calibre Bullet Being Fired from a Springfield Rifle. (From H. E. Edgerton, Massachusetts Institute of Technology; courtesy of the General Radio Company, West Concord, Mass.)

positions so that the balls appear to hang motionless at heights predicted by computations based on the acceleration of gravity and time between strobe flashes. Arrows are placed at the calculated positions below the release mechanism as a means of visual correlation with the actual spacing of the balls as seen by the students. When the Strobolume is used in this manner to study motion, the students take the place of the camera.

Fig. 24–8 Strobolume. (Courtesy of the General Radio Company, West Concord, Mass.)

24–2.2 Exposure Data

Table 24–5 contains recommended exposure guide numbers for the Strobotac. To obtain a preliminary lens aperture (f/number), the guide number must be divided by the lamp-subject distance in feet plus 1.5, because the Strobotac beam produces the same effective illumination as if it emanated from a point source approximately 18 in. behind the front of the reflector. Trial exposures should be made to obtain the best camera setting.

When a Strobolume is used, the guide number for high-intensity operation is 25 for film rated at ASA 100. The guide number should be divided by the lamp–subject distance in feet to obtain an initial f/number setting. To find a lens setting when the camera is located very close to the subject, the computed f/number must be multiplied by a correction factor. For multiplier information, see the General Radio Company *Handbook of High Speed Photography*.

Table 24–5

Recommended Guide Numbers for Type 1531-A Strobotac® Electronic Stroboscope

(Courtesy the General Radio Company)

Film Type	ASA Speed Index	Guide Number: Low Intensity (high rpm)	Medium Intensity (medium rpm)	High Intensity (low rpm)
Polaroid Type 42 and 52	200	*	15	22
Polaroid Type 44	400	*	*	30
Polaroid Type 47	3000	15	45	80
Eastman Tri-X	200	15	30	75
Eastman Royal X Pan	1600	27	62	150
Eastman Ektachrome E-3 (Daylight) CC40Y Filter	50	*	*	8
Eastman High Speed Ektachrome (EH-135) (Daylight) with CC40Y Filter	160	*	*	14
Super Anscochrome (Daylight) with CC20Y Filter	100	*	*	13

* Not recommended.

24–2.3 Triggering

The Strobotac® can be triggered by an external signal when the range switch is set on any of the external input positions. A simple contact "make" or "break" or an electrical signal of at least 6 V peak-to-peak can be used. Before connecting a signal to the input jack, make certain that the range switch is *not* set on external input. The range switch is set *after* the trigger-signal plug is inserted.

When a trigger signal is applied, the RPM dial must be set correctly since this control changes the sensitivity. The strobe can be flashed after a delay of only a few microseconds by either a positive signal or by opening a set of contacts connected to the input jack when the RPM dial is rotated to a fully *clockwise* position. If the dial is rotated *counterclockwise* from the fully clockwise position to about 90 deg, the strobe will flash once without being triggered externally. Maximum sensitivity for positive pulses is obtained when the dial is rotated back *clockwise* between 10 to 15 deg to a position just before the flash point. When the dial is shifted further counterclockwise from the flash point, the device can be flashed only by a negative input signal, or by closing a set of external contacts. The flash delay is about 20 milliseconds when the dial is in a fully counterclockwise position. This time lapse may reach as high as 300 msec at a position just counterclockwise of the flash point.

Figure 24–9 shows an inverting circuit for triggering the Strobotac by contact closure. Usually,

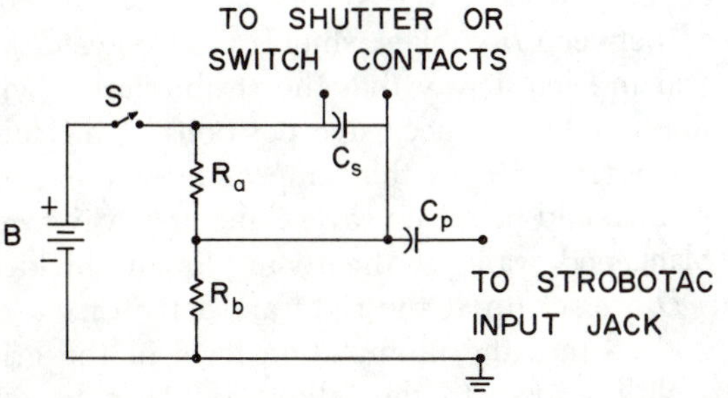

Fig. 24–9 Inverting Circuit. *B* is a 12-V (or greater) battery. R_a and R_b are carbon resistors, 1 MΩ, ½ W. *S* is a SPST on-off switch. C_s is an optional capacitor to prevent multiple triggering from contact bounce. Starting with a value of 0.01 μF and 100 V, C_s is increased until one flash is triggered from each contact closure. C_p is a capacitor rated at 0.01 μF and 100 V.

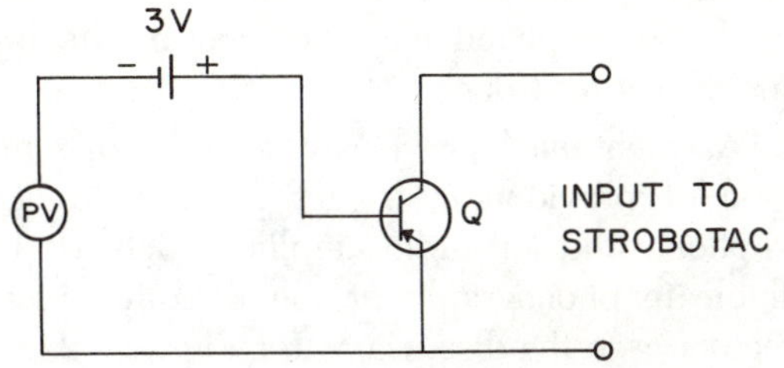

Fig. 24–10 Photocell Triggering Circuit. PV is a Photovaristor from the Heath kit Electronic Workshop 19; Q is a 2N109 transistor.

contacts in camera shutters equipped for X synchronization will operate with this circuit. When it is used, the RPM dial should be set in a fully clockwise position. Figure 24–10 shows a circuit[4] used to obtain a stroboscopic photograph of a splashing milk drop. For this photograph the RPM dial was set near the flash point, or where the triggering switches from positive to negative.

In all high speed single-flash photographic work, the strobe must be flashed at the proper time in order to stop the motion of an object in a desired position. Four methods which can be used are as follows:

1. A thin, taut wire is held tightly by supports. Leads from the wire are connected directly to the *input terminals* of the strobe, and its RPM dial is set fully clockwise. When a bullet is fired at the wire, the lamp will flash several microseconds after the wire breaks. To photograph the bullet, the camera and reflector must be aimed just past the contact wire.

2. In a similar manner, a contact sensor can be constructed by gluing two strips of metallic foil on opposite sides of a small circular hole in a thin card, and then connecting wires from the foil strips to the input terminals of strobe. A bullet fired through the card will close the circuit and flash the strobe. To keep the time delay to a minimum, the inverting circuit in Fig. 24–9 can be used.

3. Another technique is to use a photoelectric cell and a light beam.[5] In this case an electrical pulse produced by the cutting of a collimated beam of light directed at a photocell by a rapidly moving

[4] *Loc. cit.*[1]

[5] Type 1536-A Photoelectric Pickoff, available from the General Radio Company.

object is supplied to a flash-delay unit.[6] Here the pulse is amplified and then fed to the input terminals of the strobe.

4. As mentioned previously, a most convenient method of providing a trigger pulse is to use a microphone and an audio amplifier such as those available in phonographs or sound-motion picture projectors (see the discussions for Figs. 24–2, 24–4, 24–5, and 24–7).

Since the timing of the flash to stop or "freeze" the moving object in position is quite critical, the flash delay must be easily adjustable. This can usually be done by changing the distance between the firearm and the sensing device. For example, a microphone moved slightly more than a foot away from a gun will delay the light flash one thousandth of a second.

The preceding sections discuss some of the problems associated with single-flash high-speed photography and suggest various solutions. For additional information other references[7] should be consulted.

24–3 Multiple Flash Photography

Many of the strobe experiments in this book (see Table 24–1) use multiple flash techniques. In some applications the only apparatus needed is a Polaroid camera and a strobe disk. For example, in Experiment 7–2.1[8] the camera and strobe disk (300 rpm, 6 slots, $\Delta T = 1/30$ sec) are set up about 1½ meters from a vertically supported meter stick on which the 10-cm divisions are clearly marked. With the room lights on, an exposure is taken of the meter stick, which serves as a distance reference. Then with the room darkened, the strobe disk is set in motion, the camera shutter is manually opened, and a bulb attached to a 1½-V battery is dropped from the top of the meter stick into a box stuffed with rags. The camera shutter is then closed. From the trace, velocities are determined by measuring successive distances between the light flashes, which are a constant time interval apart.

Several experiments which are part of a series developed by Elisha R. Huggins[9] for use at Dartmouth College illustrate strikingly the value of multiple-flash photography. Designed for use by liberal-arts students, the strobe experiments demonstrate each of the conservation laws in mechanics. Quantitative measurements are made from the photographs.

Figure 24–11 shows equipment that is used in the projectile motion experiment. *A* is the ramp (see also Fig. 24–12) used to project a ball across the field of the 1 x 1-m grid, which is held by a framework of 2 x 4-in. wood supports. Also mounted at right angles to the bottom of the grid is a pair of 2 x 4's, which hold the grid upright. Nylon string is interlaced through holes drilled 1-cm apart in brass angles attached to the inner edges of the 2 x 4 framework. To obtain proper contrast, both white and blue (dark) nylon string are used. Every tenth string is blue.

For best results, the Strobotac *B* should be placed between *two* black shields *C* at one side of the grid in such a way that the strobe flashes will illuminate only the successive positions of the falling ball after it leaves the ramp and not the grid. The grid should be set up several meters away from the blackened walls of the room. Small shielded lights *D* are set up at the right and left in front of the grid so that the illumination level of the grid lines, with respect to the falling ball, can be adjusted.

A model 320 Polaroid camera (ASA 10,000 film) is used to photograph the ball. Although indirect triggering techniques can be used, as discussed previously, very good results are obtained by manually operating the camera with the aid of an assistant to release the ball on the ramp. Before

(6) Type 1531-P2 Flash Delay, available from the General Radio Company.

(7) For example, Harold E. Edgerton and James R. Killian, Jr., *Flash! Seeing the Unseen by Ultra High Speed Photography* (Charles T. Branford Company, Newton, Mass., 1954); Eastman Kodak Company, *Kodak Reference Handbook* (Rochester, New York), and *Schlieren Photography* (1960); George A. Jones, *High-Speed Photography, Its Principles and Applications* (Chapman and Hall, Ltd., London, Eng., 1952, John Wiley & Sons, Inc., New York, 1952).

(8) Developed by the Harvard Project Physics program under grants from the Carnegie Corporation, the Ford Foundation, the National Science Foundation, the Alfred P. Sloan Foundation, the United States Office of Education and Harvard University.

(9) Experiments discussed here were adapted with permission from Elisha R. Huggins, *Physics 1* (W. A. Benjamin, Inc., New York and Amsterdam, 1968).

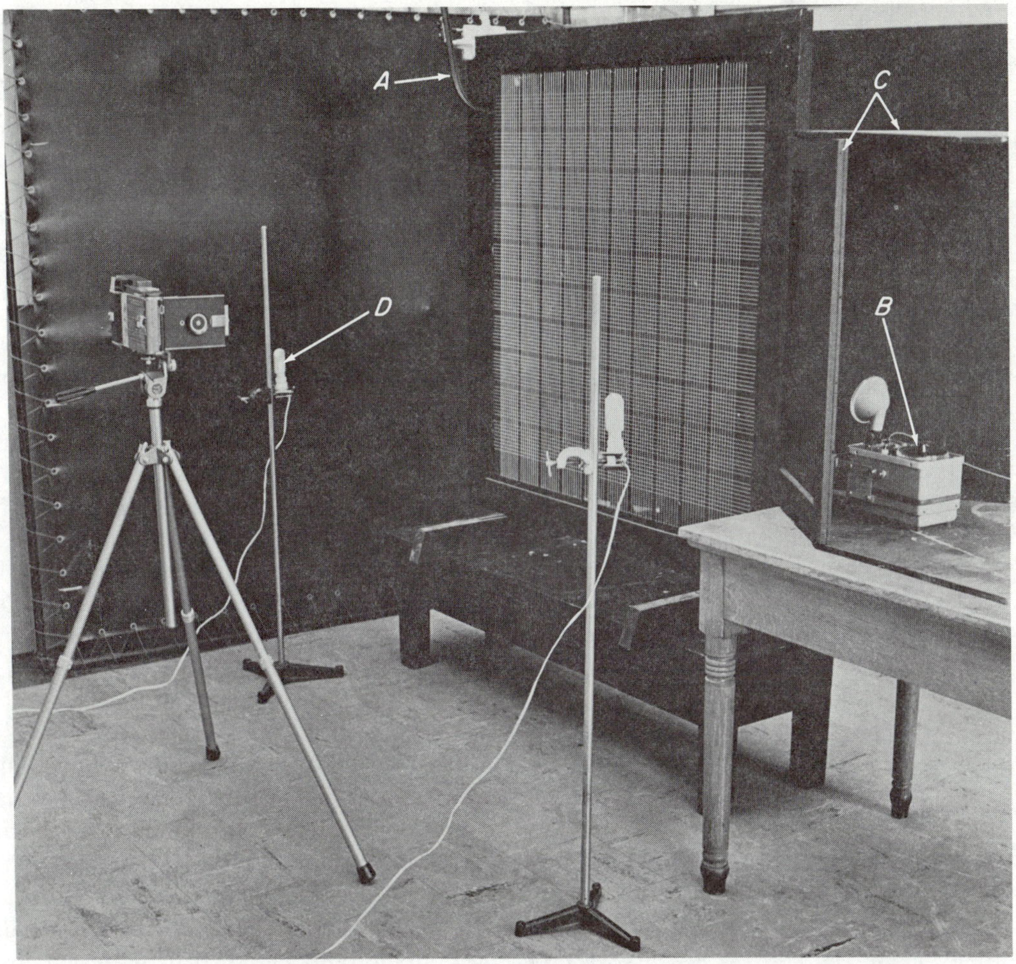

Fig. 24–11 Projectile Motion Apparatus. (Courtesy of Elisha R. Huggins.)

exposure of the film, the room lights are turned off and the strobe flashing rate is set at 600 flashes per minute ($\Delta T = 0.1$ sec). The photograph is then analyzed directly, or the data can be transferred to a slide, which can be used during lecture. In the laboratory, teams of students can quickly obtain photographs. When analyzing the results, a compass should be used to complete the ball's shape so that its center of mass can be properly located.

When the equipment is rearranged as shown in Fig. 24–13, the collision of two metal balls M of different mass (or the same mass) suspended by very long strings from the ceiling of the room can be investigated. For this experiment, the grid lights are placed beneath the grid on the front grid supports, and the camera is directed at a mirror located beneath the grid and directly below the colliding balls. To illuminate only the balls, the strobe with the same flash rate as before is positioned above the grid.

The object of this experiment is to apply the law of conservation of linear momentum to the collision of two balls. One ball is allowed to remain at rest, and the other is released in such a manner that a head-on collision is not obtained. This allows an accurate study of the vector nature of the law. To obtain data, the room is again darkened and the camera triggered manually. For a complete discussion of the results and sample data, see the publication cited in footnote (9).

Another unique stroboscopic experiment from Dartmouth can be performed with the apparatus shown in Figs. 24–14 and 24–15. It is designed for a quantitative investigation of the conservation of angular momentum, and the equipment is some-

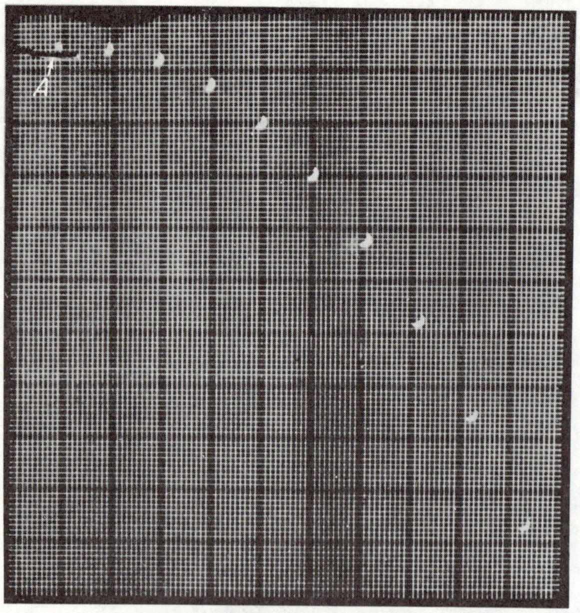

Fig. 24–12 Strobe Photograph of Projectile Motion. (Courtesy of Elisha R. Huggins.)

what more difficult to set up than that used previously.

In this experiment, a ball, approximately 2.5 cm in diam, with a mass of 240 g, is suspended below the camera on a long black nylon cord. Another black cord attached to the side of the ball is passed through a hole in a bearing *E* mounted at the top of a 3-in.-diam steel pipe, which is supported by an iron flange on a sturdy base, as illustrated in Fig. 24–15. The cord is carried through the steel pipe and under one pulley attached directly beneath the bearing in the pipe base and then under another pulley *P*. For proper performance, care must be taken to reduce the friction at the bearing and to prevent wobbling of the pipe.

To operate, pull the ball outward to some desired radius and set it moving in a circle. Then quickly shorten the radius by pulling the cord which passes through the bearing hole at the base. Since angular momentum is conserved, the ball will move much more quickly in the circle with a reduced radius.

To obtain a strobe photograph which can be analyzed, a Polaroid camera with a shutter control that automatically resets to the instantaneous position after a time exposure has been made should be used. The ball is first placed in motion in a large circle and an exposure long enough to include one orbit is taken. After the camera switches to instant, the cord is immediately pulled inward. The shutter is then clicked manually as the ball moves very rapidly in a smaller circle. Before making the exposures, several practice runs should be made in order to coordinate exposure times. During the exposures, the strobe should be set at a constant flashing rate. This rate must be high enough to obtain multiple ball images for each circle. When taking the exposures, one person places the ball in motion in the large circle and then changes the radius while an assistant uses the camera (see the publication mentioned in footnote (9) for an explanation of the measurements).

Figure 24–16 is a stroboscopic photograph of the motion of a ball bouncing on a rubber band. This experiment is used to demonstrate that Newton's second law governs the motion of the ball. Grid, strobe, camera, and the small lights which illuminate the grid are arranged as they were in Fig. 24–11. A peg is placed at the top of the grid and the ball is hung from it by a rubber band. The ball is then held to one side of the grid and released in such a way that it will complete a circuit filling most of the grid. After a few trial runs, a strobe photograph is taken, and its information is carefully transferred to graph paper or to a slide for use in lecture.

When the experiment is performed in lecture, it is suggested that calculations be made quickly for several positions of the ball. Later in the laboratory, the students working in teams can carry out a complete analysis of the photographs they have taken. Although here the analysis is more complex than for the other experiments, the results are well worth the effort. Because the total force acting on the ball at each of its strobe positions must be determined, it is necessary to obtain a calibration graph of the force needed to elongate the rubber band. This is done by hanging weights on the rubber band. From the graph, the force which the rubber band exerts on the ball at a particular position can be calculated. Since the weight of the ball is always the same, the total force acting on the ball and its direction can be determined for each position of the ball. The total force obtained by vector methods at each ball position is then compared to *ma*, which is found independently by vector techniques. In the latter case the mass *m* of the ball is measured directly, and the vector acceleration of the ball at some position can be obtained by subtracting its successive displacement vectors

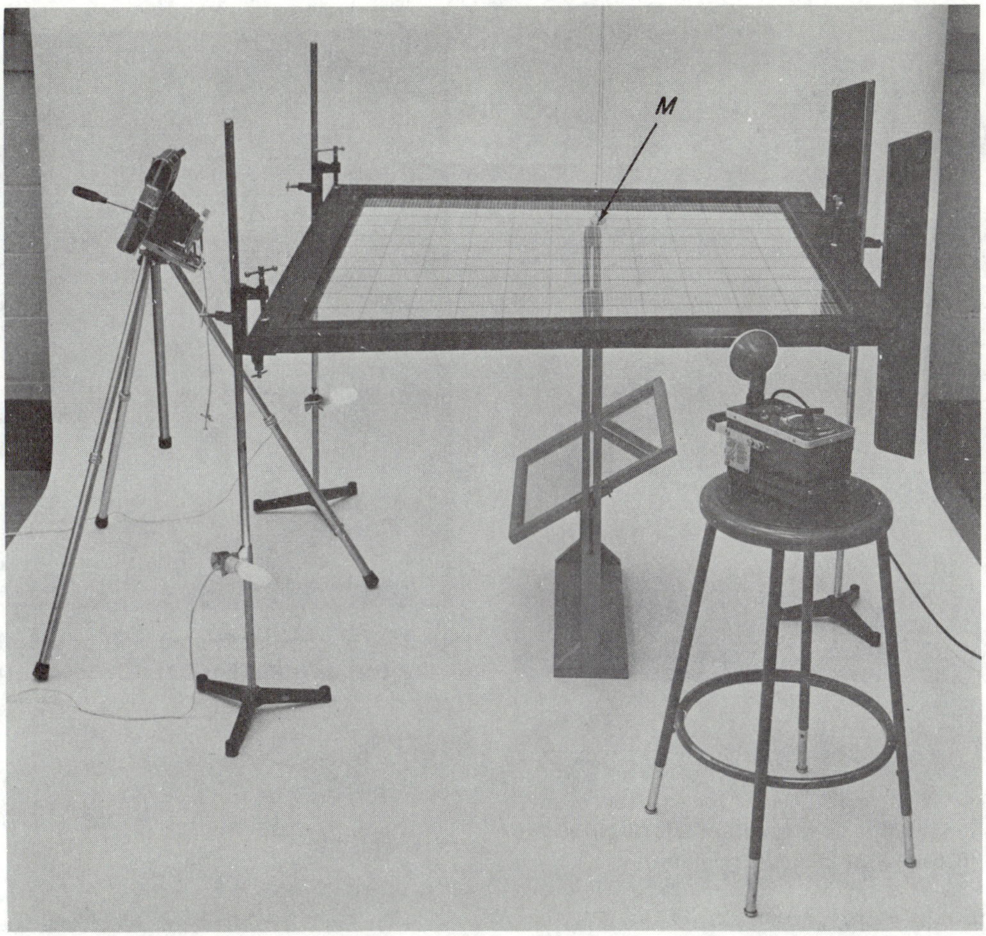

Fig. 24–13 Apparatus Used for the Experiment on Conservation of Linear Momentum. (Courtesy of Elisha R. Huggins.)

and multiplying by $1/\Delta T$ (the Huggins book provides sample data and a complete analysis).

Harvard Project Physics and the National Film Board of Canada have produced a set of stroboscopic photographs[10] which permits the detailed quantitative study of seven two-body collisions in one dimension. The student may be assigned one or more of these events as take-home problems, or as study-period (or laboratory-period) tasks. These problems are included in the *Project Physics Handbook, Unit 3, Chapters 9 and 10.* Prints and student notes are also supplied and may be kept by the student.

(10) Developed by the Harvard Project Physics program under grants from the Carnegie Foundation, the Ford Foundation, the National Science Foundation, the Alfred P. Sloan Foundation, the United States Office of Education, and Harvard University. The experiments discussed here were adapted with permission from the Project Physics *Teacher's Guide.*

In addition overhead projector transparencies of the apparatus and several events are provided to the teacher for use during lectures. With these the teacher can quickly describe the problems qualitatively to the class before he distributes the assignment to the students.

Figure 24–17 is a diagram of the apparatus used to obtain the series of photographs. Two case-hardened steel balls confined to move on circular paths in the same vertical plane are hung by a bifilar suspension from thin piano wires. The radii of these paths are the same for both balls so that the balls act as pendulums of equal periods (about 2π sec) to an excellent approximation. The balls are then released simultaneously by relays from selected initial positions and collide at the bottom of their swing a quarter period later.

The paths of the balls within the 3 x 4-ft rectangle are illuminated by four synchronized Type

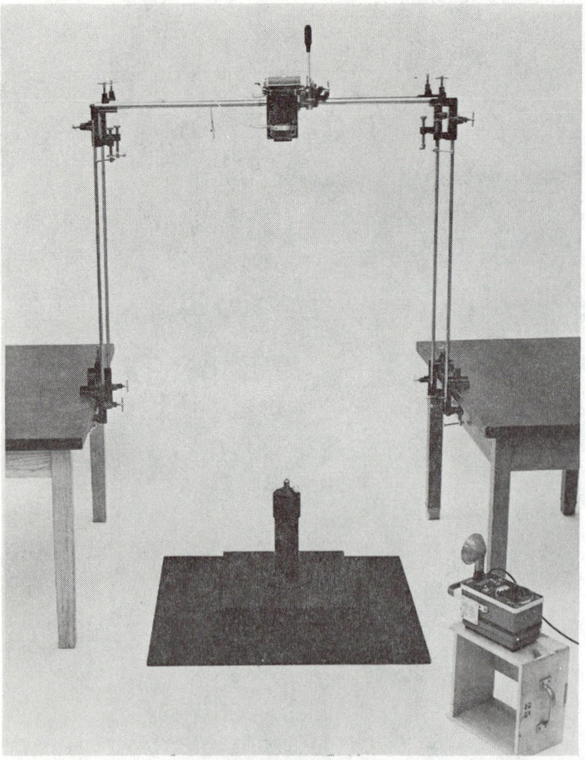

Fig. 24–14 Equipment Used for a Quantitative Investigation of the Conservation of Angular Momentum. (Courtesy of Elisha R. Huggins.)

1531-A General Radio electronic stroboscopes (not shown). A built-in calibration system in each stroboscope uses the power-line frequency for an accurate quick check and readjustment, when desired, of the flash rate.

Fig. 24–15 Steel Pipe for Angular-Momentum Experiment. (Courtesy of Elisha R. Huggins.)

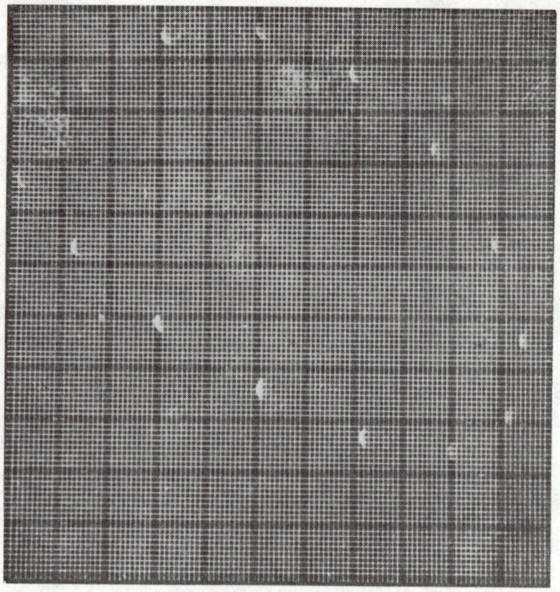

Fig. 24–16 Strobe Photograph of a Ball Bouncing on the End of a Rubber Band. (Courtesy of Elisha R. Huggins.)

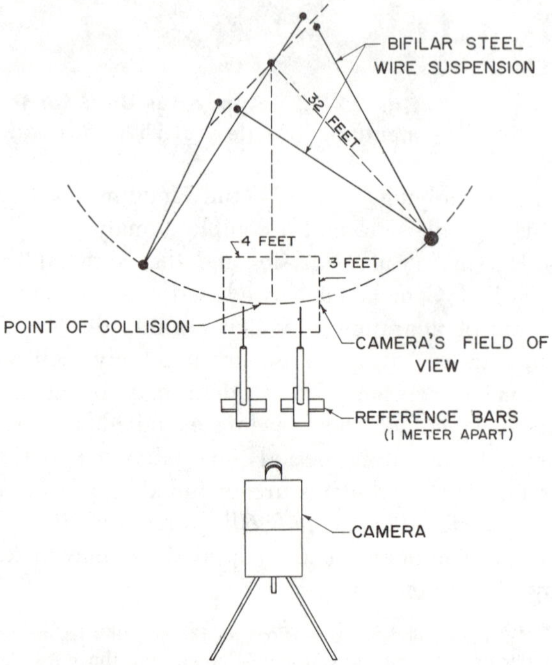

Fig. 24–17 Apparatus Used to Obtain One-Dimensional Collisions. Not to scale. (Courtesy of Harvard Project Physics.)

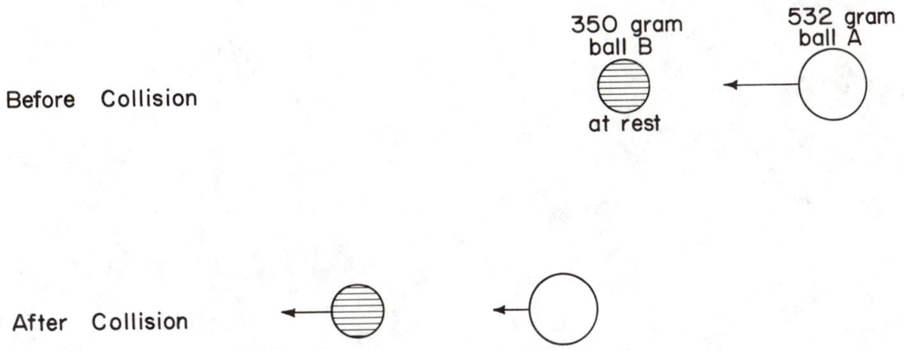

Fig. 24–18 (Diagram courtesy of Harvard Project Physics.)

Two vertical rods with their centers one meter apart[11] are also placed in the field of view. These enable the student to calculate by scaling actual distances from his measurements on the photographic prints. Although the paths of the balls are not exactly straight, the variation between the velocities calculated for a given ball as it moves toward or away from the point of collision is very small because the amplitude of swing exceeds by a sufficient factor the portion with the camera's field of view.

(11) Centers of the reference rods are 1.000±.002 m apart.

Figure 24–18 is a diagram of the conditions before and after the actual one-dimensional collision shown in Fig. 24–19.[12] It is also an example of

(12) The event illustrated in Figs. 24–18 and 24–19 is the first example in the Project Physics Film Loop *One-Dimensional Collisions:* Part I. Other one-dimensional film loops are *One-Dimensional Collisions:* Part II, *Inelastic One-Dimensional Collisions,* and *Dynamics of a Billiard Ball.* To record the events on film, a high-speed motion picture camera was substituted for the photographic camera. See Chapter 21, *Films as a Lecture Aid,* by Alfred Leitner, for a list of additional National Film Board of Canada Physics Loops, produced in consultation with Project Physics.

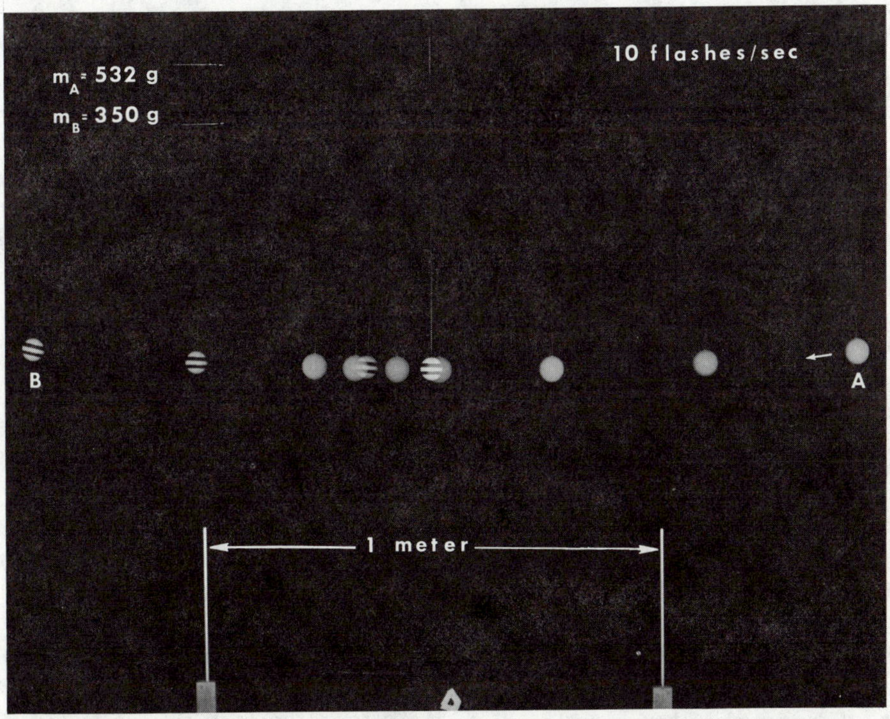

Fig. 24–19 Strobe Photograph of a One-Dimensional Collision. (Courtesy of the National Film Board of Canada.)

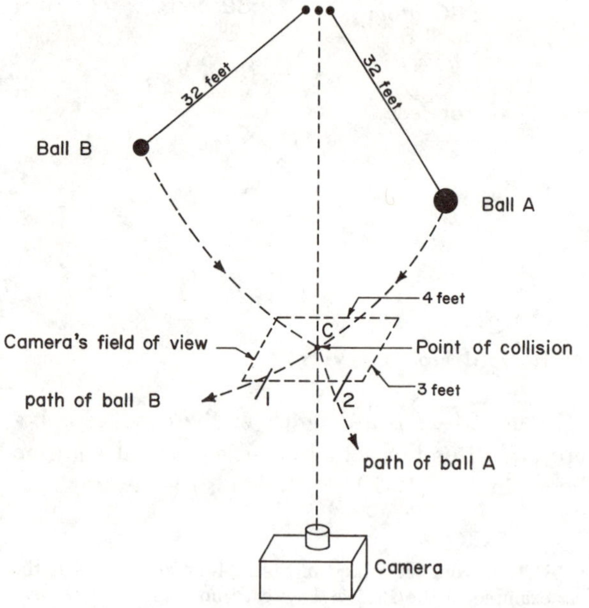

Fig. 24–20 Apparatus Used to Obtain Two-Dimensional Collisions. Not to scale. (Courtesy of Harvard Project Physics.)

the transparencies used by the teacher. Ball A moves in from the right and strikes Ball B at rest initially, then both balls move off to the left.

When the conservation of linear momentum was

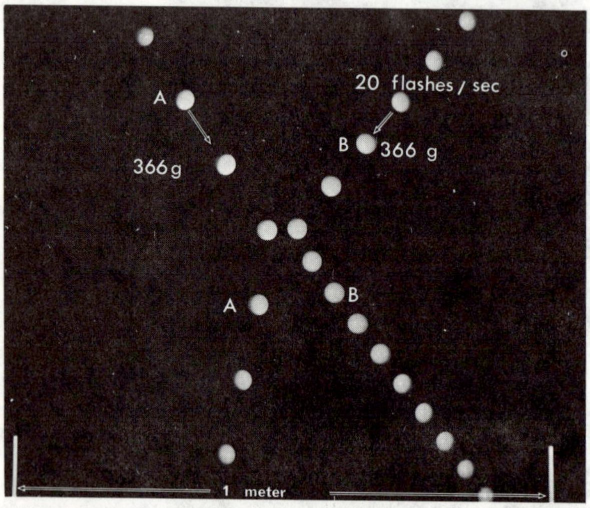

Fig. 24–21 Strobe Photograph of a Two-Dimensional Collision. (Courtesy of the National Film Board of Canada.)

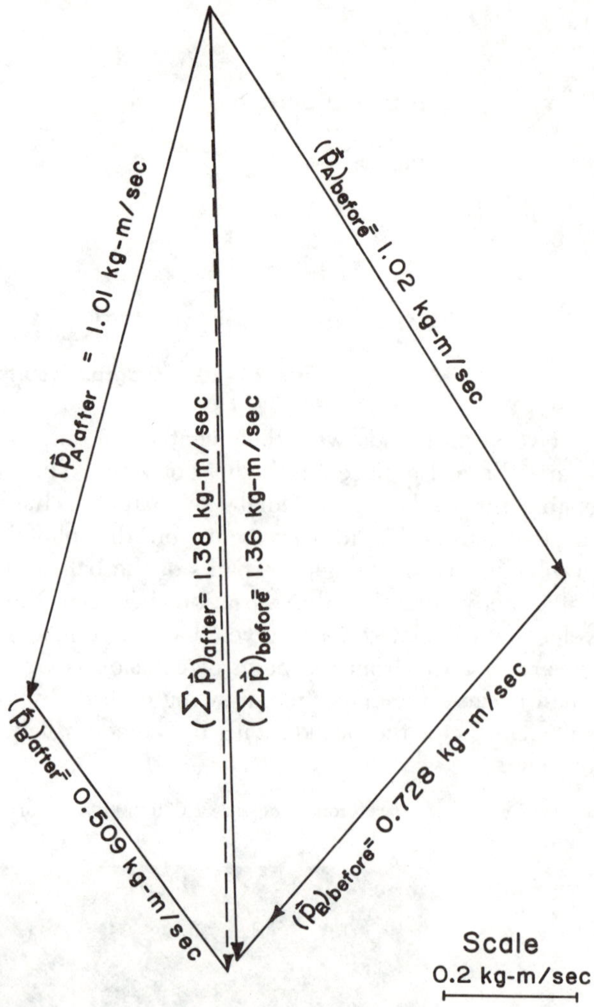

Fig. 24–22 Comparison of Total Momentum, Before and After Collision, for the event shown in Fig. 24–21. Momentum is conserved. (Courtesy of Harvard Project Physics.)

applied to the system of two balls, momentum was found to be conserved to the number of significant digits available (1.79 kg-m/sec directed to the left). Since the collision was inelastic, the kinetic energy of the system, which was 3.02 J before collision, changed to 2.64 J afterward.

In order to obtain stroboscopic photographs of two-dimensional collisions, the camera must be placed either directly below or above the point of collision C, as illustrated in Fig. 24–20. Again, four synchronized General Radio stroboscopes are used to concentrate light flashes on the 3 x 4-ft rectangle. As before, the balls were released simultaneously by electrical relays from selected initial positions.

A similar arrangement was also used for the production of film loops.[13]

When the event illustrated in Fig. 24–21 is analyzed, the total linear momentum of the system of two balls before the collision is found to be equal within the experimental error to the total linear momentum afterward. Figure 24–22 is a vector diagram of the two-dimensional collision. In this case, the total kinetic energy of the system is 2.17 J before and 1.73 J after collision.

Since two balls of equal mass are used, the instantaneous positions of the center of mass of the system can be easily located. These positions are the mid-points of the lines which join the centers of the simultaneous images of balls A and B. As is shown in Fig. 24–23, the path of the center of mass is a straight line, and its motion is uniform.

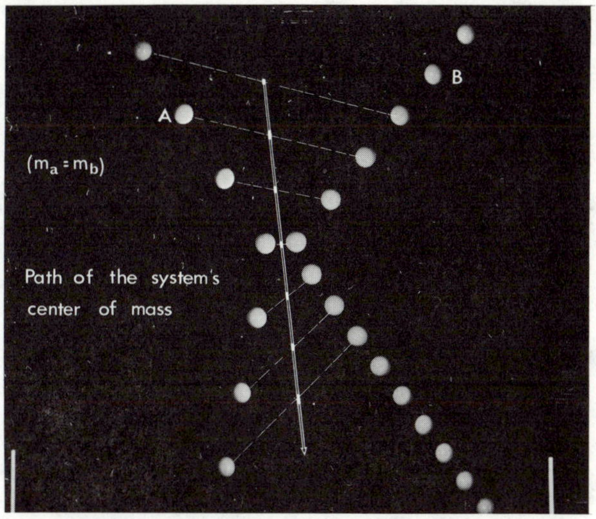

Fig. 24–23 Instantaneous Positions of the Center of Mass of the system shown in Fig. 24–22. (Courtesy of the National Film Board of Canada.)

To obtain the speed of the center of mass, the distances between successive positions of the center of mass are measured. They are then easily converted to actual distances by scaling, since the separation between the reference bars at the bottom of the strobe photograph is known.

Because the positions of the center of mass are 1/20 sec apart, its speed can be calculated ($S_{c.m.}$ = 1.89 m/sec). Since the center of mass of a system of particles can be treated as if it were a particle whose mass equals the total mass of the system and whose momentum equals the total momentum of the system, the magnitude of the momentum of the center of mass is 1.38 kg-m/sec. This value compares very well with the magnitude of the total momentum of the system found by adding vectorily individual particle momenta as shown in Fig. 24–22.

In conclusion, this chapter was prepared in response to many requests received by me from teachers of physics for a brief reference of stroboscopic techniques and experiments. Quite frankly, the use of stroboscopic techniques, whenever appropriate during lecture as well as in laboratory, helps to make physics come alive to the student. Although dramatic, these methods emphasize concepts in a manner that otherwise is not possible by example or narration in a physics textbook.

For many years I have admired the work of Edgerton. Whenever I have used techniques suggested by him in the literature during lecture, the students were both stimulated and motivated. The set-up time required was well worth the effort.

I want to thank the General Radio Company, Harvard Project Physics, Elisha Huggins and W. A. Benjamin, Inc., and D. C. Eldridge, E. M. Skinner, J. Tsepas, and Alfred Leitner for their cooperation. My service here was to piece together the good work of others with very little of my own. I hope this summary is clear and helpful.

[13] The event in Fig. 24–21 is the second example in the Project Physics Film Loop *Two-Dimensional Collisions:* Part I. Other two-dimensional film loops are *Two-Dimensional Collisions:* Part II, *Inelastic Two-Dimensional Collisions, Scattering of a Cluster of Objects,* and *Explosion of a Cluster of Objects.*

HEAT

25

TEMPERATURE

25–1 Introduction

Human subjectivity in temperature judgments has been an obstacle to understanding heat. It was not until man learned to quantify his observations of hot or cold that he could speak of heat objectively or make a science of thermodynamics.

The demonstrations in this chapter can help to acquaint students with the idea of temperature and to show them some of the methods of measurement and their limitations.

In addition to these demonstrations, some commercial apparatus may interest the lecturer.

A large (50-cm) thermometer (−50° to +160°C) with colored bands every 50° is sold by Klinger[1] (KH2003) and LaPine[2] (382 41) in the United States and by Awyco[3] (4135) in Switzerland and Leybold[4] (382 41) in West Germany.

Sets of Thermochrome temperature-sensitive crayons are available from Klinger (KH2004), LaPine (382 95), Awyco (4170), and Leybold (382 95). When an object marked with one of the crayons reaches a certain temperature, the mark changes color. Each crayon has a different reaction temperature, from 65° to 600°C. Also, temperature-sensitive paper which changes color at about 40°C is available from Awyco (4260).

A demonstration thermometer with a large circular scale (−10° to +110°C) and a separate probe on a scale may be purchased from Cenco[5] (77312). Another large thermometer is a wall type, 100 cm high (Welch,[6] 1259), with both Fahrenheit and centigrade scales (−50° to +120°F; −50° to +50°C) which can be read from up to 40 ft away.

A large (50-cm) thermometer model which may be used to demonstrate how a thermometer works may be bought from LaPine (382 02) and Leybold (382 02).

A thermometer 22 cm long (−50° to +100°C) for use with projection apparatus is sold by LaPine and Leybold (both 382 77), and Awyco (4138). Awyco's 4139 covers from −40 to +50°C.

The variation of temperature during freezing, the supercooling of water and the latent heat of fusion of ice may be demonstrated with a freezing point thermometer which consists of a 22-cm-long thermometer (−20° to +60°C) whose mercury bulb is surrounded by a sealed-on glass vessel half filled with distilled water. It is sold by LaPine and Leybold (No. 387 51 for both suppliers).

(1) Klinger Scientific Apparatus Corp.
(2) LaPine Scientific Co.
(3) Awyco AG., Lehrmittel-Verlag, Olten, Switzerland.
(4) E. Leybold, Postfach 195, Bonner Strasse 504, 5 Köln-Bayental.
(5) Central Scientific Co.
(6) Sargent-Welch Scientific Co.

A constant-pressure air thermometer which consists of a 56-cm-long graduated glass tube, sealed at one end and enclosed in a heating jacket so that a column of air over mercury in the tube can be heated to 100°C by steam and then returned to its original pressure by drawing off mercury, is distributed by Leybold and LaPine (381 56). These distributors also sell a constant-*volume* air thermometer (381 59) in which a volume of air enclosed in a sphere with a capillary tube is heated in a water bath and restored to its original volume by raising the mercury level in a manometer attached to the capillary tube.

A triple-point cell sold under the trade name of "Equiphase Cell" for establishing the temperature of the standard fixed point of thermometry can be obtained from Tran-Sonics.[7] It provides a convenient means of producing the reference temperature of 273.1600°K or 0.1600°C to an accuracy which is claimed to be greater than 5×10^{-4} degree.

The cell consists of a cylindrical Pyrex glass container with a reentrant tube, surrounded with polyurethane foam thermal insulation and a stainless steel jacket. The glass container is nearly filled with highly pure, gas-free water, and is permanently sealed. The reentrant tube has a 10-mm inner diameter and is 35 cm long. The jacket, with cover in place, has a diameter of 8 in. and is 17 in. high.

To place the cell in use, powdered Dry Ice is poured into the reentrant tube to freeze a mantle of ice around the tube. A 3/8-in.-diam rod at room temperature is then inserted into the well until the ice mantle has melted away from contact with the reentrant tube as evidenced by the free motion of the ice mantle when the jacket is rotated sharply. Alcohol or water is poured into the well to assure good thermal contact and the cell is then ready to receive the thermometer or equipment to be standardized.

The difficulty of obtaining an accurate ice point temperature with air-saturated water at standard pressure is well recognized. A triple-point cell should be of interest to elementary students even if it is not made available for their use.

25–2 Demonstrations Involving Temperature Change

25–2.1 Cubical Coefficient of Expansion

An illuminated vessel containing hot water and a thermometer are employed to demonstrate cubical coefficient of expansion via closed-circuit television. After the thermometer is inserted in the water, it is seen that the glass portion of its bulb expands first, resulting in a momentary drop in temperature reading before rising. The screen shows a temperature rise of five to ten degrees, and the initial drop noted is about one-half degree.

25–2.2 Bimetallic Thermometer

An iron rod, 1/16 in. diam x 8 in. long, is spotwelded to the end of a bimetallic strip which is mounted vertically on a brass rod. To the upper end of the iron rod, an arrowhead made from paper is fastened with pressure-sensitive tape, as shown in Fig. 25–1. When the bimetallic strip is heated with a Bunsen burner, the arrow deflects up to 45 deg. As the strip is cooled with compressed air, the arrow moves back to its original position.

[7] Trans-Sonics, Inc., Lexington, Massachusetts.

25–2.3 Heat Detection by Temperature-Sensitive Paint

Temperature-sensitive paint can be used to make qualitative demonstrations of heat phenomena, such as absorption, conduction, and specific heat. Preferential absorption is shown by coating one side of a sheet of bond paper with temperature-sensitive paint and the other side with aluminum paint on which a design is drawn with black paint. When the side containing the design is exposed to an infrared source, such as a 250-W heat lamp, the heat absorbed by the black areas raises the temperature and produces a marked color change in the temperature-sensitive coating. Filters of various kinds placed before the infrared source demonstrate different heat-transmission properties.

Note: If the temperature-sensitive paint is applied directly to aluminum or other metal surface, a reaction may take place between the metal ions and the complex silver ions. This reaction can be prevented by a protective coat of lacquer.

Relative thermal conductivity may be demonstrated by coating rods of various materials with

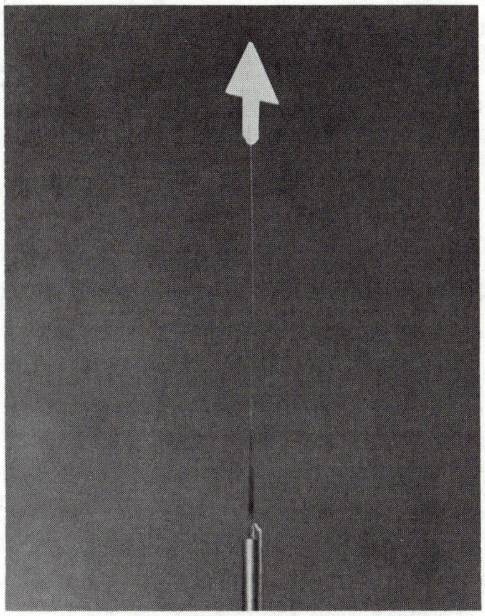

Fig. 25–1

thermosensitive paint and noting the color changes which occur when one end of the rod is heated. Test tubes coated with temperature-sensitive paint and filled with samples of various materials heated to the same temperature will show different rates of cooling, giving an indication of the relative specific heats.

DIRECTIONS FOR MAKING TEMPERATURE-SENSITIVE PAINT[8]

Make three solutions using distilled water:

Dissolve 40 g silver nitrate in 700 cc water,
60 g potassium iodide in 370 cc water, and
17 g mercuric chloride in 280 cc water.

Add the silver nitrate solution to the potassium iodide, being sure to keep the solutions completely mixed. To the resulting precipitate add the mercuric chloride solution in small quantities. The final precipitate will be yellow in color and should be filtered off and dried. This precipitate may then be ground up, mixed with lacquer, and sprayed or brushed on the surface where a change in temperature is to be indicated. At room temeprature the paint will be a chrome yellow color, while above approximately 50°C it will turn a brilliant orange. As it cools, its color will again become yellow, and it may be recycled, heated, and cooled indefinitely, always accompanied with the same color changes.

[8] This particular formula was supplied by Robert Bell, Department of Physics, University of California, Berkeley, California.

25–2.4 Gas Thermometer

Figure 25–2 shows a simple demonstration gas thermometer. The thermometer consists essentially of an inverted U-shaped Pyrex tube, with its open end *a* immersed in a dish of colored water. The tube is supported by a clamp *b*, and some of the air is pumped out through the hose *c*. In order to minimize the change in volume when the pressure changes (so that the instrument behaves nearly like a constant volume gas thermometer), the open arm contains an inner tube *d*, which is closed at the bottom. The side diameter of the outer tube is 22 mm, and the outside diameter of the inner closed tube is 16 mm. For demonstration, the bulb *e* can be warmed with hot water or with the hands and cooled with ice water. The mark on the white card indicates normal room temperature.

25–2.5 Supersensitive Thermometer[9]

A thermistor[10] and a transistor amplifier are used to make a supersensitive thermometer. Figure 25–3 is a schematic of the apparatus. The adjustment rheostat is used either in parallel or in series with the thermistor, depending on the exact parameters of the components. Connect the meter cautiously (the output current may be as much as 10 mA). If the meter is off-scale at room temperature, add a 1-MΩ rheostat in parallel. If the meter indicates zero, used a 10-kΩ rheostat in series. In either case, adjust the rheostat to obtain a full-scale meter reading at room temperature. Do not touch the thermistor during adjustment, for it responds to very small temperature changes. A 2°C increase should zero a 0.1-mA meter; a 10°C increase should zero a 1-mA meter. To show this thermometer before a lecture audience, closed-circuit television or a projection meter can be used.

One possible use of the supersensitive thermom-

[9] Developed by the Harvard Project Physics program under grants from the Carnegie Corporation, the Ford Foundation, the National Science Foundation, the Alfred P. Sloan Foundation, the United States Office of Education, and Harvard University.

[10] JA41J1, Fenwal Electronics, Inc., Framingham, Massachusetts.

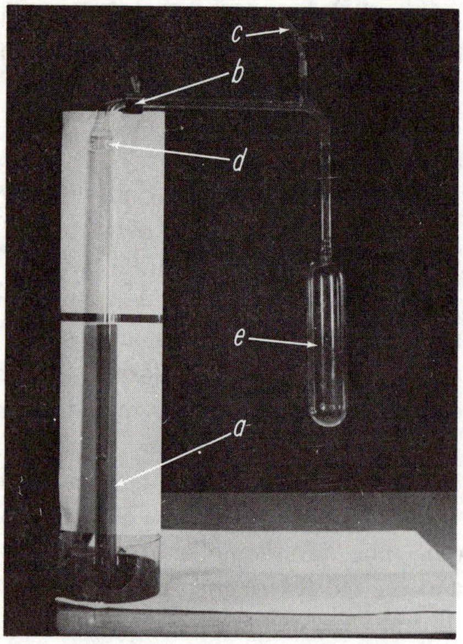

Fig. 25–2

eter is to show the kinetic energy loss of a completely inelastic collision. When two loaded carts collide, the kinetic energy absorbed by the lead bumper on one of the carts becomes heat. A thermistor soldered to the bumper indicates the heat rise by causing the milliammeter current to decrease sharply. To make the bumper, follow these steps: (1) Drill a 7/64-in. hole in a 50-g lead strip. (2) Loop one wire around the back of the thermistor and insert the thermistor in the hole. (3) Melt a drop or two of solder into the underside of the hole to hold the thermistor in place. (4) Put glue on the exposed end of the thermistor to prevent shorts with the wires. (5) Bend the lead strip into a ring, twist, and solder two long, thin wires to the thermistor wires, and tape the ring to one cart. Two PSSC carts colliding at 1 m/sec (4 J of kinetic energy) will cause about 1 cal of heat energy in the lead. The temperature of the lead should

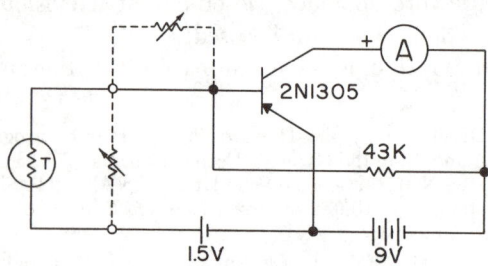

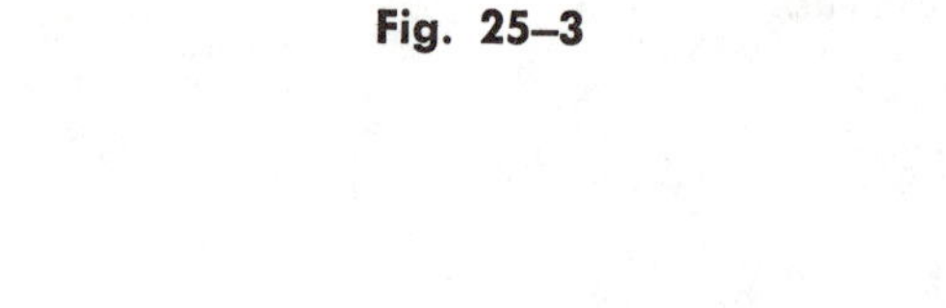

Fig. 25–3

rise by about 2–3°C. Although a 10°C thermometer (1 mA meter) will show the temperature increase, a 2°C thermometer (0.1 mA meter) shows the increase more dramatically.

25–2.6 Constant-Volume Gas Thermometer

Figure 25–4 shows five gas thermometers containing, respectively, helium, hydrogen, nitrogen, argon, and air. The bulb of each thermometer is a copper toilet-tank float to which a length of glass capillary tubing has been attached. The attachment is made by a rubber hose connection. To provide for the hose connection, a 3-in. length of copper tubing (1/8 in. i.d.) is soldered to the threaded end of the float and a 1½-in.-long brass nipple is then soldered to the copper tube. The ends of the capillary tubes are closed after a small mercury "piston" has been inserted. In use, the mercury piston should be about 6 in. from the closed end of the tube. For helium, hydrogen, and nitrogen, the capillaries should be about 29 in. long, for argon

Fig. 25–4

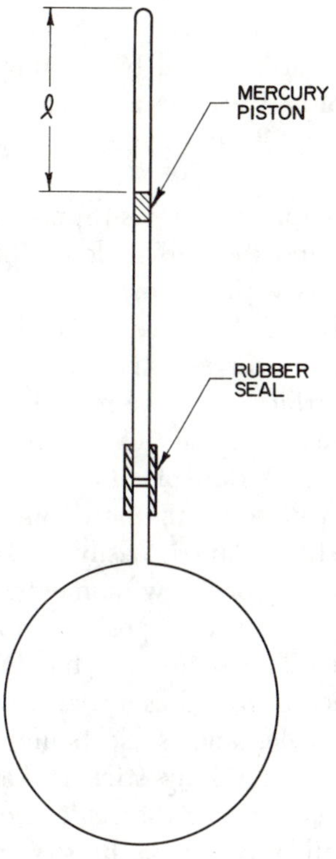

Fig. 25–5

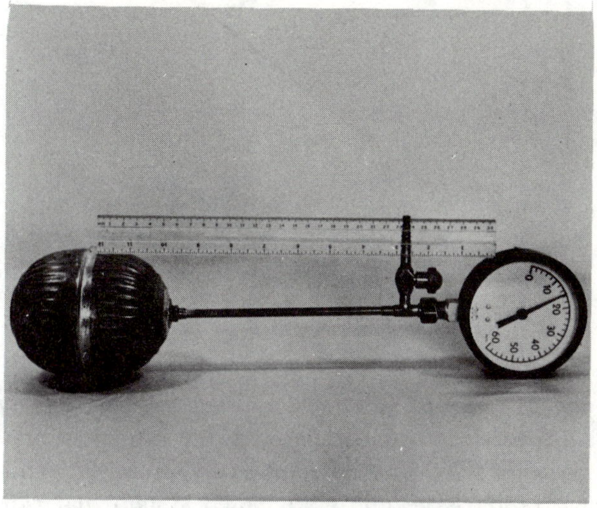

Fig. 25–6

35 in., and for air 37 in. As can be seen from Fig. 25–5, the pressure in the bulb is equal to the pressure in the capillary space above the mercury piston, if the weight of the mercury can be neglected. Since the capillary remains at room temperature, p-v changes above the piston are isothermal. Hence, the pressure p is inversely proportional to the length l. The linearity of the different thermometers can be compared over a wide range of temperature by immersing each successively in boiling water, ice water, a Dry-Ice bath, and liquid nitrogen. The readings can readily be shown by shadow projection. A Styrofoam block containing an appropriate cavity is a convenient container for the liquid nitrogen. *Caution:* Do not immerse the rubber seal in the liquid nitrogen.

25–2.7 P-T by Immersion

A helium-filled toilet tank float equipped with a Bourdon pressure gauge[11] (Fig. 25–6) is immersed in various liquids at different known temperatures to show the direct relationship between the pressure and temperature of a fixed volume of gas. The pressure is plotted against temperature as the copper float is exposed successively to boiling water, air at room temperature (~24°C), ice water, Dry Ice and acetone (−96.5°C), and liquid nitrogen (−196°C). The demonstration is then repeated using air and other gas samples in the float. Air will show pressure readings similar to those of helium, except when in the liquid nitrogen bath, where a lower pressure for air is due to the liquefaction of the oxygen in the air.

When the graph for helium is extrapolated to zero pressure, it intersects the temperature axis at about −278°C, an error of about 1.4 percent. This error is due to the small volume of gas in the gauge, valve, and stem, which cannot be immersed in the bath. Since this volume is constant, it does not affect the linearity of the *P-T* plot.

The small-bore (0.025 in.) tube[12] between the gauge and the float is about 7 in. long. Before the apparatus is assembled, the Bourdon gauge is recalibrated by removing its glass face, evacuating the gauge, and resetting the pointer to zero. To test for leaks after assembly, evacuate the instru-

[11] Welch Scientific Co., Catalog No. 1431 or comparable gauge.

[12] Commercially available from almost any metal tubing supply house.

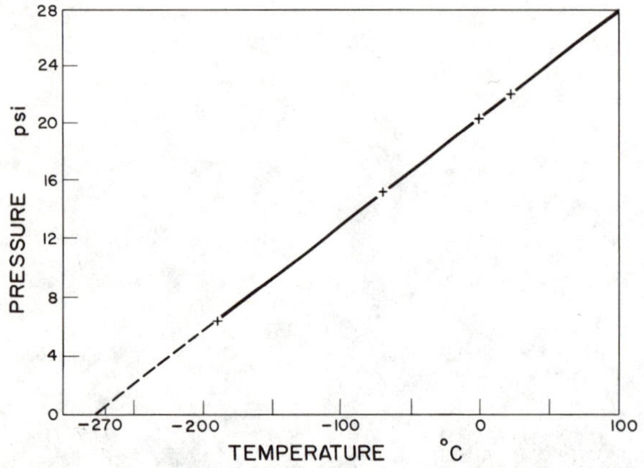

Fig. 25–7

ment, close the valve,[13] and check the gauge reading 24 hr later.

Several instruments should be available and properly filled with different gases before the demonstration is begun.[14] A closed system that can be evacuated should be used to avoid pollution of the sample gas. Before filling the bulb, *evacuate for at least 24 hr.* If air is used as one sample, it should be free of water vapor. Regardless of the gas used, the instrument should be filled to a gauge reading of about 23 when the first reading is taken at room temperature. For the rest of the readings, immerse the bulb and 4 or 5 in. of the stem in the bath. Tap the gauge before reading to insure that the pointer has settled completely.

Figure 25–7 is a typical curve for helium.

25–2.8 Constant Pressure Gas Thermometer[15]

About 6 in. of capillary tube makes a nice size gas thermometer which can be used with closed-circuit television. The dimensions of the tube are not critical, but it is very important that the bore be dry. It can be dried by heating, or by rinsing with alcohol and waving frantically—or better still, by connecting it to a vacuum pump for a few moments.

The dry capillary tube is dipped into a container of mercury and the end sealed with the finger tip as the tube is withdrawn, so that a pellet of mercury remains in the lower end of the tube. The tube is held at an angle and the end tapped gently on a hard surface until the mercury pellet slides to about the center of the tube. One end of the tube is sealed with a dab of silicone sealant; some of the sealant will go up the bore, but this is perfectly all right. The sealant is easily set by immersing it in boiling water for a few moments.

A scale now must be positioned along the completed tube. The scale will be directly over the bore if a stick is placed as a spacer next to the tube, and tube, scale, and stick bound together with rubber bands. (A long stick makes a convenient handle.) The zero of the scale should be aligned carefully with the end of the gas column—i.e., the end with the silicone seal.

In use, the thermometer should be completely immersed and the end tapped gently against the side of the container to allow the mercury to slide to its final resting place.

Other gases besides air can be used. As before, a pellet of mercury is trapped in the end of the capillary tube, but instead of the mercury's being tapped to the center, with tube laid flat the mercury is pushed to the center by gas pressure from a suitable low-pressure gas supply.

[13] Commercially available from Hoke Inc., Cresskill, New Jersey. For helium, use valve #A431 with leak-tested valve seal. For air and heavy gases, use #304.

[14] Complete assembly commercially available, see Sargent-Welch Scientific Co., Catalog No. 1602.

[15] Developed by the Harvard Project Physics program under grants from the Carnegie Corporation, the Ford Foundation, the National Science Foundation, the Alfred P. Sloan Foundation, the United States Office of Education, and Harvard University.

Demonstrations	***Contributors***
25–2.1	M. Mainardi, A. Capecelatro, Newark College of Engineering
25–2.2	Swiss Federal Institute (ETH), Zürich, Switzerland
25–2.3	A. B. Butler, Washington State University
25–2.4	Swiss Federal Institute (ETH), Zürich, Switzerland
25–2.5	Harvard Project Physics
25–2.6	University of Colorado
25–2.7	J. T. Brooks, J. M. Fowler and E. D. Lambe, Washington University
25–2.8	Harvard Project Physics

26

HEAT AND THERMODYNAMICS

26–1 Introduction

The equipment listed below will be helpful to lecturers in clearly presenting fundamental concepts of heat to their students. The list only gives an indication of the type of apparatus available from manufacturers. Teachers should consult catalogs from various companies for more complete information. These demonstrations can usually be shown to a large audience by closed-circuit TV, shadow projection, or with the aid of a demonstration thermometer.

A heat and temperature kit sold by Klinger[(1)] (KK9004) contains 50 experiments illustrating the relationship between heat and temperature. Many of the usual heat and temperature experiments are included in the kit along with an instruction manual.

Tyndall's friction cylinder for performing Count Rumford's experiment is available from LaPine[(2)] (347 78). Ether in the cylinder is made to boil from the frictional heat, and the vapor is lit at the nozzle.

One type of apparatus for determining the mechanical equivalent of heat is available from Welch[(3)] (1683) and Cenco[(4)] (77825). This type consists of a cardboard tube in which lead shot falls when the tube is repeatedly turned end over end. Measurement of the increase in temperature of the shot provides an approximate value of the mechanical equivalent of heat. A second type is sold by Ealing[(5)] (21-245), Klinger (KH2040), and LaPine (388 14). It consists of a copper drum around which a cord is wrapped several times. One end of the cord is secured, and a weight is hung on the other. When the drum is rotated, a thermometer in the drum records its rise in temperature. Ealing also offers a similar model (21-250) with a spring balance. A third type is available from Welch (1684) and Cenco (77815). This type consists of two brass cones, one inside the other, that are mounted in a calorimeter. When the outer cone is rotated, the inner cone is held stationary by a weight (Cenco) or a spring balance (Welch), and a thermometer in the calorimeter records the amount of heat generated.

Tyndall's specific-heat apparatus can be obtained from LaPine (384 01) and Welch (1623). In this apparatus, several metal weights of equal mass are heated to the same temperature and then placed simultaneously on a paraffin disk. The weights melt the paraffin at rates inversely proportional to their specific heats.

Thermal expansion and contraction in general can be demonstrated with a ball and ring sold by LaPine (381 01), Welch (1659 and 1661), and

(1) Klinger Scientific Apparatus Corp.
(2) LaPine Scientific Co.
(3) Sargent-Welch Scientific Co.
(4) Central Scientific Co.
(5) Ealing Corp.

Cenco (77450). The ball passes through the ring when it is cold, but not when it is hot. Other thermal expansion devices are ice bombs from Welch (1670), LaPine (387 61), and Cenco (77715), and a steam bomb sold by Welch (1669). The ice bombs explode when the water in them is frozen, and the steam bomb bursts when the water in it is heated.

A linear thermal expansion apparatus based on Tyndall's bar breaker is offered by Klinger (KH2021), LaPine (381 16), Cenco (77430), and Ealing (19-520). A bolt or small rod is inserted through a hole in one end of a heated bar that is mounted in a frame. When the bar cools and contracts, it breaks the bolt.

Demonstration linear thermal expansion devices are available from Welch (1640), LaPine (381 34), and Cenco (77370). One end of a tube rests on a knife edge, the other connects with a pointer. The tube expands when heated and moves the pointer. Cenco also offers a variant of the linear expansion apparatus (77411) in which the expanding tube completes an electrical circuit and lights an indicator lamp. A variety of expansion tubes made of different materials is available from all three companies. Yet another linear thermal expansion apparatus is the bimetallic strip sold by Welch (1663), LaPine (381 32), and Cenco (77455).

For change-of-state demonstrations, Wood's metal can be obtained from LaPine (387 78). This alloy, whose melting point is about 70°C, is useful for studying fusion and solidification. For studying freezing by rapid evaporation, a freezing apparatus that uses ether can be obtained from Cenco (76710). Also, a cryophorus is offered by Cenco (77705) and Welch (1667). This device consists of an upper and a lower bulb connected by a tube and partially filled with water. When the water is transferred to the upper bulb and the lower bulb is placed in a freezing mixture, the rapid condensation of the water vapor in the lower bulb causes the water in the upper bulb to evaporate so fast that it freezes.

Heat conductometers are available in a wide variety of forms. Cenco's model (77516) consists of five rods inserted around the periphery of a disk, with bits of paraffin placed in depressions on the outer ends of the rods. When the disk is heated, the paraffin bits melt at intervals proportional to the conduction rates of the rods. Cenco also sells a heat conductivity set (77531) that consists of a conductometer like the one just described plus a rod-type one made of wood partially covered with an aluminum sleeve. Welch sells a four-wire conductometer (1651) and a six-wire conductometer (1653), both of which operate like Cenco's model. Welch also sells an Igenhaus conductometer (1655) in which one end of the rods are inserted in a hot-water bath, a rod-type one (1649) like Cenco's, and another (1656) in which the rods are coated with temperature-sensitive paint and are heated with a steam tube. The paint changes color at 50°C, and the site of this temperature can be seen to move along the rods at varying rates. LaPine offers a conduction apparatus (391 27) in which three rods bent at right angles are inserted into a double thermoscope by their short arms. Heating the long arms results in the appropriate thermoscope readings. Other heat-conduction apparatus sold by LaPine includes one (389 03) similar to Welch's Igenhaus conductometer except that temperature-sensitive paint is used instead of paraffin, and one (389 10) consisting of three notched arms on which matches are placed. A conductivity-of-water apparatus is available from Welch (1647). It consists of a funnel with an air-thermometer bulb fitted in the stem and water surrounding the bulb. When ether or alcohol is poured on top of the water and lighted, only a little heat is transmitted through the water.

Apparatus for studying heat convection in liquids is available from Welch (1729), LaPine (389 18), and Cenco (77565). This apparatus consists of a closed-loop glass tube in which convection currents can be seen when one part of the loop is heated. For studying heat convection in gases, a convection box is offered by Cenco (77590) and Welch (1727). The box has two chimneys, under one of which a candle is placed. Smoke from touch paper goes in one chimney and out the other. Welch also offers a ventilation box (1727A) that has 18 small closable vents in it. Convection currents, as made visible by smoke from titanium tetrachloride, are closely controlled by the vents.

Model hot-water heaters can be obtained from Cenco (77580) and Welch (1729A). These models consist of a flask (representing the heater), a cylinder (radiator), a funnel (expansion tank), and appropriate tubing. Welch also offers a simpler model (1720).

The work performed by heat is aptly illustrated with model heat engines. Steam engines are available from Welch (1775 and 1780) and LaPine (388 16 and 388 30). Two-cycle gasoline engines are sold by LaPine (388 54) and Klinger (KH2013). Four-cycle gasoline engines can be obtained from Welch (1750 and 1753), LaPine (388 51), and Klinger (KH2011). Four-cycle diesel engines are offered by Welch (1749), LaPine (388 55), and Klinger (KH2012).

The second law of thermodynamics as it involves the Carnot cycle can be discussed conveniently with Welch's space model of the cycle (1710B). All four steps of the cycle are traced in black on the curved white surface of the molded-rubber model. One horizontal axis of the model represents volume, the other represents absolute temperature, and the vertical axis represents pressure. The traces represent an ideal gas starting as 1 mole at 546°K and 2 atm of pressure. The gas expands isothermally to become 44.8 liters at 1 atm, then expands adiabatically to 273°K, then is compressed isothermally, and finally is compressed adiabatically to its original volume.

26–2 Specific Heat

26–2.1 Specific Heat at Low Temperature

The rule of Dulong and Petit states that the molar specific heat is approximately constant for all solids. The apparatus pictured in Fig. 26–1 is designed to show that this rule does not hold at low temperatures. Two cylinders of the same size, one made of aluminum and the other of lead, are dipped into Dewar flasks filled with liquid nitrogen. The time required for each metal to heat up to a certain temperature after being removed from the nitrogen is measured.

Since both cylinders have the same surface area and the same volume, the warm-up time is proportional, ρ/A, where ρ is the density, and A is the

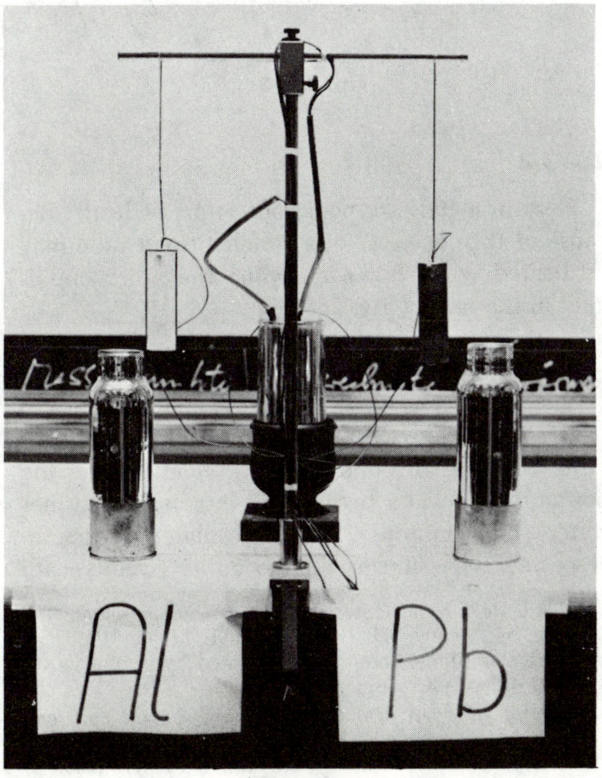

Fig. 26–1

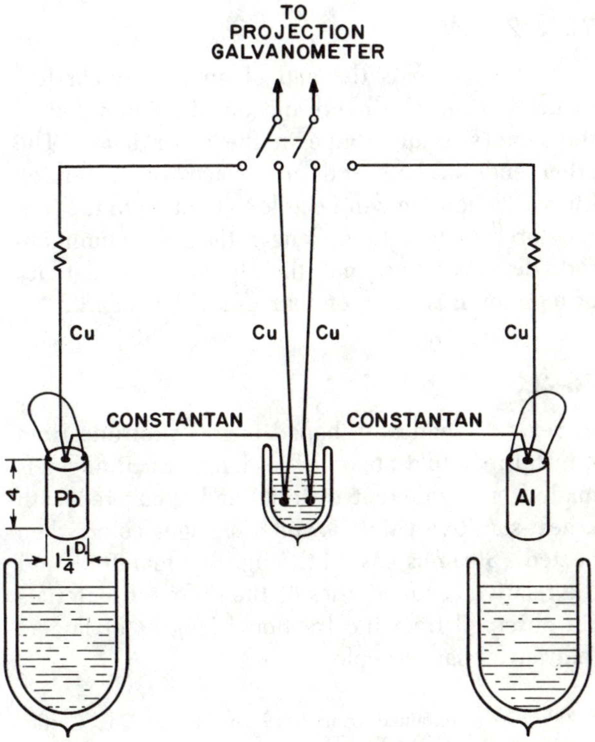

Fig. 26–2

atomic weight. Since ρ/A is approximately equal to 0.1 for aluminum and 0.05 for lead, it would take twice as long for the aluminum cylinder to heat up to a given temperature as it would for lead, if the Dulong-Petit rule held. This experiment shows, however, that both cylinders heat up at approximately the same rate, indicating that the molar specific heat is smaller for aluminum than for lead.

The temperature is measured by means of a copper-constantan thermocouple (Fig. 26–2) glued to the cylinders with epoxy resin. The reference elements are placed in a small ceramic tube and dipped in a Dewar flask filled with liquid nitrogen. An electric timer measures the time required for the cylinders to heat up to a given temperature (e.g., 30 divisions on the scape of a projection galvanometer).[6] The entire apparatus can be shown in shadow projection.

26–3 Heat Conduction

26–3.1

Two Meeker burners used to heat two plates of the same area and thickness, but of different thermal constants (aluminum or brass and asbestos slate), are laid across similar clamp rings at equal heights above the burners. Before lighting, two safety matches are placed on each plate and porcelain beads are placed on the burner screens to make the flames visible to large groups. The match on the metal plate bursts into flame first.

26–3.2

A student holds the end of an iron rod a few inches long in one hand and an aluminum rod of the same size and shape in the other hand. The other ends of both rods are placed in a Bunsen burner flame. He will be able to hold on to the iron rod three or four times longer than the aluminum rod, demonstrating that the thermal conductivity of aluminum is three or four times that of iron.

26–3.3

Several identically-shaped fingers protrude from a metal manifold about 10 in. long. Each finger is made from a different material and is covered with a heat-sensitive paint[7] which changes color when heated. Steam is passed through the manifold, and the relative conductivities of the different materials are observed from the fractional lengths of fingers that have changed color.

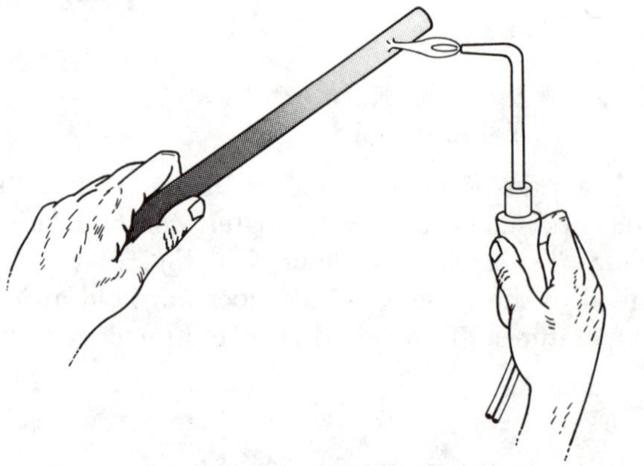

Fig. 26–3

26–3.4[8]

Certain metals are poor conductors of heat. Because of this, one end of a stainless steel tube may be heated white hot even while the other end is held in the hand (Fig. 26–3).

26–3.5[9]

If a soldering iron is dipped into a jar of water (Fig. 26–4), it is found that little heat reaches the bottom of the water because the less dense, warmer water stays on top. Isolated insulated thermometers and television or thermocouple circuits with

[6] Can be obtained from SMS Instrument Co., Castleton-on-Hudson, New York.

[7] See 25–2.3.

[8] Adapted from *University Physics, Experiment and Theory,* by George D. Freier. Copyright © 1965 by Meredith Publishing Company. Reprinted by permission of Appleton-Century-Crofts.

[9] Adapted from *University Physics, Experiment and Theory,* by George D. Freier. Copyright © 1965 by Meredith Publishing Company. Reprinted by permission of Appleton-Century-Crofts.

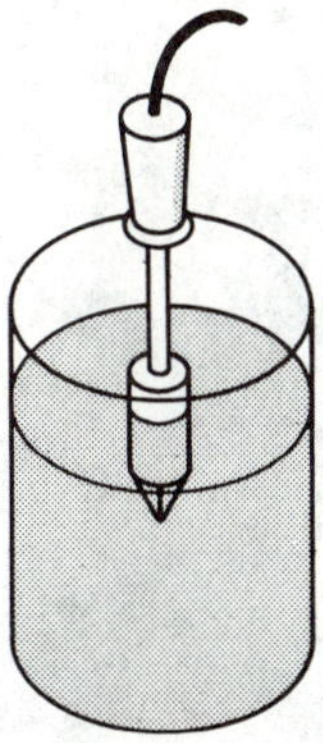

Fig. 26–4

a projected meter reading can be used to show the effect. Both **grad** p and **grad** ρ increase downward, therefore

$$\frac{dp}{dy} \times \frac{d\rho}{dy} = 0,$$

where p = pressure and ρ = density.

26–3.6 [10]

In the Thermicon[11] card, the sensitive substance is the yellow complex of mercury, silver, and iodine, $HgAg_2I_4$. When heated above 114°F it turns orange-red to cherry red in color; on cooling, it reverts to yellow. Though a variety of other "thermochromic" substances are known, mercuric silver iodide is convenient to use and has a low transition temperature and a marked change of color. Applied as a lacquer to a card material of low thermal conductivity, it becomes a flexible imaging medium, bringing within reach many colorful demonstrations in heat.

Electric Heating. Heat a soldering iron to a moderate temperature that will not burn the card. With the warm tip write a message on the front card surface, or write a hidden message on the back of the card, and it will gradually penetrate the card to appear at the front. A current-carrying iron wire and a copper wire of the same size and connected in series show different temperature changes when placed on the card.

[10] See A. Strickler, *Am. J. Phys.*, **30**, 300 (1962), and A. Strickler, *Phys. Teacher,* November 1963.

[11] Registered trademark. Kits and sensitized cards available through the Allen Strickler Co., Fullerton, California, manufactured under U.S. patent 2,945,305.

Conduction. Again touch the warm soldering iron to the back of the card, this time briefly. No immediate change occurs on the front; then a red spot appears, increasing for a while in size. It tells visibly of conduction through and across the card. Holding the card over the soldering iron or some other heat source, redden an area several inches in diameter, and press a large metal washer against the red surface. It quickly bleaches a yellow imprint against the red background. Try the same with a cork disk, which does not conduct the heat away. Again redden the card, and press an open hand against it for several seconds. The card feels warm to the touch because the hand extracts energy. Lifting the hand reveals a yellow handprint. (This experiment may not work well in hot weather.)

Convection. Explore the space around a Bunsen flame with the Thermicon card. The air beside the flame is cool; the card must be brought quite close to the flame to change its color. Holding the card vertically above the flame and tilting it slightly intercepts a hot rising current of air. The card can trace its rather narrow form upward for a foot or two.

An electric glow coil is now substituted for the flame (Fig. 26–5). With the detector card to one side of the coil, one now detects considerable heat—radiant heat—compared with that from the flame at the same position. Above the coil, as before, there is a strong rising current of warm air. If a cardboard trough is inverted over the heater, one end tilted upward, the heated air streams up through it. At the outer end of the trough, the Thermicon card detects the upside down "cascade" of emerging gases. With the V of the trough pointing down toward the heater, the warm air current divides, flowing up along the sides of the trough (Fig. 26–6).

Radiation. Turning the Thermicon card toward the sun, one uses a lens to concentrate some of the sunlight on the card surface. (To prevent overheating, use a cardboard "stop" on the lens, or avoid forming a sharp focus.) The card partly absorbs the radiation, converting its energy to heat. It is possible in this way to write with the bright spot on the card. One may also use a simple "light funnel" to concentrate the sunlight: a cone of bright aluminum foil of about 45 deg included angle,

Fig. 26–5 Demonstration of Convective Heat Flow Along Inverted Trough.

about 2½ in. diam at the upper end, and with a ¾-in.-diam opening at the bottom. Point the cone axis toward the sun, holding the card close under the funnel to catch the radiation.

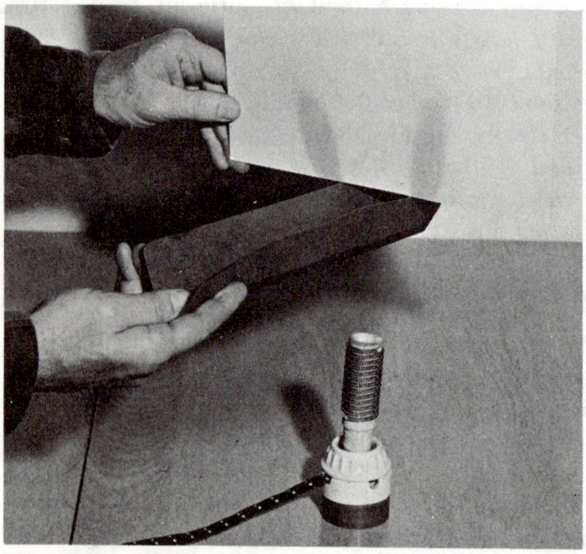

Fig. 26–6 Splitting of Convective Heat Column by Inverted Wedge.

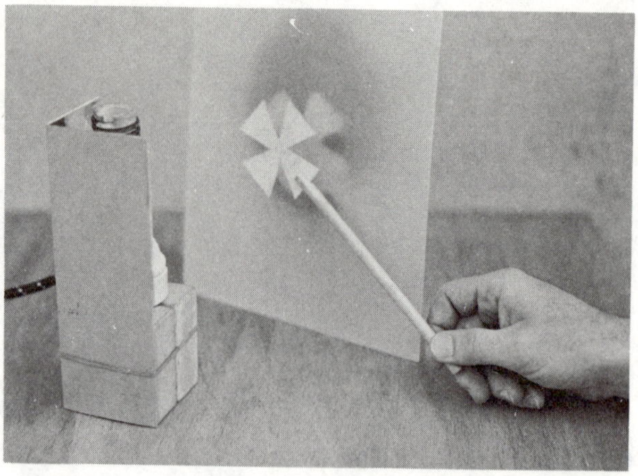

Fig. 26–7 Shadow Casting in an Infrared Beam.

Properties of thermal radiation can also be shown with these painted cards. Holding the detector card to one side of the coil and inserting an opaque object in the radiant path, one casts a yellow shadow on the card, showing that heat rays, as a form of electromagnetic radiation, travel in straight lines (Fig. 26–7). Like visible light, these rays may be *reflected* by a mirror and focused (Fig. 26–8). The mirror here is elliptical.

To cause heating, radiation must be absorbed. The radiation may be in the infrared region. As with visible rays, some surfaces reflect infrared,

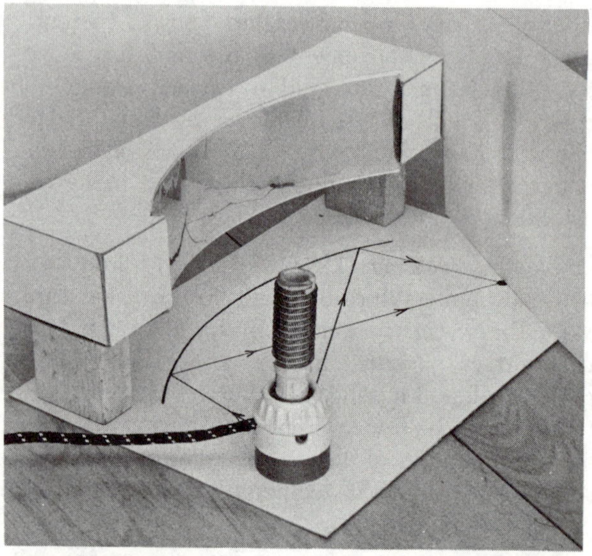

Fig. 26–8 Focusing of Thermal Radiation. Sheet aluminum reflector, shaped to elliptical contour by cardboard holder, receives radiation from a source at one focus and directs it to the detector at the other focus.

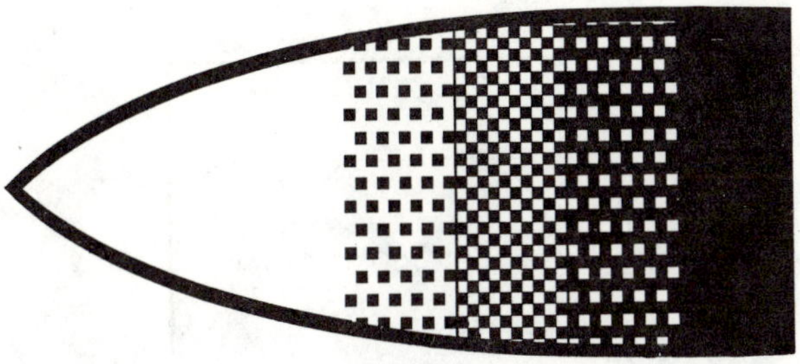

Fig. 26–9 Rocket Pattern for Demonstration of Thermal Absorption and Reflection Effects.

others highly absorb it. Surfaces highly reflective to visible light may be quite absorptive of infrared, and vice versa. Further, absorption may depend on the distribution of the infrared energy with respect to wavelength. A specially adapted Thermicon card is used to demonstrate most of these effects. The label attached to the back is aluminum foil with a printed overlay in black ink (Fig. 26–9). It pictures a space vehicle of which portions are checkered in black to adjust the solar absorption, and thereby the internal temperature.

First try a radiant source of fairly high temperature. An incandescent heating lamp such as the Amplex "Industrial Infrared" (250 W) works well. Its energy distribution as a function of wavelength is shown in Fig. 26–10. The filament appears yellow-white, but actually most of its energy is emitted in the very near infrared. Expose the labeled rear surface of the card to this source. The solid black portion of the label and the heavy black outline absorb the radiation most rapidly. This is evidenced by the red pattern appearing on the front. Continued exposure to the source shows the following descending order of absorptivity of the various surfaces toward the very near infrared radiation: solid black, checkered areas, white paper background and bare aluminum.

Repeat this experiment with infrared radiation of much longer wavelength. The source for this is a glow coil operated at rather low temperature (about 1000°K). Figure 26–11 shows its radiant energy distribution. Surprisingly, the white card surface around the pattern absorbs this radiation almost as well as the solid black areas. The bare aluminum remains, however, an excellent reflector and a poor absorber. Cut from a newspaper an area containing bold print or a line drawing. (Halftones will not work as well.) Arching the

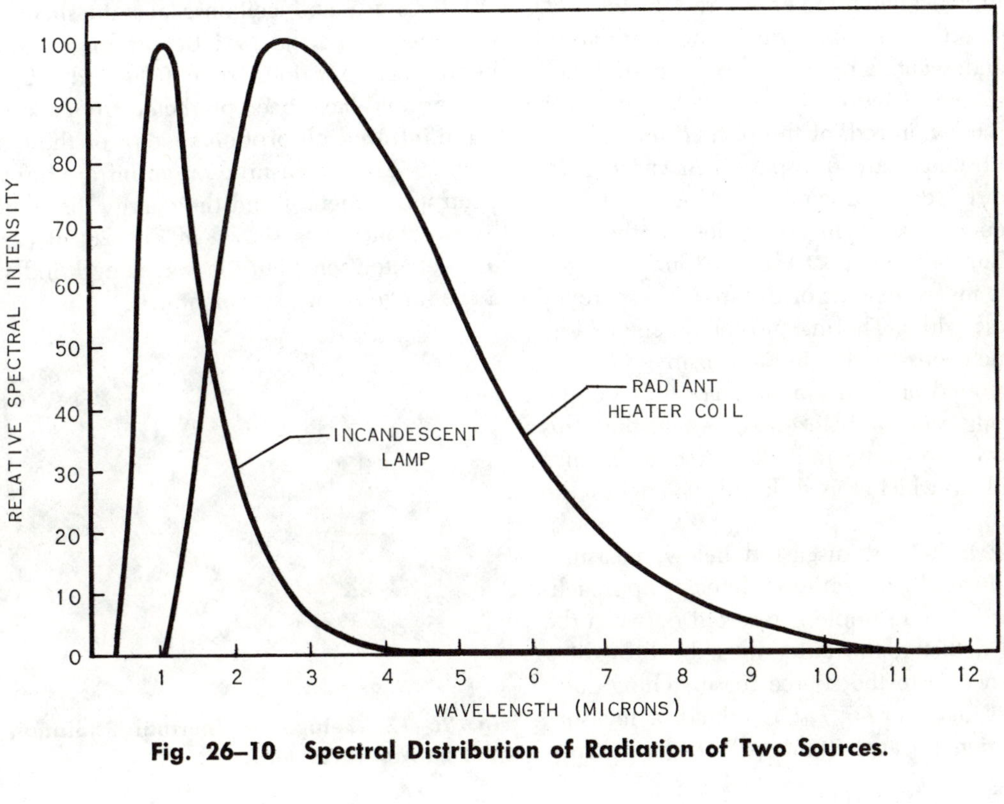

Fig. 26–10 Spectral Distribution of Radiation of Two Sources.

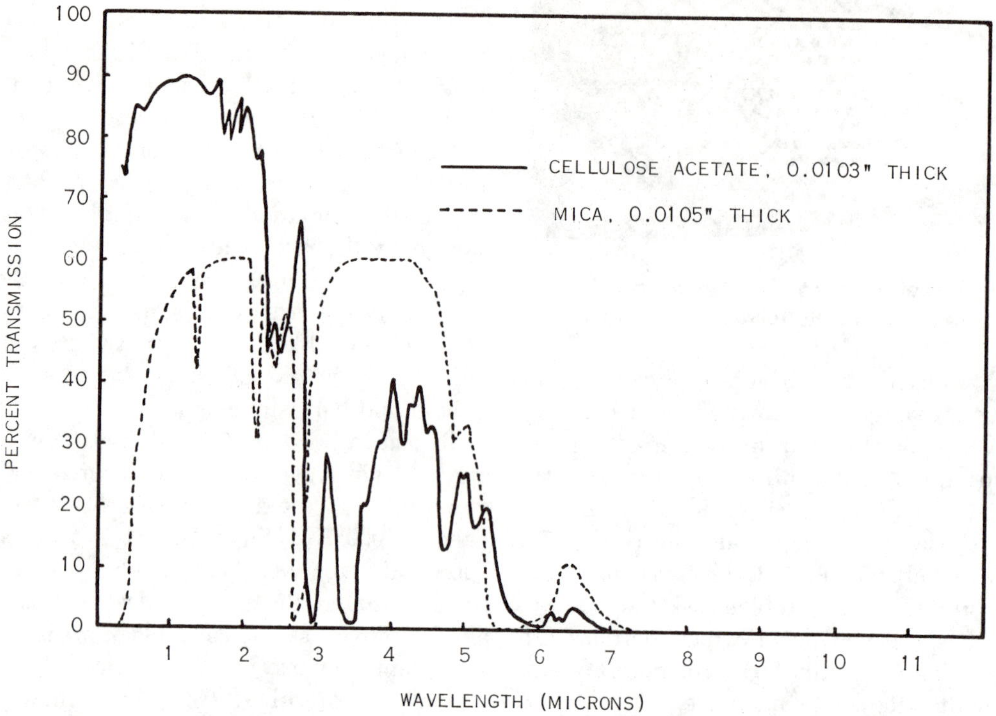

Fig. 26–11 Spectral Transmission of Cellulose Acetate and Mica Samples.

card to hold the paper in close contact with the sensitive surface (the print of interest facing outward), expose the assembly briefly to the radiation of the incandescent lamp. After a little trial-and-error adjustment of the exposure, one can obtain a fairly sharp copy, in red, of the printed matter.

Many substances are transparent in varying degree to infrared, depending on wavelength. In fact, the occurrence of resonant absorptions is the basis of infrared absorption spectroscopy. Simple, semi-quantitative measurements of infrared transparency may be made with a Thermicon-type detector (Fig. 26–12), which consists of a disk of sensitive pigment on a background of fixed orange color. The opposite, receiving side is blackened. Advancing this detector slowly toward a radiant source, a distance D_0 is found at which the disk color matches the background.

To avoid hysteresis, discussed below, measurements are always made with the detector approaching the source. If a sample is inserted between the source and detector, the latter must be brought a closer distance D to the source for matching color change, and the ratio D^2/D_0^2 is a kind of measure of the fractional transmittance T of the sample. Thus, with the incandescent lamp, $T = 0.64$ and 0.81 for mica and cellulose acetate sheet samples respectively. In the radiation from the glow coil, the respective values were 0.36 and 0.27. These values could have been predicted from a knowledge of the infrared absorption spectra of these samples (Fig. 26–11, as measured on an infrared absorption spectrophotometer) and the energy distributions of the two sources as shown in Fig. 26–10 (assuming that detector sensitivity is wavelength-independent in the range of measurement).

Fig. 26–12 Setup for Thermal Radiation Transmission Meusurement.

Surfaces that are good absorbers of a given radiation will also emit that radiation efficiently. Good reflectors correspondingly emit poorly. One demonstrates this principle by slowly heating to redness the sensitive face of the space-vehicle card bearing the patterned label on the back. Allowing the card to cool, we see it losing heat most rapidly in the areas blackened at the back, least rapidly where the rear surface is reflective. The white background radiates almost as rapidly as the black areas, because at the relatively low temperature of the card, the radiation is predominantly of rather long wavelength, peaking at about 10 μ. In this region the white and black are both strong absorbers, and both correspondingly emit well.

Energy Transformation. A portion of the Thermicon card is covered with a sheet of paper to protect its surface, and a small area of the paper is briskly rubbed with a rubber eraser. On the card below the paper is formed a bright red streak on the yellow background. Mechanical work has increased the energy of motion of the molecules on the card surface, raising their heat content and the temperature. An iron nail bent back and forth rapidly, then touched to the card, leaves a red imprint. The work of bending the nail has forced the atoms within the metal to flow past each other. Internal friction between the atoms has generated heat.

Change of State. Transformations from one to another of the familiar states of matter—gas, liquid, and solid—are accompanied by the absorption or liberation of heat, which can be illustrated with a Thermicon card. With a watercolor brush, write a message in water on the plain rear surface of the card. When the writing is exposed to a heat source such as the glow coil, the message appears in yellow against the reddened card face because evaporation has cooled the moistened areas. If we block evaporation, e.g., by covering the wetted areas with a sheet of plastic wrap, the cooling effect ceases.

Bistable Thermochromism. Warming mercuric silver iodide, we find it assumes its characteristic orange-red color at about 114° to 117°F. At higher temperature levels, there are changes to deeper tones of red. But cooling the substance again below 114°, we find the orange-red does not revert to yellow until the temperature has reached about 88°F. There is a "hysteresis loop" in the color characteristic. Within a rather broad temperature range, the color of different portions of the Thermicon card (though at identical temperatures) may remain different, depending on whether they have been warmed or cooled into this range. This "bistable" behavior is reminiscent of magnetic hysteresis.

Using a radiant heat source, advance the Thermicon card slowly toward the source until it just turns orange-red. Note the distance. Withdraw the card slowly, noting the distance at which the color fades. The card is now returned to an intermediate distance, where it equilibrates to a temperature in the bistable range. Taking now a mildly heated soldering iron, use it as a pencil to write a message in red on the yellow surface. The writing remains indefinitely, so long as the card is within the distance range corresponding to the bistable temperature limits. To erase, simply withdraw the card far enough from the source to cool it below the lower bistable limit.

On the other hand, one may preheat the card to redness and again bring it into the bistable range. Temporarily cooling any portion of it below 88°F, one stores a yellow image on the red background. If a person's skin temperature is not too high, for example, and the card is near the lower end of the bistable range, his fingerprints can be formed and retained on the card surface. Our internal body temperature is itself within the bistable limits. A Thermicon card sandwiched between the palms is raised to a temperature in this region.

26–3.7

Propagation of a heat wave in a copper rod can be illustrated by soldering a copper-constantan thermocouple into a small hole drilled 3 in. from one end of a ¼-in.-diam copper rod 1 ft long. With the flame made as sharp as possible, the opposite end of the rod is heated with a Bunsen burner for about five minutes. A mirror galvanometer connected to the thermocouple shows the output wave form to be smooth (i.e., longer) compared with the sharp (peaked) input waveform.

26–3.8

A strip of copper and a strip of iron are placed across metal bars (Fig. 26–13) to demonstrate their different heat conductivities. The strips are cov-

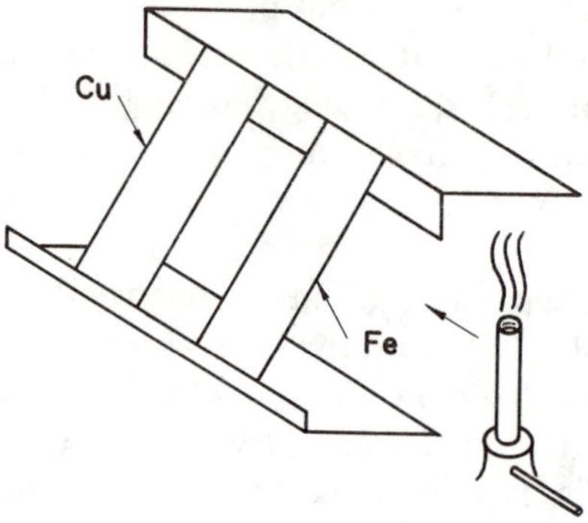

Fig. 26–13

ered with "thermal color,"[12] which is white but turns blue above 65°C. One of the metal bars is heated with a Bunsen burner, and the subsequent color change indicates the difference in heat conduction.

[12] A box of 12 thermal color pencils for temperatures between 65°C and 670°C can be obtained from Phywe AG., Göttingen, West Germany, Catalog No. 4170.

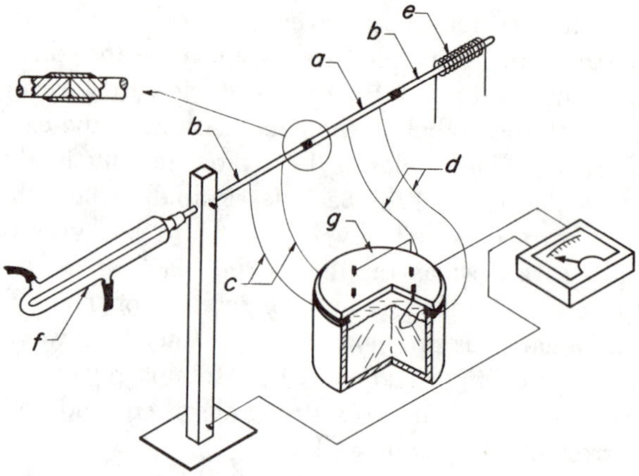

Fig. 26–14

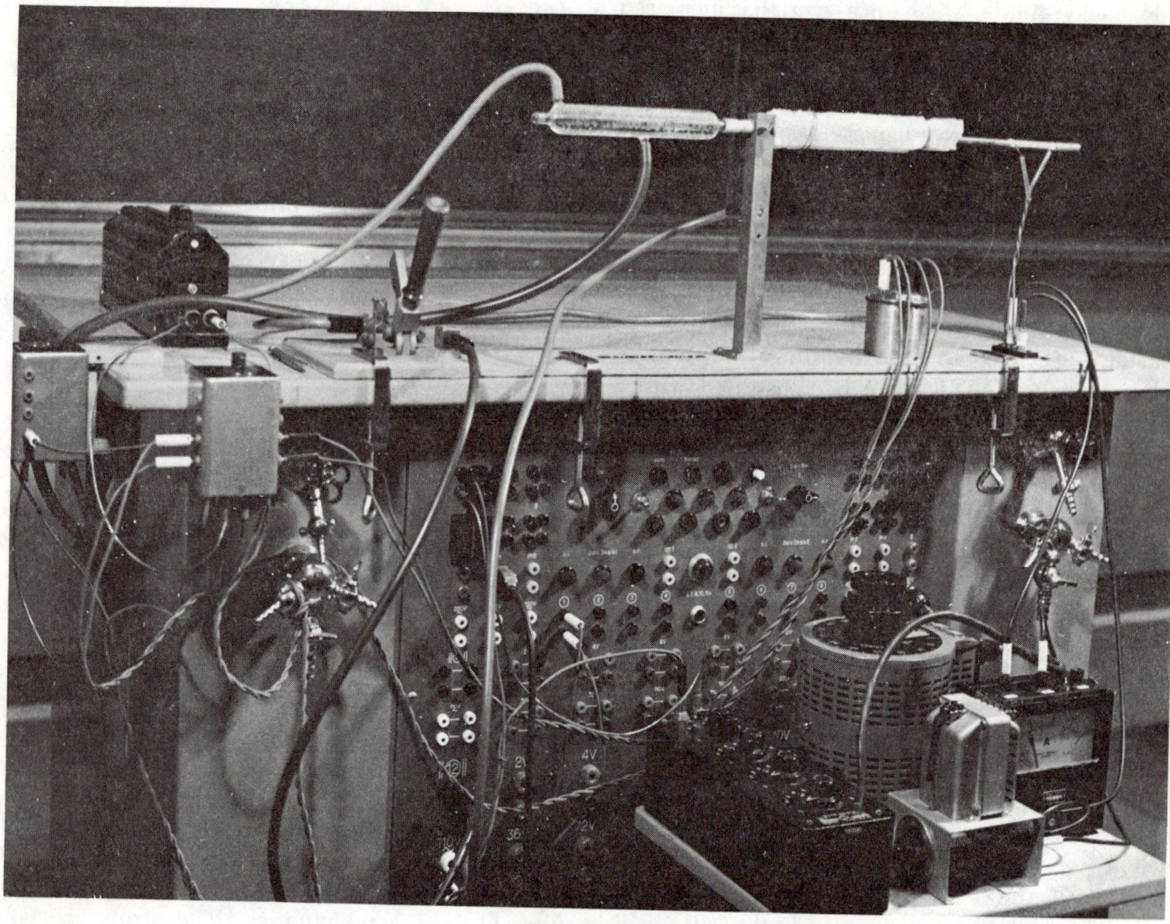

Fig. 26–15

26–3.9

A three-sectioned rod fabricated from constantan *a* and copper *b* is used to show the differences between electrical and thermal conduction (Fig. 26–14). Two 0.4-in.-diam copper rods *b*, one 8 in. long and one 24 in. long, are butted to the ends of a 0.4-in.-diam x 6-in.-long constantan rod *a*, using sleeves and solder. Two constantan wires *c* are soldered to the copper section and two copper wires *d* are soldered to the constantan section, each pair of wires 4 in. apart.

To demonstrate thermal conduction, the rod is first covered with an insulating material, as shown in Fig. 26–15. It is then heated at one end by means of a heater coil *e* and cooled at the opposite end by a water jacket *f*. The wire is wound over mica insulation and reaches a temperature of ~200°C with a potential of 6.3 V ac. The temperature of the water flowing through the cooling jacket should be 15°C or lower. A warm-up period of approximately one hour is necessary to allow the rod to reach thermal equilibrium. The ends of the wires are placed in a bath of oil g which is used as a constant reference temperature (20°C) to determine the projection galvanometer readings. These are directly proportional to the local temperature of the rod where the wires join it.

Electrical conduction is shown with the water jacket removed and the heating wire disconnected. A current of approximately 100 A is passed through the apparatus, as shown in Fig. 26–16, and the voltage drops are measured directly with a projection millivoltmeter (~60 mV range).

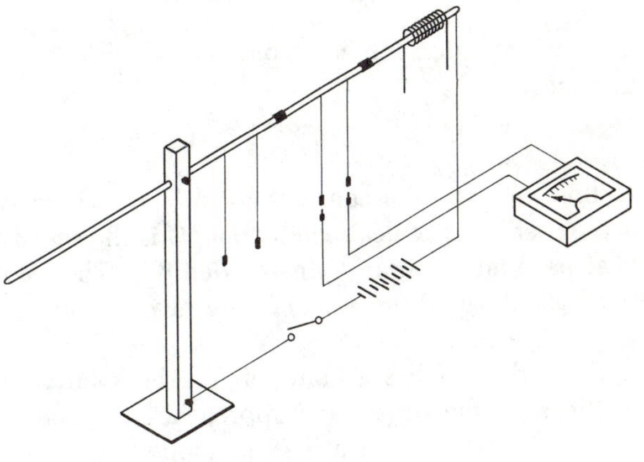

Fig. 26–16

26–3.10

Apparatus which can be used to demonstrate at different times the spreading of a "heatwave" from the center to both sides of an infinitely long thin rod is shown in Fig. 26–17.

To produce heating, the horizontal rectangular aluminum base *a* is heated at its center ($x = 0$, time $t = 0$) by a sharp flame *b*. On the plate small mirrors are mounted individually on bimetallic strips *c*. Light from an arc lamp *d* passes through a condenser lens *e*, a horizontal slit *f* and another lens *g* (f = 5 cm) to the mirrors. The mirrors are adjusted so that their images form a straight line on a screen. The heating causes the bimetallic strips to bend, forming a temperature versus distance curve on the screen.

For the idealized case of a thin rod of infinite

Fig. 26–17

length, the solution to the second order linear differential equation for heat conduction $\frac{\partial T}{\partial t} = k\frac{\partial^2 T}{\partial x^2}$ becomes

$$T(x,t) \doteq \frac{1}{2\sqrt{\pi k t}} e^{-x^2/4kt}.$$

In this equation, the constant $k = W/\rho C$, where W is the coefficient of heat conduction; C is the specific heat per unit mass, and ρ is the density. The solutions $T(x,t)$ give the temperature at each point x for any time t.

Since the rod has a finite length, the solution is modified as the heat wave approaches the end of the rod. After the heating is stopped, the area below the temperature curve is constant with time and represents the total heat energy. A plot of temperature versus distance is shown in Fig. 26–18, where the heating phase is h; i is the phase of heat propagation; and j is the final state.

The construction details are in the Appendix, page 1287.

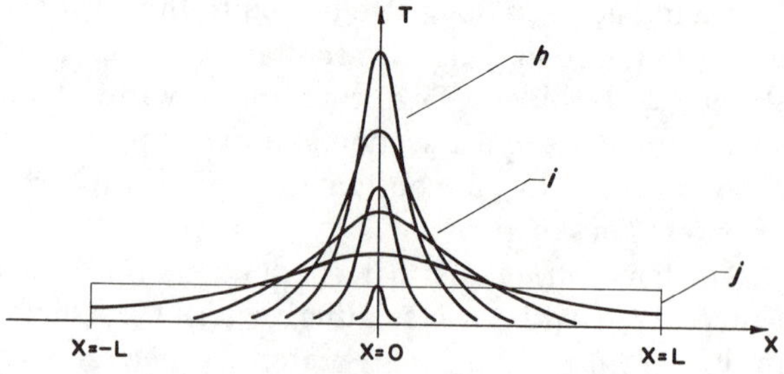

Fig. 26–18

26–4 Heat and Work

26–4.1 Rubber Band Motor[13]

Rubber bands and heat lamps provide the energy for a very simple engine. A bicycle wheel rim is joined to its hub by heavy rubber band spokes (Figs. 26–19 and 26–20). The wheel is suspended vertically and carefully balanced. Heat lamps placed on one side of the wheel cause the rubber-band spokes on that side to contract, moving the center of mass of the wheel. This causes the wheel to rotate slowly, moving the heated rubber bands away from the heat lamp and moving cold rubber bands near the lamp to continue the process.

The wheel is the rim of a 24-in. bicycle tire. The spokes of the tire are cut and formed into hooks to hold the rubber bands as shown. Each rubber band[14] measures ¼ x 3½ in. unstretched. To achieve perfect balance of the wheel small pieces of beeswax are placed at strategic positions around the rim.

Fig. 26–19

26–4.2 "J"-Value Demonstration

This experiment is not intended to give a good measurement but to show the principle of measurements, such as those of Joule, which served to establish the principle of the conservation of energy. A canvas bag containing 5 to 10 lb of small lead

(13) Richard P. Feynman, Robert B. Leighton, and Matthew Sands, *The Feynman Lectures on Physics* (Addison-Wesley Publishing Co., Inc., Reading, Mass., 1964), Vol. I, pp. 44–2.

(14) Plymouth #64, Plymouth Rubber Co., Inc., Canton, Massachusetts.

Fig. 26–20

shot is dropped from the ceiling to the floor several times in succession. The temperature of the shot is taken before the first fall and again after, say, six falls. The shot is poured into a container (preferably a paper cup, for small thermal capacity) for the temperature measurement. A projection bimetallic thermometer is immersed in the shot and the temperature read by the audience. A value of "J" is calculated from measurements of fall and temperature rise. In a typical demonstration, the bag of shot dropped from a height of 30 ft shows a temperature rise of about 3°C for about six falls. To avoid spurious warming by the room, the shot must be spread out on a tray on the lecture table for some time beforehand to bring it to room temperature.

A diagram of the projection thermometer[15] is shown in Fig. 26–21. A bimetallic thermometer equipped with a transparent scale *s* is inserted through holes drilled in the lens C_1 and in the 45-deg mirror *m* of an optical condensing system. Light from a projection lamp enters lens C_2 and is reflected through lens C_1 and the transparent scale *s*. The image of the scale is projected on a screen by the objective lens *l* and the reflecting prism *p*. The condensing system and the mounting for the objective lens are adjustable on a common support rod.

[15] H. M. Waage, *Am. J. Phys.*, **21,** 465 (1953). See also Apparatus Notes for Physics Teaching (1961), p. 99. The thermometer is a Weston bimetallic thermometer reading from −10°C to 110°C, with its opaque scale replaced by a transparent one.

26–4.3 Mechanical and Thermal Energy

The apparatus shown in Fig. 26–22 can be used to demonstrate the conversion of mechanical energy to thermal energy. A heavy steel ball (½ in.

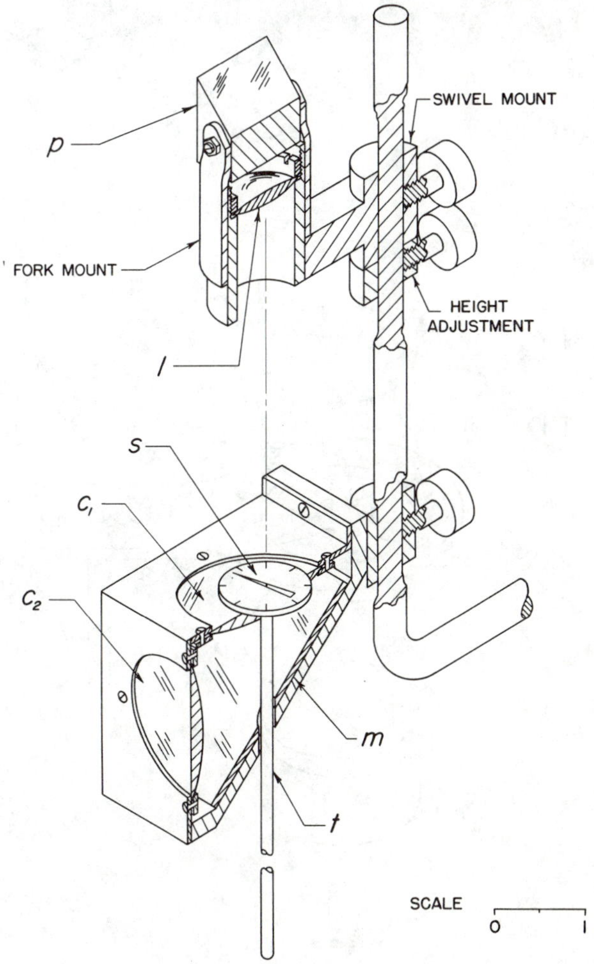

Fig. 26–21

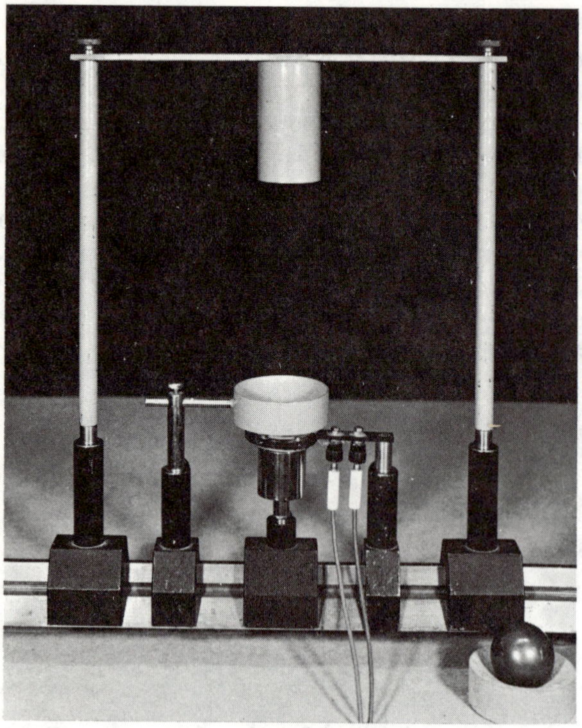

Fig. 26–22

diam) is dropped from a cylindrical chute through a brass ring *a* (Fig. 26–23) and an arrangement of thermocouples *b* onto a steel anvil *c*. Six copper-constantan thermocouples are connected in series in a star-like pattern over a 2-in.-diam hole in a paddle-shaped piece of Plexiglas. Six brass pins are pressed into the Plexiglas 1⅛ in. from the center, and another pin is placed on the rim near the "handle."

After the (0.02-in.-diam) copper and constantan wires are soldered together, an additional drop of solder is placed at the tip. The thermocouples are then soldered to the brass pins so that a copper wire and a constantan wire come together at each pin with the ends of the chain connected to sockets in the "handle." The Plexiglas ring is mounted on an optical bench with the thermocouples resting on a steel anvil covered with a fiber plate *d*, which insulates the thermocouples.

When the ball is dropped on the thermoelements, the deformation of the solder causes a temperature rise, thereby inducing a current in the thermocouples which registers on a mirror galvanometer. The apparatus is shown in shadow projection. It should be noted that although the solder on the tips of the thermocouples is soon pressed flat, the results are not influenced.

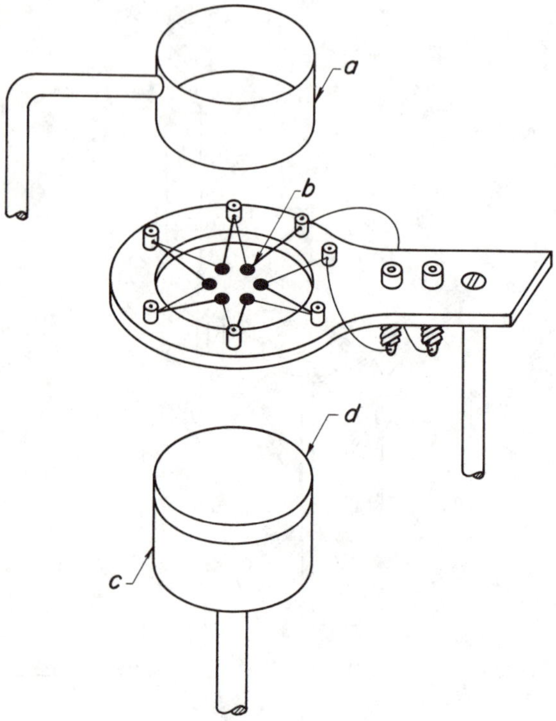

Fig. 26–23

26–4.4 Electrical Equivalent of Heat

An arbitrary mass of water in a Styrofoam container is heated by means of an immersion heater. The temperature rise is recorded by a projection thermometer made from a vapor pressure temperature indicator like that used on automobiles. Construction details are given below for the thermometer. The voltage applied and the current flowing into the immersion heater are measured with pro-

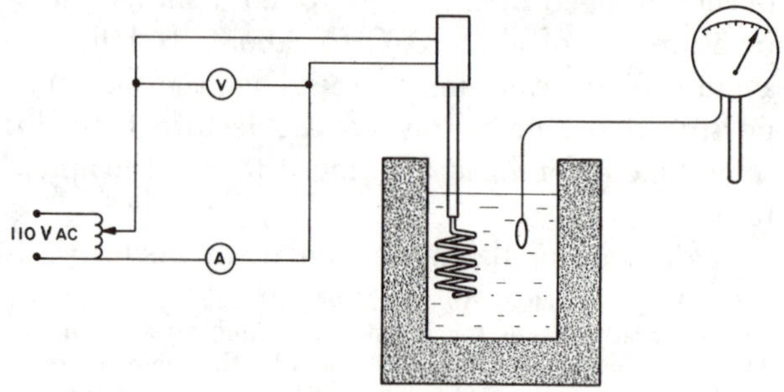

Fig. 26–24

jection instruments, as shown in Fig. 26–24. By noting the temperature rise of the water in a specific amount of time and using the proper conversion factors, one can compare the energy added to the water, which may be calculated in Btu's or calories, with the watt-hour energy fed to the immersion heater. If the heater and a stopwatch are started simultaneously and the temperature of the water is recorded at equal time intervals, the initial slope of the heating curve can be determined.

The temperature indicating device is constructed in the following manner: The indicator is removed from its original case and the pointer is lengthened by gluing onto it a very thin 3-in. length of aluminum. The aluminum may be bent to form a channel along its length to strengthen it if necessary. A case is made from 5-in.-diam aluminum pipe about 1½ in. long. Two Plexiglas disks are cut to fit the ends of the pipe. The indicator is mounted inside the pipe and the end plates are temporarily positioned. The device is calibrated using an accurate mercury thermometer. The deflections of the indicator corresponding to a given water temperature are scratched into the face of the Plexiglas and the plate is removed for engraving. For best results, the indicator should be calibrated at 20° intervals.

26–4.5 Dust Explosion

A dust explosion can be made with a paint can or any can with a tightly fitting friction lid. Punch a hole in the bottom near the side large enough to admit the stem of a small funnel inserted from inside (Fig. 26–25). Attach a length of rubber tubing

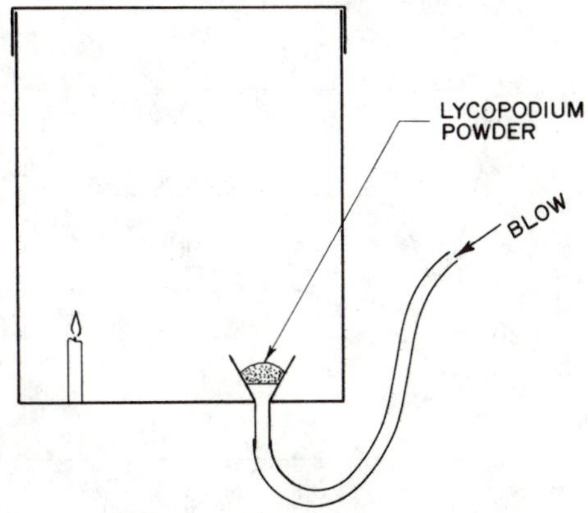

Fig. 26–25

to the extended funnel. Place a single thickness of cleansing tissue in the funnel to support a teaspoon of lycopodium powder. Place a 6-in. lighted candle on the opposite side of the can from the funnel. Close the lid firmly and give a quick puff on the rubber tube. The lid usually hits the ceiling.

26–4.6 Work from Heat

Holes (½ in. diam) are put in the top and bottom of a can which has a removable lid. The can is filled with illuminating gas and ignited at the top. At first, the flame burns low. Then it jumps through the top and the gas in the can explodes. Weights are placed on the can to show the force of the explosion. Emphasis is placed on the "unorganized" work accomplished by the heat, which is in contrast to the highly organized system of obtaining work in an automobile engine.

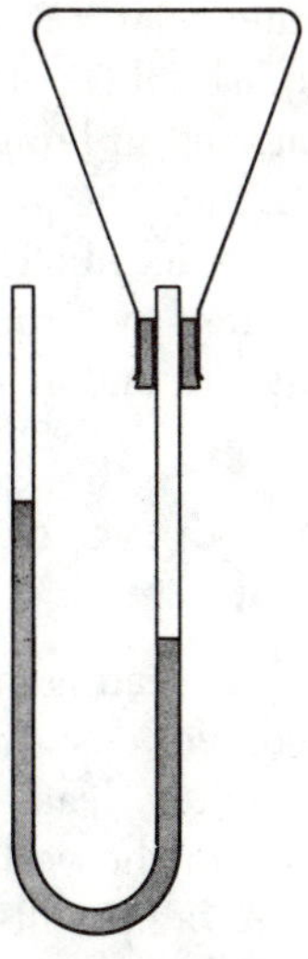

Fig. 26–26

26–4.7 Mechanical Equivalent of Heat[16]

The transformation of mechanical work into heat may be demonstrated by the forging of lead. A piece of lead is placed on an anvil and hammered with a heavy hammer. The blow is inelastic since the hammer doesn't bounce back from the lead.

[16] Translated by K. V. Robinson and H. A. Robinson (Adelphi University) from *Lecture Demonstrations in Physics*, ed. A. B. Mlodzeevskii, M. V. Lomonsov State University, Moscow, U.S.S.R.

The same piece of lead can be forged many times. If it is flattened into a plate, the plate can be bent in half and its thickness doubled.

To show that the lead becomes heated on hammering, a simple air thermoscope can be used (Fig. 26–26). A small flask is held upside down and a bent glass tube is inserted in the flask stopper to serve as a manometer. A piece of lead is first placed over the bottom of the flask and then placed on the anvil and hammered several dozen times. It is replaced over the bottom of the flask to show the rise in temperature by the change in water level in the tube. The apparatus should be shadow-projected during lecture.

26–5 Heats of Transformation

26–5.1 Heat of Fusion of Water

To demonstrate the quantity of heat per unit mass that is required to convert ice at its melting point to water, a thermocouple, a stopwatch, and a projection galvanometer are used in lecture. This instrumentation relates temperature change to time. To accomplish this quickly, the thermocouple is first immersed in liquid nitrogen and then in warm water. Around the junction a thin coating of ice forms. As the ice melts, a rough graph of galvanometer readings vs time is plotted. The curve rises, flattens out, and again continues to rise.

26–5.2 Cooling Curve of a Melted Material

A small quantity of lead is melted in a porcelain dish. A thermocouple, its junction welded, is placed into the molten lead. The thermocouple leads are connected to the vertical input terminals of an *x-y* plotter. A ramp voltage input is applied to the horizontal trace of the plotter to provide a time base for the cooling curve. Best results are obtained if the metal is allowed to cool for a short time before the thermocouple is inserted, due to the more uniform temperature distribution.

26–5.3 Two Latent Heat Experiments[17]

Both of these experiments demonstrate the latent heat of fusion when one substance loses heat to another during crystallization. In the first one, about 20 ml of light mineral oil at room temperature is placed in a wide glass test tube (3 cm i.d.), which is wrapped with insulation and placed in a foam-plastic cup. About 15 g of salol (phenyl salicylate; melting point 43°C) in an aluminum cigar tube are heated to 45–50°C by placing it in hot water. The tube is then quickly inserted into the insulated large test tube.

In order to record the temperatures of the salol and the oil at ½-min intervals, two thermometers are needed; one is used in the salol, the other is placed in the oil. Before the experiment is started, the thermometers should be fastened in a vertical position in the oil and salol.

Both temperatures-vs-time are plotted on an overhead projector or the blackboard, and the first appearance of salol crystals is noted. Stirring the salol slowly will minimize supercooling.

Figure 26–27 is a typical cooling curve; as shown, the temperature of the oil rises at the onset of freezing of the salol. The fact that the oil (already above room temperature) continues to heat up while the salol is at constant temperature indicates that the salol loses heat while freezing, although it is not changing temperature. In this example, the readings were stopped before the salol had completely frozen. By this time the temperature of the

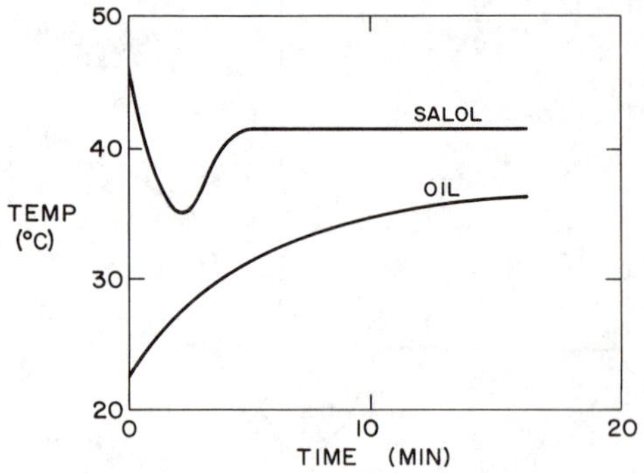

Fig. 26–27

[17] Developed by the Harvard Project Physics program under grants from the Carnegie Corporation, the Ford Foundation, the National Science Foundation, the Alfred P. Sloan Foundation, the United States Office of Education, and Harvard University.

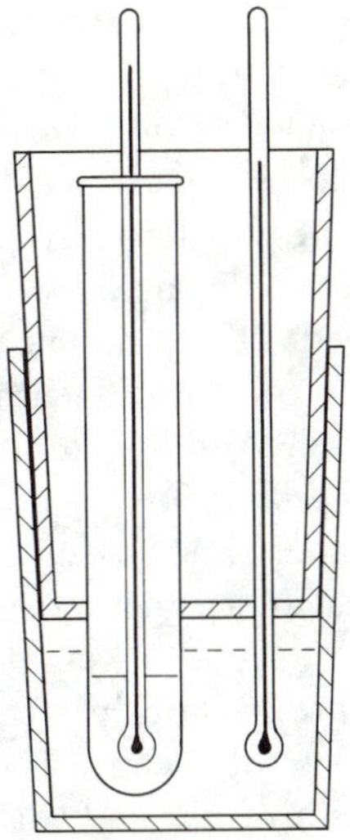

Fig. 26–28

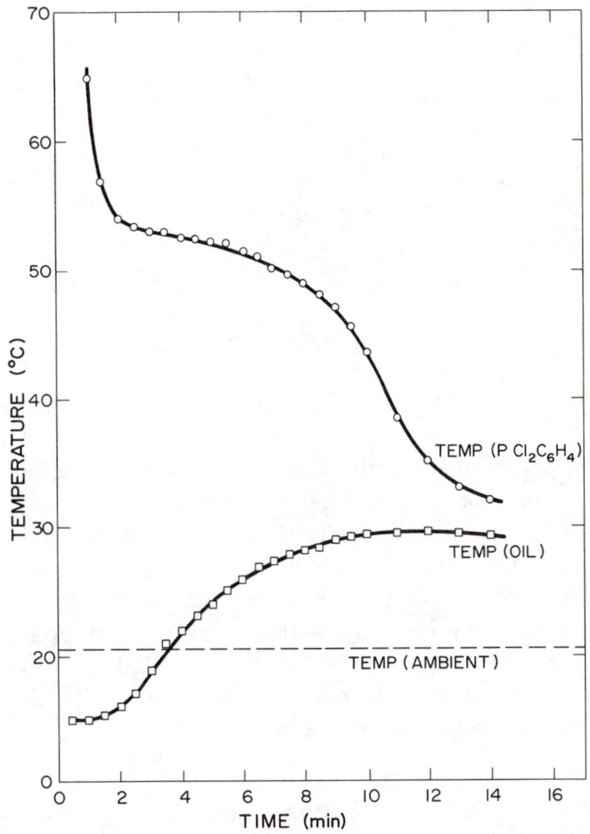

Fig. 26–29

oil was also steady; i.e., all heat being lost by the salol was being lost from the system. To obtain a more complete cooling curve showing also the loss of heat from solid salol to the oil, the oil must be cooled to ~5°C. After a period of time, the temperature of the solid salol decreases; thus the curve eventually falls.

The second experiment involves a closed system for the study of freezing *p*-dichlorobenzene. Two foam cups are needed (8 oz and 6 oz). A reference mark is made about 3 cm from the bottom of the small cup, and it is filled with light mineral oil to this mark. Two holes are then made in the bottom of the larger foam cup. Care should be taken not to make the holes any bigger than need be for an aluminum cigar tube and a thermometer to fit snugly.

As shown in Fig. 26–28, the base of the large cup is pressed into the top of the small cup so that the two cups fit tightly together and form a simple calorimeter.

After the cigar tube is half filled with crystals of *p*-dichlorobenzene, they are melted by holding the tube in hot water. A thermometer is then inserted and heating is continued until the liquid reaches 65–70°C. The outside of the tube is then wiped dry and the tube with thermometer quickly transferred to the calorimeter.

Temperature readings of the *p*-dichlorobenzene and the oil are taken at ½-min intervals. Between readings, move the calorimeter gently to keep the oil mixed. A plot of the oil temperature and the temperature of *p*-dichlorobenzene versus time is shown in Fig. 26–29. As before, data should be plotted on an overhead projector or the blackboard.

26–5.4 Heat of Melting (Crystallization)[18]

The crystallization of a super-cooled liquid can be used to demonstrate the heat of melting. A 100 to 200 cc flask with a manometer attachment, as shown in Fig. 26–30, is used for the demonstration. The manometer elbows are about 15 cm high.

About 50 gm of hypo crystals are placed in the flask and melted in a water bath. After the flask has been cooled to room temperature, the manometer-stopper is used to cork the flask. The water levels in both arms are seen to be the same. The

(18) Translated by K. V. Robinson and H. A. Robinson (Adelphi University) from *Lecture Demonstrations in Physics*, A. B. Mlodzeevskii, ed., M. V. Lomonosov State University, Moscow, U.S.S.R.

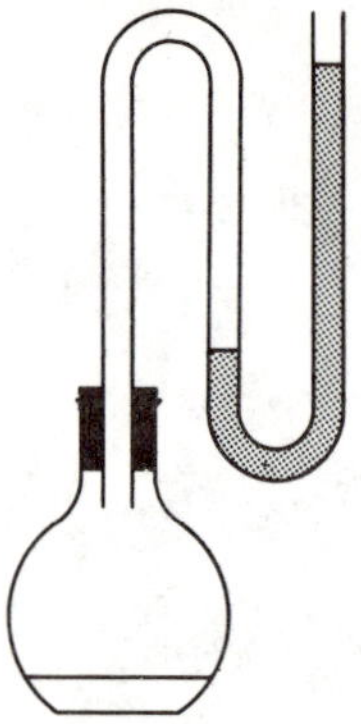

Fig. 26–30

stopper is removed, the flask content is seeded with hypo crystals, and the stopper is replaced as tightly as possible. The heat given off by the crystallization process heats the air in the flask and the manometer indicates a rise in pressure inside the flask. This rise may be so great that even if the stopper and manometer are removed from the flask several times and then replaced, the manometer will still continue to indicate increasing pressure.

26–5.5 Heat of Evaporation[19]

Cooling produced by evaporation can be demonstrated with a 200-cc flask with a rubber stopper through which passes a 40-cm-long glass tube about 7 mm in diameter. The flask is placed upside down and the lower tip of the tube is placed in a small vessel containing colored water. Several drops of ether are poured over the bottom of the flask, which may be covered with either cloth or filter paper. The level of the water will rise in the tube due to the cooling produced by the evaporation of the ether.

26–5.6 Heat of Solution[20]

The cooling which occurs when the heat of solution is absorbed can be demonstrated with simple apparatus. Pour 50 cc of water into a 250-cc flask. Then the flask is closed by a stopper which is connected to a water U-tube manometer (see Fig. 26–30). The water levels in both arms of the manometer are seen to be equal. Next, the stopper is removed and 10 g of powdered hypo or ammonium chloride are dropped into the flask and the stopper-manometer replaced. The manometer immediately shows a decrease in pressure. Heating during solution can be shown by adding a small amount of sulfuric acid instead of hypo.

A dramatic demonstration of cooling by solution consists of pouring at least 40 cc of water into a beaker and adding an equal weight of ammonium nitrate. The mixture is quickly stirred with a test tube containing 1 cc of water. The water in the test tube freezes as a result, and this can be shown by projecting onto a screen the test tube containing the water both before and after the experiment.

26–5.7 Sublimation[21]

A few ammonium chloride crystals are placed on the bottom of a clean, dry test tube, which is then heated. The ammonium chloride is evaporated without melting and forms a heavy white smoke consisting of NH_4Cl crystals separating from the vapor. A white deposit of NH_4Cl crystals forms on the relatively cold walls of the test tube while a band of clear glass remains between the deposit and the level of crystals in the test tube.

Crystallization from a vapor may be shown by means of a cylinder of liquid carbon dioxide. A small bag of heavy cloth is placed over the valve and tied securely. The valve is opened, and the gas which emerges is cooled as a result of adiabatic expansion. Because the pressure of the triple point for carbon dioxide is higher than atmospheric, carbon dioxide is condensed out of the gas as a solid. Carbon dioxide snow is deposited in the bag. A test tube of mercury placed in the snow will solidify, showing the low temperature of the snow. *Caution:* Severe burns can result from handling the snow.

26–5.8

A small amount of iodine is melted into a closed, partially evacuated glass tube which can be shown

[19] Translated by K. V. Robinson and H. A. Robinson (Adelphi University) from *Lecture Demonstrations in Physics*, A. B. Mlodzeevskii, ed., M. V. Lomonosov State University, Moscow, U.S.S.R.

[20] Translated by K. V. Robinson and H. A. Robinson (Adelphi University) from *Lecture Demonstrations in Physics*, A. B. Mlodzeevskii, ed., M. V. Lomonosov State University, Moscow, U.S.S.R.

[21] Translated by K. V. Robinson and H. A. Robinson (Adelphi University) from *Lecture Demonstrations in Physics*, A. B. Mlodzeevskii, ed., M. V. Lomonosov State University, Moscow, U.S.S.R.

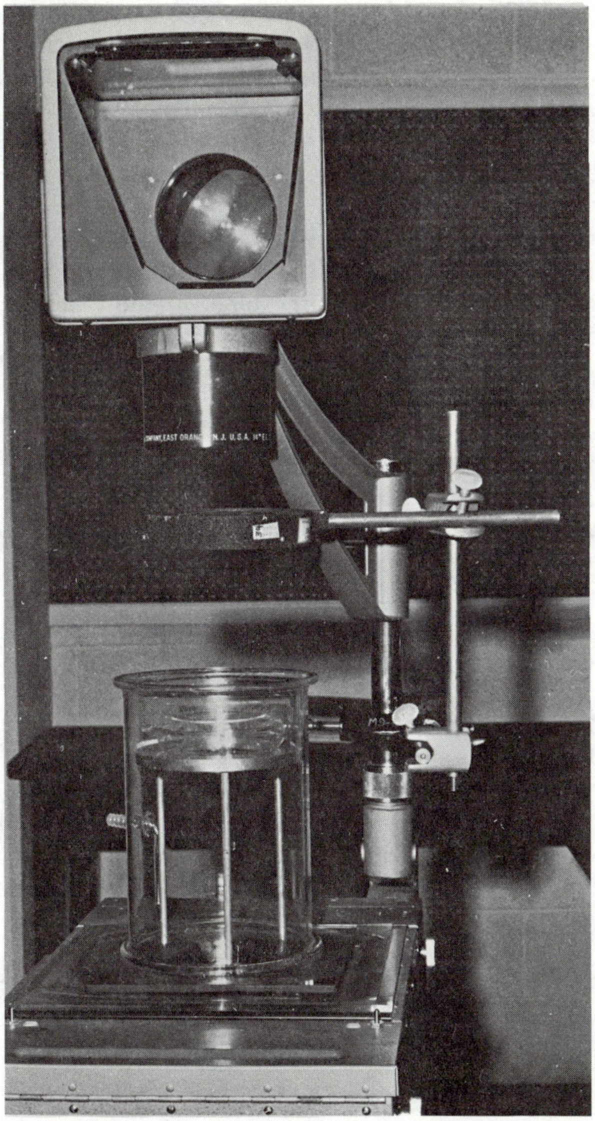

Fig. 26–31

in projection. When the tube is heated with a hair drier at the spot where the iodine crystals are melted, they are observed to disappear. Iodine crystallizes in the cold parts of the tube.

26–5.9 Cryophorus[22]

A cryophorus for use on an overhead projector is shown in Figs. 26–31 and 26–32. The vacuum chamber *a* is a glass cylinder (approximately 9 in. long, 5⅝ in. i.d.) with its ends ground flat. It is closed at the ends with glass plates sealed with vacuum grease. If desired, the lower plate can be permanently cemented to the cylinder with a vacuum-tight seal. Inside the vacuum chamber is placed a tripod *c*, consisting of an aluminum ring supported by aluminum legs. A watch glass *d* containing distilled water rests on a toroidally shaped vessel *e* containing sulfuric acid.[23] The assembly is placed on the stage of an overhead projector, and the chamber is connected to a vacuum pump. After the pump is in operation for a few minutes, the formation of ice at the triple point is observed on the screen. The transformation is rendered more dramatic when viewed by polarized light. A sheet of Polaroid *f* serves as the polarizer, and a second one at *g* is the analyzer.

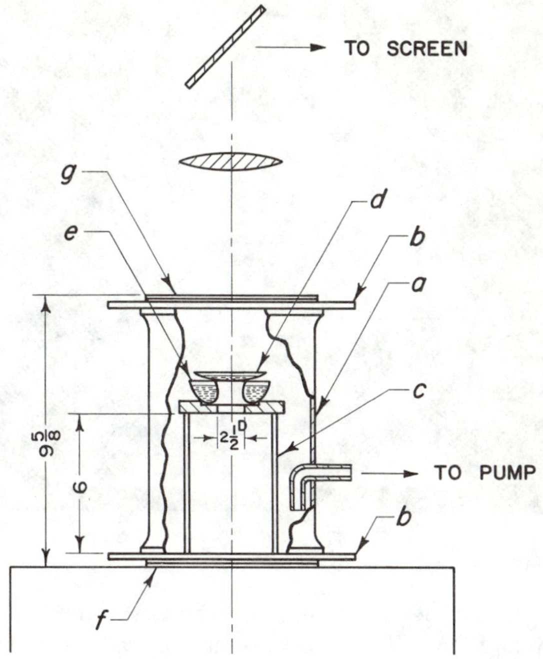

Fig. 26–32

26–5.10[24]

The apparatus shown in Fig. 26–33 is also used with an overhead projector to demonstrate to a

(22) See also R. M. Sutton, *Demonstration Experiments in Physics* (McGraw-Hill Book Co., New York, 1938), H-70, p. 215.

(23) A triangle placed over a shallow evaporation dish serves equally well.

(24) R. Resnick and H. F. Meiners, *Demonstration and Laboratory Apparatus Report of the 1960 Summer Visiting Professor Workshop at Rensselaer Polytechnic Institute.*

Fig. 26–33

large class the formation of ice crystals at the triple point. It consists of a Lucite inner cylinder *a* (Fig. 26–34), a top piece *b*, two outer cylinders *c* and *d*, a lower base *e* and an upper base *f*. Lucite is used because it simplifies construction and provides a flat surface that makes projection possible. All joints *must* be cemented tight.

A brass evacuating tube g ($\frac{3}{32}$ in. o.d., $\frac{1}{6}$ in. i.d., $2\frac{3}{4}$ in. long) is soldered into a hole drilled in a $\frac{5}{16}$-18 NC round-head brass machine screw *h* fitted with an O-ring seal *i*. Before the machine screw with tube is inserted, a small amount of distilled water is poured into the lower chamber. The water should just cover the lower base. One filling will be sufficient for a group of demonstrations separated several days apart.

After the apparatus is evacuated through the brass tube, it is sealed off by pinching, cutting, and soft-soldering the tip of the tube. If the pinch is made with a pair of *sharp* side-cutters, the seal is good enough to hold the vacuum until the pinched tip can be soldered.

When the outer annular trough is filled with a mixture of Dry Ice and acetone, the water within the evacuated chamber freezes in approximately 15 min. It can be thawed and refrozen in about 5 min. If crossed Polaroids are used with the apparatus, the growth of the ice crystals can be observed from the striking color changes that take place.

26–5.11

Liquid bromine changes state rapidly and makes a striking demonstration. The glass tube *a* ($1\frac{1}{4}$ in. diam, 22 in. long) is bent 90 deg so that one arm is about $14\frac{1}{2}$ in. long (Fig. 26–35). A small quantity of liquid bromine *b* is added after evacuation (see Demonstration 27–9.1 for a simple method of

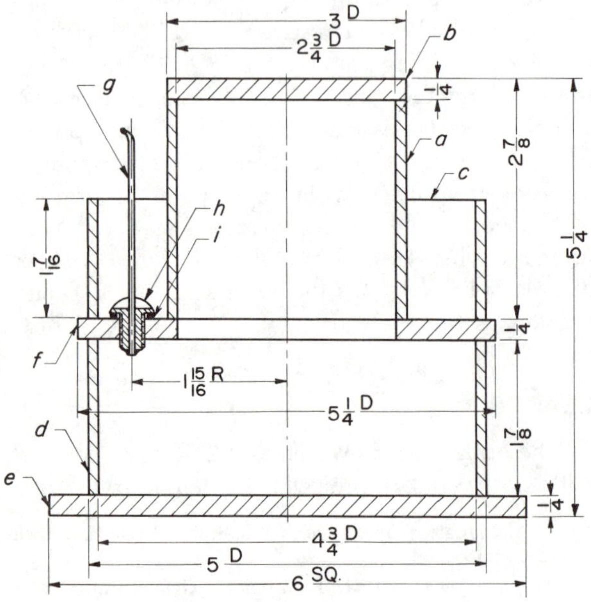

Fig. 26–34

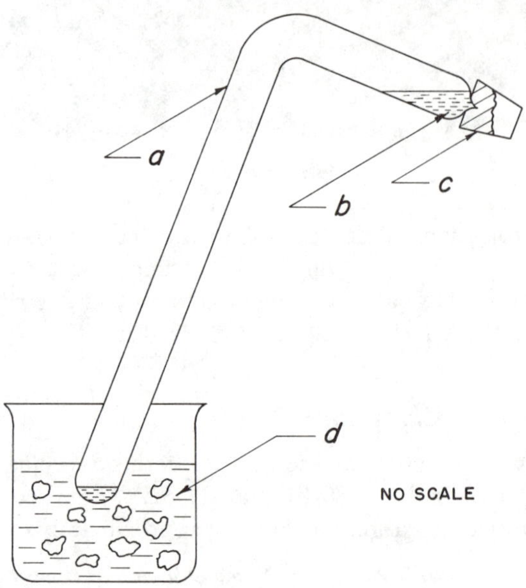

Fig. 26–35

adding a small amount of bromine after evacuation), and the tube is sealed off. A cork *c* is used to protect the tip of the tube.

As shown in Fig. 26–35, most of the liquid bromine rests in the short arm. When the long arm is immersed in a bath of Dry Ice and ethyl alcohol *d*, rapid evaporation takes place in the short arm and condensation occurs quickly at the bottom of the long arm. The rapid evaporation process subtracts enough heat from the liquid bromine in the short arm to cause it to freeze. The advantage of this apparatus over a water cryophorus is that the student can see the gas.

WARNING: Bromine is a dangerous material that attacks almost everything except glass and paraffin wax. Before attempting to prepare the bromine tube, check carefully on alternate methods for preparation.

Fig. 26–37

26–5.12 Crystallization

A wire ring with a handle is dipped into a soap solution that consists of detergent, glycerine, water, and sugar (Fig. 26–36). It is then inserted in a hollow cylinder surrounded by Dry Ice, resting on the stage of an overhead projector. Water crystals form slowly on the soap film as shown in Fig. 26–37.

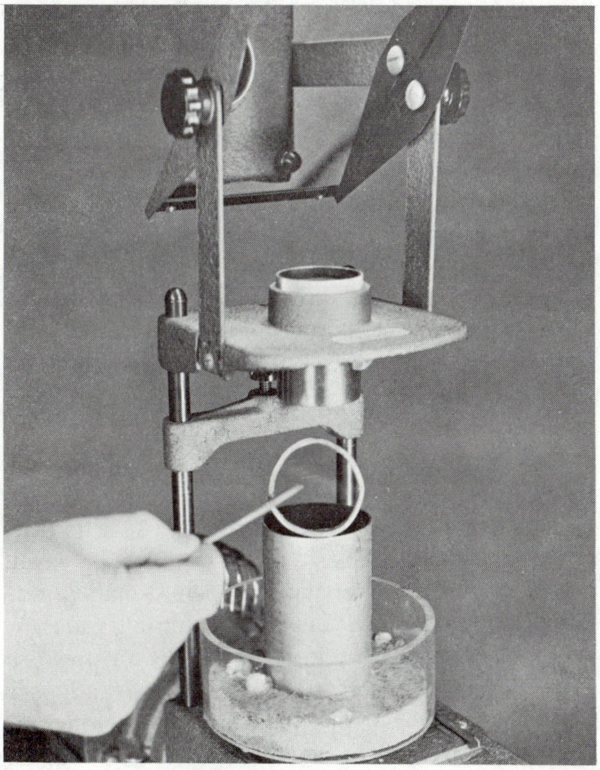

Fig. 26–36

The cylinder is a 5-in. length of brass tubing measuring 2⅜ in. i.d. and 2½ in. o.d. It rests on end within a larger plastic cylinder used to hold the Dry Ice. Both cylinders rest on a ¼-in.-thick glass plate on the projector stage. A small clamp device can be added to hold the wire frame in position, thus freeing the demonstrator's hands.

26–5.13

The crystal growth of various organic compounds can be shown with a standard lantern slide projector. Suitable compounds are DDT, dinitrochlorobenzine, naphthol, trichloroacetic acid, or recorcinol; all of these crystallize at relatively low temperatures. As shown in Fig. 26–38, two crossed Polaroids *b* are placed in the projector between the condenser *a* and the objective *d*. Between the Polaroids is slide *c* producing the image *e* on the screen. The slide is made by sandwiching the organic compound between two thin pieces of glass which are cemented together. The compounds provide colorful growth patterns under the influence of the heat from the projector lamp since the temperature gradient across the slide is not uniform. Crystal growth by evaporation can be

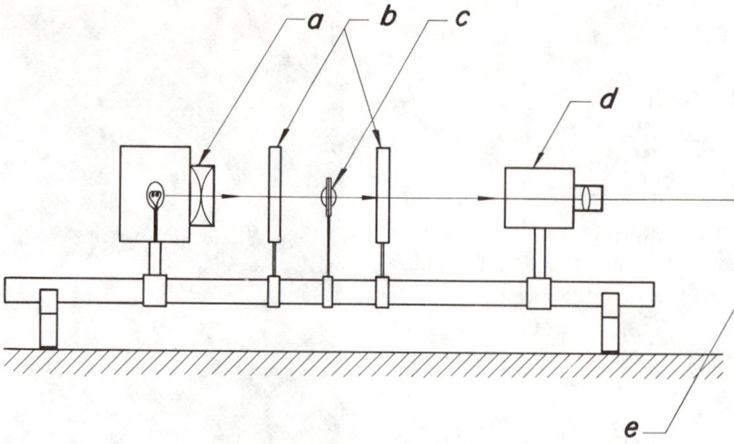

Fig. 26–38

shown if a vertical projector is used. In this case the crystals are placed on a glass plate which is inserted horizontally between cross Polaroids. A solvent which evaporates quickly is then added with a medicine dropper.

26–5.14

Crystals formed when tartaric acid and benzoic acid powders are melted together can be viewed with a vertical projector and two sheets of Polaroid. One sheet is placed on the projection surface. The acid powders are sprinkled between two glass plates, which are held in a lens holder. After the glass sandwich is held over a flame until the crystals melt, it is mounted over the projection surface. The second Polaroid sheet is positioned above the glass plates in a rotary holder. Multicolored crystals are projected when the melted powders begin to cool and crystallize.

26–5.15 The Supercooling of Liquids

Four methods are given for supercooling liquids, including the melts of a number of crystals. *Caution:* Hydrogen fluoride solution is dangerous—do not touch it.

1. Sodium thiosulfate (hypo) or sodium acetate are best handled as follows: The salt is placed in a test tube or flask which has previously been cleaned with a 10 percent solution of hydrogen fluoride and thoroughly rinsed in distilled water. The salt crystals are put into the container through a rolled-up filter paper in order to guard against contaminating the neck with crystals. The salt (mp 45–50°C) is then melted in hot water and boiled for a short period. A small amount of distilled water must be added during the boiling in order to replace vapor losses. During boiling, the neck of the flask should be stoppered with creased filter paper to keep out dust. After boiling, the flask should be closed with a two-hole rubber stopper. A glass tube (4–5 mm i.d.) with the lower end drawn down to about 1 mm is forced through one of the holes so that the tip is in the center of the flask or just above it. The flask is then cooled under running water. For the demonstration, the flask is placed in front of a projector and the image of the tip focused on a screen. It is best to have the flask immersed in a parallel-sided glass container containing water (or better yet a hypo solution) to prevent light refraction at the surface of the flask. A small crystal of hypo is dropped down the glass tube. It becomes wedged in the tip, and crystallization first starts inside the tube. As the crystallization reaches the tip, beautiful crystals begin to grow out into the tube. The flask should be slowly rotated about its vertical axis. The heat of crystallization of the hypo crystals can be easily demonstrated by inserting an open U-shaped manometer tube filled with colored water into the cork. As the supercooled liquid heats up, the air pressure in the flask will increase, causing the manometer liquid to rise. The cork may have to be removed several times to release the pressure and to prevent the manometer liquid from being ejected.

2. A few crystals of hypo, α-naphthylamine, acetamide, or salol are melted in a thin layer on a microscope slide under a cover glass over a low flame. The substance should be melted until no crystals remain on the slide and the layer of liquid melt is evenly distributed under the cover glass. (In the case of hypo, some small dehydrated crystals may not melt.) The slides are then mounted in a microprojector. It is instructive to use polarized light. Seeding is accomplished by picking up a few crystals on a needle and touching one edge boundary of the cover glass with them. Crystallization can then be followed on the screen. When salol is used, the cover glass should be dispensed with. Larger specimens, capable of being projected by ordinary projection apparatus, can be prepared by melting the crystals between ordinary lantern slide plates from which the gelatine has been removed.

3. Thymol (mp 51°C) can be melted by dropping crystals onto the surface of a crystallizing dish filled with hot water. The dish is resting on the table of an overhead projector. During the cooling process, which may be followed with thermocouples, the dish must be covered. When the dish has cooled sufficiently, crystallization follows the introduction of a seed crystal and may be watched on the screen.

4. Water may be supercooled in a flask which has been previously cleaned by a 10 percent solution of hydrogen fluoride and then rinsed with distilled water. The flask is half filled with distilled water and sealed off under vacuum (or sealed off while the water is boiling). The flask is cooled down and then set into a mixture of ice and rocksalt adjusted to give a temperature of about −9°C. The flask should remain in the mixture for about 10 min. The water remains liquid but will solidify immediately on shaking.

26–5.16 Superheating of Liquids[25]

A flask (½–¾ liter), preferably new, is washed first in a 10 percent solution of caustic soda, then in hot sulfuric acid. The latter washing can be done by pouring a little distilled water in the bottom of the flask (to a depth of about 0.5 cm) and then gradually adding sulfuric acid until the mixture of water and acid is quite hot. The flask is rinsed with this mixture, the mixture is poured out, and the flask is filled with distilled water and placed on a wire mesh over a Bunsen burner.

Since water cannot be superheated in a flask which is dirty or has internal surface scratches, this washing may have to be repeated several times until a flask is obtained which is clean and free of flaws. Consequently, the water in the flask should be boiled from time to time. As the water boils, observe whether the boiling is normal (i.e., the bubbles arise continuously in different parts of the flask) or whether superheating actually is taking place. In the latter case a few large bubbles, accompanied by a loud crackling noise, are formed over a relatively long period of time. If such boiling action is not achieved after 15–20 min of heating, the flask should be discarded and a new one tried.

The flask with the superheated water is placed on a burner side by side with a flask in which tap water is boiling normally. The difference in the boiling is immediately noticeable. In order to bring about normal boiling of the superheated water, drop a spoonful of powdered chalk in the flask. *Caution:* The boiling immediately begins with an explosion. One should not use sand in this demonstration because sand particles are likely to scratch the inside of the flask, making it unsuitable for further demonstrations.

26–5.17 Microprojector[26]

A microprojector is a useful device for demonstrating polymorphic transformations, the melting and solution of crystals, and other crystallization phenomena. Since there is greater deviation from proper form among large crystals, the shape of the crystals can best be observed when they are of microscopic dimensions.

The microprojector is made from a minerological microscope with a polarizer and analyzer. An

(25) Translated by K. V. Robinson and H. A. Robinson (Adelphi University) from *Lecture Demonstrations in Physics*, A. B. Mlodzeevskii, ed., M. V. Lomonosov State University, Moscow, U.S.S.R.

(26) Translated by K. V. Robinson and H. A. Robinson (Adelphi University) from *Lecture Demonstrations in Physics*, A. B. Mlodzeevskii, ed., M. V. Lomonosov State University, Moscow, U.S.S.R.

ordinary microscope can also be used if polarizing attachments are added.

Polaroid can be used successfully in place of a Nicol prism analyzer. Polaroid has the advantage of not limiting the field of vision, even when the microscope is used without an ocular lens. For greater stability, the Polaroid should be placed in a paper cylinder which is placed over the upper portion of the microscope.

A long-focusing objective (f ~16 mm) is used for projection. The microscope is placed in front of an arc lamp which is sufficiently bright to allow projection. The light beam is thrown onto the screen by a mirror or a totally reflecting prism, which is set on a stand directly over the analyzer.

An important additional piece of equipment for crystallization experiments is the stage heater shown in Fig. 26–39. A glass tube with an outer diameter of between 2.5 and 3 mm is bent up at the tip *e*, drawn out, and fused so that, with the heater in operation with a fully opened gas cock, the flame is about 15 mm high. The other end of the tube *d* is sealed to tube *abc*, which has an outer diameter of about 5 mm. The bottom *c* of tube *abc* is sealed off and tube *g* is attached to end *a*. Tube *g* must be of a size to allow the easy attachment of a rubber tube from the gas cock. A wide flask, weighted with sand or shot to keep it steady, is used as a stand for the burner. The end *c* of tube *abc* is inserted into a stopper in the flask. Tube *de* must be long enough so that it fits under the microscope table in such a way that the flame would be under the center of the opening in the table immediately above the condenser. The height of the burner is regulated by moving *abc* up and down in the stopper. The height of the flame and consequently the temperature of the preparation can be controlled either by the gas main cocks or by means of a screw-type clamp attached to the rubber tube carrying the gas. A 2- to 3-mm-high flame is sufficient for most experiments. If the light from the flame hinders observations, the tip of the burner may be placed somewhat eccentrically. The slide holding the preparation should be placed in such a way that the hot gases from the burner flow around the edges. The slide, therefore, should not be placed directly on the stage, but should be held slightly above it. A ring support can be made of a small tin cylinder whose diameter is greater than the width of the slide but smaller than its length.

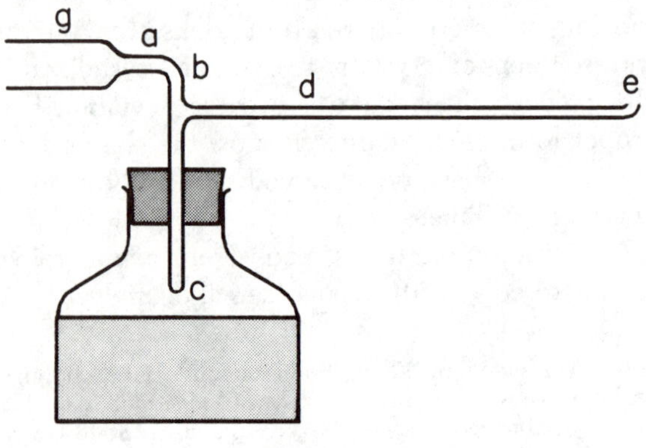

Fig. 26–39

An electric heater, shown in Fig. 26–40, may be used in place of the gas heater. It consists of a round asbestos plate with a circular opening. A Nichrome wire (0.4 to 0.5 mm in diameter and 3 to 5 cm long) is attached inside the opening. (The Nichrome wire from a 100-V ac cone heater can be used. Care should be taken in assembly to make certain that the resistance of the wire is adequate.) The ends of the wire are attached to the binding posts *h*. At the points *i* the wire is attached by fine wire hooks to the inside of the opening. Two side holes *j* are used to attach the heater to the lower side of the stage. The heater can be placed on top of the stage by attaching it by holes *j* to the tin ring mentioned previously. Neither heater should be operated without the slide in place, since this may damage the microscope objective.

26–5.18 Polymorphism[27]

One of the best examples of a polymorphic transformation is mercury iodide (HgI_2). At room temperature it is a red crystalline substance which turns yellow when heated to 126°C. The crystals revert to the red color on cooling. The yellow crystals melt at 253°C, changing into a dull-red color. To make the transformation most visible, it is best to preheat the substance until the yellow crystals melt and then coat the sides of the test tube with the liquid by tipping the tube. After the liquid solidifies, the walls of the test tube are covered by

[27] Translated by K. V. Robinson and H. A. Robinson (Adelphi University) from *Lecture Demonstrations in Physics*, A. B. Mlodzeevskii, ed., M. V. Lomonosov State University, Moscow, U.S.S.R.

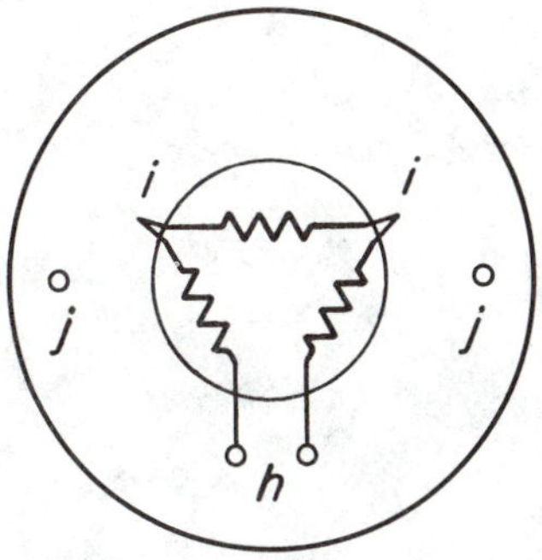

Fig. 26–40

yellow crystals. When the yellow crystals again become red, the tube is ready for the demonstration. This transformation can be speeded up by adding a red crystal to the tube. The crystals can be shown to a large audience by means of an opaque projector.

Ammonium nitrate can also be used for the demonstration. It has five solid phases at transformation temperatures of −16°C, 35°C, 83°C, and 125°C. The last phase melts at 166°C. The NH_4NO_3 transformation should be demonstrated with the microprojector with polarizer and analyzer (see Demonstration 26–5.17). Several small crystals of the substance are placed on a thin slide, covered with a cover glass, and melted over a burner. The layer between the cover glass and the slide should be as thin as possible. After the preparation has solidified and cooled to room temperature, it is placed on the microprojector stage. The image is thrown on the screen and the heater is turned on. The second phase (above −16°C) is stable at room temperature. At 35°C a sharp change in the birefringence shows the transformation into the third form. This change is associated with a change in the color of the crystal when the analyzer is crossed. At 83°C a new change in color takes place, and at 125°C the field of vision is darkened. The crystals of the last phase belong to the cubic system and do not show birefringence. The analyzer is then turned 90 deg or removed, and the crystals of the last phase are shown on the screen. Since these crystals melt at 166°C, the heater should be immediately disconnected to prevent decomposition. As soon as the crystals of the cubic phase begin to grow into dendrites, the analyzer is crossed. The gradually spreading region of the cubic phase, showing bright interference colors, will appear on the screen. The boundary of this region will move across the entire field. With further cooling all the transformations appear in reverse order, and each transformation is accompanied by a movement of the boundary of the two solid phases across the field of vision. To prevent the glass from cracking, avoid heating the crystals to the melting point.

26–6 Second Law of Thermodynamics

26–6.1 Hot Air Motor

The apparatus shown in Figs. 26–41, 26–42, and 26–43 is designed to operate as a Carnot engine, even though its thermodynamic cycle is not quite the same as the Carnot cycle. It consists of two concentric cylinders: the outer one *a* is a water jacket through which water at a constant temperature is pumped. Water enters at *b* and flows out at *c*. The inner cylinder contains two pistons whose connecting rods are connected 90 deg apart on a common crankshaft *d*. The upper piston *e* prevents escape of the air enclosed within the inner cylinder *f*, thus enabling the enclosed air to do work as it expands or contracts. The lower piston *g*, through which holes *h* have been drilled, acts as a displacement piston, passing the air within the cylinder back and forth between the upper and lower sections and acting as a heat exchanger.

The four different modes of operation which may be demonstrated are as follows:

1. *Refrigeration.* The motor is driven in a clockwise direction when viewed as in Fig. 26–41, by means of a leather string pulley which is connected through a reduction gear box to a small reversible electric motor. A thermocouple *j* which is fastened to the bottom of the inner cylinder is connected to a projection galvanometer and will show that cooling of the cylinder occurs as the motor is driven; heat is transferred from the inner cylinder to the cooling water in the outer cylinder *a*.

Fig. 26–41

2. *Heat Pump.* The motor is driven as in the preceding demonstration, but in the opposite direction. The galvanometer will indicate that the inner cylinder is being heated up as the motor is driven.

3. *Heat Engine.* The thermocouple and the pulley are removed, and the bottom of the inner cylinder is heated with a large Bunsen burner. The motor wheels will turn clockwise, the same direction as they were driven to demonstrate the refrigeration cycle.

Fig. 26–42

4. *Cold Engine.* The burner is removed and a large Dewar containing liquid nitrogen is placed so that the inner cylinder is immersed in the nitrogen. The motor wheels will turn counterclockwise, the same direction that they were driven to show the operation of a heat pump. Heat flows from the cooling water in cylinder *a* to the liquid nitrogen surrounding cylinder *f*.

26–6.2 Ratchet and Pawl[28]

Carnot's concept of a reversible engine and the second law of thermodynamics can be illustrated by a ratchet and pawl connected to a set of vanes as shown in Fig. 26–44. At first glance, this device would appear to be capable of violating the second law. If, we think, the temperature T_1 surrounding

[28] Richard P. Feynman, Robert B. Leighton, and Matthew Sands, *The Feynman Lectures on Physics* (Addison-Wesley Publishing Co., Reading, Massachusetts, 1963), Vol. I, p. 46–1.

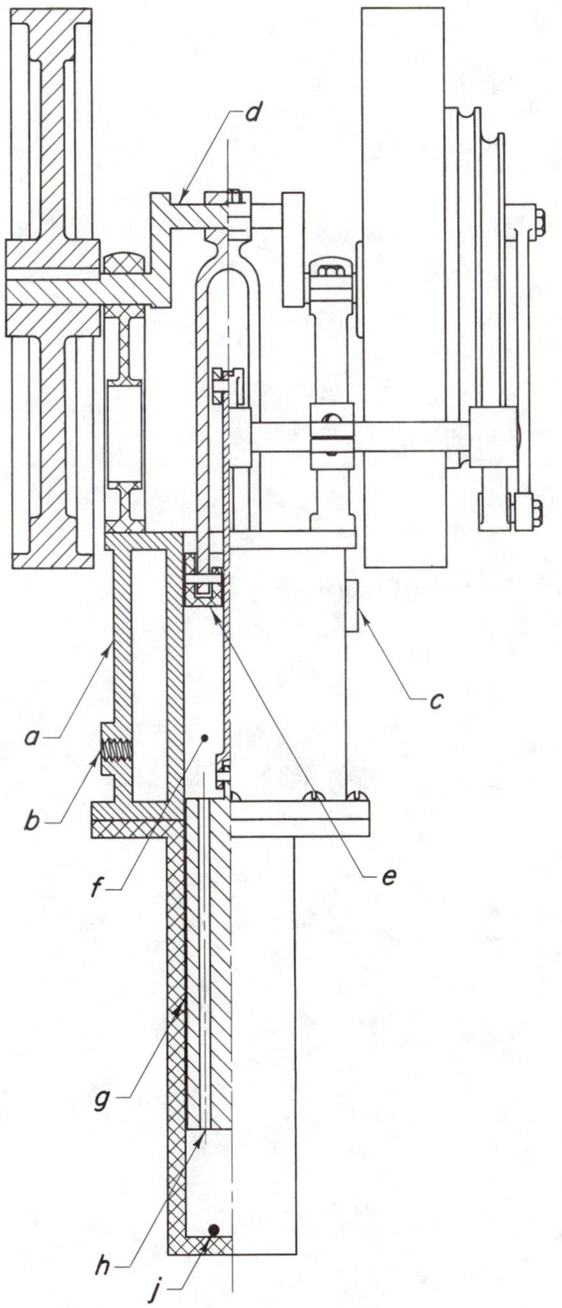

Fig. 26–43

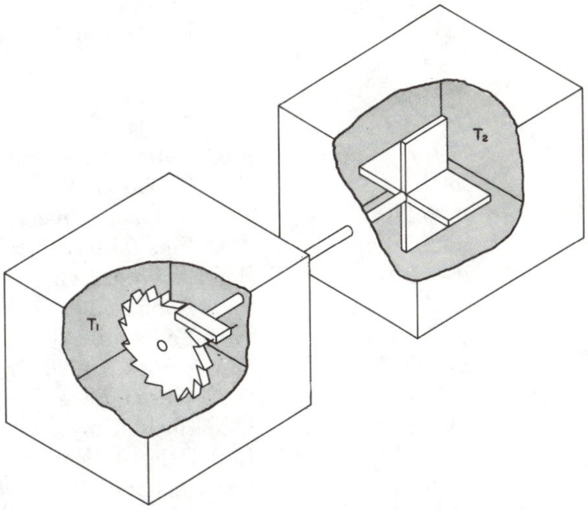

Fig. 26–44

the ratchet and pawl and the temperature T_2 surrounding the vanes are the same, gas particles will bombard both sides of the vanes. However, the pawl constrains motion except in one direction. Therefore, we erroneously conclude, the device can perform work without the presence of a temperature difference.

Actually it can be shown (see reference in the footnote) that with both vanes and ratchet and pawl at the same temperature, thermal effects on the ratchet and pawl would, on the average, cancel the thermal effects from the vanes. If the environment of the ratchet and pawl were at a lower temperature than the environment of the vanes, the device would indeed work as described. (But no longer in violation of the second law.) More interesting, however, is the fact that if the environment of the ratchet and pawl is at a higher temperature than the environment of the vanes, the effect will be to drive the shaft in the opposite direction from the direction in which the ratchet operates as a ratchet. In the demonstration, the pawl is moved by hand to simulate thermal effects and demonstrates this last phenomenon. Construction details appear in the Appendix, page 1287.

Demonstrations	***Contributors***
26–2.1	Swiss Federal Institute (ETH), Zürich, Switzerland
26–3.1	A. B. Cardwell, Kansas State University
26–3.2	L. R. Weber, Colorado State University
26–3.3 to 26–3.5	George D. Freier, University of Minnesota
26–3.6	A. Strickler, The Allen Strickler Company, Fullerton, Calif.

26–3.7 to 26–3.10	Swiss Federal Institute (ETH), Zürich, Switzerland
26–4.1	California Institute of Technology
26–4.2	Eric M. Rogers and H. M. Waage, Princeton University
26–4.3, 26–4.4	Swiss Federal Institute (ETH), Zürich, Switzerland
26–4.5	D. C. Evans, Oklahoma State University
26–4.6	R. M. Sutton, California Institute of Technology
26–4.7	H. A. Robinson, Adelphi University
26–5.1	E. M. Lyman, University of Illinois
26–5.2	Swiss Federal Institute (ETH), Zürich, Switzerland
26–5.3	Harvard Project Physics
26–5.4 to 26–5.7	H. A. Robinson, Adelphi University
26–5.8	Swiss Federal Institute (ETH), Zürich, Switzerland
26–5.9	J. W. Straley, University of North Carolina
26–5.10	H. C. Jensen, Lake Forest College
26–5.11	J. G. Likely, University of Minnesota
26–5.12	L. P. Delsasso, University of California, Los Angeles
26–5.13	Louisiana State University
26–5.14	G. Schwarz, Florida State University
26–5.15	L. Grillet and M.-T. Grillet-Amy, University of Rennes, France
26–5.16 to 26–5.18	H. A. Robinson, Adelphi University
26–6.1	Swiss Federal Institute (ETH), Zürich, Switzerland
26–6.2	California Institute of Technology

27

GASES AND KINETIC THEORY

27–1 Introduction

A limited number of demonstrations are available from manufacturers of scientific equipment which can be used during lecture demonstrations to show various properties of gases.

Welch[(1)] has three-dimensional PVT curves for water (1710) and carbon dioxide (1710A). The relationship between pressure, volume, and temperature is shown clearly on one diagram. The curves are approximately 9 x 10 x 9 in.

A Franklin's flask for showing the boiling point of water at reduced temperature and pressure is available from Cenco[(2)] (77775). The 1000-ml glass flask has a re-entrant concave bottom which holds crushed ice. It is closed by a one-hole rubber stopper that holds a thermometer.

LaPine[(3)] has a boiling flask (385 10) with a rising tube for determining the vapor pressure curve of water between the boiling point and room temperature. Water in the flask is first boiled; the open end of the rising tube is then immersed in a mercury-filled trough. As the vapor pressure of the water decreases on cooling, the mercury column will ascend in the tube.

Another vapor pressure apparatus from LaPine (385 09) measures the vapor pressure of liquids by means of a mercury column. A Torricellian tube dips into a mercury-filled trough. The tube is evacuated at the top until the mercury column indicates atmospheric pressure. When a sample of liquid is introduced into the vacuum, its vapor pressure acts upon the mercury column from above and causes it to fall by a corresponding amount.

A large demonstration pulse glass (1665) can be obtained from Welch. Two glass bulbs (10 cm diam) have been filled with a colored volatile liquid, connected by a tube, 45 cm long, and partially evacuated. When one bulb is held in the hand, the body heat is sufficient to cause rapid evaporation and increase of vapor pressure, forcing a rapid transfer of liquid to the other bulb in a bubbling pulsing manner.

Welch and Cenco have similar devices for measuring dew point. They consist of a polished cylinder with inlet and outlet tubes and an aspirator for rapidly forcing air through ether or a similar volatile liquid within the cylinder. This cooling effect reduces the temperature until dew forms on the polished surface. The cylinder of the Welch apparatus (1723B) is 70 x 30 mm, and the cylinder of Cenco's version (77015) is 28 x 54 mm.

Ealing's[(4)] dew-point apparatus (20-980) has a front chamber with two inlet tubes which hold 10 ml of air. The chamber has a glass end whose diameter is about 2 in. Through this window can be

(1) Sargent-Welch Scientific Co.

(2) Central Scientific Co.

(3) LaPine Scientific Co.

(4) The Ealing Corp.

viewed the back of the chamber, which is a stainless steel mirror 0.012 in. thick. This mirror separates the sample from a rear chamber, which is filled with a volatile coolant such as ether. The bulb of a thermometer is immersed in the coolant, and an aspirator is connected to this chamber. The sample is drawn into the front chamber, which is sealed. The aspirator is then easily worked to bubble air through the coolant and to reduce the temperature. The formation of dew on the mirror can be quickly spotted since one-half of the mirror is insulated and the contrast is sharp.

A sling psychrometer for determining relative humidity is sold by Cenco (77005). Two thermometers are mounted on a metal casting, which is provided with a wood handle and swivel.

Apparatus for demonstrating the diffusion of gases is available from LaPine, Welch, and Cenco. One Welch apparatus (0537) demonstrates the different speeds with which different gases diffuse through a porous cylinder. The porous cup is inverted and connected by a tube to a beaker filled with water. An inverted bell jar is placed over the cup, and when a low-density gas such as hydrogen is directed into the bell jar, the gas will diffuse into the cup faster than the heavier air will diffuse out. This increases the pressure, causing the mixture to bubble out at the lower cylinder. A similar apparatus from LaPine (372 43) has the increased pressure force water out of a stoppered bottom container like a fountain.

LaPine has a model of a separation column (372 45). It is used to show the thermal diffusion effect and the mode of action of separator columns for isotopic mixtures. A glass tube (80 cm long, 20 mm diam) is cooled by the ambient air from the outside. In it is a second glass tube which is coaxial with it and has a heating filament. The space between the glass tubes contains bromine vapor and air. After a few minutes, brown bromine vapor will collect in the lower part of the tube, and the colorless air will remain in the upper part. Cenco also sells a model separation column (76340) which uses a mixture of 50 percent argon and 50 percent hydrogen.

A hermetically sealed evacuated tube used to demonstrate thermal agitations of gaseous molecules is available from Cenco (77725). The tube contains a small amount of mercury, on whose surface are some particles of crushed colored glass. Upon heating, the mercury boils, and the vapor given off at high velocities carries with it particles of glass that move and collide in the space above the mercury. This can be made visible to the class by the use of a projection lantern. Welch carries a similar tube (1724).

Another diffusion device from Welch (4425) is a set of two Pyrex tubes (30 cm long, 18 mm diam), each containing a small amount of iodine crystals. The tubes are sealed off, one at moderately reduced pressure and the other at greatly reduced pressure. When the lower end of each tube is heated, the iodine vapor in the first tube remains conspicuously in the lower portion of the tube because of the slow diffusion through the remaining air. The vapor in the low-pressure tube diffuses rapidly throughout the tube because so little air is present.

Welch also has a gas diffusion apparatus (0535) which is composed of two 16-oz. bottles connected by a large-diameter brass tube and cork stopper. When smoke or a colored gas is introduced into one bottle it will be seen to diffuse rapidly into the other.

A kinetic theory model is available from LaPine (376 01). It is used to illustrate the thermal motion of molecules and atoms in a gaseous mixture exposed to the influence of the gravitational field of the earth, and to demonstrate the transformation of a gas into the liquid and, ultimately, the solid state. Colorless and colored glass beads (5–10 mm diam) represent gas molecules. The beads are set in motion in a glass cylinder (65 mm diam, 40 cm long) mounted vertically on an oscillating system with electromagnetic excitation.

27–2 Equations of State

27–2.1 Boyle's Law[5]

The pressure-volume relationship can be illustrated by shadow projection using the apparatus[6] shown in Fig. 27–1. When the stopcock is closed and weights are added to the upper plate, the volume of air in the tube decreases in an inverse proportion to the increase in pressure. As each weight is added, the internal volume of the cylinder can be read and a plot made of pressure vs volume, roughly showing the hyperbolic relationship. It is recommended that a strong arc lamp be used to shadow-project the apparatus.

27–2.2[7]

"Lab-gas" units[8] are an extremely convenient source of low-pressure gas. The unit consists of a balloon, a cylinder of the appropriate gas fitted with a plunger to break its seal, and a stopcock plug with a fitting for rubber tubing. To prepare the source, the cartridge is placed in the balloon and the stopcock plug is fitted in place. The neck of the cartridge is then rapped sharply against a table top, causing the seal to break and allowing the gas to expand and fill the balloon, which may then be connected to the experimental apparatus (Fig. 27–2).

27–2.3

A lecture demonstration of Boyle's law can be made visible to a large class with the apparatus illustrated diagrammatically in Fig. 27–3. The parts are assembled on a vertical plywood panel 48 x 60 x ¾ in. A heavy-duty Bourdon gauge *a* is mounted near the upper left-hand corner of the panel. The gauge is enclosed in a housing 26 in. square by 2¼ in. deep, which is covered by a scale *b* graduated from zero to 60 lb/in.² A large counterbalanced pointer attached to the gauge indicates the pressure. Copper tubing connects the gauge to the pressure chamber *c*, which is a vertical glass tube (2 in. diam, 3 ft long) mounted on the panel as shown. For safety, the pressure chamber is surrounded by a semicylindrical shield *d* of transparent plastic. A meter stick (dotted lines) mounted back of the pressure chamber provides a measure of the volume of air under pressure. A valve *e* at the top of the pressure chamber permits bleeding to atmospheric pressure. A drain valve *f* at the bottom provides for emptying the

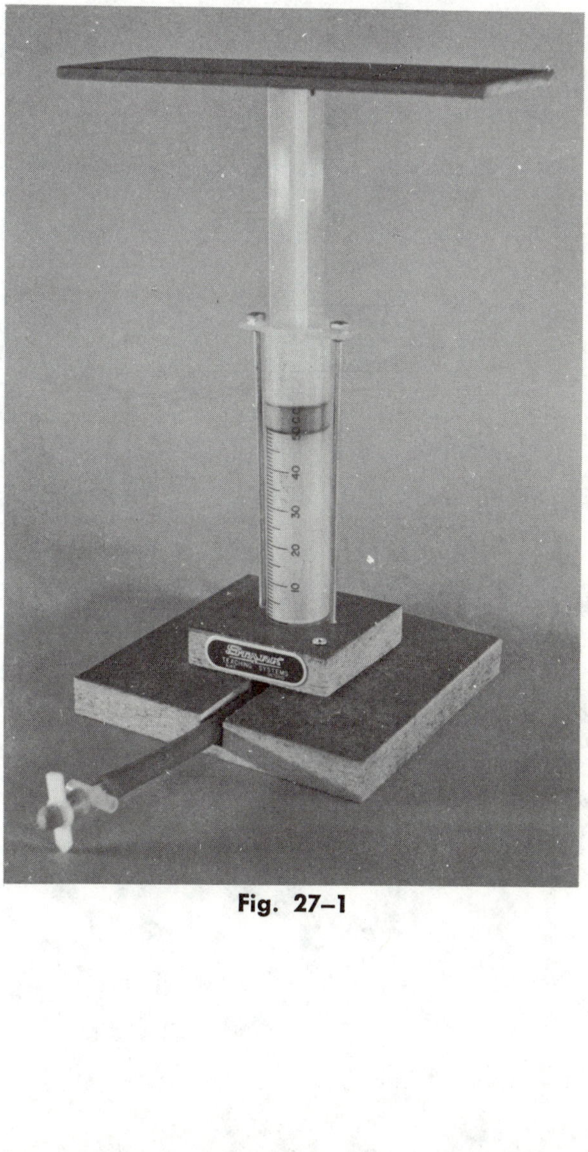

Fig. 27–1

(5) Developed by the Harvard Project Physics program under grants from the Carnegie Corporation, the Ford Foundation, the National Science Foundation, the Alfred P. Sloan Foundation, the United States Office of Education, and Harvard University.

(6) Boyle's law apparatus #C-6200. Available from Stark Electronic Instruments Ltd., Ajax, Ontario, Canada.

(7) Developed by the Harvard Project Physics program under grants from the Carnegie Corporation, the Ford Foundation, the National Science Foundation, the Alfred P. Sloan Foundation, the United States Office of Education, and Harvard University.

(8) "Lab-gas" units are available from: Simplex Products Co., Cleveland, Ohio. As of this writing, four gases are available: O_2, N_2, N_2O, and CO_2.

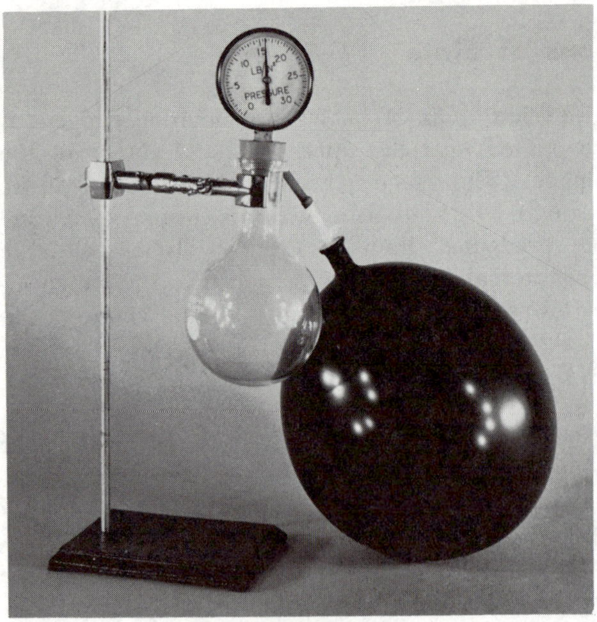

Fig. 27–2

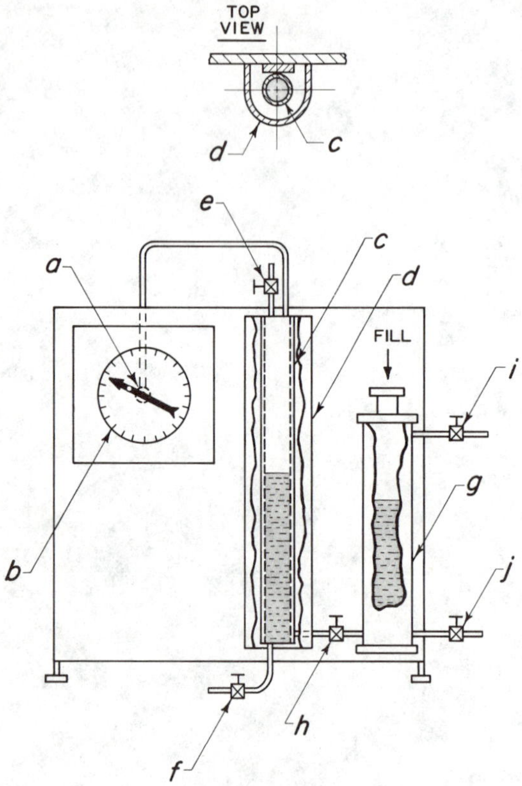

Fig. 27–3

pressure cylinder of water, which enters the pressure chamber from the reservoir *g* through the valve *h*. The reservoir is a brass tube (4 in. diam, 2 ft long) provided with a compressed air inlet valve *i* and a water drain valve *j*.

The reservoir is filled about two-thirds full of water colored with potassium permanganate. The gauge is preset to read 14.7 psi, and the bleeder valve *e* is opened to the atmosphere. After tapping the gauge lightly, close the bleeder valve. The valve *h* is opened and the colored water from the reservoir is allowed to rise to a height of approximately 10 cm in the pressure chamber and the pressure again read. By adjusting the air pressure in the reservoir and taking readings of pressure and height of water in the pressure chamber, a graph of pressure vs volume is quickly obtained.

27–2.4

Boyle's law can be simply demonstrated as shown in Fig. 27–4. An airtight glass container *a* (about 1½ in. square, 8 in. long) is connected to a water faucet *b* (Fig. 27–5) by a rubber tube *c*. (It is also possible to use compressed air if the system has been previously filled with water.) The top and bottom of the container are sealed by means of two brass plates *d* with grooves and rubber washers. Since the vessel has to withstand pressures of about 100 psi, the plates are held in place by four bolts placed at the corners. A gas valve *e*, the zero adjustment valve, is soldered in the top of the plate, and a tap for the hose is soldered in the bottom plate.

A connection is made from the faucet to a projection manometer gauge *f* (construction details in Demonstration 27–5.1), and to an outlet valve *g*. A saw-tooth scale *h* is mounted next to the pressure tank with the zero point on the level of the brass plate. The scale and the container are projected on to the wall, and the pressure gauge is projected from an optical bench.

For the demonstration, the zero adjustment valve is opened and the pressure tank is filled with water up to the position marked 10 on the scale. (The scale divisions are completely arbitrary.) The hose to the tank should be completely filled with water so that there are no air bubbles to distort the results. The pressure gauge still shows 0 corresponding to a pressure of 1 atm. Then the zero adjust-

Fig. 27–4

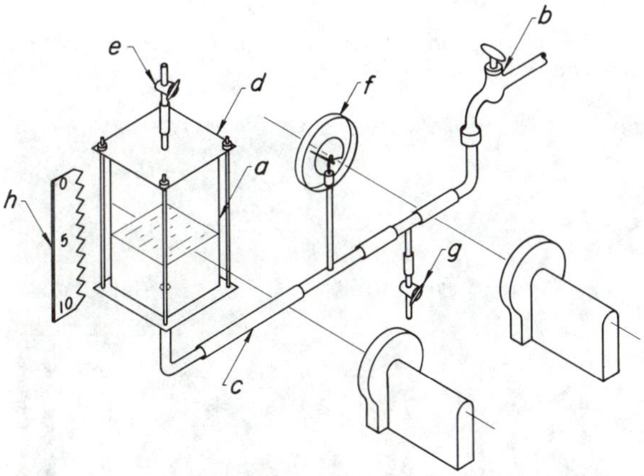

Fig. 27–5

ment valve is closed, and the water level is raised to the 5 mark on the scale, reducing the original air volume by half. The pressure gauge now shows a reading of 1, corresponding to a pressure of 2 atm. By using several different volume settings, it can be shown that the pressure is inversely proportional to the volume.

27–2.5

An arrangement for quantitatively demonstrating Boyle's law by projection is illustrated in Figs. 27–6 and 27–7. Figure 27–8 is a closeup of the water cell (*A* in Fig. 27–6) which is designed to fit into a lantern slide projector. The volume of air to be observed is confined in a section *B* of a capillary tube by a small mercury piston *C*. The thermistor probe *D*[9] is attached to a projection thermometer so that a lecture group can observe that temperature is held constant through a range of pressures. When the pressure on the mercury in the capillary tube is changed, the volume of air in the capillary tube (between the mercury column and the tube's sealed end) is changed in an inverse proportion. A commercial air pressure regulator *E* is used with a homemade pressure gauge *F* with a large dial face to enable a large group to observe the pressure applied to the mercury. There is also a small gauge on the rear for the lecturer's use. A volume

[9] Schaar & Co., Cat. No. AL 2045, 0–50°C. Yellow Springs Instrument Co., Yellow Springs, Ohio.

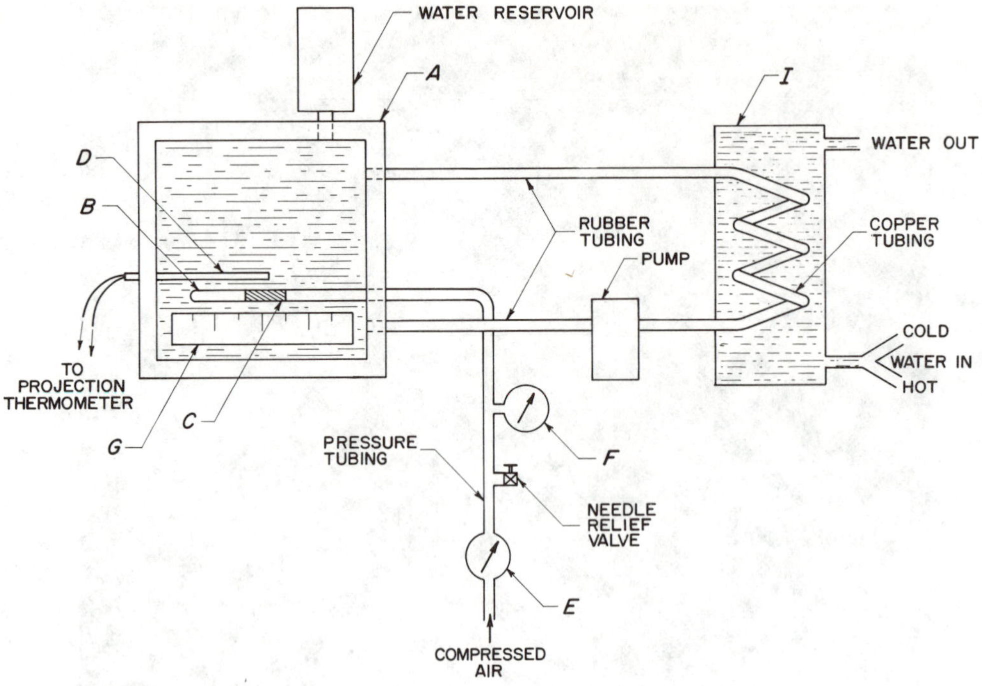

Fig. 27–6

change is noted by projecting through the cell and watching the mercury as it moves along the scale G within the cell.

The temperature of the water cell is controlled by circulating the water through a heat exchanger I, in which a mixture of hot and cold water is applied to the coil to maintain the circulating system within the cell at a constant temperature. A small pump[10] is used to provide circulation. Construction details are given in the Appendix, page 1289.

[10] Type 100, model A-1, Eastern Industries, Hamden, Connecticut.

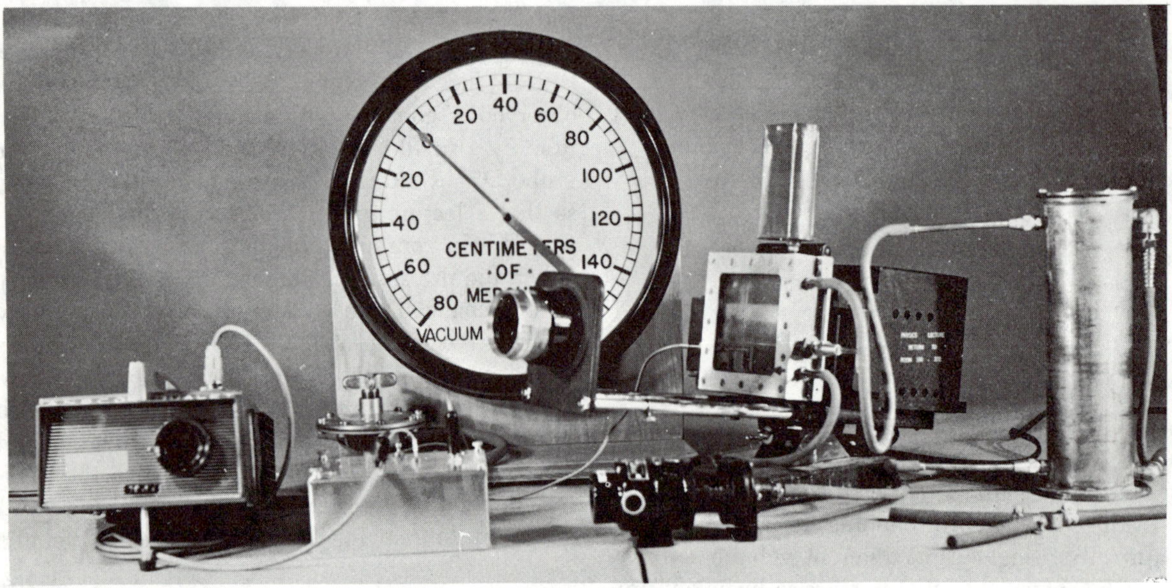

Fig. 27–7

Fig. 27–8

27–2.6[11]

The elementary study of a sample of confined air is very effectively extended by means of a device that makes use of familiar ideas. As indicated in Fig. 27–9, an all-glass stocking-shaped tube contains air in the "toe" region which is confined by mercury in the "heel" and "leg." Tipping the tube to vary the pressure is made convenient by mounting the tube on a suitably shaped board attached to the base by a single bolt *A*. This bolt serves as a horizontal axis about which the "stocking" is rotated. An angle of approximately 70 deg at the "heel" has been found to keep the confined air higher throughout the exercise than the confining mercury surface. This obviates the escape of the air past the mercury. Readings are obtained with such facility that the apparatus can be used in lecture or in the classroom to obtain data quickly for a discussion of Boyle's law. For this purpose, room temperature only is required.

By enclosing the "foot" of the "stocking" in a glass cylinder, also shown in Fig. 27–9, the device is made suitable for the study of Gay-Lussac's law for a number of different volumes. The cylinder contains a mercury thermometer whose scale is used for measuring the volumes of the air sample, in arbitrary units, as well as the temperatures. The 100° mark on the thermometer is placed at the end of the "toe," marking one end of the sample of air under study. The tubes *T* provide for the inflow and outflow of water at convenient equilibrium temperatures.

[11] M. J. Pryor, *Am. J. Phys.*, **13**, 421 (1945).

Readings are taken as follows: The equipment is initially at room temperature. Rotate the "stocking" until the mercury in the "foot" is opposite some arbitrarily selected mark on the thermometer. If the reading at this mark is *R*, the volume of the air in the uniform tube may be taken as 100 *R*. Readings of the heights *H* and *h* of the two mercury surfaces from the table top are taken by means of a meter stick. If the barometer reading is *B*, the absolute pressure P is $H - h + B$. The "stocking" is rotated to bring the mercury to each of a series of well-distributed values of *R*, and the corresponding heights *H* and *h* are observed.

Water at any fixed temperature may now be passed through the cylinder which surrounds the "foot" until an equilibrium temperature is indicated on the thermometer. The "stocking" is then rotated to duplicate the former values of *V*. When the mercury surface in the "foot" is opposite values of *R* used in the former series the corresponding values of *H* and *h* are observed. This repetition of

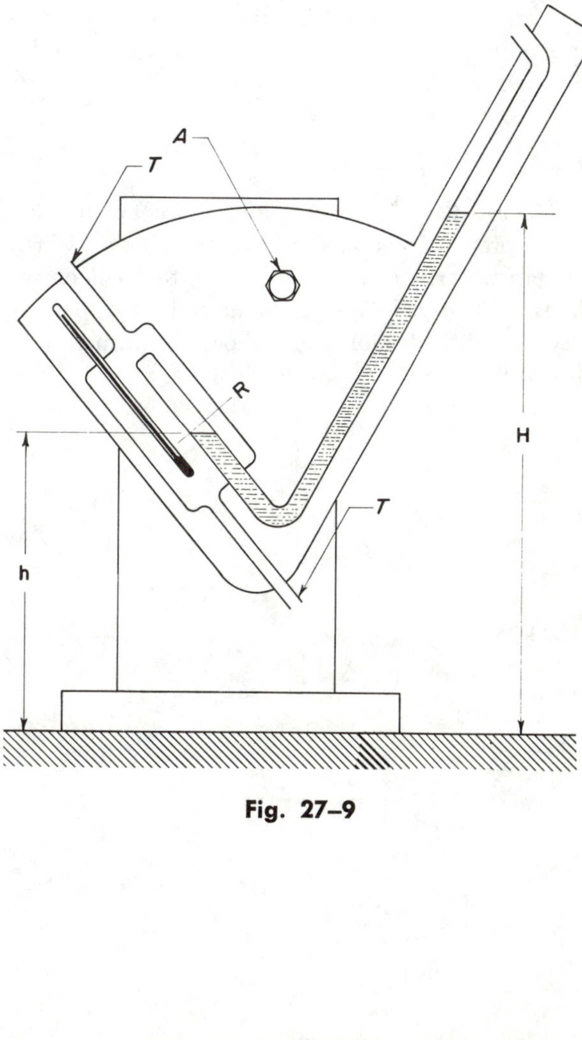

Fig. 27–9

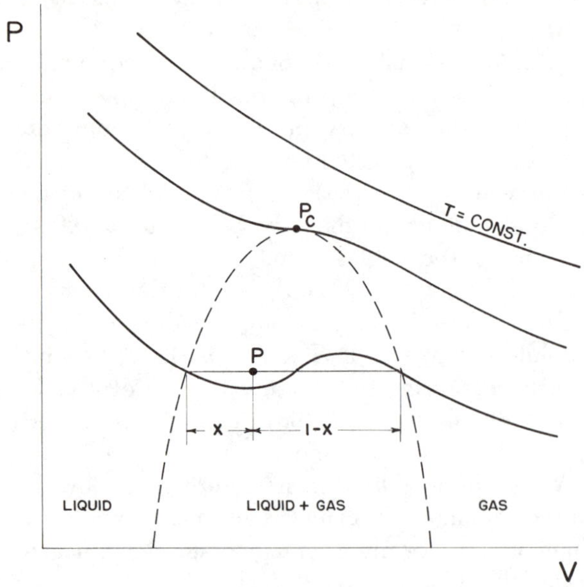

Fig. 27–10

the initial volumes is made so that, as Boyle's law data are taken for new temperatures, the data for Gay-Lussac's law are accumulated. If the base and supports are made of transparent plastic, the apparatus may be projected for observation by large classes.

27–2.7 Charles' Law

A large flask at room temperature and atmospheric pressure is capped with a toy balloon. When hot tap water is run over the flask, the balloon inflates. When the flask is immersed in a pan of Dry Ice and alcohol, the balloon is pushed into the flask.

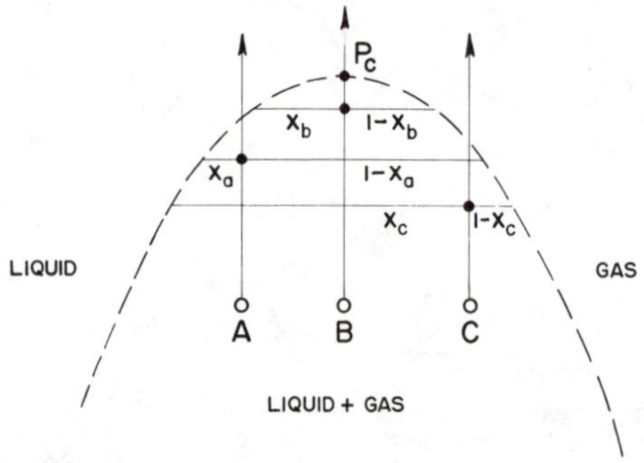

Fig. 27–11

27–2.8

A rubber balloon is partly inflated, tied to the end of a glass rod, and dipped into liquid air in a Dewar flask. When the balloon is withdrawn, it is collapsed, hard, and wrinkled. Subsequent warming to room temperature will restore its original volume and elasticity.

27–2.9 Van der Waals Gas

Gases close to liquefaction can be described by van der Waals equation

$$(P + a/V^2)\,(V - b) = RT,$$

where R is the universal gas constant, T the absolute temperature, p the pressure, and V the volume. The constant a takes into account the effect of the attractive force between molecules, and b corrects for the fact that the gas molecules have a finite volume. In the pressure-vs-volume diagram, the van der Waals equation gives third-order isotherms, as shown in Fig. 27–10. Below the critical point P_c the isotherms inside the parabola-like curve representing the two-phase (liquid + gas) region are replaced by Maxwell's straight lines. At a point P the ratio of gas to liquid will be a function of $x/(1-x)$.

In this demonstration, a given quantity of liquid CO_2 in quartz tubes is heated at constant volume. At the start there are different ratios of liquid to gas in the same volume, and then three vertical curves in the p-V diagram are followed (Fig. 27–11). The ratios of liquid to gas at each point are:

$$\text{A:}\quad \frac{x_a}{1-x_a} \to 0 \qquad \text{(Condensation)}$$

$$\text{B:}\quad \frac{x_b}{1-x_b} = \text{constant}$$

$$\text{C:}\quad \frac{x_c}{1-x_c} \to \infty \qquad \text{(Evaporation)}$$

At A the liquid rises with increasing temperature and fills the entire volume. At C the liquid evaporates gradually with increasing temperature, and only "gas" is left beyond the phase limit. In the critical case at B, the level of the liquid remains constant as the temperature rises, but the level dis-

Fig. 27–12

appears at the critical temperature since the densities (and the refractive indices) of the liquid and gaseous phases become equal.

Upon cooling, an undercooling effect is observed, and the tubes become suddenly opaque. In case A, one can imagine droplets falling out as rain, to gather as new liquid in the lower part of the tube. In case C, sudden boiling occurs and gas bubbles rise to form a gas phase above. In the critical case at B, something between A and C occurs.

The apparatus (Fig. 27–12) consists of three quartz tubes[12] filled with liquid CO_2 which are hung next to each other in a metal box *a* (12 x 2 in.) so that the level of liquid in each tube is at the same height. The box has two Plexiglas walls (3/8 in. thick) so that the tubes can be projected on an optical bench. The tube in the middle is exactly at the critical point, and the other two are above and below, respectively. An electric hair dryer *b* is attached to an easily removable 8-in.-long aluminum tube. Two of the long tubes are used; one to heat the metal box containing the three quartz tubes; the other to cool it. The tube used for cooling contains a wire mesh screen on which is placed pieces of Dry Ice. For cooling, the drier is set at the COLD position.

Caution: Very great care must be taken in handling the tubes because the pressure is about 900 psi, and rather violent explosions occur if they break. Steps should be taken to protect both the class and the demonstrator. The tubes are prepared by pumping out the tube and dipping it into liquid nitrogen to condense a given quantity of CO_2. The carbon dioxide must be pure.

[12] The tubes can be obtained in limited quantities from H. Heeb, Abt. Vorlesung, Phys. Inst., Swiss Federal Institute of Technology (ETH), Gloriastrasse 35, 8006 Zürich, Switzerland. They cannot be obtained in the U.S.A.

27–3 Vapor Pressure

27–3.1

The pressure on a volume of soda pop in an ordinary pop bottle fitted with a pressure gauge attached to a rubber stopper in the neck of the bottle as in Fig. 27–13 is released by inserting a needle between the stopper and the neck of the bottle. When the needle is removed, it is seen that the pressure in the gas over the liquid takes some time to build up again. If the bottle is shaken, producing some bubbles in the soda, the surface area across which the gas can escape from the liquid is increased substantially and the time necessary for pressure build-up is substantially reduced. The pop bottle demonstrates that the equilibrium of the gas in the liquid is a dynamic one.

Figure 27–14 shows how to connect the gauge. A hose nipple with a 1/8-in.-serrated shank is forced through the hole of a single-hole Neoprene stopper. A 1/8 to 1/4 NPT male adapter connects the hose

Fig. 27–13

nipple to a ¼-in. double male fitting. The 0-100 psi gauge[13] may be displayed over closed-circuit television.

27–3.2

A little water is boiled in a spherical flask to replace the air with water vapor. The flask is then closed with a single-hole stopper which has a long glass tube in it. The flask is inverted and the mouth of the tube is placed in a reservoir of water. As the vapor pressure in the flask drops, the water rises in the tube. It is best to measure the amount of water in advance so that all the water in the lower reservoir rises to the upper flask. Once the cold water enters the flask, the vapor pressure falls quickly and a fast-flowing fountain results.

27–3.3

Two beakers, one containing brine and the other fresh water, are put under a bell jar. Since the vapor pressure of salt water is less than that for fresh water, after a period of a few weeks the brine beaker is seen to have gained water while the other beaker has lost water. The same process occurs if alcohol and water are used. In this case the volume of liquid in the water beaker increases.

[13] Marsh Instrument Co., Skokie, Illinois.

27–3.4

When an inverted flask or graduate is dipped into a beaker containing ether, bubbles of air and ether vapor promptly emerge as the confined air pressure rises with the addition of the ether vapor pressure. Water vapor and air can be used, but the vapor pressure of water is much lower than that of ether, and thus the process is much slower.

27–3.5

A pressure cooker to which a thermometer and pressure gauge are added is used to determine how pressure affects the boiling point of water. Pressure and temperature can be read directly, and data can be obtained to construct *P-T* curves for temperatures ranging from 100° to 135°C.

27–3.6

To show that the surrounding air pressure controls the boiling point of a liquid, a flask containing a beaker of water is evacuated. Before evacuation the demonstrator first takes a sip of the water to demonstrate that it is cool. As the pressure decreases, the water will begin to boil. He then removes the beaker and takes another sip to show that the water is still cool.

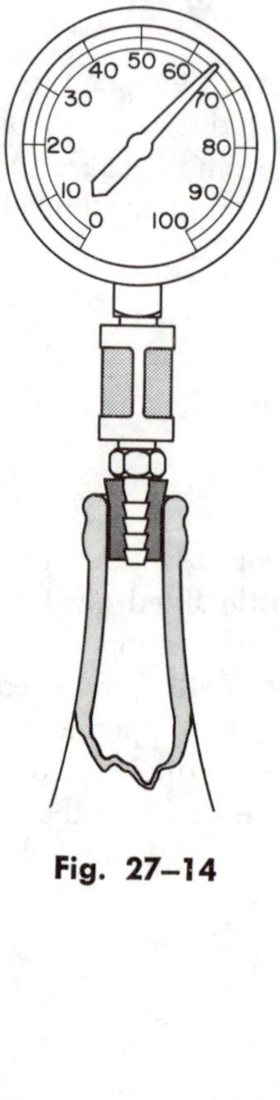

Fig. 27–14

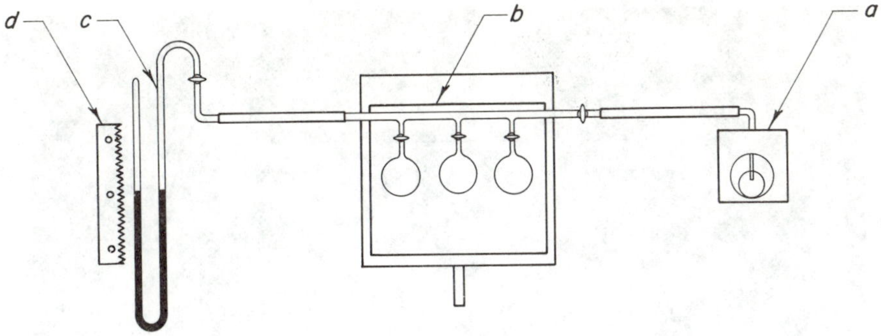

Fig. 27–15

The vacuum in the evacuated chamber is a good insulator, and when the liquid boils the heat of evacuation comes mainly from the liquid. If it is boiled long enough, it reaches the freezing point and freezes. This can also be shown by using benzine, which has a slightly higher melting point.

27–3.7

The system shown in Fig. 27–15 consists of a vacuum pump *a*, three flasks *b* filled with ether, alcohol, and water respectively, connected to a mercury manometer *c* with a saw-like projection scale *d* with teeth spaced 1 cm apart. (Small holes are drilled in the scale at 10-cm intervals.) To show the dependence of vapor pressure on temperature, pump out the whole system for a short time, close the stopcocks on the flasks, and then pump out the manometer. When the pump is turned off, open the stopcock on one of the flasks and either heat it or cool it, noting the change in scale deflection. The apparatus set up for projection is shown in Figs. 27–16 and 27–17.

27–3.8[14]

The vapor pressure curve for water may be obtained with the apparatus shown in Fig. 27–18.

(14) N. W. Goldsmith, *Am. J. Phys.*, **29**, 10, xiii–xiv (1961).

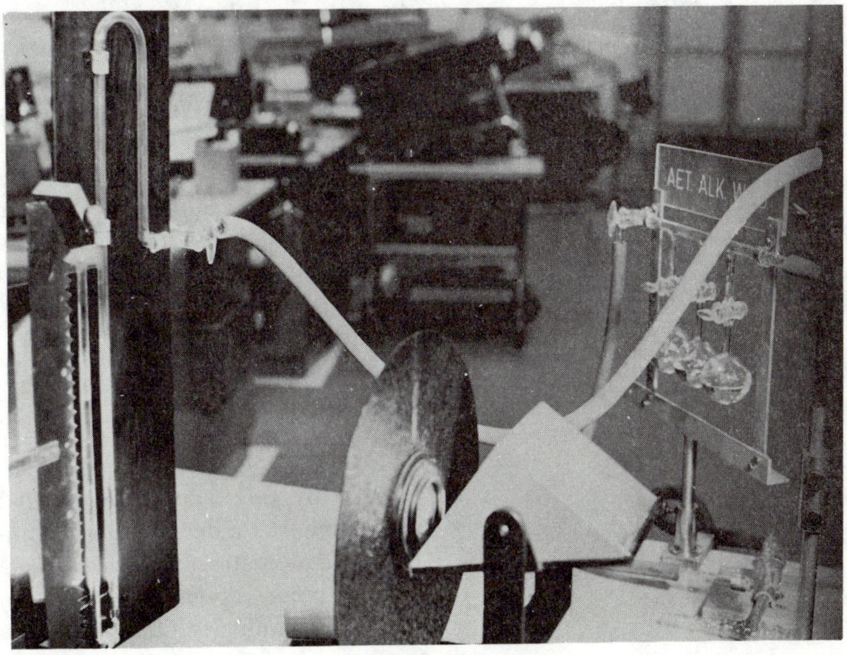

Fig. 27–16

Fig. 27–17

The stopper is left ajar and the flask is boiled to expel the air. The flame is removed, the stopper replaced firmly, and readings begun at once. The flask temperature and heights of the mercury should be read as nearly simultaneously as possible. Sets of data are made at intervals until the flask has reached room temperature. The vapor pressure is equal to the barometer reading minus the difference between the two mercury levels.

The apparatus is disassembled by heating the flask to the boiling point. When the mercury levels are equalized, the stopper can be easily removed.

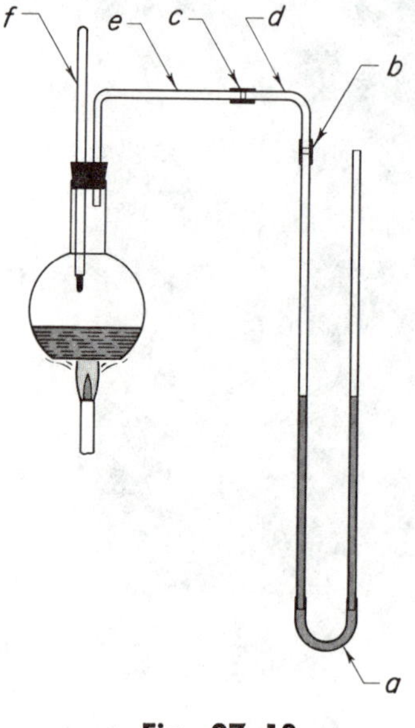

Fig. 27–18

The two vertical lengths of tubing should be about 36 in. long. They are connected by a length of thick-walled rubber tubing *a*. One tube is connected to the flask by rubber tubing (*b* and *c*) and short lengths of glass tubing (*d* and *e*). A thermometer *f* is inserted through the stopper to show the flask temperature.

The advantages of this equipment are that it is made from standard laboratory materials, is easily assembled, and offers no pressure or leakage problems since it operates at pressures less than atmospheric. The inherent disadvantage is that some air is trapped in the tubing extending from the flask to the adjacent mercury surface. The effect of the trapped air can be minimized by using tubing of fairly small bore and by using a fairly large flask only partly filled with water.

27–3.9[15]

Dalton's law states that the maximum pressure exerted by the vapor of a particular liquid in a closed space at a definite temperature depends only on that temperature, and is independent of the pressure of other vapors or gases having no chemical action (including solution) on it.

[15] J. Satterly, *Am. J. Phys.*, **13**, 50 (1945).

A common demonstration is to saturate the empty space above the mercury in a barometer tube with ether vapor, measure the depression of the mercury column, and thus obtain the saturation vapor pressure of ether at room temperature. In a companion experiment, air occupying a known volume at atmospheric pressure is saturated with ether vapor. This raises the pressure and increases the volume. The volume occupied by the air and ether vapor is then restored to its original value, and it is found that the pressure of the contents exceeds atmospheric pressure by an amount nearly equal to the saturation vapor pressure of the ether alone.

These demonstrations are well worth doing. Ether is usually selected because its vapor pressure at room temperature is on the order of 400 mm Hg so that the class can observe what is happening. One problem is that the ether is usually introduced into the tube through stopcocks, and when the volume of the gas in the tube is brought back to its original value, the ether escapes through the stopcocks because the usual tap grease is soluble in ether. The escaping ether takes air with it, and it is impossible to show that the vapor pressure of the ether in the presence of air is about 400 mm Hg at room temperature.

The apparatus shown in Fig. 27–19 is very effective in its method of introducing the ether, and it

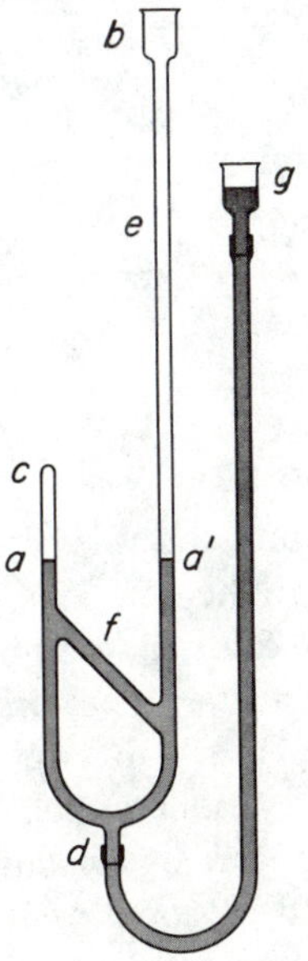

Fig. 27–19

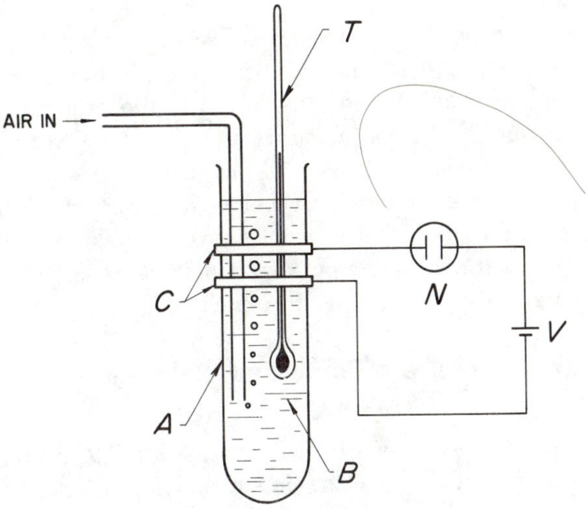

Fig. 27–20

certainly cannot leak.[16] The narrow tube f is responsible for the improvement. Tube dc is about 1 ft long, and bd is about 3 ft long. The reservoir g and the two tubes contain mercury. The tube cbd and the reservoir, both suitably supported, are manipulated so that air is moved through the mercury into dc and at atmospheric pressure has a volume of, say, tube length ca. It is quite easy to get a and a' at the same level. Point a is marked with a piece of gummed paper.

Ethyl ether is then poured in at b and a small amount of ether is run off from the a' side to the a side by lowering g and suitably inclining cdb. This process is well under control. Then, g is immediately raised to prevent a backflow of the mercury as the ether vaporizes. The ether in ac now evaporates, and in time saturates the air. If g is now raised to bring the mercury in cd back to the gummed paper, the mercury in bd rises, and the vertical distance between a and e will be about 400 mm.

The surprising thing is the length of time required by the ether to saturate the air. When the first experiment is performed (the ether alone above the mercury in a barometer tube), the process is complete in a few seconds. With the air present, however, the pressure builds up more slowly, reaching only 200 mm Hg in about 10 min, nearly 400 mm in 20 min, and its final height (400 mm at

[16] The design of this apparatus is due almost entirely to Mr. Chappell, the department glass blower at the University of Toronto.

20.0°C) in about a day. This reminds one of the delay caused by molecular collisions.

An important point is that the setting of the surface of the liquid ether in *ac* to the original air volume must be as accurate as possible. The pressure is measured from *e* to the surface of the mercury, which is now just below *a*. The correction for the liquid ether at *a* is not worth making. The temperature should be kept constant at all times, or else the tube *cd* must be jacketed.

27–3.10 Dew Point and Relative Humidity[17]

The dew point can be demonstrated to a large audience with the arrangement shown in Fig. 27–20. A test tube *A* contains a volatile liquid *B*, usually ether. Two conducting rings *C* are placed around the tube a few millimeters apart and connected in series with a neon glow lamp *N* and a voltage *V* greater than the ignition voltage of the glow lamp. When the ether is rapidly evaporated by bubbling air through it, the test tube is cooled. When the dew point is reached, moisture from the room will condense on the walls of the tube and the conductivity of the film of moisture will cause the lamp to glow. The temperature is indicated by the thermometer *T*. Since the glow from the lamp is faint, the room should be darkened or the lamp placed inside a black box open on one side.

(17) L. Grillet, *Bull Soc. Sc.* Bretagne, T. 9, 16 (1932); L. Grillet and R. Tessier, *C. R. Acad. Sc.*, T. 216, 874 (1943); R. Tessier, *Thèse d'Ingénieur-Docteur,* Paris (1943); A. Gouffe, *Journal des Usines à Gaz,* 3, 35 (1946).

27–3.11 Formation of Fog on Ions[18]

Cotton moistened with water is placed under a bell jar, and as the air is pumped out, fog forms under the bell jar. Pumping is continued until the fog is removed together with the air. Then air is fed into the bell jar through a rubber tube attached to a tap mounted in the pump plate. An ordinary glass drying tube (such as Cenco #14755) is attached to the other end of the rubber tube. The glass tube is partially filled with dust-free glass wool for filtering out dust particles.

The air is pumped out and fed back in (through the filter) several times in order to free the air in the bell jar of all dust. With every new evacuation the fog is lighter, and finally after several evacuations the fog is scarcely noticeable. Air is then once again fed under the bell jar, but a lighted burner is held close to the opening of the tube filled with glass wool. As a result, the air entering the bell jar carries with it the ions which were formed in the flame, and the subsequent evacuation of air brings about the appearance of a thick fog. If, without evacuating the fog, more air is fed into the bell jar, the fog disappears immediately as a result of the heating of the air by adiabatic compression. The demonstration should be shown in a darkened auditorium, and the bell jar should be illuminated from above with a shaded lamp for the best visibility of the fog.

(18) Translated by K. V. Robinson and H. A. Robinson (Adelphi University) from *Lecture Demonstrations in Physics,* A. B. Mlodzeevskii, ed., M. V. Lomonosov State University, Moscow, U.S.S.R.

27–4 Viscosity of a Gas

27–4.1

Illuminating gas is fed into the stem of a T made from small-bore copper tubing (approximately 1/32 in. i.d.), as shown in Fig. 27–21. The end of each arm is lighted, producing flames of equal size. A Bunsen burner set under one arm will make that flame smaller, this indicates the fact that the viscosity of the gas increases when heated and that the gas flow rate is controlled by the viscosity.

27–4.2

A demonstration of the viscosity of air as a function of density can be made with the apparatus shown in Fig. 27–22. An aluminum disk is suspended axially by a fine steel wire so that it lies in a horizontal plane between two brass disks. The assembly is placed under a bell jar (Fig. 27–23) so that the system can be evacuated. The aluminum disk is set into torsional oscillation by rotating the bell jar plate back and forth in time with the disk until the amplitude of the disk is between 3

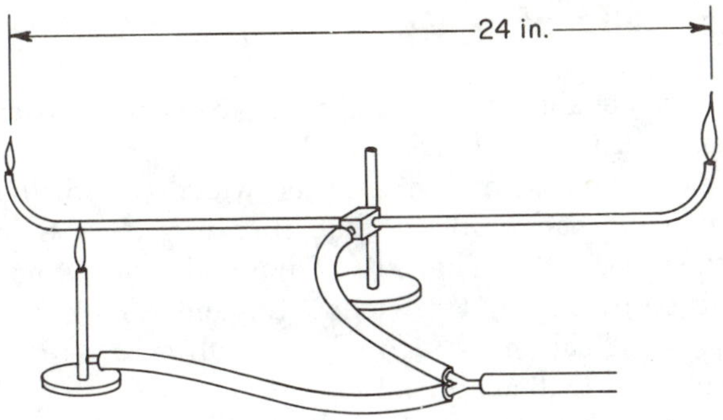

Fig. 27–21

Fig. 27–22

and 4 in., as indicated by a scale on one of the brass disks. The oscillations are damped by the viscous effect of the layer of air between the brass plates. The viscosity is determined by measuring the time for the amplitude to decay to $1/e$ of its initial value. Since the retarding torque is inversely proportional to the separation of the brass plates, the decay time is determined for a measured small separation and for a large separation (negligible viscous torque). Comparison of decay times for various pressures show that the viscosity of a gas is independent of density over a wide range.

The amplitude of oscillation may be shown to a class by means of closed-circuit television if a dot of paint or ink is placed on the disk to make the oscillations visible. Alternatively, a very small mirror may be suspended on the wire. When a narrow arc-lamp beam is focused on the mirror, the mirror reflects the beam, producing an oscillating light spot on the wall. Construction notes and theory may be found in the Appendix, page 1290.

Fig. 27–23

27–5 Thermal Conductivity of a Gas

27–5.1

In this demonstration, current is passed through platinum filaments inside a pair of Pyrex tubes. The current is of such a magnitude that if the tube contains air at atmospheric pressure, the filament glows brightly. One of the tubes is arranged so that it can be continuously flushed with hydrogen. In this case the hydrogen gas conducts heat away from the filament so rapidly that it does not get hot enough to glow. If now the hydrogen supply is turned off and the tube closed and evacuated, the filament does not begin to glow until the mean free path of the hydrogen molecules becomes comparable with the dimensions of the tube. When the vacuum reaches about 10^{-3} mm, the filament will glow more brightly than the one in the air-filled tube, showing the thermal conductivity of air.

Figure 27–24 shows the apparatus, and Fig. 27–25 is a diagram of the same equipment. The filament current is supplied by a Variac *a*. When the right-hand tube is being flushed with hydrogen from the tank, the gas is burned off at *b*. The vacuum pump *c* is an ordinary centrifugal force pump, connected via a three-way valve *d* to the hydrogen-filled tube. A projection-type pressure gauge *e* allows the audience to read the pressure during evacuation. The details of this gauge are shown in Fig. 27–26. The movement is from a commercial gauge and the housing consists of a section of aluminum tubing about 15 cm in diameter and 5 cm long, fitted with Plexiglas windows. A scale is etched onto one of the windows with the numbers backwards so that after projection they show correctly. The zero of the scale can be adjusted by turning the window containing the scale. The original pointer is replaced with a larger one cut from 1/10-mm metal foil.

Figure 27–27 shows one of the Pyrex tubes, which are about 25 cm long and 4 cm in diameter. The 1.5 x 0.1-mm platinum filament is spot-welded to its supporting electrodes, which are sealed into the glass. Both tubes are held by aluminum brackets mounted on a plywood base.

Caution: Extreme care must be used to avoid an explosive mixture of air and hydrogen in a tube containing a heated filament. Turn off the filament current before flushing the tube with hydrogen. Allow the hydrogen to flow for several seconds before attempting to light the burn-off tube.

27–5.2[19]

The apparatus shown in Fig. 27–28 is used to compare the thermal conductivity of different gases.

[19] Translated by K. V. Robinson and H. A. Robinson (Adelphi University) from *Lecture Demonstrations in Physics*, A. B. Mlodzeevskii, ed., M. V. Lomonosov State University, Moscow, U.S.S.R.

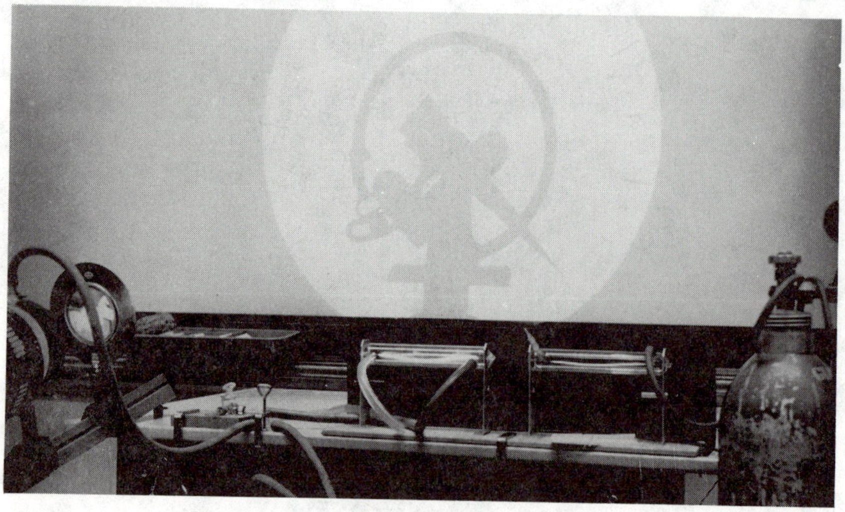

Fig. 27–24

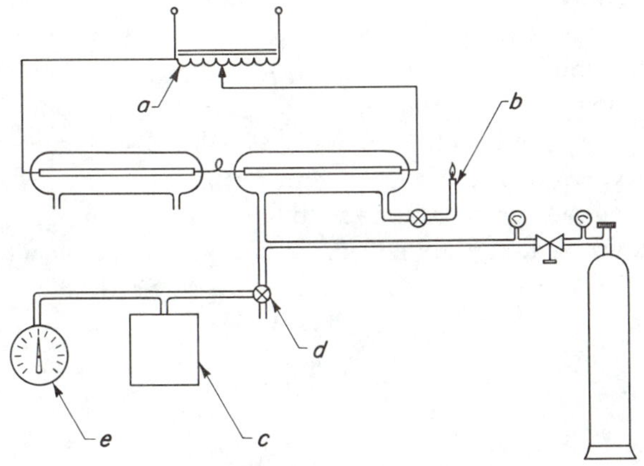

Fig. 27–25

Fig. 27–26

A glass tube *c* approximately 1 in. in diameter is hermetically sealed at both ends with brass caps *b* by means of an appropriate cement. A Nichrome wire *f* about 0.016 in. in diameter is stretched between two springs *g* and runs along the axis of the tube. In the middle section of the tube, the wire passes through a drilled brass rod *e* that is inserted through a glass stopper *d*. The stopper and the rod are coated with cement, and the wire is soldered to the left end of the rod to seal the canal and to keep the gases in the two halves of the tube completely separate. The length of the free portions of the wire in the left and right halves of the tube is about 6 to 7 in. A 5-A current is used between the binding posts *a*.

First both sides of the tube are filled with air, and the wire in both sections is heated equally. Then the current is shut off, and either illuminating gas or hydrogen is fed through one of the openings in the right side of the tube. This forces the air out through the other opening; both openings are then closed with stoppers. Current is once again supplied to the wire, and only the left portion of the wire is heated, as the right portion loses its

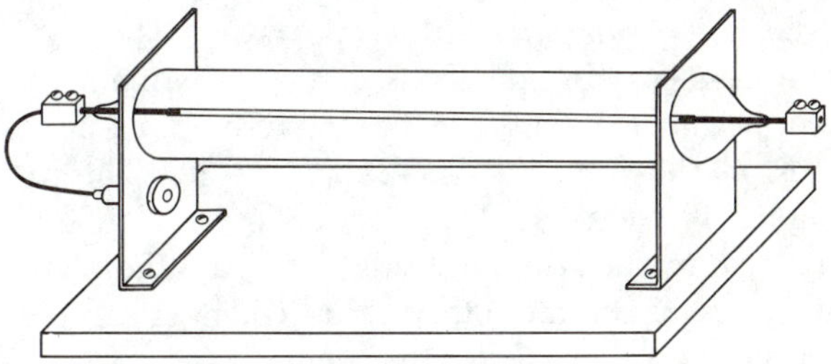

Fig. 27–27

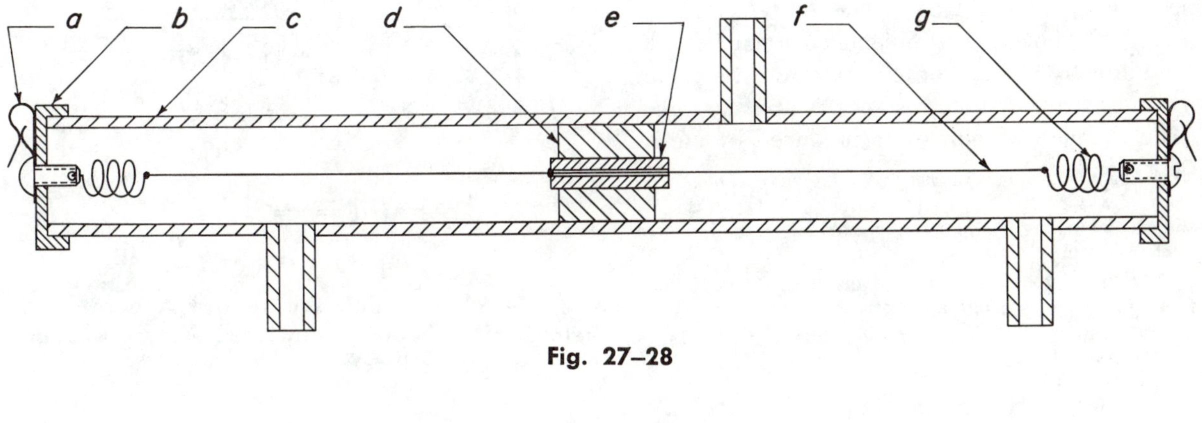

Fig. 27–28

heat due to the greater thermal conductivity of the gas.

The fact that the thermal conductivity of gases is essentially independent of pressure can be shown on the same apparatus. In this case, the right-hand portion of the tube is under atmospheric pressure and the left-hand side is evacuated by a pump. This evacuation should not take place too rapidly; otherwise the air in the left portion of the tube cools as a result of adiabatic expansion. Therefore, the left portion of the tube is connected by means of rubber tubing to the pump through a balloon having a capacity of several liters. Another tube from the balloon is connected to the pump through a T-joint. Also, attached to the T-joint is an open mercury manometer 1 m high, half filled with mercury. A current is once again fed to the wire and evacuation is begun; even though the manometer shows a considerable decrease in pressure, the heating of the wire is the same in both sections of the tube.

27–6 Adiabatic Processes

27–6.1[20]

A match placed within a closed tube fitted with a plunger may be lighted by rapidly pushing the plunger down. The resulting compression of the air liberates enough heat to ignite the match. This principle of sudden high compression is used for fuel ignition in diesel engines.

27–6.2

The temperature change of a fixed amount of gas undergoing an adiabatic compression is measured. Figure 27–29 is a line drawing and Fig. 27–30 is a photograph of the apparatus. The temperature-measuring device is a 600-Ω platinum resistor *a* wound with 0.1-mm-diam wire on a Lucite block, as shown in Figs. 27–31 and 27–32. The resistance is measured by using an external 600-Ω resistor *b* in one arm of a bridge[21] *c* and the platinum resistor in the other arm. Changes in the platinum resistance throw the bridge out of balance, and the out-of-balance current can be read by the audience on a projection-type milliammeter[22] *d*. Since the temperature coefficient of resistance of platinum is known, the milliammeter can be calibrated directly in terms of temperature if the meter current, as a function of resistance, is first measured.

The changes in volume are produced by means of a large syringe *e* connected through flexible tubing to the 1½-liter bottle *f*. Different gases may be used (for example, air, carbon dioxide, and argon), provided the bottle is rinsed well with the gas to be measured. From measurements of the initial and final volume and initial and final temperature, and assuming the ideal gas law, the ratio c_p/c_v can be determined.

[20] Adapted from *University Physics, Experiment and Theory,* by George D. Freier. Copyright © 1965 by Meredith Publishing Company. Reprinted by permission of Appleton-Century-Crofts.

[21] Phillips Type GM 5536 or equivalent; direct indicating bridge for strain gauge applications.

[22] With the Phillips bridge, a millivoltmeter with a range 5 mV/10 Ω can be used.

27–6.3

A thermocouple and a galvanometer are used to indicate a decrease in thermal energy of air as it is compressed and expanded adiabatically. A No. 5 test tube (150 mm long) is used as a pressure vessel, Fig. 27–33 (a). Within the tube, a 1/16-in. copper rod *a* is inserted into a one-hole rubber stopper *b*, which supports the hot junction of a thermocouple, consisting of the copper wire *c* and the constantan wire *d*. The junction *e* is attached to a short length of twisted copper foil to increase the effective area.

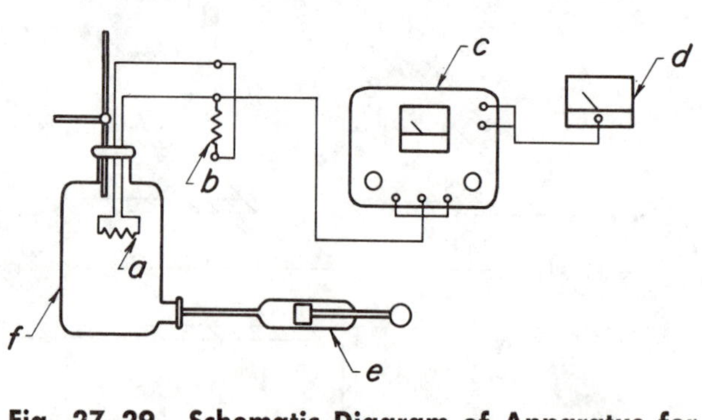

Fig. 27–29 Schematic Diagram of Apparatus for Demonstrating Volume and Temperature Changes in an Adiabatic Process.

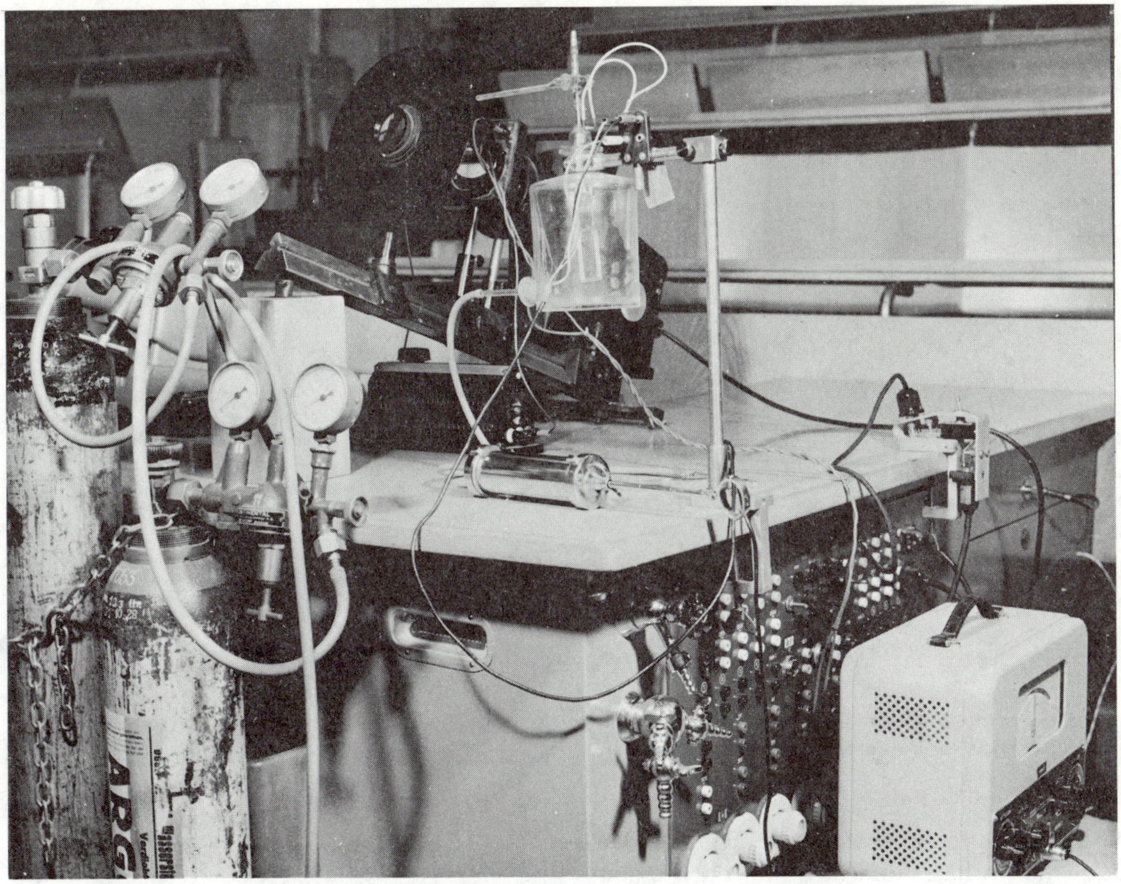

Fig. 27–30 Photograph of Actual Arrangement, showing cylinders of carbon dioxide and argon and optical projection system.

The cold junction f is contained in a circulating water bath as shown in Fig. 27–33 (b). The lead wires of the hot junction are fine enough to pass through the same hole as a small section of brass tube on which is fitted the connecting tube from a hand pump. A tight seal is needed around the stopper. Also, the hand pump must have a suitable valve arrangement to allow a slow, controlled compression and/or expansion of air in the pressure vessel, thereby maintaining approximate thermal equilibrium. Thermal energy changes are observed as deflections by a lecture μA-galvanometer.

27–6.4 Joule-Kelvin Coefficients

The outlets of two equally large tanks of N_2 and H_2 (H_2 is on the left in Fig. 27–34) are placed in opposite ends of a loosely fitting glass tube (1¼ in. diam, 1 ft long). A copper-constantan thermocouple is held by a split cork inserted into a hole of roughly ¼ in. diam which has been blown at the midpoint of the tube and is connected to a projection galvanometer. When the N_2 tank valve is opened, the galvanometer will deflect, showing that the gas cools upon expansion. (It may be necessary to direct the stream of N_2 gas at the thermocouple to obtain a large enough galvanometer deflection.) When the H_2 tank valve is opened, the galvanometer will deflect in the opposite direction, indicating that the gas heats up. It is important that the gas tanks be placed in the lecture hall at least 24 hours before the scheduled demonstration so that they will be at the same temperature.

27–6.5 Rüchhardt's Method for γ

A photoelectric pressure detector (Figs. 27–35 and 27–36) can be used with thermodynamic con-

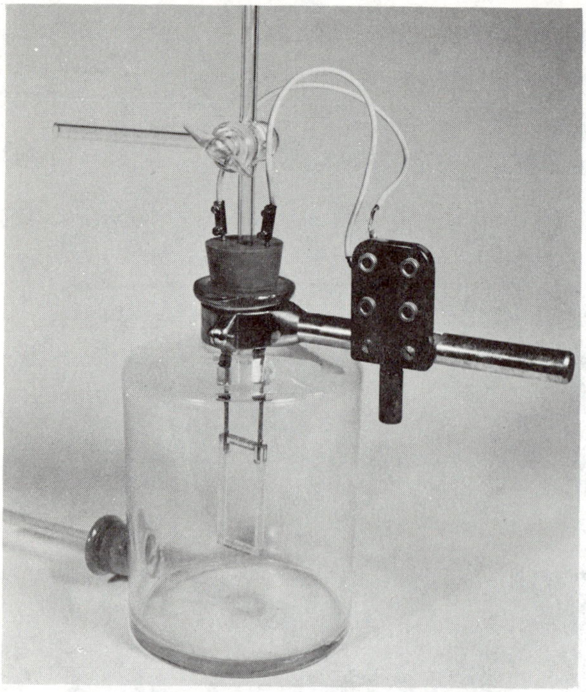

Fig. 27–31 The 1½-Liter Flask Containing the Platinum Resistor for rapid measurement of temperature.

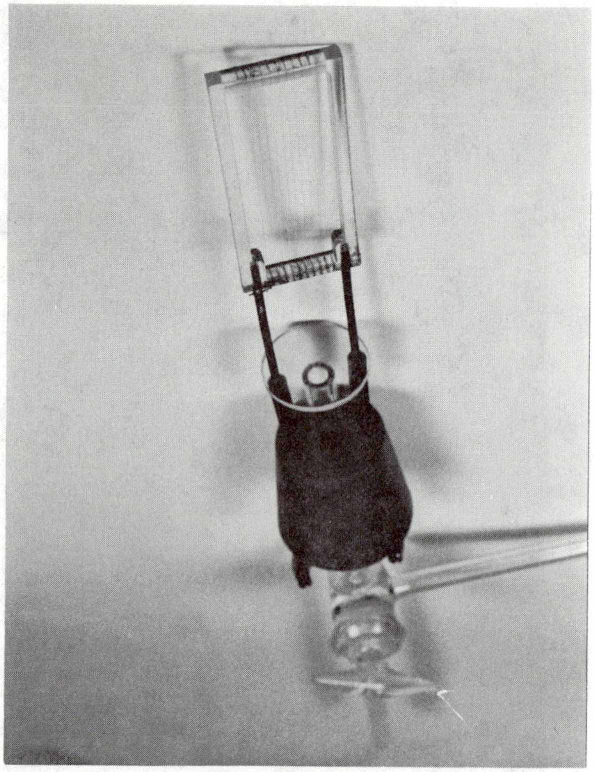

Fig. 27–32 Platinum-Wire Resistance Thermometer.

stant apparatus[23] to demonstrate damped harmonic motion with the aid of a projection oscilloscope[24] or closed-circuit TV. Pressure bulb *a* (Fig. 27–36) pumps a steel ball up the glass tube, where it is held by an electromagnet *b* (115 V ac). Stop-wire *c* prevents the ball from falling to the bottom of the glass flask. When the ball is released and oscillates, the changes in pressure cause the

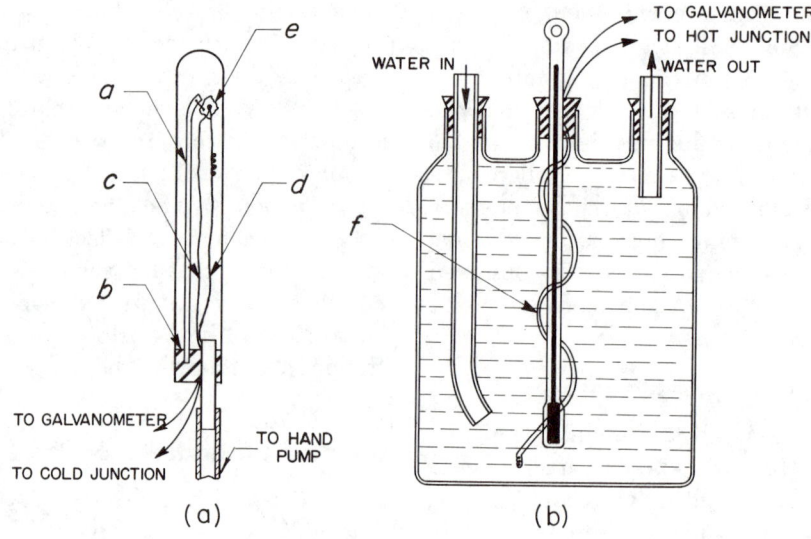

[23] Klinger Scientific Apparatus Corp., Aspirator, KH-2001; and precision glass tube with ball, KH2002.

[24] H. F. Meiners and G. Huse, *Am. J. Phys.*, 27, 680 (1959).

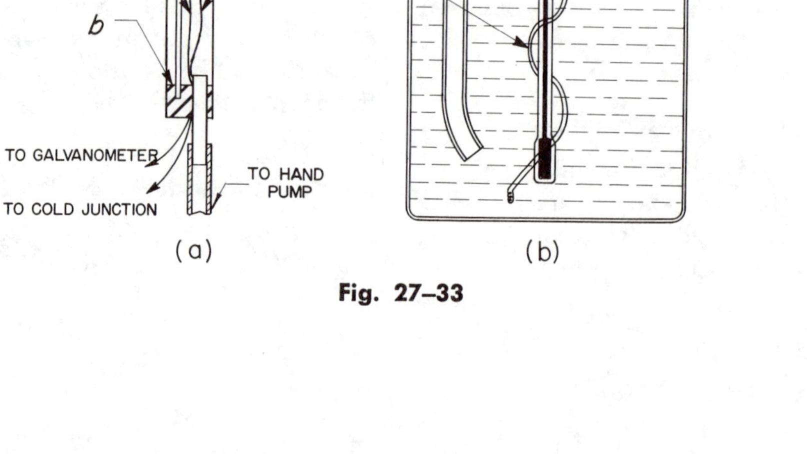

Fig. 27–33

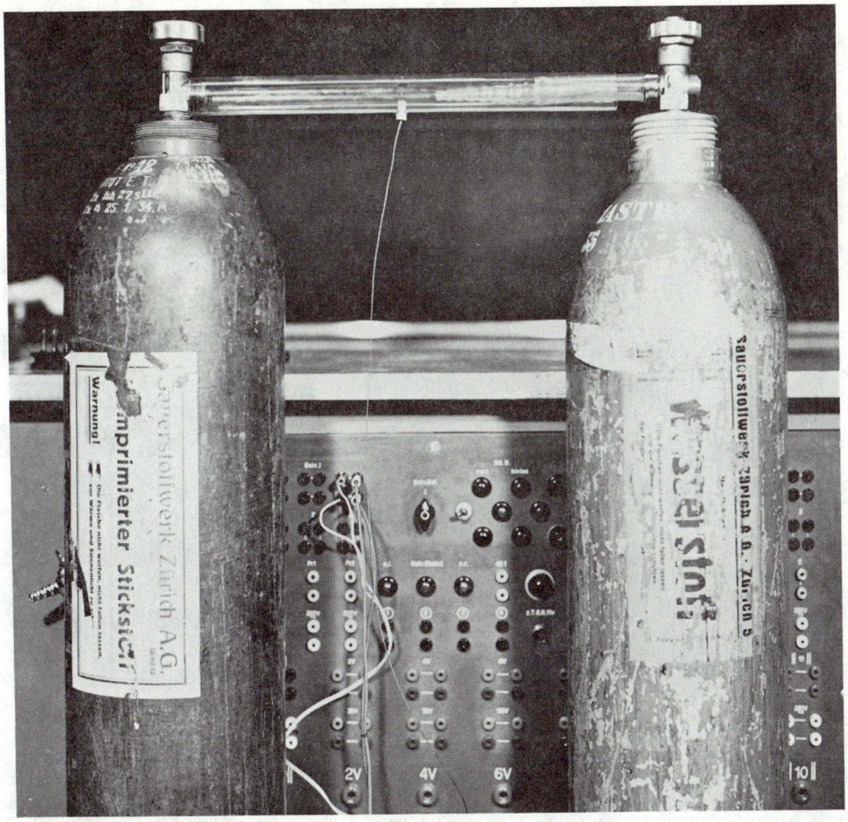

Fig. 27–34

$2\frac{1}{2}$-in.-diam x $2\frac{3}{4}$-in.-long bellows d to respond correspondingly. Closed-circuit television shows the bouncing ball. Aperture plate e (Fig. 27–37) is set so that approximately half of the light from bulb f (GE, #325) falls on the photocell g (CdS, 600 Ω, Polaris MAJ-I) when the system is in equilibrium.

As the bellows (Fig. 27–38) moves in response to pressure changes, the plate moves, thereby varying the amount of light falling on the photocell. As the amount increases, the resistance of the cell decreases, and conversely, as the amount decreases, its resistance increases. This variation causes the bridge to unbalance, giving rise to a voltage variation proportional to the pressure changes. The potential across the bridge is connected to an oscilloscope; therefore, the trace will be a function of pressure variations in the thermodynamic apparatus, that is, damped harmonic motion. The 50-μF capacitor is used to help reduce 60-cycle hum in the system. As a result, this circuit will not respond to fast variations in pressure. By very careful construction, use of a very well filtered power supply, and shielded leads, this capacitor could be eliminated. The only limitation to the frequency response would then be the response of the photocell and the bellows.

When the average period of oscillation is determined, the value of the ratio $\gamma = C_p/C_v$ can be determined from the formula

$$\gamma = 64mV/d^4pt^2,$$

where m is the mass of the ball, V the volume of the flask, p the pressure in the flask, d the diameter of the tube, and t the average period of oscillation.

27–6.6[25,26]

If the oscillating ball in Rüchhardt's apparatus is driven by a slow flow of gas and is loaded by adding additional mass as shown in Fig. 27–39, one may obtain the period as a function of the oscillat-

(25) H. Jensen and B. Donnally, *Am. J. Phys.*, **32**, 4, xv–xvi (1964).

(26) See also Chapter 22, Corridor Demonstrations.

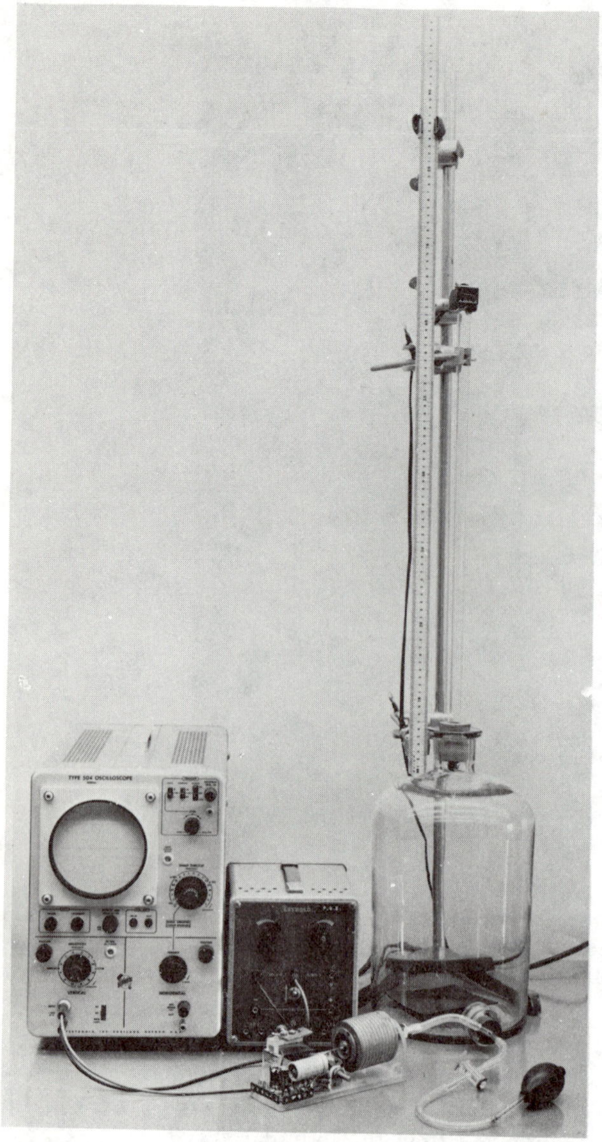

Fig. 27–35

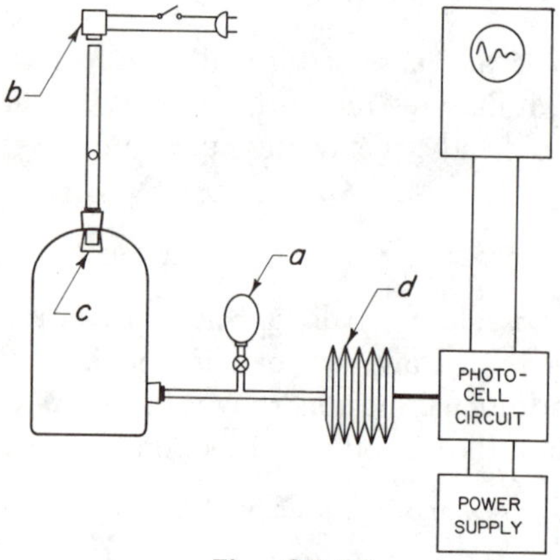

Fig. 27–36

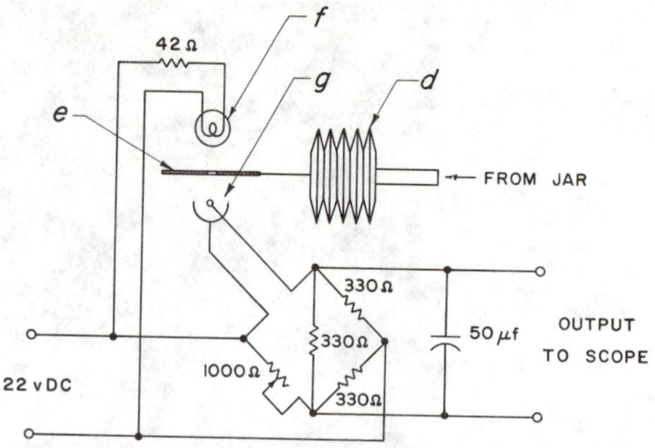

Fig. 27–37

ing mass. The masses are $\frac{3}{8}$-in.-diam washers with $\frac{3}{16}$-in.-diam holes. The $\frac{1}{2}$-in.-diam rod is tapped, and a 2-in.-long, $\frac{3}{16}$-in.-diam rod which supports the washers is inserted.

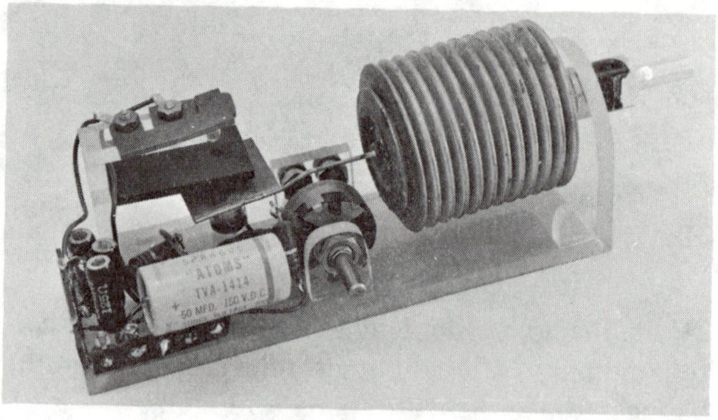

Fig. 27–38

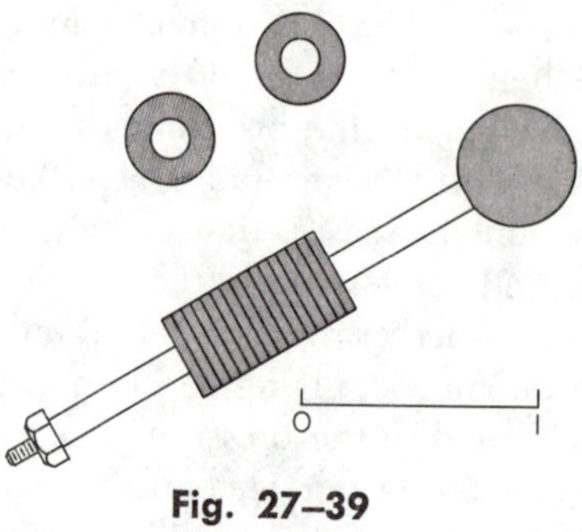

Fig. 27–39

A plot of the square of the period against the mass of a loaded ball is shown in Fig. 27–40. Note that the line has an intercept very nearly at the origin, indicating that this is the equivalent of a nearly massless spring.

27–7 Kinetic Theory Models

27–7.1 A Dynamic Hard Sphere Model[27]

Models which simulate many atoms interacting together can be useful for teaching and for giving insight on what assumptions may be satisfactory to treat multiatomic interactions.

In the model shown in Fig. 27–41, many uniform spheres are fenced in on a round horizontal glass plate and agitated by the vibrations of a wooden board, suspended from springs, on which the plate is mounted. The board is vibrated by a rotating eccentric weight which is attached to the shaft of an electric motor mounted on the underside of the board. With this excitation the spheres move continually and apparently at random. The motion of the spheres is both lateral and vertical, but a motor speed can be selected so that the vertical component is very small. At low sphere density the model exhibits "gas-like" behavior. As the density is increased the behavior becomes more "liquid-like" and then "crystallization" occurs. Many dynamic atomic phenomena, believed to occur in a gas, liquid, or solid state, are illustrated by the model.

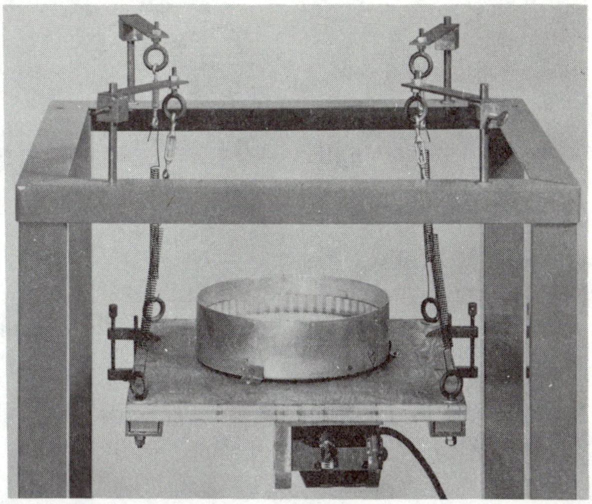

Fig. 27–41 Photograph of Hard Sphere Model.

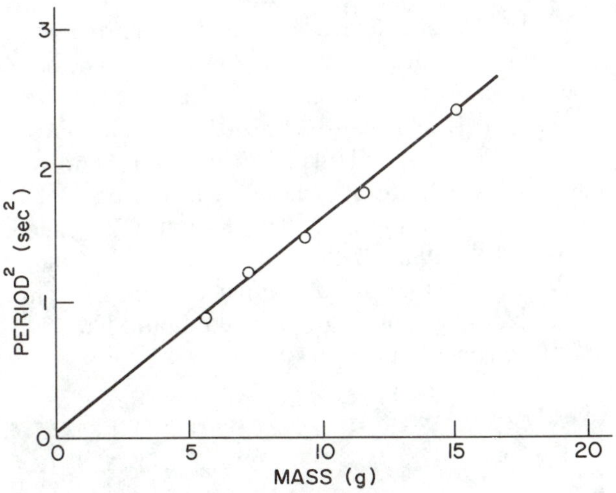

Fig. 27–40

Behavior of Model. Excepting where otherwise noted, all the observations were made with mutually unattracting glass spheres, 4 mm diam. Several high-speed photographs (Figs. 27–42 through 27–49) were made of the glass spheres in motion. For these photographs only 5-mm spheres were used instead of the 4-mm spheres. The principal variable governing the behavior of the model is the "free-space ratio" which is defined to be the ratio of the area occupied by the spheres at rest in a perfect close-packed arrangement to the total area of the model.

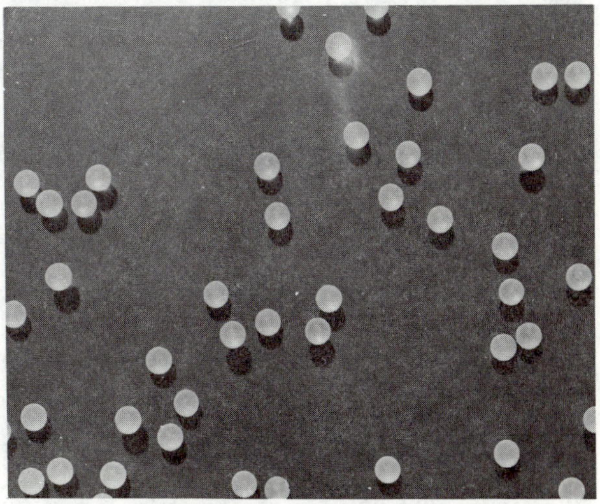

Fig. 27–42 Photograph of Dynamic Model in "Gas-like" State (small free-space ratio).

[27] D. Turnbull and R. L. Cormia, *J. Appl. Phys.*, 31, 674 (1960).

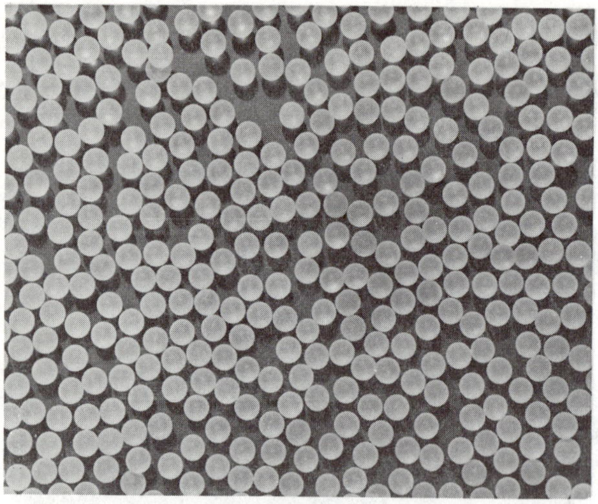

Fig. 27–43 Photograph of Dynamic Model in "Liquid-like" State (free-space ratio ~0.70).

Transition from Gas-Like to Liquid-Like to Solid-Like Behavior. In this experiment the density of glass spheres is gradually increased from zero while energy is supplied continually. The sequence of events is very like those which would occur in a material initially in the gaseous state if the pressure were increased gradually. At first the spheres are very far apart and move distances many times their diameter before colliding with another sphere. A photograph of the model in this "gas-like" state is shown in Fig. 27–42.

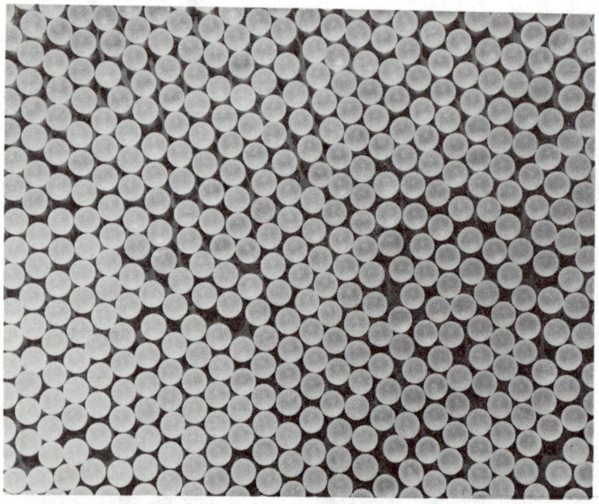

Fig. 27–44 Photograph of Dynamic Model at Transition to "Crystal-like" State (free-space ratio ~0.82).

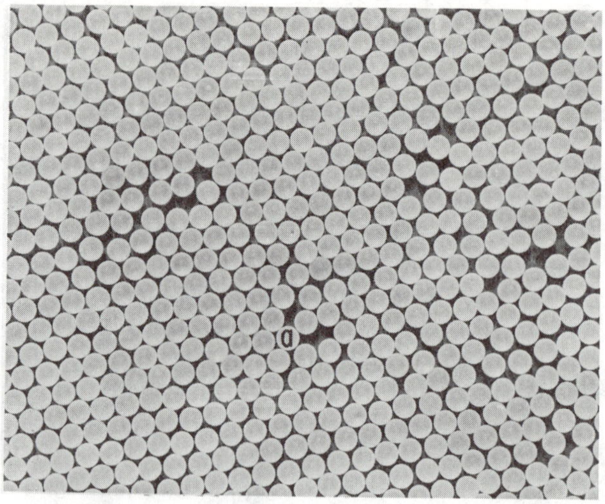

Fig. 27–45 Photograph of Dynamic Model Showing Monovacancy at point *a* in transition from one position to adjacent one (free-space ratio ~0.90).

As the density increases, the distance between collisions becomes less and less until it is no more than the diameter of a sphere. So long as the free-space ratio is less than 0.75 to 0.80, no long-range order is apparent, and a sphere added to the model easily finds a place anywhere. The behavior at a free-space ratio between 0.65 and 0.80 is called "liquid-like" although the state is, of course, more analogous to that of a highly compressed gas. Figure 27–43 is a photograph of the liquid-like state at a free space ratio of 0.70.

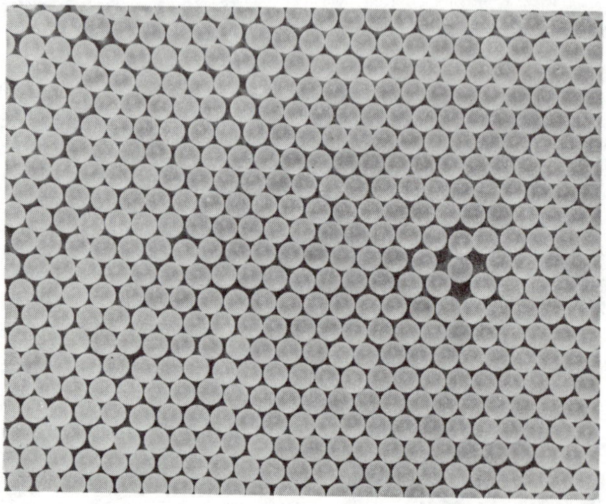

Fig. 27–46 Photograph of Dynamic Model Showing a Divacancy (free-space ratio ~0.94).

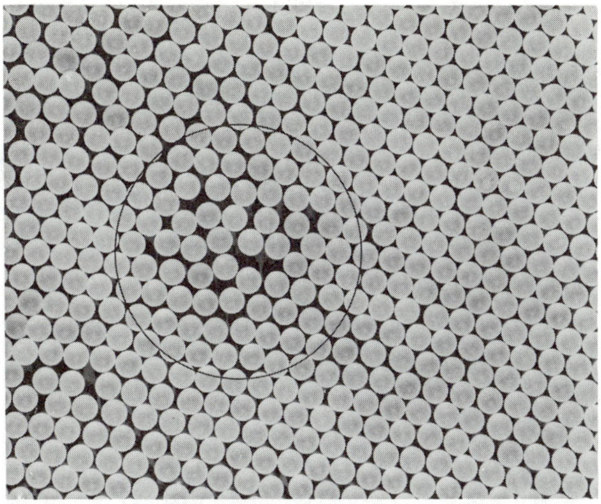

Fig. 27–47 Photograph of Dynamic Model Showing Pentavacancy within circled area (free-space ratio ~0.94).

As the free-space ratio is increased above 0.80, small domains exhibiting long-range order seem to appear quite suddenly and grow larger as the free space ratio is increased further. In this crystal-like state the spheres are confined most of the time to a single position, and the spheres added to the model flit over the surface until they encounter a defect and are incorporated into the crystal. Figure

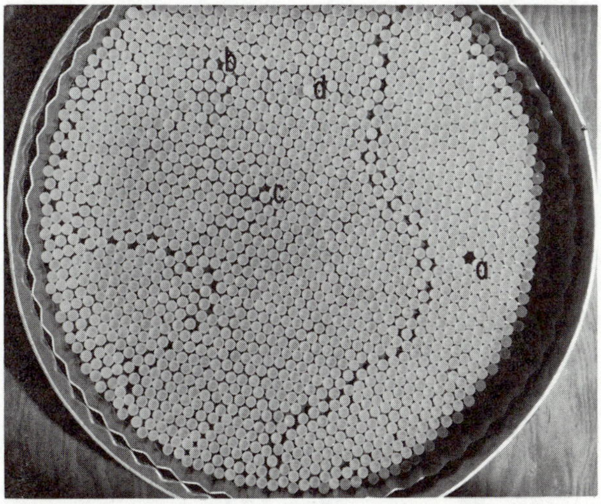

Fig. 27–48 Photograph of Dynamic Model Showing Grain Boundaries and Other Imperfections. A vacancy is in its equilibrium position at point *a*. There are dislocations centering at points *b* and *c*. A surface sphere is at point *d* (free-space ratio ~0.94).

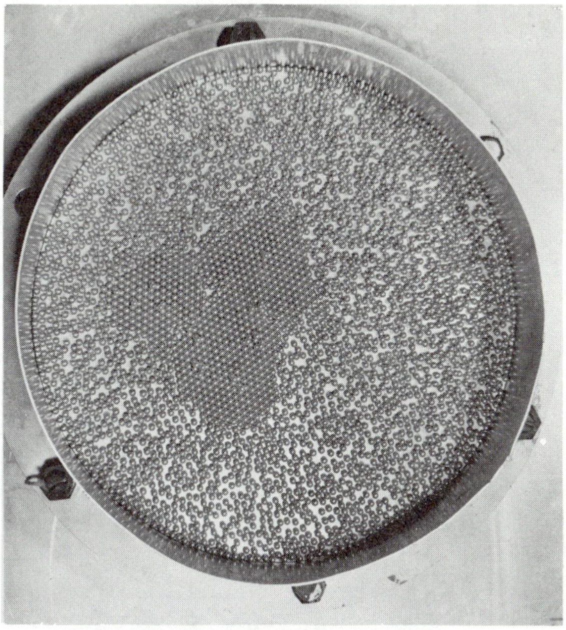

Fig. 27–49 Photograph of Nucleation produced by dropping three drops of ethyl alcohol into the operating model (using aluminum spheres and plate and the MB vibration tester).

27–44 shows the model just after long-range order has developed at a free-space ratio of 0.82. Apparently the onset of long-range order occurs at nearly the same free-space ratio at all values of the average sphere energy.

When the free space ratio is more than 0.83, the crystal-like structure of the model is well defined and various crystal defects and their dynamic behavior are clearly seen.

Liquid-Like Behavior. High-speed motion pictures were made of the liquid-like state at a free space ratio of 0.70 to 0.74. From these it appeared that most diffusive displacements result from density fluctuations and ranged in magnitude from a small fraction to, in rare instances, 2- or 3-sphere diameters. Quite a number of displacements were of the order of a 1-sphere diameter, but no quantitative assessment has been made of the relative contribution of the different displacements to the diffusion constant.

Defects in the Crystal-Like State. The dynamic behavior of monovacancies and multivacancies, dislocations and grain boundaries in the crystal-like state have been observed. At a free space ratio a

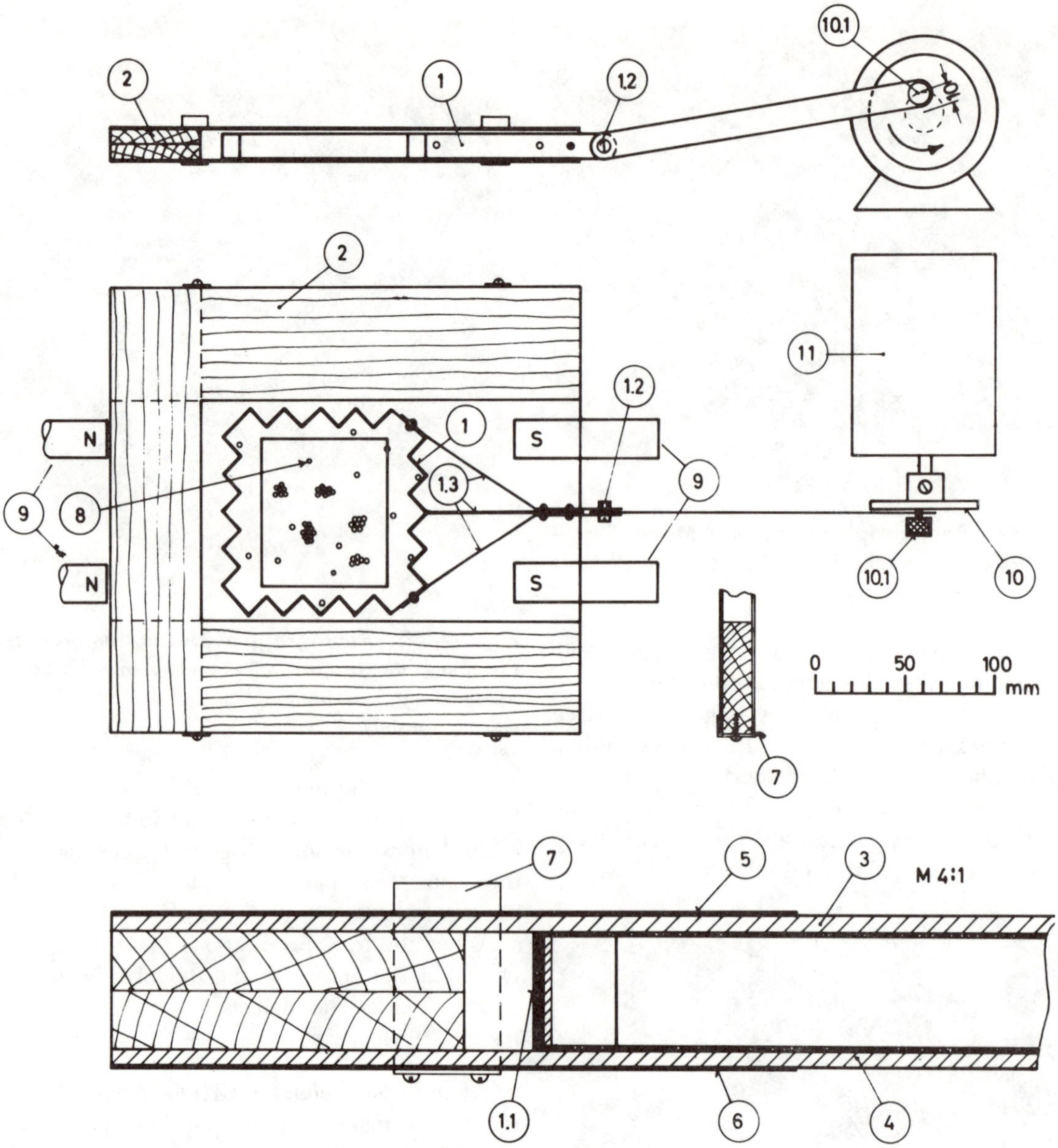

Fig. 27–50

little greater than that necessary for crystallization, 0.82 to 0.87, monovacancies form spontaneously and move quite rapidly. (Occasionally divacancies are seen to dissociate into a pair of monovacancies.) Figure 27–45 shows a monovacancy in the middle of the act of jumping from one position to an adjacent one.

When the free-space ratio is increased to about 0.93, the monovacancies become quite immobile. However, divacancies or multivacancies injected into the model at this density are very mobile and do not dissociate. Figure 27–46 shows a divacancy and Fig. 27–47 a pentavacancy. Multivacancies have an appearance somewhat resembling the configuration of the spheres in the liquid-like state. Also, it has been observed that at small free-space ratios, divacancies in 4-mm spheres are pinned by 5-mm "impurity" spheres.

There are dislocations in the model after crystallization, and they also form when the spheres are stirred. Mono- and multivacancies are annihilated when they encounter dislocations. This results in

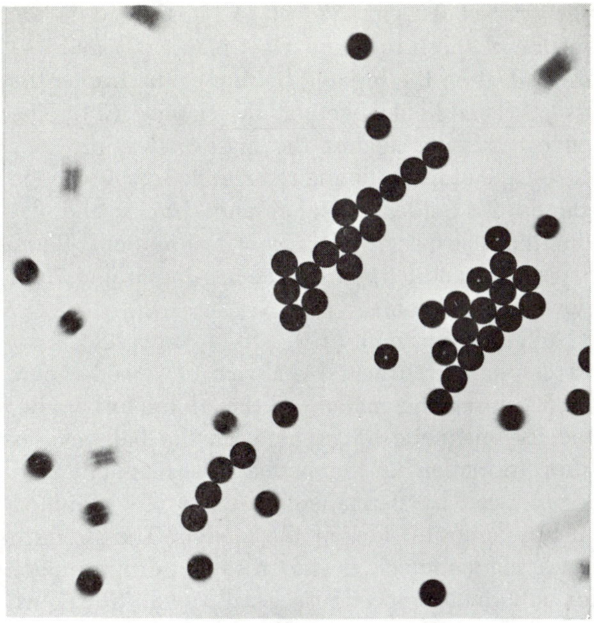

Fig. 27–51

climb of the dislocation. Climb in the opposite direction occurs when a sphere on the crystal surface enters the crystal at the dislocation.

Grain boundaries are present immediately after crystallization. Also, they can be introduced by stirring the spheres. They are seen to be sources and sinks for vacancies and surface spheres. Grain growth occurs rapidly at free space ratios of about 0.85, and sometimes at free-space ratios as high as 0.94. In time all grain boundaries and dislocations usually anneal out of the crystal except small areas at the fence. Figure 27–48 shows grain boundaries and various other imperfections a, b, c, and d.

Surface Diffusion. Spheres added to the crystal flit over its surface in what seems to be a random path. It is seen that the lighter the sphere, the greater the mean velocity of motion. High-speed motion pictures revealed that the spheres move almost always along the depressions in the surface formed by the underlying spheres.

Nucleation and Growth of Crystals. When the spheres are made attracting, either by magnetizing them or coating them with a thin film of liquid (for example, ethylene glycol), crystallization occurs by nucleation and growth. Generally, crystallization begins at the edge of the model, proceeds very rapidly, then stops abruptly, presumably when the density of the spheres in the disordered state reaches the equilibrium value for crystallization. Crystallization is promoted by decreasing either the average energy of the spheres or increasing the free space ratio. Under most conditions the "crystallites" formed by the attracting spheres drift slowly toward one side of the model. The drift could not be eliminated consistently, except as shown in Fig. 27–49, because the model is not uniformly vibrated. This drift does not occur, however, when the unit is mounted in a vibration tester.[28]

F. P. Price and W. DeSorbo of General Electric suggested the static models consisting of glass

[28] Motion pictures of various features of the model behavior have been obtained. Prints of the films may be borrowed or purchased. Requests to borrow prints (a 5-min or a 16-min film) should be sent to the General Electric Co., Metallurgy and Ceramics Research Department, Research Laboratory, Schenectady, New York. The 16-min film may be purchased, with the permission of the General Electric Research Laboratory, from Precision Film Laboratories, Inc., New York, New York.

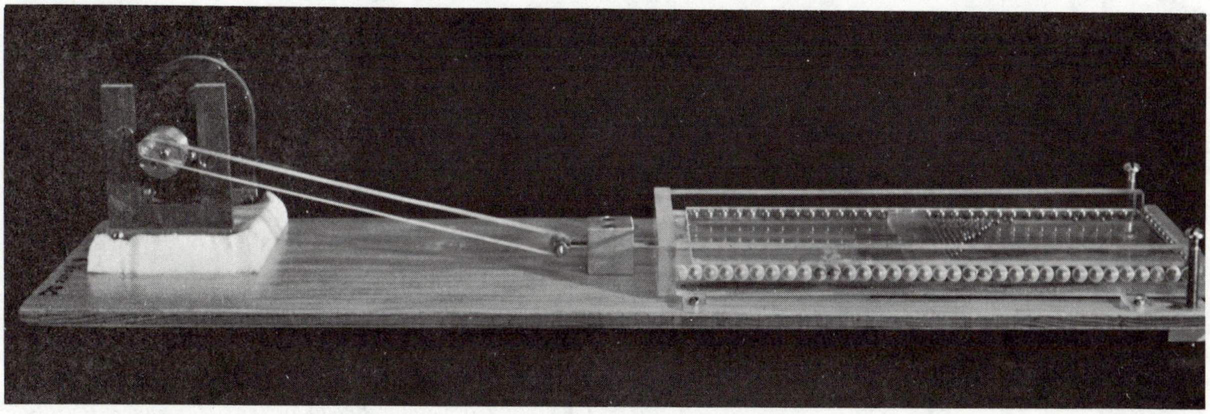

Fig. 27–52

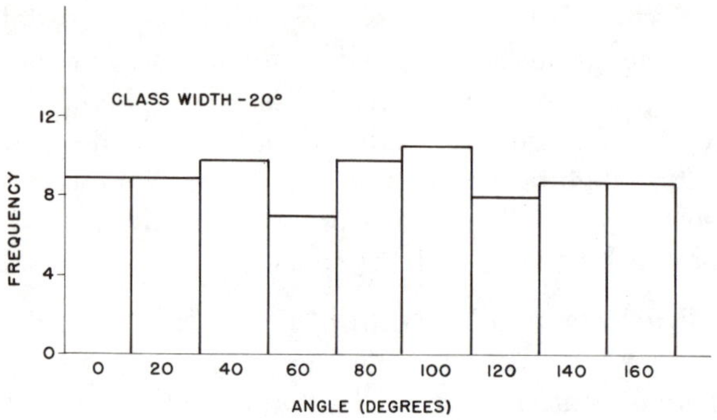

Fig. 27–53

spheres. J. W. Mitchell of the University of Virginia suggested the spring suspension in the dynamic model.

The apparatus can be used with closed-circuit TV. Construction details for the apparatus are in the Appendix, page 1292.

27–7.2 Model for Drop Formation

The model in Fig. 27–50 is used with an overhead projector to represent drop formation at condensation as well as the thermal motion of gas molecules. The $\frac{1}{8}$-in.-diam steel balls (8) are enclosed in a zig-zag aluminum frame (1) with a wall thickness of 0.02 in. Velvet (1.1) is glued to the outside of the frame, and glass plates (3) and (4) are placed on the top and bottom of the frame; the assembly is held together by clamps (7). The velvet, pressing against the glass plates, prevents the escape of the balls and reduces noise. A wooden frame (2) guides the aluminum frame, and flat aluminum strips (1.3) transmit the motion. Black paper (5) and (6) hides the zig-zag frame from view in projection.

An eccentric crank (10), (10.1), and (1.2), 1 cm off center and driven by a variable-speed motor (11), causes the motion of the aluminum frame, and the magnetic forces between the balls causes "drop formation" in the model. The magnetic field is produced by permanent magnets (9) mounted on two opposite sides of the model. The strength of the magnetic field is chosen so that drops appear for low motor speeds (several rpm); at higher speeds the drops disappear because the momentum between the spheres is high. As the speed is decreased from a high speed, a critical speed will be reached where the spheres begin to come to rest —in molecular terms, the critical temperature is reached where drops begin to condense. In order to have a few single molecules at low speeds, glass spheres, equal in size to the steel balls, are added. Figure 27–51 shows how the model illustrates disordered heat motion and the condensation and evaporation of single molecules.

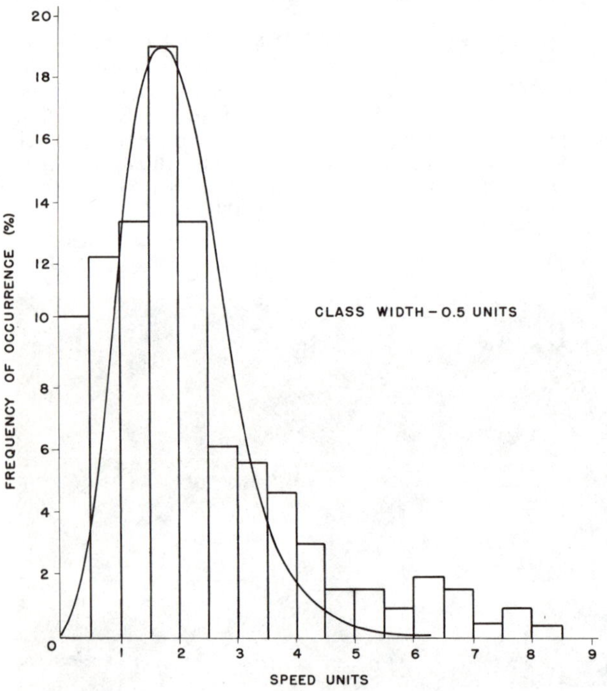

Fig. 27–54

27–7.3 Two-Dimensional Kinetic Theory Model[29]

The mechanical oscillator shown in Fig. 27–52 is used with an overhead projector. Molecules are simulated by 150 steel ball bearings ($\frac{9}{64}$ in. diam). Energy is given to the balls by the reciprocating motion of a push bar driven by a variable-speed 6 V dc motor. The Lucite enclosure walls are studded with 40 steel ball bearings ($\frac{5}{16}$ in. diam) evenly countersunk and spaced $\frac{5}{16}$ in. apart. These larger balls randomize the motion of the smaller balls, as shown in Fig. 27–53.

To determine the efficiency of the model, information about the distribution, speeds, and free paths of the spheres is determined photographically.

[29] R. Resnick and H. F. Meiners, *Demonstration and Laboratory Apparatus Report of the 1960 Summer Visiting Professor Workshop at Rensselaer Polytechnic Institute.*

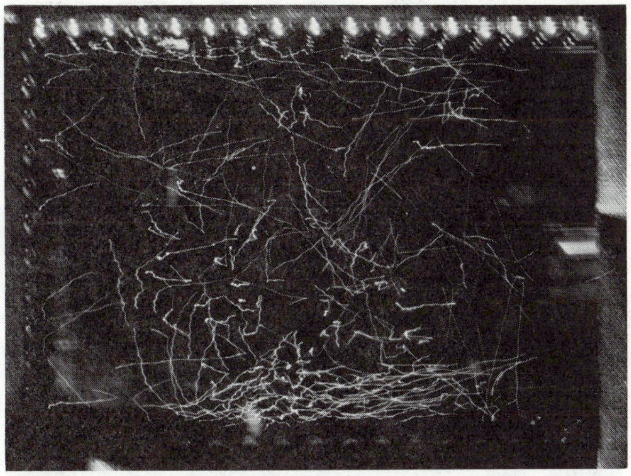

Fig. 27–55

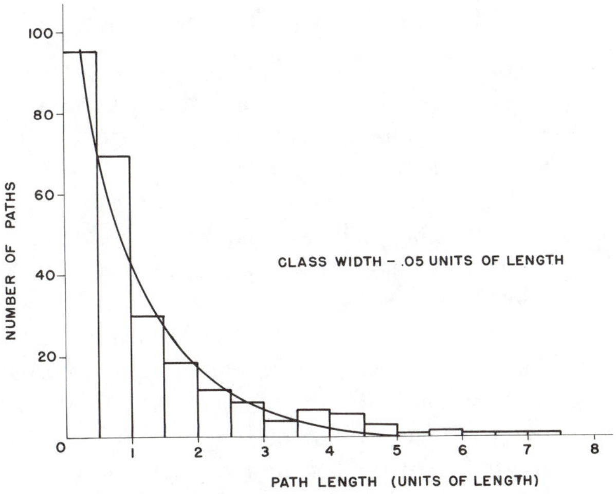

Fig. 27–56

The speed distribution was determined from the analysis of nine photographs taken at random with a Polaroid camera (set at 1/12 sec using type 47 film at a speed of 3000), and the spheres were in vigorous agitation under the stroboscopic illumination of a General Radio "Strobotac" lamp flashing 240 times per second. Measurements (in units of $\frac{1}{120}$ in.) of the distance traveled in the interim between stroboscopic flashes at 1/240 sec was used to determine the speeds of the spheres. Paths were measured well away from the enclosure walls where the spheres tend to congregate. Figure 27–54 is a histogram of the data obtained. The Maxwell-Boltzman distribution curve with the same most probable sphere velocity is shown for comparison. The limit of error of measurement in the most probable velocity is estimated to be about ± 6 percent. The free path analysis was obtained from a series of nine random photographs of the spheres under continuous illumination by a Graflex speed graphic camera set at 1/3 sec (Fig. 27–55). Only paths between two successive collisions were measured. Figure 27–56 is a histogram of the data together with a theoretical plot. Details of the apparatus are given in the Appendix, page 1294.

27–7.4 Gaseous Diffusion and Brownian Motion[30]

The process of gaseous diffusion in the mixing of two gases and Brownian motion can be illustrated with the apparatus shown in Fig. 27–57. Kinetic energy is supplied to the balls by a vibrating table. For diffusion, $\frac{1}{8}$-in.-diam plastic balls of two different colors are initially separated by a Lucite bar, and when the bar is removed they diffuse together. The random motion of a larger plastic sphere ($\frac{1}{4}$ in. diam), when subjected to continuous bombardment by the small spheres, is Brownian in the sense that the larger sphere moves, at any instant, under the action of an unbalanced force, since the number of balls striking various sides of the larger sphere at any instant is a matter of chance and may not be equal. The direction of this force is random, and its magnitude is obtained from the several impulses the larger sphere receives from the spheres which represent the gas. For best operation, the device should be carefully leveled to minimize the effect of gravitation.

Construction details are given in the Appendix, page 1295.

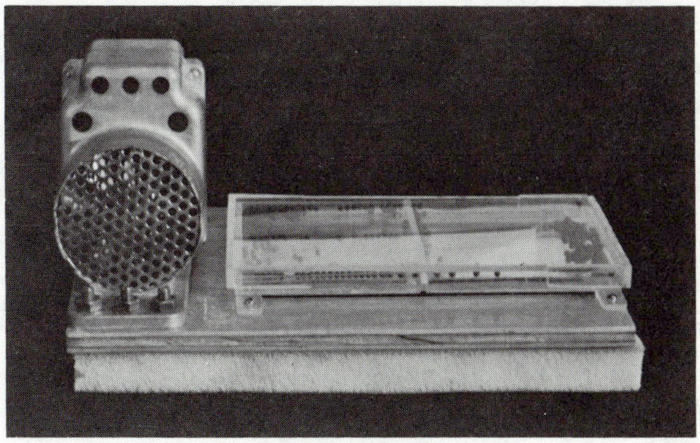

Fig. 27–57

[30] R. Resnick and H. F. Meiners, *Demonstration and Laboratory Apparatus Report of the 1960 Summer Visiting Professor Workshop at Rensselaer Polytechnic Institute.*

27–7.5 Kinetic Theory Demonstration

The apparatus in Fig. 27–58 shows the random motion of molecules of gas and the general behavior of a gaseous system with $\frac{1}{4}$-in. steel balls free to move in two dimensions within a thin rectangular enclosure. The balls are driven by a square steel rotor, and three sides of the chamber are studded with $\frac{3}{8}$-in. steel balls to randomize the motion of the free balls. In the photograph the end opposite the rotor has been removed. A fixed steel rotor bar, also removed in the photograph, absorbs the shock of the balls which sometimes are given impulses out of the plane of free motion. Except for this rotor bar, the top and bottom surfaces of the chamber are of $\frac{1}{4}$-in. plate glass, which allows projection of the motion on a screen.

When used in a vertical position, with the rotor at the bottom, the apparatus shows the general behavior of a gas (Fig. 27–59). A closed gaseous system with a movable boundary (piston) is illustrated by the use of a Styrofoam block weighted with imbedded steel balls (Fig. 27–60).

When used in an almost horizontal position, the

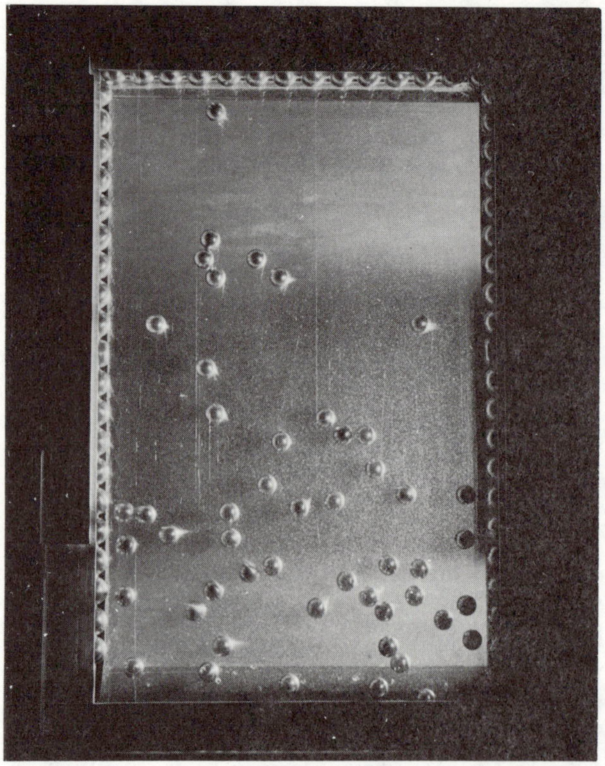

Fig. 27–59

Fig. 27–58

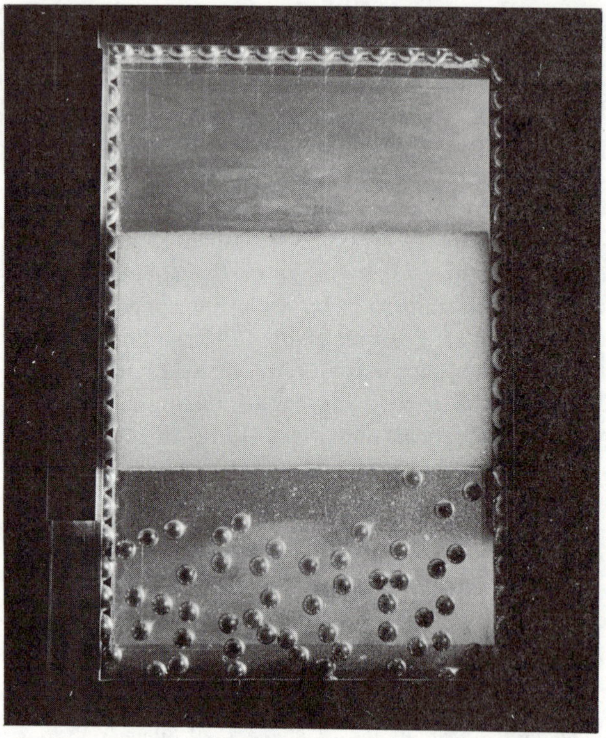

Fig. 27–60

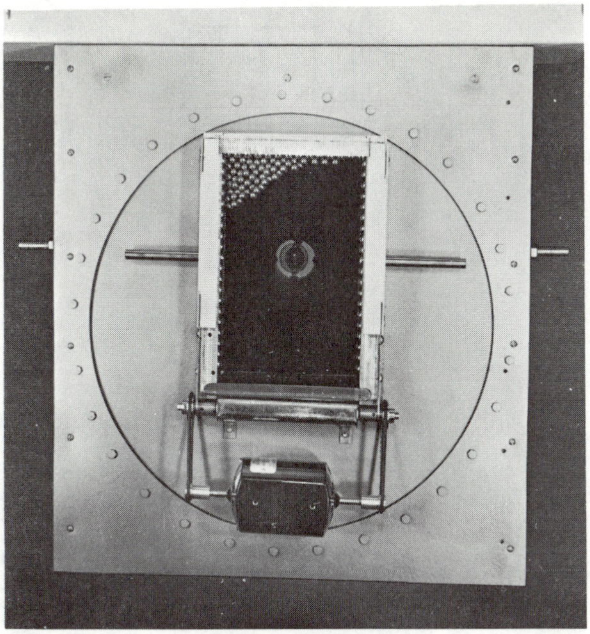

Fig. 27–61

apparatus shows Brownian motion. The rods inserted in the sides of the apparatus (Fig. 27–61) can be used in demonstrating diffusion. For this purpose, the apparatus is mounted in the same cart as that used for Demonstration 18–6.2, and the projection of motion is reflected by the mirror to the screen. A smoke particle for use in Brownian motion demonstrations consists of a cap riding on several steel balls.

For construction notes, see the Appendix, page 1295.

27–7.6 Brownian Movement Analog

Brownian movement, discovered in 1828,[31] is a direct consequence of molecular agitation in a gas or liquid. A simplified theory of this phenomenon,[32] first developed by Einstein,[33] is suitable for introductory atomic physics courses. An illustration of the phenomenon is appropriate in con-

Fig. 27–62

nection with simple kinetic theory derivations of the perfect gas law. The apparatus described here is used with an overhead projector.

A simple aluminum frame steel-ball model illustrating random molecular collisions in a gas is shown in Fig. 27–62 (top view) and Fig. 27–63 (bottom view). The ¼-in. steel balls are kept between two ¼-in.-thick glass plates and agitated by means of four rotating steel springs (coiled).

If an aluminum disk (1 in. diam, ¼ in. thick) is inserted in the space between the glass plates, it is moved around in an irregular, Brownian-movement-like fashion by the bombardment of the steel balls. The friction between the disk and the bottom

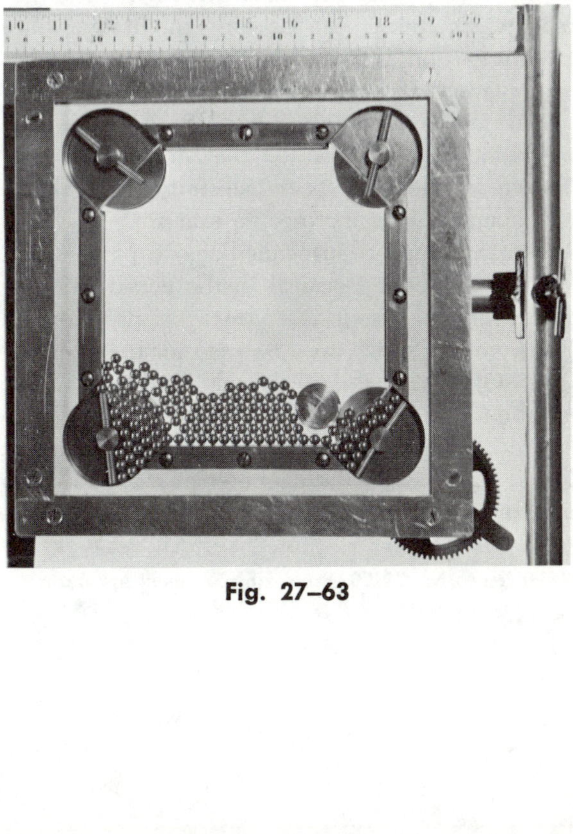

Fig. 27–63

(31) R. Brown, *Phil. Mag.*, **4**, 161 (1828).

(32) See G. P. Harnwell and J. J. Livingood, *Experimental Atomic Physics* (McGraw-Hill Book Co., New York, N.Y., 1933), 1st ed., pp. 92 ff., or E. H. Kennard, *Kinetic Theory of Gases* (McGraw-Hill Book Co., New York, N.Y., 1938), pp. 281 ff.

(33) A. Einstein, *Ann. Physik*, **17**, 549 (1905); **19**, 371 (1906).

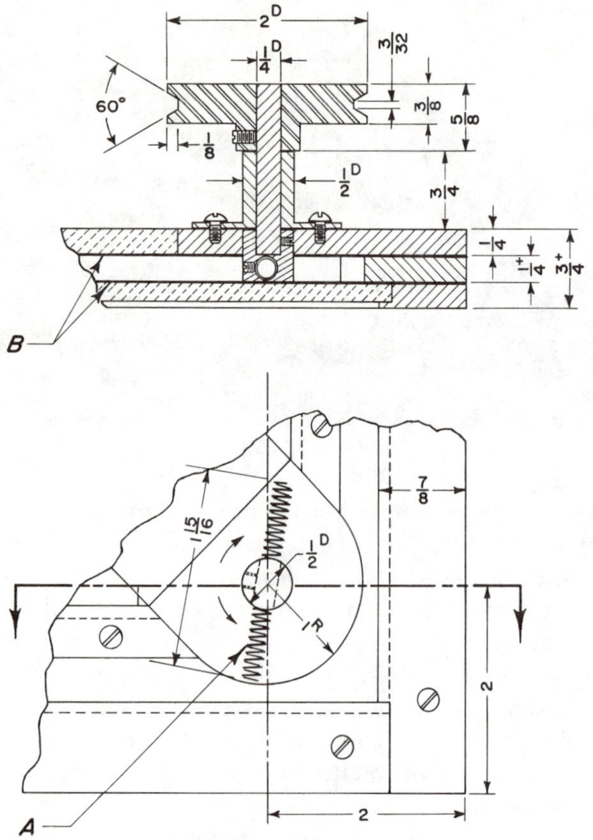

BOTTOM VIEW

Fig. 27–64

glass surface is the analog of the viscosity effect of the medium in the Einstein derivation. The analogy, of course, can be only qualitative since kinetic friction is not velocity-dependent in the same manner as viscous forces.

Half-minute demonstrations of the apparatus are sufficient to illustrate the "molecular" and "Brownian" motions. These illustrations should be accompanied with microscopic examination of real Brownian motion of suspended smoke particles.(34)

There is nothing critical in the construction of the apparatus, except that coiled springs *A* (Fig. 27–64, bottom view) must be used for the agitating arms. Otherwise, the steel balls get stuck in the agitation space. The spacing between the ¼-in.-thick glass plates *B* (Fig. 27–64) should be about 20 mils larger than the diameter of the steel balls and coiled springs and the thickness of the disk.

(34) Cenco No. 71270, Brownian movement apparatus.

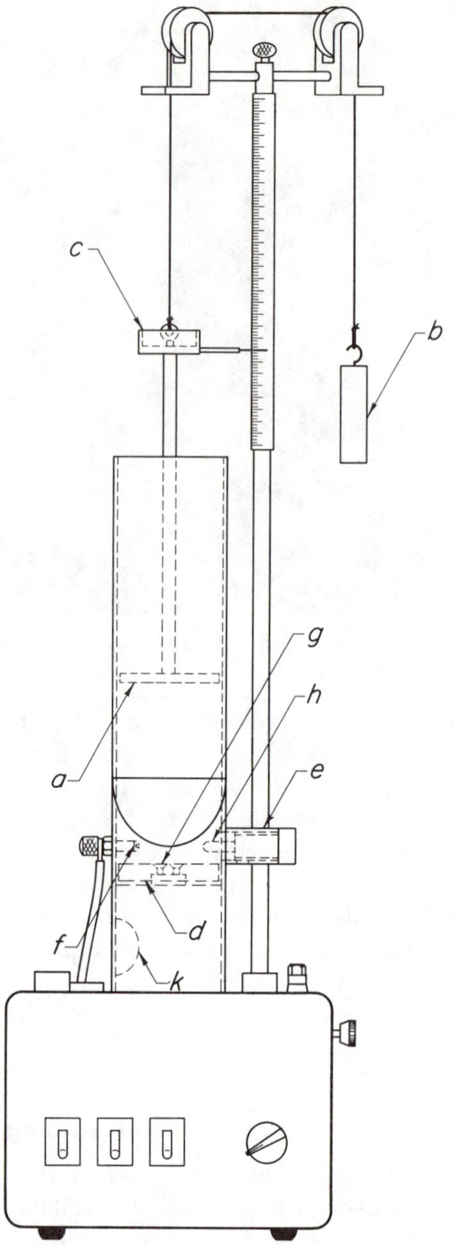

Fig. 27–65

27–7.7 Kinetic Theory Apparatus

The kinetic theory apparatus(35) used in the laboratory to study the gas laws and the kinetic theory of gases can be modified to allow shadow projection in a lecture demonstration, as shown in Fig. 27–65. The commercial apparatus consists of

(35) Central Scientific Company, Catalog No. 77722.

a round, clear plastic chamber containing $\frac{1}{8}$-in.-diam plastic balls which serve as the "molecules" of the plastic gas. The chamber is closed at the top by a movable piston *a*, which is counterbalanced by a weight *b* attached to the piston by a cord which passes over a pair of pulleys. On the piston is a weight holder *c* which carries the weights used to vary the pressure. The bottom of the chamber is closed by an impeller *d*, which is driven by a well-regulated electric motor with controllable speed. Attached to the chamber is a drain valve *e* which allows "molecules" to be bled off to change the concentration *n*. The commercial version also has a copper plate (not shown) attached to the piston and extending up through the chamber. A magnet, supported by a rod, is mounted with its poles on either side of the plate. This arrangement tends to cancel out the small oscillations of the piston due to the fact that the collisions of the balls with the piston are not quite random.

Because the plastic molecules moving at high speed become electrostatically charged when hitting each other, they will cohere to the plastic chamber and piston. To eliminate electrostatic charges, a high-voltage coil is connected between the electrode *f* and the impeller mounting screws *g* in the base of the impeller. Plastic diffusers *h* convert the rotary motion of the molecules into random motion. A window *k* in the side of the impeller-shaft housing allows the impeller speed to be measured with a Strobotac.

The plastic "molecules" are supplied kinetic energy by the rotation of the impeller, and the square of the impeller's angular speed is analogous to the temperature of a real gas. The "molecules" impinge on the piston and the walls of the chamber and exert a force on them. The pressure of the "gas" within the chamber is a function of the mass placed on the weight holder. The height of the piston above the impeller is proportional to the volume *V* of the chamber at a particular temperature *T* and pressure *P*.

Plots of *P* vs *V* (*T* and *n* constant), *P* vs *T* (*V* and *n* constant), *P* vs *n* (*T* and *V* constant), *V* vs *T* (*P* and *n* constant), and *V* vs *n* (*P* and *T* constant) can be made from the data taken with the apparatus. From the plot of volume vs the square of the impeller speed (*P* and *n* constant), a value for the diameter of the plastic balls can be found after excluded volume corrections are made. These corrections consider that a molecule can approach a wall only within a distance equal to its radius, and also that the centers of the two molecules can approach only to the diameter of one molecule. For greater accuracy a correction is made for the decrease in the concentration of molecules with height above the impeller. A very accurate value for the diameter of a molecule can be

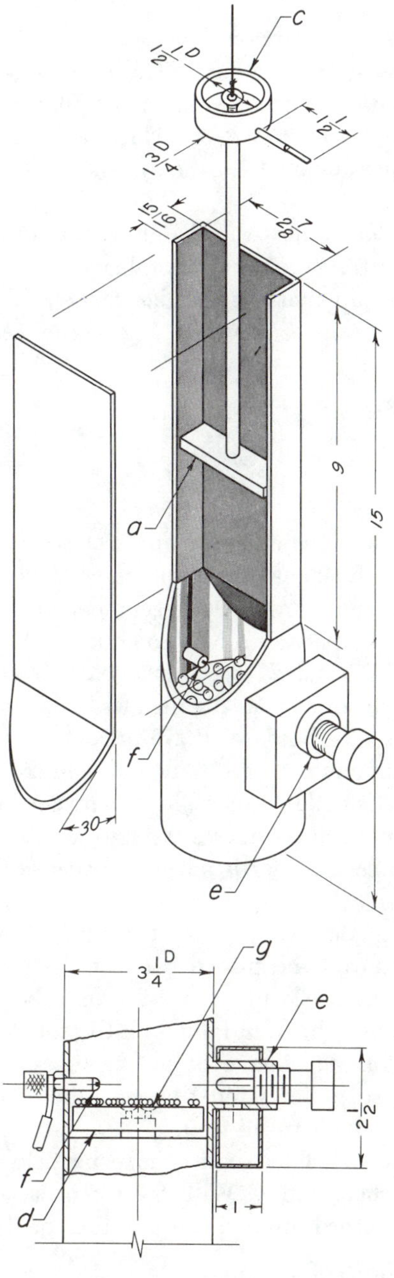

Fig. 27–66

obtained directly from the high-pressure region of the plot of pressure vs volume after corrections for excluded volume and decrease in concentration with height are made.

To obtain quantitative results with this apparatus, an expression is used which is similar to, but not the same as, van der Waal's equation, since the "plastic gas" is not a real gas. The expression is

$$(P + a/v^2)(v - b) = k,$$

where k is the random kinetic energy of the molecules in the "plastic gas." The term b is a correction for the effective volume of the plastic molecules, and the term a/v^2, which is obtained from direct measurements of the system, is added to the pressure.

In order to show the chamber by shadow projection in lecture, the round chamber which is supplied with the apparatus must be replaced by a rectangular chamber shown in Fig. 27–66. A cylinder of clear plastic, obtainable from most plastic suppliers, is cut off and the edges angled as shown in the figure. The sloping sides of the chamber must be high enough above the impeller to allow free movement of the balls, but not so high as to cut down the number of balls in the section being projected. The piston in the commercial apparatus is replaced by a rectangular piston which fits into the flat-sided part of the chamber. An arc lamp is used to project the motion of the plastic "molecules" in the rectangular chamber.

27–7.8 Equipartition of Energy Model

Two sizes of marbles are jostled by hand in a heavy tray of glass or metal to show the difference in velocities for the two sizes. A large plane mirror set at 45 deg is employed to allow the class to observe the behavior.

27–8 Brownian Motion and Mean Free Path

27–8.1

Brownian motion is shown on closed-circuit television to complement a discussion of kinetic theory. This method allows a large number of students to observe the phenomenon at one time. As shown in Figs. 27–67 and 27–68, smoke particles contained in a commercially available glass cell and illuminated by a carbon arc[36] are projected by a microscope directly onto the television camera's Vidicon tube. A simple convex lens between the light source and cell condenses the light on the particles. A microscope ocular is not used because the objective acts as a projection lens for the system. A 2-in.-long thin wall brass tube, machined to fit the lens turret and placed between the prism (over the microscope) and the Vidicon tube, prevents extraneous light from falling on the tube and interfering with the image of the particles. The light shield also protects the Vidicon tube, since direct light may be harmful to it.

The cell is filled with smoke particles from an extinguished match. With the room darkened, the lamp is turned on, and the positions of the convex lens and cell are adjusted to give maximum clarity and contrast of the smoke particles on the TV monitors. These adjustments are critical. Heat may cause excessive particle drift on the monitor screen. To correct this, the concentration of the $CuSO_4$ solution in the water cell is increased. The image can be improved by adjusting the camera controls according to the operating manual. A prism may or may not be used. If the remaining components are low enough, the camera can be angled vertically to sight directly into the microscope. See Appendix, page 1297, for materials list and demonstration setup.

27–8.2[37]

Figure 27–69 shows an optical system for exhibiting Brownian motion to a large audience via closed-circuit TV. The medium a is a water suspension of rutile (TiO_2), which is made by grinding it in a mortar, pouring the powder into water, and letting it stand for a few minutes until the larger particles have settled. If a few glass balls are added to the

(36) A point source of light may be substituted for the carbon arc lamp.

(37) A. Ehringhaus, *Die Naturwiss.*, **42** (1923) and **149** (1934); *Kolloid Zeitschrift*, XXXII, **19** (1923).

Fig. 27–67

storage vessel to allow complete mixing by shaking, the preparation remains usable for several years. The medium is viewed under dark field illumination which is provided by an arc lamp *b*, a 7.5-cm focal length condensing lens *c*, a water chamber *d* for cooling, a mirror *e*, and a dark-field condenser *f*. A drop of immersion oil is placed on the condenser, and a glass slide *g*, containing a drop of the rutile suspension, is placed directly over and in contact with the oil. The slide can be moved with the stage of the microscope. The microscope *h* consists of a 90× objective, which is dipped into the suspension, and a 6× ocular. A prism *i* deflects the final real image onto the photocathode of a TV camera tube *j*. No objective lens is used in the TV camera. In order to search for best illumination, the camera should be replaced by a screen and the final image focused on the screen. With this arrangement the entire width of the TV monitor screen corresponds to about 0.03 mm. In order to reduce the amount of stray light from the illuminating system a box can be built around the mirror, or a black cloth can be used. The apparatus is shown in Fig. 27–70.

27–8.3

The small polystyrene spheres in the monodisperse polystyrene lattices made by the Dow Chemical Co. are excellent for showing Brownian motion, and they can be used to illustrate this motion in a corridor demonstration or on closed-circuit television. Although most of the spheres appear separately in a suspension, there are also a few clumps of two, three, or even four spheres. These clumps clearly show both random rotational and translational motion, with the larger clumps undergoing lesser displacements than the smaller ones. The smaller displacements are due to the larger masses and moments of inertia, as well as to the statistical nature of the situation. Figure 27–71 is a photomicrograph of 0.80 μ spheres magnified 480 times. The clumping here is much more than average because the liquid was evaporated before the photograph was taken.

The concentration of the spheres in the available latex solution[38] is too great for use. To ob-

[38] Spheres are supplied by the Dow Chemical Co. in sizes of 0.25 μ, 0.52 μ, 0.80 μ, and 1.20 μ. These occur as the approximately 10% solid in a polystyrene latex.

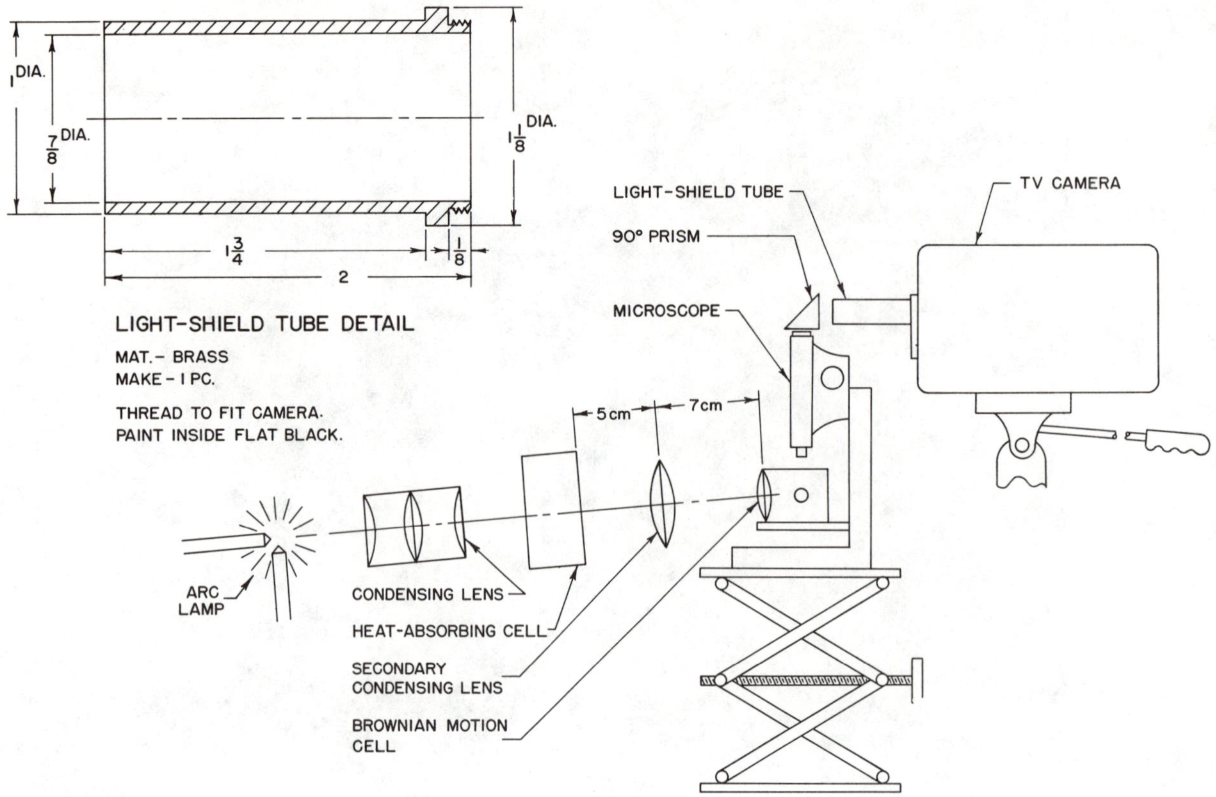

Fig. 27–68

tain a proper solution, one drop of the latex is suspended by shaking in 20 ml of distilled water. For use, one drop of this stock solution is suspended

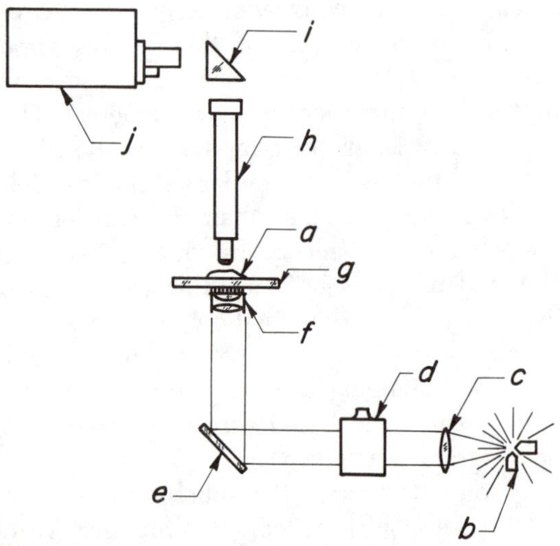

Fig. 27–69 Optical System for Closed-Circuit TV Projection of Brownian Motion.

in 25 ml or more of water. In order that the random movement be rapid, the particles in suspension should be as small as can be observed readily with a microscope. A magnification of 400× to 500× is enough to show the 0.80 μ particles as spheres, but it is not large enough for those with diameters of 0.52 μ. Higher magnifications require the use of an oil immersion objective. The Brownian movement of the particles can be seen easily in bright field, but dark field is even better. For the latter, a paraboloid condenser may be preferable to either the cardioid type or a dark field stop.

Samples of the dilute suspension can be conveniently stored in a depression or microculture slide, which is filled with the solution and sealed with a slide cover glass of the correct thickness for the microscope used. Some patience and manipulation are required to get the cover glass in place without leaving air bubbles in the liquid. Carefully dry the area around the cover glass before applying sealant. Colored fingernail lacquer is a satisfactory sealant if it is allowed to become tacky

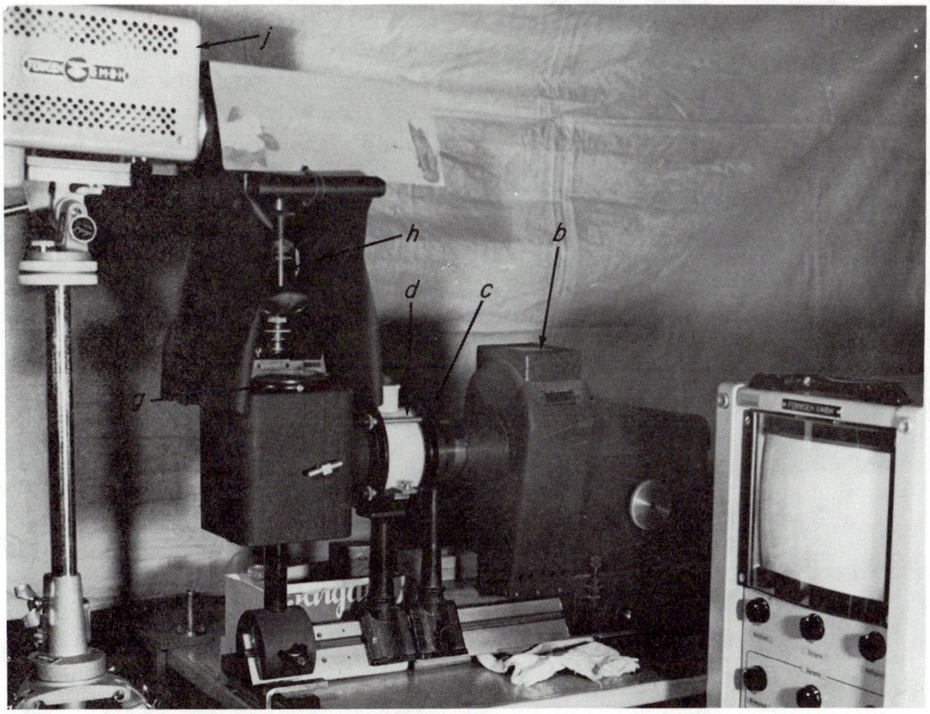

Fig. 27–70

before application.[39] Because of the color, it is easy to see if any of the sealant has "bled" under the cover glass far enough to contaminate the suspension. With care these depression slides should last for years.

27–8.4[40,41]

This corridor demonstration, contained in a reasonably student-proof box, permits quantitative observations for a computation of Avogadro's number.

Brownian movement of polystyrene latex spheres (1.3 μ diam) suspended in water can be seen by means of a 1900-power microscope with a 6-in. projection screen. Particle position can be read on the screen from a projected grid at fixed intervals indicated by an audible 5-sec signal. External focus adjustment allows the observer to keep a particular particle in view as it rises or falls in the water sample. Equipment for cooling the sample and temperature measurement is contained in the microscope console.

(39) A commercial sealant is Permount, Catalog No. 12–568, Fisher Scientific Co., Pittsburgh, Pennsylvania.

(40) *Am. J. Phys.*, **32**, 7, v–vi (1964).

(41) Prototype built for the Science Teaching Center (Massachusetts Institute of Technology) by Courtland Randall at Technical Marketing Associates, Inc., Concord, Massachusetts. Dr. James Burkhardt at the Science Teaching Center was in charge of development.

27–8.5

The formation of flat-sided crystals of lead carbonate in a glass cell is projected on a screen. A demonstration similar to this is described by R. M. Sutton,[42] but the proportions of the constituents used here are somewhat different. To make the cell solution, 1 g of lead acetate crystals is dissolved in 150 cc of water. The same thing is then done with 1 g of potassium carbonate crystals. A cubic centimeter of each of these solutions is added to 300 cc of water, first adding the potassium carbonate and then the lead acetate solution with the cell in a lantern slide projector. Care must be taken to add the lead acetate very slowly while

(42) R. M. Sutton, *Demonstration Experiments in Physics* (McGraw-Hill Book Company, Inc., New York, 1938), No. A-50, p. 463.

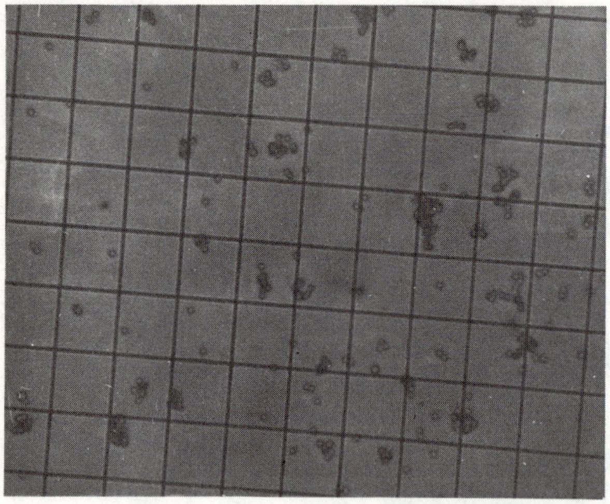

Fig. 27–71

stirring. A suitable cell is 11.5 x 11 x 3 cm made of Pyrex glass.[43]

27–8.6 Mean Free Path

The concept of mean free path is demonstrated quantitatively with steel balls rolling across the face of a pinboard elevated at one end. Using the pinboard as a simple two-dimensional analogy of a gas, the lecturer can show the relationship of mean free path to molecular size and the density of matter.

[43] Cenco No. 87790 or equivalent.

A convenient size for the pinboard is 24 x 36 in. Ninety-six holes are arranged in 12 rows and 8 columns. Eight pins, one in each column, are inserted randomly across the board. The pins are ⅛ in. in diameter and the steel balls are 1⅛ in. in diameter.

The following equation is verified experimentally:

$$\lambda = \frac{A}{dN} = \frac{(24)(36)}{(1\frac{1}{4})(8)} = 86.4 \text{ in.},$$

where λ is the mean free path, A the area of the pinboard, d the effective cross section of the pins and balls ("molecular size"), and N the number of pins ("density of matter"). The effective cross section of a pin and a ball is the sum of their diameters, 1¼ in.

The theoretical value for λ is shown to be accurate in the following way:

1. $\text{Probability of a collision} = \frac{\text{Total effective cross section}}{\text{Total width}} = \frac{10}{24} = \frac{5}{12}$
2. For 12 balls, the probability is 12 times as great: $12\,(5/12) = 5$ collisions.
3. Since 12 balls rolling through the 26-in. length of the pinboard experience 5 collisions, the distance one ball will roll before a collision (i.e., its mean free path) will be $\frac{12 \times 36}{5} = 86.4 \text{ in.} = \lambda.$

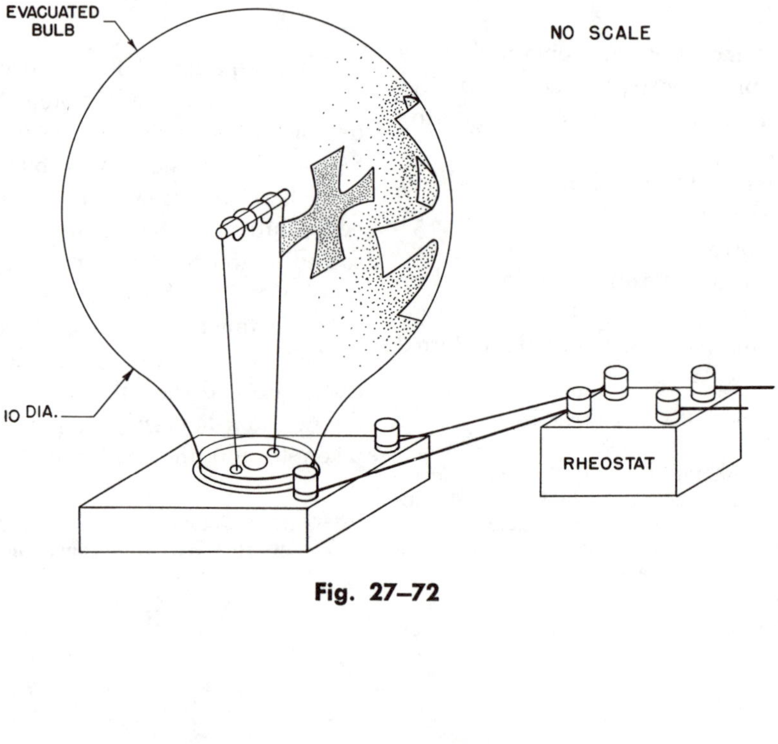

Fig. 27–72

The mean free path can be repeatedly verified if a blindfolded student starts a ball through the board a number of times, and the number of collisions is counted. The length of the mean free path is reduced by (1) increasing the "density of the gas" (using more pins) or (2) increasing the "molecular size" (using larger steel balls).

27–8.7

The long, straight line trajectories of atoms at low pressure are shown by using a glass bulb containing a tungsten filament and a Maltese cross. The nickel cross is mounted between the filament and the inner wall of the bulb by a support attached to the base of the bulb (not shown). A small piece of aluminum foil is inserted in the filament as shown in Fig. 27–72. The bulb is placed on a vacuum pump[44] and evacuated to a pressure less than 10^{-5} mm Hg. At this point the aluminum atoms will travel to the glass wall when the filament is brought up to a temperature causing the aluminum to melt and evaporate. The "shadow" of the Maltese cross is thus formed on the inner surface of the bulb.

27–9 Diffusion

27–9.1 Speed of Molecule Demonstration and a Rough Estimate of Avogadro's Number

Bromine vapor is allowed to diffuse within two large, tall glass tubes, one evacuated and one filled with air. A translucent screen illuminated from behind makes the progress of the bromine vapor easy to see.

The slow diffusion of bromine in air is first shown, and the lecturer suggests that its motion is obstructed by air molecules. Then the very fast diffusion of bromine in vacuum is shown as a qualitative demonstration of the high speed of molecules. The diffusion of bromine in air (for ~500 sec) is then repeated, and macroscopic measurements of the time and the extent of diffusion are made; to obtain the measurements a centimeter scale is placed beside the tube and the audience decides where the vapor seems "half brown." Individual votes usually range around 10 cm, with some at 9 and 11 and a few outside that range. From the votes a rough estimate of the mean free path of bromine molecules in air is made, and the mean free path of air molecules in air is assumed to be approximately the same.

Using kinetic theory and a knowledge of the densities of air as a gas and as a liquid, one can make an estimate of the diameter of an air molecule. Then an estimate of Avogadro's number is obtained. This is a rough order of magnitude estimate, not one within 5 to 10 percent range of accuracy, both because of simplifications in kinetic theory treatment and because the estimate made by the audience affects the result strongly.

Bromine is now available in small glass 1-cc ampules.[45] The ampule is easily broken to release the bromine. (Other methods of providing a small amount of bromine to be released in the main tube are quite dangerous because they involve the handling of liquid bromine from an open bottle. The ampules are much safer.)

WARNING: Bromine is a dangerous material that attacks almost everything except glass and paraffin wax. It is best to do the cleaning out of doors.

Each of the main diffusion tubes (Fig. 27–73) has a neck at the side near the bottom which leads to a glass stopcock of wide bore (e.g., 8 mm, preferably spring held). A glass "cap tube" (½ in. i.d.) to hold the ampule, closed at the far end, is connected to the stopcock by a short piece of

[44] See Movable Vacuum Pump, Demonstration 16–6.1.

[45] Ampules may be obtained from Matheson, Coleman & Bell, East Rutherford, New Jersey.

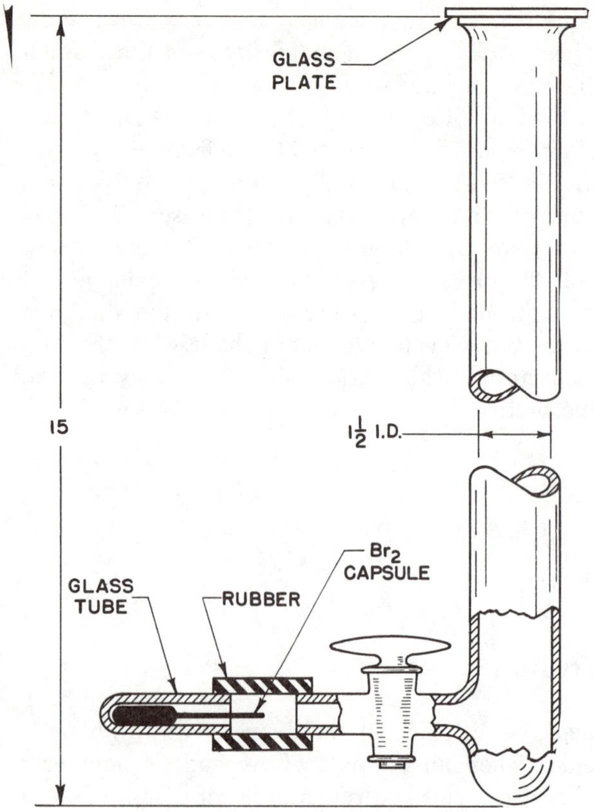

Fig. 27–73

Fig. 27–74

rubber tube with a fairly thin wall. The ampule is placed in the cap tube. To release the bromine, the lecturer tilts the cap tube and taps it so that the ampule slides in the rubber tube. The ampule is then crushed inside the rubber tube with a pair of pliers. The stopcock is kept turned off while the bromine is being released. Then the stopcock is opened to admit a small quantity of liquid bromine to the bottom of the main tube.

When the diffusion is to be shown in a vacuum, the main tube is first evacuated by attaching the stopcock to the pressure tubing from a pump instead of the cap tube and ampule (Fig. 27–74). Then the stopcock is turned off, the tube from the pump removed, and the cap tube and ampule are attached with rubber tubing as before (Fig. 27–75). The ampule is crushed; then the stopcock is opened (Figs. 27–76, 27–77, and 27–78). Although the thin rubber tube collapses, the demonstration is not affected.

The apparatus is made safe for cleaning by

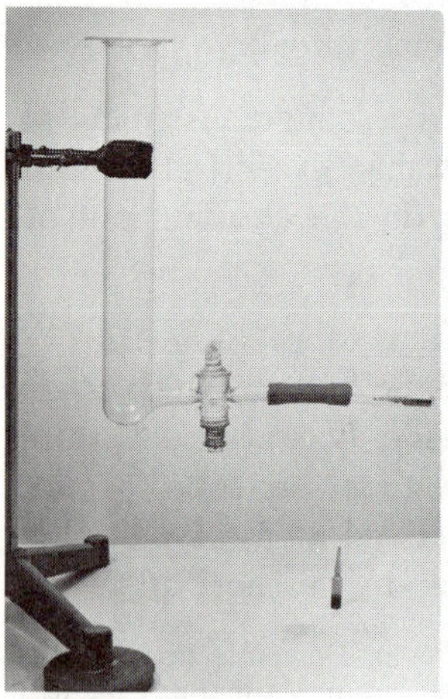

Fig. 27–75

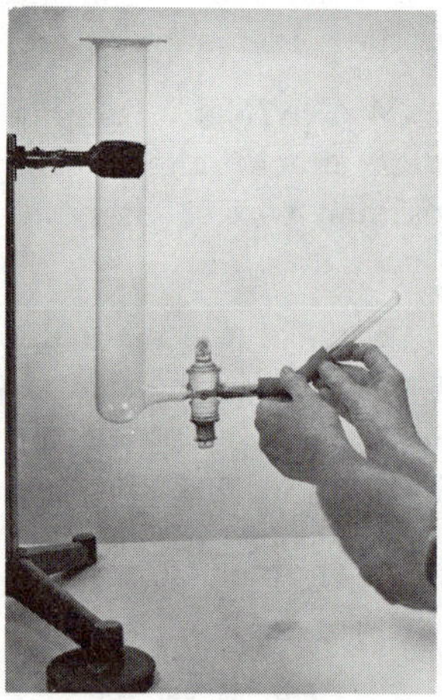

Fig. 27–76

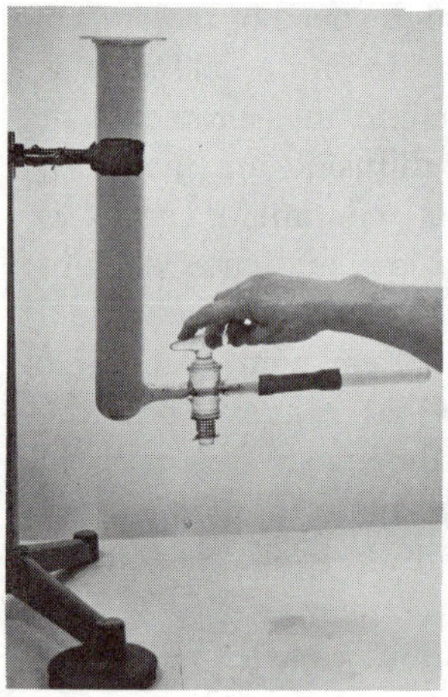

Fig. 27–77

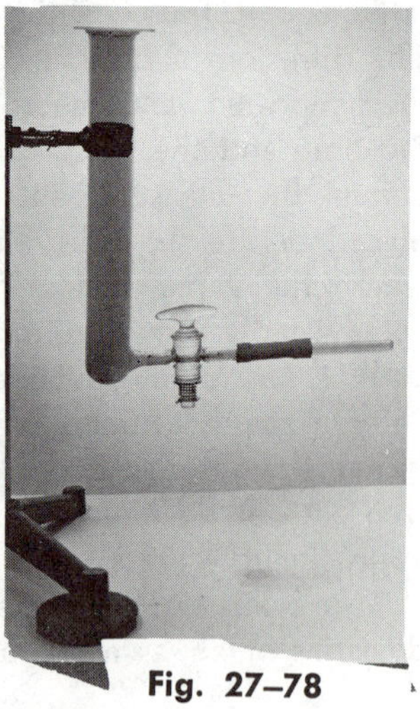

Fig. 27–78

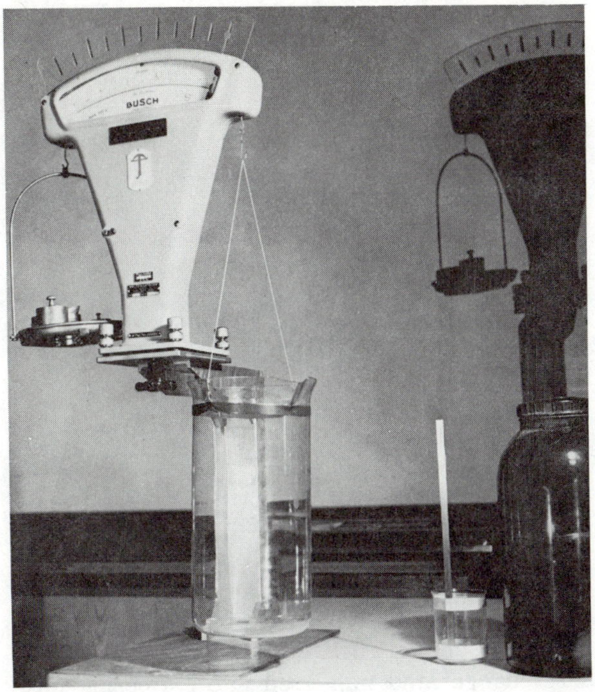

Fig. 27–79

being immersed in a strong ammonia solution—which should also be available during the experiment in case of accidents. (A 0.880 ammonia solution diluted with an equal volume of water is suitable.) A solution of photographic hypo is not as safe.

27–9.2 Apparent Surface of Gas

A balance is mounted on a rack high enough so that a 1-gal beaker can be hung from it as shown in Fig. 27–79. After the arrangement has been balanced, ether vapor is poured from a wide-mouth 1-gal bottle (containing about 100 cc of ether vapor and a little cotton wool) into the beaker. *Caution:* Care must be taken, since ether is very flammable.

The ether vapor is visible in shadow projection and behaves like a liquid. The heavy gas in the beaker is registered on the scale. This liquid-like behavior is demonstrated by the formation of a nonequilibrium apparent surface before diffusion sets in.

The scale is modified for shadow projection by elongating the indicator and mounting a Plexiglas scale on the top. Gases lighter than air can be weighed by hanging the beaker upside down. A glass bulb with a stopcock can be used for weighing the air. First it is weighed with air, then reweighed after it has been evacuated.

27–10 Liquefaction of a Gas

27–10.1

A conical container made from very thin copper sheeting is supported by a Styrofoam ring mounted on a ring stand. When the cone is filled with liquid nitrogen, oxygen from the atmosphere will condense on the cone's outer surface and drip from the tip into a receiving flask placed beneath it. The presence of oxygen in the flask can be demonstrated by placing a burning splint over the mouth of the flask.

27–10.2

The difference in behavior between a gas (air) and a vapor (sulfur dioxide) under compression is shown. Equal pressures are applied to the two samples by dual mercury pistons connected to a mercury reservoir to which air pressure can be applied. Beginning with equal volumes, an increase of 1 atm compresses each of half the original volume. The compressibility of SO_2 seems to coincide with that of air, which follows Boyle's law. As pressure increases, the SO_2 continues to match the air sample until about 2½ atm, when SO_2 occupies a somewhat smaller volume. At still higher pressures, SO_2 suddenly collapses to a drop of liquid (invisibly small). The pressure needed for the liquefaction of the SO_2 depends on the temperature. If the available pressure is not sufficient to liquefy the SO_2 at room temperature, the apparatus may be enclosed in a Lucite box with chips of Dry Ice in the bottom and a current of air fed through the box around the apparatus; or cold (gaseous) air may be allowed to pour down over the apparatus from an open bottle of liquid air.

The apparatus (Fig. 27–80) consists of a glass U-tube containing air in one arm and SO_2 in the

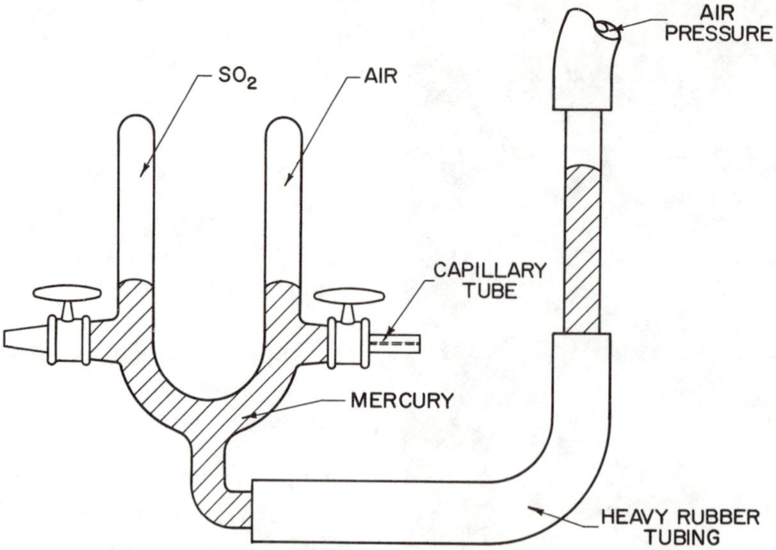

Fig. 27–80

other, but under equal pressures provided by a common line containing mercury. Each arm has a stopcock for the introduction of its gas. A rubber hose connects the center of the bottom of the U-tube to a reservoir tube of mercury, which is attached to a compressed air line. The demonstration is shown by lantern projection, and the physical size of the apparatus is determined by the lantern used.

The following procedure should be followed in filling the U-tube: (1) Pump out both sides to a good vacuum. (2) Introduce SO_2 on one side, air on the other—*slowly*—keeping them matched.

It would be desirable to be able to observe the condensed SO_2 liquid, but the vapor capacity of the system would not yield an observable quantity of the liquid.

27–10.3

The Joule-Thomson effect and an easily constructed heat exchanger are used to liquefy oxygen. The system consists of a 40-liter tank of oxygen, an intermediate cooler, the heat exchanger, a needle valve, and an unsilvered Dewar flask. The complete heat exchanger is shown in Fig. 27–81. See the Appendix, page 1297, for construction details and testing procedure.

To fill the spaces caused by the uneven wall of the flask, a cloth is wrapped around the exchanger before it is placed in the flask. The intermediate cooler is connected between the oxygen tank and the exchanger and immersed into a freezing mixture of ice and rock salt. The tank valve should be

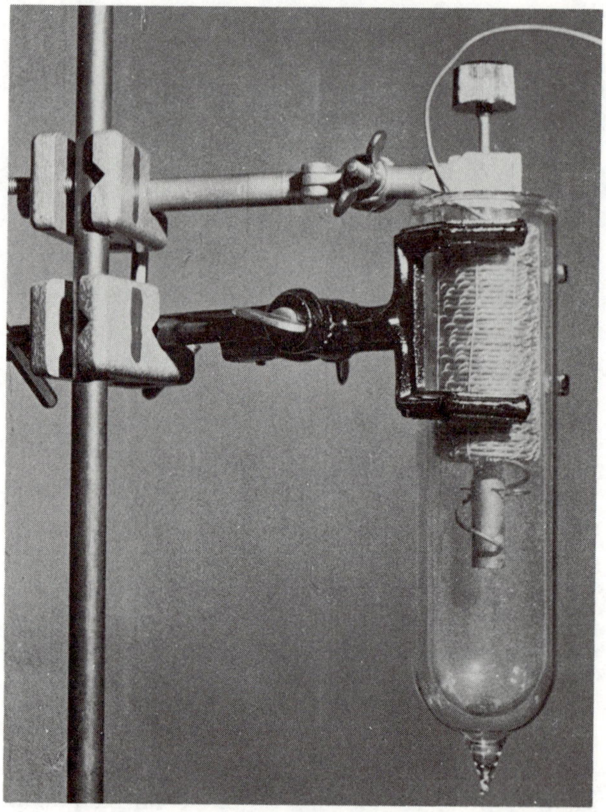

Fig. 27–81

completely open to avoid throttling at this point, and the needle valve is permanently open to avoid its becoming blocked. With an initial pressure of 130 atm, a jet of liquid oxygen from the needle valve is visible after about 5 min. This jet can be projected, using a $CuSO_4$ solution in a flat-walled container as a heat filter. Within about 10 min, a few cc of liquid oxygen may be seen at the bottom of the flask. During one demonstration, the pressure in the tank drops about 10 atm.

Demonstrations	***Contributors***
27–2.1, 27–2.2	Harvard Project Physics
27–2.3	E. G. Ebbighausen, University of Oregon
27–2.4	Swiss Federal Institute (ETH), Zürich, Switzerland
27–2.5	Stanford University
27–2.6	M. J. Pryor, State University of New York at Albany
27–2.7	University of Minnesota
27–2.8	University of Illinois
27–2.9	Swiss Federal Institute (ETH), Zürich, Switzerland
27–3.1	California Institute of Technology
27–3.2 to 27–3.4	R. M. Sutton, California Institute of Technology
27–3.5	R. Katz, Kansas State University
27–3.6	H. R. Dursch, Skagit Valley College
27–3.7	Swiss Federal Institute (ETH), Zürich, Switzerland
27–3.8	N. W. Goldsmith, State University of New York at Oswego
27–3.9	J. Satterly, University of Toronto
27–3.10	L. Grillet and M.-T. Grillet-Amy, University of Rennes, France
27–3.11	H. A. Robinson, Adelphi University
27–4.1	University of Minnesota
27–4.2	F. Oppenheimer, University of Colorado
27–5.1	Swiss Federal Institute (ETH), Zürich, Switzerland
27–5.2	H. A. Robinson, Adelphi University
27–6.1	University of Minnesota
27–6.2	Swiss Federal Institute (ETH), Zürich, Switzerland
27–6.3	Duke University
27–6.4	Swiss Federal Institute (ETH), Zürich, Switzerland
27–6.5	H. F. Meiners and F. Stiteler, Rensselaer Polytechnic Institute
27–6.6	B. L. Donnally and H. C. Jensen, Lake Forest College
27–7.1	D. Turnbull, Harvard University, and R. L. Cormia, General Electric Research Laboratory, Schenectady, New York
27–7.2	P. Huber, University of Basel, Basel, Switzerland
27–7.3, 27–7.4	E. O. Cook, State Teachers College, Frostburg, Md.
27–7.5	J. T. Brooks, Washington University
27–7.6	W. E. Meyerhof, Stanford University
27–7.7	H. F. Meiners, Rensselaer Polytechnic Institute
27–7.8	Eric M. Rogers, Princeton University
27–8.1	P. Whiting and J. Carr, University of Minnesota
27–8.2	Swiss Federal Institute (ETH), Zürich, Switzerland
27–8.3	M. R. Wehr, Drexel Institute of Technology
27–8.4	J. G. King, Massachusetts Institute of Technology
27–8.5	Massachusetts Institute of Technology
27–8.6	E. M. Lyman, University of Illinois
27–8.7	G. D. Freier, University of Minnesota
27–9.1	H. D. Smyth and Eric M. Rogers, Princeton University
27–9.2, 27–10.1	Swiss Federal Institute (ETH), Zürich, Switzerland
27–10.2	Eric M. Rogers and H. M. Waage, Princeton University
27–10.3	H. Werheit and F. Günneberg, University of Köln, West Germany

28

LOW-TEMPERATURE PHENOMENA

28–1 Basic Apparatus

A number of demonstrations[1] of very-low-temperature phenomena and of the unusual properties of liquid helium can be made with the apparatus shown in Fig. 28–1. Since the complete assembly is quite complex and the requirements very specific, a detailed parts list and construction details are given in the Appendix, page 1305. The Appendix also contains a detailed sequence of techniques and procedures for the handling of liquid helium.

It is no longer necessary for a physics department to have its own cryogenic facilities. Liquid helium is now available commercially in rented storage containers at a reasonable price per liter. Liquid nitrogen is also generally available from welding suppliers at a cost much less than that of liquid helium.

Reduced to its barest fundamentals, the apparatus consists of (1) a transparent low-temperature chamber in which the experiments are performed, (2) a source of liquid helium with a convenient means of transfer, (3) a vacuum system that includes appropriate pump, valves, and pressure gauge, and (4) suitable means for introducing and manipulating the components inside the vacuum chamber. The latter is an unsilvered Dewar flask, shown in Fig. 28–2. It is a commercial double-Dewar of single-unit construction. The flask has a special neck to accommodate gasketed flanges which provide for a cover containing ports for evacuating the Dewar, for filling it with helium, for attaching the pressure gauge, and for manipulating the experimental equipment inside the chamber. Specific details and instructions are given in the Appendix, page 1300.

The apparatus can be used to show the "floating magnet" and the "persistent current." Both of these experiments illustrate the phenomenon of superconductivity.[2] In addition, the equipment can be

[1] The demonstrations using the transparent low-temperature chamber, 28–2.1 through 28–2.6, were contributed by A. Leitner and R. Au of Michigan State University. These experiments were verified and extended in the physics work shop at Rensselaer Polytechnic Institute during the summer of 1964. They are now part of the regular lecture sequence at Rensselaer, as well as at Michigan State University. Experience has shown these experiments to be well adapted to projection techniques and to presentation by closed-circuit television. See also G. Salinger and A. Leitner, *Am. J. Phys.*, **34**, 692 (1966).

[2] Alfred Leitner has produced two excellent 16-mm sound films on liquid helium which are available for purchase or rental from the Audio Visual Center of Michigan State University. Their titles are *Liquid Helium II, the Superfluid,* and *Introduction to Superconductivity.*

used to perform other experiments which demonstrate the extraordinary properties of liquid superfluid helium II, such as "superleak," the "fountain effect," and the "Rollin creeping film." This list is not exhaustive, and still other demonstrations can be developed by the ingenious experimenter. (Note: A judicious choice of lecture scheduling is necessary because of the expense of liquid helium and the preparation time involved. A two-hour session with an intermission would be quite advantageous. The intermission is used for a second transfer of liquid helium.)

The Dewar is prepared and precooled before the lecture. All the required apparatus should be already hung from three nickel-silver tubes in the movable seals of the vacuum lid. The first part of the demonstration includes the transfer of liquid helium and the experiments performed at atmospheric pressure.

In the following descriptions, *the symbols in parentheses refer to details in the parts list in the Appendix.*

28–2 Demonstrations with Liquid Helium

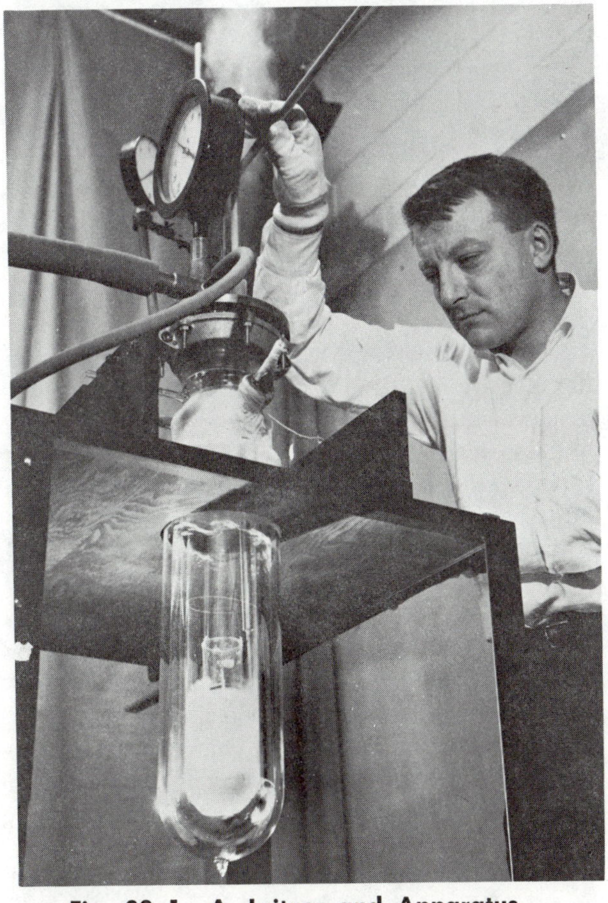

Fig. 28–1 A. Leitner and Apparatus.

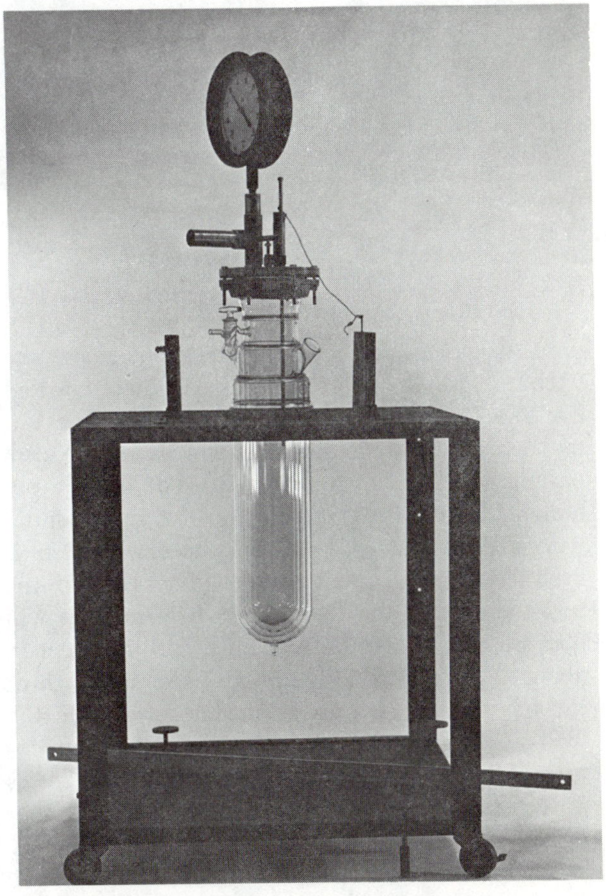

Fig. 28–2 Unsilvered Glass Double Dewar on Mounting Cart. The cart can stand on casters, or fixed-bolt-type legs, or on the bottom board if casters and legs are taken off. Vacuum lid is mounted on the Dewar with pipe flange.

28–2.1 The Floating Magnet Experiment

A Styrofoam hemisphere is fitted to the inside bottom of the Dewar. Next, a lead disk (10c, page 1305) is laid on the Styrofoam with its concave face

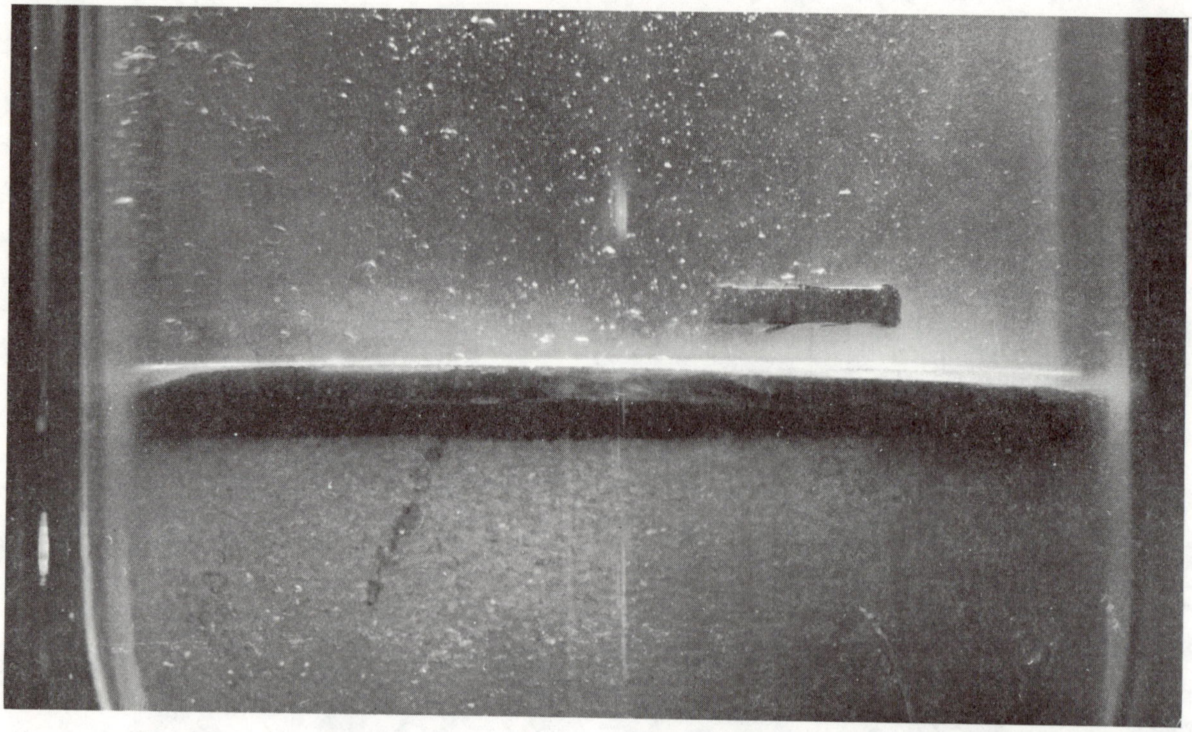

Fig. 28–3 Bar Magnet Floating Above a Lead Plate in a bath of liquid helium at 4.2°K.

upward and leveled. A short length of glass tube (10b) (with fire-polished ends) is then lowered into the Dewar and onto the disk. Its purpose is to keep the floating magnet away from the disk periphery, where it would tend to slide down a potential hill. A stirrup (10e) made of light copper wire and suspended by a fine necklace chain is used to lower the Alnico magnet (10a) through the transfer port in the Dewar lid. The magnet will float off just above the lead disk and will remain about ¼ in. over it (Fig. 28–3). The approach of the bar magnet sets up persistent currents in the lead disk, which is superconducting at the temperature of liquid helium. In accordance with Lenz's law, these currents set up a magnetic field that opposes the approach of the magnet.

28–2.2 The Persistent Current

The coil assembly (11) is shown in Fig. 28–4. It is attached to a suspension tube. The leads pass through this tube to the niobium coil (11e), which will remain superconducting in liquid helium at 4.2°K for currents up to about 5 A. The maximum flux density of the coil is about 360 G/A. A short length of tantalum wire (11f) is shunted across the niobium leads to the coil. The wires must be in superconducting contact (Appendix) with each other below the transition temperature for tantalum, which is about 4.3°K (just a little above the normal boiling point of helium). The external circuit is shown in Fig. 28–5. The tantalum shunt acts as a switch, that is, as the shunt is moved above or below the liquid helium surface, it turns the persistent current off or on. The demonstration can be done in two different ways. One way is to submerge the coil in the bath, leaving the tantalum shunt above the surface. Connect to the battery, and adjust the current to one or two amperes. The niobium coil is now superconducting, but the shunt is not. When the shunt is submerged and the battery subsequently removed, the current persists without the applied emf, as shown by the magnetic action of the coil on a nearby magnet or other magnetic material. A second method of inducing a persistent current in the coil requires no external

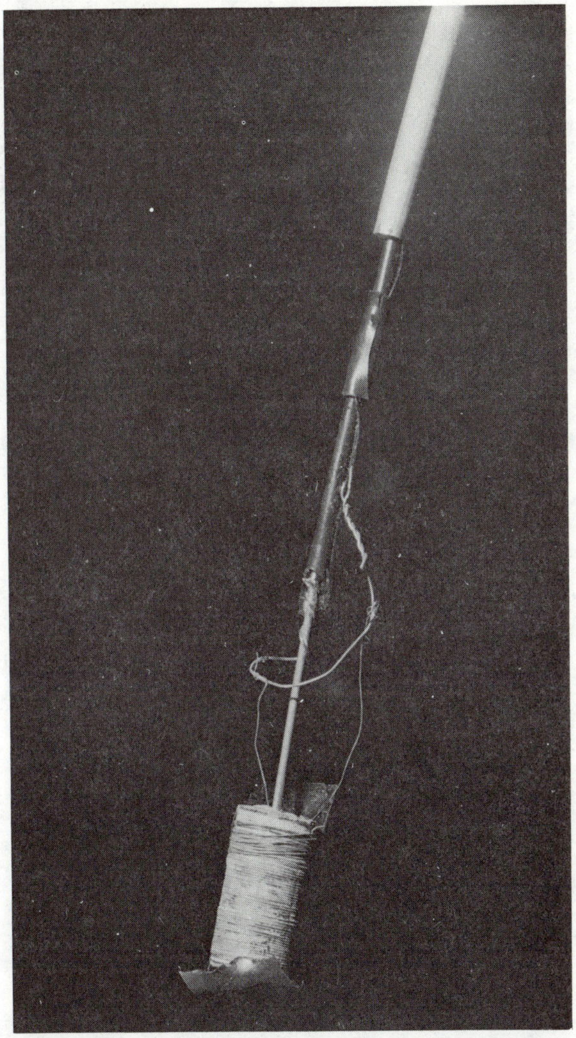

Fig. 28–4 **Small 10-Mil Niobium Wire Coil** with tantalum superconducting shunt. Mounted with thin brass rod on a piece of nickel-silver tubing. Copper leads through this tube allow electrical access to the coil from outside the Dewar.

leads. The coil is immersed, shunt and all, into the liquid helium, which is in the field of a strong external magnet. Care must be taken not to exceed the critical fields of tantalum and niobium. The current produced by removing the magnet will be observed to persist. In either method the coil's persistent flux will be sufficient to pick up iron filings. It will also attract the (already floating) magnet. When the shunt is lifted above the bath surface, the current decays, and the iron filings or the magnet is released.

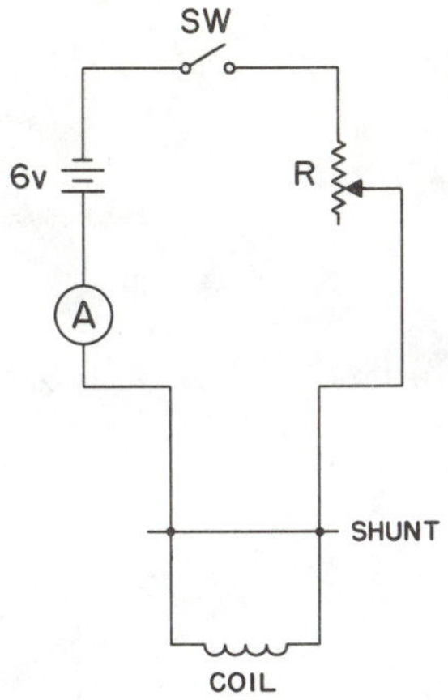

Fig. 28–5

28–2.3 λ-Point Transition

The next four demonstrations are designed to show differences in the physical behavior of liquid helium I and II. Since the transition point between helium I and helium II is at 2.17°K (the λ-point), it is necessary to cool a sizable quantity of helium I by evaporation until it reaches the transition temperature. This transition can be shown as follows.

The Dewar is sealed, the pump turned on, and the valves opened slowly (needle valve 4a first). As the liquid cools, boiling will be visible, exhibiting larger bubbles than at 4.2°K. Because of its low thermal capacity, at least one-third of the liquid helium will be pumped away. The boiling becomes violent just before reaching the λ-point and suddenly stops as the λ-point is passed. This is clearly visible on a projection screen.

28–2.4 Superleak

This experiment is begun with liquid helium below the λ-point. The valves are now closed. The heat of the projector (and/or a heat lamp) will soon raise the bath temperature slightly above the λ-point. The modified Buechner funnel (12) is now partly filled with liquid helium and immediately raised; no leakage through its fritted disk can be observed, Fig. 28–6 (a). The valves are now reopened rapidly. The transition will be clearly visible in the funnel. Following this, a rapid leakage of liquid helium through the fritted disk is observed, Fig. 28–6 (b).

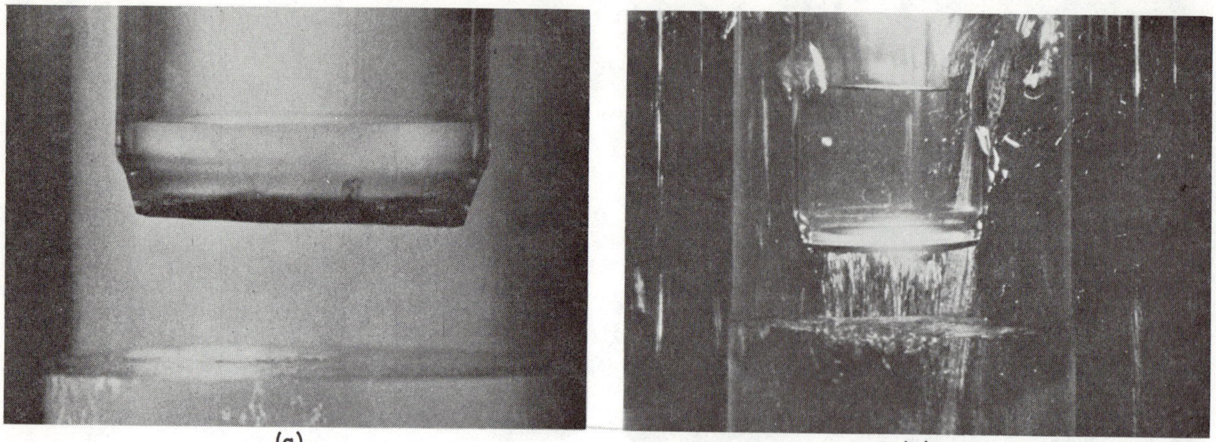

(a) (b)

Fig. 28–6 (a) Above λ Point; (b) Below λ Point.

28–2.5 The Fountain Effect

The static fountain effect accessory (13), *a* in Fig. 28–7 (a), and the dynamic fountain effect accessory (14) (*b* in the figure) are used to demonstrate the fountain effect, first statically, then dynamically. The liquid helium level will rise into the filter tube (13), Fig. 28–7 (b), and will spout from the capillary as a fountain (14) (Fig. 28–8), if an unfiltered light beam is focused on their respective rouge-filled

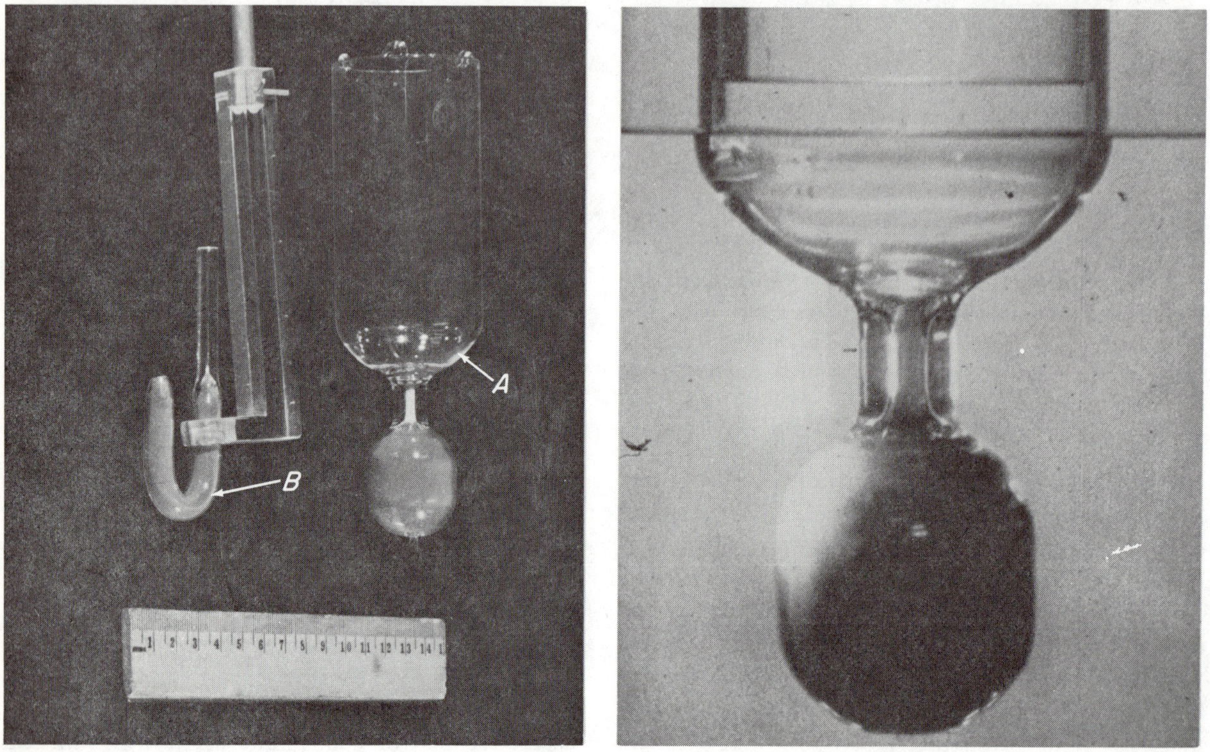

(a) (b)

Fig. 28–7 (a) "Static" Fountain. (b) "Dynamic" Fountain in Lucite holder, which is fixed by cotter pin to a tube.

Fig. 28–8 **Dynamic Fountain.**

portions. The temperature gradient produced by the light determines the heights of rise. The light beam from the projector may be sufficient to produce these effects.

28–2.6 The Rollin Creeping Film

A shallow glass dish made from the bottom of a test tube (15) is partly filled with liquid helium below the λ-point. The film rises along the inside surface and "creeps" over the lip and down along the outside, collecting in drops at the bottom, Fig. 28-9. The thickness of this creeping film is only

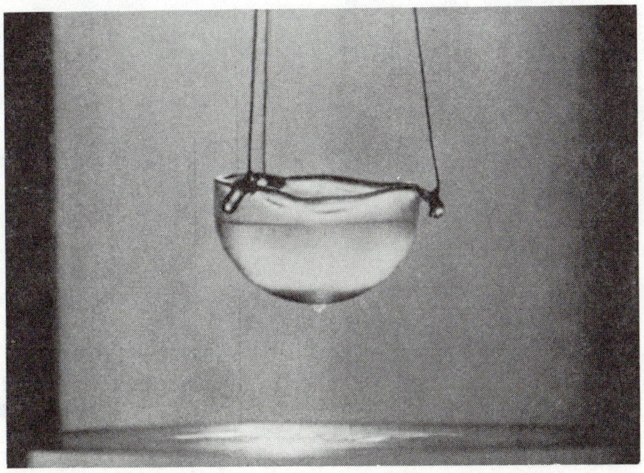

Fig. 28–9 **The Rollin Creeping Film.**

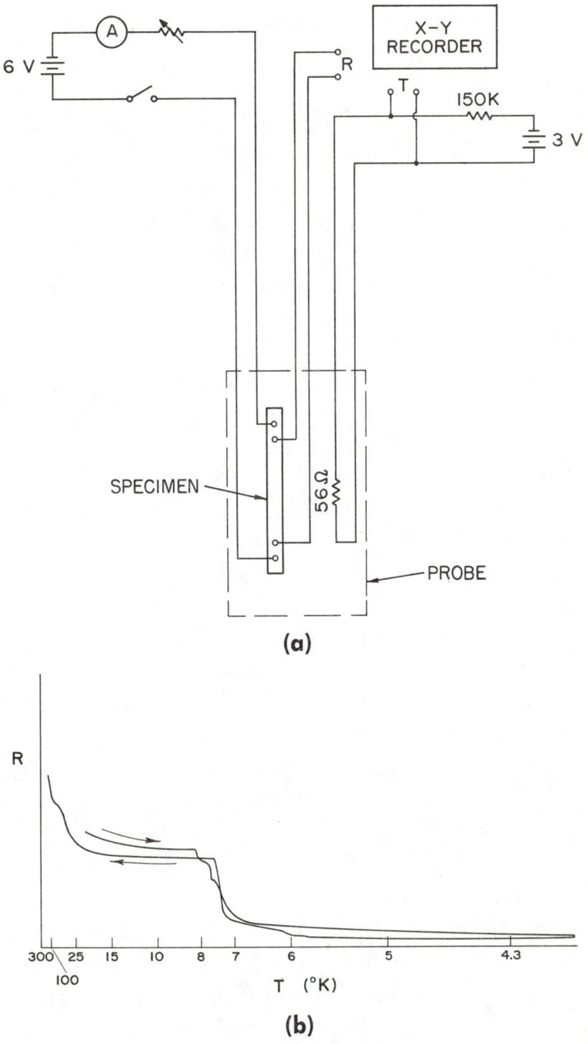

Fig. 28–10

a small fraction of one micron, and of the order of two to three hundred atomic layers.

This is the most difficult of the demonstrations to perform using projection techniques. Even though the light is filtered, its heat leakage tends to evaporate the film before drops can collect. For all demonstrations of the superfluid properties of liquid helium, the temperature should be as far below the λ-point as possible.

28–2.7 Resistance versus Temperature[3]

The circuit shown in Fig. 28–10 (a) can be used to demonstrate superconductivity in lecture. A

(3) This demonstration was contributed by R. W. Shaw, Rensselaer Polytechnic Institute.

sample of the material to be tested is placed in a container filled with liquid helium. When the switch is closed in the circuit, current passes through the sample. The voltage drop across the sample, which is proportional to its resistance, is fed into the *Y* input of an *X-Y* recorder (1 mV/in. sensitivity in each direction). The temperature is measured by the voltage across a 56-Ω Ohmite ¼-W carbon resistor. This is fed into the *X* input of the recorder.

Figure 28–10 (b) is a typical example of a graph obtained during a lecture demonstration. The drop in the superconductivity state is not as sharp as it might be, due to the temperature gradients present. At the low temperature the current was turned on and off several times and the lack of a signal noted. This was also done at high temperatures to convince the audience that the signal really depends on the current.

This apparatus can also be used to illustrate the resistance vs temperature characteristics of other superconductors, nonsuperconducting metals and alloys, as well as semiconductor specimens.

The probe consists of a thin-walled stainless steel or CuNi tube, ⅜ in. o.d. and 3 ft long, with a copper chamber silver-soldered to the end. The chamber is ½ in. o.d., ⅜ in. i.d., and 3 in. long. It is split lengthwise so that one half can be removed to change specimens. The specimen and the carbon resistor are placed inside the chamber, and the movable half is secured for the experiment with transparent cellophane tape. The diameters are chosen so that the probe will fit down the neck of a Supairco or Standard Air 25-liter liquid helium container. The amount of liquid used in carrying out the demonstration is extremely small.

The specimen can be a length of any superconducting wire which has a transition above 4.2°K. Approximately 4 in. of 20 percent indium-in-lead wire, 0.020 in. diam, yields good results. A pure lead sample would probably show a sharper transition, but because of lower resistance at low temperatures, a much longer and/or finer specimen would be required to achieve a usable signal. Niobium is not recommended because of the difficulty in attaching leads to it.

A very satisfactory means of providing demountable electrical connections is to use Winchester M-4 (Monoblock) plugs and receptacles ground down to fit into the ⅜-in. thin-walled tube. The specimens can be permanently mounted to such plugs, and a receptacle permanently fixed at the top of the copper chamber so that specimens are easily changed. No. 30 enamel-insulated copper wire encased in Fiberglas sleeving in the thin-walled tubing is used for the electrical leads.

The extreme temperature sensitivity of the resistance of the carbon resistor in the low temperature range makes it useful as a temperature-measuring device. This leads to an extremely nonlinear scale, but one which is large in the range of interest. Of course, it would be possible to drive the T axis with a thermocouple, if the sensitivity at low temperature were not desired.

ELECTRICITY AND MAGNETISM

29

ELECTROSTATICS

29–1 Electric Charge

29–1.1

The Braun electroscope shown in Fig. 29–1 has a vane 2 ft long and 1 in. wide. The fixed rods are of metal, 1 x 3/8 in. in cross section, the lower rod being insulated from the base by a rod of polystyrene or polyethylene. The vane is a piece of aluminum, 0.012 in. thick, 2 ft long, and 1¼ in. wide, with ⅛-in. bends along the long sides to provide stiffness. The details of the bearings used to support the vane are shown in Fig. 29–2. Hardened steel screws *a*, machined to conical points, support a threaded, hardened steel pin *b*, to which the vane *c* is fastened by means of two locknuts *d*. Two more locknuts *e* maintain the position of the screws with respect to the fixed arms *f*. The proper amount of restoring torque can be provided by placing a piece of soft wax somewhere along the lower half of the vane.

29–1.2

The electroscope in Fig. 29–3 is constructed from nylon cord 150 cm long, with an aluminum-painted Ping-Pong ball at each end. The cord is wound at its middle point around a wooden or plastic dowel, which serves as a support. The balls are charged by a small Van de Graaff generator and after charging are touched together to equalize the charge. This is easily accomplished without grounding since the nylon is nonconducting. This electroscope differs from the conventional inasmuch as the supports for the balls are nonconducting.

29–1.3

Figure 29–4 shows a large-scale demonstration electroscope, which consists of two table tennis balls attached to separate rods and suspended from a stand so that the balls just touch when hanging freely. The rods, 18-in. lengths of 1/16-in.-diam aluminum welding rod, are pushed through the balls and secured with glue. The table tennis balls have a thin coating of aluminum paint to make their surfaces conducting. The upper ends of the rods are attached to small rings fastened to the metal cross bar of the support stand. The cross bar is attached to the 2-ft-long, ¾-in.-diam vertical Lucite rod of the support stand. The balls are thus electrically insulated from the support stand and hold a charge for a long period of time. This electroscope is sufficiently sensitive that a friction-charged glass or ebonite rod touched several times to the top of the electroscope results in a deflection of 2–3 in. (Fig. 29–4). As shown in Fig. 29–5, the classical Faraday ice pail experiment can be made very impressive with the aid of this electroscope.

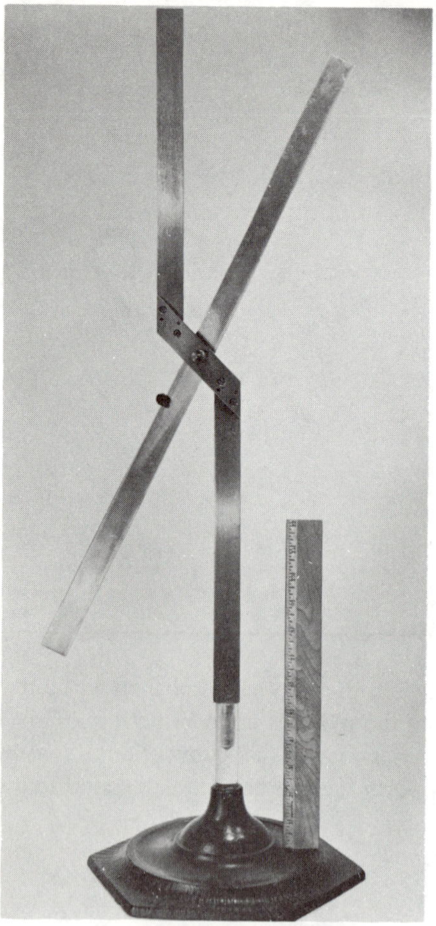

Fig. 29–1

29–1.4

An overhead projector is used to demonstrate electrostatic attraction and repulsion. Two balls of pith or plastic foam coated with Aquadag, are bifilarly suspended side by side in a clear plastic framework. Monofilament thread, such as spun

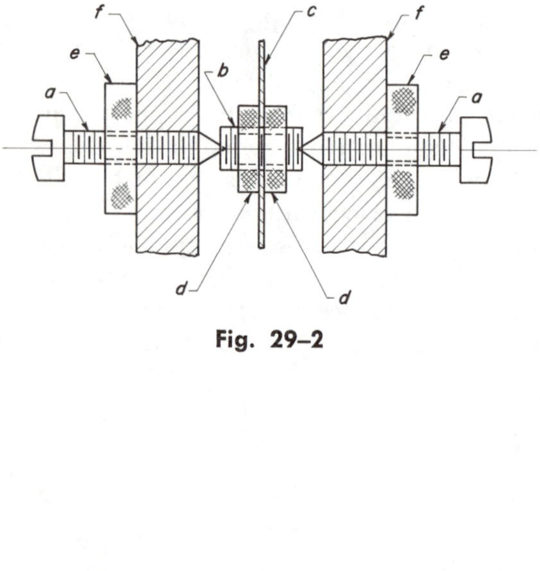

Fig. 29–2

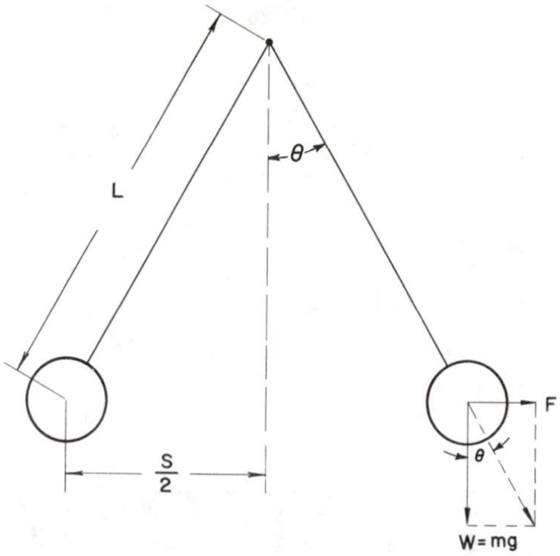

$$F = W \tan\theta = \frac{q^2}{4\epsilon S^2}$$

$$q = 2S\sqrt{\epsilon W \tan\theta}$$

Fig. 29–3

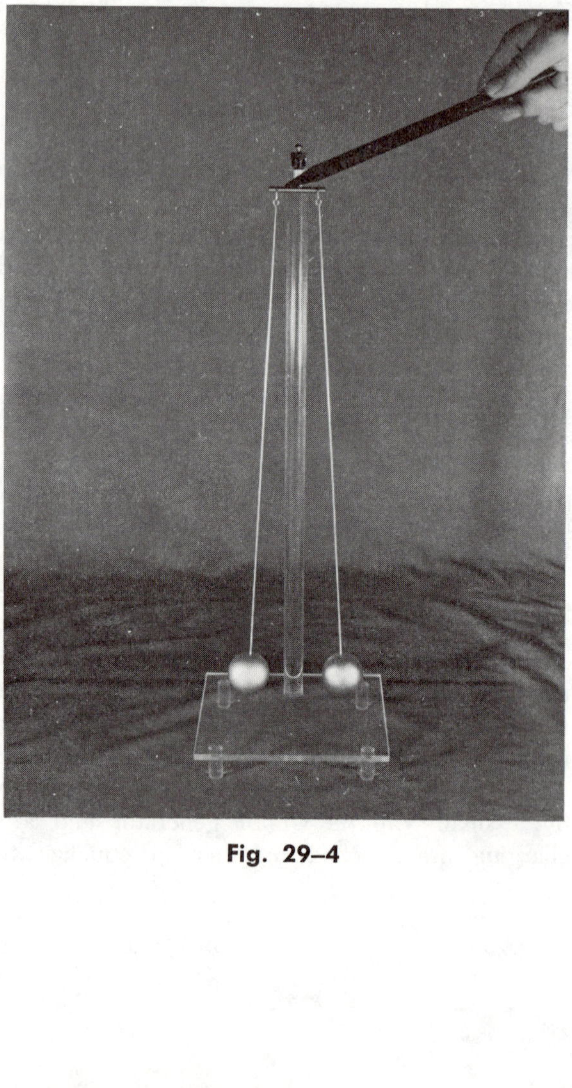

Fig. 29–4

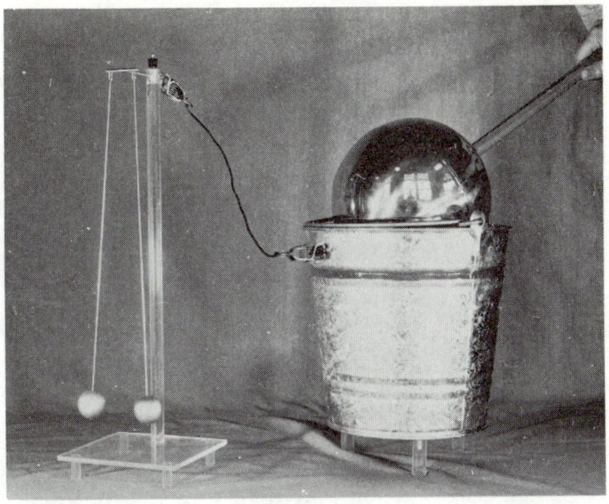

Fig. 29–5

nylon, makes excellent suspensions. The spheres are statically charged from a hard rubber rod stroked with cat's fur.

29–1.5

Two square sheets of aluminum foil 15 x 15 cm are suspended from the same horizontal glass rod by two thin wires about 1 m long. When the sheets are charged, they stand apart.

29–1.6

A large-scale demonstration electroscope (Fig. 29–6) consists of a long strip of metallized Mylar film *T* attached at one end to a brass strip S that is suspended from a Lucite rod *L* by means of the hook *H* formed by bending the brass strip as shown. The Mylar tape[1] is 1½ in. wide and about 15 in. long. When the electroscope is charged in the usual manner, the metallized tape is deflected from the brass strip, behaving like the gold leaf in a conventional electroscope. Obviously, larger electric charges are required for demonstration. A convenient source of charge is an electrophorus made from a Lucite disk and an aluminum plate fitted with a Lucite handle (Fig. 29–7). The Lucite disk is charged by friction and the aluminum plate by induction from the disk.

(1) Mylar tape, ¼ mil thick and metallized on both sides, can be obtained from Hastings Aluminized Films, Philadelphia, Pennsylvania. Half-mil Mylar tape, metallized on one side only, is available from Gomar Manufacturing Inc., Linden, New Jersey.

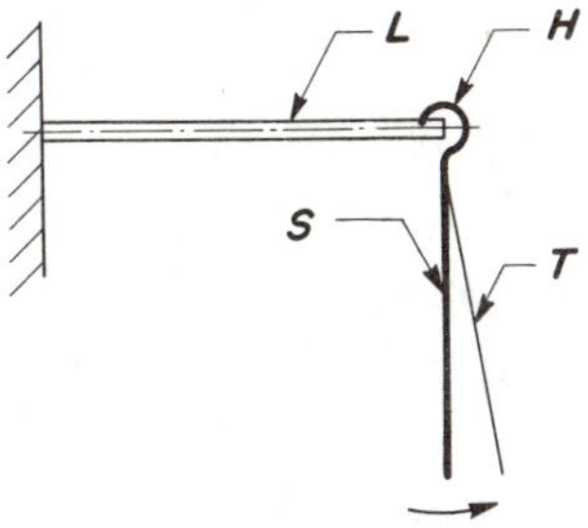

Fig. 29–6

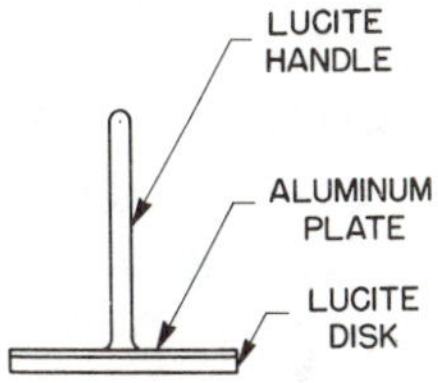

Fig. 29–7

29–1.7

An aluminum foil electroscope *a* is attached to the plate connection of a rectifier diode tube[2] *b* that is connected to a filament transformer[3] *c* as shown in Fig. 29–8. The electroscope is charged positively with an ebonite rod. When the filament power is turned on, the electroscope discharges.

29–1.8

Aluminum foil of the standard household thickness can be used as an excellent substitute for pith balls. The aluminum foil is made into a crude sphere by wrapping about a marble. After the marble is removed, the foil is suspended by a silk thread or some other nonconductor. Such hollow foil balls are as light as pith balls and are more visible due to their shiny surface. Balls as large as Ping-Pong balls can be made.

29–1.9

A Van de Graaff generator and a balloon sprayed with aluminum paint are used to show that like charges repel each other. A broomstick is screwed into a hole in the top of the generator, and the

(2) Raytheon Jan CRP-72 or equivalent.

(3) Thordarson-Meissner Mfg. Co., Mt. Carmel, Illinois, Catalog No. T21F00.

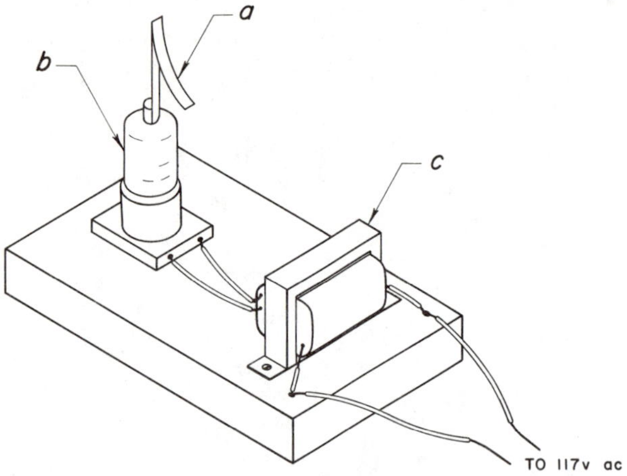

Fig. 29–8

balloon is hung against the Van de Graaff globe from a string tied to the tip of the broomstick. The broomstick and string are also coated with aluminum paint. When the generator is turned on, the balloon will stand out away from its charged metal sphere.

29–1.10

Charged balloons may be used with a hand-held brass rod to illustrate the discharging effects of points.[4] Three balloons,[5] inflated by a blast-suction pump[6] with illuminating gas to about 20 cm diam and sealed with rubber bands, are suspended by threads from a common point on the ceiling of the room. Helium-nitrogen mixtures may be preferable to illuminating gas in order to avoid the danger of fire. In this condition, the balloons weigh less than 1 g, and when charged by rubbing with a piece of fur, the balloons will stand apart some 30 cm from the vertical. A $\frac{3}{8}$-in. brass rod 35 in. long is pointed at one end. The other end is threaded and a $1\frac{1}{2}$-in.-diam metal sphere is drilled and screwed onto the rod. When the rod is held in the hand, sphere uppermost as in Fig. 29–9, the only effect observed is a slight attraction of the balloons to the sphere due to the induced charge on it, but when the pointed end is held

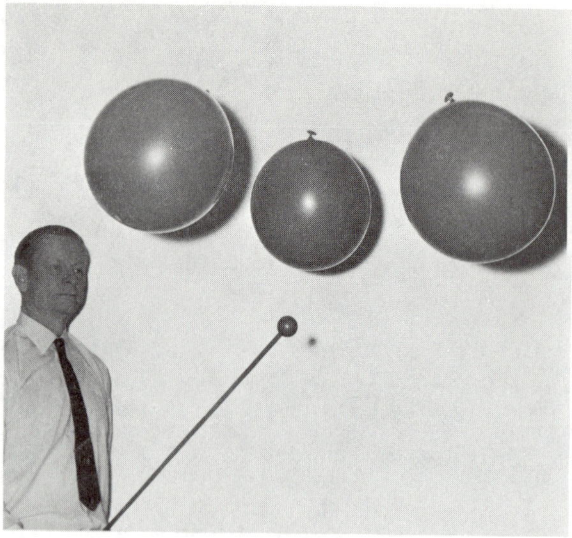

Fig. 29–9

uppermost as in Fig. 29–10, the balloons very quickly lose their charge and fall together. Assuming a negative charge on the balloons, the negative ions drift to the point, neutralizing its charge, and the positive ions drift to the balloons, neutralizing their charges.

Fig. 29–10

Assuming that the charges are uniformly distributed over the balloons' surfaces, and knowing the effective weight of the balloons and their configuration, fairly accurate quantitative calculations can be made. For example, the amount of charge,

(4) P. Rood, *Am. J. Phys.*, 8, 320 (1940).

(5) Available from Welch Scientific Corp., Catalog No. 1538.

(6) Available from Eberbach Scientific Corp., Ann Arbor, Michigan, Catalog No. 72–610.

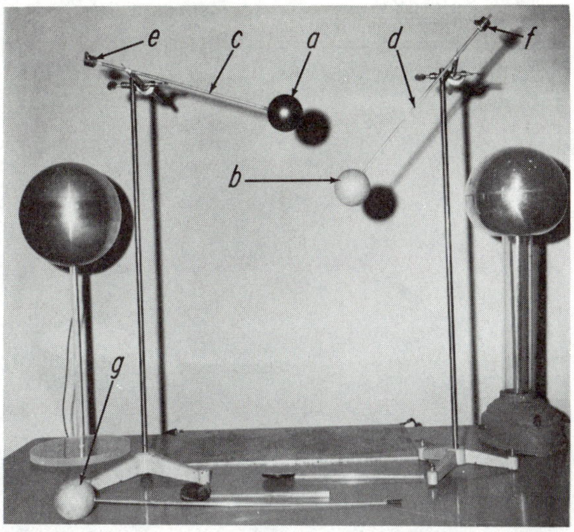

Fig. 29–11

the potential, or the average ion current flowing during the discharge can be found.[7]

Unlike other demonstrations in electrostatics, this demonstration is reliable in spite of rather large changes in the humidity of the surrounding air.

29–1.11[8]

The forces of attraction or repulsion between charged bodies can be demonstrated in a striking way with the simple apparatus shown in Fig. 29–11. This apparatus serves as a substitute for the traditional pith balls or inflated balloons.

Two polystyrene spheres[9] *a* and *b*, 3 in. diam, are cemented at the ends of Lucite rods *c* and *d* (¼ in. diam, 18 in. long). Counterweights *e* and *f*, consisting of pieces of brass rod, are threaded onto the other ends of the Lucite rods. Each rod is pivoted on a sewing needle, which fits loosely into a hole drilled 4⅛ in. from the threaded end and which is pressed into the end of another Lucite rod (½ in. diam, 6 in. long) mounted on a stand. To reduce friction, the needles are coated with Krylon silicone spray (No. 1325A). The counterweights are adjusted so that, when uncharged, one ball is above its pivot and the other below, the rods being vertical. The center of gravity, in each case, is only a short distance below the pivot, and the gravitational restoring torque, when the systems are displaced from the vertical, is thus very small.

The total weight of the system can be reduced by cutting the balls in half, hollowing them out with a penknife, and cementing them together again. They should be made conducting by painting their surfaces with Aquadag or conducting silver paint.[10] A fairly long brass screw threaded transversely into the counterweight serves as a setscrew, and (perhaps with the addition of one or two nuts) can be oriented to correct for lack of symmetry of the system about the axis of the rod.

A ball g of the same size or larger, made conducting and mounted on a Lucite rod, can be used to charge the pivoted balls. It is best to charge this ball from an influence machine or a small Van de Graaff generator.

Figure 29–12 shows how the pivoted balls can be arranged to demonstrate the force of attraction between unlike charges and the force of repulsion between like charges.

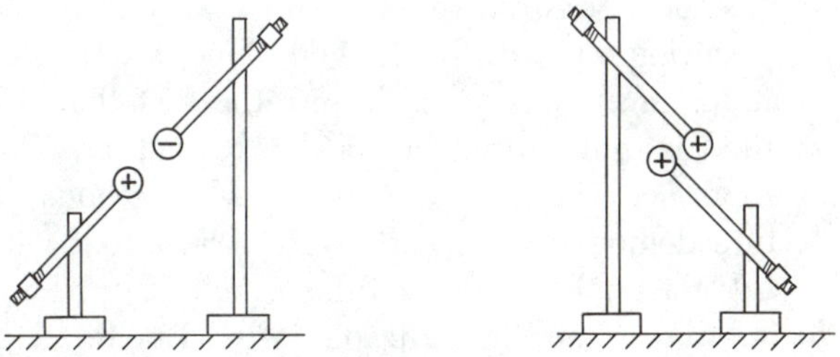

Fig. 29–12

The image force between a single charged ball and a conducting plate can be shown in an interesting way by holding a bright aluminum cookie sheet so that the class can see both the ball and its optical image, formed by reflection in the cookie sheet. With proper manipulation of the plate, the ball can be held in equilibrium with its support rod at an angle of 45 deg or so with the vertical, and the class can see the optical image of the ball at the same position as its electrical image. If a plus sign is painted on the side of the ball facing the class, and a minus sign on the side away from them, the optical image has the same sign, as well as the same position, as the electrical image.

(7) P. Rood, *Am. J. Phys.*, **14**, 445 (1946).

(8) F. W. Sears, *Phys. Teach.*, **11**, 225 (1963).

(9) Available from Plasteel Corp., Inkster, Michigan.

(10) Available from Lectronic Research Laboratories, Inc., Philadelphia, Pennsylvania.

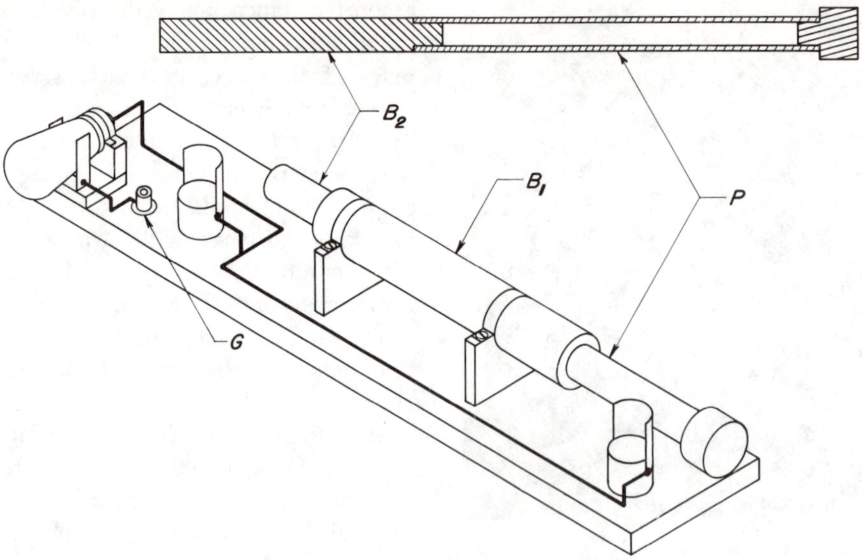

Fig. 29–13

29–1.12[11]

Nearly all elementary physics textbooks suggest the use of a glass rod rubbed with silk as a method for securing a positively charged object for demonstrations in electrostatics, despite the fact that this method is quite unreliable in damp weather.

In an endeavor to find more reliable methods for lecture demonstrations, tests were carried out with a soft glass tube, a Pyrex tube (each 12 in. long, 1 in. diam), and a strip of ¼-in. Lucite plate (12 in. x ¾ in.). These were rubbed with silk and a piece of Neoprene rubber sheeting, respectively, and tested by inserting them in a Faraday pail connected to a Braun electrometer. The capacitance of the combination was of the order of 25 pF and each division of the electrometer scale corresponded to a potential of 100 V.

These tests showed an increase in the amount of charge, in humid weather, of up to 50 percent when Neoprene was used in place of silk. In view of this, only the results obtained with Neoprene are given. With the relative humidity in the range from 50 to 60 percent, the electrometer deflections for the foregoing samples of soft glass, Pyrex glass, and Lucite–rubbed with Neoprene–averaged approximately 4, 14, and 35, respectively. With the relative humidity much higher than usual, the corresponding deflections were 0, 0, and 16.

Another item, concerning which physics textbooks show little variation, is the electrophorus. Nearly all specify soles of ebonite or resin and suggest the use of the finger as a means of detecting electrification. Either Lucite or polystyrene is a better material for the electrophorus sole. The action of the electrophorus can be demonstrated to a group of moderate size by the use of a neon lamp held in the hand as a means of grounding the metal plate and testing the plate for a charge after it has been removed.

A type of electrophorus employing cylindrical elements is shown in Fig. 29–13. This consists essentially of a stationary brass tube B_1, supported on insulators, and a movable tube for which the details are shown in the upper part of the drawing. This part consists of a polystyrene tube P (1 in. diam, 12 in. long) and a brass extension B_2. To operate this electrophorus, the polystyrene section of the movable tube is charged by rubbing it with fur. It is now inserted into the stationary tube B_1 and moved back and forth. At each end of the movement the tube B_1 is electrically connected, through the brass section B_2 and one of the contacts, to the central terminal of a neon lamp supported on an insulated stand. The circuit is com-

[11] D. S. Ainslie, *Am. J. Phys.*, **26**, 50(A), 549 (1958).

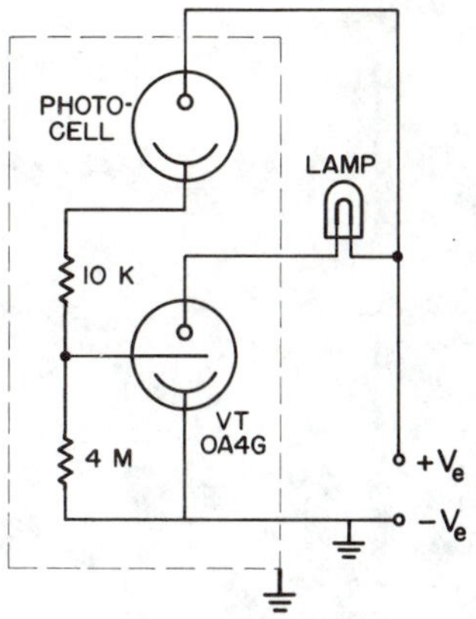

Fig. 29–14

pleted by means of a metal sheath touching the glass envelope of the lamp and connected to a grounded terminal *G*. With this arrangement a momentary flash is produced in the lamp at each end of the movement. One interesting feature of this electrophorus is its ability to operate indefinitely without recharging the polystyrene section provided it is left in the closed position when idle.

The following demonstration experiment was devised using electronic equipment to illustrate the importance of shielding sensitive electrical equipment. The circuit for this experiment, as shown in Fig. 29–14, is one of the conventional types using a photocell as the control unit for a cold cathode thyratron, OA4G. This equipment is usually operated on a 110-V ac circuit and the main current is started by illuminating the cell.

As a variation to the foregoing procedure, the apparatus was used with a dc voltage, in the range of 200 V, applied to the main circuit. It was started by bringing a charged rod near either the thyratron or the circuit components between the photocell and the grid of the thyratron.

To show the shielding action of a closed conductor, the sections inside the dotted line were enclosed in a shield consisting of a metal base with removable sides constructed of wire screening and a top consisting of a metal plate with three circular openings to allow for the insertion of a charged rod. With the shield in place and properly grounded, the circuit proved to be immune to electrostatic effects. In this connection it is important to have the negative terminal of the dc power supply grounded. If this precaution is overlooked, the action of the circuit is most erratic. The OA4G was used in place of a hot cathode thyratron because the cathode glow accompanying the main discharge is visible for a reasonable distance.

29–1.13

If the two parallel, circular metal plates shown in Fig. 29–15 are held at a high potential difference and a Ping-Pong ball with a conducting surface is inserted between them, the ball will oscillate between the plates, showing the repulsive forces caused by like charges. The parallel plates *a* are sheet aluminum (1/8 in. thick, 12 in. diam) and are insulated from one another by two plastic posts *b* (5/8 in. diam, 4½ in. long). To prevent the ball *c* from escaping, short plastic dowels *d* (½ in. diam, 2¼ in. long) are fixed at regular intervals around the edge of the lower plate. The entire assembly rests on three plastic legs (5/8 in. diam, 5 in. long). A Winshurst electrostatic generator can be used to charge the plates.

A small pith ball *e*, anchored with a light thread *f* to a 5-g weight *g* on the lower plate (insert), will "float" like a balloon because of the electrostatic forces.

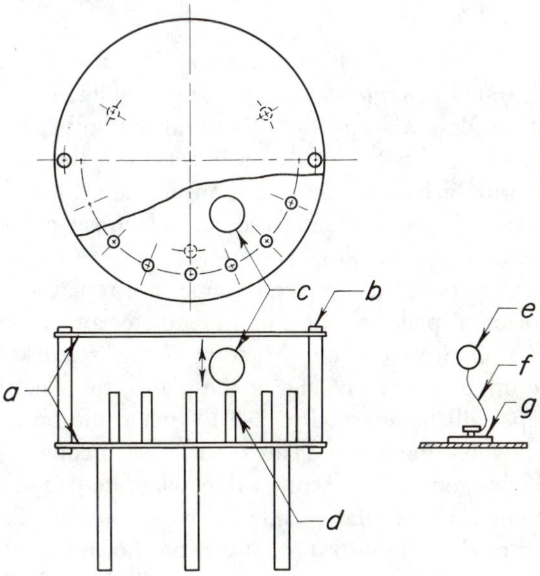

Fig. 29–15

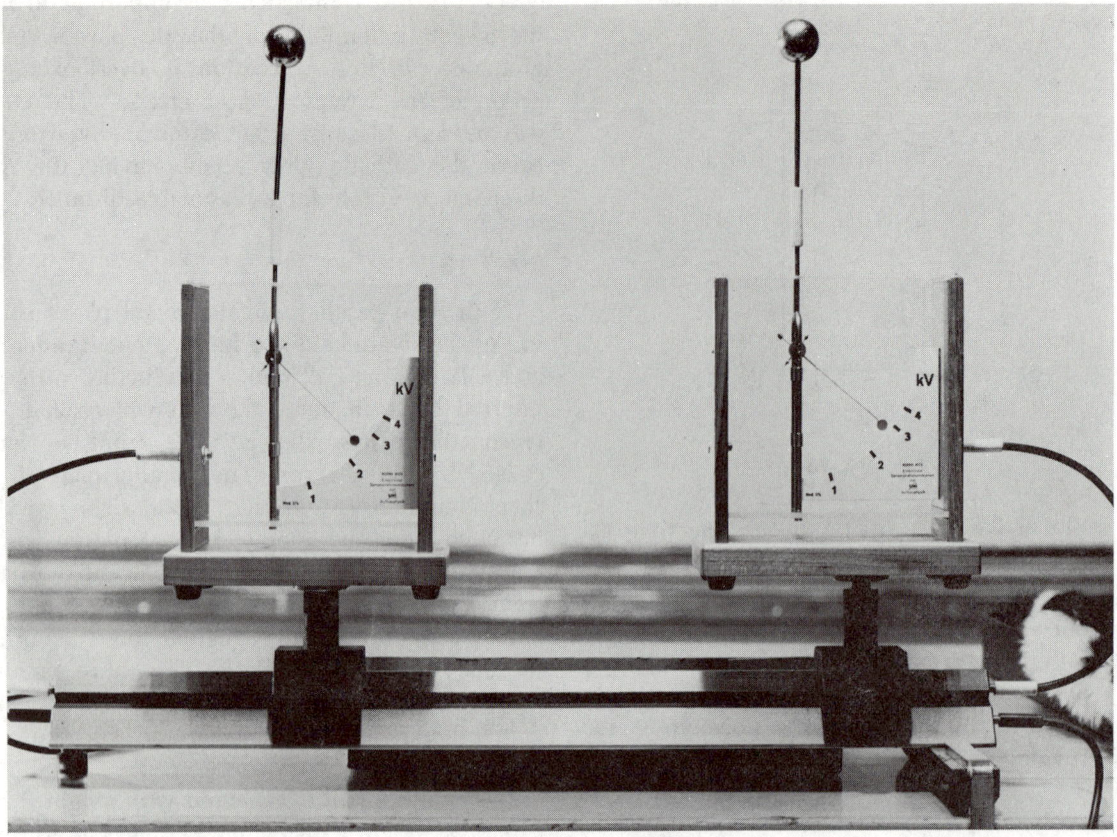

Fig. 29–16

29–1.14

Two identical electroscopes, Fig. 29–16, are oppositely charged to the same voltage, one with a glass rod and the other with an amber rod. The charge on one can be transferred to the other by using a ball or plate with an insulating handle. As the charge is transferred, the electroscope deflection drops to zero.

If the balls on the electroscopes are replaced by horizontal plates about 2 in. in diameter and a 4-in. piece of Scotch Magic Mending Tape is stuck to the upper surface of one of the plates, the electroscope will be charged to 2–3 kV when the tape is pulled off the plate. The other electroscope will be charged to the same value when the tape is touched to its plate, and, by repeating the first part of the demonstration, it can be shown that the charges are equal and opposite. To avoid the effects caused by rubber-soled shoes, the demonstrator should wear grounded handcuffs.

29–1.15[12]

A Ping-Pong ball painted with Aqua Dac silver paint is suspended from a Plexiglas stand by means of a 0.1-mm piano wire and is connected to a high-voltage electrostatic generator (50–100 kV) through the wire (Fig. 29–17). A grounded aluminum sheet (2 x 2 ft x 1/16 in.) held vertically by two 2 x 4's with slits, is placed about 4 in. away from the ball. The ball's original position is marked with an arrow. As the voltage applied to the ball is increased, the ball moves toward the aluminum plate. This experiment may be performed using an ordinary high-voltage generator, but a safety re-

[12] The following demonstration can also be performed: Have the supporting string insulated and place the voltage on another large insulated sphere on the machine itself. The ball is attracted until it touches and then springs away by repulsion. A lighted cigarette near the ball will cause it to lose charge so it will again be attracted to the machine. One can play Ping Pong with the machine. Suggested by G. Freier, University of Minnesota.

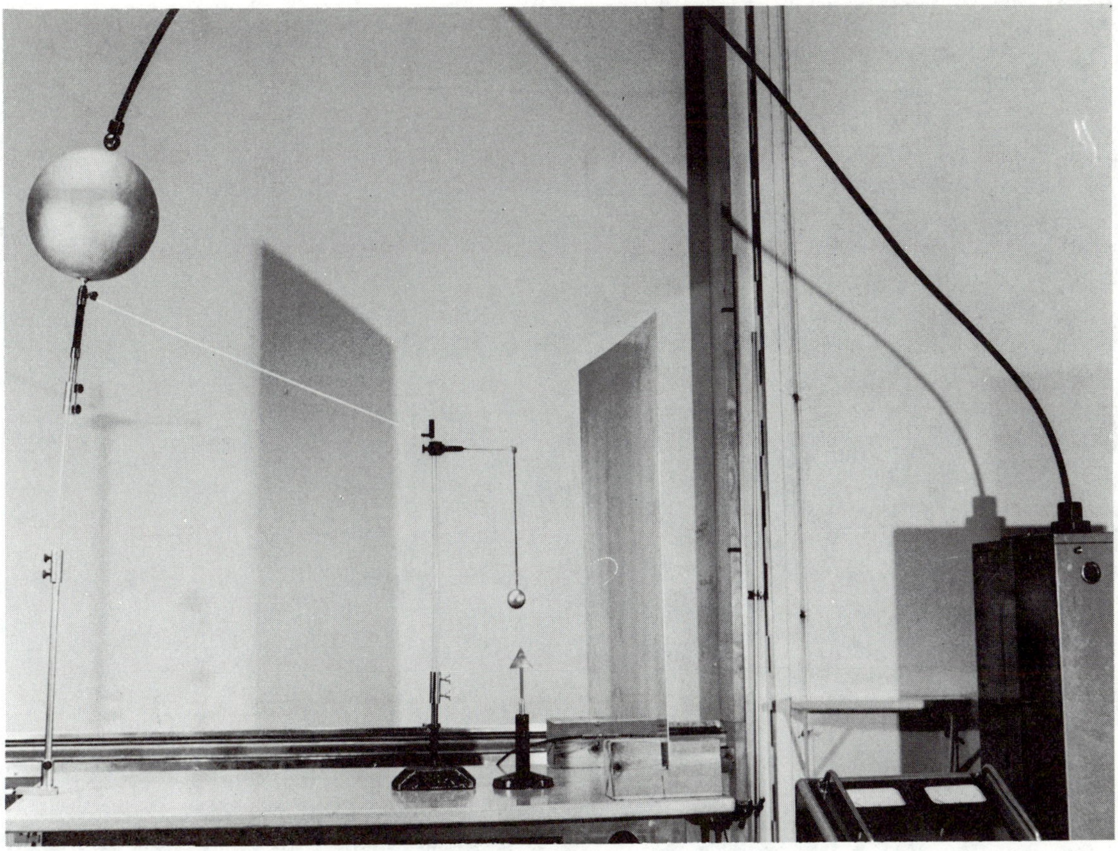

Fig. 29–17

sistor must be added to prevent an accidental short circuit.

29–1.16

The electrostatic breakdown of the surface tension of a liquid is demonstrated with a Van de Graaff generator and carbon tetrachloride.

A small circular dam of sealing wax is formed by hand on the top of the generator globe and filled with the carbon tetrachloride (Fig. 29–18). When the generator is turned on, a fountain of tiny droplets sprays upward to a height of approximately 24 in.

29–1.17

Electrostatic pressure causes water to disperse into fine droplets. This is demonstrated by allowing water to drip from a pipette *a* (Fig. 29–19) which is at a high potential (about 5,000 V). The pipette is supported by a glass rod *b*. The droplets are made visible by light shining from a light source *c* through a lens *d* (focal length 20 cm) onto a screen.

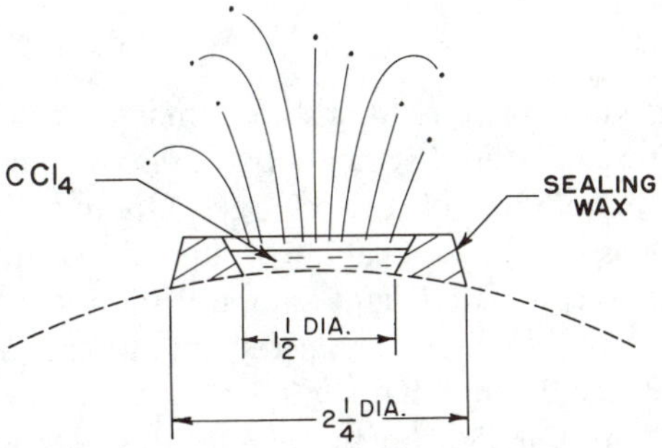

Fig. 29–18

29–1.18

If the end of a nylon clothesline is split with a razor blade, the loose strands will fan out by mutual

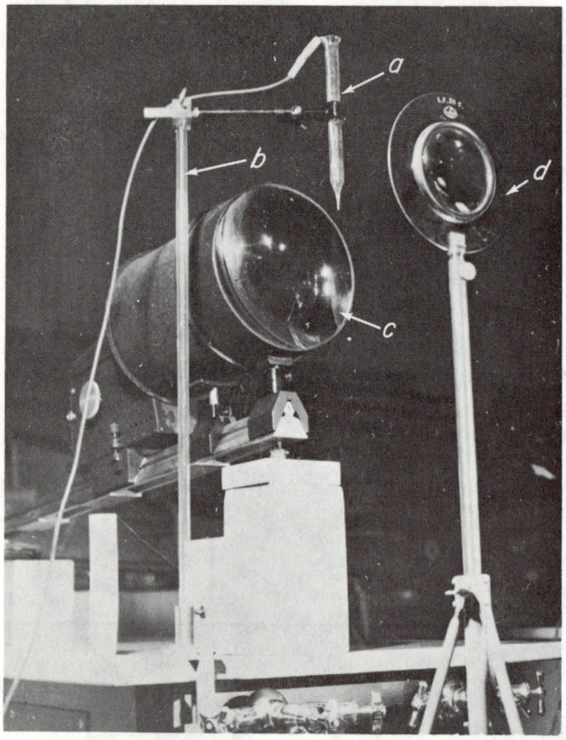

Fig. 29–19

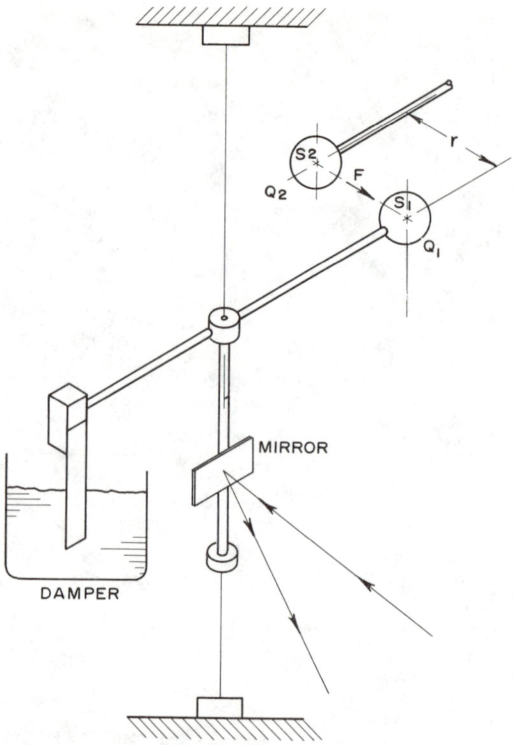

Fig. 29–20

repulsion when charged by an electrostatic machine, simulating the effect of human hair standing on end when it is stroked with a hard rubber comb.

29–1.19

The functional relationships among an elctrostatic force F, the distance r between two charges, and the quantity of two charges Q_1 and Q_2 are shown using the optical lever illustrated in Fig. 29–20. A collimated light source[13] is focused on a mirror and the light is reflected onto a scale on the wall of the lecture hall. Figure 29–21 shows the change in the light path due to the rotation of the optical lever. The distance L from the mirror to the wall combined with the deflection of the light spot determines the angle 2θ. Knowing this angle and the torsion constant of the wire allows the torque produced by the electrostatic force to be calculated.

The spheres are both charged in the same manner and then touched together. Deflection measurements are made with S_1 and S_2 separated by 40, 30, 20, 15 and 10 cm.[14] These measurements should be made as quickly as possible to avoid charge leakage. The force existing between the charges in each position is determined and compared with those expected from an inverse square variation. After placing the spheres 15 cm apart, the variation of force with charge is demonstrated by touching first S_2, and then S_1, with an uncharged auxiliary sphere. Each time, the data is taken and the forces are calculated.

The spheres or the insulating rods supporting them should not be handled. Carefully cleaned surfaces are necessary to retain charge on the spheres. The apparatus should be free of drafts or vibrations of any kind and the damping vane must be free to move in the oil container. Construction details are given in the Appendix, page 1311.

(13) *Editor's Note:* A small, low-output laser might be used as the light source since it produces a bright, low-dispersion beam.

(14) *Editor's Note:* In the laboratory, the separation between the spheres is measured by the use of a telescope sliding on a scale. For a lecture, however, the mount for the sphere S_2 should contain a scale and sliding pointer in order to read the separation of the spheres. The spheres should be set at about 1 mm separation and the pointer at 4.1 cm before charging. The pointer will then read the center-to-center separation of the balls directly.

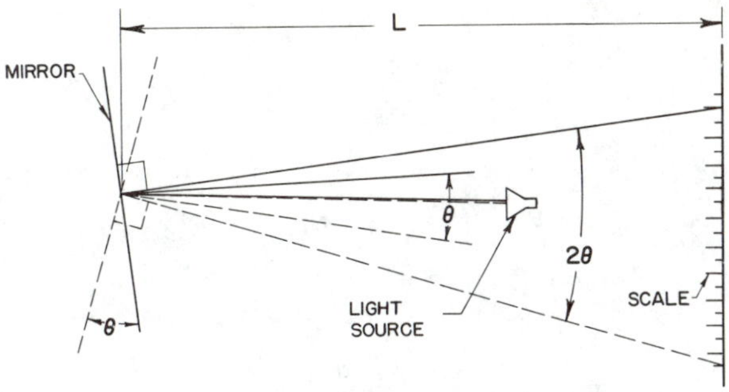

Fig. 29–21

29–1.20

The apparatus shown in Fig. 29–22 can be used to demonstrate electrostatic forces and Coulomb's law on an overhead projector. Two Ping-Pong balls *a* and *b* are coated with aluminum spray paint. Ping-Pong ball *a* is mounted on a clear acrylic plastic base *c*, and *b* is suspended by a nylon thread from as high a point as possible. A small Van de Graaff generator is used to charge the conducting spheres, and the effects of electrostatic forces are then easily visible to large audiences. Antistatic polish[15] is sprayed on the plastic supports to reduce conduction.

The geometry of the setup for the Coulomb's law demonstration is shown in Fig. 29–23. The equilibrium position *P* of the suspended sphere *b* should be noted prior to starting by marking the position of its shadow on the wall or blackboard. The deflection *D* of *b* will be measured from *P* in each instance. First, sphere *a* is placed near the equilibrium position *P*, resulting in a large deflection *D* of *b*. As *a* is moved to the right, the distance between the spheres *R* is increased, and *D* is therefore decreased. The decrease in the deflection *D* should be made by a factor of two in each measurement.

The deflection *D* of *b* is assumed proportional to the force on *b*, which is reasonable if a long nylon thread is used, since both spheres will then lie in nearly the same horizontal plane. From Coulomb's law the force on *b* is also proportional to $1/R^2$. It is then easily shown that the product DR^2 must be equal to a constant. The values of *D* and *R* collected from this experiment are used to verify this relationship.

The mounted sphere must be moved slowly so that the suspended one does not oscillate, but the entire run must be made in about three minutes to minimize leakage effects.

[15] Available from Park Chemical Co.

29–1.21

Coulomb's law can be demonstrated with the apparatus shown in Fig. 29–24, which consists of an insulated conducting sphere *b*, 15 cm in diameter and mounted on a movable stand, and a small pith ball *f* suspended bifilarly by nylon threads. The zero position of the pith ball is *e*. Positions are measured by projecting the shadows of the spheres on a screen. When the spheres have like charges, they repel each other. This force is measured by comparing it with the gravitational force,

$$F = mg \sin \theta,$$

where *m* is the mass of the pith ball, *g* is the acceleration due to gravity, and θ is the angle from the vertical through which the pith ball is deflected. For small angles, θ is approximately proportional to x/l, where *x* is the chord of the arc through which the ball is deflected and *l* is the length of the sus-

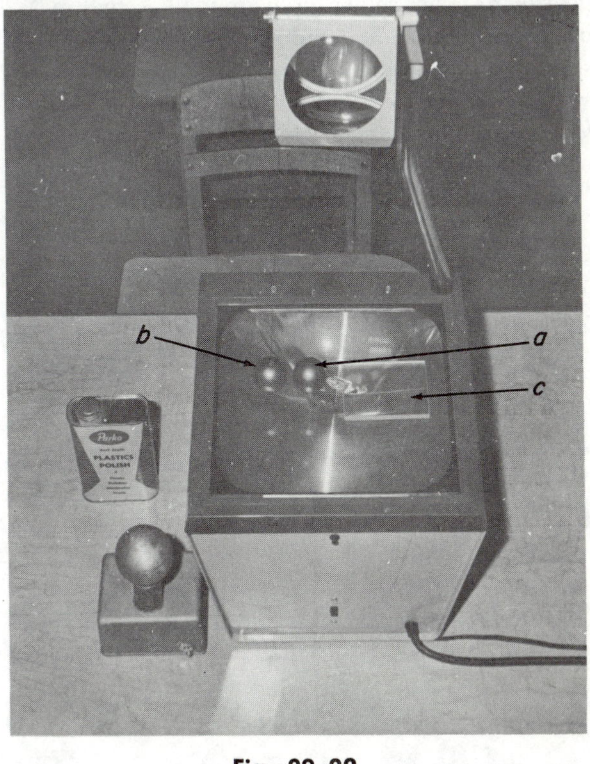

Fig. 29–22

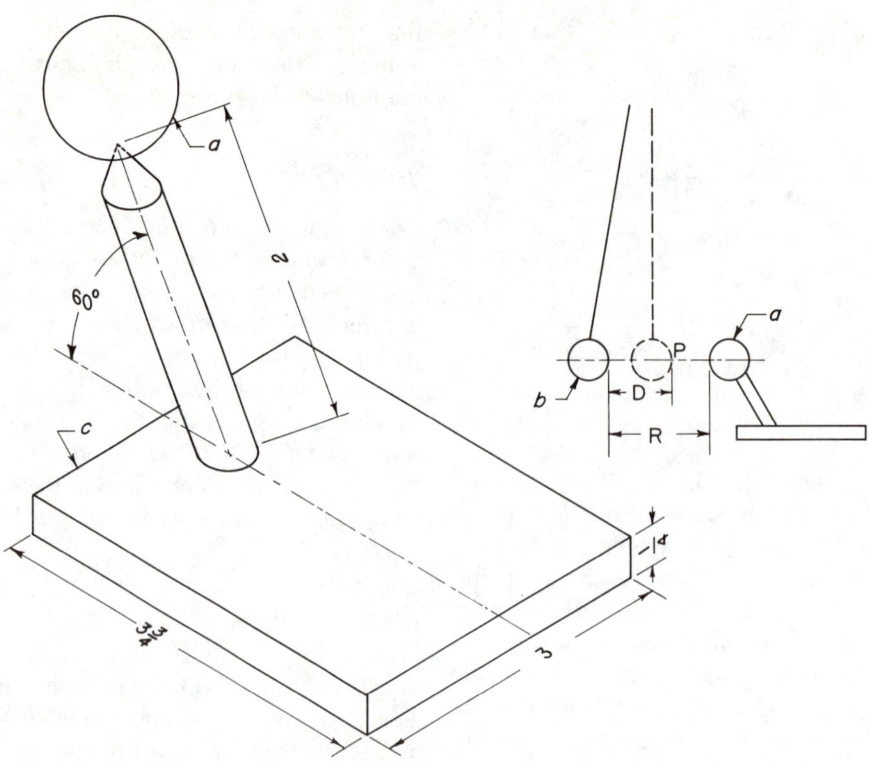

Fig. 29–23

pension (l is not shown in diagram). Actual measurements are made from the projected shadows which are larger than the true distances by a constant scaling factor. This affects only the numerical value of the constant $1/4\pi\epsilon_0$ deduced from such an experiment.

To demonstrate Coulomb's law, turn on the light *a* and discharge the pith ball and sphere by touching them. Mark the position of the edge of the shadow of the pith ball. Charge the larger sphere with a Van de Graaff generator by leaving the sphere in contact with the generator until after it is turned off. Since there will be some leakage of charge into the insulating rod, let the sphere stand a few minutes

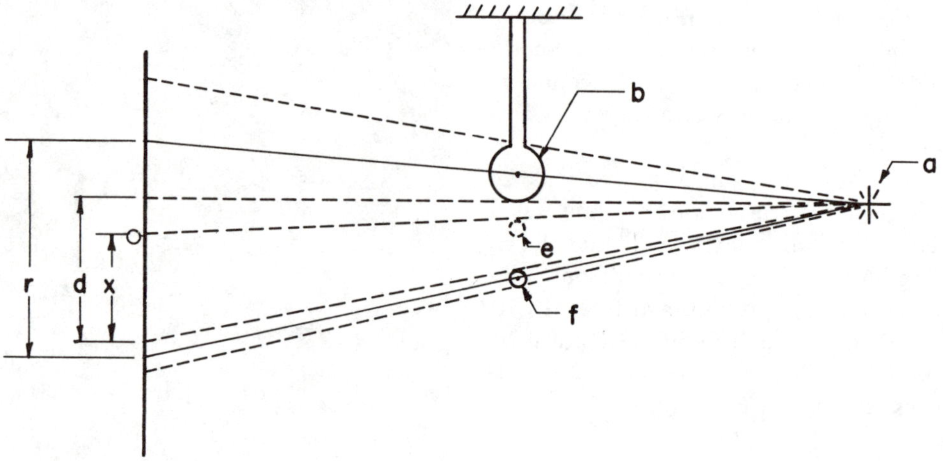

Fig. 29–24

before using it. Then move the charged sphere sufficiently close to the pith ball so that the induced polarization is enough to bring the pith sphere in contact with the larger sphere. The pith ball will then be repelled after picking up an amount of charge q_1/q_2 which is proportional to the ratio of the radii of the two spheres. Mark the edge of the shadows on the screen to measure the deflections. Measure the diameters of the shadows of the spheres to obtain the value r as indicated in the figure. Once the large sphere has been charged, work as rapidly as possible so that the loss of charge during the experiment will not be appreciable. The product of x and r^2 should be constant as the distance d between the spheres is changed.

The charge on the large sphere can be halved by sharing the charge with a second initially uncharged sphere of equal radius. To get the same deflection with a half-charged sphere as was obtained with a fully charged sphere, r is reduced by the factor of $1/\sqrt{2}$.

29–1.22

In this demonstration, separation of charge is accomplished by allowing water to drip from a valve a (Fig. 29–25), through a brass funnel b coated with paraffin, and into a brass cup c. The cup and the funnel are electrically connected to separate gold leaf electroscopes d. When the drops of water run through the funnel, they acquire a charge at the expense of the paraffin, and the charge on the water drops is transferred to the cup. The surface charge on the paraffin induces a like charge on the brass funnel so that the leaves of the two electroscopes become oppositely charged. The leaf deflections are shadow projected by means of lamps e, diaphragms f, and inverting prisms g.

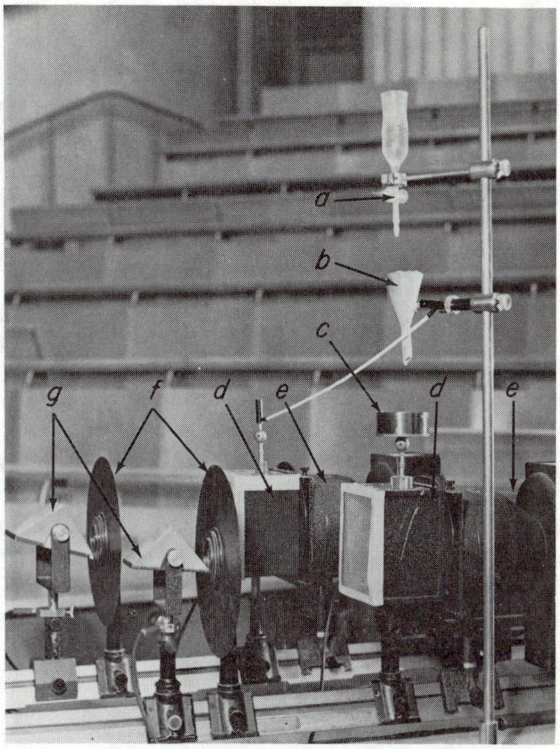

Fig. 29–25

29–1.23

About a century ago Lord Kelvin invented the water drop charge generator (Fig. 29–26), a simple but fascinating device. Two smooth water jets flowing down from a common tank break into droplets within the metal collars, and the drops fall into metal cans. Each collar is electrically connected to a can, but instead of a collar being connected to the can below it, it is connected to the other can. The cans are electrically connected to

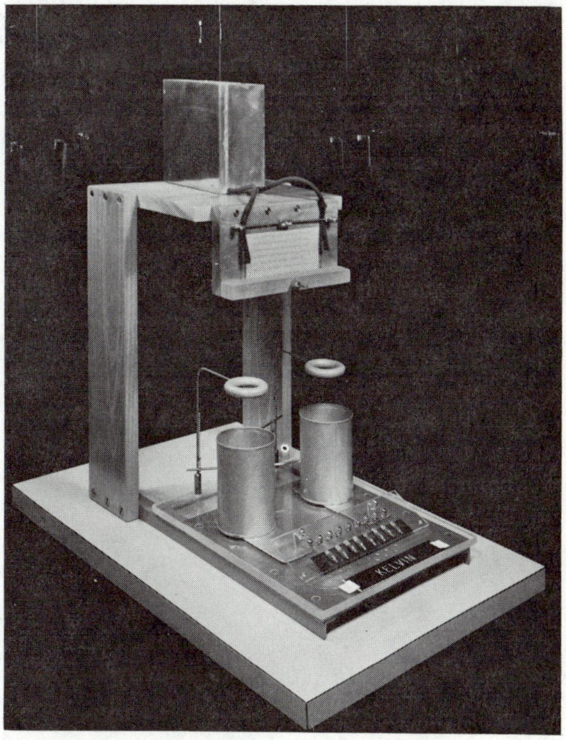

Fig. 29–26

their respective collars, but the cans and base plates rest on a Plexiglas plate; the cans are therefore insulated from each other. The two cans, electrically isolated, have the same potential—if they are connected with a wire. But remove the wire, and the cans soon drift apart in potential. For one thing, differing effects of ionized air drifting past the cans will insure a difference of potential. Suppose the can at the right becomes positive; then its collar is also positive. The can at the left, and its collar, will be negative. The jets, together with the water conducted to the nozzles, and the supply-tank water, form a continuous conductor. The electrical charges on the collar will induce opposite charges on the ends of the jets. In the right half of the assembly, the negative collar induces positive charge on the jet; positively charged drops break off and fall into the positive can, thereby making it still more positive. At the left side, negative drops make the negative can still more negative. Once started, charges build up very rapidly.

One terminal of a neon lamp is connected to one base plate and the other is bent close to the second plate so that an air gap (1/8 to 1/4 in.) exists between the terminal and the plate. If the gap is properly adjusted, the potential difference builds up to 5000 V or more, the gap breaks down, and one electrode flashes in the bulb. Then the cycle starts over again, and there is another flash in two or three seconds. It is the bulb's negative electrode that flashes. Which electrode will be negative is a matter of chance, but polarity can be imposed by touching the leads from a 45-V B battery to the cans. If

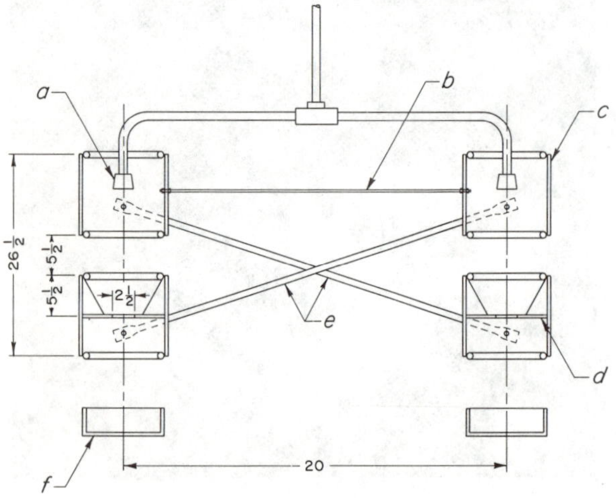

Fig. 29–27

the gap is made too wide for breakdown, the voltage rises so high that water drops are deflected by being attracted toward the collars and repelled by the cans. Water will be sprayed around and wet the surroundings.

Incidentally, these high voltages do *not* cause any hazard whatever. The capacity of the system is too small to store enough energy to do any harm. Much of the generator's capacitance is between the two cans, acting as capacitor plates.

Two extra cans set on the base plates an inch or so apart and connected, each to a different generator can, will increase the capacitance of the system. By adjusting separation of the additional cans, one can easily double the capacitance, double the time between flashes, and get a bigger flash. A more dramatic flasher is made with several neon lamps: mount them in series on their own piece of Plexiglas, make a small air gap at each end terminal, and connect to the generator. They will all flash at once.

Other demonstrations can be performed with the extra cans on the base plates. Drill a small hole horizontally into a small Styrofoam block that rests on the cans, and push a length of 1/8-in. metal rod into the hole, letting an inch or so stick out. Take a plastic soda straw, cut off a 1/2-in. length of it, and slide it over the rod. For a clapper, use several inches of rod. Attach its upper end to the soda straw sleeve by sticky tape. Start the generator. When it has built a potential, swing the clapper over to touch a can. Thereafter, it will clap back and forth, banging on the cans, and demanding attention. Or replace the rod clapper with a length of pull-chain—the kind used on lamp sockets. It also will slap back and forth. This is a sure attention-getter.

These cans can also be used for a leakage test. Fasten a few inches of fine, lightweight chain to a metal plate, remove the clapper mechanism from the cans, and lay the plate on top of one can, so that the chain hangs between the cans. Place it perhaps 1/4 in. from its own can and an inch or so from the other. Start the generator. As the potential builds up, the chain stiffly starts to bend toward the other can. With still more build-up it will reach over, touch, discharge, and repeat. This chain makes a fine test for leakage. Raise the voltage until the chain is reaching for the other can, but not touching, then turn off the generator. If it soon relaxes and hangs straight, there is leakage somewhere.

A capacitor is made of two lengths of copper ribbon, each bent into an L-shape. The ribbon is ½-in. wide, quite thin (about 0.005 in.), and flexible. The pieces stand up like bookends, facing each other, 5 in. high, and spaced about ⅛-in. apart, with the feet weighted down to the base plates. The lamp gap is opened too wide for breakdown to occur. When the jets are turned on, potential builds up, the condenser charge rises, the strips attract each other and are drawn together. On touching, the system is discharged and the strips spring apart, vibrating as they come apart. If a 3 x 5-in. card is stuck down into the condenser gap with generator running, the strips close on the card, clamp tightly to it, and press tightly enough to call for a surprising amount of pull to get the card out. Construction details of this apparatus appear in the Appendix, page 1311.

29–1.24

The cans are connected electrically as shown in Fig. 29–27. One pair of cans is connected to an electroscope with its case grounded.

The apparatus consists of two ordinary shower heads *a* enclosed in cans *c*. The cans are 10¼ in. in diameter, 10½ in. long, and are made from 1/64-in.-thick aluminum. The lightest possible weight aluminum tubing is rolled on the top and bottom of each can to prevent corona losses. Four cans are required. The upper two cans in Fig. 29–27 are connected by a ¾-in.-diam clear acrylic plastic rod

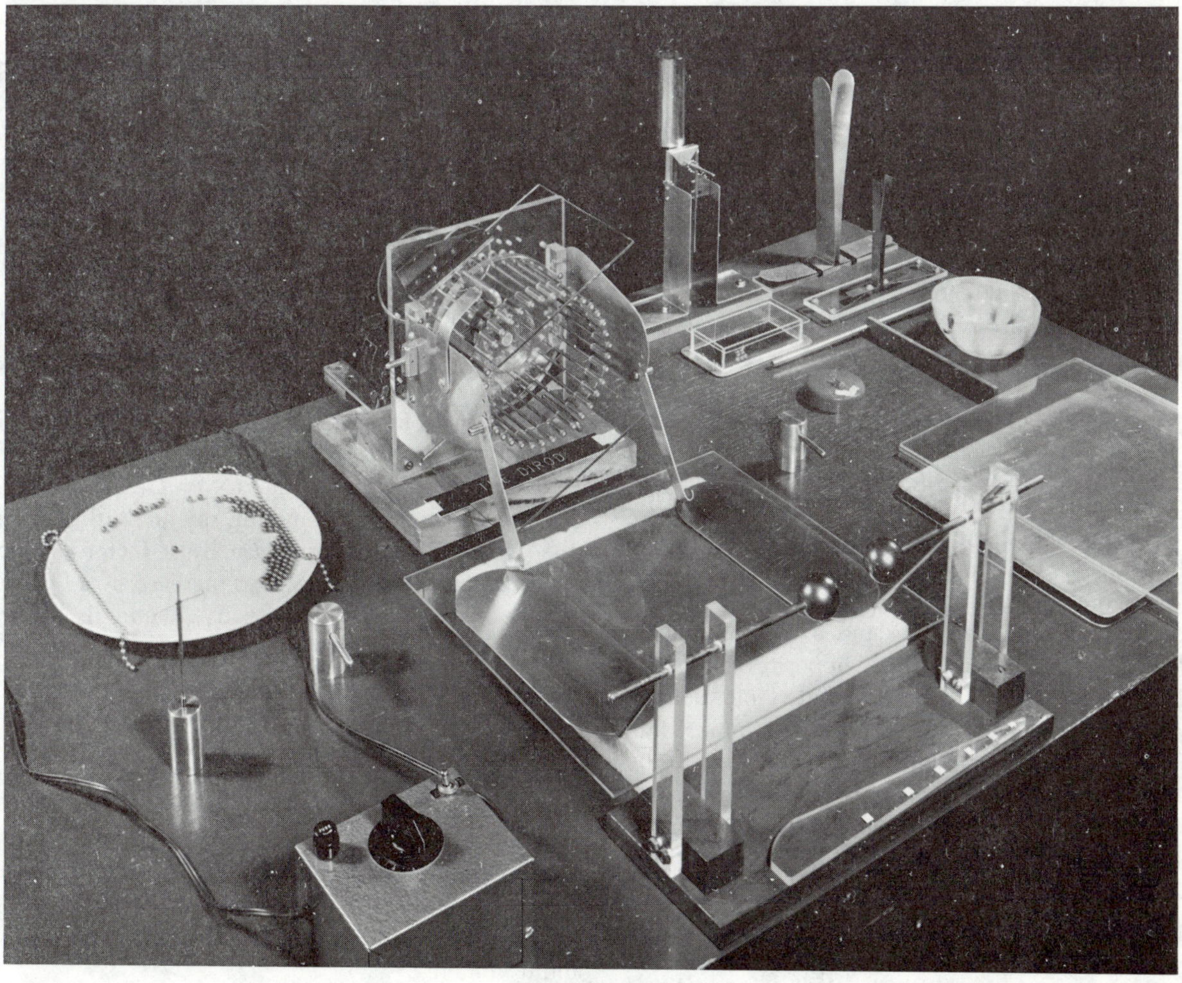

Fig. 29–28

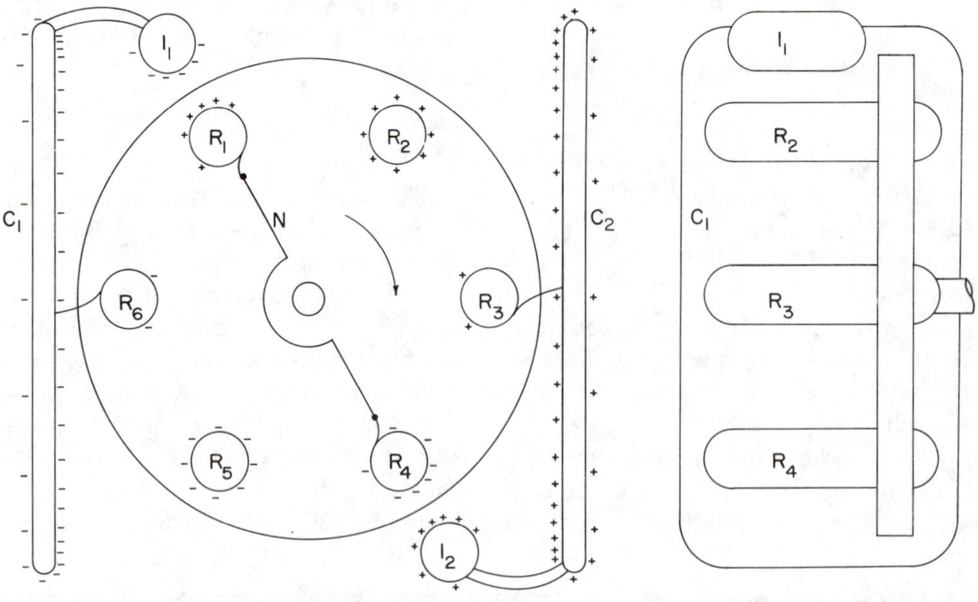

Fig. 29–29

b. The lower cans each contain a shelf *d,* held in place with heavy spring wire. The center part of the shelf is perforated to allow the water to pass. The upper and lower cans are cross-connected with ½-in.-o.d. light aluminum tubing *e* bent as required. The ends are flattened and riveted to the cans. Catch basins *f* are provided to catch the water that has passed through the system.

29–1.25[16]

Figure 29–28 shows a rotating electrostatic generator called a dirod (DISK and ROD). Figure 29–29 is a simplified diagram which illustrates how the dirod works. Disk *a* is made of Plexiglas. The rods R_1, R_2, etc. are metal (see Appendix, page 1314), as are the collector plates C_1, and C_2. Assuming that the collectors already have opposite charges on them, the charge will be distributed on their surfaces, as indicated by the plus and minus signs. Two metal inductors I_1 and I_2 are connected to their respective collectors. A neutral wire *N* has conducting brushes on each end which are touching R_1 and R_4 in the diagram.

If the collectors and inductors are initially uncharged, no charges appear on the R_1-N-R_4 combination. Since unlike charges attract, the negative charges on I_1 attract positive charges causing them to appear on rod R_1; and the positive charges on I_2 cause negative charges to appear on R_4. The charges on R_1 and R_4 are induced charges, and the rods have become charge carriers.

When the machine is rotated clockwise, R_1 and R_4 move ahead carrying their captured charges with them. The rods then touch the conducting brushes which stick out from the collectors and deposit about three-fourths of their charge on the collectors.

Every rod passing either C_2 or C_1 increases its total charge by a fixed amount, say one-fifth. When the collector charges increase by a factor of one-fifth, the inductor charges also increase by a fifth, as do the induced charges on rods R_1 and R_4. If we represent the charge on C_2 by the number 1, when it collects from the first rod it goes to 1.2, next 1.2×1.2, or $(1.2)^2$, next $(1.2)^3$, etc. It builds up according to the compound interest law.

There are 36 rods in the dirod shown in Fig. 29–28. If each rod increases its charge by only 4 percent, one machine revolution will increase the charge by four times; two revolutions, 16 times; three, 64 times. Charge can therefore be built up very rapidly. The corona effect will limit the maximum voltage.

Since the Plexiglas parts of the dirod have been

[16] From *Electrostatics,* A. D. Moore (copyright © 1968, Doubleday & Company, Inc., New York, New York). The dirod is available from Macalaster Scientific Co.

handled frequently during construction and assembly, there is usually enough charge on the surface to induce charges on the collector which is sufficient to start the charge build-up. In certain cases, when the weather is hot and humid and the Plexiglas parts have not been cleaned recently, it may be difficult to build up a charge. In this case, allowing the rods to run briefly in contact with a limp plastic bag will build up the charge. The dirod is a very rugged and reliable open-air electrostatic generator.

Although the dirod of Fig. 29–29 has only six rods, if the proper mechanical parts were added to hold it together and make it run, it would certainly generate. In fact, it would generate if there were only two opposite rods.

Figure 29–30 shows a major improvement, a machine with 24 rods. If each rod carries a certain charge, the more rods there are, the faster the charge will build up on the collector plates. This was tested for as follows: First, one opposite pair of rods was put in, and the machine was run at a certain speed. The machine was connected to a capacitor and a sphere gap, and the number of sparks per minute was counted. A second pair of rods was then added, and the test was repeated. This was done until all 24 rods were installed in their holes in the disk. The number of sparks per minute was found to be directly proportional to the number of rods. For example, 16 rods gave twice the sparking rate of 8 rods.

This poses the question. "If 24 rods are good, why not use 48?" Although 48 rods might double the build-up rate, two problems might be encountered. First, that many rods might weaken the disk so much that rod break-out would occur, causing an accident. Also, the rods might be too crowded electrically.

The dirod of Fig. 29–29 would rapidly build up its voltage until it sparked over internally. A spark would jump between I_1 and R_1, with another jumping between I_2 and R_4 at the same instant. Nearly all of the separated positive and negative charges would neutralize each other, and almost all of the charge would disappear. The machine would then start to build up charge again.

To increase the voltage before spark-over takes place, the inductors I_1 and I_2 could be moved farther away. If this were done, however, less charge would be induced on the rods, and the build-up rate would be slower.

In Fig. 29–30 the inductors have been moved in and the rods out to increase the induced charges. This shortens the gaps, and would reduce the spark-over voltage if it were not for the spark shields S_1 and S_2. These shields are glass or Plexiglas sheets which must be thick enough so that they will not fail by being punctured, and they must extend far

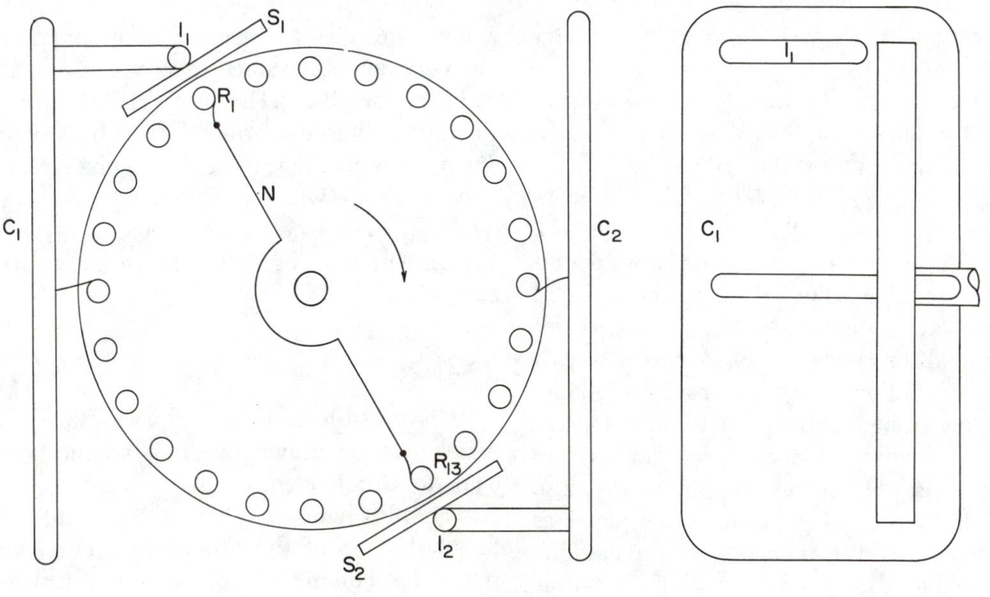

Fig. 29–30

Fig. 29–31

enough to prevent the spark from going around them.

A dirod with a 5-in. disk and ¼-in. rods would spark over at about 20 or 30 kV without shields. With shields, it might go to 60 or 70 kV.

Radial dirods (Fig. 29–31) can also be constructed with the rods sticking out from the disk like spokes of a wheel. Figure 29–32 shows a simpler, easier-to-build version of the radial dirod shown in the left of Fig. 29–31. A main object of building a radial dirod was to see if corona shields could be eliminated by using encapsulated inductors. Also, this design permits each side to have double inductors, thereby increasing the charges put on the rods. The collector plates are also double. With the rod going between two plates as the charge is collected, it is virtually in a Faraday cage and should give up practically all of its charge.

One surprising property of the radial dirod is that if it is shorted, it reverses. Both radial machines run very smoothly since the rods stick out on the plane of the disk. The large radial dirod has the advantage that its terminals can be swung up or down through a large arc. This change of level is often welcome when making connections to various pieces of demonstration equipment.

Construction details are given in the Appendix, page 1312.

29–1.26[17]

It is possible to double the length of the spark which can be drawn off from an inexpensive toy Van de Graaff electrostratic generator by making two simple changes. First, add a ¾-in. plastic collar to the top of the insulating column which extends this column up into the spherical electrode.

(17) J. Fowler, *Am. J. Phys.*, **32**, No. 5, xii (May 1964).

Then, replace the simple wires used for adding electrons to the belt or removing them from it by teeth cut with scissors from 0.005-in. brass shim stock.

29–1.27

A motor is operated by electrostatic charges drawn from a Van de Graaff generator (300,000 V output) (Fig. 29–33). A nonconducting support stand *a* holds a long aluminum tube *b* so that it touches the sphere *c* of the generator. On the same support is a cantilever-mounted Styrofoam wheel *d* which rotates on ball bearings. It lies in the same plane as the tube and can be adjusted vertically. Embedded in the wheel (15¾ in. diam, 1 in. thick) are 12 Aquadag-coated Ping-Pong balls *e*, equally spaced on a 13-in.-diam circle. This wheel assembly must be dynamically balanced, which can be done by adding pieces of tape to appropriate points on its periphery. A length of No. 8 gauge copper wire *f* (positive electrode) is coiled around the tube so that it hangs in the shape of a pair of ice tongs around the circle of balls. The points are sharpened, spread to a gap of 1¾ in., and positioned about 2 in. off the wheel's vertical centerline. Resting on a grounded (to the generator) aluminum plate *g*, another heavy copper wire *h* is shaped to encompass the circle of balls in the same manner as previously described, but with its electrode tips on the opposite side of the wheel's vertical centerline.

29–1.28

The motion of a charged body in an electrostatic field is demonstrated with the arrangement shown in Fig. 29–34. It consists of a Dry Ice puck *P* sur-

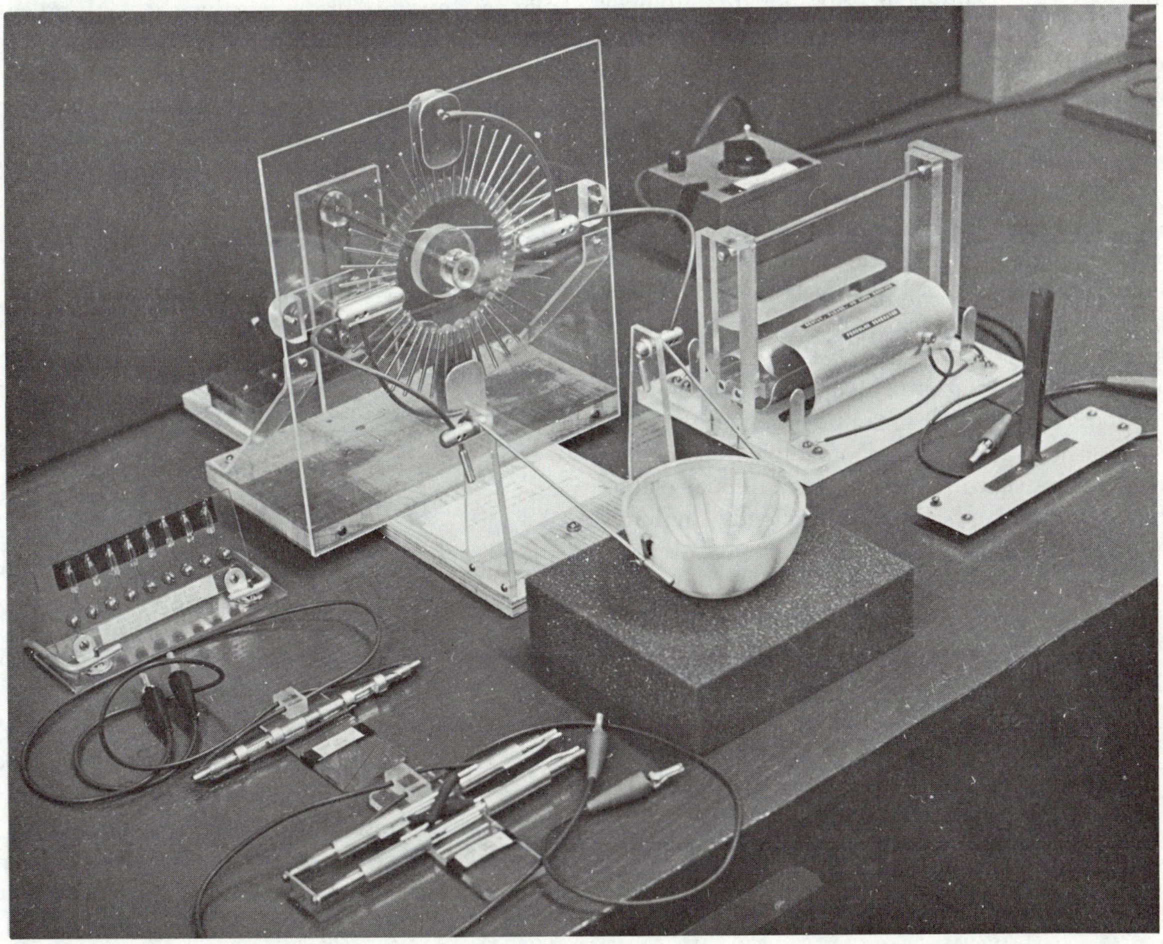

Fig. 29–32

Fig. 29–33

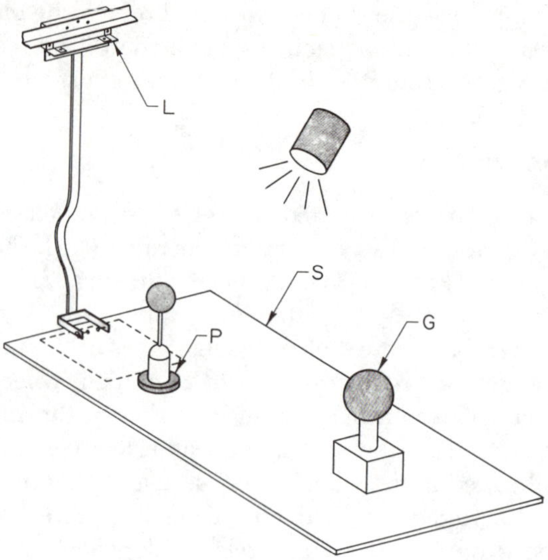

Fig. 29–34

mounted by a conducting sphere, a small Van de Graaff generator *G*,[18] a puck launcher *L*, and a plate glass riding surface S. Additional equipment includes a shielded strobe light, two aluminum sheets (12 x 30 x 1/32 in.) and a 10- to 20-kV dc power supply.

The apparatus demonstrates the mechanics of electrically charged particles. For example, repulsive scattering occurs when the small conducting-surface sphere on top of the puck has been charged from the sphere of the Van de Graaff generator and the puck launched so that it travels slowly towards the generator. The puck gently veers away in a hyperbolic path. The action can be photographed from above by using stroboscopic illumination for quantitative analysis, or it can be observed directly. Also, by using stroboscopic photography techniques, orbits can be shown. If desired, additional elaborate equipment can be employed to study Coulomb scattering in detail.

When the two aluminum sheets are set up vertically and parallel to one another, about 20 in. apart, and connected to the 10- to 20-kV power supply, the movement of a charged particle in an electric field can be shown (Fig. 29–35). If the charged puck is launched from one end of the passageway formed by the sheets while there is no field, its path shows no deflection (with a level table). Similar results are obtained if an uncharged puck is launched in a constant electric field. When a positively charged puck is launched through the field, it follows a parabolic path away from the positive sheet. If a negatively charged puck is directed from the same point on the positively charged plate side, it follows a parabolic path toward the plate. Stroboscopic photography techniques again can be used to obtain quantitative data to show the electrical forces involved.

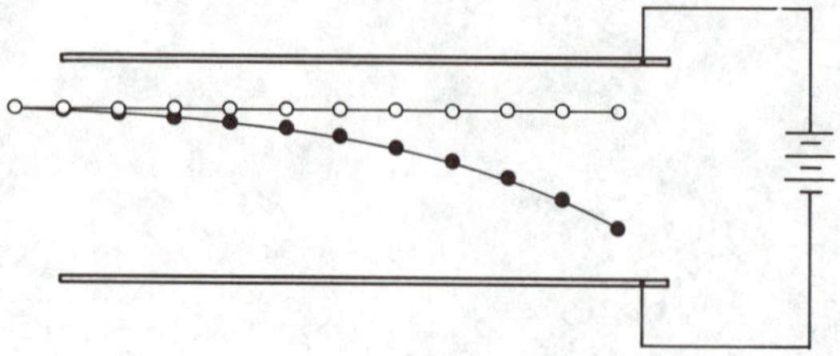

Fig. 29–35

Construction of the puck is described in the Appendix, page 1318.

29–1.29

A "Faraday cage," Fig. 29–36, can be used with a very sensitive dc VTVM electrometer[19] to show charge transfer from plastic strips. The cage consists of an electrostatic shield approximately 5 x 4 x 3½ in. The base and two sides (5 x 3½ in.) are 1/32-in.-thick brass, and the sides have 3¼ x 1⅜-in.

[18] Atomotron High Voltage Generator Cat. No. 71846 or 71848, available from Central Scientific Co.

[19] Available from Madison Associates, Inc., Madison, New Jersey.

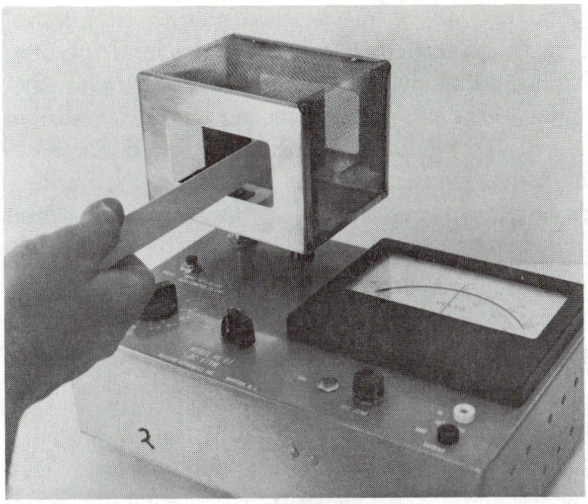

Fig. 29–36

apertures for access to the interior. The other three sides are copper window screening.

Inside the cage are two brass plates, each 2 x 2 x 1/32 in., spaced 2¼ in. apart. These plates have extensions by which they are mounted to banana plugs outside the cage. One plate is grounded, the other is insulated with Teflon. The banana plug spacing will depend on the electrometer used.

The dc VTVM electrometer should be able to read at least 1 V each side of center-zero, with shunts for 10^{-7}, 10^{-8}, and 10^{-9} A. Two strips (10 x 1 x 1/32 in.), one each of cellulose acetate and Vinylite, are needed in this demonstration.

In another demonstration, the "grounded" plate is also insulated from ground and a single pole normally closed jack connects the plate to ground. A plug opens the jack and applies 250 V between plate and ground. A ¾-in.-diam brass ball, or cylinder, with an insulated handle is needed.

The cage is plugged (with the insulated plate to the input) into the VTVM, the shunt at 10^{-8} A, and the needle set at center-zero.

A D battery held across the plates will indicate +1.5 V when its plus terminal is touching the insulated plate. Reversal of the battery causes a −1.5 V reading. The high-resistance shunt does not change the response of the instrument as a voltmeter.

When the two plastic strips are rubbed against each other, the Vinylite strip is left with an excess of electrons and the cellulose strip is left positively charged.

When the positively charged strip is held motionless or moved parallel to the plates, the VTVM reading remains zero. While the strip is moved to the left toward the input plate, the needle of the VTVM will read the conventional current direction for positive charges. Movement of the strip to the

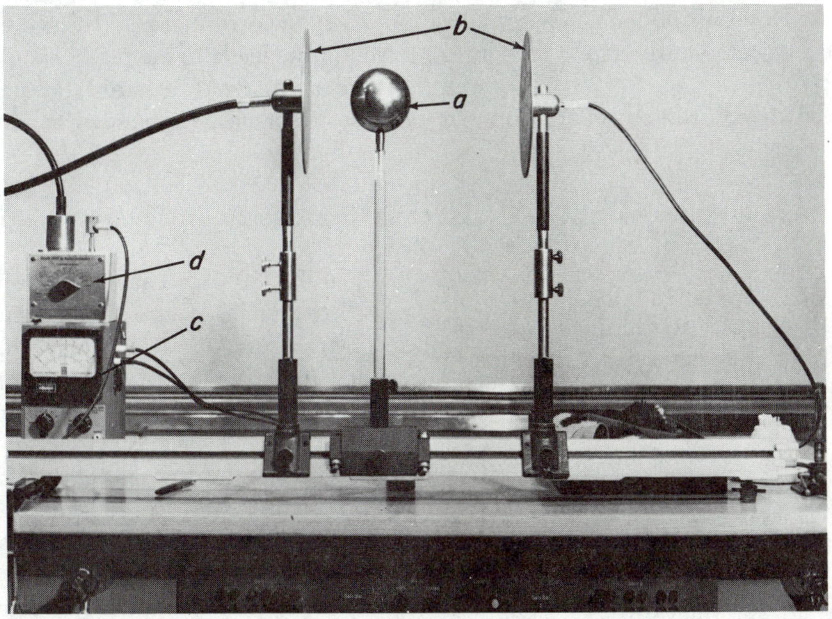

Fig. 29–37

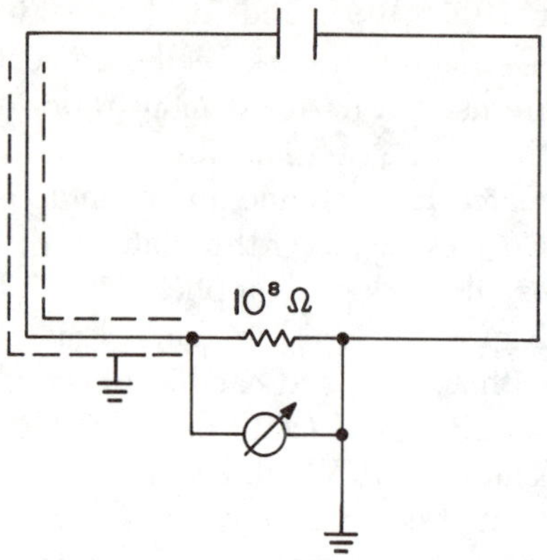

Fig. 29–38

right will reverse the VTVM, also indicating a conventional current. When the negatively charged strip is used, an opposite current is indicated.

In a second demonstration, with 250 V at the jack and the shunt set to indicate a current of 10^{-9} A, a conducting sphere at the end of a rod can be used to carry electrons from one plate to the other, with "holes" being carried in the opposite direction. The faster the conducting sphere is moved back and forth between plates, the higher the current reading. It should be demonstrated that reversing the battery polarity will reverse the current.

Protect the battery, or power supply, with a 10- to 20-kΩ resistor in each of the leads. When the sphere touches the plate connected to the battery, it charges up fast. However, when it charges by touching the plate connected to the input, it charges up slowly.

As in the second demonstration, charge up the conducting sphere at one of the plates. Then move it back and forth between the plates, and the first demonstration will be duplicated. Charging the sphere at the other plate reverses the charge on the sphere, and this can be shown by again duplicating the first demonstration. Reverse the battery polarity and repeat as above to show reversed charge polarity on the sphere.

29–1.30

A charged metal sphere *a* moving between the plates of a capacitor *b* (Fig. 29–37) induces a current in the external circuit (Fig. 29–38). The sphere can be charged positively or negatively, and the reading on the electrometer[20] *c* is proportional to the current. A 100-MΩ shunt *d* is used for an 8-V electrometer, although these values depend on the distance between the plates. A projection instrument can be used to show the electrometer reading. This reading is shown to depend on the amount of charge on the sphere (which is proportional to the voltage used to charge it up), on the velocity of the sphere, and on the sign of the charge.

The vertical plates are 6 in. in diameter; one is grounded, and the other is insulated with amber or Plexiglas. Ball bearings on the sphere support enable it to roll on the assembly bench. Care must be taken to make sure that the Plexiglas insulation does not carry away any charge. Furthermore, the insulation of the leads to the plates should be grounded. In order to avoid electrostatic perturbation, the demonstrator wears grounded handcuffs.

29–2 Electric Fields

29–2.1

The patterns of electric fields can be demonstrated with small bits of material,[21] suspended in oil, which will align themselves with the applied electric field. The apparatus is illustrated schematically in Fig. 29–39. The electric terminals are connected to a small electrostatic generator or a 10-kV dc power source, and fine, black rayon fibers suspended in the oil trace the resulting electric field associated with a specific pole arrangement. Variously shaped electrodes (Fig. 29–40), made of 1/16-in.-diam brass rod, fit into a Lucite ring (Fig. 29–41) placed over the edge of a 3-in.-deep glass dish. The dish contains approximately 3/4 in. of castor oil mixed with a small quantity of the rayon fibers. The demonstration is conveniently projected on a screen with an overhead projector.

[20] The Keithley electrometer with decade shunt can be obtained from Keithley Instruments, Inc., Cleveland, Ohio.

[21] Fine clippings of rayon fibers, commercially known as "velveteens," are used for producing a velvet surface on shoes. They are available in plastic containers in most department stores in the United States. When suspended in oil, these tiny fibers act as sensitive dipoles which align themselves with an applied electric field.

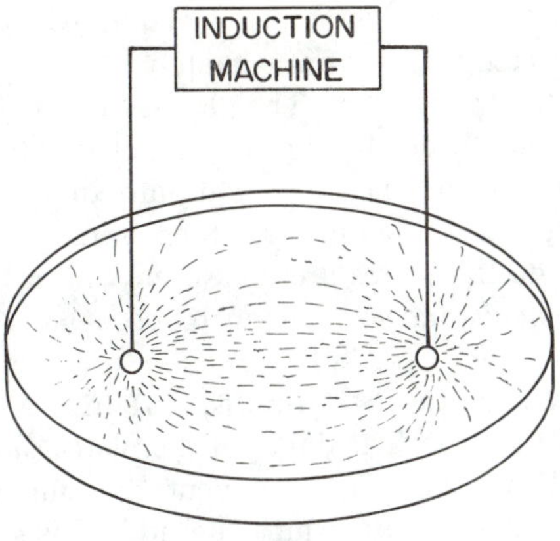

Fig. 29–39

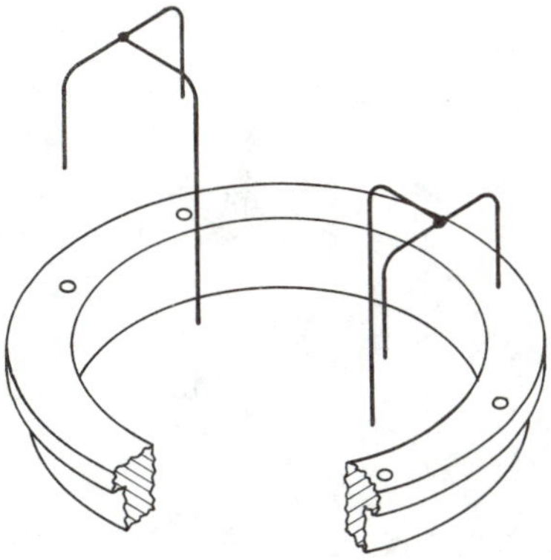

Fig. 29–41

29–2.2

Air bubbles in an inhomogeneous electric field are repelled by the region of high electric field. The air bubbles and the electric field are created in a bath of paraffin oil or silicone oil.

A pointed electrode *a*, a brass plate *b*, and a nozzle *c* made from glass tubing are mounted in a square-walled glass container *d* (4 x 4 x 1 in.) filled with paraffin oil or silicone oil as shown in Figs. 29–42, 29–43 and 29–44. The electrodes, which have rounded corners to avoid breakdown, are held to the container by small brass clamps *e*. The pressure of the compressed air is adjusted by means of a reducing valve. If the nozzle is in the oil before the demonstration starts, the air is kept under pressure to prevent oil from getting into the tubing. The presence of oil in the tubing causes an irregular production of bubbles. When the output of a Van de Graaff generator *f* is connected to the electrodes, a stream of air bubbles *g* emerging from the nozzle will be deflected as it passes the pointed electrode.

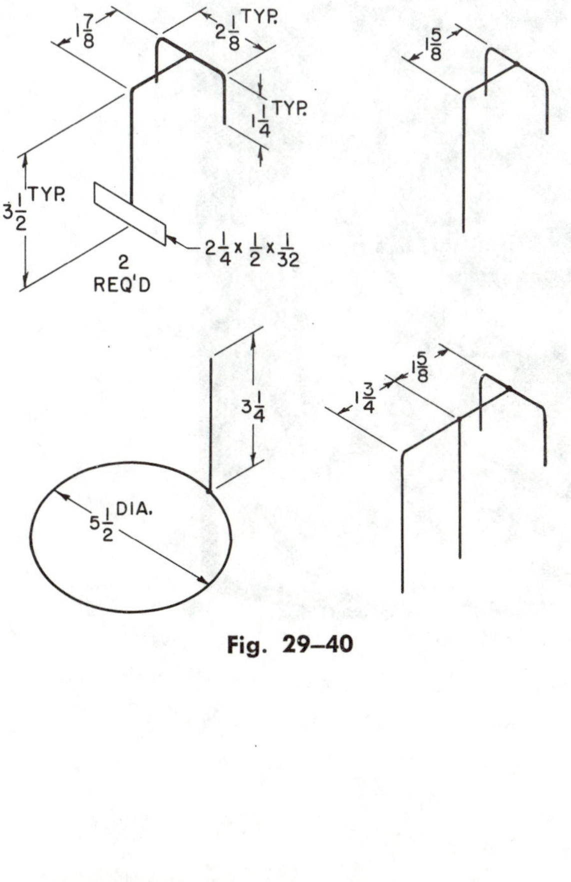

Fig. 29–40

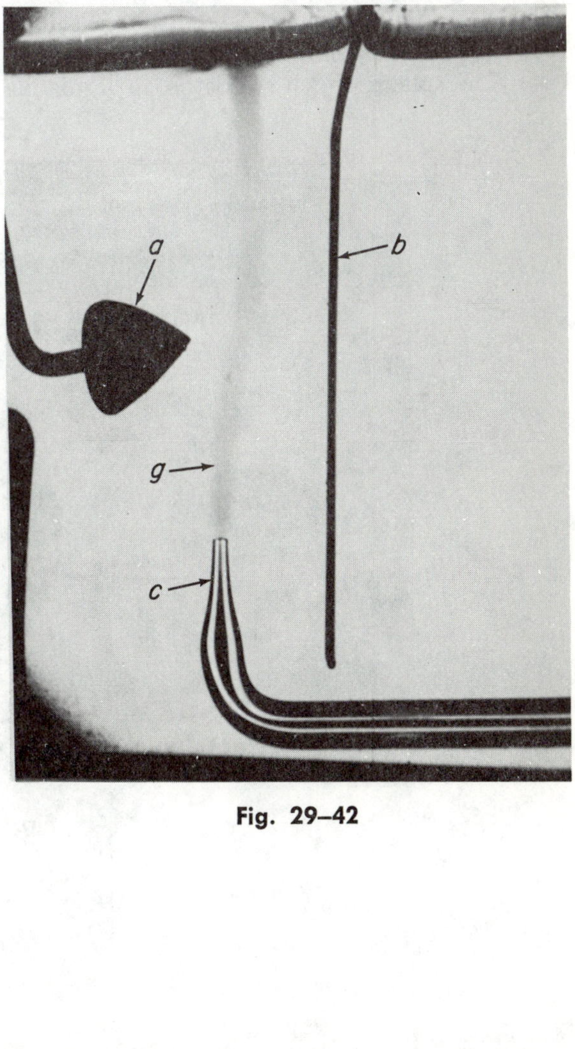

Fig. 29–42

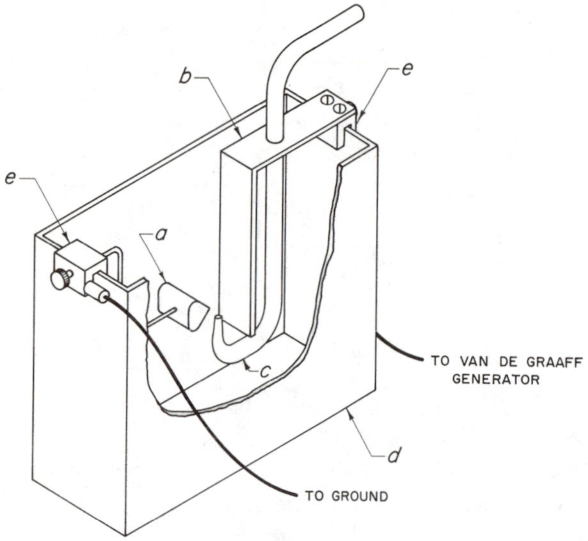

Fig. 29–43

This phenomenon can be shadow projected very effectively.

29–2.3

The electric field around a large sphere can be measured with double brass plates as shown in Fig. 29–45. The sphere, which stands on an insulating stand or is suspended from the ceiling with a nylon cord, is charged to approximately 80 kV with an electrostatic generator. The plates, 4 in. in diameter and mounted on Plexiglas handles 20 in. long, are touched together, brought into the field surrounding the sphere, and then separated. The charges on the plates are measured with a ballistic galvanometer and are shown to be equal and opposite.

To achieve the best results, certain precautions should be taken. The edges of the plates should be rounded and smooth to prevent sparking effects. A high-voltage resistor must be added in series to the charging circuit if the generator produces more than a few μA. The demonstrator should wear a grounded handcuff to avoid body effects.

29–2.4

A water droplet model demonstrates the motion of a stream of charged particles in an electric field. A pulsed stream of water breaks up into droplets on emerging from a jet, and the motion of the stream of droplets is shadow-projected and viewed stroboscopically, as illustrated schematically in Fig. 29–46. Reservoir *a* supplies water to the jet *f*. The flow is regulated by the clamp *h* and pulsed at 120 drops/sec by the vibrator *g*. The illuminating sys-

Fig. 29–44

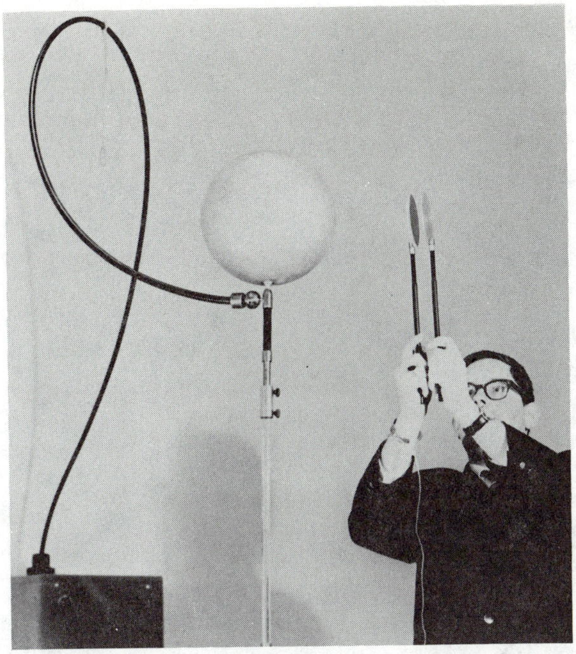

Fig. 29–45

tem, consisting of an arc *e* and a stroboscopic disk *b*, driven by a 60-cycle, 1800-rpm synchronous motor *c*, projects the shadow of the droplets upon a screen *i*. The vibrator[22] consists of a small electromagnet run on ac, driving a metal reed. The rubber tubing connecting the reservoir to the jet passes between the reed and a stationary block so that it is squeezed as the reed vibrates.

In an improved version the water reservoir is dispensed with and the vibrator and rubber tube are modified to form a simple pump by the addition of two valves. The water is caught and recycled continuously.

The droplets are given an electric charge, and they subsequently pass through an electric field between two plates. The induced charge on the drops can be increased by a metal ring placed near

[22] Such vibrators are available from several scientific apparatus manufacturers (as apparatus intended for Melde experiments). A complete assembly is also available from Central Scientific Co., water drop parabola demonstrator, Catalog No. 76030.

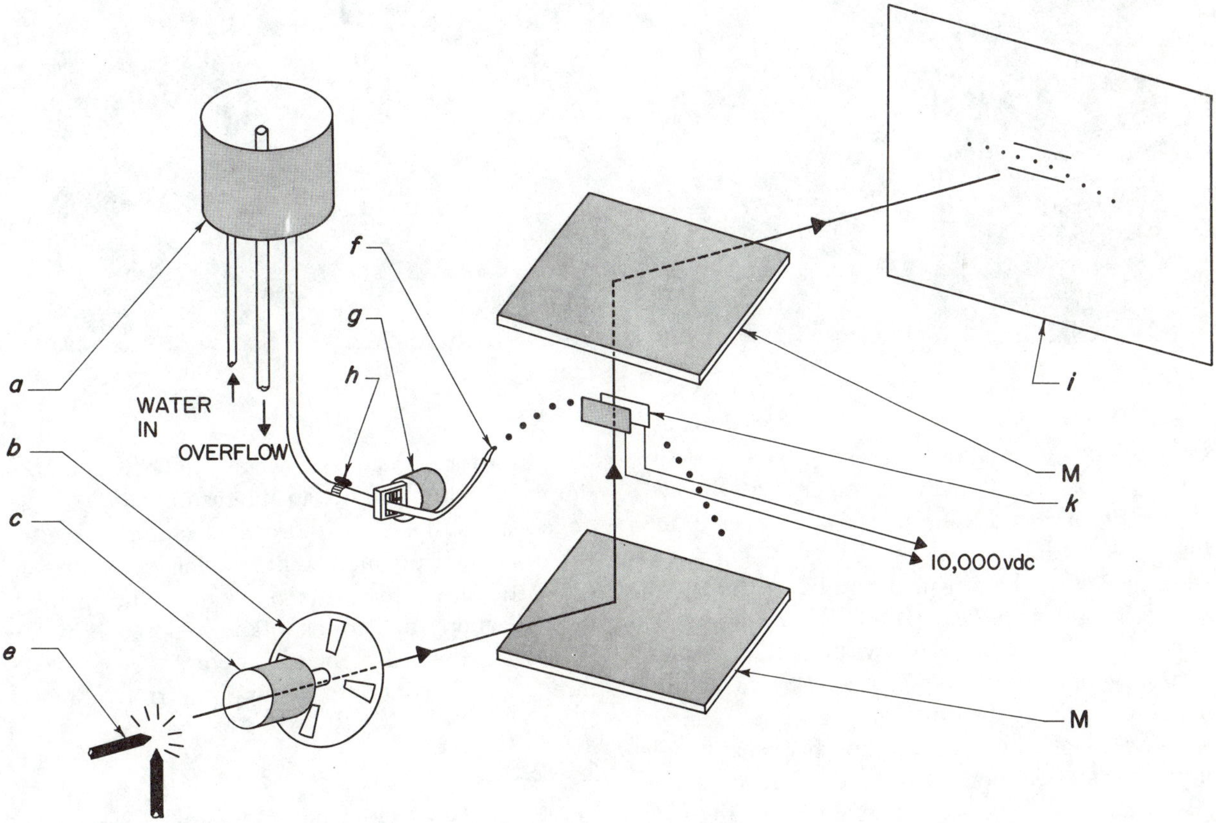

Fig. 29–46

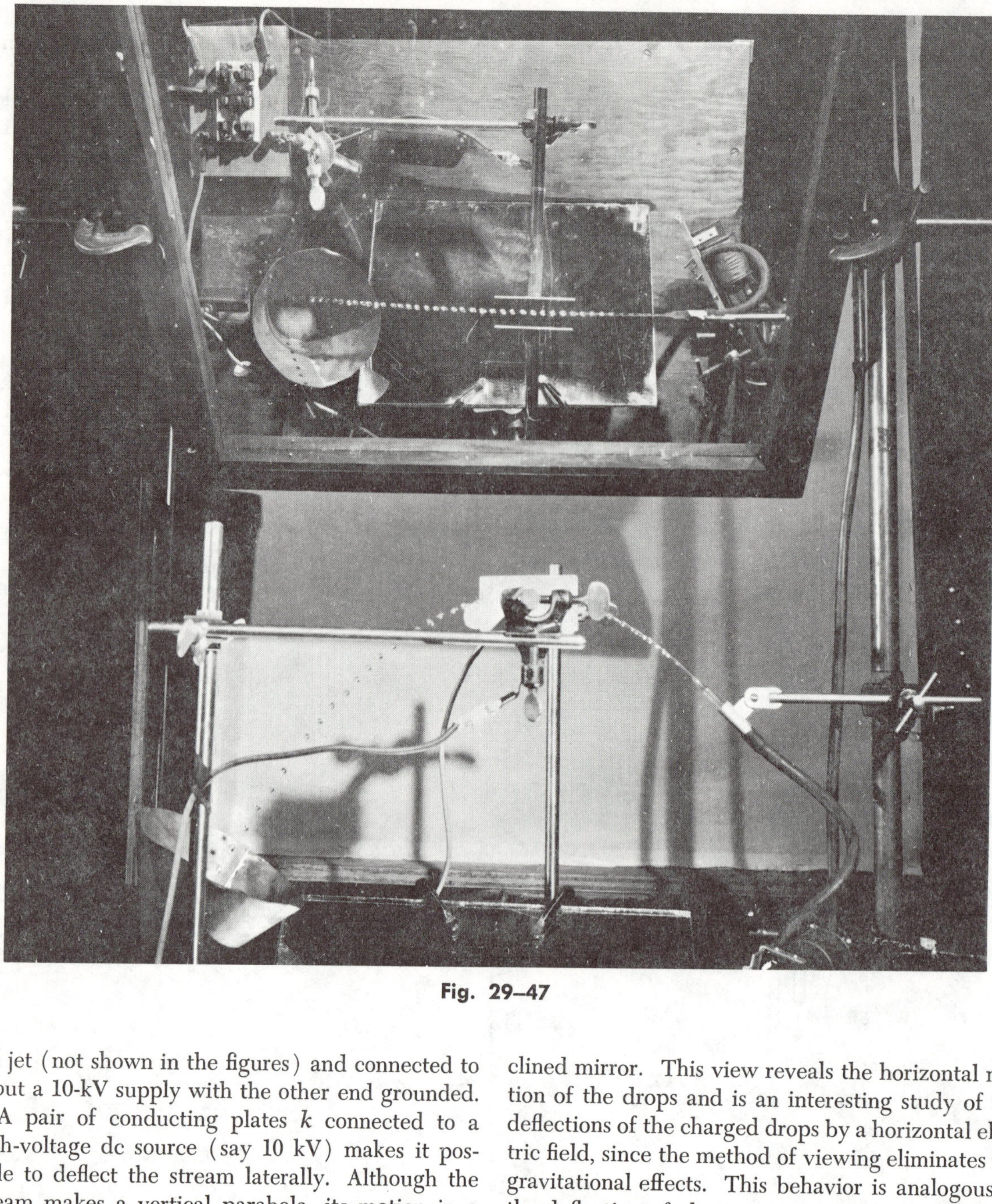

Fig. 29–47

the jet (not shown in the figures) and connected to about a 10-kV supply with the other end grounded.

A pair of conducting plates *k* connected to a high-voltage dc source (say 10 kV) makes it possible to deflect the stream laterally. Although the stream makes a vertical parabola, its motion in a vertical plane is concealed by an optical arrangement: two mirrors *M* and *M*, set at 45 deg with the horizontal, direct the stroboscopic light beam upwards through the stream drops between the deflecting plates *k* and thence to the screen. Figure 29–47 shows the complete assembly with the parabolic jet and the top view as seen in the upper inclined mirror. This view reveals the horizontal motion of the drops and is an interesting study of the deflections of the charged drops by a horizontal electric field, since the method of viewing eliminates the gravitational effects. This behavior is analogous to the deflection of electrons in an oscilloscope by the horizontal deflecting plates. Figure 29–48 shows three views looking into the upper mirror.

29–2.5

The model of Millikan's oil drop experiment described in this demonstration levitates tiny glass

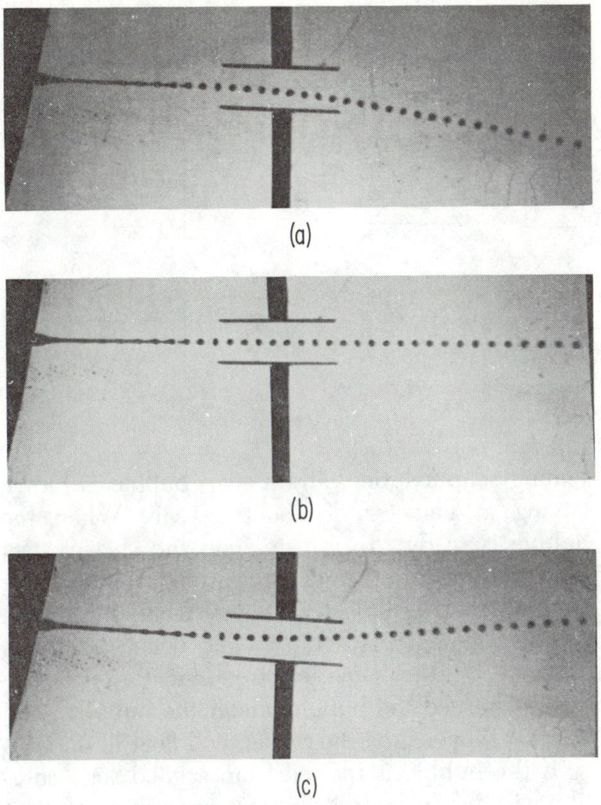

Fig. 29–48

balls in oil by the force exerted by an inhomogeneous electric field on the induced electric dipole moment in the balls. Despite this important difference from the Millikan experiment, some of the latter's essential features can be demonstrated to a large class with this model.

The glass balls, representing the oil droplets in Millikan's experiment, are used in an electrode cell (Fig. 29–49). The cell is made of $\frac{3}{32}$-in. glass plates cemented together as shown. The cover plate (Fig. 29–50) with the electrode leads *m* and *n* passing through it rests loosely on top of the cell. The electrode leads are #8 threaded brass rods, 3 in. long, whose positions can be adjusted by the locking nuts *l*. To the lower end of the rod *m* is soldered a brass plate *p* in which a small cavity *q* has been formed so as to lie immediately below the end of the rod *n*, which terminates in a brass disk *r*.

Fig. 29–49

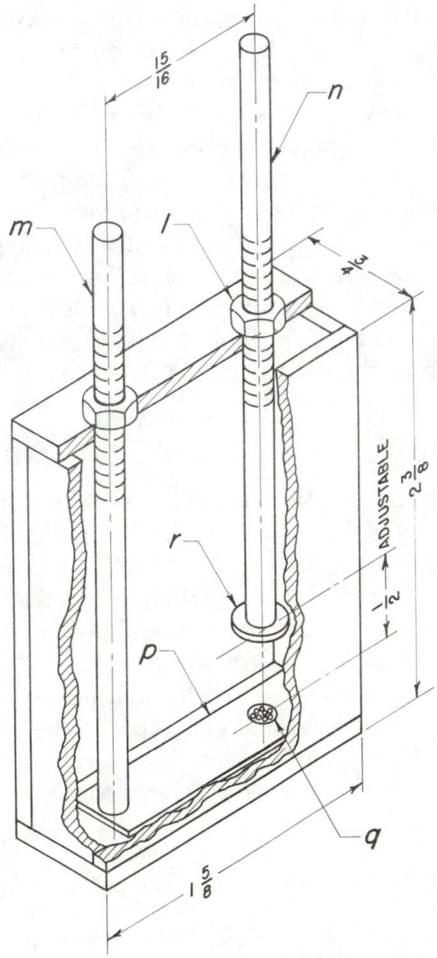

Fig. 29–50

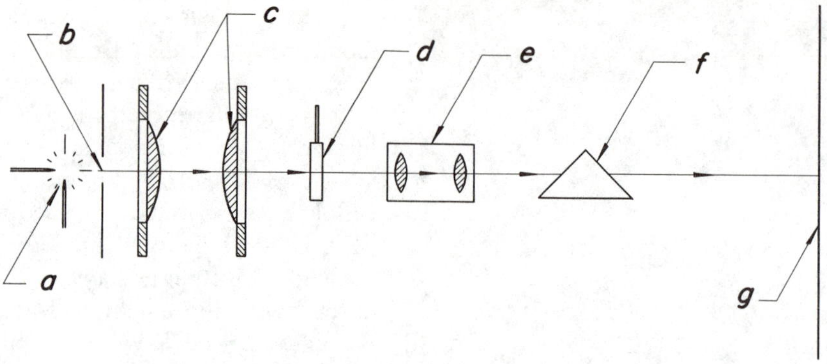

Fig. 29–51

The electric field between the electrodes q and r is produced by applying a dc potential to the terminals m and n. A number of glass spheres[23] 0.02 in. in diameter are placed in the cavity q, and the cell is filled with clear mineral oil. By manipulating the voltage up or down to a maximum of 8 kV, the glass balls can be made to rise or fall. The optical arrangement is shown in Fig. 29–51. The projection light source is an arc lamp a. After passing through the aperture b, the beam passes through two condensing lenses c, the cell d, a projection lens system e, a prism f to invert the image, and then on to a screen g. All of these components except the screen are mounted in line on an optical bench. A lantern slide projector with a projection voltmeter allows the applied voltage readings to be viewed on the same screen as the action in the cell.

29–2.6[24]

A racket, similar to a tennis racket but somewhat smaller and having an insulated handle, is made from metal screening c (Fig. 29–52). One end of a glass tube, bent at a right angle, terminates in a metal sleeve b. To the other end a rubber tube a, about 14 in. long, is connected. The glass tube is set up on an insulated stand, and a wire is attached to the metal sleeve and connected to the electrode of an electrostatic machine.

The tip of the sleeve is submerged in a cup containing soap solution, and a soap bubble is blown having a diameter of about 1½ in. When the bubble is ready to separate from the sleeve, start the rotation of the electrostatic machine. The charged soap bubble tears away from the sleeve. At this moment place the racket, which has been charged by the same electrode of the machine which charged the bubble, under the bubble. The bubble, repulsed by the racket, will float in mid-air.

If the bubble is formed from a good soap solution with no drops on its bottom surface, it can float this way for half a minute or longer. By moving the racket under the bubble, one can force the bubble upwards, making it float about the auditorium. This demonstration, however, requires considerable preliminary training. If the air in the auditorium has been contaminated with smoke, the demonstration is often not very successful.

29–2.7

The apparatus shown in Fig. 29–53 is used to show the basic theory involved in the famous Millikan oil-drop experiment and to illustrate electrostatic force. In the classical experiment a small drop of oil is acted upon by three forces: an upward force produced by a vertical electrostatic field, a downward gravitational force, and a viscous force acting in a direction opposite to the motion of the drop. By measuring the velocity of the drop before and after the electric field is applied, the charge on the drop can be determined, and it is always found to be an integral multiple of a unit charge, which is the charge on an electron.

A small Wimshurst machine is used to create the

[23] Available from Kimble Glass Co., Boston, Massachusetts.

[24] Translated by K. V. Robinson and H. A. Robinson (Adelphi University) from *Lecture Demonstrations in Physics*, A. B. Mlodzeevskii, ed., M. V. Lomonosov State University, Moscow, U.S.S.R.

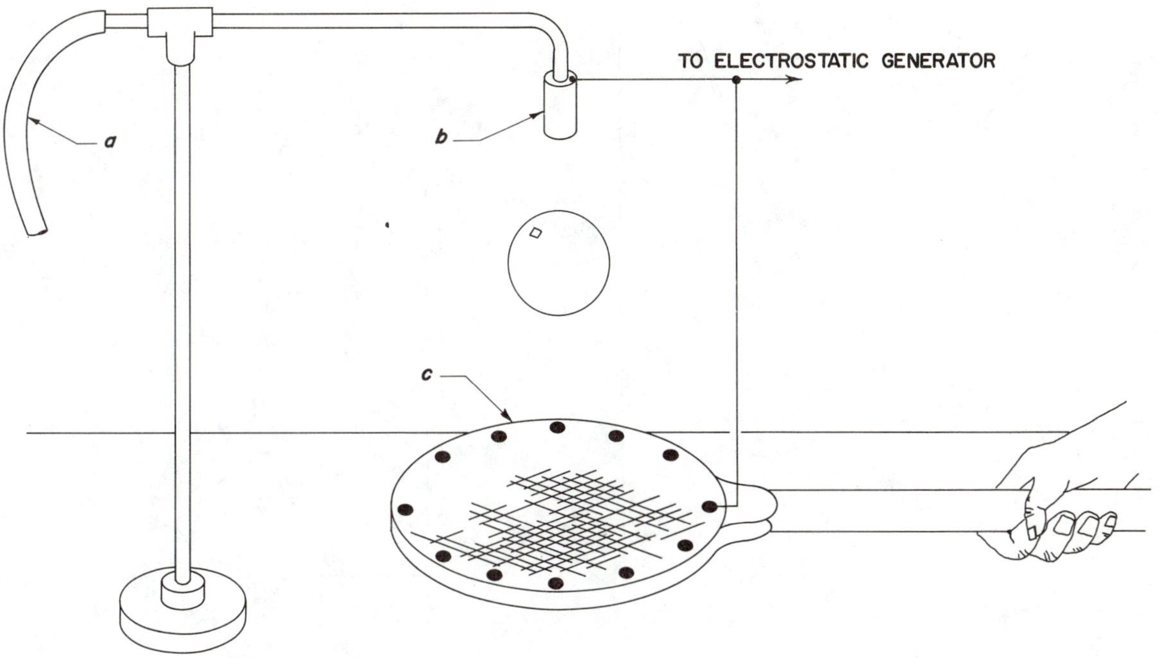

Fig. 29–52

electrostatic potential applied to the circular plates. A light foam plastic ball replaces the charged oil-drop of the classical experiment. Since the terminal velocity in air of a body this size is too large for a reasonable demonstration, viscous effects cannot be considered. Instead the charge on the ball is computed by balancing the effects of gravity and the upward electrostatic force. The condition for maintaining the charged sphere in a stationary position between the plates is

$$qE = mg,$$

where q is the total charge on the sphere, E the electric field, m the mass of the sphere, and g the acceleration due to gravity. If we neglect fringing effects near the edge of the plate, E is equal to V/l, where V is the applied potential and l is the separation between the plates. The charge on the ball is then calculated from

$$q = mgl/V.$$

It is necessary to use either an electrostatic voltmeter with the proper range or a spark gap in order to measure V, which can vary between 15,000 and 30,000 V. The graph of spark gap distance vs

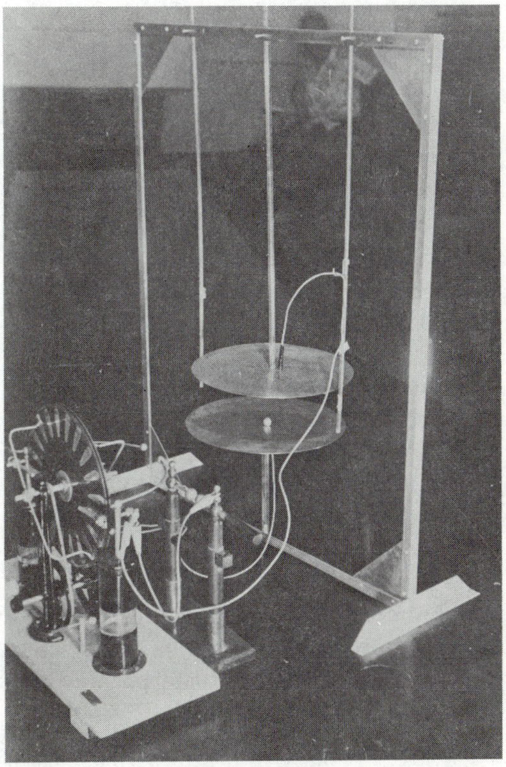

Fig. 29–53

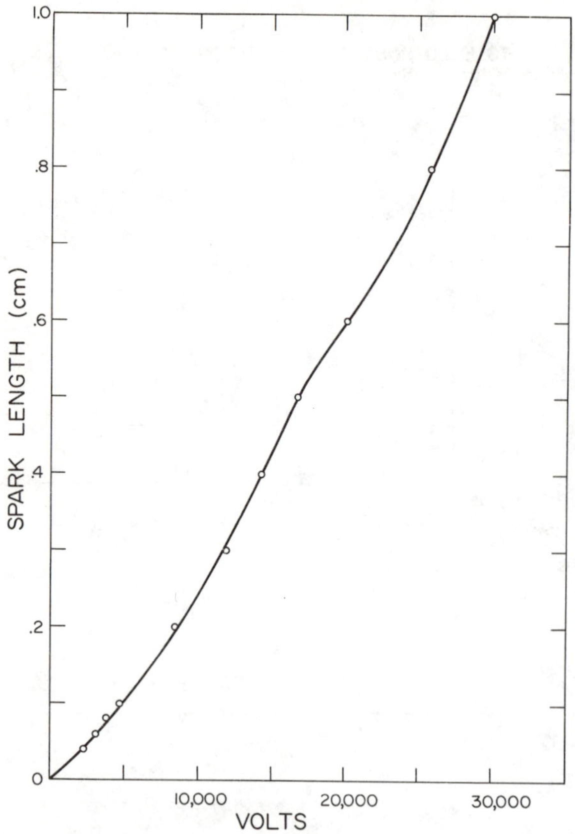

Fig. 29–54

steady potential difference for ball electrodes 4 cm in diameter is shown in Fig. 29–54.[25] The values from the graph are for initial breakdown voltage and not the potential for continuous sparking, which is considerably lower. When measuring the voltage with a spark gap setup, the Wimshurst machine should be driven by a motor and a variable transformer at an adjustable low speed. (A different electrostatic generator with an easily controlled current or voltage output might also be used.) The spark gap is opened wide, and while the system is maintained in the condition needed to hold the ball suspended, the gap is reduced until a spark occurs. Figure 29–54 is then used to convert the gap setting to a voltage reading.

The design of the apparatus is illustrated in Fig. 29–55. The supporting rods *a*, *b*, *c*, and *d* are made of Plexiglas. The very light foam plastic ball *e* is

[25] Plotted from tables in *Smithsonian Physical Tables*, W. E. Forsythe, ed., 9th revised edition (Smithsonian Institute, Washington, D.C., 1954), p. 421.

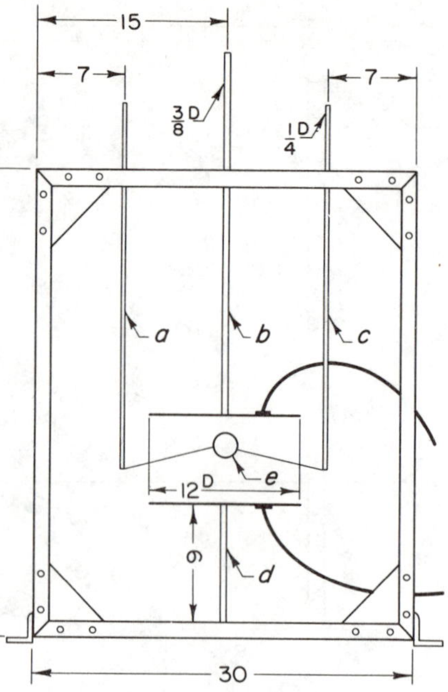

Fig. 29–55

an ordinary Christmas tree ornament, which is made conducting by gluing electroscope-leaf aluminum foil to most of the surface. It is necessary to peel off all the loose foil to eliminate rapid dissipation of the charge. Including the foil, the ball should weigh about 18 mg and have an average diameter of 1 cm. In order to maintain the ball in the field, a synthetic fiber runs through the ball horizontally and is tied to *a* and *c* at each end. The ball is allowed to touch the lower plate but is held back an inch or more from the upper plate to keep it from discharging.

29–2.8

It can be shown that surface charge density on a conductor is proportional to the electric field strength (Fig. 29–56). A can *a* is connected to a 150-V dc source. The charge on the surface is demonstrated by touching one lead from a grounded 25-W (150-V) light bulb *c* to the edge of the can. Can *b* is connected to an electrostatic projection voltmeter[26] *d* (range 0–250 V), which consumes no current.

The first part of the demonstration consists of moving a metal ball on the end of a quartz handle

[26] Can be obtained from Trüb, Täuber and Co., 8634 Hombrechtikon, Zurich, Switzerland.

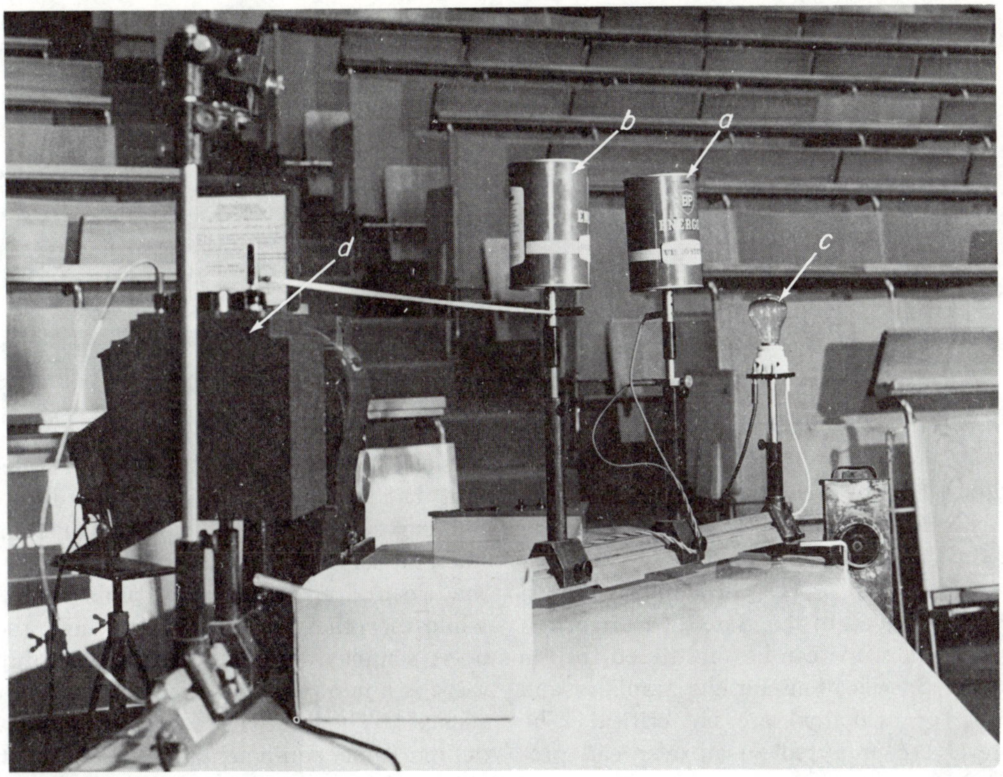

Fig. 29–56

back and forth between the edge of can *a* and the edge of can *b*. The voltage on can *b*, which is insulated from its support, rises exponentially to a value $V_0 = 150$ V.

With can *b* at V_0 the charge is then transferred from the edge of can *a* to the inside bottom of can *b* without touching the walls of the can. The voltage will keep rising on the voltmeter beyond V_0 since the field is zero inside the can, thus proving the principle that the charge is completely transferred from the sphere to the can no matter how high the potential is. The last part of the demonstration involves touching the bottom of can *b* with the metal sphere and then transferring this charge to the bottom of can *a*. The voltage on *b* drops off while the voltage on *a* remains constant.

29–3 Electric Potential

29–3.1

Water in a gravitational field flows down a hill at right angles to contour lines, each of which is at a particular altitude (potential) above sea level. Similarly, there are lines of flow in an electric field that are everywhere perpendicular to equipotential lines; that is, they form mutually perpendicular, or orthogonal, sets of lines. Hence, if equipotential lines can be found, lines of flow and lines of force can be located.

Apparatus for doing this is shown in Figs. 29–57 and 29–58. A shallow, glass-bottomed tray is filled with tap water to a depth of about ¼ in. In the bottom of the tray is placed a glass plate (9¾ x 13¾ x ⅛ in.) ruled with a coordinate grid. Each grid line is numbered or lettered as shown. Two electrodes are placed in the water at each side of the tray and are connected to an audio frequency oscillator set at 1000 cps. If direct current is used in this circuit, hydrogen gas (an insulator) will form at the negative electrode, eventually cutting off the

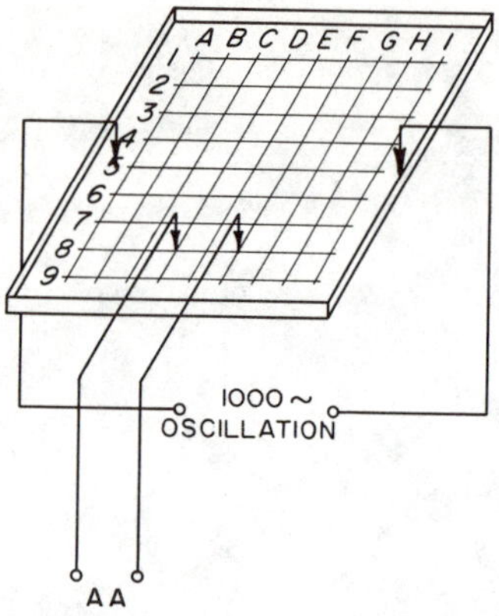

Fig. 29–57

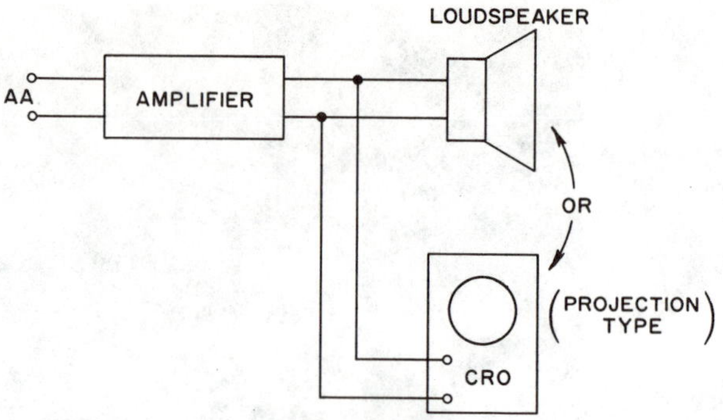

Fig. 29–58

current flow. To avoid this problem, alternating current is used. With the audio oscillator set at 1000 cps, the gas does not have time to collect at either electrode. Two probes, an amplifier, and a loudspeaker are added to the system (a projection cathode ray oscilloscope can be substituted for the loudspeaker). Specifications for the amplifier and loudspeaker (or oscillator) are not critical. The tray and grid are projected on an overhead projector.

In the water different electrodes are paired to produce different field configurations as shown in Fig. 29–59. Other objects can be added (as shown) to modify the fields. As the current passes through the water, the conducting ions move along the electric lines of force. This is so because the ions travel so slowly and consequently have so little momentum that they do not overshoot the lines of force, but curve with them. To find the equipotential lines, immerse the two probes in the water between the electrodes. One of the probes is kept fixed while the other is moved about until there is no noise coming from the loudspeaker. (In practice, there is a minimum of sound rather than complete silence.) The locations of the two probes are points on the same equipotential line. When the movable probe is not on the same equipotential line, the current flowing through the water is diverted through the loudspeaker and can be heard as a 1000-cycle note. To locate more points on the same equipotential line, the fixed probe is kept stationary and the movable probe is repeatedly shifted closer to or farther from it and again moved about until

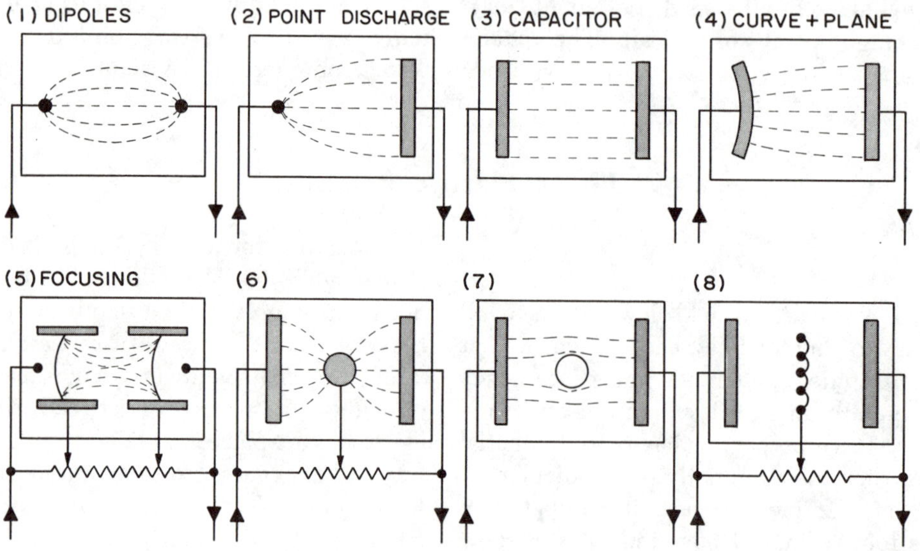

Fig. 29–59

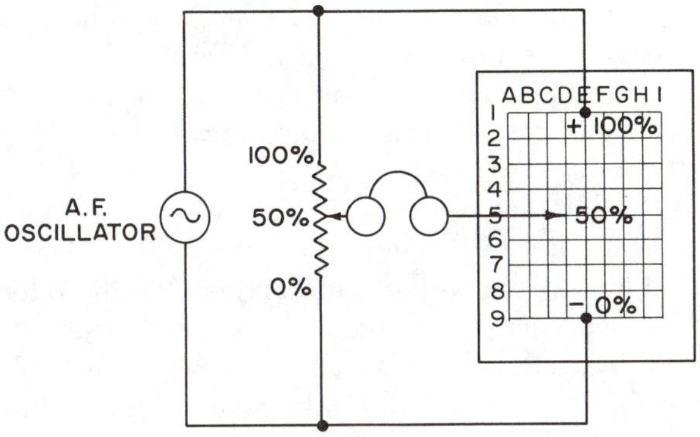

Fig. 29–60

the noise is at a minimum. To locate a second equipotential line, the fixed probe is shifted and the entire process is repeated. A whole family of equipotential lines can be plotted in this manner.

A "one-probe" system can be used which makes it possible to tag each equipotential line in terms of a percentage of the total voltage drop. As is shown in Fig. 29–60, ten equal resistances are connected in series. With the fixed probe connected in the middle of the resistances, the other probe is moved about as previously described. The plotted equipotential line has a potential of 50 percent of the total drop between the electrodes. Other equipotential lines can be located with this "one-probe" system by connecting the fixed probe to other points between the resistances.

For class participation during lecture, one student operates the probes and calls out the coordinates. A second student writes them on the blackboard in tabular form and in crude graphical form. The other students plot the equipotential lines at their seats. The electric lines of force are then drawn in. They will be perpendicular to the equipotential lines and to the electrodes.

29–3.2

The potential gradient is measured by using the two-pointed probe, Fig. 29–61 (a). The probe consists of two cylindrical metal wires fixed diametrically opposite in grooves on a plastic cylinder so that there are pointed projections at one end. At the other end, the wires are connected to a galvanometer. The pointed ends are dropped into electrolyte, and the potential difference between them is measured by the galvanometer.

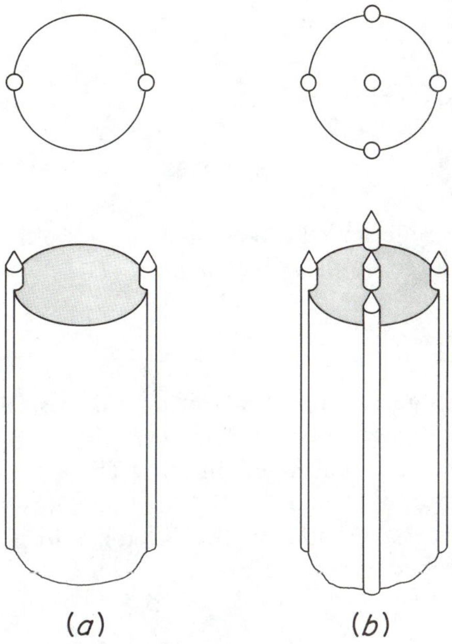

Fig. 29–61

To measure the second derivative of potential, the five-pointed probe, Fig. 29–61 (b), is used. Four of the wires are placed symmetrically on the outer surface of the cylinder and the fifth is placed along the axis. The potentials measured by the outer four points are added by a summing circuit and then divided by four. The difference between

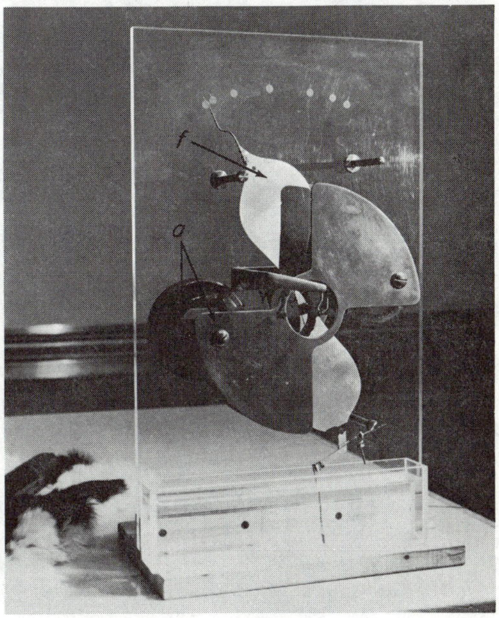

Fig. 29–62

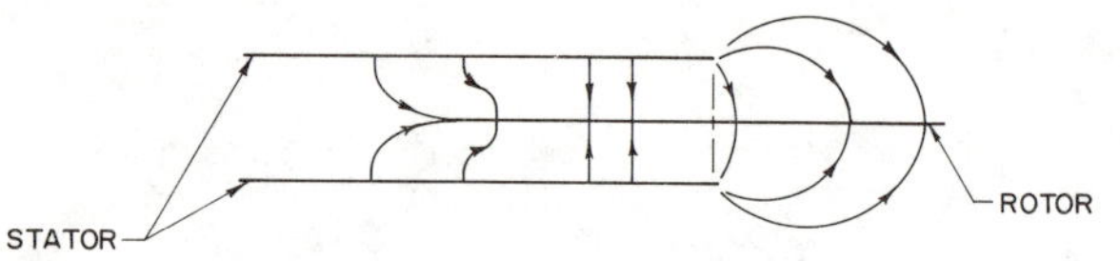

Fig. 29–63

this average and the reading for the central point is measured by the galvanometer.

29–3.3

The stator *a* and rotor *f* of an electrostatic voltmeter (Fig. 29–62) are charged to different potentials by any convenient means. The equilibrium position of the rotor relative to the stator is then taken as an indication of the difference in potential existing between them.

The principle of the electrostatic voltmeter is attraction or repulsion of conductors at different potentials. The electric field established by the stator and rotor appears roughly as shown in Fig. 29–63.

With the stator assumed to be positively charged and the rotor to be negatively charged, the electric field lines begin on the stator and end on the rotor. The field energy decreases as the rotor penetrates into the stator; therefore an attractive force is obtained. The force is also attractive for reverse polarity; thus the response is quadratic. Since an exact theoretical expression for the potential in terms of the position of the rotor is difficult to obtain, the apparatus must be calibrated experimentally. The construction details are given in the Appendix, page 1320.

29–4 Capacitors and Dielectrics

29–4.1

A demonstration capacitor large enough to be seen at a distance of 30 ft is shown in Fig. 29–64. The potential difference between the plates can be read to an order of magnitude and changes in the potential difference of the order of 10 percent.

A vertical Zeiss optical bench bar supports two

Fig. 29–64

6 x 8 x ¼-in. aluminum plates *a* (Fig. 29–65) so that they lie flat, one above the other, in horizontal planes. The upper plate is insulated and the lower is grounded. The lower one can be raised or lowered to change spacing and/or can be swiveled about its central normal axis through 90 deg to vary the plate area. A 6 x 8 x 1-in. Lucite plate *b* is also mounted on the bench bar so that it can swing freely in and out of the space between the aluminum plates on pivot pin *c*. The Lucite plate is counterbalanced by a lead counterweight *d*. Potential difference readings are projected from a calibrated electroscope.

Several demonstration experiments can be conducted with this apparatus:

1. *Dependence of capacitance on area at constant charge.* The lower plate is rotated about its normal axis. Changes in potential difference are projected on a screen.

2. *Dependence of potential difference on vertical spacing of plates at constant charge.* The capacitor is charged to approximately 2000 V and the spacing is changed by manipulating the lower plate while observing the potential difference on the electroscope.

3. *Dependence of potential difference on dielectric medium at constant charge.* The capacitor is charged to approximately 2000 V and the dielectric material introduced into the space between the plates while the potential difference is observed on the electroscope.

4. *Measurement of capacitance.* With the electroscope attached to the upper plate, the capacitor is charged to a potential difference of approximately 1000 V. The electroscope is next disconnected and the capacitor is recharged using the same combination of batteries. The capacitors are subsequently discharged through a ballistic galvanometer.[27] The ratio Q/V is quickly determined. The unchanging ratio is noted.

5. *Result of charge growth.* A physically large radio capacitor is placed in parallel with the plates, and the capacitor is charged to 1000 V. While the potential difference is observed on the electroscope, the effective area of the radio capacitor is slowly decreased. The potential difference increases in an impressive way, that is, it goes up as high as 3000 V, whereupon a spark forms, discharging the capacitor.

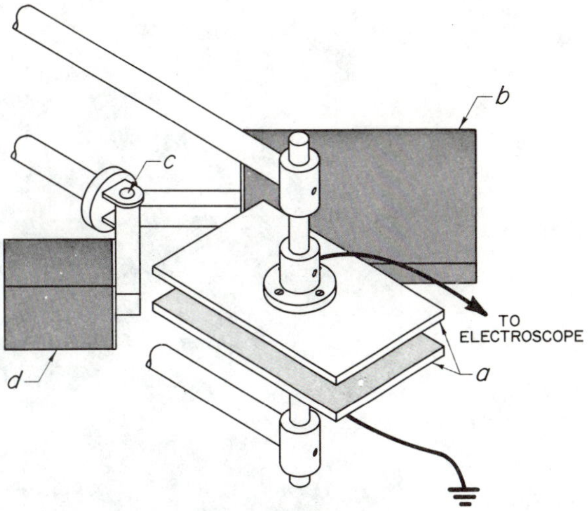

Fig. 29–65

6. *Charged sphere experiment.* A large conducting sphere approximately three feet in diameter is attached by means of a nylon thread to the ceiling and suspended at least six feet from water pipes, radiators, etc. It is charged by means of the batteries to approximately 2000 V and discharged through a ballistic galvanometer. In this manner the capacitance (and therefore the radius) of an isolated spherical conductor is measured.

29–4.2

The charging and discharging of a capacitor, and the storage of energy within it, is demonstrated by the apparatus in Fig. 29–66. A large capacitor *a* (300 μF, 250 V ac) is charged through a light bulb *b* (220 V, 10 W) from a potential of 300 V dc, as shown in Fig. 29–67. Bulb *b* lights up while the capacitor is being charged; bulb *d* is disconnected by switch *c* during the charging. When the capacitor is discharged by switch *c* through bulb *d*, it will light up momentarily to the same brightness as did bulb *b* during charging.

When the demonstration is repeated with 220 V ac, bulb *b* will continue to be brightly illuminated as long as the capacitor is in the circuit. The in-

[27] The ballistic galvanometer used in this experiment was a Leeds and Northrup Model 1429. It has a charge sensitivity of 10^{-8} C/cm at a scale distance of 125 cm. Deflections are easily observable on a translucent screen.

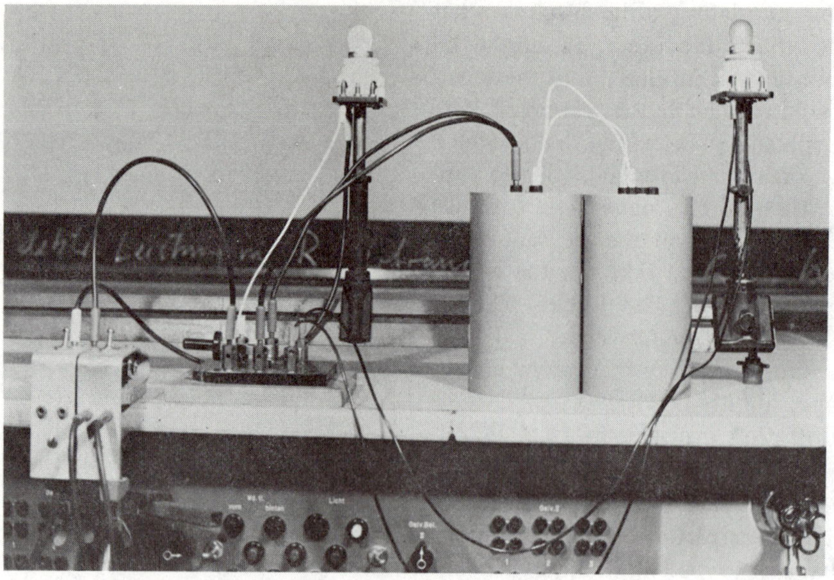

Fig. 29–66

stant switch c is thrown to connect bulb d across the capacitor, bulb b goes out; bulb d does not light up, showing that it is not possible to store charge in the capacitor in the same manner as the dc instance. A steady charge does not accumulate on a capacitor with ac charging.

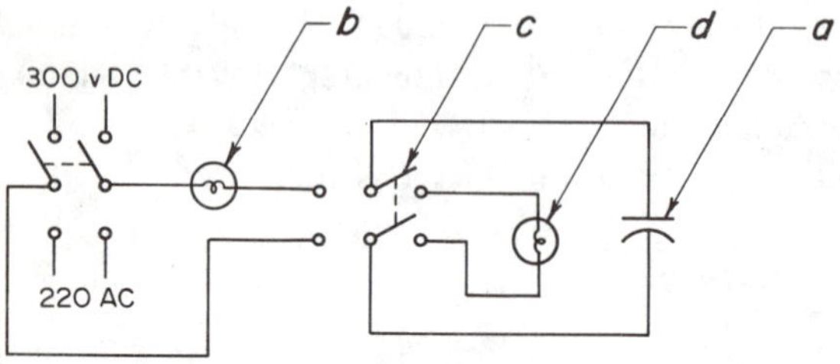

Fig. 29–67

29–4.3

A gas-filled bulb can be used to detect the rise in potential between two sheets of aluminum foil sandwiched between two boards and separated by a dielectric. The circuit diagram is given in Fig. 29–68. A 90-V battery is used as a potential source,

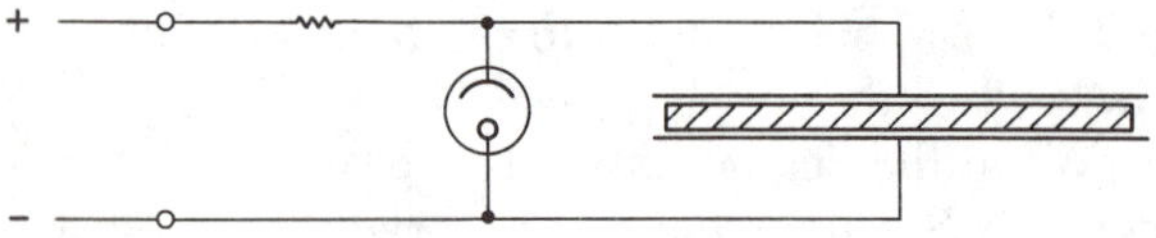

Fig. 29–68

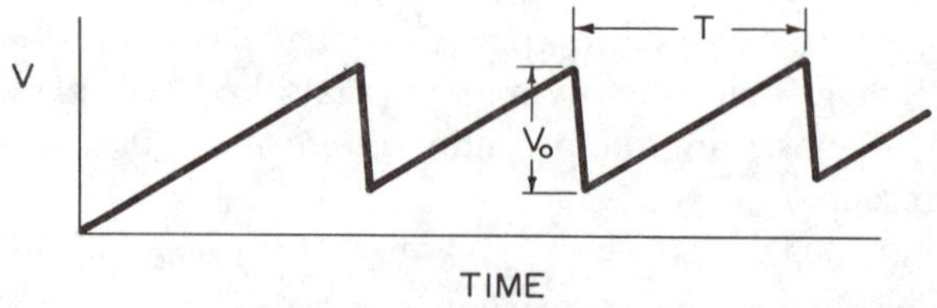

Fig. 29–69

and a 22-MΩ resistor is used. When the potential across the bulb reaches a fixed maximum value, the gas begins to conduct, and most of the charge on the capacitor goes through the bulb. The potential across the bulb and capacitor drops suddenly. The gas then ceases to conduct, and the capacitor begins to recharge (Fig. 29–69).

The light flashes of the bulb are timed, and the capacitance can be computed even though the current and potential values cannot be measured directly. They are characteristic of the circuit. From the relationship

$$q_0 = iT = CV_0,$$

where

$$C = (i/V_0)T,$$

the value of i/V_0 can be determined by placing a capacitance of known value in the circuit and measuring the light flash intervals. The current is considered constant for simplicity. Once the value of i/V_0 is known, the dielectric constant for various

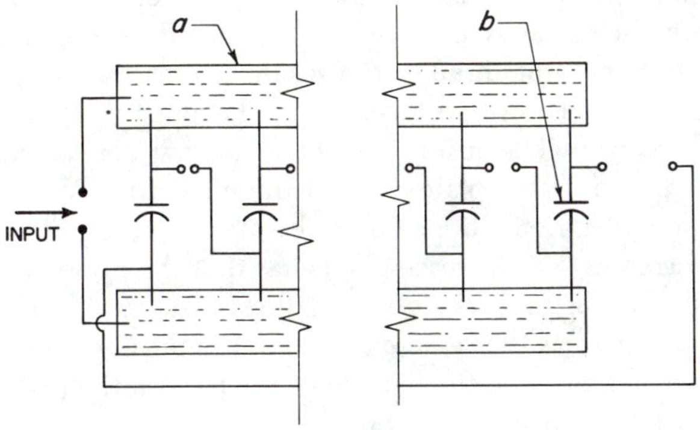

Fig. 29–70

materials, such as a Mylar sheet,[28] can then be found by simply measuring T. See Appendix, p. 1321.

29–4.4[29]

The Arkad'ev capacitor-bank transformer makes it possible to obtain very high voltages by automatic electric switching of charged capacitors from a parallel to a series connection. This principle is used to obtain short-lived high voltages for the acceleration of ions.

The principle may be shown with the circuit in Fig. 29–70. The two resistances a each total 240 kΩ and can be either ceramic or liquid resistances. One of the simplest liquid resistances consists of a molybdenum glass tube with sealed-in molybdenum wires filled with a copper sulfate solution (concentration of 0.4 g per liter of water). Eight to ten Leyden jars b, each with a capacitance of about 1000 C, are connected between the series of resistances. The inner lining of each jar is connected through a spark gap to the outer lining of the adjoining jar. The first gap is adjusted in relation to the voltage feeding the setup, and each successive gap is larger than the preceding one. The gaps should have spheres with a diameter of 1.5–2 cm. Distances between the spheres should be adjustable, preferably with a micrometer, and are selected as the demonstration is set up.

When the difference in the potentials between the linings of the jars reaches the disruptive voltage for the first gap, the gap will break down, connecting (by means of the spark) the inside lining of the first jar to the outside lining of the second. As a result, the potential of the second jar will increase, and the second gap will break down, etc. The breakdown of the last gap will discharge the entire battery of jars in a spark 10–20 cm long. The resistances of the two series must be high enough to prevent the series-connected jars from discharging during the breakdown of the gaps. At the same time, however, the jars must be simultaneously charged to an equal potential through these resistances. A high-voltage power supply or an electrostatic machine can be used to charge the jars.

29–4.5[30]

A corrugated paper Chinese lantern which has been covered with aluminum paint is used to demonstrate how the capacitance is changed when the dimensions of a capacitor are varied. The lantern is folded like an accordion and placed on a disk connected to an electrometer. A lead weight is placed in the bottom of the lantern so that it will sit tightly on the disk. Charge the lantern with an electrified rod and observe the deflection of the electrometer pointer. Hook an ebonite rod into the lantern handle and stretch out the lantern, thus changing its dimensions and its capacitance. The potential will rapidly decrease and the electrometer pointer will decrease its deflection. Fold the lantern again and observe the increase in the deflection of the electrometer pointer.

29–4.6

It is well known that, due to polarization, repeated charges can be drawn from a Leyden jar. Furthermore if the jar is demountable, touching the inner and outer conducting surfaces after charging will result in no spark if the dielectric is removed. The spark will subsequently appear, however, after the jar has been put back together with the dielectric in position.

The effect of dielectric polarization can be shown by connecting one of the electrodes of the Leyden

[28] Available from Cadillac Plastic and Chemical Co., Detroit, Michigan.

[29] Translated by K. V. Robinson and H. A. Robinson (Adelphi University) from *Lecture Demonstrations in Physics*, A. B. Mlodzeevskii, ed., M. V. Lomonosov State University, Moscow, U.S.S.R.

[30] Translated by K. V. Robinson and H. A. Robinson (Adelphi University) from *Lecture Demonstrations in Physics*, A. B. Mlodzeevskii, ed., M. V. Lomonosov State University, Moscow, U.S.S.R.

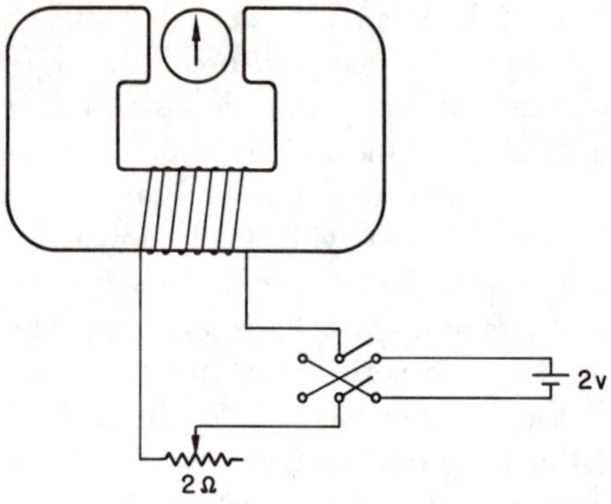

Fig. 29–71

jar to one of the electrodes of a neon discharge tube. The correct neon tube to use is the one designed for use as an indicator on a 115-V ac line. The normal pair of discharge tongs is then used to complete the circuit between the other electrode of the Leyden jar and the other electrode of the discharge tube. Note, however, that the Leyden jar must first be discharged once before it is connected to the discharge tube, or else the tube may burn out. One hundred or more successive discharges can be shown by this method.

It is also possible to show the polarity of the charges on the two surfaces of dielectric. In this connection it is noted that, when the grid of a triode is made positive, the anode current increases, and conversely, when the grid of a triode is made negative, it decreases. In using this behavior the Leyden jar is held in one hand and charged with the center knob (inside shell) positive. The inside shell is now removed with insulated tongs and the outside shell and dielectric set on an insulated metal plate connected to the grid of a triode. When the dielectric is removed by insulated tongs, the anode current through the triode decreases due to removal of positive charges on the outside of the dielectric. The inside electrode is now connected to the insulated plate without the dielectric. When the dielectric is slipped over the electrode, the plate current again decreases, showing that the approach of the negative charges (on the inside of the dielectric) has the same effect as the removal of positive charges (on the outside of the jar) on the plate current.

29–4.7

A small, magnetized needle on a pivot is placed in a Plexiglas box filled with gasoline. The box is sealed and then placed between the poles of an electromagnet which has 20 turns of wire connected through a variable resistor to a polarity changing device (Fig. 29–71). The polarity changer is detailed in Fig. 29–72. It is made from thin brass strips set into a Plexiglas cylinder attached to a variable-speed motor whose speed can be varied from 0 to 1000 rpm. Four input brushes feed strips half-encircle the cylinder, and two output brushes contact strips completely encircle the cylinder. The strips are connected as shown in the schematic drawing of the "unwrapped" surface of the cylinder (Fig. 29–73).

In operation, the needle is aligned with a small magnet which is then removed. The motor is started and run at a low speed. While the polarity is changing slowly, the magnet needle follows every

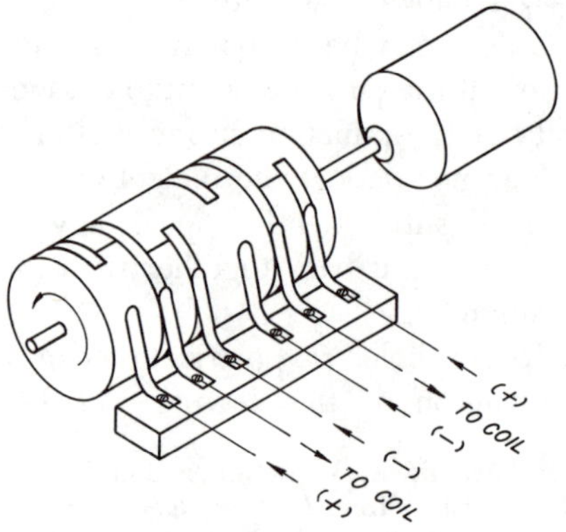

Fig. 29–72

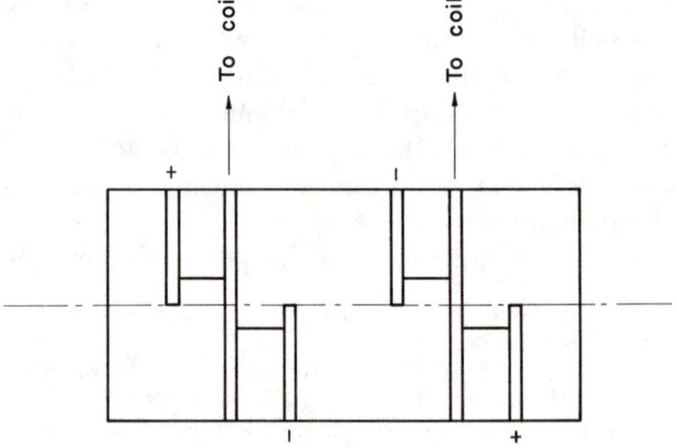

Fig. 29–73

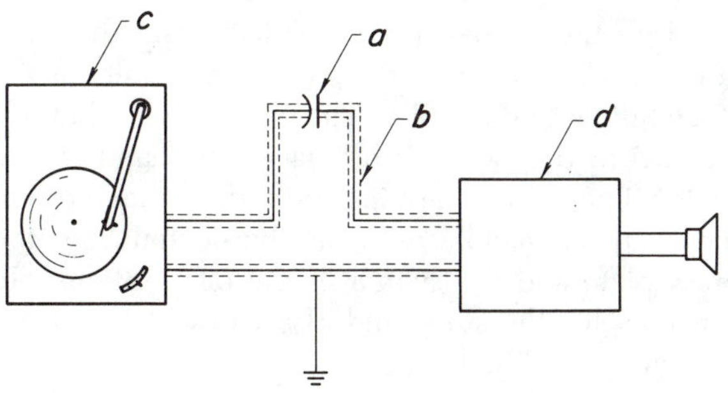

Fig. 29–74

change of the magnet field in the gap. However, when the frequency of polarity change is stepped up, the needle is unable to keep up with the change in the magnetic field, and will take a fixed position perpendicular to the magnetic field.

29–4.8

A brass parallel plate capacitor *a* (1½ in. diam) is placed in series with the input to an amplifier as shown in Fig. 29–74. Shielded line *b* is used between a phonograph *c* and the amplifier *d*. With a dielectric of air, glass, or Plexiglas, practically no sound is produced by the system. If a disk of $BaTiO_3$ is placed between the plates, the music is heard clearly, due to the very low impedance of the capacitor with a high dielectric constant. Figure 29–75 is a photograph of the apparatus.

29–4.9

Two pith balls, similarly charged, are supported from a rigid upright as two pendulums. One pendulum is set in oscillation. As it slows down, the other pendulum starts to oscillate. If the pith balls are not charged, this oscillation transfer does not occur.

29–4.10

A 45-V dc motor *a* (Fig. 29–76) is fitted with a pulley, a length of light cord, and a small weight. An electrolytic capacitor (525 μF) is charged to a higher potential than the rated voltage of the motor and is then discharged through the motor. The work done in lifting the weight may be compared to the energy stored in the capacitor. The vertical distance the weight moves is measured with a meter stick. Corrections are made for energy lost in the resistor.

29–4.11

A 2000-μF, 15-V capacitor *a* is mounted on an insulating base equipped with two prongs which allow it to be hung in a rack as shown on the left in Fig. 29–77. The rack has two built-in resistors *b* and is connected across the poles of a 12-V battery (not shown). The right rack is connected to a dc motor *c* with high impedance and low rpm (a dc watt-hour meter), which is used to discharge the capacitor. Motor motion can best be shown by shadow projection of a small marker on its shaft. With a sufficiently high motor impedance, discharge time can reach about 20 sec.

29–4.12

A flat, ground lithographic stone plate (3 x 5 x 1½ in.) and an exactly fitting brass plate (Fig.

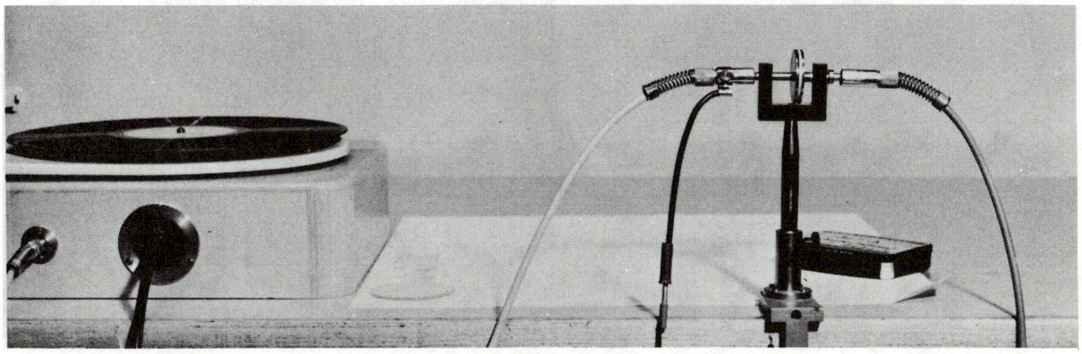

Fig. 29–75

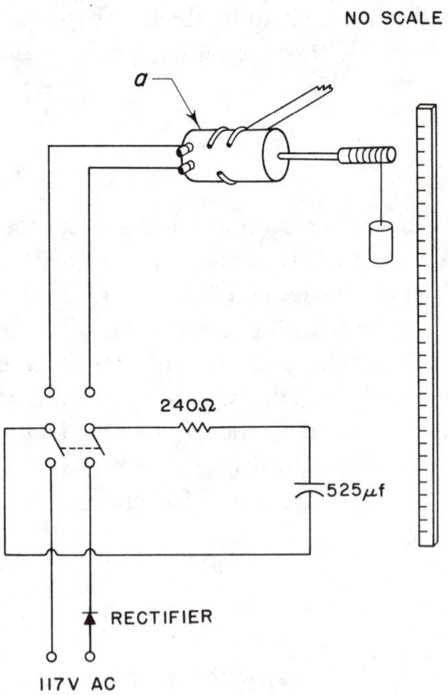

Fig. 29–76

29–78) are both connected to a 300-V dc source (200 V may do) by means of pressed-in plugs. The source line to the stone contains a 5-MΩ safety resistor, and the other line contains a switch.

With the switch turned on, the stone can be lifted off the padded table by picking up the insulated handle. Turning the switch off results in the stone falling to the table. The experimenter himself may act as the switch if he places one hand on the switch line source terminal and the other around the insulated handle with his thumb touching the brass plate and the switch in the off position. He can now lift the stone and also cause it to fall by removing his thumb.

29–4.13

Apparatus used at the University of Colorado to measure the force of attraction between charged parallel plates and the dielectric constant ϵ of a piece of material inserted between the plates is shown in Fig. 29–79. Two charged parallel plates are mounted on a triple beam balance so that the force between them can be measured. The forces are measured for separations of 3 and 6 mm at potentials of 2 and 4 kV. At a separation of 6 mm, a 6-in.-square by 3-mm-thick slab of cloth impregnated with epoxy is inserted between the plates, and the force is measured at a potential difference of 4 kV. The theoretical forces involved are calculated and compared with those measured. From the measured and theoretical forces, the dielectric

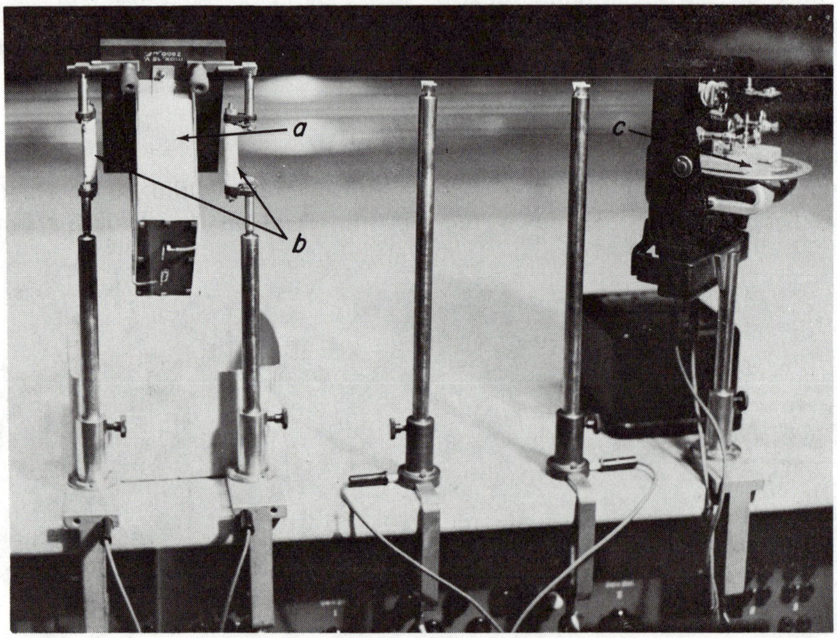

Fig. 29–77

Fig. 29–78

constant of the material inserted between the plates can be determined.

Similar apparatus, used at the Swiss Federal Institute (ETH), is shown in Fig. 29–80. Again two charged plates are attached to a balance so that the force between them can be measured. In addition, the potential difference between the plates is measured with a projection voltmeter. Construction details are given in the Appendix, page 1322.

29–4.14[31]

The pulling of solid and liquid dielectrics into homogenous electric fields may be shown in projection as follows. An elongated ellipsoid made of a solid dielectric, such as paraffin, is suspended between two flat electrodes connected to the poles of an electrostatic machine. The ellipsoid is suspended horizontally so that its long axis is parallel to the planes of the electrodes. When the electric field is turned on, the ellipsoid turns so that its long axis is in line with the lines of force. When the electric field is switched off, the ellipsoid returns to its original position.

A liquid dielectric, such as kerosene, is poured into a container with parallel sides. Two parallel electrodes are put into the kerosene and connected to a source of high voltage, such as a tube rectifier of 2000–3000 V. The distance between the electrodes should be 0.08–0.12 in. When the voltage is turned on, the liquid is drawn into the capacitor and the rise of the liquid between the plates is shown by projection.

In a nonhomogeneous electric field, a body whose dielectric susceptibility is greater than the susceptibility of the surrounding medium will be drawn into the region of the field having the highest intensity. On the other hand, if the dielectric susceptibility of the body is less than the susceptibility of the surrounding medium, then the body will be pushed out of the regions of the field having the highest intensity.

(31) Translated by K. V. Robinson and H. A. Robinson (Adelphi University) from *Lecture Demonstrations in Physics*, A. B. Mlodzeevskii, ed., M. V. Lomonosov State University, Moscow, U.S.S.R.

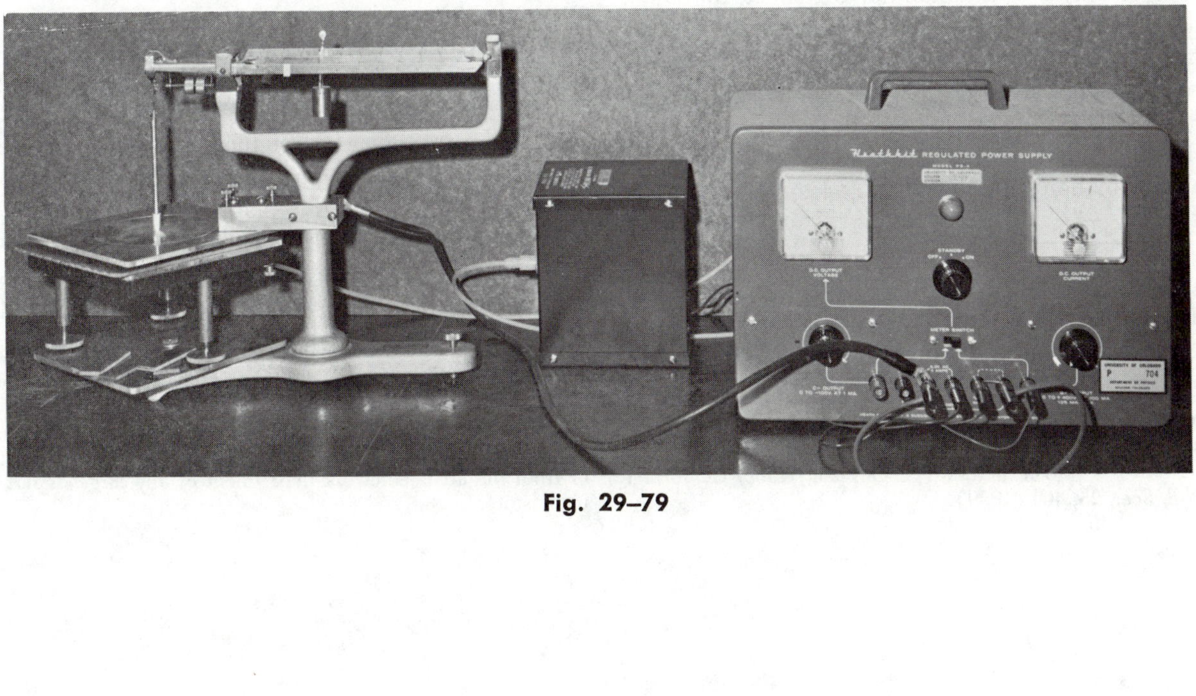

Fig. 29–79

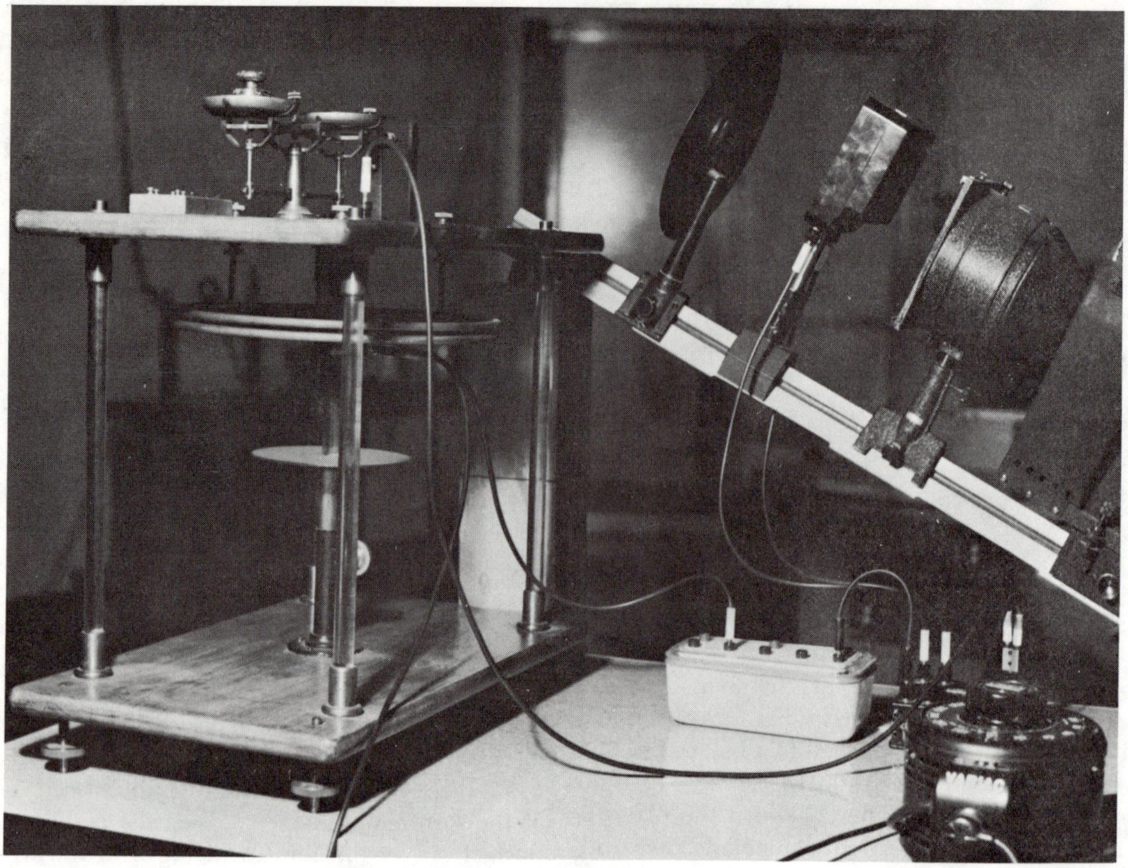

Fig. 29–80

29–5 Analog of Laplace's Equation

29–5.1

If a stretched sheet of rubber tissue is deformed at right angles to the plane of the sheet, the equation for the distorted surface is of the same form as Laplace's equation for two dimensions.(32) Specifically, if the undistorted plane is the x-y plane, and the displacements are in the z-direction, it has been shown(33) that the surface is described to a good approximation by

$$\frac{\partial^2 z}{\partial x^2} + \frac{\partial^2 z}{\partial y^2} = 0,$$

if the slope of the deformed surface is so small that $\tan\theta \approx \sin\theta \approx \theta$. The angle between the x-y plane and the tangential plane is θ. This principle is readily employed in analog demonstrations in electrostatics, magnetostatics, and steady state of heat and fluid flow.

A convenient arrangement is obtained by stretching rubber tissue(34) over a 23-in.-diam quilting hoop.(35) The best method of mounting the sheet involves two persons stretching a square of the material over the inner frame while a third person slides the outer frame into place and tightens it.

A mounted rubber sheet with no distortions applied is shown in Fig. 29–81, and Fig. 29–82 shows a number of electrodes used in demonstrations of

(32) The contributor first saw this analog in 1935 during a trip through Dr. Zworykin's section of the research laboratories at the RCA Company in Camden, New Jersey.

(33) P. B. Moon, and M. L. Oliphant, *Proc. Cambridge Phil. Soc.*, **25**, 460 (1929).

(34) Rubber tissue (dental dam), 0.012 in. thick. Available from Arthur H. Thomas Co., Philadelphia, Pennsylvania.

(35) Quilting frames of various shapes and sizes can be obtained from Sears Roebuck Co., Montgomery Ward Co., or from the art needlework department of any large department store.

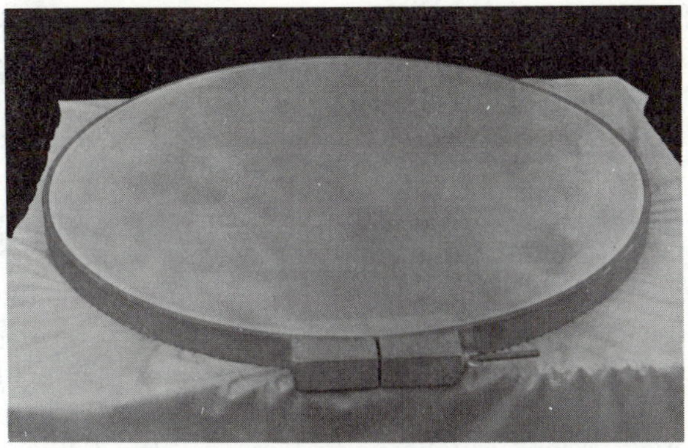

Fig. 29–81 Mounted Rubber Sheet with no distortions applied.

electrostatic fields. In use, the electrodes are mounted on a piece of plywood and the mounted rubber tissue is pressed down over the electrodes. The voltage of the electrodes is represented by the height of the electrode, and the analog voltage may be changed by varying the number or thickness of wood supports. A small steel ball may be used with each of these electrodes to show the paths taken by electrons in these fields. Figure 29–83 demonstrates a simple way of measuring the amount of distortion from the x-y plane (undistorted plane), here employed on an analog of a point electrode or line electrode in a coaxial cylindrical arrangement.

Contour lines may be projected on this surface in the following way. Strips of black paper are glued onto a clear $3\frac{1}{4}$ x $4\frac{1}{4}$-in. lantern slide in such a way that there are a series of clear 2-mm-wide horizontal slits extending across the slide. They are spaced 1 cm apart, center to center, vertically. Then this slide pattern is projected horizontally as

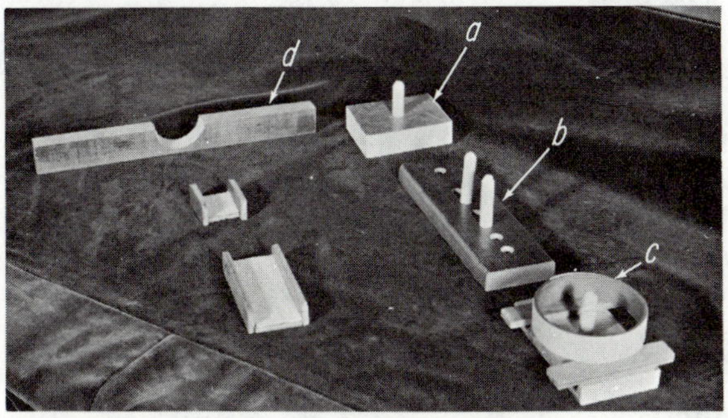

Fig. 29–82 Electrodes: (*a*) Point electrode (or point charge); (*b*) two-point electrodes (or point charges); (*c*) cathode and grid of a triode of cylindrical geometry; (*d*) electron gun, consisting of cathode and two parallel-plate focusing electrodes.

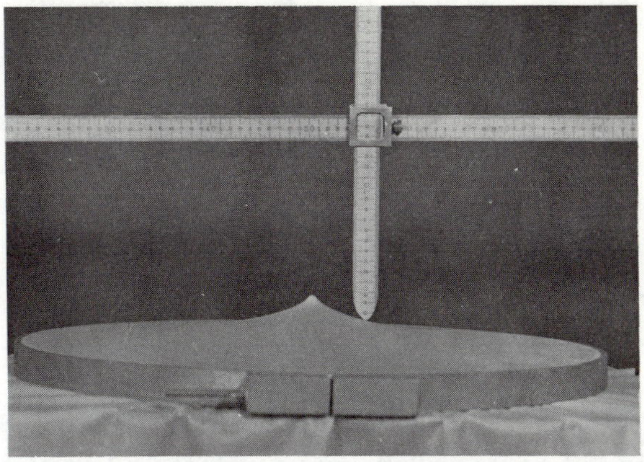

Fig. 29–83 Analog of Point Electrode or Line Electrode in Cyclindrical Configuration. Meter stick arrangement provides easy way of quantitative analysis.

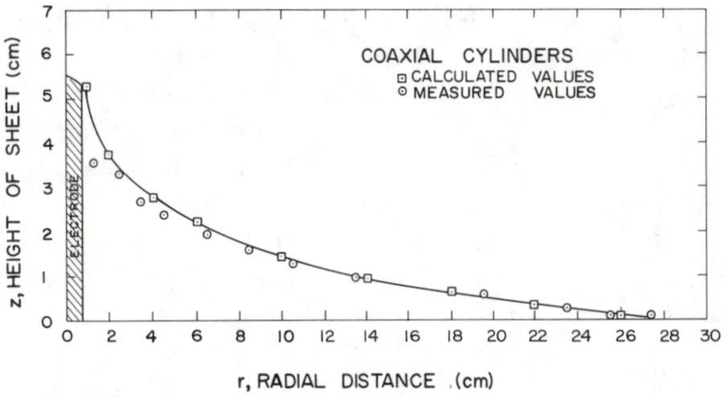

Fig. 29–84

a parallel beam which falls upon the distorted rubber sheet. The bright contour lines show fairly effectively where the slope of the sheet is greatest. However, the mapped field is rather small and, of course, shows only on the side of the sheet nearest the projector. The graph in Fig. 29–84 shows the agreement between the calculated and measured values of z obtained for this arrangement.

In the cylindrical analog, shown in Fig. 29–85, the grid is positive with respect to the cathode.

Figure 29–86 shows an arrangement whereby a depression can be made in the membrane. This is equivalent to reversing the polarity, and it has a simple and obvious application in the field between

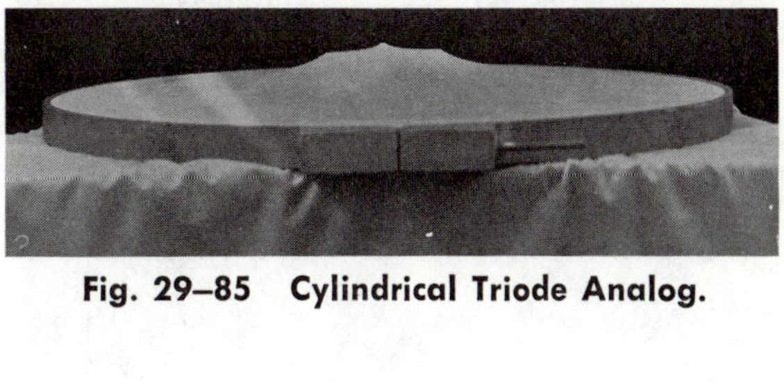

Fig. 29–85 Cylindrical Triode Analog.

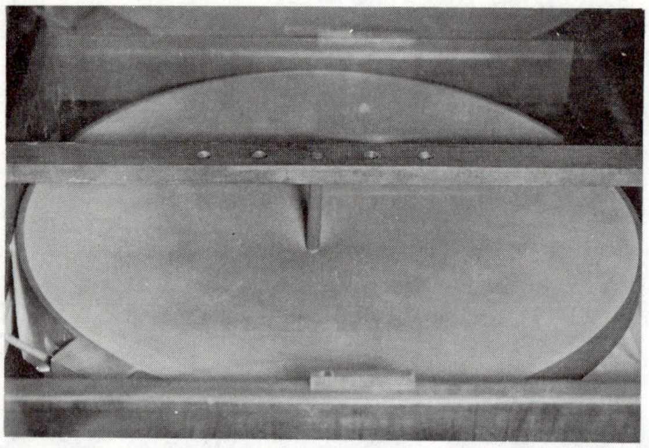

Fig. 29–86 Auxiliary Framework for "Reversing Polarity" in Electrical Analogs.

the opposite charges. It can also be used in demonstrating orbits by rolling a small steel ball along the surface of the sheet. Depending on the initial direction of motion, circular or elliptical orbits may be generated, with friction gradually reducing the ball's speed until it is "captured" and comes to rest at the center. To illustrate the positive-negative "charge" situation, one peg pushes upward on the underside while another pushes downward on the top face. For this case the hoops must be held down with a wooden frame arrangement. These configurations can be picked up on closed-circuit television to be shown to large groups of students or may be built to a larger scale.

Demonstrations	*Contributors*
29–1.1	W. F. C. Ferguson, New York University
29–1.2	H. A. Robinson and D. E. Albert, Adelphi University
29–1.3	H. W. Dosso and R. H. Vidal, Victoria College, Victoria, British Columbia, Canada
29–1.4	P. Sprawls, Emory University
29–1.5	R. M. Sutton, California Institute of Technology
29–1.6	H. M. Waage, Princeton University
29–1.7	University of Nevada
29–1.8	W. R. Mellen, Lowell Technological Institute
29–1.9	E. B. Nelson, State University of Iowa
29–1.10	P. Rood, Western Michigan University
29–1.11	F. W. Sears, Dartmouth College
29–1.12	D. S. Ainslie, University of Toronto, Toronto, Ontario, Canada
29–1.13	R. W. Kenworthy, University of Washington
29–1.14	Swiss Federal Institute (ETH), Zürich, Switzerland
29–1.15	Swiss Federal Institute (ETH), Zürich, Switzerland
29–1.16	G. Likely, University of Minnesota
29–1.17	Swiss Federal Institute (ETH), Zürich, Switzerland
29–1.18	L. W. Morris, Louisiana State University
29–1.19	J. D. Barnett, Brigham Young University
29–1.20	C. Francis, St. Augustine High School, Brooklyn, New York
29–1.21	F. Oppenheimer, University of Colorado
29–1.22	Swiss Federal Institute (ETH), Zürich, Switzerland
29–1.23	A. D. Moore, University of Michigan
29–1.24	California Institute of Technology
29–1.25	A. D. Moore, University of Michigan
29–1.26	J. T. Brooks, Washington University
29–1.27	J. H. Howey, Georgia Institute of Technology
29–1.28	E. D. Lambe and J. M. Fowler, Washington University
29–1.29	A. Capecelatro, Newark College of Engineering
29–1.30	Swiss Federal Institute (ETH), Zürich, Switzerland
29–2.1	H. M. Waage, Princeton University
29–2.2	Swiss Federal Institute (ETH), Zürich, Switzerland
29–2.3	Swiss Federal Institute (ETH), Zürich, Switzerland
29–2.4	H. M. Waage, Princeton University
29–2.5	H. E. Anderson, Massachusetts Institute of Technology

Demonstrations	*Contributors*
29–2.6	H. A. Robinson, Adelphi University
29–2.7	T. E. Wade, Jr., University of Nebraska
29–2.8	Swiss Federal Institute (ETH), Zürich, Switzerland
29–3.1	J. B. Hoag, University of Florida
29–3.2	California Institute of Technology
29–3.3	Swiss Federal Institute (ETH), Zürich, Switzerland
29–4.1	J. W. Straley, University of North Carolina
29–4.2	Swiss Federal Institute (ETH), Zürich, Switzerland
29–4.3	J. D. Barnett, Brigham Young University
29–4.4	H. A. Robinson, Adelphi University
29–4.5	H. A. Robinson, Adelphi University
29–4.6	L. Grillet and M.-T. Grillet-Amy, University of Rennes, France
29–4.7	Swiss Federal Institute (ETH), Zürich, Switzerland
29–4.8	Swiss Federal Institute (ETH), Zürich, Switzerland
29–4.9	R. Gray, University of Massachusetts
29–4.10	G. Freier, University of Minnesota
29–4.11	Swiss Federal Institute (ETH), Zürich, Switzerland
29–4.12	Swiss Federal Institute (ETH), Zürich, Switzerland
29–4.13	F. Oppenheimer, University of Colorado
29–4.14	H. A. Robinson, Adelphi University
29–5.1	M. R. Wehr, Drexel Institute of Technology

30

ELECTRICAL PROPERTIES OF MATTER

30–1 Current and Resistance

30–1.1 [1]

An apparatus which enables large classes to read resistance settings is the decade resistor box shown in Figs. 30–1 (front) and 30–2 (back). The design depends on the size and shape of the two resistance decades taken from an old resistance box. The device (Figs. 30–3, 30–4) is constructed from two pieces of ⅛-in. aluminum plate, 11¼ x 21½ in. and 6 x 21½ in., respectively, bolted together by means of a ⅛-in. aluminum angle, 1 x 1 x 21½ in. The two resistance decades *c* and dial faces *a* are supported by means of two ⅛-in. aluminum plates, each 4 x 7½ in., and attached to the base by an aluminum angle. The dial faces *a* are each cut to size from circular plates of Lucite (⅛ in. thick, 10 in. diam) and screwed to the knobs. Number decals are located on the front, and translucent paper is glued to the back of each of the Lucite sheets. Stopscrews *e* limit the rotation of the dials. The illuminator *b* is attached using a suitably bent aluminum plate as a reflector and mount. Low-resistance wires *d* connect the decades to the binding posts *h*. Figure 30–4 shows the spacing of the four 1½-in. holes *f*, the positioning of the binding posts *h*, and the spacing of the dial plate *a* with regard to the front panel and the ⅜ x 3⅞ x 3¼-in. Bakelite plate *g*.

[1] R. Resnick and H. F. Meiners, *Demonstration and Laboratory Apparatus Report of the 1960 Summer Visiting Professor Workshop at Rensselaer Polytechnic Institute.*

30–1.2

When a solution of potassium iodide is electrolyzed, iodine is liberated at the anode. Ordinary starch turns a dark violet when it reacts with iodine. These two reactions can be combined to demonstrate the continuous nature of direct current and the fluctuations of alternating current.

A tinned iron sheet 32 in. square is mounted on a piece of plywood 3 ft square and covered, except for one corner, with a tightly stretched piece of white cloth tacked onto the wood (Fig. 30–5). The cloth is painted over with a 50–50 mixture of a saturated solution of starch and a 75 g/100 cc of water solution of potassium iodide. The mixed solution is unstable and cannot be stored. Two copper electrodes are connected to a 12-V ac or dc source through a changeover switch. The electrodes are short pieces of ¼-in.-diam copper rod attached to insulating handles. A connecting wire

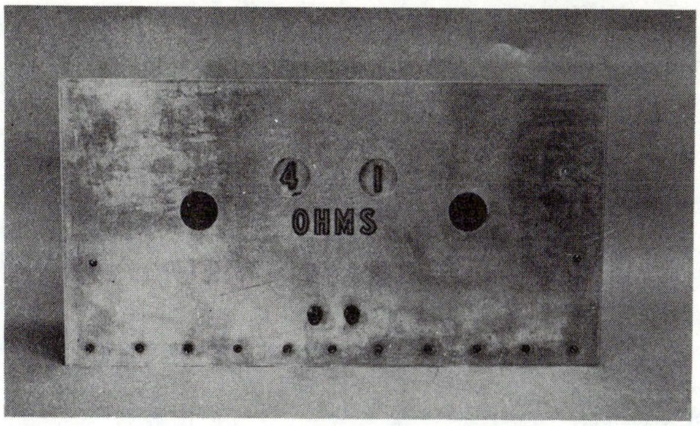

Fig. 30–1

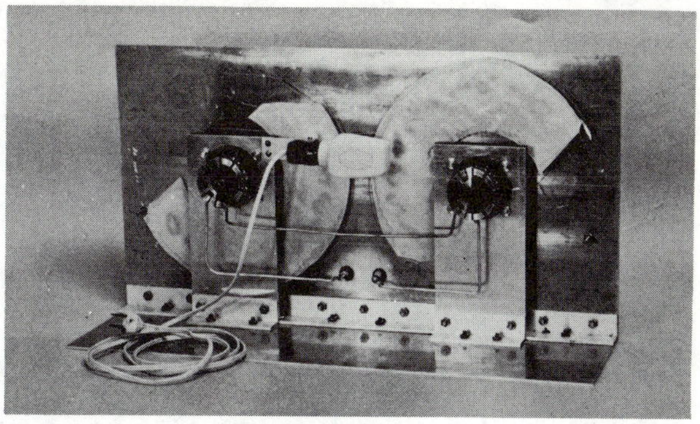

Fig. 30–2

runs from the copper rod, through the handle, to the electrical supply. For improved contact, the contact electrode can have a piece of copper strip attached to it by spring clips.

With the negative electrode in contact with a corner of the iron plate, it is possible to draw on the board with the other. With dc a continuous black line appears, and with ac, a striped band. If a timed sweep is made and the number of stripes counted, the frequency of the ac supply can be determined. The coloration is caused by the liberation of free iodine by electrolysis and its subsequent reaction with the starch.

30–1.3[2]

A piece of soft glass capillary tubing (2 cm long, 6 mm o.d., 1 mm i.d.) has nickel wire electrodes sealed into its ends. The other ends of the nickel wire are screw-clamped into the ends of two vertical brass rods. The rods are screwed into a nonconducting base and are of such a length that the glass is at the correct height to be heated with a Bunsen burner. The electrodes are connected to a 110-V ac line through an ammeter and a solenoid with a movable magnetic core. After heating with the Bunsen flame, the ammeter will indicate an increase in the conductivity. About 0.8 A is sufficient to maintain the temperature in the glass without the flame. This current may be set by adjusting the magnetic core. Since the luminosity of the glass is so high, dark glasses or some direct shielding is advisable. Such a unit is a convenient spectrographic source for sodium D lines. The capillary tubing acts like a vacuum tube with sodium vapor, and it is possible to observe a bright yellow discharge.

(2) L. Grillet, *Bull. Union des Physiciens,* No. 296, 24 (1936).

30–1.4

A simple device for showing that the resistivity of a metal increases with increasing temperature is shown in Fig. 30–6. Forty turns of 0.02-in.-diam iron wire are wound on a frame of ceramic material which is supported by two copper wires in a stand. This resistance is connected in series with a light bulb (6 V, 25 W) and a 35-V ac source from a Variac or a transformer. The bulb burns normally. When the resistance is heated to redness with a large Bunsen burner, the light goes out.

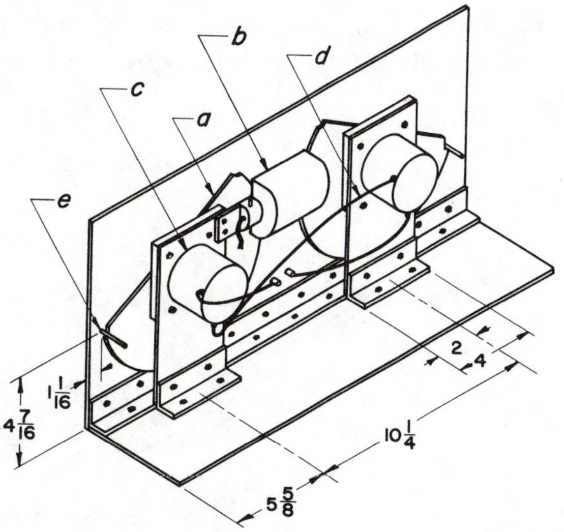

Fig. 30–3

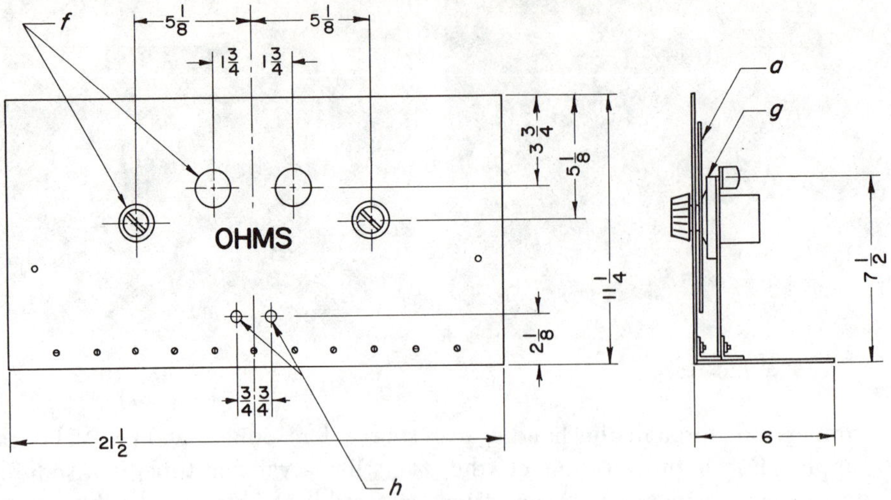

Fig. 30–4

30–1.5

An electric arc is struck between a carbon rod *a* (Fig. 30–7) on an insulated handle and a solution of $ZnSO_4$ *b* in a shallow dish. They are connected through a 50-Ω resistance to a 600-V, 10-A dc source. An aluminum plate may be substituted for the $ZnSO_4$ solution. If the applied voltage is reversed, the arc will go out, demonstrating that a hot, electron-emitting cathode is necessary for the arc to be sustained, as in Fig. 30–8 and that the temperature of the anode has negligible effect.

30–1.6

Eighteen turns of heavy duty electrical cable are wound around a 300-turn coil connected to a 220-V, 30-A source. A short length of zinc-coated 2.5-mm-diam iron wire is connected to the ends of the heavy duty cable. When the primary coil is turned on, approximately 500 A will surge through the secondary coil, vaporizing the wire in a brilliant flash (Fig. 30–9).

30–1.7

The face of a standard meter can be projected on a screen with the simple arrangement shown in Fig. 30–10. A front-silvered mirror *M* is fixed at the appropriate angle as shown. The dial *D* is illuminated with a standard slide projector, and the image is projected on the screen by the objective lens *L* removed from the projector and placed

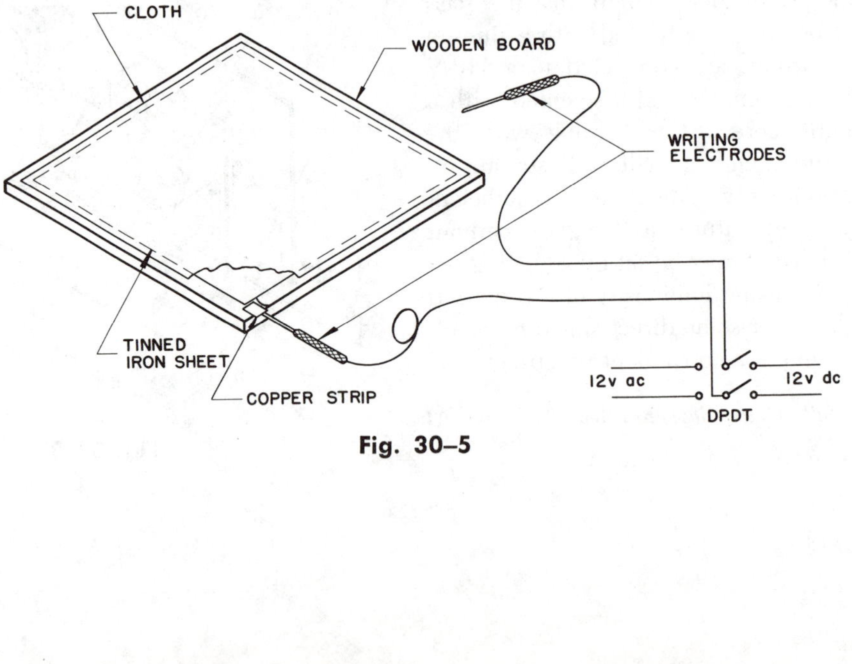

Fig. 30–5

Fig. 30–6

to receive light reflected from the meter face and the mirror.

30–1.8

A Cenco projection meter can be used in a Delineascope Lantern Slide Projector[3] (Model D, 500 W, with a 750-W bulb) if it is cooled. The

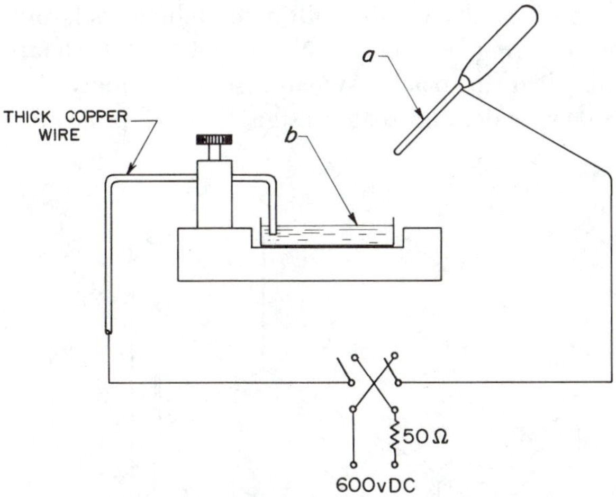

Fig. 30–7

water cooling cell and mount described below were designed for the Cenco meter, but can be modified to hold others intended for projection.

The cooling cell is shown in Fig. 30–11. The two face plates are aluminum with $2\frac{7}{8} \times 3\frac{1}{2}$-in. windows. The $\frac{1}{8}$-in.-thick glass plates for the windows are slightly larger than the openings and are held in place by a U-shaped rubber bracket. Dovetail brackets are screwed to the rear face so that the cell, after it is filled with water, may be mounted in the lantern. Two holes in the front face of the cell are threaded for attachment of the meter holder.

Fig. 30–8

The holder (Fig. 30–12) is also made of aluminum, with two copper spring clips to hold the meter in place. All of the pieces are screwed together, as in the cooling cell.

30–1.9

A projection meter mount can be made for use in a commercial slide projector. A standard Ansco Dualet (Model JN234) projector,[4] shown in Fig.

[3] Spencer Lens Co., Buffalo, New York.

[4] Ansco, Binghamton, New York.

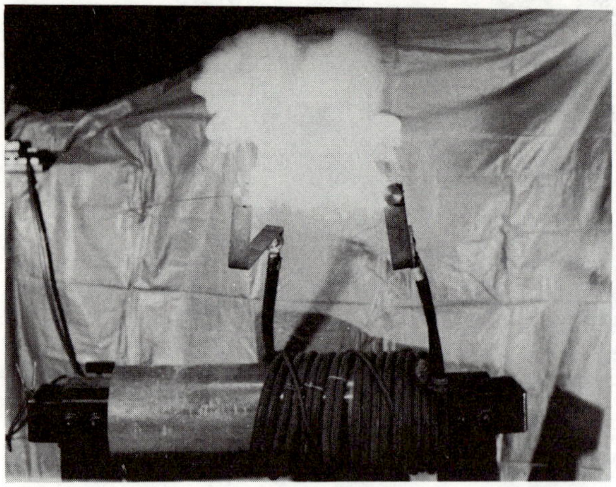

Fig. 30–9

30–13, can easily be modified to house special meter movement mounts. This projector has a 300-W lamp, a short focal length lens, and cool running qualities. Electrically, the only modification is the addition of an on-off switch. Very few mechanical changes need be made to locate a mount. The bottom is cut back to the first lens, a connector hole is cut in the top, and the slideholder is removed. Only two wing screws need be tightened down after positioning the mount in the projector from the bottom.

Back and bottom views of the meter mount without the meter movement are shown in Fig. 30–14, along with the bracket for attachment in the projector. Note that positions of the mounting holes

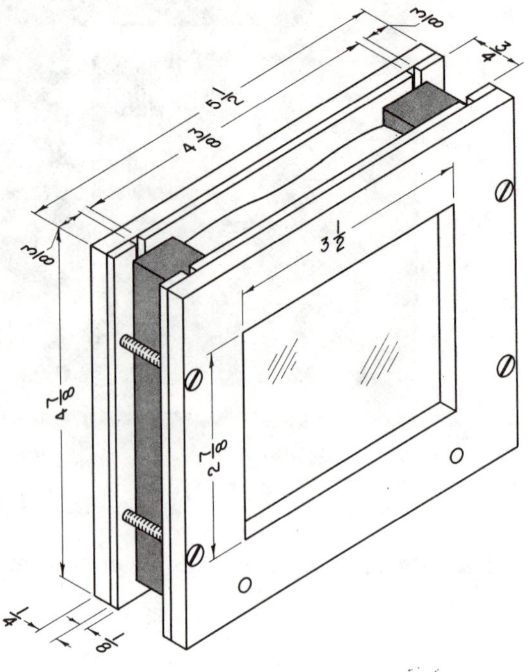

Fig. 30–11

are for Simpson meter movements only. Except for the two thin Lucite windows, all pieces are aluminum. The divisions are engine divided on the back of the Lucite with a scratching tool, and the lettering is done with a hot metal stamp and filled in black. When assembled, only the pointer, scale, and scale designation are projected.

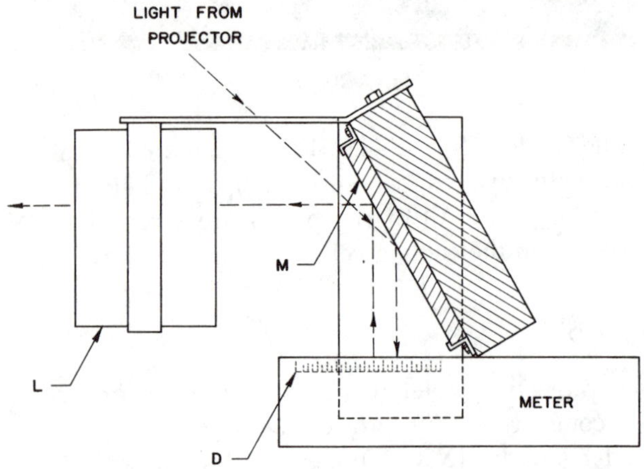

Fig. 30–10

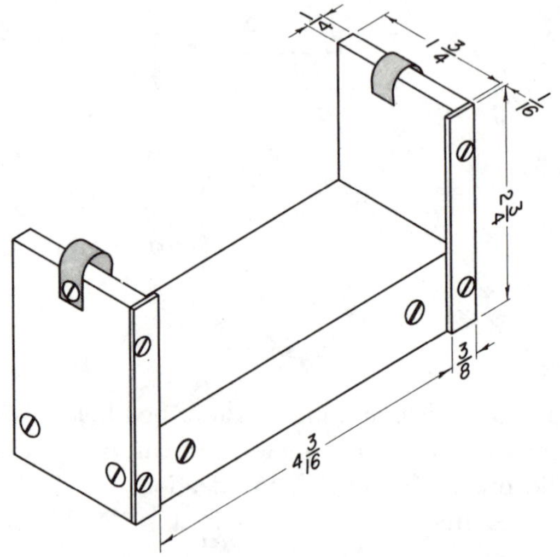

Fig. 30–12

A female Amphenol No. VG-103/V connector *a*, which fits through the projector casing, is used so that a connection can be made from above when in the operating position.

Fig. 30–13

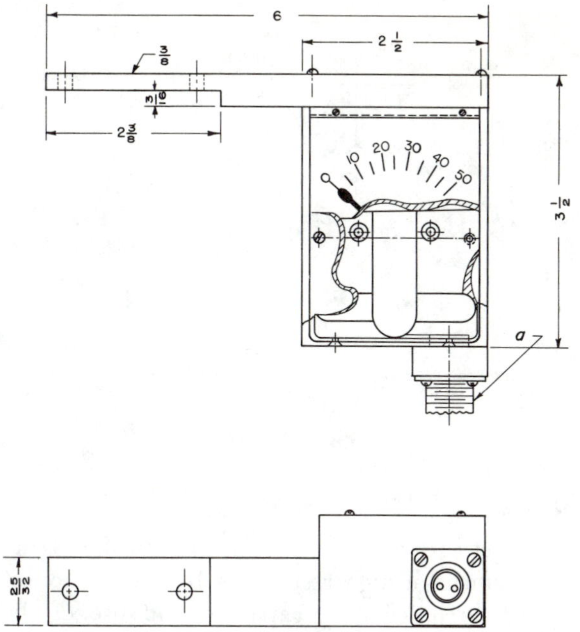

Fig. 30–14

30–2 EMF and Circuits

30–2.1

The apparatus in Fig. 30–15 is used to illustrate by analogy voltage drops across resistances in a parallel or series dc circuit. A small circulating pump *a* is used to feed water from a photographic tray *b* to an inverted flask *c*, which serves as a constant-level reservoir. A piece of rubber tubing with a clamp is analogous to the internal resistance of a voltage source. This is placed in the connection from the flask to the external circuit. The vertical tubes are used as "voltmeters" to measure the potential at different places in the circuit. Pieces of rubber tubing with clamps affixed are inserted at various positions in the circuit. The open end of the tubing is placed in a small beaker, which is analogous to a coulombmeter. A small paddle-wheel "ammeter" can be used in place of the beaker.

The potential drop across the resistances is shown by the difference in the water level in any two vertical pipes. The vertical pipe nearest the flask illustrates that the voltage output is less than that of an ideal source which would have no internal resistance. Transients can be illustrated, but it should be noted that the time relations of the transients are misrepresented and the resistances depend on geometrical factors.

All "wires" are made of glass tubing, ¼ in. o.d., and the resistances are 3-in. lengths of rubber tubing with hose clamps to vary the resistance.

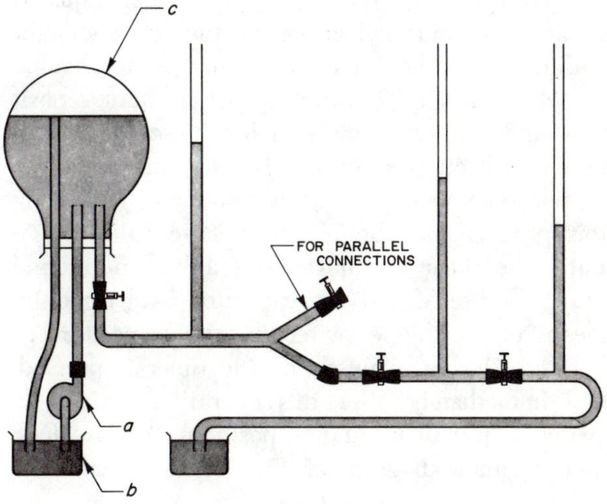

Fig. 30–15

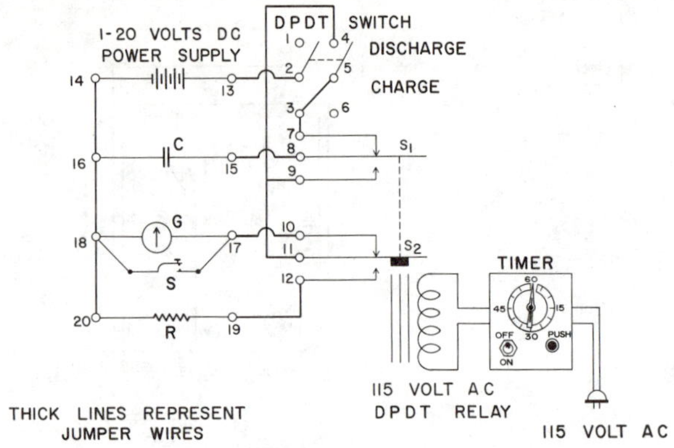

Fig. 30–16

30–2.2[5]

The circuit diagram in Fig. 30–16 allows the measurement of high resistances by observation of rate of discharge of a capacitor and use of the equation

$$q = q_0 \exp(-t/RC),$$

where q is the residual charge, q_0 the initial change on the capacitor, t the time, R the resistance, and C the capacitance. Use of an electric timer wired into the circuit allows semiautomatic operation and eliminates the usual errors from human reaction time. The circuit is drawn on a 12 x 12 x ¼-in. piece of Lucite, and 20 five-way binding posts are mounted in the positions shown in Fig. 30–16. A timer rests on a plywood board, which also supports the Lucite sheet in a vertical position.

The experimental procedure is straightforward. The voltage of the power supply is first adjusted so as to obtain deflections of the galvanometer within scale. The capacitor is charged to a value q_0, when the DPDT switch is put in CHARGE position, and returning this switch to DISCHARGE position a deflection θ_0 of the ballistic galvanometer proportional to charge q_0 is obtained. Repeating this process one obtains an average value of θ_0. Later the timer is adjusted for a known interval (say, $t = 5$ sec), and the capacitor is charged by means of the toggle switch in CHARGE position as before. The push button on the timer is pressed, and immediately after this operation the toggle switch is put in DISCHARGE position. A close look at the circuit diagram of Fig. 30–16 reveals that in this process the capacitor is first discharged through the high resistance R for a preset time (of $t = 5$ sec), after which it is connected to the galvanometer by the deactivated relay (when it is deactivated), producing a deflection θ of the galvanometer proportional to the residual charge q (after $t = 5$ sec). An average value of θ is obtained by repeating the process several times. Similarly, different values of θ corresponding to different values of t are obtained, from which the values of log (θ_0/θ) are computed.

A linear plot is obtained when values of log θ_0/θ are plotted against t. The slope of this line is used in computing the effective value R_e of the high resistance under investigation. This is not the true value of the high resistance; it represents the resistance of the parallel combination of the true value of the high resistance R and the gap resistance R_g,

$$1/R_e = 1/R_g + 1/R.$$

The true value of R can easily be evaluted from this equation. In order to find the gap resistance R_g, the experiment should be repeated without the high resistance in the circuit (i.e., across binding posts No. 19 and 20). This correction is necessary to account for any leakage of charge across the binding posts due to accumulated dust or other factors.

30–2.3

As shown in Fig. 30–17, four light bulbs are mounted in a rectangular network of pipes, with one bulb for each arm of the bridge. A projection ammeter a replaces the conventional galvanometer. A current transformer b protects the projection meter (110 V ac is used).

When four similarly rated bulbs are used, the meter shows no deflection. But if one bulb is removed or replaced by a bulb of different rating, the meter will show the bridge to be unbalanced.

The line voltage and the line current can be measured with a voltmeter across e and f and an ammeter across c and d.

30–2.4[6]

Thevenin's theorem states that any two terminals of a general network of linear components can be

[5] B. A. Syed, *Am. J. Phys.*, **33**, 850 (1965).

[6] L. Grillet, *Bull. Soc. Sci. Bretagne*, **38**, 35 (1963).

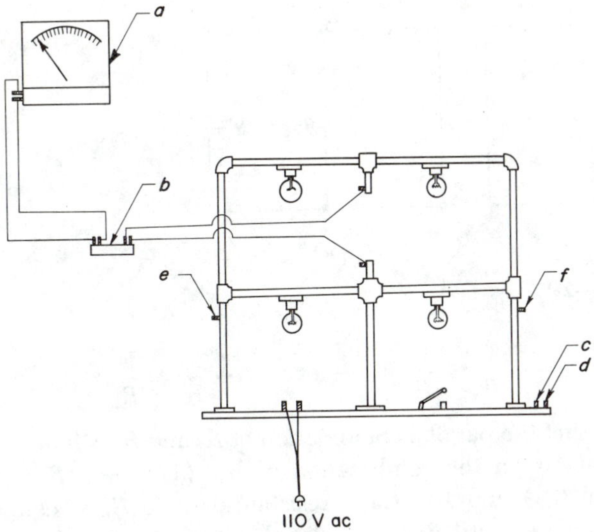

Fig. 30–17

considered to be the terminals of a single linear component consisting of a single voltage source E_0 in series with a single impedance Z_0. This theorem can be shown to be valid in the simple case of an unbalanced Wheatstone bridge. For this problem the superiority of an analysis utilizing Thevenin's theorem over other conventional methods will be immediately apparent.

A circuit is constructed as shown in Fig. 30–18. An overhead projector can be used to show this principle in a lecture demonstration. The circuit is laid out on clear plastic with a projection ammeter across the bridge and with the voltage source V and the values of the resistances clearly labeled. An ordinary 1.5-V dry cell is the voltage source,

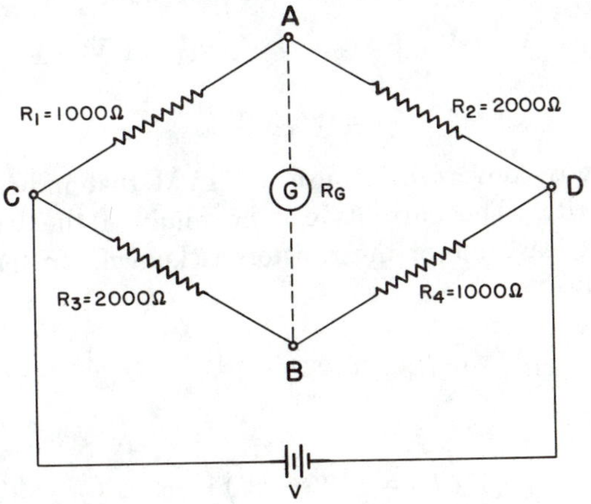

Fig. 30–18

and the current is measured with a 100-μA meter G, which has an internal resistance R_G of 5000 Ω. According to Thevenin's theorem, the resulting equation for the current I_G through the detector will be

$$I_g = \frac{E_0}{Z_0 + R_G},$$

where E_0 is the potential difference between terminals A and B when the detector is removed, and Z_0 is the impedance of the whole circuit as seen from the terminals A and B (here a simple resistance). In the new circuit V is replaced by its internal resistance (considered negligible here).

Under these conditions,

$$E_0 = \frac{R_3}{R_3 + R_4} V - \frac{R_1}{R_1 + R_2} V = \frac{V}{3} \quad \text{volts,}$$

$$Z_0 = \frac{R_1 R_2}{R_1 + R_2} + \frac{R_3 R_4}{R_3 + R_4} = \frac{4000}{3} \quad \text{ohms.}$$

It follows at once that $I_G = 79$ μA.

Terminals A, B, C, and D are fitted with banana plug jacks for ease in inserting and removing the detector and shorting out V. With G removed, a high-impedance voltmeter will read $V_{AB} = V/3$ volts and directly, with G inserted, the ammeter will read 79 μA as a direct reading, thus giving an illustration of the validity of the procedure in this case. The impedance Z_0 may also be checked experimentally by replacing V with a short circuit and measuring the network resistance.

30–2.5

The circuit shown in Fig. 30–19 can be used to show the basic principle of an analog computer—

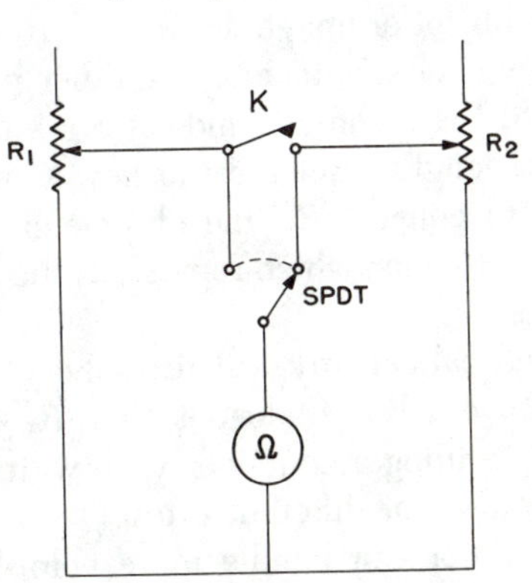

Fig. 30–19

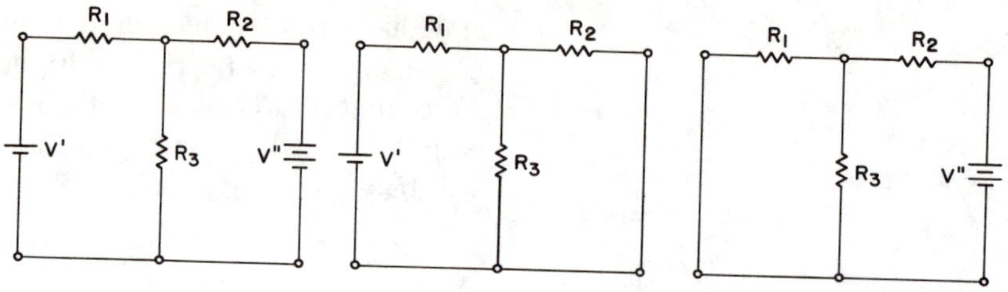

Fig. 30–20

the representation of a physical problem in an electrical analog, finding the electrical solution and thereby finding the solution to the given problem.

In Fig. 30–19 the resistances obey the law,

$$\frac{1}{R_1} + \frac{1}{R_2} = \frac{1}{R_t}.$$

Since in optics, for example, we have a similar law,

$$\frac{1}{S} + \frac{1}{S'} = \frac{1}{f},$$

we can use the computer to calculate the focal length f, given the object distance S and the image distance S'.

To find the focal length, first open key K and throw the switch to connect R_1. Turn potentiometer R_1 until the ohmmeter indicates the value of S. Then throw the switch to connect R_2. Turn potentiometer R_2 until the ohmmeter indicates the value of S'. The computer is now programmed. Next, connect R_1 and R_2 in parallel by closing key K and read on the ohmmeter the total resistance that represents the value of the focal length f in the analog.

If the focal length and either the object or image distance is given, the other can be found by representing the object or image distance as R_1 and then, with the key K closed, turning the other potentiometer R_2 until the ohmmeter indicates a value equal to the focal length. Release the key K and throw the switch to connect R_2; the ohmmeter will read the value of R_2 alone which represents the unknown optical distance.

This electrical network will also solve other problems that have a law analogous to $1/R_1 + 1/R_2 = 1/R_t$, such as filling a container with water coming from two faucets at different rates.

The computer can handle more complex problems of the form

$$\frac{1}{R_1} + \frac{1}{R_2} + \frac{1}{R_3} + \ldots + \frac{1}{R_n} = \frac{1}{R_t}.$$

First the parallel combination of R_1 and R_2 is found, and then the combination of R_{12} $(1/R_1 + 1/R_2 = 1/R_{12})$ and R_3. Each resistance up to R_n is considered until R_t is calculated.

The accuracy of the answer depends on the accuracy of the ohmmeter. The choice of potentiometers is not critical, but they should both have similar values.

30–2.6[7]

The superposition theorem states that each emf in a linear dc network produces a current in any given branch independent of the action of other emf's, and that the resultant total current in any branch is the algebraic sum of the contribution of current due to each of the emf's. This can be verified by measuring the voltages across the resistors R_1, R_2 and R_3 in the circuits of Fig. 30–20. Measurements are made of V_1, V_2 and V_3 in (a); V'_1, V'_2 and V'_3 in (b); and V''_1, V''_2 and V''_3 in (c). If the superposition theorem holds, then

$$V_1 = V'_1 + V''_1, \qquad V_2 = V'_2 + V''_2$$

and

$$V_3 = V'_3 + V''_3.$$

It is convenient to use a VTVM that indicates polarity. The currents can be found if the value of at least one of the resistors is known. In terms of R_3,

$$R_1 = (V' - v')\frac{R_3}{v'}$$

and

$$R_2 = (V'' - v'')\frac{R_3}{v''},$$

[7] L. Grillet, *Bull. Soc. Sci. Bretagne,* **36,** 341 (1961).

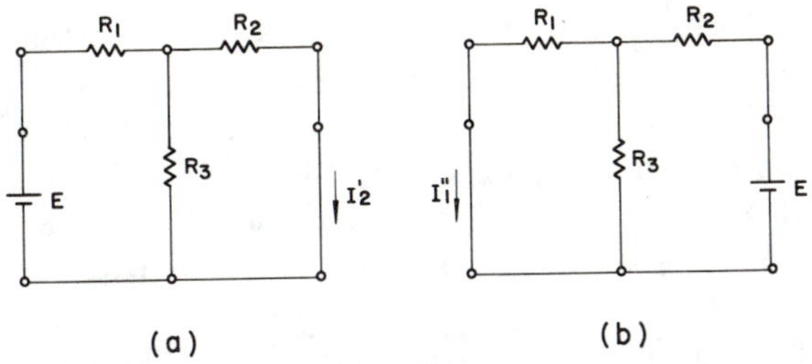

Fig. 30–21

where v' is the voltage drop across R_3 when the branch in circuit (b) containing R_2 is open-circuited, and v'' is the voltage drop across R_3 when the branch in circuit (c) containing R_1 is open-circuited. To facilitate demonstration in lectures, a plastic plate can be laid out with banana plug sockets in the six positions represented by very small circles in Fig. 30–20.

Resistors are soldered between the sockets with the following convenient values: $R_1 = 2\ \text{k}\Omega$, $R_2 = 5\ \text{k}\Omega$, and $R_3 = 4\ \text{k}\Omega$. Ordinary 1.5-V dry cells are used as voltage sources. Overhead projection apparatus is used with projection ammeters and voltmeters to present the results to the class.

30–2.7[8]

The reciprocity theorem states that if an emf E_0 is placed in one branch of an electrical circuit, giving rise to a current I in another branch, the same current I is obtained in the original branch when the emf E_0 is transferred to the branch where the current I originally arose. A demonstration of the validity of this theorem can be performed quite easily with the aid of the circuits shown in Fig. 30–21.

Banana plug sockets are mounted on a sheet of plastic in the positions denoted by the eight very small circles. Resistances are soldered between these sockets with the following convenient values: $R_1 = 2\ \text{k}\Omega$, $R_2 = 5\ \text{k}\Omega$, and $R_3 = 3\ \text{k}\Omega$. An ordinary 1.5-V dry cell can serve as a convenient voltage source.

The voltage drops V_i across the resistances R_i can be measured with a high-impedance projection voltmeter equipped with plugs. The various voltage drop measurements are denoted as follows: V'_1, $V'_2 = V'_3$ for (a) and V''_1, $V''_2 = V''_3$ for (b). Thus $I'_2 = V'_2/R_2$ and $I''_1 = V''_1/R_1$. The statement of the theorem requires that $I''_1 = I'_2$, and the readings taken at this point are sufficient for verification. A more general proof can be obtained by measuring the ratio

$$\frac{R_2}{R_1} = \left(\frac{V - V'}{V'}\right)\left(\frac{V''}{V - V''}\right),$$

where V' is the measured voltage across R_3 in (a) when R_2 is removed and the circuit left open; V'' is the measured voltage across R_3 in (b) when R_1 is removed and the circuit left open; and V is the applied voltage.

(8) L. Grillet, *Bull. Soc. Sci. Bretagne,* **36,** 343 (1961).

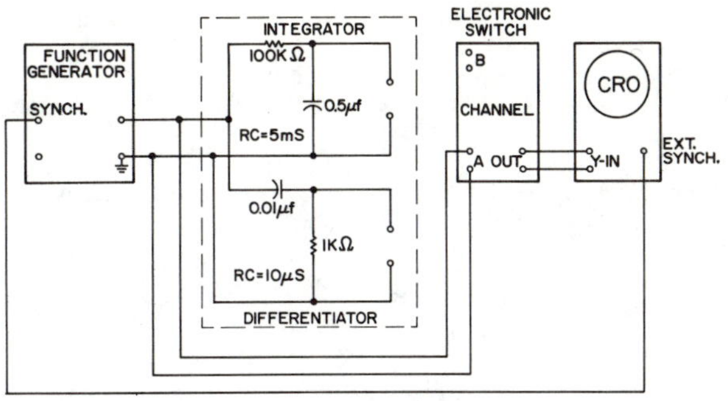

Fig. 30–22

30–2.8

Some of the properties of RC differentiating and integrating networks as well as electrical transmission (delay) lines can be examined in conjunction with lectures on linear systems.

The integrator and differentiator are mounted on the same 9 x 6 x ¼-in. acrylic plastic plate to facilitate shadow projection, if this is desired. They are wired, as shown in Fig. 30–22, so that the input waveform (a square, triangle, or sine wave obtained from a Hewlett-Packard[9] Model 202A function generator) is applied simultaneously to both. Normally an electronic switch (Heathkit[10] Model 1D22 or Dumont[11] Model 185A) is used to provide a dual trace presentation on the oscilloscope, thus permitting a comparison of the input waveform with either output, and a measurement of phase shift when the oscilloscope sweep is triggered by the generator output. With the function generator operating at about 1 kc/sec, the component

(9) Hewlett-Packard Co., Palo Alto, California.
(10) Heath Co., Benton Harbor, Michigan.
(11) Dumont Laboratories, Clifton, New Jersey.

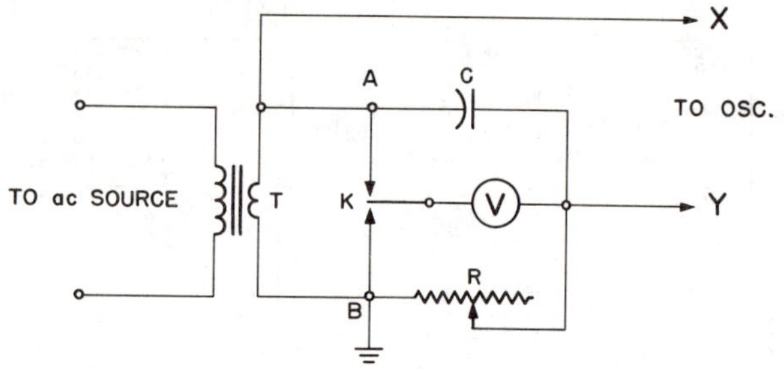

Fig. 30–23

values shown will provide accurate integrals and differentials of the input provided that the integrator is not shunted seriously, i.e., works into an impedance of $>2\,M\Omega$ shunted by <50 pF.

For a square wave input, the integrator gives an accurate triangular wave output whose peak amplitude is $\sim 1/100$ of the input, while the differentiator produces the appropriate spiked wave form whose peak amplitude is twice that of the input. Because of the large attenuation of the integrator, it is advantageous to insert a preamplifier between the network and the electronic switch. This permits both the switch and CRO to be operated at less than maximum gain, thus minimizing drift and noise. A suitable low cost amplifier is the G.E.[12] UPX-003C microphone-phonograph preamplifier, which has a gain of ~ 30 db and an integral power supply. It is, of course, used in its "flat" mode, i.e., selector switch set to MICROPHONE.

30–2.9[13]

The RC series circuit shown in Fig. 30–23 is connected to the 10-V secondary of a step-down transformer T. Here V is a projection voltmeter used to measure either the voltage drop V_C across the capacitor C or the voltage drop V_R across the resistance R, depending on the position of switch K. If the capacitance is kept constant, the impedance depends only on R. When the resistance R and the capacitive reactance $X_C = 1/(2\pi fC)$ are equal, V_R and V_C are also equal. For $V = 10$ V and $C = 10\,\mu$F, $X_C = 265\,\Omega$ for 60 cycles.

Curves are plotted of V_R vs R and V_C vs R. The intersection of the two curves gives the real value of X_C from which C can be computed. For some values of R, diagrams can be drawn of the vectors corresponding to V, V_R and V_C. Point M, the intersection of V_R and V_C lies on a semicircle of diameter V. The argument ϕ can be measured on the diagrams.

If V and V_R are respectively applied to the x and y plates of a cathode ray oscilloscope, a Lissajous ellipse appears which varies with R. The measurement of ϕ can be carried out by means of the ellipse. If V_R and V_C are applied to the x and y plates respectively, the ellipse will degenerate into a circle when $V_R = V_C$. The axes of the ellipse coincide with the x and y axes.

For a different demonstration the positioning switch K is replaced by another potentiometer between A and B in Fig. 30–23. Banana plug jacks can be used, so that if both demonstrations are to be done in the same lecture, the position switch can be quickly and easily replaced by the potentiometer.

When the cursor of the potentiometer is at A or B, then the voltmeter is reading V_C or V_R, respectively (as in the first demonstration). But as the cursor is gradually moved from A to B, the reading on the voltmeter V varies and has a minimum somewhere in this range.

Figure 30–24 is a vector diagram depicting the vectors V, V_R, V_C, and v. The minimum value v_m of the variable reading occurs when vector V is perpendicular to vector v. When $V_r = V_c$ the minimum v_m occurs when the cursor is in the middle of the potentiometer. In this case, v_m reaches its maximum value with respect to cursor position with respect to its relative values. The argument ϕ is again measurable from the diagram.

Again with C kept constant, measurements are taken of V_R, V_C, and v_m for several values of R lower and higher than X_X (constant capacitive reactance of C). With this data, the diagram can be

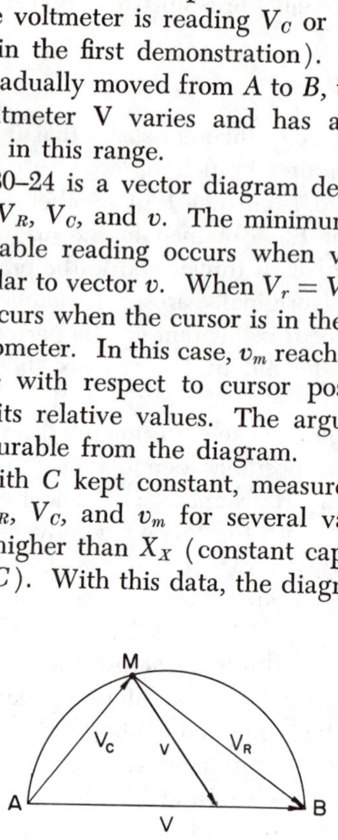

Fig. 30–24

[12] General Electric Co. Apparatus Sales, Schenectady, New York.

[13] L. Grillet, *Bull. Soc. Sci. Bretagne,* **24,** 50 (1949) and **37,** 209 (1962).

drawn for $V_R = V_C$ and the curve $V = f(X)$, where X is the position of the cursor given by the dial of the potentiometer. The purpose of this setting is to show the principle of a low-frequency impedometer.

30–2.10

The apparatus shown in Fig. 30–25 can be used to demonstrate that the back emf developed by an electric motor when running with no load is almost equal to the driving voltage. In addition, the behavior of an electric motor running as a dynamo can be shown.

To demonstrate for a large audience, assemble the apparatus on a sheet of plywood, 4 x 3 ft, and display it vertically. The wiring, switches, and lamp should be mounted on the front surface of the board, so that the audience can easily follow the circuit. Meters *A*, *B*, and *C* are large, center-zero current meters, with a full-scale deflection exceeding the no-load current of the motor.

The lamp and meter *C* are in the circuit only when switch S_2 is closed. Meter *A* measures the total current in the circuit, and meter *B* measures current taken by the motor. The motor should be fitted with a flywheel, so that a braking force can be applied.

With the circuit connected to a suitable power supply, switch S_2 is opened and switch S_1 closed. The motor starts, and meters *A* and *B* give large deflections. When the motor reaches its normal running speed, both meters will have dropped to a much lower value of current for the no-load condition. If a braking force is now applied to the flywheel by means of a block of wood (motor loaded condition), both meters show an increase in current.

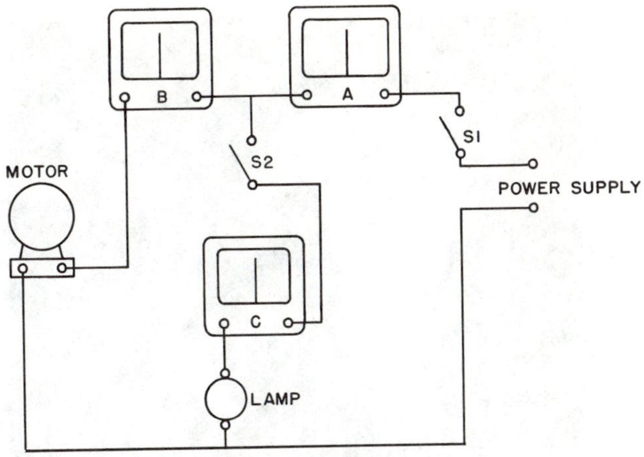

Fig. 30–25

With the motor running with no load and switch S_2 closed, the lamp lights and meter *C* shows the current drawn by the motor; meter *A* will show the sum of the current drawn by the motor and lamp. If switch S_1 is now opened, the power supply is cut and the reading on meter *A* falls to zero. Meter *C* still shows a current, and the lamp continues to glow, but meter *B* shows a current in the opposite direction. Both meters and lamp fall to zero as the motor slows and stops. Since the back emf is the emf induced in the armature coils of the motor by their rotation in the magnetic field, this emf will continue to be generated when switch S_1 is opened and the armature continues to rotate. This produces the reverse current observed in meter *B*.

30–3 Electrolytic Conduction

30–3.1

Faraday's first law of electrolysis states that in a given electrolytic reaction, the amount of material deposited (or released) is proportional to the electric charge which has passed through the electrolyte. This may be stated mathematically as

$$m = \frac{MJt}{F},$$

where m is the mass deposited (or released), M is the electrochemical equivalent (atomic weight/valence), J is the current, and t is the total elapsed time. The proportionality constant F is Faraday's number, whose accepted value is $F = 9.652 \times 10^4$ C/mole.

The silver coulombmeter shown in Fig. 30–26 provides a visible demonstration of Faraday's law of electrolysis and can also be used to determine Faraday's number. The scale is used to read continuously the weight of a pure silver anode suspended in an electrolytic solution of silver nitrate and water. Figure 30–27 illustrates the electrical details of the experiment. The silver anode *a* is

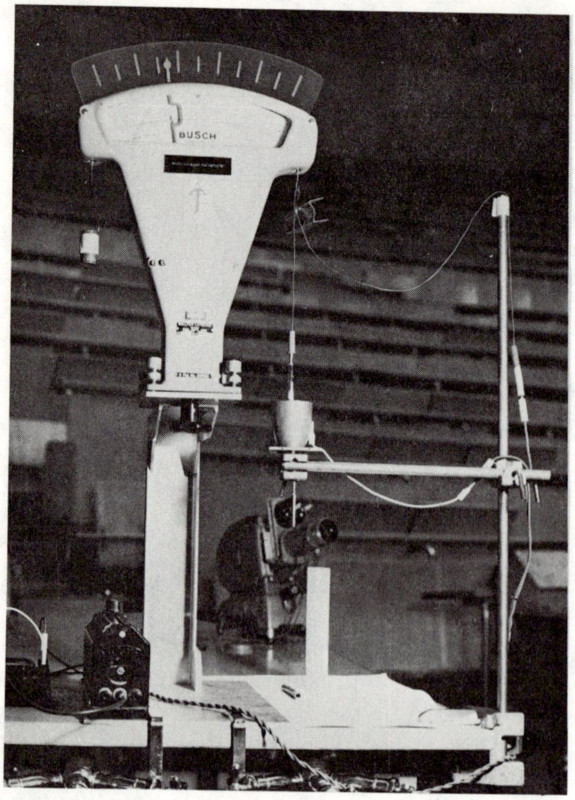

Fig. 30–26

immersed in an electrolytic solution b of $AgNO_3 + H_2O$, which also contains a metal cathode c. In series with the electrolytic cell is a variable resistance d (0–35 Ω, 50 W) for controlling the current, a 40-V dc regulated power supply e, and a projection ammeter f (1 A dc). A small weight (~1 g) should be included between the scale and anode as in Fig. 30–26 to overcome the sideways pull of the electrical wires.

Passing a 1-A dc current through the cell for 1000 sec causes a change in weight of the anode of approximately 1 g, which can be readily detected by the scale.

30–3.2

An arrangement for demonstrating the migration of permanganate ions in an electric field utilizes a thin electrolytic cell[14] placed on the stage of an overhead projector. The cell is very small, measur-

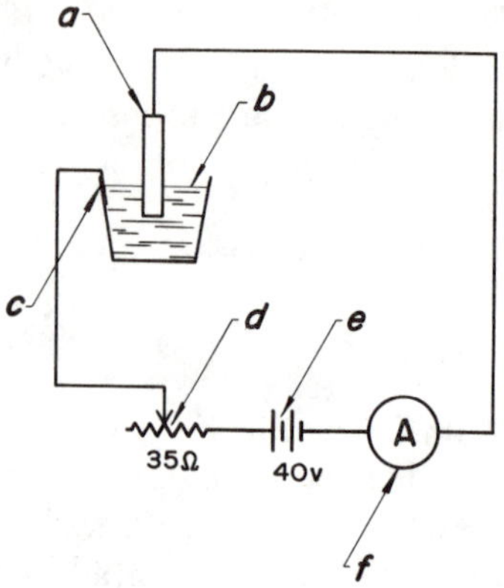

Fig. 30–27

ing 5/8 in. wide by about 1 7/8 in. long with a depth equal to the thickness of rubber tape (approximately 1/32 in.). The cell assembly, of laminated glass construction, is shown in Fig. 30–28. The base plate p is a plate of glass 4 x 7 x 1/8 in. To this plate are cemented two strips of glass g 1 in. apart. Two strips of plastic electrical tape t 3/16 in. wide, attached to the base plate p alongside the plates g, form the longitudinal boundaries of the cell and determine the depth of the liquid. The ends of the cell are closed by two lead electrodes e made of 1/16-in. lead strips 1 in. wide. The electrodes are connected to a 110-V dc source through a rheostat, a milliammeter, and a reversing switch. A glass cover plate c is shorter than the distance between the electrodes by 3/32 in. to provide an opening s through which permanganate can be introduced. With the cover off, the cell is filled to the level of the plastic strips t with a very dilute solution of KNO_3. The cover is then replaced, and one drop of a saturated salution of $KMnO_3$ is added through

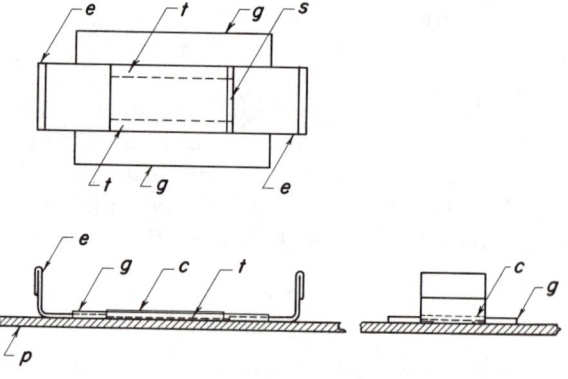

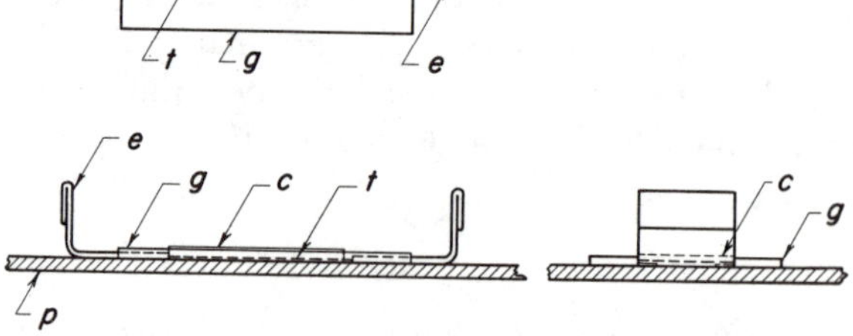

Fig. 30–28

[14] See also R. W. Pohl, *Elektrizitätslehre* (Springer Verlag, Berlin, Germany, 1957), 16th ed., p. 194.

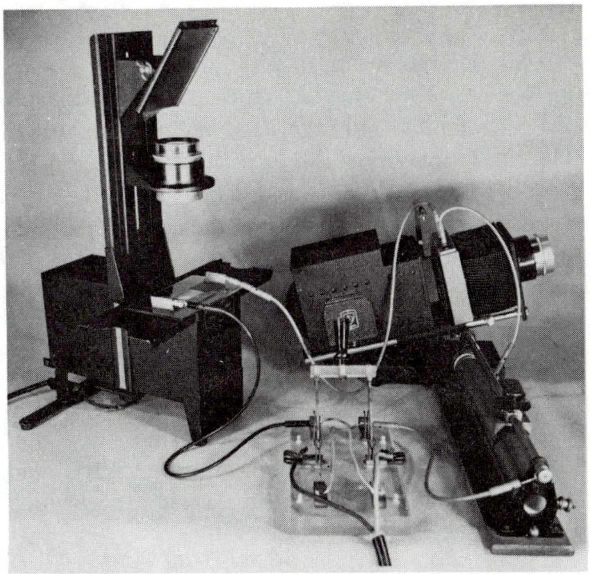

Fig. 30–29

the aperture *s*. A fairly well-defined red front drifts from the negative to the positive electrode, and the effect of reversing the field is readily observable. The current reading, about 0.50 mA, can be projected by a zero-centered projection galvanometer. Figure 30–29 shows the complete assembly.

30–3.3[15]

A piece of filter paper is saturated with a solution of table salt and phenolphthalein and wrapped around a glass tube 10–15 cm long and 3–4 cm in diameter. The ends of the paper are fastened by several twists of bare copper wire which serve as electrodes. A 110- or 220-V dc supply is connected to the electrodes after a 15–25-W incandescent lamp has been inserted into the circuit. As the current passes through the paper, a reddish-colored band of ions is formed around the cathode and begins to move in the direction of the anode at a speed of several millimeters per minute. The colored band is formed by the interaction of the phenolphthalein with the caustic soda produced by the electrolysis.

[15] Translated by K. V. Robinson and H. A. Robinson (Adelphi University) from *Lecture Demonstrations in Physics*, A. B. Mlodzeevskii, ed., M. V. Lomonosov State University, Moscow, U.S.S.R.

The rate of motion of colored MnO_4 ions is shown in projection with the apparatus illustrated in Fig. 30–30. A stopcock with a small opening is sealed to the bottom of a U-shaped glass tube of suitable size for projection. A glass tube with a funnel is then sealed to the stopcock. The funnel tube is filled, up to the stopcock, with a solution of 0.5 g of potassium permanganate and 50 g of urea in 1 liter of water. A solution of 0.3 g potassium nitrate to 1 liter of water is poured into the arms of the U-shaped tube to about one-third of its height. Platinum electrodes are inserted in the tube, and the stopcock is very carefully opened, allowing the solution from the funnel tube to enter the U-shaped tube. The boundary between the two solutions should reach about one-third the height of the U-shaped tube; the electrodes must be covered. When an 0.1–0.3-A current from a storage battery passes through the solutions, the MnO_4 ions will move and the boundary between the solutions will be shifted.

30–3.4

Theory indicates that when current flows through an electrolyte of sodium sulfate, sulfuric acid is formed at one electrode and sodium hydroxide at

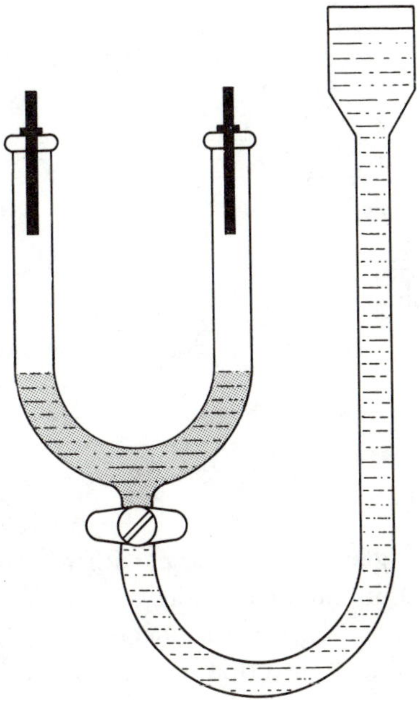

Fig. 30–30

the other. With the proper indicator, one should be able to show that around one electrode the solution is acidic and near the other electrode it is basic, while much of the electrolyte remains neutral. Purple cabbage may be used to demonstrate this quite strikingly. Remove the leaves from a head of purple cabbage, place them in a pan, cover with water, and press and agitate them to remove the dye. Add sodium sulfate (sodium chloride may be substituted) and place in a tank. When current flows through this tank, one end of the tank turns pink (the acid end) and the other end becomes green (the basic end) while the central section remains purple. For the colors to show to the best advantage, the tank should be about an inch thick. Either metal or carbon rods may be used for electrodes.

30–3.5

Two pointed strips of zinc and copper hooked up to a galvanometer are stuck into various fruits and vegetables (orange, lemon, potato, onion, celery), or may just be separated by a crushed green leaf or blade of grass. The qualitative effect is readily shown. A large deflection may be achieved on a zero-centered 500-μA galvanometer by placing the electrodes in the mouth.

30–3.6

Demonstrate oxidation of ferrous iron to ferric iron by putting some ferrous salt in hot water and adding a small amount of concentrated nitric acid. Notice the color of the solution after heating for a few minutes.

30–4 Conduction in Gases

30–4.1

Paschen's law of gas discharge states that the voltage across a gas discharge tube is proportional to the product of the pressure and the distance. This can be demonstrated by the arrangement shown in Fig. 30–31, consisting of two 3-in.-diam

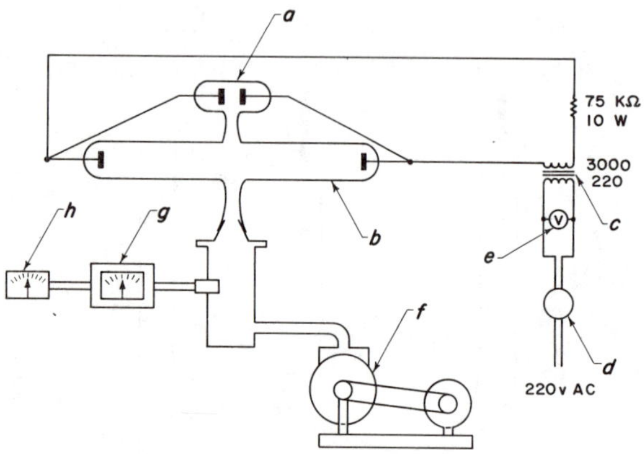

Fig. 30–31

horizontal glass tubes, one *a* with a gap of ¾ in. between aluminum electrodes and the other *b* with a 20-in. gap. They are connected in parallel to a 3000:220-V ac transformer *c* and a 75-kΩ protective resistance of adequate power rating. A Variac *d* regulates the voltage between 0 and 3000 V. This voltage is measured on the primary side with a projection voltmeter *e*.[16]

The tubes are connected to a vacuum pump *f*, and the pressure is measured by a vacuum gauge *g* coupled to a projection meter *h*. The projection meter is made of a meter-face identical to that in the vacuum gauge, using a transparent photocopy of the scale, and mounted in a Plexiglas replica of the original opaque case.

As the pressure decreases, with the voltage kept constant (at about 2000 V), the tube with the short gap fires first, and then, at a lower pressure (10^{-2} torr) the other tube fires. At still lower pressures the discharge is observed only in the long tube; finally, none. A quantitative picture of the phenomenon can be obtained by plotting the ignition voltages at different pressures. The result is called a Paschen curve.

30–4.2[17]

The mode of discharge in a gas is highly dependent on the cathode temperature. As long as the cathode surface is kept cold, electrons are pro-

(16) EMA Multimeter 4S, obtainable from SMS Instrument Co., Castleton-on-Hudson, New York.

(17) Engel, Seeliger, and Steenbeck, Z. *Physik.*, **85**, 144 (1933), and Grotrian, *Ann. Physik.*, **47**, 194 (1915).

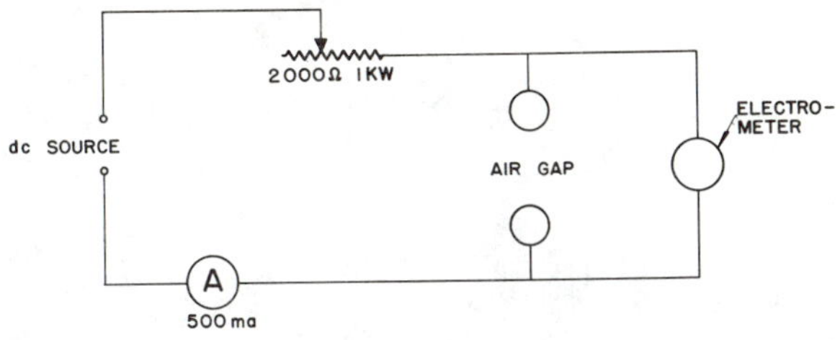

Fig. 30–32

duced by the glow discharge mechanism in the cathode drop area, with cathode glow and dark space. If the cathode is allowed to warm up, electrons are produced by the hot material, without a cathode drop. This is a typical arc discharge. The processes at the anode change correspondingly.

A high-voltage supply producing at least 1200 V and 200 mA is needed. (The more current and voltage, the better.) Figure 30–32 shows the electrical circuit. Two electrodes, consisting of hollow copper hemispheres of 0.2-mm wall thickness and at least 3-cm curvature radius are water-cooled. The water runs through rubber hose connections (3 m long and 0.2 cm i.d.) to avoid short-circuiting the electrodes. As shown in Fig. 30–33, the electrodes are mounted on a vertical optical bench by means of sliders. The vertical arrangement is necessary for the aerodynamic stability of the discharge column.

The discharge is started by bringing the electrodes in contact and then pulling them apart. As long as the water flow is high and the current is kept below 200 mA by a resistor, a typical glow discharge occurs exhibiting the spectrum of air. The cathode dark space is 0.2 mm from the cathode surface. When the water flow is reduced, the discharge mode suddenly changes. The voltage between the electrodes falls by about 400 V. As the cathode drop disappears, the current consequently rises (beware of overloading the power supply), the structure of the column alters, and the copper spectrum appears as bright green radiation from all parts of the arc.

The length to which the column or arc can be extended depends upon the power supply voltage and the aerodynamic stability of the column. With a 1300-V supply, the glow discharge burns stably up to a length of 12 mm, and up to 20 mm for average times of 5 sec. A 500-mA arc burns stably up to a length of 20 mm, and for 5 sec up to 50 mm.

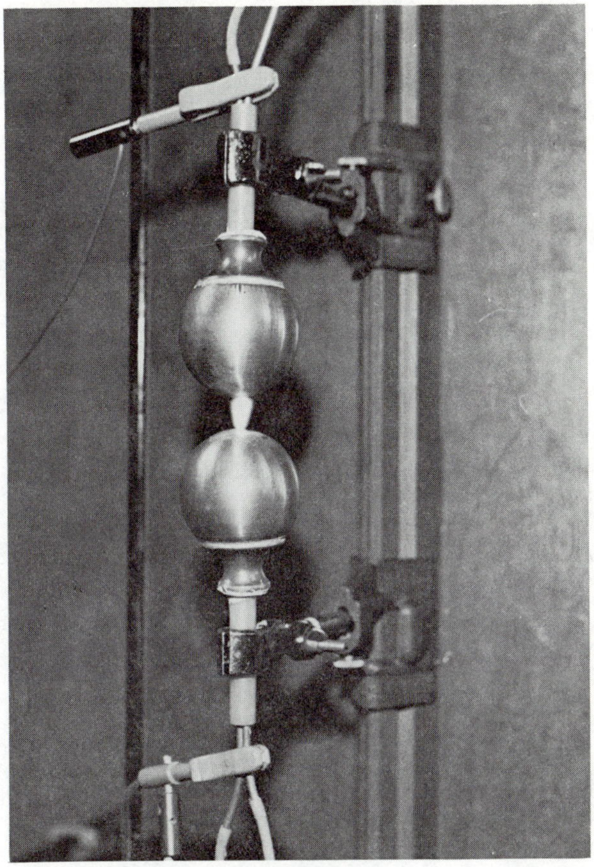

Fig. 30–33

30–4.3

In this experiment, use is made of an electrometer[18] (micromicro- or picoammeter) to measure ionization currents and to exhibit them by driving an ordinary demonstration voltmeter. The conductivity of air containing ions is demonstrated when a lighted match is held between two 6 x 8-in. copper plates with an electric field maintained between them. The plates, separated by an air gap, are connected in series with a 1200-V battery and the input of the electrometer. The output of the electrometer is connected to a large wall voltmeter (3-V scale) which registers a voltage to show that a tiny current is passing between the plates. Here, the flame provides ions in the gap.

A large alpha particle source is then held in or near the gap, and the ions produced carry current. When the source is held farther away, the voltmeter reads zero; there is no current, showing that

[18] Keithley Model #610.

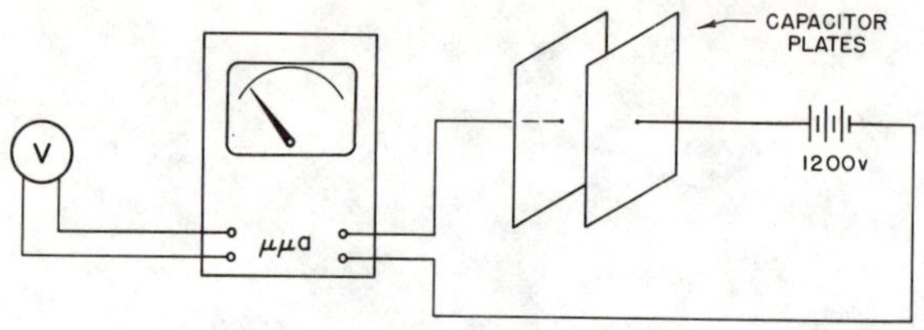

Fig. 30–34

the alpha particles have a short range. But if the source is held some distance away in line with the gap and the lecturer blows across it towards the gap, a reading is again obtained. Ions made by the alpha particles are carried by the lecturer's breath to the gap. The source is removed, and the last experiment is repeated to show that the lecturer's breath is not radioactive or ionized. The circuit is shown in Fig. 30–34.

30–4.4

A 2-cm length of 0.5-mm diam constantan wire is mounted on an insulated stand approximately 1 cm from the charging sphere of a gold leaf projection electroscope. The electroscope is charged, and the wire is heated to red heat by passing a current through it. As soon as the wire turns red, the electroscope begins discharging, due to thermally induced ionization.

30–4.5

A simple Cottrell precipitator is shown in Fig. 30–35. It consists of a glass or plastic tube *a* fitted with a rubber stopper *b* at each end. Two heavy copper wire (#12) electrodes *c* and two glass tubes *d* for letting smoke in and air out fit into the stoppers. The tube is filled with smoke. When the electrodes are connected to a Van de Graaff generator, an electrostatic generator, or other high-voltage source, the smoke particles become charged and are attracted to the electrodes.

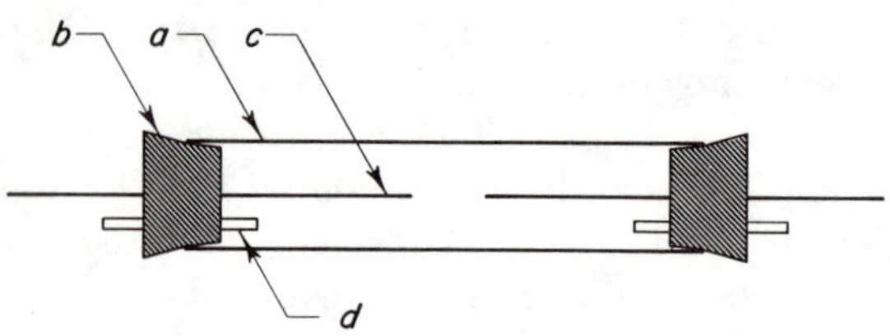

Fig. 30–35

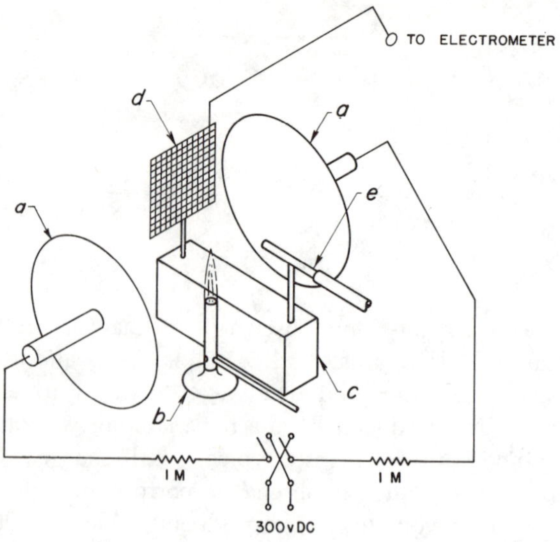

Fig. 30–36

30–4.6

A 300-V potential is applied through two 1-MΩ resistors to a capacitor made from two 6-in.-diam brass plates *a* (Fig. 30–36) mounted 3 in. apart. A Bunsen burner *b* is mounted so that its flame burns at a distance of 1 in. in front of one of the plates. A Plexiglas block *c* with a wire mesh *d* at one end and a short length of copper tubing *e* at the other is placed in front of the other plate so that when the copper tubing is connected to a compressed air source, the air will blow past the surface of the second plate. The wire mesh is connected to a zero-center electrometer, which will show the charge on the mesh. With the flame burning and the air stream blowing past the plate, the electrometer will deflect for a given polarity of the applied voltage. If the polarity of the voltage is reversed, the electrometer will deflect in the opposite direction.

30–4.7

An open Pyrex tube *a* (4 ft long, 1 in. i.d.) is sealed with electrodes *b* at its bottom, middle, and top (Figs. 30–37 and 30–38). In order to obtain maximum ionization on humid days, a heater consisting of a 0.3-mm-diam constantin wire *c* is wound externally around the tube and connected to terminals *d*. When the heater is used, a potential of 30–40 V is needed to keep the tube several degrees above the room temperature.

Fig. 30–37

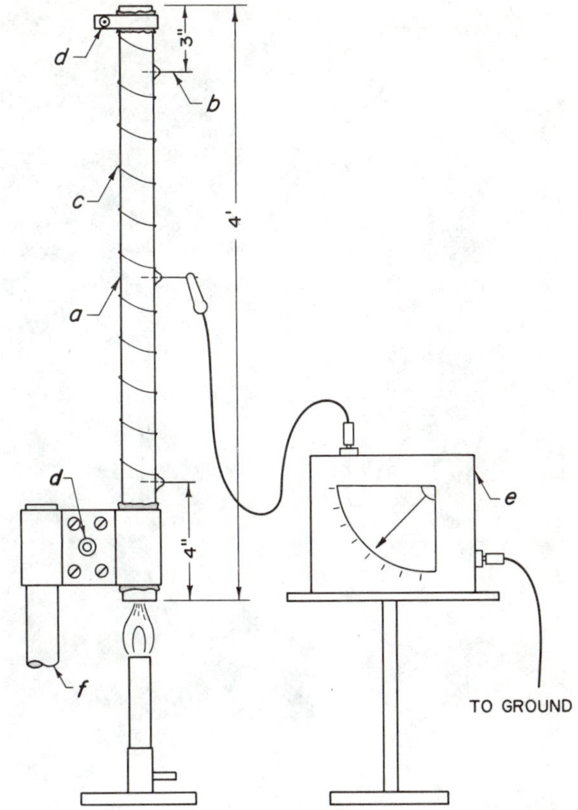

Fig. 30–38

The tube is supported vertically by rod f, and a Bunsen burner with about a 1-in. flame is placed under it. Electrometer e is clipped to the three electrodes in turn and is charged to 2000–3000 V.

Ions are produced by the Bunsen flame at the lower end of the tube. Rapidity of discharge of the electrometer indicates the number of ions present. Discharge is very fast at the lower end, slow in the middle, and hardly perceptible at the top, showing rapid loss of ions due to recombination. The experiment should also be shown without the flame in order to indicate that the system is insulating.

30–4.8

The Cerberus smoke detector[19] pictured in Fig. 30–39 uses an ionization chamber to detect smoke produced in the early stages of development of a fire.[20] An ionization chamber is used because the conductivity of the air in the chamber is a function of its chemical and physical composition.

As shown in Fig. 30–40, a battery a applies a potential across plates b and c of the chamber. A radium source d emits alpha particles, which ionize the air in the chamber. The ions move as shown. A galvanometer e measures the current, which depends on the strength of the source and, to a certain extent, the battery voltage. This relationship between the ionized current and the applied potential can be explained by the fact that, at low ionizing potentials, only part of the ions reach the electrodes. The remainder collide on the way with electrons, causing a recombination or neutralization of the charge to take place. Only when the potential reaches a certain limit do all the ions formed reach the electrodes. This is known as the saturation

(19) Cerberus Smoke Detector, in the United States called Pyr-a-larm, can be obtained from Pyrotronics, Union, New Jersey.

(20) E. Meili, "Ionisations-Feuermelder," Bulletin No. 23 of the Swiss Electrotechnical Assn., 1952.

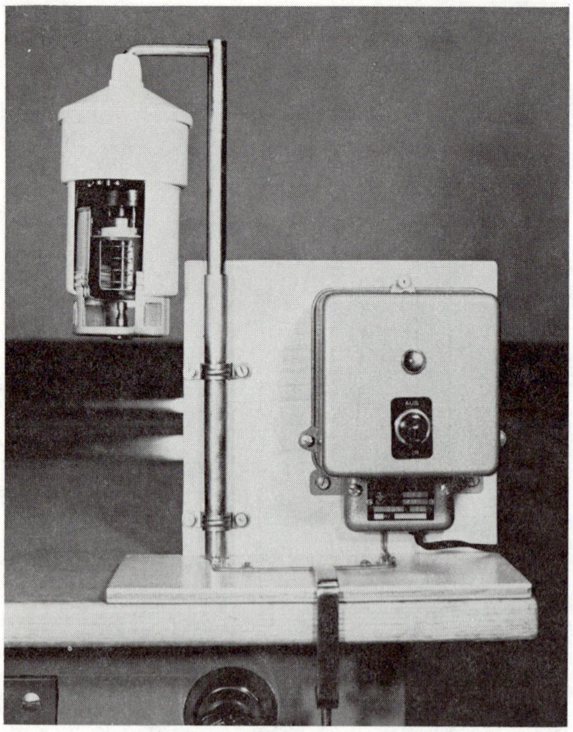

Fig. 30–39

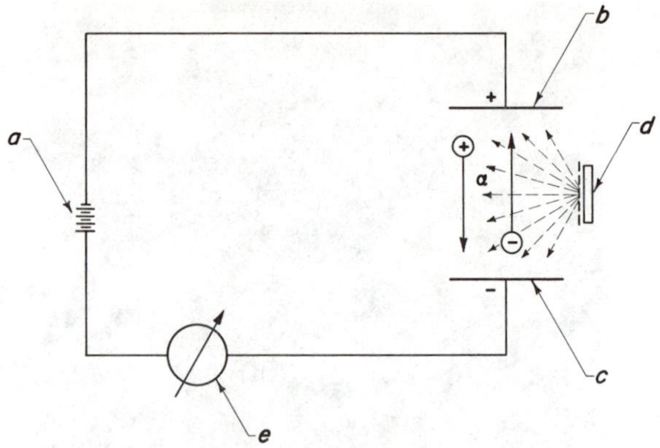

Fig. 30–40

point. Beyond this, the current remains essentially constant with increasing potential.

Aside from these factors, the current in the ionization chamber depends on the composition of the gas between the electrodes. More or fewer ions will be produced, depending on the number and size of the gas molecules. Furthermore, the rate of drift of the ions is closely related to their size and mass, but there is little difference between gases.

Conditions are quite different when combustion products enter the ionization chamber; a sharp drop in current results. Almost every combustion produces particles suspended in high concentration in the air, some of which are visible as smoke. Most of these particles, however, are too small to be seen but still may be a thousand times bigger than a gas molecule.

These particles hinder the production of ions by increasing absorption of the alpha particles and also hinder ion transport in the electric field, thus decreasing the current in two ways. When being ionized themselves, these heavy particles are much less mobile than the air molecules, and as a result their chances of being neutralized by free electrons before reaching one of the electrodes is nearly 100 percent. Since the equivalent circuit of an ionization chamber is a variable, nonlinear resistor, combustion gases entering it increase its resistance value, making it possible to detect a fire from its combustion products rather than from its flame.

30–5 Thermoelectric Effects

30–5.1

When a current is passed through a circuit composed of two dissimilar metals, one of the junctions between the metals becomes warmer while the other becomes cooler; this is known as the Peltier effect.

In the apparatus shown in Fig. 30–41, two counterpole GeAs elements[21] *a* are connected in series between three copper blocks *d* and *c*. A vapor pressure thermometer is inserted into the hole in the larger copper block *c*. The two smaller blocks *d* are glued together with resin so that they are electrically insulated from each other. A hole is drilled through each of the smaller blocks so that water for cooling supplied by the two copper pipes *e* can flow through both blocks.

For insulation, the Peltier junctions shown in the upper part of the figure are enclosed in a Styrofoam box. The counterpole elements are composed

(21) Peltier elements can be obtained from Philips AG, Holland.

of couples made of two types of semiconducting materials, known as n-type and p-type. The n-type material has free negative carriers, or electrons, and will, when heated at one end, show a negative potential at its cold end with respect to its heated end. The p-type material has positive carriers or holes and will, when heated at one end, show a positive potential at its cold end with respect to its heated end. Connecting the p- and n-type materials by copper blocks is a convenient way to combine these substances to obtain maximum results. The term counterpole is used because of the arrangement of the materials. The p-type is connected to the large copper block *c*, which represents the cold side of the junction in such a manner that its positive potential is at the *cold* end. The n-type is connected so that its negative potential will be at the *cold* end. The negative potential of the p-type and the positive potential of the n-type are connected to the small copper blocks *d*, which are at the *hot* side of the junction.

Water and the current (~75 A dc) are both provided to the small copper blocks *d* by the two copper pipes *e*. The current passing through the system causes one junction to increase in temperature and the other to become cooler. Heat is also generated due to the I^2R law. In addition, the temperature difference produced between the hot and cold junctions results in heat transfer by thermal conduction back from the hot to the cold junction. The net heat transfer and temperature changes depend on these effects. The applied voltage must be high enough to overcome the resistance of the system and the back Seebeck voltage caused by the temperature difference between the cold and the hot junctions.

For demonstrations, the apparatus is shown in shadow projection and the thermometer is projected from an optical bench. The current supply must be adjustable so that the maximum permissible current for the element is not exceeded. The cold junction will cool to about −20°C.

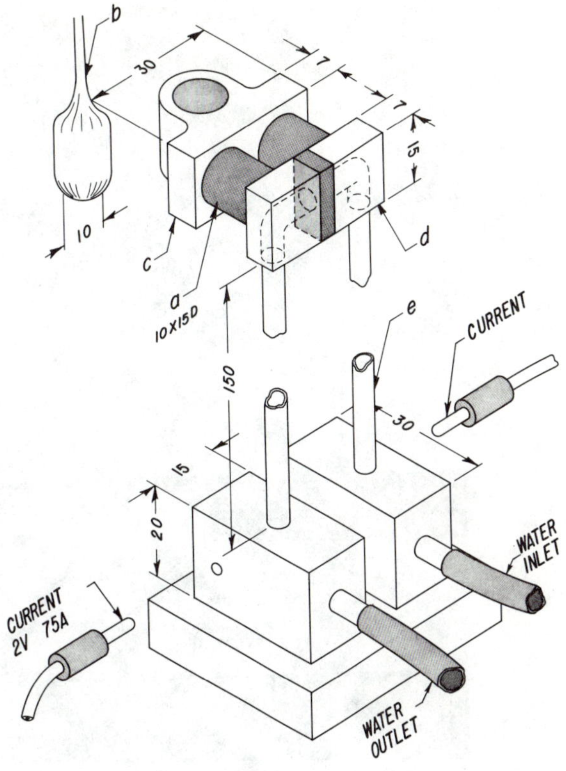

Fig. 30–41

30–5.2

The effect of temperature changes on the polarization of a pyroelectric crystal is demonstrated for two crystals, tourmaline and $BaTiO_3$. These crystals possess a permanent electric polarization which is temperature-dependent; hence, a change of the polarization with temperature can be observed. The electrode faces of the crystals are covered with silver paint. An alternative method of obtaining a conducting surface is to evaporate silver in a vacuum onto the faces.

The $BaTiO_3$ crystal, having been soldered directly to the end of a grounded cable, is protected against undesirable radiation by an aluminum can (1 in. diam, 3 in. long). The top of the can *a* in Fig. 30–42 is mounted on the cable, close to the crystal. The crystal can be heated through a hole in the bottom of the can (e.g., by the heat radiation of a flashlight or of the hand). The tourmaline is held between two plates (2¾ in. diam) by two springs *b*. It is heated by a hair drier *c*. The voltages produced with both crystals are measured by an electrometer[22] *d*, which can be connected to a projection scale.[23] The decade shunt *e* shown in the photo is left on OPEN (i.e., it is really not necessary). The measuring range depends on the heating of the crystals.

[22] Model 210, Keithley Instruments, Inc., Cleveland, Ohio, or equivalent.

[23] MB5, Keithley Instruments, Inc., Cleveland, Ohio.

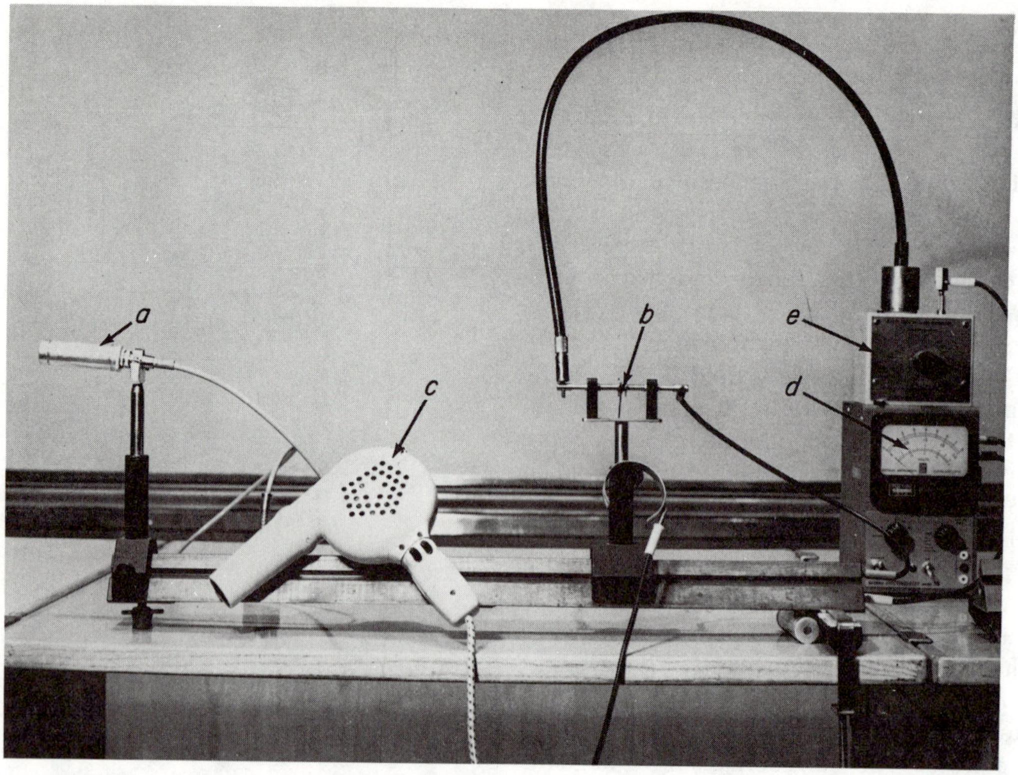

Fig. 30–42

The so-called pyroelectric crystals possess a permanent electric polarization along a polar direction. This is any direction of which the two ends are not related by any symmetry element of the point group of the crystal. Among the 32 point groups there are 21 noncentrosymmetrical groups. But only the following 10 of them may theoretically show pyroelectricity and are called the polar classes:

1, 2, 3, 4, 6, *m*, *2mm*, *3m*, *4mm*, *6mm*.

The permanent electric moment is usually not evident since the corresponding surface charges become neutralized by imperfect insulation. However, the electric moment is, in principle, temperature-dependent, and hence a change of the polarization is observed when the temperature of the crystal is varied uniformly. For a small temperature variation ΔT,

$$\Delta \mathbf{P} = (\mathbf{p})\, \Delta T,$$

where the vector **p** is called the pyroelectric coefficient. Its nonzero components are restricted by point group symmetry; e.g., in a tourmaline crystal

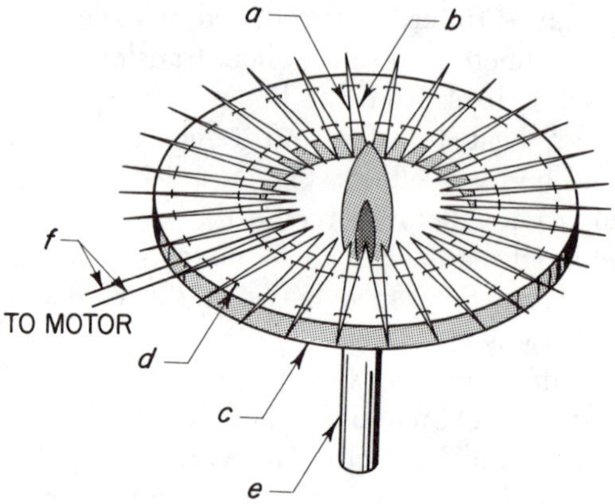

Fig. 30–43

(3 m) there is only one, nonvanishing component, the one parallel to the polar axis.[24]

Ferroelectrics such as $BaTiO_3$ form a subgroup of pyroelectrics in which the spontaneous polarization can be forced to change the direction by 180 deg or 90 deg by the application of an electric

[24] For further details see J. F. Nye, *Physical Properties of Crystals,* Clarendon Press, Oxford, 1960.

field exceeding the coercive field or by a mechanical stress. In ferroelectrics the pyroelectric coefficient can easily be two orders of magnitude higher than in tourmaline. A change of 0.001°C can be measured without difficulty.

30–5.3

The conversion of heat to electricity can be demonstrated with the apparatus shown in Fig. 30–43. Sixty 2-in. lengths of #14 copper wire *a* and #14 iron wire *b* are arranged alternately around a 3/8-in.-thick 6-in.-diam Transite disk *c* with a 3-in.-diam center hole. The ends of the wires are cleaned, twisted four times, and soldered. The wires are fastened to the disk by nylon strings *d* threaded through holes drilled in the disk. If the center of the disk is heated with a Bunsen burner *e*, a milliammeter connected to the wires at *f* will register a current of about 90 mA. The current can be used to run a Micromax motor.

30–6 Piezoelectric Effects

30–6.1

In a piezoelectric crystal, compression stress produces an electric polarization which shows up as surface charges on the sides of the crystal. The apparatus in Fig. 30–44 is designed to show this effect. It is also used to show that the voltage is proportional to the applied stress, and that when

Fig. 30–44

the crystal is stressed for a long period of time, the voltage slowly diminishes due to gradual cancellation of the surface charges by conduction through the crystal and by ions attracted from the air.

As shown in Fig. 30–45, a Seignette salt crystal[25] *a* is loaded by placing weight *d* on the lever arm *b*, which was cut from 3/16-in. aluminum stock. A flat steel disk *c* the size of a U.S. nickel should be placed on the top surface of the crystal to insure that it is uniformly stressed. The leads from the crystal are connected to a projection electrostatic voltmeter[26] with a 200-V full-scale range. As more weights *d* are added to the lever, the voltage increases, and if they are all removed at once, it will

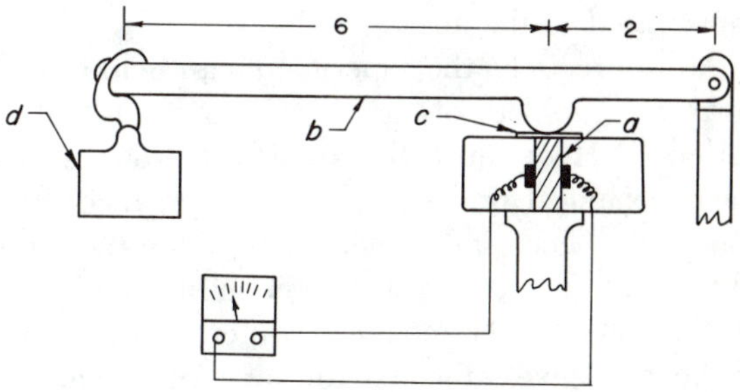

Fig. 30–45

be seen that a reverse polarity is set up due to the relaxation of the applied stresses.

30–6.2

After showing the production of an emf by pressure applied to a Seignette salt crystal, the reverse effect is demonstrated with the apparatus in Fig. 30–46. A flat, openly mounted Seignette crystal is fed with an audio voltage from a record player and amplifier. The sound is barely audible and all treble. By touching the bottom of a drawer with the crystal, the full sound is heard. The effect is used

(25) Leybold Catalog No. 58725. Available through LaPine Scientific Co.

(26) Trüb, Täuber and Co., 220 V, full scale.

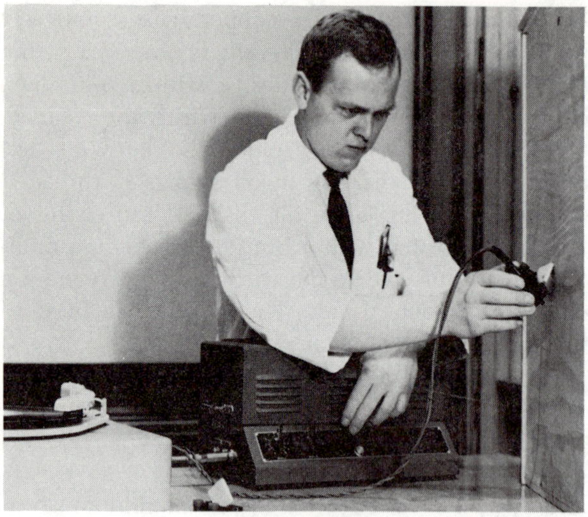

Fig. 30–46

to discuss qualitatively the acoustical impedance and necessary size of a sounding board. For example, if the sounding board is small compared with the wavelength of sound, the pressures on the front equalize according to the laws of hydrostatics, and no net force occurs, and therefore no power is transmitted to the air.

More precisely, the experiment can help to make clear the problem of matching the acoustical impedance. How must the transformation ratio of the mechanical transformer be chosen in order to transmit all energy (no reflections) from the crystal to the air? The theory of reflection of sound waves at the interface of media of different mechanical properties gives the condition: $Au\zeta = \text{constant}$, where A is the area of piston, u is the velocity of sound, and ζ is the density of the material.

The crystal used is a Seignette-salt crystal sold by Leybold (type 58725) which is enclosed in an air-tight transparent housing. A light pressure on the places designated produces easily a potential of several hundred volts, which can be measured with an electrostatic voltmeter, electroscope, or electronic electrometer. To transmit music to the sounding board, a mechanical link, e.g., a screw glued on the housing with double-coated plastic tape, must be provided as shown in Fig. 30–47. The amplifier should have an output impedance of at least 500 Ω. If it does not, a suitable audio transformer can be inserted. Take care not to load the crystal electrically for too long a time since it may melt.

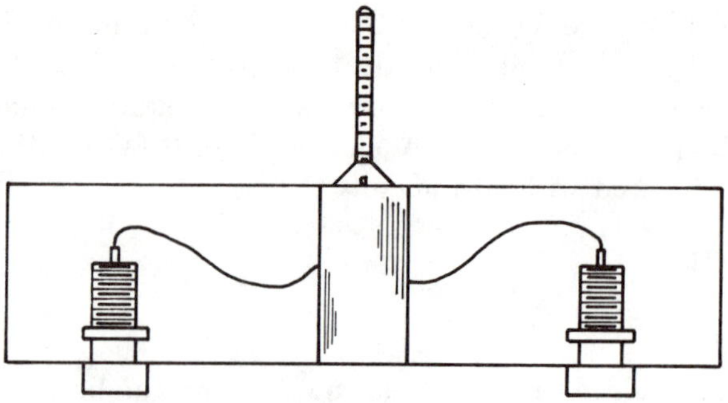

Fig. 30–47

30–6.3[27]

The phenomenon of ferroelectricity can be demonstrated with the apparatus represented by Fig. 30–48. A specimen of Rochelle salt K is contained within a multipurpose holder, which is described in the Appendix, page 1322, along with the preparation of the specimen. An audio generator S is connected in series with the specimen K and an oscilloscope O. The horizontal input to the scope

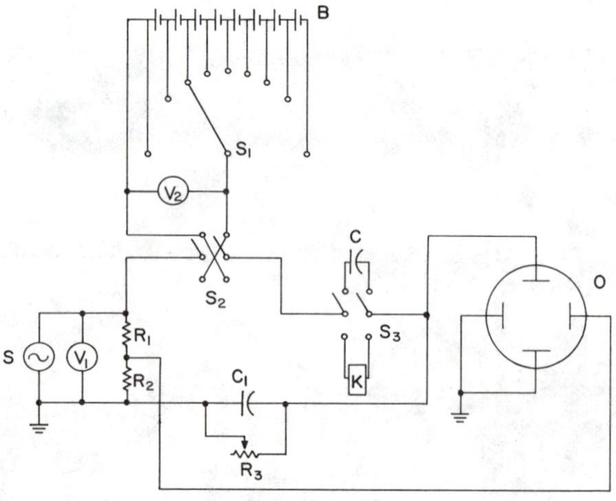

Fig. 30–48

is taken from a voltage divider arrangement R_1 and R_2. In order to minimize the phase shift at the oscilloscope, the value of C_1 and shunt resistor R_3 should be adjusted so that

$$\frac{C_K}{C_1} = \frac{R_3}{R_K},$$

where C_K is the capacitance of the crystal and R_K is its resistance. Since R_K will vary with the humid-

[27] Translated from K. Karmen, *Usp. Fiz. Nauk,* **63,** No. 4, pp. 819–823 (1957).

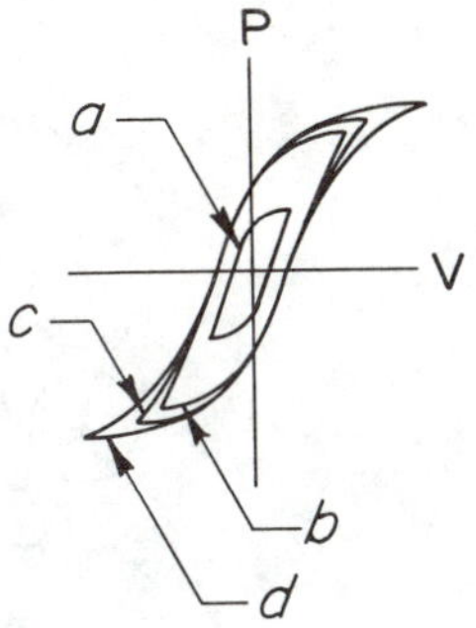

Fig. 30–49

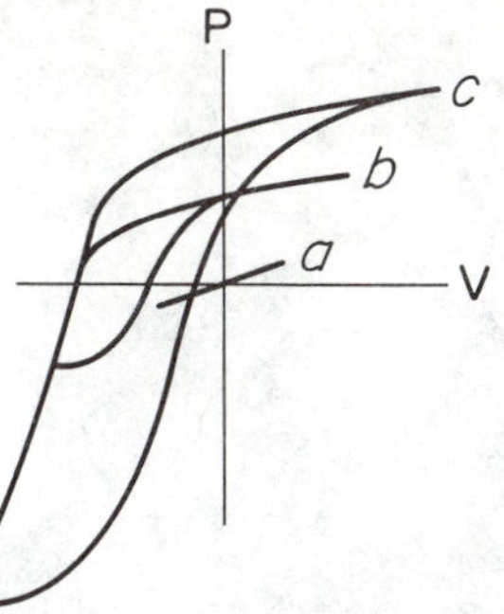

Fig. 30–50

ity, stress, and the applied frequency, further adjustment of R_3 will be necessary. A constant applied voltage may be obtained with the DPDT switch S_2, battery B and slider S_1 combination. The DPDT switch S_3 allows substitution of a mica capacitor box C for K.

To demonstrate hysteresis, battery B is removed from the circuit by S_1, and the audio generator is set at 50 cps. The oscilloscope patterns at 18°C for various generator voltages are shown in Fig. 30–49. Curves *a*, *b*, *c*, and *d* correspond to 50 V, 100 V, 150 V, and 185 V, respectively. To evaluate the capacitance of the specimen and for the purposes of comparison, S_3 is used to replace K with C.

The existence of the upper Curie-point is easily demonstrated by removing the specimen from its holder and setting the middle of the small side parallel to the filament of a 75-W lamp at a distance of 5 to 8 cm. Within a few minutes this will cause a transition to the nonferroelectric state, which is indicated by a straight line on the oscilloscope. After the lamp is switched off, it takes approximately the same amount of time for the hysteresis curve to be gradually reestablished.

To demonstrate the influence of a constant displacing field on the polarization process, S_2 is used to connect the battery B in the circuit. Figure 30–50 shows the curves obtained with a constant 80 V, a generator frequency of 50 cps, and different alternating voltages. Curves *a*, *b*, and *c* correspond to generator voltages of 50 V, 100 V, and 185 V, respectively.

The curves obtained with an applied frequency of 10^4 cps are shown in Fig. 30–51. Curve *a* is the original state, and curves *b*, *c*, and *d* are 30 sec, 2 min, and 5 min, respectively, after switching to the higher frequency. These phenomena can be explained by the hysteresis losses heating the specimen.

The direct piezoelectric effect in the static condition can be demonstrated using the specimen and holder, with the holder's terminals directly connected to a ballistic galvanometer. Placing a load on the plates of the holder will cause a sudden displacement of the galvanometer to one side, while removing the load will have the opposite effect. By loading stepwise from 0.5 to 5 kg, the saturation of the piezoelectric polarization can be observed.

The piezoelectric effect under dynamic conditions is conveniently shown by connecting the terminals of the holder directly to an oscilloscope. A load of about 1.5 to 2 kg is placed on the plate of the holder and then stroked with a small wooden hammer. This produces vibrations which will cause periodic variations of the piezoelectric polarization. With some practice it is possible to obtain damped sine waves on the oscilloscope.

The converse effect can be demonstrated by connecting the specimen under the same constant load to the output of the audio generator. At 200 to 400 cps a distinct sound can be heard, which grows much louder as resonance is approached.

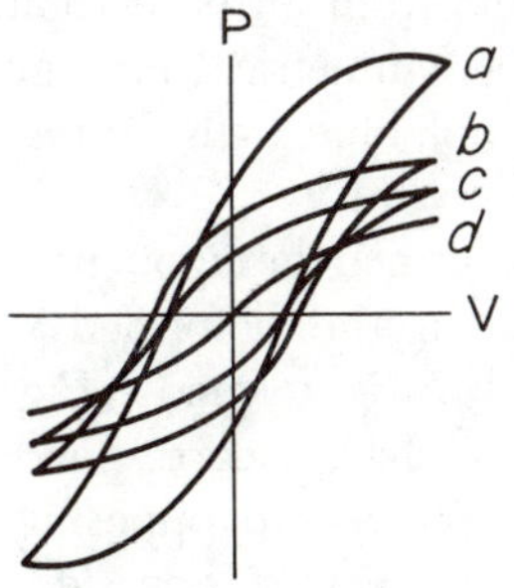

Fig. 30–51

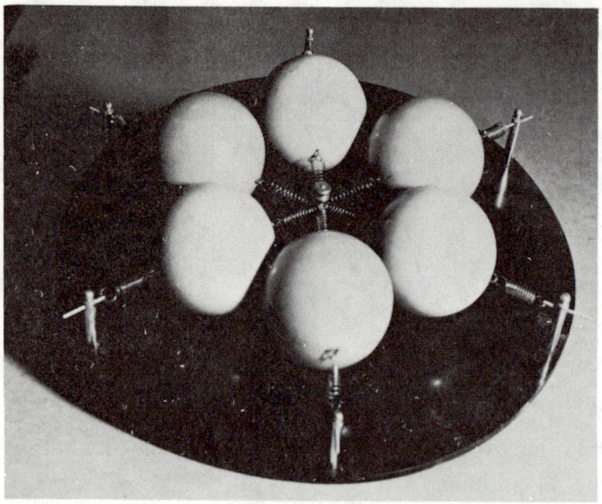

Fig. 30–52

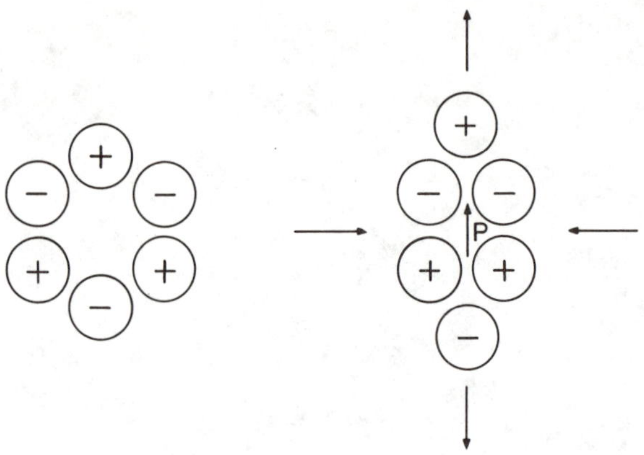

Fig. 30–53

30–6.4

Piezoelectricity is the formation of an electric polarization by mechanical stress. Even if a crystal is built up of ions, the symmetry of the lattice provides cancellation of positive and negative charges (total charge equals zero) and the centers of gravity of positive and negative charges coincide (no total electric dipole moment). Certain crystals will produce a polarization when the lattice is deformed by stress.

To show that even simple lattices can produce this effect, a flat model of threefold symmetry is built up by red and blue balls fastened by springs as shown in Fig. 30–52. Seven bolts (about 1¾ in. long, $\frac{3}{16}$ in. diam) are fastened to a plywood board, one at the center and the others in a hexagonal pattern, about 6 in. from the center. Three 3-in.-diam red balls representing positive charges, and three similar blue balls representing negative charges, are held by two springs, one from the center bolt and one from an outer bolt. The spheres are placed in a symmetrical arrangement, with alternating red and blue balls corresponding to the neutral quartz crystal.

In the undeformed hexagon no dipole moment exists, whereas in the deformed cell an electric dipole moment P is formed (Fig. 30–53). The movement of the charges corresponds to a displacement current. When two opposing balls are displaced toward the center, one side becomes more positive (red) and the other side becomes more negative (blue). By addition of many cells to a crystal, it is easily seen that inside the displaced charges cancel. Therefore, the surface charges create a macroscopic dipole.

30–6.5[28]

To demonstrate hysteresis in ferroelectric crystals, a convenient material to use is one of the perovskites, barium titanate, which is available in the form of a polycrystalline ceramic. The specimen in the circuit of Fig. 30–54 is a type HKY capacitor,[29] 0.01 μF, 50 V.

The polarizing field, which is proportional to the generator voltage, is applied to the x-plates of the cathode ray oscilloscope. The resulting instantaneous state of polarization of the specimen is proportional to the charge and therefore to the voltage on the 1-μF capacitor. This instantaneous voltage is applied to the y-plates of the oscilloscope.

Figure 30–55 shows the effects of an alternating electric field of sufficient magnitude to reverse the direction of polarization; the electric equivalents of remanence and coercive force appear. The magnitudes of these effects are measurable with the energy loss per cycle computed from the loop plot.

Figure 30–56 shows a modification of the circuit for use with a x-y plotter. The entire hysteresis curve can be traced, starting from the state of zero polarization.

(28) R. Resnick and H. F. Meiners, *Demonstration and Laboratory Apparatus Report of the 1960 Summer Visiting Professor Workshop at Rensselaer Polytechnic Institute.*

(29) Sprague Catalog No. TGS 10.

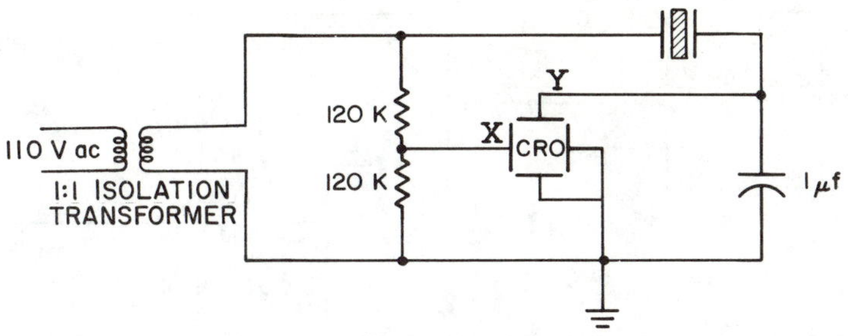

Fig. 30–54

30–6.6

Domains of electric polarization in tiny $BaTiO_3$ single crystals held on a heatable glass slide (made by putting a transparent conducting layer of tin oxide on glass) can be easily seen with the apparatus shown in Figs. 30–57 and 30–58. An arc lamp *a* provides unpolarized light, which passes through a condenser lens *b*, water cooler *c*, polarizing Nicol prism *d*, substage mirror *e*, an Abbe condenser *f*, heatable glass slide *g*, microscope *h* (60x), and a second Nicol prism *i*. A right-angle prism *j* is used to project the image onto a wall.

The glass slide is covered with a layer of $BaTiO_3$ single crystals. On each end of the slide a 1-cm-wide strip of gold is applied (painted or evaporated). A Variac connected directly to the slide by two insulated spring clips is used to heat the ~1200-Ω slide. At the critical Curie temperature, the domains disappear suddenly. In cooling, they reappear in different patterns.

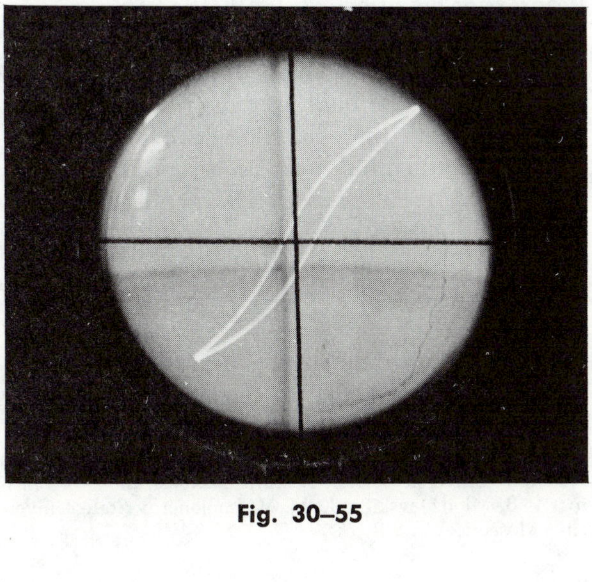

Fig. 30–55

30–6.7[30]

In ferroelectric materials the mechanical strain generated by the application of an electric field depends on the direction of the field, causing expansion and contraction as the field is reversed. This phenomenon, known as the piezoelectric effect, can be demonstrated by connecting a capacitor containing a high percentage of $BaTiO_3$ to an audio oscillator having an output of about 20 V peak to peak. A type HYK capacitor,[31] 0.01 μF, 50 V, can be heard to resonate at a number of frequencies in the audible range. Since the amplitude of the resonance is low, amplification through a standard microphone and PA system or visual presentation on an oscilloscope is required for classroom demonstration.

30–6.8[32]

In most cases polycrystalline sheets of piezoelectric crystals, such as Rochelle salt, do not show direct or inverse piezoelectric effects because of the random orientation of the crystallites. It is possible, however, to create polycrystalline plate textures which show these effects.

Fine crystals of Rochelle salt are placed in a beaker in a water bath. The salt melts at 63°C, but it is best to prepare the plates by heating the melt to 100°C. A small amount of the melt is picked up on a stiff cylindrical paintbrush about

(30) R. Resnick and H. F. Meiners, *Demonstration and Laboratory Apparatus Report of the 1960 Summer Visiting Professor Workshop at Rensselaer Polytechnic Institute.*

(31) Sprague Catalog No. TGS 10.

(32) Translated by K. V. Robinson and H. A. Robinson (Adelphi University) from *Lecture Demonstrations in Physics*, A. B. Mlodzeevskii, ed., M. V. Lomonosov State University, Moscow, U.S.S.R.

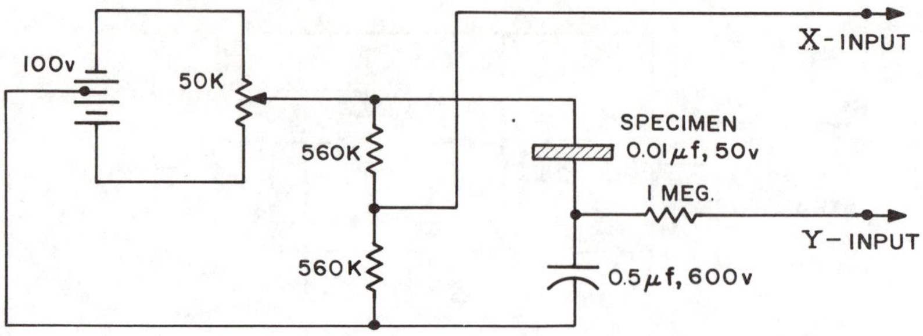

Fig. 30–56

½ in. in diameter. The material is then brushed onto a clean metal or glass plate of the desired size. Brush in one direction only (say, from left to right). When the salt begins to crystallize (glitter), a second brushful is transferred by brushing in the opposite direction. Between brush strokes the brush should be replaced in the beaker so that all crystals adhering will be remelted. Plates of any desired thickness can be built up by this alternate direction brushing technique. Plates prepared in this manner will not show the piezo effects immediately, but they will do so after several days, during which time the conductivity decreases.

Electrodes are easily attached to such plates by carefully moistening the surface and pressing aluminum or lead foils onto them by cotton pads, even though these plates are somewhat brittle. Alternatively, the electrodes may be applied to the lower surface of the plate during the process of preparing the textured plate: place a smooth piece of foil on the plate and brush the salt melt onto the foil. Leads can be attached to the foils by contact springs.

Fig. 30–57

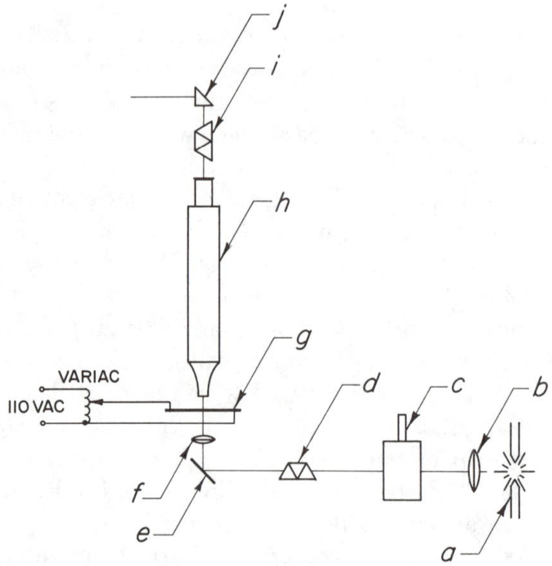

Fig. 30–58

Such textures have many applications. For example, they can be built up directly on tuning fork prongs, close to the stem. The emf produced by such a vibrating fork can be demonstrated on a cathode ray oscilloscope, or if amplified, fed directly to a loudspeaker.

30–6.9[33]

To demonstrate the inverse piezo effect is a crystal of Rochelle salt, that is, the deformation of the

[33] Translated by K. V. Robinson and H. A. Robinson (Adelphi University) from *Lecture Demonstrations in Physics*, A. B. Mlodzeevskii, ed., M. V. Lomonosov State University, Moscow, U.S.S.R.

crystal when affected by an electric field applied across its plates, obtain a 45 deg x-cut of a single salt crystal. (Cenco's Piezoelectric Demonstration Apparatus No. 81335 could be used.) The crystal plate, 8 x 25 x 40 mm, is placed between two wide spring brass contacts. The size of the contacts is such that the upper edge of the crystal is several millimeters above the ends of the contacts. The wide surfaces of the crystal are connected by means of the contacts to an audio oscillator. Since the audio oscillator should apply a voltage of 100–120 V, an audio frequency amplifier must be set up between the crystal and the oscillator.

If the voltage supplied to the plates is sufficient, the crystal starts vibrating in a vertical direction, emitting a sound corresponding to the frequency of the alternating voltage applied to it. To amplify the sounds, set the apparatus on a wooden board which resonates to the oscillations of the crystal. If the amplified output from a radio receiver is applied to the crystal (in the same manner as the audio oscillator), it will act as a loudspeaker.

Demonstrations	***Contributors***
30–1.1	H. C. Jensen, Lake Forest College
30–1.2	Royal Institution of Great Britain, London
30–1.3	L. Grillet and M.-T. Grillet-Amy, University of Rennes, France
30–1.4 to 30–1.6	Swiss Federal Institute (ETH), Zürich, Switzerland
30–1.7	M. Mainardi, Newark College of Engineering
30–1.8	J. W. Straley, University of North Carolina
30–1.9	Stanford University
30–2.1	M. Iona, Jr., University of Denver
30–2.2	Bashir A. Syed, University of Saskatchewan, Canada
30–2.3	E. B. Nelson, Iowa State University of Science and Technology
30–2.4	L. Grillet and M.-T. Grillet-Amy, University of Rennes, France
30–2.5	J. D. McGovern, Brother Rice High School, Birmingham, Michigan
30–2.6, 30–2.7	L. Grillet and M.-T. Grillet-Amy, University of Rennes, France
30–2.8	California Institute of Technology
30–2.9	L. Grillet and M.-T. Grillet-Amy, University of Rennes, France
30–2.10	Royal Institution of Great Britain, London
30–3.1	Swiss Federal Institute (ETH), Zürich, Switzerland
30–3.2	Massachusetts Institute of Technology
30–3.3	H. A. Robinson, Adelphi University
30–3.4	V. E. Eaton, Wesleyan University, Conn.
30–3.5	J. G. Black, Eastern Kentucky University
30–3.6	D. C. Evans, Oklahoma State University
30–4.1	Swiss Federal Institute (ETH), Zürich, Switzerland
30–4.2	University of Cologne, Cologne, Germany
30–4.3	J. Brault, Eric M. Rogers, Princeton University
30–4.4	Swiss Federal Institute (ETH), Zürich, Switzerland
30–4.5	L. Hirsch, California State College, Los Angeles
30–4.6 to 30–4.8	Swiss Federal Institute (ETH), Zürich, Switzerland
30–5.1, 30–5.2	Swiss Federal Institute (ETH), Zürich, Switzerland
30–5.3	L. Hirsch, California State College, Los Angeles
30–6.1, 30–6.2	Swiss Federal Institute (ETH), Zürich, Switzerland
30–6.3	J. R. Taylor, Materials Laboratory, General Electric Co., St. Petersburg, Florida
30–6.4	Swiss Federal Institute (ETH), Zürich, Switzerland
30–6.5	E. O. Cook, Frostburg State College, Md.
30–6.6	Swiss Federal Institute (ETH), Zürich, Switzerland
30–6.7	E. O. Cook, Frostburg State College, Md.
30–6.8, 30–6.9	H. A. Robinson, Adelphi University

31

MAGNETISM

31–1 Ampere's Law and the Magnetic Field

31–1.1

The interaction between a current-carrying conductor and a magnetic field can be demonstrated by the apparatus in Fig. 31–1. The moving element is a U-shaped wire swing supported by flexible brass strips from two horizontal brass rods. These brass rods are mounted on rigid uprights attached to a wooden base. The U-swing, of any convenient size, is of $\frac{1}{16}$-in. aluminum welding rod. The supporting springs are brass strips 2 in. long x $\frac{1}{4}$ in. wide x .005 in. thick. The horizontal bars to which the swing is attached are brass rods bolted to 1-ft-long and $\frac{1}{2}$-in.-diam brass uprights. A Lucite brace adds to the rigidity of the system. The uprights are electrically connected to a reversing switch on the baseboard, and a tap key is connected across one set of terminals of the switch. The force on a current-carrying conductor in a permanent magnetic field of about 2000 G is illustrated in Fig. 31–1. The principle of the electric motor is also demonstrated. The inverse situation, the generation of an induced emf by moving a conductor across a magnetic field, is shown in Fig. 31–2. An induced current of 1 mA is readily obtained. This is the principle of the generator. Alternating currents are demonstrated by making the swing oscillate in the field. Electromagnetic damping can be demonstrated by short circuiting the oscillating spring with the tap key. The mutual force between two parallel conductors carrying current is illustrated in Fig. 31–3. For this demonstration a second swing, electrically connected to the first by terminals that can be connected for parallel or antiparallel flow, must be added. A current of 40 or 50 A is required to produce a reasonable deflection.

31–1.2

A triangular loop of wire, as shown in Fig. 31–4, is hung on a large spring balance with one leg of the triangle in the field of a large electromagnet. Experiment shows that the electromagnetic force on a wire in a given magnetic field varies with the current in the wire, as shown by a large ammeter.

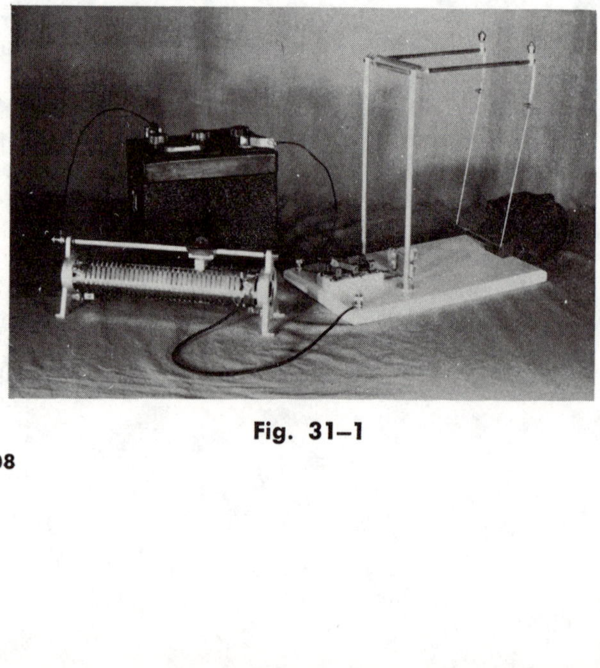

Fig. 31–1

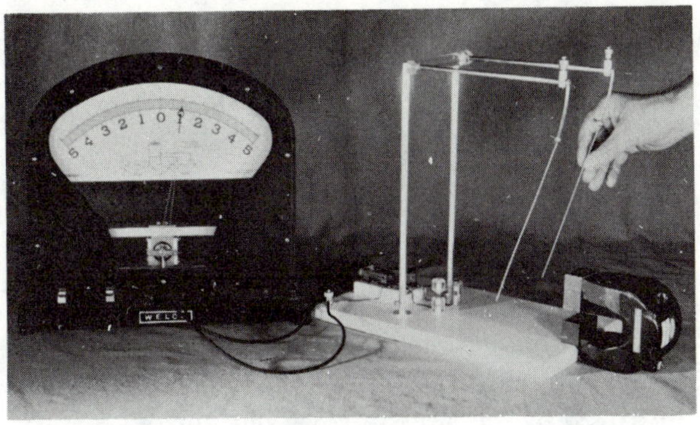

Fig. 31–2

A tall, narrow, rectangular coil, hung from a balance so that one of its ends is at the center of a large, vertical, circular coil, can also be used to show electromagnetic force when currents are sent through both coils.

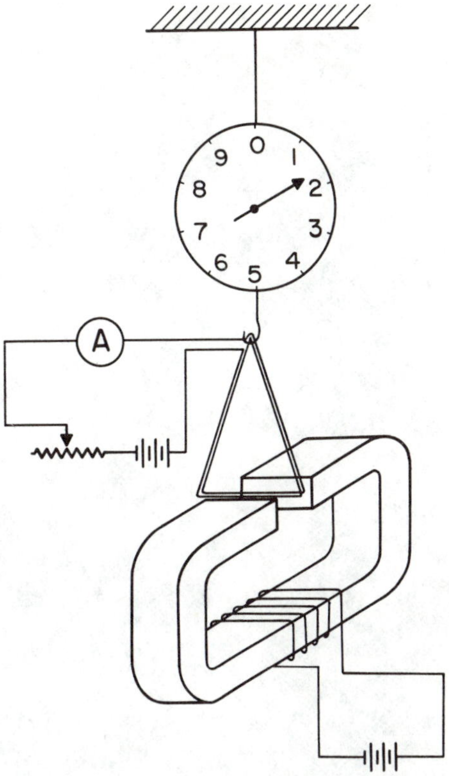

Fig. 31–4

31–1.3

An interesting form of Ampere's motor apparatus is shown in Fig. 31–5. Three large permanent ring magnets *m* are arranged in line between two long flat steel plates *p*. Attached to the long edges of the top plate are two copper side rails *r* which are electrically insulated from the plate. When a cylindrical, non-magnetic, electrically conducting rod *c* is placed across the rails at one end, the current through the rod (via the rails) causes it to roll rapidly along the rails. The direction of motion depends upon the direction of the current through the conductor. The magnetized nails in the figure show the direction of the magnetic field.

In one form of the equipment, the magnets were supplied from surplus loudspeakers. Each has a 6-in. o.d. and a height of 3 in. If ring magnets are not available, electromagnets can be used. The steel plates each measure 6 x 24 x $\frac{3}{8}$ in. Each side rail is made from copper, $\frac{3}{4}$-in. wide x $\frac{1}{8}$-in. thick and 25 in. long, allowing an extension of 1 in. beyond the plate for ease of connection to the power source. These plate and rail dimensions vary with the size of the magnet used. Direct current power is supplied to the rails by a 6-V automobile battery. The conductor rod *c* is a copper-coated carbon rod ($\frac{1}{4}$ in. diam and 8 in. long). These are commercially available under the trade name "National Suprex Positive," No. 688. Other rods of various sizes and materials are used in place of the copper-coated carbon rod to show the effects these variables have on the motor effect.

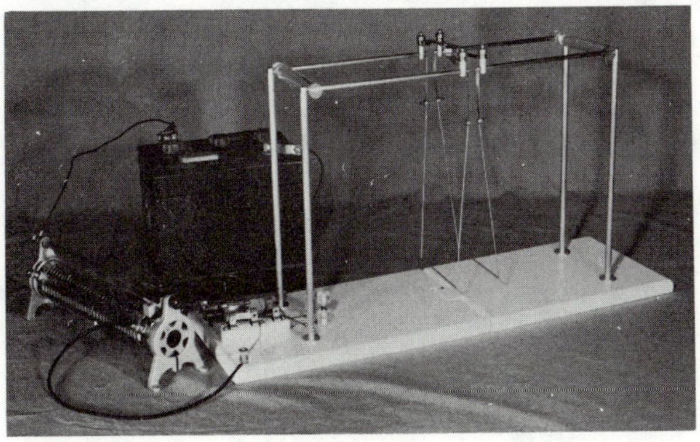

Fig. 31-3

31–1.4

Large magnets are connected so that similar poles (north poles) are at the top (see Fig. 31–6). The vertical magnetic field should be about 2000 G. A pair of rails *a*, made of 2 in. brass angles (15 x

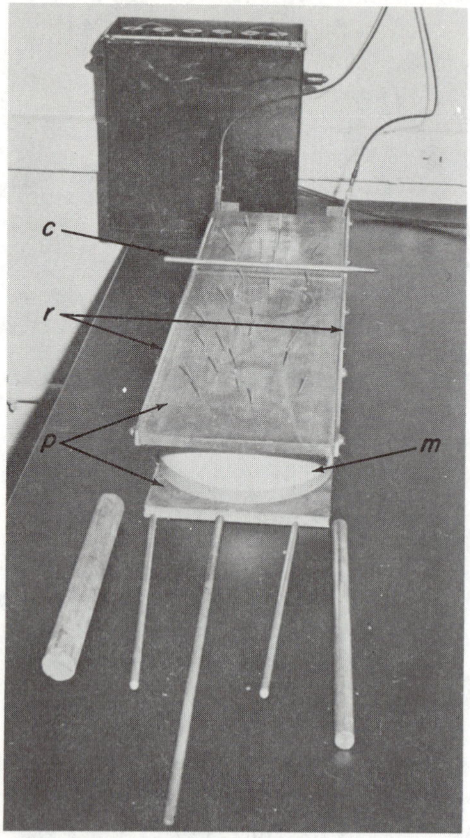

Fig. 31–5

15 x 2 mm), with vertex upward, is supported with two 6 in. long insulating bands *b*. At the ends of the rails are brass blocks *c* which serve as bumpers; the bumpers at one end have holes for electrical

Fig. 31–6

connectors *d*. On the rails runs a pair of wheels *e* with a brass axle. The 3-in.-diam x 3/8-in.-thick wheels have a V-shaped flange which fits over the rails. There are two demonstrations:

1. A projection galvanometer (5 mV, 10 Ω, zero in the middle) is connected across the rails. The galvanometer deflects when the wheels are pushed back and forth by the demonstrator. It is important to have the circuit open when the magnet is switched on or off or else the galvanometer will be destroyed by the large induced current.

2. A current source which can supply 10 to 15 A is connected to the rails. The wheels roll in a direction which depends on the direction of flow of the current.

31–1.5

A stainless steel wheel 8 in. in diam and ~.04 in. thick is mounted with a ball bearing on a 1/4-in.-diam steel shaft. This shaft is fastened in a ring stand so that the lower edge of the wheel dips into a mercury-filled, milled groove in a piece of Plexiglas. A 12-V potential is applied between the hub of the wheel and the mercury bath, and a strong dc electromagnet (approximately 1500 G) is placed so that the current flowing in the wheel must travel through the magnetic field (Fig. 31–7). With the current flowing the wheel will rotate, and if the polarity of the applied potential is reversed, the wheel will reverse direction. A series of equally spaced small holes may be drilled in the periphery of the wheel to enable the rotation to be shown in shadow projection.

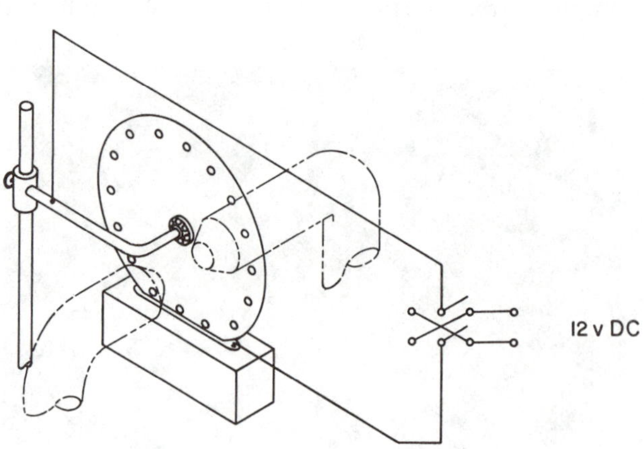

Fig. 31–7

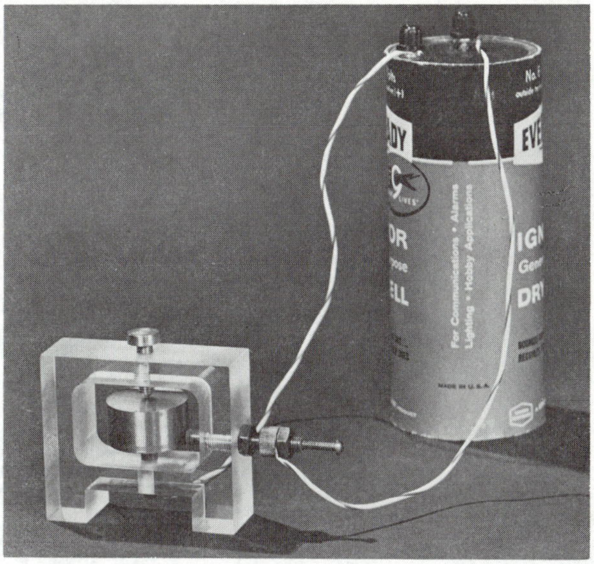

Fig. 31-8

31-1.6[1]

The familiar Barlow's wheel is one example of a unipolar motor. It consists of a copper disk which can be made to rotate on its axis when it is placed in an external magnetic field (perpendicular to the plane of the disk) and when current is flowing in the disk from the axis to the rim. A consideration of the torques involved in this device leads one to question whether the disk would rotate if the external magnet were attached to it, or alternatively, if the disk itself produced the magnetic field.

That such a motor would rotate has been noted by several authors[2] and a small model which illustrates the principles involved in such a device provides an interesting "corridor type" demonstration.

Figure 31-8 shows a model of this type in use. It consists of a short cylindrical magnet, magnetized in a direction parallel to its axis, arranged between two pivots so that it can rotate freely on its axis. Electrical connection is made through the bottom pivot and through a mercury contact on the rim.

If the apparatus is connected to a dc source such as a No. 6 dry cell, it can be made to rotate as a motor provided there is very little friction at the pivots. Because of the rather heavy current drain on the battery, and the fact that at high speeds it tends to throw mercury off at the rim contact, it is not recommended that the motor be run for more than a few seconds at a time. The device can also be used as a unipolar generator[3] by replacing the battery with a sensitive galvanometer and spinning the magnet by hand.

Variations of the basic design shown in Fig. 31-8 are almost unlimited. The model illustrated consists of an Alnico magnet, 1¼ in. in diam by ⅝ in. long, but similar magnets can be obtained up to 3 in. in diam.[4] The material is very hard but small depressions can be made at the center of each face of the magnet with a diamond cutter or drill. The magnet is then mounted between two pointed brass (or bronze) pivots, held in place in a Lucite frame.

Electrical contact is made along the axis of the magnet by connecting a wire to the lower pivot. The contact at the rim can be made through a threaded, hollow, brass rod attached through the side of the frame. If this rod is filled with mercury, and a smaller screw is threaded into the outer end of the rod, then a small bubble of mercury can be forced out to make contact with the magnet.

31-1.7

This experiment is related to the operation of Barlow's wheel. Two brass plates are connected by a spring, Fig. 31-9, (a) and (b). One plate has a slot with a screw and spring for adjustment in approximately the center, and the other plate has a screw in the center. Both plates have standard terminals at the top. The plates are $\frac{1}{16}$ in. thick. The slotted plate is 2¾ in. wide by 6¾ in. long and has a slightly curved edge, while the other one is 3 in. by 7 in. They are mounted on a 0.35-in.-thick plastic plate and connected to a galvanometer. The plates are inserted between the poles of an 1800-G magnet.

The plates are used to show that some of the ideas about flux linkages are difficult to comprehend

(1) Developed while the contributor was an N.S.F. Faculty Fellow at the University of Wisconsin.

(2) L. Page and N. I. Adams, *Principles of Electricity* (D. Van Nostrand Company, Inc., Princeton, New Jersey), 3rd ed., p. 267; A. K. Das Gupta, *Am. J. Phys.*, **31**, 428 (1963); and T. D. Strickler, *Am. J. Phys.*, **29**, 635 (1961); R. L. Burling, *Am. J. Phys.*, **31**, 217 (1963).

(3) See W. Panofsky and M. Phillips, *Classical Electricity and Magnetism* (Addison-Wesley Publishing Company, Inc., Reading, Massachusetts, 1962), 2d ed., pp. 166, 338; and J. W. Then, *Am. J. Phys.*, **28**, 557 (1960).

(4) Empire Magnetics, Inc., Maumee, Ohio.

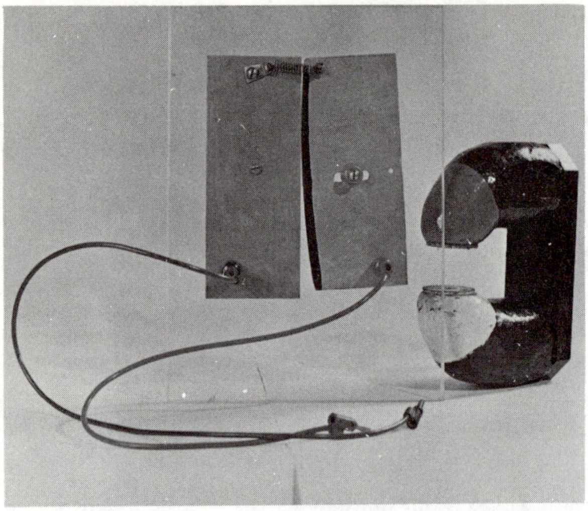

(a)

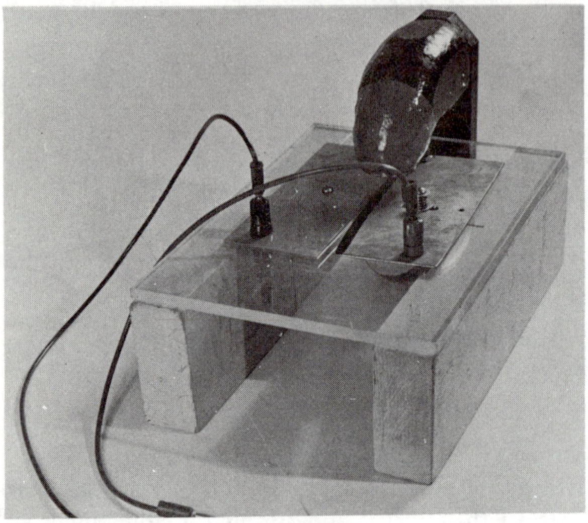

(b)

Fig. 31–9

when using extended circuits, that is, large sheets of metal. When just wires are used, it is possible to define what is meant by flux linkages and by the generation of an emf. With sheet conductors, as in this experiment, the conductors can be rocked back and forth. The rocking plates produce what seems like very large changes in flux, but these changes do not generate an emf.

The main point of this experiment is to show that the correct physics is given by

$$\mathbf{F} = q\mathbf{E} + q\mathbf{V} \times \mathbf{B}$$

and that other things deduced from the experiment

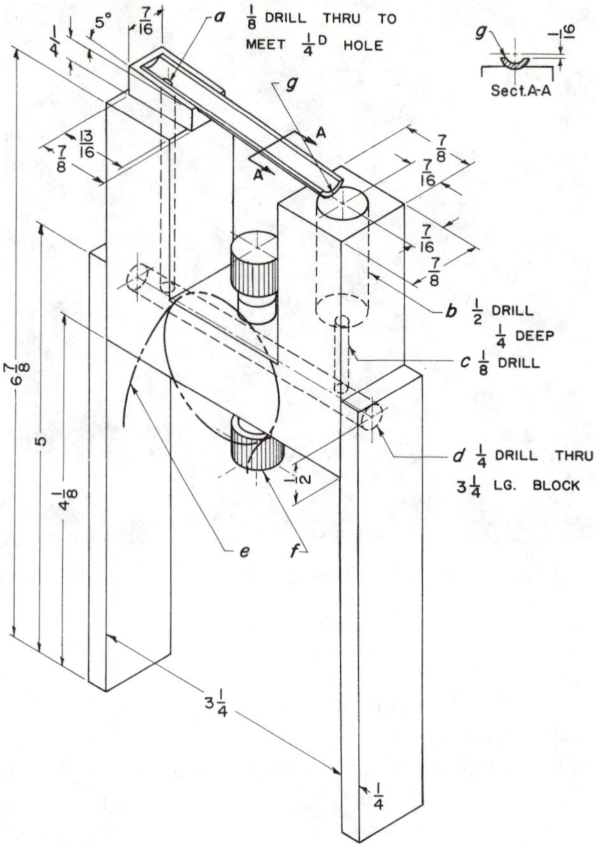

Fig. 31–10

may or may not be true, but that the basic force law is fundamental and is always true.

31–1.8

A dc arc will bend and finally blow out as a strong permanent bar magnet is brought closer and closer to it. This demonstration is observed by a lecture group by projection of the arc on a blue screen and by direct observation through a large sheet of red glass. The electrodes are set up vertically and 20 V dc is applied through adequate resistances.

31–1.9[5]

The action of the electromagnetic pump in Fig. 31–10 can be shown by projection apparatus. With the exception of two standard binding posts *f* and

(5) R. Resnick and H. F. Meiners, *Demonstration and Laboratory Apparatus Report of 1960 Summer Visiting Professor Workshop at Rensselaer Polytechnic Institute.*

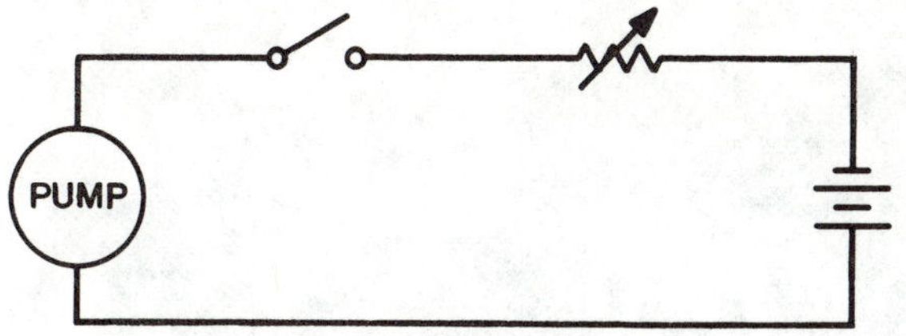

Fig. 31–11

a horseshoe magnet (~1800 G), all parts are made of Lucite and are cemented together. The terminal of the binding post shown in the upper part of Fig. 31–10 protrudes slightly downward into the drilled channel *d*. The terminal of the binding post shown in the lower part protrudes slightly upward into the channel.

Drilled channels *a, b, c,* and *d* are completely filled with mercury. The magnet (not shown in Fig. 31–10) is positioned according to outline *e*. The direction of the current brought to the binding posts and the polarity of the magnet should be arranged so that the mercury flows from right to left in channel *d* and up channel *a* to overflow into trough *g*. Due to gravity, the mercury flows down the trough and through *b* and *c* into channel *d* to complete the cycle. Figure 31–11 is a wiring diagram for the apparatus.

In Fig. 31–12 is shown the upper details of a paddle-wheel pump with slightly changed dimensions. The paddle wheel enables the circulation of the "pumped" mercury to be easily observed. A steel pin (1/8 in. diam, 1¾ in. long) is used for an axle.

31–1.10

The electromagnetic pump shown in Fig. 31–13 is a sealed, closed circuit system made from Pyrex glass tubing and filled approximately two-thirds full with mercury. A 50-A current passes through the mercury at the bottom of the vessel by means of two banks of six electrodes *a* (Fig. 31–14) made from Kovar, which has the same thermal expansion properties as Pyrex. Each coil of the electromagnet *b* is wound with 600 turns of 1.8-mm-diam copper wire and connected in series with a 25-V, 12-A source. The pole pieces *d* are separated by two aluminum or brass bars *e*. With the transverse current flowing through the mercury and the electromagnet on, the mercury will rise in one outer tube and fall into the center tube, the direction of flow dependent on the direction of the current and the polarity of the magnet. This action is shown by projecting the indicated part of the apparatus onto

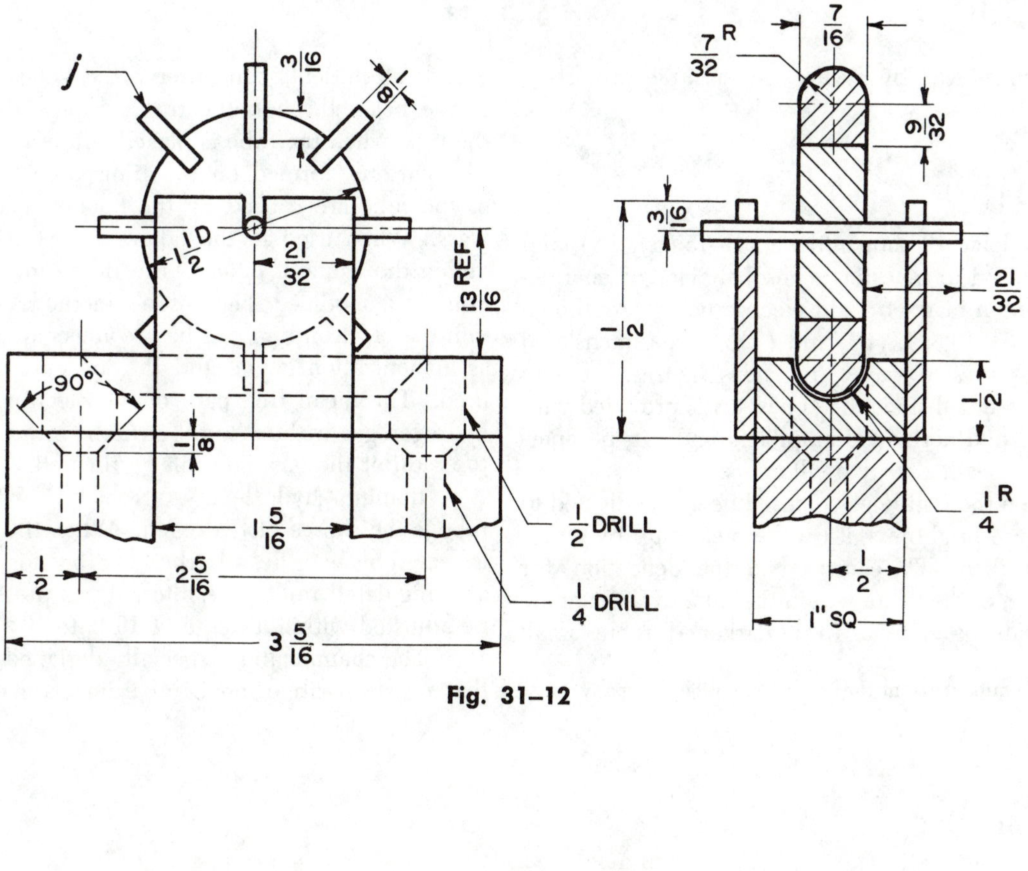

Fig. 31–12

Fig. 31–13

a screen or wall by means of a suitable projection apparatus.

31–1.11

The device[6] shown in Fig. 31–15 (a), (b), and (c) is used to investigate the behavior of charged particles in magnetic and electric fields. As shown in Fig. 31–16 (a), (b), and (c), three-dimensional patterns of sine waves, Lissajous figures, etc. can also be studied since the apparatus is provided with vertical and horizontal amplifiers and can be operated as a cathode ray oscilloscope.

When the visible gas focused beam is adjusted to maximum brightness, it can be picked up by a sensitive vidicon TV camera and the deflection, for example, of the beam within the tube by a magnetic field can be shown. In a darkened room small groups of students can observe the helical motion of the beam through the transparent end face of the tube when the tube is placed within a solenoid. The patterns formed on the transparent end face of the tube are caused by the fluorescence of the glass. When used in the laboratory, students can obtain the value of e/m by the Busch method and also by a modified Thompson's method. In addition the electron gun can be reconnected and used as an ionization gauge and the ion current measured. The mean free path of the electrons in the hydrogen gas within the tube can be estimated and the radii of the gas molecules estimated.

A titanium hydride crystal channel within the tube is the source of hydrogen. When the channel is heated by ac or dc, adsorbed hydrogen is released at a regulated rate. The internal gas pressure can be adjusted within a range of 10^{-6} to 10^{-1} mm of Hg. The channel and a specially designed cathode that resists positive ion bombardment enables the

[6] Obtainable from the Welch Scientific Company.

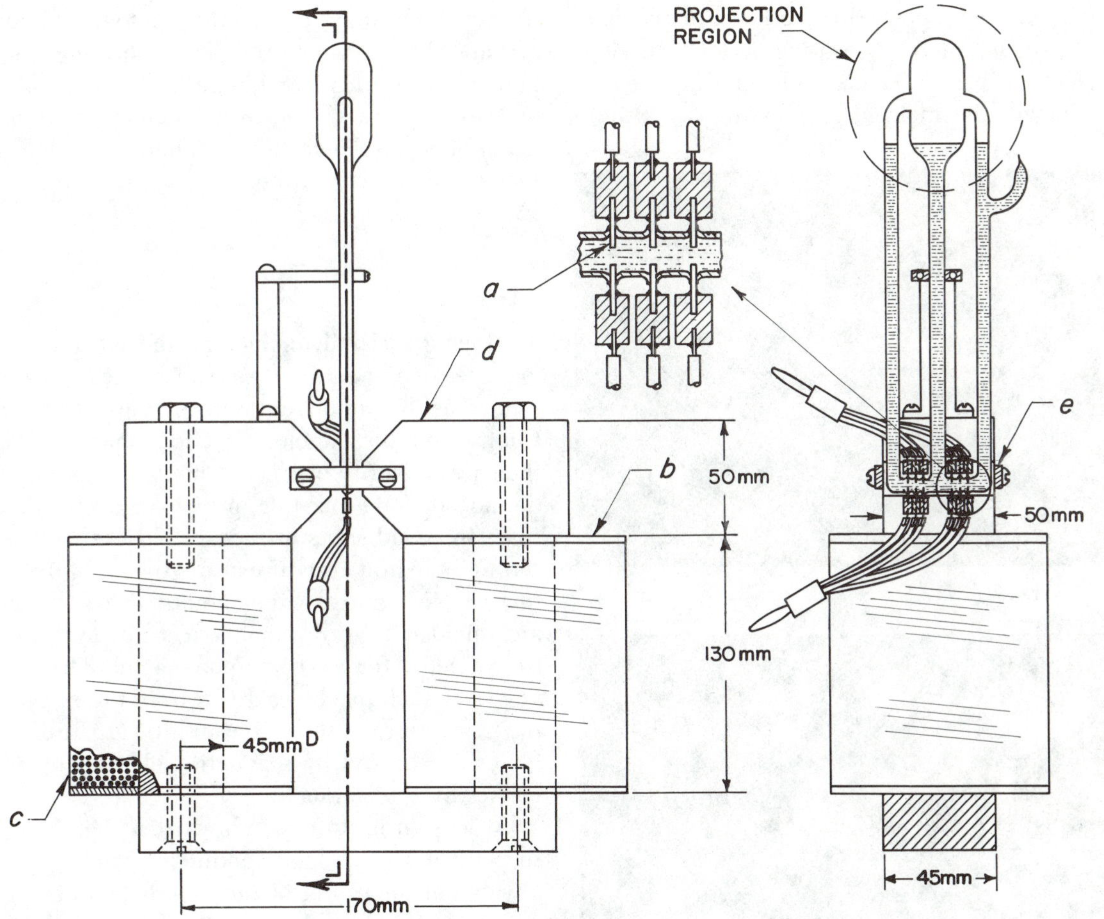

Fig. 31–14

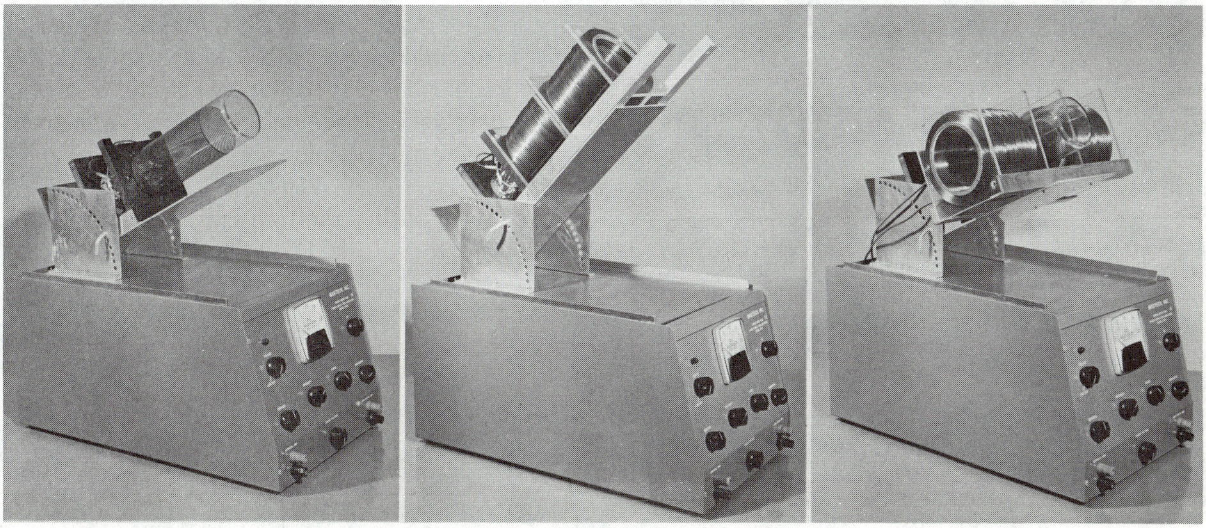

(a) (b) (c)

Fig. 31–15

tube without being opened to be filled many times with hydrogen, thereby insuring a long lifetime. The electron gun produces a beam of electrons which ionizes the molecules of hydrogen. Brightness of the beam at accelerating potentials between 200 and 2000 V is controlled by adjusting the bias potentials on grids 1 and 2 and the focus. All parts of the tube are visible and the movement of the beam between the vertical and horizontal deflecting plates can be seen. The tube is 3 in. in diam and 13½ in. long.

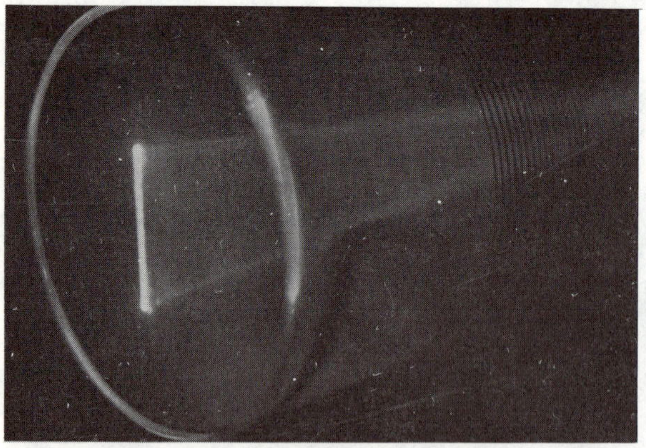

(a)

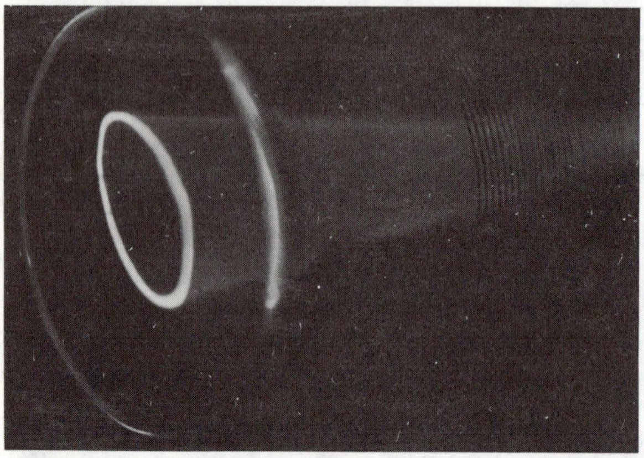

(b)

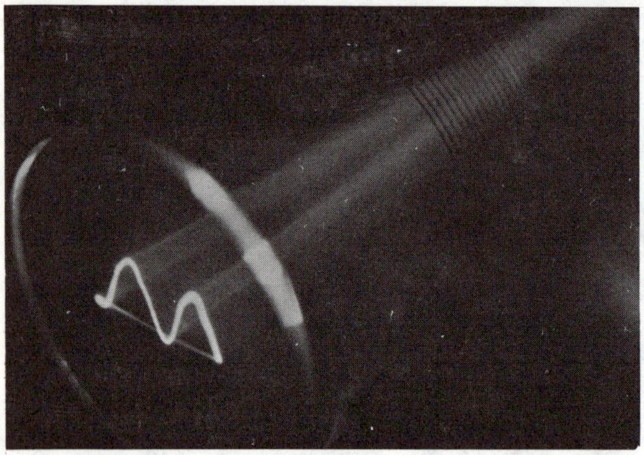

(c)

Fig. 31–16

31–1.12

Moving coil galvanometers and electrodynamic loudspeakers operate as a result of electromagnetic forces acting upon current-carrying conductors. Contrary to the statements made in many textbooks, the great majority of electric motors and generators do not operate on this principle. In most such cases the conductors are located in channels bored through the iron armature or in grooves in the armature, as in squirrel cage motors. The conductors are shielded from the magnetic fields and no electromagnetic forces act upon them. The motive power of such machines arises from forces between the iron masses of the stator and the armature. The forces are caused by magnetic fields set up in the masses by the conductors.

A long iron tube *f*, shown in Fig. 31–17, is suspended by two long conducting wires *g* in such a way that the center of the tube is in a homogeneous, one-dimensional magnetic field (see Fig. 31–18). The homogeneous field is produced by two small Leybold transformer coils *a* mounted on a U-shaped core *b* (Leybold 56211). Each arm of the U is extended by a yoke *c* (Leybold 56211A). A wooden block *d* fits freely between the free ends of the two yokes; the block is about 2 percent shorter than the normal separation of the two yokes.

A weight *e* pushes the free ends of the yoke together onto the block. The diameter of the iron

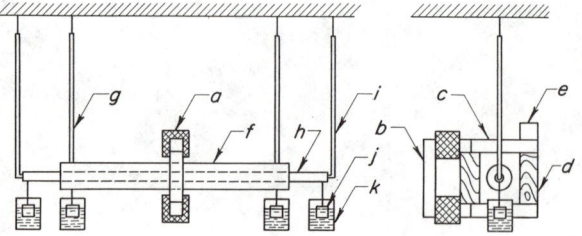

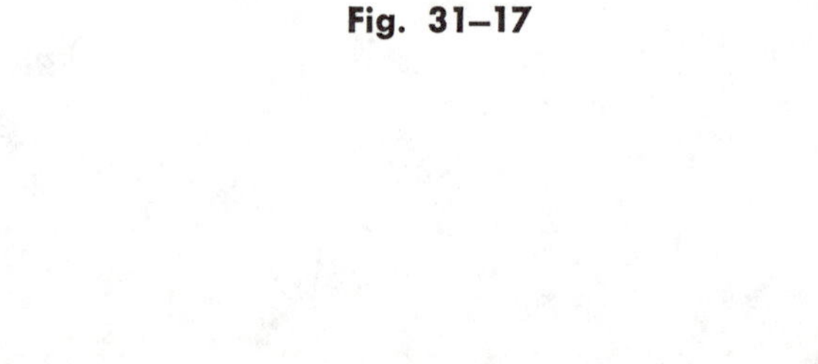

Fig. 31–17

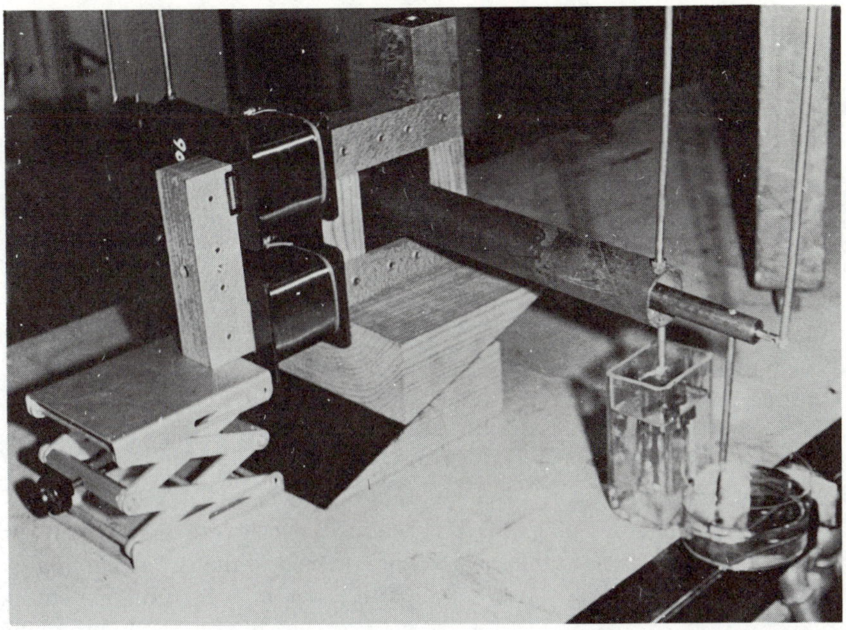

Fig. 31–18

tube is one half the separation between the yokes; its length should be at least three times that distance. The wall thickness of the tube must be such that it is not oversaturated by the magnetic field (i.e., around 3 mm). A conducting nonferrous rod *h*, 30 percent longer than the tube and exactly the same weight, is hung coaxially with the tube by two further conducting suspensions *i*, the same length as suspensions *g*. Both the rod and the tube have soldered to them brass vanes *j* which dip into separate glycerine filled dashpots *k*. The homogeneity of the field is adjusted by changing the separation between the free ends of the added yokes. Adjustment is achieved if the tube does not move when the magnetic field current is switched on. All electrical connections should be made with ba-

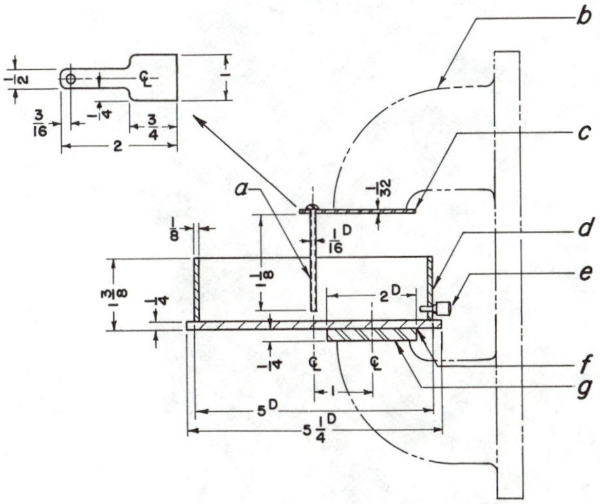

Fig. 31–19(a)

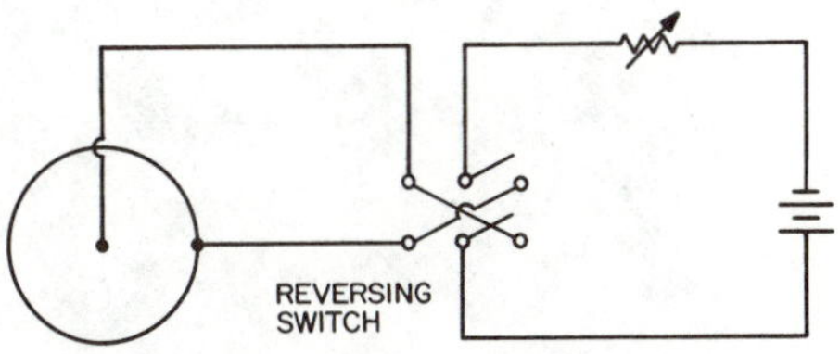

Fig. 31–19(b)

nana plugs for ease in mounting and dismounting.

First, the rod is hung in the magnetic field and the current through the rod is adjusted so that the rod's deflection is somewhat less than one half the inner diameter of the tube. (The deflection can be shown to large groups by projecting the shadows of two pointers fixed to the suspension wires.) The rod is then suspended coaxially within the iron tube, and the same current is passed through the rod. In this case the rod remains in position and the tube deflects by the same amount as the rod did previously. When the tube is fastened in its normal position and current passed through the rod, neither the tube nor rod move. Since the suspension is essentially a force measuring device, it is clear that no force acts on the conductor when it is shielded by the tube, and that the magnetized iron tube experiences an identical force to that arising on an unshielded conductor.

31–1.13[7]

To allow projection on an overhead projector, the ion motor shown in Fig. 31–19 (a) is made from Lucite. The magnet *b* (about 1800 G) is used to support the electrode holder and the motor dish. A copper center electrode *a* is soldered to the steel plate *c*. The side wall *d*, which is made from a 5-in.-diam Lucite cylinder, is cemented to the base plate *f* which in turn is cemented to the steel support disk *g*. The second electrode *e*, a standard binding post, is mounted through the side wall.

When copper sulfate solution with cork dust is put in the Lucite dish and the current, Fig. 31–19 (b), is turned on, the particles swirl in a circular motion due to the crossed electric and magnetic fields. The direction of motion is determined by the polarity of the electric field.

31–1.14

The purpose of this demonstration is to illustrate some of the basic principles involved in the operation of a cyclotron. A long trough-shaped magnet produces a magnetic field which corresponds to the electric field in the gap of the real accelerator. The particle is represented by one end of a long magnet hung vertically as a pendulum above the horizontal gap magnet as shown in Fig. 31–20. A reversing switch controls the current for the large gap magnet, while the pendulum is supplied with continuous current. Since only one pole of the pendulum is close enough to be affected by the gap field, the pendulum is accelerated each time it crosses the gap.

It should be noted that in an actual cyclotron

Fig. 31–20

Fig. 31–21

the particles move in a strong vertical magnetic field and experience a force, $F = q(\mathbf{v} \times \mathbf{B})$, which causes circular motion. In this demonstration the pendulum magnet also experiences a restoring force, but this is due to gravity. Both of these forces will cause periodic motion, but the pendulum must be given a lateral displacement in order to get open loops, and due to friction these loops will eventually degenerate into motion only in the plane perpendicular to the long axis of the trough magnet.

In operating the system the current in the gap magnet is reversed each half cycle with a knife switch. Trial runs were made with the apparatus using 10 A of current for both magnets. The amplitude measured at the end of the pendulum was about 15 cm for one cycle and reached a maximum of about 31 cm in 8 or 9 cycles.

For construction details, see the Appendix, page 1323.

31–1.15[7]

In Fig. 31–21 is shown a model cyclotron that produces a gravitational acceleration of a steel ball which is analogous to the electrical acceleration of particles in an actual cyclotron. The model is designed so that the action of the ball can be shown on an overhead projector.

The dees which form the circular part of the path are connected by short straight sections of accelerating "hills." One dee is fixed while the other reciprocates above and below to give acceleration. The acceleration is limited by the size of the track cut in the plastic. Synchronization of the revolution of the ball with the motion of the dee is accomplished by adjusting the variable-speed motor driving the dee. Leveling of the whole device is most important. In this model a $\frac{3}{8}$-in. or $\frac{1}{2}$-in. steel ball bearing is accelerated in such a manner as to illustrate approximately equal periods for half orbits of increasing radii for 5 revolutions. Therefore, the speed of the driving motor can remain constant. Construction details are given in the Appendix, page 1324.

31–1.16

The pendulum shown in Fig. 31–22 can be made to oscillate in either its normal position or in an inverted position (Fig. 31–23) if its point of support is oscillated at a proper frequency and amplitude.

[7] R. Resnick and H. F. Meiners, *Demonstration and Laboratory Apparatus Report of the 1960 Summer Visiting Professor Workshop at Rensselaer Polytechnic Institute.*

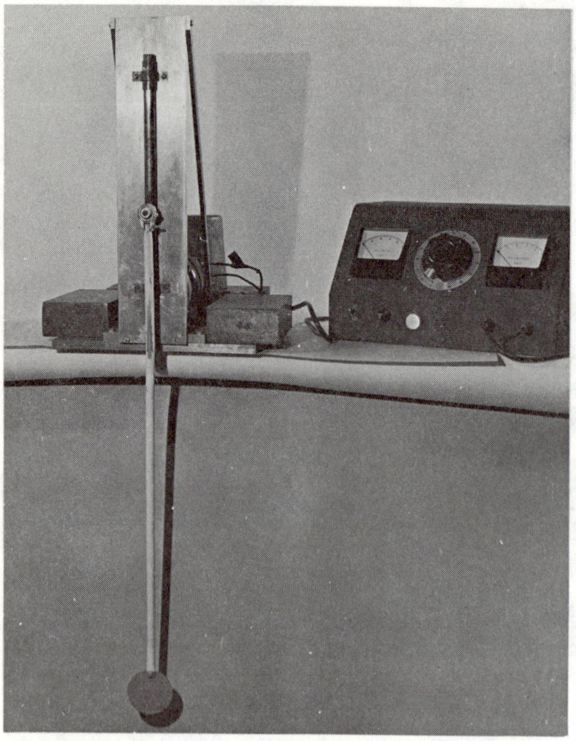

Fig. 31–22

Fig. 31–23

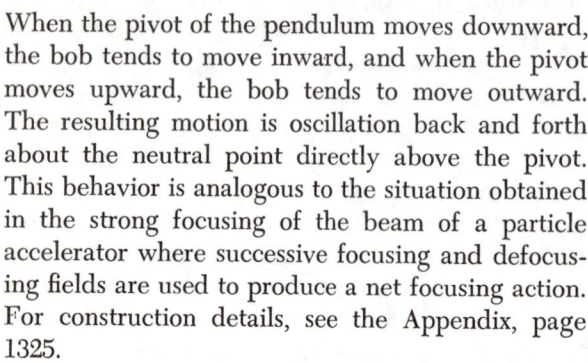

When the pivot of the pendulum moves downward, the bob tends to move inward, and when the pivot moves upward, the bob tends to move outward. The resulting motion is oscillation back and forth about the neutral point directly above the pivot. This behavior is analogous to the situation obtained in the strong focusing of the beam of a particle accelerator where successive focusing and defocusing fields are used to produce a net focusing action. For construction details, see the Appendix, page 1325.

To obtain effective radial focusing in an accelerator, a large positive magnetic field gradient is necessary, although the vertical forces would then be strongly defocusing. A large negative gradient would produce strong vertical forces but would also cause radial defocusing. It was discovered, however, that a force that alternates between strong focusing and strong defocusing can still have a net focusing force.

This can best be explained by considering the case of two "quadrupole lenses," which are actually four-pole magnets, that are based on the same principle. A positive particle entering a horizontal focusing field from the right or the left will be deflected towards the center while a particle entering from the top or bottom will be deflected away from the center. If the poles are reversed, one obtains a vertical focusing lens which has the reverse effect.

If these two lenses are placed in series and a particle enters with some horizontal displacement from the axis, it will be deflected towards the axis by the first lens. When it reaches the second lens it will be closer to the axis, and the outward force will be less. This produces a net horizontal focusing. If the particle enters in the vertical direction, it will first be directed away from the axis. When it reaches the second lens with this larger displacement, it will experience a stronger force causing it to move towards the axis. In either case there is a net focusing.[8]

31–1.17

With the apparatus diagramed in Fig. 31–24 one can show the magnetic field lines which sur-

[8] See R. P. Feynman, R. B. Leighton, M. Sands, *The Feynman Lectures in Physics* (Addison-Wesley Publishing Co., Inc., Reading, Mass.; Palo Alto; London; 1964), Vol. II, pp. 29–4 to 29–8.

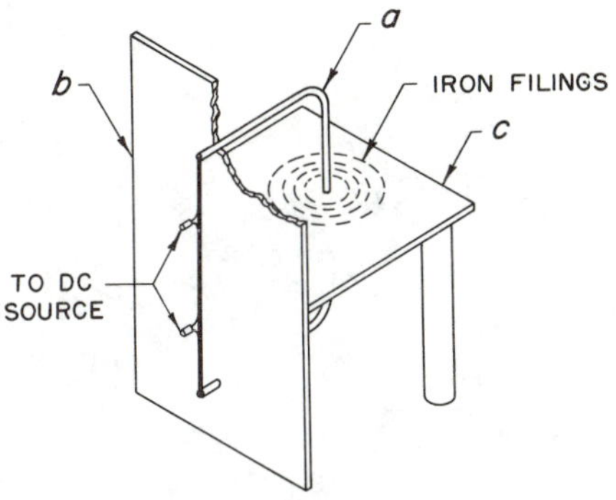

Fig. 31-24

round a straight section of a current-carrying conductor. If the experiment is to be used with an overhead projector, the deck *c* and supports should be made from Lucite, while the backboard *b* may be constructed of any suitable material. Several lengths of small enameled wire are placed through a straight length of copper or plastic tubing *a*, which is inserted through the deck at the desired position. After drilling two holes in the proper place in the backboard, the tube is bent and mounted in place. The wires are now soldered together in series, insulated from each other, and connected to a dry cell. When iron filings are sprinkled on the deck, the field pattern will appear.

31-1.18

The apparatus shown in Fig. 31-25, used on the stage of an overhead projector, demonstrates the orientation of a magnetic field surrounding a single current-carrying conductor. A ¼-in.-diam x 4½-in.-long copper rod *ab* is pressed through a hole in the center of a 10 x ¼ x 12-in. Lucite sheet. About 3¾ in. of the rod are allowed to project above the plate. Four compass needles *c* are mounted on short pivot pins pressed through the sheet, 90° apart in a circle of about 3½ in. diam. Current is

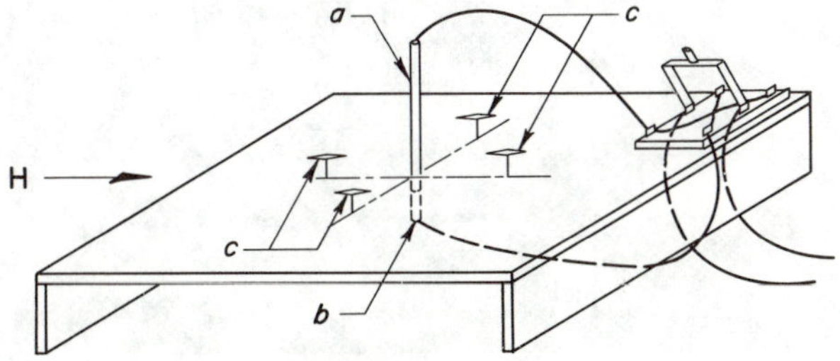

Fig. 31-25

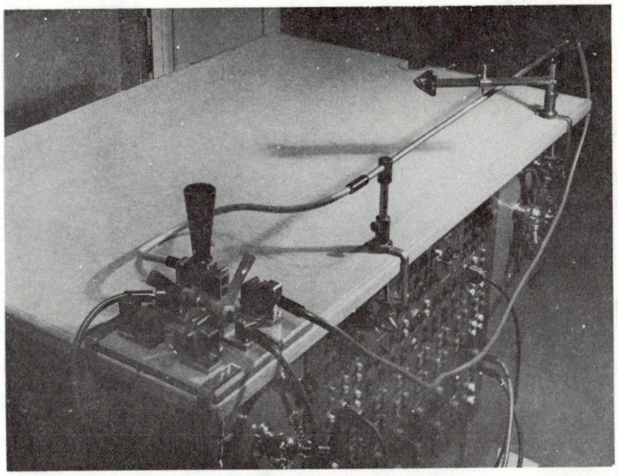

Fig. 31-26

supplied to the conductor *ab* by a Heathkit battery eliminator (Model BE-4) through a reversing switch. The compass needles align themselves with the field about the current-carrying conductor. When the current is reversed, the needles swing through 180 deg. Another method is to use small transparent compasses, which can be moved at will and removed for storage.

31-1.19

The magnetic field produced by a current-carrying conductor can be detected by a large compass pointer supported on a sewing needle attached to the conductor. The field is shown to be present and to be independent of the nature of the current-carrying conductor.

Three types of conductors are used, a metal rod, an electrolyte, and a gas discharge tube. Each conductor is three feet long. The rod in Fig. 31-26 is made of ⅜-in.-diam brass. A sulphuric acid solution in a ½-in.-i.d. glass tube is used as an electrolyte (Fig. 31-27). The electrodes (copper or platinum) project into the solution at each open end of the tube. The tube ends are bent upward to hold the electrolyte. The gas discharge tube is constructed from 1-in.-i.d. Pyrex glass (Fig. 31-28). After about 25 to 30 cc of Hg are placed in the tube, it is evacuated and sealed. Kovar inserts are used at the ends. To ignite, the discharge tube is tilted.

Since the compass needle should be parallel to the conductor when the current is off, magnets are placed beneath each tube on the table in order to compensate for the earth's magnetic field.

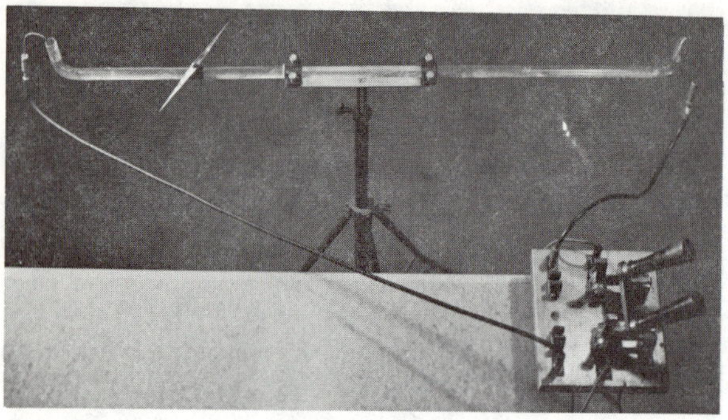

Fig. 31–27

A current of about 100 A is used for the brass rod. Ten amperes are needed for the electrolyte and the discharge tube. Because the gas discharge tube is overloaded with a current of 10 A, it is used only for short periods of time.

31–1.20

The magnetic field pattern surrounding the current in a coil can be shown with the apparatus illustrated in Fig. 31–29. If the experiment is to be used with an overhead projector, the platform *a* should be a piece of 4 x 6 x ¼-in. Lucite, while the supports may be made from any suitable material. A piece of 1-in.-diam tubing or rod, which has been split in half lengthwise, is used as a form for winding the coil *b*. With the paper still attached, the Lucite is cut to the proper dimensions, and the holes for the coil are marked and drilled, making sure that the diameter of the coil is slightly larger than the forms. After the paper has been removed and the platform mounted on the supports, the coil is wound around the forms with 14 or 16-gauge enameled wire, and the tubing is then removed. When a battery is con-

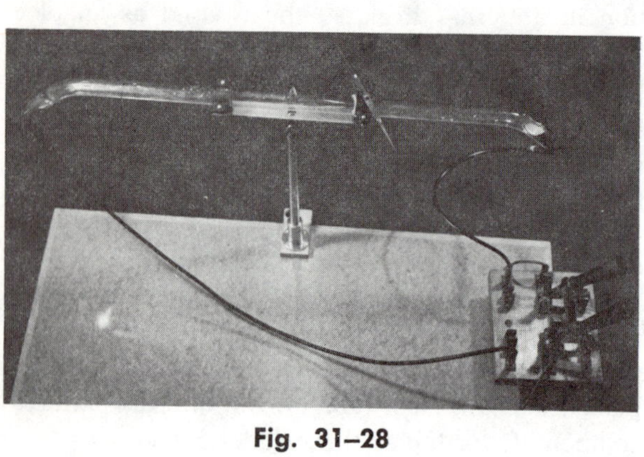

Fig. 31–28

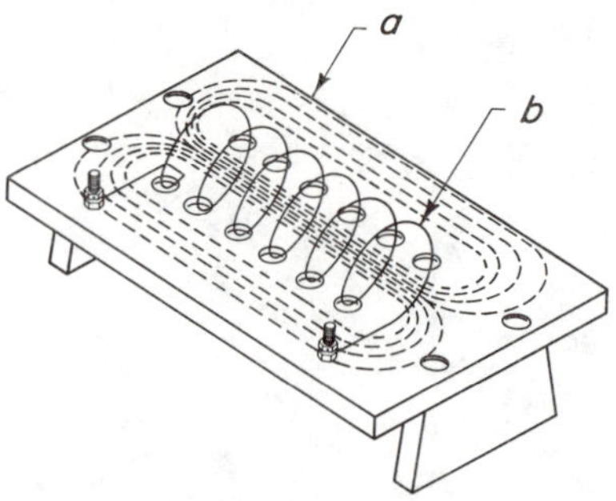

Fig. 31–29

nected through a small rheostat and iron filings are sprinkled around the coil, the field lines will appear.

31–1.21

The magnetic field at a point on the axis of a solenoid *a* is determined with a magnetometer *b* (Fig. 31–30) as well as a 100-turn induction coil *c* (Fig. 31–31). The magnetic field on the axis of a solenoid depends on the length of the coil according to the relation

$$B = \frac{\mu_0 NI}{2l}(\cos\beta_1 - \cos\beta_2),$$

where N is the number of turns, I is the current through the solenoid, l is the length of the solenoid, β_1 is the angle between the axis and a line running from the point on the axis to a point on the circumference of the coil at one end, and β_2 is the angle between the axis and a line running from the point on the axis to a point on the circumference of the coil at the other end. The coil is constructed so that the windings can be slid closer together or

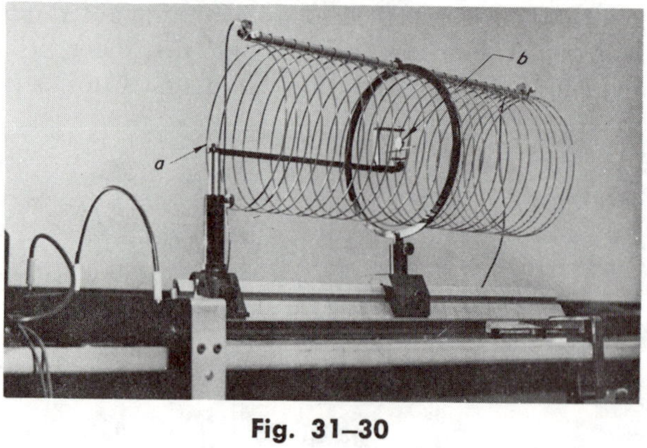

Fig. 31–30

Fig. 31–31

further apart (see Fig. 31–32), thereby allowing the length of the solenoid to be changed.

The support for the solenoid consists of a metal ring *d* and a hard paper tube *e*. The solenoid, which has 25 windings, is made from steel wire 2 mm in diameter. Brass rings *f*, which vary in inside diameter from 11 to 14 mm, are used to suspend the windings from paper tube *e*. The rings

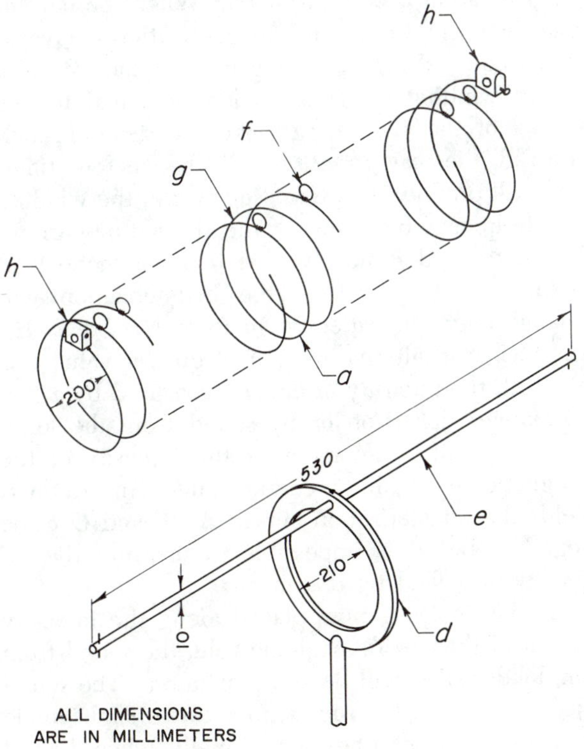

Fig. 31–32

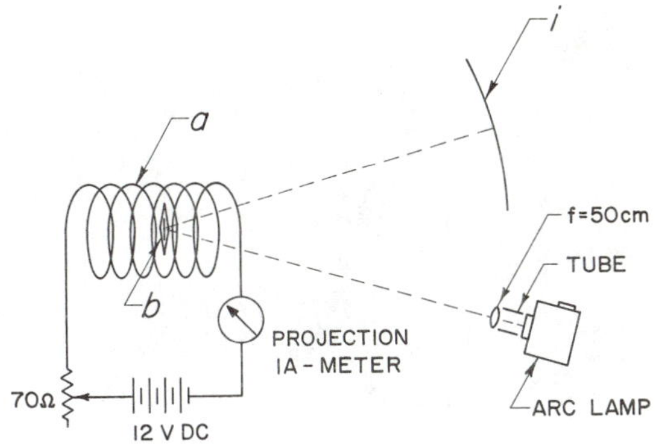

Fig. 31–33

are hard-soldered onto all of the windings except the first, the last, and the thirteenth *g*. The thirteenth winding is connected directly to the metal ring *d*. The first and last windings have brass thumbscrew devices *h* soldered to them in order to allow the length of the solenoid to be fixed at any desired value.

The magnetometer *b*, which consists of a compass needle, a spring, and a mirror, is placed on the axis of the solenoid. A current is passed through the solenoid, thereby establishing a magnetic field. The compass needle comes to equilibrium under the influence of the spring and the magnetic field. The orientation of the compass needle is determined by reflecting a light beam off of the mirror attached to the compass needle onto a scale *i*, which is attached to a wall. (See Fig. 31–33.) The demonstration is repeated after having changed the length of the solenoid.

The demonstration is also performed with the use of an induction coil. The induction coil is placed on the axis of the solenoid. The primary circuit (see Fig. 31–34) is opened or closed, thereby inducing a voltage in the induction coil. The induced voltage is observed with the aid of a mirror galvanometer *j* attached to the lecture hall wall. The length of the solenoid is changed, and the demonstration is repeated.

31–1.22[9]

A weak-field magnetometer may be made by winding 5000–7000 turns *d* of 0.004-in. well-insu-

(9) Translated by K. V. Robinson and H. A. Robinson (Adelphi University) from *Lecture Demonstrations in Physics*, A. B. Mlodzeevskii, ed., M. V. Lomonosov State University, Moscow, U.S.S.R.

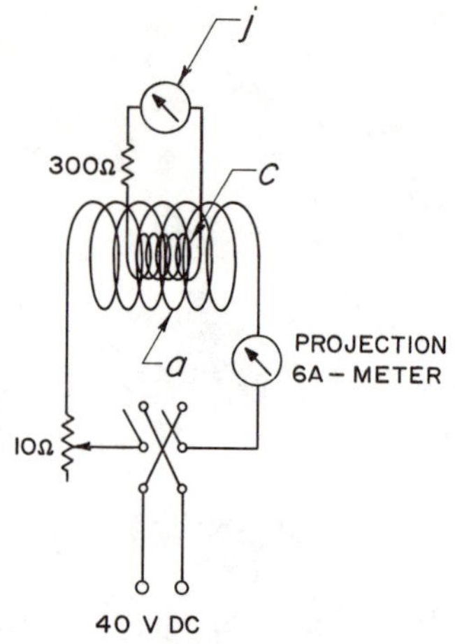

Fig. 31–34

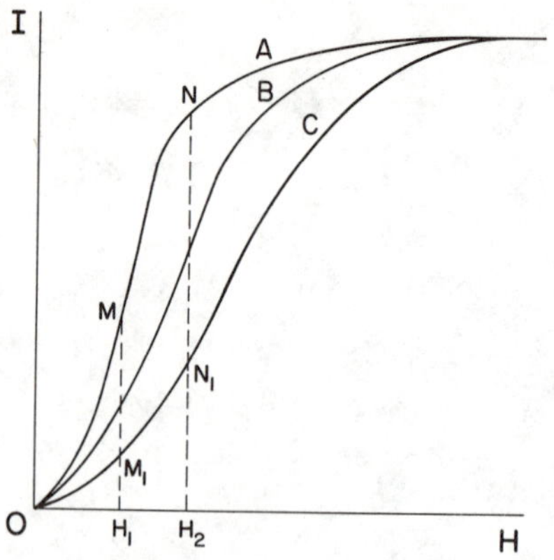

Fig. 31–36

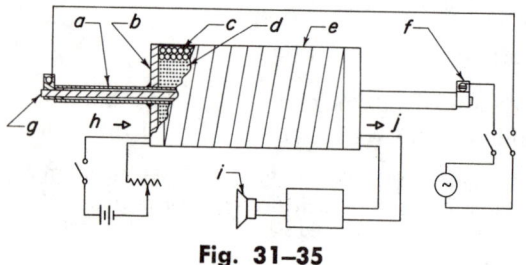

lated copper wire around a glass tube *a,* 10 in. long, 0.35 in. diam (Fig. 31–35). Lucite end pieces *b* hold the windings in place. A compensating winding *c* of 0.008-in. wire is wound over the main winding and the windings are wrapped in tape *e*. A thin, tempered permalloy strip *g* is inserted in the tube and held in place by metal clamps *f*. The main winding *d* is connected either to an audio amplifier and loudspeaker *i* or to a rectifier and sensitive galvanometer (not shown). The compensating winding *c* is connected to a battery, rheostat, and switch, as shown, and the permalloy rod is connected to a 120-V ac circuit capable of supplying 1–3 A.

If a ferromagnetic rod, e.g., permalloy, is placed parallel to the magnetic field direction in a homogeneous field of sufficient strength, all the different regions of spontaneous magnetization will turn along the rod (longitudinal magnetization). The curve of this magnetization corresponds to curve *A* in Fig. 31–36. If a direct current were passed through the rod, the longitudinal magnetization curve would be lowered in the direction of curve *C* (the degree of lowering increasing with the current,

Fig. 31–35

but not affected by the polarity). If, instead, an ac current is passed through the rod, curve *A* will be the curve of longitudinal magnetization corresponding to the time when the current equals zero, and curve *C* will correspond to the time when the current is at its maximum.

When the permalloy strip (with an ac current passing through it) is placed parallel to a magnetic field of intensity H_1 (see Fig. 31–36), then when the current curves pass through zero the magnetization is equal to MH_1. When the current reaches its maximum, the magnetization decreases to a value of M_1H_1, resulting in an inductive pulse in the winding *d*. Thus, in a time equal to one period of the alternating current passing through the rod, the magnetization will change four times and will produce an alternating emf in the winding at a frequency twice that of the alternating current. When the rod is next placed in a magnetic field with intensity H_2, then the magnetization will change over the wider range from NH_2 to N_1H_2, producing an alternating emf of greater value. As a result, the intensity of emf *j* (measured by a galvanometer deflection or by sound from the loudspeaker) will be an index of the intensity of the magnetic field within a certain range. In saturated fields the magnetization curves *A*, *B*, and *C* come together and it is impossible to measure the intensity of a field by this means.

If the device is now placed along the intensity vector of the earth's magnetic field, the sound from the loudspeaker will be at a maximum. The sound decreases when the rod is tipped at a slight angle or is in a horizontal position. Finally, when the rod is placed in the horizontal plane, perpendicular to the horizontal component of intensity of the earth's field, the sound in the loudspeaker will stop. If a permanent magnet is brought close to the magnetometer, the loudspeaker will sound louder and louder. When the magnet is very close to the magnetometer (0.5–1 in.), that is, when a fairly large longitudinal field acts on the device, the sound will stop, indicating that the permalloy rod has been magnetized to the point of saturation. (The experiment will show these effects only if the earth's field H coincides in direction with the outer field of the permanent magnet.)

If the permanent magnet is brought toward the magnetometer so that the South Pole is up and the North Pole is down (opposite to the field of the earth), then at a certain distance the field of the earth and the average field of the permanent magnet will be mutually compensated and the sound from the loudspeaker will cease. When the magnet is brought still closer, the magnetometer will once more be affected by the external field and the speaker will emit sound. It should be noted that the magnetometer is not sensitive to the sign of the field, so that if the meter is turned 180 deg the sound will remain the same. However, if a dc current h from a battery is applied to the compensating winding, the demonstrator can easily pick the direction and value of the current which will stop the sound in the loudspeaker. Knowing the direction and strength of the current and the number of turns in the compensating winding, it is easy to determine the direction and average value of the field required to eliminate the field being measured. Such a magnetometer is, of course, useful only for measuring relatively weak magnetic fields.

31–1.23

Ampere's law describing the magnetic field produced by electric currents states that

$$\int_C \mathbf{H}\cdot ds = \iint_A \mathbf{J}\cdot d\mathbf{A} = I,$$

where $\mathbf{H}$ is the magnetic field intensity, ds is a differential distance along curve C, $\mathbf{J}$ is the current density, A is the area bounded by curve C, and I is the total current passing through the area A. This experiment helps to demonstrate Ampere's law by a direct measurement of

$$\int \mathbf{B}\cdot ds$$

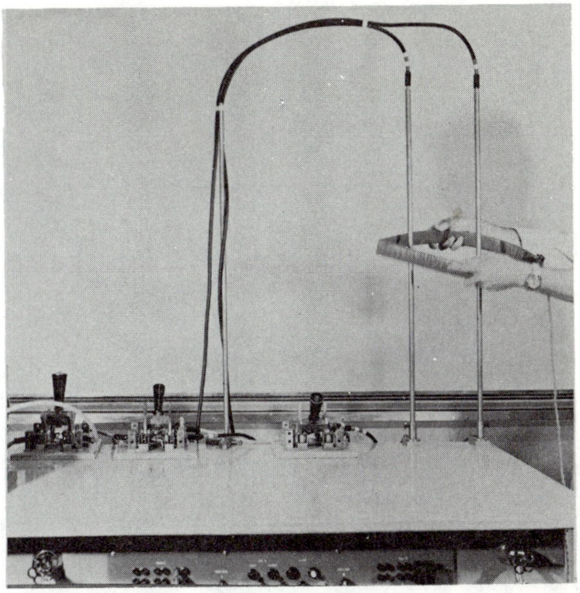

Fig. 31–37

between two points in space, where $\mathbf{B} = \mu\mathbf{H}$ is the magnetic induction. Thus, a direct measurement of magnetic potential differences is accomplished.

The coil shown in Fig. 31–37 was proposed by Rogowski.(10) It is connected to a ballistic galvanometer and the induction voltage integral is measured when the currents are switched on and off.

For a coil of length L with n turns per unit length, the magnetic flux through one turn of the coil is

$$d\phi = \mathbf{B}\cdot d\mathbf{a},$$

where $d\mathbf{a}$ is the area of one turn of the coil. If the total number of turns is $N = nL$, then the total flux linked by the coil is

$$\phi = \int \mathbf{B}\cdot d\mathbf{a} = K\int \mathbf{B}\cdot ds,$$

where K is a constant. Using Lenz' law, the voltage impulse seen by the ballistic galvanometer when the flux is changed is

$$\int V dt = -\int \frac{d\phi}{dt}\, dt = K' \int \mathbf{B}\cdot ds.$$

Thus the galvanometer reading is proportional to the magnetic potential.

Two brass rods (3 ft long, ½ in. diam) are mounted 6 in. apart on a table (Fig. 31–37). The rods are capable of carrying about 40 A. With

(10) Construction details are given in Dem. 31–1.24.

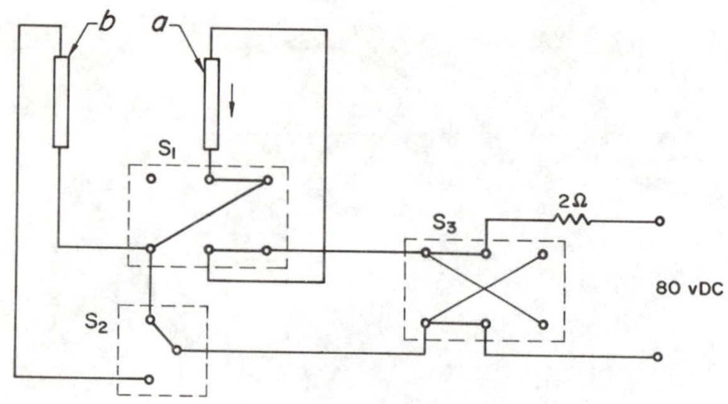

Fig. 31–38

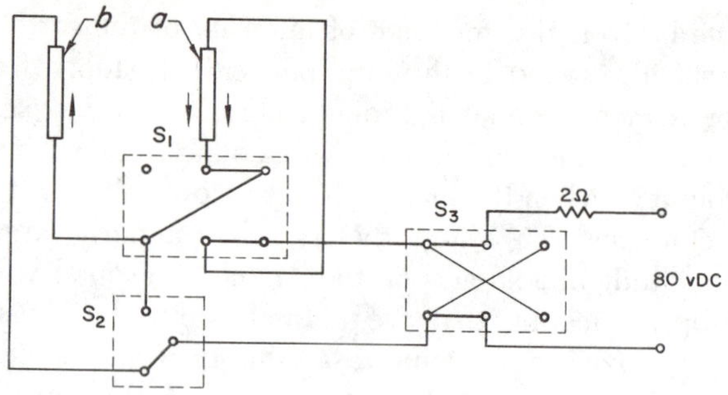

Fig. 31–40

switches S_1 and S_2 set as in Fig. 31–38, the current flows only in rod *a*. With the switches set as in Fig. 31–39, the current flows in both *a* and *b* in the same direction. With the switches set as in Fig. 31–40, current flows in opposite directions in *a* and *b*. S_3 serves to change the polarity, and the 2 Ω resistor is used to keep the current constant.

The Rogowski coil is now connected to a ballistic galvanometer and held around the rods. If S_1 and S_2 are now set as in Fig. 31–38 and S_3 is turned on, there will be a deflection on the galvanometer. If S_1 and S_2 are set according to Fig. 31–39 and S_3 is turned on, the deflection on the galvanometer will be doubled. When S_1 and S_2 are set as in Fig. 31–40, there is no deflection. Reversing S_3 reverses the direction of the deflection.

Also, the Rogowski coil may be distorted and the experiment repeated to show that the magnetic potential is independent of the path. (See also Dem. 31–1.24.)

31–1.24

With the Rogowski coil, direct measurement of magnetic potential between two points can be made. Thus, it can be used to effectively demonstrate Ampere's law. The coil consists of a flexible nonmagnetic core made of leather or plastic tightly wound with thin wire. It is connected to a ballistic galvanometer which will indicate any induced current.

If n is the number of windings per unit length, $n = N/L$, and A is the cross-sectional area of the coil's core, then the flux is given by

$$\phi = nA \int_0^L \mathbf{B} \cdot ds.$$

nA may be taken outside the integral since it is constant for any particular coil (see Fig. 31–41).

As the field is switched on and off, the deflection of the galvanometer is proportional to the flux through the coil. Thus the magnetic potential difference between two points can be determined.

Figure 31–42 shows how the coil can be used to demonstrate Ampere's law around a single conductor. The potential between two points is independent of the path unless the conductor is crossed. The potential difference if it is crossed is a function of the current in the wire and the number of times it is circled. The field of a solenoid can also be investigated as shown in Fig. 31–43. The results are shown to be:

1. Potential between the ends is large when the path is inside the coil.
2. The field outside the coil is almost zero.

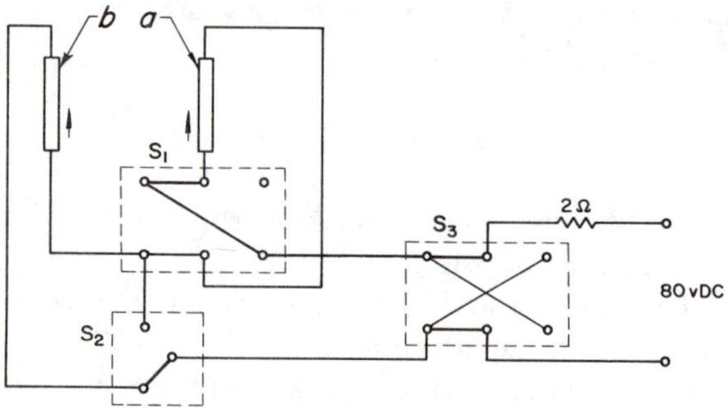

Fig. 31–39

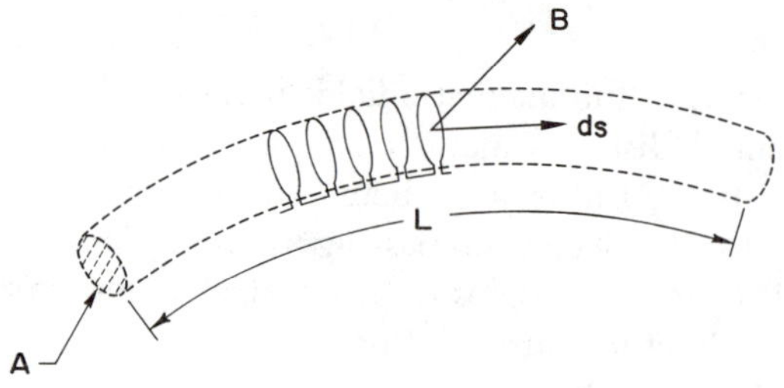

Fig. 31–41

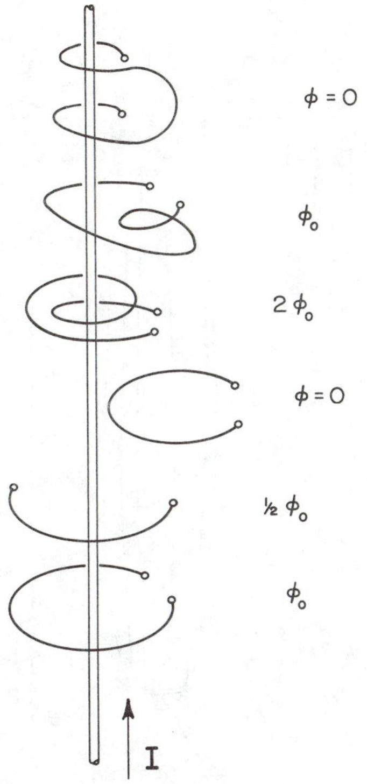

Fig. 31–42

3. The effect of an iron tube inside the solenoid is significant.

The core is a leather belt (or some similar plastic material) 50 in. long with a cross-sectional area of 1½ x 3/16 in. It is wound with 9600 turns of .012-in.-diam Cu wire with silk insulation. The winding is started from the middle and wound tightly towards the ends with the return leads brought back to the middle to a small insulating end plate with two banana plugs. The whole coil is then covered with electrical tape.[11] The construction of the large solenoid coil is described in Demonstration 31–2.15.

31–1.25

Very large currents can be used in the demonstration of magnetic fields around current-carrying conductors if a flat, braided brass cable (½ in. wide x ~1/32 in. thick) is used instead of the usual copper wires.

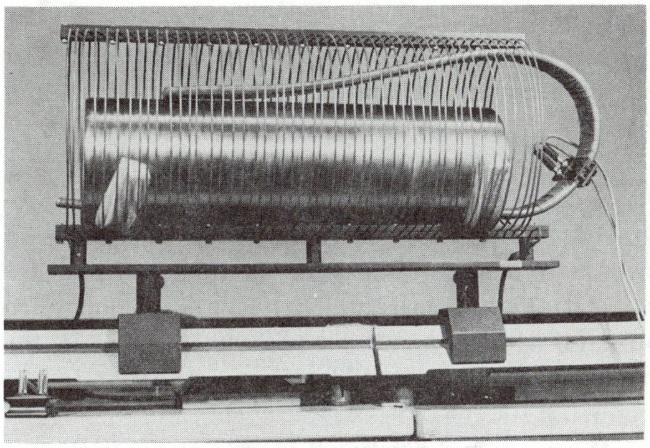

Fig. 31–43

31–1.26

The force between two wires carrying a current can be demonstrated using the apparatus shown in Fig. 31–44. A plane loop *a* of wire is suspended from a suitable support *b* by a single wire *c* so that the loop is free to oscillate about the axis of suspension. Current is fed to one end of the loop through a cup of mercury *d* and exits through another. A reversing switch *e* enables the direction of the current in the loop to be changed. The same current from power supply *j* runs through both the

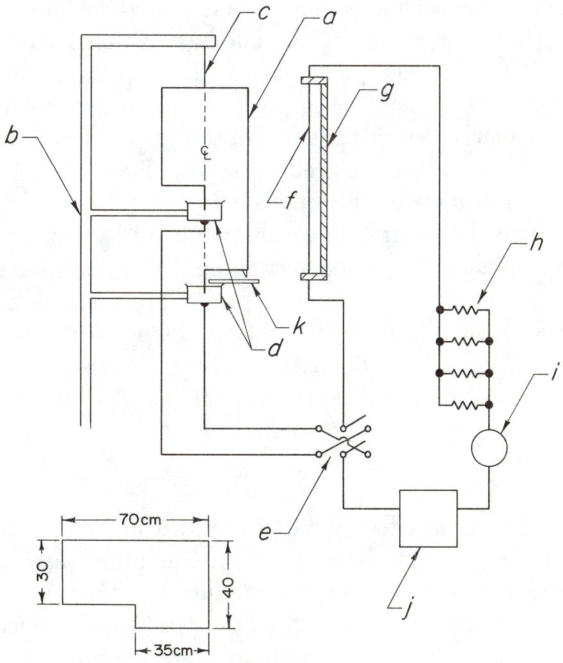

Fig. 31–44

[11] A similar model can be bought from Spindler and Hoyer K.G., Göttingen, Germany, No. 120703, or from Jacob Klinger Scientific Apparatus Co., Jamaica, New York.

loop and vertical wire *f*. The current can be determined by measuring the voltage across four 0.3-Ω resistors *h* in the circuit or by using an ammeter *i*.

The single long vertical wire *f* with brace *g* is located near the edge of the loop. Protractor *k* is mounted so that angular deflections of the loop can be measured. A ruler is used for measuring the separation between the loop edge and the wire. When this measurement is made, the plane of the loop should be perpendicular to the ruler.

The mass and rotational inertia of the loop can be calculated, since its material and physical dimensions are known. Then, by obtaining the period of the rod, the torque on the suspension and its force constant can be calculated. Using the torque constant and the angular displacement of the loop, the force exerted by one conductor on the other can be determined. The plot of θ vs I^2/D, where θ is the angular displacement of the loop, I is current, and D is the separation of wire and loop, should theoretically yield a straight line. A discussion of causes of deviation from the expected straight line may be included. The effect of reversing the current may be examined.

To account for the earth's magnetic field, the vertical wire *f* is removed from the vicinity of the loop. The current is then turned on and rotation of the loop will be observed. The base is then rotated until the magnetic moment of the loop is parallel to the earth's field. The pointer is used as a reference marker for this position of the loop. The vertical wire is then placed near the loop and the demonstration conducted.

This demonstration can be modified by employing shadow projection techniques. A light source could be directed at a clear plastic protractor from below and the deflections of the loop on its scale could be projected on the ceiling, on an angled screen, or on a wall (via mirrors).

31–1.27

The pinch effect apparatus shown in Fig. 31–45 can be used to show the magnetic forces associated with current-carrying conductors. Six rubber-covered multiconductor No. 14 copper wires *a*, each 23½ in. long (less connections), are loosely strung vertically from corresponding peripheral points of two brass disks *b* spaced 22 in. apart. Each disk is 1⅛ in. in diam, 1/16 in. thick, with six holes drilled equally spaced around the circumference. The wires are soldered to the disks which are attached to a 5-in.-long, ½-in.-diam aluminum support rod *d*. The stand *c* is a ½-in.-diam aluminum rod mounted on an 8-in.-diam x ¾-in.-thick plywood base with rubber feet.

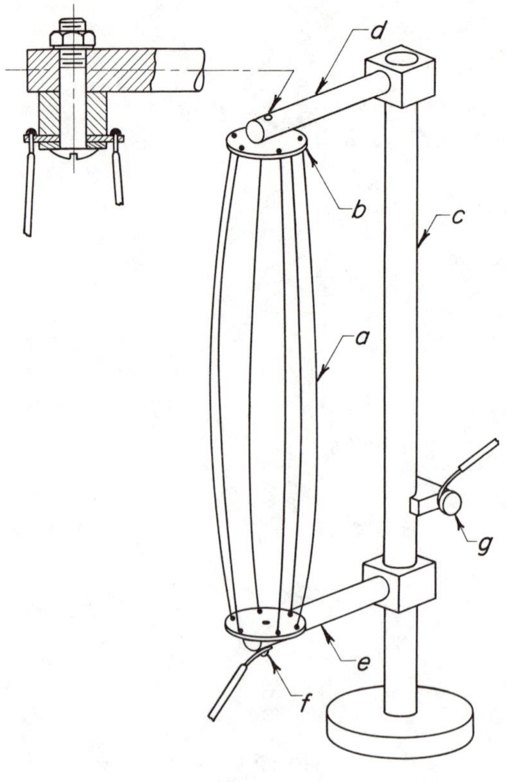

Fig. 31–45

One lead *f* is attached to the bottom disk and a second lead is attached to the aluminum stand at point *g*. A support arm *e* of Lucite insulates the bottom disk from the stand rod. The pinching effect is caused by the current flowing in only one direction through all six wires. Current is supplied by an ac step-down transformer.

31–1.28

Another demonstration illustrating the attractive magnetic forces between conductors carrying currents in the same direction uses the setup shown in Fig. 31–46. A 6 x 6-in. piece of aluminum foil is cut into narrow strips along its midsection, as in Fig.

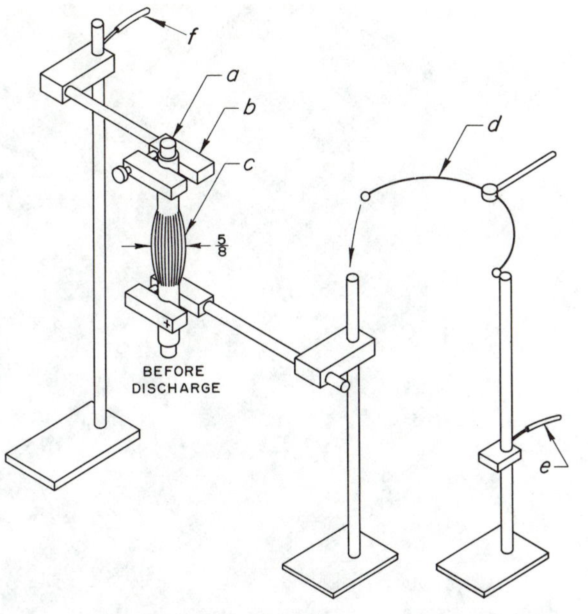

Fig. 31-46

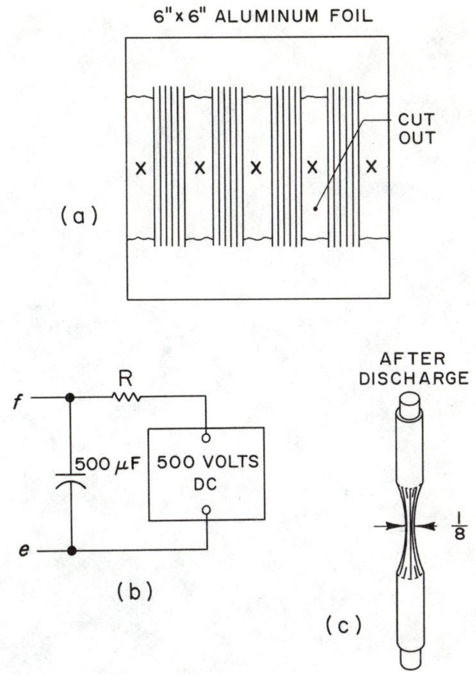

Fig. 31-47

31-47 (a), with ¼-in sections (marked x) cut out completely. The foil is then formed into a cylinder by wrapping the upper and lower (unsliced) parts of the foil around short lengths of ½-in.-diam rod *a*, Fig. 31-46, and held to the rods with a pair of parallel bar clamps *b*. The sliced section *c* is fluffed out for increased "before-after" effect. A large capacitor (e.g., 500 μF) charged to 500 V, Fig. 31-47 (b), is then discharged through the foil with a capacitor discharge rod *d* or a well-insulated screwdriver. A capacitor is connected across *ef*, and a resistance R is put in series with the voltage source to limit the current within the current capacity of the source. The electromagnetic forces will constrict the sliced part of the foil greatly, Fig. 31-47 (c). Precautions should be taken when using a high voltage with a large capacitance as indicated here.

31-1.29[12]

A cylindrical coil *d* 12-15 in. long and about 4 in. in diam is mounted in a horizontal position on a stand *c*, shown in Fig. 31-48. The coil remains stationary and is wound with one layer of insulated wire 0.04-0.06 in. in diameter. A second coil *e* is made of 0.03-0.05-in.-diam wire in durable silk or paper insulation. Enameled wire cannot be used because this movable coil strikes the stationary one during the demonstration and would very quickly damage enameled insulation. The second coil consists of about 100 turns and is wound on a frame

(12) Translated by K. V. Robinson and H. A. Robinson (Adelphi University) from *Lecture Demonstrations in Physics*, A. B. Mlodzeevskii, ed., M. V. Lomonosov State University, Moscow, U.S.S.R.

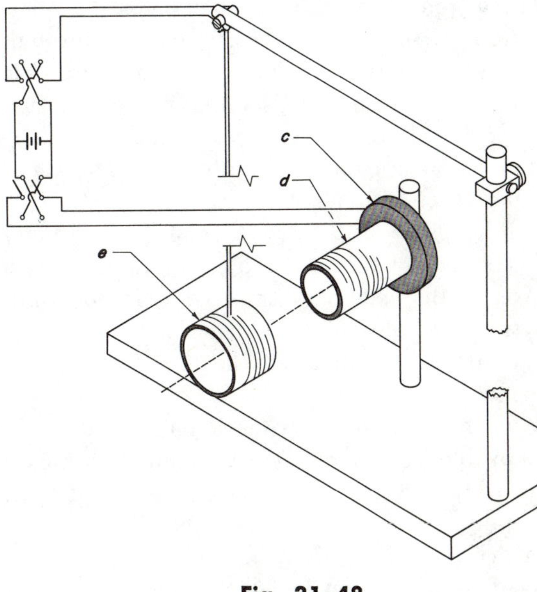

Fig. 31-48

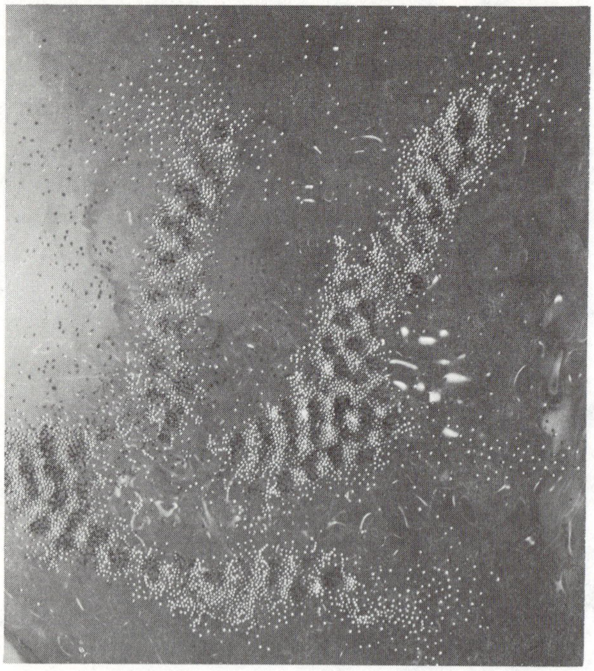

Fig. 31–49 The Magic Carpet on Water, in free raft form. A dynamic carpet of magnetized balls driven by a vertical ac magnetic field.

Fig. 31–50 More Compact Raft, at lower field strength.

whose inner diameter is about 1 in. larger than the outer diameter of the first coil. The winding is of a honeycomb type and is permeated with Bakelite or other varnish to give it the necessary rigidity. The coil should be covered at the top and sides with a thin layer of insulating material to protect the wire insulation from damage by the blows.

Current is supplied to the second coil through wires about 0.01 sq in. in cross section and 3–4 ft long, the wires themselves serving as a suspension for the coil. Current-reversing switches should be connected in the circuits of both coils. The stand with the second (movable) coil is set up so that the coil hangs about 2 in. away from the stationary coil. The axis of the movable coil is aligned with the axis of the stationary coil, but it is not absolutely necessary to have the plane of the movable coil parallel to the plane of the turns on the stationary coil.

When the current is switched on in both coils, the movable coil is either immediately attracted to the stationary one and slips over it, or it is at first repulsed, then turns 180 deg and slips over the stationary coil. If the direction of the current is changed in one of the coils, the movable coil is forced off the stationary one, turns around, and once again slips over the stationary one, this time from the other direction. To differentiate between the ends of the coils, they should be painted different colors. To demonstrate the effect of magnetic permeability, a core of iron rods may be inserted in the stationary coil; the attraction or repulsion of the movable coil will then be much stronger.

31–1.30[13]

Thousands of small magnetized balls, floating freely on the surface of water, form standing waves when excited by an ac magnetic field. When the balls are unbounded, they form a free form raft as shown in Figs. 31–49 and 31–50. When the water surface is bounded, however, the balls form a regular pattern of hills and hollows as shown in Fig. 31–51.

The mechanism of production of the standing waves is the magnetic interaction of the balls. If they are poured onto the surface without a field, a random pattern of chaotically interlocking chains is formed. When the field is turned on, repulsion

[13] A. D. Moore, "The Magic Carpet," *IEEE Conference Paper No. CP 63–106* (1963).

starts to break up the chains and they begin to seesaw. Each group starts a small component train of traveling waves. At some time, a stronger pair of wave trains is produced by a group of chains oriented in parallel. As this stronger wave moves along, the chains it approaches may be oriented at random. They will swing around, however, and point up and down the greatest slope of the wave because the wave has the same frequency as they have. The chains become slope-oriented and add their energy to accentuate the wave train.

Major build-up of the waves comes from the square (or rectangular) frame. Traveling wave components going diagonally will strike the wall at 45 deg, be reflected at the same angle, and continue to be reflected at each wall. If these are in phase, they will accentuate each other. Waves parallel to two sides of the frame will build up by back and forth reflections between the sides. Parallel waves don't last very long, however, because they cannot correct for deficiencies in the wave trains; whereas in the diagonal mode, the repeated reflection of the waves will tend to shuffle the balls into organization. Even in the unframed raft, however, the diagonal pattern predominates because it is the only pattern which allows a stable waveform.

The wavelength of the pattern is approximately constant for frame dimensions from 1–4 cm and water depths from ⅛–½ in. This shows that the wavelength is not described by water wave theory, since water waves are dependent on both frame size and depth of the water. The chain length of about five balls is the determining factor of the wavelength and the water is in forced oscillation at this wavelength.

The balls used in the demonstration are hard cast iron shot meant for shot-peening and air-blast cleaning of castings. The diameter is close to $\frac{1}{80}$ in. Careful selection of the shot is necessary to eliminate imperfect spheres. The balls are copper-plated and burnished to prevent iron to iron contact. They are magnetized by moving an Alnico U-magnet underneath a plastic bowl containing the balls. Care must be taken to produce the correct amount of magnetization because if overmagnetized the chains are too strong and the field needed to produce elasticity is too large. If undermagnetized, the balls will separate and become inactive. A pair of Helmholtz coils furnish the vertical field. The rms field density should be about 54.5 G for best results.

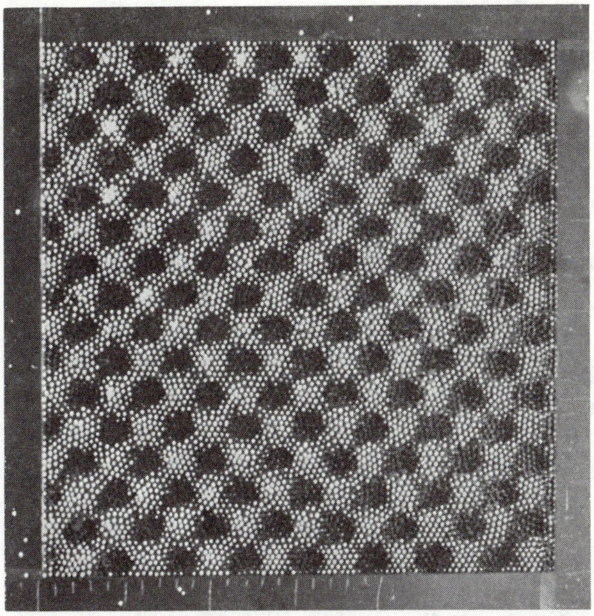

Fig. 31–51 Checkerboard Standing Waves Pattern of Carpet in 3-cm square. Bright areas are hills; dark areas, hollows. The array is reversed in the next half-cycle.

To pour the balls close to the surface and direct them to remaining bare spots, a feeder is made from a waxed paper soda straw. The straw is bent at right angles about an inch from the end and the top of the boot slit open. The balls will cling together until the field is turned on. A medicine dropper inserted through the carpet may be used for accurate adjustment of the water surface to frame top level. The frame, itself, is made of four pieces of Plexiglas, held to plate glass by modeling clay. If the carpet shivers with full field, more balls are needed. After operation, the balls are collected on the end of a smooth Alnico rod magnet. Since this overmagnetizes them, the balls are put in the bowl and the bottom is moved in contact with the upper coil in which high momentary current is used.

By using a General Radio Strobotac at line frequency, the striking beauty of the magic carpet is revealed. An even more striking effect can be seen when the Strobotac frequency is changed to near 60 cycles, putting the carpet into slow motion. This can be demonstrated to large audiences by closed circuit television.

31–2 Faraday's Law

31–2.1

A voltage can be induced in a loop by moving a rod magnet in and out. This is shown by two equally large loops, one turn made from copper wire and the other turn from cotton string (wound around a Plexiglas tube), successively connected to a projection millivoltmeter (10 Ω impedance) and a Keithley electrometer[14] (10^4 Ω impedance) whose output is connected to a projection microammeter. Shielded connection wires are necessary to prevent the electrometer from being influenced by induction.

When the magnet is passed through the copper loop, the low impedance millivoltmeter deflects, as does the high-impedance electrometer. When the magnet is passed through the string loop, there is a deflection by the electrometer but not the millivoltmeter, showing that it is voltage and not current that is induced.

31–2.2

This apparatus is designed to demonstrate induction effects in a secondary coil when the current is changed in the primary, so that the factors affecting the amount of induction are visible. In particular, it can be used to show that the mutual inductance depends on the degree of magnetic coupling (relative positions of the coils) and the approximate variation of flux along the axis of a current-carrying solenoid of finite length.

The primary and secondary coils are arranged as shown in Fig. 31–52, the former consisting of a single layer solenoid mounted permanently on a wooden base, while the latter is a honeycomb wound multi-layer coil which can be moved up and down along the axis of the primary. The secondary assembly moves along a vertical post which is ruled at centimeter intervals so that the (up and down) displacement of the secondary coil from the center of the primary is known at all positions. When the secondary is off the primary, its center can be rotated in a horizontal circle about the post, and the axis of the secondary may be rotated vertically through a full circle as well.

Fig. 31–52

The circuit (Fig. 31–53) in which the apparatus is used consists of a 32-V power supply connected in series with a rheostat for current control and (through a reversing switch) the primary coil. A knife switch and ballistic galvanometer are placed in series with the secondary.

When the apparatus is used for classroom demonstration it is necessary to make the ballistic galvanometer deflection visible by shining a beam of light at the mirror and picking up the reflection on a screen or a large scale prepared for this purpose. A projection ammeter is used to read current.

The procedure consists of the following steps: The reversing switch is thrown one way and the

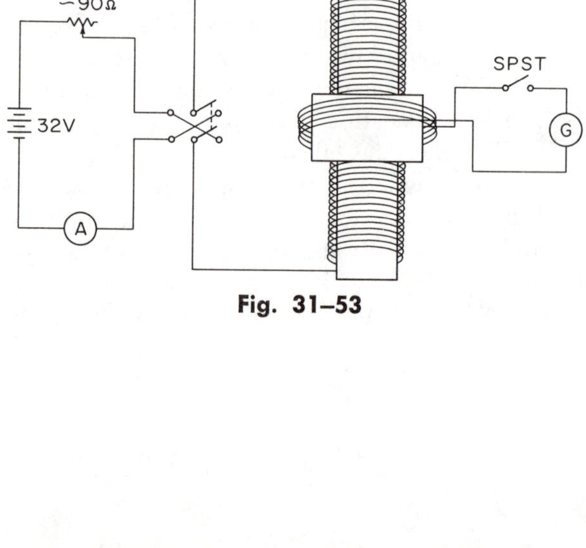

Fig. 31–53

[14] Keithley electrometer Model 210 and projection microammeter MB5 from Keithley Instruments, Cleveland, Ohio.

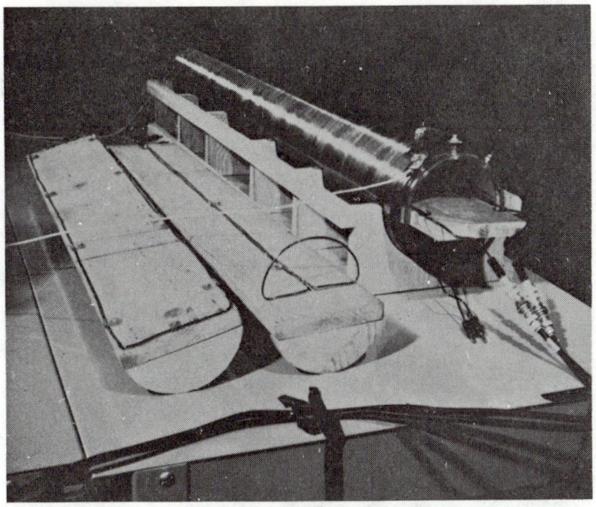

Fig. 31–54

current adjusted to about 2 A. The knife switch in series with the secondary is closed and the current in the primary is reversed ($\Delta I = 4$ A), galvanometer deflection being noted. The dependence of the effect on the direction and magnitude of the change in the flux through the secondary can be shown by noting the galvanometer deflection as the current is reversed (switch thrown directly to the opposite side) and then repeating the whole process with a current of 1 A. After each galvanometer reading, the knife switch is opened briefly to allow the undamped galvanometer to swing back toward zero a little faster.

Finally with $I = 2$ A, the deflection is noted when the primary current is reversed with the secondary in various positions of interest with respect to the primary and/or with a small piece of iron inside the primary. Conclusions are then made concerning the amount of magnetic coupling under the various conditions.

In quantitative work the mutual inductance of the coils or the total flux through the secondary can be calculated for all of the experimental conditions above. However, the galvanometer must first be calibrated with the secondary coil across its terminals since it is used in this (highly damped) fashion. A suggested method is to discharge a capacitor, having known values of charge on its plates, through the parallel combination of galvanometer and secondary coil and note the deflections for the fraction of the charge which passes through the galvanometer. When care is taken with the calibration and measurements, the probable error in the results can be held to less than 5 percent.

When using the apparatus make sure to (1) turn off the primary current when the circuit is not in use, since 2 A is almost the current limitation of the primary coil and may cause overheating over long periods of time, and (2) take care not to get the secondary coil upside down, since the graduations on the post along which the secondary assembly moves tell where the coil is with respect to the center of the primary only when it is upright.

Construction details are given in the Appendix, p. 1325.

31–2.3

Faraday's law of induction states that a changing magnetic flux through a loop or coil will induce an electromotive force. A relatively uniform magnetic field can be produced by applying a constant voltage V to a long solenoid. If this applied voltage is suddenly reduced to zero, there will be an emf induced in the coil given by

$$\epsilon = \frac{S'V}{Anl},$$

where n are the turns per unit length. The cross-sectional area of the coil is A; the length is l, and S' is the area of the loop which is perpendicular to the magnetic field. The experimental arrangement is shown in Fig. 31–54.

The schematic of the apparatus is shown in Fig. 31–55. The solenoid is constructed in three layers L_1, L_2, and L_3 which are wired in parallel, and switch S_3 is used to change from one layer to three. Several different coil configurations with various values of S' are arranged on wooden frames and can be inserted into the solenoid. By varying R_2 the voltage V across L can be maintained at a constant 18 volts with either one or three layers connected. When S_2 is closed, V immediately becomes

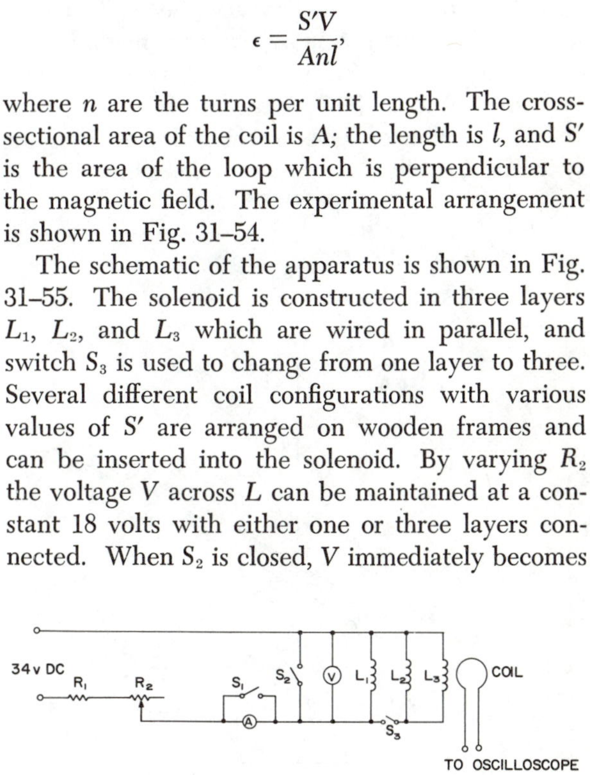

Fig. 31–55

zero, causing the magnetic field to collapse and inducing a voltage ϵ in the coil. The emf's are displayed and measured on an oscilloscope and compared with the calculated values. In order to protect the ammeter *A*, S_1 should always be closed before S_2 is used to short out the solenoid.

Opening and closing S_3 will verify the fact that ϵ does not depend on the number of layers since *n*, the number of turns per unit length, is the same for one, two, or three layers connected in parallel. However, this does change the resistance of the solenoid, which alters the time constant of the decay. This can also be confirmed on the oscilloscope. For construction details, see Appendix, page 1327.

31–2.4

Two Helmholtz coils *a*, Fig. 31–56, each 225 turns (15 layers of 15 turns each), 3 ft in diam, and approximately 15 Ω, are placed close together. One coil is connected to a galvanometer *G*, and the other is connected in series with an ammeter *A*, a rheostat *R*, and the output of a linear variable potential divider. The meters may be wall or projection type. The potential divider is made from a Nichrome strip *e*, 1 in. wide and 20 ft long, with a resistance of 2 Ω. The voltage across the strip is 16 V. A contactor *f* has spring contact with the strip and is pulled along it at a constant speed in the range of ¼–½ m/sec by a cord *d* connected to a drum *c* (4½ in. diam) driven by a motor *b*. The motor is a ⅛-hp, series-wound, 120-V dc universal motor. In order to have a wide range of speeds, it is connected as a shunt motor with a rheostat in series with the field. It is set for about 3000 rpm so that the drum turns about 1 rev/sec. A 48:1 speed reduction gear (not shown in the figure) is used with the motor.

While the ammeter reading steadily increases as the potential difference tapped off from the Nichrome strip increases, the galvanometer reading remains fixed, indicating that the induced current does not depend upon the primary current, but upon the rate of change of the primary current. Switch S_1 must be closed after S_2 is closed, or opened before S_2 is opened in order to protect the galvanometer.

31–2.5

An experiment using solid and laminated iron cores suspended in a magnetic field may be used to demonstrate Lenz's law. A coil turning in a magnetic field will have an electric current induced in it. This induced current will produce a magnetic field which according to Lenz's law is in a direction so as to oppose the change in the field through the coil. The result is that the coil experiences a torque which opposes its motion. This is illustrated in Fig. 31–57 where the ribbon, from

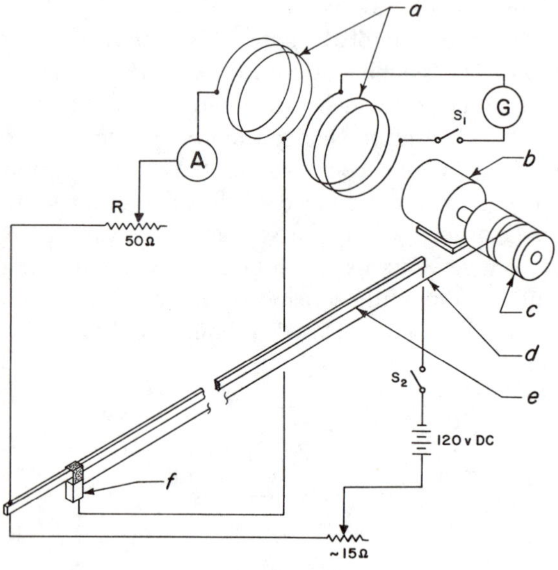

Fig. 31–56

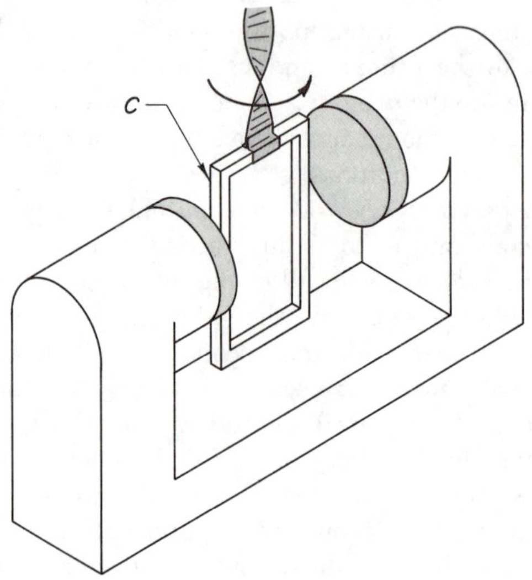

Fig. 31–57

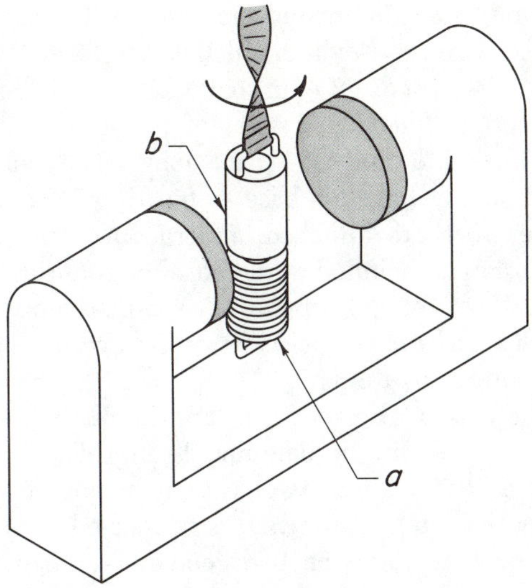

Fig. 31–58

which the coil *c* is suspended, is twisted, due to the torque on the coil.

The same effect can be seen using a solid iron core *b*. Eddy currents will be set up in the rotating core, and again the ribbon will experience a resistance to its motion as shown in Fig. 31–58. However, if a laminated core *a* is used as in Fig. 31–59 the ribbon turns freely. Eddy currents are greatly reduced due to the lamination, with the result that the motion is unopposed. If desired, the permanent magnet may be replaced with an electromagnet in order to demonstrate the magnitude of the effect.

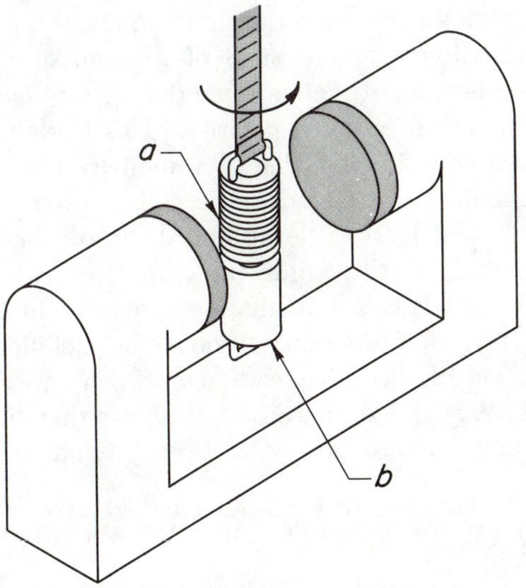

Fig. 31–59

31–2.6

This device demonstrates the forces produced by eddy currents in two aluminum disks. Each disk is 6 in. in diam and 1/8 in. thick. They rotate on bearings (3 3/8 in. apart), in a rigid Y-shaped hand-held frame so that they overlap. The disks are mounted in parallel planes and are placed 1/16 in. apart. If the overlapping portion of the disks is positioned vertically between the flat poles of a Leybold transformer magnet, they will spin rapidly in opposite directions. The Y-shaped frame should be mounted on a track so that it can be moved by external controls on the display case. This experiment makes a very good corridor demonstration.

31–2.7

A large horseshoe magnet *a* (Fig. 31–60) is rotated around an aluminum cylinder *b* to induce eddy currents, thus creating a torque on the cylinder, causing it to turn in the same direction. The thin-walled cylinder is suspended by a thin thread vertically on its long axis between the poles of the magnet. The magnet is mounted on the spindle of a hand rotater *c*.

31–2.8

The apparatus shown in Fig. 31–61 can be used to demonstrate the effects of a changing magnetic

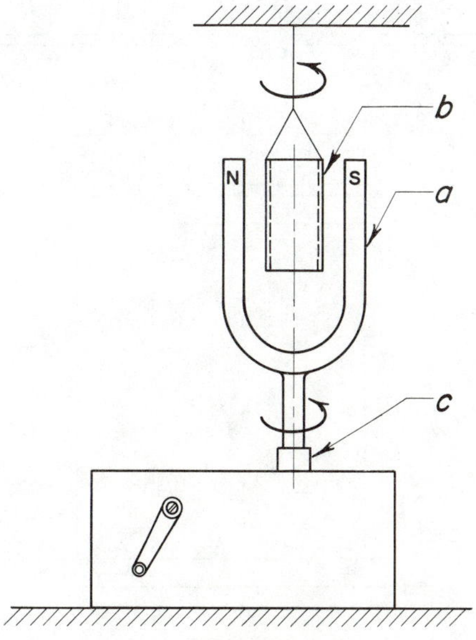

Fig. 31–60

Fig. 31–61

field. A large, vertical, iron core solenoid supplied with 120 V ac is used to produce the field. When an aluminum disk is held in a horizontal plane above the core, eddy currents are induced in the metal. If the disk is not located symmetrically above the solenoid, the interaction of the time varying magnetic field with the eddy currents will cause the disk to rotate rapidly. When a ⅛-in.-thick aluminum plate is introduced between the solenoid and disk, the rotation ceases, showing the shielding action of the metal.

The solenoid shown in Fig. 31–62 is made by winding three 156-turn layers *e* of No. 14 enameled copper wire on a cylindrical Bakelite form *f* measuring 2¼ in. i.d. x 2½ in. o.d. x 11⅛ in. in length. The core is filled with strips of transformer iron *g* 9⁄16 in. wide and 11¼ in. long. The solenoid stands on a Bakelite base *h* 6 x 6½ x ¼ in., and rubber feet are attached underneath. Two binding posts are included at the base for connections to a 120 V ac voltage supply. Two 5-in.-diam x ⅜-in. Bakelite disks *d* are held with setscrews to the top and bottom of the coil.

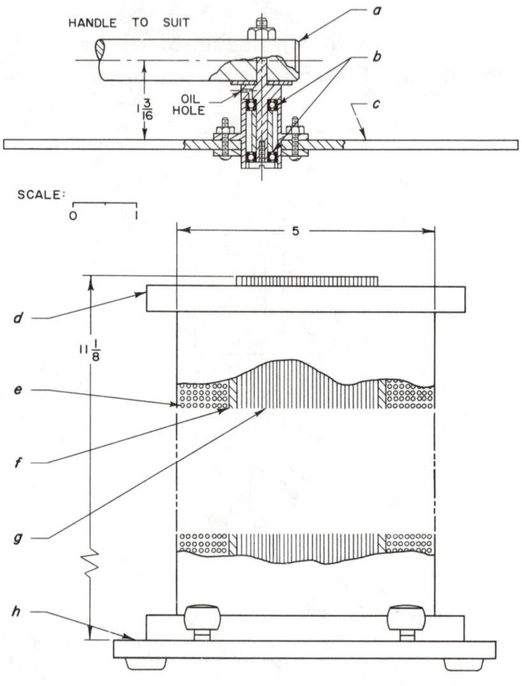

Fig. 31–62

The disk *c*, also shown in Fig. 31–62, is made of aluminum 8¼ in. in diam and ⅛ in. thick, and the front is divided into twelve sectors painted alternately black and orange. It is connected to an axle shaft which rotates on two good sets of ball bearings *b* and is attached to a ⅝-in.-diam aluminum handle *a*.

31–2.9

Some interesting points associated with Lenz's law formulation of electromagnetic induction can be extracted from this experiment. A solenoid with an iron core protruding a few inches from one end is placed in a vertical position, and an aluminum or copper ring is placed over the core to rest on the coil end. The observation is made that when the coil is energized from an ac source, the ring is always repelled from the coil and thrown off if the current is strong enough. An indiscriminate application of Lenz's law to explain this phenomenon will predict a repulsion during any part of a period of increasing magnetic field of a given direction, but predicts an attractive force during the decreasing strength of a given polarity. This leads to the seeming contradiction that in a number of tries the ring should be attracted to the coil as often as repelled, or at least rattle up and down on the core. The fallacy of this argument stems from the fact that Lenz's law is a qualitative corollary to Faraday's law of induction, without the detailed information of the time relation between cause and effect. A detailed analysis[15,16] shows that the repulsion here is an averaged effect stemming from

(15) L. Page and N. I. Adams: *Principles of Electricity*, 3rd ed. (D. Van Nostrand Co., Inc., Princeton, N.J., 1958), p. 329.

(16) S. G. Starling: *Electricity and Magnetism* (Longmans, Green & Co., Inc., New York, 1912), p. 369.

the phase difference between inducing and induced currents. If a soft iron ring is used, it will always be attracted. This may serve as a stepping stone for discussion of diamagnetism vs para- and ferromagnetism.

31–2.10

An electromotive force is induced in a coil inserted in the plane of the large nonconducting disk shown in Fig. 31–63 as the disk is rotated by hand through the field on an ~1800-G permanent magnet. The effect is demonstrated qualitatively in lecture by using the induced emf to light a flashlight bulb mounted on the disk. The greater the speed of rotation, the brighter the lamp flashes. The ½-in.-thick disk, 14½ in. in diam, is cut from two ¼-in. Masonite sheets. A 5-in.-diam hole for the coil is cut in the wheel on a 4⅛-in. radius. The coil consists of approximately 200 turns of No. 20 insulated copper bell wire. Its terminals are connected to the flashlight bulb which is mounted diametrically opposite the coil near the periphery of the disk. The disk is balanced dynamically by a lead counterweight. The Masonite disk is held by setscrews to a 12 x ¼-in.-diam steel shaft, which rotates on ball bearing assemblies at each end. These assemblies are bolted to 10½ x 2½ x ¾-in. plywood supports. An aluminum handwheel, 4 in. in diam and ½ in. thick, is attached at one end of the steel rod so the disk can be rotated by hand or a pulley and motor system. The main plywood supports are bolted to a 11½ x 9 x ¾-in. base and braced with two 9 x 2½ x ¾-in. boards, also made of plywood.

Fig. 31–63

Fig. 31–64

31–2.11

A device for producing a uniform magnetic field within a volume visible to a lecture hall audience is shown in Fig. 31–64. It consists of two large, open Helmholtz coils suspended about a central axis. Details for construction of the coils are given in the Appendix, page 1328.

Various demonstrations showing the effect of a magnetic field on solenoids, streams of charged particles, and a coil of wire can be performed with the device. A solenoid with a battery attached can be suspended in the field by a cord, as in Fig. 31–64, and the alignment with the field shown. To show the effect of a magnetic field on a stream of charged particles, a cathode-ray tube mounted in

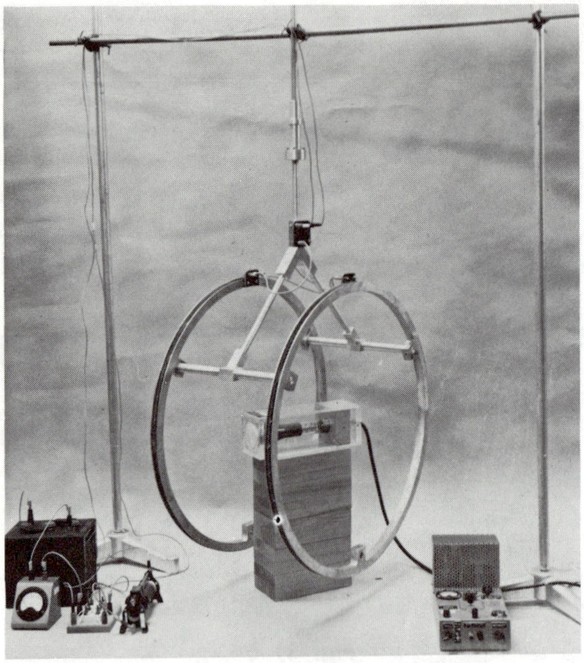

Fig. 31–65

a plastic box (Fig. 31–65) can be placed in the field and the coils rotated around it to show the deflection of the electron beam. A frame of wire mounted on a Lucite support and connected to a ballistic galvanometer is shown in Fig. 31–66. It consists of a center-tapped coil made of 20 turns

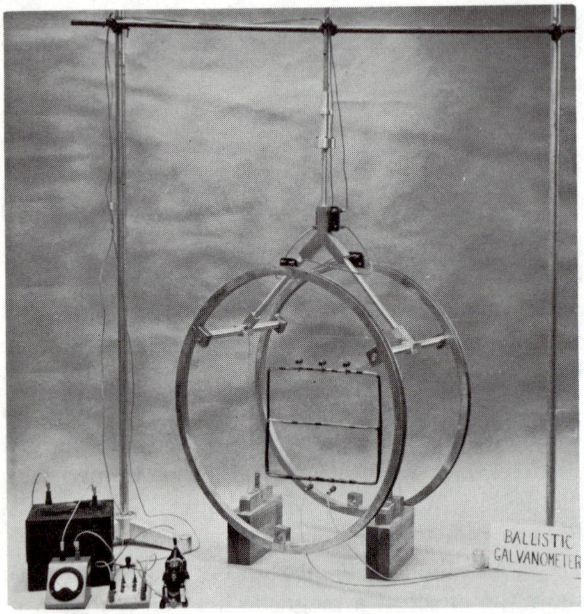

Fig. 31–66

Fig. 31–67

of wire encompassing 1 sq ft and a similar coil encompassing ½ sq ft, which can be used to show the dependence of an induced emf on coil dimensions. A flip coil, shown in Fig. 31–67, can be used to demonstrate that the total charge that flows in the circuit is independent of the speed with which the magnetic lines of force are cut.

31–2.12

The relation between electric and magnetic fields is demonstrated by the homopolar generator in Figs. 31–68 and 31–69. Brass cylinder *a* is rotated at a constant speed with the permanent magnet *b* (1½ in. diam, 1½ in. high) at rest in the brass holder *i*. A millivoltmeter is connected to the sliding contacts *c* mounted on a Plexiglas strip, and shows a voltage *V* which is due to the Lorentz force on the electrons in the cylinder.

When the magnet is rotated at the same speed as before with the cylinder at rest, no voltage is observed. This is because the voltage due to the Lorentz force on the electrons of the cylinder exactly cancels the voltage due to the Lorentz force in the rest of the circuit.

If the magnet is coupled to the cylinder and rotated, the same voltage is noted as when the cylinder alone was rotated. The entire voltage is

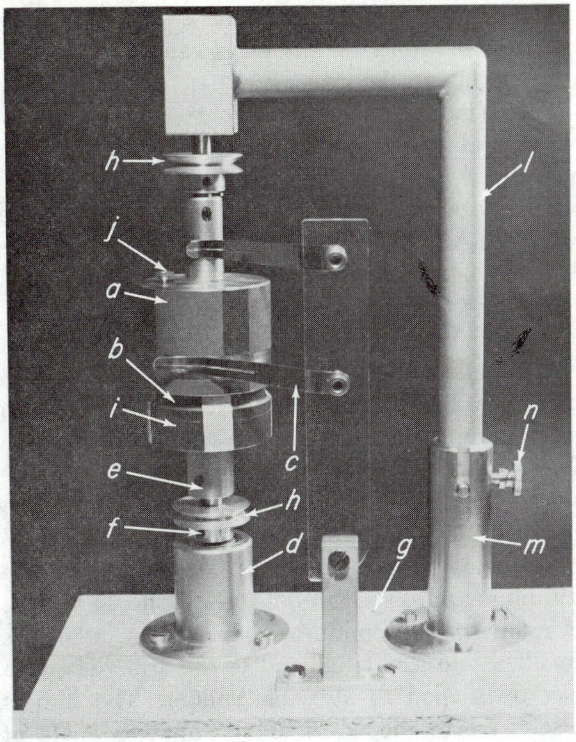

Fig. 31–68

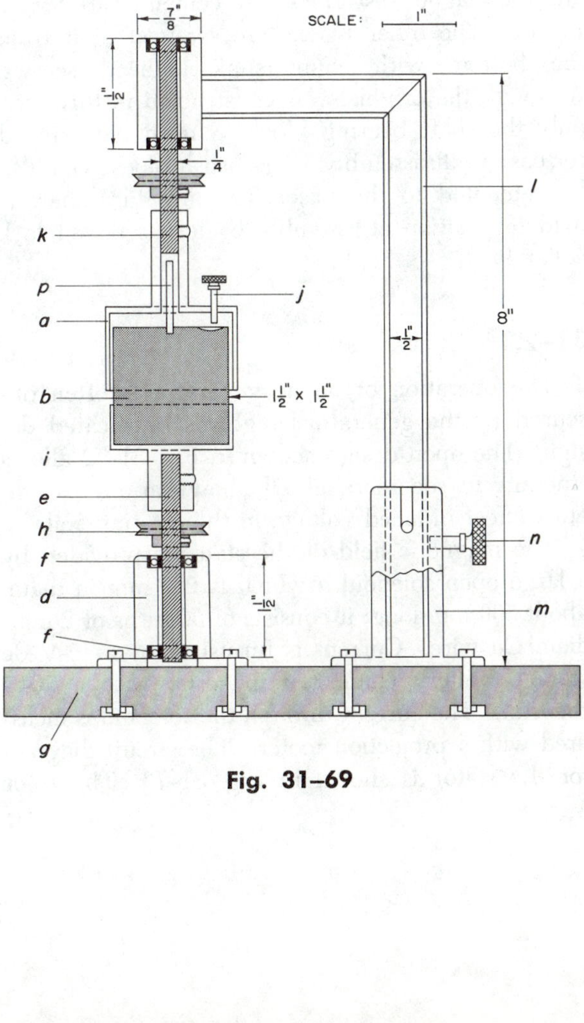

Fig. 31–69

due to the Lorentz force in the part of the circuit which is at rest. Since the magnet is a conductor, the cylinder is really unnecessary. The upper part of the machine can be removed allowing the contacts to touch directly on the magnet and on the pin *p* about which the magnet rotates.

To construct the homopolar generator, a brass socket *d* (1½ in. high, 1 in. diam), inside of which a steel shaft *e* (2¾ in. long, ¼ in. diam) rotates on ball bearings *f* on the top and bottom of the socket, is screwed onto the wooden base *g*. Aluminum pulleys *h* are attached to the steel shaft by setscrews. Cylinder *a* can be attached to the magnet by a screw *j*. Note that the magnet is mounted so that its magnetic field is symmetric when it is rotated in the indicated way. A pulley and ball bearing socket arrangement like the one on the bottom is mounted on another steel shaft *k* and attached by a setscrew onto cylinder *a*. The upper socket is supported by a brass rod *l* (8 in. high, ½ in. diam) mounted in a brass socket *m* and can be raised and held in place by a setscrew *n*.

31–2.13

A coil of insulated bell wire *a* (½ x ¾-in. cross section) is taped 4½ in. from one end of a long nonconducting rotor shaft *b* as shown in Fig. 31–70. Outside coil dimensions are 1¾ x 5½ in. The rotor shaft is ⅜-in.-diam, 11-in.-long cloth-based Phenolic. A brass ferrule on the bottom (with a shoulder for thrust) is snug-fit on a ball bearing *c*

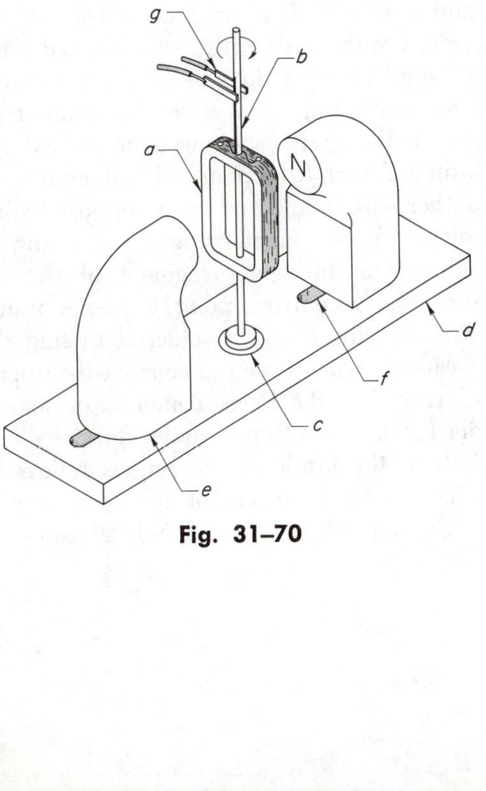

Fig. 31–70

nested in a steel base *d*, 3 x 14 x ¾ in. The poles *e* of a magnetron magnet have been removed from their base and remounted in slots *f* on each side of the rotor bearing (for adjustment).

Power from a surplus 2-V dc wet cell is supplied to the armature by externally supported brushes *g*. The winding lead tips are stripped, run up opposite sides of the shaft and taped to it. An emf is supplied (or taken from) the bared ends of the tips by the brushes.

To demonstrate the operation of a generator, another nearly identical shaft is used. It differs in that the wires running up the shaft lead to slip rings (each shaft is removable from the bearings). The current is carried off by brushes to a projection meter. Only hand rotation of the armature is required to produce a meter deflection as the winding cuts the field of the magnet (about 1800 G). To make the rotation of the armature more visible, alternate sides of the winding can be covered with black and white paper (or with reflecting foil).

31–2.14

In Fig. 31–71 is a generator designed for use on an overhead projector. A projection meter is connected to the brushes *3*. Most of the components are made of clear acrylic plastic for good electrical insulation, transparency, and ease of construction.

The commutator segments for the armature *1* can be made from either brass tubing or from solid brass stock drilled out for a smooth fit over the shaft. The tube is ¼ in. i.d. x ⅜ in. o.d. x $^{3}/_{16}$ in. long and is split with a band saw. The segments are screwed to the shaft with small flathead screws.

Slip rings, identical in size to the commutator ring before splitting, can either be cemented in position on the shaft or screwed or pinned to it. The insulated wire from point *a* is soldered to ring *c*. Another wire is soldered to the opposite side of the ring and passes uninterrupted under ring *d* to be soldered to the upper segment of the commutator. The wire from point *b* passes uninterrupted under ring *c* and is soldered to ring *d*. A short insulated wire is used to connect the opposite side of ring *d* to the lower commutator segment. In order for the wires to pass under the rings, either the shaft or the inside of the ring is relieved, or both. The original version of the generator employs ⅜-in.-o.d. slip rings and No. 20 connecting wire. If finer wire and rings of a larger diameter are used, the relief for the wires can be made with a rat-tail file on the ring.

The winding passes through the shaft in three places and is bent to shape. The dimensions are not critical except for the width of 1⅜ in. This is necessary because of the need for clearance as the winding is rotated between the magnet poles. The distance between the poles of the magnet used determines the aforementioned dimension.

The slip ring brushes *2* are attached to one block. The threaded brass studs *e* can be No. 4-40 N.C. and about ½ in. long. They are threaded into bottom tapped holes in the slip ring brush block *5*. Two smaller screws provide brush pressure on the rings. The commutator brushes *3* are screwed to the insides of their respective blocks *4*.

The crank arm *7* consists of a hub soldered to a thin brass arm. The crank handle *6* should be free to rotate on a roundhead brass screw which is threaded into a nut soldered to the arm. The arm may be shaped to suit the builder. The hub is setscrewed to the ¼-in.-diam armature shaft at assembly.

The clear acrylic plastic base *10* is cut to shape and the magnet rest blocks are cemented to it with Plexite. The other blocks are screwed to it from the bottom with countersunk flathead screws. Although the original was constructed in this way, only the shaft bearing blocks *8* need be screwed for ease in disassembly. The brush blocks can also be cemented to the base. The armature shaft is held in position at assembly by a setscrewed lock collar *9*.

31–2.15

The operation of an ac generator is often obscured by the generator's highly sophisticated design. The open design shown in Fig. 31–72 allows students to see more clearly how various parameters affect induced voltage in the rotating coil.

The magnetic field of the stator is provided by a large open solenoid *a* which is 200 mm in diam, about 500 mm long; it consists of 50 turns of 2-mm-diam Cu wire. Current is furnished by a 12-V dc storage battery connected in series with a 20-Ω rheostat. The current through the solenoid is measured with a projection meter. The circuit diagram for the stator is shown in Fig. 31–73. The rotor

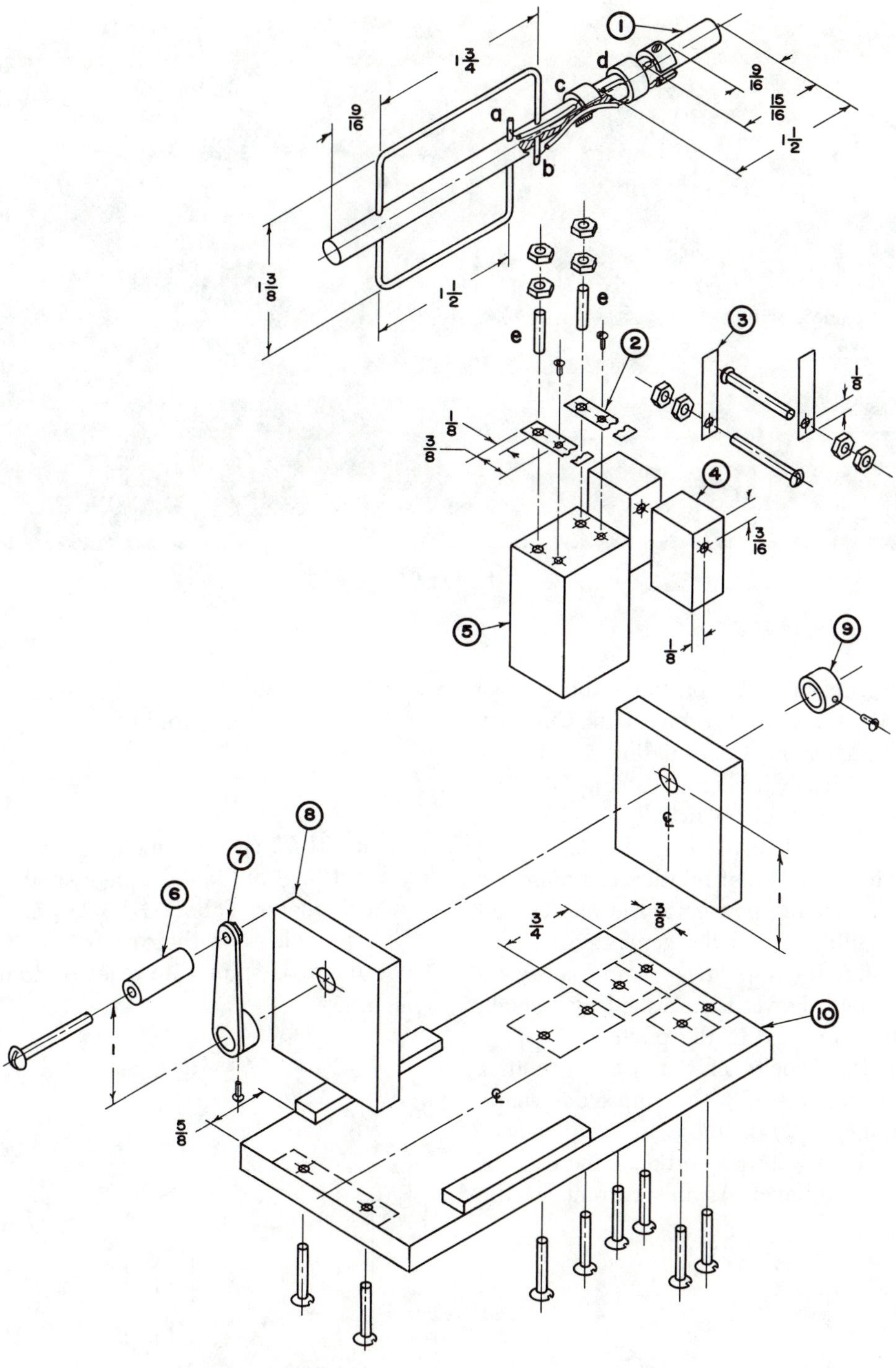

— — — — INDICATES BLOCK ARRANGEMENT ON BASE

SCALE:

Fig. 31-71

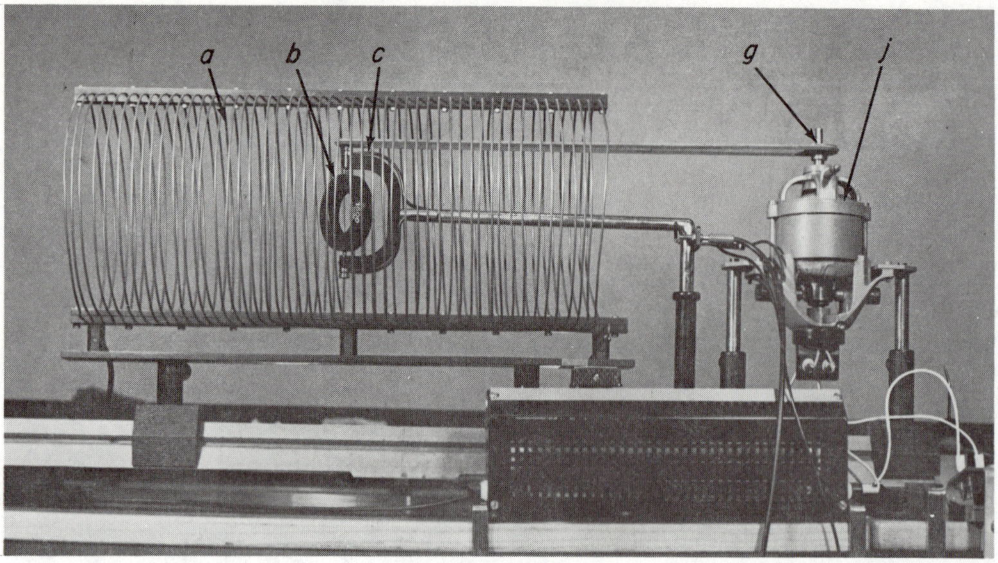

Fig. 31–72

consists of a plastic, Plexiglas, or hard-rubber spool *b* holding 1600 turns of 0.1-mm-diam Cu wire mounted on a brass-stemmed insulating Y *c* of plastic, Plexiglas, or hard rubber (see Fig. 31–74). The rotor's steel axle *d* rotates in cylindrical bronze bearings *e* and is held in place by bronze spring washers *f*. The rotor's axis of rotation is a diameter of the solenoid. Driving power for the rotor is fed by a belt and pulley *g*, and the generated voltage is brought out through slip rings *h* and lead wires *i* to an oscilloscope. Power is furnished by a small, variable speed dc motor *j* in the photograph.

As the angular velocity of the rotating coil is increased, one observes (1) the sinusoidal shape of the voltage signal, (2) the increased frequency of the voltage, and (3) the proportional increase of the peak-to-peak voltage. As the current in the stator is increased, the ac voltage shows a proportional increase in amplitude.

31–2.16

Figure 31–75 shows a motor-generator combination consisting of two coupled small permanent-magnet dc motors (about 1½ x 1½ cm) with plastic disk propellers on their shafts. If the propeller of one motor is spun with a jet of compressed air

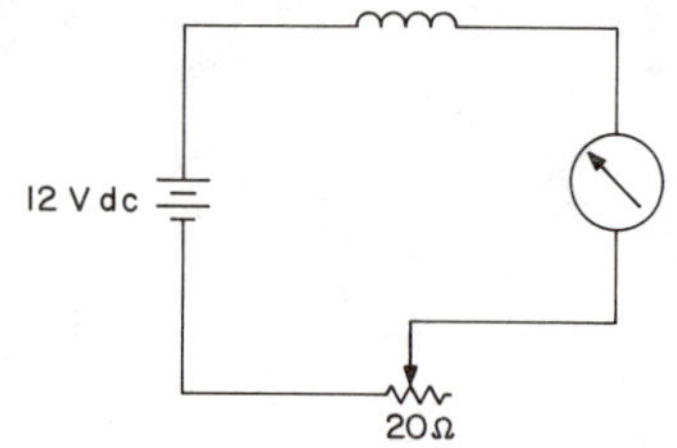

Fig. 31–73

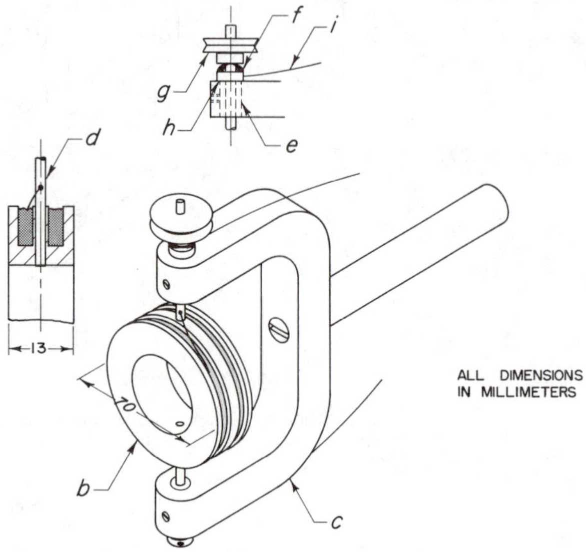

Fig. 31–74

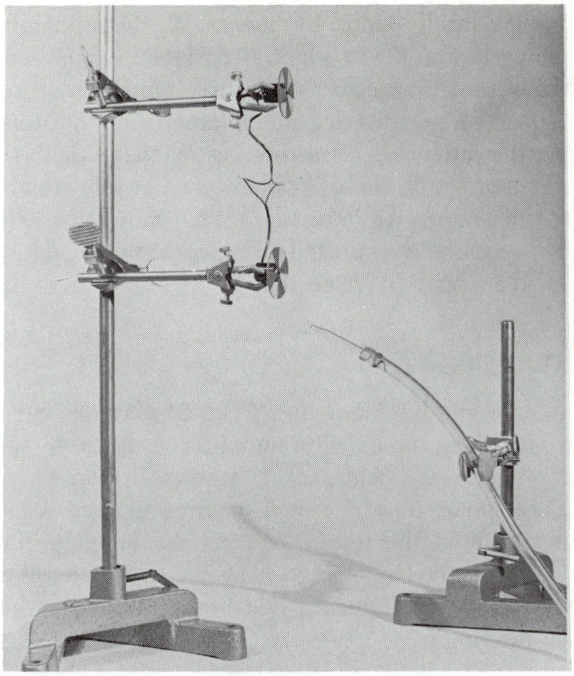

Fig. 31–75

supplied through a Tygon hose, the second propeller will also spin.

31–2.17

Many students are surprised to learn that there are no electrical connections to the rotor of an induction motor. A model of a two-phase induction motor is shown in Fig. 31–76. Two coils, one horizontal and one vertical, carry alternating current. Placing a capacitor in one of the lines produces a phase difference of almost 90 deg and a rotating magnetic field. If a coffee can mounted on a spindle is held in the field it will rotate like the rotor in an induction motor.

31–2.18

A hollow copper sphere *a* in a beaker of water provides a simple demonstration of a spinning magnetic field. A 110-V ac source is connected to a focusing coil *c* from a TV set or a cathode ray oscilloscope in series with a decade capacitor[17] (Fig. 31–77) to tune the circuit. On top of the focusing coil and under the beaker is a ¼-in.-thick circular

[17] Cornell-Dubilier Decade Capacitor, 0–10 μF, Model CDC, Serial 9091.

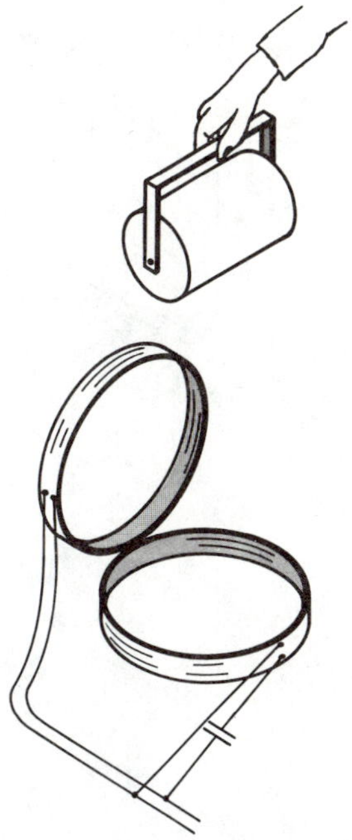

Fig. 31–76

disk *b*, one semicircle of which is aluminum and the other Lucite. The fields from the induced eddy currents and the coil are out of phase, causing a rotating magnetic field. This will cause the sphere to spin.

31–2.19

Figure 31–78 shows an eddy current motor which can be assembled from commonly available items.

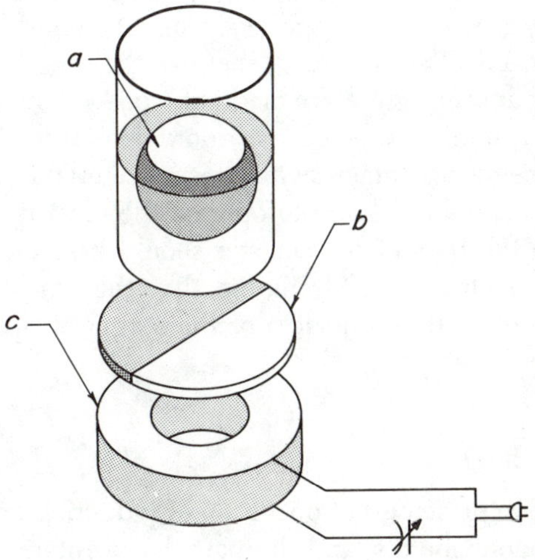

Fig. 31–77

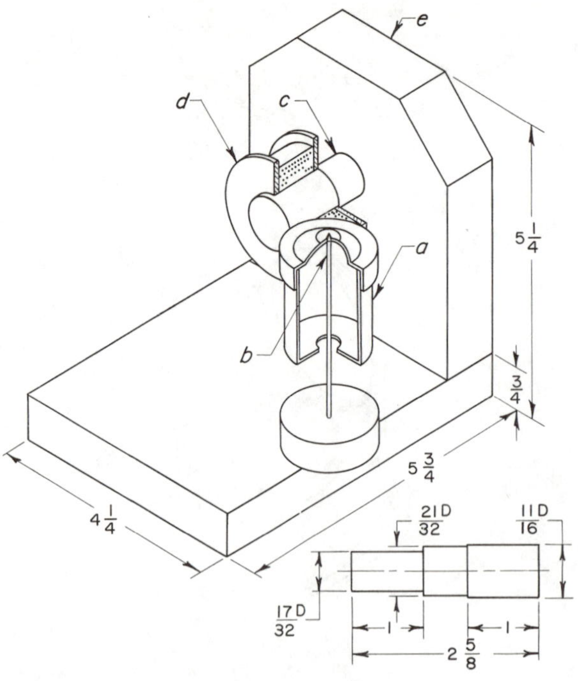

Fig. 31–78

A 35-mm film canister *a* is supported on its axis by a pointed pivot post *b*, $\frac{1}{16}$ in. in diam. The top is dimpled at the center for a bearing. If the can is placed near, but off to one side of an electromagnet, the can will spin rapidly when the coil of the magnet is energized. If the can is then moved to the other side of the tip, it will spin in the opposite direction.

The magnet core *c* is made from a piece of $\frac{11}{16}$-in.-diam soft iron, $2\frac{5}{8}$ in. long. Around the core is a dynamic speaker magnet coil *d*. Sixty-cycle 110 V ac is used to energize the magnet. The base and upright *e* are made of wood. This experiment makes a good corridor demonstration. A momentary contact switch fastened to the display table should be used to operate the electromagnet. Also the base of the canister should be attached to the electromagnet frame so that the canister can be moved from one side of the magnet's tip to the other.

31–2.20

The principles of operation of induction and synchronous motors, and the part that a rotating magnetic field plays in each type of motor, can be demonstrated easily by means of the apparatus shown in Fig. 31–79 which is designed for use with an overhead projector. The three pairs of coils are connected to a 360-deg potentiometer, so that turning the latter creates a rotating magnetic field. In the photograph the permanent magnet rotor rotates at the center. An induction rotor (aluminum cylinder) can be substituted. For construction details see the Appendix, page 1329.

31–2.21

A large aluminum ring (Fig. 31–80), resting in a glass dish on Bakelite supports, is made to spin rapidly in the center of a horizontal iron toroid. Three-phase ac is supplied to three separate windings around the toroid, causing the magnetic flux in the center to change continuously in a rotary fashion, and producing eddy currents and torque on the ring.

Fig. 31–79

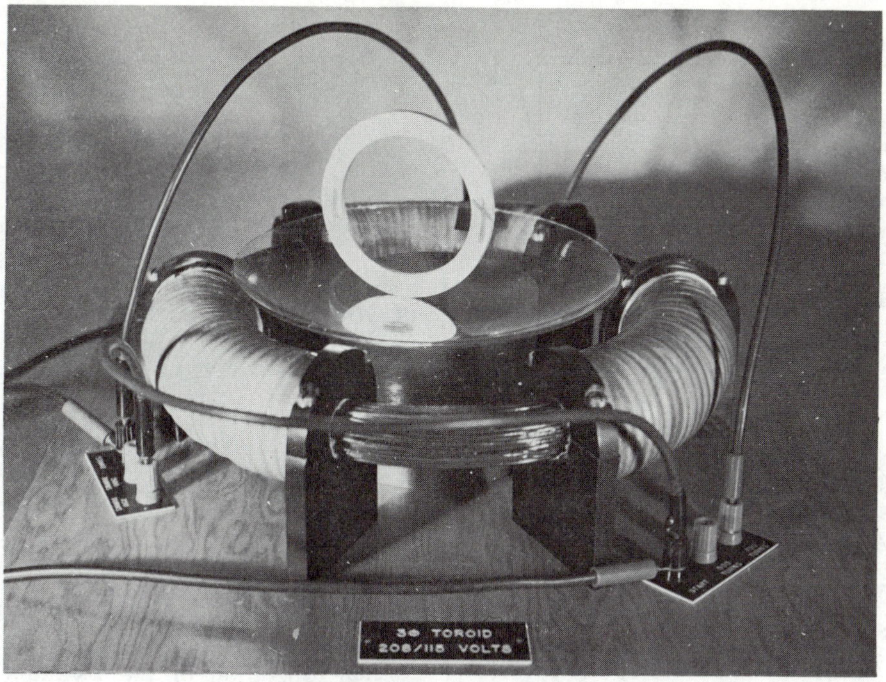

Fig. 31–80

The device is essentially a modified Rowland ring with a mean diameter of 12 in. The toroidal core *a* (Figs. 31–81 and 31–82) is formed from 500 ft of No. 12 lacquered iron wire (baling wire). The core diameter is ~1¼ in.

Instead of a continuous coil, three identical but separate coils *b* are spaced symmetrically around the core. Each occupies about 60 deg, leaving about 50 deg of uncovered core between coils. The remaining portion of the circle is occupied by nonconducting support brackets *c*.

The coils, each consisting of 750 turns of enameled copper wire with a tap at 500 turns, are delta-connected to a 208-V three-phase source *d*. Formvar fiberglass insulation is used between layers. Electrical connections are made via banana plug jacks, which are arranged in three groups of three each. A group is positioned outside each coil on a 20 x 20-in. board *e* and is labeled "Start," "500," and "750."

The operating current is approximately 13 A, or 4.5 A per coil. The recommended time limit for continuous operation is 5 min.

The effect of the field on other objects of differing size and composition can be demonstrated. A half dollar, for example, or a hollow copper ball floating in water, will also spin when placed within the toroid. The following rings are used with different results:

1. Aluminum: 3⅜ in. i.d. x 4⅜ in. o.d. x ¼ in. thick.
2. Same as (1), but ⅛ in. thick.
3. Same as (2), but ring is slit.
4. Same as (1), but of iron; also eighteen ⅛-in.-diam holes equally spaced around the ring.

31–2.22[18,19]

A levitator has been designed which can be constructed without much difficulty in the average physics department shop; it weighs only 100 lb and can be operated from the 110-V ac line at a power consumption of about 400 W. If capacitors are used which have a voltage rating of 440 V, the levitator can also be operated from a 220-V ac line.

The fact that the large electrical capacitance needed for the device can be made up from easily available and inexpensive surplus capacitors re-

(18) This research and development has been made possible through grants from the Fund for the Advancement of Education of the Ford Foundation, and the U.S. Department of Health, Education, and Welfare, Office of Education.

(19) H. E. White and H. Weltin, *Am. J. Phys.*, **31**, 925 (1963).

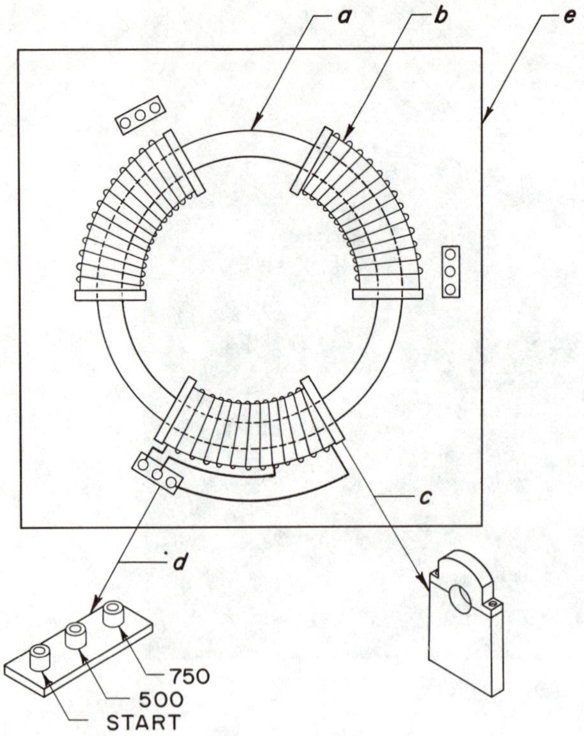

Fig. 31–81

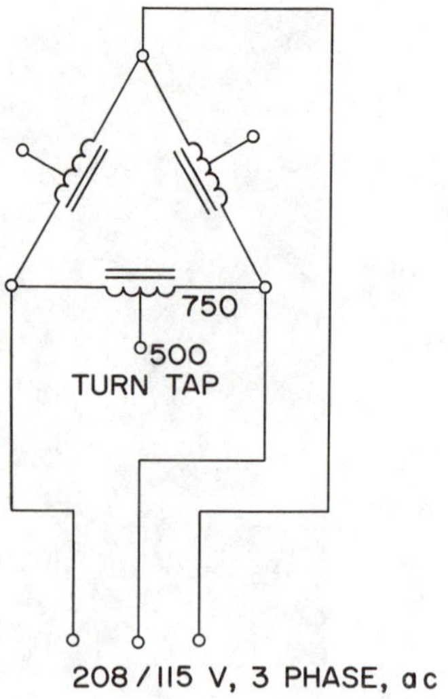

Fig. 31–82

duces the cost sufficiently to place the device within reach of any physics department interested in adding a levitator to already existing electromagnetic demonstration apparatus. Wire for the coils may also be obtained from Federal surplus sources.[20]

The three essential parts of the levitator are (1) a bank of capacitors, (2) an electromagnet consisting of four concentric coils, and (3) a laminated iron core assembled from six C-shaped and twelve L-shaped pieces that are placed around a circle in a six-arm star array, each arm of the array made up of one C-shape and of two L-shapes as shown in Fig. 31–83. Figure 31–84 shows the overall design of the magnet with the coils in place.

It is important to note that the coils are wound in the same sense and are numbered consecutively from the center toward the periphery. Coil (1) and coil (2), and coil (3) and coil (4), respectively, are wound upon one another. The magnet rests on top of a rectangular box in which the bank of capacitors is mounted and electrical connections are concealed.

The operation of the levitator is based upon Lenz's law.[21] A sketch of the magnet and of the coils is shown in Fig. 31–85. Because the pole pieces reverse their polarity periodically with the alternating current in the coils, the changing magnetic field induces strong eddy currents in the conducting aluminum pan. In accordance with Lenz's law, the eddy currents give rise to opposing fields and the aluminum pan is continually subject to an upward force in the central region of the magnet.

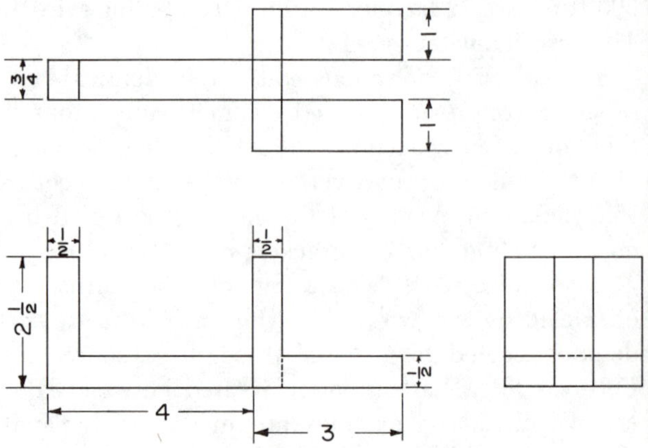

Fig. 31–83

(20) Available through the Federal Surplus Property Donation Program, Operated by the Department of Health, Education, and Welfare. Contact your State Agency for Surplus Property for details and your requirements.

(21) Harvey E. White, *Modern College Physics* (D. Van Nostrand Company, Inc., Princeton, N.J., 1962), 4th ed., p. 473.

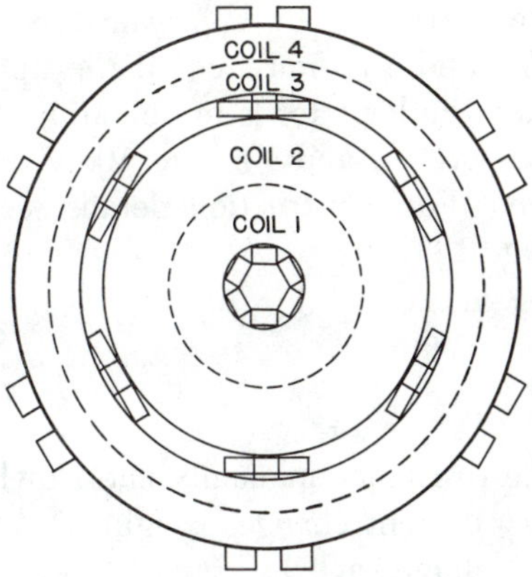

Fig. 31–84

The field along the circumference of the magnet provides the needed stability. When the pan is displaced from its equilibrium position above the center of the magnet, the stronger eddy currents induced in that side of the pan nearest the periphery of the magnet increase the repulsive force components acting on the pan in a direction opposite to the pan's displacement.

The electrical connections of the coils and of the capacitors are shown in Fig. 31–86. The numbers within the circles denote the coil numbers, and the letters S and *F* signify start and finish, respectively, of each coil. From Figs. 31–84 and 31–85 it is clear that the electron currents in the two inner coils (1) and (2) and in the two outer coils (3) and (4), respectively, must have the same direction, and that the electron currents in these two pairs of coils must oppose each other if the magnetic induction shown in Fig. 31–85 is to be achieved.

It has already been stated that all four coils are wound in the same sense. If one now assumes that the bottom line of Fig. 31–86 is momentarily negative, the currents in coil (1) and coil (2) appear to be in opposite directions, because the electrons flow from S to *F* in coil (1) and from *F* to S in coil (2). The same seems true of the currents in coils (3) and (4) in which electrons appear to flow from *F* to S in coil (3) and from S to *F* in coil (4).

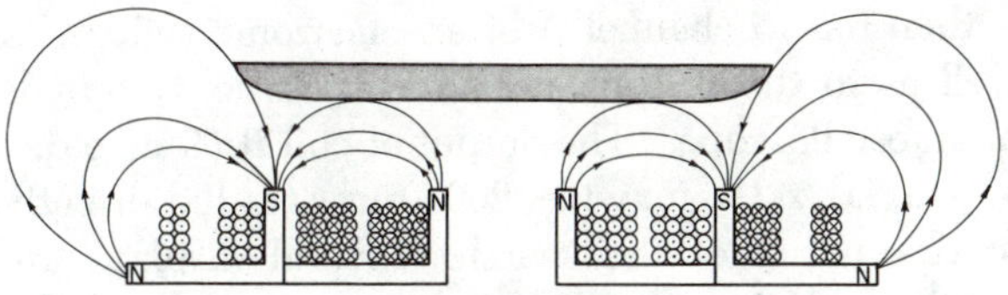

Fig. 31–85

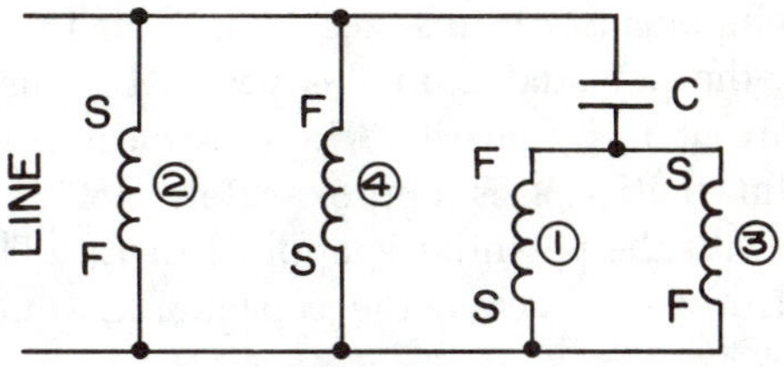

Fig. 31–86

In actuality, however, the voltage across the pair of coils (2) and (4) is approximately 180° out of phase with the voltage across the pair of coils (1) and (3) because of the action of the capacitor, as can readily be seen from the following considerations.

The circuit of Fig. 31–86 can be simplified by combining impedances as shown in Fig. 31–87.

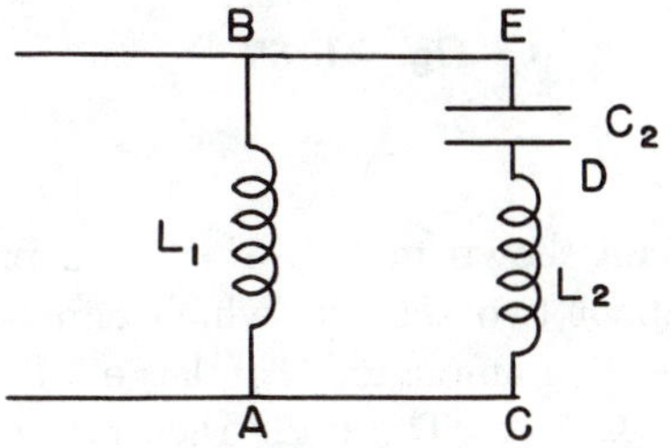

Fig. 31–87

The resulting circuit is a parallel resonant circuit. By increasing the capacity sufficiently, the circuit is brought into resonance. Under this condition branch *CE* is capacitive and branch *AB* is inductive. If *V* is the applied voltage, the potential drops across *AB* and *CE* are also *V*. Since the resistances of the coils are small, the current I_L in L_1 lags behind *V* by nearly 90°, and the current I_C in branch *CE* leads the voltage *V* by nearly 90° as shown in the vector diagram, Fig. 31–88. Now the current I_C is the same in both C_2 and L_2 since these

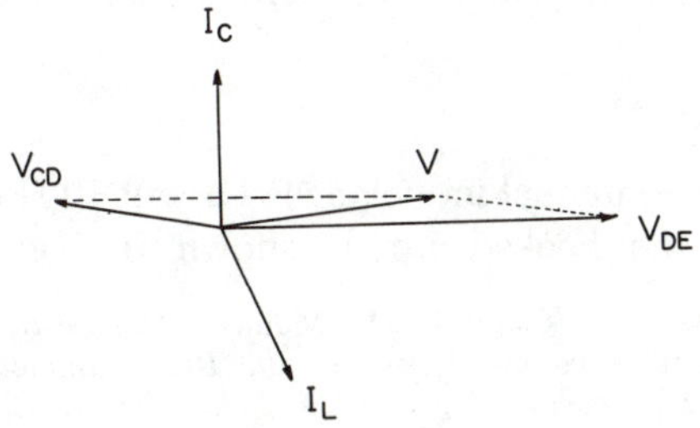

Fig. 31–88

two components are in series. But L_2 is inductive; hence, the potential drop V_{CD} across L_2 must lead the current I_C by nearly 90°. It is clear from the diagram that V_{CD} is approximately 180° out of phase with the potential V applied to L_1. The potential drop V_{DE} across the capacitor C_2 must lag behind the current I_C by approximately 90°. The sum of V_{CD} and V_{DE} must equal the applied voltage. When the levitator is in operation, the measured values of V_{CD} and V_{DE} are 100 V and 215 V, respectively. For construction details see the Appendix, page 1332.

31–3 Inductance

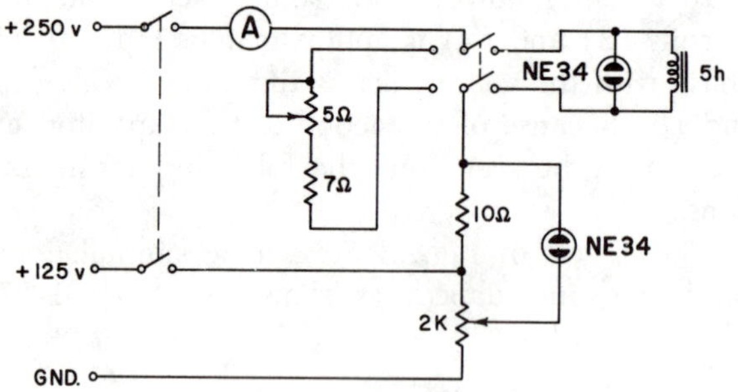

Fig. 31–89

31–3.1 [22]

The circuit shown in Fig. 31–89 can produce a delay of about two seconds which can be shown with a projection ammeter. The large 5-H choke is rated at 10 Ω, 5 A. The 2-kΩ rheostat is rated at 5 W and the 5-Ω adjustable resistor at 200 W. The 7-Ω and 10-Ω resistors are cone type, toaster-heater elements, 110 V, 600 W (modified to the required resistances).

31–3.2

A compass needle and pivot are located at the middle of a clear glass plate on a vertical projector lantern as shown in Fig. 31–90. The needle is deflected by the current in a single copper wire which is located just above the needle. Among the demonstrations which can be done with this indicator are Oersted's discovery of electromagnetism and Faraday's discovery of electromagnetic induction.

31–3.3 [23]

A large mutual inductance of about 10 H or more should be hooked up as shown in Fig. 31–91. Change the resistance manually and slowly so that the primary current changes roughly 2 A/sec, and read the voltage at the same time. Calculate $M = E/(dI/dt)$, where M should be checked against the known value.

Large mutual inductances are sometimes hard to find. The model transformers used in many laboratories are about 1 H. A high-voltage (secondary 8,000–25,000 V) transformer is usually well over 10 H. When used, the primary current should be decreased gradually to zero (rather than suddenly stopped) after increasing to a maximum, so as not to damage the voltmeter. An electromagnet can be used if it has two coils—one can be used as primary.

If none of the above mutual inductances are available, another method is to use 60 cycle ac on a power supply transformer of a radio set, or any power transformer whose primary is for 120 V. Measure the no-load primary current (in mA) and secondary voltage. The ratio of voltage to current is 2π times the frequency times M. The higher the frequency, the less the need for a large mutual inductance, so a 1000-cycle electronic generator would be advantageous.

31–3.4

Mutual inductance is demonstrated qualitatively in a straightforward way as shown in Fig. 31–92. Coil A (9 in. o.d. x 4½ in. i.d. x 3½ in. wide, 2000 turns) is connected to an audio frequency oscillator through a 100-Ω resistor. The voltage across this resistor, proportional to the current through coil A, is then fed to channel A of an electronic switch, as well as to the x-input and external sync. terminals of an oscilloscope. The output of coil B (9 in. o.d. x 4½ in. i.d. x 3½ in. wide, 2000 turns) is fed directly to channel B of the switch. Several displays are possible. Using the linear sweep of the oscillo-

[22] R. Resnick and H. F. Meiners, *Demonstration and Laboratory Apparatus Report of the 1960 Summer Visiting Professor Workshop at Rensselaer Polytechnic Institute.*

[23] E. M. Little, *Am. J. Phys.*, **25**, 491–92 (1957).

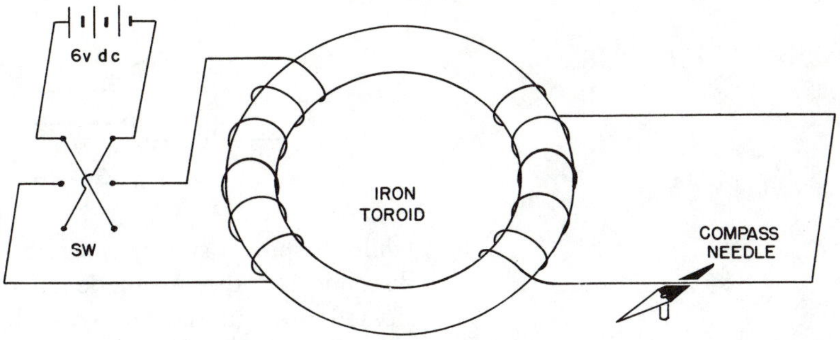

Fig. 31–90

scope, synchronized to the current through coil *A*, the phase relationship (in terms of time) between the current in *A* and the voltage induced by it in *B* can be shown. Applying the signal from R_A directly to the *x*-input of the CRO (sweep off) will display the same information in the form of a Lissajous figure. In the latter case it is usual to feed only channel *B* of the switch to the *y*-input of the CRO.

The effect of geometry on the magnitude of the mutual inductance between the coils is demonstrated by moving them closer or further apart when they are coaxial, or by rotating one about a vertical axis until coupling is a minimum, i.e., until their axes are at 90 deg to each other. A further demonstration with the coils coaxial consists of inserting a 1-in.-diam x 6-in.-long bar of solid iron and a stack of transformer core laminations alternately through the centers of both coils. The relative influence of the two forms of iron on the coupling between the two coils is graphically illustrated.

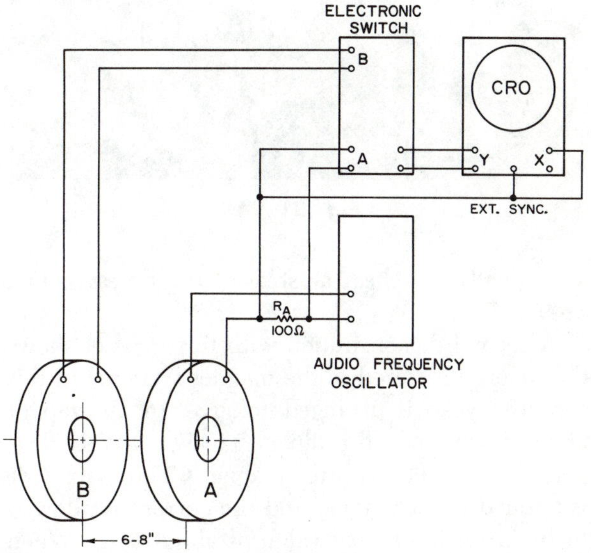

Fig. 31–92

31–3.5

When a steady voltage is suddenly applied to a large inductance, the current rises slowly due to the back emf produced by the changing current. Here, two automobile lamps (6 V, 25 W) are used to indicate the voltage across and current through a large electromagnet after a 28-V dc supply is connected. The magnet used had an inductance of 36 H with a yoke across the poles, and 7 H without the yoke. The resistance of the magnet's windings is 6 Ω. Figure 31–93 shows the circuit diagram. A rheostat (20 Ω) is used in series with the voltage-sensing lamp so as not to burn it out. When the switch is closed, the current-sensing lamp in series with the magnet begins to glow after about 4 sec

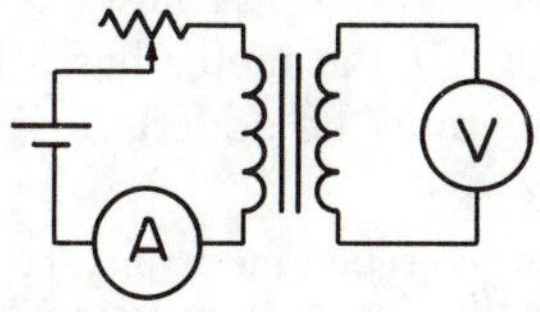

Fig. 31–91

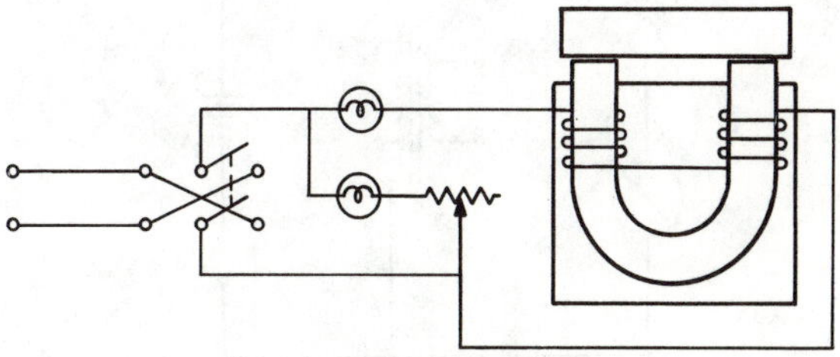

Fig. 31–93

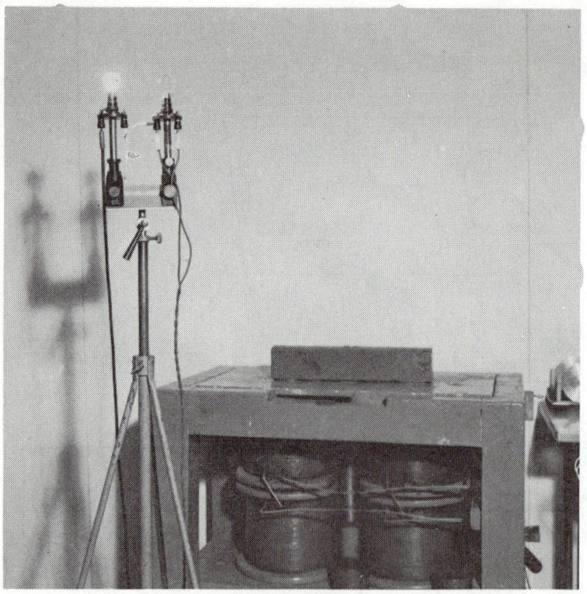

Fig. 31–94

and finally reaches a steady brightness (Fig. 31–94).

A second demonstration with this magnet shows that energy is stored in the magnet system. In this case the yoke is arranged to give an air gap of about ¼ in. and an ordinary 110-V, 40-W bulb is placed in parallel with the magnet. Thirty-six V dc is applied to the magnet and the current is allowed to build up to its final value of about 6 A. When the power supply switch is opened, the entire 6 A is switched into the lamp, which burns out in a spectacular flash. In an alternative demonstration a 6-V, 25-W lamp is connected in series with a rectifier across the magnet terminals. The rectifier prevents current from flowing through the lamp

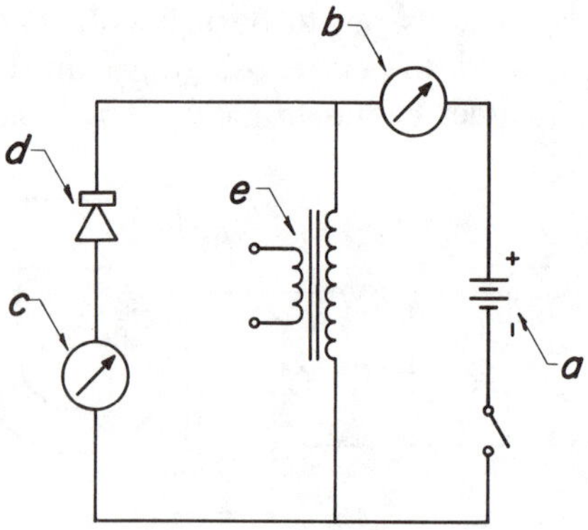

Fig. 31–95

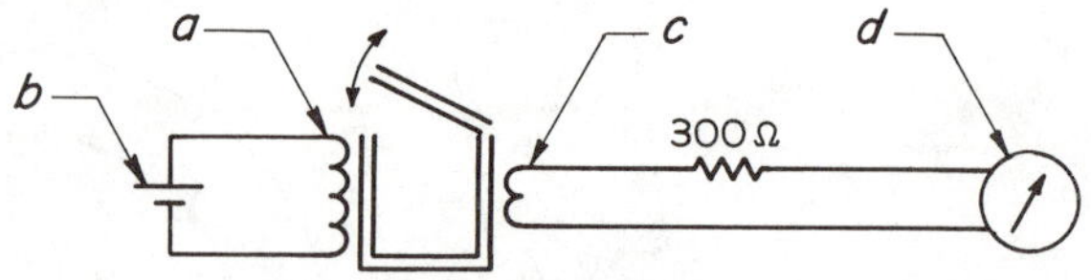

Fig. 31–96

while the power supply is connected, but when it is disconnected the magnetic energy is slowly dissipated through the lamp which glows for a few seconds and then dims.

31–3.6

A large transformer (220:30; 1 kVA) is used as a self inductance. The transformer e is magnetized by a 6-V battery, a as shown in Fig. 31–95. The current in the primary is shown by a projection meter[24] b (about 15 A). A second meter c is connected in series through a diode d (Thomson 13 R-2) and in parallel with the transformer. The diode is connected so as to stop the primary current flow.

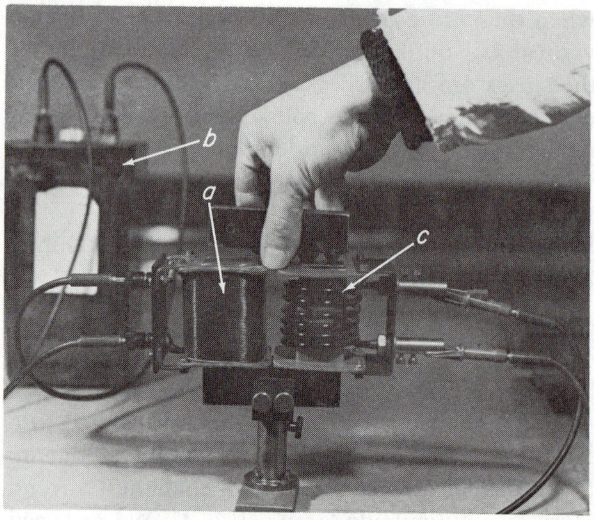

Fig. 31–97

When this primary current is cut off, an induction current of about 1.5 A flows through the second instrument. If the meter has a low resistance, a series resistor must be added.

[24] A suitable projection meter may be purchased from: the Central Scientific Company or from EMA AG., Meilen/Zürich, Switzerland.

31–3.7

Changing the air gap in a magnetic circuit (Fig. 31–96) will induce a voltage in the coil. This changes the reluctance and hence the flux, so there is an induced voltage in any circuit enclosing the flux. A coil *a* of 600 turns[25] is connected to a 2 V dc source *b* to produce the flux. A second, six turn coil *c* surrounds the iron core. The coil *c* is connected to a mirror galvanometer *d* which measures the current. The demonstrator holds a bar across the top of the two cores (Fig. 31–97) and than varies the air gap between the bar and the cores. The size of the air gap is shown in shadow projection with an arc lamp.

(25) Can be obtained from Phywe AG., 34 Göttingen, Post Box 665, West Germany.

Demonstrations	Contributors
31–1.1	H. W. Dosso and R. H. Vidal, Victoria College, Canada
31–1.2	Eric M. Rogers, Princeton University
31–1.3	Louisiana State University
31–1.4, 31–1.5	Swiss Federal Institute (ETH), Zürich, Switzerland
31–1.6	T. D. Strickler, Berea College
31–1.7	California Institute of Technology
31–1.8	University of Illinois
31–1.9	H. C. Jensen, Lake Forest College
31–1.10	Swiss Federal Institute (ETH), Zürich, Switzerland
31–1.11	H. F. Meiners, Rensselaer Polytechnic Institute
31–1.12	J. Jaumann, H. Werheit, and F. Günneberg, University of Köln, West Germany
31–1.13	H. C. Jensen, Lake Forest College
31–1.14	T. E. Wade, Jr., University of Nebraska
31–1.15	E. J. Eastmond, Brigham Young University
31–1.16	California Institute of Technology
31–1.17	L. Hirsch, California State College, Los Angeles
31–1.18	P. Sprawls, Emory University
31–1.19	Swiss Federal Institute (ETH), Zürich, Switzerland
31–1.20	L. Hirsch, California State College, Los Angeles
31–1.21	Swiss Federal Institute (ETH), Zürich, Switzerland
31–1.22	H. A. Robinson, Adelphi University
31–1.23, 31–1.24	Swiss Federal Institute (ETH), Zürich, Switzerland
31–1.25	Michigan State University
31–1.26	F. Oppenheimer, University of Colorado
31–1.27	University of Washington
31–1.28	P. Younger, St. Cloud State College, Minn.
31–1.29	H. A. Robinson, Adelphi University
31–1.30	A. D. Moore, University of Michigan
31–2.1	Swiss Federal Institute (ETH), Zürich, Switzerland
31–2.2	P. R. Mason and J. L. Teegarden, Rose Polytechnic Institute
31–2.3	F. Oppenheimer, University of Colorado
31–2.4	Eric M. Rogers, Princeton University
31–2.5	M. J. Pryor, State University of New York at Albany
31–2.6	A. A. Bartlett, University of Colorado
31–2.7	University of Illinois
31–2.8	University of Washington
31–2.9	J. B. Hart, Xavier University
31–2.10	H. F. Richards, Florida State University
31–2.11	J. W. Straley, University of North Carolina
31–2.12	Swiss Federal Institute (ETH), Zürich, Switzerland
31–2.13	H. C. Jensen, Lake Forest College
31–2.14	W. Eppenstein, Rensselaer Polytechnic Institute

Demonstrations	*Contributors*
31–2.15	Swiss Federal Institute (ETH), Zürich, Switzerland
31–2.16	California Institute of Technology
31–2.17	V. E. Eaton, Wesleyan University, Conn.
31–2.18	Harvard Project Physics
31–2.19	H. C. Jensen, Lake Forest College
31–2.20	Swiss Federal Institute (ETH), Zürich, Switzerland
31–2.21	E. M. Lyman, University of Illinois
31–2.22	H. E. White and H. Weltin, University of California, Berkeley
31–3.1	D. K. Berkey, Colgate University
31–3.2	R. M. Sutton, California Institute of Technology
31–3.3	E. M. Little, U.S. Navy Electronics Laboratory, San Diego
31–3.4	California Institute of Technology
31–3.5, 31–3.6, 31–3.7	Swiss Federal Institute (ETH), Zürich, Switzerland

32

MAGNETIC PROPERTIES OF MATTER

32–1 Poles and Dipoles

32–1.1

Several properties of the magnetic field can be demonstrated on an overhead projector with the aid of a box with a transparent top and bottom, filled with a transparent liquid and some iron chips. The box is shaken, inverted, and then placed over or near the magnets. The chips, still floating downward, arrange themselves in the appropriate configuration. The viscosity of the liquid, size of the chips, and strength of the magnets are adjusted so that the formation of the field configuration can be followed conveniently for 2 or 3 sec.

The box is made from 9-cm-diam knurled aluminum tubing and has a Plexiglas top and bottom. It is filled with liquid through an opening in the tubing, which is then closed by a screw and rubber washer. The liquid is a mixture of 75 percent glycerol and 25 percent alcohol.

For demonstration of various magnetic field configurations, a large Plexiglas plate is slipped over two pins on the projector stage, covering the whole stage. The sheet has a central 12-cm-diam hole, into which fit circular disks which serve as mounting bases for studies of a variety of magnet arrangements.

Figure 32–1 shows the field configuration obtained with a single bar magnet. Magnetic fields produced by two bar magnets placed either parallel or antiparallel can be studied. The effect of different materials on the magnetic field can also be

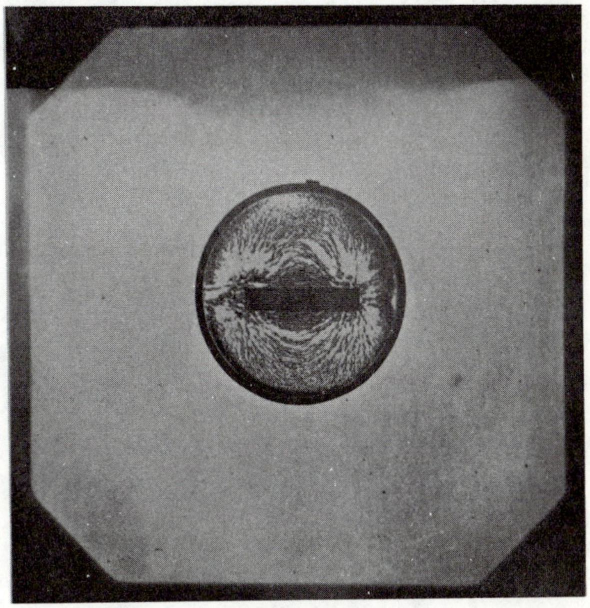

Fig. 32–1

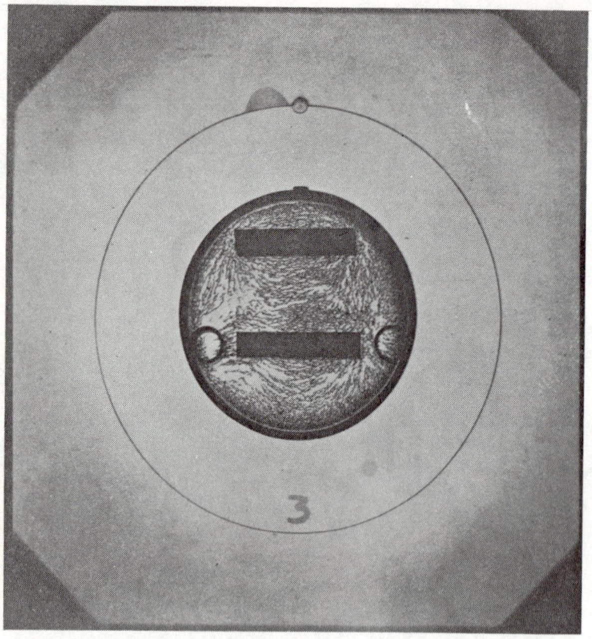

Fig. 32–2

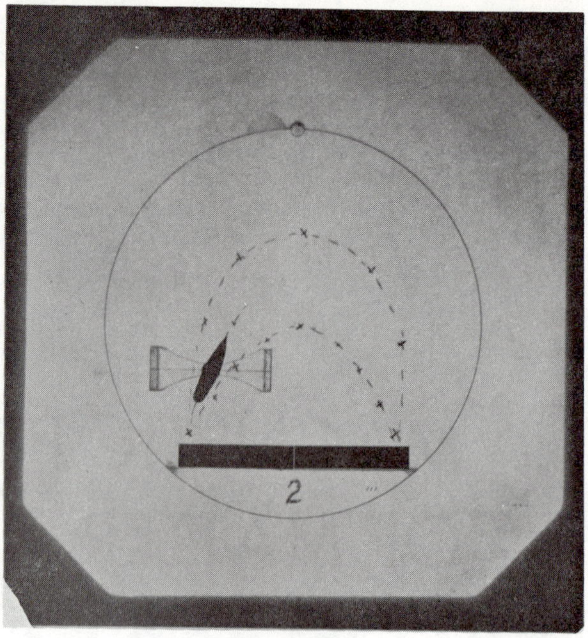

Fig. 32–4

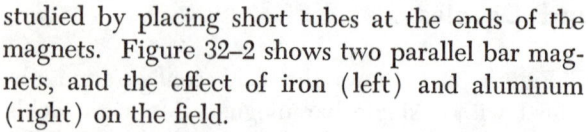

studied by placing short tubes at the ends of the magnets. Figure 32–2 shows two parallel bar magnets, and the effect of iron (left) and aluminum (right) on the field.

Plotting of field lines can be done with the compass-needle bridge shown in Fig. 32–3. A good magnetic needle is mounted on a standard compass-needle pivot (with jewel bearing). The field lines are traced as shown in Fig. 32–4. The compass needle is moved in a stepwise manner so that at each consecutive position its north end is where its south end was at the previous step. The position of the end of the needle is marked with a grease pencil for each step.

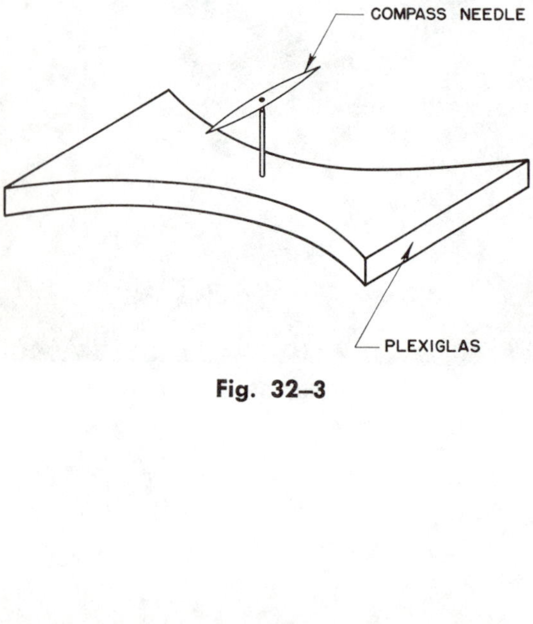

Fig. 32–3

The field lines may turn out to be asymmetric with respect to a perpendicular through the center of the magnet, due either to the influence of the magnetic field of the earth, or to the magnets being of unequal strength. In order to check the cause of the asymmetry, the positions of the magnets can be exchanged.

Magnetic screening can be demonstrated by constructing a screen from a 2.5-cm-wide silicon iron strip which is bent into a 10-cm-diam cylinder. When the field demonstration box is placed inside the screen, a bar magnet on the outside has no effect on it. The screen is not perfect, as can be shown by replacing the box with a magnetic needle. The oscillations of the needle will change from slow to violent and to slow again if the screen is first put in place, removed, and then put back again.

The variation of magnetic field strength with distance can be demonstrated using a coordinate system drawn on a sheet of Mylar. The magnet is placed in the middle of a margin as shown in Fig. 32–5. The compass needle is moved along a chosen coordinate line, and the frequency of oscillation ν is noted.

The magnetometer (Fig. 32–6) is a standard 9-cm-diam magnetometer box with a transparent bottom which is graduated in degrees along the circumference, with every fifth degree marked.

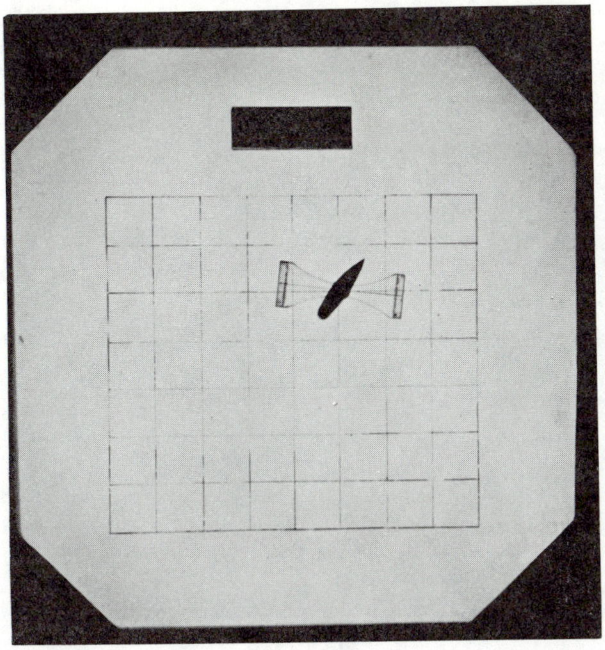

Fig. 32–5

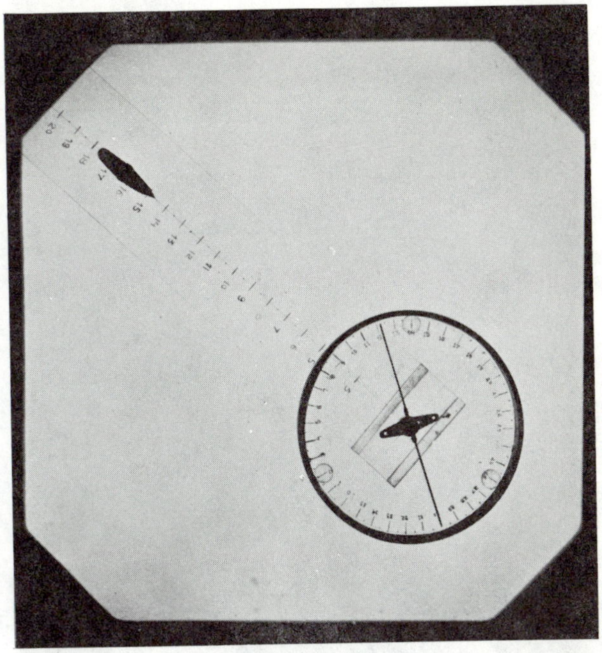

Fig. 32–6

Three stubs, placed at 20 deg, 140 deg, and 290 deg along the circumference and one under the center, serve as feet.

The variation of magnetometer deflection with distance can be measured. Since the bar magnet would be much too strong, it is replaced by a compass needle which is placed in the east-west direction. The magnetometer is moved along a coordinate line, and the deflection as a function of distance is noted.

Figure 32–6 shows the arrangement for determining the magnetic moment of a bar magnet. A special transparent ruler has a hole inside a slot. The ruler is attached to the center foot of the magnetometer; the bar magnet is moved along it, and the deflection of the needle is noted. The ruler lies perpendicular to the local magnetic meridian. It should always be possible to turn the magnetometer so that the ruler does not have to meet one of the foot stubs.

The magnetic moment can also be determined by the oscillation method. The bar magnet is placed in the slot of the ruler as shown in Fig. 32–7, and the compass bridge is moved along the ruler.

In a similar manner, the horizontal component of the magnetic field of the earth can be determined. The compass needle bridge is placed in the east-west direction where the magnet was on the Mylar sheet, and its frequency of oscillation is determined. The needle is then removed from the bridge and placed on the sheet at the same place, and the deflection of the magnetometer caused by the magnetic needle is noted.

Properties of a solenoid can be easily demon-

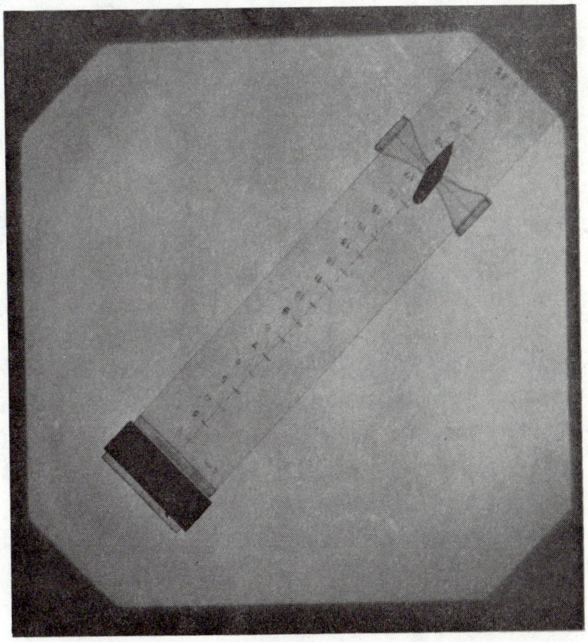

Fig. 32–7

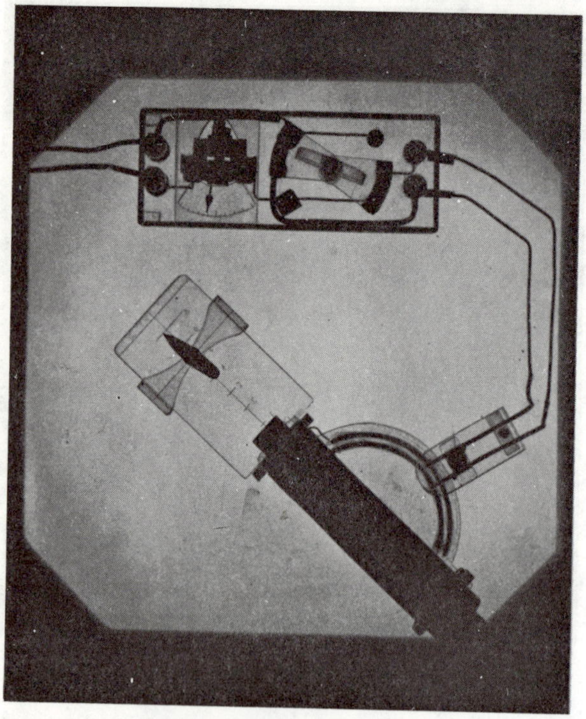

Fig. 32–8

strated with the overhead projector. The solenoid core can be made from a brass tube filled with milled chips of transformer iron. The solenoid itself can be turned to coincide with the local field. Either the compass bridge or the magnetometer is laid on the attached short ruler (Fig. 32–8) and the deflection is noted as a function of the current in the solenoid, the penetration of the core, and the distance of the magnetic needle from the solenoid.

With the core completely inserted, the current is turned off, and the presence of the iron core is shown to have little influence on the deflection. The demonstration shows the augmented deflection with the core inserted, but also proves that the iron core itself is not the main cause of the deflection.

The field of a solenoid can be mapped using the same technique as was described before. A long solenoid with several taps is used (Fig. 32–9), with auxiliary circuitry consisting of an ammeter and a double-pole, double-throw switch for reversal of the current.

32–1.2

The equipment in Fig. 32–10 is used to show that the force between two small dipoles whose

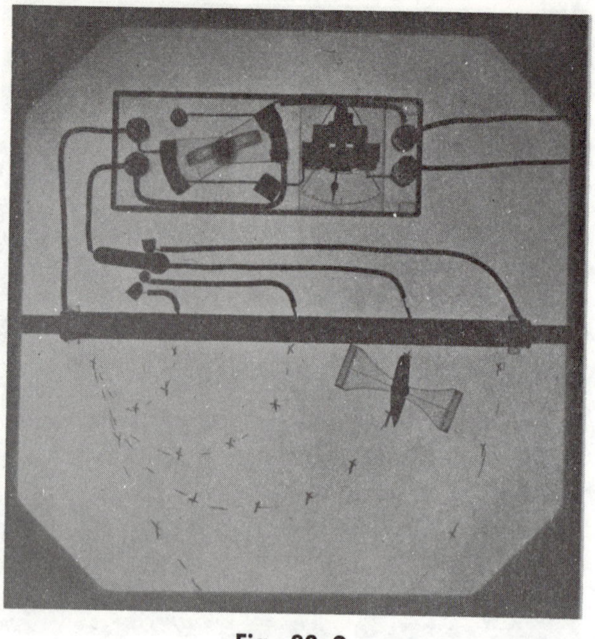

Fig. 32–9

moments are in the same straight line varies as the inverse fourth power of their separation. A thin-walled brass tube *a* (Fig. 32–11) 75 cm long and 0.5 cm in diam at its upper end, is soldered to a knitting needle *b* which serves as a crossbar. This needle rests in two oppositely-bored holes in a U-shaped aluminum sheet *c*, 15 cm wide. Motion along the supporting axis is prevented by rubber stops on the needle. The movement of this plane pendulum is measured on a vertical scale by a light beam deflected by a small mirror *d* fastened to the turning axis (the needle) of the pendulum.

A core magnet[1] (a magnet with a hole in its center) *e* is fastened at about 10 cm above the lower end of the brass tube with its magnetic moment in the pendulum plane and perpendicular to the tube. To damp out oscillations, a cross made from sheet metal strips is fastened to the lower end of the tube and dipped into a dashpot containing glycerin. The lower part of the complete pendulum is located in a square glass container to prevent wind disturbances. A strong loudspeaker magnet *f* is placed at different distances from the pendulum magnet by sliding it along a non-ferrous

[1] Very handy short magnets with a central bore are available as core magnets for moving coil instruments from Krupp Widia Fabrik, Essen, Germany. The magnets are ellipsoid, with the ends rounded and the sides flattened. The two round ends are the poles.

Fig. 32–10

angle trough g which is horizontally mounted off to one side but in the plane of the pendulum so that the two magnetic moments line up. The deflection of the pendulum is measured as a function of the separation between the magnets. The results plotted in log–log coordinates produce a straight line with a slope of -4. A slide image of log–log coordinates on the blackboard is helpful.

With this apparatus one can also investigate the interaction of perpendicular magnetic dipoles.

32–1.3

The force between a magnet and a piece of soft iron varies as $1/r^7$, where r is the distance between them. The apparatus for measuring this force is shown in Fig. 32–12. A glass tube a about 50 cm long and 3 cm in diam, is used to support a suspension and to shield it from effects of wind. Two heavy-gauge (1-mm) wires b pass through the rubber plug c and are attached to the fine filaments d ($\sim$50 cm long) as shown. The lower ends of the filaments are attached to a non-ferrous cross bar e which extends 5 mm beyond the filament joints. On one side of the cross bar a piece of demagnetized soft iron f is hung, and on the other extension a brass balance weight g is fastened and to this balance weight is attached a dampening

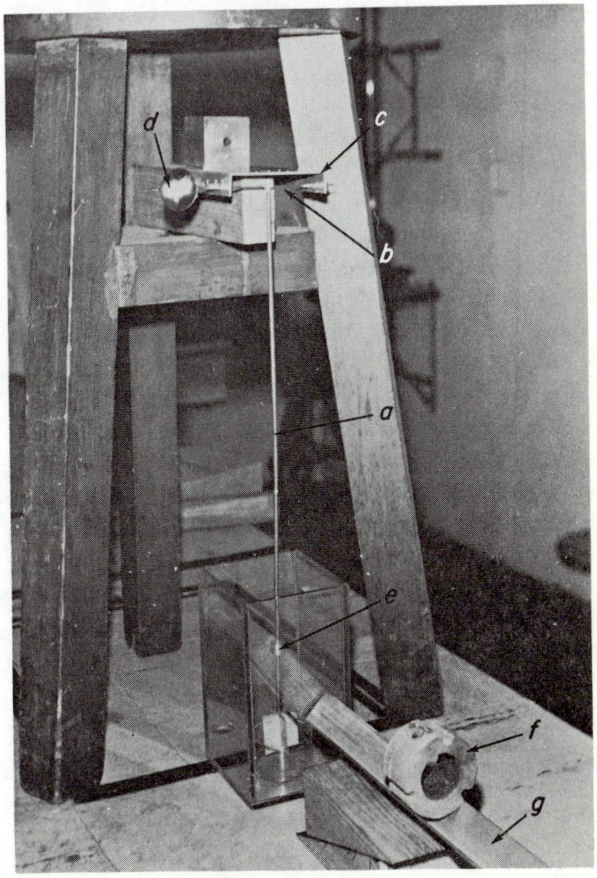

Fig. 32–11

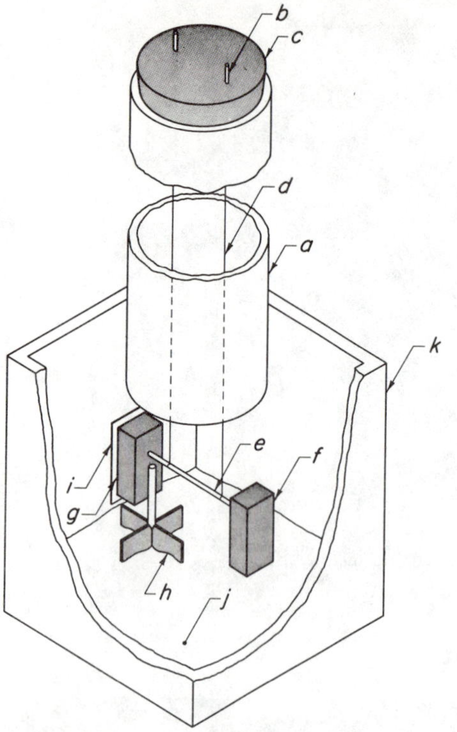

Fig. 32–12

vane h and a mirror i. The vane hangs in a pool of glycerin j. By reflecting a beam of light from the mirror, the deflection of the suspension can be measured. The magnet in the horizontally mounted aluminum angle (see Fig. 32–13) is moved along the normal to the plane of the suspension. The glass box k is used to shield the suspension and hold the glycerin.

The $1/r^7$ dependence can be shown by suspending a piece of soft iron on a long string and measuring the deflection caused by a magnet. A log–log plot of the soft iron deflection (measured by the reflection of a light beam from a mirror attached to it) vs the magnet-iron distance r results in a straight line of slope-7.

The force on a dipole in an inhomogenous field H is the product of the dipole strength M and the space derivative of the field,

$$F = M \frac{\partial H}{\partial R}.$$

The magnetization M is proportional to the field strength H

$$M = \chi H,$$

Fig. 32–13

where χ is the susceptibility of the soft iron; it is independent of H.

Using this, the force becomes

$$F = \chi H \frac{\partial H}{\partial R},$$

$$F = \frac{\chi}{2} \operatorname{grad}_R (H^2).$$

The strength of the field of a dipole is given by

$$H \propto \frac{1}{R^3}.$$

Therefore,

$$F = \frac{\chi}{2} \operatorname{grad}_R \frac{1}{R^6},$$

which can be rewritten as

$$F \propto \frac{1}{R^7}.$$

32–2 Diamagnetism and Paramagnetism

32–2.1

When samples with magnetic properties are placed in a magnetic field, they tend to orient themselves so as to minimize their potential energy. A paramagnetic sample in an inhomogeneous magnetic field will have an induced moment μ parallel to the field direction, and will move toward a region of greater field strength. A diamagnetic sample has the induced moment antiparallel to the field direction, and will move toward a region of weaker field strength.

Figure 32–14 shows a large magnet (about 2 ft long) capable of producing fields of 10^4 G over a few square inches of area when the poles are separated about ½ in. and the applied potential is 36 V. The inhomogeneous magnetic field is produced by tapering one of the pole pieces, while the other pole face is kept flat.

Samples of diamagnetic material such as bismuth or glass (10-mm-diam balls) will move toward the flat face when the magnet is energized, whereas a paramagnetic substance such as aluminum will progress toward the stronger field at the tapered tip. Magnetic properties of fluids can be studied by placing them in thin-walled vials small enough to fit between the pole faces.

32–2.2[2]

A magnetic field strength of 4000–6000 Oe acts on paramagnetic and diamagnetic bodies suspended in a paramagnetic solution. A properly constructed electromagnet with 30,000–40,000 amp-turns will produce the necessary field if the poles are kept close together. This requires, however, that the samples be small and that the electromagnet gap be shown in projection. The paramagnetic body which is used is a small glass tube about 0.3 in. in diam and about 1.2 in. long, filled with a weak solution of ferric chloride (Fe_2Cl_6). The diamagnetic body is a similar tube filled with a strong solution of bismuth nitrate. (Note: both solutions have very strong oxidizing properties and will darken the surface of many metals.) If a sedi-

[2] Translated by K. V. Robinson and H. A. Robinson (Adelphi University) from *Lecture Demonstrations in Physics*, A. B. Mlodzeevskii, ed., M. V. Lomonosov State University, Moscow, U.S.S.R.

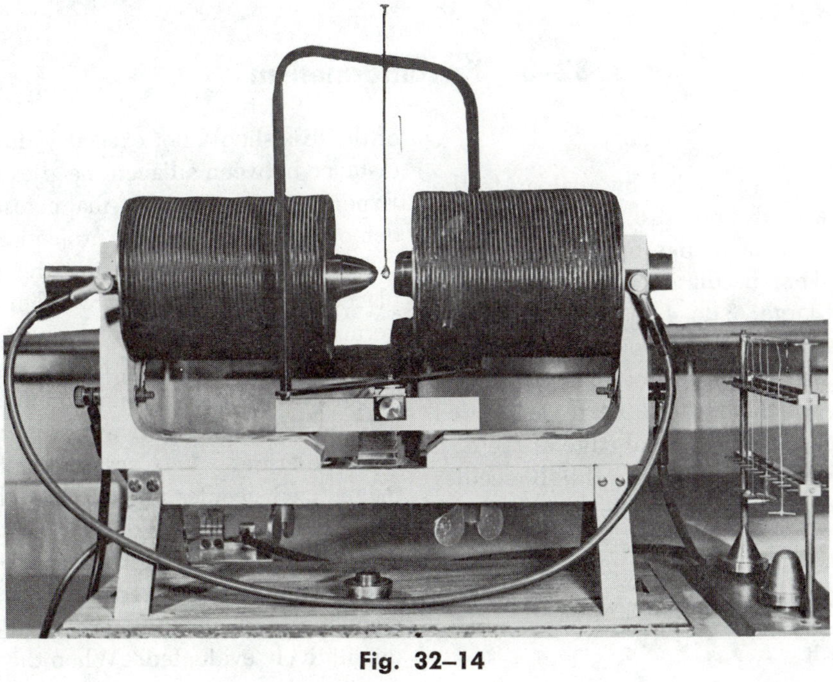

Fig. 32–14

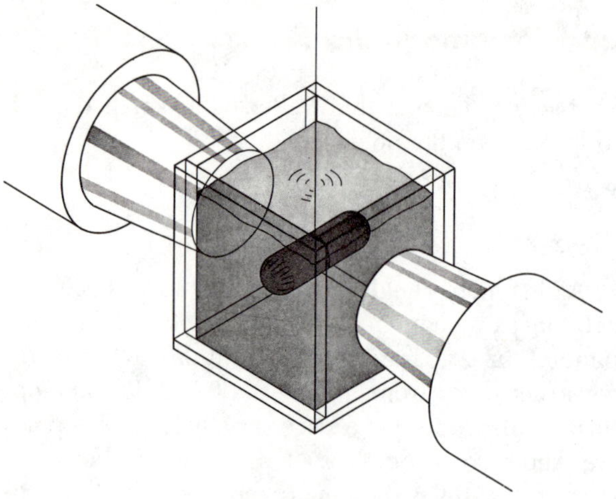

Fig. 32–15

ment appears when crystalline bismuth nitrate is dissolved, a few drops of nitric acid should be added to the solution.

The tube with the ferric chloride (paramagnetic body) is suspended on a thread in the space between the poles of the electromagnet as in Fig. 32–15. The tube in each of these experiments must be centered in the space without touching any part of the apparatus. When the electromagnet is connected to dc, the tube becomes stabilized in a position coinciding with the direction of the magnetic field at the position of strongest field intensity. If the electromagnet is disconnected, the tube returns to its original position. A cubic glass vessel containing a strong solution of ferric chloride is now placed between the poles. The tube is put in the solution, between the poles. When the electromagnet is again connected, the medium (the strong ferric chloride) surrounding the tube will be magnetized more strongly than the substance filling the tube (weak ferric chloride). The tube will therefore be ejected from the strong field region. Thus a weak paramagnetic substance in a strong paramagnetic medium behaves like a diamagnetic body upon magnetization.

To demonstrate the behavior of diamagnetic bodies, the tube with the bismuth nitrate solution is suspended by a thread between the poles of the electromagnet. When a sufficiently strong field is applied, the tube will move to a region of weaker field strength. A rheostat is placed in the circuit and the intensity of the magnetic field between the poles is reduced to such an extent that the tube no longer reacts when the electromagnet is switched on. Now, without changing the current setting, the container of ferric chloride is placed between the poles and the tube is suspended in the solution. When the electromagnet is energized, the medium, which is capable of stronger magnetization, ejects the tube containing the bismuth nitrate from the strong field region.

32–3 Ferromagnetism

32–3.1[(3)]

The magnetoscope (so called by analogy to the electroscope) is a small brass disk with 10–12 steel needles suspended from hooks on its under side (Fig. 32–16). The needles (darning needles) should be fairly large: 3 to 4 in. long and 0.04–0.06 in. in diam at their thickest point. They should be identical in size and made of the same steel. The hooks from which the needles are suspended should be set tangent to the lateral edge of the disk so that the needles can move freely. If 9–10 needles are used to make the magnetoscope, the diameter of the disk should not exceed ½ in. Otherwise, the distance between adjacent needles will be too great. Experiments using the magnetoscope should be demonstrated in shadow projection.

The brass rod supporting the magnetoscope is placed in the holder of a wooden stand. A strong permanent magnet is slowly brought from below to within close range of the points of the needles. The needles will become magnetized in the external field of the magnet and will acquire identical magnetic polarities. As a result, the free ends of the magnetized needles will repel each other, forming a conical bunch. The degree of magnetization can be evaluated by noting the angle of divergence between neighboring needles. The intensity of the magnetic field affecting the needles may thus be qualitatively evaluated. When the permanent mag-

(3) Translated by K. V. Robinson and H. A. Robinson (Adelphi University) from *Lecture Demonstrations in Physics*, A. B. Mlodzeevskii, ed., M. V. Lomonosov State University, Moscow, U.S.S.R.

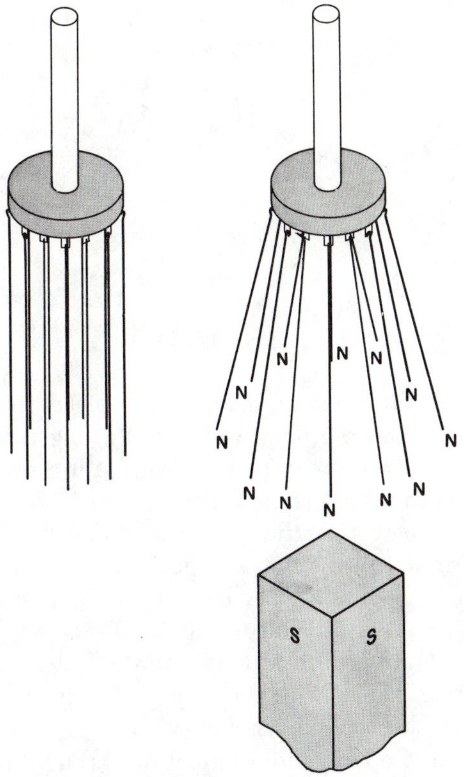

Fig. 32–16

net is removed, the needles will not return to their original vertical position because the steel retains a certain amount of residual magnetism.

The permanent magnet is turned 180 deg and again slowly brought within close range of the needles. Because of the change in the direction of the magnetic field, the needles will initially become demagnetized, losing their residual magnetism so that the angle of deflection between the needles decreases. At a certain distance between the pole of the permanent magnet and the tips of the needles, the needles will assume a completely vertical position. This position of the needles shows the total demagnetization of the steel, and at the given position of the permanent magnet, the average intensity of its external field in the area of the steel needles is approximatly equal to the coercive force of the steel from which the needles are made. As the permanent magnet is brought closer, the steel needles become magnetized in the opposite direction and a repulsive force is again apparent between the ends of the needles.

In another version of this experiment, a permanent magnet is brought very close to the tips of the needles (so close that they almost touch). In this case the attraction between the steel needles and the pole of the magnet will be considerably greater than the repulsive force between the individual needles; hence, the needles will assume a vertical position. If the magnet is moved away from the needles very slowly, the attraction between the tips of the needles and the pole of the permanent magnet is decreased. The steel needles, which have been magnetized to the same degree, now repel each other, forming a conical bundle.

32–3.2[4]

A glass tube 14–16 in. long, with an interior diameter of ½ in., is sealed on one end and filled ⅚ full of magnetically soft, pure iron filings. The tube is then fitted with a cork or stopper which can be pushed completely into the tube to compress the filings. It is best to attach a handle or small copper rod to the cork so that it may be easily moved in and out of the tube.

A tube which has just been filled with filings may be compared to a demagnetized iron rod. If the tube (with the filings compressed) is brought close to a magnetic needle, the needle reacts very feebly, showing that the mass of filings is not magnetized as a whole. With the filings compressed by the cork, as above, the tube is placed inside a dc coil to magnetize the filings. When the tube is again brought near the magnetic needle, the sharp deflections show that the mass of filings now has an effect like that of a solid bar magnet; that is, the filings have acquired an external magnetic field around the tube. When the audience is convinced that the tube of filings acts as a true magnet, the cork may be slightly loosened and the tube shaken vigorously to thoroughly agitate the filings. The filings are compressed again and the tube is brought close to the magnetic needle, producing a very feeble deflection. The audience is shown, then, that as a result of shaking the tube, the filings were stirred and neutralized each other, destroying the external magnetic field.

[4] Translated by K. V. Robinson and H. A. Robinson (Adelphi University) from *Lecture Demonstration in Physics*, A. B. Mlodzeevskii, ed., M. V. Lomonosov State University, Moscow, U.S.S.R.

32–3.3[5]

A simplified model of a magnet may be made by pouring transparent castor oil into a flat, glass container, sprinkling iron filings into the oil, and stirring the two thoroughly. The oil should spread over the bottom of the container in a homogeneous layer no thicker than 2–3 mm. When the container is shown in horizontal projection, the audience will see on the screen only chaotically scattered filings with small, random agglomerations of filings. When these small groups of filings are affected by a magnetic field, however, they reorganize and stretch out to form chains along the field. A viscous castor oil must be used in this demonstration so that the magnetic field will only force the filings to turn in the oil.

32–3.4

The historic experiment by Gilbert consists of aligning a small, soft iron bar with the earth's magnetic field (determined with a compass). The bar is then hit repeatedly with a hammer, and the bar is found to be magnetized in one direction. When the bar is reversed in the earth's field and hit again, the compass indicates that the bar is magnetized in the opposite direction. A bar of permalloy changes polarity without any hammering.

32–3.5

The object of this demonstration is to clinch a simple discussion of theory which predicts that a ring of steel could be "magnetized" and yet show no poles. A ring is made of oil-hardened die stock 3½ in. i.d. x 6 in. o.d. x 1⁄16 in. thick. After the ring is made, it is heated and made glass-hard by quenching in oil. It is magnetized before the demonstration. The ring is tested with iron filings in front of a translucent screen illuminated from behind. It shows no poles, but it is suggested that the ring may be magnetized circularly, in which case there would be no poles, because the lines of force form the closed loops inside the ring. The ring is then broken into two parts by a sharp blow with a non-magnetic mallet. The pieces are again tested with iron filings and show strong poles. (A smaller version may be arranged with a small ring placed in a projection lantern.)

[5] Translated by K. V. Robinson and H. A. Robinson (Adelphi University) from *Lecture Demonstrations in Physics*, A. B. Mlodzeevskii, ed., M. V. Lomonosov State University, Moscow, U.S.S.R.

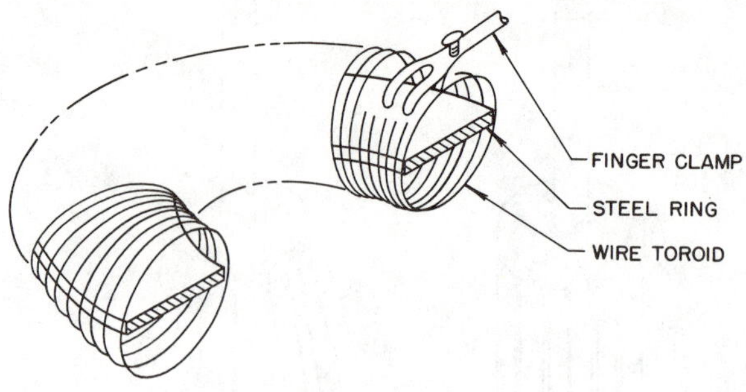

Fig. 32–17

To magnetize the ring, the magnetic field of a current in a long straight wire is employed. Since a current that is large enough for this ring (~1000 A) is difficult to obtain, the current is run up several wires in a bundle through the center of the ring. Far above the ring each wire runs well out to the side, down below the ring, and in to become the next wire running up through the ring. This array of wire is, therefore, like a large toroidal winding, but all parts of it except the sections that run up through the center of the wire are kept far away from the ring and are distributed fairly symmetrically around it. An alternate method uses a toroidal winding as illustrated in Fig. 32–17. A 120-V dc supply is applied to the coil for a very short time. It is not necessary to maintain the current for long, and the insulation burns off dramatically as shown by the photographs (a) through (d) in Fig. 32–18. The coil must be wound by hand before use and cut away with shears after the current has been passed. A considerable number of rings can be threaded and magnetized at the same time.

32–3.6

A short horizontal permanent magnet *a* (Fig. 32–19) is suspended by a thread to which a mirror *b* is attached. The magnet turns inside a small brass-sided pot *c* with an aluminum bottom. Eddy currents set up in the pot provide damping so that the magnet points stably in a North–South direction. The pot is set inside a square four-sided plate-glass container *d* to minimize wind disturb-

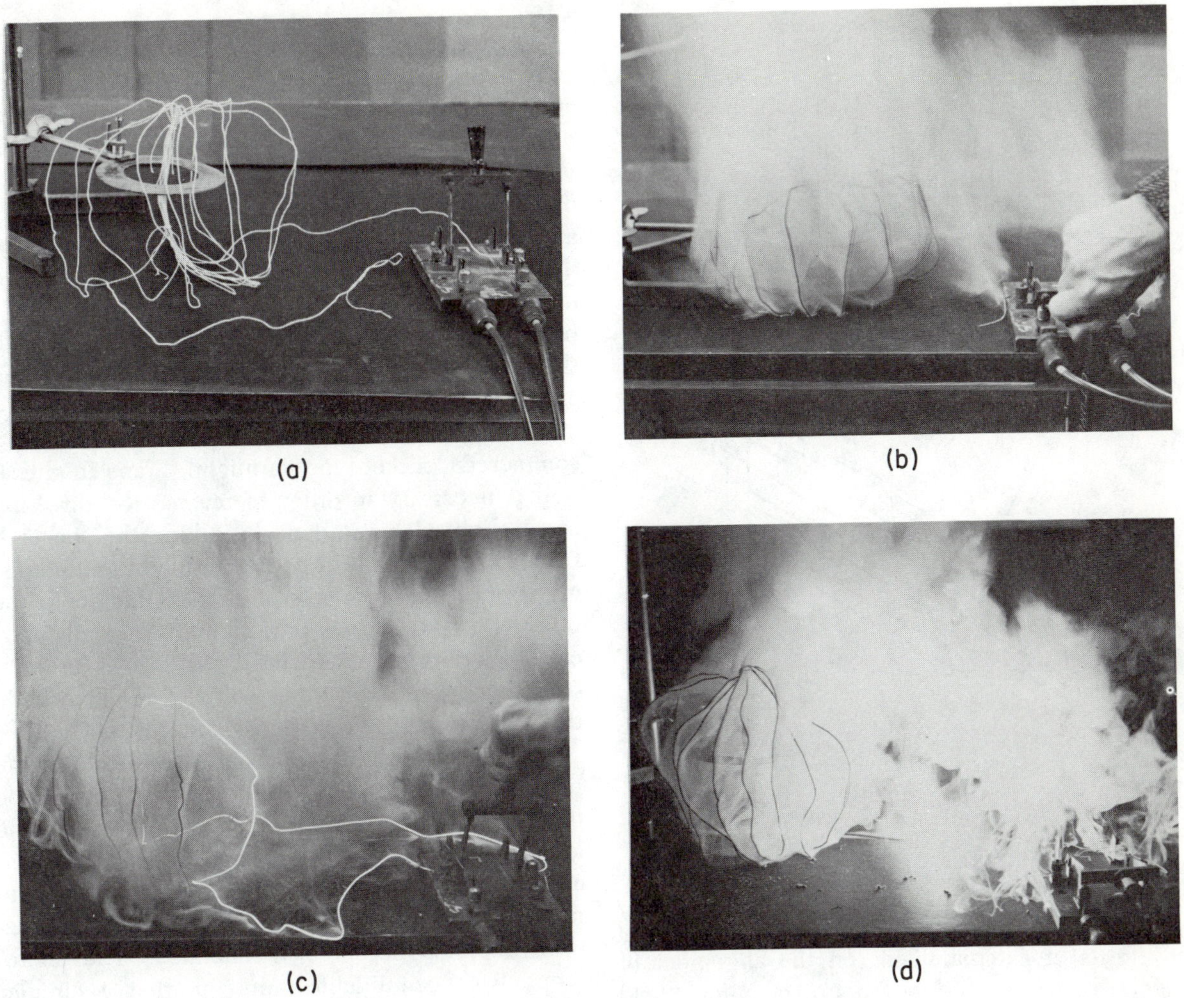

Fig. 32–18

ances. When a second short magnet (not shown in Fig. 32–19), pointing in an East–West direction, is placed below the first magnet, the latter will show a deflection. The deflection vanishes when a 100-cm-long, 5-cm-diam, soft iron tube *e* is pushed over the second magnet, due to the shielding properties of the tube. A single pole of a very long permanent magnet cannot be so shielded, as pushing the iron tube over one end of a long magnetized needle will show. To demonstrate the effects of shielding magnetic fields produced by current-carrying conductors, a long wooden lath *f*, capable of being fitted into the iron tube, is now prepared by having a narrow loop conductor *g* mounted on one end (the loop should be less than 100 cm long) and a straight conductor *h* mounted along its whole length. The leads to the loop conductor should be twisted. The straight and looped conductors can be alternately connected through rheostats to electrical supply lines, taking care to keep such circuits as far away from the measuring magnet as possible. The lath is now placed on a non-ferrous angle strip *i* (to enable easy shifting of the iron tube) and placed under the measuring device. When the loop circuit carries a current, the measuring magnet will show a deflection which drops markedly (but not entirely) when the iron tube is shifted over the loop for shielding purposes. When the straight wire carries the current, the iron tube is far less successful as a shielding device. The deflection may in fact increase somewhat due to the partial screening of the earth's field by the tube. Figure 32–20 is a photograph of the experimental arrangement.

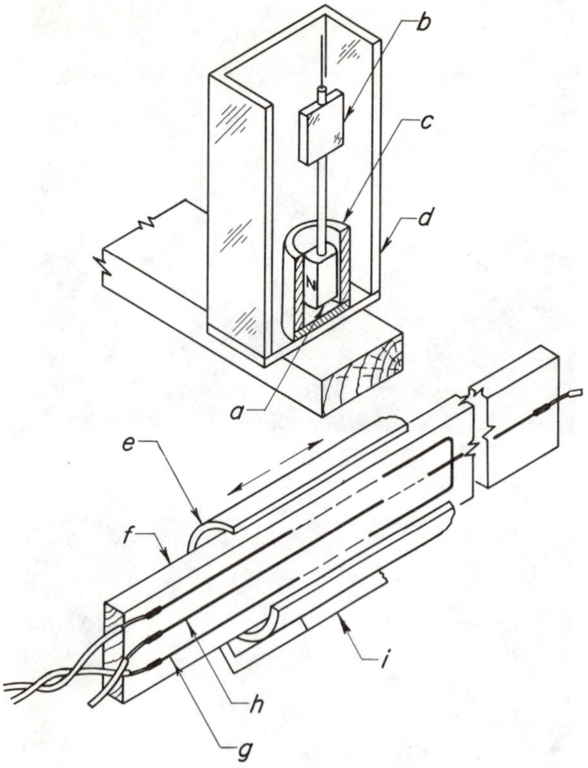

Fig. 32–19

32–3.7

The apparatus shown in Fig. 32–21 demonstrates the effects of external fields on the alignment of magnet "domains." When the strong uniform field of an electromagnet is applied, the domains all align. When the relatively weaker field from a bar magnet is applied, the order is disrupted and the array is left in random order.

The domains are magnetized segments of blued spring steel cut from 0.015-in.-thick x 1/4-in.-wide strip stock as shown in Fig. 32–22. The ends of each of the 72 domains are pointed, and simple brass pivots are pressed through the centers. The domains swing on the points of common pins which are pressed through holes in a 20 x 26 x 1/2-in. clear plastic board and arranged in a 9 x 8 coordinate grid of 2 1/8-in. squares.

For viewing with an angled mirror, like poles of each magnet are painted white, while the opposite poles are painted black. To facilitate use of shadow projection, like poles of each magnet have holes drilled in them. The opposite poles are left solid for differentiation.

32–3.8[6]

A completely self-contained demonstration device, consisting of a light source, a mounted specimen and a polarizing microscope, is shown in Fig. 32–23. This unit is used to demonstrate the domain structure in a thin ferromagnet garnet and is helpful in connection with discussions of domains, domain wall, domain configuration, and magnetostatic energy. The domains are made visible through the polarizing effect associated with the orientation of the domains relative to the line of sight.

The light source *a* in Fig. 32–23 is made from a commercial pocket-type flashlight. Some modification is necessary in order to reverse the direction of the light. This is done by taking the socket *b* from a second identical pocket flashlight and soldering it onto the brass crosspiece *c*. Soldered to the other end of this brass piece is the threaded top *d* of the battery case. An insulated wire is brought from the center of this threaded cap through a slot in the crosspiece to the center contact of the bulb.[7]

The specimen is then affixed over a hole in the blackened metal plate *e*. The pocket microscope used is not the cheapest on the market, but the less expensive models are not suitable because their apertures are very small and their focusing mechanism is usually inadequate. The focusing mechanism depicted here is self-contained. The microscope has been modified only in that a circular disk of Polaroid material *f* has been placed below the specimen, and a second circular disk *g* has been placed beneath the eyepiece retaining ring. The microscope as a whole fits snugly into a tube *h* which is in turn soldered to the battery box. It is essential for the purposes of the demonstration that it be possible to rotate either the analyzer or the polarizer. In the instance of the microscope shown in the drawing, this was accomplished by rotating the microscope body within the bracket by which it is held to the battery case. In another version, the polarizer between the light and the sample was mounted within an annular ring which was free to rotate within the bracket. This could be turned through the requisite angle by means of

[6] J. F. Dillon and H. E. Earl, *Am. J. Phys.*, **27**, 201 (1959).

[7] For best results the bulb used is of the prefocused variety.

Fig. 32–20

a small screw which protruded through a slot. A microscope with a magnification of perhaps 40× to 60× seems most appropriate. The most desirable magnification depends on the scale of the domain structure, and this in turn depends on the thickness of the crystal as well as its magnetization and its mechanical preparation. Specimens of about a thousandth of an inch in thickness are the most appropriate. In addition to showing the existence of domains, the apparatus can easily show the effects of temperature and a magnetic field on the sample.

Single crystal garnets with the formula $M_3Fe_2(FeO_4)_3$, where M is a rare earth with an atomic number of 62 or more, give good results. The single crystal must be prepared, polished, and handled very carefully; and because of the difficulty in making them, it is advised that one buy commercially prepared crystals.[8]

[8] $Gd_3Fe_2(FeO_4)_3$ crystals can be obtained from Semi-Elements Inc., Saxonburg, Pennsylvania, or Linde Co., Division of UCC, Speedway Laboratories, Indianapolis, Indiana.

The plane-polarized light, passing through the crystal, undergoes a nonreciprocal rotation. That is, if the magnetization is parallel to the line of sight, the plane of polarization is rotated some angle θ. If M_s is antiparallel to the line of sight, the angle is $-\theta$, and lastly if M_s is perpendicular to the line of sight, there is no rotation. The angle

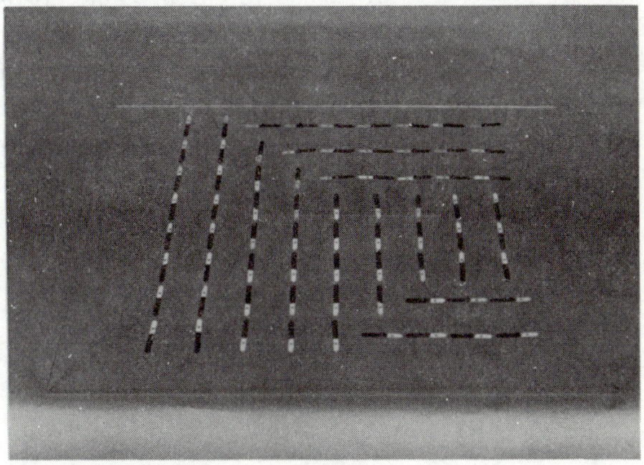

Fig. 32–21

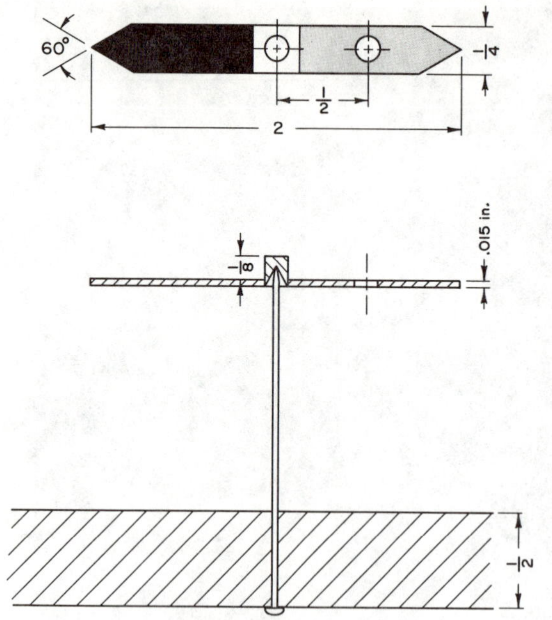

Fig. 32–22

of rotation varies as the cosine of the angle between the line of sight and the magnetization. The rotation is, of course, proportional to the thickness, and it varies with color.

Figure 32–24 illustrates the observation of magnetic domains in one of these transparent crystals. When the crystal is introduced into the optical path between the polarizers, we must consider separately the light through each of the three domains, *A*, *B*, and *C*. The magnetization vectors in these, M_A, M_B, and M_C, are, respectively, antiparallel, parallel, and perpendicular to the optical path. For light which passed through domain *B*, the plane of polarization of light of some particular wavelength will be rotated through an angle θ. The corresponding angle for domain *A* will be $-\theta$, and for domain *C* it will be zero. Thus depending on whether we set the analyzer at θ, 0, or $-\theta$, we will see the domain structure in one of three states of

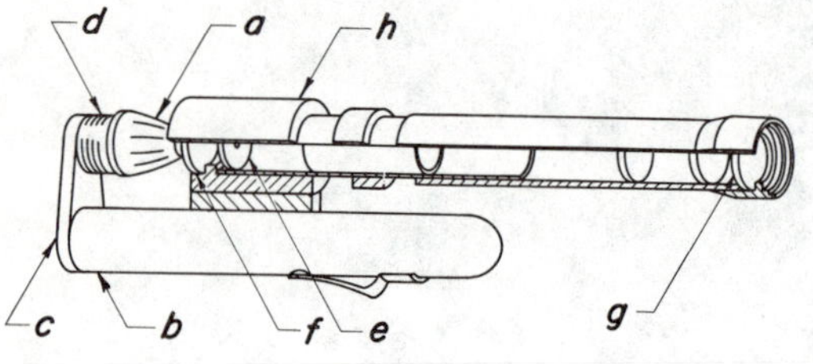

Fig. 32–23

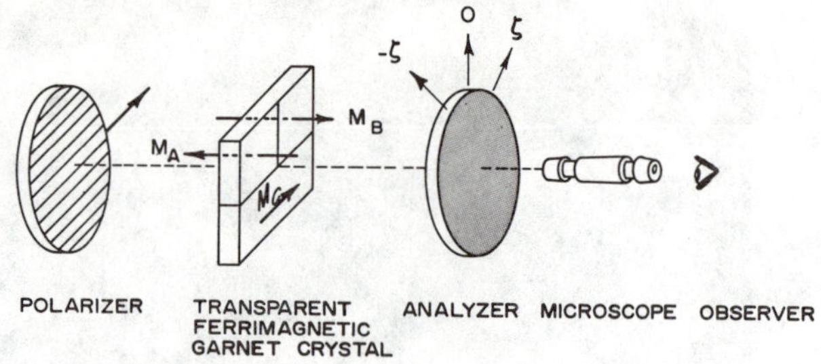

Fig. 32–24

contrast. These are indicated in Fig. 32–25 as (a), (b), and (c), respectively. Between domain *A* and *B* there is a domain wall in which a thin sheet of magnetization lies perpendicular to the line of sight. This sheet will not rotate the plane of polarization. Thus, in *b*, light passing through *A* and *B* will be rotated away from the original extinction position, but we see the domain wall as a black line since light passing through it is not rotated.

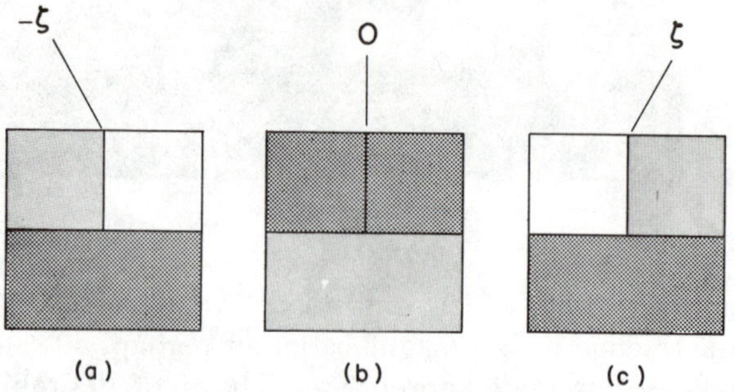

Fig. 32–25

32–3.9

Weiss domains are regions of a crystal which exhibit spontaneous, complete polarization of their magnetic moments. They can be shown by the apparatus pictured in Fig. 32–26. The crystal structure determines possible orientations, and the domain size decreases with increasing temperature until the domains finally disappear at the Curie point temperature.

A transparent chip of Gadolinium–Iron–Garnet[9] ($Gd_3Fe_2(FeO_4)_3$) crystal with a surface area of approximately 2 sq mm is mounted in a polarizing microscope *a* for optical display. When the crystal

[9] Can be obtained from Semi-Elements Inc., Saxonburg, Pennsylvania, or Linde Co., Division of UCC, Speedway Laboratories, Indianapolis, Indiana.

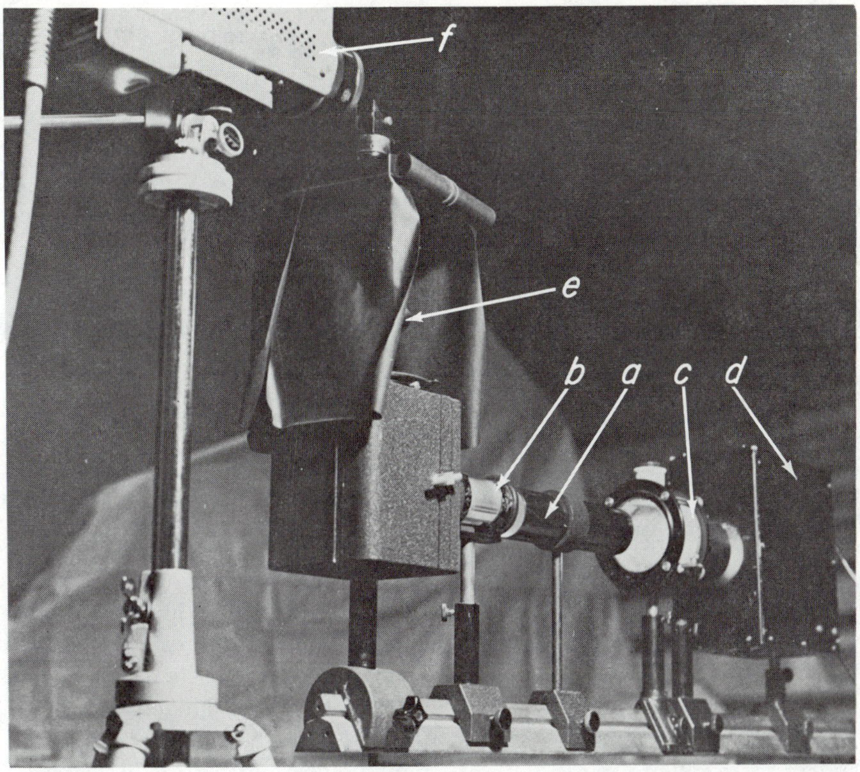

Fig. 32–26

is initially magnetized, a rotation of the plane of polarization occurs. The brightness may now be adjusted by the polarizers *b* (Nicol prisms) so that domains magnetized upward become bright and those magnetized downward become dark. In the unmagnetized initial condition, the bright and dark domains (Fig. 32–27) have the same total area. When an external vertical field is applied, one set of bands grows at the expense of the others.

The effect of temperature on the domains can be shown by taking the heat filter *c* ($CuSO_4$ solution) off the illuminating lamp *d* (xenon, 350 W) and letting the crystal heat up until the Curie point is reached and the bands disappear. If the crystal is allowed to cool by turning off the light, the domains will reform in a different pattern and can then be viewed by turning the light back on and inserting the heat shield. If the heating is discontinued before the Curie point is reached, the pattern will be preserved although the domains will become smaller and smaller.

The crystal image is magnified with a microscope *e* (150×). The image is projected directly onto the photo-cathode of a vidicon TV tube. For specific details on the theory, construction details, and crystal preparation, see the article by Dillon and Earl.[10]

32–3.10

The Barkhausen effect is audibly demonstrated with the apparatus in Fig. 32–28. Two permanent bar magnets *m* (3/8 in. diam x 5 in.) are mounted north pole to south pole on a horizontal brass bar *b*, which can revolve about a central vertical axis. Attached to the top of the post on which the horizontal bar revolves is a small, cubical Lucite box containing a coil *c* of about 5000 turns of fine enameled copper wire. All sides of the box except the front are painted white. Various core materials can be inserted in the coil. The ends of the winding are connected to an audio amplifier and loudspeaker. When the magnet bar is spun by hand, the relative motion between the moving magnets and the fixed coil induces an emf in the coil, intermittently producing an audible sound from the

[10] J. F. Dillon and H. E. Earl, *Am. J. Phys.*, **27**, 201 (1959).

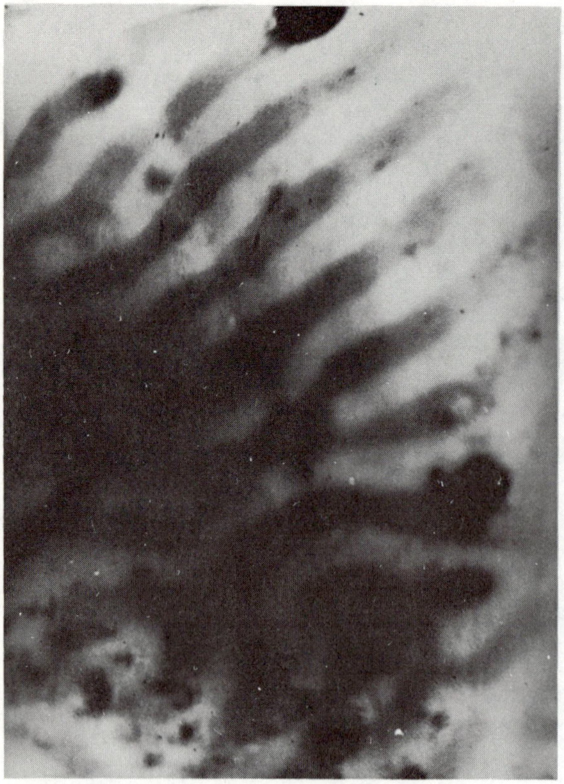

Fig. 32–27

speaker by virtue of the sudden alignment and realignment of the magnetic domains within the core. Sound differences can be heard when different magnetic materials are used for cores, and no sound at all is heard when an aluminum core is used.

32–3.11[11]

A variation of the demonstration of the Barkhausen effect may be carried out using an iron-nickel alloy (15 percent Ni and 85 percent Fe). When a sample of this alloy is strained by stretching, the hysteresis loop becomes rectangular. When the alloy is then remagnetized, irreversible processes of slip at domain boundaries (Barkhausen increments) simultaneously encompass the total volume of the sample.

A brass bracket *c* has a pickup coil *b* mounted

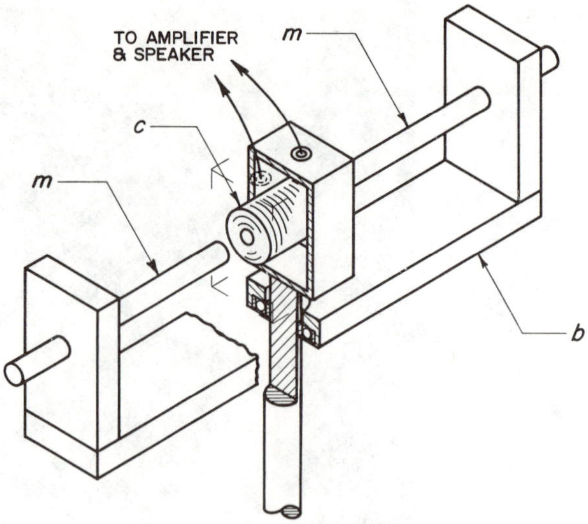

Fig. 32–28

on it as shown in Fig. 32–29. The coil consists of about 10,000 turns of 0.004–0.005-in. copper wire and is set so that its axis coincides with the axis of the two wire holders mounted in the bracket. A wire *a* of the iron-nickel alloy is attached to the left holder which is held in place by a setscrew. The wire is passed through the coil and attached to the right holder which has a threaded rod. The rod is keyed, as shown, so that when the nut is tightened only pure stretch can result in the wire. By turning the nut, one can stretch the wire so tightly that the hysteresis loop of this alloy will become rectangular. The two leads from the pickup coil are connected to a two or three tube amplifier which drives a large loudspeaker. The resistance of the coil must match the impedance of the amplifier's input transformer.

A large permanent magnet is slowly brought

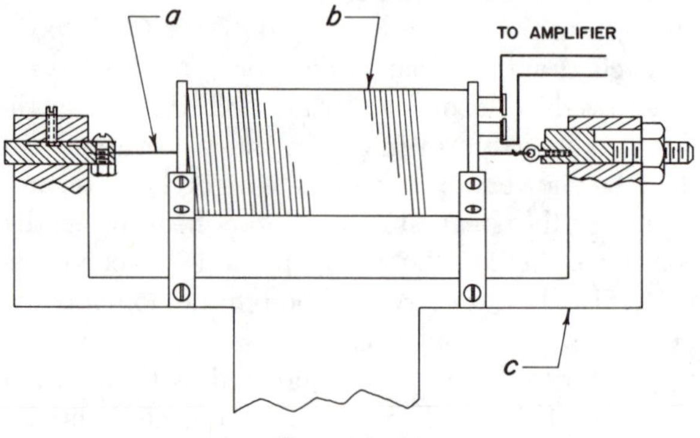

Fig. 32–29

[11] Translated by K. V. Robinson and H. A. Robinson (Adelphi University) from *Lecture Demonstrations in Physics*, A. B. Mlodzeevskii, ed., M. V. Lomonosov State University, Moscow, U.S.S.R.

close to the coil. (The distance between the poles, whether on a horseshoe or bar magnet, should be approximately equal to the length of the wire.) When the magnet is a certain distance from the wire, a sudden, simultaneous magnetization of the total mass of the ferromagnetic wire takes place, resulting in a sharp clear sound in the loudspeaker. If the magnet is brought closer to the wire, no further noise will result showing that the magnetization of the wire no longer changes. The magnet is next withdrawn, turned 180 deg, and slowly brought close to the wire once more. Again, at a similar distance, the sharp clear noise will be produced, resulting from the sudden, simultaneous remagnetization of the entire wire. This effect may also be produced by holding the magnet at the distance which produces the magnetization and slowly turning the magnet through 180 deg. At a precise moment in the turn, the remagnetization and resulting noise will suddenly take place.

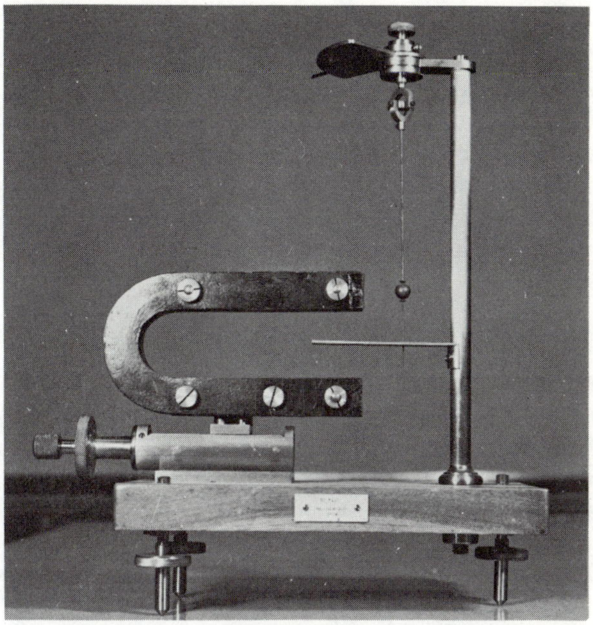

Fig. 32–30

32–3.12

A small sphere of Pyrothit (a mineral of composition between FeS and Fe_6S_7) is suspended on a thin brass rod from a sensitive universal joint which may be turned about a vertical axis with the knob and indicator at the top of the stand (Fig. 32–30). A large magnet is placed so that one of its poles lies in the same horizontal plane as the sphere. As the knob is turned, there will be certain orientations where the sphere will not be attracted to the magnet and other orientations where the sphere will be attracted to the magnet, showing paramagnetic and ferromagnetic behavior respectively. A small pointer suspended from the bottom of the sphere serves as a position marker.

32–3.13[12]

A search coil connected to a ballistic galvanometer is placed over a sample, which is then placed between the poles of an electromagnet. The electromagnet *a* (Fig. 32–31) should be capable of producing a field strength of 800–1000 Oe between the poles (in a space of 250–500 cc). The direct current in the windings of the electromagnet must be varied over a wide range from very low currents (fractions of an ampere) to high currents (several amperes). Since rheostats with such a range are not generally available, two or three rheostats with different resistances are connected in series.

The sample *b* must be a completely demagnetized piece of steel or other material, long enough to fit between the widely separated poles of the electromagnet. The search coil *c* which is slipped over the sample must be constructed on a rigid in-

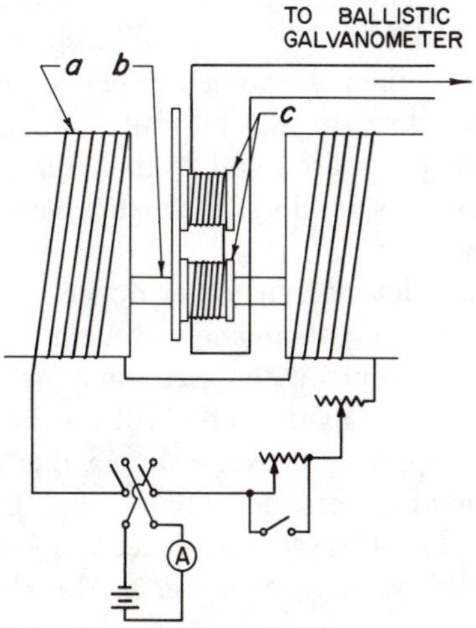

Fig. 32–31

[12] Translated by K. V. Robinson and H. A. Robinson (Adelphi University) from *Lecture Demonstrations in Physics,* A. B. Mlodzeevskii, ed., M. V. Lomonosov State University, Moscow, U.S.S.R.

sulating frame whose cross section is slightly larger than that of the sample. The length of the coil should be about one-fifth the length of the sample. For the greatest sensitivity, the resistance of the windings on the coil should be equal to the internal resistance of the galvanometer to which it is connected. The galvanometer must be such that its deflections can be shown or projected for a large group.

The best results are obtained if the search coil actually consists of two identical coils mounted rigidly on a common base and connected in series opposed. When the sample is placed in one of the two coils and the entire apparatus is set in the homogeneous portion of the magnetic field, the induced emf will be proportional, not to the total change in the magnetic induction B, but only to the change in the intensity of magnetization I of the sample. That is, with the arrangement of coils described, the magnitude of the magnetic field strength H which acts on the sample is taken into account as follows:

$$\text{Galvanometer deflection} = \underbrace{(H + 4\pi I)\,S_n}_{\text{Flux through coil with sample}} + \underbrace{(-HS_n)}_{\text{Flux through coil with no sample}}$$

With the apparatus set up as shown in Fig. 32–31, the direct current is switched into the windings of the electromagnet, thus magnetizing the sample and increasing its inductance. The emf generated in the double search coil will result from the change in inductance, but it will be proportional to the intensity of magnetization I, so that the galvanometer deflection will also be proportional to I. If the galvanometer deflections are calibrated according to the current induced in the windings of the search coil c, then the galvanometer becomes essentially a fluxmeter, and the reading is equal to I.

When the first galvanometer deflection has been recorded and the galvanometer returns to the zero position, the circuit of the electromagnet windings is opened. The galvanometer will immediately deflect in the opposite direction to a value not equal to the value of the first deflection (when the current was applied). This second deflection corresponds to the residual magnetization I_r in the sample when the field is removed. Next, the current in the windings is reversed, using the reversing switch, but is maintained at the same level as before. The galvanometer deflection in this case corresponds to the magnetization of the sample in the opposite direction. Finally, the circuit is opened and the deflection corresponding to the residual magnetization (in the opposite direction) is recorded.

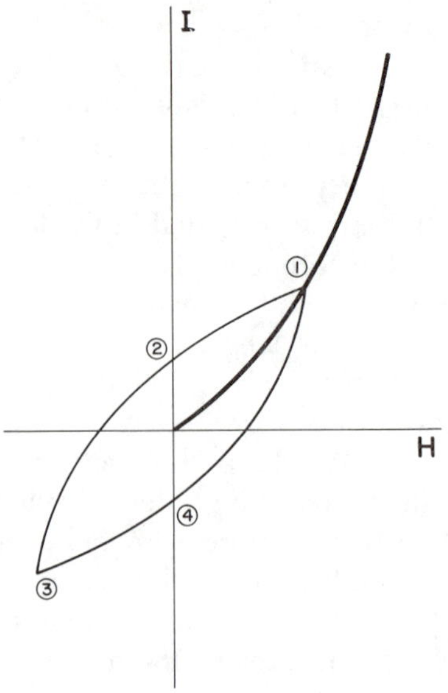

Fig. 32–32

The process of switching the current in the electromagnet and recording the deflection of the galvanometer is actually the process of changing the magnetic field from H to $-H$ and recording the resulting magnetization. The sample is brought from 0 (demagnetized) to point 1 (Fig. 32–32) or I, by field H_1. Point 2 is I_r; point 3 is $-I$, produced by field $-H_1$, and point 4 is $-I_r$. The plot of the points indicates the hysteresis curve for the initial field strength. The process of determining points should be continued at increasing field strengths until the saturation point (I_s) is reached, where further increases in the intensity of the magnetic field produce practically no change in the magnetization. With a few such hysteresis curves plotted, it is possible to plot the magnetization curve $I = f(H)$ as in Fig. 32–32 and to indicate I_s and I_r for the sample.

32–3.14

Hysteresis in a magnetic sample is demonstrated by tracing the hysteresis loop on the screen of a modified television tube.[13] The apparatus consists of an old television tube with its electron-gun circuitry. Only the electron gun is employed; the yoke coils are removed. The arrangement is illustrated schematically in Fig. 32–33. Two identical magnetizing coils *A* and *B* are mounted horizontally, one on each side of the neck of the tube, and another pair *C* and *D* are mounted vertically above and below the neck.

The magnetization of the sample is shown by vertical deflection of the electron beam by the sample's own magnetic field. The sample, a thin rod of iron or steel, is placed in magnetizing coil *A*. The opposite coil *B* carries the same current in the reverse direction, for compensation, so that the field due to currents in the two coils alone causes no deflection. The pair of coils *C* and *D*, mounted above and below the tube, deflect the electron beam horizontally when current passes through them. The current through *C* and *D* is the magnetizing current through *A* (or a fraction of it, in phase); hence, the horizontal deflection is proportional to the magnetizing field.

When an alternating source sends current through *A*, *B*, *C*, *D* in series, the beam traces a hysteresis loop on the screen. When a dc supply is used, with a soft iron sample, the beam traces the initial magnetization curve from zero to saturation as the current is increased by means of an external rheostat.

The magnetizing coils *A* and *B* each consist of eight layers of 120 turns each, wound on a fiber tube 6 in. long x 1 in. in diam. The current-sample coils *C* and *D* each consist of ten layers of 70 turns each, in a coil of about 2¾ in. i.d., about 4 in. o.d. The soft iron sample is a small bundle of strips of transformer stamping, each about 6 in. long. The steel sample is a triangular or a rat-tail file.

The coils *A* and *B* are placed on a shelf so that they can be positioned with respect to the neck of the tube. They are placed close to the neck, and one is moved until a large current through both (empty) gives no vertical deflection to the beam. The coils *C* and *D* are placed so that a large alternating current through both coils spreads the spot to a fairly straight horizontal line on the screen.

[13] This method is not so neat as the usual one with an integrating circuit, but the principle is simpler, and easier for beginners to understand. Also, this scheme can be used with dc to plot the initial magnetization curve.

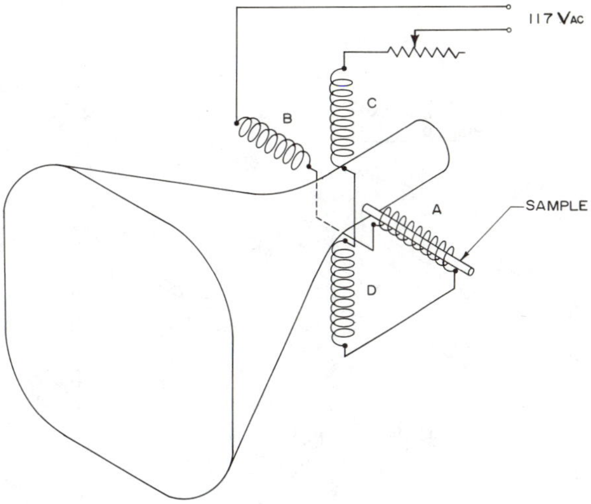

Fig. 32–33

32–3.15[14]

A simple and relatively quick way to demonstrate hysteresis in a commercially important sample of iron is to use a dc generator. The current for the stator of the generator is supplied by a constant voltage source in series with a rheostat. A reversing switch allows the direction of the current in the stator to be changed. The current *I* is measused by means of a projection ammeter. An ac motor turns the rotor of the dc motor at a constant rate. A projection voltmeter, connected across the rotor, displays the output voltage *V*. The curve of *V* vs *I* is proportional to the normally obtained *B* vs *H* curve.

32–3.16[15]

The apparatus shown in Fig. 32–34 can measure magnetic flux simply by determining the induced emf in conductors moving in the field. A small dc series motor[16] *n*, used as a generator, with its field coil assembly removed, is placed between two coils[17] *f*. It is turned by a constant speed driving

[14] L. Grillet and M.-T. Grillet-Amy, University of Rennes, France.

[15] R. Resnick and H. F. Meiners, *Demonstration and Laboratory Apparatus Report of the 1960 Summer Visiting Professor Workshop at Rensselaer Polytechnic Institute.*

[16] Type C-2A-1B, KS5829–0; John Oster Mfg. Co., Ft. Lauderdale, Florida; 27-V, 7-A used, although any small dc motor can be used.

[17] 1200 turns of No. 18 enameled copper wire on each, approximately 3100 feet required; wrap with electrical tape.

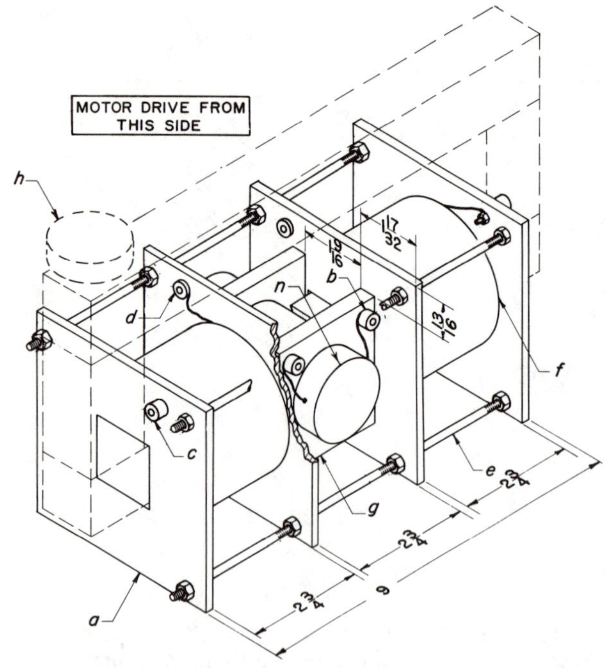

Fig. 32–34

motor (not shown). When current is applied to the coils through the standard binding posts *c* and the telephone type jacks *d*, the strength of the resultant magnetic field is measured by noting the voltage induced in the generator across the universal binding posts *b*. In the absence of iron, the magnetic field is found to be proportional to the applied current.

The four framing pieces *a* are Masonite 4⅝ in. square by 3⁄16 in. thick with a 1 9⁄16 in. sq hole in the center. The framing rods *e* are brass, No. 10–24 N.C. threaded. The generator mounting *g* is 2¾ x 3 x ⅜-in. Lucite.

An iron path (*j*, *k*, and *m*) is now formed which is closed except for the small space occupied by the generator (Fig. 32–35). The iron is 1½ in. square and is held together by ⅜–16 N.C. socket head cap screws, 2 in. long; except for one corner, which is tapped with a ⅜–16-in. N.C. thread and held together by the brass screw *h*. Starting with unmagnetized iron, the magnetization can be traced from zero to saturation. The current can be decreased or reversed.

The magnetization can be carried out slowly; it can be stopped at will, and the significance of any point discussed. It can be shown that to demagnetize the iron, one must form ever-decreasing cycles

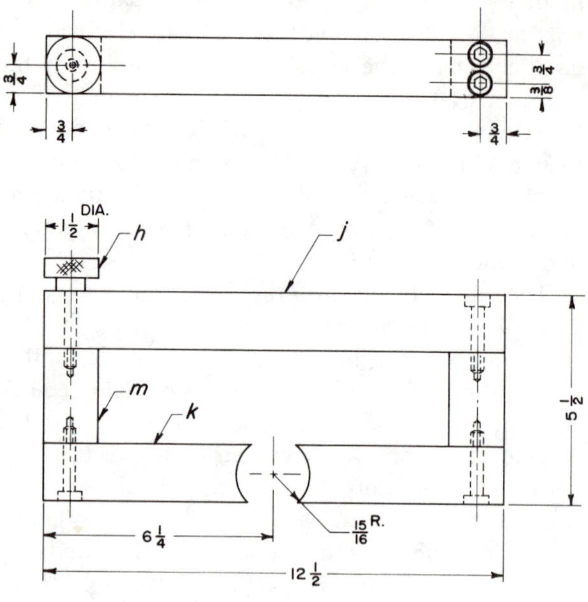

Fig. 32–35

and that only in this way can a condition be reached in which the generator detects no flux.

32–3.17[18]

The circuit shown in Fig. 32–36 makes use of readily available equipment to show a hysteresis

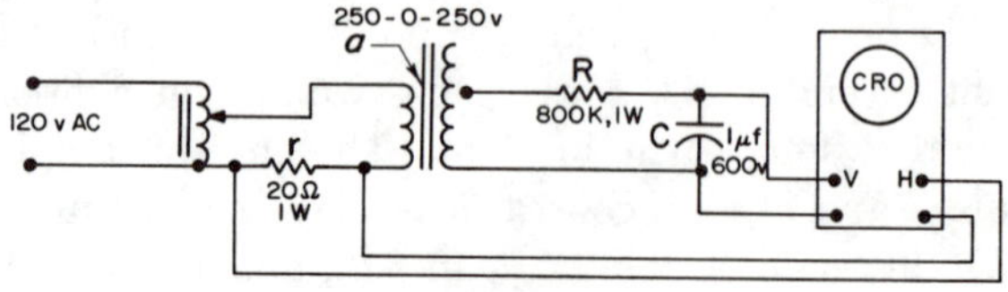

Fig. 32–36

curve on a cathode-ray oscilloscope. The 20-Ω resistor *r* should be noninductive and capable of carrying the current in the transformer *a*. A 1-W radio-type composition resistor proves satisfactory. One side of the center-tapped high-voltage secondary gives an ample output. The *RC* circuit makes the potential across the capacitor proportional to the *B*-field of the transformer core. This output is applied to the vertical deflection amplifier of the oscilloscope. The voltage across the resistance is

[18] R. Resnick and H. F. Meiners, *Demonstration and Laboratory Apparatus Report of the 1960 Summer Visiting Professor Workshop at Rensselaer Polytechnic Institute.*

Fig. 32–37

proportional to the current in the primary and corresponds to the magnetic field strength H. When this is applied to the horizontal deflection amplifier of the oscilloscope, the familiar B–H curve shown in Fig. 32–37 (a) results. If the display appears as in Fig. 32–37 (b), the oscilloscope leads should be reversed.

The time constant RC of the circuit should be large compared to one cycle of the alternating current. The potential difference across the capacitor must be small compared to that across R. Therefore, the reactance $1/2\pi fC$ must be large compared to R. The secondary must supply a potential difference sufficiently high so that the potential difference across C is enough to operate the CRO. The circuit in Fig. 32–36 produces about 0.3 V.

T. B. Brown's[(19)] use of a coil instead of a transformer was tried and found successful. The circuit shown in Fig. 32–36 enables the teacher to explain easily what goes on in the primary (magnetization of the core) and to explain the induced emf in the secondary ($V = k(dB/dt)$), as given by Faraday's law.

If the circuit is modified as shown in Fig. 32–38,

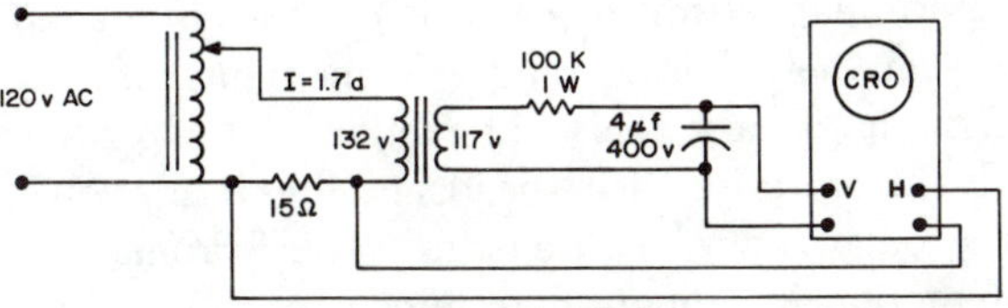

Fig. 32–38

the properties of various core materials can be studied. Figure 32–39 is a diagram of the dissectable transformer used with this circuit. Cores and coil turns which were used are shown in Fig. 32–40. Characteristics of the coils are as follows:

	Primary	*Secondary*
Turns	3700	3700
Wire gauge	No. 22	No. 36
Resistance	28 Ω	1000 Ω

(19) T. B. Brown, *Electronics* (John Wiley & Sons, Inc., New York, London, Sydney, 1954), Chapter 11.

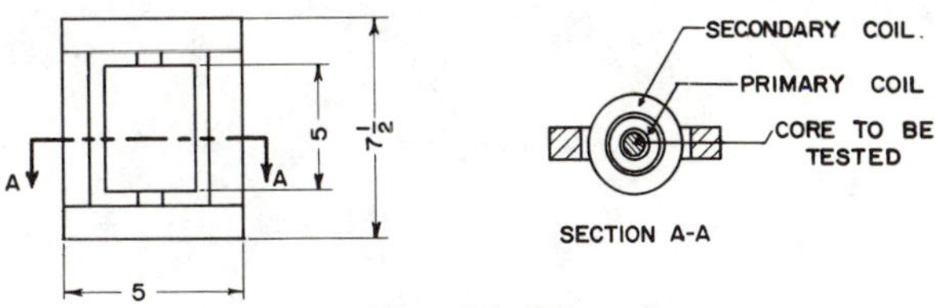

Fig. 32–39

If solid iron and steel rods (¼–1 in. diam) are used, the hysteresis curves will be distorted from the classical shape, probably due to eddy currents. A satisfactory core can be made from 17 hacksaw blades taped together. Larger cores can not be saturated without resorting to currents higher than those available with this unit.

32–3.18

A ferromagnetic substance is one having the following two properties which set it apart from other substances: first, the permeability is very high; and second, a net magnetization remains after an external field is removed, and the familiar hysteresis loop occurs in the plot of B vs H. The net magnetization results because some of the domain wall motion that occurs as the external field H is increased is irreversible. The existence of these spontaneously magnetized domains is explained as arising from exchange energy, i.e., a change in the electrostatic energy caused by the Pauli exclusion principle, between electrons in a partially filled energy band. Distortion of the wave functions by this exchange interaction results for certain materials; namely, those which are ferromagnetic, in a lower energy in the magnetized state than in the unmagnetized state. However, the energy difference between the magnetized and unmagnetized states is characteristically only a fraction of an electron-volt and, as one might expect, increasing the temperature will destroy the spontaneous mag-

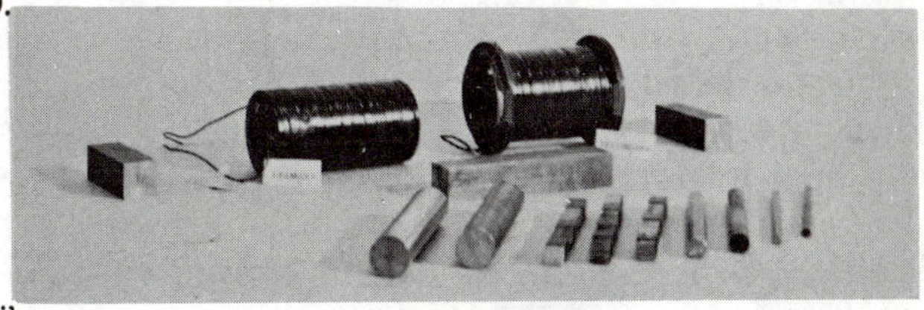

Fig. 32–40

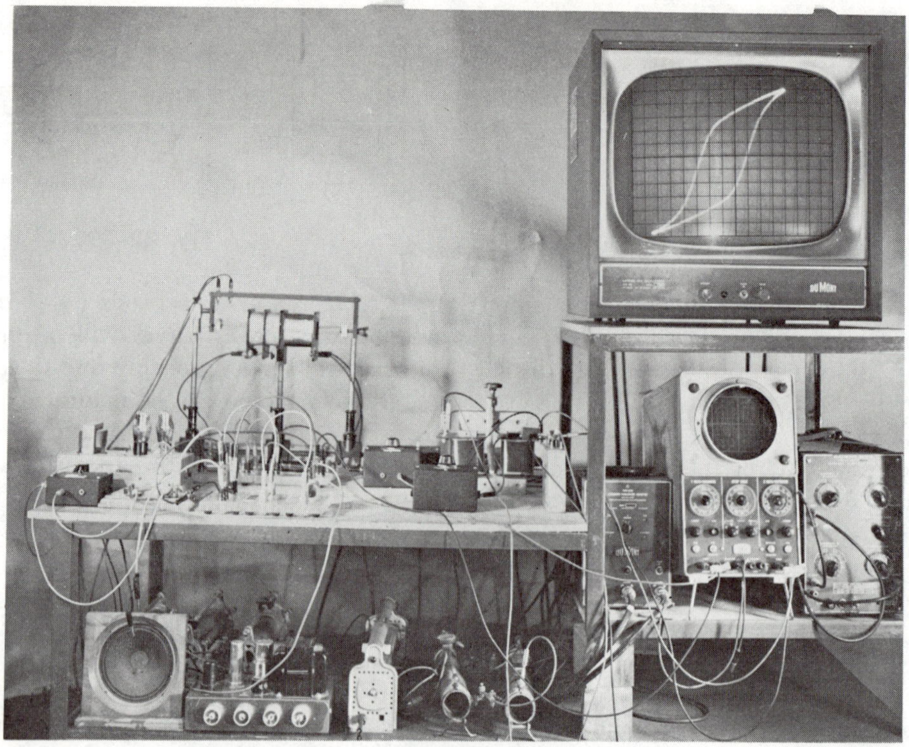

Fig. 32–41

netization and, hence, the ferromagnetism. This destruction is accompanied by a breakdown of the hysteresis loop. The temperature at which this occurs is characteristic of the particular substance and is called the Curie temperature. Above the Curie temperature the substance is paramagnetic.

The apparatus shown in Fig. 32–41 is designed to illustrate the breakdown of the hysteresis loop at the Curie temperature. The external field is supplied by an ac-driven transformer primary *a;* the driving signal is also applied to the *x*-axis of an oscilloscope *b* as shown in Fig. 32–42. The output of the transformer secondary *c* is applied to the *y*-axis, and the combination of the two signals produces the hysteresis loop of the core of the transformer. A 110-V dc Nichrome heater *d,* supported by the two porcelain tube sockets *e,* is arranged along the coil axis. On heating the coils the hysteresis loop disappears.

For use in a large classroom, it is best to use a Dumont 401A oscilloscope which is connected via a cathode follower adapter to a Dumont model 345 slave oscilloscope *f.* A similar setup, omitting the heater, can be used to illustrate hysteresis. The modified circuit for this application is shown in Fig. 32–43. The same oscilloscope and slave monitor is used. Coil *1* has 250 turns and coil *2* has 1000 turns. For a materials list, see Appendix, page 1333.

32–3.19

Ferromagnetism is caused by the interaction between neighboring magnetic moments of paramagnetic atoms. This interaction, which occurs only in crystal lattices, leads to the formation of fully polarized Weiss domains. The ferromagnetic interaction is strongly determined by the structure of the lattice and by the temperature. Above a certain temperature, the Curie temperature, thermal energy prevents formation of polarized domains and the substance can only be paramagnetic. Also, the ferromagnetic interaction produces mechanical forces within the lattice, thereby determining lattice constants and influencing phase transitions.

This experiment demonstrates two phenomena occurring in iron at the Curie point temperature (770°C): the disappearance of ferromagnetic susceptibility and the change of phase from α-iron (2.86 Å) to β-iron (2.90 Å).

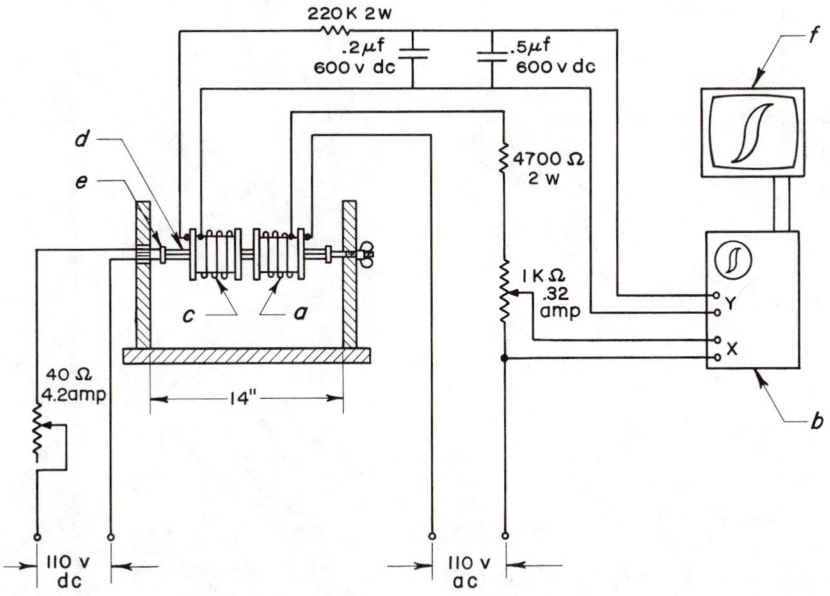

Fig. 32–42

The experiment is set up as illustrated in Fig. 32–44. A long iron wire *a* (length, ~8 m; diam, 2 mm) is hung by its extremities with a slight sag at room temperature. The wire can be heated to about 900°C by passing a large current through it (110 V ac, ~50 A). Due to the heat capacity of the wire, the heating requires about 10 sec. There will be a noticeable sagging of the wire due to thermal expansion. The glowing wire is more easily visible if the lecture room lights are dimmed.

To observe the disappearance of ferromagnetic susceptibility, a ferrite ring magnet *b*, is mounted as shown so as to produce a field parallel to the wire and magnetize it. Thus the flux passing through a 10,000 turn ring (~0.1-mm-diam wire), coil *c*, mounted next to the magnet and connected to a mirror galvanometer, is partly due to the ferromagnetic susceptibility of the wire. When, during heating, the Curie point is reached, this contribution is suddenly lost, resulting in a strong induction

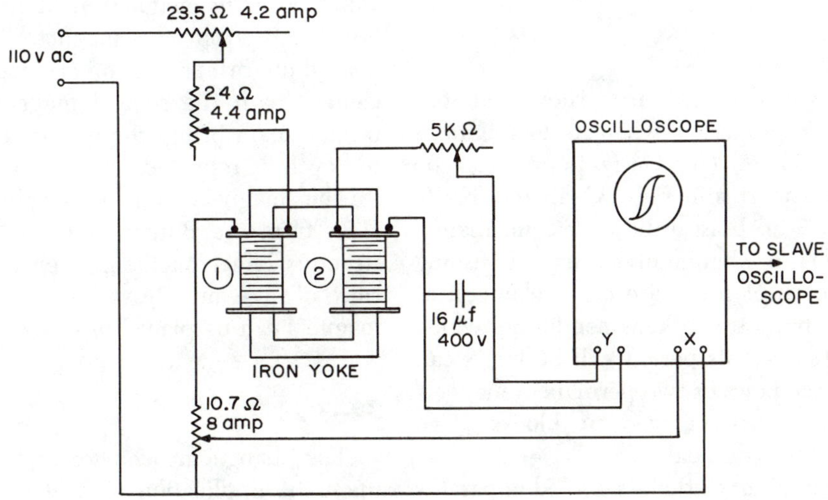

Fig. 32–43

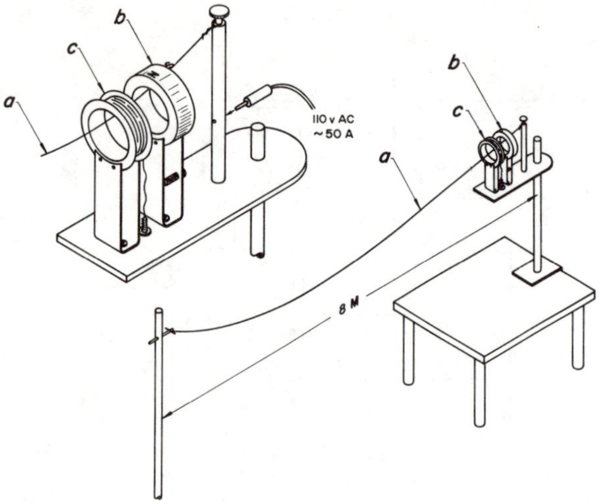

Fig. 32–44

pulse of the galvanometer. Passing the Curie point from the other direction gives an opposite pulse due to sudden remagnetization. However, in cooling, the Curie point is more clearly detected because there is no ac current to disturb the galvanometer.

The change in lattice constant occurs at practically the same moment and manifests itself very strikingly. Upon heating, the wire expands and its sag increases. At the phase transition the expansion stops, a short recontraction occurs, and then expansion sets in again as the temperature keeps rising. Again, this phenomenon is more easily detected upon cooling as the temperature is then more homogeneous.

32–3.20

An interesting demonstration to show that iron loses its magnetic properties when its temperature is raised above 770°C (its Curie point) can be shown with the appartus in Figs. 32–45 and 32–46. It consists of a 7-in. brass tube d ($\frac{3}{8}$ in. diam), suspended from an aluminum gantry by fishing line a. Short loops of iron wire e are clamped at each end of the brass tube. This pendulum, which should be long so that its period will be large, can swing between the poles of two permanent magnets which are mounted onto wooden blocks. Two Bunsen burners provide heat.

When the pendulum of the assembled apparatus is displaced to one side, the iron wire is held to the pole face by magnetic attraction. The flame from the Bunsen burners should just touch the pole face of the magnets. The flame heats the iron wire, which loses its magnetism at its Curie point and falls away from the magnet. The swing of the pendulum brings the other loop of iron wire in contact with the second magnet. Heat from the burner again brings the iron to its Curie point, and the cycle is repeated.

Fig. 32–45

The magnets can be surplus radar magnets. Heat from the Bunsen burners has very little destructive effect on them, even after demonstration runs of five minutes or more. If necessary, the magnets can be placed in water baths.

32–3.21

The change in behavior of a magnetic material upon the application of heat can be shown in an interesting manner by using a spoked wheel, the

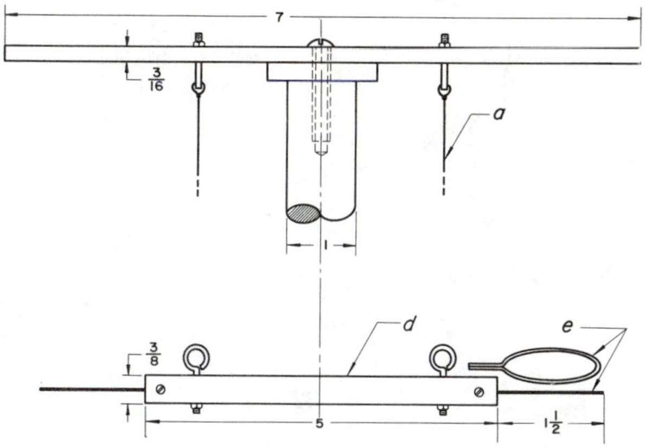

Fig. 32–46

rim of which is constructed of Monel metal tape[20] (Fig. 32–47), and a permanent magnet. Monel metal is an alloy of nickel, copper, aluminum, plus traces of other elements. Its Curie point is about 140°F. The magnet can be a surplus magnetron magnet mounted on a brass rod (Fig. 32–48).

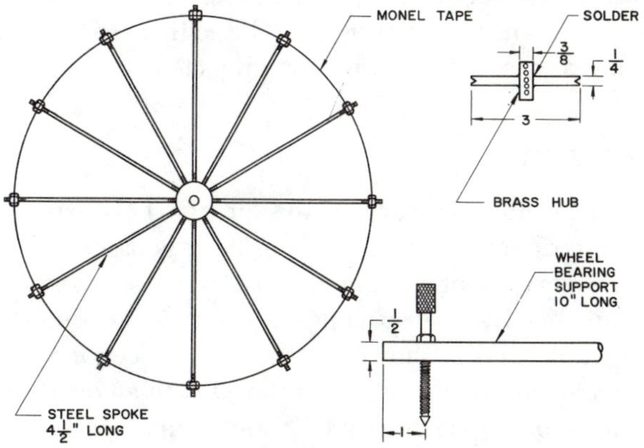

Fig. 32–47

The hub of the wheel is turned from brass, and a ¼-in. hole drilled through the center. In this hole is soldered a steel spindle 3 in. long and ¼ in. in diam, the ends of which are center-drilled to give a bearing surface. Around the circumference of the hub, twelve equally-spaced holes are drilled and tapped. Twelve spokes are cut from steel rod. These should be 4½ in. long and have a thread cut for about ½ in. along each end. One end of each spoke is then screwed into the hub. A 32-in.-long strip of Monel tape has twelve holes drilled along its length, spaced to fit the spokes. A small nut is screwed over the ends of each spoke, and the drilled tape fitted over the spokes. Another nut is fitted to each spoke, and the tape is pinched into position.

Fig. 32–48

The two wheel bearing supports are cut from brass rods, ½ in. in diam and 10 in. long. About 1 in. in from one end of the rods, a hole is drilled and tapped. The ends of two steel screws are turned to a point to form a bearing with the spindle. Locking nuts are placed on the screws, and the screws assembled in the rods.

With the Monel tape rim just inside the poles of the magnet, and the wheel adjusted on its bearings to turn quite freely, heat the tape, with a match or cigarette lighter, slightly to one side of the magnet. The wheel starts to turn. If the heat is withdrawn and applied to the tape at the other side of the magnet, the wheel stops and rotates in the other direction.

[20] Supplied by Henry Wiggin and Co., Birmingham, England.

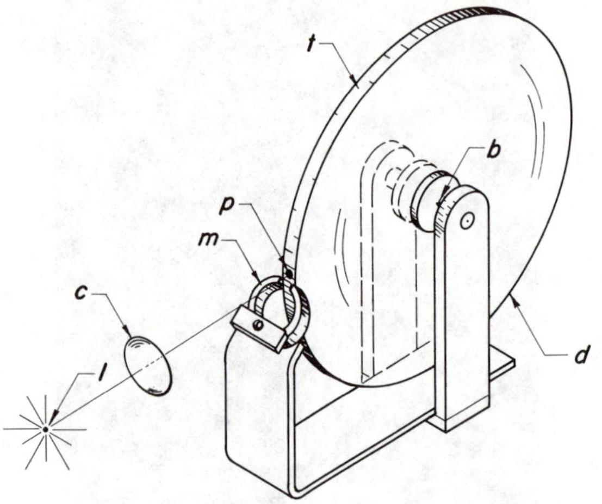

Fig. 32–49

32–3.22

The thermomagnetic motor (Fig. 32–49) is a device for demonstrating motion resulting from a change in magnetization caused by local heating.[21] The motor consists of a carefully balanced plastic disk *d*, 6 in. diam x 1/8 in. thick, mounted on low-friction, needle-point bearings *b*. Accurate balancing may be accomplished by drilling holes in the plastic disk. Cemented to the rim of the disk is a strip of permalloy tape *t* .010 in. thick. An Alnico horseshoe magnet *m* is so located as to produce an intense transverse field in the permalloy tape. Heat from a 6-V automobile headlight lamp *l* is focused by the lens *c* on the tape at a point *p* slightly above the line joining the poles of the magnet. The permeability of the tape is reduced in the region of the focal spot, allowing the more highly magnetized portion of the tape to be attracted into the field, thus causing rotation. The motor is contained in a glass-walled box to eliminate the effect of air currents. The source of power could be concealed by using infra-red light filters over the source. This apparatus is well adapted to corridor displays.

A modification of the device, devised by T. E. Nicholson, is illustrated in Fig. 32–50.[22] A ring of 24-gauge nickel wire is supported on a vertical

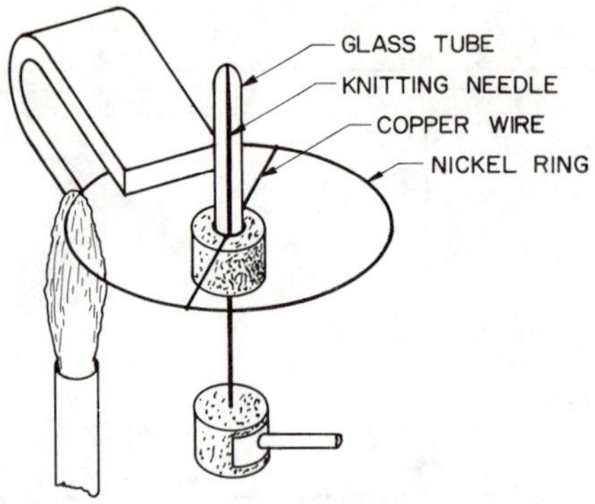

Fig. 32–50

axis by means of a small test tube axially attached to the ring by a copper wire along a diameter of the ring. The test tube slips over a vertically mounted knitting needle, thus making a low-friction bearing. A permanent horseshoe magnet provides a magnetic field across a small section of the nickel ring. When the nickel wire is heated by a Bunsen burner located a little to one side of the magnet, the disk rotates. Moving the burner to the other side of the magnet reverses the direction of rotation. Apparatus can be either shown on closed circuit TV or shadow projected.

32–3.23

The effect of varying the air gap between two magnets can be studied by using a gaussmeter.[23] The equipment for this demonstration is shown in Fig. 32–51. The spacing between the magnets is variable and is shown by shadow projection. The gaussmeter readings are displayed with an overhead projector. Spacings of 1, 2, and 3 mm are recommended.

32–3.24

Closing the gap of the magnetic circuit shown in Fig. 32–52 results in a transient drop of the current in the energizing coil. A 600-turn coil[24] *a* is connected in series to a 2-V dc source *b* and then to a galvanometer *c* which is shunted with a 1-Ω resistor.

The demonstrator varies the air gap *d* of the

[21] R. M. Sutton, *Demonstration Experiments in Physics* (McGraw-Hill, New York, 1938), p. 294; also J. Mills, *Am. Phys. Teacher*, **5**, 40 (1937).

[22] Welch, *Physics and Chemistry Digest*, No. 34, p. 49.

[23] Type G1 (Range 25,000 G), The British Thomson Houston Co. Ltd., Rugby, England.

[24] Can be obtained from Phywe AG., 34 Göttingen, Post Box 665, West Germany.

Fig. 32–51

iron core (Fig. 32–53) and notes any change in current while changing the circuit. Varying the air gap changes the reluctance of the path, producing a changing magnetic induction in the coil which in turn induces a change in current. If the air gap is held constant, then the current will not change.

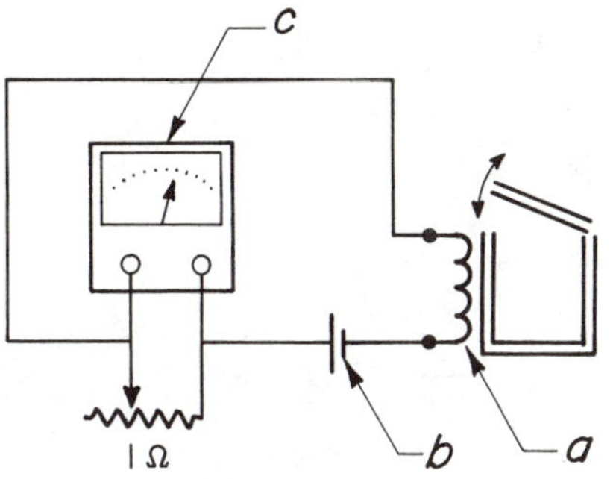

Fig. 32–52

The size of the air gap can be projected by the use of an arc lamp.

32–3.25

The equipment illustrated schematically in Fig. 32–54 (a) and represented pictorially in Fig. 32–54 (b) includes a 10-in.-sq iron magnetic path of 1-in.-sq cross section, supported by wooden dowels on a 16½ x 18 x ¾-in. plywood board. One section *m–n* of the path is removable. A coil *c* with approximately 1200 turns of No. 18 copper wire supplies the magnetomotive force, and a variable 0–15-V dc power supply with a 3-A capacity is used to energize the coil. The current in the coil can be reversed by the switch *s*. The flux is measured by a D'Arsonval flux meter *g*, con-

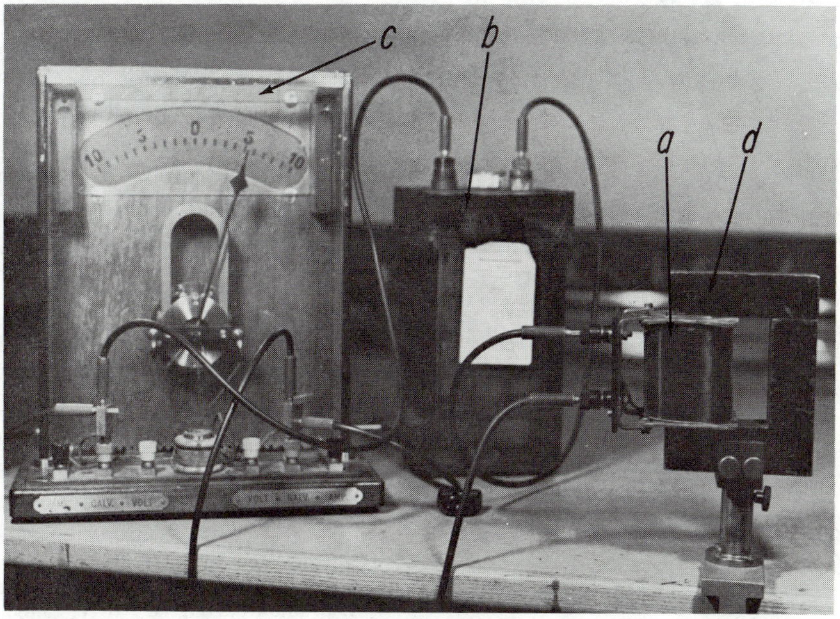

Fig. 32–53

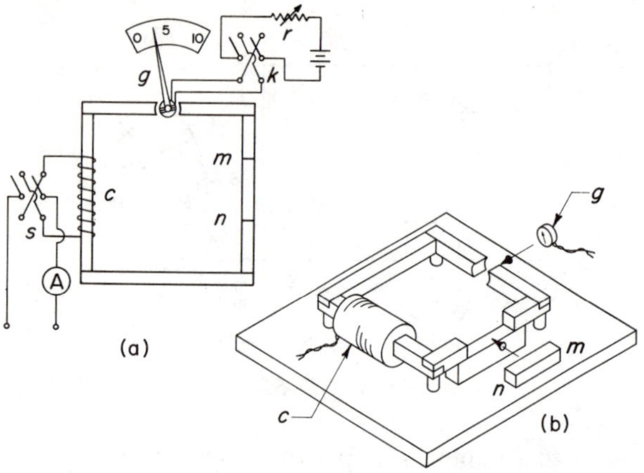

Fig. 32–54

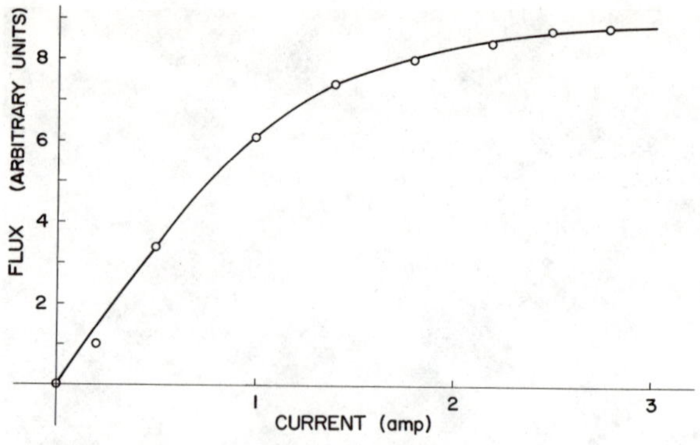

Fig. 32–56

structed by removing the field magnet from a 1-mA or 10-mA meter and inserting the movement in the magnetic path so that the deflection is proportional to the field in the loop. The movement of the flux meter is held in place by a little cement. The range of the meter is changed by varying the resistance r of the order of 1000 Ω. Two dry cells supply constant current to the coil of the flux meter. A reversing switch k in series with the galvanometer coil permits deflection in either direction. A complete hysteresis loop can be run by increasing the current in steps and reading the flux. Figures 32–55 and 32–56 show, respectively, a hysteresis curve and magnetization curve for data taken with the apparatus. The magnetic properties of various materials can be compared by replacing the removable link m–n with the same size links of different materials, such as brass and aluminum. The apparatus may be adapted to projection by mounting the magnetic circuit on a plastic base for overhead projection and using projection meters. Alternatively, the meter readings can be shown by closed-circuit television.

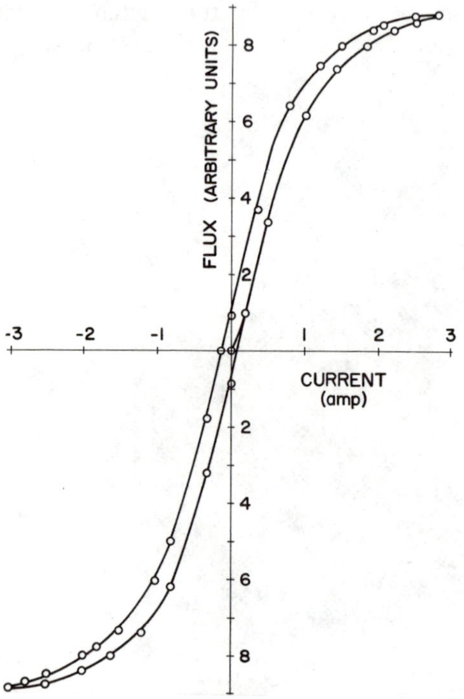

Fig. 32–55

32–3.26[25]

Two cylindrical electromagnets are attached in a vertical position, parallel to each other, on an aluminum frame d as in Fig. 32–57. The height of the electromagnets is $5\frac{5}{8}$ in., the diameter of the core is 1 in., and the distance between the electromagnet axes is 5 in. There are about 500 turns of .02-in.-diam wire on each electromagnet. The electromagnet cores protrude over the windings by $\frac{7}{8}$ in. on each end, and can be closed above and below by passive plates (yokes) of soft iron a and e. The entire apparatus is placed above a demonstration table on a firm stand. The electromagnet windings c are connected in series, and an ammeter and a switch are connected through a rheostat to a dc source. When a relatively strong direct current is switched on, the electromagnet cores become so strongly magnetized that the lower

(25) Translated by K. V. Robinson and H. A. Robinson (Adelphi University) from *Lecture Demonstrations in Physics*, A. B. Mlodzeevskii, ed., M. V. Lomonosov State University, Moscow, U.S.S.R.

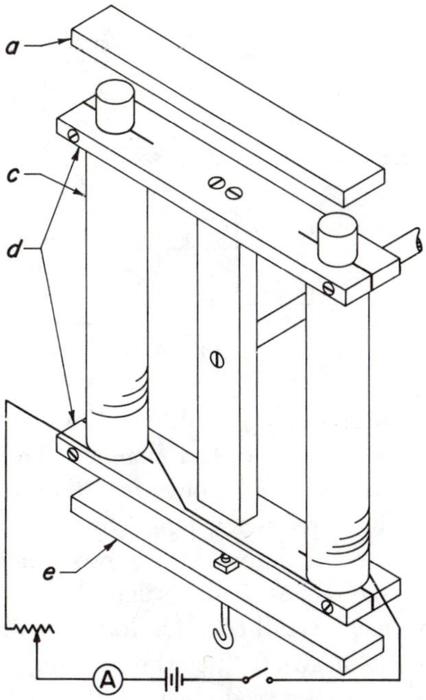

Fig. 32–57

plate (yoke) lifts and holds a considerable weight. Now if the upper yoke is removed from the electromagnet, the bottom plate with the weight drops. This occurs because when the upper yoke is removed, the reluctance of the whole circuit is suddenly increased, producing a decrease in the flux across the iron cores, which in turn produces a decrease in the lifting power of the apparatus causing the weight to drop.

The bottom plate is replaced with the upper plate still removed. The strength of the current is increased in the electromagnet winding until the lower yoke, together with its load, is held by the electromagnet. The current strength in the winding of the electromagnet should be compared for both versions of this experiment. Using plates of different thickness for the top plate provides other interesting demonstrations.

A series of other experiments can also be demonstrated using this apparatus. Maintaining a constant current in the electromagnet circuit (constant magnetomotive force of the system), close the upper poles of the magnet with iron plates of equal area but different thickness. Given a considerable difference in the thickness of the upper plates, it is easy to show that the flux at the lower poles of the electromagnets will be higher when the upper plate is thicker.

Control in this comparative experiment is the amount of load held by the electromagnet. A number of ferromagnetic plates of equal area and equal thickness but having different permeability are prepared and are then used to cover the upper poles of the electromagnets. The experiment will show a difference in the lifting power of the electromagnets even though the magnetomotive force of the system remains the same. Soft iron may be used as the material tested; steel which is relatively soft in a magnetic sense may also be used, as well as other products. In addition, plates having the same shape, but made of brass, copper, and aluminum, may be used. It can be shown that the permeability of these materials is nearly equal to one.

This experiment may also be shown using two permanent magnets set in a double parallel bar yoke; the yoke is made of nonferromagnetic material. The apparatus using permanent magnets may be used to show practically all of the demonstrations described above with the exception of the variation requiring a change in the magnetomotive force of the system.

32–4 Magnetostriction and Magnetoresistance

32–4.1 [26]

A ferromagnetic rod in an alternating magnetic field periodically changes its length with a frequency twice that of the applied alternating current. The amplitude of these longitudinal, mechanical oscillations of the rod will reach a maximum when the frequency of the forced oscillations of the rod coincides with its natural frequency. In other words, the first maximum arises when one half an elastic wave $\lambda/2$ is established along the length of the rod. Since the velocity of sound in nickel is about 4.76×10^5 cm/sec, a 40-cm nickel rod will show a first maximum at

[26] Translated by K. V. Robinson and H. A. Robinson (Adelphi University) from *Lecture Demonstrations in Physics*, A. B. Mlodzeevskii, ed., M. V. Lomonosov State University, Moscow, U.S.S.R.

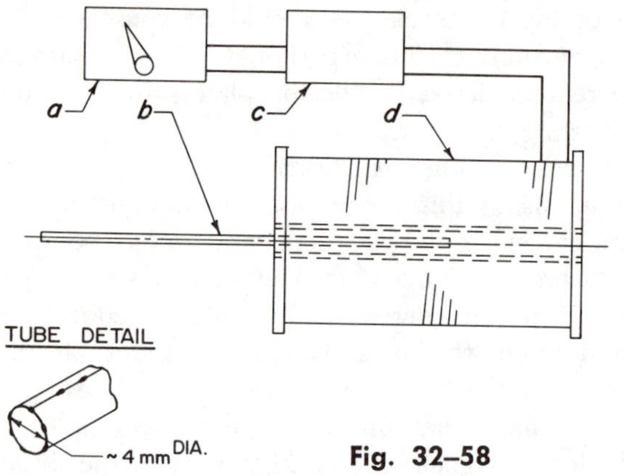

Fig. 32–58

$$V = \frac{\nu}{\lambda} = \frac{\nu}{2l} = \frac{4.8 \times 10^5}{2 \times 40} \approx 6000 \text{ cycles.}$$

Thus in order to obtain maximum vibrations of a nickel rod 40 cm long, an alternating current with a frequency of 3000 cycles must be applied to the winding of the magnetizing coil.

If the nickel rod is magnetized by an ac field with a frequency two to three times higher than that calculated above, two, three, or more half-waves will be set up in the rod, i.e., the rod will be excited to one of its harmonics. Normally magnetostrictive vibrations occur at the fundamental mode of the rod. To obtain higher frequencies from the magnetostrictive oscillator, decrease correspondingly the dimensions of the rod.

To set up the apparatus, connect an audio oscillator *a* (up to 20,000 cycles) to a low-frequency amplifier *c* (see Fig. 32–58). Voltage from the amplifier (100–120 V) is applied to a horizontal magnetizing coil *d*. A nickel or permalloy rod, thin strip, or hollow tube *b* can be placed inside the coil. The rod should be about 40 cm long and about 3–4 mm in diam, or if a strip is used, it should be as long as the rod with a cross section of about 2 × 10 mm. The magnetizing coil should be as long as the rod or a half inch shorter, and the intensity of the alternating magnetic field in the coil canal should reach 100–150 Oe.

The experiment is most successful if a hollow tube of thin sheet nickel is used. The seam of the tube need not be soldered all the way; two or three points of attachment are sufficient. The tube will be less heated by eddy currents than will the solid rod or strip. Small disks may be soldered to the tube at one or both ends to intensify the emitted sound.

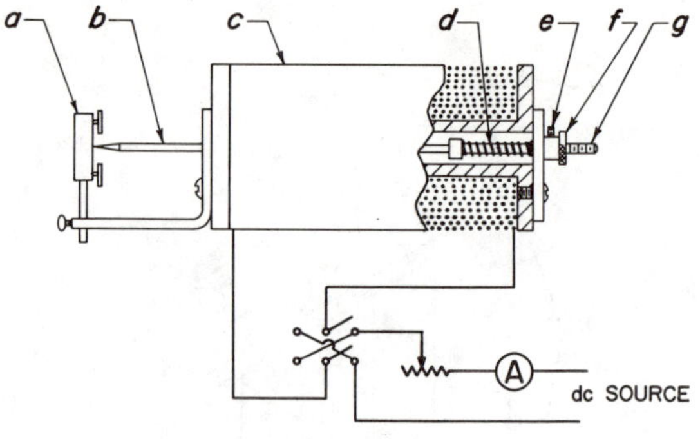

Fig. 32–59

When the frequency of the audio oscillator is changed smoothly, the ferromagnetic rod starts oscillating, and its oscillations are accompanied by a sound with a frequency equal to twice the frequency of the oscillator. If the rod (having the same cross section over all its length) is well annealed and has a small coercive force, its excitation takes place at a strictly determined frequency. A small change in the frequency of the oscillator to one side or another brings about a sharp decrease in the sound emitted by the rod. If another rod of a different length is introduced into the coil, its excitation will occur at a different frequency.

When the ferromagnetic rod is vibrating sharply, remove the rod from the coil, noting the cessation of the sound. If a brass or copper rod of the same dimensions as the ferromagnetic rod is now placed inside the coil, no sound will be emitted. If during the resonance oscillations of the nickel or permalloy rod, the rod is magnetized to saturation in a constant field by bringing a permanent magnet close to it, the effect of resonance, calculated by the above formula, disappears. In this instance, resonance of the ferromagnetic rod sets in when the natural frequency of the rod coincides with the frequency of the alternating current.

32–4.2[27]

The apparatus (Fig. 32–59) consists of a magnetizing coil *c*, a ferromagnetic rod *b*, and an optical system for the demonstration of Newton's rings *a* (such as Cenco's Newton's Rings Apparatus No. 87460). The coil should be at least 10 in. long,

[27] Translated by K. V. Robinson and H. A. Robinson (Adelphi University) from *Lecture Demonstrations in Physics*, A. B. Mlodzeevskii, ed., M. V. Lomonosov State University, Moscow, U.S.S.R.

with about 1200 windings, and should have a magnetic field strength of about 1500 Oe. The rod, about 11 in. long and 0.2 in. in diam can be made from cobalt steel, which possesses positive magnetostriction. Magnetostriction of iron, nickel, and other ferromagnetics is also fairly strong.

Brass plates, having openings along the axis of the coil, are attached to the end walls of the wooden frame of the coil. The ferromagnetic rod, passing through these holes, is held along the axis of the magnetizing coil. A bent brass holder is attached to the left flange of the frame; the vertical rod of the optical system is held in the holder. By such an arrangement of the apparatus, the sharpened tip of the rod can be brought to the back surface of the optical system, and the other end can be held by a screw.

Because it is difficult to work cobalt steel, a small, threaded iron rod *g* is soldered to the right-hand end of the cobalt steel rod; the iron rod is about 2 in. long and has the same diameter as the rod of cobalt steel. By means of a knurled nut *f* and spring *d,* the ferromagnetic may be moved from one side to another and locked in place by setscrew *e*.

The apparatus is set on a wooden stand at such a height that the parallel light beam from the projection lantern falls on the front pane of the optical system. To prevent possible heating up of the system by light rays, place a plane, parallel container, filled with water, at the cone of the projection lantern.

Even the slightest increase in the length of the rod will make the tip push against the back wall of the optical system and produce a change in the thickness of the air layer between the convex lens and the flat plate. The change will immediately influence the conditions of light reflection from the air layer, and the coloration of Newton's rings in the reflected light will change.

By manipulating the regulating screws, usually provided in the system, the demonstrator can produce Newton's rings in the central part of the field and project them onto a screen with a lens. The experiment is most successful when a light green spot can be shown in the center of the rings.

When the current is switched on in the windings, the magnetic field produced along the axis of the coil will cause a lengthening of the ferromagnetic rod, which will bring about a change in the thickness of the air layer. As the conditions of reflection of the parallel light beam change, the colored image of the rings on the screen will become slightly wider, and the spot in the center of the rings will change color. When the current in the winding of the coil is switched off, the rod will largely lose its magnetization, returning nearly to its initial length; the ring system will contract and resume its former coloration.

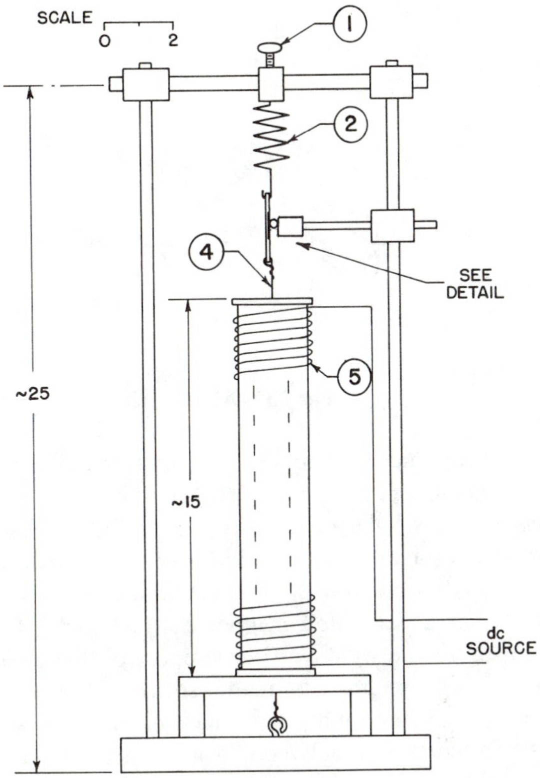

Fig. 32–60

32–4.3[28]

The magnetostriction of a nickel wire in a constant magnetic field is shown with the apparatus in Fig. 32–60. A long cylindrical coil 5 is set up vertically on a stable wooden base. The length of the coil is 15 in., and the magnetic field in the coil is about 200–250 Oe. A nickel wire *4,* 0.020–0.028 in. in diam, is set up along the axis of a coil. The lower end of the wire is attached to a stationary

[28] Translated by K. V. Robinson and H. A. Robinson (Adelphi University) *from Lecture Demonstrations in Physics,* A. B. Mlodzeevskii, ed., M. V. Lomonosov State University, Moscow, U.S.S.R.

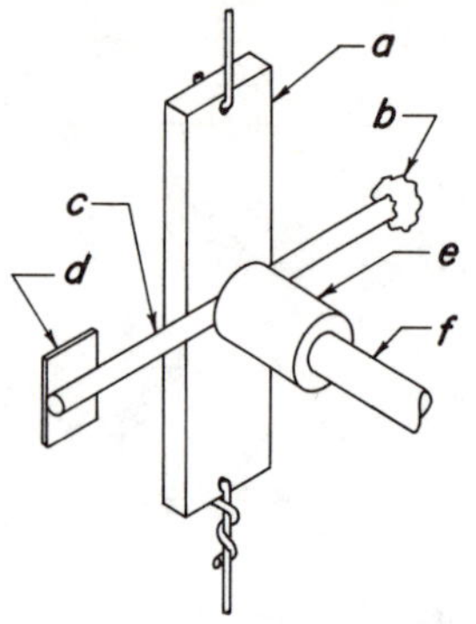

Fig. 32–61

screw-eye, and the upper end to a small metal plate *a* (see Fig. 32–61). The metal plate (and with it the whole nickel wire) is pulled up by a spring *2* attached to the middle of the horizontal frame of the apparatus. A horizontal metal rod *f*, which has a flat-ended magnet *e* attached to it, is clamped to the right vertical support of the apparatus. Adjustments are made so that the small metal plate *a* and magnet *e* touch each other with their flat surfaces. Between these surfaces is placed a small cylindrical iron rod *c* about 2 in. long, to which is attached at one end a small mirror *d*. A small counterweight of wax *b* is attached to the opposite end of the rod. This cylindrical rod is held between the surfaces of the plate and magnet by friction and by magnetic attraction.

A beam of light is directed at the small mirror *d*. The position of the reflected light beam on the screen shows the original length of the nickel wire. The coil is connected to dc and the nickel wire is magnetized. As a result of its magnetostrictive properties, the wire shortens and produces a corresponding lowering of the metal plate *a*, which turns the mirror through a small angle and lowers the light beam. When the current is turned off, the nickel wire returns to its original length and the light beam returns to its original position on the screen. (The residual magnetization of the wire is disregarded and also residual magnetostriction.)

The shortening of the nickel wire when magnetized is not very great. Therefore, in order to produce a noticeable shift in the light beam reflected on the auditorium screen: (1) use a sufficiently long wire, (2) magnetize it to a considerable degree, (3) use a perfectly cylindrical rod of about 0.008 in. diam to support the mirror, and (4) set up the apparatus at least 13–15 ft away from the screen.

In order to obtain the greatest possible relative decrease in the length of the nickel wire, heat it in a hydrogen atmosphere at a temperature of 800–900°C for 2–3 hr. If it is not possible to subject the nickel wire to such a treatment in the physics laboratory, it may be subjected to a very strong electric current for a period of several minutes to bring it to red heat. In both cases the wire should be straightened by stretching before the treatment. All the parts of the apparatus not otherwise specified (including the spring) are made of non-ferromagnetic material.

When the length of the wire changes as a result of its magnetization, the metal plate *a* moves up or down, thus rolling the cylindrical rod (which carries the mirror) over the side surface of the magnet *e*. This method of recording magnetostriction is convenient since the friction which arises between the movable parts of the apparatus is considerably decreased. The forces of friction will be constant if the rolling of the cylindrical rod *c* always takes place in a homogeneous part of the magnet's field. The outer magnetic field of the magnet will slightly magnetize the nickel wire, but this will not change the demonstrated effect to any great extent. The clamp *1* that supports the spring and nickel wire is used for regulating the initial degree of tension in the wire.

32–4.4

When mechanical stress is applied to a ferromagnetic material in a magnetic field, a change of magnetization generally occurs. The magnitude of the change depends on the total intensity of magnetization and is zero at saturation. This indicates that stress may change the orientation of the domains. The resulting strain can be considered to have two components; one is purely elastic in origin, and the second is due to the magnetic-mechanical process. Thus the values of the elastic moduli of a ferromagnetic material depend on the state of magnetization. In particular, the variation of Young's modulus E with intensity of magnetization is known as the ΔE effect.[29] That is, when

[29] L. F. Bates, *Modern Magnetism* (Cambridge Univ. Press, Cambridge, 1963), p. 424.

Fig. 32–62

the magnetostrictive effect is superimposed on the ordinary elastic elongation, the measured Young's modulus is smaller than normal. The presence of a large magnetic field (specimen saturation) prevents orientation of the magnetization by stress and so raises Young's modulus to its purely elastic value.

A 3-mm-diam, 50-cm-long rod of pure nickel is carefully annealed by heating in an evacuated quartz tube for 5 hr at 800°C. The rod (Fig. 32–62) is allowed to cool down overnight and must be handled with great care since any strain influences the results. The rod is suspended in two nylon cord slings set 12.5 cm from either end as in Fig. 32–63. A coil, with an impedance matched to that of an audio oscillator at about 10 kc, is placed around the rod as close to one of the suspensions as possible. The rod thus can be set in longitudinal vibration as a result of the Joule magnetostrictive effect. The energy input from the oscillator should be kept as low as possible; 0.05 mW should be sufficient. (Above 2 mW distortion occurs.) A pickup coil (1 cm long and 1 cm i.d. with 10,000 windings of 0.02-mm-diam wire) is placed around the rod close to the other suspension and is connected to an oscilloscope.

To eliminate direct pickup from the first coil, a third (compensating) coil must be placed in the pickup circuit in series with the pickup coil (Fig. 32–64). This third coil should be placed on the opposite side of the first exciting coil, and its position should be adjusted to give an essentially zero signal when the rod is not at resonance. The first and third coils need not be identical but should be comparable in the product of the area and the number of turns. Complete compensation is not possible, in general, but the visibility of the resonances is greatly improved. All leads should be shielded. The suspended rod with its coils should be mounted in an 80-cm-long glass or clear plastic tube, 8 cm in diam. A magnetizing coil consisting of one winding of 1.5-mm-diam wire per cm is

Fig. 32–63

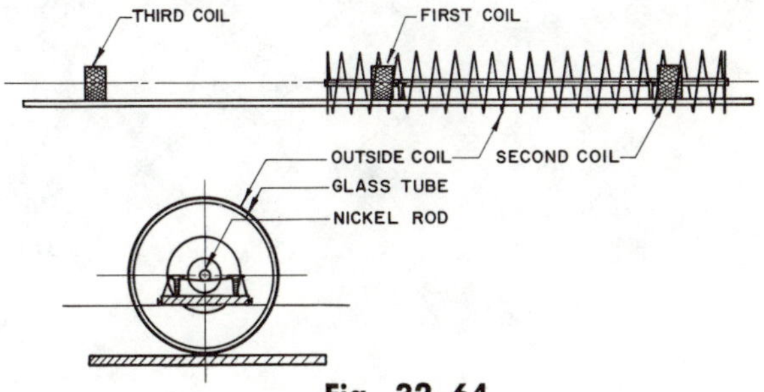

Fig. 32–64

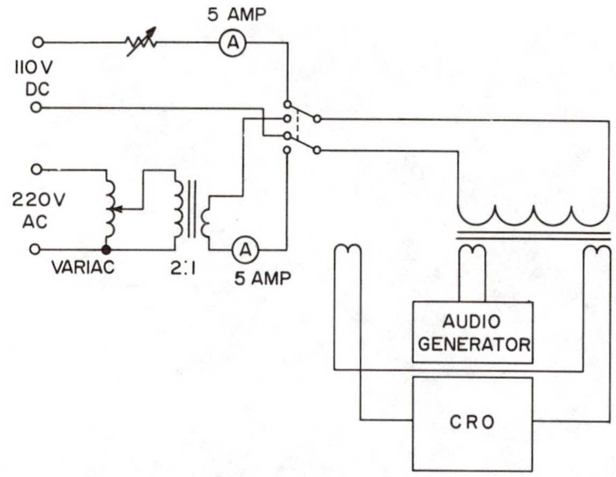

Fig. 32–65

wound around the cylinder and cemented in place. The cylinder and coil can be fastened to another board to prevent rolling. A circuit diagram is shown in Fig. 32–65.

In demonstrating the effect, the nickel rod first is demagnetized by gradually reducing 10 A ac in the magnetizing coil to zero. (Disconnect the oscilloscope and oscillator during this process.) When the oscillator is reconnected, no resonance due to damping by magnetic hysteresis is found (see Dem. 32–4.5), and as a result, E cannot be measured in the annealed state. A 0.5-A dc current is now applied for a short time through the outer coil. After switching this current off, a resonance is found in the vicinity of 9.4 kc. The signal can be demonstrated on the oscilloscope or made audible by an amplifier and loudspeaker. A direct current of 5 or more amps is next fed through the outer coil. The resonance now occurs at a frequency which is 5–7 percent greater than in the previous case. Since $E = 4\rho f^2 l^2$, where ρ is the density, f is the frequency, and l is the specimen length, this corresponds to an increase in Young's modulus of around 10–15 percent. Due to the decrease in damping, the peak signal is several times greater than it is in the absence of a field. Harmonic resonances also can be induced with this apparatus when the audio energy input is increased to several watts. That is, when the oscillator produces 2.35 kc, rod resonances are observed at 9.4 kc.

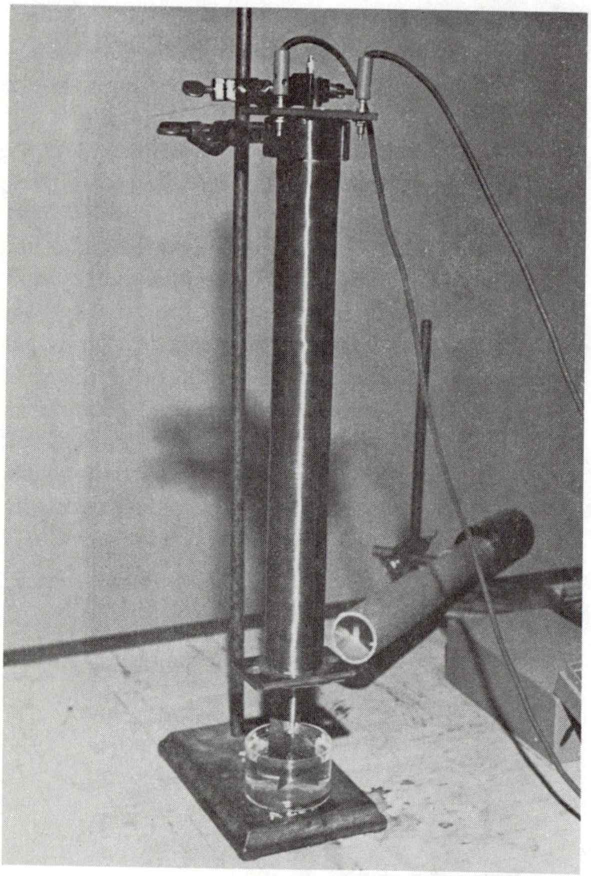

Fig. 32–66

32–4.5[30]

The inverse magnetostrictive effect causes displacement of magnetic domain boundaries when a magnetic material is subjected to tension. These displacements are only partly reversible. Two observations are made: (1) deformations show hysteresis, and (2) oscillatory deformations are damped by this hysteresis.

A nickel wire, 40 cm long and 1 mm in diameter, is carefully annealed. (See Dem. 32–4.4 for the annealing process.) Do not try to anneal in open air; the resulting oxide surface spoils the effect.

A coil capable of producing an ac magnetic field of 500 Oe (Fig. 32–66) is wound from enameled copper wire (1.3 mm diam) in ten layers on a 40-cm-long, 2.5-cm-diam brass tube, which is slit along its entire length to avoid eddy currents. The tube has plastic end plates cemented with Araldite. The

[30] See R. Becker and M. Kornetzki, Z. *Physik.*, **88**, 634 (1934), and *Encyclopedic Dictionary of Physics* (Pergamon Press, New York, 1961), vol. 4, p. 456.

coil requires about 110 V ac (60 cps) for a current of 5 A. The electrical circuit is shown in Fig. 32-67.

The annealed wire is suspended vertically along the axis of this coil. The upper end is fastened in a banana plug that is firmly gripped by a clamp. A damping vane is fastened onto the lower end of the wire. The vane consists of two 5 x 5-cm crossed brass sheets, soldered to a banana plug and dipped into a dashpot of oil. A small mirror is fastened to the vane and directs a beam from a lamp to a wall 5 m away.

First the wire is demagnetized by decreasing the alternating current through the coil from 5 A down to zero. The zero position of the light spot is marked on the wall. Now a torque is applied to the lower end of the wire, so that the light spot moves 1 m, at a distance of 5 m (not more, or else plastic deformation would result, making a new annealing procedure necessary). When the torque is released, the light spot comes to a standstill at about 5-10 percent of the distance of the initial deviation. The domains that accommodated their position to fit into the twisted crystal structure now partially retain this deformation. When the ac current is turned on again, the domain boundaries are shaken; the initial magnetic disorder is re-established, and the light spot instantaneously returns to the zero mark. The same effect can be shown by orienting the domains by a dc field and 100 Oe before and after the torque.

Remove the dashpot and replace the damping vane with an aluminum strip, 20 cm long with two 10-g brass screws at the ends. This construction forms a torsion pendulum with high rotational inertia and low weight to keep the longitudinal stress low on the wire. On the wall mark the zero point and points at 0.5 m and 1 m (at 5 m distance) from zero. The wire is demagnetized by the Variac and coil. Then the pendulum is caused to oscillate, beginning with an amplitude slightly greater than 1 m. Measure the time for the amplitude to decay to its half value (about 30 sec). Then a dc field of 500 Oe is produced in the coil, and the experiment repeated. The time to decay to half-value is five times as great. If only 100 Oe can be produced, the time is three times as great. The dc field keeps the domains in a fixed position, so that damping by domain boundary displacement hysteresis is suppressed.

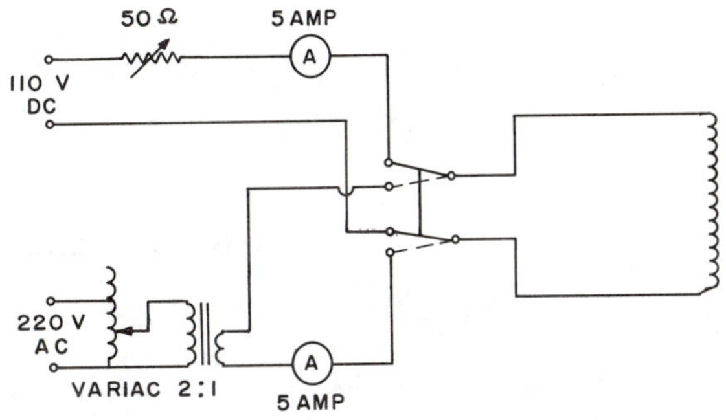

Fig. 32-67

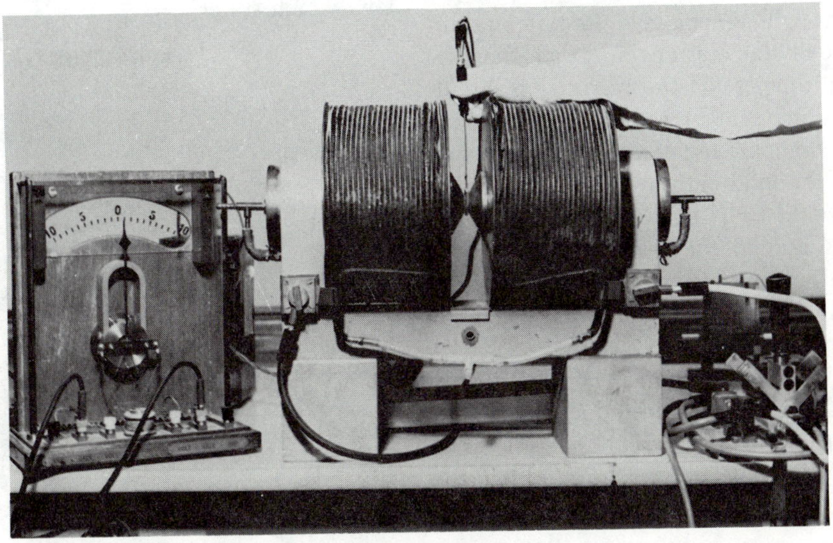

Fig. 32-68

32–4.6

The apparatus consists of a Bi-spiral[31] a in a 25-kG magnetic field (see Figs. 32–68 and 32–69). A current of about 0.1 A is sent through the spiral by a 2-V dc battery c. As the magnetic field is changed, a change in the resistance of a is observed by a 300-mV demonstration voltmeter d.

32–5 Adiabatic Demagnetization

32–5.1

The effect of adiabatic demagnetization in a piece of gadolinium is demonstrated at room temperature. The result of this effect, which is most apparent in materials of high paramagnetic susceptibility, can be derived from basic thermodynamic considerations.

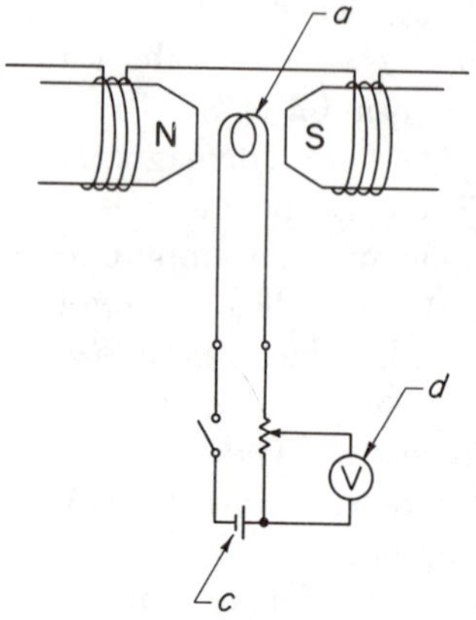

Fig. 32–69

When a material of high magnetic susceptibility is placed in a strong magnetic field, its temperature rises as a result of the rearrangement of the material's magnetic dipoles. If the material is simultaneously (or subsequently) cooled so that it remains at the original temperature (isothermal magnetization), the total entropy of the material is lower than it would be if no magnetic field were present. This is due to the resulting increase in order with the alignment of the dipoles. When the field is then reduced to zero adiabatically, there is a resulting temperature drop.

Although the demagnetization is not completely reversible, no great error is introduced by considering it to be so; thus $\Delta S = 0$. The process is shown on the TS diagram in Fig. 32–70. When the material is magnetized isothermally (1-2), it goes to a lower entropy state. Subsequent adiabatic demagnetization (2–3) results in a large temperature drop. The effect is especially noticeable at room temperature since Gd has a ferromagnetic Curie point at 289°K, which is just below room temperature. Thus the paramagnetic susceptibility (χ) has

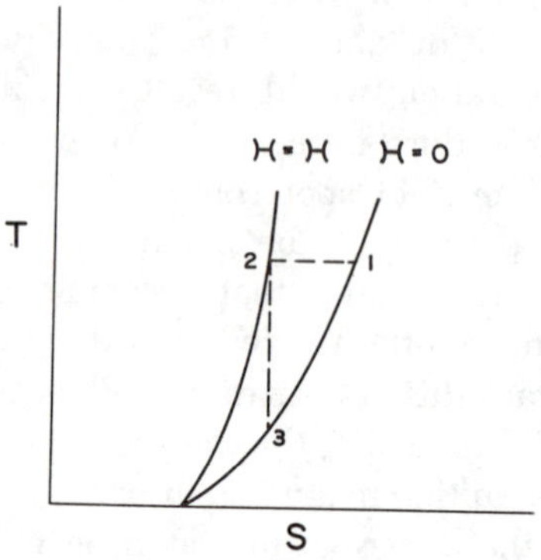

Fig. 32–70

a large negative temperature coefficient giving rise to a large magnetocaloric effect. The change in the magnetic susceptibility is given by the Curie–Weiss law, which states

$$\chi = \text{constant}/T.$$

Fig. 32–71

[31] Available from Hartmann and Braun, Frankfurt a/M, West Germany.

Successive adiabatic demagnetizations have been used to reach extremely low temperatures, approaching absolute zero.

A thermocouple is built into the Gd, which is mounted in a Plexiglas tube. The piece fits exactly into the gap of a strong magnet (25 kG) and is supported by nonmagnetic clamps. The thermocouple is connected directly to a lecture-hall mirror galvanometer. Turning the magnet on produces a strong deflection as the temperature of the Gd rises. The material is then cooled to room temperature by blowing compressed air into the gap. This is continued until the galvanometer reads zero once again. When the magnet is now turned off, the galvanometer is deflected through the same distance as before except in the opposite direction, indicating a cooling process. The apparatus is shown in Fig. 32–71. The experiment should be carried out as quickly as possible to avoid overheating the electromagnet. For construction details see the Appendix, page 1334.

Demonstrations	***Contributors***
32–1.1	D. Nahshol, Tel Aviv University, Israel
32–1.2, 32–1.3	J. Jaumann, H. Werheit, and F. Günneberg, University of Köln, West Germany
32–2.1	Swiss Federal Institute (ETH), Zürich, Switzerland
32–2.2, 32–3.1, 32–3.2, 32–3.3	H. A. Robinson, Adelphi University
32–3.4	R. M. Sutton, California Institute of Technology
32–3.5	Eric M. Rogers, Princeton University
32–3.6	J. Jaumann, H. Werheit, and F. Günneberg, University of Köln, West Germany
32–3.7	Massachusetts Institute of Technology
32–3.8	J. F. Dillon and H. E. Earl, Bell Telephone Laboratories
32–3.9	Swiss Federal Institute (ETH), Zürich, Switzerland
32–3.10	G. Schwarz, Florida State University
32–3.11	H. A. Robinson, Adelphi University
32–3.12	Swiss Federal Institute (ETH), Zürich, Switzerland
32–3.13	H. A. Robinson, Adelphi University
32–3.14	Eric M. Rogers, Princeton University
32–3.15	L. Grillet and M.-T. Grillet-Amy, University of Rennes, France
32–3.16	J. W. Straley, University of North Carolina
32–3.17	D. K. Berkey, Colgate University
32–3.18	R. A. Llewellyn, Purdue University
32–3.19	Swiss Federal Institute (ETH), Zürich, Switzerland
32–3.20, 32–3.21	Royal Institution of Great Britain, London
32–3.22	J. H. Howey, Georgia Institute of Technology
32–3.23, 32–3.24	Swiss Federal Institute (ETH), Zürich, Switzerland
32–3.25	P. Sprawls, Emory University
32–3.26, 32–4.1, 32–4.2, 32–4.3	H. A. Robinson, Adelphi University
32–4.4, 32–4.5	J. Jaumann, H. Werheit, and F. Günneberg, University of Köln, West Germany
32–4.6, 32–5.1	Swiss Federal Institute (ETH), Zürich, Switzerland

33

ELECTROMAGNETIC OSCILLATIONS

33–1 RLC Oscillations

33–1.1

Various aspects of an RLC circuit can be demonstrated with the circuit shown in Fig. 33–1. A motor driven commutator S permits alternately charging a capacitor C and discharging through an inductor L and resistor R connected in series. The resulting damped oscillations are viewed on an oscilloscope connected across the capacitor, as illustrated in Fig. 33–2.

In the actual demonstration a number of different combinations of resistance, inductance, and capacitance can be used, as shown in Fig. 33–1, and therefore a variety of effects can be shown. Since the resistance is kept small, the frequency f of the oscillation is given by:

$$f = \frac{1}{2\pi\sqrt{LC}}.$$

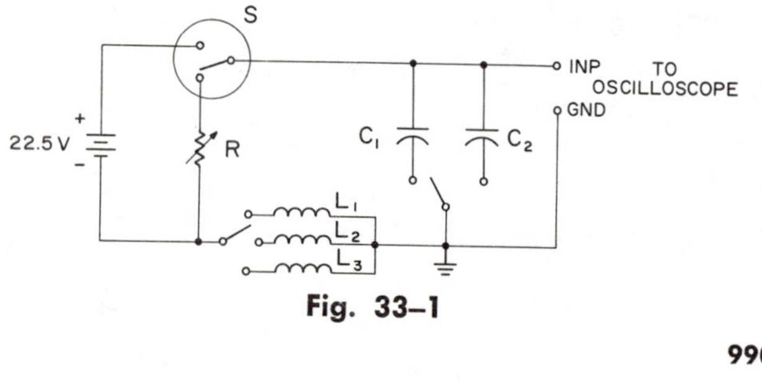

Fig. 33–1

By interchanging values of capacitance and inductance, one can demonstrate the change of frequency. The values of R may be calculated by measuring the time required for the amplitude to decrease to ½ or 1/e of its initial value. The voltage V on the capacitor is given by:

$$V = V_0\, e^{-Rt/2L}\,(\sin 2\pi ft),$$

and from this the value of an unknown inductance may be calculated. This value may be compared with the inductance computed from

$$L = \mu N^2 Al,$$

where l is the length of the coil, A is the cross-sectional area, N is the number of turns per meter and μ is the permeability of the core. Using first an air core and then a ferrite core, the ratio of the permeability of iron to that of air can be calculated. The ratio of the energy stored to the energy lost per cycle, or the Q value of the circuit, can be compared for a long solenoid and a ferrite core inductance. Finally, the difference in the oscillations with a soft iron rod, a ferrite rod, or a test tube full of water can be discussed. Ferrite is made of a powdered magnetic oxide imbedded in clay and has very small eddy current and hysteresis losses.

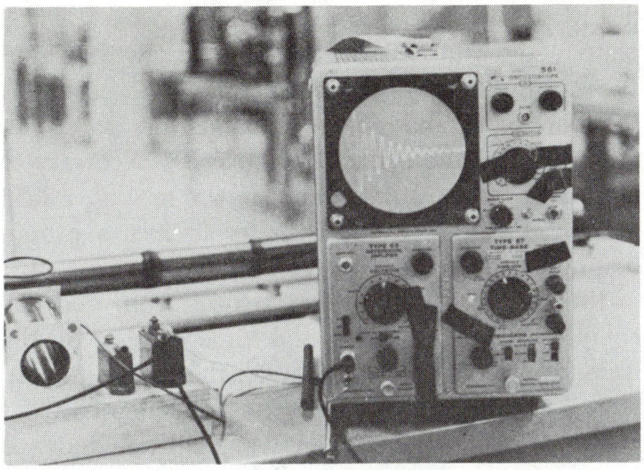

Fig. 33–2

For construction details of the commutator and a list of materials, see the Appendix, page 1334.

33–1.2

The relaxation oscillator shown schematically in Fig. 33–3 demonstrates the relationship between the period of the oscillation and the components, a capacitor C and resistor R. When the switch is thrown, the capacitor is charged up by the current through the resistor. The neon bulb I will remain "OFF" until the voltage across it and the capacitor is enough to fire it. When this occurs, the bulb flashes as the capacitor discharges through it. After discharge, the process repeats as the capacitor starts recharging. The period T of the flashes is proportional to the product of R and C, where T is in seconds, R is in ohms, and C in farads. It is most convenient to have C in μF (10^{-6} F) and R in MΩ ($10^6 \Omega$) so that T is between 1 and 10 sec. If R or C is doubled, the period is doubled.

For easy demonstration of the proportionality, R and C can be placed by decade resistors and capacitors, respectively. If variable components are not used, binding posts will allow easy switching of individual resistors and capacitors. It is important that the voltage V_0 be at least enough to fire the neon bulb (usually ~70 V) and that the bulb be neon.[(1)] A supply of about 90 V is best.[(2)] The components and battery are mounted on a small square of plywood and wired according to the schematic.

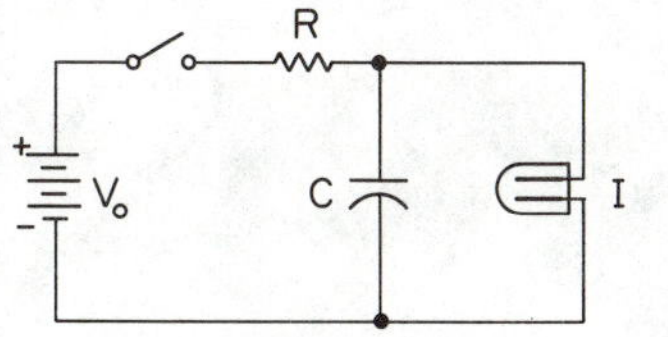

Fig. 33–3

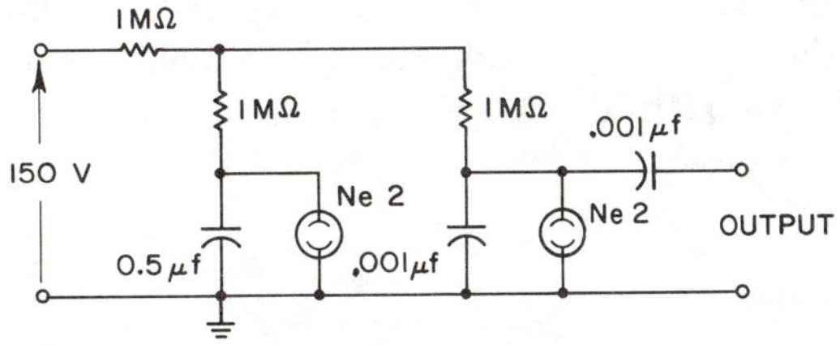

Fig. 33–4

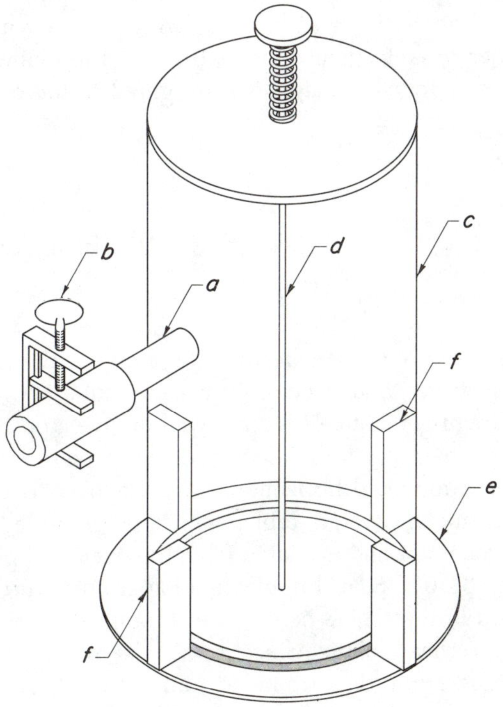

Fig. 33–5

33–1.3

The relaxation siren oscillator in Fig. 33–4 will give a signal which is a saw tooth wave of a few hundred cps modulated by a saw tooth of very low frequency. When it is connected to earphones, it

(1) Bayonet base neon lamp NE51.

(2) 2 No. XX30 Burgess 45 V batteries.

sounds much like a siren. The circuit is composed of three 1-MΩ resistors, one 0.5-μF, 200-V capacitor, two 0.001-μF, 200-V capacitors, and two Ne 2 neon bulbs.

33–1.4

The apparatus shown in Fig. 33–5 may be used to illustrate the concept of relaxation oscillations in an electrical circuit. A ½-in.-o.d. glass inlet tube *a* is connected to a water faucet with a short length of rubber hose. A hose clamp *b* in the line acts as a "resistor." The inlet tube feeds the base of a 4-in.-o.d. x 14-in.-long glass cylinder *c*, which acts as a "capacitor." The top of the cylinder is sealed, and a clearance hole for a ⅛-in.-diam brass rod *d* is drilled in the seal plate. The rod is spring-supported at the top and holds an iron plate *e* with a rubber gasket against the bottom of the cylinder. Three or four bar magnets *f* are glued to the cylinder to provide the main support force on the iron plate while the cylinder is filling up with water. The spring should only exert enough force to lift the iron plate and gasket after all the water has flowed out of the cylinder. It may be necessary to put a centering guide (not shown in Fig. 33–5) near the bottom of the cylinder to insure proper seating of the bottom plate.

In operation, the water is turned on, and the cylinder will fill up until the force exerted by the water is strong enough to break the magnetic contact of the iron plate. The water will then flow out the bottom of the cylinder. When the cylinder is nearly empty, the spring will lift the bottom plate until magnetic contact is made. This cycle will repeat itself as long as the water is left on. The exponential rise of potential in the "capacitor" can be observed, and the frequency can be adjusted by varying the clamp pressure on the rubber hose "resistor."

33–2 Electronics and AC Circuits

33–2.1

Soap bubbles can be used to simulate the flow of electrons in an electronic vacuum tube. A Van de Graaff generator[3] is employed to give a positive charge to a flat toroidal-shaped metal plate. This plate is mounted horizontally on two posts over a large, metal-covered table, which is grounded to the generator and insulated from the overhead plate (Fig. 33–6). Soap bubbles issuing from a vertical orifice in the table's center rise through the hole in the overhead plate. Since they take on negative charges from the grounded table top, they are attracted to the overhead plate's top surface. Two other vertical, parallel metal plates, connected by two horizontal rods, can be set at various positions between the grounded table top and the overhead plate. When the vertical plates are charged negatively, the bubble stream is kept in a central column. This accessory is used to help cancel the effects of disturbing air currents. For construction details see the Appendix, page 1336.

(3) Model 21, American Electrostatic Co., Tulsa, Oklahoma.

33–2.2

The circuit shown in Fig. 33–7 is used to demonstrate the current-voltage characteristics of various electrical components, which can be switched on individually, each one completing a current path

Fig. 33–6

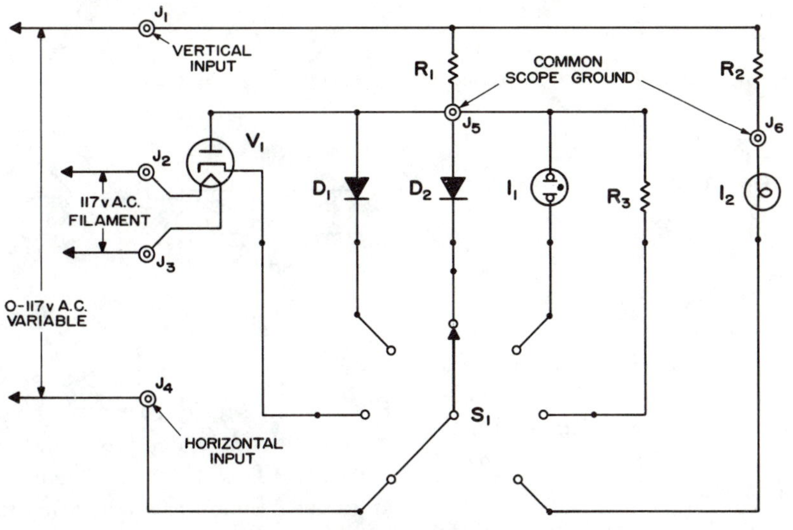

CURRENT - VOLTAGE CHARACTERISTICS

Fig. 33–7

through R_1 and the component in series. The voltage measured from J_1 to J_5 is proportional to the current through R_1, and the same amount of current must flow through the component. This reading is fed into the vertical input of the oscilloscope. The voltage across the component is measured from J_5 to J_4 and is fed into the horizontal input of the scope. On the scope will appear a plot of voltage vs current as the voltage changes from 0–110 V. Using the resistor component R_3, for example, the plot will be a straight line since the voltage is proportional to the current. To measure the characteristics of the incandescent lamp I_2, the scope connections at J_5 are moved to J_6. In order to show the circuit clearly on an overhead projector, it is best to build it on a 12 x 12 x ¼-in. sheet of clear acrylic plastic and wire as shown in the schematic. See Appendix, page 1337, for a list of materials.

33–2.3

The electronic circuit shown in Fig. 33–8[4] is used to demonstrate the phase relationships between currents in resistive, inductive, and capacitive circuits. The circuit, driven by a square wave

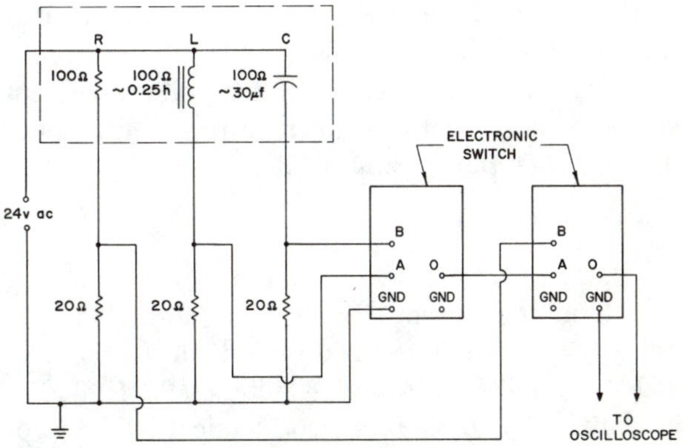

Fig. 33–8

generator, is connected by two electronic switches[5] to the vertical axis of a Welch demonstration oscilloscope[6] which in turn is connected to a 6-V external synchronizer.

If the current in the resistance branch of the circuit is first displayed on the scope, followed by a display of the inductor and capacitor branch currents, the capacitive current will be shown to lead the resistive current, while the inductive current will lag. The impedance values shown in Fig. 33–8 are for 60 cps.

(4) The part of the circuit enclosed by dotted lines is No. 80375 Choke Coil and Resonance Apparatus, Central Scientific Company.

(5) Electronic switch and square wave generator from Allen B. Dumont Laboratories, Inc., Passaic, New Jersey, Type 185-A.

(6) No. 2139, Welch Scientific Company.

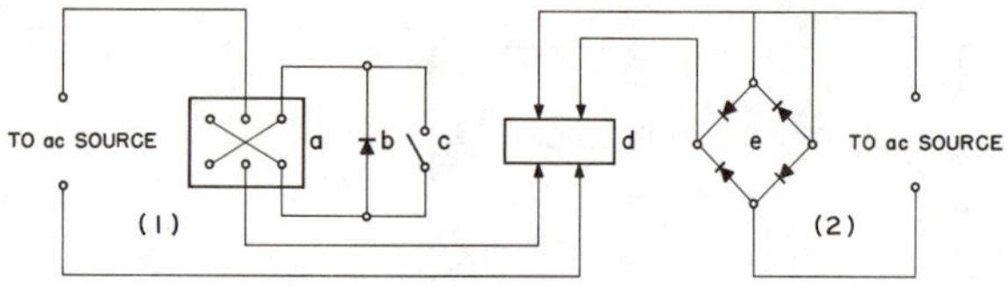

Fig. 33–9

33–2.4

The circuit shown in Fig. 33–9 can be used in demonstrating half-wave and full-wave rectification. A pendulum with a horseshoe-magnet bob which swings into and out of a 10,000-turn coil is used as a very-low-frequency ac source. In the circuit *a* is a reversing switch, *b* is a germanium-diode rectifier, *c* is a shorting switch, *d* is a recording galvanometer, and *e* is a germanium-diode bridge. With the coil connected to the terminals at (1) and *c* closed, the current is not rectified at all. But with *c* open, there is half-wave rectification; the choice of eliminating the top or bottom half is made by reversing *a*. When the coil is connected to the terminals at (2), there is full-wave rectification. Because of the voltage–current characteristics of the rectifier, the oscillations are damped, not purely sinusoidal.

33–2.5

Alternating-current output of a normal amplifier (1000 cps) *a* is rectified by a diode *b* (50 mA/150-V) and filtered by an L–C filter *c* (0.1 H, 3 μF), as shown in Fig. 33–10. Two identical 30-mA projection meters[7] *d*, placed before and after the filter, measure the current, which must be the same on both sides. With a second amplifier *e* and loudspeaker *f*, the value of the ac will be shown before and after the filter. A loud sound will be heard when the amplifier is connected at point *A*, and no sound when connected at *B*. The difference of the filter effect with high and low frequencies may be demonstrated in the range of 10–10,000 cps.

33–2.6

The functional relationship between resistance *R*, capacitance *C*, and inductance *L* can be effectively demonstrated by the apparatus shown in Fig. 33–11. Music from a phonograph *a* is fed into the amplifier *b*, which has a high output impedance (500 Ω). This amplifier is connected to a second amplifier *c* by either a coil[8] *d* (1200 turns, 35 mH, 12 Ω), a decade capacitor[9] *e* with 0.1-μF steps, or a decade resistor *f* with 1-, 10-, and 100-Ω steps. The three components can be switched into the circuit individually.

The capacitance and the inductance must be adjusted so that, for intermediate frequencies, the speaker loudness is roughly the same for each component switched into the circuit. When the resistor is switched into the circuit, essentially the full range of frequencies is transmitted to the speaker. With the capacitor in the circuit, only the higher frequencies can be heard, while with the inductor, only the lower frequencies pass through to the speaker. With both of them switched into the circuit, the full sound spectrum is transmitted. More quantitative effects may be demonstrated by using an audio oscillator instead of the phonograph.

[7] Projection meter is EMA Type Multimeter 4S with Projection Instrument Type PL4; made by EMA AG., Meilen/Zürich, Switzerland.

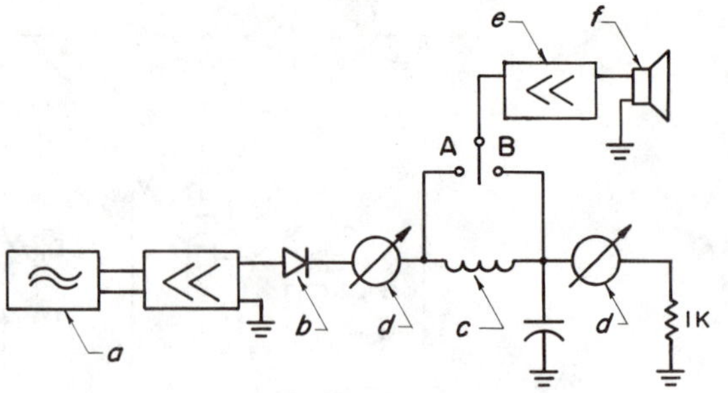

Fig. 33–10

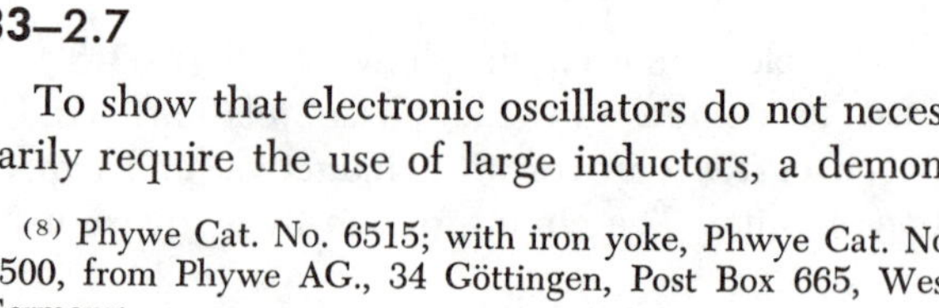

Fig. 33–11

33–2.7

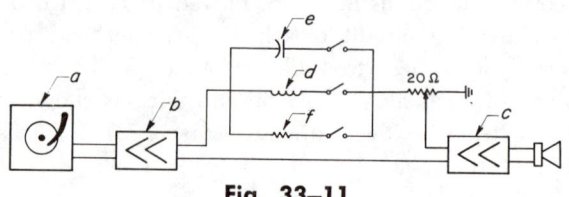

To show that electronic oscillators do not necessarily require the use of large inductors, a demon-

[8] Phywe Cat. No. 6515; with iron yoke, Phwye Cat. No. 6500, from Phywe AG., 34 Göttingen, Post Box 665, West Germany.

[9] Can be obtained from General Radio Co., West Concord, Massachusetts.

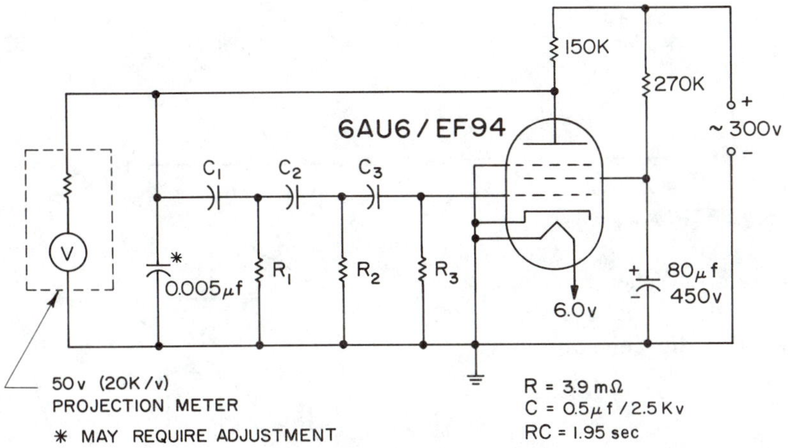

Fig. 33–12

stration is built around a simple RC phase-shift oscillator. To facilitate a discussion of the component parts, these have been mounted on an acrylic plastic plate (9 x 6 x ¼ in.), with the parts placed in the same relative positions as their symbols on the circuit diagram (Fig. 33–12). The unit is large enough to be easily seen when placed on the lecture hall bench, but it can also be fitted to the slide stage of a standard 10 x 10-in. overhead projector, thus permitting considerable magnification by shadow projection.

The circuit is quite standard, and is described in F. E. Terman, *Radio Engineers Handbook* (McGraw-Hill Book Co., Inc., New York, 1943), p. 505, as well as elsewhere. The period of oscillation is about 20 sec. The signal at the plate is displayed on a 50-V (20 kΩ/V) projection meter and varies between 8 and 42 V. The EF 94/6AU6 should be selected for low grid current (I_g) and operated at slightly lower than nominal filament voltage (5.9–6.0 V) so that I_g will be reduced further. The capacitors for the phase-shift network should have very low leakage currents if the oscillator is to be satisfactorily stable and predictable at very low frequencies. The economical solution adopted was to use 2.5-kV oil-paper, hermetically sealed, military grade capacitors obtained on the surplus market. Ideally the time constants in the phase-shift network (R_1C_1, R_2C_2, R_3C_3) should be the same. Good results are obtainable, without resorting to expensive close-tolerance components, by carefully selecting resistors appropriate to the available capacitors. The 0.005-μF capacitor, shown between plate and cathode, was used to suppress high frequency oscillations in the prototype, but may not be essential. Occasionally, oscillations will not start spontaneously and must be stimulated by varying the plate supply ±25 V at a rate approximately equal to the natural period of the circuit. Once some variation in plate current has been obtained, the B+ is set at 300 V. The circuit will then stabilize in a few minutes, producing an output signal of constant amplitude and frequency.

33–2.8

The circuit shown in Fig. 33–13 is designed to show the effect of the addition of waveforms of different frequencies. The output of an audio oscillator,(10) fed through a step-up transformer, is

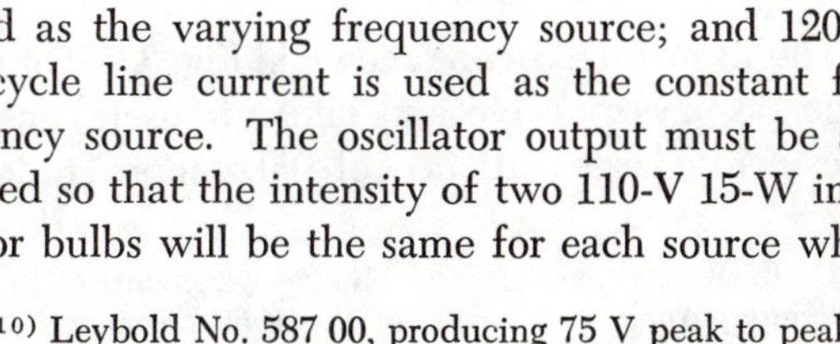

Fig. 33–13

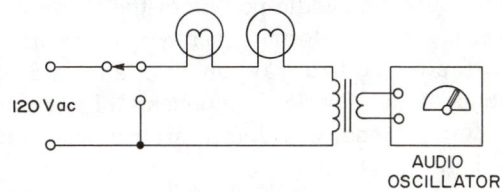

used as the varying frequency source; and 120-V, 60-cycle line current is used as the constant frequency source. The oscillator output must be adjusted so that the intensity of two 110-V 15-W indicator bulbs will be the same for each source when

(10) Leybold No. 587 00, producing 75 V peak to peak, is especially suitable. Available through LaPine Scientific Co.

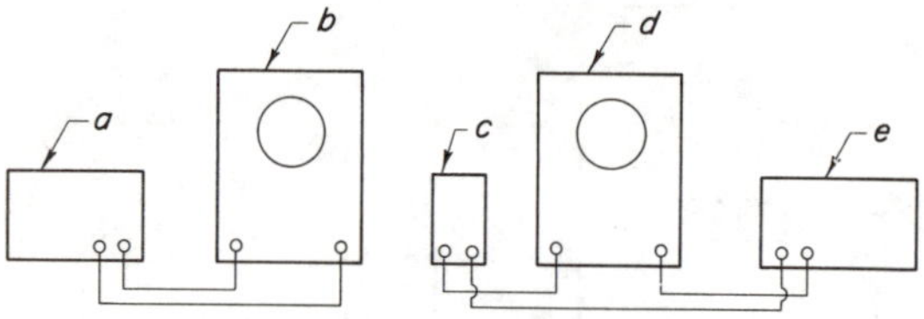

Fig. 33–14

they are applied separately. When the sources are applied simultaneously, the bulbs will show beats as the frequency of the audio generator is varied.

33–2.9

Waveforms of any desired shape can be produced by varying the light intensity reaching a photocell or photomultiplier. The variation of intensity with time is easily accomplished using light from the face of an ordinary oscilloscope. A 1-Mc/sec sine wave from generator *a* in Fig. 33–14 is applied to the vertical input of the scope *b*, which is set for a sweep frequency of about 50 cps. This will result in a vertical stripe which progresses across the scope face. For best results, the persistence time of the phosphor should be as small as possible. A cardboard mask in the shape of the desired wave is then placed over the scope face, and the photocell *c* is placed in front of the scope face without any intervening optical devices. The signal from the photocell has the shape of the cutout with a repetition rate equal to the repetition rate of the low-frequency sweep, as can be seen on a second oscilloscope *d*. In order that the intensity of the vertical stripe be uniform over the height of the cutout, only the middle portion of the stripe should be used. The waveforms generated by this arrangement can be fed to a wave-analyzer *e* for a demonstration of the Fourier components of a complex waveform. Some waveforms are shown in Fig. 33–15.

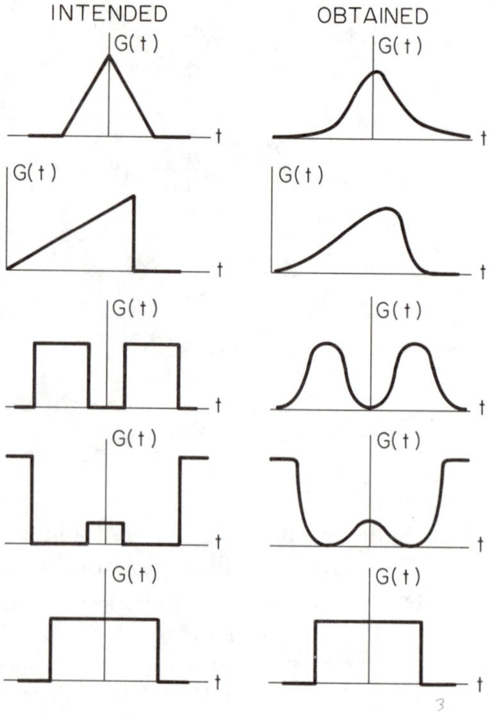

Fig. 33–15

33–2.10[11]

The 12-in. oscilloscope shown in Fig. 33–16 makes large displays possible and hence is useful in lectures before large audiences of 100 or more students. Since it is designed primarily for demonstration work, it has been possible to bring the cost of the instrument down by omitting extra circuits needed in quantitative, high-sensitivity, or high-frequency work. The amplifiers are linear well beyond the audio range, and the input required is about 2 V. The electronic details of the scope are given in the Appendix, page 1337.

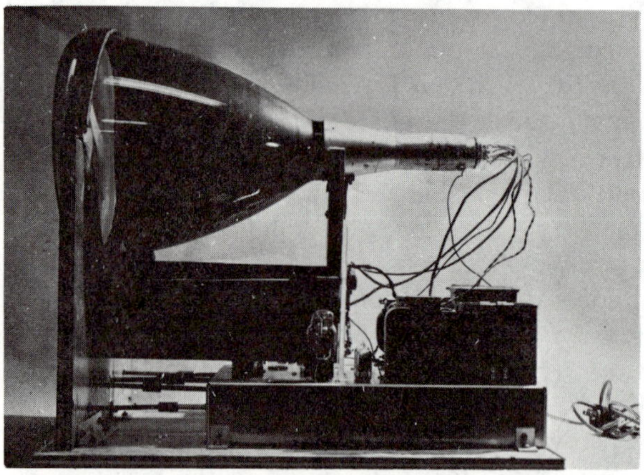

Fig. 33–16

[11] R. Resnick and H. F. Meiners, *Demonstration and Laboratory Apparatus Report of the 1960 Summer Visiting Professor Workshop at Rensselaer Polytechnic Institute.*

33–3 Forced Oscillations and Resonance

33–3.1

A mechanical analog[12] of an electric filter circuit demonstrates parallel resonance. The pulley and cord arrangement and the circuit it represents are shown in Fig. 33–17. The external circuit can be motor-crank operated, or better, manipulated by hand until a flexible meter stick (representing a capacitor) and a loaded truck (which represents a coil) attain parallel resonance. At this frequency, both the meter stick and the truck move with large amplitude in opposite directions. The hand on the external circuit moves very little, but feels considerable force. The pulleys must be good ones with little friction, and the strings must be taut.

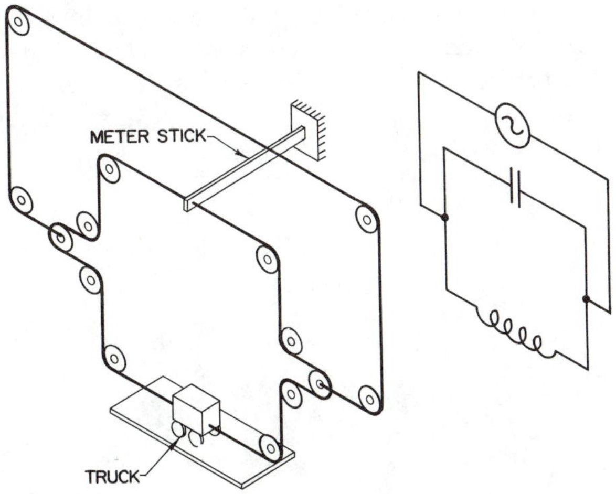

Fig. 33–17

33–3.2[13]

The oscillation circuit consists of a Leyden jar and a large rectangular loop of thick brass or copper wire (60–80 cm long x 25–30 cm high). One end of the loop is in contact with the outside coating of the jar, and the other end forms a spark gap with the center terminal. (See Welch resonant Leyden jar apparatus #2629.) The circuit is supplied by a medium-sized induction coil. The resonating circuit is similar except that one end of the loop is open. A movable cross-connector of copper wire forms a bridge between the two long edges of the rectangle, making it possible to adjust the size of the loop. In addition, a small Geissler tube is connected across the plates of the second Leyden jar.

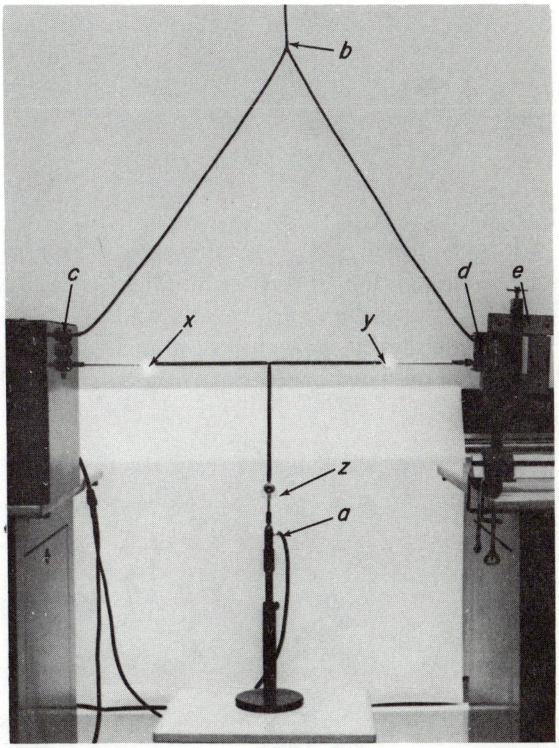

Fig. 33–18

Set up the two circuits so that the loops are parallel and are about twenty centimeters apart. Connect the induction coil and adjust the second circuit into resonance with the first by moving the cross-connector. At the moment resonance is reached, the voltage across the plates of the resonating Leyden jar increases considerably, and the Geissler tube glows. By changing the position of the cross-connector, it can be shown that the tube will not glow in the absence of resonance. The auditorium should be partially darkened, and the spark gap should be covered with a cardboard box.

(12) E. M. Rogers, "Mechanical Analogs of Electric Circuits." *Am. J. Phys.*, **14**, 318 (1946).

(13) Translated by K. V. Robinson and H. A. Robinson (Adelphi University) from *Lecture Demonstrations in Physics*, A. B. Mlodzeevskii, ed., M. V. Lomonosov State University, Moscow, U.S.S.R.

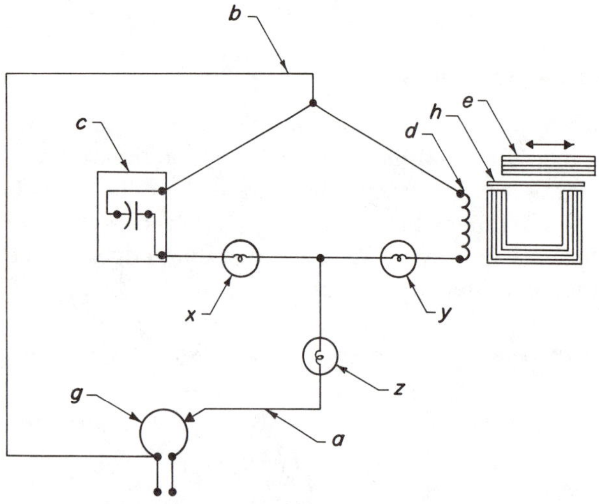

Fig. 33–19

33–3.3

Currents flowing in a parallel LC circuit are easily demonstrated to a large audience with the apparatus shown in Figs. 33–18 and 33–19. A Variac g supplies 60-cycle current to points a and b. Part of this current flows through an 80-μF, 250-V ac oil capacitor c, and part flows through a variable inductor d, which consists of 500 windings on a U-shaped iron core.[14] Tuning is accomplished by sliding an iron bar e to and fro across the top of the U. A 5-mm-thick plastic spacer h supports the tuning bar. Three 6-V 25-W automobile lamps indicate the total current drawn from the Variac and the current through the inductor and capacitor. At resonance the currents in the two branches are nearly 180 deg out of phase and add up to give a very small current from the source. Lamps x and y glow brightly while lamp z stays dark. Removing each lamp separately and changing lamp z shows qualitatively the current distribution.

33–3.4

Next to the coupling coil of a HF generator (30 Mc, 500 W), a series circuit is set up on a tripod with a three edged rail, as shown in Fig. 33–20. The circuit consists of a loop b (one 8-in.-diam turn of ⅛-in.-diam copper wire); insulated,

[14] The U-core and metal bar are available from LaPine Scientific Co. (Cat. No. 56211). The coil, which slips onto the U, is available separately (Cat. No. 56214).

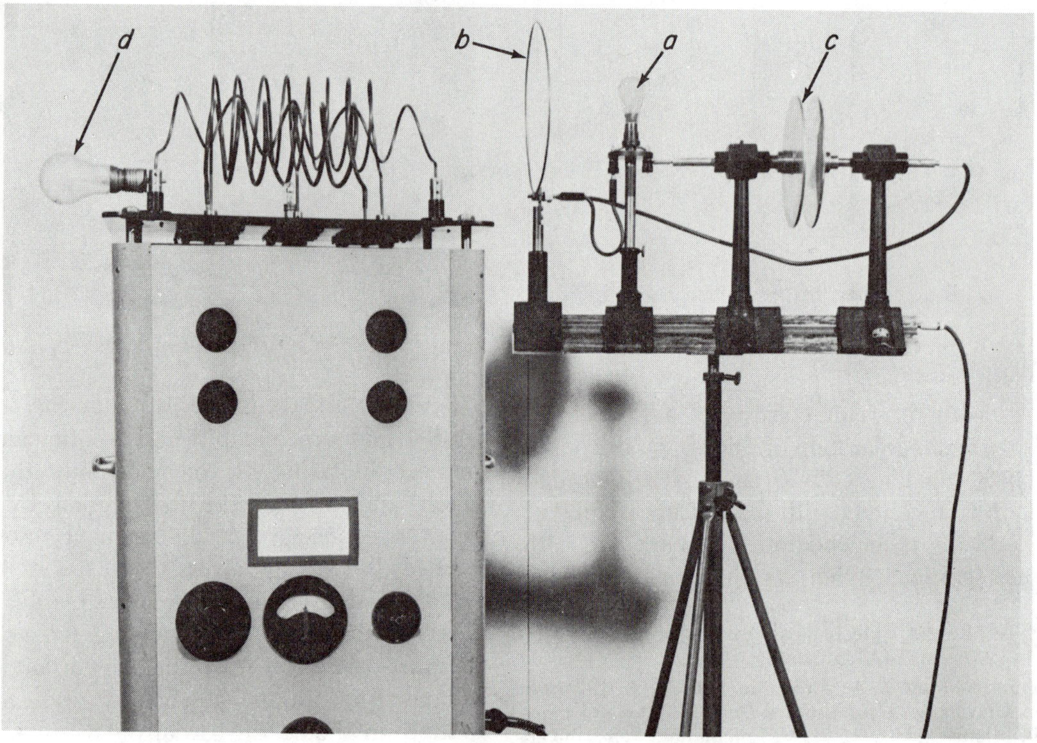

Fig. 33–20

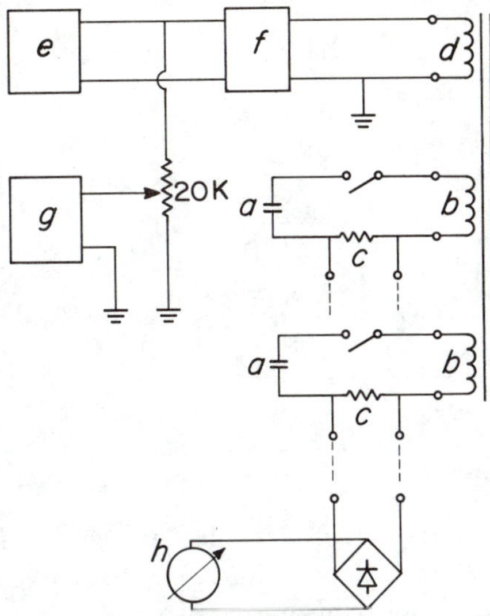

Fig. 33–21

vertical capacitor plates c (6-in.-diam brass); and a bulb a (6 V, 25 W) as a current indicator. In setting up the apparatus, care must be taken that each individual element can be seen when the setup is projected on a screen.

When the generator is turned on, an incandescent lamp d, connected with one terminal to the coupling coil, will indicate the existing H-F field. In changing the distance between the capacitor plates, the secondary circuit can be adjusted until the bulb a burns bright, indicating resonance. The distance between generator and circuit may have to be increased so the lamp will not burn out. It is important to note that the resonance of the circuit is very sharp, and therefore one of the capacitor plates must be moved slowly. For an easier determination of the resonance point, one can use the incandescent lamp d. The bulb used as a current indicator (6 V, 25 W) should have very little inductance.

33–3.5

Two identical oscillating LRC circuits (Fig. 33–21)—each composed of a 0.5-μF capacitor a, a 120-turn inductor coil b, and a 1-Ω resistor c in series—are so positioned (Fig. 33–22) that the two inductor coils can be "connected" by an iron core running through them. The core is long enough so that a 600-turn coupling coil d may also be "connected."[15]

The coupling coil d is fed by a sound generator e (1–10-kc/sec range) in conjunction with a power amplifier f (100 W, 150 Ω). The applied frequency is monitored by a suitable frequency meter g and

[15] The coils are available from Phywe AG., 34 Göttingen, West Germany. Similar coils with a different number of turns (Cenco-Dissectible Transformer, PG 4305) may be purchased from Central Scientific Co.

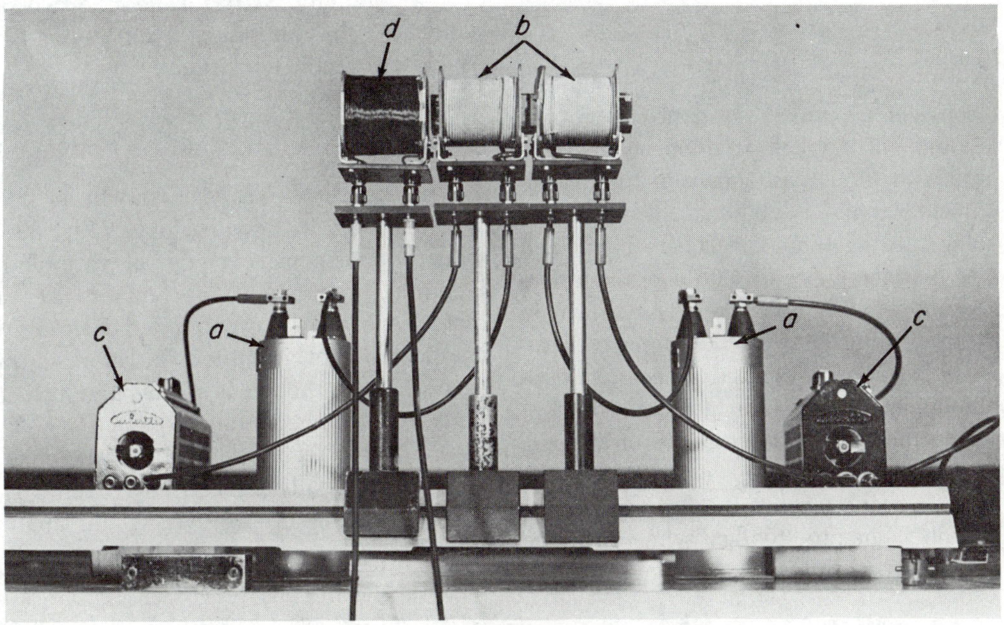

Fig. 33–22

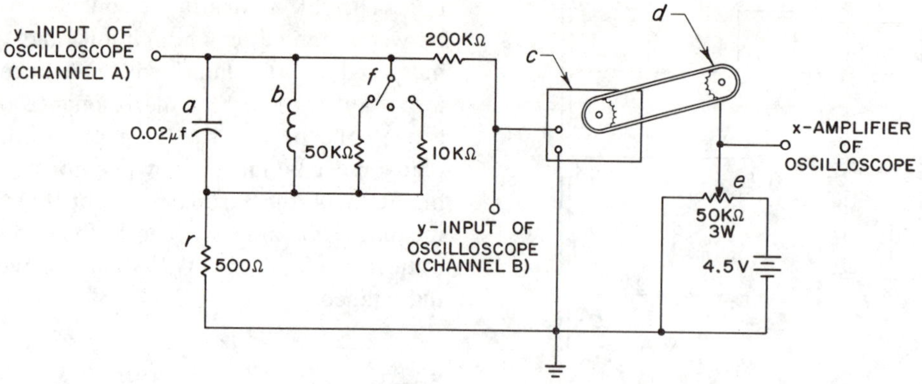

Fig. 33–23

projection unit. A second projection system *h* for alternating voltages, for example, a projection milliammeter or a VTVM, is used to measure the induced emf's in the LRC circuits.

The effect of the coupling is shown as follows: the voltmeter *h* is connected across the resistor *c* of one of the LRC circuits. The other circuit is left open. The applied frequency is varied until resonance occurs (maximum voltage). This will occur at approximately 4.5 kc/sec. The procedure is repeated with the roles of the LRC circuits reversed. The two circuits should have the same resonance frequency. Both circuits are then operated simultaneously, and two resonance frequencies are found; one at 2.7 kc/sec and the other at 6.2 kc/sec.

33–3.6

The response of a parallel resonant circuit to a continuous range of applied frequencies may be demonstrated with the circuit shown in Fig. 33–23. A .02-μF capacitor *a* and a 1200-turn solenoid[16] *b*, with an inductance of about 2 mH, are connected in parallel to produce a circuit with a resonant frequency of approximately 25 kc/sec. As shown in Fig. 33–24, the tuning knob of the signal generator[17] *c* is replaced by a toothed wheel. A wire-wound potentiometer *e* with an identical wheel *d*, is connected to the generator by means of a chain. The *x*-input of the oscilloscope is connected to the potentiometer, and the *y*-input is connected to the resonant circuit. When the toothed wheel is turned rapidly by means of a knob (not shown in the picture), the envelope of the resonance curve of the circuit will appear on the oscilloscope. The shape of the curve is changed with the addition of various resistances by means of switch *f*. By inserting ferrite cores in the solenoid *b*, the resonant frequency can be changed.

The phase relationship between current and voltage may also be shown. One input to a dual trace oscilloscope[18] is the applied voltage; the other input is the voltage across the resistor *r*, the latter being proportional to the current. At resonance a 180-deg phase-shift is observed between the two signals. With small damping the shift is very sudden, whereas with large damping the shift occurs gradually. At resonance the two signals are in phase. The experiment is shown to a large group by closed circuit television.

33–3.7[19]

The harmonic analyzer shown in Fig. 33–25, together with its power supply (Fig. 33–26), provides an inexpensive model of variable-frequency laboratory equipment in common use. This harmonic analyzer, however, demonstrates harmonic composition of complex waveforms quite adequately, even though it is only designed to operate on a fixed fundamental frequency.

A positive feedback loop increases the effective Q of the basically high-Q, L–C circuit, L_1 and C_1, C_2, etc., in Fig. 33–25, making this analyzer quite

(16) Available from Phywe AG., 34 Göttingen, West Germany.

(17) Philips, type GM 2616, 3–30 kc/sec.

(18) Tektronix, type 545A, dual trace may be used.

(19) R. Resnick and H. F. Meiners, *Demonstration and Laboratory Apparatus Report of the 1960 Summer Visiting Professor Workshop at Rensselaer Polytechnic Institute.*

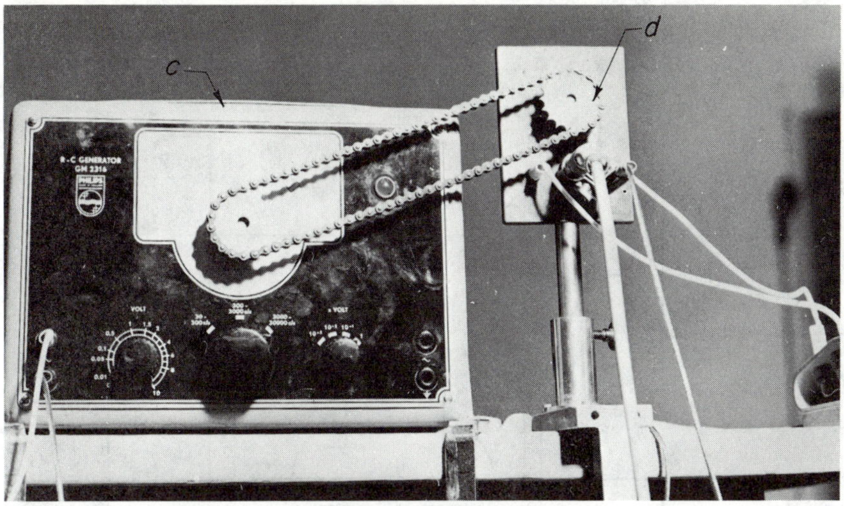

Fig. 33–24

selective. Ten circuits L_1 in parallel with C_1, C_2, etc., activated individually by a ten-position switch, S_1, have component values such that they are selective to ten harmonics of the analyzer's fundamental frequency of 235 cps. As S_1 is rotated through its ten positions, the analyzer output indicates magnitudes of corresponding frequency components of a complex input waveform.

A cathode-ray oscilloscope and an audio oscillator are used to adjust or tune the harmonic analyzer. The ten rheostats designated R_1 and R_3 are set to a minimum value and output potentiometer R_{15} is set in mid-range. After a 15-min instrument warm-up period, the audio oscillator and CRO are connected to the analyzer. With the oscillator set to zero output, each rheostat (R_1 and R_3) is so adjusted that feedback is maximum (without oscillation).

The audio oscillator output is then brought up to a 10-V (p–p) level, at a frequency ten times the fundamental. With switch S_1 in position *10*, the output potentiometer R_{15} and the CRO are so adjusted that the CRO plot amplitude is about 20 percent greater than a main division line on the scope mask. The corresponding position-10 feedback rheostat is adjusted until the CRO plot amplitude decreases to just fit within the main division line on the scope mask. The analyzer is thus calibrated for a tenth harmonic. A similar procedure is used to calibrate the analyzer for the remaining harmonics. When this is done, outputs will be of the same amplitude for all harmonic inputs of the same amplitude.

When calibrating, it is important that the sensitivity of both the CRO and the oscillator be independent of frequency within the range of operation.

When in use, the analyzer operates with a 10-V (p–p) input waveform. Output can be observed with either a CRO or a high-impedance ac voltmeter.

Although of complex waveform, the input signal must have a fundamental frequency of 235 cps (±10 percent) to operate this analyzer. Fine frequency adjustment for maximum output improves results. Harmonic components are read from the output individually, as selected by switch S_1. The list of components of the circuit is given in the Appendix, page 1337.

33–3.8

The spark-excited Tesla high-frequency transformer is now so old that it has never been seen by many present generation students. It provides such an impressive demonstration of electrical resonance that it makes an ideal lecture experiment, especially when a 30-in. discharge is drawn off to a hand-held fluorescent or incandescent light. Mass produced 60- or 120-mA, 15,000-V neon-sign transformers with magnetic shunts have voltage-drop characteristics that make a rotary spark gap unnecessary. Today, the use of signal generator and oscilloscope makes initial adjustments simpler than in Tesla's time. Following are some suggestions it is hoped will be of assistance to anyone setting up a Tesla

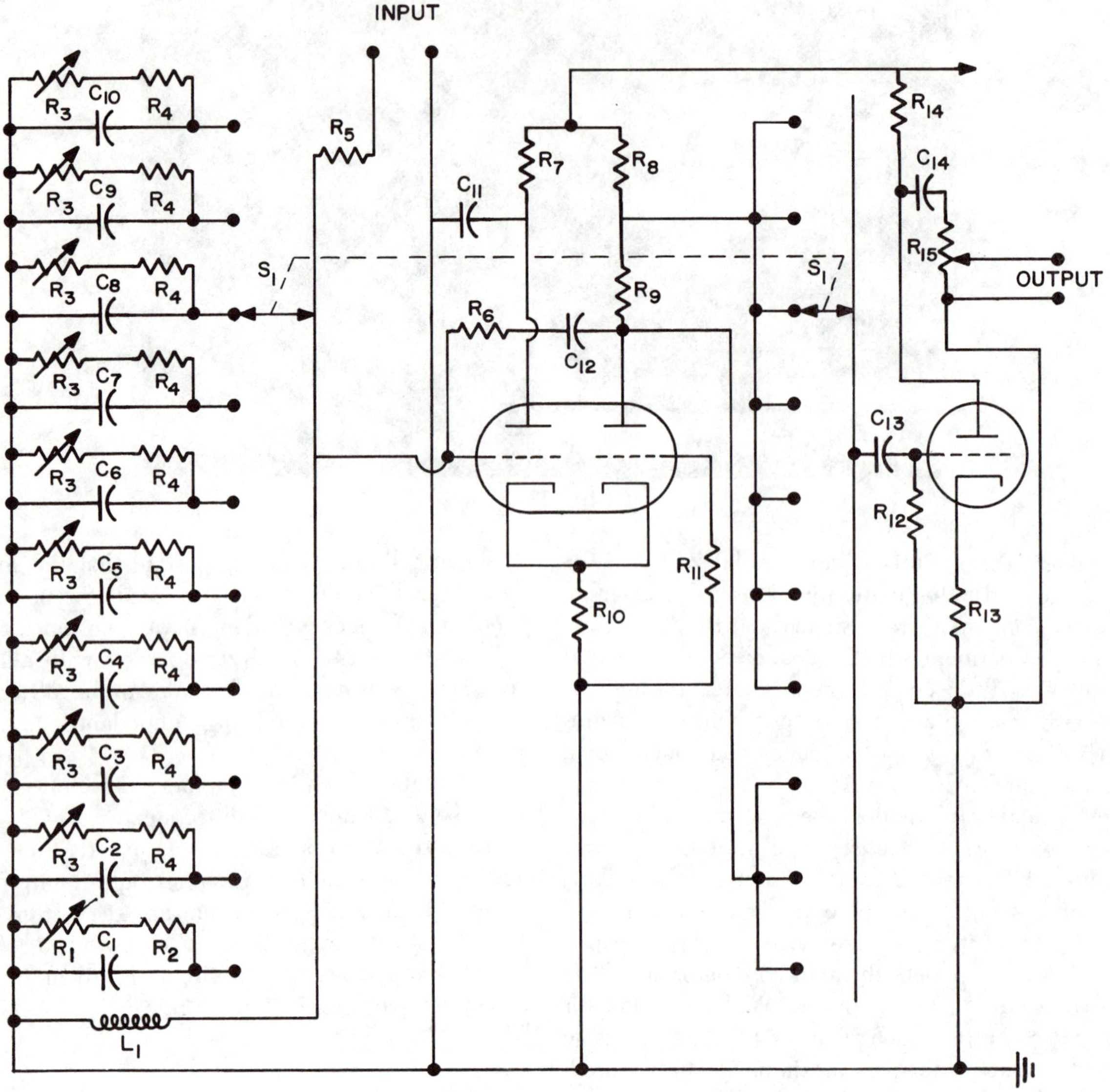

Fig. 33–25

coil (or to be precise an Oudin coil when one end of the secondary is grounded).

A 7-in.-i.d., 40-in.-long Transite (asbestos) pipe has its outer surface trued to a cylinder in a lathe[20] and is threaded, 20 threads to the inch. The lathe bed is protected by papers, and while slowly rotating, the form is given several coats of shellac, allowing time to dry between coats. The grooved form is then wound with a continuous length of No. 26 or 28 Formex insulated copper wire, by allowing the wire unwinding from the reel to slip between two pieces of fiber clamped to the tool rest with its drive adjusted to wind 20 turns/in. Finally it is given two coats of shellac to keep the wire in place and to prevent discharge between adjacent turns. While having lower loss, Fiberglas is carbonized by an arc, whereas the asbestos is not.

An 8-in.-diam gong from an electric bell (discarded because of burnt-up breaker points) makes a suitable top electrode. The bottom of the secondary is wrapped in polyethylene or placed in a

[20] If a lathe is not available, a 6-in.-diam, 30-in.-long, fiberglass tube already wound can be obtained from Polyfibre Ltd., Renfrew, Ont., Canada (the quarter-million-volt oblate for Van de Graaff kits from Morris and Lee, Buffalo, New York, fits the Polyfibre secondary).

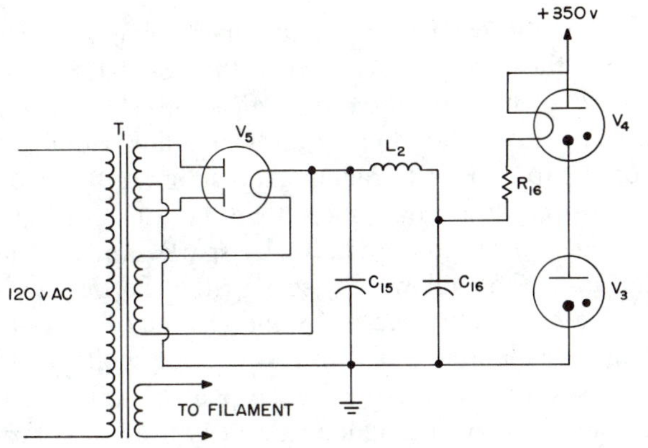

Fig. 33–26

polyethylene waste basket to prevent sparking to the primary, which must be close enough to insure at least critical coupling. (When the primary was excited by an oscillator having its frequency swept in synchronism with the time base of the oscilloscope used to detect resonance in the secondary, the appearance of the characteristic double hump from overcoupling was not observed). The primary diameters are 18 in. and 24 in., respectively, wound with copper shielding for cables or with battery strap or No. 6 stranded wire on wooden dowels held in place by Scotch electrical tape or grooves.

A large top electrode reduces the L/C ratio and hence the voltage, since the Q-value for the system at resonance is

$$Q = \frac{2\pi f L}{R} = \frac{1}{2\pi f C R} = \frac{1}{R}\left(\frac{L}{C}\right)^{1/2}.$$

Around the bottom end of the secondary, place a few turns of wire fed from a signal generator covering 150–450 kc, and connect the top electrode, through a 1-MΩ isolating resistor, to the vertical input of an oscilloscope with the lower end of the secondary to the ground terminal of the oscilloscope. The increases in signal amplitude at the fundamental and harmonics of the secondary's self-resonant frequency are easily observed; the fundamental is 200 kc for the Transite secondary, and 380 kc for the Polyfibre secondary.

The primary must be tuned to the secondary's self-resonant frequency. If maximum energy transfer is to take place from the neon transformer's secondary to the capacitors, their impedance must be equal, so the transformer rating determines the approximate size of the primary capacitors; and this capacitor size with the self-resonant frequency of the secondary then fixes the primary inductance. A 120-mA 15,000-V transformer implies an impedance of 125,000 Ω requiring C = 0.02 μF for 60-cycle matching.

Because neon transformers have the centertap of their secondary grounded to the case, a symmetrical primary circuit balanced to ground is required, so two 0.04-μF capacitors in series are needed. Type G mica are best, but two 16,000-V, 0.05-μF oil-filled capacitors in series from surplus airborne radar modulators are used for short periods with the 60-mA transformer and the Polyfibre secondary. Capacitors can be made from polyethylene or glass, although if made of glass and the glass cracks from uneven heating, it may ruin the transformer.

A quenched spark gap of 4 parallel copper plates 2 in. in diam separated by mica rings, or 8 gaps 5/16 in. in diam in series, is used with the Polyfibre and Transite coils, respectively.[21]

To keep the high-frequency current out of the neon-transformer secondary, where it could break down the insulation between turns, pi-wound radio-frequency chokes from a surplus radio transmitter are used. Alternatively, the chokes may be made with cotton covered wire wound on spools turned in a 2-in.-diam wooden dowel.

The spark gap is shorted to put the capacitors in parallel with the primary, and the length of the primary each side of its grounded centertap is adjusted by sliding a knife-switch jaw along the copper braid until an oscilloscope across the primary (no isolating resistor needed) shows resonance at the secondary's self-resonant frequency when inductively excited from the signal generator. The secondary must not be in the vicinity during this adjustment.

A Variac must be used to bring up the primary of the neon transformer because, if switched directly onto the line at the wrong part of a cycle, the switching transient can double the voltage and break down the capacitors and may destroy the transformer secondary with them.

The bottom of the secondary, the center tap of the primary, and the transformer case with its centertap connection are well grounded. This guar-

[21] A subdivided spark gap (581 35) is available from LaPine Scientific Company.

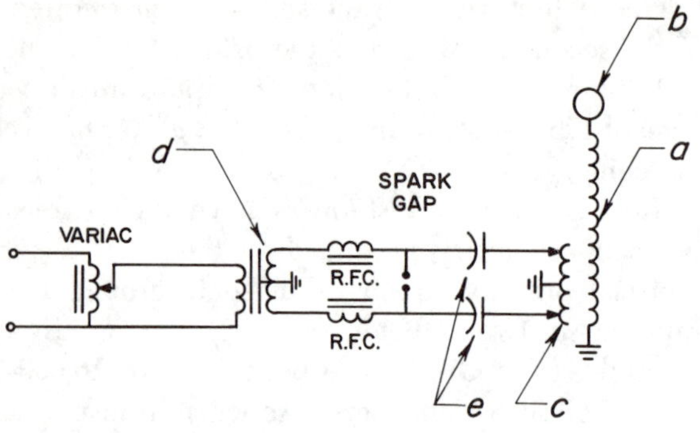

Fig. 33–27

antees that no lethal 60-cycle voltage can exist between the secondary's top terminal and ground, because the reactance of the secondary at 60 cycles is practically zero. One should stand on an insulated stool to prevent sparks through the shoes to the floor connecting the body to ground. When drawing off the high-frequency discharge, a fluorescent lamp or incandescent lamp or metal should be held in the hand to protect the skin from spark burns.

In the circuit of Fig. 33–27, the transformer *d* charges the capacitors *e* until their voltage breaks down the spark gap allowing the oscillatory discharge of the capacitors through the Tesla primary *c*. The magnetic shunt causes the flux from the spark gap current to reduce the voltage of the neon transformer secondary more than its primary's back voltage, so the ionization at the spark gap is extinguished. This allows no escape from the primary's oscillatory energy except by the secondary's *a* resonant oscillation generating sufficient voltage to cause corona and spark discharge at *b*. If a transformer without a magnetic shunt is used, a rotary spark gap must be employed to prevent the spark developing into a power arc. A resistor in series with the primary can also prevent this. The 30-in. Polyfibre coil with a 60-mA transformer in Fig. 33–28 produces 18-in. sparks, and the 40-in. Transite coil with a 120-mA transformer in Fig. 33–29 produces 30-in. sparks. Do not operate without the resonant secondary because the voltage rise on the primary due to its Q may blow the capacitors. Fine adjustment of the primary to ⅛ of a turn leads to improved performance.

In order to minimize the effect of the high-voltage discharge on local radio reception, the coil should be operated in a shielded room or one with metal lath in the walls and ceiling.

The secondary may be regarded as a transmission

Fig. 33–28

Fig. 33–29

line with voltage rise produced by the Ferranti effect. Thus, in the early days, if the length of arc drawn as one went up the secondary passed through a maximum, the frequency was known to be too high and was reduced until the top electrode was ¼ wavelength from the grounded end.

The *Electrical Journal* for May, 1931, described a coil with a primary of 8 turns of 1-in. copper pipe, 48 in. in diam, spaced 3 in. between centers; and a secondary of 600 turns of No. 18 wire, 100 in. long, and 24 in. in diam. The top of the secondary was protected by a guard ring 30 in. in diam, above which extended a 36-in. rod terminating in an 8-in.-diam sphere. The capacitor used 24–30 plates of ⅛-in.-thick glass, 18 x 24 in., coated with tinfoil 12 x 18 in., in series with the high end of the primary and charged by a 10-kVA, 60-cycle, 220–12,000-V transformer. A spark gap of 12 electrodes on a 12-in.-diam Bakelite disk driven by a ½-hp, 1800-rpm motor was used with two fixed electrodes. It produced 10-ft sparks.

John J. O'Neill[22] tells how in 1899 sparks over 135-ft long came from a 3-ft sphere, 200 ft above the earth, excited by a 75-ft-diam primary, before the overload burnt out the alternator of the Colorado Springs Electric Company.

33–4 Displacement Current

33–4.1[23]

With the use of a dielectric, such as barium titanate, which has a large relative dielectric constant $\epsilon_r = 2800$, the displacement current can be demonstrated directly.

The arrangement is shown in Fig. 33–30. The current from the signal generator is passed through the barium titanate capacitor C, a piece of straight wire W, and a resistor R of 20 Ω. The capacitor C, as well as the straight wire W, are surrounded by two transformer cores T_1 and T_2. These cores of 4–79 Mo Permalloy (Arnold Engineering Co. No. 5515-P2) have an i.d. of 0.650 in., an o.d. of 0.900 in., and a thickness of 0.125 in. They are encased in a nylon case with a free opening of about 0.5 in. The cores carry a winding of $N = 100$ turns of No. 40 copper wire. The winding covers only a small portion of the core so that it is possible to surround the winding with a shield S which is connected to the grounded side of the output.

The barium titanate capacitor consists of a piece of barium titanate cut from a large capacitor. The piece has an o.d. of 0.5 in. and a thickness of 9/32 in. The electrodes are painted on with silver paint, and the connecting wires are similarly attached with silver paint. One of the transformers thus has the real current as the primary, while the other has the displacement current as the primary. Since both currents are the same, both output voltages are the same.

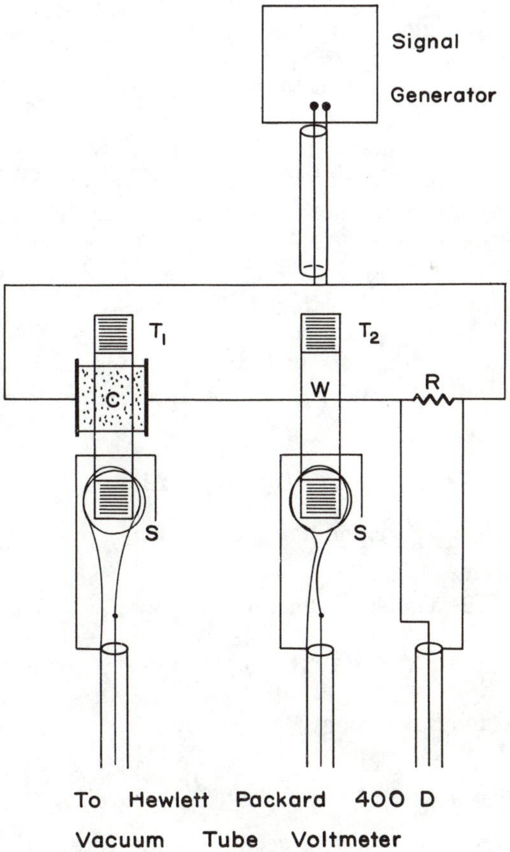

Fig. 33–30

[22] John J. O'Neill, *Prodigal Genius, The Life of Nikola Tesla* (Ives Washburn Inc., New York), ch. 11. See also H. S. Norris, *Induction Coils—How to Make, Use, and Repair Them* (Spon and Chamberlain, 1909), pp. 34–36, and George F. Haller and E. T. Cunningham, *Tesla High Frequency Coil—Its Construction and Uses* (D. Van Nostrand Co., New York, 1910), for further information about Tesla coils.

[23] H. Meissner, *Am. J. Phys.*, **32**, 916 (1964).

The output voltages are measured with a Hewlett Packard 400 D vacuum-tube voltmeter. This voltmeter has an output which gives, for full scale deflection, 0.4 V peak-to-peak with a dc bias voltage of 0.4 V. This requires an intermediate amplifier, such as a so-called "public address amplifier," with an output of 70 V, before a projection meter can be driven satisfactorily.

We give here the simplest theory necessary to show the expected frequency dependence. This derivation is necessary to prove the correctness of the measurement. When the shields (see Fig. 33–30) are left off, a signal is also obtained. This signal, however, does not have the correct frequency dependence.

For classroom discussions other derivations starting with the displacement current are more appropriate.

The current through the system is very nearly given by (V_0 = rms supply voltage)

$$i = \omega C V_0, \tag{1}$$

with (neglecting edge effects)

$$C = \epsilon_0 \epsilon_r \pi r^2 / d, \tag{2}$$

where r is the radius, d the thickness, and ϵ_r the relative permittivity of the piece of barium titanate.

At the mean radius r' of the core, this current sets up a magnetic field H of

$$H = i/2\pi r', \tag{3}$$

where i is the current through the wire for toroid T_2; but for toroid T_1, i is the displacement current i_D:

$$i_D = \epsilon_0 \epsilon_r \int dE/dt \, dA. \tag{4}$$

Since the toroid T_1 encloses all of the displacement current, $i = i_D$.

The magnetic induction in the core is given by

$$B = \mu_0 \mu_r H \tag{5}$$

and the flux through N turns by

$$\Phi = N \mu_0 \mu_r H A, \tag{6}$$

where A is the cross section of the core and μ_r is the relative permeability of the core material.

If V_0 varies sinusoidally with a frequency ω, the output voltage V is given by

$$V = d\phi/dt = \omega\phi. \tag{7}$$

The insertion of Eqs. (1–3 and 6) yields

$$V = NA\mu_0\mu_r\omega^2\epsilon_0\epsilon_r r^2 V_0/2r'd. \tag{8}$$

Alternatively, Eq. (8) can be derived by calculating the electric field across the capacitor and from this the displacement current Eq. (4), etc. The arrangement allows the following measurements:

1. The direct measurement of the current by measuring the potential difference across the resistor R. From Eq. (1) the capacitance can be calculated and the relative permittivity ϵ_r can be determined separately using Eq. (2).

2. The output voltages of both transformers can be measured. They are found to be equal (as they should be, and between 500 cps and 5 kc/sec they vary with the square of the frequency as given by Eq. (8).

3. The self inductance of the transformer winding can be measured. This self inductance is given by

$$L = N^2\mu_0\mu_r A/2\pi r'. \tag{9}$$

This allows an independent determination of the relative permeability of the core material μ_r. (Care must be taken to measure L with extremely small voltages since μ_r depends on the value of H.)

The results are shown in Figs. 33–31 and 33–32. Figure 33–31 shows the voltage across the resistor R

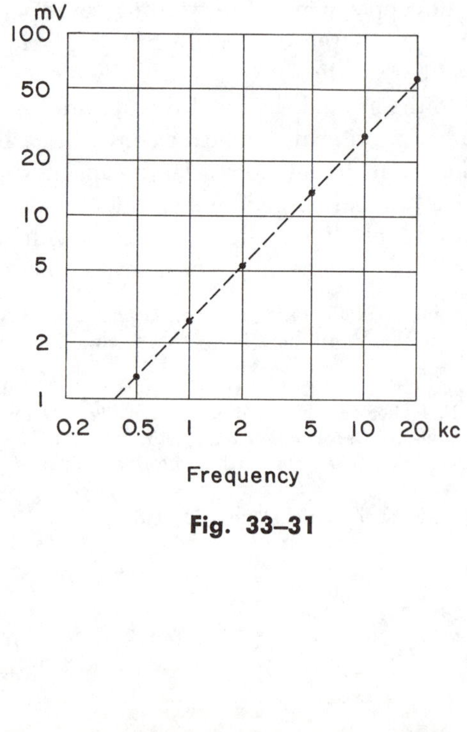

Fig. 33–31

as a function of the frequency. With $R = 20\ \Omega$, and $V_0 = 50$ V, one obtains

$$C = 4.5 \times 10^{-10}\ \text{F}$$
$$\epsilon_r = 2.8 \times 10^3.$$

Figure 33–32 shows that the voltages from the two transformers are very closely the same. In the range between 500 cps to 5 kc/sec, they vary as the square of the frequency. At 10 kc/sec and above the sensitivity of the transformer decreases so that deviations from the square law are observed.

The independent determination of the self-inductance gave

$$L = 4.3 \times 10^{-2}\ H,\ \mu_r = 2.1 \times 10^4.$$

From the values of μ_r, ϵ_r and the dimensions, one obtains at $f = 2$ kc/sec with $V_0 = 50$ V from Eq. (8),

$$V = 1.5 \times 10^{-3}\ \text{V},$$

while the actual measurement gave

$$V = 1.5 \times 10^{-3}\ \text{V}.$$

The agreement proves that the setup actually performs as expected.

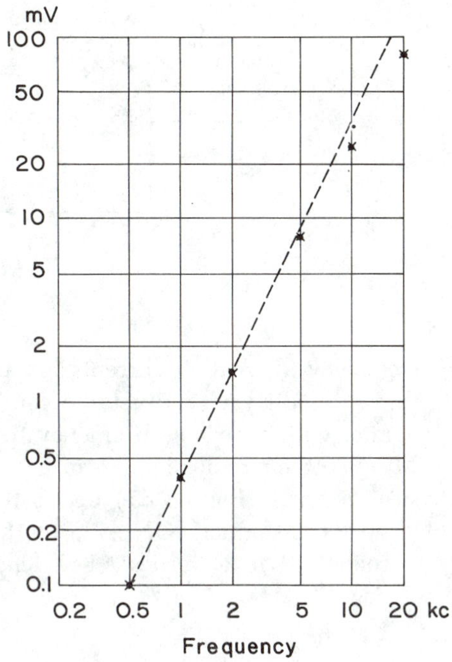

Fig. 33–32

33–5 High-Frequency Oscillations

33–5.1

An incandescent bulb I (40 W, 110 V) receives its power either from a high-frequency ac generator (30 Mc, 500 W) through a 5-turn coupling coil c, or from a 200-V dc source in series with a variable resistor R (300 Ω, 40 W). The schematic is shown in Fig. 33–33. If a suitable dc supply is not available, a 220-V ac, 60-cycle source can be used. The resistor should be so adjusted that the bulb glows equally brightly with either the dc or the high-frequency ac. Note that with 30-Mc ac, the length and arrangement of the lead wires from the coupling coil play an important role. Best results were obtained with 5-ft-long free-hanging cables.

A 10-ft-long cable is looped into a 4-turn induction coil (8 in. diam) and is connected at points A and B in parallel with the bulb. When the dc power source is used, the coil has a lower impedance than the bulb and therefore short-circuits it (the bulb does not burn). When the high-frequency ac source is used, however, the coil has a higher impedance than the bulb and thus acts as an open circuit (the bulb burns). The opposite effect can be demonstrated by replacing the induction coil with a 0.022-μF capacitor in parallel with the bulb. The capacitor acts as an open circuit with dc (the bulb burns), but as a short circuit with high-frequency ac (the bulb does not burn).

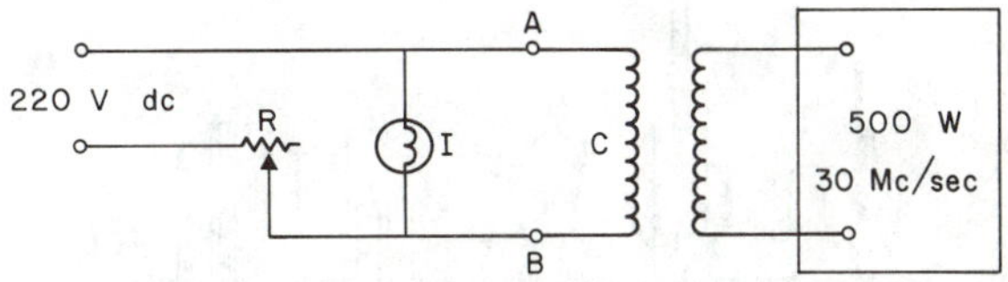

Fig. 33–33

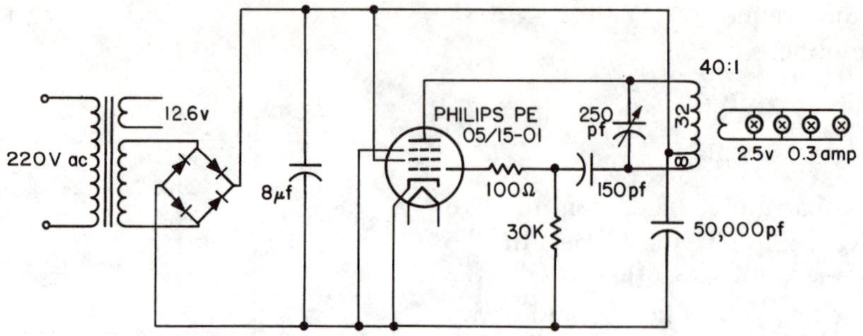

Fig. 33–34

33–5.2

High-frequency alternating currents are carried along a thin surface layer of a metallic conductor. An rf generator with a variable frequency between 1 and 4 Mc (see the circuit diagram in Fig. 33–34 and physical layout in Fig. 33–35) drives the primary of a 40:1 transformer. One side of the secondary is connected to a vertical, 1-m length of brass angle; the other side is connected to each of four concentric conductors, about 1 m long. The inner conductor is 0.5-mm-diam wire, the outer ones are brass tubes with inside diameters of 1, 2, and 4 mm, and outside diameters of 1.5, 3, and 5 mm, respectively. Each conductor carries the high-frequency current to the center terminal of one of four flashlight bulbs, all of whose sockets are connected electrically to the brass angle. Details of the bulb assembly and connections from the four conductors to the bulbs can be seen in Fig. 33–36. The four

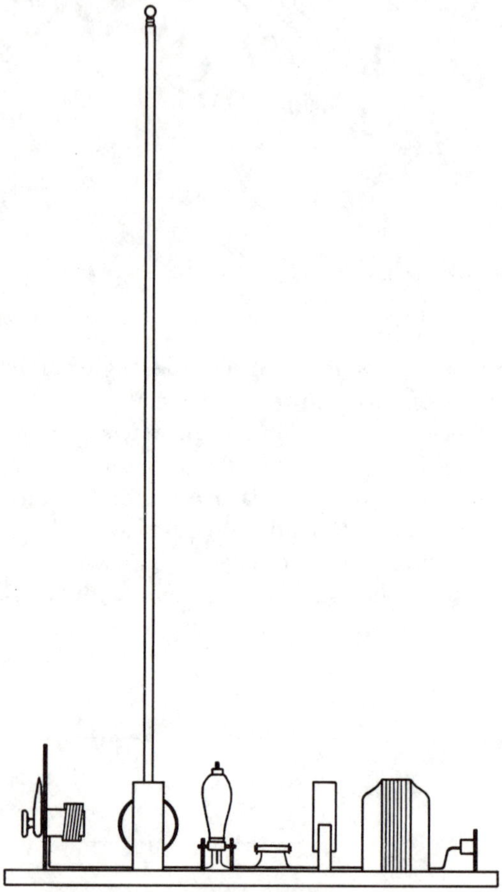

Fig. 33–35

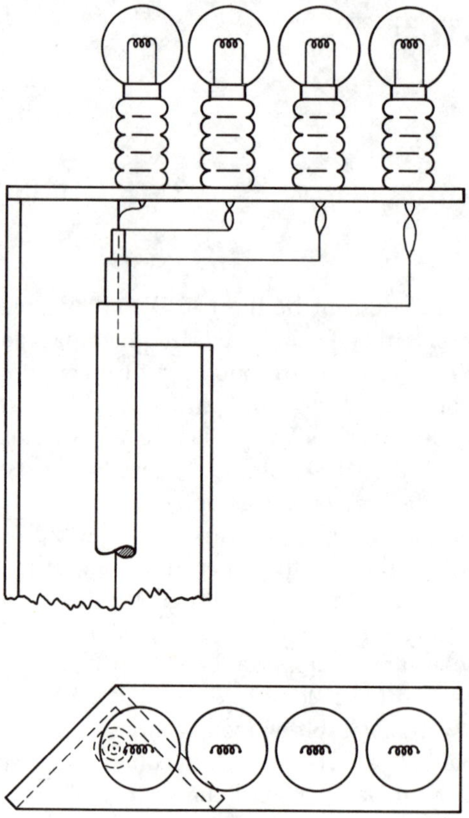

Fig. 33–36

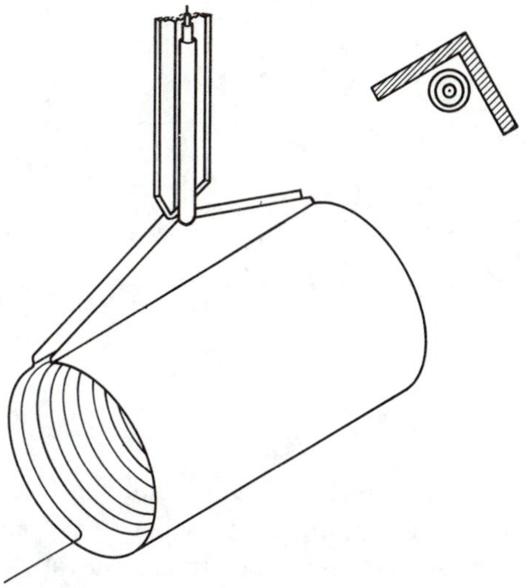

Fig. 33–37

conductors are insulated from one another along their entire length (except at the bottom) by insulating varnish or PVC tape. The primary of the transformer consists of 40 turns of copper wire; the secondary is a 0.5-mm-thick copper sleeve, which fits around the primary coil and is insulated from it by PVC tape (Fig. 33–37). The frequency of the generator is controlled by a variable air capacitor.

In operation all four lamps are lit at the lower end of the frequency range. As the frequency is increased, the lamps fed by the smaller conductors will dim because of the high resistance presented by the thin conducting layer around the outer circumference of the tubes. At 4 Mc, the conductance of the largest tube is still large enough for its lamp to light.

33–6 Transmission Lines

33–6.1

The behavior of transmission lines when excited with both transient and steady (repetitive) waves is illustrated with a line made of sixty sections of series inductors and shunt capacitors. Emphasis is placed on the amplitude, phase, and time relationships between the input and output. An analysis based on low-pass multisection filter theory may also be made.

The transmission line used is of the lumped parameter type and quite standard in design.[(24–26)] Characteristics of the line are shown in Figs. 33–38 and 33–39 and Table 33–1. They can be summarized as follows: number of sections $n = 60$, $L = 10$ mH, $C = 2200$ pF, total delay time $T_d = 281$ μsec, rise time $T_r = 19.6$ μsec, cut-off frequency $f_c = 68.8$ kc/sec, and characteristic impedance $Z = 2130$ Ω.

A generator[(27)] with a variable repetition rate is almost essential. The period of the input square wave should be set to be somewhat in excess of twice the maximum delay time so as to avoid a confusing display. Accurate measurements of delay times, and phase relationships when using a sinusoidal input, are possible only if the oscilloscope sweep is triggered from the input waveform or, if available, the trigger pulse from the signal generator. It is assumed that the two channels of the electronic switch have closely matched pass bands, gains, and delay characteristics. Best results have been obtained when the switching rate of the electronic switch is set low enough and in the correct time relationship with the generator repetition rate so that the oscilloscope display is of the "Alternate

Table 33–1

$L = 10^{-2}$ H $\pm$ 2% ($Q = 60$), Dust Core Toroid
(Surplus–Part No. GH-3104–2)
$C = 2.2 \times 10^{-9}$ F $\pm$ 1%, Silver Mica (Surplus)
$n = 60$–59 π sections + $\pi/2$ matching section at each end
(61 inductors)

$$Z = 2130\ \Omega = \frac{(R_1 + R_G)R_2}{R_1 + R_G + R_2} = \sqrt{\frac{L}{C}}$$

$T_D = n\sqrt{LC} = 4.69$ μsec (1 section)
$= 28.1$ μsec (6 sections)
$= 281$ μsec (60 sections)

$T_R = 1.07\, n^{1/3}\sqrt{LC} = 5.02$ μsec (1 section)
$= 9.14$ μsec (6 sections)
$= 19.6$ μsec (60 sections)

$f_c = \sqrt{LC} = 68{,}800$ cps

(24) J. Millman and H. Taub, *Pulse and Digital Circuits* (McGraw-Hill Book Company, Inc., New York, 1956), chap. 10.

(25) J. F. Blackburn, *Components Handbook–M.I.T. Rad. Lab. Series*, vol. 17 (McGraw-Hill Book Company, Inc., New York, 1949), chap. 6.

(26) W. C. Elmore and M. Sands, *Electronics: Experimental Techniques*, (McGraw-Hill Book Company, Inc., New York, 1949) sec. 2.4.

(27) Hewlett-Packard 202A Function Generator, Hewlett-Packard Co., Palo Alto, California.

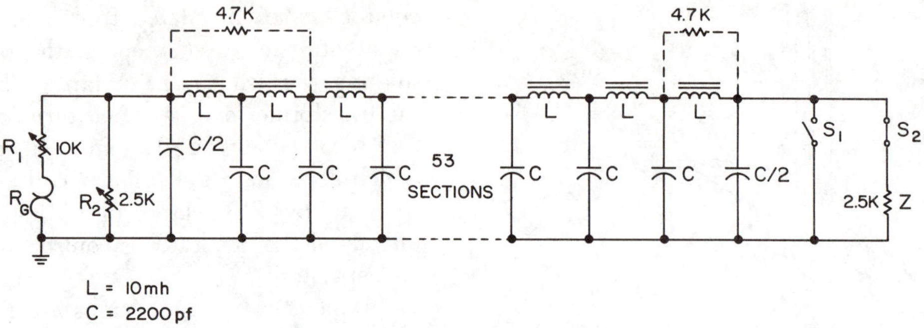

Fig. 33–38

Trace" type (see Tektronix[28] catalog description of Dual Trace Plug-Ins). Most electronic switches do not have adequately high switching rates (7200 kc/sec) to permit their use for a "chopped" display with a delay time of this magnitude.

The line in its present form provides input terminals and the associated matching resistors R_1 and R_2 at one end. The other end of the mounting board contains a set of output terminals, another set of terminals connected in parallel with the input, the terminating resistor Z, two toggle switches S_1 and S_2, and an 11-position rotary selector switch S_3. Proper use of S_1 and S_2 permits the end of the line to be open, terminated, or short-circuited. The selector switch S_3 permits examination of the wave at various distances down the line from the input if it is connected to the oscilloscope. If, however, the oscilloscope is connected to the auxiliary input terminals, and the desired termination connected across the "output" terminals, this termination can be placed at various points along the line and its effect on the deflected wave at the input displayed.

Construction is straightforward, though a few precautions are desirable if best results are to be obtained. Both the inductors and capacitors were obtained as surplus. In spite of the small tolerance on their sizes, their values were measured accurately and assembled with the smallest inductor mated with the largest capacitor, and so on down the line so that a minimal discontinuity was obtained between adjacent sections. The excellent results obtained more than justify the few hours time required for this operation. (The ripple at the top of the waveform is about 20 percent of that obtained when the components were assembled randomly.) The two 4.7-kΩ resistors (shown with dotted lines) were found empirically to reduce the amplitude of the ripple even further (~50 percent). The use of a wire-wound potentiometer for Z, rather than a noninductive carbon unit, also assisted in correcting the phase distortion that produces the ripple. The function of the potentiometers R_1 and R_2 is to permit driving the line from its characteristic impedance while simultaneously presenting the function generator with its optimum load, i.e., 4000 Ω. Adjustment of these is most expeditiously carried out with the line shorted at its far end, and with the oscilloscope set up to observe the waveform on both sides of R_1. Use of other generators, designed to operate into other load resistances, would require rearrangement of the positions of R_1 and R_2, and perhaps an adjustment in their values.

The inductors are fixed to the mounting board with brass wood screws. Plastic shoulder washers prevent chafing of the wires. Tie points for the inductor and capacitor leads consist of insulated standoffs with their studs screwed into aluminum bars. The leads from the selector switch S_3 to the appropriate points on the line, and the leads connecting the input to the auxiliary input, are ordinary vinyl-insulated, stranded, hook-up wire, cabled for neatness and tucked down between the aluminum bars that carry the standoff insulators. Undesirable coupling between sections of the line, due to cabling these leads, is not measurable on this unit.

[28] Tektronix Inc., Beaverton, Oregon.

33–6.2

In connection with discussions concerning the propagation of signals along wires, the following questions arise: When the switch is closed connecting one end of the wire to a constant voltage source V_0, does the voltage V_0 *immediately* appear at the

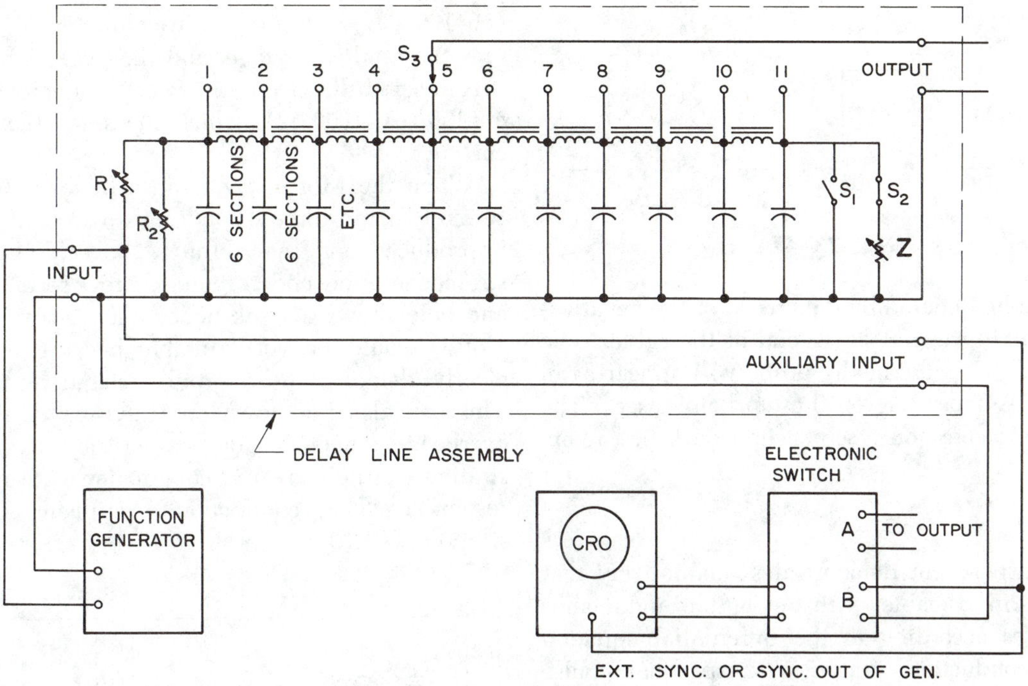

Fig. 33–39

other end of the wire? If not, how long does it take? Analysis shows that to charge a coaxial line very fast requires a very large current in the center conductor. This current produces a magnetic field, and this energy must be supplied by the source. From knowledge of the rate at which energy is supplied by the source and the capacitance and inductance of the cable, the velocity of propagation of the signal can be computed. That this velocity may be considerably smaller than 3×10^8 m/sec and, in fact, is readily measurable comes as a surprise to many students. This demonstration is designed to show that a finite time is required for a signal to reach the end of a cable of even moderate length and to measure the velocity of propagation approximately.

The coaxial cable used is made of two 20-ft sections of RG65/u, a high impedance (950 ohm) commercially available coax. The two sections are joined by an Amphenol type CPH-49190, M-158 tee fitting, and wound on any convenient nonconduction drum as shown in Fig. 33–40. The cable is secured to the drum with standard cable clamps. Additional tee fittings are placed at the ends of the two attached sections and a single tee one foot from the free end of the lower section. This arrangement provides an input connection (at the free end of the lower section) and three taps, 1 ft (a), 20 ft (b), and 40 ft (c) from the lower end.

The input signal must be provided by a good quality square wave generator, e.g., a Hewlett-Packard model 211-A. This signal is applied to the lower end of the cable, and from the same tee a line is run to the external trigger input of a Tektronix 541A oscilloscope with a type 53/54 fast rise preamplifier and fitted with a projection lens. The oscilloscope probe (y-input) is plugged into the desired tap, and the probe ground connection is clipped to the tee shield. With the scope trigger mode in the negative, external position at AC Slow and the stability and triggering level near zero, the

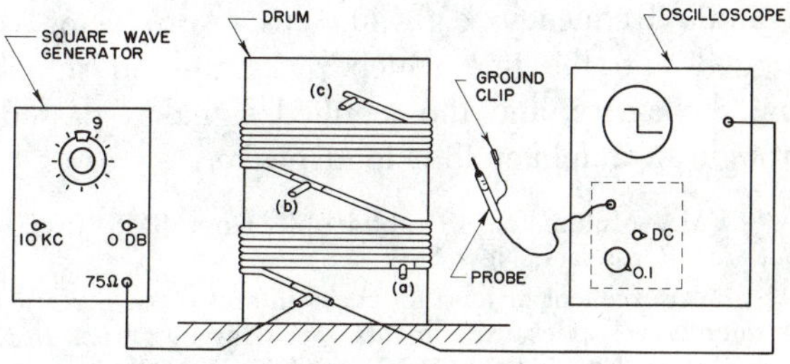

Fig. 33–40

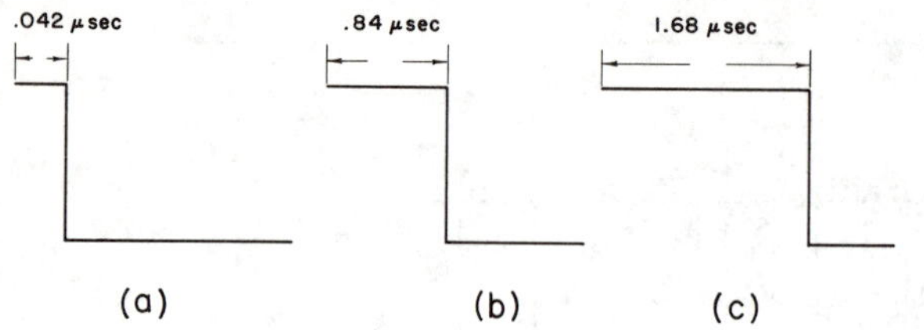

Fig. 33–41

trace will be initiated by the arrival of the negative-going square wave at lower end of the cable. The arrival of the signal at the probe will appear as in Fig. 33–41, depending on the tap being used. The trace is projected on a screen in a darkened room.

33–6.3

This experiment demonstrates qualitatively that a signal in a cable with negligible inductance propagates according to the differential equation of heat conduction: that is, a sharp square pulse at the input gives rise to a delayed flat peak at the output. The artificial cable consists of 151 resistors (1 kΩ) and 150 paper capacitors (2 μF) arranged in an oak box *a* as shown in Fig. 33–42. It is connected over a Morse key *b* to a battery *c* (50–100 V). A dc projection voltmeter *d*[29] measures the applied voltage, and the end of the cable is connected directly to a projection microammeter *e* (150 μA, 1000 Ω), which measures the output current.

When the Morse key is pushed down (for between two and three seconds) a square wave signal is produced and sent through the cable. The instrument at the end responds a few seconds later and only shows a weak deflection. A two-channel plotter would be very suitable because it would clearly demonstrate the pulse shape and delay. This can also be shown on television or with an overhead projector. For comparison this demonstration can be shown in conjunction with the corresponding heat conduction experiment, Demonstration 26–3.10.

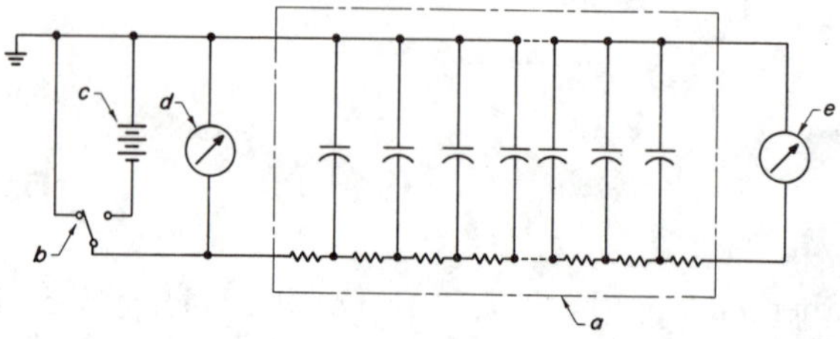

Fig. 33–42

33–7 Microwaves[30]

33–7.1

Microwave resonance can be exhibited with the apparatus shown in Fig. 33–43. A Hewlett-Packard 616A UHF Signal Generator *a* is used as a source. The output of cavity *b* is run through a Dynakit Preamplifier *c* and amplifier *d* to an AR-2 speaker *e*.

A beer can may be used as a cavity, as may a 6-in.-diam, 5-in.-high brass cylinder, fitted with sockets for coaxial lines. The generator frequency used is less than 4000 Mc/sec and is modulated at an audio frequency, e.g., 400 cps. A detector is plugged into the line which picks up the signal from the cavity, and the rectified signal is passed through an amplifier to a loudspeaker.

Small wire loops are fitted to the ends of the lines (Fig. 33–44a) so that the input produces an oscillating magnetic field perpendicular to the plane of the loop. The pickup lead is excited by perpendicular fields. Action of the pickup can be demonstrated by bringing two loops together, giving an audible signal.

With the leads plugged into opposite sides of the cavity and the loops in the vertical plane, the first resonance is found at around 1990 Mc/sec for the cylinder and 3550 Mc/sec for the beer can. Once achieved, resonance can be destroyed by twisting the loops or by poking a stiff wire down

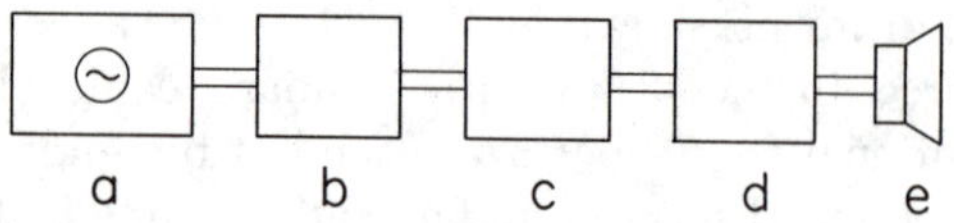

Fig. 33–43

[29] EMA Multimeter 4S, obtainable from SMS Instrument Co., Rensselaer, New York.

[30] Two excellent articles for those interested in performing microwave optics experiments are: Harvey, *Proc. Inst. Elect. Engrs.*, **106**, 141–51, 1959), and G. F. Hull, Jr., *Am. J. Phys.*, **23**, 559 (1949).

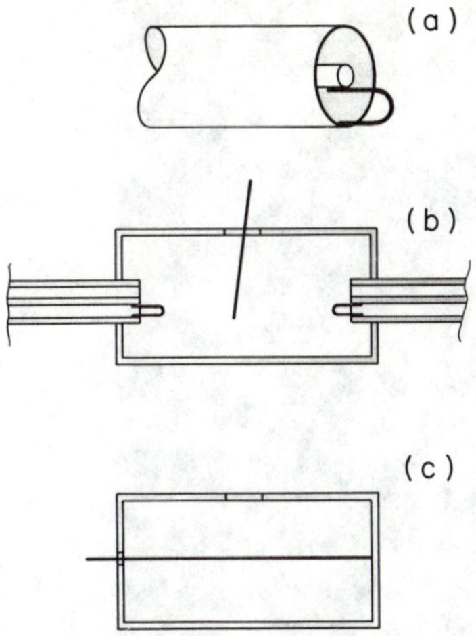

Fig. 33–44

the axis of the cavity (Fig. 33–44b). Poking the wire in sideways (through a small hole drilled for this purpose), as in Fig. 33–44c, shifts the resonant frequency rather than destroying it. Generally if the wire is pushed down only partially, or if a metal plug is inserted into the top of the cavity, the resonant frequency decreases.

33–7.2

Two accessories for demonstrating the optical properties of 3-cm microwaves can easily be constructed from aluminum sheet stock. One is a concave microwave lens, and the other is a prism. The concave lens acts as a convex glass lens would act with light waves, i.e., the emerging waves converge. The concave lens is made up of a series of seventeen or eighteen properly contoured, $\frac{1}{32}$-in.-thick metal sections which are uniformly separated by $\frac{5}{8}$-in. spacers on $\frac{1}{8}$-in.-diam rods (see Fig. 33–45a). The proper separation is just over one half the 3-cm wavelength. The height of the lens is 22 in., its width is 8 in., and its overall length is about 12 in. A microwave prism is shown in Fig. 33–45b. The assembly of the prism is similar to that of the lens but uses twenty-six sections. Each section is an isosceles triangle with a 12-in. base and a $23\frac{5}{8}$-in. height. If the lens and prism are to be used on other microwaves, the dimensions must be changed.

33–7.3

A 4 x 4 x 1-in. Plexiglas box constructed of $\frac{1}{16}$-in.-thick pieces is placed between the transmitter and receiver of a 3-cm-microwave apparatus as in Fig. 33–46. The empty box can be considered to be transparent to both light and microwaves. When the box is filled with water, it is still transparent to light but is opaque to microwaves. A glass box is unsatisfactory because the glass is not totally transparent to microwaves.

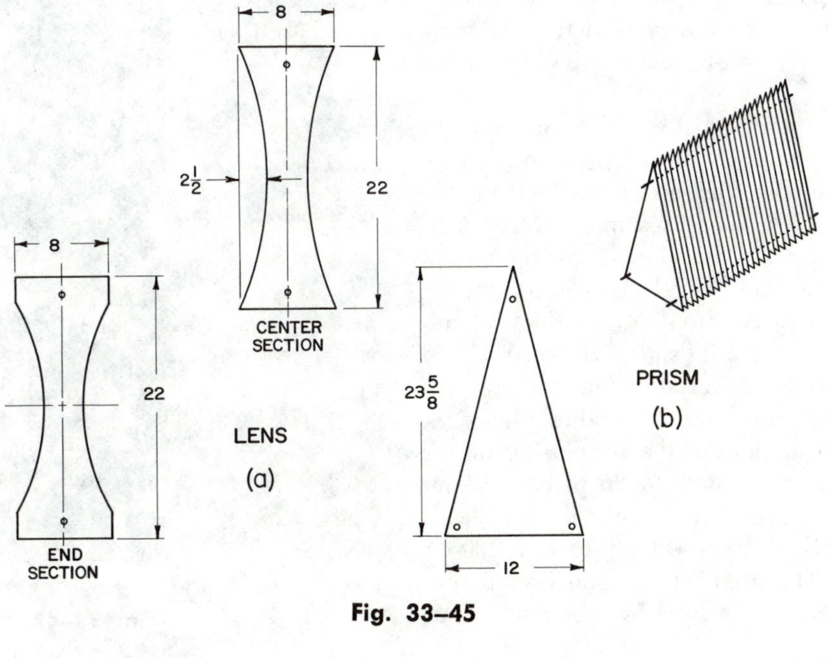

Fig. 33–45

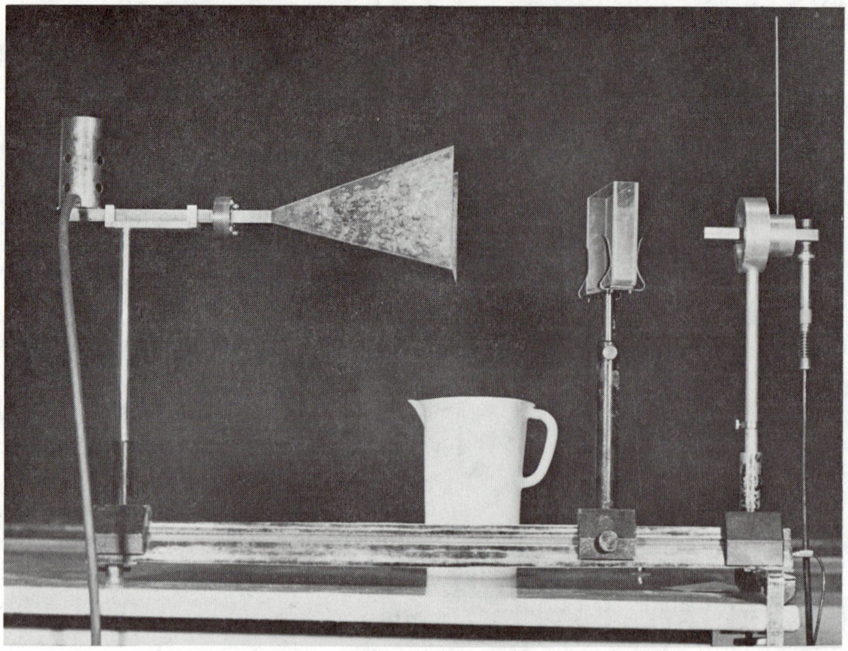

Fig. 33–46

33–7.4

Three microwave interferometers (1) Lloyd's mirror (Fig. 33–47), (2) Michelson's interferometer (Fig. 33–48), and (3) a grid-detection interferometer (Fig. 33–49) serve as a brief introduction to optical interferometers. The purpose of these demonstrations is to shorten the introduction to interference of light by use of hand-sized wavelengths of microwaves.

Each of the interferometers is mounted on a board ⅜ x 8½ x 11 in. The oscillator, power supply, and crystal detector are described in the Appendix, page 1338. The fixed mirror of Michelson's interferometer and the supports for the crystal detectors of Lloyd's mirror and Michelson's interferometer are made alike from $\frac{1}{16}$-in.-thick aluminum with mirror faces 2 in. sq. A rip cord of two leads is soldered to the crystal detector on each side of the glass capsule and drawn through a tight fitting slot in the middle of the supporting mirror so as to hold the crystal detector in place at about a quarter wavelength from the mirror. The leads from the crystal detectors extend to a 0–100-μA microammeter. The fixed mirrors and crystal supports are fastened to the base by wood screws as

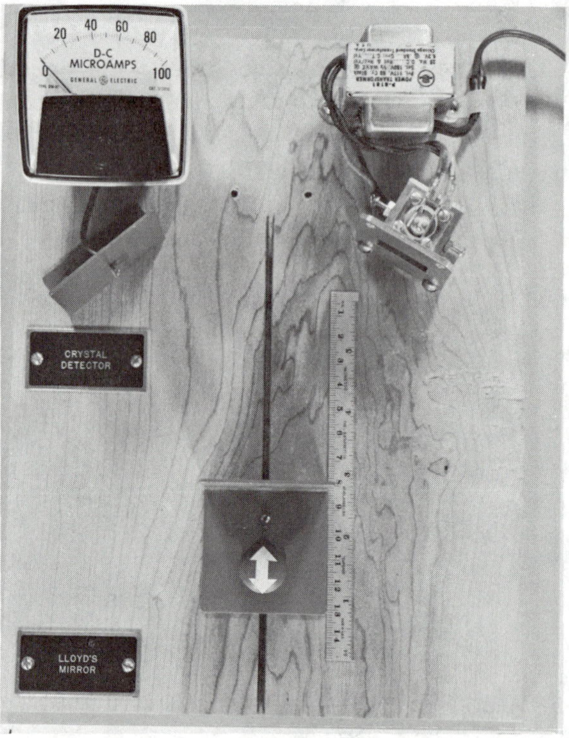

Fig. 33–47

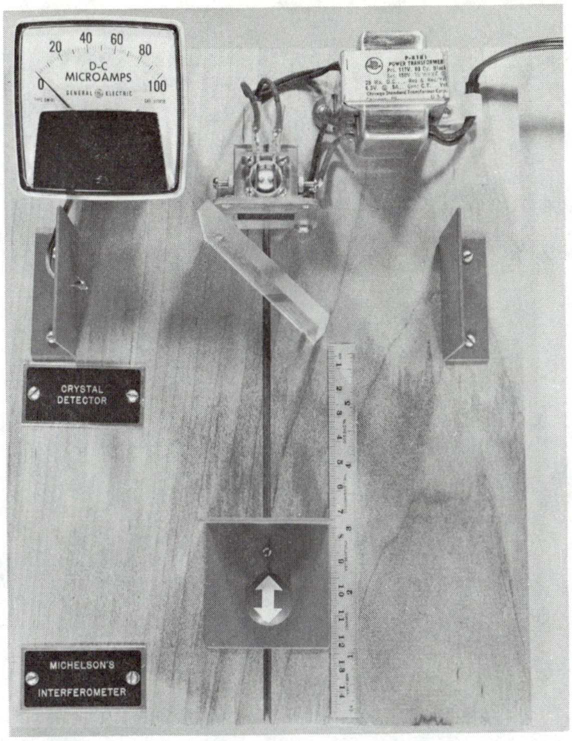

Fig. 33–48

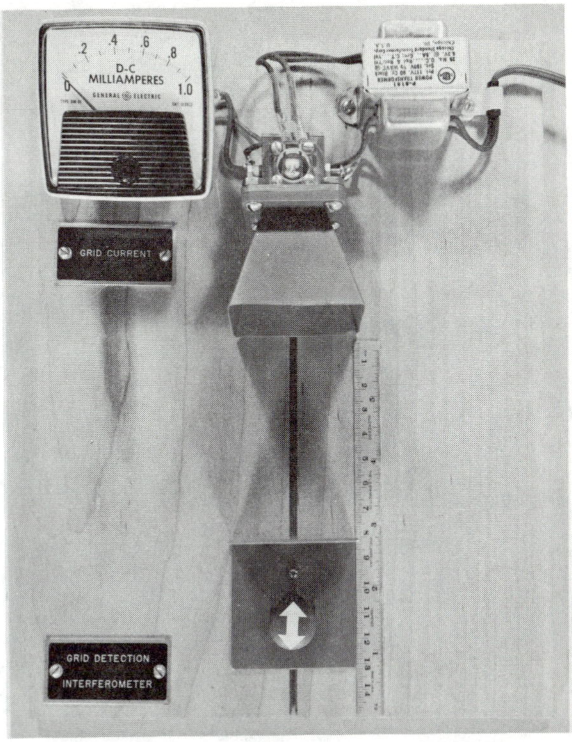

Fig. 33–49

shown in Fig. 33–48. Angle brackets[31] $\frac{5}{8}$ x $\frac{5}{8}$ in. are used to support the oscillators and meters. By bending the supporting brackets, the face of the meter can be set at the desired angle. To serve as a guide for the movable mirrors, a groove $\frac{1}{8}$ in. deep and $\frac{1}{8}$ in. wide extends from the end for 8 in. along the middle of each board. The movable mirrors are made by bending $\frac{1}{16}$-in.-thick aluminum to give two surfaces each 2 in. sq at right angles to each other. To give a more finished appearance, the mirror can be made from 2-in. lengths of 2 x 2-in. angle iron[32] or aluminum angle bar.[33] A metal bar $\frac{1}{8}$ x $\frac{1}{8}$ x $\frac{1}{2}$ in. long is held by a No. 4–40 screw to the bottom of each movable mirror, and slides in the groove of the base board. A wooden handle is fastened to the base of each mirror by a flathead wood screw in the bottom. The arrows on the tops of the handles are cut from white, plastic, gummed tape.[34] The half reflecting mirror for Michelson's interferometer is a sheet of Lucite 2 in. high x 3 in. long. The thickness of Lucite for half reflection at 45 deg incidence was found to be $\frac{1}{4}$ in. The centimeter rulers[35] are glued to the board so that the right hand edges of the mirrors slide along them.

The grid lead of the grid-detection interferometer is brought through an insulator in the left hand wall of the oscillator in the same manner as the plate lead in Fig. 33–48 of the preceding section. A mica washer between a metal washer and the inner wall of the waveguide serves to prevent the microwave energy from leaking out of the waveguide by way of the grid lead. The grid current is indicated by a 1-mA ammeter. If one wishes to make the current vary over a greater range as the mirror is moved away from the source, a sheet metal horn can be added to the oscillator. Figure 33–50 indicates the dimensions of a 0.020-in. sheet

(31) Angle brackets. Cat. No. 3201, Amaton Electronic Hardware Co., New Rochelle, New York.

(32) Angle iron. Edgecomb Steel of New England, Nashua, New Hampshire.

(33) Aluminum Angle Bar, Eastern Metals Wholesale, Inc., Albany, New York.

(34) White Plastic Tape, Labelon Tape Co., Inc., Rochester, New York.

(35) Steel rulers, No. CME 300. The L. B. Starrett Co., Athol, Massachusetts.

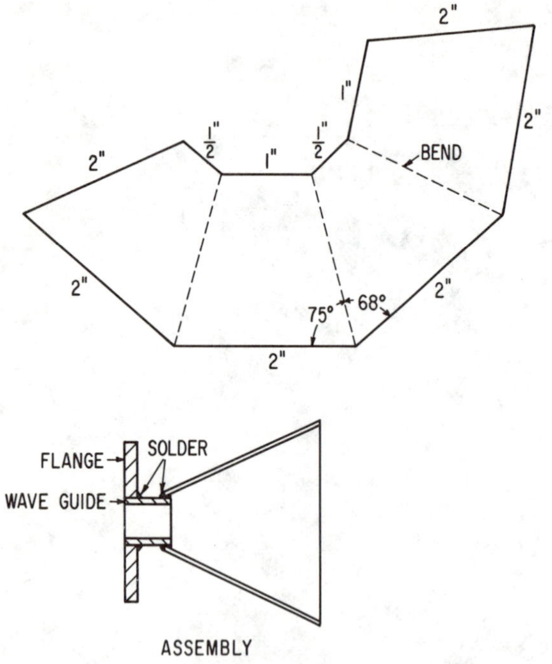

Fig. 33–50

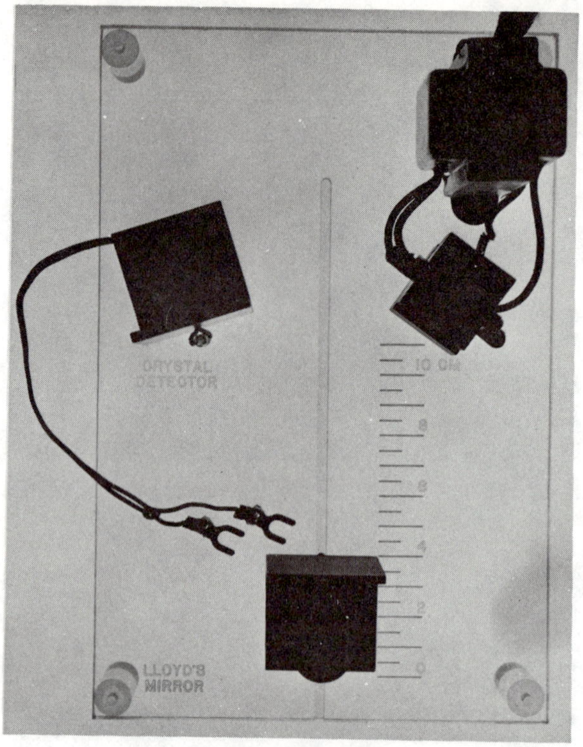

Fig. 33–51

of brass from which to fold a rectangular horn. Beneath it is shown the assembly of an *X*-band waveguide flange, a half-inch length of *X*-band waveguide and the horn soldered together.

If the microwave interferometers are to be demonstrated to large audiences, they can be mounted on sheets of Lucite ¼ x 6 x 9 in., instead of large wooden bases, and projected from an overhead projector. Figure 33–51 indicates the Lloyd's mirror as projected on a screen. The leads from the crystal detector may extend to a Cenco projection meter[36] for which there is room on the horizontal illuminated platform of the overhead projector.

Figure 33–52 shows a projection of Michelson's interferometer. The outline of a miniature projection meter is seen on the base of the crystal holder. The needle extends to a scale made with ink marks in the upper left-hand side of the figure. This projection meter was made by removing the plastic case from a 50-μA AM tuner meter.[37] It is exposed to dust. Commercial projection meters are more dependable.

Since the glass platform of the projector has an index of refraction of 2.5 for microwaves, it also serves as a Lloyd's mirror. The maximum intensity received at the crystal detector was multiplied several fold by mounting the lucite plate on 1-in. legs made of Lucite dowel.

In the Lloyd's mirror demonstration, the direct beam from the source and the reflected beam from the movable mirror interfere at the crystal detector. Since the reflected beam undergoes 180-deg phase change upon reflection, the meter will indicate a minimum signal when the path difference between the direct and reflected waves reaching the crystal is an integer number of wavelengths including zero.

In the demonstration of Michelson's interferometer, four minima in intensity are observed as the mirror is moved the length of the scale. Thus a distance of two wavelengths may be measured on the ruler.

If the same triode tube is used as both source and grid detector, the 45-deg mirror and fixed mirror of Michelson's interferometer are not needed. The incident wave and the reflected wave from the movable mirror form a standing wave that extends back into the resonance cavities of the oscillator to interact with the electron beam. As the mirror

(36) Projection Meter 82550. Central Scientific Co., Chicago, Illinois.

(37) Tuning meter, 50 μA, TM12AM, Lafayette Radio Co., Natick, Massachusetts.

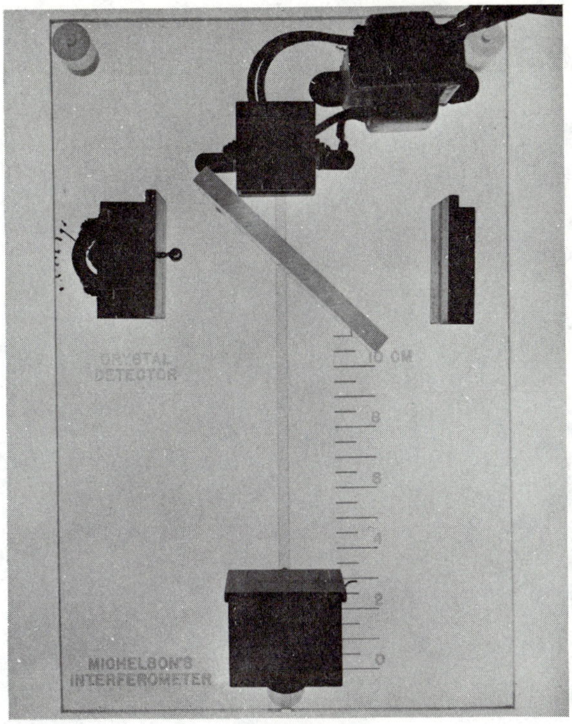

Fig. 33–52

of the grid detection interferometer is moved the grid current undergoes the same changes as that of the crystal detector in Michelson's interferometer.

33–7.5

Microwaves of 3 cm wavelength are particularly well suited for the quantitative investigation of the optical properties of electromagnetic radiation because the radiation is monochromatic, plane polarized, and coherent. Since the wavelength is longer than that of light, both laboratory experiments and lecture demonstrations in physical optics can be performed easily on a convenient observable scale with simple accessories. Bragg's law of diffraction, for example, can be demonstrated on a macroscopic scale. Although macroscopic radiation is not identical to that of light, the use of microwaves will enable the student to understand wave phenomena which are fundamentally the same for each form of radiation.

The eleven microwave optics demonstrations[38] (33–7.6 through 33–7.16) which follow use basically the same apparatus shown in Fig. 33–53 set up for a polarization experiment. Three-cm microwaves are generated by the reflex Klystron *K*, which rests on a transmitter horn *T*. The receiver horn *R* is attached to a crystal mount containing a 1N23B silicon diode. Both horns can be supported on 80-cm booms *B* which are free to rotate. The spectrometer *S* supports the booms and permits measurement of the angular position of the transmitter and receiver horns to 1°.

Since most microwave optics experiments can be performed without external modulation of the Klystron power supply, a galvanometer with a sensitivity of less than 0.01 μA per division can be used as a detector.[39] When a very sensitive galvanometer is used, a shunt may be necessary. If an ac detector, such as a standing wave indicator or a sensitive VTVM, is to be used, modulation may consist of a 1000-cps square wave, a signal from any audiogenerator, or the 60-cycle output of an autotransformer. An ac detector[40] with a sensitivity of about .002 V/division is recommended. For large audiences an amplifier and loudspeaker or an oscilloscope with a closed circuit TV system can be used. The basic apparatus is available commercially.[41]

33–7.6

Michelson's interferometer can be duplicated with microwaves, using the arrangement shown in Fig. 33–54. The transmitter horn *1* and a movable sheet metal mirror *2*, 30.5 x 30.5 cm, are mounted at opposite ends of a double rod optical bench. A metal mesh mirror *3*, with round holes approximately 9 mm in diam spaced 4.5 mm apart, is placed on the bench at an angle of 45°, 25 cm from the transmitter horn *1*. The receiver *4* and another sheet metal mirror *6*, 30.5 x 30.5 cm, are mounted directly opposite each other on a line perpendicular to the bench. Both the mirror *6* and the receiver *4* are stationary and stand 40 cm from the metal

(38) From H. F. Meiners, Walter Eppenstein, Kenneth H. Moore, *Laboratory Physics*, preliminary ed. (John Wiley & Sons, Inc., New York, 1969).

(39) Rubicon, Honeywell, or D'Arsonval type galvanometers, for example.

(40) Such as Heathkit AV-2, VTVM.

(41) The particular equipment used in these experiments was purchased from Welch Scientific Co., Cat. No. 2641. Equivalent apparatus from other companies will work equally well.

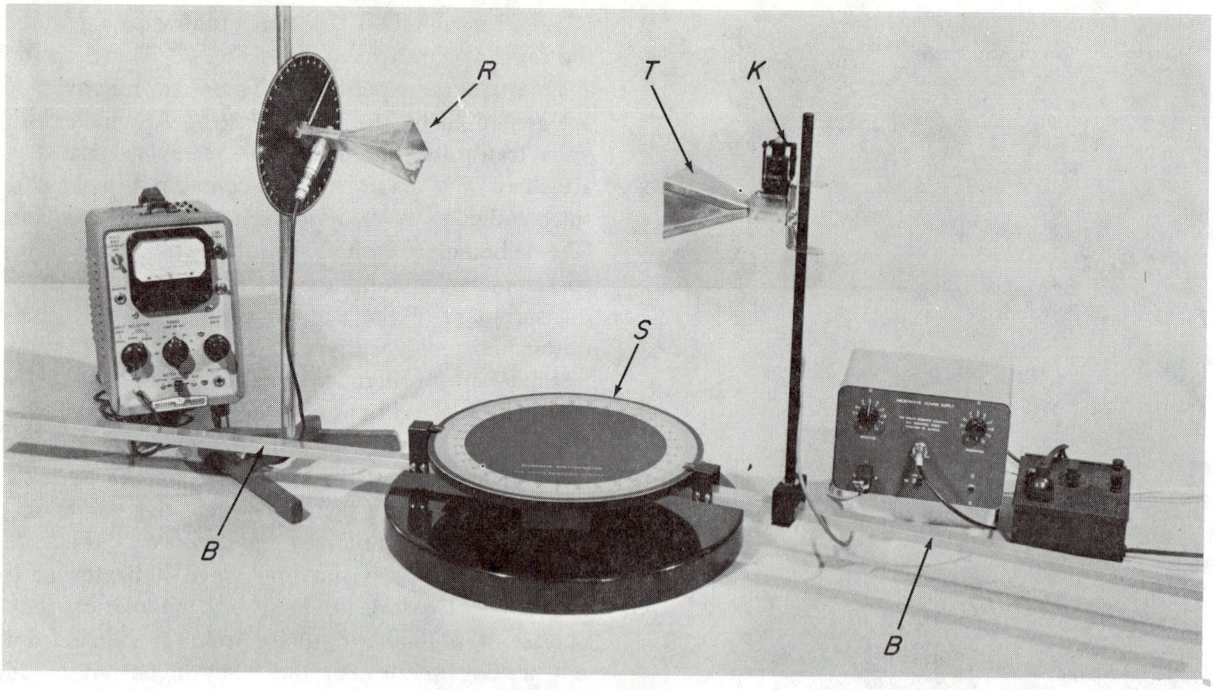

Fig. 33–53

mesh mirror 3. This arrangement can be used to measure the wavelength of the microwaves or to calculate the index of refraction of a dielectric.

The path *ABDBE* is constant throughout the experiment, while the length of path *ABCBE* can be varied by moving mirror 2. When

$$ABCBE = ABDBE + n\,\lambda,$$

where n is an integer and λ is the wavelength, there will be constructive interference and a maximum intensity at the detector. If this path difference is an integral number of half wavelengths, there is destructive interference and a minimum intensity. The wavelength can be calculated by measuring the distance moved by mirror 2 when passing a known number of maxima. First, one minimum N_i is located. Then mirror 2 is moved until twenty maxima are counted, and the position of the final minima N_f is recorded. Since the microwave radiation must traverse the path from B to C twice, we have the following relationship for the wavelength:

$$\lambda = \frac{N_f - N_i}{10}.$$

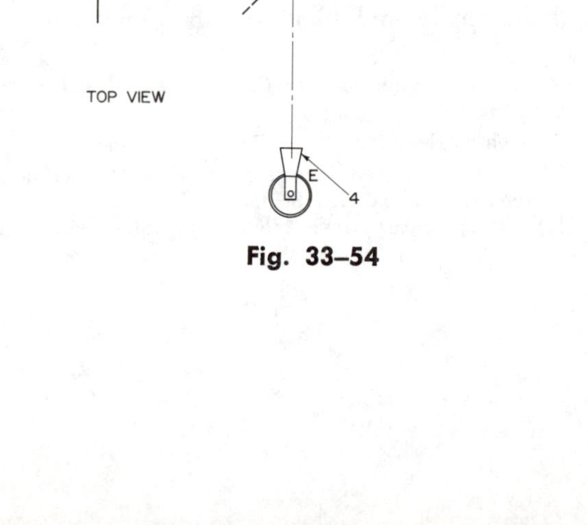

Fig. 33–54

The index of refraction of low-loss materials in thin sheet form (e.g., polyethylene, polystyrene, paraffin wax, Teflon) can also be computed with the interferometer. First, a maximum is located on the detector. The dielectric is then placed between points B and C, parallel to mirror 2, which changes the optical path length. Mirror 2 is then moved a measured distance r in order to restore the original maximum. It can be shown that the index of refraction n is then given by

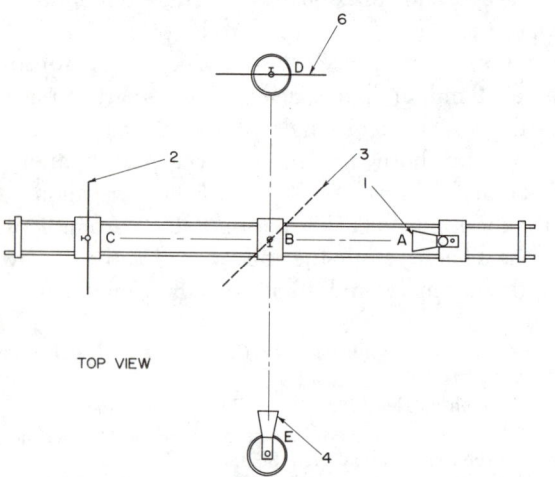

$$n = 1 + \frac{r}{d},$$

Fig. 33–55

where d is the thickness of the dielectric. There will be minor errors in this calculation due to multiple reflections within the material.

The microwave interferometer shown in Fig. 33–55 can be used to show how a platinum–iridium bar(42) f, or a meter stick, can be calibrated with microwaves. Using a known wavelength, the distance required to produce a certain number of maxima is determined and used for calibration. This is similar to the method used to calibrate a standard meter bar using the orange light from Kr^{86} and an optical interferometer.

The interferometer principle is the same as for the optical case, but the operation can be shown in daylight for a large group. Microwaves are radiated from the transmitter A to the partially reflecting, partially transmitting grid B which splits the beam into two components. One component is reflected from the fixed mirror D (aluminum plate) back to the receiver E, while the other component is reflected from the movable mirror C and back to the receiver. The output of E is fed into an amplifier and loudspeaker. As C is moved along the optical bench, the two components of the microwave beam are alternately in phase and out of phase, producing sound maxima and minima which are easily heard by the audience.

(42) The platinum–iridium bar shown is a copy of the standard meter bar kept at the National Bureau of Standards, Washington, D.C. The standard meter is the distance between two scratches on the bar which are located by the white paper arrows.

33–7.7

In each of the microwave optics experiments which follow, it is necessary to have an accurate value of the wavelength of the microwave radiation. The purpose of this demonstration is to determine this wavelength using standing waves.

The experimental arrangement is shown in Fig. 33–56. The Klystron transmitter a and a 30 x 30-cm steel mirror b are mounted at oppposite ends of an optical bench. An appropriate detector or a loudspeaker is connected to the receiver c, which is capable of moving between the transmitter and mirror. By producing standing waves between the transmitter and mirror and measuring the distance between a known number of nodes, we can compute the wavelength λ. As can be seen from Fig.

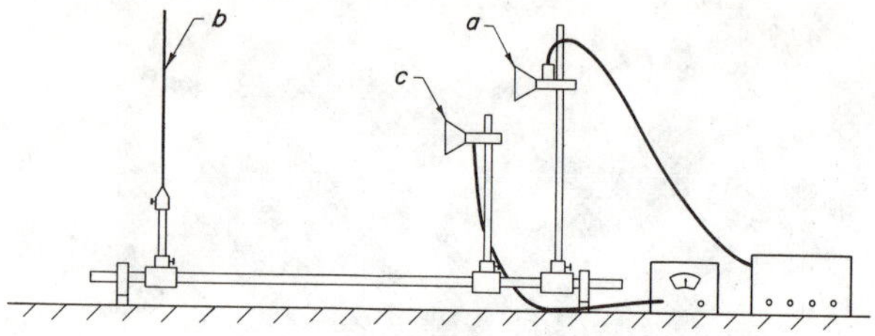

Fig. 33–56

33–57, the distance between two adjacent nodes n or antinodes a is $\lambda/2$. One minimum node is located carefully, and its position N_i is recorded. The receiver is then slid along the optical bench slowly until twenty maxima or antinodes have been passed, and the position of the last minimum N_f is recorded. The wavelength is then given by

$$\lambda = \frac{N_f - N_i}{10}.$$

In a particular sample run, using a standing wave indicator[43] as a detector, this arrangement yielded a value of 3.37 ±0.1 percent cm for the wavelength.

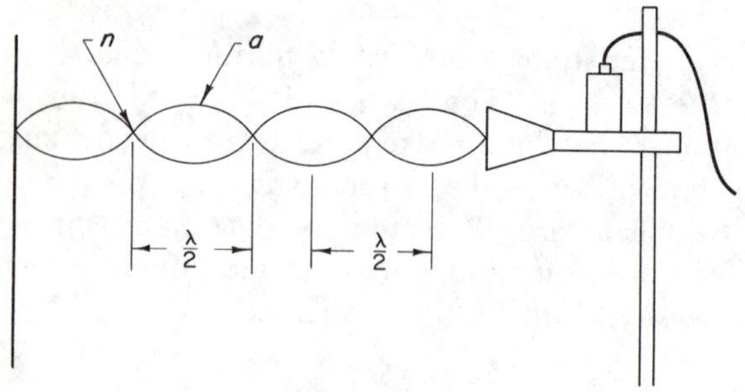

Fig. 33–57

33–7.8

Figure 33–58 shows apparatus that can be used to determine with microwaves the index of refraction of a liquid. A double rod optical bench supports a Plexiglas tank which contains the liquid to be investigated. Suspended in the tank is a reflector c made of sheet steel attached to a board. A scale is mounted on the top of the tank so that the distance moved by the mirror can be recorded. The Klystron transmitter a and receiver horn b are supported by a stand which is located a short distance from the liquid. Standing waves are set up in the liquid, and by measuring the distance between a known number of minima, the wavelength in the dielectric λ_n can be computed. First, the position of one minimum N_i is recorded. Then the reflector is moved until eight maxima have been passed, and the position of the last minimum N_f is located. The wavelength in the liquid λ_n is then given by

$$\lambda_n = \frac{N_f - N_i}{4}.$$

Since the wavelength of the microwaves in air λ is known, the index of refraction n may be calculated from

$$n = \lambda/\lambda_n.$$

With this arrangement a value of 1.45 was obtained for the index of refraction of transformer oil. This agreed very well with the tabulated value of 1.46.

[43] Hewlett–Packard standing wave indicator, Model 415B.

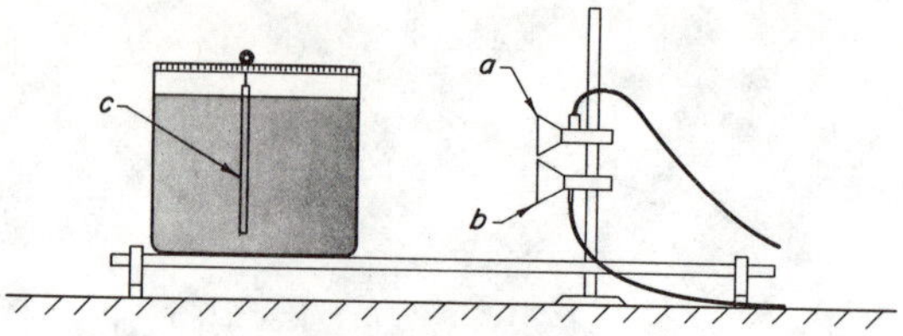

Fig. 33–58

33–7.9

To demonstrate double slit diffraction with microwaves, the apparatus in Figs. 33–59 and 33–60 can be used. A Klystron transmitter a is mounted on a stationary boom 16.5 cm directly behind the double slit assembly b. The receiver horn c is supported by a movable boom which is free to rotate 90° in either direction from its position 51 cm in front of the slits. Both horns are situated 42 cm above the base d which supports the booms. The apparatus provides for measurement of the angular position of the receiver system to 1°. Figure 33–61 shows the layout of the double slit plate which is constructed from cardboard covered with aluminum foil. To some extent the dimensions will depend on the wavelength λ; but with 3.37-cm microwaves, the following values will give good results: the slit width a is 1.6 cm, slit length e is 5.5 cm,

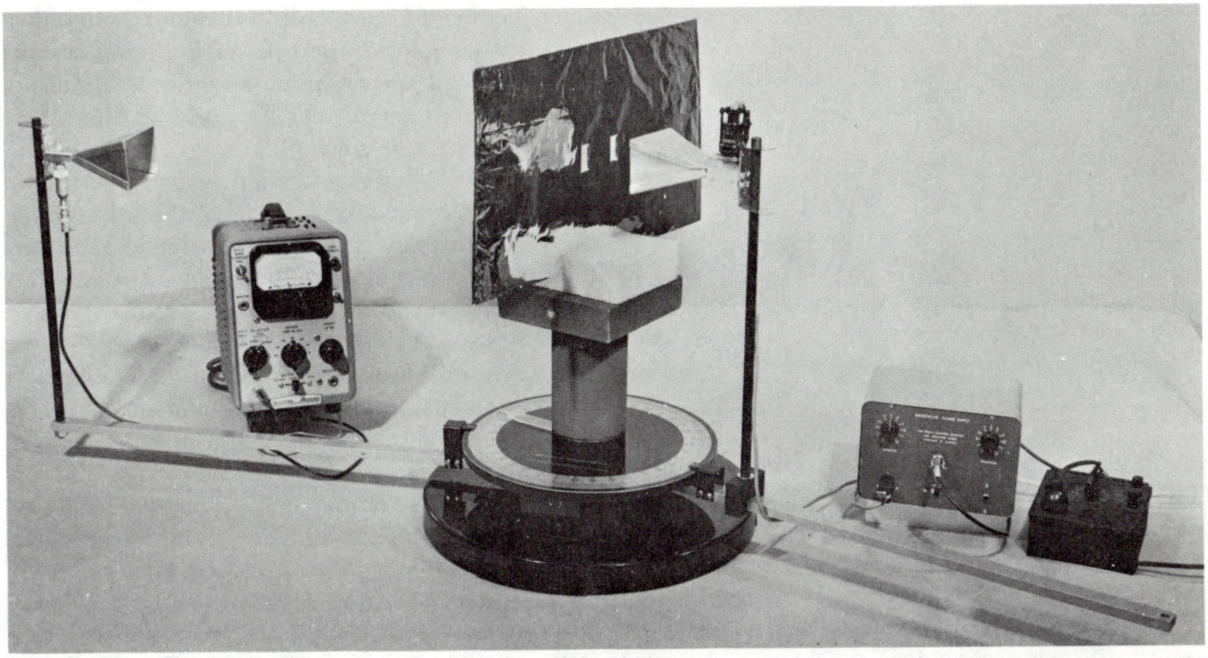

Fig. 33–59

slit separation d is 6.6 cm, and the cardboard is 22 x 30 cm.

The procedure is simple. With an appropriate detector, the intensity is measured every two degrees as the receiver boom is moved from $\theta = -90°$ to $\theta = +90°$, as shown in Fig. 33–60. The theory for Fraunhofer double slit interference predicts that minima will occur when θ satisfies

$$(n + 1/2)\,\lambda = d \sin\theta \qquad n = 0, 1, 2, \ldots$$

and maxima will appear when θ satisfies

$$n\lambda = d \sin\theta. \qquad n = 0, 1, 2 \ldots$$

Therefore, if λ is 3.37 cm, and the slit separation d is 6.6 cm, maxima are predicted at 0° and ±30° 44′, whereas minima should occur at ±14° 50′ and ±50° 10′. The results from a sample run using a sensitive VTVM as a detector are plotted in Fig. 33–62, which shows good agreement with the theory.

If the slit width a is much less than the wave-

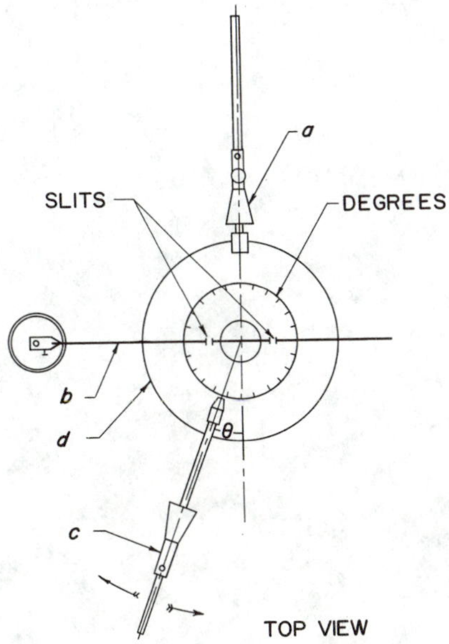

Fig. 33–60

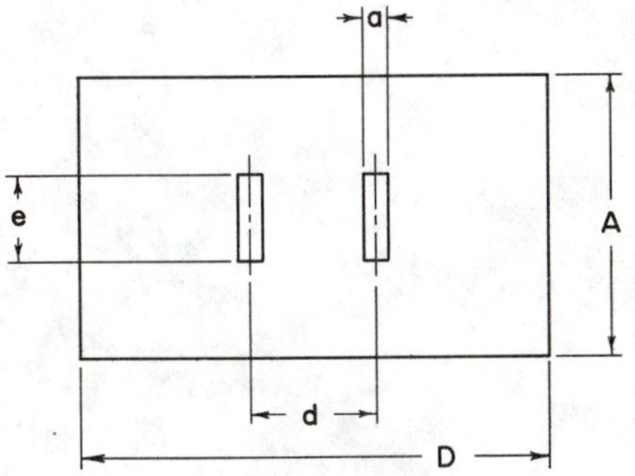

Fig. 33–61

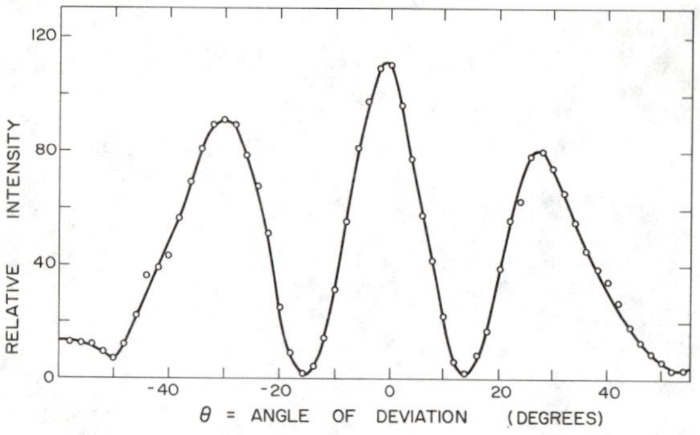

Fig. 33–62

length λ, all maxima will have the same intensity. This requirement is not met in this experiment since a is nearly equal to $\lambda/2$. The result is the appearance of a single slit diffraction envelope which modifies the double slit interference pattern. This is illustrated in Fig. 33–62 where the second order maxima are reduced in intensity.

33–7.10

The index of refraction of a prism can be determined with the equipment in Figs. 33–63 and 33–64, with the addition of a plastic or paraffin prism *4*. Construction of the paraffin prism requires special mention. Hot liquid paraffin is poured in layers ½ in. thick into a Plexiglas mold with the dimensions specified in Fig. 33–65. Each layer must be allowed about 2-hr for complete cooling and hardening before the next layer is poured. After this process is completed, the sides of the prism can be smoothed with a warm iron. In order to eliminate reflection from nearly metallic surfaces, the prism is mounted on a Styrofoam base *3*.

Before the index of refraction n can be measured, one must determine the angle A of the prism using the procedure illustrated in Fig. 33–66. With the transmitter *1* located opposite angle A, move the receiver *5* and record the angular positions B and C from scale *2* of the maximum reflection from either side of the prism. It can be shown that the angle A is equal to one-half the difference between B and C. Using the prism shown in Fig. 33–65, A should be close to 60°.

Next, the angle of minimum deviation D must be determined. With the Klystron *1* aimed as shown in Fig. 33–64, the receiver is moved to find a maximum. Then the prism is rotated so as to achieve the greatest reading. The difference between this setting and the position of the undeviated beam without the prism in place is the angle of minimum deviation D. It can be shown that the index of refraction n is given by

$$n = \frac{(\sin A + D)/2}{\sin(A/2)}.$$

Using a sensitive VTVM as a detector, a value of 35.5° was obtained for D. With A equal to 60°, the calculated value of n is 1.48, which agrees very well with the tabulated value of 1.47 for the index of refraction of paraffin.

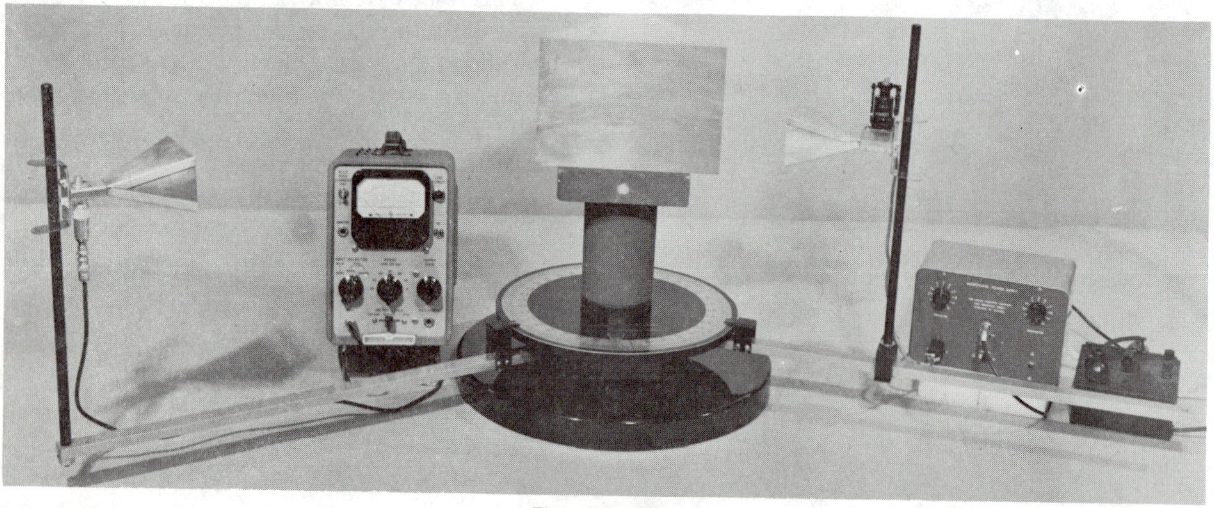

Fig. 33–63

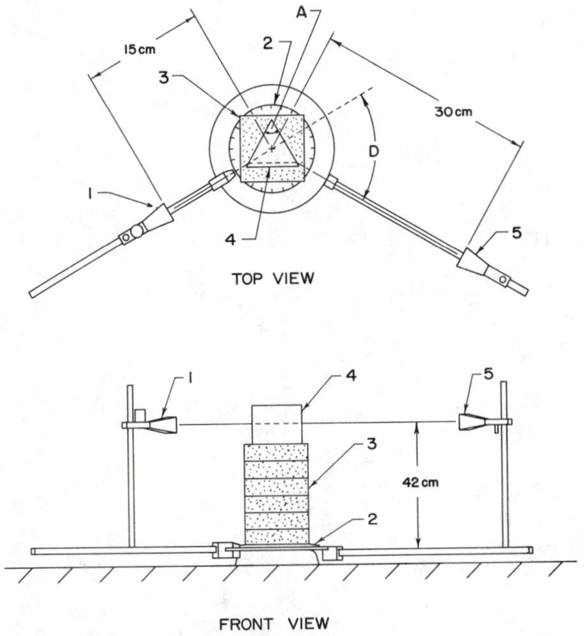

Fig. 33–64

33–7.11

Polarization of microwaves can be demonstrated with the apparatus shown in Fig. 33–53. The radiation from the Klystron transmitter will be polarized in the direction of the antenna, which is vertical for the particular equipment used here. With the receiver placed opposite the transmitter, the intensity should vary from a maximum to a minimum as the receiver horn is rotated 90° from the vertical to a horizontal position.[44] Fig. 33–68 shows the results for a sample run, which follow closely the predicted formula

$$I = I_0 \, (\cos \theta)^2.$$

Polarization by a grating can be shown with the arrangement diagramed in Fig. 33–69 (a). The grating *3* has the dimensions shown in Fig. 33–69 (b) and is constructed with wires 1.3 mm in diam and placed 1.8 cm apart. First, the transmitter *1* and receiver *2* are placed 20 cm from the grating and so positioned that the angle of incidence equals the angle intercepted by the receiver. With the

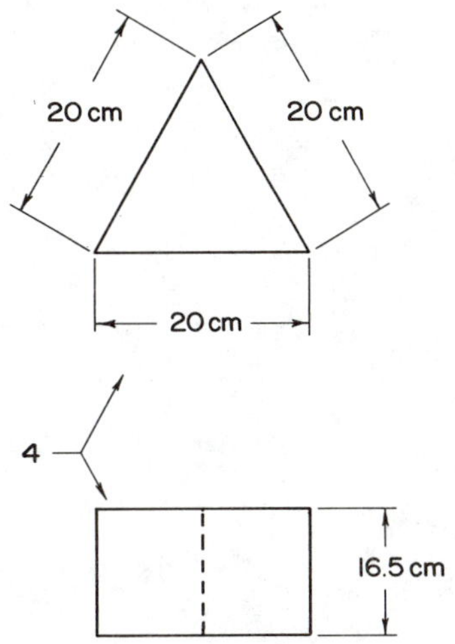

Fig. 33–65

grating wires horizontal there will not be a reflected wave, since the vertically polarized microwaves are transmitted. When the wires are vertical there should be complete reflection from the grating. This can be verified by moving the receiver opposite the transmitter with the grating in place. When the wires are horizontal there will be a maximum reading. Similarly, when the wires are in a vertical position, the intensity is nearly zero or at least greatly reduced, showing that a grating will polarize microwaves.

33–7.12

The index of refraction of a dielectric may be determined using the apparatus shown in Fig. 33–70.

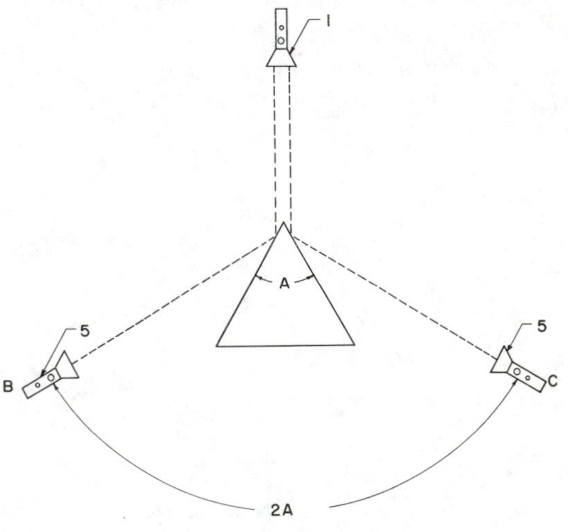

Fig. 33–66

[44] The rotatable receiver horn should be assembled as shown in Fig. 33–67. All parts except the plastic calibrated scale are aluminum. Part *a* attaches to the receiver support, and the receiver horn is held by part *b*. The dimensions are not critical.

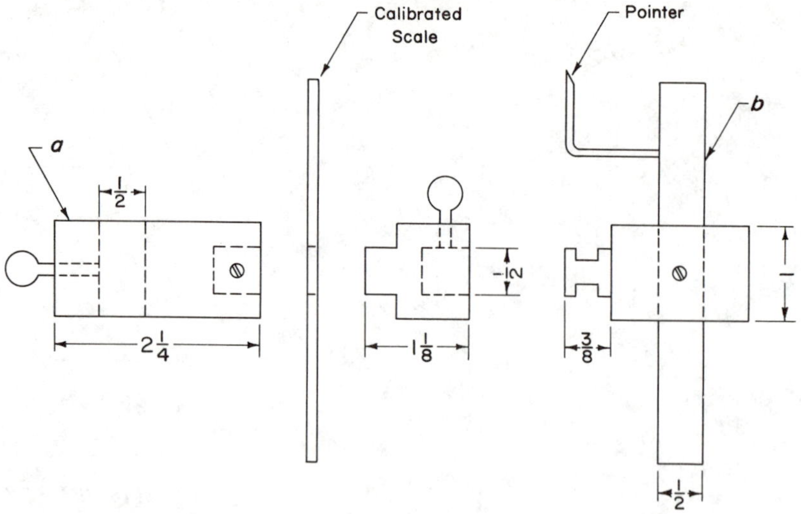

Fig. 33–67

It is known that electromagnetic radiation is polarized perpendicular to the plane of incidence upon reflection from a surface. This polarization of the reflected wave is a maximum at Brewster's angle θ_B. When the Klystron transmitter *1* is used with the antenna in the vertical direction, the microwaves are already polarized in the plane of incidence. Therefore, at Brewster's angle there will be no reflection of this wave, and the transmitted component picked up by the receiver *3* will be a maximum. Since the index of refraction n is equal to $\tan \theta_B$, n can be computed once the angle of maximum transmitted intensity is found.

As shown in Fig. 33–70, a slab of paraffin *2*, 30 x 30 x 4 cm, rests on a base of Styrofoam *7*. A protractor *4* is attached to the slab, while a weight suspended by a string measures the angle of inclination of the paraffin. From Fig. 33–71 it can

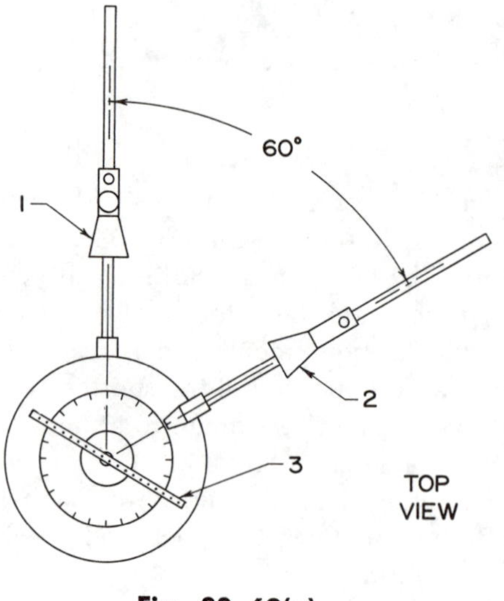

Fig. 33–69(a)

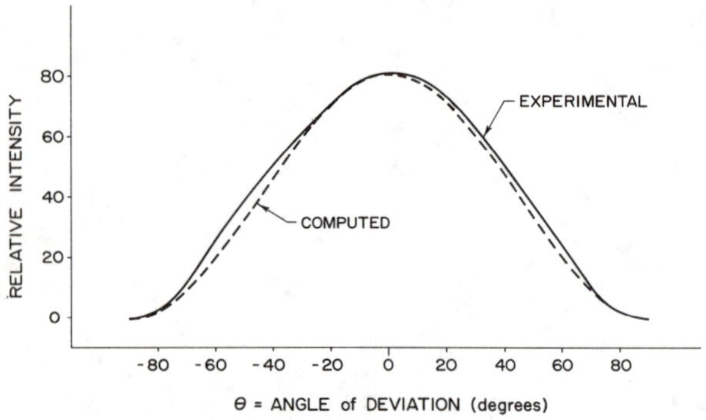

Fig. 33–68

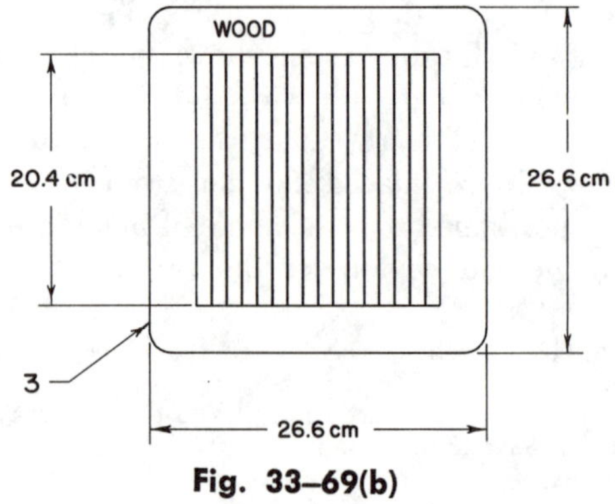

Fig. 33–69(b)

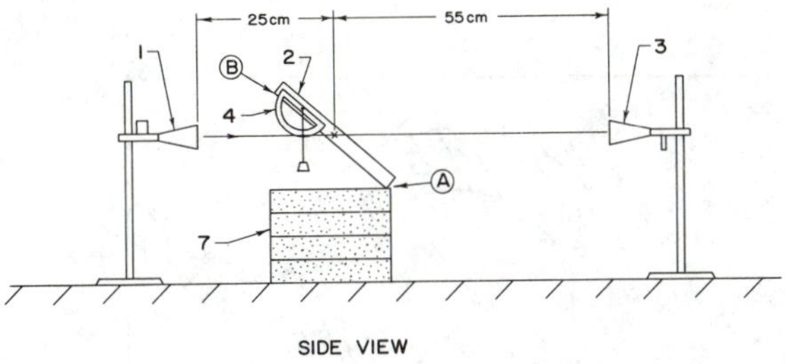

Fig. 33–70

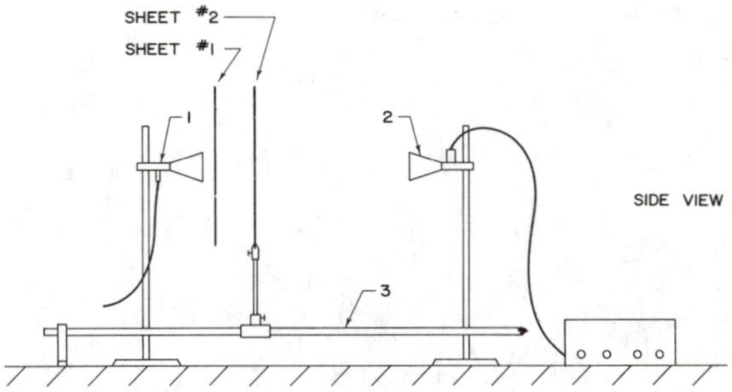

Fig. 33–72

be proved that the angle shown by the protractor θ' is equal to the angle of incidence θ. The slab is rotated about point A until the angle of maximum transmitted intensity is found, which is Brewster's angle θ_B. In order to eliminate error in the mounting of the protractor, the slab is also rotated about point B, and the average of these two readings is used. The index of refraction n is then computed from the formula

$$n = \tan \theta_B.$$

Using the arrangement shown in Fig. 33–70, a value of 55.9° was obtained for θ_B which yields 1.48 for the index of refraction n. This agrees quite well with the tabulated value of 1.47.

33–7.13

Interference from thin films can be demonstrated using the setup shown in Fig. 33–72. In this case the film is air with a thickness of the order of centimeters corresponding to the longer wavelength of microwave radiation. The apparatus consists of two 20.5 x 20.5-cm ground glass sheets about 2.5 mm thick, a receiver *1*, Klystron transmitter *2*, and an optical bench *3*. Sheet No. 1 is mounted on a stationary stand, while sheet No. 2 is free to move on the bench. The receiver *1* is placed about 30 cm from the transmitter 2 and about 1 cm from sheet No. 1. At the beginning the sheets are placed close together and made parallel. Their separation and the position of sheet No. 2 on the optical bench are recorded. Intensity readings are then taken as sheet No. 2 is moved away from sheet No. 1. Theoretically, there should be a maximum when

$$d = \frac{n\lambda}{2}, \qquad n = 0, 1, 2, \ldots .$$

where d is the separation of the sheets, λ is the wavelength, and n is the order of the maximum.

Interference from thin films by reflection can be shown with the arrangement in Fig. 33–73. The receiver *1* and transmitter *2* are mounted on the same stand, and intensity readings are taken as sheet No. 2 is moved away from sheet No. 1. For this case maxima are predicted when

$$d = \left(\frac{2n+1}{4}\right)\lambda, \qquad n = 0, 1, 2, \ldots .$$

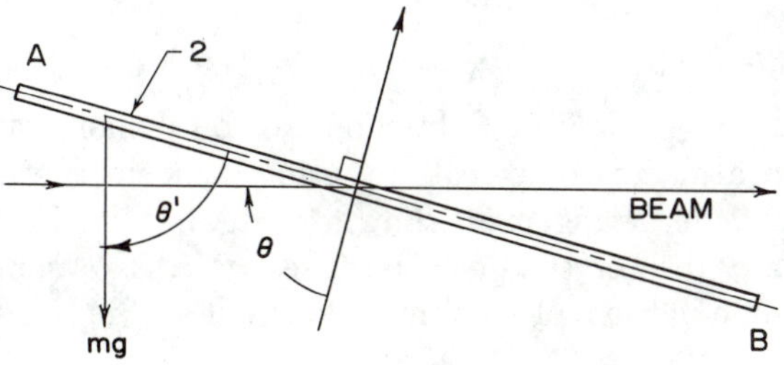

Fig. 33–71

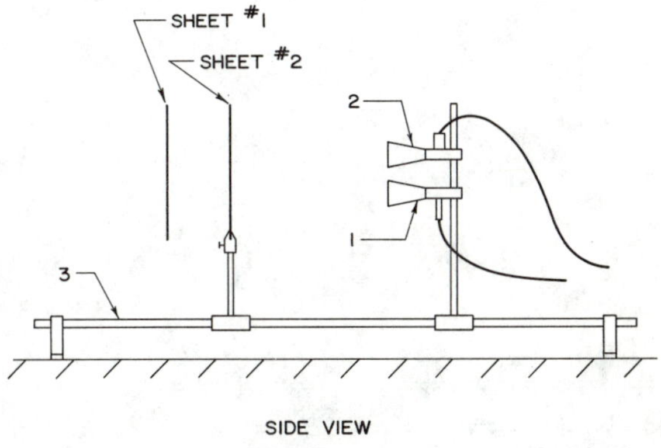

Fig. 33–73

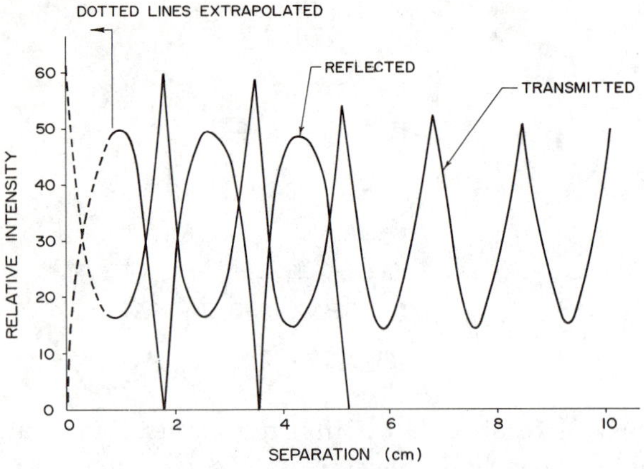

Fig. 33–74

Figure 33–74 is a plot of the transmitted and reflected waves using 3.37-cm microwaves and a VTVM for detection. The results agree quite accurately with the theory.

33–7.14

The optical phenomenon of Fresnel diffraction can also be demonstrated with microwaves, using the arrangement shown in Fig. 33–75. The apparatus includes a Klystron transmitter *1* and a movable receiver horn *3,* supported on a double rod optical bench *4.* A 60.5 x 45.6-cm piece of cardboard *2* is placed a distance $d = 70$ cm from the transmitter *1.* The foil-covered cardboard shown in Fig. 33–76 has a circular opening C 13.75 cm in radius (R_2) in its center. Two circular cardboard inserts, also 13.75 cm in radius, can be placed in this

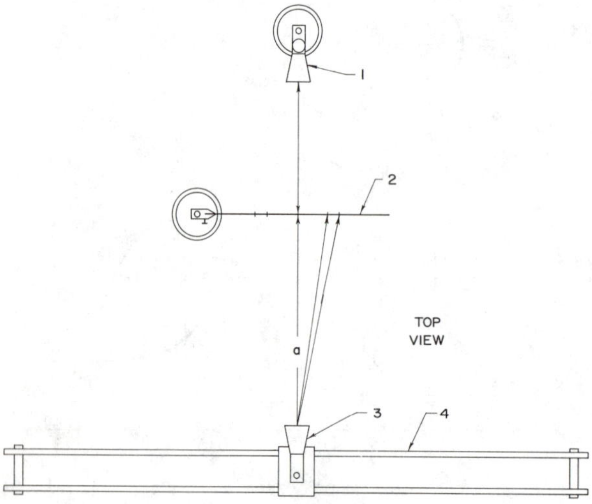

Fig. 33–75

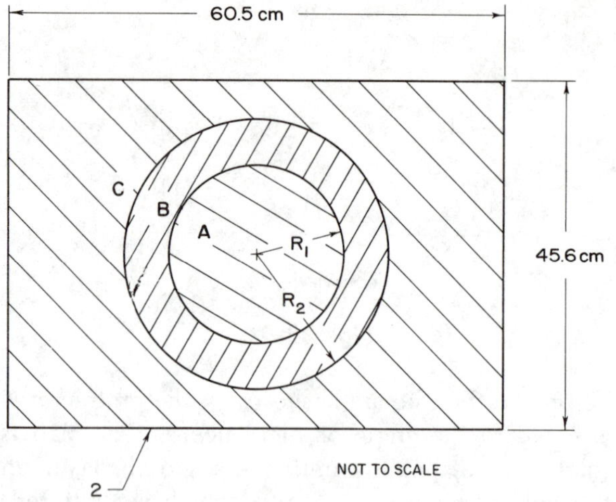

Fig. 33–76

aperture. A circle of aluminum foil 9.6 cm in radius (R_1) is centered on insert A, while a ring of foil with an inside radius (R_1) of 9.6 cm and an outside radius (R_2) of 13.75 cm is glued to insert B. Using the two different inserts, it is possible to simulate the zone plates of Fresnel's theory of diffraction for a circular aperture.[45]

The procedure is simple. With insert B in place, the receiver is moved slowly along the optical bench, which is located at a distance $a = 35$ cm from the cardboard. Intensity readings are then taken. Figure 33–77 is a plot of the results for insert B in position (case *1*), no insert (case *2*), and insert A in position (case *3*). The theory predicts a maximum intensity at the center position when an odd number of consecutive zones are exposed which is represented by case *1;* a minimum intensity for an even number of consecutive zones which is case *2;* and a maximum when the first zone is blocked and the second zone is exposed, case *3.* It can be seen from Fig. 33–77 that the results follow the predictions of the theory.

33–7.15

Bragg's law of diffraction can be demonstrated on a macroscopic scale by using 3-cm microwaves and the arrangement shown in Fig. 33–78. In this case, the "crystal" *c* to be investigated is a simple cubic lattice of 10-mm metal spheres imbedded

[45] See, e.g., F. A. Jenkins and H. E. White, *Fundamentals of Physical Optics* (McGraw-Hill Book Company, Inc., New York, 1950), p. 175.

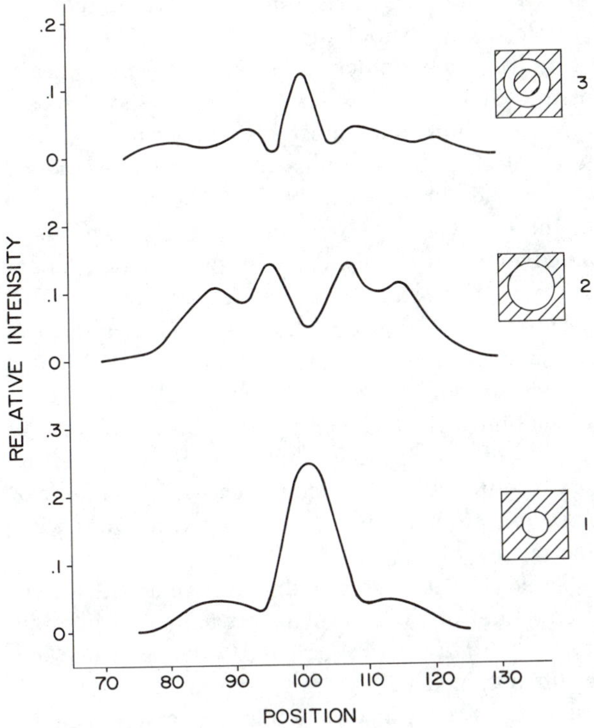

Fig. 33–77

about 4.25 cm apart in an 18 x 18 x 18-cm foam plastic matrix. The plastic is transparent to the 3-cm radiation and is built in layers so that the actual separation of the spheres may be measured after the experiment has been completed. A column of foam plastic *a* at the center of the spectrometer *s* supports the matrix in order to reduce reflections from the base. In lecture demonstrations an oscilloscope, loudspeaker, or projection galvanometer can be used for detection. The support column and foam plastic matrix are available commercially.[46]

Two conditions must be satisfied in order to detect Bragg diffraction. First, the angle of the detector must be equal to the angle of incidence of the microwave radiation. Second, Bragg's law

$$2d \sin \theta = n\lambda, \qquad n = 0, 1, 2 \ldots$$

must be satisfied, where d is the spacing between the parallel planes of spheres, θ is the angle the radiation makes with the planes, λ is the wavelength, and n is the order of the diffraction. Using a known wavelength λ and measuring the angle θ of maximum reflected intensity, the spacing d between planes can be computed.

The experiment is begun with the receiver and transmitter horns on a line directly opposite each other, and the crystal lattice is so oriented that the particular set of planes to be investigated lie along this line. Then, the booms supporting the receiver and transmitter are moved toward each other in

[46] Welch Scientific Co., Cat. No. 2641 A.

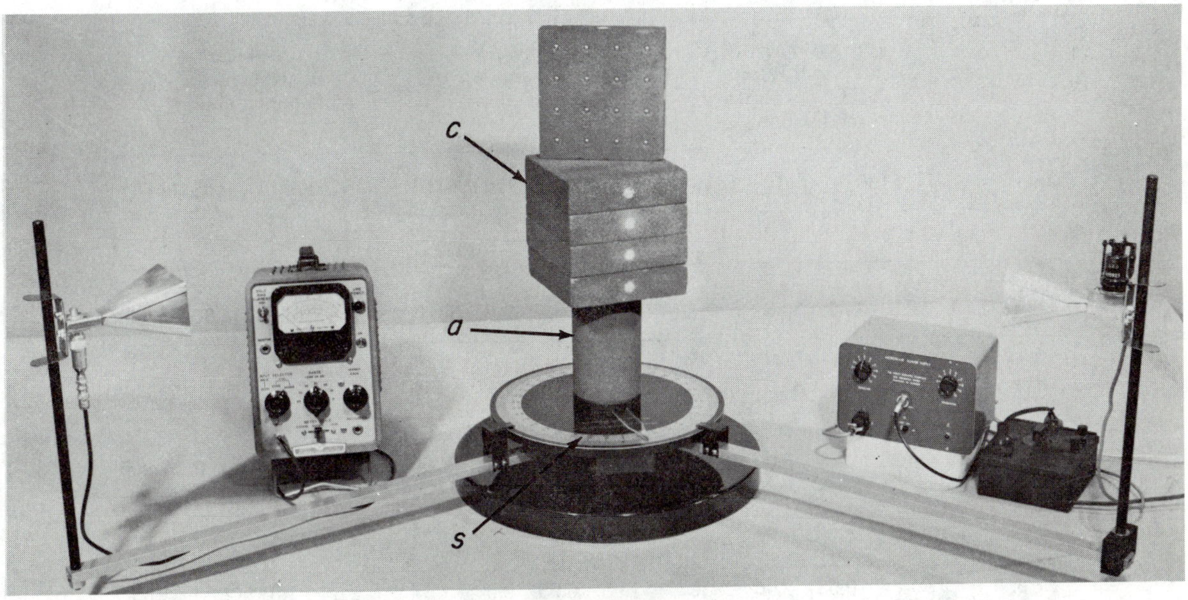

Fig. 33–78

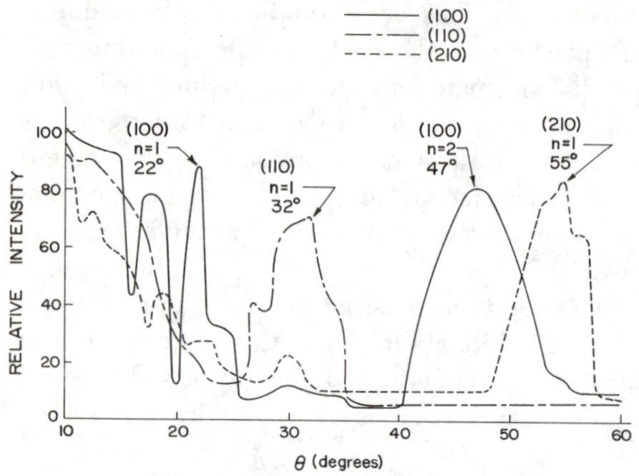

Fig. 33–79

equal steps, and intensity readings are taken every 1° or at whatever interval is suitable.

Graphs of the relative reflection intensity vs θ for the (100), (110), and (210) families of planes obtained with a sensitive galvanometer[47] are shown in Fig. 33–79. The interplanar spacing d_0 of the crystal used in this investigation was 4.25 cm, and the wavelength λ was equal to 3.2 cm. The apparatus was arranged in the center of the lecture table, and all the other equipment which could have produced spurious reflections was removed. Detector, power supply, etc. were placed below the level of the spectrometer. Because the $n = 1$ peak for the (100) planes is always near the direct feed-through region, data must be taken carefully between 0–20°. Best results are obtained in rooms with plaster or concrete walls. The $n = 1$ peaks for the (110) and (210) families of planes are obtained with little difficulty. If spurious reflections are very troublesome, sheets of microwave absorbent material[48] should be used. This absorber is a mat of spun animal hair impregnated with rubber containing carbon black which provides a conduction loss of rf energy. It absorbs about 99 percent of the incident energy. The absorber is very useful when precision microwave optics experiments are performed, particularly in rooms with steel walls.

Table 33–2 contains both the actual and calculated values of the interplanar spacings for a cubic crystal with $d_0 = 4.25$ cm and $\lambda = 3.24$ cm. The results obtained with several types of detectors and different forms of modulation are compared. Other crystal lattices can be constructed and analyzed.

For a simple cubic structure which has the same type of atom at each lattice site, there is a decrease

(47) Honeywell galvanometer, Model 3125, sensitivity .0035 MA/MM.

(48) B. F. Goodrich Co., Sponge Rubber Products Divison, Shelton, Conn.; (24 x 24 x 2-in. sheet) Type 12 CM.

Table 33–2

	Hewlett–Packard 415B S.W.I. (1000-cps sq wave modulation)				Heathkit AV-3 (60-cps sine wave modulation)				Rubicon Galvanometer (no modulation)			
Planes	(100)		(110)	(210)	(100)		(110)	(210)	(100)		(110)	(210)
n	1	2	1	1	1	2	1	1	1	2	1	1
0° Theoretical	22.5	49.9	32.8	58.6	22.5	49.9	32.8	58.6	22.5	49.9	32.8	58.6
0° Observed	22	49	32	56	22	49	32	58	23	51	32	59
d(cm) Actual	4.25	4.25	3.00	1.905	4.25	4.25	3.00	1.905	4.25	4.25	3.00	1.905
d(cm) Calculated	4.32	4.29	3.06	1.95	4.32	4.29	3.06	1.93	4.15	4.17	3.06	1.89
% Error	1.65	0.95	2.00	2.36	1.65	0.95	2.00	1.31	2.35	1.88	2.00	1.05

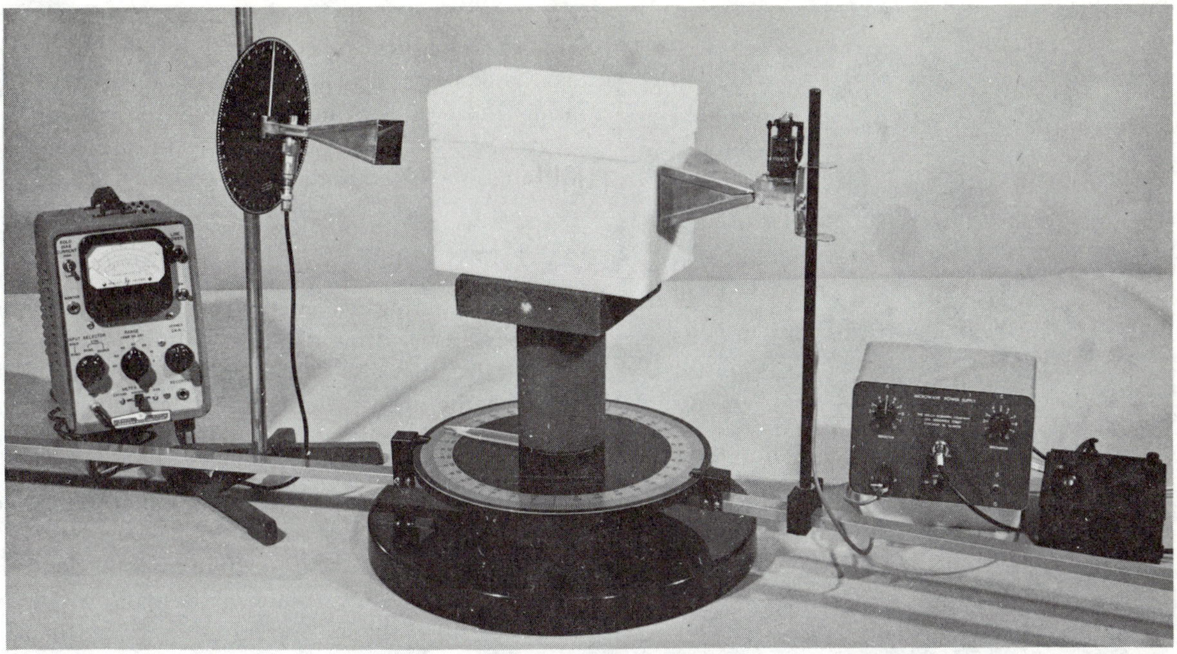

Fig. 33–80

in the number of atoms per plane associated with a decrease in *d*. This implies that the maxima are strongest for the 100 planes, weaker for the 110 planes, and greatly reduced for the 120 planes, which can also be verified with this experiment.

33–7.16

Three-cm microwaves can be used to demonstrate on a macroscopic scale the ability of certain sugars to rotate the plane of polarization of light.[49] The $5\frac{1}{2} \times 9 \times 12$-in. Styrofoam box in Fig. 33–80 contains approximately 1200 2-turn coils. Each coil is wound $\frac{1}{4}$ in. in diameter with 22-gauge enameled wire and wrapped with tissue paper. It is important that each coil is wound in the same sense, that is, all right-hand or all left-hand coils. After shaking the box so that the coils assume random orientation, it is placed directly in front of the microwave transmitter horn. The microwaves from the transmitter are already polarized in a particular direction. By rotating the receiver horn, it will be found that the plane of polarization has been rotated after passing through the Styrofoam box.

Plane polarized radiation is composed of right and left circularly polarized waves. The coils accelerate one component and hinder the other. After passing through the Styrofoam box, the two components add with a resulting shift in the plane of polarization. Right hand coils will rotate this plane in a direction opposite to that of a box of left hand coils. If a box of right hand coils and an identical box of left hand coils are used, the rotation effect will be cancelled.

The rotation of the plane of polarization of microwaves depends on the length and diameter of the coils. Two Styrofoam matrices (Fig. 33–81), $8\frac{1}{2} \times 7\frac{1}{4} \times 1\frac{3}{4}$ in. and $9\frac{1}{2} \times 9\frac{1}{2} \times 2$ in., contain 36-turn, copper-wire (22-gauge) coils, arranged in a symmetrical array, 13×15 and 11×11 rows, re-

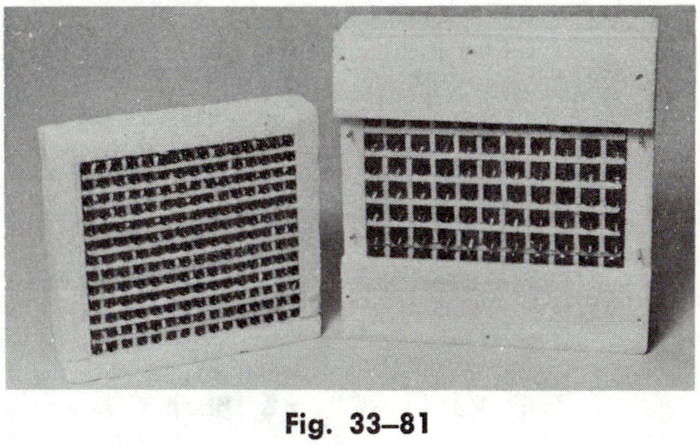

Fig. 33–81

[49] See A. Sommerfeld, *Optics* (Academic Press, Inc., New York, 1954), p. 165.

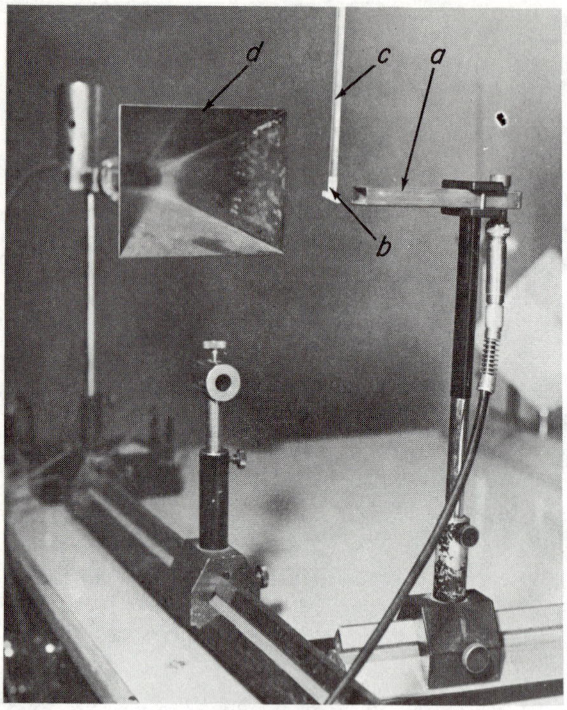

Fig. 33–82

spectively. The helices for the larger array are 1 in. long by ½ in. in diam and for the smaller, 1 in. long by ¼ in. in diam. Each row is separated by a layer of styrofoam.

Two aspects of the rotation can be shown.[50] When the Klystron transmitter horn is aligned with the axes of the coils, the rotation is a maximum (about 25 deg for the smaller coils) and decreases as the horn is moved from this orientation. By using the two Styrofoam matrices described, the degree of rotation can also be shown to depend on the cross-sectioned area of the coils.

[50] Investigated by V. Dionne at Rensselaer Polytechnic Institute.

33–7.17

The scattering of microwaves by a dielectric dipole can be demonstrated with the apparatus shown in Fig. 33–82. Microwaves with 3-cm wavelengths are emitted through a waveguide *a*. They are scattered by a dipole *b* mounted at the end of a Lucite rod *c*. The waves are scattered to the detector *d* which is perpendicular to the wave guide and about a foot away.

The intensity of the scattered wave depends upon the dielectric constant of the dielectric body and upon its shape. A rod of barium titanate ($\epsilon \approx 10^3$) scatters almost like a copper rod of the same dimensions, while the scattering due to a Teflon rod of the same size is barely detectable. The polarization of the scattered wave can be studied by varying the orientation of the rods.

In this demonstration copper dipoles 1 mm in diam with lengths of 15, 30, 50 and 8 mm are used. Maximum scattering is observed when the length is an integral number of half wavelengths (15 mm, 30 mm, etc.). The barium titanate rod is 2.7 x 2.7 x 30 mm, and the Teflon rod is 3 mm in diam and 30 mm long.

Editor's note: Apparatus using ultrasonic waves to obtain Bragg diffraction in Sec. 33–7.15 was developed by Dr. V. N. Bindal at the National Physical Laboratory, New Delhi, India.

A macroscopic analog of a crystal lattice is made from 2.5 mm diam plastic spheres cemented 1.0 cm apart on a 0.3 mm diam wire which passes through them. Eight such wires are fixed 1.0 cm apart in a vertical position to a rectangular frame of 1.0 mm diam wire. A number of frames are placed one behind the other at a distance of 1.0 cm to form a lattice. The back, top and the bottom of the lattice are surrounded with foam rubber to suppress spurious reflections from the walls of the room. Ultrasonic waves of about 7.5 mm wavelength are used. Two matched ultrasonic transducers with horns are used as transmitter and detector. Transducers are mounted on each of two booms which can be adjusted at various angles with respect to the lattice. With the help of a filter in the receiver and care in obtaining uniform spacing in the lattice, an accuracy of 10% can be easily obtained. A 9 V dc power supply is used.

Demonstrations	***Contributors***
33–1.1	F. Oppenheimer, University of Colorado
33–1.2	Oak Ridge Institute of Nuclear Studies
33–1.3	L. Hirsch, California State College, Los Angeles

Demonstrations	***Contributors***
33–1.4	M. Iona Jr., University of Denver
33–2.1	L. P. Delsasso, University of California, Los Angeles
33–2.2	W. Eppenstein, Rensselaer Polytechnic Institute
33–2.3	J. I. Swigart, University of Utah
33–2.4	L. Grillet and M.-T. Grillet-Amy, University of Rennes, France
33–2.5, 33–2.6	Swiss Federal Institute (ETH), Zürich, Switzerland
33–2.7	California Institute of Technology
33–2.8	Massachusetts Institute of Technology
33–2.9	J. A. Soules, New Mexico State University
33–2.10	H. C. Jensen, Lake Forest College
33–3.1	Eric M. Rogers, Princeton University
33–3.2	H. Robinson, Adelphi University
33–3.3 to 33–3.6	Swiss Federal Institute (ETH), Zürich, Switzerland
33–3.7	H. C. Jensen, Lake Forest College
33–3.8	W. K. Allen, Western Canada High School, Calgary, Alberta, Canada
33–4.1	H. Meissner, Stevens Institute of Technology
33–5.1, 33–5.2	Swiss Federal Institute (ETH), Zürich, Switzerland
33–6.1	California Institute of Technology
33–6.2	R. A. Llewellyn, Purdue University
33–6.3	Swiss Federal Institute (ETH), Zürich, Switzerland
33–7.1	California Institute of Technology
33–7.2	J. H. Howey, Georgia Institute of Technology
33–7.3	Swiss Federal Institute (ETH), Zürich, Switzerland
33–7.4	C. L. Andrews, General Electric Co., Schenectady, New York
33–7.5 to 33–7.16	H. F. Meiners, Rensselaer Polytechnic Institute
33–7-17	Swiss Federal Institute (ETH), Zürich, Switzerland

OPTICS

34

GEOMETRICAL OPTICS

34–1 Reflection and Refraction

34–1.1

Mount two front surfaced mirrors so that they can be rotated about the line of the intersection of their planes. If an object, such as a small light bulb, is then placed between the mirrors, many images of the object will be seen due to the possibility of multiple reflections. The number of images depends upon the angle between the mirrors. An image of the light bulb can be seen in each mirror (provided, of course, that you are looking in the right direction). These images are produced by rays which leave the light bulb and are reflected directly to the eye. Other rays leave the light bulb and are reflected to the eye and one can see an additional image of the light bulb in the second mirror. This image can be considered to be the image of an image. If M_1, M_2, M_3, etc. are the images seen in mirror M (Fig. 34–1) and those in mirror m, m_1, m_2, m_3, etc., M_1 is the image of the bulb O in M, and m_1 is the image of the bulb in m. M_2 is the image of m_1 in M and m_2 is the image of M_1 in m; M_3 is the image of m_2 in M, etc. The image of a point in plane mirror will lie behind the mirror on a line drawn from the point perpendicular to the plane of the mirror and the image will be the same distance behind the mirror that the object is in front of it. Figure 34–1 shows the construction of the images for two mirrors making an angle of 75°.

Investigate the number of images as a function of angle θ between the mirrors and observe what the maximum number of images can be if $360/\theta$ is an integer. Also, notice what is seen as you move your head from side to side while looking into the mirrors along a line perpendicular to their intersection when the mirrors make an angle of exactly 90°.

34–1.2[1]

The signaling mirror is a rather simple, but ingenious device demonstrating the reflective properties of a plane surface. This plane mirror reflects on both sides and has a small, unsilvered area in the center. Figure 34–2 illustrates the principle of the device. To use it, the sender sights at the target through the unsilvered aperture. At the same time, the sunlight shines on the mirror and a small portion passes through the aperture forming a spot of light on the sender's hand or a card. The sender, looking in the mirror, can see this spot by reflection, and, if he tilts the mirror to align this reflected image with the transmitted light from the

[1] C. Harvey Palmer, *Optics Experiments and Demonstrations* (The Johns Hopkins Press, Baltimore, Md., c. 1962), Experiment A4, pp. 14–16. The editor wishes to thank C. Harvey Palmer and The John Hopkins Press for granting permission to use the following experiments in Chapters 34, 35, and 38: 34–1.2, 34–1.23, 34–1.25, 34–1.27, 35–2.8, 35–5.1, 35–5.2, 35–7.8, 38–5.13.

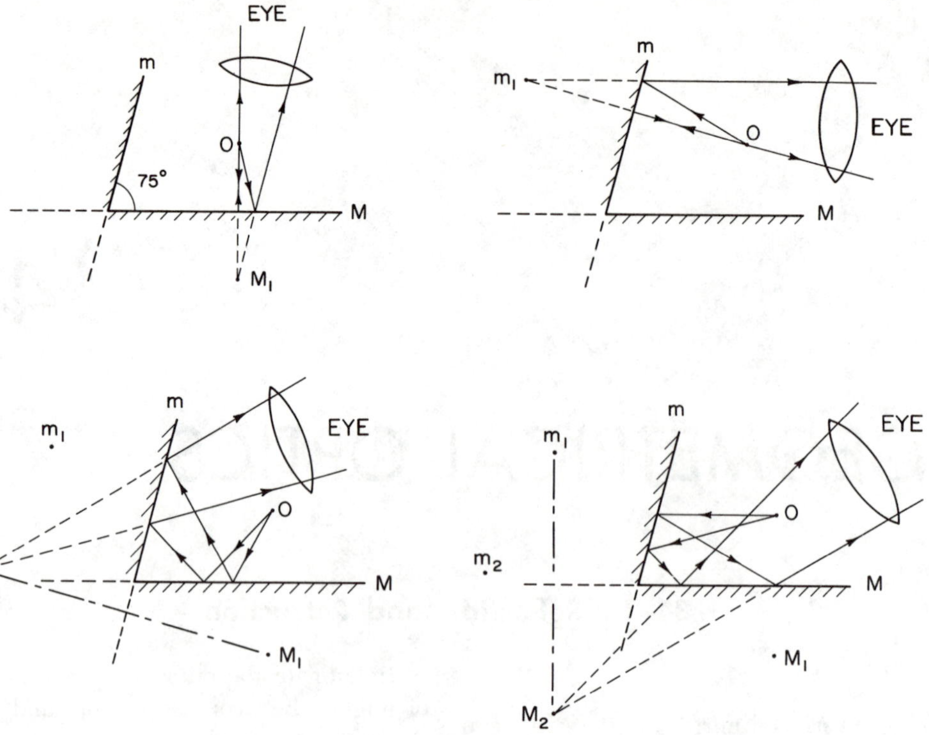

Fig. 34–1

target seen through the aperture, he can be sure that the mirror is reflecting sunlight toward the distant target.

Such a device is easily made from an inexpensive pocket mirror from which the backing paint has been removed with paint remover to provide reflecting surfaces on both sides. Kleenex should be used to wipe off the paint so as not to scratch the reflecting surface. A pencil eraser will remove the silvering where desired to form the aperture—a cross with arms about 1/8 in. wide is convenient.

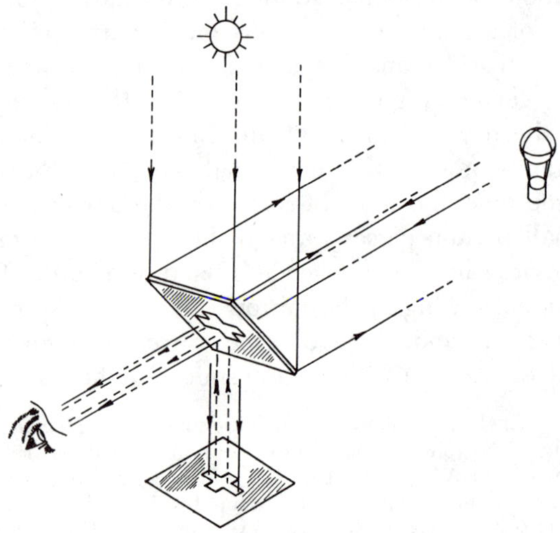

Fig. 34–2

Unsilvered or partially silvered mirrors provide both transmission and reflection. They may, therefore, be used either to combine two light beams, or fields of view, or else split one light beam into two parts. The thickness of the metal coating depends on the desired use of the mirror.

34–1.3[2]

When an object is at the center of curvature of a concave mirror, an inverted real image is formed on the same side of the mirror as the object. As the object is displaced toward the principal focus, a larger inverted image appears, and as the object is displaced away from the principal focus, a smaller inverted image appears. A convenient object is an oscillating frosted incandescent bulb suspended at the end of a 2-m-long cord, with its plane of oscillation containing both the center of curvature

[2] H. Dezouche, *Bull. Union des Physiciens,* No. 341–343, 13 (1945).

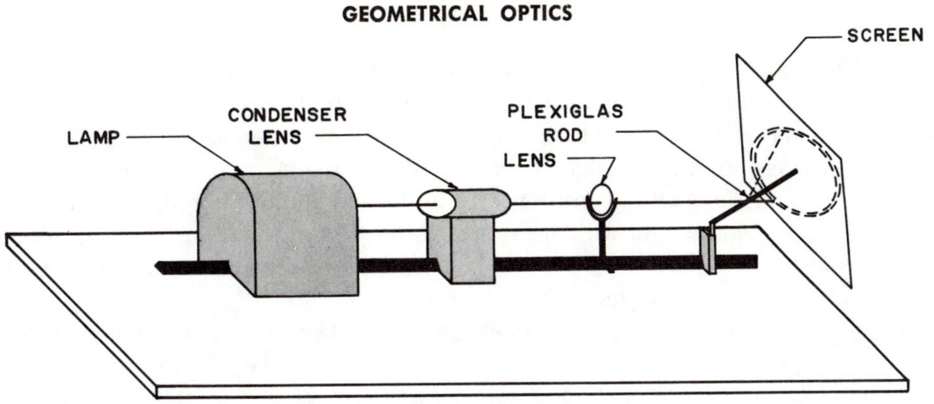

Fig. 34–3

and the principal focus of the mirror. The changing position and size of the real image should be viewed along the principal axis of the mirror as the bulb oscillates.

34–1.4[3]

A carbon arc projector with condenser (Fig. 34–3) emits a narrow beam of light. A white cardboard screen is cemented to the end of a round glass rod a foot or so in length. The light from the arc is allowed to fall on the glass rod at some angle. After reflection, the locus of the light rays forms a hollow cone which will illuminate a narrow ring on the cardboard. The half angle of the cone will be equal to the angle of incidence of the light on the glass rod. The apparatus can be used for a corridor demonstration.

34–1.5[4]

To study reflection from a mat surface, the light from a projection lantern is collected by a condenser onto a slit and the image of the slit is projected by an objective lens onto a screen. Then a frosted piece of glass is placed in the path of the light directly in front of the objective lens so that its plane is parallel to the slit and the frosted surface is directed towards the incident light. If the light is reflected approximately at right angles by the frosted glass, the light falling on the screen, which has been placed in the path of the reflected light, has been scattered by the frosted surface and does not give a sharp image of the slit. When the frosted glass is rotated about an axis parallel to the slit, the image becomes clearer and clearer as the angle of incidence is increased and with glancing incidence, i.e., when the angle of incidence is close to a right angle, a clear image of the slit is apparent on a vertically placed screen.

It is interesting to substitute a piece of smooth paper or smoked glass plate for the frosted glass. It should be noted that in the case of frosted glass a second clear image will be seen; this image is produced by the reflection of light from the back, smooth surface of the glass. This image is always white. In order to extinguish it, cover the back surface of the glass with black lacquer.

34–1.6

A thin sheet of light passing through a semi-circular cylinder of glass mounted at the center of a disk that can rotate about the axis of the semi-circle can be used to demonstrate Snell's law. The sheet of light is wide enough that the path of the light can be seen on the surface of this disk.

The periphery of the disk is marked in degrees and two perpendicular diameters are drawn on the disk. The semi-cylinder of glass must have its diameter accurately aligned with one of these lines on the disk and the center of the semi-circle must coincide with the center of the disk. In addition the sheet of light must closely follow the diameter of the disk.

Under these conditions the light will always enter or leave the circular surface of the glass cylinder radially. It will thus be parallel to the normal of the glass and not be deflected as it

(3) Robert W. Pohl, *Optik und Atomphysik* (Springer Verlag, Berlin, 1967), 12th ed., p. 9.

(4) Translated by K. V. Robinson and H. A. Robinson (Adelphi University) from *Lecture Demonstrations in Physics*, A. B. Mlodzeevskii, ed., M. V. Lomonosov State University, Moscow, U.S.S.R.

enters the glass. Thus any angular deflection of the sheet of light will occur at the flat surface which forms the diameter of the semi-cylinder of glass. The degree markings around the edge of the disk enable measurement of the angle of incidence and the angle of refraction of the light as it passes through this flat surface.

The disk is oriented so that light falls on the flat surface of the glass, and the angle of refraction is measured at intervals of 5- or 10-deg angles of incidence. If the angles on either side of the normal differ by more than a degree or two, the disk should be realigned. The angle of incidence is plotted against the angle of refraction, and the index of refraction is computed.

The experiment is repeated, allowing the light to enter the glass along its circular surface. Total internal reflection is observed, and the critical angle for this to happen is observed.

Colors will be seen if a white card is placed in the path of the sheet of refracted light at the outer edge of the disk. The critical angle for the blue light will be somewhat smaller than the critical angle for the red. The difference in the index of refraction of the glass for red and blue light can then be measured.

34–1.7

The Abbe refractometer is used to measure the refractive index of liquids from the critical angle. Transmitted light is used to determine the critical angle because it gives a brighter image. The illumination transmitted through P_2 from a light source must fall at grazing incidence on the hypotenuse of the prism P_1 in Fig. 34–4. Grazing incidence is obtained by the prism P_2. A drop of the liquid being tested forms a thin film between the two prisms.

The cross hairs of a telescope are lined up with the dark edge along axis I–I′. Usually the telescope is mounted on an arm attached to a scale calibrated to read the refractive index of the liquid for yellow lines directly. When axis I–I′ is normal to the output surface of P_1, the critical angle C_0 has the same value as the angle A of prism P_1.

When the Abbe refractometer is used without source and without liquid, the index of refraction N for the yellow lines can be found by measurement. The angles of prism P_1 are known; usually A is 60 deg.

The optical axis of the telescope is set perpendicular to face AB of prism P_1. On the scale, corresponding to this position, we obtain a value n_0 for the index of refraction.

If c_0 is the value of the critical angle,

$$\sin c_0 = \frac{n_0}{N},$$

where N is the value of the index of refraction of prism P_1. Since $c_0 = A$, where A is known,

$$N = n_0/\sin A.$$

This is the value for the yellow lines.

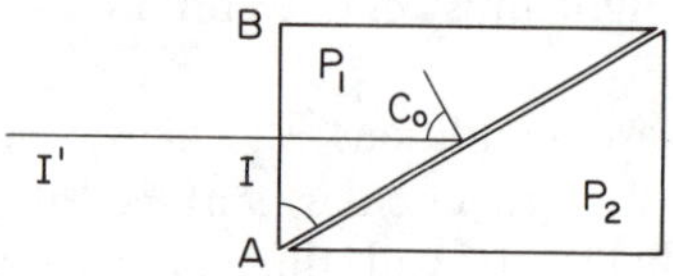

Fig. 34–4

34–1.8

A telescope is fixed on an optical bench with its axis horizontal and its focus about 15 cm. A thick (4 cm) transparent plate of Plexiglas is so mounted that it can be moved back and forth on the bench. By the displacement of this plate, the telescope is

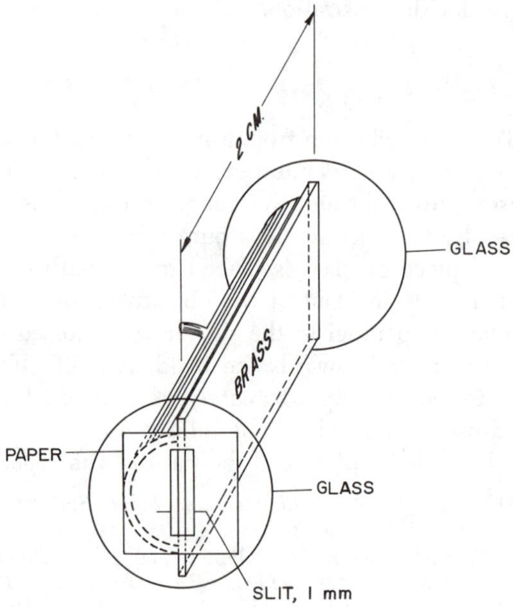

Fig. 34–5

focused on the front face, then the back face. The index of refraction is found from the relation $n = d/(d - a)$, where d is the plate thickness and a is the displacement of the plate. To focus the two faces, both surfaces are dusted with lycopodium powder; the powder is removed from the left half of the front and from the top half of the back. This same method can be used for liquids held between a thin-walled, parallel-face container. The light source is an opal lamp with a red filter (to illuminate the back face of the plate).

Using a high-intensity mercury lamp with a yellow-green filter, a lens (instead of the telescope) focuses the front face, then the back face on a screen.

34–1.9[5]

The Raleigh refractometer[6] provides an accurate method of measuring the index of refraction of air. However, there are several practical difficulties in obtaining results with the usual form of the apparatus. Because of the large separation of the slits, the interference fringes are close together and are difficult to see and count. Also, when long tubes are used and the air removed, the fringes are no longer visible, so that it is necessary to remove only a part of the air and calculate the index by proportion.

These difficulties are overcome in the apparatus shown in Fig. 34–5. With the construction shown, the two slits are very close together and large, clear fringes are easily obtained. Also by using a short tube (2-cm), all the air may be removed and the fringes may still be seen.

The apparatus consists of a brass or copper tube 2 cm long, cut in half and soldered to a flat piece of brass about 0.25 mm thick. A side tube for connection to a vacuum pump is soldered to the tube. A thin piece of glass is cemented to each end with Cenco-Sealstix cement. On the outside of one of the pieces of glass is glued a piece of black paper with a slit about 1 mm wide cut in it, in such a position that the end of the brass plate bisects the slit in the paper, thereby forming two slits.

The equipment is set up as shown in Fig. 34–6.

[5] P. S. DeLaup, *Am. J. Phys.*, **14**, 383 (1946).

[6] J. K. Robertson, *Introduction to Physical Optics* (Van Nostrand, 1935), p. 199; Wood, *Physical Optics* (Macmillan, 1934), p. 173.

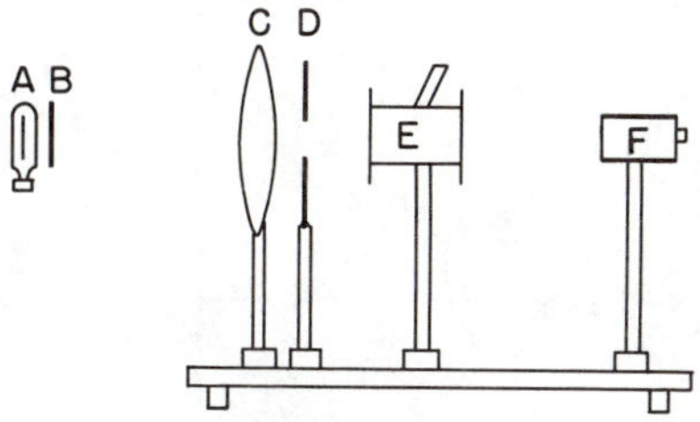

Fig. 34–6

The slit end of apparatus E is toward the lens C. A straight-filament lamp A is placed about 6 ft away with a red glass filter B mounted in front of it. The light passes through lens C with a focal length of about 50 cm, through screen D, and through apparatus E to eyepiece with cross hairs F. The system can be lined up by removing the filter and forming the image of the two slits on a piece of paper. It is important that the slit be parallel to the filament if the clearest fringes are to be obtained.

The apparatus E is connected to a vacuum pump by a rubber tube. As the air is removed, the fringes will shift. After the best possible vacuum is obtained, the pump is disconnected, air is slowly admitted by adjusting a pinch clamp on a rubber tube, and the number of fringes passing the cross hair on the eyepiece is counted. Fractions of a fringe may be estimated. If difficulty is found in admitting the air slowly enough, a large bottle may be placed in the vacuum system.

The index of refraction μ can be calculated by means of the equation

$$\mu = 1 + \left(\frac{\lambda N}{l}\right),$$

where N is the number of fringes shifted, l is the length of path, λ' is the wavelength in the air, and λ is the wavelength in vacuum. To obtain the above equation, note that N is the difference between the number of waves in the air and vacuum paths; hence

$$N = \frac{l}{\lambda'} - \frac{l}{\lambda} = \frac{l}{\lambda}\left(\frac{\lambda}{\lambda'} - 1\right) = \frac{l}{\lambda}(\mu - 1).$$

The wavelength can be taken as the average of the wavelength range transmitted by the filter. More accurate results can be obtained by replacing the lamp and filter by a strong source of monochromatic light.

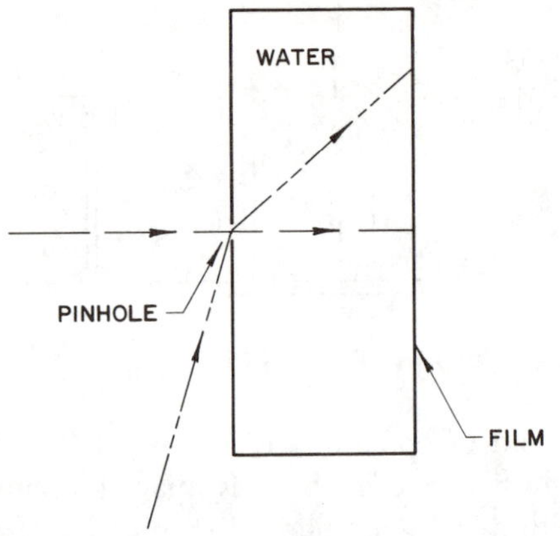

Fig. 34–7 Schematic Side View of Fish-Eye Camera, showing typical light rays.

34–1.10

A pinhole image can be shown by closed-circuit TV by removing the TV-camera lens and replacing it with a short length of brass tubing, one end of which is closed by a sheet of aluminum foil. The "pinhole" is formed by puncturing the foil with a needle. The sharpness of the pseudo image depends upon the size of the hole. A clear incandescent lamp bulb makes a satisfactory object.

34–1.11

The refraction of light rays by plane surfaces can be demonstrated easily by using R. W. Wood's "Fish-eye Camera"[7,8] adapted with a Polaroid camera back for rapid viewing. A sketch of the apparatus is shown in Fig. 34–7. The device consists of a water-tight pinhole camera with a clear Lucite back, against which the film is pressed. The pinhole is covered with Lucite, and the camera is filled with distilled water. Further construction notes are given in the Appendix, page 1345. Alternately, the camera might be constructed of a solid block of Lucite; this type of construction is preferable to the water-filled camera. An identical air-filled pinhole camera is used for comparison purposes.

The class is instructed to sit very still, and with normal lecture hall illumination a 20-sec exposure of Polaroid A.S.A. 3000[9] film is made. The camera, mounted on a tripod, is located a foot or so in front of the first row of people, and the exposed film may be projected almost immediately with an opaque projector. The result (Fig. 34–8) is a picture of the class as would be seen by a fish in an aquarium. This photograph demonstrates that the rays gathered from a nearly 2π solid angle have been compressed into a cone of much smaller angle and that the light passing through the pinhole actually forms an image. For comparison, a second picture is taken with the air-filled pinhole camera. Figure 34–9 shows a photograph taken with a solid Lucite fish-eye camera, which has about the same dimensions and the same size pinhole as the water-filled camera. This view corresponds to that of a biological specimen mounted in clear plastic.

34–1.12

A light beam can be bent by passing it through a small tank containing a medium of varying index of refraction. To construct the tank, a bottom plate *a* 1 x 24 in. and a back plate *b* 6 x 24 in. are cut from 1/16-in.-thick brass and a front plate *c* 6 x 24 in. and two end plates *d* 1 x 6 in. are cut from glass. The tank is glued together, as shown in Fig. 34–10. The brass bottom plate is drilled in its center to receive a T-type stopcock *e*, which is cemented in place. A series of depth marks *f* ½ in. apart is etched in the front plate to facilitate the filling operation. The refracting medium consists of ten layers of varying concentrations of a mixture of benzol and CS_2. The top layer is 100 percent benzol, the second layer is 90 percent benzol and 10 percent CS_2, and in each successive layer, the concentrations of benzol and CS_2 are decreased by 10 percent and increased by 10 percent, respectively.

To fill the tank, the following procedure is used: Ten aspirator bottles, each one containing one of the solutions of benzol and CS_2, are fitted with short lengths of rubber tubing and numbered from 1 to 10, beginning with the solution of 100 percent benzol, *proceeding in order of decreasing benzol concentration,* and ending with the solution of 100 percent CS_2. The tube from bottom No. 1 is fitted

(7) R. W. Wood, *Phil. Mag.,* **12,** 159 (1906); No. 68; see also his book *Physical Optics,* 3rd ed. (The Macmillan Co., New York, 1936), p. 67.

(8) Not to be confused with J. C. Maxwell's "Fish-eye." See, for example, M. Born and E. Wolf, *Principles of Optics* (Pergamon Press, New York, 1959), p. 146.

(9) Type 57, 4 x 5-in. Polaroid film packet.

Fig. 34–8 **Fish-Eye View of Class,** normal room illumination, 20-sec exposure.

to the bottom pipe of the stopcock and air is bled from the line with the stopcock in the position shown in Fig. 34–11. As soon as fluid flows from the horizontal pipe of the stopcock, it is turned to the shutoff position. The stopcock is then slowly turned to the fill position, until the fluid just begins to flow into the bottom of the tank. When the fluid reaches the first depth mark, the stopcock is shut off, the tube is removed and the tube from bottle No. 2 is put on the stopcock pipe. The operation is repeated for all the bottles, making sure to bleed the air from the line each time, each bottle adding enough fluid to bring the level to the next higher depth mark. During the filling operation, it is most

Fig. 34–9 **Photograph Taken with Solid Lucite Fish-Eye Camera,** 25-sec exposure.

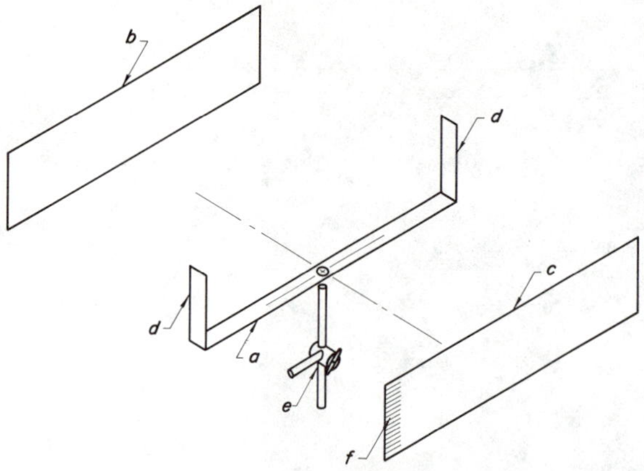

Fig. 34–10

important that the fluid level in the bottle be higher than the fluid level in the tank in order to prevent backflow. It is also important that each successive fluid layer be added very slowly to prevent mixing as much as possible.

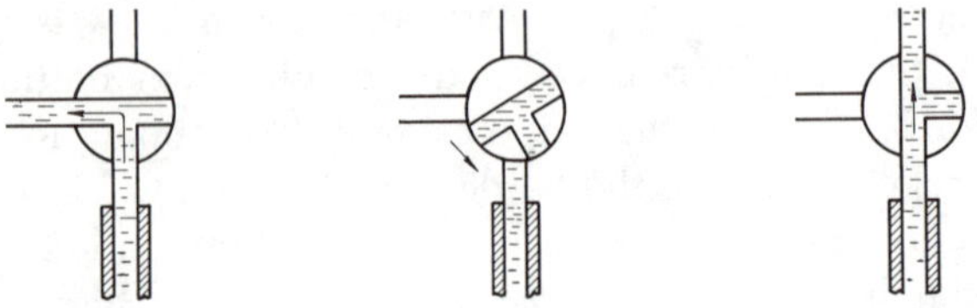

Fig. 34–11

In operation, a beam of light from an arc lamp is directed through a horizontal slit lengthwise through the tank, at a small angle to the horizontal and at a small angle to the back plate, as shown in Fig. 34–12.

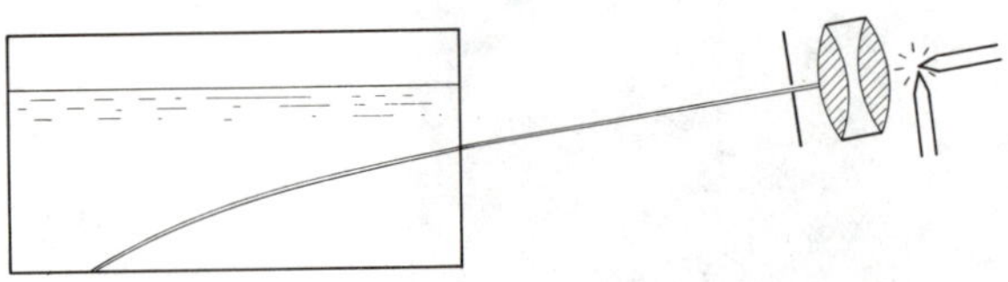

Fig. 34–12

34–1.13

A corridor type demonstration illustrating single-surface refraction is shown with a thermometer immersed in a water-filled 5-liter flask. The effect is observed clearly if the flask is wrapped with a

Fig. 34–13

single layer of ½-in. plastic tape. See Fig. 34–13. A thermometer is a convenient object to show the "spherical lens" effect or to use at the center of curvature for comparison with an object at the far side.

34–1.14

Figure 34–14 shows a method of locating a real image by parallax.[10] The screens are made by folding red and green graph paper so as to show lines on both sides, and tacking them to small boards. The positive lens (or two identical lenses) may have a 15- or 25-cm focal length. The 10-W lamp has a clear bulb and a C-shape filament in a plane. The tops of the screens and the centers of lens and lamp should all be at the same height. The distances of separation are adjusted until a sharp image of about half the filament appears on the image screen; the screen that has green lines. Now, get far back in line with the system and look straight toward the lamp. You should see the part

[10] M. J. Pryor, *School Science and Mathematics* (May 1959).

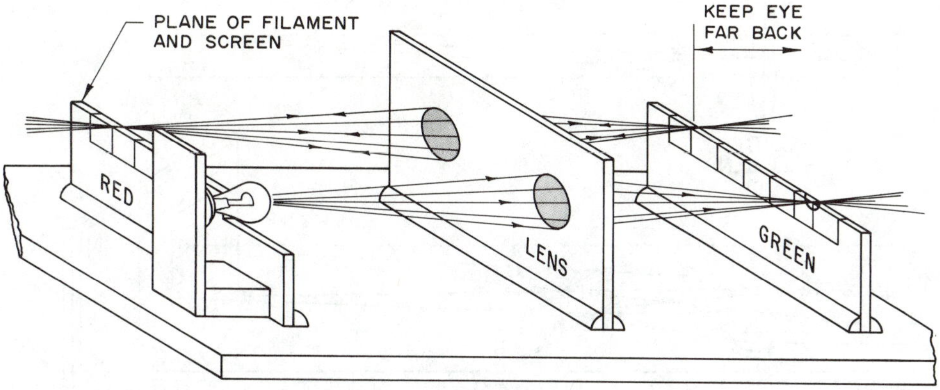

Fig. 34–14

of the image which is above the screen. In Fig. 34–14 it is the down-turned images of the ends of the filament that in the lamp are pointing up. You could now study parallax, but the light is too bright. Exchange the lamp filament for the screen with red lines. (Just move to one side if you use two lenses.) Images of part of the red lines are made of the green screen with part extending above it as drawn for the lamp. However, the image on the screen can not be seen because the lines are too faint. Look as if along the rays represented in Fig. 34–14 and you should see red lines above the green screen. It is helpful to stand back and vary your eye level. A test is made by a slight sidewise eye motion. The red lines of the image should not move relative to the lines on the green screen. Now go behind the red screen at the other end. You should see images of the green lines above the red lines. By this alteration you have exchanged object and image. Now move the green screen away from you toward the lens. Then, as you move to the right or left the green lines will seem to move with you past the red lines. This is because the green lines are farther away from you than the red-line images are.

If the screen is placed too close to you, the nearer green lines go past the red ones in a direction opposite to your motion. (Since the lens is far beyond both sets of lines you are looking at, you can see the lines move rapidly past the lens.) Concentrate on the lines, now, and move the screen to the position which again shows no parallax between the images of red lines and the actual green lines. Check the accuracy of your work by looking from the other end. This arrangement serves to locate real images—images that could be seen on a screen were they bright enough. You may replace the lamp for a re-check, but after a few trials you will be perfectly satisfied with the accuracy of the parallax method of using the two screens.

The arrangement in Fig. 34–15 is used to locate a virtual image by parallax. Reflected light from each point on the red screen *R*, which is closer than the principal focal point, converges enough that it seems to come from a more distant point. The dotted lines on Fig. 34–15 show rays from the same distance as the rays from the half-lens. Equipment like that in Fig. 34–14 may be used, except that you should run a glass cutter along a diameter of the lens and tap it gently until you have two half-lenses. Eye-center, half-lens-top, and the top of the red screen *R* should be in a plane that cuts the green screen *G*. You see an enlarged image of the red lines by light that comes through the lens and enters the lower half of the pupil of your eye, and a green image of light that misses the lens and enters the upper half of the pupil of the same eye.

Ordinarily you will see a motion of the red lines relative to the lines on the green screen when your eye is moved parallel to the lens' edge. The more distant lines move the same way your eye does, relative to the nearer lines. Slide the screen nearest you until there is no relative motion. As related to the half-lens, the red screen in the object and the green screen is at the position of the red image. By this procedure a virtual image is located by parallax for a positive lens.

Figure 34–16 shows a method of locating a virtual image formed by a negative lens. The half-lens makes it possible to see both sets of lines with the same eye. The large image screen is the nearer one so it needs a slot at the lens level to admit light

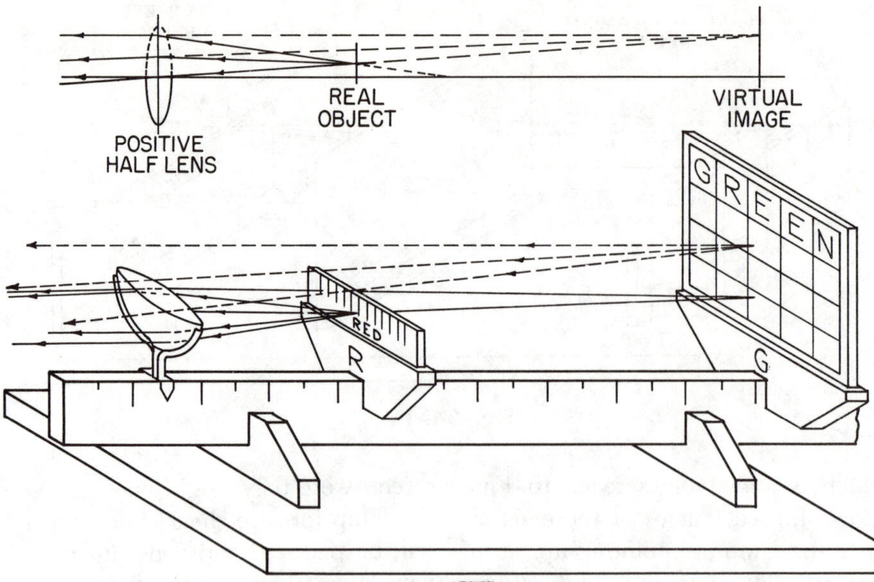

Fig. 34–15

from the red screen. You must stay back at least 10 in. (25 cm) from the image screen in order to see it clearly. The procedure regarding Fig. 34–15 may be followed, as given above, in other respects.

Virtual objects can easily be represented with the equipment discussed. In Fig. 34–17 the apparatus of Fig. 34–15 is assumed to be in adjustment so that there is no parallax between images of the red lines and the green lines. The positive half-lens is removed or covered. The lamp and another positive lens are so placed that light passes through the lens holder and strikes both the red and green screen. Adjust the positions of the lamp and whole lens until a focused image is on the green screen. Replace the half-lens. Now a good image is focused on the red screen. The light from a point of the filament is incident on the half-lens but is converging toward a point on the green screen. Note

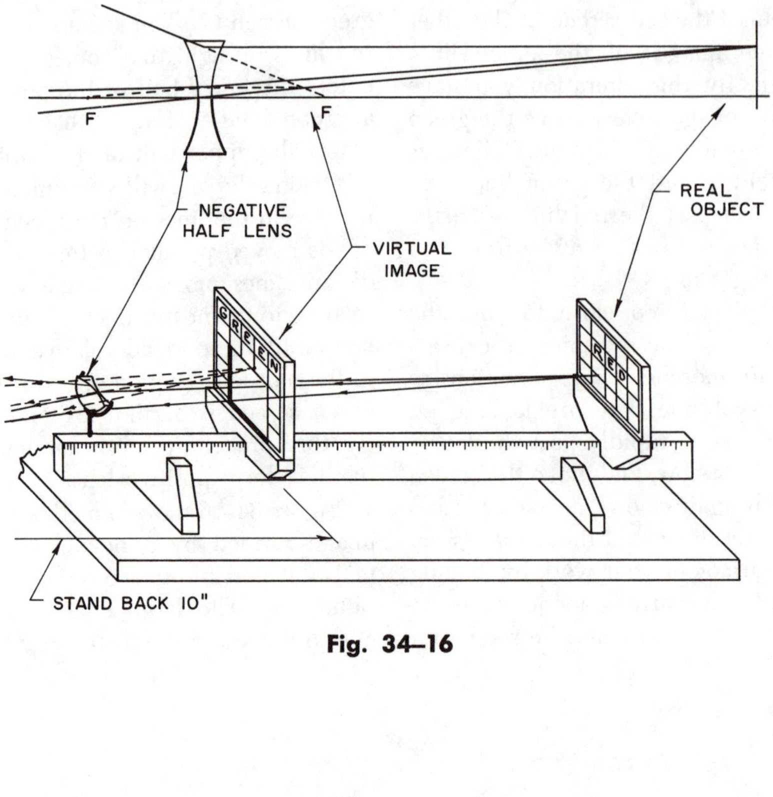

Fig. 34–16

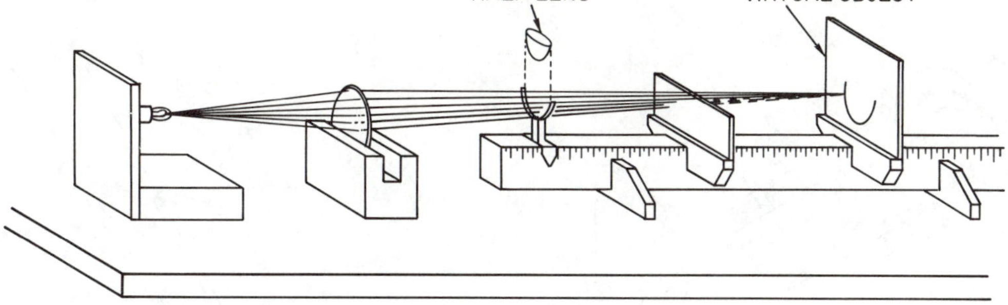

Fig. 34–17

the difference. Light spreads from a point to the whole lens. The lamp is a real object. Light incident on a lens is related to the object. Light incident on the half-lens is not spreading, but is converging. A point to which this light is converging is a point on the object. This type of object is spoken of as a virtual object. The light which was already converging is converged still more by the positive lens and so focuses at a closer position —right on the red screen as originally set by the parallax method. You now have a real image on the red screen where the real object was originally and a virtual object on the green screen. Light that still misses the half-lens still shows the virtual object while light passing through the lens makes the real image.

By this procedure object and image have been exchanged. You can represent a virtual object and exchange object and image for the negative lens shown in Fig. 34–16 by a like procedure.

34–1.15

A slide projector is set up with its optical axis coinciding with the long axis of a 6-in. length of brass channel as shown in Figs. 34–18 and 34–19. The 1-in.-wide channel is mounted in a convenient position by a laboratory support stand. An opaque slide, with a small triangular aperture .04 in. on a side, is inserted in the projector. When the channel

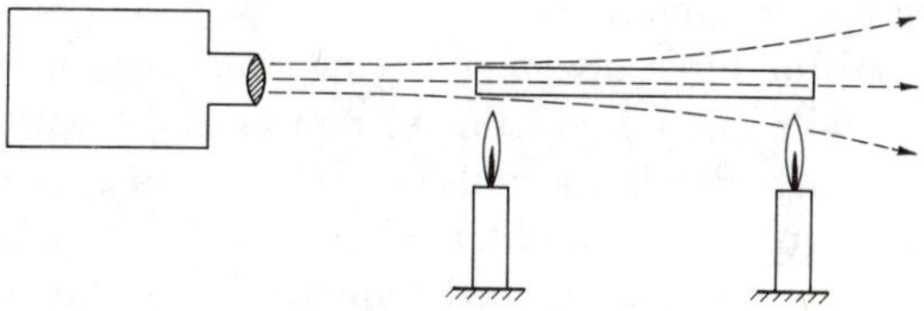

Fig. 34–18

is cold, only a small bright triangle appears on a screen. When heated rapidly at both ends, the triangle enlarges and separates since the light path is refracted upwards above the heated bar and downward below it.

34–1.16

Sometimes a chalkboard diagram is difficult to understand, whereas a three dimensional model makes a point quite obvious. Such an example is

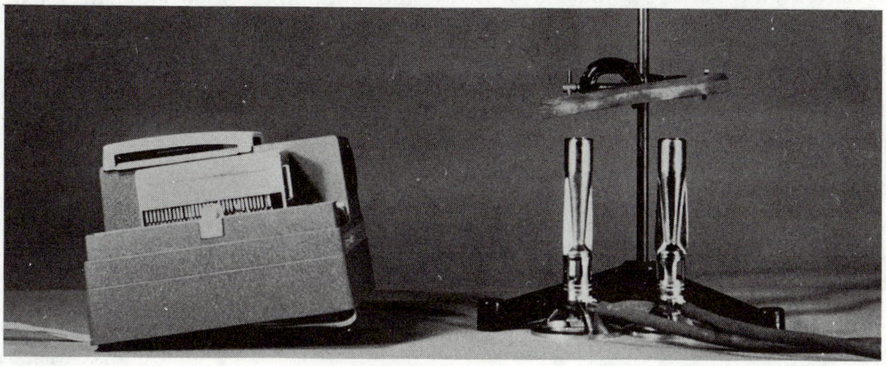

Fig. 34–19

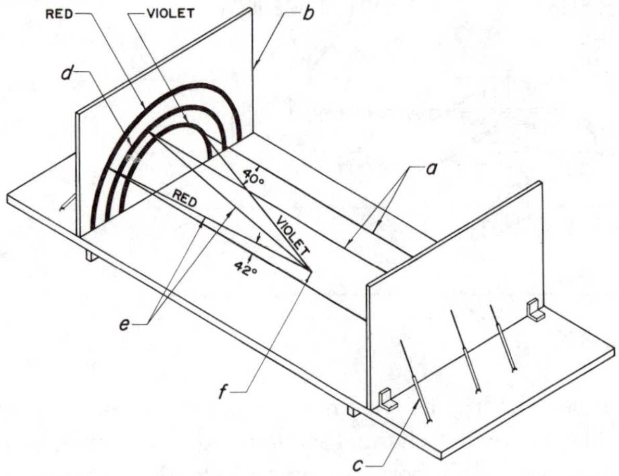

Fig. 34–20

the model shown in Fig. 34–20 which explains the origin of a rainbow.

Two or three dozen skeins of violet, green, red and white embroidery thread represent the various light rays. Parallel strands of white thread *a*, representing incident sunlight, are stretched between two vertical sections of a wood framework *b*. The ends of the thread pass through holes and are anchored to the baseboard with rubber bands *c*.

The other ends go to various sections on the other board depicting the rainbow spectrum *d*. Appropriately colored strands *e* pass through a hole *f*, which represents the eye in the 5-ft-long baseboard, and are stapled to the underside.

34–1.17

A mechanical model for demonstrating the formation of a rainbow is shown in Fig. 34–21. A rigid assembly of 1-in.-diam brass balls connected by brass rods is mounted on a hardwood base (8 x 11 x 1 in.) so the assembly can swing in two mutually perpendicular planes. Three parallel 3⁄16-in.-diam brass rods *a*, supported in the same plane by a brass tube *b* (1⁄4 in. o.d., 3⁄16 in. i.d.) and a 3⁄16-in.-diam brass rod *d*, represent light rays coming from the sun. A bi-axial swivel joint *c* connecting the rod assembly to the base allows the assembly to be rotated about a vertical axis and also about an axis parallel to the simulated light rays. Thus, the model shows why the rainbow is produced and why its size depends upon the time of day.

The balls are mounted 1½ in. apart on the lowest "light ray," 2¼ in. apart on the middle ray, and 3 in. apart on the top ray. The rays are spaced 5½ in. apart vertically. The horizontal tubing is painted white, while the left-hand (as seen in the diagram) vertical member is painted red, and the right-hand one, blue.

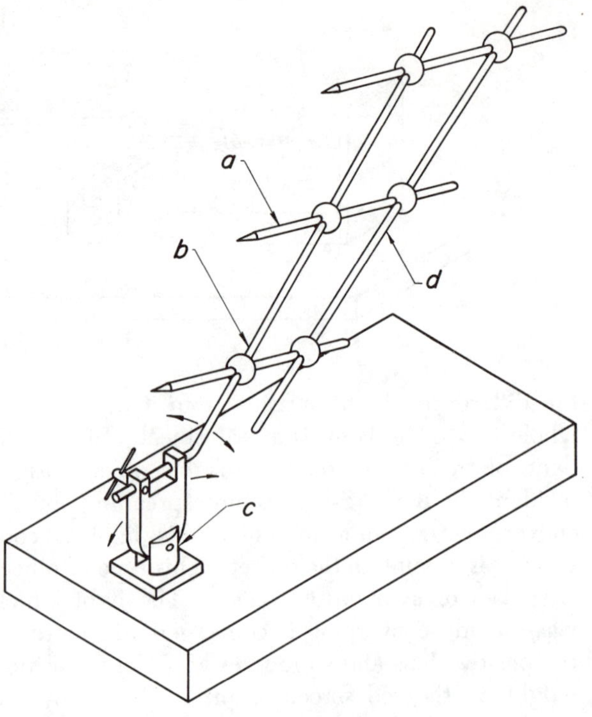

Fig. 34–21

34–1.18

The Hartl optical disk is a well-known and useful device, but the size of the audience which can see demonstrations with the disk is limited. A 3 x 4-ft translucent screen,[11] a standard slide projector,[12] and a number of large, flat, Plexiglas lenses and prisms allow the demonstrations to be seen by a large audience. The demonstrator stands behind the screen and holds a lens or prism against the back side of the screen as in Fig. 34–22. The light rays from the projector strike the lens and are refracted.

The screen, Fig. 34–23 (a), is stretched over a wooden frame and is wound on a ½-in.-diam alumi-

[11] "Roscolene Matte" comes in 100-ft rolls, 40 in. wide, 0.003 in. thick, Rosco Laboratories, Brooklyn, New York.

[12] Delineascope Model MC, available from American Optical Co., Instrument Div., Buffalo, New York.

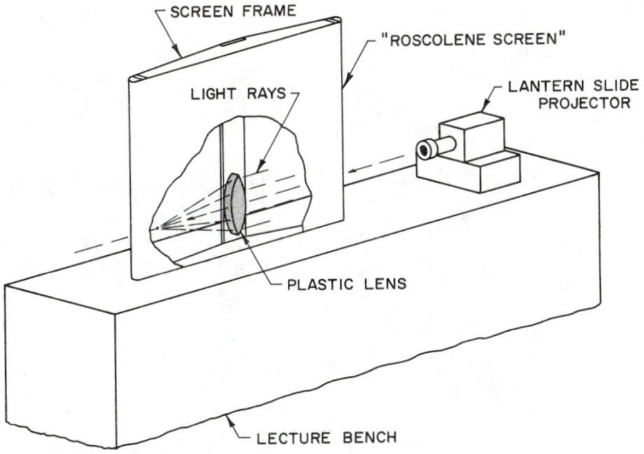

Fig. 34–22

num rod. A setscrew in the rod maintains tension on the screen material. Slits ⅞ in. long cut into a 2 x 2-in. brass plate break the light up into rays, Fig. 34–23 (b). A .08-in. slitting saw is used to cut the slits and the plates can then be placed in the projector as though they were slides.

Two 1-in.-thick lenses, one concave and one convex (Fig. 34–24 (a)), with apertures of over 1 ft, are shaped from Plexiglas. The flat surfaces of the lenses are left unpolished and the curved surfaces are sawed to shape and filed using #320 paper on the file. Final polishing is done with Tri-X diamond buffing compound.

A large 1-in.-thick Dove prism is used to demonstrate refraction at three surfaces. A brass slide

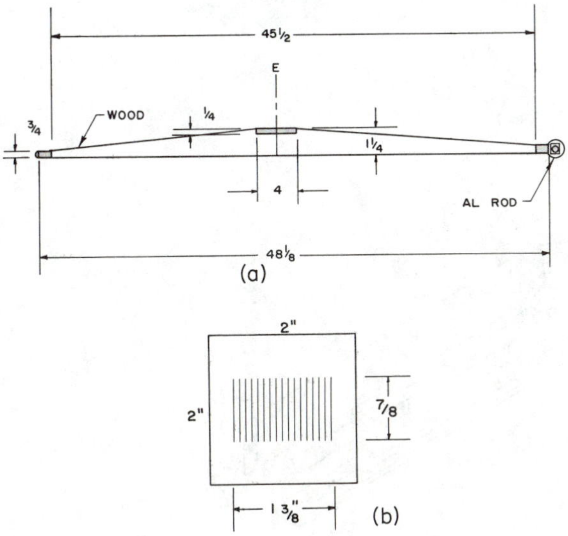

Fig. 34–23

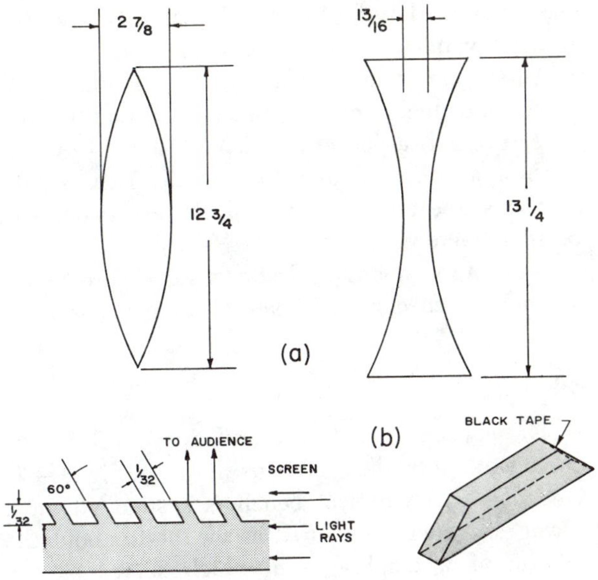

Fig. 34–24

with two parallel slits is used to limit the number of rays, and one slit is covered with a red gelatin lantern slide filter for differentiation between the two ray paths as they enter and leave the prism. The side of the prism nearest the audience is serrated with a 1/32-in. slitting saw as shown in Fig. 34–23 (b) to refract part of the light toward the audience. The sawed edges are filed and polished.

The rear leading edge of the prism is covered with black tape to prevent parallel rays.

34–1.19

Developments in manufacturing have made plastic lenses available in a variety of sizes and shapes.[13,14] Because of the improved quality of plastic lenses, they can replace glass lenses in many optical systems.

Plastic lenses have several advantages over glass lenses besides being of lower cost. A plastic lens of optical grade acrylic plastic is substantially lighter in weight than its glass counterpart. Plastic lenses can be machined or drilled for mounting with little danger of chipping or cracking. They can be made relatively shockproof and are free of striations, bubbles, and other distortions. Light transmission in plastic lenses is slightly higher than in

(13) Bolsey Research and Development.

(14) Fostoria Corp., Horsham, Pennsylvania.

glass lenses. Plastic Fresnel lenses combine the above advantages with compactness.

Among the types of plastic lenses available are plano- and double convex, plano- and double concave, Fresnel, aspheric and axicon lenses. A few of these are illustrated in Fig. 34–25. Lenses with special shape and transmission characteristics can be manufactured on request.

A few applications of plastic lenses are projectors, magnifiers, reflectors, collimators, and concentrators.

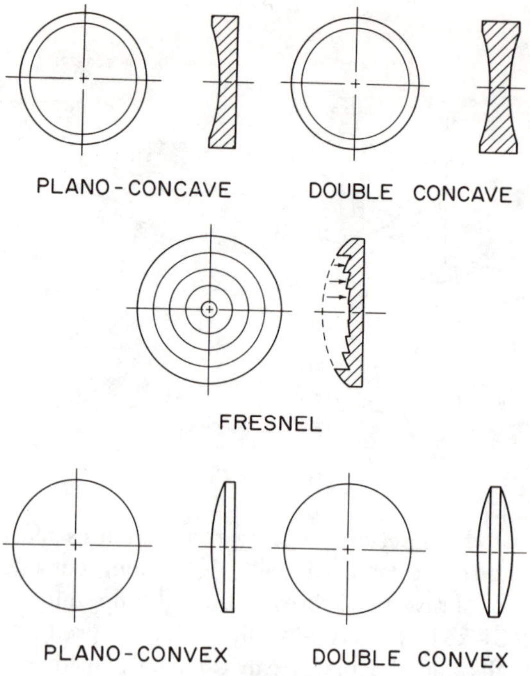

Fig. 34–25

34–1.20[15]

The concept of entrance and exit pupil can be demonstrated by the apparatus shown in Fig. 34–26. One half of an optical bench is suspended on a spring *c* in such a way that it can rotate about the middle of a diaphragm *d* which serves as the entrance pupil. The object *a* is a small opening illuminated from the rear by a carbon arc and condenser. A lens is mounted against a card in such a way that the beam in both the object and image spaces is made visible (The card is set into and at a slight angle to the beam axis.) Under these conditions the object *a* can be made to vibrate up and down. The light rays in both object and image space will also vibrate up and down with the object. However, both the entrance pupil *b* and the exit pupil *B′B′* will remain stationary. The upper half of the entrance pupil can be covered

[15] Robert W. Pohl, *Optik und Atomphysik* (Springer Verlag, Berlin, 1967), 12th ed., pp. 31–34.

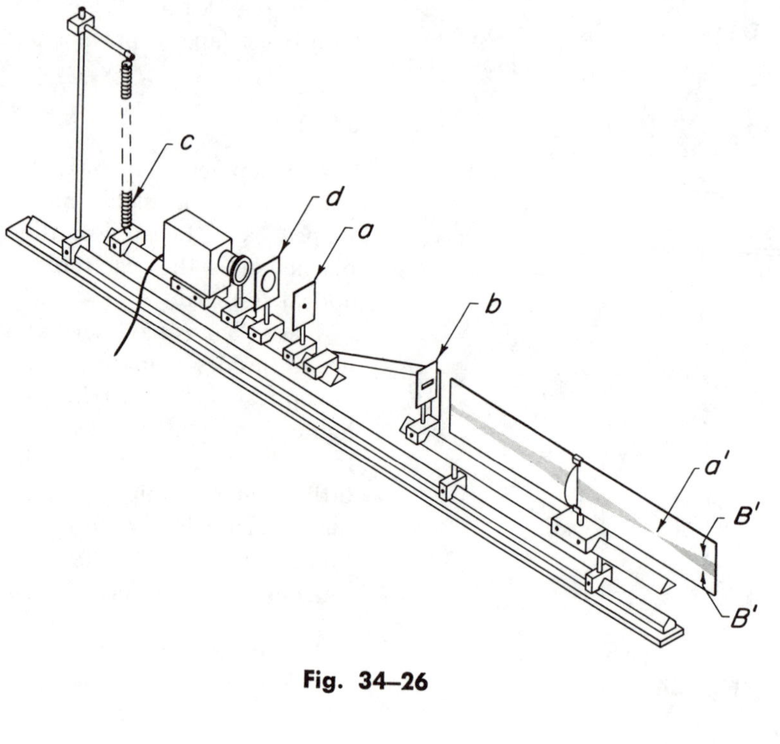

Fig. 34–26

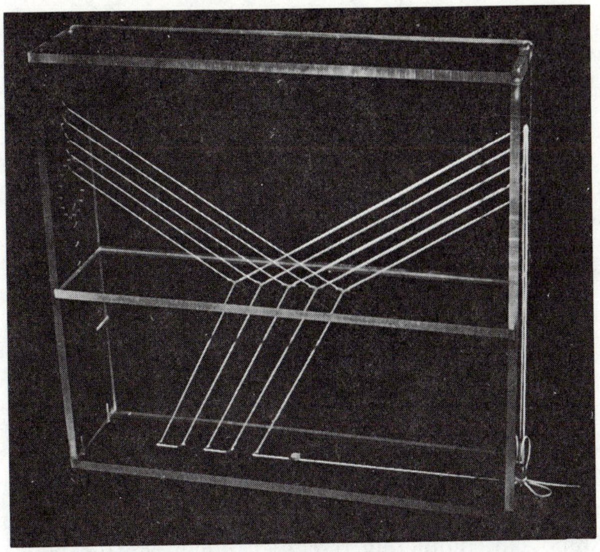

Fig. 34–27

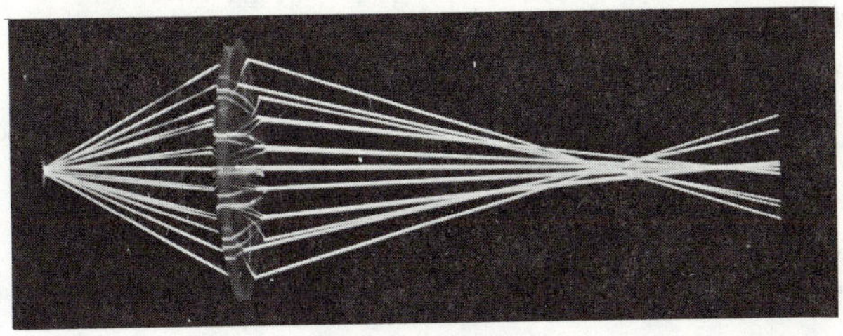

Fig. 34–28

with a red filter, the lower half with a green one. Under these conditions the lower half of the exit pupil will be red, the upper green. The vibrating focal point a' will be colorless since it is illuminated by equal amounts of light of complementary colors.

34–1.21[(16)]

Figures 34–27 to 34–31 are photographs taken from models used to demonstrate basic principles of image formation by refraction through simple lenses and to show and differentiate between some of the defects in the images formed. Elastic fluorescent strings represent light rays. In lecture ultraviolet light is used.

Reflection and refraction of light rays is shown in Fig. 34–27. Light incident from air onto the plane surface of a denser transparent medium such as glass is, in general, partly reflected back into the air and partly refracted into the glass. The incident and reflected rays make equal angles with the surface (or surface normal). The rays refracted into the glass are bent away from the surface (toward the surface normal). The angles of incidence and refraction are related by Snell's law, the ratio of the sine of the angle-of-incidence to the sine of the angle-of-refraction being equal to the index of refraction of the glass relative to air.

(16) R. Resnick and H. F. Meiners, *Demonstration and Laboratory Apparatus Report of the 1960 Summer Visiting Professor Workshop at Rensselaer Polytechnic Institute.*

When light rays are refracted through a simple glass lens with convex surfaces, they converge to form an image. Parallel rays incident on the lens form an image at the focal point of the lens. The distance from the center of the lens to the focal point is called the focal length of the lens. In general, lenses of reasonable aperture allow large cones of rays to pass through, and the images formed show certain defects known as aberrations. Even with monochromatic light, spherical aberration, coma, and astigmatism can occur in the images.

Spherical aberration (Fig. 34–28) occurs when monochromatic rays are refracted at convex spherical surfaces. To visualize this, consider rays coming through various zones in the lens (a zone being a ring on the lens all points of which are equidistant from the axis). Rays from an object point on the axis refracted through a lens zone of large radius form an image closer to the lens than rays refracted through a zone of smaller radius. The resultant effect is a linear separation of these image points along the axis. For all zones of the lens active, the best image produced on a screen perpendicular to the axis will be at the point at which the diameter of the circular cross section of the rays is least. This best image is called the circle of least confusion.

The aberration called coma (Fig. 34–29) occurs for rays that come from object points off the axis of the lens. The principal ray (i.e., that passing through the center of the lens) from the off-axis

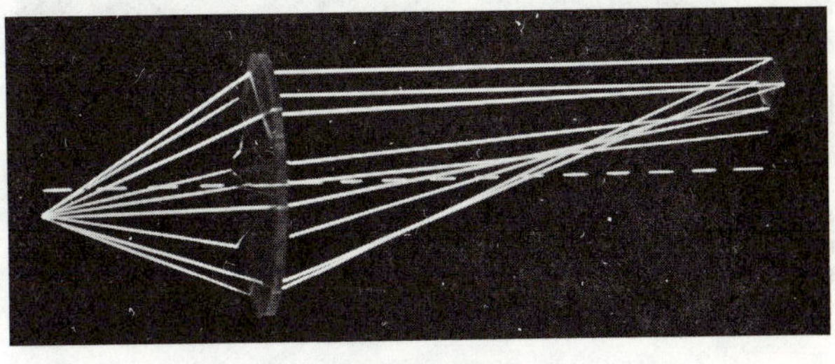

Fig. 34–29

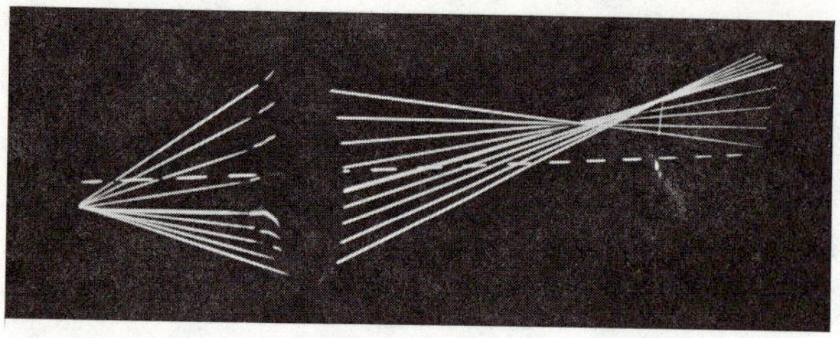

Fig. 34–30

object point forms a point image on an image screen perpendicular to the lens axis. Cones of rays refracted through lens zones of different radii form a series of comatic circles, displaced laterally from the principal ray spot and overlapping one another to form a comet-shaped image. Hence, the name coma.

Astigmatism (Fig. 34–30) occurs for off axis object points. The primary or tangential plane is defined by the lens axis and the oblique principal ray. The secondary (sagittal) plane is perpendicular to this. Rays from the off-axis object-point, forming a fan in the sagittal plane, come to a focus at a point a different distance from the lens. The series of point images from all tangential fans form a line image called the tangential focus. The series of point images from all sagittal fans form a line image called the sagittal focus. These two line images from a given object point are perpendicular to one another.

Dispersion by a prism is shown in Fig. 34–31. With use of polychromatic light, the amount of angular refraction at a plane surface depends on the wavelength of the light. Two plane surfaces at an angle to one another form a prism. Light refracted through such a prism is deflected from the original incident direction. The refraction for blue (shorter) wavelengths is greater than for red

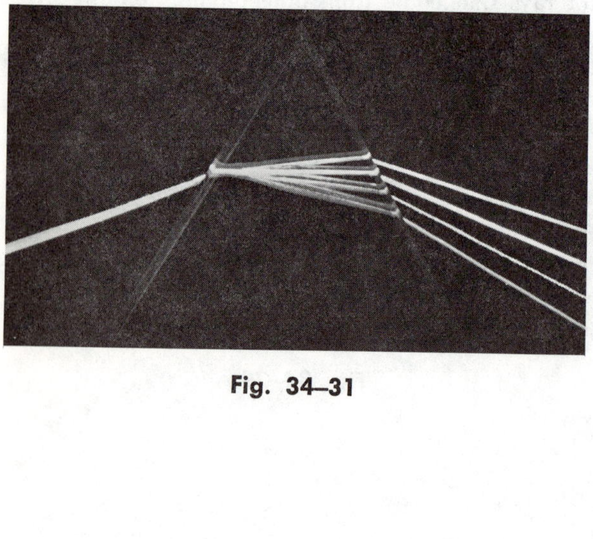

Fig. 34–31

(longer) wavelength rays. The blue is thus deviated more than the red with a resultant angular dispersion of the rays.

Such dispersion gives rise to chromatic aberration, a defect of simple lenses apparent when polychromatic light is used. On refraction through the lens, blue rays are deviated more than red; thus the blue image is formed closer to the lens than the red. A plane image screen at right angles to the axis, even at the position of best focus, shows a circle of least confusion which is colored at the edges.

Each of the lens aberrations has been treated individually assuming no others simultaneously present. In actual practice, several aberrations may occur at once. While in general these defects cannot be completely removed, much can be done by suitable control of lens surfaces and with compound lenses to improve the quality of the image produced by a lens system. Construction details of the models are given in the Appendix, page 1345.

34–1.22[17]

A strong incandescent lamp with a straight filament is mounted inside a black box having a hole in one side covered with a filter which will pass both red and ultraviolet light. A convenient filter is made from a solution of 10 mg nitrosodimethyl analine in 100 cc of water.[18] An ordinary plano-convex glass lens having a focal length of 30-cm for visible light is now placed 35-cm from the filament. The red image will be focused on a white screen at about 200 cm. The ultraviolet image will be focused at about 150-cm and is best shown on a fluorescent screen. Since the focal length f for a plano-convex lens is given by $f = \frac{R}{n-1}$, where R is the radius of curvature and n is the refractive index of the glass, the latter can be calculated for the two wavelength regions if f and R are measured It is convenient in this demonstration to use two half screens, one mounted above the other but separated by the correct distance in order to observe both sharp images simultaneously.

[17] L. Grillet, *Bull. Union des Physiciens,* No. 266, 37 (1933).

[18] A number of other suitable filters are discussed in N. M. Mohler and J. R. Loofbourow, *Am. J. Phys.,* **20,** 499 (1952).

34–1.23[19]

This demonstration is intended to show the aplanatic properties of a sphere. Aplanatic systems are lens or mirror combinations which have the property of introducing no spherical aberration or coma for some special position of the object and image. Such systems may use aspherical surfaces or, in a few cases, spherical surfaces. These few spherical systems are the exceptions to the usual rule that spherical surfaces produce spherical aberration. Probably the most important example is that of the oil immersion microscope objective of short focus. Since one of the aplanatic points of a sphere lies within its surface, the oil immersion objective is made of a hemisphere of glass and the object is placed in a drop of oil of the same refractive index as the glass at the aplanatic point.[20]

Figure 34–32 illustrates the cross section of a sphere with the aplanatic points, *P* (real object point) and *Q* (virtual image point). That is, if an object is placed at point *P*, a virtual image of it is formed at point *Q*. The distance between the center of the sphere *C* and the object *P* is r/n, where *r* is the radius of the sphere and *n* is its refractive index. The image distance *CQ* is equal to rn. This relation between object and image points is rigorously true even for very large angles, and spherical aberration and coma are absent.

There are two alternative procedures for the demonstration. The straightforward one is often used in elementary courses to trace rays through a prism or lens, and it is convenient for locating a virtual image. The alternative method reverses the object and image positions and allows the convergence of rays to be observed by means of fluorescence.

In the straightforward demonstration the apparatus required is simply a container with a cylindrical wall; this represents a cross section of the sphere. It may be a goldfish bowl with a cylindrical side, a large diameter beaker, or a flat-bottom petri crystallizing dish.

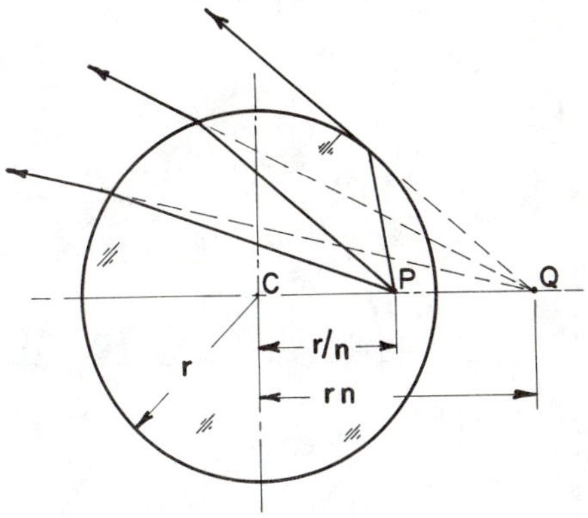

Fig. 34–32

The container is filled with water, and the object, a straight heavy wire supported vertically, is held at the calculated aplanatic object point. Pins are used to mark, on a large piece of paper, the position of the virtual image as seen from various directions. After the position of both the container and object are marked on the paper, the container and object are removed and lines drawn to locate the position of the virtual image. If the experiment is carefully done, all the lines will cross very nearly at one point. Then put the object just outside the container and repeat the experiment. The spherical aberration will now be quite apparent.

In the alternate method, a dish of fluorescein is used to observe the convergence of the rays. Since the image formed by the spherical lens *c* (or its two dimensional cylindrical equivalent) is virtual, the aplanatic properties are best demonstrated in reverse. That is, convergent light rays are directed toward the previous virtual image point *Q* (Fig. 34–33), and upon striking the curved refracting surface, are converged toward the former real object point *P*. In this demonstration *P* becomes the real image point, and *Q* the virtual object point. The ray directions may be made visible in a fluorescein solution.

Figure 34–33 illustrates how the demonstration may be set up directly on a bench top. An auto lamp *a* and a small collimating lens *b* are clamped as close to the table top as practical. The lens may be held in place with wax if desired; it is set to give a parallel beam at right angles to the desired optical axis centerline. Five small plane mirrors each having a reflecting surface of about ⅛ x 1 in. are cut from an aluminized microscope slide or from an inexpensive compact mirror. These narrow

[19] C. Harvey Palmer, *Optics Experiments and Demonstrations* (The Johns Hopkins Press, Baltimore, Maryland, 1962), Experiment A11, pp. 53–56; see also J. Strong, *Concepts of Classical Optics* (W. H. Freeman and Co., San Francisco, 1958), sec. 14.5, pp. 314–315; sec. 15.4, pp. 332–336.

[20] See F. A. Jenkins and H. E. White, *Fundamentals of Optics*, 3rd ed. (McGraw-Hill, New York, 1957), pp. 146–147.

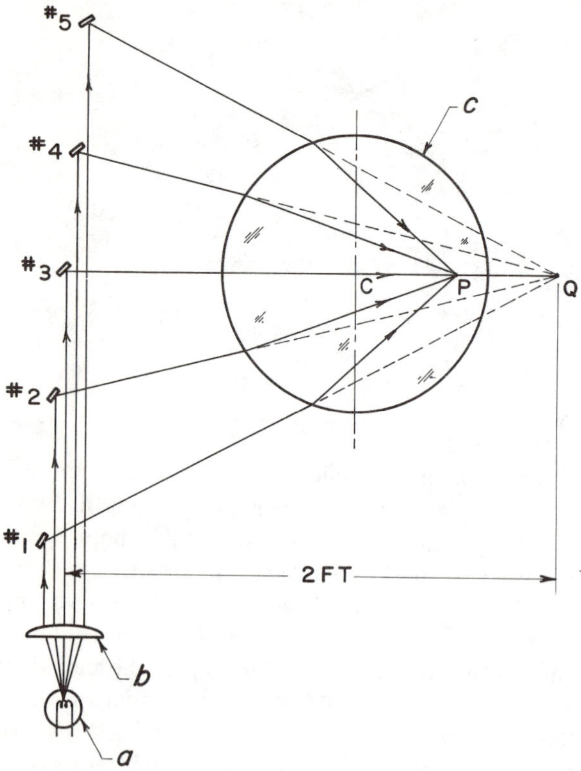

Fig. 34–33

Fig. 34–34

mirrors (#1, . . . , #5) are arranged to tap off portions of the collimated beam and to direct narrow pencil beams toward a point Q. (The dish with fluorescein is not used yet.) The point Q should be marked on paper or on the table top and mirror #3 placed so that it throws rays along the chosen optical axis. It should be tipped very slightly downward so that the path is clearly visible on the table top near Q. The other four mirrors are then arranged to tap off other portions of the collimated beam to throw light through the point Q at angles of 20 to 40 deg with the axis.

Calculate the aplanatic points for the system and the distance between the surface of the container and the point Q. Prepare the fluorescein solution by dissolving enough fluorescein powder in the water to render light beams traversing it readily visible. Place the cylindrical container (such as those mentioned previously), at the calculated distance from the original image point Q, making sure that the center of the dish lies on the path of ray #3. This ray should not be deviated by the dish. If the calculations are correct and the surface of the container is smooth and regular, five rays should be seen in the water, and they should intersect very nearly at one point. A piece of black paper below the container will improve the visibility of light beams through it. Extraneous room light should be kept at a minimum to improve the demonstration.

If the results are satisfactory, the water "lens" should be moved to other positions to show the appearance of spherical aberration when the aplanatic points are not used. If the center of the water "lens" is placed over point Q, then there is no spherical aberration either; however, there is also no magnification. For a large group the above results could be shown with closed-circuit TV.

34–1.24[21]

When two periodic arrays in which the orientations or the "wavelengths" of the periodicities are not quite the same are superposed, their periodic patterns will be alternately in and out of phase. The resulting large scale periodic pattern is known as a moiré pattern.[22] These moiré patterns recently have stimulated renewed interest because they can be seen in an electron microscope when two imperfectly matching crystal sheets are superposed.

[21] E. A. Wood, *Am. J. Phys.*, **30**, 381 (1962).

[22] Editor's Note: Many varieties of moiré pattern can be obtained in black and white and color kits from Edmund Scientific Co., Barrington, New Jersey. An experiment manual, *The Science of Moiré Patterns* by Dr. G. Oster, Brooklyn Polytechnic Institute, can also be obtained from Edmund.

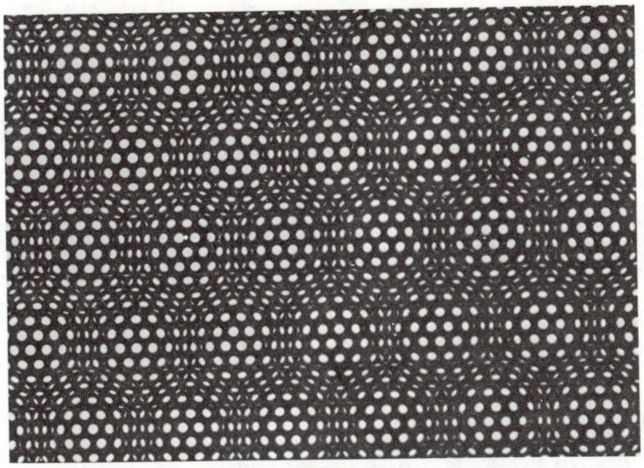

Fig. 34–35

Moiré patterns can be quite simply shown by using the cover of an electronic circuit chassis: Figs. 34–34, 34–35 and 34–36 illustrate a moiré pattern resulting from a slight difference in "wavelength" of two periodic patterns. Since the back of the box is farther from the camera than the front of the box, its projected length is shorter, and the repeat distance or "wavelength" of the far pattern appears less than that of the near pattern. Therefore, the patterns are alternately in and out of phase, in two dimensions, as seen in Fig. 34–34. For Figs. 34–35 and 34–36 the camera was progressively defocused to simulate, in Fig. 34–36, the kind of pattern observed in an electron microscope. From the resolvable periodicity of such patterns, one deduces the much smaller scale periodicity of the array of atoms which cannot be resolved by the microscope.

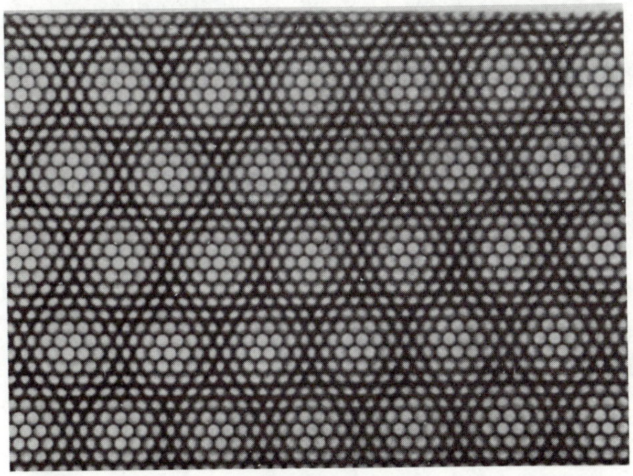

Fig. 34–36

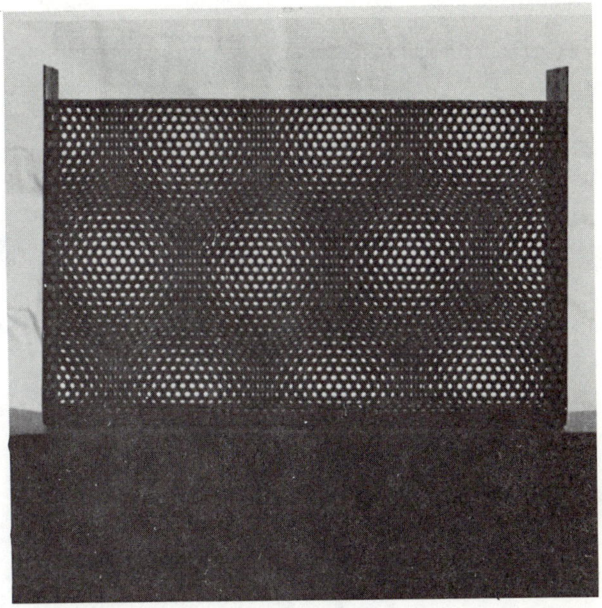

Fig. 34–37

If one moves the camera farther away from the chassis box (Fig. 34–37), the angle subtended by the back of the box becomes more nearly equal to that subtended by the front, so that the "wavelengths" of their periodic units, or holes between successive out-of-phase (dark) nodes of the moiré pattern, become larger.

34–1.25[23]

This is another demonstration concerning moiré fringes, a purely ray optic phenomenon, and shows how they may be utilized in making measurements of extraordinary delicacy. In precision apparatus moiré fringes can reveal mirror rotations of 10^{-10} radians or linear displacements of conceivably 10^{-12} cm—smaller than the diameter of an atom! One well known very sensitive infrared detector, the Golay cell, makes use of moiré fringes.

Moiré fringes are produced by the overlapping of two grid patterns or of one grid and the image

[23] C. Harvey Palmer, *Optics Experiments and Demonstrations* (The Johns Hopkins Press, Baltimore, Md., 1962), Experiment A20, pp. 108–119; see also R. V. Jones and J. C. S. Richards, "Recording Optical Lever," *J. Sci. Instr.*, **36**, 90–94 (1959); M. B. Stout, *Basic Electrical Measurements*, 2nd ed. (Prentice-Hall, Englewood Cliffs, N.J., 1960), secs. 4.10, 4.32, 5.13; Gareth Thomas, *Transmission Electron Microscopy of Metals* (John Wiley & Sons, Inc., New York, London, Sydney, 1962), ch. 2, pp. 51–57.

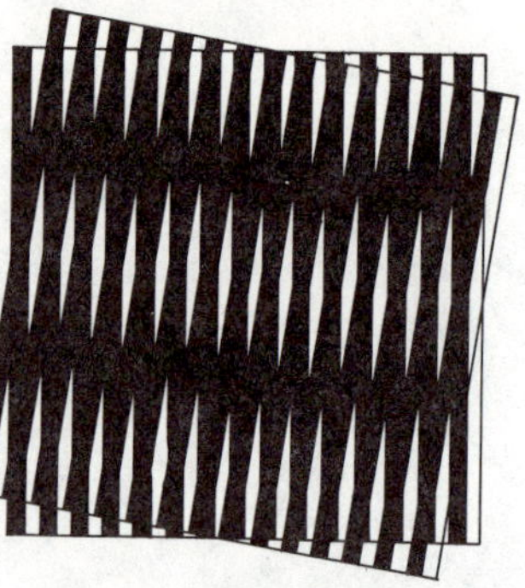

Fig. 34–38

of a grid. The grids themselves ideally consist of straight opaque stripes separated by transparent stripes of equal width sometimes called Ronchi rulings or Levey screens. They are made with high precision either by a photographic process or by a ruling process. Figure 34–38 shows the appearance of two such overlapping grids slightly tipped with respect to each other. The grids in the illustration are very coarse so that the structure of the broad, horizontal moiré fringes is evident. The black moiré fringes are separated by more or less transparent bands for which the transmission rises to a maximum of 50 percent (which is, of course, the transmission of a single grid). It is evident, then, that these fringes are a purely ray optic phenomenon having to do with the geometry of the grids. It is true that interference and diffraction may also be involved, especially when the grids have many lines per inch, but that aspect is not involved in the present discussion. The width of the moiré bands depends upon the angle made between the two grids and the number of stripes per unit width of the grid. The smaller the angle and the fewer stripes per unit width, the wider the moiré fringes will be. If one of the grids is translated in a direction at right angles to its stripes, the moiré fringes move up or down.

Suppose now that the two precision grids are accurately parallel. The transmission then obviously depends upon the "phase" between the two grids, that is, the relative degree of overlap of the sets of stripes. If the black stripes are superposed, the transmission will be a maximum, that of a single grid, or 50 percent on the average. If the black stripes of one grid overlie the transparent stripes of the other, the transmission is obviously zero. Figure 34–39 (A) is an end view of two superposed parallel grids. The upper grid is displaced a distance x with respect to the lower grid. The grid stripes have a periodic spacing of a, and parallel light incident from below is transmitted through the clear spaces of width b, common to the two grids. The transmission through the grids is $T = b/a = (a/2 - x)/a = 1/2 - x/a$ for values $0 \leq x \leq a/2$. For larger values of x, between $a/2$ and a, the transparent region is determined by the left edge of the upper opaque stripes and the right edge of the lower stripes. Here $T = b/a = (x - a/2)/a = x/a - 1/2$. For the complete period, $T = 1/2 - x/a$ for $0 \leq x \leq a$. The theoretical transmission as a function of x is shown in Fig. 34–39 (B).

A pair of precision grids can be used to measure distances of translation accurately. Suppose, for example, one wishes to measure accurately the translation of a lathe carriage along its longitudinal ways. With the aid of the grids, a few simple optical parts, and some electronic components, distances say of 10 in. can be measured with an error of less than 0.001 in. using grids with 100 lines/in. Figure 34–40 is a simplified diagram of how this measurement might be affected. The light source a, lens system b consisting of two plano-convex lenses, photocell c, and a short grid d are fixed to the lathe carriage e. A second grid f is fixed to the lathe ways g, parallel to the first grid and the ways. An amplifier increases the signal of the photocell for a counter and a meter. The counter records the integral number of fringes moved, and the meter records the fraction of a fringe. Since two positions of the grids give the same transmission, one must note whether the photocell signal is increasing or decreasing during the fractional fringes at the

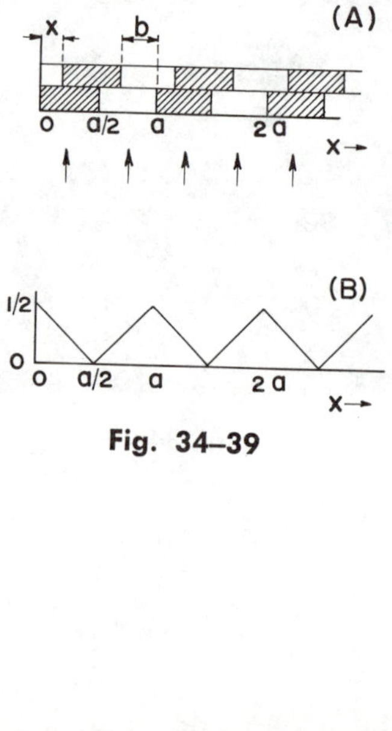

Fig. 34–39

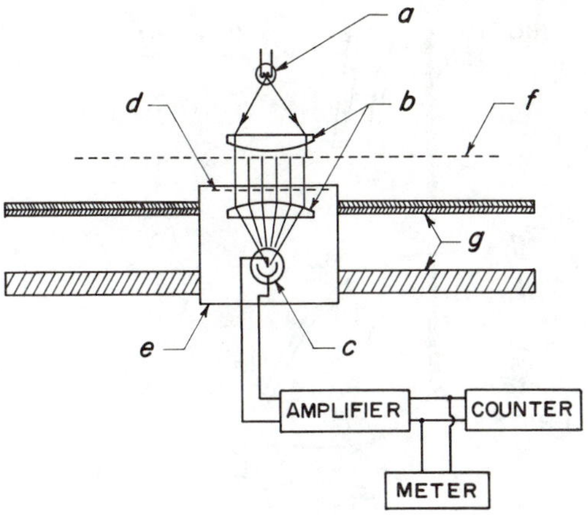

Fig. 34–40

beginning and end of the translation. This scheme with suitable refinements can be used for automatic machine systems.

So far, perfect grids have been assumed; in practice, of course, errors occur. Small random errors in some of the lines have little effect. For instance, suppose one of the grids in Fig. 34–40 had one line displaced $a/2$ from its proper position but no other errors. If the smaller grid d on the movable carriage had 100 lines/in. and were 2 in. long, the positional error would be 1/2 percent or 0.0005 at most. Such gross errors do not occur, but smaller ones are inevitable. The measurement of translation can be no more accurate than the average accuracy of the grids.

Figure 34–41 shows the basic principle involved in measuring small angular deflections. Suppose that an illuminated grid h is placed at the focus f of a lens i behind which is a mirror j. The light striking the mirror consists of a set of parallel beams, one from each point of the object grid. The reflected light, passing again through the lens, is

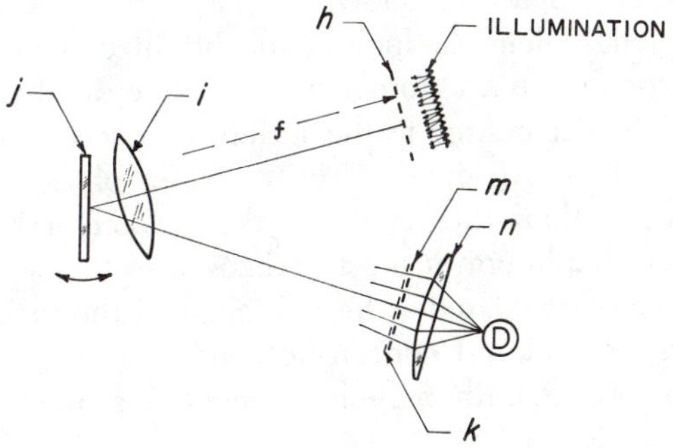

Fig. 34–41

imaged in the same plane as the grid itself and is the same size. If the illuminated grid is moved to one side of the axis so that the grid image k is formed on the other, then a second grid (or extension of the first) m will or will not transmit light according to the phase of the image with respect to the second grid. The transmitted light is then focused by a plano-convex lens n to a detector D. A small rotation of the mirror will change light to dark.

If there are n lines per unit length on the grid and the focal length of the lens is f, a mirror rotation of $\Delta\theta = (1/2) \times (1/2n) \times (1/f)$ will cause the transmission to change from maximum to minimum or vice versa. The first factor of 1/2 occurs since the mirror doubles the angular deflection of the beam. The other two factors express the radian measure of the required displacement of the image. This scheme is very sensitive to small mirror rotations, but it is also very sensitive to small changes in illumination and detector sensitivity.

Considerable improvement in stability and also in sensitivity results from the use of two detectors arranged behind two grids which are out of phase so that if one grid transmits the other does not and vice versa. Figure 34–42 shows the complete system. Light from the source S is focused by lenses L_1 on M_1. An image of the grid is formed by L_2 partly on the rest of grid 1 and partly on grid 2, which is out of phase with grid 1. Light passing out of the lower part of grid 1 is reflected by mirror M_2 through lenses L_3 to photocell P_2; light passing through grid 2 is reflected by M_3 through L_3 to photocell P_1. Since the image of the grid formed by L_2 has arbitrary phase with respect to the remainder of grid 1 and grid 2, a pair of compensating plates C_1 and C_2 are used to restore the phase. These plates are simply located in the beam near the grid unit where the incoming and return beams do not overlap. The two photocells form two arms of a Wheatstone bridge shown within the dotted lines in the lower part of the figure. Mirror M_1 is attached to anything whose rotation is to be measured. The magnetostrictive device (MD) shown at the left of the figure is suitable. The optical lever–electrical bridge circuit apparatus is capable of giving full-scale galvanometer deflection for very small magnetization of the iron. By direct measurement with conventional telescope and scale, the rotation of the mirror is all but unobservable.

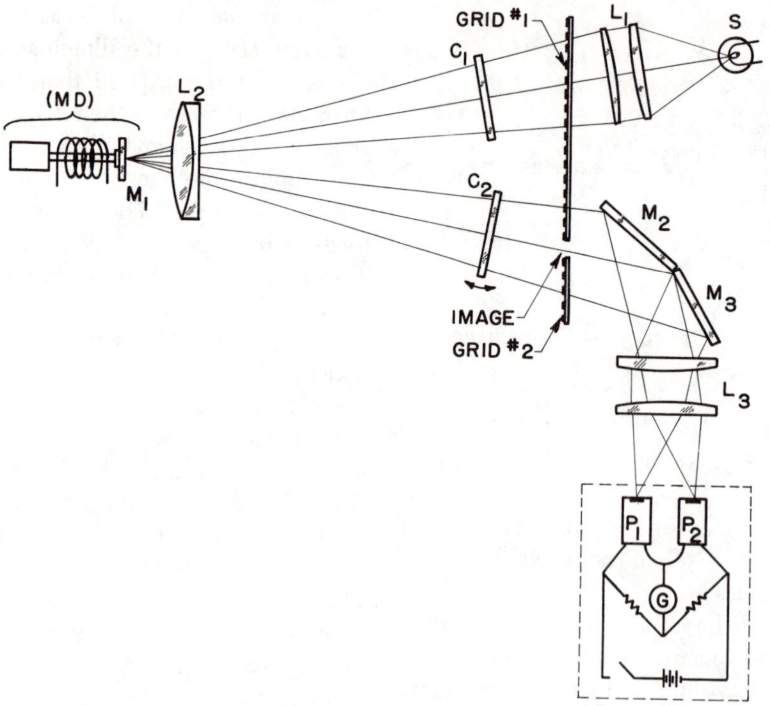

Fig. 34–42

It is also possible to feed the output of the photocells into individual projection meters so that a large group can view the results.

The demonstration could clearly be made quantitative if desired. One would have to calibrate the optical lever in terms of the signal corresponding to one full grid displacement using low sensitivity of the bridge and making sure the deflection was proportional to displacement—that is assuming that the bridge was not too far off balance. The magnetostrictive device would require calibration—the magnetic field within the iron wire would have to be determined, and the length and spacing of the wires ascertained so that the change in length of the iron wire could be determined from the measured mirror rotation. The calculations, in short, are not very straightforward. For the construction details and material list of the optical lever system see Appendix, page 1346.

34–1.26[24]

In many cases light when traversing a given "object" will undergo such small changes in phase or such small deviations in direction that the overall energy change is slight and the object remains unperceived by the eye. Such "objects" may be a stream of flowing gas or the inhomogeneities inside a polished glass plate. In the case where the light is slightly deviated—either by refraction or diffraction or both—the object may be rendered visible by placing a point source of light behind it and allowing the intercepted light to fall on a screen. The screen will show shadows or "schlieren" of light and dark areas. The incident light has been deflected from the dark areas and falls in the light areas, thus proving the presence of the deflected beam.

A more sophisticated experiment depending on this same effect can be carried out with the Toepler Schlieren apparatus shown in Fig. 34–43. A 1-m focal length lens L_1 focuses the bright end of the + carbon of an arc onto a small opaque diaphragm d which is cemented to the surface of a second lens L_2 of 2-m focal length. The diameter of L_2 must be larger than the image of the carbon and the opaque diaphragm d must be chosen so that it just (but no more) catches the full image of the carbon. If desired, colored concentric zones cut from red, green, etc. gelatin may be cemented around the

[24] R. W. Pohl, *Optik und Atomphysik* (Springer Verlag, Berlin, 1967), 12th ed., p. 96.

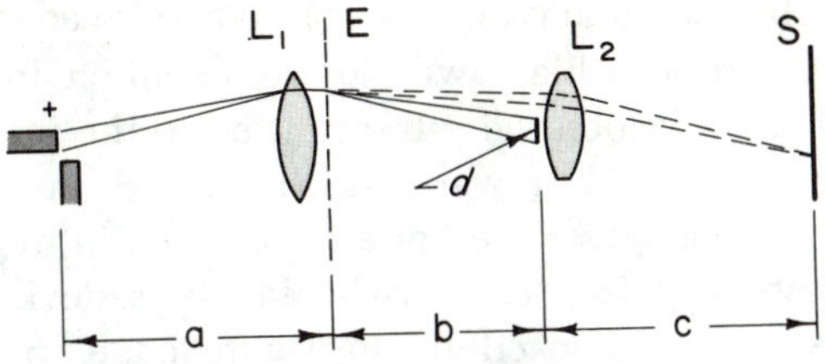

Fig. 34–43

stop. With these lenses, the distance a will be 1.5 m, and the distances $b = c = 2f = 4$ m. The diameter of lens L_1 should be about 12 cm. (Note that this is essentially a projection lantern with E the plane of the slide and with the small diaphragm stop added.) Under these conditions the only light falling on screen S will be those rays which have been deviated by one of the "objects" being investigated. Light which is deviated slightly will be pinkish; light showing greater deviations will be green, etc., as a result of the action of the colored gelatin zones. The principle used here is essentially that of dark field illumination in a microscope.

34–1.27[25]

This demonstration illustrates two ray optic methods by which very small changes in the refractive index of air may be made evident. A third method, closely related to these two, but depending on interference is mentioned briefly, but not demonstrated. The three methods, direct shadow, Schlieren (German for streaks or striae), and interferometric, all depend on the optical effects resulting from local variations of refractive index associated with density changes in air. Though they are all dependent on refractive index changes, the methods give different information and supplement each other. The relationship between the methods can be illustrated by Fig. 34–44. A ray of light a incident from the left would ordinarily strike the screen b at point P_1. If a model is placed in the testing region c near the origin of coordinates, the airflow will be disturbed, and as a result there will be local changes in density and therefore in refractive index. The incident ray is deviated through a very small angle $\Delta\theta$, and will then strike the screen

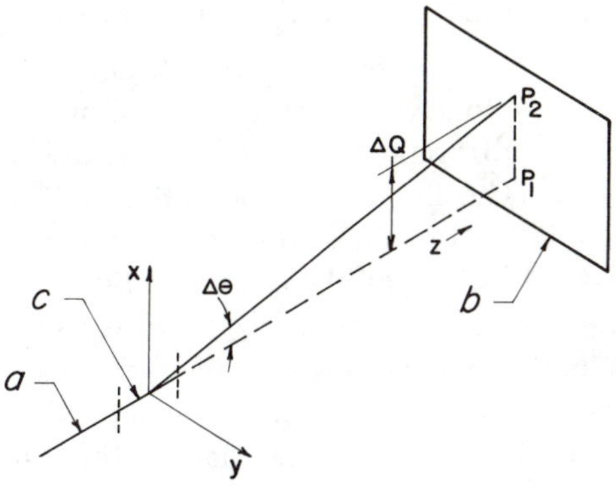

Fig. 34–44

at the point P_2, a distance ΔQ from P_1. The direct shadow technique gives a measure of ΔQ; the Schlieren method gives a measure of $\Delta\theta$; and the interferometric method a measure (in fringe shift) of the change in optical path $\Delta(nd)$. The change of n in the test region causes the path difference $\Delta(nd)$, deflection $\Delta\theta$, and displacement ΔQ. In the expression $\Delta(nd)$, n is the refractive index, and d the geometric distance between an arbitrary point ahead of the working section and the screen; both quantities change, and it is the difference in the product which is measured by the interferometric method.

The interferometric method usually requires the use of high quality, expensive optical components, and is not considered further.

Two arrangements for the direct shadow method are illustrated in Fig. 34–45. Light from a small bright flash of short duration d passes through the

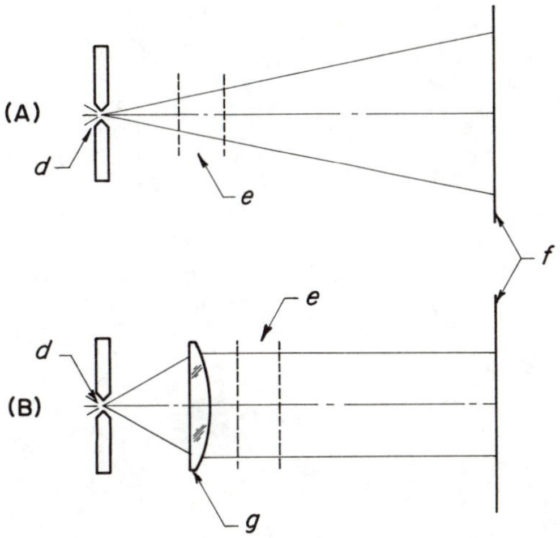

Fig. 34–45

[25] C. Harvey Palmer, *Optics Experiments and Demonstrations* (The Johns Hopkins Press, Baltimore, Maryland, 1962), Experiment A21, pp. 123–129.

test area e, either as a divergent beam (A) or as a parallel beam (B), and falls on a distant screen f. It is desirable to have the screen 20 ft or more away from the small bright source d—carbon arc or the equivalent. The illumination is better if a mirror or lens g is used to make the light beam parallel. It is best to block off all the unwanted light in order to make the pattern readily visible. Changes in density and refractive index in the test area result in corresponding changes in illumination on the screen. Though the interpretation of the patterns obtained may not be the simplest, the apparatus surely is.

For photographic purposes, a flash device is used, and if the room is completely darkened, the camera aimed at the area on the screen to be illuminated may have its shutter left open for a few seconds prior to the flash. In this way, synchronization is not needed; and indeed, with a flash tube, the synchronization problem might be difficult. Unless the flash has both very small dimensions (or a small hole is used to restrict its size) and is of very short duration (less than a millisecond), the photographs will show very little detail.

The principle of the Schlieren method is illustrated in Fig. 34–46. With this technique, light from an illuminated adjustable slit h is rendered parallel by a plano-convex lens i (4–6 in. diam; focal length f should be 50 cm or more). After the parallel beam j has traversed the test area, it is converged by the same type of lens k to a knife edge m (a mounted razor blade will do) set to cut off about half of the rays otherwise transmitted to the screen n. Rays deviated upwards—as indicated in Fig. 34–46 (A) pass over the knife edge and add to the illumination of the screen, whereas, rays deviated downward are blocked by the knife edge. Since the plane o indicated by the vertical dotted line is in focus on the screen n, an object placed at that point in the beam produces a sharp image on the screen. Changes in refractive index near the object deviate the rays and cause changes in illumination at the corresponding point in the image. The image is thus surrounded by a shaded pattern indicating air density changes.

For visual observation of Schlieren patterns the optical system should be set up so that the light travels several meters from the test area to the knife edge. A straight filament lamp p (6 V) should be used as the source. It is placed at the focus of the mirror q (focal length should be 50 cm or more), a little to one side of the projected beam. About 3 or 4 m away, the second mirror r is arranged to reflect the parallel beam to a focus on the other side of it. The patterns are directly observable at the knife edge s or they may be projected on a ground glass screen (not shown). In the latter case, the position of the test region t must be such that it is in focus on the screen. The necessary conditions are illustrated in Fig. 34–46 (B).

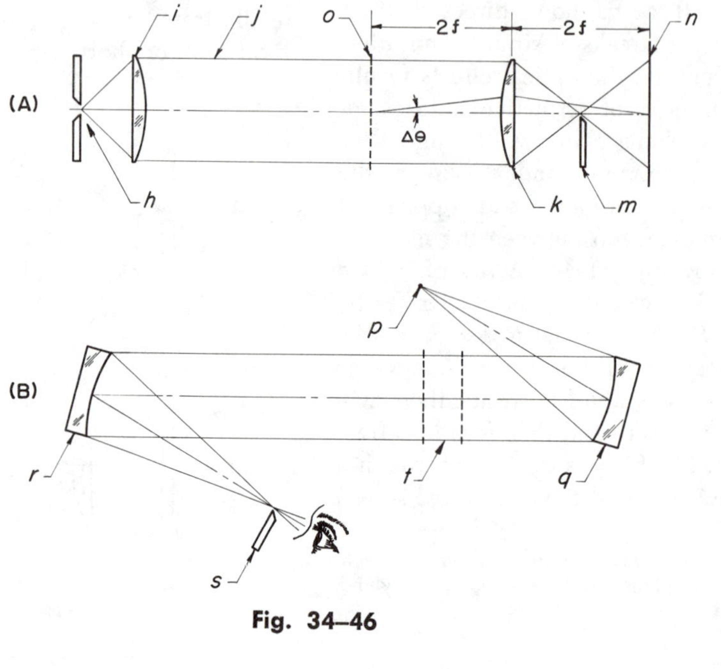

Fig. 34–46

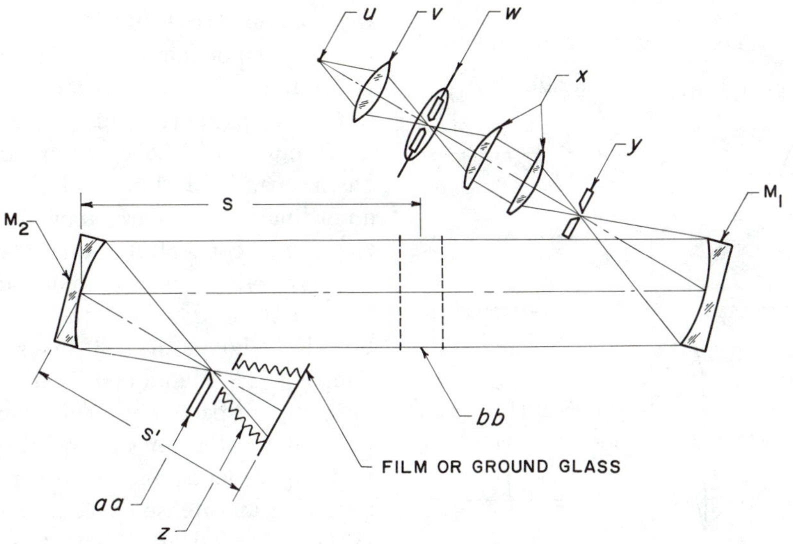

Fig. 34–47

A convenient arrangement of apparatus to photograph Schlieren patterns is shown in Fig. 34–47. The straight filament lamp *u* is used for alignment purposes; a simple lens *v* focuses the continuous source at a spot between the electrodes of the flash lamp *w* or at the spot where any other flash source would be put. A pair of condenser lenses *x* are needed to focus the flash tube light on the slit *y* for Schlieren pictures. The adjustable slit is needed to restrict the light rays to a narrow line. A camera *z* with a 4 x 5-in. view is excellent for the photography, though almost any camera could probably be used. Very fast film should be used. The lens board of the camera is removed if possible, and the image focused on a ground glass plate at the position of the film holder. Black cloth or paper is used to cut out all unwanted illumination. The knife edge *aa* is carefully adjusted to give the optimum pattern of the object in the test region *bb*, and the filament lamp then turned off. The object distance is *s*, the image distance is *s′*, and M_1 and M_2 are concave mirrors.

At this point, a cardboard shutter should be held over the opening of the camera and the film holder opened. A couple seconds before the flash takes place the shutter is removed. After the flash the shutter is replaced and the film holder covered.

It has been shown that the direct shadow and the Schlieren patterns can be demonstrated directly, or photographed if a suitable short duration flash is available. With either arrangement, the sensitivity is great enough to allow the air currents to be seen rising above one's hand because of its warmth. The air currents rising from a warm electric soldering gun give a spectacular demonstration of either smooth or turbulent air flow. With a little more elaboration, one could observe shock waves from a suitably shaped nozzle attached to the compressed air line. A small enclosed test section with glass or Plexiglas windows might be very convenient for observing airflow over a model wing section, etc. One might also study the standing wave pattern of a reflected sound wave as produced in a rectangular pipe fitted with Plexiglas walls. For more detail of the flash tube and power supply, see Appendix, page 1352.

34–2 Optical Instruments

34–2.1

The simple optics of the human eye can be demonstrated with a 22-liter Pyrex flask filled with water containing a little fluorescein, a source of light, and a simple iris-lens system. In Fig. 34–48, the flask of water represents the aqueous humor, and the fluorescein makes the path of the light beam

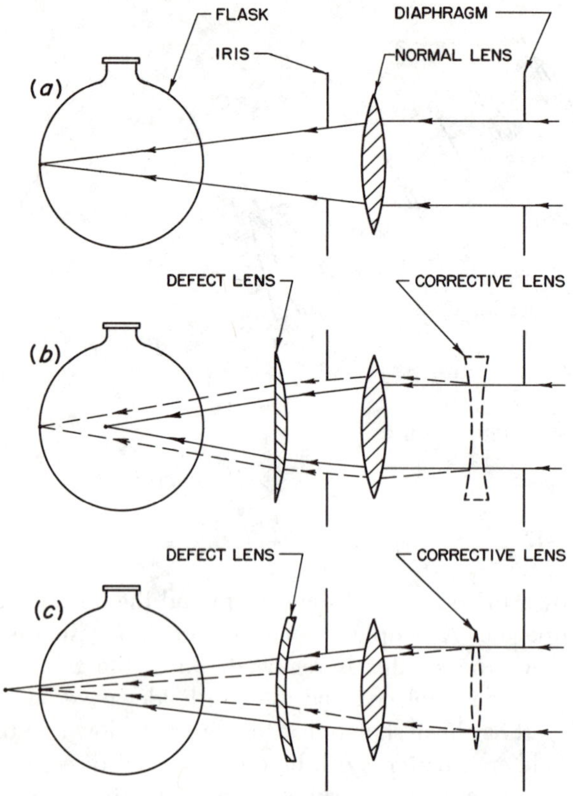

Fig. 34–48

visible. Part (a) represents the normal eye, in which the convergent lens (+34.5 cm f) forms an image of the source (an arc) on the "retina." Part (b) illustrates a myopic, or nearsighted, eye in which the "defect" is produced by interposing a convergent lens (+32 cm f) between the iris and the flask. The image is formed in front of the iris, and a negative corrective lens (−30 cm f) restores the image on the retina. Part (c) represents a hypermetropic, or farsighted eye produced by a divergent defect lens (−30 cm f). The image is formed back of the retina and a positive corrective lens (+34 cm f) converges the image upon the retina. Astigmatism can also be shown by the use of cylindrical lenses.

34–2.2

There are many requirements in physics for a very strong point source of light. One such light source of exceptionally high brightness is the Osram HBO–109,[26] a small, high-pressure mercury-vapor lamp with a fused-quartz envelope. It produces a stable arc constricted to 0.3 mm and has an average surface brightness of 140,000 cp/cm^2, an intensity of 125 cp. For comparison, the sun has a surface brightness of about 170,000 cp/cm^2, an ordinary carbon arc, about 20,000 cp/cm^2, and a zirconium concentrated-arc, about 4,500 cp/cm^2. The HBO–109 has a luminous area slightly smaller than that of the 10-W zirconium lamp, but the total light flux is over 20 times greater. Zirconium lamps of 100 W and 300 W are rated at 100 cp and 275 cp, respectively. This lamp should be of particular value for shadow projection, for diffraction experiments, and for other optical procedures requiring an intense concentrated light beam. The HBO–109 spectrum is characterized by much broadened mercury lines superposed on a continuous background. The infrared content is low, but the ultraviolet emission is particularly intense and extends to the short-wave region. A completely enclosed lamp housing is needed with a window that gives adequate ultraviolet protection. The lamp requires a minimum potential of 50 V dc and a series resistor to limit the current to about 5 A. It is necessary to adjust the current to a precise value specified with each individual lamp. The operating potential drop is between 16 and 24 V. When rectified ac power is employed, the filtering should be such that the minimum current is at least 60 percent of the maximum current. An ignition pulse is needed to start the arc. A Tesla coil can be used for this purpose if a complete power supply unit is not purchased.

34–2.3

Figure 34–49 illustrates the construction of a convenient overhead projector. A box shaped cart c houses a 1000-W projection lamp[27] l with a large aluminum-foil reflector r behind it. Between the lamp and a large plastic plano-convex lens p is a Masonite board b covered with aluminum foil. This board is adjustable up and down. The plastic plano-convex lens is 6 in. in diam and has a focal

[26] Available from George W. Gates & Co., Inc., Franklin Square, Long Island, New York. See H. P. Stabler, "Apparatus Notes," *Am. J. Phys.* (September 1960).

[27] General Electric No. T20 with a C-13 filament.

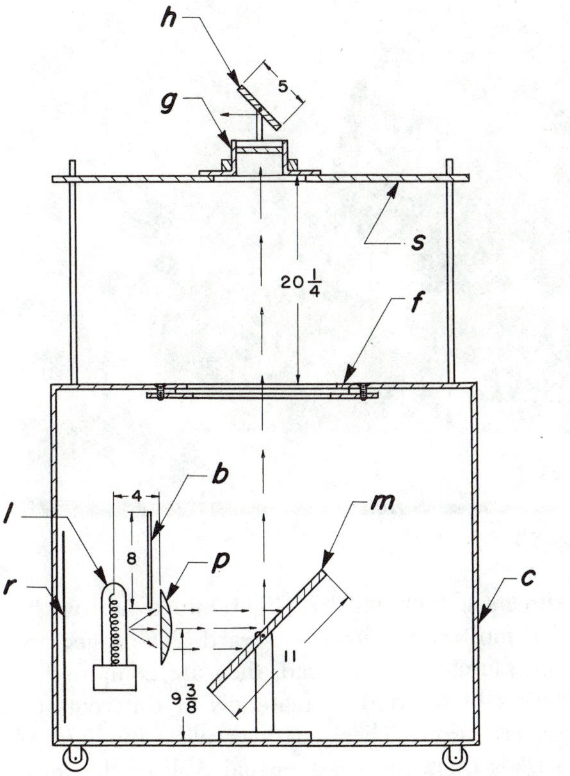

Fig. 34–49

length of about 4 in.[28] A plate glass mirror *m*, 9 x 11 in., is mounted on a tilting stand located on the vertical centerline of the cart. The mirror reflects the available light vertically up through a large plastic Fresnel lens *f* fitted to the cart's top surface. The lens is 15 in. in diam, has a focal length of 12½ in., and is 3/32 in. thick. Its circularly serrated surface has 120 lines per inch of radius. The lens is over-corrected for spherical aberration. It is sandwiched between a Masonite frame and a slightly smaller opening in the cart's top surface, which is subsequently fitted with a clear sheet of Lucite. This sheet protects the serrated side of the Fresnel lens. Two vertical rods attached to opposite corners of the cart support a condensing lens *g*, a light shield *s*, and a front-surface mirror *h*. The mirror is cemented to a horizontal axle rod, which pivots in a simple bracket mounted on the lens housing. The top surface of the cart is large enough to provide work space, and the entire unit can be wheeled to and from the lecture area. It is of wood and Masonite board construction with one side removable. Overall dimensions are 24 in. wide, 36 in. long, and 31 in. high. The interior is ventilated by a large number of ½ in. holes. No forced air system is used to cool the lamp. Apparatus with modification could be adapted for vertical field projection.

[28] The Fostoria Corp., Horsham, Pennsylvania.

34–2.4

An excellent translucent projection screen can be made by stretching drafting linen on a frame. Such a screen is useful not only for both front and rear projection, but also an illuminated background against which instruments, such as barometers and manometers, can easily be seen in silhouette.

Drafting linen scatters light over wide angles. Therefore, it is especially suitable for a screen of this kind where an audience seated in a wide room needs to see the screen equally bright when viewed from the end of a row as from the center. Ground glass is fragile and more expensive and usually scatters over much smaller angles, so it is less suitable. Opal glass is suitable.

34–2.5

A standard straight-filament showcase bulb can be used in demonstrations that require a straight line light source.

34–2.6

Convenient for use in optics demonstrations to show spectra and diffraction patterns in large lecture halls is the high-intensity movie theater arc lamp (60 V dc, 50 A) shown in Fig. 34–50. It features an automatic advancing mechanism for the electrodes. The slits and diaphragms must be water-cooled. To demonstrate diffraction by a grating, the direct beam (order zero) is darkened by a strip of black velvet hung on the projecting wall. Be careful to screen all stray light from the optics so that high orders can be seen.

34–2.7

A lightweight, cheap projection system for use by schools with limited budgets is shown in Fig.

Fig. 34–50

34–51. The light source is a 110 V ac 300-W GE PAR 56/2NSP narrow spot, cool beam lamp. The $6\frac{1}{2}$-in.-diam lamp has a special reflector which directs a high level of light into a narrow beam, but most of the heat backward. The reflectors are not aluminized. Instead, they are composed of 17 layers of alternating materials of different indices of refraction. These coatings reflect the light which strikes them, but pass virtually all of the infrared. As a result, the main beam is much cooler than that of a standard reflector lamp. The cool-beam lamp can be used directly with a projection meter that has a plastic meter face, but as an additional safeguard ordinary heat absorbing window glass is used as a protective filter. Blocks of wood or polystyrene can be used to support the apparatus. A double convex lens with about the same diameter as the cool beam lamp and a focal length of about 15 in. can be used for a projection lens. These lamps are also recommended for applications where a high level of illumination is desired, but the work area must be kept as cool as possible.

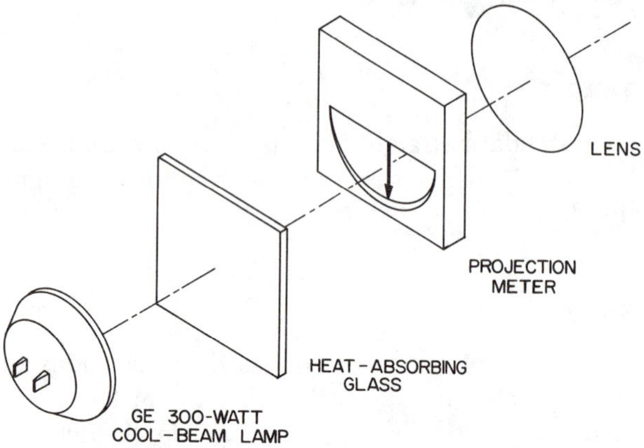

Fig. 34–51

Demonstrations	***Contributors***
34–1.1	F. Oppenheimer, University of Colorado
34–1.2	C. H. Palmer, The Johns Hopkins University
34–1.3	L. Grillet and M.-T. Grillet-Amy, University of Rennes, France
34–1.4	R. W. Pohl, University of Göttingen, Göttingen, West Germany
34–1.5	H. A. Robinson, Adelphi University
34–1.6	F. Oppenheimer, University of Colorado
34–1.7, 34–1.8	L. Grillet and M.-T. Grillet-Amy, University of Rennes, France
34–1.9	P. S. Delaup, Southwestern Louisiana Institute
34–1.10	M. Mainardi and A. Capecelatro, Newark College of Engineering
34–1.11	H. C. Bryant and C. E. Williams, University of New Mexico

Demonstrations	*Contributors*
34–1.12	Swiss Federal Institute (ETH), Zürich, Switzerland
34–1.13	M. Mainardi and A. Capecelatro, Newark College of Engineering
34–1.14	M. J. Pryor, State University of New York at Albany
34–1.15	M. Mainardi and A. Capecelatro, Newark College of Engineering
34–1.16	B. G. Saunders, University of Indonesia; now at Radiation Laboratory, University of California
34–1.17, 34–1.18	V. E. Eaton, Wesleyan University
34–1.19	Bolsey Research and Development; Fostoria Corporation, Horsham, Pennsylvania
34–1.20	R. W. Pohl, University of Göttingen, Göttingen, West Germany
34–1.21	E. J. Eastmond, Brigham Young University
34–1.22	L. Grillet and M.-T. Grillet-Amy, University of Rennes, France
34–1.23	C. H. Palmer, The Johns Hopkins University
34–1.24	E. A. Wood, Bell Telephone Laboratories, Murray Hill, New Jersey
34–1.25	C. H. Palmer, The Johns Hopkins University
34–1.26	R. W. Pohl, University of Göttingen, Göttingen, West Germany
34–1.27	C. H. Palmer, The Johns Hopkins University
34–2.1	L. Carver and H. M. Waage, Princeton University
34–2.2	A. Hudson, Occidental College
34–2.3	L. W. Morris, Louisiana State University
34–2.4	E. M. Rogers, Princeton University
34–2.5	L. W. Morris, Louisiana State University
34–2.6	Swiss Federal Institute (ETH), Zürich, Switzerland
34–2.7	D. M. Glassman and H. F. Meiners, Rensselaer Polytechnic Institute

35

PHYSICAL OPTICS

35–1 Nature and Propagation of Light

35–1.1

In the rotating mirror apparatus shown in Fig. 35–1, a light ray (1.2) from an illuminated slit (6) falls on a rotating mirror (8) where it is reflected to the plane mirror (14) through the lens (13). By means of a partially reflecting mirror (7), lens (13) also produces (not shown in Fig. 35–1) a second image of slit (6) on the entrance slit (16) of a photomultiplier (20). If the rotating mirror revolves clockwise (light ray 1.1), the beam falls to the right of slit (16); if counterclockwise (light ray 1.3) it falls to the left.

The speed of light can be determined if the following are known: a, the total deflection between clockwise and counterclockwise produced light beams; D, the distance between the rotating mirror (8) and the plane mirror (14); l, the distance between the rotating mirror and the slit (16); f_1 and f_2, the frequencies of rotation of the rotating mirror in the clockwise and counterclockwise directions, respectively. The speed of light is calculated from the relation:

$$c = \frac{8\pi (f_1 + f_2) Dl}{a}$$

This determination can be carried out in a large lecture hall if the distance a between the slit images is measured by means of a photomultiplier. The multiplier with its entrance slit has a pointer (18) rigidly coupled to it, and both are fastened to a track in such a way that a screw (15) can move them back and forth along the track. The pointer (18) and a fixed millimeter scale (19) are projected on the wall so that the position of the slit (16) can be read on the scale (19). When the image of the beam from the moving mirror falls precisely on the entrance slit (16), a maximum reading will be given by a scaler. Both positions of slit (16) can therefore be read directly from the position of the projected pointer (18).

The two frequencies f_1 and f_2 can also be measured with the same photomultiplier and a proper scaler. For this purpose the arc lamp (1) is not suitable as a light source, since its intensity varies and produces additional impulses in the photomultiplier. An auxiliary 12-V dc light source (12) is used, and its illuminated slit (10) is focused on the entrance slit (16) of the photomultiplier with lens (9).

The demonstration is carried out as follows. The arc lamp, photomultiplier, and scaler are turned on. The mirror is rotated in the clockwise direction. The screw (15) is turned until a maximum counting rate is obtained on the scalar. The position of the pointer on the scale (19) is recorded. The procedure is repeated for a counter-clockwise rotation. The difference between the two pointer positions gives the distance a. The dc light source (12) is

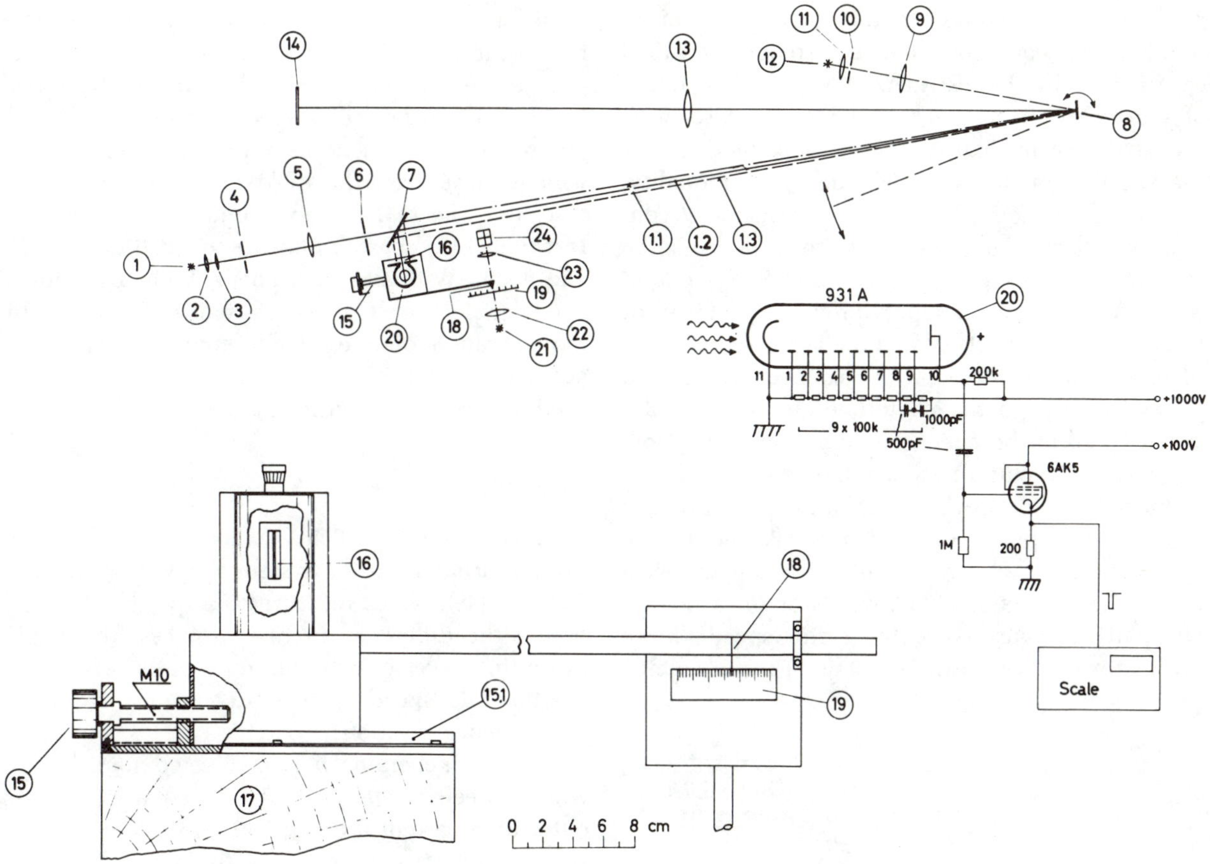

Fig. 35–1

connected and the frequencies of rotation, f_1 and f_2, are determined by means of the scaler. The distances, D and l, are measured with a tape. See the Appendix, page 1353, for a list of materials.

35–1.2[1]

In the directions accompanying the Leybold apparatus for measuring the speed of light, it is stated that the rotational speed of the motor "is found by comparing the noise of the motor with a tuning fork of known frequency." While this method may be very satisfactory for someone with an ear for music, it proves impossible when tried, and so other means are presented for making this measurement.

The first method, and one which has proved quite successful, makes use of a guitar microphone taped to the motor housing. The output of this microphone is applied directly to the vertical deflecting plates of a G.E. type ST–2A oscilloscope. With a Hewlett–Packard audio oscillator connected to the horizontal plates of this oscilloscope, the rotational speed of the motor can be found from the resulting Lissajous pattern. If the frequency of the oscillator is adjusted so that the one-to-one Lissajous pattern is formed on the oscilloscope screen, the oscillator frequency will give the rotary speed of the motor directly in rps. In order to improve the accuracy, the oscillator is first calibrated by means of the Lissajous figures which its output forms with tuning forks of frequencies 320, 480, and 512 cps.

A second method employs a crystal microphone set very close to (but not touching) the motor housing and connected to an amplifier. The output of the amplifier, in this case the signal from the voice coil of a 12-in. Jenson speaker, is applied to the vertical plates of the oscilloscope and a one-to-one Lissajous pattern is formed with the aid of the oscillator. In the particular arrangement used, the

[1] H. N. Maxwell, *Am. J. Phys.*, **29**, 711 (1961).

microphone is set close to the rotating mirror to reduce extraneous noise and the frequency (1090 cps) is found to be just twice the speed of rotation of the motor. More background noise exists with this arrangement than for the previous one, so that the Lissajous patterns are of much poorer quality.

A third method which works satisfactorily consists in reflecting a beam of white light from the rotating mirror onto the cathode of a 930-A phototube. A Kepco power supply provides the 90 V for the phototube circuit. When the vertical deflecting plates of the oscilloscope are connected across a 4.7-mΩ resistor in the phototube circuit, the rotational speed of the motor can again be found from the Lissajous pattern. Since the rotating mirror supplied with the Leybold apparatus possesses two surfaces, two pulses of light are reflected to the phototube for each revolution of the motor. As a result, the frequency of the oscillator for a one-to-one pattern is just twice the motor speed. More satisfactory patterns can be obtained if the voltage pulses across the 4.7-mΩ resistor are amplified before being applied to the oscilloscope.

If the equipment is available, the best method for measuring the motor speed is to use the 930-A phototube circuit mentioned above, but to feed the voltage pulses from the 4.7-mΩ resistor directly into a scaler. The scaler used is one made by Atomic Instrument Company with scales of 16, 64, 256, 1024, and 4096. With this instrument, both the counter and electric timer can be started and stopped by the operation of one switch, thus increasing the accuracy of the measurement. The 1024 scale is used and counts taken for 100 sec.

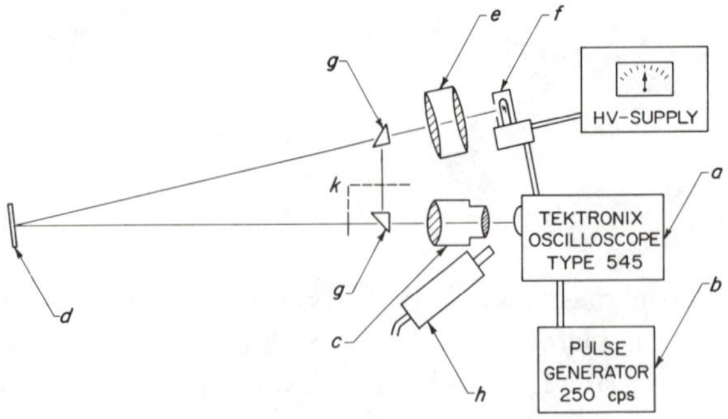

Fig. 35–2 (a)

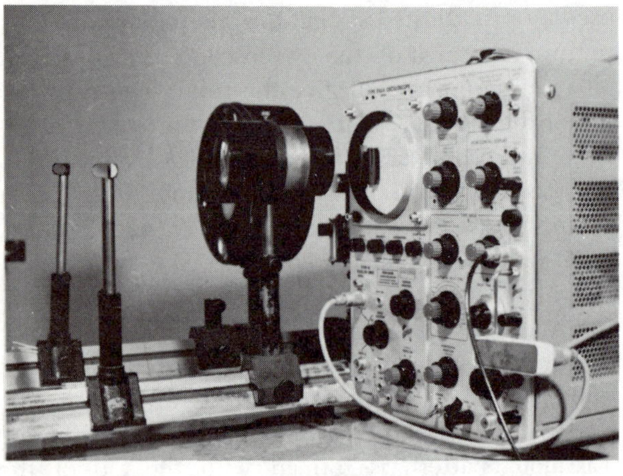

Fig. 35–2 (b)

35–1.3

A Tektronix type 545 Oscilloscope *a* is the primary instrument in the system shown in Fig. 35–2, (a) and (b), which measures the speed of light by noting the difference in time it takes for a light beam to traverse two different paths of known length. The speed is then determined from the time distance relation.

A pulse generator *b* is connected to the horizontal sweep trigger so that the scope will sweep out a time base of 0.1 μsec/cm on its screen at the rate of 250 sweeps/sec. At the far left of the screen, a piece of Scotch tape with a small hole in its center is placed so that when the sweep begins, the hole will be illuminated from behind (Fig. 35–3). The light that comes through this hole is picked up by a projection lens *c* ($f = 150$ mm) and collimated so that it is reflected off a mirror *d* approximately 15 m away, falls into a condenser lens (220 mm in diam) *e*, and is focused on a photomultiplier *f*. Two small prisms *g* are placed in the light beam to divert a small part of it on a much shorter path and into the photomultiplier.

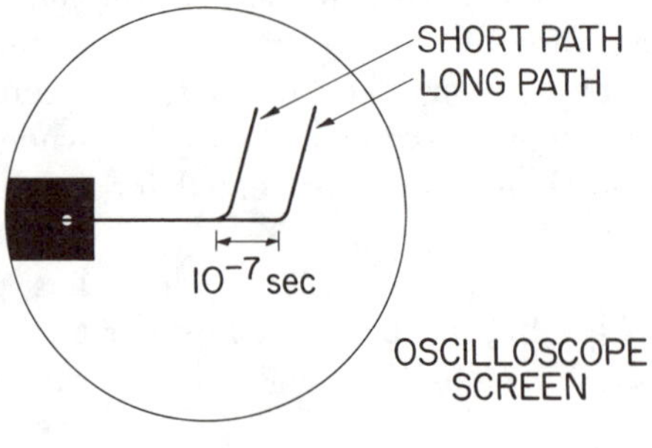

Fig. 35–3

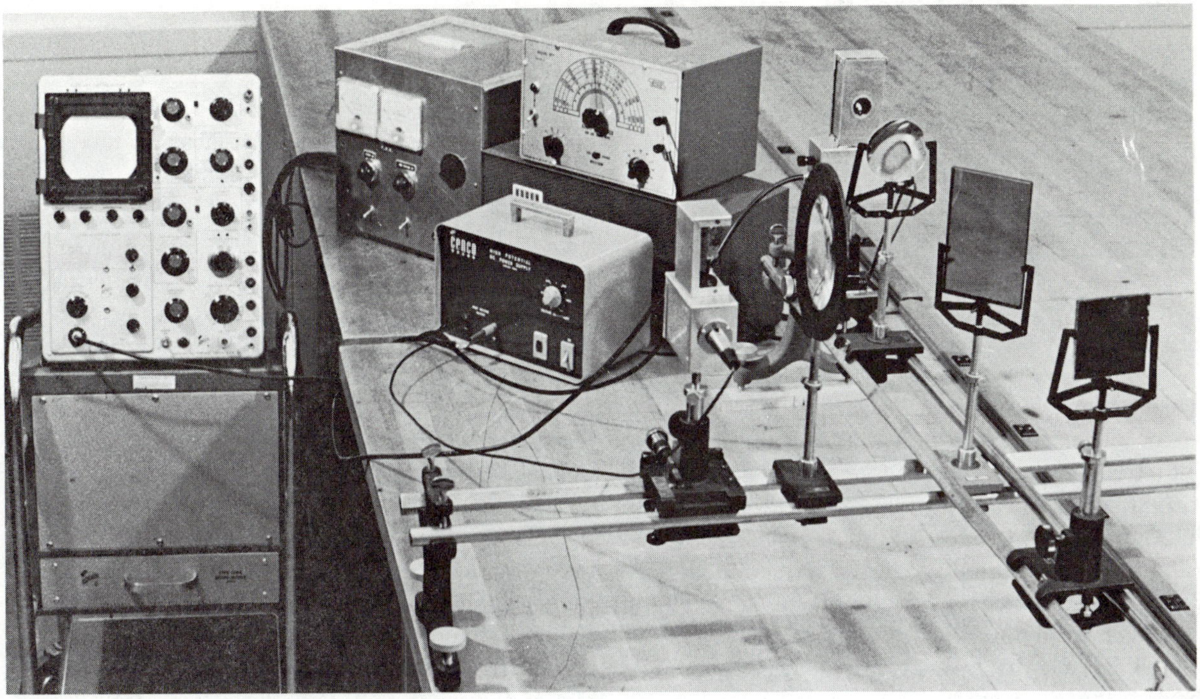

Fig. 35–4

The signal of the multiplier is connected to the y input of the scope without amplification. Screens k are used to block unwanted light paths. It is extremely important that the light traveling both paths is focused on the same part of the photomultiplier with approximately the same intensity. The simplest way to achieve this condition is to set a point source of light in the photomultiplier location and adjust each path separately, using screens, so that the light falls exactly within the hole in the tape on the scope screen. A TV camera h is so mounted that it will pick up the scope screen and not interfere with the experiment. In operation, the scope should be allowed to warm up for about one hour before the experiment is to be performed. With the room darkened, the pulse generator is turned on and the photomultiplier pickup causes two vertical displacements on the screen as shown in Fig. 35–3. The distance between these can be measured and converted to time units through the scope sweep rate. The path difference is the distance from the first prism to the mirror plus the distance from the mirror to the second prism minus the separation distance of the two prisms, all distances measured from reflecting surfaces.

Caution: Cover the photomultiplier before turning on the lights. Electroluminescent nightlights can be used to illuminate the apparatus that must be operated.

35–1.4[2]

Basically, a measurement of a speed v involves the measurement of a distance s and a time t in which the distance is covered: $v = s/t$. Since the speed of light c is very large, early attempts to determine c either were unsuccessful (Galileo) or required extremely long distances (Roemer) and long waiting. Fast electronics now allow the measurement of c, with an accuracy of 2 percent, in a medium-sized lecture hall or laboratory.

A 1-nsec light flash is produced by a light source.[3] As is shown in Figs. 35–4 and 35–5, the light is focused with a long-focal-length, large-

(2) W. Eppenstein and R. Resnick, *Report on the Workshop for the Development of Apparatus for College Physics at Rensselaer Polytechnic Institute, 1964–1965,* March 1966, pp. 61–66.

(3) K. Nygaard and S. Malmskog, *Nucl. Instr. and Methods,* **22**, 363 (1963); J. T. D'Alessio, P. K. Ludwig and M. Burton, *Rev. Sci. Instr.,* **35**, 1015 (1964); also *Chem. Eng. News,* Aug. 10, 1964 (p. 35), and "Radiation Laboratory Counting Handbook," UCRL 3307. A simple light source is described in the Appendix, page 1353.

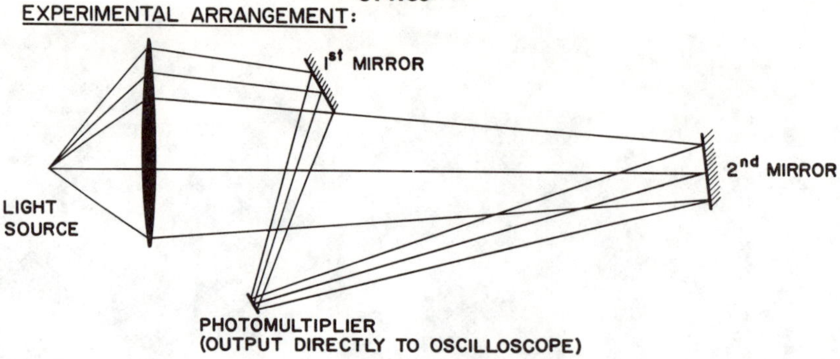

Fig. 35–5

aperture lens (in one setup, a surplus aerial camera lens, f = 6 in., f:6.3 was used). In Fig. 33–4, a beam splitter is mounted where the optical benches cross at right angles to each other. Part of the light from the source in the left foreground is reflected by the beam splitter to the mirror in the foreground. It is then reflected back through the splitter to a photomultiplier located on the optical bench in the background behind the beam splitter. Therefore, at a short distance s from the lens, part of the almost parallel light rays are reflected into a photomultiplier (1P28 or 931A; see Fig. 35–6) while most of light rays are transmitted by the beam splitter and travel over a longer distance d and are reflected by a second mirror located about 10 m to the right (not shown) back into the same detector. The light flash from the first mirror is used to trigger the sweep of a fast oscilloscope (Tektronix 585, 545, or 541), and the separation between the two flashes gives directly the time difference t. The result gives $c = 2(d - s)/t = x/t$, with an accuracy of about 2 percent, given by the accuracy of the reading on the oscilloscope. In addition, the calibration of the time scale of the oscilloscope must be correct. It can be calibrated by displaying a known frequency, say 20 or 40 Mc/sec, on the screen.

The signals from the photomultiplier are negative because ordinarily the anode is at ground potential and the electrons produced by each light pulse reduce the potential to below ground potential. The voltage applied to the cathode should be chosen to give a sufficient signal. Also, while the rise of the pulse on the oscilloscope is fairly clean, there is some after-pulsing and sometimes some afterglow occurring after about 30 nsec. The first mirror should be adjusted to produce a first pulse equal in height to the second pulse. The rise time of the pulses (about 4 nsec on the 585, and between 20 and 25 nsec on the 545 and 541) is mainly determined by the rise time of the amplifier of the oscillator. A small increase in the rise time is caused by the spread in arrival time of the electrons at the photomultiplier anode and by the duration of the light flash.

The experiment is now being performed by all junior physics majors at Rensselaer Polytechnic Institute with very good results. The difficulties of alignment have been overcome by many students since a portable gas laser beam has become available to them. A few student photographs of the scope face are shown in Figs. 35–7, 35–8, and 35–9.

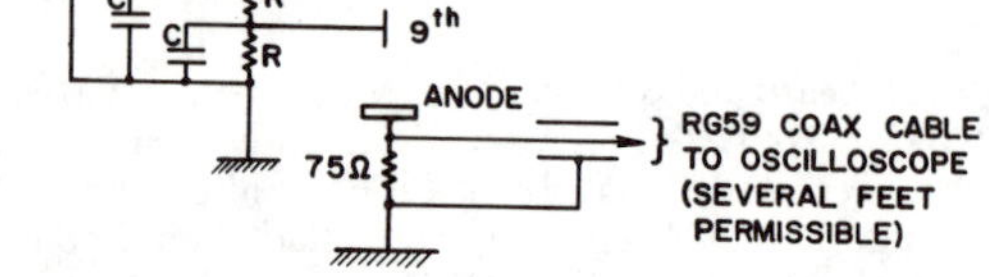

Fig. 35–6

35–1.5

The speed of light in a vacuum (c_0) is the best known universal constant. Its value is presently given as $(2.99792 \pm 0.000003) \times 10^{10}$ cm/sec. One method for determining the speed of light is the direct method;[4] the time needed for a pulse of

[4] Eric Berstrand, "Determination of the Velocity of Light," *Encyclopedia of Physics,* Vol. XXIV (Springer Verlag, Berlin, Göttingen, Heidelberg, 1956).

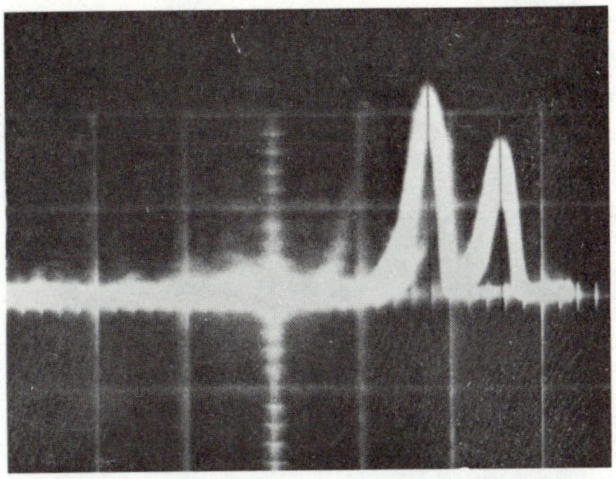

Fig. 35–7 x: 0.05 μsec/cm. y: 0.5 V/cm.

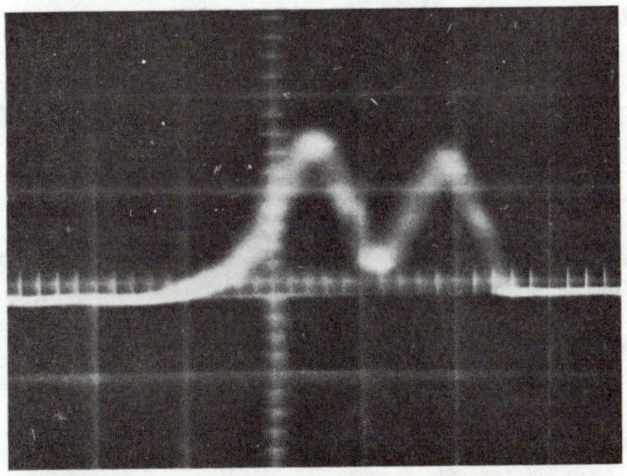

Fig. 35–8 x: 0.1 μsec/cm. y: 1.0 V/cm.

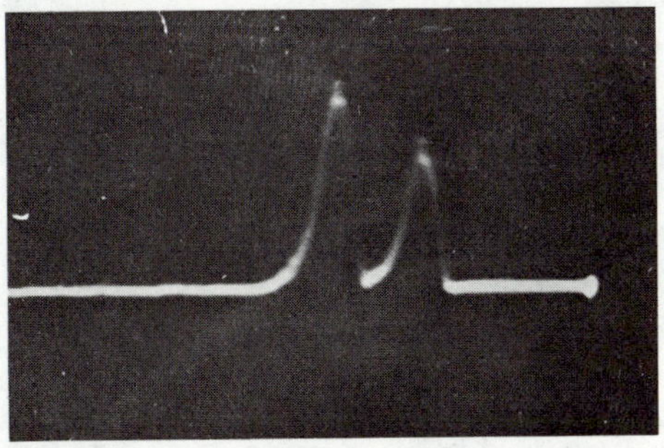

Fig. 35–9 x: 0.05 μsec/cm. y: 0.5 V/cm.

light to cover a known distance is measured. This is probably best for lecture demonstrations as it is easily understandable. Most experiments using the direct method for precision measurement of light speed require distances of 40 m to 40 km for the optical light path. To produce the short light pulses, revolving mirrors, toothed wheels, optical Kerr cells, or ultrasonic quartzes are used as optical shutters. The shorter the light pulses, the smaller the apparatus may become. Using quartz crystals, the chopping frequency can be as high as 300 Mc/sec. If a measurement of the light speed has to be demonstrated in a lecture hall, one must use a short light path and a high pulse frequency. However due to the shortness of the light path, the demonstrator must forgo extreme accuracy for compactness. Lindberg[5] and Kollath[6] have both dedescribed demonstration methods. Lindberg used Foucault's method with a revolving mirror, while Kollath worked with an ultrasonic quartz crystal using Houston's method.[7] The following method makes it possible to determine the speed of light without difficult adjustments and long light paths.[8]

In this experiment, a radio frequency oscillator generates vibrations of a quartz crystal in a high harmonic. Light modulated by this crystal strikes the photomultiplier after having covered a certain distance. The radio frequency output of the photomultiplier and the RF signal of the generator have a phase difference which depends on the length of the light path. If the length of the light path is changed, the phase difference will change correspondingly. Both signals are fed into a phase detector; if the length of the light path is changed in such a way that the phase detector indicates a phase difference of 2π, then this change must have been one wavelength (λ) of the modulation. If the modulation frequency (ν) is known, the speed of light can be found from $c = \lambda\nu$.

The apparatus (see Fig. 35–10) consists of four main parts: (1) transmitting optics with the light source *LS*, lenses L_1 and L_2, slits S_1 and S_2, quartz crystal *Q*, and radio frequency generator RFG; (2) variable light path; (3) receiving optics with lens L_3 and L_4 and photomultiplier (P.M.); (4) electronics with amplifiers, phase detector and a projection voltmeter (Instr.). All optical devices, light source, and RFG are mounted on optical benches (Fig. 35–11). The lens L_1 produces an image of the slit S_1 at S_2. The crystal *Q* is excited to a high harmonic by a frequency $\nu = 49.6$ Mc/sec and repre-

(5) Lindberg, *Praxis*, 1955/4 S. 85ff.

(6) R. Kollath, "Messung Der Lichtgeschwindigkeit," *Physikalische Verhandlungen*, 1961 4/6 S. 52.

(7) Houston, *Proc. Roy. Soc.*, Edinburgh A, **61**, 102 (1941), **63**, 95 (1950).

(8) A good reference for the electronics described in this article is Samuel Seely, *Electron Tube Circuits* (McGraw-Hill Book Co., Inc., New York, Toronto, London). Light modulation and quartz crystals are described in Bergmann, *Der Ultraschall* (Hirzel Verlag, Stuttgart, 1954).

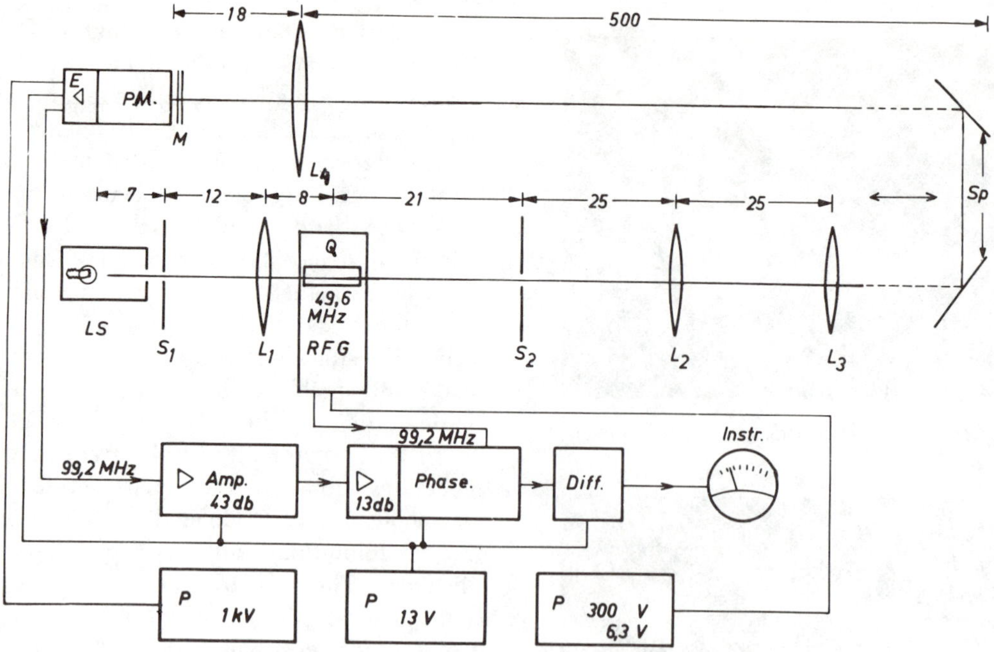

Fig. 35–10

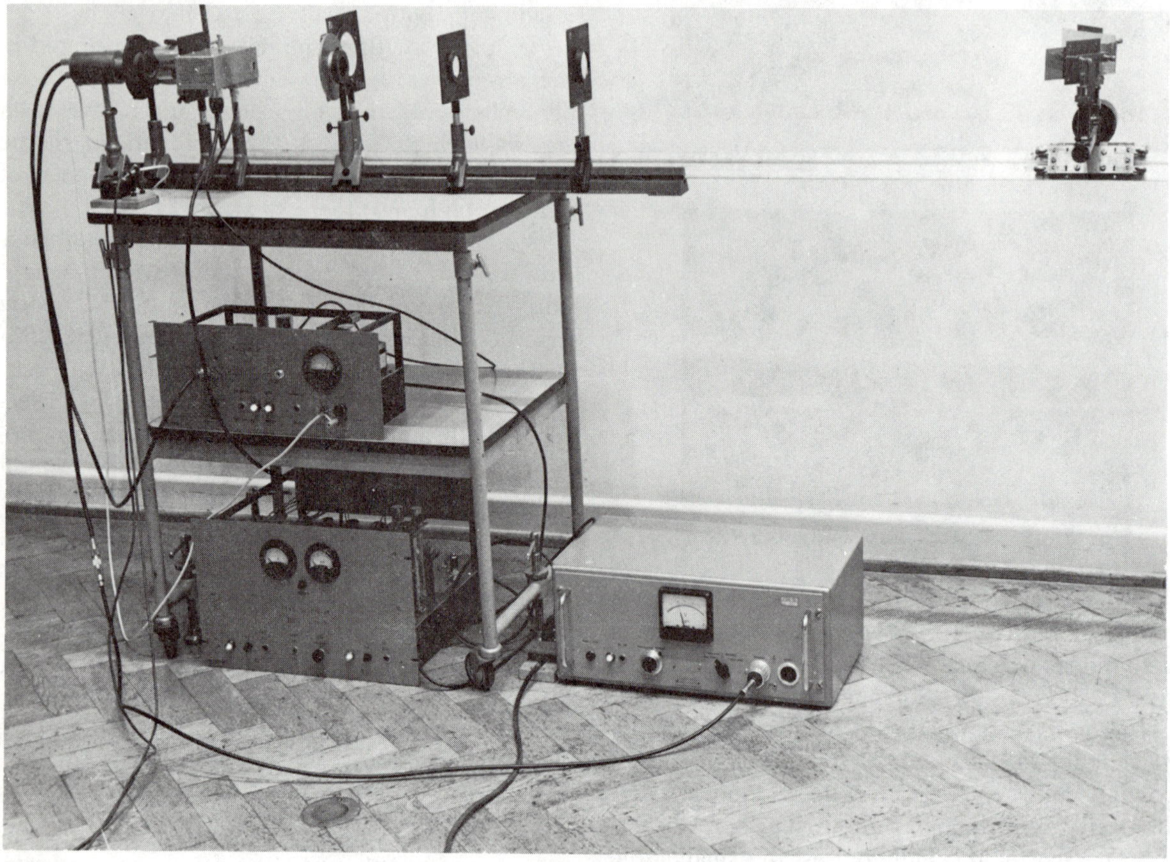

Fig. 35–11

sents an optical diffraction grating which appears and disappears with the frequency $2\nu = 99.2$ Mc/sec. The apparatus is arranged so that only the first order of the diffraction pattern passes through slit S_2. The light rays coming from slit S_2 are made parallel by lenses L_2 and L_3 and arrive at L_4 after being reflected by a pair of mirrors S_P. The photocathode of the photomultiplier lies in the focus of L_4. There are two ground-glass screens M in front of the photocathode. The output signal of the photomultiplier is amplified and fed into the phase detector. The second harmonic of the 49.6-Mc/sec generator is used as a 99.2-Mc/sec reference signal. The output of the phase detector is connected to a difference amplifier which drives the low-resistance projection voltmeter.

After having applied all supply voltages to the apparatus, a sheet of paper is hung over slit S_2 to adjust the vibration of the crystal. The third order of the diffraction pattern should be visible on the paper. If necessary, the frequency of the RFG may be varied by adjusting the fine tuning screw. If the paper is then removed, the modulated light beam will hit the ground-glass screens. The window of the photomultiplier housing is opened, and a reading appears on the projection meter. The pair of mirrors is movable on an optical bench parallel to the light path. If the pair of mirrors is shifted by ΔL, the change in length of the light path will be $2\Delta L$. The phase detector output will be zero if the phase difference of the input signals is 90, 270, 540 deg, etc. The pair of mirrors is set so that the phase detector output reads zero, and this position is marked. Then the pair of mirrors is moved until the phase detector again reads zero. Between these two zero points, a phase shift of 180 deg occurs. This represents a shift in length of the light path of $\lambda/2$ (λ being the wavelength of the modulation). If ΔL is the distance between two zero marks, then $\lambda/2 = 2\Delta L$; hence $\lambda = 4\Delta L$.

If the modulated light beam is not exactly parallel to the optical bench on which the mirror is mounted, it will not hit the same spot of the photocathode when the mirror is displaced along the optical bench. The photocathode does not have the same sensitivity over its entire surface. Therefore two ground-glass screens are fixed in front of the photomultiplier to scatter the light over a greater region. This reduces the variation of the multiplier output to ± 2.5 percent if the pair of mirrors is moved over a distance of 4 m. The 99.2-Mc/sec frequency has a wavelength of $\lambda = 3.022$ m. A $\lambda/2$ variation of the length of light path is equivalent to a 75.6-cm displacement of the mirrors when using the described apparatus. If the multiplier output signal is held constant within $\pm$ 2.5 percent, the total error of each measurement will be within 3 percent. This is equal to a ± 2.2-cm uncertainty in mirror displacement.

For easier installation of apparatus, the optical

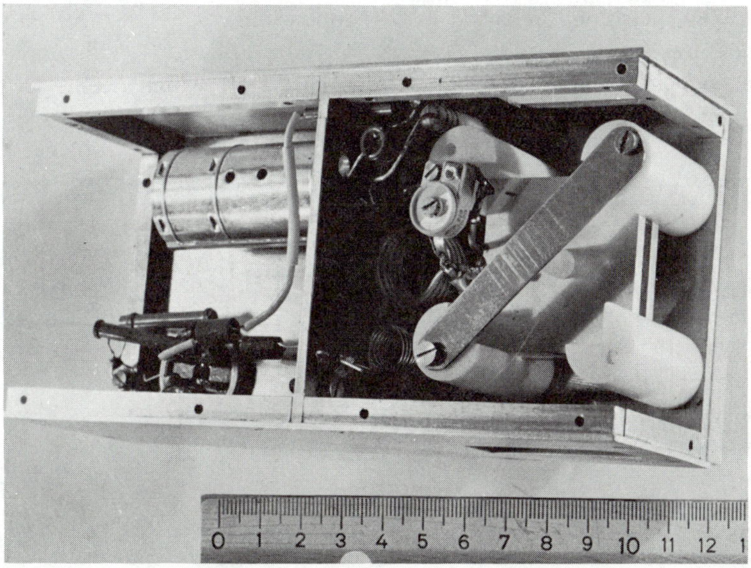

Fig. 35–12

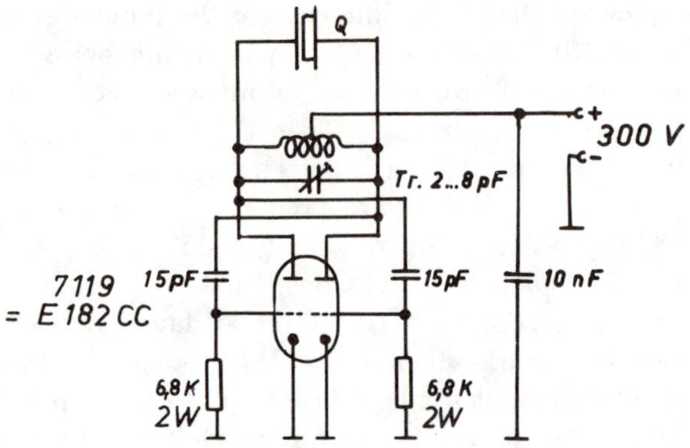

Fig. 35–13

bench for the mirror displacement should not extend more than 5 m. A low-voltage incandescent filament lamp (12 V, 2.2 A) with no reflector was used as the light source.

The quartz crystal and the radio frequency generator are mounted on the same chassis (Fig. 35–12). A simple push-pull oscillator with a 7119 tube was used (Schematic Fig. 35–13). The tube is operated as class C. Main tuning is done by varying the trimmer capacitor *Tr.*, while fine tuning is accomplished by varying the short circuit turn. This turn consists merely of a silver plated brass screw 4 mm in diam and 40 mm long, screwed into the center of the resonant circuit coil. The quartz is held between two silver plated brass plates of 4-mm thickness directly soldered to the resonant circuit coil. The circuit capacitance is mainly due to these plates with the quartz as dielectric material. The quartz measures 4 x 40 x 40 mm. It is not silver plated, and receives an rf voltage of 600 V p–p. Its fundamental frequency is approximately 712.5 kc/sec and it is excited at the 69th harmonic.

After an initial warm-up period of 30 min, the generator frequency at which the quartz has its maximum vibration amplitude was measured. This frequency was 49.598 Mc/sec. The frequency shift during three hours was ± 0.003 Mc/sec. The greatest shift during a single measurement was 200 cps.

The modulation frequency of the light passing the crystal was measured as 99.2098 ± 0.0009 Mc/sec during three hours. All frequency measurements were made with an electronic counter whose accuracy is greater than $\pm 5 \times 10^{-8}$.

The photomultiplier (FS 9-A, Fernseh GMB H, Darmstadt, Germany) is used as a light detector. The photomultiplier output is loaded with a 100-Mc/sec parallel resonant circuit, and the input impedance of a grounded-collector amplifier stage. This stage is used to match the 50-Ω line to the resonant circuit (Fig. 35–14).

The phase detector is fed by the RFG with a 99.2-Mc/sec signal of 1.2 V rms. For maximum sensitivity of the phase detector, the photomultiplier output signal must be amplified to the same voltage. The largest part of the required amplification (approximately 50 dB) is provided by a four stage, transistorized amplifier (Fig. 35–15). It is built as a narrow band amplifier which provides a high gain per stage. Each stage is protected against parasitic oscillations by neutralization. The rf voltage at the output is rectified to monitor the operation.

The apparatus will work correctly only if the phase shift which occurs to both of the rf signals

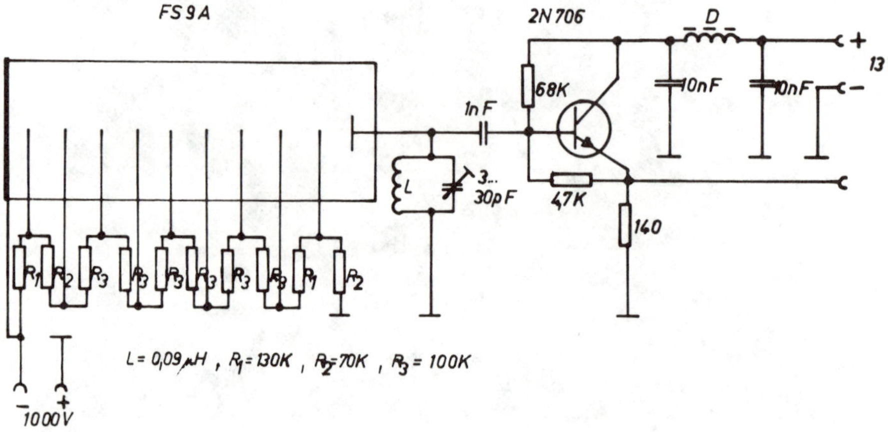

Fig. 35–14

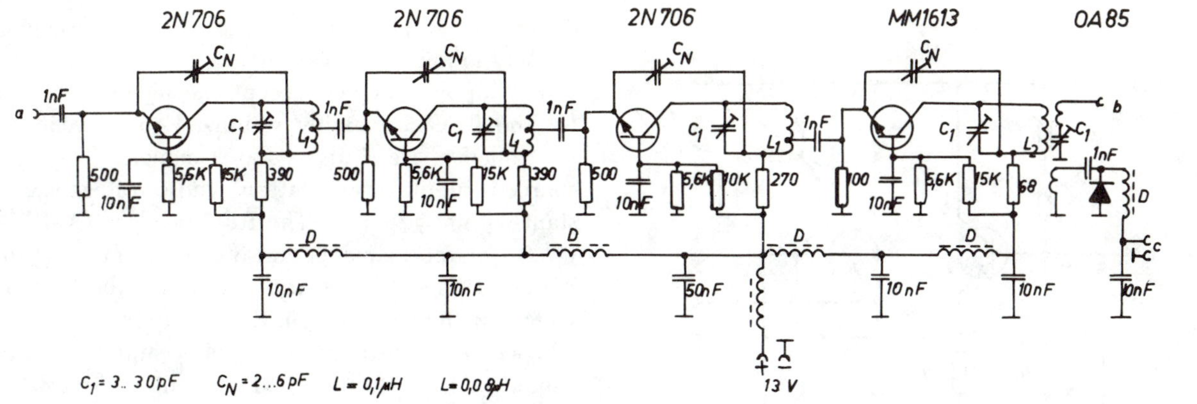

Fig. 35–15

is constant during the time of measurement. The phase shift of an amplifier depends on temperature supply voltage, and on the applied input voltage. It is evident that at this point extreme care must be taken to avoid changes in the phase shift. Therefore, the supply voltage is stabilized and a constant input signal is fed to the amplifier. Also, if the measurement is carried out in a short time, the influence of temperature will be negligible.

A constant input signal, the phase shift of the amplifier used was less than 0.25 deg during 10 min. This corresponds with a 5-mV variation of the phase detector output (which, in turn, is equivalent to a 0.2-cm uncertainty in length measurement).

The phase detector (Fig. 35–16) consists of a single stage amplifier and a phase-sensitive rectifier stage. The input and output terminals of the stage are mismatched. This results in better stability and broader bandwidth. The phase sensitive circuit is fed with two voltages, E_1 (the phase reference signal, coming from the RFG) and $2E_2$ (from the multiplier). A high resistance voltmeter V is connected to the terminals a and a'. If only voltage E_1 is applied to the circuit, a zero reading appears on the voltmeter V. The same result is obtained if only $2E_2$ is applied. Now, if the circuit is fed with both E_1 and $2E_2$, one diode will get $E_1 + E_2$ and the other $E_1 - E_2$. Assume further that the amplitudes of the two signals are equal and their phase differences are zero. Then diode D_1 will get a maximum signal and D_2 will receive none; a deflection will appear on the voltmeter V. If the phase difference is 180 deg, D_2 will get the maximum signal, D_1 will get none, and the meter will indicate a deflection of the opposite polarity. If the signals have a phase difference of 90 deg, both diodes will get the same voltage, and a zero reading will appear on the meter. The mathematical analysis of the circuit gives

$$E_1 = E_s \sin \omega t \quad ; \quad E_2 = E_s \sin(\omega t + \alpha). \qquad (1)$$

The ac voltage arriving at D_1 is given by:

$$E_a = E_1 + E_2,$$

and at D_2

$$E_b = E_1 - E_2.$$

By use of Eq. (1), it follows that

$$E_a = E_s(\sin \omega t + \sin(\omega t + \alpha))$$

$$E_a = 2E_s \cos\frac{\alpha}{2} \sin\left(\omega t + \frac{\alpha}{2}\right)$$

$$E_b = E_s(\sin \omega t - \sin(\omega t + \alpha))$$

$$E_b = -2E_s \sin\frac{\alpha}{2} \cos\left(\omega t + \frac{\alpha}{2}\right)$$

The resulting dc voltages are given by the expressions

$$E_{a'} = 2\,\mu\,E_s \cos\frac{\alpha}{2} \quad ; \quad E_{b'} = 2\,\mu\,E_s \sin\frac{\alpha}{2},$$

where μ is the efficiency of the rectifiers. The voltmeter V reads the resulting voltage

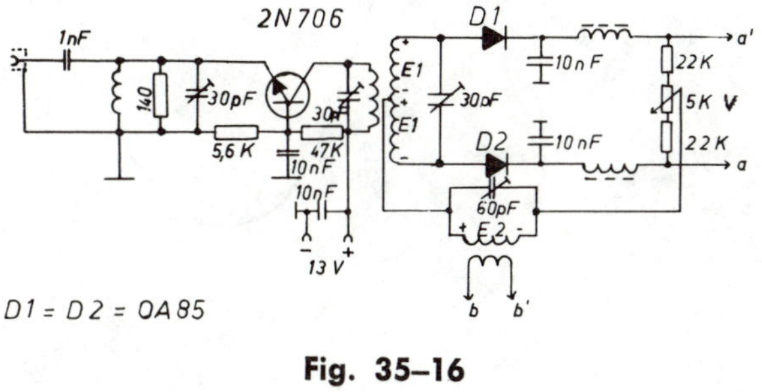

Fig. 35–16

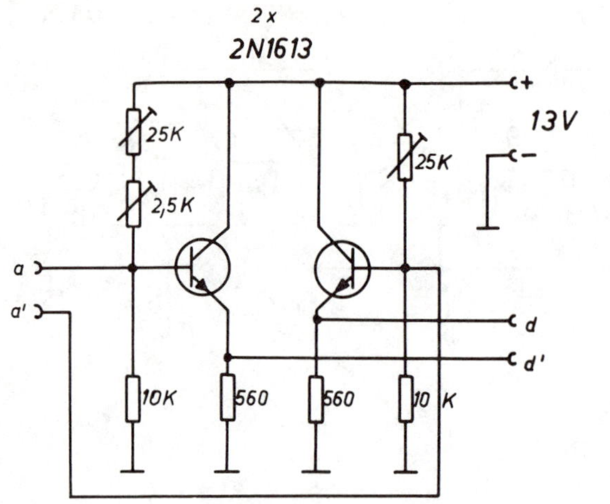

Fig. 35–17

$$E = E_{a'} - E_{b'} = 2E_s\left(\left|\cos\frac{\alpha}{2}\right| - \left|\sin\frac{\alpha}{2}\right|\right).$$

It is important that the output voltmeter read zero for phase angles 180 deg apart. A zero reading is easily reproduced and, by using a dc amplifier, the accuracy of the zero indication can be increased. For the demonstration, the phase detector's output is applied to a difference amplifier (Fig. 35–17) rather than the voltmeter. This amplifier has a voltage gain of approximately unity, and is used to feed a low-resistance projection voltmeter. Figure 35–18 is a graph of the output of the phasemeter versus the mirror displacement.

All supply voltages, except the filament voltage for the RFG are highly stabilized to prevent uncontrolled phase shifts of the rf signals. The light source is supplied by a battery whose voltage is constant within ±25 mV. The RFG needs 6.3 V RMS, 0.7 A, for the filaments and 300 V, 50 mA, for the plates. It is not necessary to stabilize the filament voltage if the variations in the ac supply are within 10 percent. However, the plate supply voltage should be kept within ±0.01 percent. A transistorized unit (Fig. 35–19) is used to supply the amplifiers and the phase detector. The power supply will deliver currents up to 3.5 A if a heat sink is used with the AD 103 transistor. This gives the possibility of feeding the light source with this power supply. With the help of the 5-kΩ trimmer potentiometer the inner resistance can be suppressed far below 1 MΩ. The output voltage depends on temperature. This is caused by the variation of Zener voltage of the reference element and the rapid increase of the temperature depending on what current passes through the transistors. However, this can be compensated for by use of germanium diodes in series with the reference element.

If the supply voltage of the multiplier is varied

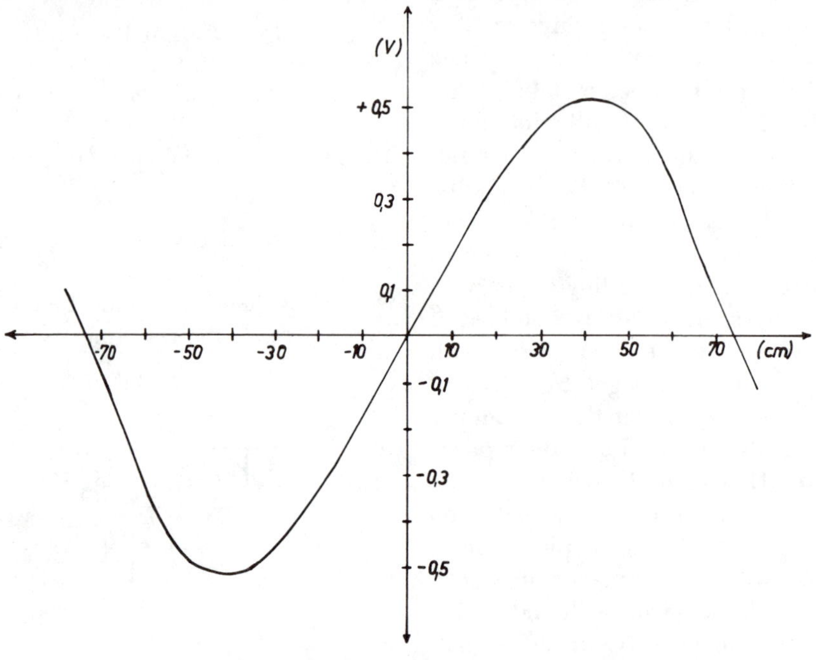

Fig. 35–18

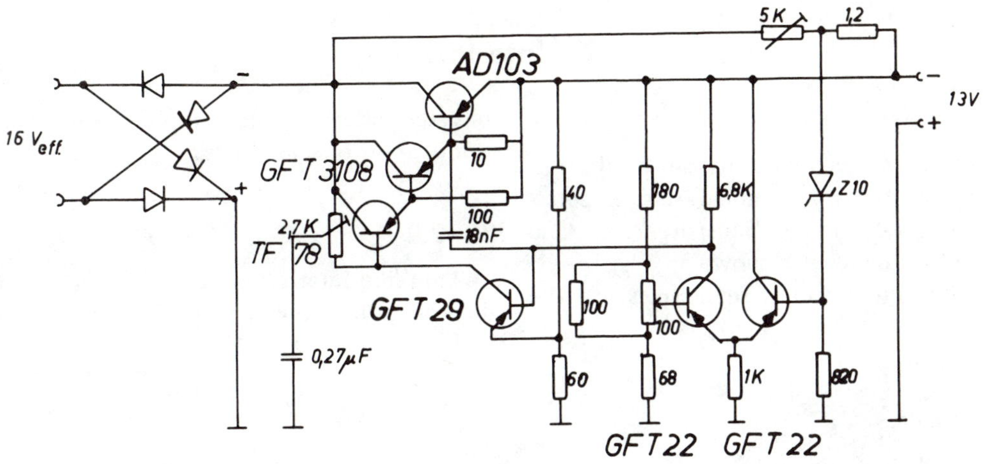

Fig. 35–19

by 10 percent, the gain will vary approximately 100 percent. Therefore the high voltage for the multiplier must be stabilized. A factory built supply, Nucletron NK 103 was used. It has the following technical data: output is 1 to 3 kV variable, 20 mA; an ac supply voltage change of 10 percent causes a ±0.05 percent variation in output voltage.

Replacement information can be found in the Appendix, page 1353.

35–1.6

In a darkened room, a collimated beam of light from an arc lamp *a* falls on a grey board *b* approximately 6 ft away, producing a white spot. The intensity of the beam can be measured using a solar cell and a projection ammeter of suitable range. A half-spectrum[9] filter *c* and mirror *d* arrangement is placed into the beam as in Fig. 35–20. The filter reflects green, blue etc., and transmits red, yellow, etc. The mirror is adjusted so that the two separate beams join at the surface of the board. When the solar cell is placed in the spot of white light produced by the addition of the two beams, the ammeter reading will be the same as that noted for the single beam of white light. A sheet of white cardboard *e* may be placed behind the beams so that they can be more easily seen.

[9] Available from Balzers AG., Balzers, Liechtenstein.

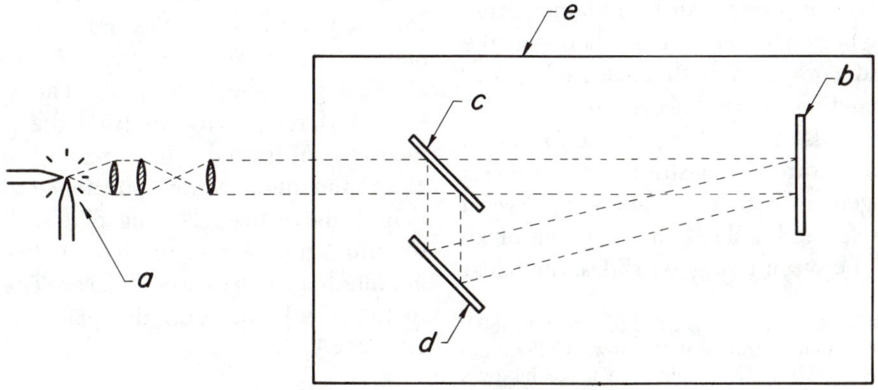

Fig. 35–20

35–2 Interference

35–2.1[10]

Double slit interference patterns can be demonstrated by means of two identical sets of semicircles on two photographic transparencies. One such pattern of semicircles is shown in Fig. 35–21. The spaces between the dark semicircles are the

Fig. 35–21

same width as the semicircles themselves. To demonstrate double slit interferences the two transparencies are placed so that their centers are a small distance d apart. The angles θ at which constructive interference occurs can be obtained from $n\lambda = d \sin \theta$ (where λ is the spacing between the semicircles) and compared with measured angles at which constructive interference occurs.

The device can also be made up in the form of a lantern slide for lecture demonstration purposes: a sandwich arrangement with a sheet-film negative of a set of semicircles and a sheet-film negative of an angle indicator, between two glass slides, one clear and the other a photographic plate bearing an identical set of semicircles.

(10) E. L. Cleveland, *Am. J. Phys.*, 27, 521–522 (1959). This device was first demonstrated at the Iowa Colloquium of College Physics in 1953. The reader can make his own transparencies or obtain them from a distributor of scientific apparatus.

35–2.2

Thin film interference can be demonstrated with the apparatus shown in Fig. 35–22. Light from two

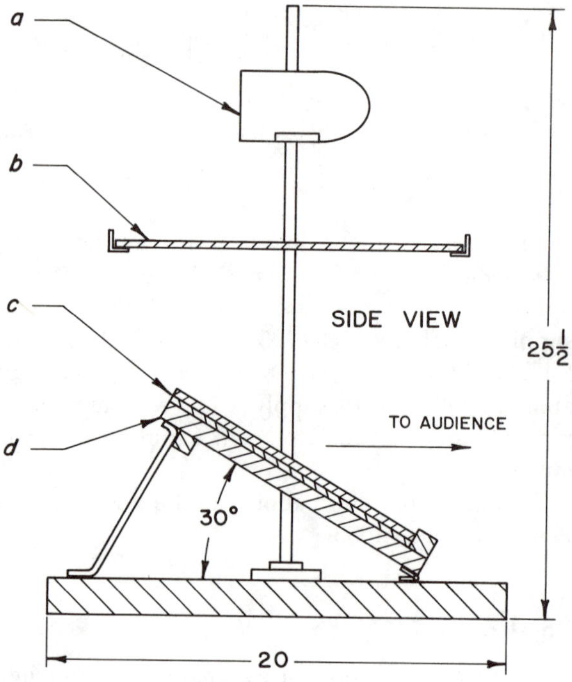

Fig. 35–22

sodium lamps[11] *a* placed about 20 in. apart impinges on a horizontal, frosted glass plate *b*. The frosted plate diffuses the light to two plane glass plates *c*, which rest on a piece of plywood *d* angled at 30 deg to the horizontal. The audience views the interference patterns from the plain glass surfaces. The frosted plate, resting in a frame of aluminum angle, can be positioned at an adjustable height above the reflecting plates. The frame and sodium lamps are supported by two vertical rods mounted on a hardwood base. The frosted plate is 15 x 36 x $\frac{1}{8}$ in., and the plane glass plates are 12 x 36 x $\frac{1}{4}$ in.

(11) Commercially available from George W. Gates & Co., Franklin Square, New York.

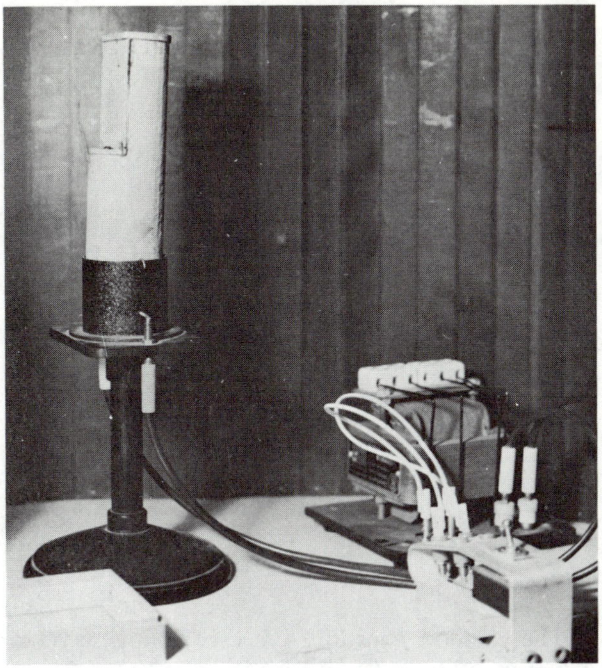

Fig. 35–23

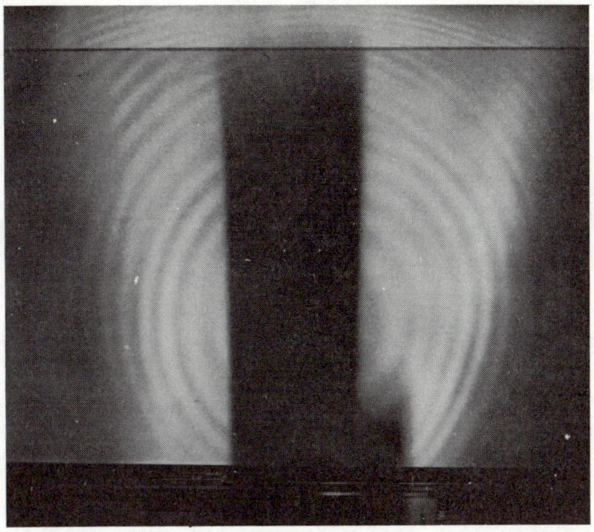

Fig. 35–24

35–2.3

The apparatus, as shown in Fig. 35–23, is used to demonstrate the interference of two reflected, coherent images. The front and back surfaces of a thin mica sheet act as two mirrors producing two coherent sources able to interfere with each other. With two point sources, the loci of constructive interferences are rotation hyperbolae, as these are the surfaces where the difference of the distances to the two points is a constant number of full wave lengths. What is observed on the wall is the cut of these hyperboloids by a plane, i.e. a series of concentric circles. This is shown in Fig. 35–24. The interference rings are observed even if the source is of finite size, as long as it is smaller than the ring spacing on the wall. The progression of radii of the rings (see Fig. 35–25) as a function of order of interference can be derived in the following manner. Interference maxima are found when

$$\Delta_b - \Delta_f = \frac{\lambda}{2} + m\,\lambda.$$

The term $\lambda/2$ comes from the phase jump of π occurring at the reflection from the thinner to the denser medium at the front surface.

It can be assumed that $d << S << D$. Therefore the radius of the rings is

$$r = D \tan\phi.$$

On this assumption the interfering rays are parallel and the phase differences can be calculated. Thus the front reflected ray (see Fig. 35–26) travels the optical distance

$$A \rightarrow B_f = 2\,d \tan\psi \sin\phi;$$

the back one

$$A \rightarrow C \rightarrow B_b = \frac{2d}{\cos\psi}\,n.$$

From the law of refraction,

$$\sin\phi = n \sin\psi.$$

After eliminating ψ, the path difference is

$$\Delta = 2dn\sqrt{1 - \frac{\sin^2\phi}{n^2}}.$$

Considering the minima and calling $\Delta_0 = 2dn$, the optical path back and forth at $\phi = 0$, then

$$\Delta = \Delta_0 - m\,\lambda.$$

Substituting the expressions above and squaring, we have

$$\sin^2\phi = \frac{n\,\lambda}{d}\,m - \frac{\lambda^2}{4d^2}\,m^2;$$

for small order m the linear term prevails, therefore

$$\sin\phi \sim r \sim \sqrt{m}.$$

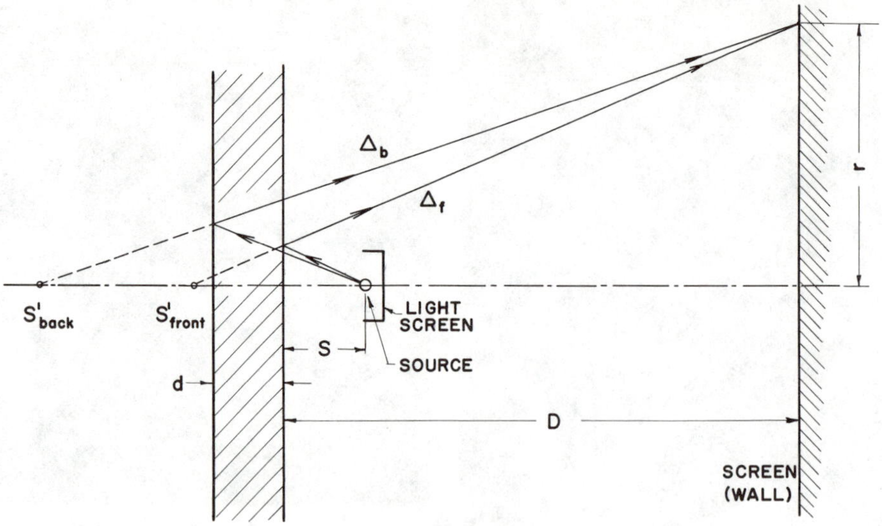

Fig. 35–25

The rings are widely spaced in the center and are more closely spaced as ϕ increases.

The quadratic term is never very important, as can be shown by calculating the value of m corresponding to the maximum of $\sin^2 \phi$. This is independent of λ/d and equal to n^2. It is therefore well beyond the real values of ϕ. At bigger ϕ, the ring spacing increases again because the radius increases as $\tan \phi$. As the spacing depends on λ, the ring systems for spectral lines of different colors start overlapping when the order increases. They overlap from the start, as the number of wavelengths on the fixed path difference Δ_0 differs. With Hg-light, the rings stay clearly visible, as the strong spectral lines are few and well separated. The picture can be improved by selecting one line with an appropriate filter, but it becomes less luminous.

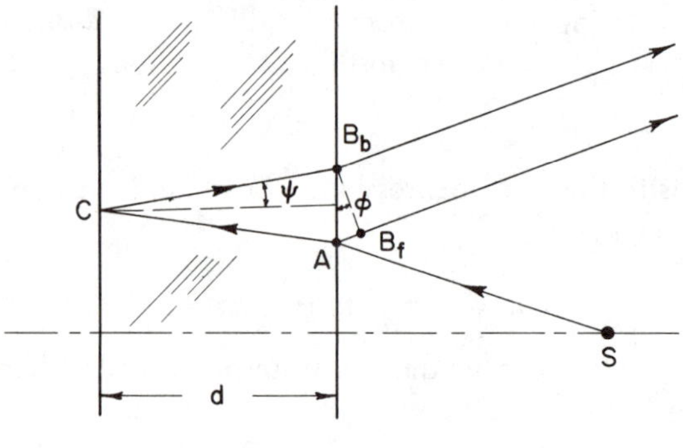

Fig. 35–26

The method constitutes a very simple way to determine thicknesses of flat transparent foils, as no optics whatsoever are needed.

The beam of a mercury lamp is directed backwards on a mica sheet held behind it. This produces a beautiful interference ring system on the opposite wall. The lamp is set up with the slit towards the audience and screened. The mica plates of about 4 x 4 in. are put behind the lamp so that the light is reflected toward the wall of the lecture room (the distance is unimportant, typically about 5 m). By moving the plate, the best part (that with the most constant thickness) can be selected. This results in a very clear picture of the interference rings. Mica plates of different thicknesses give interference patterns with variable ring spacing.

The mercury lamp can be constructed from a commercial mercury high pressure lamp *a* of approximately 125 W (e.g., Philips HP 125; Osram HQA 500, 125 W, or similar). These lamps have to be connected with a series-choke. The outer glass is removed by cutting along the base of the lamp with a glass knife and carefully breaking it. The new light-tight housing *b* must resist heat and can be built from Eternit (asbestos-cement plates) which can be glued with a mixture of kaolin and water glass (sodium silicate). It has to be as narrow as possible so that its shadow will not obstruct the picture. The assembly is shown in Fig. 35–27 on next page.

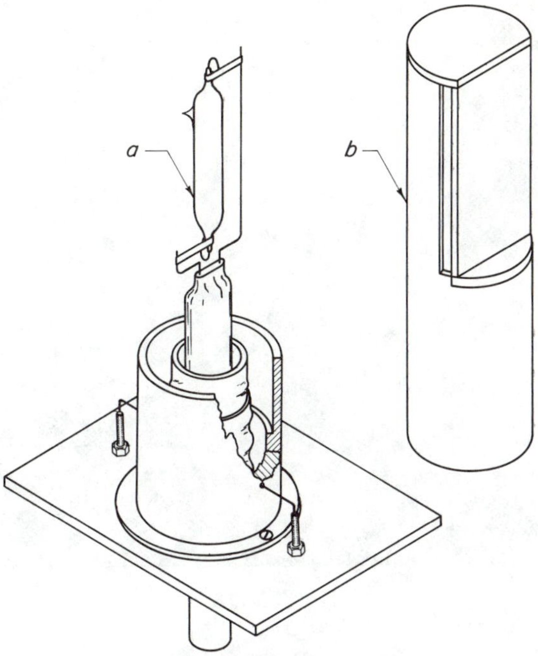

Fig. 35–27

that the shorter the focal length of the lens, the greater the difference in the focus of the various sections of the water surface; consequently, the incident angle should be small. A bit of paper or a match should be placed on the surface of the water to simplify the focusing.

The water should be drawn directly from the faucet, and the container must be thoroughly washed in running water. If at first the film gives off white light of the first order, more drops of turpentine should be added until bright colors are produced. As the turpentine evaporates the color of the film gradually changes, and, if the original film was very thick, colors appear on evaporation. Evaporation can be hastened by blowing on the film. Very handsome patterns may be shown on the screen if the water is set in motion by blowing or stirring with a small rod. If the experiment is to be repeated, the container should be washed out and fresh water used, otherwise the film will not spread.

35–2.4[12]

Beautiful interference colors can be obtained with a turpentine film on water. Water is poured into a horizontal container or a crystallizing dish 4 in. in diam with a sheet of black paper on the bottom to eliminate light reflection. A turpentine drop is shaken from a glass rod onto the surface of the water. The surface of the water is illuminated with an arc lantern having a condenser producing a slightly converging beam; the light is directed towards the container at an angle from above by means of a flat mirror. The angle of incidence of the light to the surface of the water must be approximately 45–60 deg. The light, which is reflected from the water, falls on a lens with a focal length of 15–20 cm, which throws the image of the water surface onto the ceiling or onto a screen by means of a second mirror. The arc lantern should be adjusted so that the converging beam of light, reflected from the surface of the water, will fall completely within the projecting lens. When choosing the incident angle of the light, one should remember

35–2.5[13]

"Temper colors" on the surface of a metal can be obtained when the metal is heated and a thin film of oxide has been formed on its surface. The surface of an iron or steel plate is polished with a fine file and emery paper until it has the desired gloss. Then the plate is heated over a flame, either at the center or at one edge, and projected onto a screen in reflected light in the same way as done for a turpentine film or other similar films (see Demonstration 35–2.4). It is even better to heat the plate during the projection, having first adjusted one of its edges in a stand to obtain a bright image of it on the screen. It is very easy to show the formation of color bands and their gradual shift as the temperature rises. The thicker the plate and the slower its heating the slower will be the rate of formation and the motion of the temper colors.

Beautiful interference colors can also be produced from other materials. For example, interference colors from some samples of mother-of-pearl can be demonstrated on a screen in reflected light. If the sample is turned a little bit during the

[12] Translated by K. V. Robinson and H. A. Robinson (Adelphi University) from *Lecture Demonstrations in Physics*, A. B. Mlodzeevskii, ed., M. V. Lomonosov State University, Moscow, U.S.S.R.

[13] Translated by K. V. Robinson and H. A. Robinson (Adelphi University) from *Lecture Demonstrations in Physics*, A. B. Mlodzeevskii, ed., M. V. Lomonosov State University, Moscow, U.S.S.R.

demonstration, a handsome interplay of colors can be seen. Another example is the thin, insoluble film sometimes found on the inside walls of glass containers due to some unknown chemical reaction. Such a film produces beautiful interference colors which can also be demonstrated on a screen. The film is occasionally formed on the inside of a jar which has contained Vaseline for a considerable length of time.

35–2.6[14]

An interference filter consists of an optically thin film of transparent material[15] coated on both sides with either a reflective coating or a coating with a different index of refraction. A narrow band pass filter may contain several such layers. The transmitted band is sensitive to the angle of the incident light. It is convenient to demonstrate this behavior in a filter with a reflective coating built to pass the 5460-Å green line of mercury in the second order at normal incidence. The filter can be illuminated either with white light (frosted incandescent bulb) or a mercury arc. When white light is used, the transmitted wavelength can be measured on a spectrometer and the progressive color change plotted against the angle of incidence. At angles of approximately 40 deg the transmitter beam is split into two, each beam polarized at 90 deg to the other. This can be demonstrated using Polaroids. When using a mercury lamp, the blue 4916-Å line will be transmitted, when the angle of incidence is around 45 deg. If a neon glow lamp is used, the transmitted light is green at normal incidence, blue at higher angles and red when the angle is around 60 deg. The specific type of behavior will of course depend on the specific filter chosen.

35–2.7

A commercial laboratory interferometer[16] (Figs. 35–28 and 35–29) can be used to project interference fringes from both a white light source and a monochromatic light source, such as a sodium lamp. Figure 35–29 illustrates the arrangement of components necessary to perform this demonstration for white light. The frosted glass, which acts as a diffuser, is not necessary for a monochromatic source. The patterns are projected through a 3-in. double convex lens.

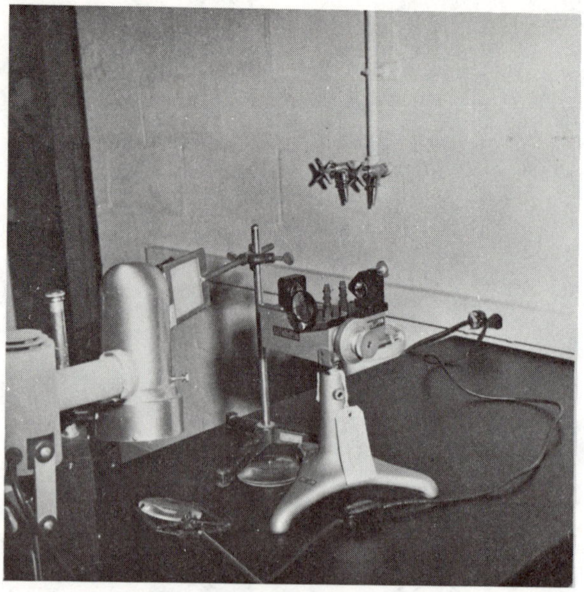

Fig. 35–28

35–2.8[17]

The triangular interferometer shown in Fig. 35–30 can be used to produce white light fringes or colored fringes which can be projected on a white screen some distance away from the instrument.

The beam splitter in Fig. 35–30 is symmetrical and it gives rise to a black or white central fringe. An unsymmetrical beam splitter, coated on one side, gives rise to a colored central fringe.

Light from a diffuse source *a*, either white, such as from an auto lamp, or monochromatic, is incident on a beam splitter *b* about a foot away. The

[14] L. Grillet and J. Bouyer, *Bull. Soc. Sci. Bretagne,* **35,** 141 (1960).

[15] See N. M. Mohler and J. R. Loofbourow, *Am. J. Phys.,* **20,** 583 (1952).

[16] Available from Gaertner Scientific Corp., Chicago, Illinois, model I–1010.

[17] C. Harvey Palmer, *Optics Experiments and Demonstrations* (The Johns Hopkins Press, Baltimore, Md., 1962), pp. 160–165; see also J. Strong, *Concepts of Classical Optics* (W. H. Freeman and Co., San Francisco, Cal., 1958), Appendix A (W. E. Williams) and Appendix B (J. Dyson), pp. 373–392; P. Hariharan and D. Sen, "New Gauge Interferometer," *J. Opt. Soc. Am.,* **49,** 232–234 (1959); "Triangular Path Macro-Interferometer," *ibid.,* pp. 1105–1106; and P. Hariharan and R. G. Singh, "Achromatic Fringes Formed in a Triangular Path Interferometer," *ibid.,* p. 732.

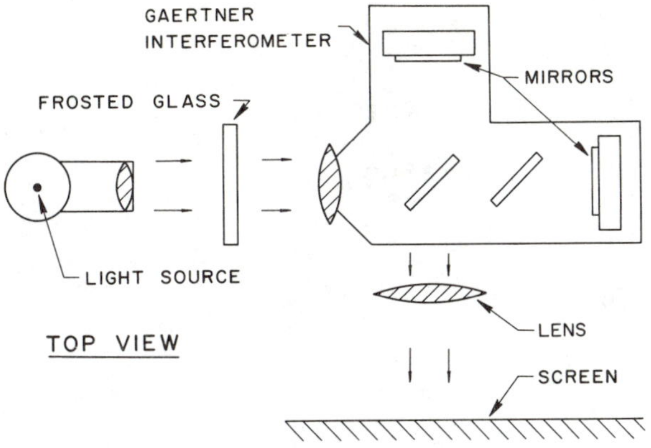

Fig. 35–29

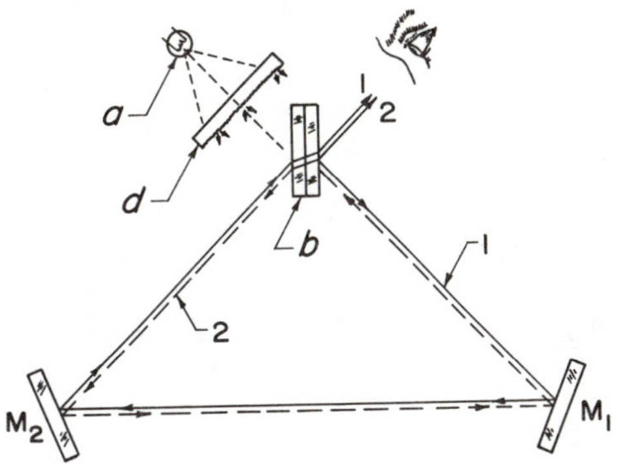

Fig. 35–30

beam splitter, consisting of a partly silvered surface sandwiched between two ¼-in.-thick glass plates,[18] divides the incident ray such that ray No. 1 (solid line) is transmitted clockwise as shown, and ray No. 2 (dashed line) is reflected in the equivalent counterclockwise path. A beam splitter composed of plates which reflect about one-third of the incident light is preferable. One of these or a pair with their reflecting sides in contact can be used.

The system is aligned by placing a finely ground glass screen or a card *d* with a small cross cut in it between the beam splitter and the source. The glass screen is 4 in. square and ground fine enough so that if an auto lamp is used as the source, the lamp filament is distinguishable through the screen.

In Fig. 35–30, the rays have been sheared for clarity, but actually for the symmetrical case shown, the rays coincide at all points—there is neither shear nor path difference.

The more general case is illustrated in Fig. 35–31 which has been simplified by the omission of the glass on either side of the beam splitter. Here the successive images of the virtual source S_0 (where the beam splitter divides the incident ray) produced by the various reflections are indicated. Mirror M_1, front-silvered plate glass about 1½ x 3 in. and ¼ in. thick, has been tipped and the amplitude-divided rays from S_0 appear to come from S_1'' and S_2'''. These two virtual (coherent) sources are separated both laterally (sheared) and in depth. The latter path difference is responsible for the interference fringes. Simple considerations show that the emerging rays which originate from a common source point are parallel at an angle with the plane of the beam splitter *B* as indicated. Skew rays, out of the plane of Fig. 35–31, are not shown, but their effect can be inferred from the diagram.

During alignment, two sets of images are seen (generally), the number in each set depending on whether the beam splitter is symmetrical or not. When the brightness of each set is brought into coincidence by tilting or rotating either of the identical mirrors M_1 or M_2, white light fringes should be seen; if they are not, the mirrors should be adjusted slightly until color is noted. The fringe system can be made circular or nearly straight depending on the adjustment of the mirrors.

The ground-glass screen can be moved to the exit beam and an achromatic lens substituted to produce collimated light incident on the beam splitter for observation of fringes on the glass. The ground glass can be removed and the colored fringes projected on a white screen some distance away from the instrument.

Further interesting effects can be observed with either ground glass used in the exit beam and parallel light incident, or in the entering beam to diffuse the incident light. First, one of the mirrors (M_1 or M_2) is rotated to obtain quite fine and vertical fringes. Then a thick glass plate *e*, about 2 in. square, ½ in. thick, is placed between the mirrors as shown in Fig. 35–32 and rotated slowly to restore the fringe pattern to its central position. Reasonably good achromatized fringes may be obtained by proper adjustment of the plate, or still more colorful fringes by rotating the plate in the other direction.

With the glass plate *e* removed, obtain good vertical fringes, with about five fringes in the field of view. Then monochromatic light is substituted. A 1–5 cm cell with plate glass ends or at least flat

[18] Available from Edmund Scientific Corp.

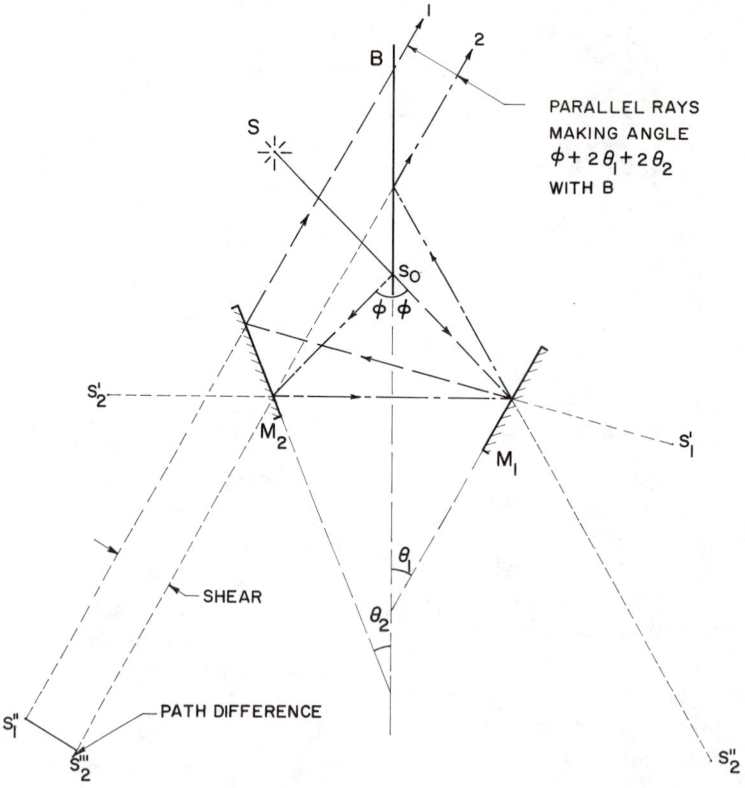

Fig. 35–31

ends, is filled with water and placed between the mirrors. If two or three drops of acetone are mixed into the water, the fringe pattern readily shows the small changes in refractive index during the mixing.

For interferometer construction details see Appendix, page 1353.

35–2.9

A mercury arc lamp[19] with an adjustable slit housing *a,* two condensing lenses *b,* a projection lens system *c* (these are mounted in that order on an optical bench), and two identical mirrors *d* arranged on a massive rigid base are used to display interference fringes. This homemade Jamin interferometer is shown in Fig. 35–33. The mirrors are critical components of the apparatus, and the reflecting surfaces of each should be as optically flat and parallel as practicable. Only one side of each is silvered. A mirror thickness of ¾ in. is required for prevention of distortion due to temperature fluctuations and/or physical stresses. The mirror housings are constructed as shown in Fig. 35–34. The assembly is mounted on a long steel channel on a heavy wooden base so that one mirror can be adjusted while the other one remains fixed with relation to it. The distance between mirror pivots is not critical although symmetry about the channel's center is.

(19) General Electric No. H-4 mercury arc lamp powered by a General Electric autotransformer, Catalog No. 59G16.

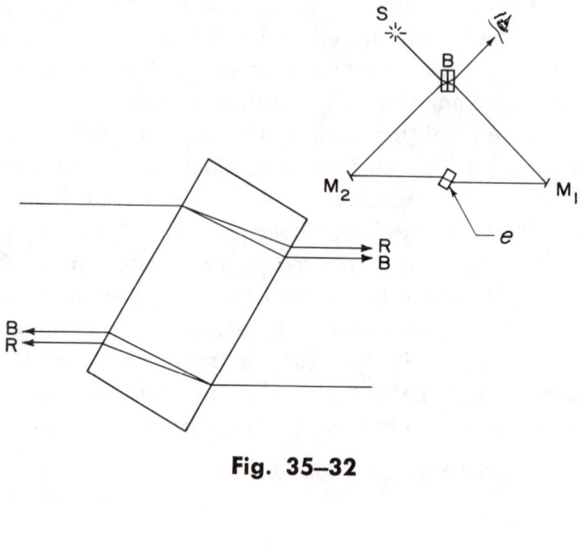

Fig. 35–32

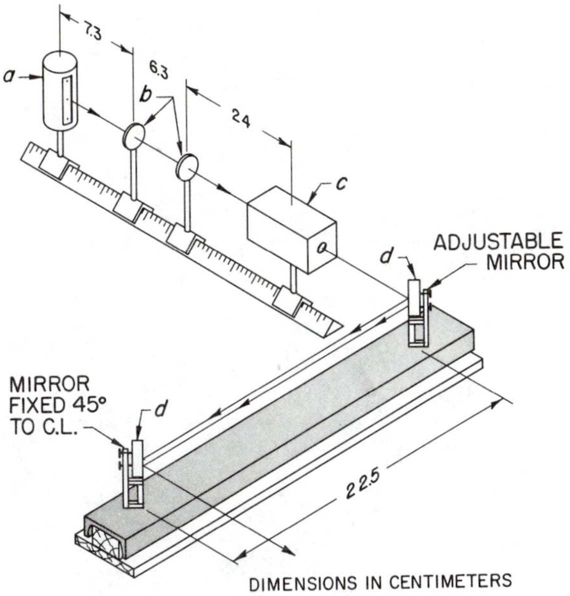

Fig. 35–33

The image of the vertical housing slit is focused sharply so that it falls on the first mirror some distance off its center. The slit is adjusted to give an image width of about ¼ in. Part of the beam is reflected from the frontal surface and part from the silvered surface, resulting in two parallel beams. The flatter and more parallel the mirror surfaces are, the more parallel the emerging paths. (This is very critical if the apparatus is to function at all.) These are caught by the second mirror (identical to the first) and reflected towards a projection screen, forming interference fringes.

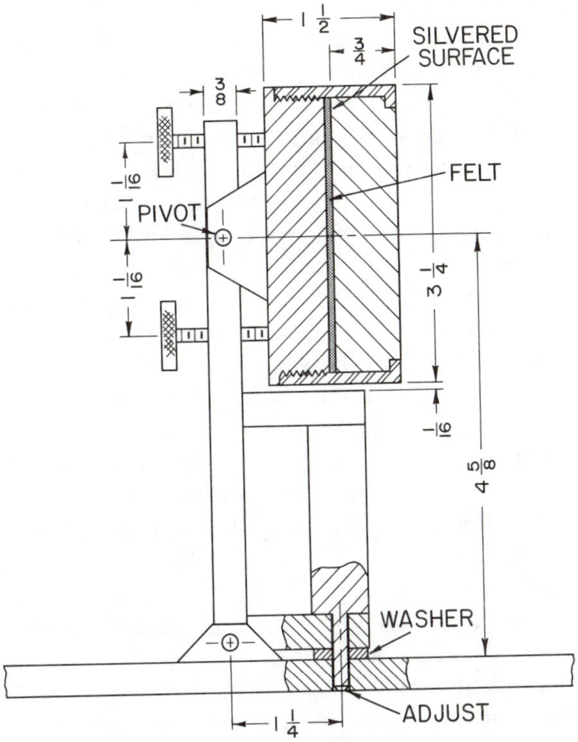

Fig. 35–34

35–2.10[20]

A Jamin apparatus shown in Fig. 35–35 provides far brighter interference fringes than those obtained with a biprism or with the Fresnel mirror. A plane parallel glass plate *c* divides the incident ray into two beams which are once again converged into one by means of a second similar plate *d*. As long as both plates have the same thickness and are placed parallel one to the other, the light beams will have a zero path difference, i.e., will appear to emerge from the same virtual source *e* produced by reflection in both plates and shown in the diagram by the dotted line. If one of the plates is deflected slightly from a parallel position, this imaginary image bifurcates so that two coherent light sources appear, as in the case of the Fresnel mirrors. The distance between these two light sources may be very easily adjusted by turning one of the plates by

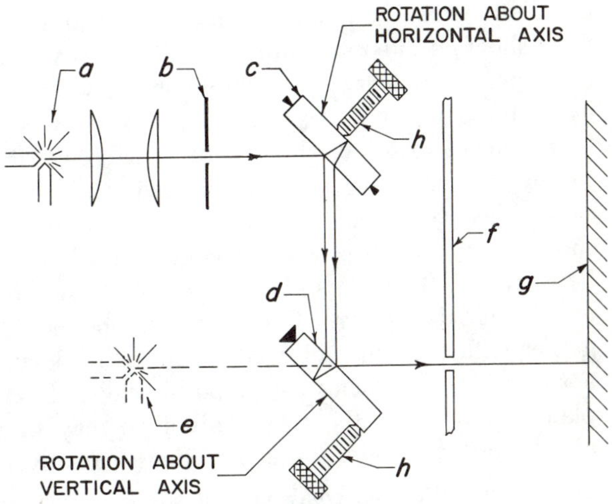

Fig. 35–35

[20] Translated by K. V. Robinson and H. A. Robinson (Adelphi University) from *Lecture Demonstrations in Physics*, A. B. Mlodzeevskii, ed., M. V. Lomonosov State University, Moscow, U.S.S.R.

means of a micrometer screw *h;* this makes it possible to change the breadth of the fringes to a considerable degree.

Plates *c* and *d* are placed in vertical planes. A diverging beam of light from a lantern *a* passes through a narrow diaphragm *b* and falls on plate *c*. The light which is reflected from plate *d* falls onto the screen *g* through a window which has been cut out of a cardboard sheet *f* in such a way that the rays, reflected from *d*, but not taking part in the interference, do not fall upon the screen. Micrometer screws make it possible to turn the plates about mutually perpendicular axes: plate *c* about a horizontal axis and plate *d* about a vertical axis. By means of these screws it is possible both to control the width of the fringe and to move the fringes in a direction perpendicular to their length, thus placing the median white fringe in the middle of the field of vision. The colored interference fringes in this demonstration are very bright and almost horizontal. A light filter may be used to show the dependence of the fringe width on the wavelength; this is impossible using the Fresnel biprism because of the insufficient light.

35–2.11[21]

Two separate Fresnel mirror experiments are described here. The first of these uses plane front silvered mirrors set at a small angle one to the other. Adjustment with this setup is difficult and the coherency requirement necessitates the use of a narrow slit which considerably cuts down on the amount of light available. Furthermore, in adjusting the mirrors, the point of contact between them must show *no* difference in height since this would correspond to a definite path difference between the two interfering rays. As a result the setup would be restricted to the production of interference between relatively long wave trains and white light cannot be used. The second setup is very much simpler and consists of an air wedge of small angle between two thick glass plates 7–8 cm on a side set one behind the other. Two edges are bound together with tape and the wedge formed by laying a strip of tinfoil along the opposite edge of one plate but between the two. This edge may then also be taped. Under these conditions a relatively broad light source such as a carbon arc can be used to project the fringes. Interference arises between rays reflected from the two front surfaces of the glass plates.

[21] Robert W. Pohl, *Optik und Atomphysik* (Springer Verlag, Berlin, 1963), 11th ed., p. 73.

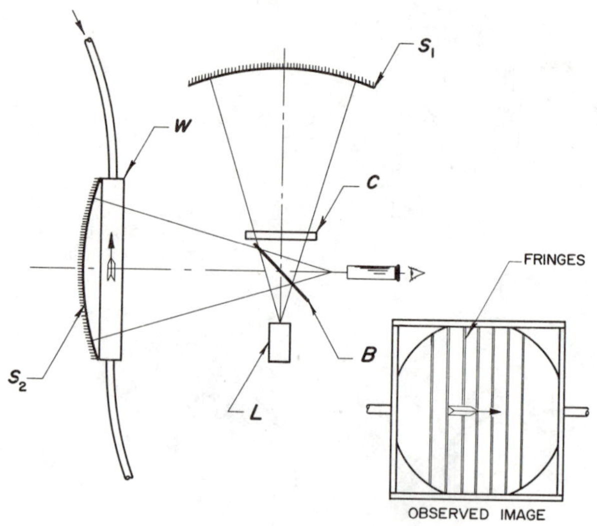

Fig. 35–36

35–2.12

The interferometer is designed here to show in color wind currents around objects, heat currents, and strain effects. The apparatus, Fig. 35–36, consists of two spherical mirrors S_1 and S_2; a beam splitter *B*, a window tunnel *W*, a light source *L*, and a magnifier. S_1 and S_2 are oriented at 90 deg to each other and their common centers of radii of curvature pass through the beam splitter. These mirrors need not be identical; however, adjustment is facilitated by their being so. Each mirror should be fitted with four adjustment screws. Mirror S_2 should be mounted on a focusing slide so that the fringes can be adjusted straight and parallel when both images of the mirror are at their precise radii of curvature and are superimposed.

A 50/50 partially aluminized or silvered microscope cover glass serves as a beam splitter. It is mounted at a 45-deg angle on a rod with lab wax; the aluminized side faces one spherical mirror and the light source. The returning beam from the spherical mirror S_2 should pass through the center of the beam splitter. Block off S_1 with a cardboard and locate the returning image from S_2. The window tunnel consists of two plates of glass in a

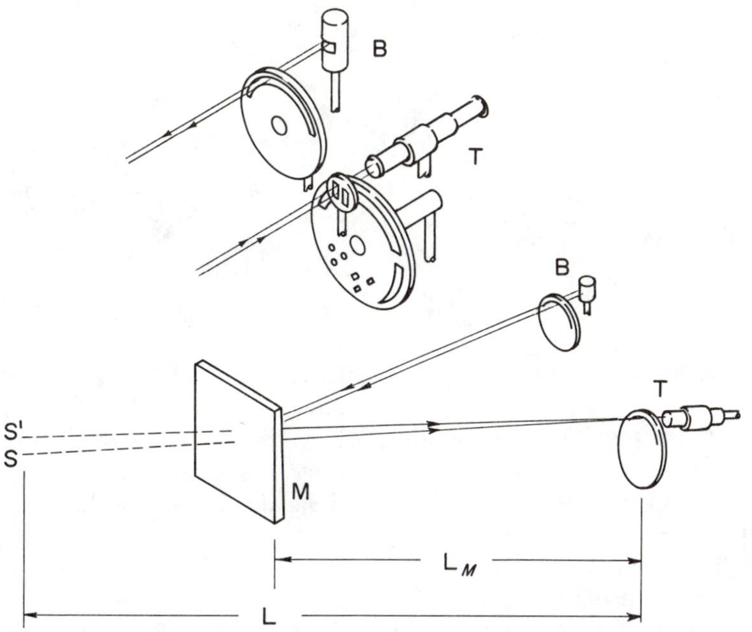

Fig. 35–37

wooden frame with an air inlet (vacuum sweeper hose) at one end and an outlet at the other. An effective light source is a mercury lamp with a source aperture of about .01 in. The two images near the beam splitter are located with a 5× magnifier.

The interferometer is aligned by first adjusting each mirror separately until the returning beam passes through the beam splitter. Both images should be equal in size. If they are not, adjust the sliding mirror and at the same time superimpose both images. If the fringes are circular, adjust the mirror until the fringes are parallel and ¼ in. or more apart.

Besides the above mentioned experiments, one can measure refractive variations in glasses (a compensating piece *c* of approximately the same thickness should be placed in the other beam, very close to the beam splitter).

Because it is an interferometer, the apparatus should be set up on a solid bench.

35–2.13[22]

The measurement of stellar diameters by the Michelson method can be modified to measure the angular separation of two artificial stars by relatively simple equipment. The model uses a common laboratory telescope[23] *T* of only 1-in. aperture. Its position relative to other parts is shown in Fig. 35–37. The disks are made of heavy paper that can be cut with clean edges. The one at the telescope serves to produce, in the position illustrated, two small apertures whose separation can be varied continuously by a simple rotation produced by the observer's fingers. The variable single slit[24] in front of the disk forms the vertical boundaries of the apertures. The components for the artificial stars are reached easily by the observer at the telescope so that he can make adjustments while watching for pattern changes. The "stars" are small areas of a ribbon-type filament projection lamp,[25] housed in a brass cylinder *B*. A narrow slit in the housing exposes the filament. A disk, mounted to rotate in front of this lamp, is made to expose either one or two small sections of the filament to a front-surface mirror *M* several meters away. The artificial stars, then, are the distant images S and S′, formed by the mirror.

The forms adopted for the disks are shown more completely in Fig. 35–38. The drawing of the

(22) M. J. Pryor, *Am. J. Phys.*, **27**, 101–103 (1959).

(23) Gaertner Scientific Corp., No. M522.

(24) Cenco No. 86140.

(25) Cenco No. 86606.

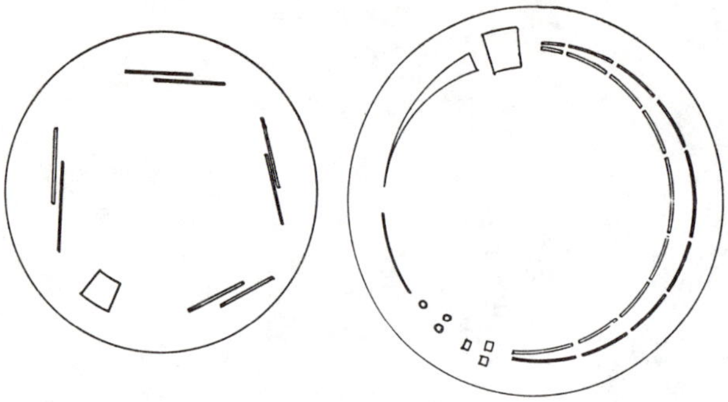

Fig. 35–38

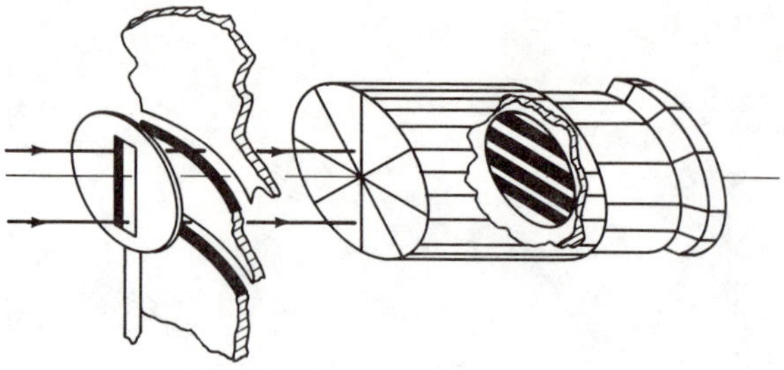

Fig. 35–39

larger disk shows that the provision for variable separation is achieved by cutting slits about 1 mm wide along eccentric circles with separation varying from as close as possible to about 4 mm between slit centers. The parts left uncut, for support, are at nonsymmetrical positions on the two sides so all values of separation may be had as one turns the disk in front of the telescope. The large openings are for preliminary alignment. Moving the tapered opening brings out the effects of aperture size. With the large section in front of the telescope a focused image of a "star" and its surroundings are seen. With narrower sections, fringes appear and for the narrowest part these fringes are quite broad. The round and square holes show that the virtually square holes, which are inherent in the device, give essentially the same patterns as round holes.

In the smaller disk, the separation of the different parallel slits provides for stars of different separation. By cutting the slits as each single pair is shown, one may "expose" either star or both stars. The purpose of this feature follows:

Place the disks in the positions shown in Fig. 35–37 with the double part of a pair of lines in front of the lamp. Then turn the large disk which varies aperture separation slowly until the dark lines across the pattern, which persist most of the time, disappear. The entire area is bright. Then the small disk is turned to expose but one star. The pattern of light and dark lines should come out clearly. Repeated trials with both stars and with single stars result in a good setting for measurement.

The principles of operation are represented in Figs. 35–39 and 35–40, which duplicates in three drawings the relative positions of apertures, telescope lens of the focal length F, and interference pattern. Figure 35–39 shows the relationship to the drawings of Fig. 35–37. In Fig. 35–40, top, it is shown that d, the half-distance between interference fringes, may be varied by changing the separation of the slits D as expressed in the equation

$$\frac{d}{F} = \frac{\lambda/2}{D} = \beta_1 \quad \text{or} \quad \frac{d}{F} = \frac{3\lambda}{2D} = \beta_2.$$

The angle β is the angle between destructively interfering wave fronts and the angular separation of the midpoints of dark and light fringes. The lower drawing in Fig. 35–40 shows that the separation of the centers of the geometric images of two selected stars of angular separation α is given by the equation

$$\frac{d}{F} = \frac{SS'}{L} = \frac{SS'}{2L_M} = \alpha.$$

The distance to the artificial stars L is twice the distance to the mirror L_M. For a selected pair of stars, the aperture disk is turned until the bright

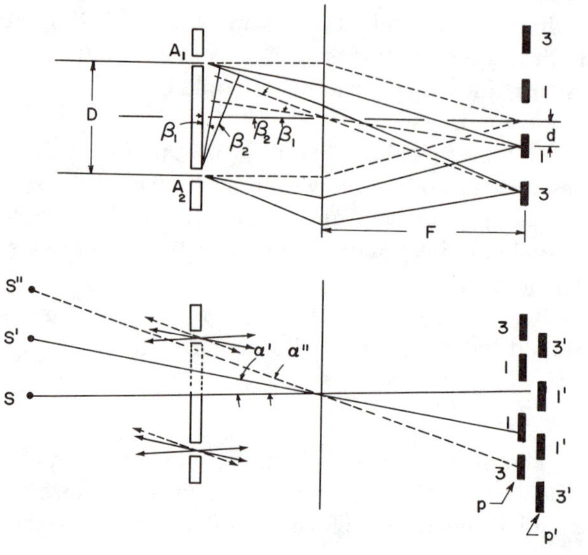

Fig. 35–40

bands of the pattern for one star fall on the positions of dark bands for the other star. For this adjustment, $\beta = \alpha$ and

$$SS' = L_M \frac{\lambda}{D} \quad \text{or} \quad SS' = L_M \frac{3\lambda}{D},$$

depending on the separation of the stars. The interference pattern of the star S′ is lower than that of S by the separation of a dark and bright line, while that of S″ is three times that distance.

Some sample measurements are represented in Table 35–1. 5500 Å was arbitrarily taken to be an effective value of the wavelength. All lengths are given in centimeters. Note that the entries in the line SS'' are equal to 3×0.144 and 3×0.098, respectively. The factor "3" is used when "1" fails for the amount of pattern shift in fringe widths cannot be observed and counted. The values opposite S in this table are direct vernier measurement of separation of the slits giving double stars.

Table 35–1
Measurements of the Separation of Artificial Double Stars (in centimeters)

D	0.20	0.38	0.23	0.28	0.42
L_M	773	2400	773	737	741
SS'	0.21	0.35	0.18		
SS''				0.43	0.29
S	0.20	0.37	0.17	0.42	0.28

35–3 Diffraction

35–3.1[26]

A narrow (~1 mm) slit is illuminated by a carbon arc having a condenser of 7.5-cm focal length (Fig. 35–41). The slit is focused on a distant screen with an achromatic lens of 30-cm focal length. A second slit is placed at the focal point of the achromatic lens. As the second slit is narrowed to less than 1 mm, the diffraction pattern appears on the screen.

35–3.2

Diffraction and interference patterns produced by single slits, double slits, and gratings with various spacing can be demonstrated to a large class using Cornell plates.[27] A good lecture-table arrangement is illustrated schematically in Fig. 35–42. A light source, consisting of an unfrosted line-filament lamp, is mounted vertically on the lecture table and viewed by the class, individually provided with Cornell plates. A meter scale is mounted horizontally above the lamp approximately at right angles to the student's line of sight. To facilitate locating the angle of the diffracted image, a second line-filament lamp is located just above the scale so that it can be moved to the right or left along the scale. The apparatus is shown in Fig. 35–43.

Using only the stationary light source, the characteristic diffraction patterns of single, double, and multiple slits are observed. With the single slit, the students should note that the central maximum is twice as wide as the side maxima, the intensity decreases with the order and red light is deviated more than blue. In the higher orders the maxima of different colors will overlap, ultimately resulting in elimination of the pattern. With the double slit, it will be noted that the maxima are uniformly spaced, and the closer the slits are together, the greater will be the distance between maxima. It is also possible to observe both the single slit and double slit patterns simultaneously. With the multiple slit, the characteristics of the diffraction grating are readily demonstrated. Using the movable lamp to locate the various orders, a qualitative measurement of the wavelength of light is made by the

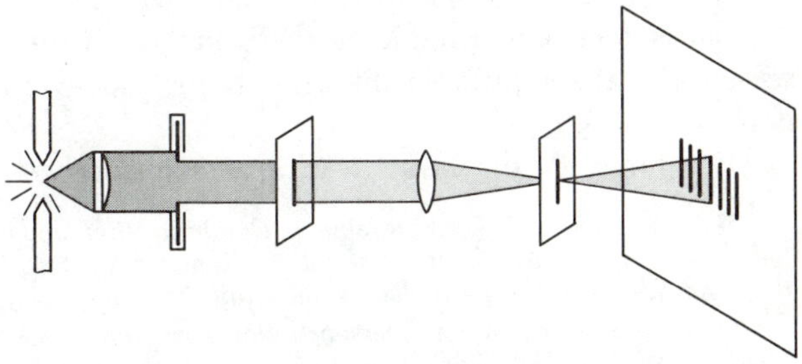

Fig. 35–41

[26] Published with permission from Hermann Daniel, *Demonstrations Versuche aus der Optik* (Musterschmidt-Verlag, Göttingen, 1960), 2nd ed., p. 40.

[27] Available from the National Press, Palo Alto, California.

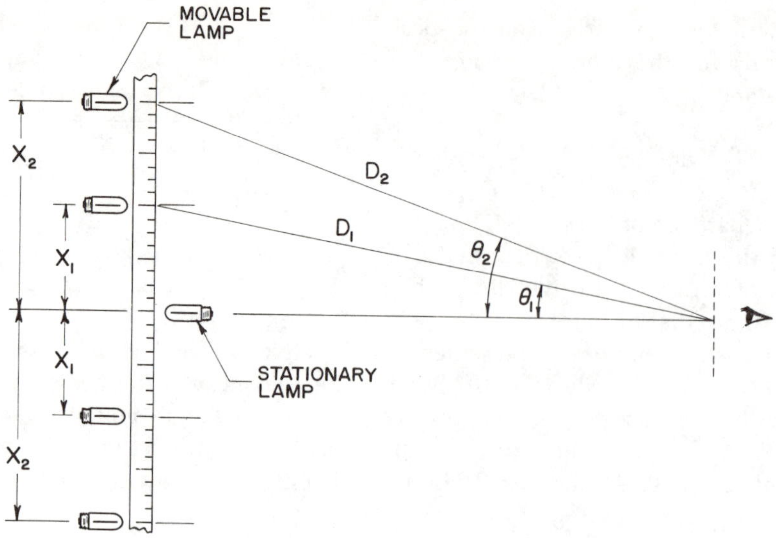

Fig. 35–42

student. The grating spacing d is specified on the Cornell plates. The formula $n\lambda = d \sin \theta$ can be approximated by $n\lambda = d \cdot (x/D)$ where D $(= D_1 = D_2)$ is the distance from the observer to the image on the scale. Hence, for the first order, $\lambda = (x_1/D_1) \cdot d$ and for the second, $2\lambda = (x_2/D_2) \cdot d$. It is also interesting to observe the effect of rotating the grating about an axis along the line of sight and about a vertical axis.

35–3.3[28]

The apparatus shown in Fig. 35–44 allows interference patterns of single or double slits to be displayed with an intensity much higher than normal. This intensity is achieved by superimposing patterns produced by many sets of single and double slits.[29] In the setup, the condenser lens b (f = 25 cm, d = 18 cm) must produce an image of the arc a on the adjustable slit c and be large enough to fill the projection lens d (Bausch & Lomb lens from an opaque projector: f = 45 cm, d = 10 cm) with light. The latter lens must produce a sharp image on the screen e of the adjustable slit when the slit array f is removed.

High-intensity displays of various interference patterns were obtained by the following procedure: From a table of random numbers, a random set of 60 positions was selected in the range 0–60. Parallel lines 18 in. long were ruled on a sheet of acetate (see Fig. 35–45) with "Chart Pak," a technique which rolls onto the sheet a narrow strip of tinted "Scotch" tape. Each of the 60 lines was $\frac{3}{16}$ in. wide and its position was dictated by the set of random numbers. This display was then duplicated on a second sheet of acetate which was superimposed slightly offset on the first sheet. The resulting display consisted of 60 identical pairs of lines randomly spaced. This composite display was then photographed using high-contrast "Kodalith" film. The negative thus acquired became the slit array f in Fig. 35–44. Single-slit displays were made by turning one acetate sheet end-for-end before superimposing it on the other sheet, thus giving 120 randomly spaced lines.

(28) R. Resnick and H. F. Meiners, *Demonstration and Laboratory Apparatus Report of the 1960 Summer Visiting Professor Workshop at Rensselaer Polytechnic Institute.*

(29) Pointed out by Sutton, *Demonstration Experiments in Physics* (McGraw-Hill Book Co., Inc., New York, 1938), p. 402.

35–3.4[30]

While the pattern cannot be made as large as desirable, the effect of the diffracting edge of a stop on the resulting image can be demonstrated at least to small groups. A small two dimensional point lattice is constructed by piercing a thin foil in a square array of about 25 small holes (5 holes on a side). The holes should be about 0.2 mm in diam-

(30) Robert W. Pohl, *Optik und Atomphysik* (Springer Verlag, Berlin, 1963), 11th ed., pp. 24–26.

Fig. 35–43

eter and spaced roughly 0.7 mm apart. Each side of the square array will thus be around 3 mm long. The lattice a is intensively illuminated from behind using a carbon arc b as in Fig. 35–46 and focused on a screen c with a 5 cm diam, 70 cm focal length lens d. The condensing lens e must focus the image of the arc b onto d in such a way that d is somewhat overfilled with light. This ensures that the rim of the lens is in fact the exit pupil. A color filter should be inserted into the system (monochromatic light). Under these conditions a clear image of the 25-point array should be visible on the screen. An auxiliary diaphragm f_1, with a variable circular opening, is then inserted near a and stopped down so as to allow only the image of a single point to be seen. The image is still clear. A perpendicular variable-width slit f_2 is then inserted just in front of the lens d and closed to a width of about .3 mm. The single spot spreads out into a horizontal diffraction pattern. It is clear that the image of a system is in fact the diffraction pattern of the edge of the stop. Since the light through the small central hole is in fact parallel, the first minimum of the diffraction pattern will occur at angle α where $\sin \alpha = \lambda/B$, with λ the wavelength and B the (f_2) slit width; f_2 can of course be replaced by a small round opening, in which case a circular diffraction pattern will appear.

The diaphragm f_1 is now opened so that all 25 points would appear on the screen were it not for

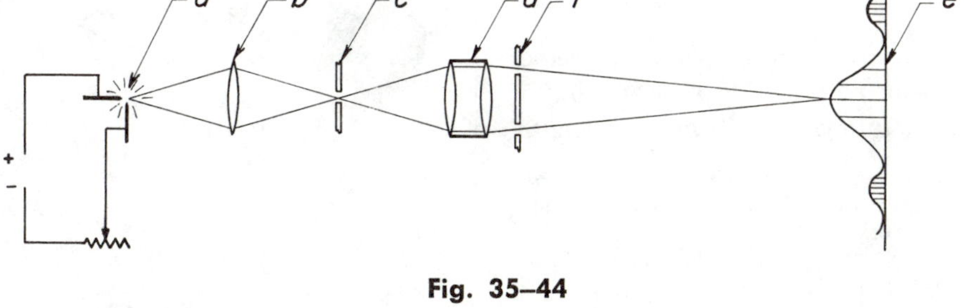

Fig. 35–44

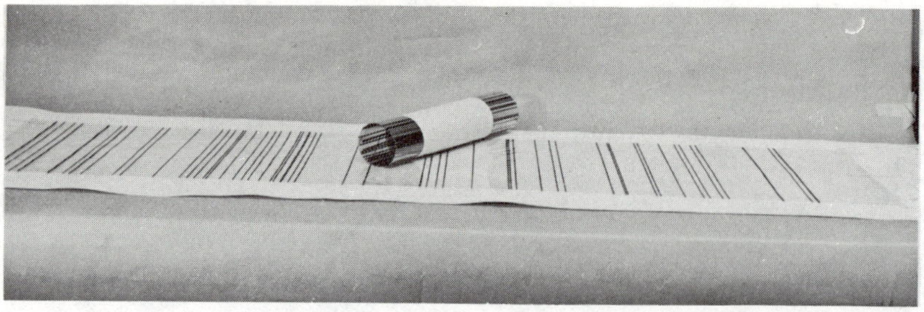

Fig. 35–45

slit f_2. The images will consist of 5 horizontal lines (f_2 is still vertical) resulting from the overlapping of the fine individual diffraction patterns. If the slit f_2 is now rotated to 45 deg the pattern will consist of fine parallel lines of light also at 45 deg. A horizontal slit will show a pattern of 5 vertical lines, etc. Other shapes of pattern can be shown by replacing the slit with openings of various shapes. As an example, a six pointed star will arise from a small triangular opening.

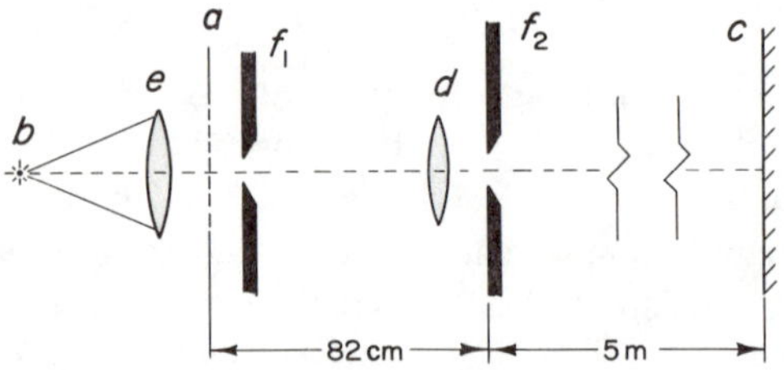

Fig. 35–46

35–3.5

A point source of light casts a shadow of a small disk (Fig. 35–47). With an arc lamp, this requires a translucent screen, and students must line up behind the screen to see the shadow. With the new Osram mercury lamp,[31] the shadow is just visible on a white screen with students standing in front. See also Chapter 4, Vol. I., Demonstration 4–2.3(m).

35–3.6[32]

A beam of light from an arc lamp a passes through a diaphragm b with a 1½ mm opening, is collimated by a lens c having a 30-cm focal length, passes through a glass plate d lightly dusted with lycopodium powder (small opaque spheres about 30 μ in diameter) and the resulting diffraction pattern falls on a wall at least 60 ft from the lens, as shown in Fig. 35–48. To enable higher order maxima to be more easily seen, a piece of black velvet is placed on the wall so that the central maximum will fall on it. In this way, the glare associated with the central maxima is eliminated. The first order maximum should be about 50 cm from the second one. It is possible to compute the size of the particles from the diffraction pattern.

35–3.7

The phenomena of diffraction is usually associated with light interacting with a physical object having a size or structural dimension comparable to its wavelength. In this case, however, diffraction can be made to appear with a very coarse grating whose rulings are 4 mm apart. Thus, the wavelength of the light can be measured directly by mechanical length standards.

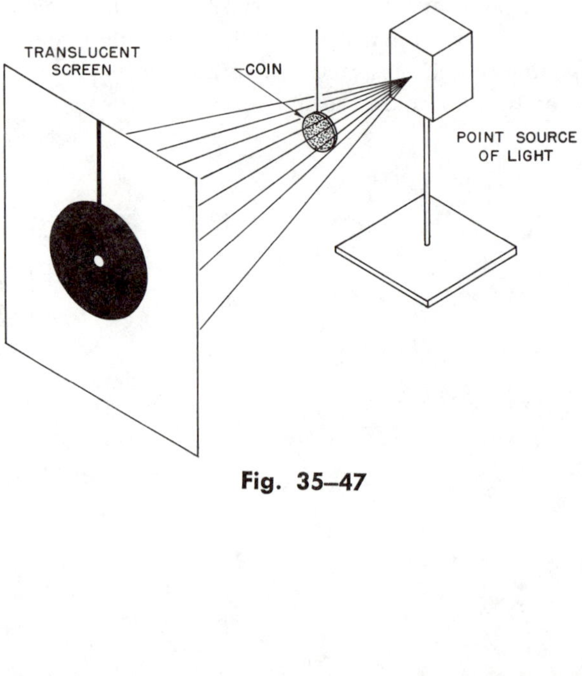

Fig. 35–47

[31] Commercially available either from PEK Laboratories, Sunnyvale, California, or from George W. Gates & Co., Franklin Square, New York.

[32] See also R. W. Pohl, *Optik und Atomphysik* (Springer Verlag, Berlin, 1963), pp. 87–88.

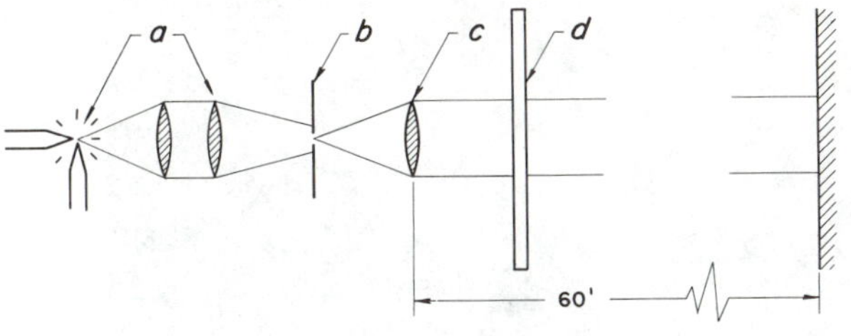

Fig. 35–48

Grazing incidence can be used to produce minute path differences for rays emerging at various angles. By applying Huygen's principle, the path difference becomes

$$\Delta = \delta_0 - \delta_n = d\,(\cos\theta_0 - \cos\theta_n)$$
$$= \frac{d}{2}(\theta^2{}_n - \theta^2{}_0)\,(\text{for small }\theta\text{ —grazing incidence}),$$

where $\theta = \theta_0$ (see Fig. 35–49). The condition $\Delta = n\lambda$, which insures a maximum intensity in the diffraction pattern, yields

$$\theta_n - \theta_0 = \frac{2n\lambda}{d(\theta_n + \theta_0)}.$$

The angular distance between orders can be made arbitrarily large as rays are made more grazing. Only a limited number of maxima of negative order occur, although the number increases with increasing angle of incidence.

The diffraction pattern is shown in Fig. 35–50. A red filter helps keep the higher order maxima separated. When white light is used, all maxima

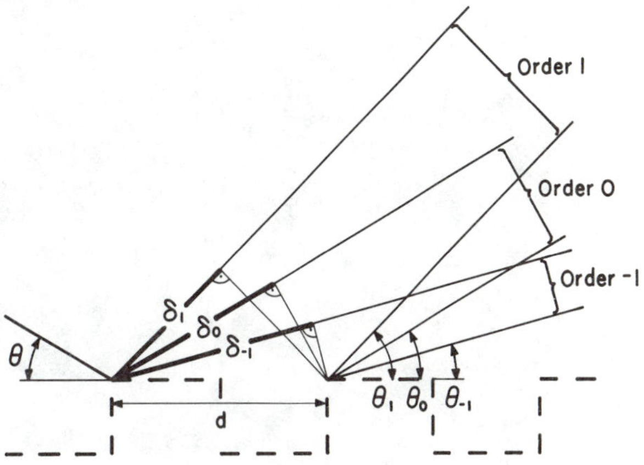

Fig. 35–49

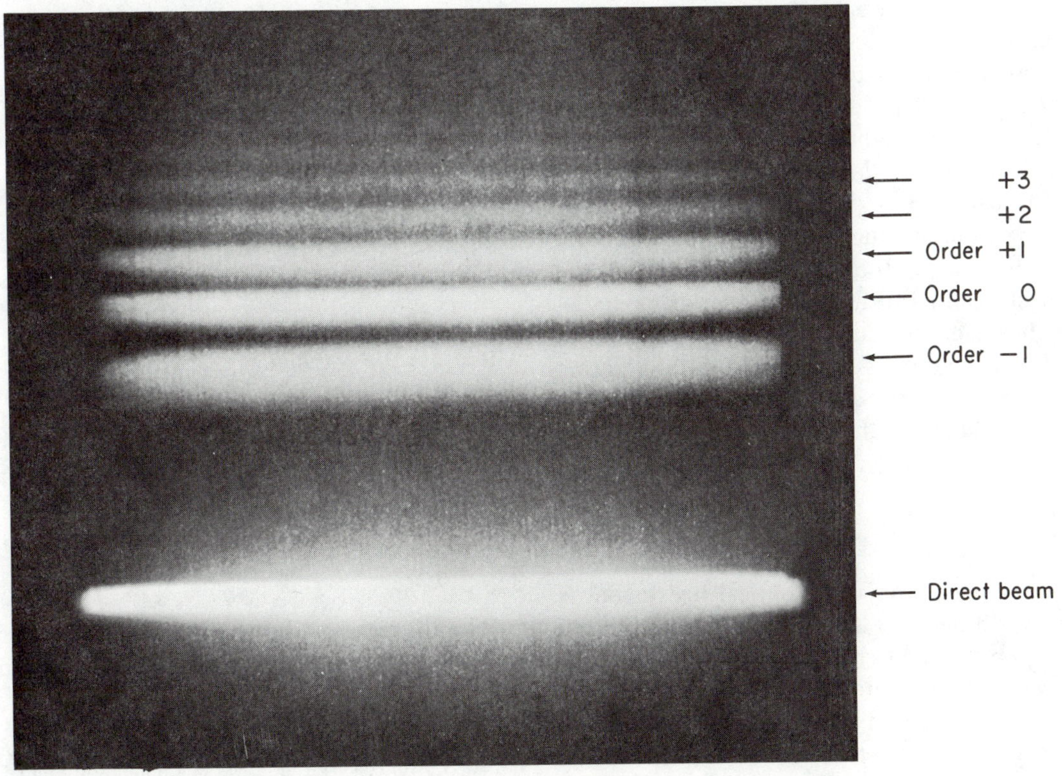

Fig. 35–50

Fig. 35–51

except the one of order zero (and the direct beam) appear as spectra.

The apparatus is set up as shown in Fig. 35–51. An arc lamp *a* and condenser lens *b* illuminate a horizontal slit *c* as brightly as possible. An achromatic lens *d* focuses the image of the slit on a screen. A coarse grating *e* is placed in the path of the light beam at grazing incidence. The red filter is inserted between the condenser lens and the slit. The slit can be turned around an axis parallel to the light beam, to insure that it is parallel to the plane of the grating. A setscrew on the grating makes adjustment of the angle of incidence possible. Construction details of the grating are given in Fig. 35–52.

35–3.8

The resolving power of an optical instrument is limited by diffraction. This can be demonstrated by the apparatus pictured in Figs. 35–53 and 35–54. Light from an arc lamp *a* (Fig. 35–53) passes through a filter *b* and a condenser lens *c*, through a narrow slit *d* and another lens *e*. This resulting beam passes through a diaphragm with an opening of variable width *f* (Fig. 35–54), through a telescope *g* (magnification of about 50×) to a television camera *h*.

The observed diffraction pattern is drawn in Fig. 35–55. The width of the central maximum 2ϕ is shown to increase with decreasing diaphragm width *d* according to the relation

$$\sin \phi_{\min} = \frac{\lambda}{d},$$

where λ is the wavelength of the incident light. To demonstrate the side-maximum, it is best to place a black strip of tape over the image on the photocathode of the TV tube, in order to avoid stray light.

35–3.9[33]

The diagram of the apparatus used to study resolving power of a microscope is shown in Fig.

[33] Translated by K. V. Robinson and H. A. Robinson (Adelphi University) from *Lecture Demonstrations in Physics*, A. B. Mlodzeevskii, ed., M. V. Lomonosov State University, Moscow, U.S.S.R.

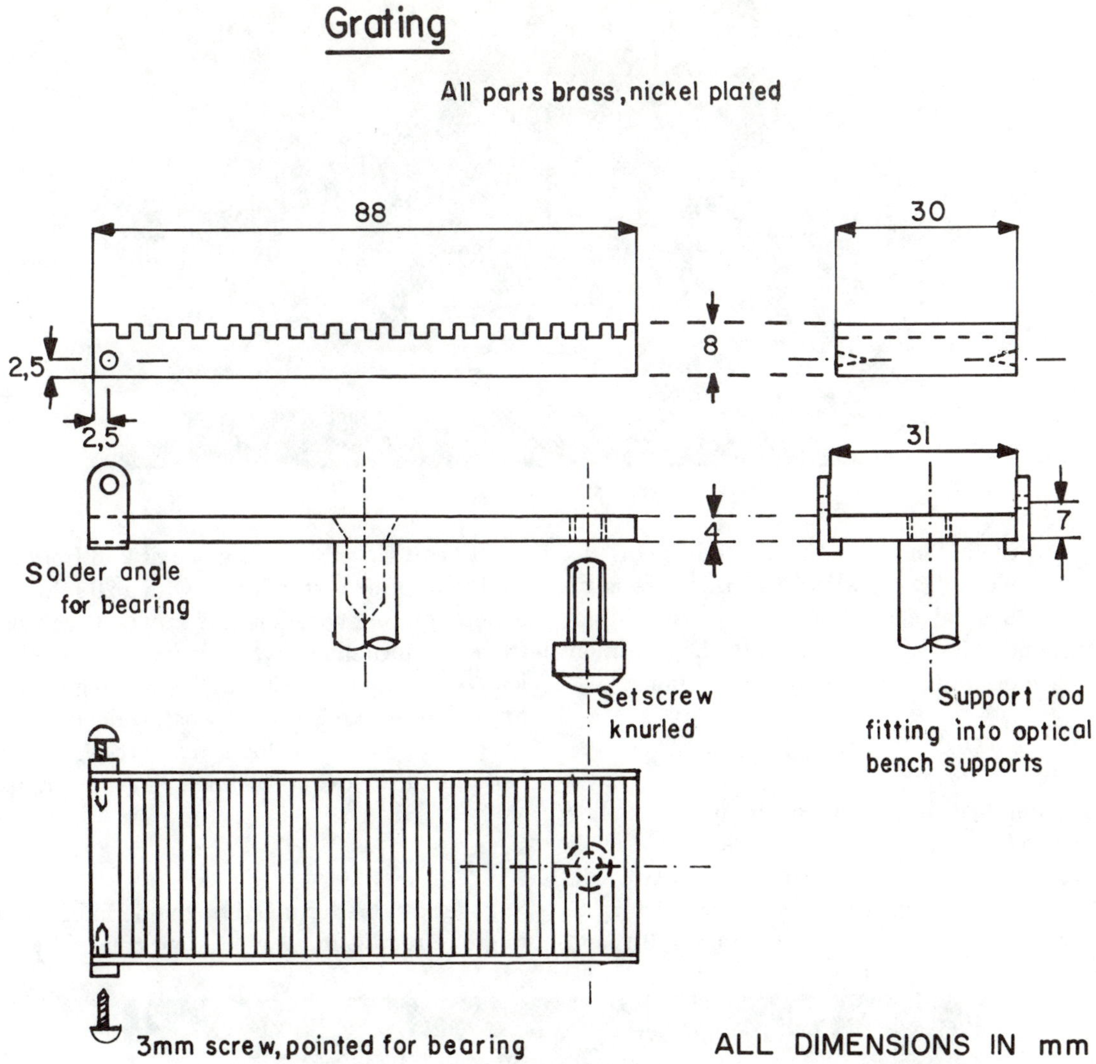

Fig. 35–52

35–56. Slit b is placed in the converging beam of light from the condenser lens a of a projection lantern. The grating d is placed directly in front of lens c and is in turn projected by objective lens e. A second slit f is placed in the plane in which the image of slit b is formed. If slit f is wide enough to let through not only the image of slit b but also the first pair of diffraction maxima, the image of the grating is clearly seen on the screen. If the slit is narrowed so that it lets through only the central image of b, the image of the grating disappears.

This effect is more complex than it seems at first sight: when slit f is very narrow, the image of the grating should disappear for another reason; diffraction in slit f broadens the images of the individual slits of the grating. Therefore, if this demonstration is to be convincing, considerable care must be exerted. With a given grating constant, if the width of slit f is too great, it does not stop the diffraction maxima; and if it is too narrow, diffraction occurs in the slit itself. To avoid this diffraction, the slit must be wider than the light beam passing through it and the edges of the slit must not interfere with the passage of the light. In other words, slit f must have a width D approximately twice as large as the width of the image of slit b formed on it ($D = 2s$). On the other hand, in order to shield the first order diffraction bands, the width of the slit must be less

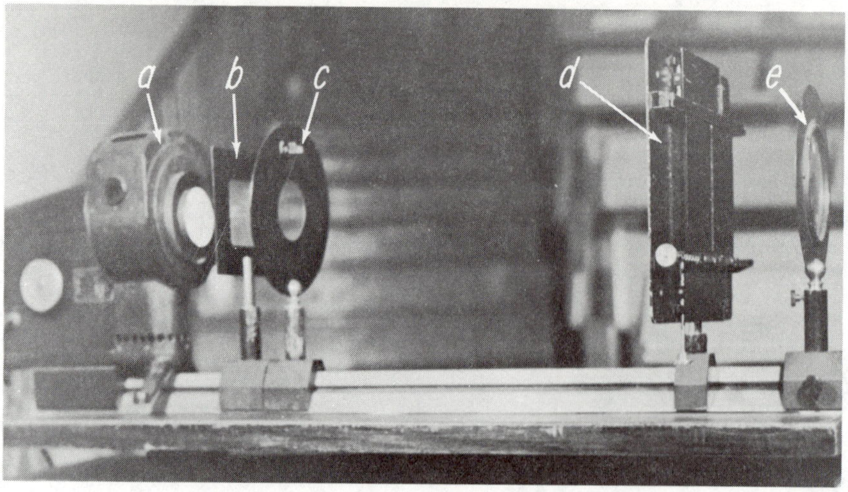

Fig. 35–53

than the distance between these bands. This distance is equal to $2f\lambda/p$, where f in this case is the focal length of objective lens e, p is the grating spacing, and λ the wavelength. Thus the conditions of the demonstration are set by the relation

$$D < \frac{2f\lambda}{p} \quad \text{or} \quad p < \frac{2f\lambda}{D}.$$

Assuming that the width s of the image of slit b is equal to 0.5 mm (brightness decreases if the slit is narrower), i.e., $D = 1$ mm, and that $\lambda = 0.6\mu$, and $f = 10$ cm, then the greatest possible value for p is 0.12 mm. A grating with a constant of 0.1 mm, $f = 10$ cm, placed at a distance of 2 m from the objective, appears on the screen with a 20-fold magnification (grating constant of 2 mm). It can be seen from the formulae given above that an objective lens with a short focal length cannot give a larger image, because p is proportionately decreased. The 2-mm-wide bands on the screen are visible only at close hand, unless an inclined screen is used.

35–3.10

Light can be diffracted by the periodic variations in the refractive index of a liquid caused by density

Fig. 35–54

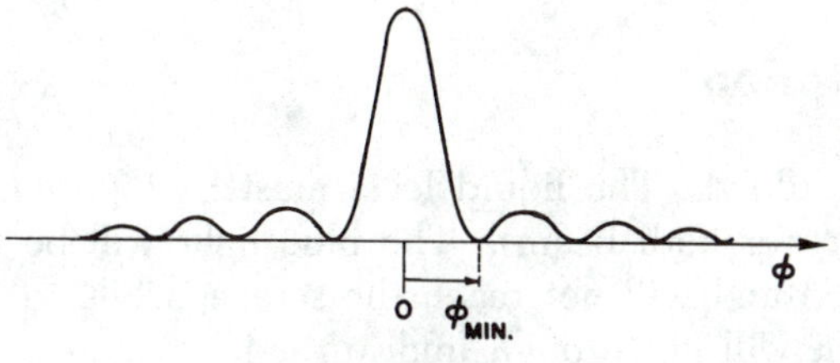

Fig. 35–55

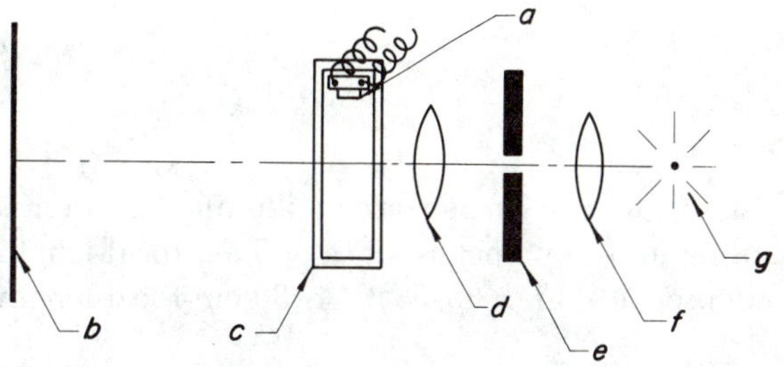

Fig. 35–57

variations occurring when an ultrasonic wave generates compressions and rarefactions. The compressions and rarefactions diffract the light in much the same manner as an ordinary grating. Provided that the time required for the light to pass through the liquid is much less than the period of the ultrasonic vibrations, the results are independent of whether the ultrasonic waves are traveling or standing.

In the simplest case, a plane monochromatic light wave is incident at right angles to the ultrasonic wave. Then the diffraction maxima are governed by the usual law:

$$n\lambda = d \sin \theta_n,$$

where n is the diffraction order, λ the wavelength of the light, θ_n the angle between the direction of the incident and diffracted light waves for nth order, and d the "grating" spacing, which is the wavelength of the ultrasonic wave. Since θ_n is very small, $n\lambda = d\theta_n$, or $\lambda = d(\theta_n - \theta_{n-1})$. If this formula is used, only the distance between adjacent images is needed. The distance d is found to be the wavelength of sound in the liquid. A simple qualitative description and a more complicated mathematical theory of this phenomenon is found in Born and Wolf.[34]

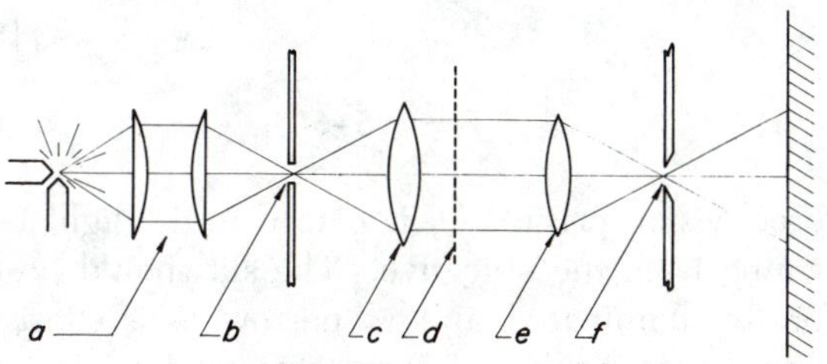

Fig. 35–56

The equipment is shown in Fig. 35–57. Light from a strong source[35] g is focused on an adjustable slit e by means of lens f. Light from e, rendered parallel by lens d, passes through a transparent "optical cell"[36] c, containing the liquid, and on to screen b, which is located approximately 15 ft from the rest of the apparatus. In addition to the liquid, the cell contains a piezoelectric crystal in a suitable holder[37] a, which is excited at a frequency of a few megacycles by a power oscillator (range ~1 to 10 Mc/sec). After the equipment is aligned, the power oscillator is adjusted until a resonant frequency of the crystal is obtained, at which point the diffraction pattern appears. The pattern then can be improved by slight adjustments. If quantitative results are desired, the frequency must be carefully determined, for example, with a wavemeter. If white light is used, colored fringes are obtained; to obtain a monochromatic wave, an approximately monochromatic filter[38] can be inserted in the light beam. Toluol and such liquids have been used in the cell, but a silicone fluid such as Dow–Corning 200 is preferable.

(34) M. Born and E. Wolf, *Principles of Optics* (Pergamon Press, London, 1959), pp. 590*ff*. See also L. Bergmann, *Der Ultraschall* (S. Hirzel Verlag, Stuttgart, 1954), pp. 248*ff*., and J. Blitz, *Fundamentals of Ultrasonics* (Butterworth and Co., Ltd, London, 1963), pp. 138–40.

(35) A convenient source is a microscope illuminator.

(36) Approximately 1¼ x 2½ x 3½ in.

(37) Available from J. Klinger Scientific Apparatus.

(38) A metal-dielectric interference filter is available from Grubb Parsons, Newcastle upon Tyne, England.

35–4 Absorption

35–4.1[39]

Light from a carbon arc *a* illuminates an iris diaphragm *b* by means of a 7.5-cm-focal-length condenser *c* (Fig. 35–58). A 30-cm-focal-length achromatic lens *d* focuses the iris onto a screen *e*. A plane glass trough *f*, about 10 x 10 x 3 cm, filled with 250 cc of water and 25 cc of a concentrated hypo solution (sodium thiosulfate) is placed between the diaphragm and the lens. A 10 percent solution of HC1 is slowly dropped into the glass trough. The reaction liberates colloidal sulfur. The increasingly cloudy solution absorbs the blue light first; the screen image is first orange, then red. It may be necessary to shield the trough on all sides as the solution scatters a good deal of light.

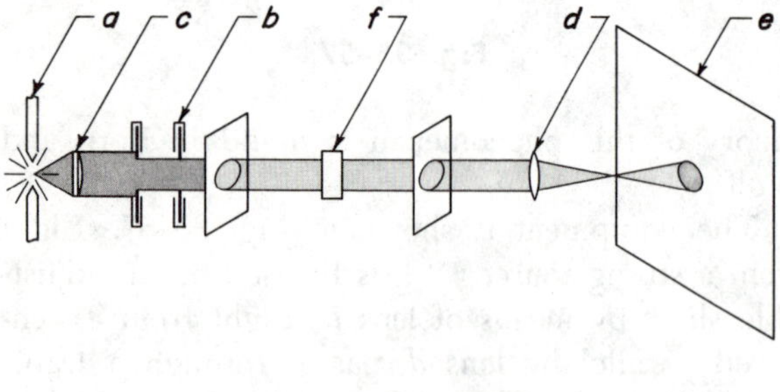

Fig. 35–58

35–4.2[40]

Collimated light from a carbon arc *c* (Fig. 35–59) illuminates a double iris diaphragm *c*, which consists of two circular openings one over the other, through a condenser *b* of 10-cm focal length. A 30-cm-focal-length achromatic lens *d* focuses both openings onto a screen *f*. The upper opening is covered with a red filter, the lower with a blue one. A large plane glass trough *e* (20 x 8 x 6 cm) is placed in the beams with the long sides parallel to the beams. It is filled with distilled water to which a few drops of an alcoholic solution of gum mastic are added. The liquid level must be high enough to cover both beams. The blue light will be scattered and will not reach the screen, while the red light will go through undisturbed.

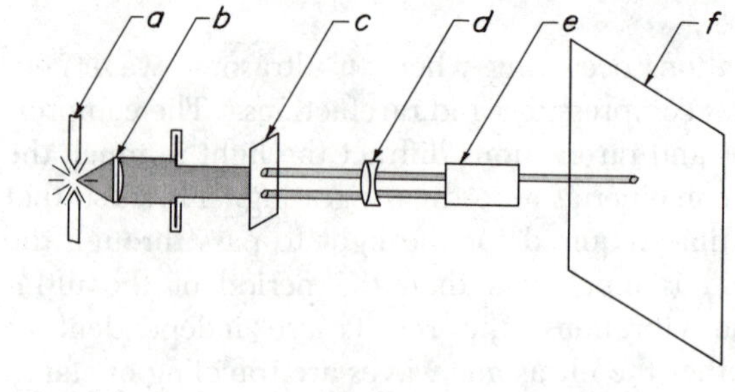

Fig. 35–59

35–4.3[41]

Any colored salt or solution of a colored salt can be used to demonstrate absorption in the visible spectrum. The simplest light source is a vertical slit *a* (Fig. 35–60) illuminated by the converging beam of light from condenser lens *b* of a carbon arc *c*. The image of the slit is focused on a screen *d* by a 30-cm-focal-length achromatic lens *e*. A large direct-vision prism[42] *f* is placed in the light beam issuing from the objective. The slit should be uniformly illuminated and as narrow as is consistent with a reasonably bright spectrum. A second projection system can be used to superimpose a bright scale on the spectrum. By this means, the position

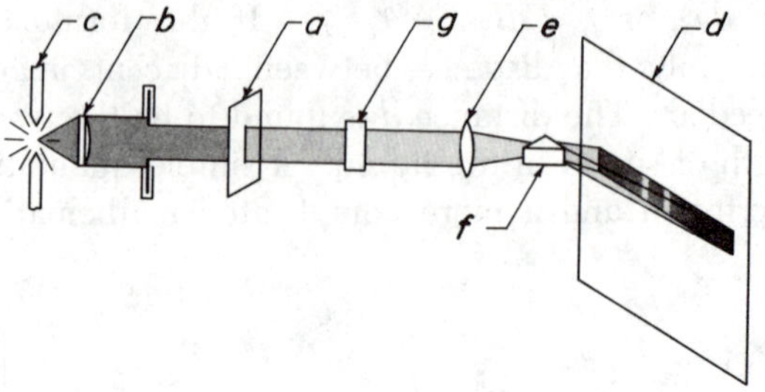

Fig. 35–60

[39] Hermann Daniel, *Demonstrations Versuche aus der Optik* (Musterschmidt-Verlag, Göttingen, 1960), 2nd ed., p. 128.

[40] Hermann Daniel, *Demonstrations Versuche aus der Optik* (Musterschmidt-Verlag, Göttingen, 1960), 2nd ed., p. 130.

[41] Hermann Daniel, *Demonstrations Versuche aus der Optik* (Musterschmidt-Verlag, Göttingen, 1960), 2nd ed., p. 124.

[42] Klinger Scientific Apparatus Company.

of the absorption edge can be determined. The solution should be placed in a plane-sided glass absorption cell g placed in front of the slit. Potassium permanganate shows characteristic absorption bands in the green region.

In operation, one places clear water in the cell and adds the colorant drop by drop until a suitable absorption is seen. A weak solution of permanganate differs very little in color from didymium glass, yet the absorption is significantly different, as can be demonstrated. Solutions of dyes such as cyanin (yellow-orange absorption) and fuchsin (wide-band absorption between red and violet) are interesting. A solution of chlorophyll prepared by infusing dried grass or leaves in alcohol or ether is often shown. Of interest is the absorption of fresh animal blood diluted with a .1 percent sodium carbonate solution (in the ratio 1 to 100) and well shaken with air. There are two absorption bands, one in the yellow and the other in the yellow green. The oxyhemoglobin in fresh blood may be reduced to hemoglobin by adding a small amount of ammonium sulfide. Both of the former absorption bands disappear and are replaced by a broad absorption in the green. Hemoglobin can be destroyed by adding a few drops of HC1 to a more concentrated blood solution. The latter turns brown. Two broad absorption bands appear in the green and an additional one in the red. If CO (illuminating gas) is bubbled through the weak sodium carbonate blood solution described above, the solution turns bright red and the absorption bands shift toward shorter wavelengths (carbon monoxide poisoning).

35–5 Absorption Anomalies and Dispersion

35–5.1[43]

Dispersion is called normal if the refractive index decreases smoothly and continuously as the wavelength increases. Colorless, transparent substances such as water or glass exhibit normal dispersion (in the visible spectrum). On the other hand, it is possible for the refractive index to decrease at first, then to rise rather abruptly over a short wavelength span, and then to decrease smoothly again. A substance for which this is true is said to exhibit anomalous dispersion. The rapid rise in index, which makes the dispersion anomalous, occurs in a band of wavelengths which the substance absorbs strongly. One such substance is an alcoholic solution of the dye fuchsin, and it absorbs strongly in the yellow green part of the spectrum.

The names "normal dispersion" and "anomalous dispersion" are retained only for historical reasons, for if the infrared and ultraviolet regions are included, then all substances show absorption bands and thus "anomalous dispersion." "Anomalous dispersion" of quartz, for instance, occurs at 8 μ, and may be used to separate long infrared radiation from the shorter wavelengths by Wood's method of focal isolation.

Perhaps the easiest demonstration of anomalous dispersion is based on the experiments of Christiansen (1870). A solution of ½ g of powdered fuchsin dye is dissolved in about 5 g (or 6¼ cc) of ethyl alcohol. If the solution is much stronger, it is too opaque.

The prism is readily made by clamping together two 1½-in.-sq pieces of plate glass about ⅛ in. thick, separated along one edge by a strip of brass 0.020–0.030 in. thick. The edges opposite the wedge should make contact. The glass plates are held together either with a large battery clip or a rubber band and the prism is filled, when needed, with a few drops of solution from a medicine dropper. Surface tension keeps the liquid in the prism. Paper towels should be kept handy, for the dye stains everything within reach.

The chief difficulty in the demonstration arises from chromatic aberration. Thus a small student spectrometer, although equipped with achromatic lenses, still has too much residual chromatic aberration (secondary spectrum). There are several ways to set up the experiment satisfactorily. One way

[43] C. Harvey Palmer, *Optics Experiments and Demonstrations* (The Johns Hopkins Press, Baltimore, Md., 1962), Experiment A7, pp. 33–35. See also J. Strong, *Procedures in Experimental Physics* (Prentice-Hall, Englewood Cliffs, N.J., 1939), pp. 97–98; R. W. Wood, *Physical Optics* (Macmillan Co., N.Y., 1934), 3rd ed., pp. 118–121, and T. Preston, *The Theory of Light* (Macmillan Co., N.Y., 1928), 5th ed., pp. 513–516.

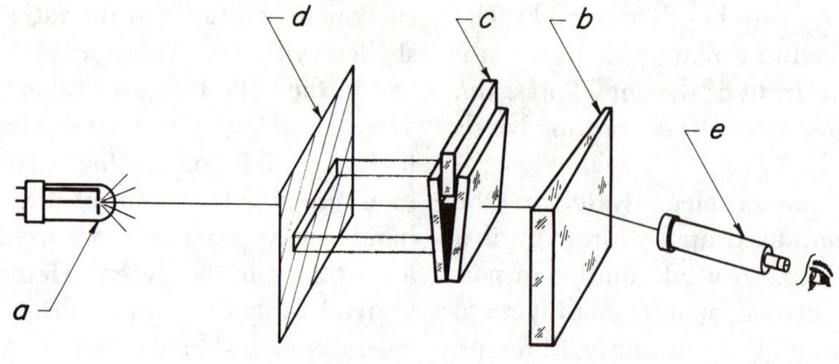

Fig. 35–61

is to use a high quality spectrometer (such as a Gaertner precision research instrument). Another less expensive way is to omit the collimator altogether.

Figure 35–61 shows how the demonstration may be set up without the use of a collimator. A point light source *a* is placed at the right height 3 or 4 ft away from a small student spectrometer of moderate quality. On the prism table, mount a thin glass prism *b* having a fairly small refracting angle—(20–30 deg), so that a short horizontal spectrum is obtained. It is important to be able to see the whole visible spectrum at the same time so that the red and blue can be compared. The fuchsin dye prism *c* should now be mounted just in front of the glass prism with its refracting edge horizontal. A diaphragm *d* with a horizontal slit, approximately ⅛ in. wide x 1 in. long, is placed just ahead of this prism to restrict the light rays to those passing through the dye prism, getting rays as close to the edge as possible, but blocking rays not passing through the dye prism. The spectrum now seen through the telescope *e* should consist of two sections—two lines which slant in such a way as to indicate that yellow is deviated more than red and red more than blue. The green is strongly absorbed and absent unless the dye film is ultra thin and the source particularly intense.

Figure 35–62 shows how the demonstration may be set up using a good spectrometer *f* and a mercury arc source *g*. If the arc source is used, it is possible to dispense with the small angle glass prism needed in the first arrangement. The collimator *h*, fuchsin prism *i*, telescope *j*, horizontal slit *k*, and a focusing lens *m* are arranged as shown. In this case, one sees only a few spectral lines, and the demonstration is perhaps somewhat less dramatic.

This experiment requires viewing through a telescope, and is appropriate as a corridor demonstration.

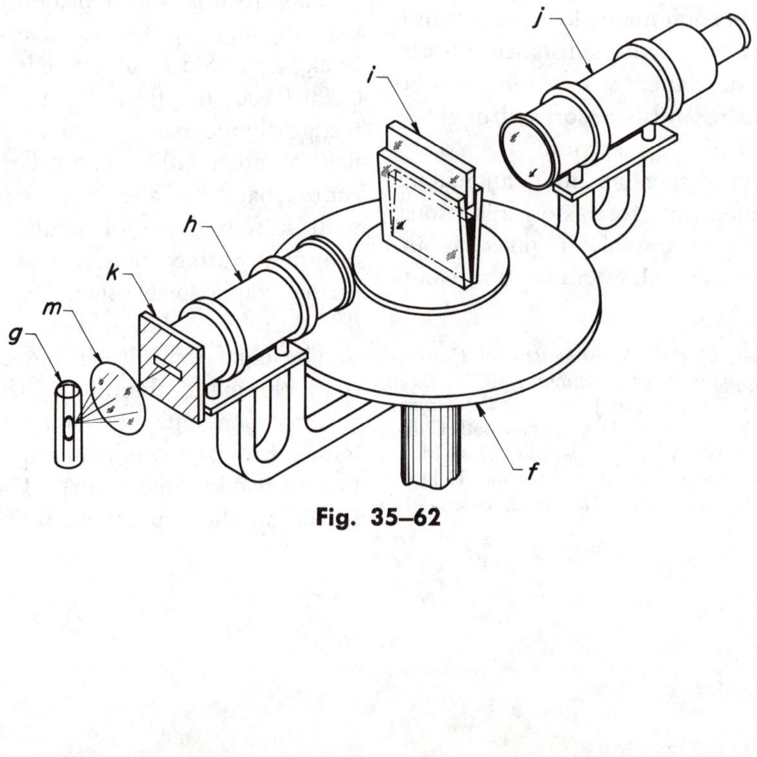

Fig. 35–62

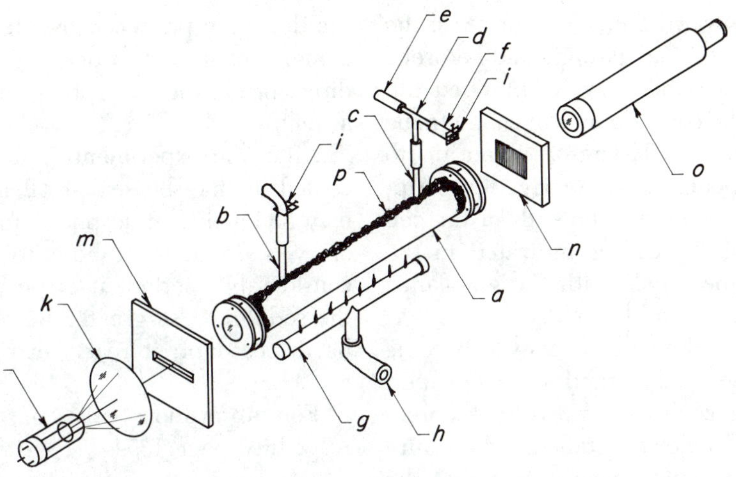

Fig. 35–63

35–5.2[44]

The fuchsin dye solution which was used in Demonstration 35–5.1 has, characteristically, a broad absorption band. Therefore, the anomalous dispersion curve is such that the whole visible spectrum needs to be observed to see the effect. Sodium vapor, on the other hand, being a low-pressure gas, has very narrow absorption lines (D lines) in the yellow whose two wavelengths differ by only one part in a thousand (5890 and 5896 Å). Thus, in this case, the refractive index differs very little from unity except in the yellow very near the absorption lines. The basic phenomenon, the anomalous rise in refractive index is, however, much the same in the two experiments.

The arrangement shown in Fig. 35–63 enables the observation of the anomalous dispersion of sodium. Basically it consists of an absorption-tube system, a heater, and optical apparatus. The absorption tube *a* has a vacuum-sealed window on each end and two 6-in. side tubes silver-soldered into the main tube. One of the side tubes *b* serves as a flushing tube leading through good rubber or plastic tubing (clamped when not in use) to a specified pressurizing gas. The other tube *c* is used in evacuation, and by means of a T-shaped piece *d*, one outlet *e* leads to a Hg U-tube manometer (reading from 1 mm to 1 atm) and the other outlet *f* leads to a Dry Ice trap and a small vacuum forepump capable of reducing the pressure in the cell to less than 1 mm. Use of a Dry Ice vapor trap between the absorption tube and the vacuum pump is highly desirable to prevent sodium vapor from passing through the pump. The absorption tube can be heated by a row of gas jets, as shown in Fig. 35–63. The heater *g* is made of a piece of pipe (e.g., ½-in. iron pipe) into which a row of $\frac{1}{16}$-in. holes are drilled about 1 in. apart. The length of the series of gas jets should be about 4 in. shorter than the absorption tube so that the ends of the cell may be cooled rather than heated. A fitting is provided to allow the heater pipe to be connected by rubber tubing *h* to the gas line. The ends of the heater tube must be capped. At all outlets using rubber or plastic tubing, use of glass stopcocks or good tubing clamps *i* may be desirable.

If the absorption tube is vacuum tight (see Appendix, page 1354), it is securely clamped into position with one window removed. About a dozen pea-size pieces of sodium metal are cut from a larger chunk submerged in a dish of kerosene. *Use protective goggles.* The pieces are removed with tweezers and the excess kerosene blotted off with *dry* paper towels (*dry hands*). These lumps are then rinsed in a beaker of gasoline (which vaporizes at a lower temperature than kerosene), again blotted as dry as possible, and put in the absorption cell. They may be distributed along the length of the cell using a rod to push them. The lumps should be kept several inches from either window. The missing window is now replaced and the cell evacuated. After pumping for about 15 min, turn on the

[44] C. Harvey Palmer, *Optics Experiments and Demonstrations* (The Johns Hopkins Press, Baltimore, Md., 1962), Experiment A8, pp. 36–41; see also R. W. Wood, *Physical Optics* (Macmillan, New York, 1934), 3rd ed., pp. 492–7.

gas flame and gently warm the cell to increase the vapor pressure of the residual gasoline and kerosene so that as much as possible may be pumped off.

After evacuating the cell, add hydrogen, methane (illuminating gas) or, as a last resort, even air, to give a pressure of about 3 or 4 cm Hg. Put dripping wet cotton wads *p* around the ends of the cell and along the top of the cell as indicated in the figure. Warm the tube slowly with the gas flame, keeping the cotton wet at all times.

If (unfortunately) the cell windows become fogged with kerosene or deposited sodium vapor, then the cell must be cooled, air pressure restored, and the extra set of windows installed. Make sure there are no wet cotton wads nearby and that the cell is thoroughly dry when changing the windows.

After the kerosene and gasoline have been removed in the above manner, the demonstration may be set up. Light from a point source *j* is collimated with a short focus achromat *k* and directed through the cell. A cardboard mask *m* with a horizontal slit should be used to restrict the beam to a horizontal sheet—otherwise, the nonuniformity of the sodium "prism" may spoil the observations.

The light emerging from the cell should be horizontally dispersed by the diffraction grating *n* and viewed through a telescope *o*. A student spectrometer with the collimator turned out of the way is convenient.

With a grating, several orders of spectra should be observable. The spectrum, at first, should be simply a long thin line. As the sodium melts and vaporizes, the sodium prism forms and the spectrum develops a break in the yellow corresponding to a sharp change in refractive index. The blue side of the spectrum bends in the direction corresponding to an index less than unity and the red side the other way corresponding to an index greater than unity.

To observe the deviation between the two D lines, high quality optical parts will be needed and considerable care exercised in setting up the demonstration.

The progressive changes in the density of the sodium "prism" resulting from the heating is evidenced by the gradually increasing deviations of the yellow region of the spectrum.

It is also of interest to turn the grating so that its rulings are horizontal rather than vertical. In this orientation, the grating adds to or subtracts from the deviation produced by the sodium vapor prism alone. For grating orders on the other side of the direct beam, the absorption lines may appear to be absent.

After the experiment is finished, and the tube cooled, it may be sealed off at low pressure or it may be filled to atmospheric pressure with nitrogen or even air. If the tube is to remain unused for a considerable period, it is perhaps best to remove the sodium. Use great caution if the sodium is to be disposed of; it reacts extremely violently with water!

For absorption cell construction details see Appendix, page 1354.

35–5.3[45]

A continuous spectrum is projected onto a screen by a direct vision prism in order to demonstrate anomalous dispersion. The vertical slit which is used should be as short as possible, so that the spectrum on the screen appears as a narrow band (two metal or cardboard strips may be attached at right angles to the slit or a diaphragm can be placed next to the slit). The flame of a gas burner, tinted by sodium vapor, serves as a deflecting prism, for the density of this vapor decreases rapidly from the bottom to the top of the flame. The gas burner is placed between the slit and the lantern and an iron spoon with metallic sodium is held in the flame. The spoon with the sodium must be held as close as possible to the bottom of the slit without blocking the slit.

When the sodium vapor in the flame becomes sufficiently thick, the dark absorption line on the screen widens and the edges of the spectrum adjoining it bend. The edge closer to the red end of the spectrum bends down and the edge closer to the violet end of the spectrum bends up, thus forming the characteristic curve of anomalous dispersion. It should be noted that these directions of spectrum bending are the exact opposite of what actually takes place, as the lantern objective produces a reversed image. The closer the spoon with the sodium is to the lower edge of the slit, the more distinct is the dispersion; i.e., the spectrum spreads

[45] Translated by K. V. Robinson and H. A. Robinson (Adelphi University) from *Lecture Demonstrations in Physics*, A. B. Mlodzeevskii, ed., M. V. Lomonosov State University, Moscow, U.S.S.R.

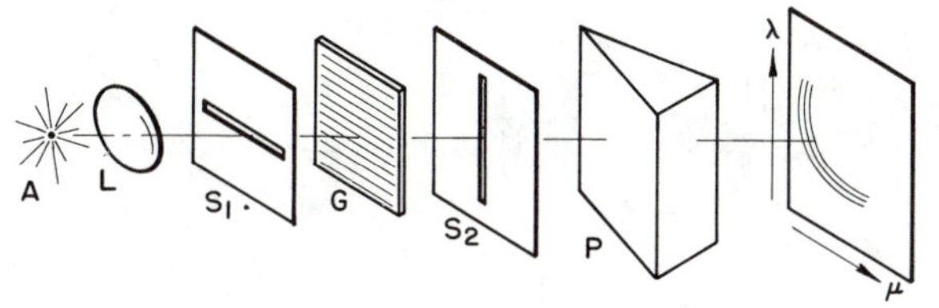

Fig. 35–64

out from the widening band of absorption, bending in opposite directions.

35–5.4

The dispersion curve of a prism can be shown in color by the arrangement shown in Fig. 35–64. Light from an arc source *A* passes through a lens *L* and a horizontal slit S_1, and is dispersed vertically by the transmission grating *G*. The vertical slit S_2 is so located that it is illuminated by a continuous spectrum varying from violet at the bottom to red at the top. The prism *P* deviates the colors in the inverse order of their wavelengths, thereby producing a dispersion curve in color on the screen.

35–5.5[46]

A carbon arc *a* (Fig. 35–65) with a 7.5-cm-focal-length condenser lens *b* illuminates a slit *c* which is

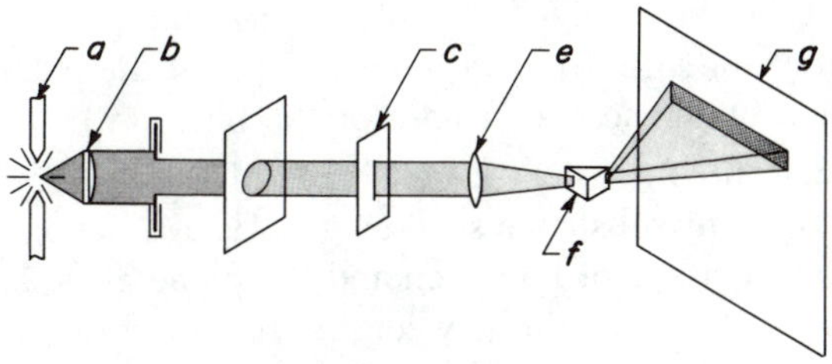

Fig. 35–65

focused on a screen *g* by a 30-cm-focal-length achromatic lens *e*. A 60-deg prism *f* is placed on a table so that it intercepts the light beam from the slit. The 60-deg prism is adjusted by rotation for minimum deviation. The beam will, of course, be bent away from its original direction. Constant deviation on other prisms can be demonstrated in the same simple setup and thus dispersion compared.

(46) Hermann Daniel, *Demonstration Versuche aus der Optik* (Musterschmidt-Verlag, Göttingen, West Germany, 1960), 2nd ed., pp. 90 and 100.

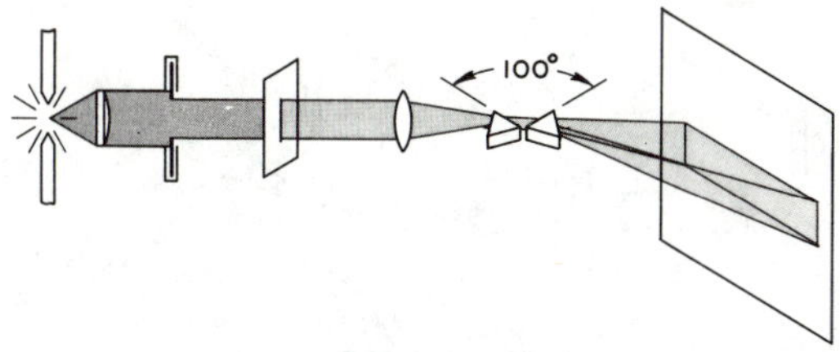

Fig. 35–66

It is also possible to obtain prisms in sets of different glasses[47] such that when put together enable an undeviated but dispersed beam to be projected.

The dispersion can be increased after setting the first prism for minimum deviation by laying an identical prism on the same table just after the first prism, so that the angle between neighboring faces of the two prisms is about 100 deg (see Fig. 35–66). If, however, one interposes the second identical prism in such a way that the dispersing edge is traversed by the beam in a reverse direction, it is possible to recombine the rays and reproduce the slit in white light. If the two prisms are separated by a small amount, it is possible to catch the light which has traversed the first prism and is reflected off one side of the second prism on to a white card or transparent screen and demonstrate that it is still actually colored before passing through the second prism (Fig. 35–67).

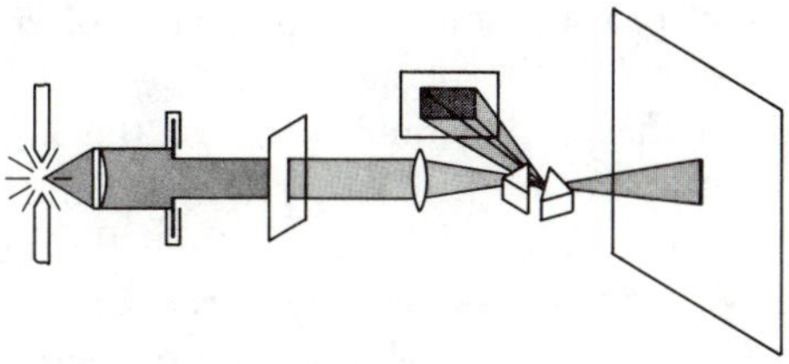

Fig. 35–67

Light dispersed through a prism can also be recombined in a number of other ways to reform white light as follows: (1) An iris diaphragm *i* (Fig. 35–68) is set close to the prism in such a way that the beam is intercepted. It will be noticed that the image is a light spot surrounded by colored edges. The size of the image can be adjusted by moving the condenser lens slightly. A second 30-cm lens, large enough in diameter to take the total area of the beam (including the color), focuses the

(47) Klinger Scientific Apparatus Corp.

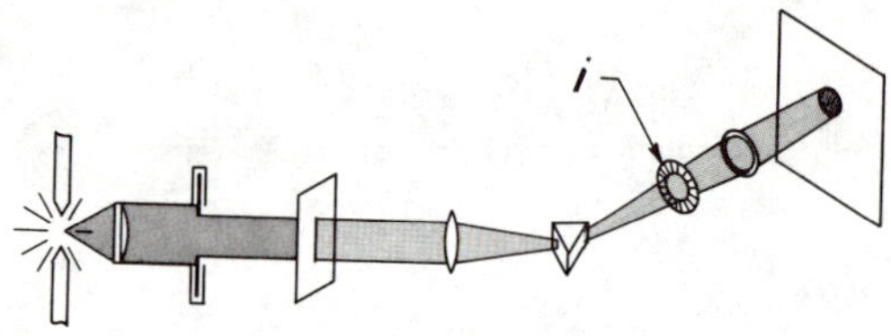

Fig. 35–68

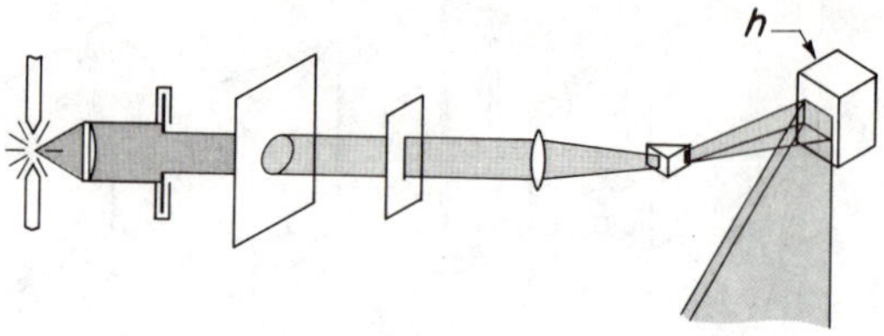

Fig. 35–69

iris diaphragm onto the screen. The final image will be white. A 30-cm concave mirror can be used in place of the last lens with the same result. (2) A somewhat more instructive method is to allow the dispersed light from the prism to fall onto a several sided rotatable mirror[48] h (Fig. 35–69) before imaging on the screen. Rotating the mirror by hand will cause the spectrum to spread out into a colorless white band.

To demonstrate that colored monochromatic light cannot be further dispersed by additional prisms, it is necessary to add further components to the simple setup as follows: The deviated dispersed light from the prism passes through an iris diaphragm closed down far enough to cut out the colored edges of the light beam (see (1) above). A 30-cm achromatic lens focuses the colorless light onto a slit mounted in such a manner that its position is adjustable to the right and the left. The spectrum appears on the slit, and by moving the slit (right or left) particular monochromatic beams may be selected. This slit is now focused on the screen with a third 30-cm-focal-length lens, giving a sharp monochromatic slit image. A second prism set in the beam at minimum deviation will give rise to further deviation only and no further dispersion.

35–6 Polarization

35–6.1

Light is a transverse wave, that is, in a vacuum the electric field vector is perpendicular to the direction of propagation. The plane of polarization can be defined as the plane containing the electric vector and the propagation vector of the light. If there is any way of predicting the orientation of this plane as a function of time, then the light is said to be polarized. If no such prediction is possible, so that the plane of polarization varies randomly during times comparable to the period of the light (or of the time of observation), then the light is said to be unpolarized. When the orientation of the plane of polarization remains constant for the light arriving at some point, then the light is said to be plane polarized. If this plane rotates once per period of the light, then the light is either circularly or elliptically polarized, depending on whether or not the magnitude of the electric vector is a function of the orientation of the plane of polarization.

[48] Catalog No. 41 521a or 50 496a, LaPine Scientific Company.

There are several practical methods of producing plane polarized light:

1. The vibrating charges which give rise to the light can be confined to a plane (e.g., the polarization of the light from a dipole antenna of the polarization of the scattered light of the blue sky).

2. The intensity of reflection of light from a transparent substance such as glass is a function of both the angle of reflection and the plane of polarization. If the angle between the refracted ray and the incident ray is 90 deg, then the reflected ray is plane polarized perpendicular to the plane of incident. The angle for which this occurs, called Brewster's angle, is an angle of incidence, θ, such that $\tan \theta = n$, where n is the index of refraction. One can make use of the polarized light at Brewster's angle to find the polarizing axis of a polarizer such as Polaroid sheet, or a Nicol Prism.

3. In many crystals, the index of refraction depends on the orientation of the plane of polarization with respect to the crystal axes. There are several specially constructed combinations of crystals which transmit only one plane of polarization.

In the Nicol Prism one plane of polarization is transmitted and the perpendicular plane of polarization is totally internally reflected at an interface between two crystals that has been cut at a carefully selected angle.

4. Some chemical compounds have molecules that absorb one plane of polarization more than another. Such substances are called *dichroic* compounds. If the molecules of such a substance can be laid down in a sheet with all the molecules having the same orientation, then this sheet will transmit only one plane or polarization. "Polaroid" sheet is made up of such oriented dichroic molecules.

A plane polarized vibration can theoretically be constructed in a variety of ways. It can be made by two in-phase vibrations at right angles to each other. The resultant vibration will be plane polarized. Alternatively a plane polarized wave can be constructed from two circularly polarized waves whose plane of polarization is rotating in opposite directions. Thus, although one experiment may lead to the conclusion that the light that passes through a sheet of Polaroid is in a unique state of polarization, a different experiment on the same beam of light may indicate that it consists of the superposition of two different states of polarization.

When plane polarized light passes through a solution containing complex asymmetric molecules, its behavior can be explained by considering the plane polarized beam to be constructed from two circularly polarized beams. The clockwise polarization travels at a different velocity through the solution than the counterclockwise polarization. Therefore, the phase relationship between the two circularly polarized beams changes as the light moves through the solution. When the emergent light is analyzed by a Polaroid disk, the plane of polarization will not be the same as that of the light entering the solution. The plane will have been rotated either to the right or the left, depending on which sense of rotation has the higher velocity in the solution.

In many crystals, the velocity of light will be different for two selected planes of polarization; in particular the planes parallel to the optic axis and plane perpendicular to the optic axis. If a thin section of a crystal is cut with its optic axis parallel to the surface, it can affect the plane of polarization of the incident light.

35–6.2

A horizontal beam of plane polarized light (horizontally oriented) is projected onto a piece of window glass so as to reflect from the glass to the ceiling of the room. As the glass is tilted to move the reflection across the ceiling, at one point, the oncoming light will strike the glass at Brewster's angle (approximately 57 deg for glass), and the reflection will disappear as virtually all of the incident light is transmitted through the glass.

35–6.3

Two pieces of plate glass *A* and *B* are placed parallel to one another and so arranged that a beam of light from a source *S* is incident on *A* at Brewster's angle (Fig. 35–70). The beam is again reflected at the same angle by the second glass plate *B*. As *B* is revolved around a vertical axis, the intensity of the reflected beam decreases and finally disappears every 90 deg. If the entire apparatus is enclosed in a glass box with only the

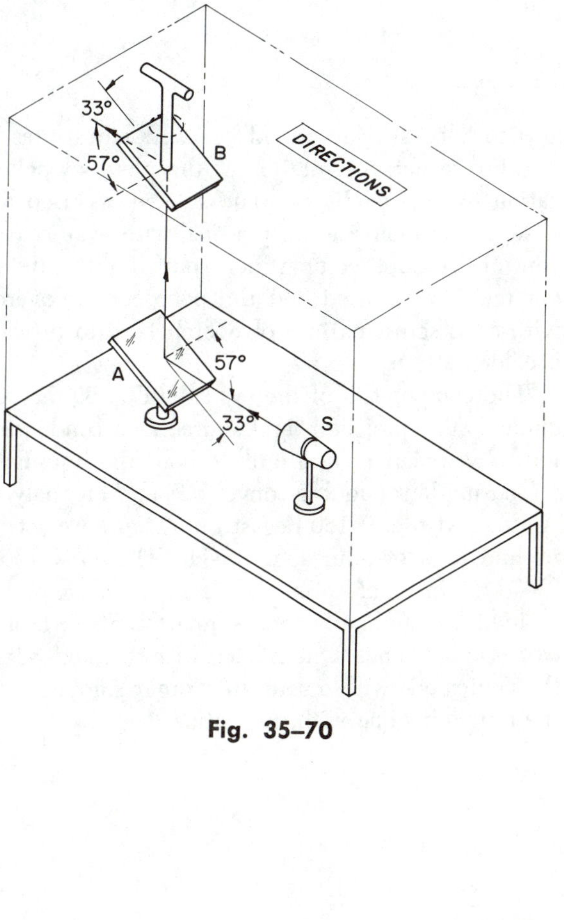

Fig. 35–70

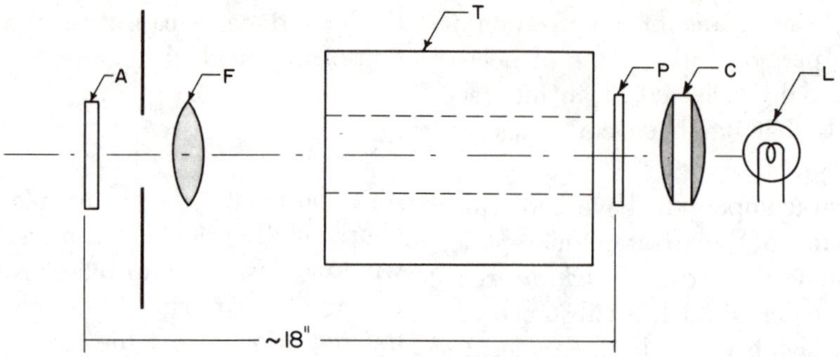

Fig. 35–71

handle for turning the glass plate protruding as shown in the figure, the device makes an excellent corridor demonstration.

35–6.4

A fixed and a rotatable disk of Polaroid film are mounted on a common axis in a wooden frame. With a desk lamp behind the turning disk and a double refracting material between the two Polaroids, the polarizing effect is demonstrated by a color change every 90 deg of rotation. The rotatable disk is driven at constant speed by a continuous duty dc motor. Suitable double refracting materials are Scotch tape on a piece of acrylic plastic or cellophane.

35–6.5

The apparatus described below demonstrates to a lecture group the rotation of the plane of polarization by two mediums. The substances used are a sugar solution and turpentine with water as a control. A three compartment tank for the different mediums is used and the polarizer is covered with a crisscross pattern of Scotch tape to provide a color pattern.

The components of the system (Fig. 35–71) include a slide projector light source *L*, a condensing lens *C*, a polarizer *P*, a tank to hold the liquids *T*, a focusing lens (double-convex) *F*, and an analyzer *A* attached to a 0–180 deg scale. These are set up in line as shown in Fig. 35–71. The tank (Fig. 35–72) is made of ⅛ in. clear acrylic plastic and is divided longitudinally into compartments a, b, and c of equal volume. One is filled with distilled water, the center one with a saturated sugar solution, and the remaining one with turpentine.

To take quantitative measurements, the tank is moved into a position such that the light beam passes longitudinally through the water compartment to focus on a screen and to establish a scale zero point. Then the sugar solution is moved into this position and the analyzer is rotated to match the colors on the screen. The turpentine is treated similarly.

35–6.6

A glass tube *a*, 1½ in. diam x 4 ft long, is sealed at one end with a flat glass plate and mounted vertically. It is filled with a sugar solution, which is made by combining three parts by weight of sugar with two parts by weight of water, boiling the mixture and then filtering it while hot. Allow the solution to stand at least 24 hours before using. The tube is illuminated from below with polarized light by means of an arc lamp *b*, prism *c*, and

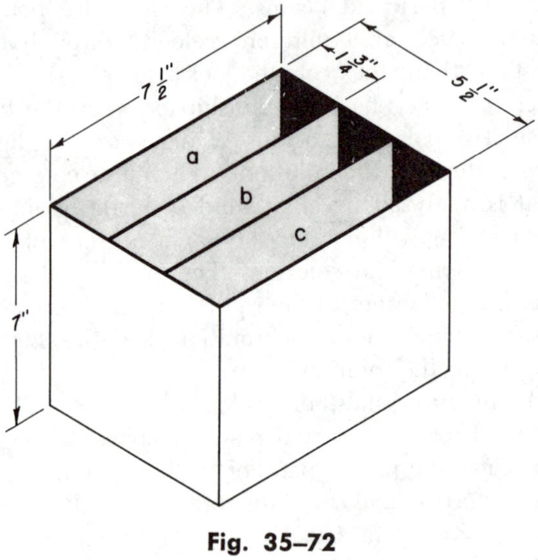

Fig. 35–72

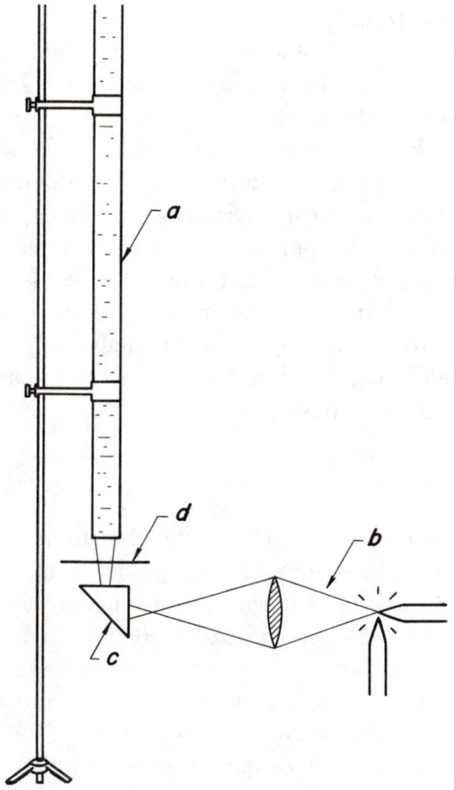

Fig. 35–73

Polaroid filter *d,* as shown in Fig. 35–73. A helix of light is seen within the tube due to the circular polarization effect of the sugar solution. The helix is turned when the polarizer is rotated. Toward the top of the tube, colors are seen within the helix due to the fact that the polarization effect depends on the wavelength. (Absinthe can be used as a diffusive element.)

35–6.7[49]

A cylindrical container about 4 in. in diam and about 3½ ft high, made from a glass or Plexiglas tube with a bottom cemented in, is used to demonstrate scattering of light in a cloudy medium. The cementing should be strong to withstand the weight of the water filling the container. There are several methods of preparing the cloudy emulsion. One of the best methods consists of adding a rosin solution in alcohol to water. The solution is prepared beforehand and is added drop by drop to the column of distilled water directly before the demonstration.

(49) Translated by K. V. Robinson and H. A. Robinson (Adelphi University) from *Lecture Demonstrations in Physics,* A. B. Mlodzeevskii, ed., M. V. Lomonosov State University, Moscow, U.S.S.R.

The cylindrical container is placed on a table or bench over a circular hole whose diameter is the same as the diameter of the cylinder. A mirror is placed under the table to direct the beam of light from the projector upward into the cylinder. A parallel beam of light is obtained in the cylinder by placing a lens having a long focal length under the bottom of the container. By changing the distance to the lantern and by placing cardboard diaphragms of different diameters under the cylinder, one can select an optimum width of the light beam so that the whole container will glow evenly.

The intensity of the scattered light is inversely proportional to the fourth power of the wave length. Hence the container seems to be tinted a deep blue. The upper portion is tinted yellow or orange, because here the basic hue is determined not by the scattered light but by the light rays emerging from the emulsion below. A reddish spot is projected on the ceiling by the transmitted light.

The experiment may be changed slightly by placing a Polaroid *between* the mirror and the bottom of the container. A parallel beam of plane-polarized light is then directed into the container. The width of the light beam is selected as explained in the preceding experiment. When the polarizer is rotated, the polarization plane of the beam changes and the container will appear to the audience alternately light and dark. When the polarizer is stationary, the illumination of the container will appear different from different points in the auditorium.

If a Porro prism or mirror is used in this version, particular care must be taken that the general picture is not complicated because of the polarized light reflected from the face of the prism (mirror).

35–6.8[50]

The Tyndall experiment requires a cube-shaped glass container with a metal casing. The glass cube may be made of plate glass held together with an appropriate cement; the glass should be roughened

(50) Translated by K. V. Robinson and H. A. Robinson (Adelphi University) from *Lecture Demonstrations in Physics,* A. B. Mlodzeevskii, ed., M. V. Lomonosov State University, Moscow, U.S.S.R.

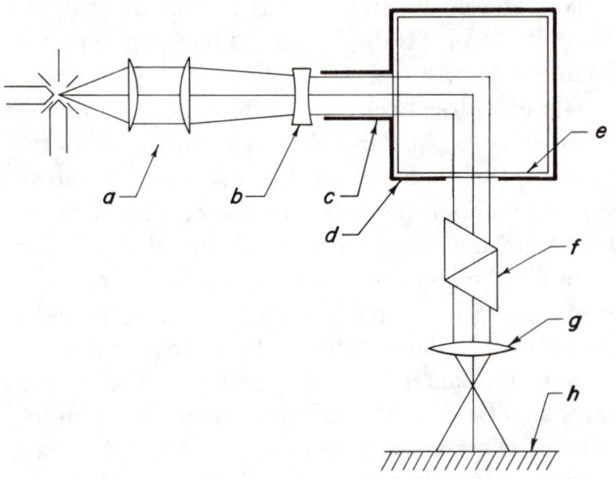

Fig. 35–74

with fine emery at the place of junction. The shape of the metal casing *d* is shown in Fig. 35–74; it is a cube open at the top and having circular openings centered in two of the mutually perpendicular sides. A metal or cardboard tube or shield *c* is attached to one opening. The cloudy medium is prepared by dissolving rosin in alcohol and adding this solution drop by drop to water until the latter appears slightly milky. The solution of rosin and alcohol should be prepared in advance but should be added to the water just before the demonstration.

The beam of light from the condenser lens *a* of the arc lamp passes through the diverging lens *b* and falls on the glass cube *e* through the tube *c*. Extraneous light should be removed and may be screened between the lantern and the end of the tube by a camera bellows or tube of black paper.

The light is scattered by the cloudy medium and emerges from the second opening. The light is projected by an objective lens *g* to get maximum illumination of the screen *h*, which has been set to one side of the container. If the light is not sufficiently bright on the screen, the screen may be replaced by a mirror turned towards the audience and placed as near as possible to the apparatus. A Nicol or Polaroid analyzer *f* is then placed between the container and the objective lens. The scattered light emerging from the second opening is polarized. Hence, when the analyzer is rotated, the field of vision on the screen is observed to darken at two positions.

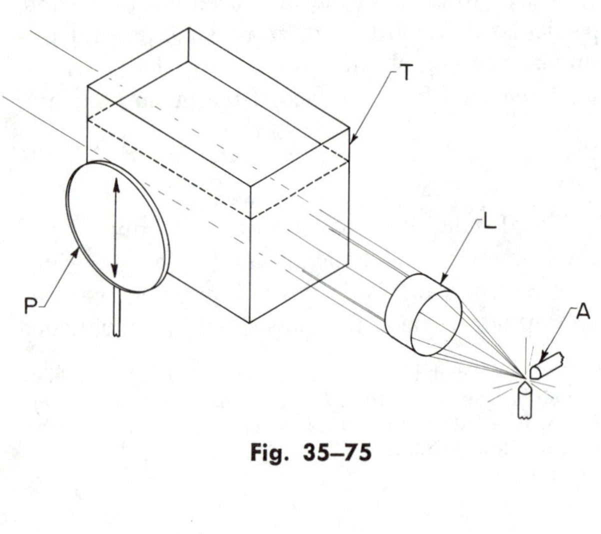

Fig. 35–75

35–6.9

The polarization of the light from an unpolarized beam that is scattered by a suspension of fine particles can be shown with the arrangement of Figs. 35–75 and 35–76. The scattered light is polarized with the polarization vector perpendicular to the axis of the incident beam. The light from an arc, or other strong source, *A* is collimated by the lens *L*. The plastic tank *T* contains a milky solution of soap and water. The light scattered to the side is analyzed with a large Polaroid sheet *P* and found to be polarized at right angles to the incident beam. The light scattered upward can be observed with a mirror inclined at 45 deg with the horizontal.

35–6.10[51]

Depolarization of light by diffuse reflection from a rough surface can be demonstrated by means of the setup shown in Fig. 35–77. The fresh surface of a piece of chalk, a chalk plate, or a magnesium oxide surface[52] is used as a reflecting surface. The chalk surface must not be handled and should be sanded with fine grade sandpaper before every demonstration. A beam of light from the projector lantern *a* passes through a diaphragm *b* and a polarizer *c*. It is then directed by the objective lens *d*, which has a short focal length, to the chalk surface *e*, set at a 50 deg angle to the incident ray. By moving the objective *d* and changing slightly

(51) Translated by K. V. Robinson and H. A. Robinson (Adelphi University) from *Lecture Demonstrations in Physics*, A. B. Mlodzeevskii, ed., M. V. Lomonosov State University, Moscow, U.S.S.R.

(52) Supplied with GE–Hardy Spectrophotometer.

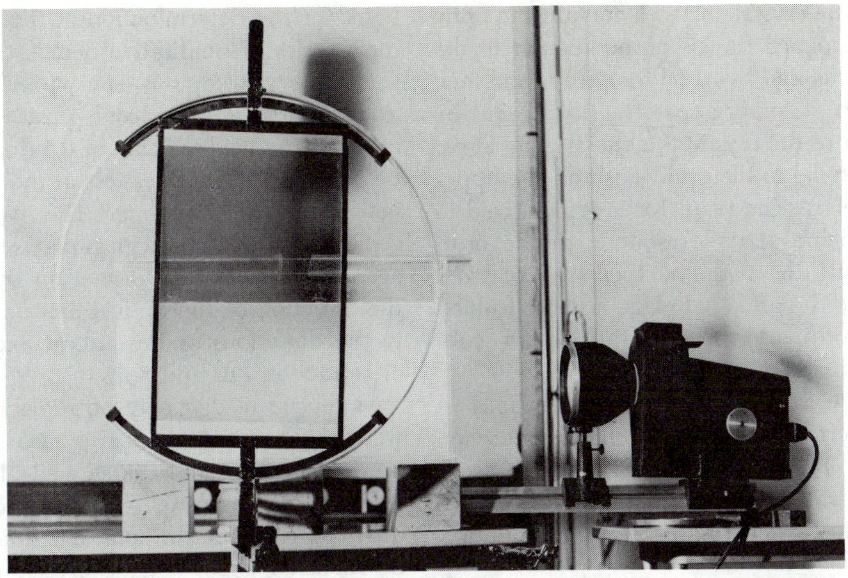

Fig. 35–76

the inclination of the plate, the image of the diaphragm aperture is focused on the chalk surface. Objective lens *g* (a biconvex lens), with long focal length, projects this image onto screen *h* set to one side. An analyzer *f* is placed between the objective *g* and the chalk surface. When the analyzer is rotated, it can be shown that due to the depolarization accompanying diffuse scattering, the light spot fails to disappear regardless of the positions of the analyzer and polarizer.

35–6.11

A good but little known polarized light experiment verifies the existence and form of the Huygens wave surfaces for a doubly refracting uniaxial crystal and demonstrates polarization of both the ordinary and extraordinary waves by the crystal.[53] Since the crystal surfaces must be viewed with a microscope, this experiment is appropriate for only small groups.[54] Details of the experiment can be found in the reference.

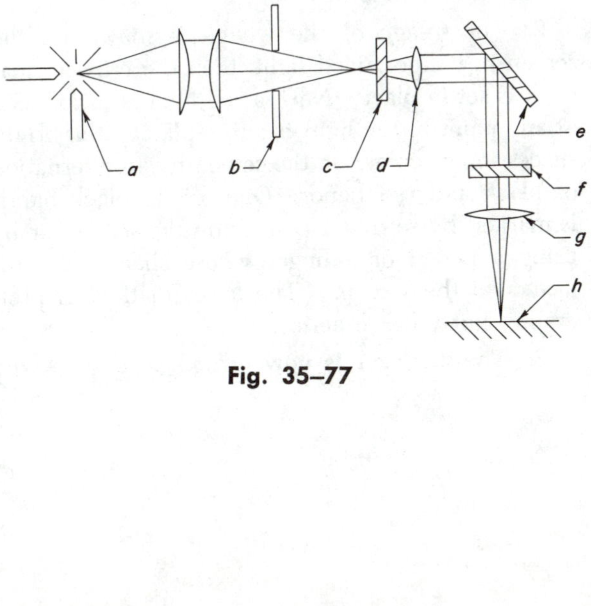

Fig. 35–77

A pinhole source of light, three plane, parallel slabs of calcite crystal, and a low power microscope are needed. The crystals are cut with their surfaces: (1) perpendicular to the optic axis, (2) parallel to the optic axis, and (3) along the cleavage faces. The illuminated pinhole is placed in contact with the lower surface of the crystal, and the image of the pinhole is focused on the cross hairs of the microscope. The pinhole acts as a point source from which waves spread as in the Huygens construction. The various images produced by the cut surfaces are observed:

CASE 1. When the crystal faces are cut perpendicular to the optic axis so that the pencil of rays travels along the axis, two positions of the microscope give distinct images of the source. The separation of these positions is about 15 percent of the thickness of the crystal. Both images are unpolarized and essentially achromatic.

[53] W. R. Wright and M. W. Garrett, *J. Opt. Soc. Am.*, **26**, 360 (1936). The editor wishes to thank Prof. Wehr for calling attention to the article by W. R. Wright and M. W. Garrett. The brief summary which follows was prepared from this article by the editor.

[54] For projection techniques, see Dem. 26–5.17, Microprojector.

CASE 2. If the crystal surfaces contain the optic axis so that the pencil travels perpendicular to the axis, a point focus and two line foci are found, one lying below and the other above the point focus by about 7 percent of the crystal thickness. The lower line focus is parallel to the optic axis and the upper one perpendicular. The point focus is polarized in the principal plane and corresponds to the ordinary wave, while the two line foci are polarized perpendicular to the principal plane and are formed from the extraordinary wave. All three are substantially achromatic.

Case 1 and Case 2 comprise the directions in which double refraction appears to be absent when judged by eye. It is often assumed that in Case 1 there is no separation into ordinary and extraordinary waves. This is not true if the pencil subtends a finite angle.

CASE 3. The general case, in which the faces of the crystal are cleavage faces, is more complicated. There is now double refraction in the usual sense, i.e., a separation of the images can be detected by the eye. With a microscope, three images are again formed, but the line foci are displaced laterally from the point focus in the principal plane. The lower line focus lies in the principal plane and the upper one perpendicular to it. When white light is used, both line foci are colored, since the displacement in the principal plane is accompanied by dispersion. The upper line focus in this case has the appearance of a small prismatic spectrum since the dispersion is transverse to the line images. A reasonably sharp line spectrum is produced with a source such as a mercury arc.

35–6.12[55]

The interference colors of a crystalline plate in polarized light depend on the path difference, that is, on the thickness of the plate and on the wavelength of the light. This can be observed particularly well in a quartz wedge.[56] Wedges may also be used to determine the thickness of crystalline plates by observing their color in polarized light, for the determination of the crystal sign, and for a series of qualitative demonstrations.

A quartz wedge is cut parallel to the optical axis (quartz is a uniaxial, positive crystal). The angle of the wedge is about 0.5 deg, its length may be 4–5 cm, and its thickness at the wide end should not exceed 0.2–0.3 mm. The wedge should be cemented between two glass plates to prevent breakage. Arrows may be drawn on the glass showing the direction of the optical axis and corresponding to the directions of the largest and smallest index of refraction (n_g and n_p).

A quartz wedge may be replaced by a gypsum wedge. These are easier to make; however, they are less durable and under white light the succession of colors is usually not as smooth. In certain qualitative demonstrations the quartz wedge may be replaced by one made of cellophane, but here again the color succession is not smooth. Sheets of cellophane are stacked in a step formation and arranged to produce a succession of colors, corresponding to the color succession of a quartz wedge.

For the demonstration, the quartz wedge is placed at f between lenses e and g in a parallel beam of light from the projection lantern a and lens c (Fig. 35–78). In monochromatic polarized light the wedge seems to be covered by a series of approximately equidistant light and dark bands. With crossed polarizers d the wedge should be placed in a diagonal position for the greatest color intensity. Dark bands coincide with path differences equal to a whole number of waves. The experiment is carried out as follows:

1. The projection lantern is turned on and adjusted to produce the most even and bright illumination of the screen h. The polarizer may be left in for the entire experiment since it does not change the illumination to any degree.

2. The image of the wedge is shown on the screen in nonpolarized light, then the crossed analyzer is set in place. Next, a red filter is introduced at any point in the light beam, replacing the bright succession of colors on the screen by an alternation of black and red bands. One of the black bands is marked by setting a pointer on the screen, or by using a pencil or a finger whose shadow is projected on the screen. (The latter method is preferable, but takes practice.)

3. The analyzer is now turned 90 deg. A red

[55] Translated by K. V. Robinson and H. A. Robinson (Adelphi University) from *Lecture Demonstrations in Physics*, A. B. Mlodzeevskii, ed., M. V. Lomonosov State University, Moscow, U.S.S.R.

[56] Quartz wedges are available from Bausch & Lomb, Inc.

band appears at the place the black band had been marked, because a 90-deg turn of the analyzer corresponds to a change of π in the phase difference, or a path difference of half a wave.

4. Without moving the pointer from the original position, replace the red filter with a green filter. The position of the black bands changes because the path differences depend not only on the thickness of the wedge but on the wave length. As done for red light, demonstrate the change in the light and dark bands when the analyzer is turned 90 deg.

5. If the red and green filters have evenly cut edges, both filters may be arranged so that the boundary between them cuts the wedge lengthwise, i.e., so that the wedge will be simultaneously in both green and red light. It can be seen that the black bands in the two colors do not coincide and that the bands are closer together in the green section than in the red.

6. Rotate the wedge and note its fourfold extinction over 90 deg. It is advisable to mark the ends of the wedge in some manner.

7. Remove the filters and show the wedge in white light. Mark one of the bands by some means and then turn the analyzer 90 deg, showing the change of color to the complementary one. For example, the band which is pink when the Nicols are crossed becomes green when they are parallel; a yellow band becomes blue. Call the attention of the audience to the fact that when the Nicols are crossed, the field around the wedge is black, and when the Nicols are parallel, the field is light.

8. With the wedge still in white light, turn the analyzer 45 deg. Note that the colors of the wedge gradually grow dimmer and turn white.

35–6.13[57]

A spectral analysis of the colors of a quartz wedge may be shown on the setup in Fig. 35–78. Place a slit 3–5 mm wide at *f* in the parallel beam of light and adjust the light source *a* so that the slit is evenly illuminated. Clamp the quartz wedge directly to the slit in a vertical position. The polarizer should be rotated so that the direction of its oscillations is at a 45-deg angle to the direction of oscillation in the wedge. The analyzer is set in front of the objective g and crossed with the polarizer. Using the objective, project the image of the slit onto a screen *h;* the image will be tinted by the successive colors of the wedge. Set a direct vision prism directly behind the analyzer (and in front of the objective *g*) to produce a horizontal spectrum. The continuous spectrum which appears will be cut by a series of curved black bands.

To explain the composition of the spectral colors, the demonstrator determines with the audience what color should be produced at a given horizontal cross section when the colors are combined at any given spot of the spectrum crossed by the absorption bands. It is best to determine in advance the spots of the most simple combinations of colors. Having determined the color which should be produced at a given spot, remove the prism without removing the pointer or marker, and note that the color of the image of the slit in each spot is the one expected.

[57] Translated by K. V. Robinson and H. A. Robinson (Adelphi University) from *Lecture Demonstrations in Physics,* A. B. Mlodzeevskii, ed., M. V. Lomonosov State University, Moscow, U.S.S.R.

35–6.14[58]

To show interference effects of polarized light in a plane parallel crystalline plate, one should have a number of plates of different thicknesses. These should be cut parallel to the optical axis (in the case of a uniaxial crystal) or parallel to the plane of the optical axis (in the case of a biaxial crystal). Gypsum or mica plates, split off along the cleavage planes, are particularly appropriate. The plates are mounted in rotating frames and are placed at *f* in the parallel beam of light in the setup shown in Fig. 35–78. Focusing of the image of the plate edges is done in nonpolarized light.

1. Crossing the Nicols *d* results in total darkness on the screen *h.* A crystalline plate is then set at *f* and a bright interference color appears in the field of vision.

2. The analyzer is turned 90 deg, and the color of the plate changes to the complementary one. Colors disappear with a 45-deg turn.

[58] Translated by K. V. Robinson and H. A. Robinson (Adelphi University) from *Lecture Demonstrations in Physics,* A. B. Mlodzeevskii, ed., M. V. Lomonosov State University, Moscow, U.S.S.R.

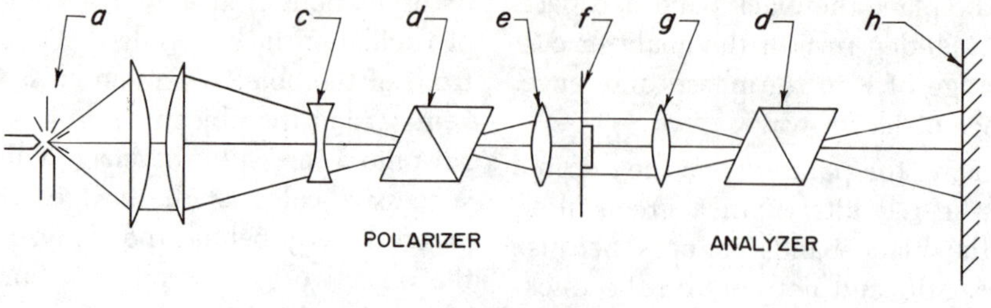

Fig. 35–78

3. The plate is turned through a full circle between crossed Nicols and the fourfold extinction alternating with positions of maximum intensity is noted.

4. Experiments 1, 2, and 3 are repeated, using plates of different thickness (different colors). These demonstrations are particularly handsome if the plate used is not uniformly thick. Gypsum plates, roughly split off along the cleavage planes, and placed between crossed Nicols, will give an extremely beautiful pattern of colors corresponding to the different thicknesses of the plate.

5. It is interesting to use as an analyzer a crystal of Iceland spar or a birefringent polarizing prism. Two separate images should be visible on the screen, the colors being complementary (due to the fact that the directions of the oscillations are mutually perpendicular). When the crystal is rotated, one of these images (the extraordinary ray) passes around the second one (the ordinary ray) and the images change color. If the crystal is such that the two images partly cover each other, the common field remains white all the time.

6. A crystalline plate is set between crossed Nicols in the position giving the maximum color intensity. The plate is then slowly tipped in relation to the beam. Due to a change in the path difference of the beam, the color of the plate gradually changes and will reverse if the plate is tipped in the other direction. This step of the experiment should be shown when preparing to demonstrate the effects of interference in a converging beam of polarized light. It illustrates the change in the path difference when the angle between the light beam and the crystalline plate is changed.

7. Particularly effective demonstrations can be done with mosaics made from sheets of mica, gypsum, and ordinary cellophane of different thicknesses. For example three rhombic plates of gypsum or mica having different thicknesses may be glued together; in polarized light they appear as a cube having different colored faces. The color of the faces changes when the analyzer is rotated.

Glowing etchings are also extremely beautiful in polarized light. They are produced by scraping away or by partially dissolving parts of the transparent gypsum plates to form a picture. Due to the different thicknesses of the birefringent material, the various parts of the plate take on different colors in polarized light. Again, the colors will change when the analyzer is rotated.

35–6.15[59]

This experiment serves as a graphic demonstration of the function of a half-wave plate. The plate should not have a frame or else should be cemented between two glass plates in such a manner that its edge will separate the field on the screen *h* (Fig. 35–78) in half. The edge of the plate should be cut parallel to the direction of the oscillations which pass through it.

A quartz wedge is placed at *f* between a crossed polarizer and analyzer *d* and projected on the screen *h*. A filter is placed in the light path and the half-wave plate is mounted in contact with the wedge so that it covers half the width of the wedge. The light bands which then appear on the image of the uncovered half of the wedge will be directly across from the dark bands appearing on the image of the covered half. When the analyzer is turned from a crossed to a parallel position, the light and

[59] Translated by K. V. Robinson and H. A. Robinson (Adelphi University) from *Lecture Demonstrations in Physics,* A. B. Mlodzeevskii, ed., M. V. Lomonosov State University, Moscow, U.S.S.R.

dark bands from both sides of the wedge will change places.

A thin sheet of cellophane may be used as a half-wave plate if the edge of the cellophane is cut parallel to the principal cross section of the sheet. The cross section may be checked by examining the sheet between a crossed analyzer and polarizer. It is easy to choose a sheet which will produce the effect described above under either a red or a green filter. When cellophane is used, it can be shown graphically that the half-wave plate functions only for a determined wavelength. That is, if the dark and light bands from the two halves of the wedge coincide more or less accurately with one filter, then with another filter the correspondence will not hold.

If cellophane is folded in half along a line parallel to its principal cross section, the resulting double sheet will have a path difference of one wavelength. By setting such a plate against the wedge so that the folded line divides the width of the wedge in half, it can be shown that the light and dark bands from both halves of the wedge are continuations of one another. A change of filters in this case produces an even greater discrepancy of the bands than was the case with the half-wave plates. (This fact may be put to use in choosing a suitable sheet of cellophane.)

35–6.16[60]

Quarter-wave and half-wave plates are usually matched to the wavelength of sodium light. Therefore it is best to show this demonstration in monochromatic light, either yellow or close to yellow. A heavy yellow filter is quite satisfactory. The position of the yellow filter is relatively unimportant. The plates are attached to rotating frames and are inserted at f in the beam of parallel light shown in Fig. 35–78.

The plates are sometimes made in the form of squares with the sides parallel to the direction of the oscillations. This is convenient, for if the squares are completely visible on the screen, the rotation of the plates may be observed directly. If the edges of the plate are not seen on the screen h, the plate must have an arrow attached to it. Likewise, the analyzer should have a pointer for all experiments with these plates.

The demonstration of a half-wave plate is as follows:

1. The analyzer is first set for extinction, then the yellow filter is inserted and the half-wave plate is placed in the beam at f in a diagonal position which produces the greatest illumination on the screen h. Extinction occurs when the analyzer is then turned at right angles.

2. The plate is removed and the analyzer is once again set for darkness. The plate is inserted and set at the position of extinction (its principal cross section either parallel or perpendicular to the polarizer oscillations). Rotate the plate to a point considerably less than 45 deg, so that a noticeable lightening is apparent. It is then possible to show that the analyzer must be turned at an angle twice that of the plate rotation if darkness is to be produced; this is due to the fact that the oscillations entering the plate and leaving it are symmetrical in relation to its principal cross section.

To demonstrate a quarter-wave plate, place the analyzer and polarizer in crossed positions, adjust the yellow filter, and then set the quarter-wave plate in position at f.

1. Turn the plate any given angle (except 45 deg) to the incident oscillation. When the analyzer is rotated, periodic maxima and minima of illumination appear on the screen. These are due to elliptical polarization and none of the positions of the analyzer will produce darkness.

2. When the analyzer is crossed, the plate is set for a maximum of light at a 45 deg angle to the incident oscillations. In this case circular polarization is produced and rotation of the analyzer does not bring any changes in illumination.

35–6.17[61]

The sign of crystals is determined by observing them in converging polarized light with a sensitive

[60] Translated by K. V. Robinson and H. A. Robinson (Adelphi University) from *Lecture Demonstrations in Physics*, A. B. Mlodzeevskii, ed., M. V. Lomonosov State University, Moscow, U.S.S.R.

[61] Translated by K. V. Robinson and H. A. Robinson (Adelphi University) from *Lecture Demonstrations in Physics*, A. B. Mlodzeevskii, ed., M. V. Lomonosov State University, Moscow, U.S.S.R.

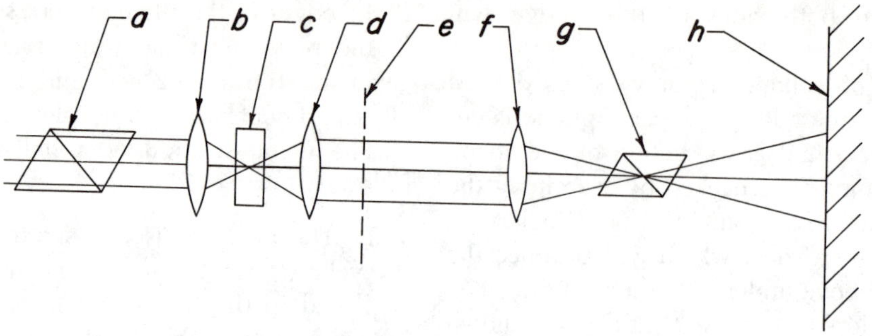

Fig. 35–79

tint plate or a quartz wedge as in Fig. 35–79. A light beam from a projector passes through a polarizer a and is changed into a strongly converging beam by lens system b with a short focal length (such as a microscopic Abbe condenser). The crystal c to be studied is placed on a rotating table between this lens system and a symmetrical system d which makes the beam parallel.

In general, the larger the aperture of the b and d lens systems, the larger the field of vision, i.e., the more interference bands on the screen. The maximum field of vision with this setup is obtained when the entire light beam enters aperture d.

The quartz wedge or sensitive plate e is placed in the parallel beam and an objective lens f is set up to focus the focal plane of the second lens system on the screen. A rotating analyzer g is placed between the objective lens f and the screen h. The plate or wedge must be set at such a position that the direction of its oscillations will make a 45 deg angle with the direction of the oscillations in the polarizer.

In order to determine the sign of uniaxial crystals, the crystal used must be cut perpendicular to the optical axis. A black cross and concentric rings will then be visible on the screen when the polarizers are crossed. When the wedge is inserted, the rings shrink in two opposite quadrants, as if they were pulled towards the center and expanded into the opposite quadrants. If the crystal is positive, the expansion takes place in the quadrant lying along the optical axis of the wedge, and if the crystal is negative the widening takes place in the opposite pair of quadrants. The sign of biaxial crystals may be determined on any section crossing the optical axis, although the determination works best in sections perpendicular to the acute bisector. When the wedge is inserted, a similar narrowing and widening of the rings is seen. The sign may also be determined by means of a sensitive tint plate, inserted in the same way as the wedge. When the sensitive tint plate is used, the color of the conoscopic picture changes: two opposite quadrants close to the center become blue and two others become yellow. If the blue coloration coincides with the direction of the optical axis in the sensitive tint plate, the crystal is negative; in the opposite case, the crystal is positive.

35–6.18

When linearly polarized light is passed through CS_2 in a glass cell inside a solenoid, the plane of polarization is rotated by the longitudinal magnetic field. The angle through which the plane of polarization is rotated is given by

$$\alpha = HlV,$$

where H is the magnetic field strength, l is the path length in the magnetic field, and V is a constant for a particular medium. The demonstration of this effect can be performed with readily available equipment.

A beam of light from an 8-V, 50-W, movie projector lamp a (Fig. 35–80) is directed through a lens b ($f = 1$ in.) which causes the beam to converge slightly. The light bulb is parabolic in the rear and spherical in the front. The entire bulb is silvered except for one point on the center of the spherical side. The beam then travels through a Nicol prism c and into a 2 x 2 x 6-in. glass container d filled with CS_2 which is positioned inside a large solenoid e. The coil, shown in detail in Fig. 35–81, is wound on a 6-in. length of insulated pipe with about 1000 turns of 1-mm-diam copper wire. The beam emerges from the fluid, passes through

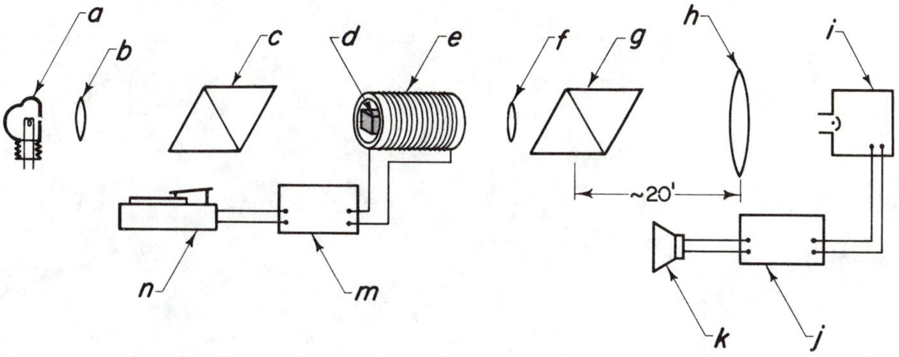

Fig. 35–80

another lens f (f = 8 in.), a second Nicol prism g, and falls on a large condenser lens h 10–20 ft away. The condenser lens focuses the light beam on the face of a photocell and integrated pre-amplifier i which is connected to an amplifier j and a loudspeaker k. The solenoid surrounding the CS_2 solution is fed by the output of a 20-W amplifier m which is driven by the pickup crystal of a record player n.

In operation, the Nicol prisms are arranged so that a minimum of light reaches the photocell when the solenoid is turned off. When the solenoid is put into operation, the Faraday effect causes the plane of polarization of the light to rotate as it passes through the magnetic field. This causes the intensity of the light falling on the photocell to fluctuate with variations in the magnetic field, and the record being played on the record player will be heard on the loudspeaker which is driven by the photocell amplifier.

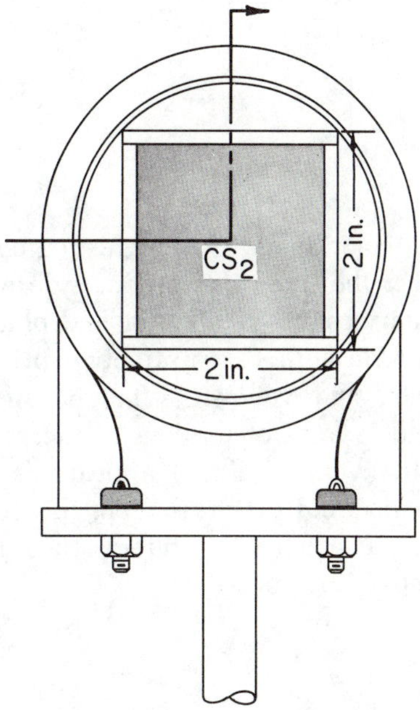

Fig. 35–81

35–7 Color

35–7.1

A standard 3¼ x 4-in. slide projector can be used to demonstrate the mixing of primary colors. The projector with its lens and bellows removed is shown schematically in Fig. 35–82. The two plates *A* and *B* are made of cardboard or metal. The holes in plate *A* are about 3 cm in diam, with hole centers at the points of a triangle 5 cm on each side. A plano-convex lens of ~15 cm focal length[62] is taped over each hole. A small cardboard shim under the outer edge of each lens tilts them inward so that the images overlap on the viewing screen. The holes in plate *B* are 2 cm in diam and are also 5 cm apart. With suitable filters[63] mounted over the holes in plate *B* (such as Eastman Kodak filters No. 29, 61, and 49) a good quality of white, cyan, magenta and yellow is obtained in the familiar overlap pattern.

35–7.2

A special lantern slide has seven ⅛-in. slits uniformly spaced. All except the center slit are covered

[62] This is suitable for a medium size screen not too far away from the lecture table. For longer throws select a longer focal length. These lenses were purchased from Edmund Scientific Co., Barrington, N.J.

[63] Available from Eastman Kodak, Rochester, Corning Glass Works, Corning, and Charles Products Co., New York.

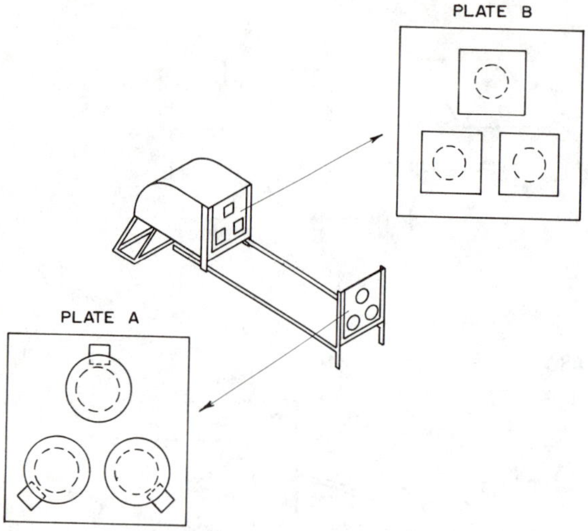

Fig. 35–82

with different color filters. An interesting combination is to cover the three slits on one side with the additive primary colors—red, green, and blue; those on the other side with the subtractive primaries—cyan, magenta, and yellow. When a diffraction grating is held in front of the lens of the projector, the center slit (with no filter) appears as a horizontal color spectrum on the screen, the colors of which are matched by the filters, thus demonstrating selective absorption.

35–7.3

A row of colored skeins of yarn is illuminated by light which is passed through different filters. The filters could be located on a rotating disk, which would enable the observer to change the filter simply by rotating the disk.

35–7.4

The projection arrangement represented in Fig. 35–83 is convenient for demonstrating various color phenomena, including the subjective nature of color, selective absorption, and additive and subtractive color reproduction processes. The basic equipment consists of a 2 x 2-in. slide projector ($f = 7$ in.) set up on a rotatable table at the rear of the lecture hall and directed toward the rear wall on which are mounted three 6-in.-diam, ball-joint mirrors. This arrangement eliminates the need for three

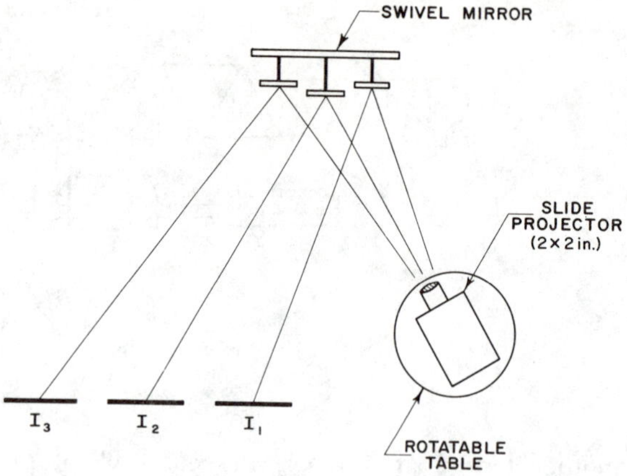

Fig. 35–83

projectors and the difficulty of manipulating them. The demonstrations utilize several specially constructed slides. The subjective color slide (Fig. 35–84) is an opaque sheet of metal (i.e., shim brass) with two ½-in. holes in it, one of which is covered with a Kodak No. 33 green filter. Both sides of the slide are covered with $\frac{1}{16}$-in. glass. An additive 2 x 2-in. color slide (Fig. 35–85) has three holes (1 in., center to center) arranged in a triangular array and covered with blue, green, and red filters, respectively. Both sides of this slide are also covered with $\frac{1}{16}$-in. glass. A third 2 x 2-in. slide (Fig. 35–86) has seven rectangular holes ($\frac{1}{8}$ in. wide, $\frac{3}{16}$ in. high, and separated by $\frac{1}{32}$ in.) in a row, the middle one of which is clear. This is a template. A photo is obtained. The negative is black with clear holes which are carefully covered with tiny pieces of Kodak filters. Double-sided Scotch tape

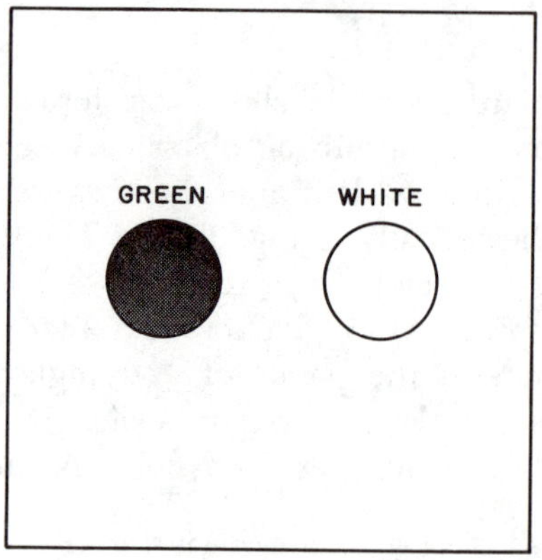

SUBJECTIVE COLOR SLIDE

Fig. 35–84

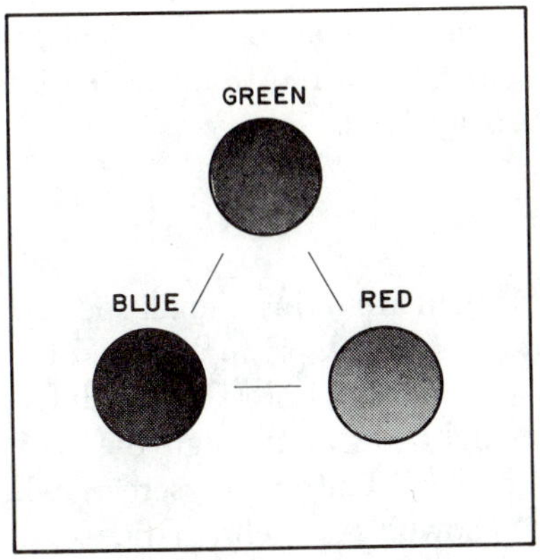

PRIMARY COLOR SLIDE

Fig. 35–85

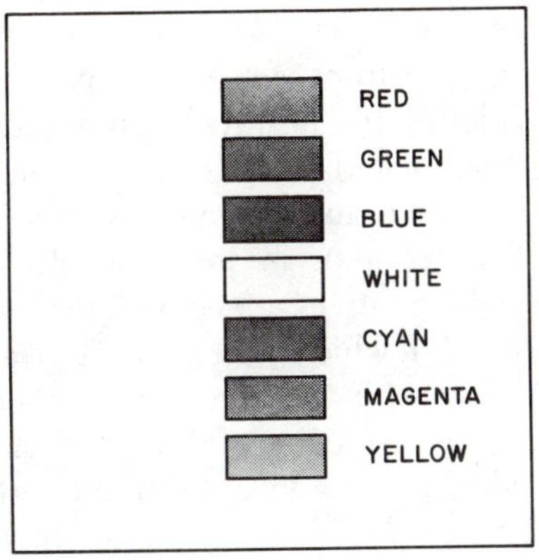

SLIDE OF PRIMARIES AND THEIR COMPLEMENTS

Fig. 35–86

is used on each side of the row of holes. Then each side of the negative is covered with 1/16-in. glass. The three holes on one side of center are covered with filters of the three additive primary colors (red, green, and blue) and those on the other side of center are covered with the complements of the additive primaries, the so-called subtractive primaries (cyan, magenta, and yellow).(64) For demonstrating the additive color process, there is a pair of slides, each containing three positive images of a colored object, each positive having been made from a negative taken with one of the additive primary filters. One of this pair of triple-image slides is left black-and-white. The other has each image covered with a filter of the same color as that used in making the corresponding negative.

For demonstrating the subtractive color process, a second set of three black-and-white positive transparencies is made from the same set of negatives. Each positive is dyed with the complement of the color with which the negative was made, and the three colored positives are bound in register in a "triple-deck" sandwich. Other auxiliary equipment includes a double refracting prism and one or more diffraction gratings.

Subjective Nature of Color Perception. With the subjective color slide in the projector, project the green and white images side by side. Adjust the mirrors until the images partially overlap and note that the outside white appears pink. Cover each mirror separately with a piece of cardboard and note the effect.

Dependence of Color on Illumination. The fact that the color of an object depends upon the character of the incident light can be demonstrated by illuminating bits of colored paper and cloth with a sodium lamp, a mercury vapor lamp, and the standard room lights.

Selective Absorption. Using a slide with a single aperture (a slit 1/8 to 1/4 in. wide), place a diffraction grating in front of the projector lens and project the continuous spectrum on the screen (directly or by reflection). Introduce various colored transparent substances in the beam and note the elimination of certain portions of the continuous spectrum. Didymium, for example, absorbs a band in the yellow-orange region. Various colored glasses and solutions can be shown to exhibit selective absorption.

Additive Primary Colors. With the additive color slide in the projector, produce 3 colored images on the screen. Manipulate the mirrors until the three primary colors overlap, producing white light; note that where any two of them overlap, the resulting color is of a spectral hue lying between the two, i.e., the complement of the third primary.

Additive Color Process. The basis of all three-color reproduction processes is that, with a suitable choice of three colors—a red, a green, and a blue—nearly all of the hues found in nature can be reproduced. Any three-color photographic process involves four steps: (1) making three separation negatives with light of the three additive primaries,

(64) Eastman filters: Additive primaries—red "A" No. 29; green "B" No. 61; blue "C5" No. 47. Subtractive primaries—cyan (minus red) No. 44A; magenta (minus green) No. 32; yellow (minus blue) No. 12.

using appropriate filters; (2) making three positive transparencies, one from each of the separation negatives; (3) coloring the positives appropriately; (4) combining the colored positives appropriately.

In both additive and subtractive color photography, the first two steps are the same. In the additive process, the positives are colored with the color of the light with which the respective negatives were made, i.e., with the additive primaries. The three images are then combined in parallel, i.e., the light from all three is added to produce the composite image. In this demonstration, three black-and-white positive images are grouped on a single slide. When this slide is projected, the three separate black-and-white images are seen to have certain tonal differences. When the black-and-white slide is replaced by an identical one with the additive primary colors added, three images are produced—a red, a green, and a blue. By manipulating the mirrors, the images can be superimposed in register and a reasonably good color picture is obtained.

Subtractive Color Process. The subtractive process differs from the additive process only in the last two steps. For the demonstration, three positive transparencies on film have been dyed with the complements of the colors with which the negatives were made. When projected individually, the separate images are cyan (minus red), magenta (minus green), and yellow (minus blue). Three such transparencies bound together in register produces a colored picture by subtracting the appropriate colors from a beam of white light.

Color and Polarization. Interference effects in bi-refracting crystals can be shown in many ways. It is instructive first to show that the two images produced by a bi-refracting prism are polarized at right angles to each other. A titanium dioxide (rutile) prism[65] with its optic axis parallel to the refracting edge is mounted in front of the projector lens and set for minimum deviation. With a slit source, two images (spectra) are seen. Rotating a polarizing screen (Polaroid) in the beam alternately extinguishes the images.

Interference colors in polarized light can be shown by introducing thin sheets of bi-refracting materials, such as mica or cellophane, between crossed polarizing plates. Pieces of Plexiglas stressed in the beam illustrate the basis of photoelastic stress analysis.

35–7.5

Light from an arc forms a small spectrum. Small wedges of glass (weak prisms) placed in that spectrum deflect some colors to slightly different directions. The light passes through the region of the spectrum to a distant white screen where color mixing is shown. An achromatic lens, suitably placed, forms a large image on the screen of a diaphragm through which all the light passed on its way to the spectrum. Without any glass wedges in the spectrum, light of all colors is gathered to make a pure white patch on the screen. With the deflecting wedges, some colors are deflected to make patches nearby. Where all the patches overlap, perfect white is seen. Thus, this arrangement demonstrates complementary colors with any selections from the spectrum.

The optical arrangement is shown in Fig. 35–87. The carbon arc *a* is well housed to avoid stray light. The arc itself takes the place of a slit in the formation of the spectrum. A large good lens *b* forms an image of the arc at infinity. Just beyond the lens *b* a large prism *c* is placed, roughly at minimum deviation. Just beyond the prism a diaphragm *d* with a circular aperture in it is placed so that all of the aperture receives white light. An iris diaphragm should be used so that the size of the aperture can be adjusted to exclude colored edges. A good achromatic lens *e* is placed some distance beyond the iris diaphragm to receive the light and form a spectrum. The prism *c* and lens *e* should be chosen so that the width of the spectrum from red to violet is 2 to 4 in. Since the light striking the prism is rendered practically parallel, the spectrum is formed in the focal plane *f* of the achromatic lens. But all of the light proceeding to that lens and on to the spectrum has passed through the iris aperture. Therefore all of that light must pass through an image of the aperture formed by the achromatic lens. Since the iris is much nearer than infinity, its image will be much farther away than the spectrum. The achromatic lens *e* is placed to form that image on the distant screen *g*. Since white light passed through the iris—not yet separated noticeably to form the spectrum—the image

[65] Available from Union Carbide, Linde Division, Indianapolis, Indiana.

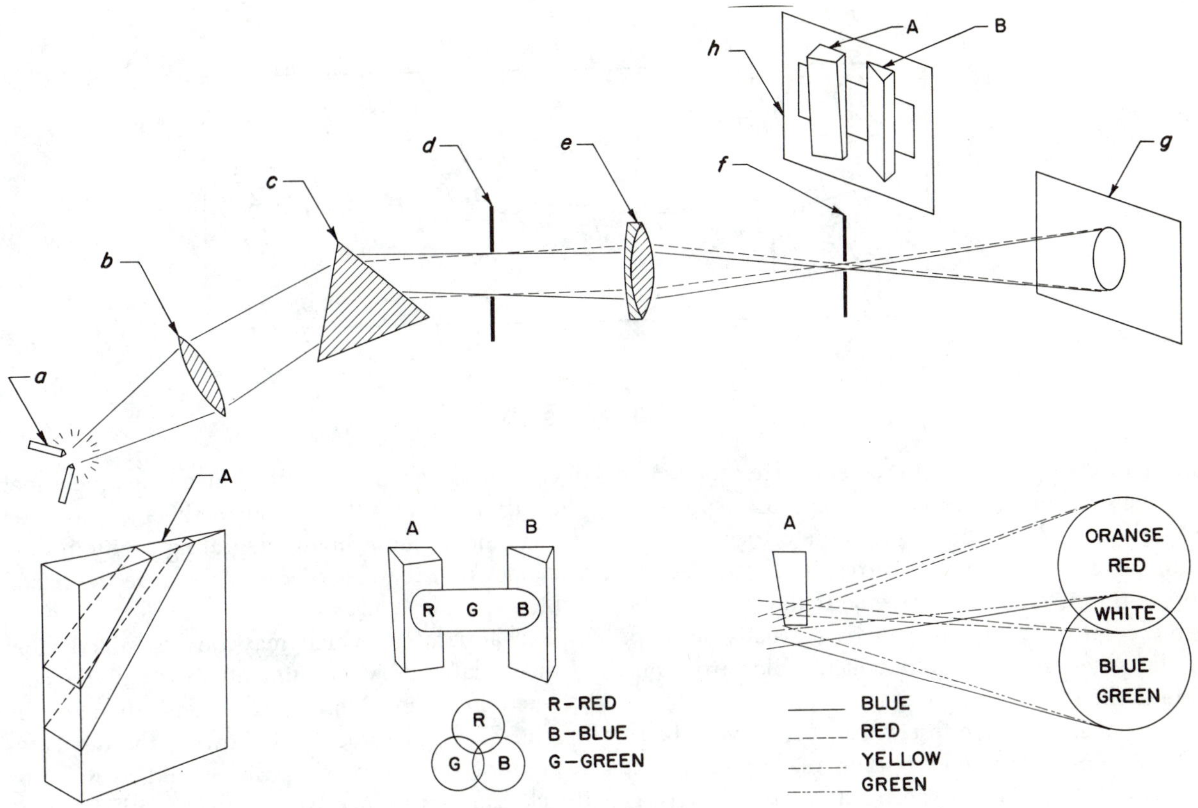

Fig. 35–87

is a white disc. Thus the achromatic lens *e* performs *two* image-forming functions at the same time.

If a small wedge of glass (Fig. 35–87 A) is placed in the spectrum at *f* to cover the red region, all red light will be deviated to a round patch to one side of the original patch on the screen. The remaining light makes a complementary patch, blue-green, in the original position. Where these two patches overlap, they make white. By using another small wedge (Fig. 35–87 B), cut on a slant as shown, we can produce three patches on the screen in three different colors, overlapping to make white.

It is important to place the wedges in the plane *f* where the spectrum is formed, and to make them easy to maneuver. To facilitate this, a sheet of metal *h* with a rectangular opening is placed there to accommodate the spectrum, and the glass wedges are suspended from the upper edge of this frame, or "spectrum gate," so that they hang vertically in the spectrum.

In showing this demonstration it is necessary to make the arrangement clear to the audience. To begin with, an extra projection lens is placed beyond the spectrum so that it forms an image of the spectrum itself on the remote screen. The audience sees the spectrum surrounded by the spectrum gate frame. Then a small opaque strip is hung in the spectrum gate in the red region. The audience sees the strip obscuring the red in the image of the spectrum on the screen and predicts that, if all the rest of the light could be collected, it would make a blue-green color. The projection lens is removed (swung out on an arm) and the blue-green patch is seen. Then the projection lens is inserted again and the opaque strip replaced by a strip of glass wedge. The audience sees that the wedge catches the red light and should deflect it to a separate spot. The projection lens is removed again and the two spots, blue-green and red, are seen. If the iris is small, the two spots are quite separate; if it is large, they overlap.

To adjust this whole arrangement, the demonstrator places the iris just beyond the prism and moves the achromatic lens until it makes a sharp

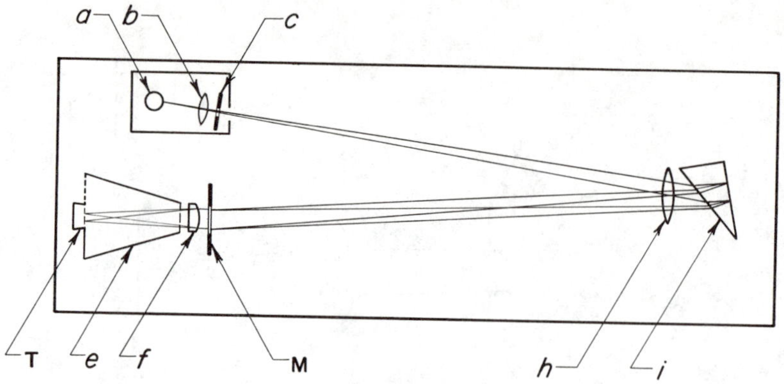

Fig. 35–88

image of the aperture on the screen. He then keeps the lens fixed there. He locates the spectrum roughly and places the spectrum gate there. Then, hanging a small opaque strip in the gate he observes the patch on the screen. If the patch is the same color all over, the gate is correctly located. If the patch shows a gradation of color from one side to the other, the gate must be moved farther or nearer. This sensitive test will be found very helpful. The projection lens is then placed so that it forms an image of the spectrum on the screen.

The glass wedges are strips cut with a diamond from opthalmic small-angle prisms (for spectacles) with a power of 1 or 2 diopters.

35–7.6[66]

In an elementary discussion of color, it is desirable to demonstrate the connection between color and the spectral energy distribution of light (including, as a special case, the color of monochromatic light). The "color synthesizer" described here enables one to demonstrate any desired energy distribution and visually to show, in at least a qualitative manner, the significance of such quantities as dominant wavelength, purity, and luminosity. In addition, the apparatus may be used to demonstrate monochromatic and heterogeneous additive primaries, and to show the influence on color of a surrounding field of white light of variable luminosity.

The function of the equipment is to disperse the light from a tungsten-filament projection lamp into a spectrum, where the spectral energy distribution may be modified in any desired manner by means of a suitably shaped opening in an opaque sheet (hereafter referred to as the modifier), and then to combine the light in an integrating cavity in such a manner that the observer may perceive the mixed light.

The apparatus, which may be assembled from ordinary laboratory equipment and may be constructed in various forms, is shown schematically in Fig. 35–88. Photographs showing the device as seen from the lecturer's position and as seen by the class are reproduced, respectively, in Figs. 35–89 and 35–90. The light source *a* is preferably a ribbon-filament projection lamp rated at 6.0 V and 108 W. A simple condensing lens *b* forms a crude image of the lamp filament in the plane of lens *h*. Although the slit *c* should, ideally, be curved, good results are obtained with a simple, straight slit 2.5–3 mm wide and 40 mm long. Lens *h* is placed at a distance from the slit equal to its focal length (f = 50 cm is recommended) and is oriented with the prism *i* to produce a spectrum at *M*; it may be

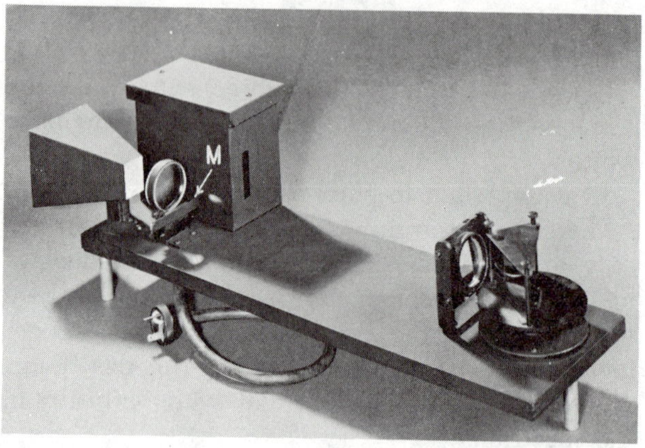

Fig. 35–89

[66] J. A. Van den Akker, *Am. J. Phys.*, **16**, 1 (1948).

Fig. 35–90

an inexpensive simple lens and may be slightly tilted about a horizontal axis if difficulty is encountered with stray light at M arising in surface reflections at h. A prism of the Littrow type (back suface silvered) was employed in order that the apparatus might be compact and readily portable; very good results may be obtained with an ordinary 60-deg prism, but equipment utilizing such a prism is awkward to carry and store because of its greater length and the "dog-leg" bend in the middle.

The adjustment of both the lens h and the prism to form a spectrum at the relative position shown in Fig. 35–88 is facilitated by placing a "monochromatic filter" over the slit and mounting a white card at M for the purpose of viewing the approximately monochromatic image of the slit; if such a filter is not readily available, the lamp may be temporarily replaced with a mercury arc, and the adjustments performed with the green and yellow spectrum lines of mercury.[(67)] The modifier support M is located where the spectrum is formed; it consists simply of a narrow slot (formed, for example, by folding a 1 x 4-in. strip of thin brass sheet about its longer central axis), and it is mounted at such elevation that the top edges of the slot are just below the lower boundary of the spectrum.

After passing through the modifier at M, the beam is condensed by the lens f on a small block of magnesium carbonate T, which diffusely reflects more than 95 percent of the incident light into the integrating cavity e. The focal length of f is chosen so that this lens forms a crude image on the target T of the front face of the prism (for light incident on the prism—in other words, the object-to-lens distance is a few centimeters greater than the distance from f to the back face of the prism). Lens f is preferably a thick, inexpensive condensing lens of the kind used in lantern slide projectors; a lens of this description (selected for "whiteness" of glass) was used in the apparatus assembled by the writer, after a local optical company reduced its diameter to about 2.5 in.

The integrating cavity, formed from fairly stiff sheet metal, has the shape of a truncated four-sided pyramid, although it is probable that other cavities, such as that which would be provided by a hemispherical shell, would function well. The dimensions of the cavity are not critical. The smaller opening should be just large enough to pass the visible beam, and the side dimension of the larger opening should be of the order of 70 percent of the distance between the openings. (It is of interest to note at this point that only about 2 percent of the light diffusely reflected from the target T escapes directly through the smaller opening.)

It is important that the inner surface of the cavity be coated with magnesium oxide rather than white paint, because the reflectance of the latter varies appreciably with wavelength, and is apt to be too low for really effective results. It is a good plan, however, to use white paint as a base for the magnesium oxide. See the reference mentioned in footnote (66) for additional details.

The instrument is arranged on the lecture table so that the class can view the large opening of the integrating cavity. During the demonstration, the room should be as dark as possible because room light will lower the purity of colored light issuing from the cavity; this is of particular importance in the demonstration of monochromatic light.

Perhaps the first test of the instrument lies in the production, with a rectangular modifier in place, of light having the yellowish-white color characteristic of tungsten-filament lamps. With the exception of surface reflection losses, slight absorption in the glass optics, and a loss at the target of the order of 3–5 percent (depending on the quality of the magnesium carbonate), all the light from the slit passing through the lens f should enter the integrating cavity.

A number of interesting, although purely qualitative, demonstrations may be performed with the uncalibrated instrument. However, it is desirable

(67) Strictly, the spectrum is formed in a slightly curved surface cutting across the beam at an angle. However, the functions of the apparatus do not require good resolution, and the support for the modifier at g (Fig. 35–88) may be arranged perpendicularly to the beam.

to have reasonably accurate information on the relationship between wavelength and distance across the beam at M. The supporting slot at M for the modifier should be furnished with a stop at its far end. This stop furnishes a convenient fiducial point for the calibration.

Good results can, of course, be obtained by replacing the projection lamp with a mercury arc and other sources of line spectra. A white card placed in the modifier support at M is moved against the stop, and the positions of known spectrum lines are carefully marked on the card with a sharply pointed pencil. Greater accuracy (in view of the wide slit) can be obtained through the technique of measuring effective wavelength.[68]

The following demonstrations can be performed:

Continuous Spectrum. If the class is not large, the students may be asked to come to the lecture table to observe the complete spectrum on a white card inserted on the modifier support or, if this is undesirable, the instrument may be rotated to show the spectrum to the seated group. The card may then be removed to demonstrate the elementary principle that a mixture of the continuum of monochromatic radiations from the source yields white light.

Monochromatic Light. A modifier consisting simply of a fairly wide slit (about 5 or 6 mm for good luminosity) is moved across the spectrum. It is suggested that white markers be placed on the black card so that the lecturer can call out the wavelengths, 400, 450, 500, . . . mμ as he slowly moves the slit through the spectrum.

Complementary Colors. Two rectangular openings, each somewhat wider than the spectrum, are cut in a black card, with a strip several millimeters wide separating the openings. This strip removes segments of the spectrum, thus presenting colors to the class that are complementary to those removed. Strikingly beautiful colors may be obtained through suitable adjustment of the width and position of the strip.

Additive Primaries. Three slits cut in a thin metal sheet, located to pass red, green, and blue-violet portions of the spectrum, and provided with vertically adjustable slides, enable the lecturer first to show the primaries alone, then in variable combination to yield an unlimited number of colors. With fairly narrow slits (a few millimeters wide), the use of monochromatic primaries may be shown. The use of heterogeneous primaries (covering a small range of possible colors) may be demonstrated by cutting wider slits in the metal sheet.

Color of Various Light Sources. In addition to the color of tungsten-filament light (rectangular modifier in position), the lecturer can create light of relative spectral energy distribution closely similar to that of any one of several other light sources. The physicist often has occasion, in theoretical work, to employ the "equal-energy spectrum"; few physicists, however, have seen light of constant energy distribution. To prepare a modifier for the equal-energy spectrum, it is only necessary to plot ordinates on a black card proportional to $1/E$, where E is the spectral energy distribution of the light emitted by the projection lamp (this, of course, neglects selective losses in the instrument).

Color Synthesis. The color synthesizer may be used to produce light of relative spectral energy distribution the same as that of light reflected from, or transmitted through, colored bodies and, hence, to reproduce the colors of bodies as seen under various illuminants. Let us consider the color of a paper viewed in tungsten-filament light. The spectral energy distribution of light entering the observer's eye after reflection from the paper is proportional to the product ER, where the two quantities are functions of wavelength. The spectral energy distribution of light produced by the color synthesizer is proportional, with fair accuracy, to the product Ey, where y is the ordinate of the opening in a modifier. Hence, if we prepare a modifier in which $y = kR$ (k being a suitable constant) and insert this modifier in the instrument, we will generate light of color similar to that seen when the paper is held under a tungsten-filament lamp.

Dominant Wavelength, Purity, and Luminosity. The three attributes of color may be demonstrated in an effective manner by means of modifiers constructed like that shown in Fig. 35–91. A horizontal opening of vertical dimension a passes a limited fraction of the whole spectrum in a nonselective manner; this light, after mixing in the cavity, should be "white" light of the same character as that emitted by the projection lamp. A vertical opening, of horizontal width b and height c, passes monochromatic light. The admixture of the white and monochramatic light simulates the

[68] J. A. Van den Akker, *J. Opt. Soc. Am.*, 33, 257 (1943).

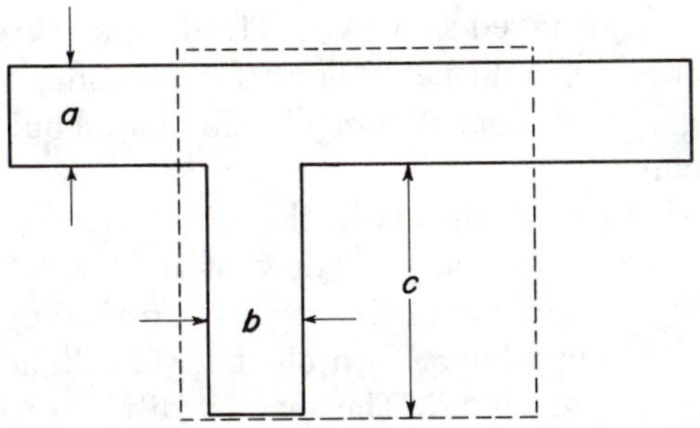

Fig. 35–91

mixture involved in the ICI monochromatic method of specifying color.[69] If the luminosities of the white and monochromatic beams are so adjusted that the color matches that of a sample, then the wavelength of the monochromatic light is the *dominant wavelength* for the sample (for the given white light source), and the luminosity of the monochromatic light relative to the total luminosity is the purity. The total luminosity may be conveniently varied by moving wire-cloth screens of different transmissions into the beam near the opening in the lamp house.

The influence on color of surrounding colors and even of a surrounding neutral field of illumination is generally appreciated. For example, a colored object viewed in the typical environment of a room will give rise to a different response from that experienced when the same object, viewed under the same illuminant, is observed in a dark room, even when allowance is made for the fact that the walls and furniture of the room modify the character of the light falling upon the colored object. Let us consider the appearance of a piece of brown paper. On constructing a modifier for the brown paper and inserting the modifier in the color synthesizer, we are surprised to note that the cavity light is not brown, but straw-yellow, depending upon the nature of the original brown color. If now we surround the cavity with an illuminated diffusing screen, and gradually increase the level of illumination of the screen, we find that the cavity light (which has remained unchanged) appears to be brown. The effect of the surroundings is striking in the case of the spectrum colors. The appearance of saturation of monochromatic light is often disappointing, because spectrum colors are usually viewed in a dark room, with a dark surrounding. The maximal saturation of spectrum colors can be appreciated when a surrounding field of white light of appropriate luminosity is provided.

[69] A. C. Hardy, *et al.*, *Handbook of Colorimetry* (Technology Press, Cambridge, 1936).

35–7.7

In Fig. 35–92 a light beam from a mercury lamp *a* passes through a condenser lens *b* and illuminates a diaphragm *c*. This diaphragm consists of an opaque surface with a normal slit in its lower half and an opening with a narrow stop in its upper half. The lens *d* focuses the image of the diaphragm.

The mercury lamp, such as Philips HP 80W, has its glass envelope removed. Both the slit and stop are 1 mm wide; the slit is 10 mm high, and the complementary opening has a total width of 32 mm.

After the image of the diaphragm is focused, a prism *e* is introduced in the light beam for color dispersion. The standard yellow, green, and violet lines of the mercury spectrum appear in the top half of the image; these lines are produced by the slit in the lower half of the diaphragm. The spectrum appearing in the lower half of the image, which is caused by the top half of the diaphragm, is called an "inverted" spectrum and appears as lines of complementary colors. For the "inverted" spectrum to appear, the color bands must overlap in the "inverted" spectrum; therefore, the prism can not have too high a dispersion.

The appearance of the complementary colors in the inverted spectrum can be understood as follows. If a single mercury line is selected (by using a green filter, for example), only this transmitted

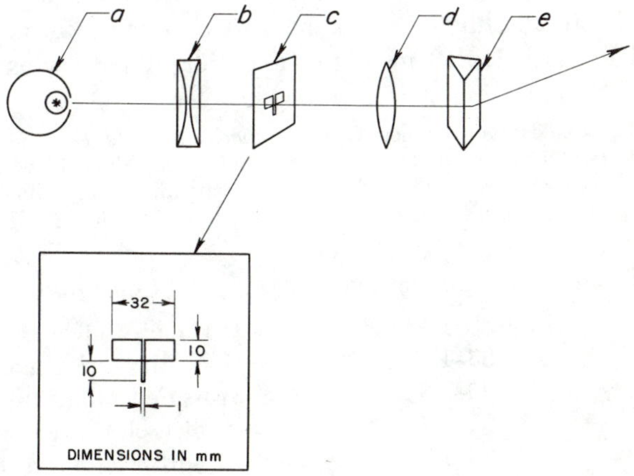

Fig. 35–92

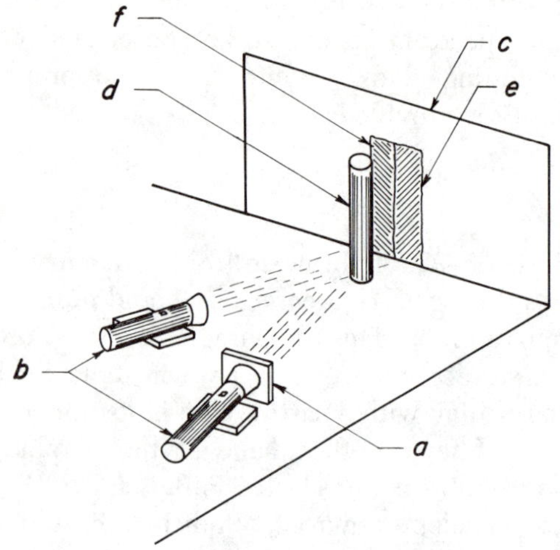

Fig. 35–93

(green) line appears in the normal spectrum. In the inverted spectrum, however, a black line is seen exactly below the normal spectrum line. This black line is framed by a band having the color of the transmitted (green) line. If all colors in the source are let through, a color arising from the superposition of all colors, except the green spectral line, appears in place of the black line.

35–7.8[70]

Color, or the sensation of color, may be perceived because of the abundance of that color, or because of the over-abundance of a complementary color. To demonstrate this, a piece of red cellophane or red glass filter *a* (Fig. 35–93) is secured to one of two flashlights *b*. A white card or paper screen *c*, 8½ x 11 in., is propped on a table top, with the flashlights and a cylinder *d*, about 1 in. in diam and 2 or 3 in. high (a flashlight battery is suitable), arranged as shown. The distances are not critical and should be adjusted for optimum effect during observation. Room illumination should be minimum.

First, turn on the flashlight not provided with a color filter. A sharp shadow of the cylinder is cast on the card, with the region surrounding the shadow being white. Turn off the white light and turn on the red light. The shadow displaced from the first one is surrounded by red light. Note that the region of the "red shadow" lies on the part of the card previously illuminated by white light, and the "white shadow" region is now illuminated by red light.

Both the white and the red light are turned on separately while the shadow regions and surrounding regions are studied carefully; no abundance of green light will be noted. With both lights on, the shadow regions are again studied. The former "white shadow" *e* is red, as expected, since the red light falls in that region. The former "red shadow" *f*, illuminated only by white light, appears green. By adjusting the lights, the shade of green can be varied. The same demonstration can be tried, substituting green cellophane or glass for the red.

A very striking demonstration of this same effect may be made by using somewhat more elaborate equipment. If two slide projectors are available, they are used in the following manner:

First, make a slide with an opaque patch in the middle. Then make a second slide opaque everywhere except for a transparent patch that corresponds in size and position to the opaque patch on the former slide. Cover the projection lens of one projector with a red filter.

Use the slide with the opaque patch in this projector. Using the other slide in the unfiltered projector, illuminate a screen with both projectors, fitting the white patch into the dark area caused by the opaque patch of the red projector. When intensities are adjusted, the white patch area appears brilliantly green.

[70] C. Harvey Palmer, *Optics Experiments and Demonstrations* (The Johns Hopkins Press, Baltimore, Md., 1962), pp. 105–107; see also E. H. Land, "Experiments in Color Vision," *Scientific American*, **200**, pp. 84–99 (1959).

Demonstrations	***Contributors***
35–1.1	P. Huber, University of Basel, Switzerland
35–1.2	H. N. Maxwell, Ohio Wesleyan University
35–1.3	Swiss Federal Institute (ETH), Zürich, Switzerland
35–1.4	H. Bichsel, University of Southern California
35–1.5	A. Karlsches and A. Scharmann, I. Physikalisches Institut der Universität, Giessen, West Germany

Demonstrations	*Contributors*
35–1.6	Swiss Federal Institute (ETH), Zürich, Switzerland
35–2.1	E. L. Cleveland, New Mexico State University
35–2.2	University of Washington
35–2.3	Swiss Federal Institute (ETH), Zürich, Switzerland
35–2.4, 34–2.5	H. A. Robinson, Adelphi University
35–2.6	L. Grillet and M.-T. Grillet-Amy, University of Rennes, France
35–2.7	L. W. Morris, Louisiana State University
35–2.8	C. H. Palmer, The Johns Hopkins University
35–2.9	H. Mueller, Massachusetts Institute of Technology
35–2.10	H. A. Robinson, Adelphi University
35–2.11	R. W. Pohl, University of Göttingen, Göttingen, West Germany
35–2.12	A. S. DeVany, Williams College
35–2.13	M. J. Pryor, State University of New York at Albany
35–3.1	H. Daniel, Musterschmidt Verlag, Göttingen, West Germany
35–3.2	F. Oppenheimer, University of Colorado
35–3.3	J. W. Straley, University of North Carolina
35–3.4	R. W. Pohl, University of Göttingen, Göttingen, West Germany
35–3.5	Eric M. Rogers, Princeton University
35–3.6, 35–3.7, 35–3.8	Swiss Federal Institute (ETH), Zürich, Switzerland
35–3.9	H. A. Robinson, Adelphi University
35–3.10	H. L. Armstrong, Queens University, Kingston, Ontario, Canada
35–4.1, 35–4.2, 35–4.3	H. Daniel, Musterschmidt Verlag, Göttingen, West Germany
35–5.1, 35–5.2	C. H. Palmer, The Johns Hopkins University
35–5.3	H. A. Robinson, Adelphi University
35–5.4	R. M. Sutton, California Institute of Technology
35–5.5	H. Daniel, Musterschmidt Verlag, Göttingen, West Germany
35–6.1	F. Oppenheimer, University of Colorado
35–6.2	California Institute of Technology
35–6.3	H. E. White, University of California, Berkeley
35–6.4	H. C. Jensen, Lake Forest College
35–6.5	J. Christie, Oak Ridge Institute of Nuclear Studies
35–6.6	Swiss Federal Institute (ETH), Zürich, Switzerland
35–6.7, 35–6.8	H. A. Robinson, Adelphi University
35–6.9	Swiss Federal Institute (ETH), Zürich, Switzerland
35–6.10	H. A. Robinson, Adelphi University
35–6.11	M. R. Wehr, Drexel Institute of Technology
35–6.12, 35–6.13, 35–6.14, 35–6.15, 35–6.16, 35–6.17	H. A. Robinson, Adelphi University
35–6.18	Swiss Federal Institute (ETH), Zürich, Switzerland
35–7.1	L. R. Weber, Colorado State University
35–7.2	G. Schwartz, Florida State University
35–7.3	H. E. White, University of California, Berkeley
35–7.4	V. E. Eaton, Wesleyan University, Conn.
35–7.5	H. M. Waage, Princeton University
35–7.6	J. A. Van den Akker, The Institute of Paper Chemistry, Appleton, Wisconsin
35–7.7	P. Huber, University of Basel, Switzerland
35–7.8	C. H. Palmer, The Johns Hopkins University

ATOMIC AND NUCLEAR PHYSICS

36

GAS LASER DEMONSTRATIONS

By Haven Whiteside
Oberlin College

36–1 Introduction

The availability of continuous visible gas lasers at a reasonable price has made possible a number of new demonstrations in optics that should be of general interest. In the first place, a laser may be treated as a high intensity monochromatic point source and may be used for demonstration of the usual interference and diffraction phenomena. In the second place, the unusual coherence of laser light may be demonstrated and used to show phenomena that do not exist for ordinary light, no matter how intense. After a discussion of the properties of laser and ordinary light sources and a brief introduction to the theory of operation of the gas laser, we will present various demonstrations that are made possible by or improved by the use of a gas laser source. These demonstrations are mostly chosen for their relevance to the subject of physical optics rather than the study of the properties of the laser itself. It is to be noted that the number of people who can see adequately varies a great deal from one demonstration to another, ranging from perhaps thirty for diffraction from a pinhole to several hundred for demonstrations using a grating. Ruby lasers have not been included because the author does not know of any useful demonstrations using them, aside from burning holes in razor blades.

36–2 Laser Light Compared with Ordinary Light

There are three important properties that we will wish to compare between a laser and an ordinary light source: coherence length, transverse spatial coherence, and maximum image brightness.

Coherence is usually defined in terms of some sort of interference experiment. In generalized terms, light from two point "sources" is brought to an observation point. The two "sources" may be quite independent, or they may be slits in the field of a given source, or they may be the real and the virtual sources formed when a real beam goes through a beam splitter. The amplitude of the optical disturbance at the observation point is a function of time obtained by adding the amplitudes due to the two waves. If there is a constant phase difference between these wave amplitudes, the light is said to be coherent, and interference fringes can be observed in the vicinity. If, on the other hand, the phase difference is rapidly fluctuating with equal probability for all possible values from zero to 2π, the light is said to be incoherent, and no interference fringes are observed. It is clear that complete coherence and complete incoherence are limiting concepts and that we usually have a situation of partial coherence which lies somewhere in between. Thus we must carefully define what is meant by the degree of coherence, and introduce a numerical measure of it.

36–2.1 Partial Coherence Function and Fringe Visibility

For simplicity, let us ignore polarization and treat the amplitude of a light wave as a scalar quantity. This is of course equivalent to the special case where the two electric vectors are parallel at the observation point, but generalization to the case of arbitrary polarization would not essentially affect any of the conclusions to be drawn here. In the vicinity of the observation points Q, the fringe visibility is defined as

$$V = \frac{I_{\max} - I_{\min}}{I_{\max} + I_{\min}}, \tag{1}$$

where $I_{\max}$ is the light intensity at the nearest interference maximum, and $I_{\min}$ is the intensity at the adjacent interference minimum. At point Q, the complex amplitudes of the waves arriving by path 1 and path 2 can be designated $\psi_1\,(Q,t)$ and $\psi_2\,(Q,t)$ so that the total amplitude is

$$\psi\,(Q,t) = \psi_1\,(Q,t) + \psi_2\,(Q,t), \tag{2}$$

where ψ is proportional to the amplitude of the electric vector, but normalized so that the instantaneous power at Q is equal to the square of the absolute value of ψ, thus

$$P(Q,t) = \psi^*\psi = |\psi_1|^2 + |\psi_2|^2 + \psi_1{}^*\psi_2 + \psi_1\psi_2{}^*, \tag{3}$$

or

$$P(Q,t) = |\psi_1|^2 + |\psi_2|^2 + 2\,Re\,(\psi_1{}^*\psi_2), \tag{4}$$

where Re stands for the real part of a complex number. But the variations of $P(Q,t)$ with time are too rapid for observation, so that only the average of $P(Q,t)$ over many cycles is observable. This time average, the observed intensity I, will be denoted by brackets thus $\langle\ \rangle$. It is immediately clear from Eq. (36–4) that the intensity due to the first wave alone is $I_1(Q) = \langle|\psi_1(Q,t)|^2\rangle$ and for the second wave alone, $I_2(Q) = \langle|\psi_2(Q,t)|^2\rangle$. At this point it is convenient to introduce the normalized partial coherence function

$$\gamma_{12} = \frac{\langle\psi_1{}^*(Q,t)\,\psi_2(Q,t)\rangle}{\langle|\psi_1(Q,t)|\,|\psi_2(Q,t)|\rangle}, \tag{5}$$

so that the total intensity at Q may be expressed as:

$$I(Q) = I_1 + I_2 + 2\sqrt{I_1 I_2}\,Re\,\gamma_{12}. \tag{6}$$

Before applying this equation, let us examine two limiting cases. First, the case of complete coherence may be represented by letting both point sources emit infinite spherical waves of the same frequency, but including a possible difference in phase and amplitude. Thus

$$\psi_1(Q,t) = \sqrt{I_1}\,e^{i\omega t} \tag{7a}$$

and

$$\psi_2(Q,t) = \sqrt{I_2}\,e^{i(\omega t + \phi)}, \tag{7b}$$

where ϕ, I_1, and I_2 are all functions of Q.

From Eq. (5) we find $\gamma_{12} = \langle e^{i\phi}\rangle = e^{i\phi}$

and

$$I(Q) = I_1 + I_2 + 2\sqrt{I_1 I_2}\cos\phi. \tag{8}$$

The case of complete incoherence, on the other hand, can be represented by the same expressions, but with ϕ a random function of time instead of a constant. This gives

$$\gamma_{12} = \langle e^{i\phi(t)}\rangle = 0 \quad \text{and} \quad I(Q) = I_1 + I_2. \tag{9}$$

In summary we see that:

$|\gamma_{12}| = 1$ implies complete coherence,

$|\gamma_{12}| = 0$ implies complete incoherence,

and

$0 < |\gamma_{12}| < 1$ implies partial coherence.

In actual practice, if $|\gamma_{12}| > 0.88$ we describe the light as "almost coherent.[1]

It is also interesting to relate γ_{12} to the fringe visibility V. To do this, write $\gamma_{12} = |\gamma_{12}|\,e^{i\alpha}$, where

[1] M. Born and E. Wolf, *Principles of Optics* (Pergamon Press, Oxford, 1964), 2nd ed., p. 511.

$|\gamma_{12}|$ is assumed to be constant in the neighborhood of the observation point although α may vary with position. This is a good approximation if Q is not too near the source so that $|\psi_1|$ and $|\psi_2|$ are nearly constant in the neighborhood in question. This leads at once, using Eq. (6), to

$$I_{\max} = I_1 + I_2 + 2\sqrt{I_1 I_2}\,|\gamma_{12}| \qquad (10a)$$

and

$$I_{\min} = I_1 + I_2 - 2\sqrt{I_1 I_2}\,|\gamma_{12}| \qquad (10b)$$

in the vicinity of Q. Using Eq. (1), the visibility is then:

$$V = \frac{2\sqrt{I_1 I_2}}{I_1 + I_2}|\gamma_{12}|. \qquad (11)$$

In the special case, $I_1 = I_2$, which reduces to the interesting result $V = |\gamma_{12}|$, thus emphasizing the intimate relation between coherence and fringe visibility.

36–2.2 Coherence Time and Coherence Length

To apply the preceding analysis to coherence time and coherence length, let a beam of light from a simple point source P be split in half, sent through some kind of interferometer, and recombined at the observation point Q. Then $\psi_2(Q,t) = \psi_1(Q, t+\tau)$, where τ is the difference in time of flight from P to Q via the two paths. If both intensities are set equal to unity, the coherence function is

$$\gamma_{12}(\tau) = \langle \psi_1{}^*(Q,t)\, \psi_1(Q, t+\tau)\rangle. \qquad (12)$$

The coherence time τ_c is defined as the root-mean-square width of the square of the magnitude of $\gamma_{12}(\tau)$:

$$\tau^2{}_c = \frac{\int_{-\infty}^{\infty} \tau^2\, |\gamma_{12}|^2\, d\tau}{\int_{-\infty}^{\infty} |\gamma_{12}|^2\, d\tau}. \qquad (13a)$$

Thus $|\gamma_{12}|$ is small and interference cannot be observed if τ is much greater than τ_c. It is also convenient to define the coherence length which is just the distance traveled by light in one coherence time,

$$L_c = c\tau_c, \qquad (13b)$$

where c is the velocity of light, not to be confused with the subscript for coherence. Therefore, to observe sharp fringes in an interferometer, one should

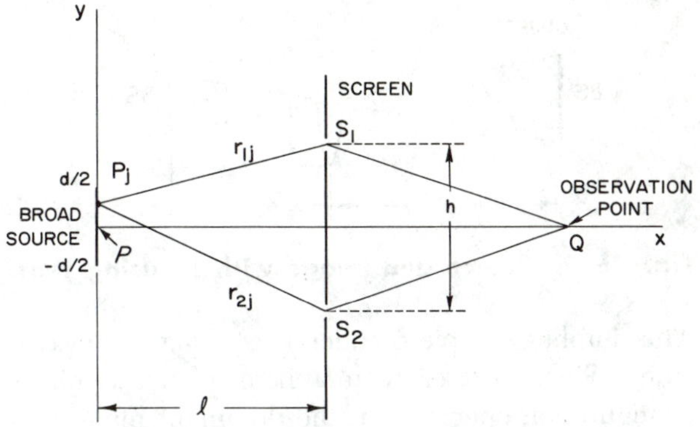

Fig. 36–1 Double-Slit Experiment.

keep the path difference less than a coherence length.

For many sources of interest the spectral density is a Gaussian function centered about a frequency ν_0 with a full linewidth $\Delta\nu$:

$$P(\nu) = P_0 \exp\left[-2\left(\frac{\nu - \nu_0}{\Delta\nu}\right)^2\right]. \qquad (14)$$

It can be shown that the coherence time is then

$$\tau_c \approx \frac{1}{2\pi\,\Delta\nu}, \qquad (15)$$

which relation can also be obtained for a general spectrum if τ_c and $\Delta\nu$ are defined in an rms sense.[(2)]

36–2.3 Spatial Coherence

The same formulation can be used again to discuss transverse spatial coherence in the context of the classic double-slit experiment. As in Fig. 36–1, a source P is set a distance l behind a screen with two holes S_1 and S_2 of separation h, and an observation point Q is chosen which is equidistant from S_1 and S_2. The visibility of fringes at Q depends upon the coherence of the waves arriving at Q, but this is the same as the coherence of the waves in the two slits. If the distances from the source to the slits are nearly equal, we can set the intensity at each slit equal to unity and then Eq. (5) can be adapted in the form

$$\gamma_{12}(S_1, S_2) = \langle \psi_1{}^*(S_1, t)\, \psi_2(S_2, t)\rangle. \qquad (16)$$

If P is a point source and S_1 and S_2 lies on a sphere centered at P, then $\psi_2(S_2, t) = \psi_1(S_1, t)$ and $|\gamma_{12}| = 1$.

(2) Born and Wolf, *op. cit.*, p. 540.

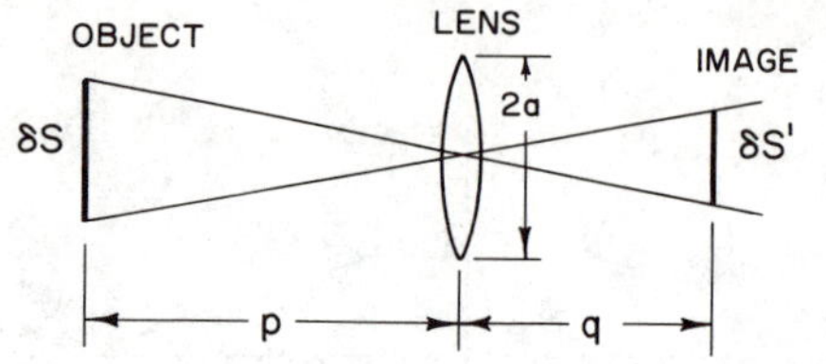

Fig. 36–2 Source and Image with Focusing Lens.

This implies complete coherence along the wavefront. Since a laser with a hemispherical mirror configuration operating in the dominant mode acts as an ideal point source with the exception that the wavefront has limited angular extent, we see that complete spatial coherence is obtained over the area of the beam.

A conventional source has finite extent, and the effect of this must be included in determining the degree of coherence. Suppose that P is a line source of length d divided into regions P_j of atomic size Δ_y, so that the amplitude at S_1 may be found by summation over the source points:

$$\psi_1(S_1, t) = \sum_j \frac{A_j}{r_{1j}} \exp\left[i(kr_{1j} - \omega t + \phi_j\right] \qquad (17)$$

and similarly for $\psi_2(S_2, t)$, where A_j are the individual source amplitudes, ϕ_j the phases, r_{1j} are the distances from the jth source point to slit S_1, and k is 2π divided by the wavelength. Then

$$\langle \psi_1{}^*(S_1, t)\, \psi_2(S_2, t)\rangle$$
$$= \left\langle \sum_{n,j} \frac{A_n A_j}{r^2} \exp\left[i\left(k\left(r_{2j} - r_{1n}\right) + \phi_j - \phi_n\right)\right]\right\rangle, \qquad (18)$$

where it has been assumed that $r_{1n} \approx r_{2j} \approx r$ in amplitudes but not in phases.

If the source is completely incoherent, the atomic radiators are turning on and off independently of each other, which implies that $\phi_j - \phi_n$ is a random function of time for $j \neq n$. Thus in the time average all terms in Eq. (18) vanish except those with $n = j$. Also if $l >> h$ and $l >> d$ then in Fig. 36–1, $(r_{2j} - r_{1j}) \approx hy/l$ and

$$\langle \psi_1{}^*(S_1, t)\, \psi_2(S_2, t)\rangle = \sum_j \frac{<A^2{}_j>}{r^2} e^{ikhy/l} \qquad (19a)$$

$$= \sum_j \int_{y_j}^{y_j + \Delta y} \frac{A^2}{r^2 \Delta y} e^{ikhy/l}\, dy \qquad (19b)$$

$$= \int_{-d/2}^{d/2} \frac{A^2}{r^2 \Delta y} e^{ikhy/l}\, dy, \qquad (19c)$$

where $A^2 = \langle A^2{}_j\rangle$ is assumed the same for all source points, and the sum has been replaced by an integral for ease in evaluation. This can be integrated and used in an expression like Eq. (5) to give[(3)]

$$|\gamma_{12}| = \frac{\sin(khd/2l)}{khd/2l}, \qquad (20)$$

where the normalizing factor is found most easily by the physical requirement that $|\gamma_{12}| = 1$ for $d = 0$. From this result it is seen that the wavefront from a line source of length d is almost coherent ($|\gamma_{12}| > 0.88$) over separations h such that

$$h < 0.27\, \lambda l/d. \qquad (21)$$

If the source is extended but now completely coherent, then $(\phi_j - \phi_n)$ is a constant with respect to time in Eq. (18), and the time average of the double sum is just the double sum itself. Likewise $|\psi_1|$ and $|\psi_2|$ are constant in time so that the numerator and denominator cancel in Eq. (5), giving $|\gamma_{12}| = 1$ for all points S_1 and S_2, thus demonstrating the perfect coherence of the field due to a coherent, but extended, source.

36–2.4 Image Brightness

A plane incoherent source may often be treated[(4)] as a plane element of constant brightness B W/cm²-steradian. Suppose a lens of radius a is placed on the axis at a distance p from the source, thus forming an image at a distance q from the lens, as shown in Fig. 36–2. What is the illumination L in W/cm² at the image plane? Assume $a << p$, $a << q$, and a source of moderate extent with an area δS. From purely geometrical considerations, the power accepted by the lens from such a source is

$$P = B \frac{\pi a^2}{p^2} \delta S, \qquad (22)$$

(3) This is a special case of the van Cittert-Zernike theorem, which relates the coherence function of an extended source to the diffraction pattern obtained by coherent illumination of an aperture with the same dimensions as the source. See Born and Wolf, *op. cit.*, p. 560.

(4) Born and Wolf, *op. cit.*, p. 182.

and at best this will all be focused into the image of area $\delta S'$, giving an illumination

$$L_1 = P/\delta S'. \tag{23}$$

However, from geometrical optics $\delta S/\delta S' = p^2/q^2$, so that the illumination of the image will at most be

$$L_1 = B\,\pi a^2/q^2 \text{ W/cm}^2. \tag{24}$$

This is ultimately limited by how large one can make a/q, and thus by the ratio of the radius to the focal length of the lens.

On the other hand, a laser with hemispherical configuration acts almost like an ideal point source, except that the output is confined to a narrow beam. Since the brightness of a point source approaches infinity, Eq. (24) would also imply infinite illumination at the image point. However, in actuality the minimum size of the image is limited by diffraction.

If the lens diameter is not as large as the incident beam diameter, the image size is limited essentially to the central maximum of the diffraction pattern, called the Airy disc.[5] The angular half width of the Airy disc is $\theta = 0.6\,\lambda/a$, and its area at an image distance q is $0.36\,\pi\lambda^2 q^2/a^2$. For a monochromatic incident plane wave, 84 percent of the incident energy lies within the central maximum so that the average illumination of the smallest image possible with a laser source of power P is approximately

$$L_2 \approx 0.7P\,a^2/\lambda^2 q^2 \text{ W/cm}^2. \tag{25}$$

If, on the other hand, the lens diameter is large compared to the incident beam, it is the transverse intensity distribution in the beam itself that limits the image size by diffraction. If the laser is operating in a single TEM_{00} mode (see Sec. 36–3.4), the intensity distribution may be written[6]

$$I(\theta) = \frac{P}{\pi\alpha^2 p^2} e^{-\theta^2/2\alpha^2}, \tag{26}$$

where P is the total power, p the distance from the point source to the observation point, θ the angle from the beam axis, and α the angular half width. The usual Kirchoff-Huygens diffraction integral gives a Gaussian intensity distribution in the image as well[7] with a half width inversely proportional to α so that narrowing the beam broadens the image and conversely. The illumination at the center of the image is

$$L_3 \approx 2\pi P\,(2\alpha p)^2/\lambda^2 q^2 \text{ W/cm}^2, \tag{27}$$

which bears a resemblance to Eq. (25) with the lens radius α replaced by a "beam radius" $2\alpha p$ and with a larger proportionality constant. (Eq. (26) shows that at this radius the beam intensity drops to $1/e^2$ of its central value.)

36–2.5 Numerical Comparisons

Now that the theoretical groundwork has been laid, let us take a typical small gas laser and compare it with a conventional source.

1. *Coherence Time.* The coherence time is determined by the linewidth according to Eq. (15), and for a single spectral line from a conventional source, the linewidth is the Doppler width $(\Delta\nu)_D$ (see Eq. 37). For 6328-Å light emitted from a neon source at room temperature, $(\Delta\nu)_D = 1830$ Mc/sec and the coherence time is

$$\tau_c = 0.26 \text{ nsec (conventional source)}. \tag{28a}$$

The effective linewidth of a laser source in single-mode operation depends on various engineering variables such as stability of the mirrors against vibration. For lasers of research grade this can be as low as 20 cps, depending on the mounting, but for demonstration quality lasers, it may be much higher. It must of course be much less than the cavity width (see Eq. 38) to get adequate laser action. Assuming a cavity width of $(\Delta\nu)_c = 1.6$ Mc/sec, we estimate a maximum laser linewidth $(\Delta\nu)_L \approx 2 \times 10^4$ cps which implies

$$\tau_c = 8 \text{ microseconds (laser)}. \tag{28b}$$

2. *Coherence Length.* As for coherence length, these coherence times give

$$L_c = 3 \text{ cm} \quad \text{(conventional source)} \tag{29a}$$

and

$$L_c = 2.4 \text{ km} \quad \text{(laser)}. \tag{29b}$$

3. *Transverse Spatial Coherence.* As an example

[5] Born and Wolf, *op. cit.*, p. 395*ff*.

[6] G. D. Boyd and J. P. Gordon, *Bell System Tech. J.*, **40**, 489 (1961).

[7] Robert C. Rempel, "Optical Properties of Lasers as Compared to Conventional Radiators," *Spectra-Physics Laser Technical Bulletin*, No. 1, June, 1963.

of an incoherent source, consider a line source 1 mm long radiating at a wavelength 6328 Å. Equation (20) predicts that the wavefront at a distance of 1 m will be almost coherent over a distance

$$h_c \approx 0.02 \text{ cm} \quad \text{(conventional source)}. \qquad (30)$$

A laser operating in a single mode is essentially a perfectly coherent source, and thus spatial coherence is maintained over the entire wavefront. Typically the full-angular spread of the wave from a small laser with hemispherical configuration may be 0.01 radian which at 1 meter gives a spatially coherent wavefront of width

$$h_c \approx 1 \text{ cm} \quad \text{(laser)}. \qquad (31)$$

4. *Image Brightness.* A 1000-W mercury arc, for example a General Electric AH6, has a brightness of about 10 W/cm^2-steradian in the 5461-Å line,[8] while a small gas laser radiates about 0.5 mW with α, the half angle of the Gaussian wavefront, equal to about 0.005 radian. Equations (24) and (25) predict that, with the use of a given lens and source position, the ratio of the power per square centimeter arriving at the image of the laser source to that for the conventional source when lens diffraction limits the size of the laser image is

$$\frac{L_2}{L_1} = 2.9 \times 10^3. \qquad (32)$$

When beam diffraction determines the size of the laser image, Eqs. (24) and (27) give

$$\frac{L_3}{L_1} = 1.0 \times 10^5 \quad \text{(laser/conventional source)}. \qquad (33)$$

In summary we see that typically a small laser source is better than a conventional source by a factor of 10^5 in coherence length, 10^2 in transverse coherence, and 10^3 to 10^5 in brightness. These qualities make it exceedingly useful for optics demonstrations.

36–3 Theory of Operation of the Gas Laser

Although it is beyond the scope of this work to give a detailed treatment of the theory of the gas laser, a short qualitative discussion may be of value as background for the demonstration use of a laser.

36–3.1 Emission of Radiation

The actual emission process may best be discussed in terms of the energy level diagram[9] in Fig. 36–3. The neon atom in the ground state can be thought of as a core of 9 electrons in a configuration $1s^2\ 2s^2\ 2p^5$ and a 10th electron which is also in a $2p$ level. Excited states of the atom are formed when this 10th electron goes into a higher energy level, such as one of the levels labeled $5s$ on the chart. The numbering indicates the actual n and l levels for the electron.[10] The excited electron may then go to a lower level such as a $3p$ level, emitting the excess energy as light. Such a transition is allowed by the usual selection rule $\Delta l = 1$. The $5s$ state is actually made up of 4 substates, and the $3p$ state is made up of 10. The 6328-Å visible red line of neon comes from a transition of the $5s \rightarrow 3p$ type between a particular pair of substates. The radiative decay of the atom then proceeds via $3p \rightarrow 3s \rightarrow$ ground, thus effectively emptying the $3p$ state.

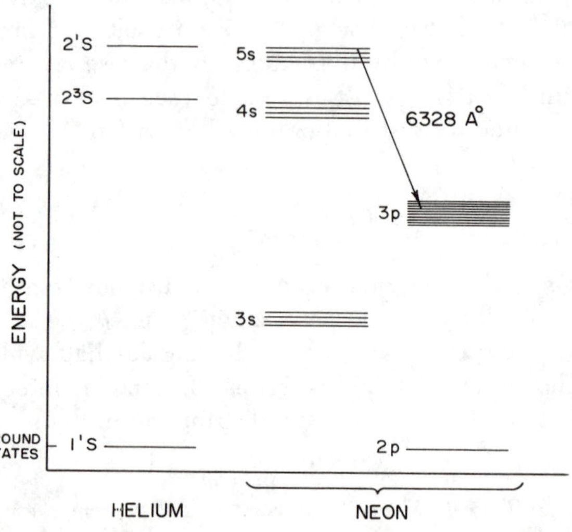

Fig. 36–3 **Energy Levels in Helium and Neon.**

[8] D. Dutton, M. P. Givens, and R. E. Hopkins, *Am. J. Phys.*, **32**, 355 (1964).

[9] A more detailed discussion of the energy levels may be found in B. A. Lengyel, *Introduction to Laser Physics* (John Wiley & Sons, Inc., New York, 1966), p. 177*ff*.

[10] Note that this is not the customary Paschen notation.

At thermal equilibrium the population of the upper state is just enough less than that of the lower state to prevent any net emission of radiation. This can only be achieved if an external mechanism is used to upset equilibrium and over-populate the upper state; that is to say, a population inversion must be achieved. This is usually accomplished by the collisions of atoms with electrons in a high voltage gas discharge. The energy required is about 20 eV, and it turns out to be more efficient (but not essential) for the electron collisions to excite helium atoms, which are much lighter than neon. Fortunately helium has an excited state (designated 2 S in Fig. 36–3 using *L–S* notation) at almost the same excitation energy as the 5*s* level in neon. Subsequent collision of the excited helium atom in this state with a neon atom in the ground state is very likely to transfer the energy to the neon atom and excite it to the 5*s* level as desired.

36–3.2 Stimulated and Spontaneous Emission

Let us now consider a simplified two level atomic system and its interaction with radiation. The total probability that an atom in state n will undergo a transition to a lower state m is:

$$P_{n\to m} = A_{n\to m} + Nh\nu_{nm}B_{n\to m}, \tag{34}$$

where $h\nu_{nm}$ is the energy difference between levels n and m and N is the number of photons per unit volume with that energy. The first term is the probability of spontaneous emission, and the second is the probability of stimulated emission, to be discussed below. Also, the total probability for an atom to absorb a photon of energy $h\nu_{nm}$ is

$$P_{m\to n} = Nh\nu_{nm}\,B_{m\to n}. \tag{35}$$

Relations between $A_{n\to m}$, $B_{n\to m}$ and $B_{m\to n}$, the so-called Einstein coefficients, may be found from simple statistical mechanics,[11] but the value of one of them must be determined by the usual methods of time-dependent perturbation theory in quantum mechanics.[12] The main difficulty involved is the complication of multi-electron wave functions.

(11) O. S. Heavens, *Optical Masers* (Methuen and Co. Ltd, London, 1964), chap. 1.

(12) R. B. Leighton, *Principles of Modern Physics* (McGraw-Hill Book Company, Inc., New York, 1959), Chap. 6.

The term $Nh\nu_{nm}B_{n\to m}$ which describes stimulated emission is the one of interest to us in obtaining laser action. It is called a stimulated emission term because of its proportionality to the number of photons already present in the radiation field at the transition energy.

36–3.3 Laser Action

The state of a photon can be completely represented by its wave vector **k** and its spin σ, and the remarkable thing about stimulated emission is that the emitted photon has the same values of **k** and σ as that portion of the radiation field that caused the stimulating perturbation. Thus, if we had a box of atoms in the upper state, a photon emitted spontaneously by the first atom could stimulate other atoms to emit like photons. These in turn increase the probability of stimulated emission by other atoms, so that a tremendous buildup of photons with a particular **k** and σ could be obtained provided that

1. The population of the upper state does not become too depleted; and
2. The emitted photons are not lost to the system, but are kept in the region of the emitting atom.

The first condition can be achieved by continually exciting atoms from the ground state to the upper state, using collisions in a gas discharge as previously described in Sec. 36–3.1. Condition (2) is achieved by placing the discharge tube in a cavity with reflecting end walls (Fabry-Perot cavity).

36–3.4 The Fabry-Perot Cavity

The cavity, of cylindrical shape, confines those photons with **k** parallel to the axis, so that they get reflected back and forth about a hundred times before finally escaping into the external beam. This is achieved by making the ends of the cavity of multi-layer dielectric mirrors with a reflection coefficient of about 99 percent. This provides a radiation density in the cavity high enough to cause stimulated emission of at least one photon with the correct **k** and σ for every photon that is lost out the ends, thus allowing sustained oscillation. It is not necessary to enclose the sides of the cavity, and leaving them open allows the escape of many of

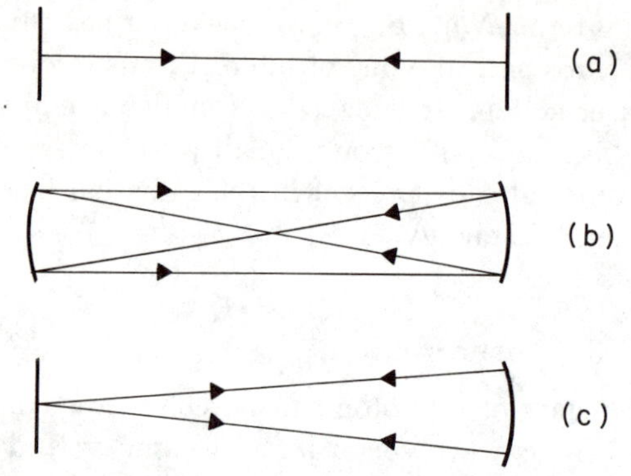

Fig. 36–4 Mirror Configurations: (a) plane parallel, (b) confocal, (c) hemispherical. The path of a typical ray is shown from a starting point until it returns to that point.

the photons that are spontaneously emitted with unwanted values of **k**.

There are three useful configurations of end mirrors, as shown in Fig. 36–4. The plane parallel configuration conveniently produces a plane wave output, but is quite difficult to align. The confocal configuration, so-called because the focal points of the two mirrors[13] lie at the center of the cavity, is easy to align, but the oscillations are likely to occur in high order transverse modes. The hemispherical configuration is the most popular, since it is easy to align and operates well in a single TEM_{00} mode (see below). The output is a spherical wavefront originating at a point on the plane mirror which is at the center of curvature of the hemispherical mirror.

As with a microwave cavity, the laser cavity has certain normal modes of oscillation. For the modes of interest the electric and magnetic field vectors are mostly perpendicular to the axis; so they are designated TEM_{rm}, where m is half the number of zeros in the angular distribution and r is the number of internal zeros in the radial distribution. TEM refers to the transverse electric and magnetic field.[14] The electric field configurations for several modes are illustrated in Fig. 36–5.

For a given transverse mode there is a set of allowed longitudinal modes, which determine the

[13] The focal length of a mirror is half its radius of curvature.

[14] A. G. Fox and T. Li, *Bell System Tech. J.*, **40**, 453 (1961), have a discussion of mode theory; and W. W. Rigrod, *Appl. Phys. Letters*, **2**, 51 (1963), has a series of photographs of various modes.

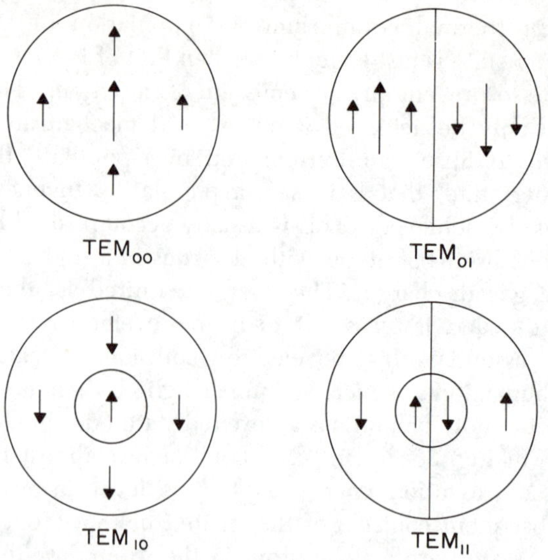

Fig. 36–5 Electric Field Configurations in Various Modes of a Resonator with Circular Mirrors.

frequency according to the cavity resonance condition $2l = n\lambda$ or

$$\nu = nc/2l, \qquad (36)$$

where l is the cavity length and n is any integer. At resonance, standing waves are set up with $n + 1$ nodes along the length of the cavity. For a typical laser, n is very large, perhaps 10^6. At low power levels the laser oscillates in a single longitudinal mode, while at higher power levels several modes may be present at once. This can be important in demonstration use, as different modes are not in general coherent with one another. It is therefore necessary to ensure that only a single mode is present or that a given mode always interferes with itself.

36–3.5 Polarization

As described so far, a laser could oscillate with any or all possible polarizations of the electric field, and that is indeed the case for lasers with the mirrors sealed within the discharge tube. However, many lasers are constructed with external mirrors as indicated in Fig. 36–6 both to increase the life of the mirrors by removing them from the discharge region and also to allow access to the beam within the Fabry–Perot cavity. In order to bring the beam out of the plasma tube to the mirrors with negligible loss, end windows oriented

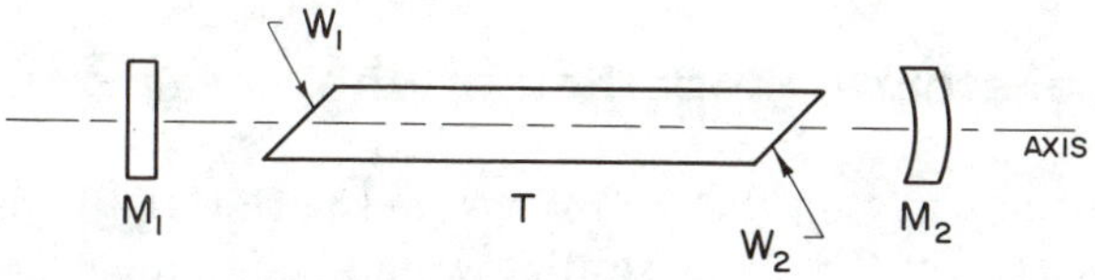

Fig. 36–6 Schematic Diagram of Laser with External Mirrors. T is the discharge tube, W_1 and W_2 are windows set at Brewster's angle, and M_1 and M_2 are the mirrors that form the Fabry-Perot cavity.

parallel to each other and at Brewster's angle to the beam are used. Since lossless transmission at the window occurs only when the electric vector is polarized in the plane of incidence, the oscillations that build up in the laser will be linearly polarized and so will the external beam.

36–3.6 Linewidth

Finally, it is necessary to discuss linewidth. For a conventional gas discharge tube at moderate pressures, it is possible by the use of filters to select a single spectral line. However, the fact that the emitting atoms have a distribution of thermal velocities causes a spread in the emitted frequencies. The calculated spectrum is Gaussian with a full linewidth[15]

$$(\Delta\nu)_D = 2\nu_0 \sqrt{\frac{2kT}{Mc^2}\, ln\, 2}, \qquad (37)$$

where ν_0 is the central frequency of the line, k is Boltzmann's constant, T the absolute temperature of the gas, and M the mass of the gas atom.

If now the source is placed in a Fabry–Perot cavity as described above, the external beam will only contain frequencies corresponding to cavity modes allowed by Eq. (36). Of course these cavity modes are not infinitely sharp but have a full width at half-maximum power of[16]

$$(\Delta\nu)_c = cf/2\pi l, \qquad (38)$$

where f is the fractional energy loss per transit. Thus the output of this device will consist of several lines of width $(\Delta\nu)_c$ and spacing $(\delta\nu)_m = c/2l$, contained within the Doppler envelope as shown in Fig. 36–7. Typical values of these parameters for

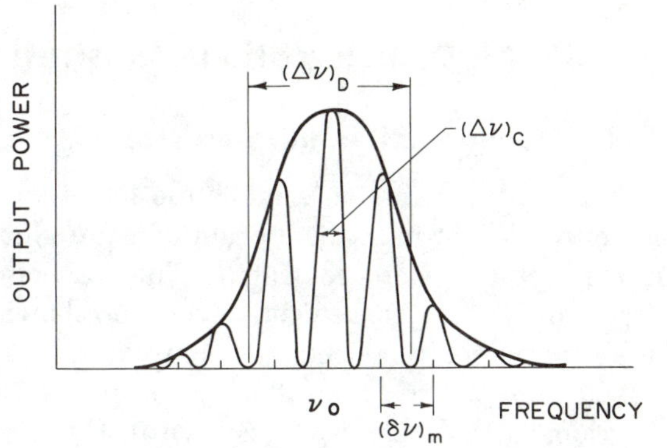

Fig. 36–7 Output Spectrum of a Gas Discharge Source in a Cavity.

visible light in a 30-cm cavity are: $\nu_0 = 0.5 \times 10^9$ Mc/sec, $(\Delta\nu)_D = 1500$ Mc/sec, $(\Delta\nu)_c = 1.5$ Mc/sec, and $(\delta\nu)_m = 500$ Mc/sec.

So far laser action has not been included although the effect of the cavity has. When the former is also present, the stimulated emission process tends to build up the output at a very particular frequency within a cavity mode at the expense of neighboring frequencies, thus narrowing each line still further. The ultimate limit on the linewidth in a single mode is theoretically[17]

$$\Delta\nu = \frac{8\pi h\nu\,(\Delta\nu)^2{}_c}{P},$$

where P is the optical power. For a 1-mW laser this gives $\Delta\nu_a = 10^{-3}$ cps. Values obtainable by experiment have not been nearly this low, mainly because of additional noise effects due to such things as acoustic vibrations of the end mirrors. However, there is experimental evidence for laser linewidths as low as 20 cps.[18]

Finally, if the excitation of the discharge is low near the threshold for laser action, oscillation occurs only in the dominant cavity mode. This is the method used in practice to provide a light source with maximum monochromaticity and coherence. At higher levels of excitation other cavity modes will also be present with a consequent broadening of the observed spectrum and loss of longitudinal coherence.

(15) B. A. Lengyel, *op. cit.*, p. 44.

(16) W. R. Bennett, Jr., *Applied Optics, Supplement on Optical Masers* (1962), p. 26.

(17) C. H. Townes, *Advances in Quantum Electronics*, ed. J. Singer (Columbia University Press, New York, 1961), p. 3.

(18) T. S. Jaseja, A. Javan and C. H. Townes, *Phys. Rev. Letters*, **10**, 165 (1963).

36–4 Demonstrations Illustrating Coherence Properties of the Laser

36–4.1 Long Path Interferometer

The first demonstration exhibits the longitudinal coherence of light in a single beam and introduces the concept of coherence length. The Michelson interferometer is a convenient device for showing this. Fringes are produced in the usual way, except that an external movable mirror is used to allow long path differences, as shown in Fig. 36–8. In addition, the high intensity of the laser source allows the fringes to be viewed on a screen so that about two dozen people or so can see them at the same time. Visible circular fringes can be observed with the movable mirror at least 6 m farther than the fixed mirror from the beam splitter, thus giving a coherence length of at least 12 m, in striking contrast with a conventional source. Since the beams do travel greatly different distances, it is desirable to use a converging lens in the long beam to make the two spots about the same size and obtain good contrast. It also may be useful to use a diverging lens system in front of the laser to make sure that the beam fills the aperture of the interferometer. The main difficulty encountered is the relative vibration of the distant mirror and the interferometer mount which causes the fringes to appear and disappear rapidly. It can easily be demonstrated by varying the mirror distance that this effect depends only on the mountings and not on the light itself, but the demonstration would be improved by using vibration-free mountings if they are available.

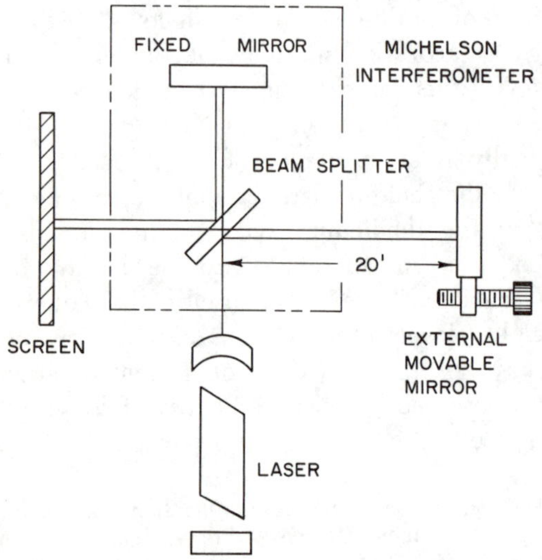

Fig. 36–8 Michelson Interferometer, Modified for Long Path Difference.

36–4.2 Long Path Interferometer, Second Version

A second version of this demonstration that gives an audible presentation that is suitable for a large lecture class is shown in Fig. 36–9. A long-path Michelson interferometer is used again, but the output beam is brought to a small silicon photovoltaic cell such as the Texas Instruments LS–222 rather than to a screen. The optics are adjusted so that a single interference fringe fills the aperture of the photocell, and the output of the photocell is connected to an audio amplifier which has a loudspeaker. As one mirror is moved, the path difference between the two beams changes so that bright and dark fringes will alternately appear at the photodetector. If the mirror velocity is v, then the output voltage of the photocell will vary at a frequency $f = 2v/\lambda$ cps so that a velocity of 1 mm/sec produces a conveniently audible output frequency of 3.2 kc/sec. This result can also be interpreted as a Doppler shift of the frequency of the light reflected from the moving mirror which causes beats to be produced in the photocell current at the difference frequency between the two optical frequencies.

The apparatus is illustrated using cube-corner reflectors instead of plane mirrors since the latter require very precise tilt adjustments while the

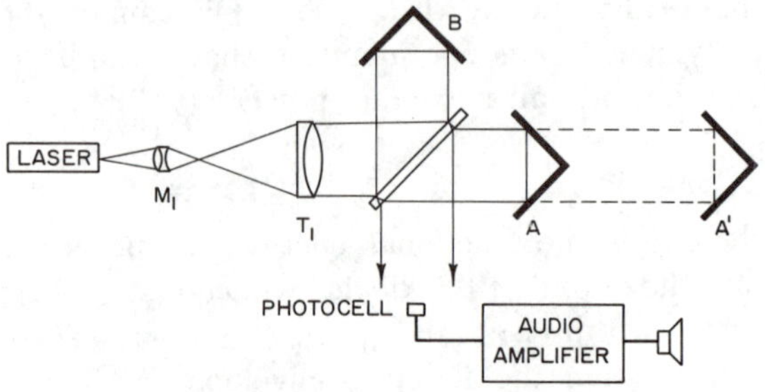

Fig. 36–9 Arrangement for Audible Demonstration of the Michelson Interferometer. M_1 is a microscope objective, T_1 is a telescope objective, B is a fixed cube-corner reflector, and A is a movable cube reflector. (Courtesy of D. Dutton.)

former require only an approximate lateral alignment. However, the cube-corners are rather expensive, and are not essential unless the very longest paths are desired. If cube-corners are to be used at all, it would be well to get two because, unlike plane mirrors, they cause light from the top of the input beam to come out from the bottom of the output beam. If the laser is oscillating with anything but a uniphase wavefront, as it may well be if maximum power is desired, then use of like mirrors is essential to ensure that corresponding parts of the wavefront are brought together in the interference pattern. Otherwise interference may not be observed!

The fact that true interference is being observed can be demonstrated by blocking first one, then the other beam, causing most of the audio signal to vanish. There is a slight rumble remaining due to a portion of the beam that is reflected back to the laser where it undergoes an additional reflection, re-emerging to interfere with the primary beam.

36–4.3 Double Ended Interferometer

This demonstration exhibits an effect that is not visible at all with a conventional light source—the coherence of light beams emitted from opposite sides of the source. A simple interferometer, Figs. 36–10, 36–11, and 36–12, can be constructed to recombine the beams from opposite ends of the laser, and good quality circular fringes can be observed on a screen by a small group in a semi-darkened room (Fig. 36–13). Because the interferometer is a single unit, the fringes persist for ten seconds or so between washouts so that vibration is not a serious problem in this demonstration. In building the interferometer, a plywood base and inexpensive optics can be used. However, the mirror mounts must be reasonably rigid and allow for rotation of each mirror about both a horizontal and a vertical axis in order to permit correct beam alignment (Fig. 36–12). A removable centering pin is very useful for the initial adjustments but is not needed for the minor re-adjustments required later. It is necessary to keep the optical surfaces clean, since small dust particles produce diffraction patterns of their own. The centering pin (also shown in Fig. 36–12) is a small rod with a pointed top and fixed height which fits in the mounting holes

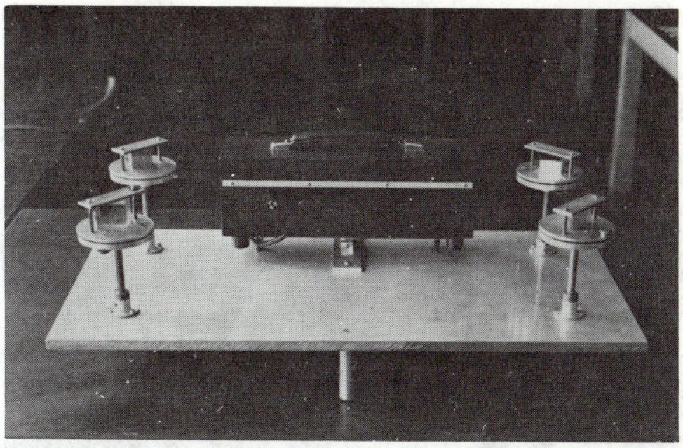

Fig. 36–11 Interferometer of Fig. 36–10. The laser is clamped to the table to maintain alignment. It acts as an ideal point source at the plane mirror (at right end of laser).

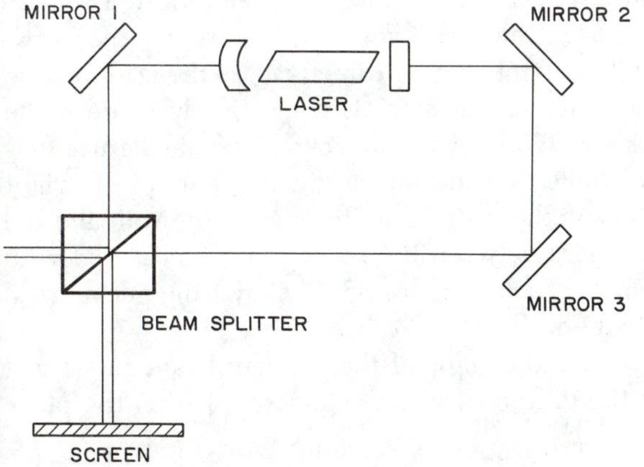

Fig. 36–10 Interferometer for Combining Beams Emitted from Opposite Ends of the Laser.

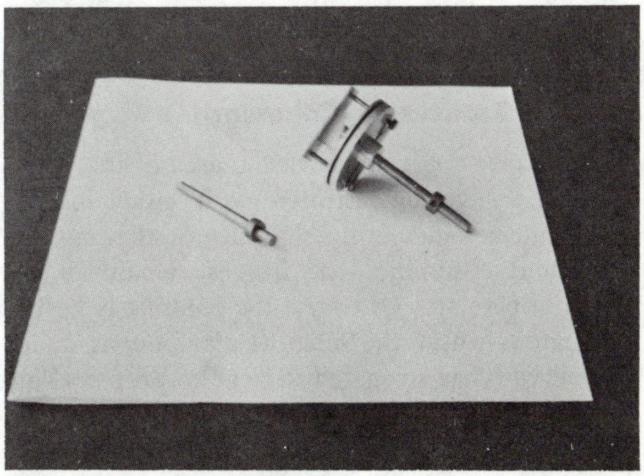

Fig. 36–12 Interferometer Details: centering pin (*left*) and mirror mount (*right*).

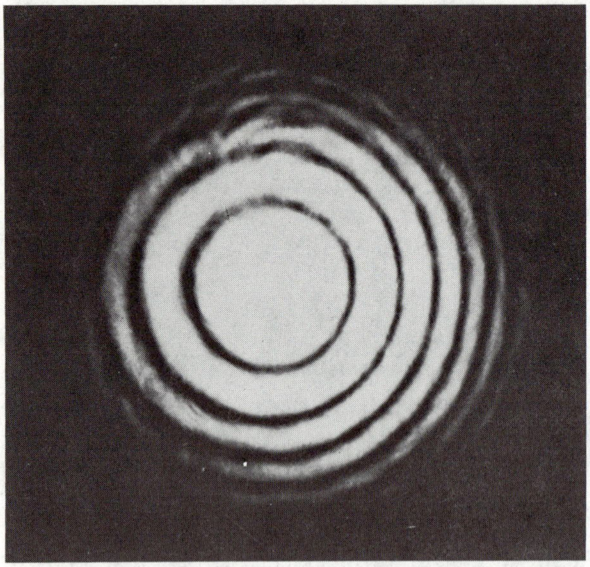

Fig. 36–13 Fringes Observed with Interferometer of Fig. 36–10, as seen on a screen.

in the base when the mirror mounts are removed. In order to align the system all the mirrors and the beam splitter are removed and the pin is placed alternately in positions Nos. 1 and 2 to allow adjustment of the height and orientation of the laser until the beams reaching these positions each arrive at the point of the pin. Then mirror No. 2 is replaced and adjusted to the correct tilt and rotation with the pin in position No. 3. Next, mirrors Nos. 1 and 3 are replaced and adjusted with the pin in the beam splitter position. Finally, the beam splitter can be replaced and adjusted to give concentric spots on the screen from the two beams. Fringes should then be observable, and final adjustments can be made to optimize them.

36–4.4 Transverse Coherence

Transverse coherence over the area of the wavefront is easily demonstrated with either of the above interferometers by slightly mis-aligning one beam and observing that fringes remain in the region of overlap. Of course the contrast is best in those areas where the intensities are about equal. To one who has struggled to get the proper alignment of an interferometer before seeing anything, it is both interesting and refreshing to have a source that produces visible interference fringes in the usual setup whenever the two beams overlap at all.

36–4.5 Thick Reflecting Plate

A demonstration involving both longitudinal and transverse coherence uses interference by reflection from a thick glass plate. Since the beam from the back reflecting surface is displaced laterally with respect to that from the front, this effect is limited with ordinary light to relatively thin films. With the laser, however, very good straight fringes can be observed using a flat glass plate of considerable thickness. The dependence of the fringe spacing and orientation on the reflection angle can be easily illustrated by moving the plate. A good piece of thick flat glass is often available as a compensator plate for a Michelson interferometer, but a piece of ordinary plate glass will do if that is not available.

In a classroom situation one trick that is useful for this or another demonstration that cannot be shown to a large class at one time is to have the pattern projected on the wall near the door, where the students can see it as they come in. For safety's sake the entire laser beam should be kept well below eye level. The room does not have to be completely darkened to make the pattern visible.

36–4.6 Measuring the Wavelength of Light with a Ruler[19]

This is a demonstration with a provocative title that is excellent for even a very large audience, as are most demonstrations that use some sort of diffraction grating. The arrangement is shown in Fig. 36–14. A steel machinist's scale with rulings of $\frac{1}{64}$ or $\frac{1}{100}$ in. is placed horizontally in front of the laser. The laser beam is then adjusted so that it grazes the last two inches or so on the scale, while part of it goes directly past the scale to the wall. This adjustment is conveniently made if the laser rests on two Lab-jacks. The steel ruler acts as a reflection grating producing a series of bright spots on the wall. The distance to the wall and the distances between the spots can be measured with a meter stick. Figure 36–15 shows the geometrical situation.

If α is the angle of the incident beam and β that of the diffracted beam as measured from the plane

[19] This experiment was first developed by A. L. Schawlow and published under this title in *Am. J. Phys.*, 33, 922 (1965). Thanks are extended to him for allowing it to be included here and for supplying the original figures.

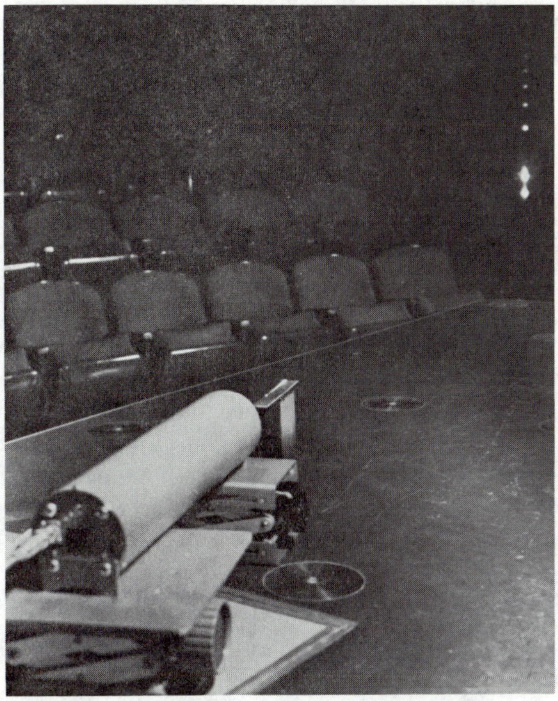

Fig. 36–14 Measuring the Wavelength of Light with a Ruler. The laser in the foreground shines on the rulings of a ruler, producing the diffraction pattern on the far wall. (Courtesy of A. L. Schawlow.)

of the "grating," the condition for the nth order maximum may be written

$$n\lambda = d(\cos\alpha - \cos\beta_n), \qquad (39)$$

where n is an integer, λ is the wavelength, and d is the spacing between rulings. In the experiment the distance to the screen x_0 and the distance between spots on the screen are measured. Since $\alpha = \beta_0$ (the zeroth order is specularly reflected), the intersection of the plane of the grating with the screen lies half-way between the spots of the direct beam and the zeroth order diffracted beam. Taking this point as the origin for measuring distances along the screen, the intersection of the direct beam is at $-y_0$ and those of the diffracted beams are at y_0, y_1, y_2, etc. Now,

$$\cos\beta_n = [1 + (y_n/x_0)^2]^{-1/2}, \qquad (40)$$

and if $y_n << x_0$ this may be approximated by

$$\cos\beta_n \approx 1 - y^2{}_n/2x^2{}_0.$$

Also,

$$\cos\alpha = \cos\beta_0.$$

Therefore,

$$n\lambda \approx \frac{d}{2}\left[\frac{y^2{}_n - y^2{}_0}{x^2{}_0}\right], \qquad (41)$$

which allows the calculation of λ from the data. Since the approximation used is best for the first order diffracted beam, $n = 1$ is expected to give the best value of λ. However, in a typical case, the value of λ calculated from the 6th order diffracted spot was only 1 percent greater than that from the 1st order. Even if the experiment is performed without special care, an accuracy of 1–2 percent is easily possible.

An additional point that can be readily illustrated with this apparatus is the dependence of the diffraction pattern on grating spacing. On the usual machinist's scale, longer markings are used to set off every second or fifth division of the fine scale. A slight change in the alignment of the laser beam allows it to be reflected from this coarser grating, and a more closely spaced diffraction pattern appears on the wall.

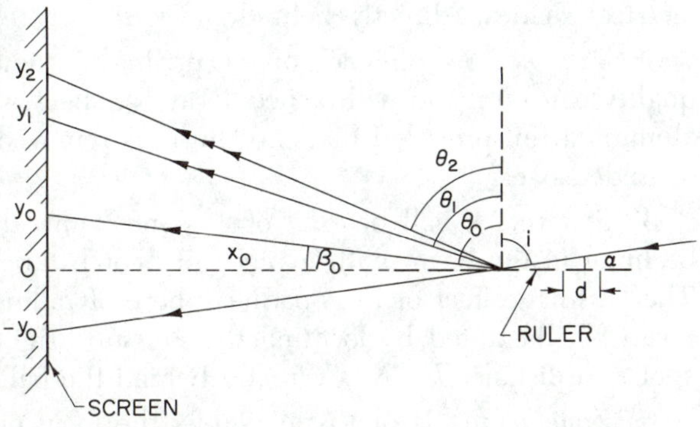

Fig. 36–15 Angles and Distances for Calculating Wavelength from the diffraction pattern of a reflection grating. (Courtesy of A. L. Schawlow.)

36–4.7 Sparkling Spots and Random Diffraction[20]

An interesting and unexpected effect can be observed by simply shining the laser beam on the rough surface of a wall. This works best if the spot diameter is about 2 in. and the viewers are reasonably near. One sees a general sparkling

(20) Articles describing this phenomenon may be found in *Proceedings of the IEEE*, **50**, 2367 (1962) and **51**, 220 (1963).

within the illuminated region, an effect that is a bit hard to describe on paper, but that is unforgettable when it has once been seen. This is due to the formation of an interference pattern in three dimensions in the region near the spot as the coherent light scattered from the randomly oriented surface elements combines to make a null here and a bright region there.

36–5 Demonstrations of Interference and Diffraction

36–5.1 General Comments

One of the most important properties of laser light from a teaching point of view is its high intensity, which allows the direct formation on a screen of the usual diffraction and interference patterns that formerly required individual viewing. For a class of thirty or forty, all of these patterns may simply be shown on a screen at the front of the room, as shown in Fig. 36–16. To make them all visible to a larger class requires a bit of ingenuity. One approach is to hang a sheet of paper among the seats in the classroom for use as a screen. The paper is translucent enough to allow the students behind it to see the pattern by transmission, while those in front see it by reflection as usual. To ensure safety it is essential to lay out the apparatus in such a way that it is physically impossible for anyone to look accidentally into the direct beam. Finally, for a really large class, one can set the apparatus on the lecture table and then use a mirror to project the desired patterns successively on each of several such screens, placed strategically about the room. A final word of caution to the lecturer is to stand as far from the screen as the farthest student to make sure that the phenomena he is supposedly illustrating are truly visible to the students as well as to the instructor.

One great advantage of the setup indicated in Fig. 36–16 is its extreme simplicity. The absence of additional lenses for focusing and collimation is important in allowing the audience to concentrate on the diffraction phenomenon itself, and the formation of the fringes on a screen seems to give them a greater objective reality for the viewer than they would have if seen through a telescope or by holding slits up to the eye.

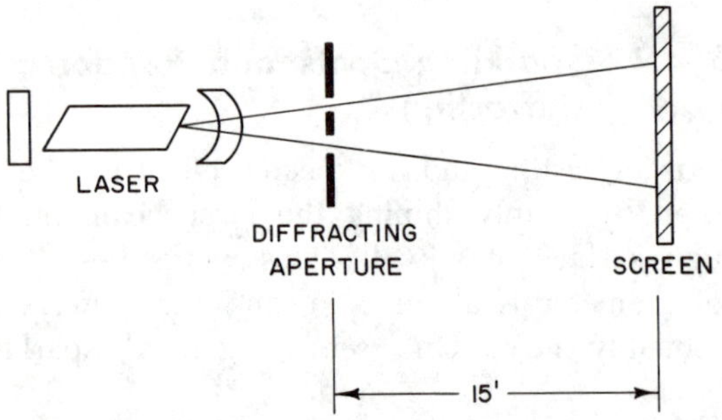

Fig. 36–16 Diffraction Demonstrations: experimental layout.

36–5.2 Suggested Diffracting Objects[21]

1. *Straight edge.* A razor blade makes a straight edge of good quality. Or, if an adjustable single slit is available, it may be opened wide and one of its jaws used. Notice the diffraction pattern in the region where light is not allowed to go by geometrical optics.

2. *Single slit* (Fig. 36–17). A slit with adjustable width is very convenient, especially if it opens and closes symmetrically. Several manufacturers also make glass slides with slits of various widths in them.

3. *Single wire.* This is approximately the complement to the single slit. A piece of piano wire with a diameter of the order of 0.15 in. has an interesting diffraction pattern in that it is possible to obtain a bright line in the middle of the geometrical shadow, directly behind the wire.

4. *Pinhole.* A pinhole of surprisingly good quality can be made with a needle and a sheet of aluminum foil provided that the "burr" is removed or folded back.

5. *Sphere.* A ball bearing of $\frac{1}{16}$ in. diam can be hung in the beam with a piece of Scotch tape. The lens-like effect of the opaque sphere of radius r can be illustrated by locating the Poisson bright spot at a distance $l = r^2/\lambda$ directly behind the ball.

6. *Zone plate.* If one is available, the focusing

[21] No discussion of diffraction theory is included here, since it is readily available in any standard optics text.

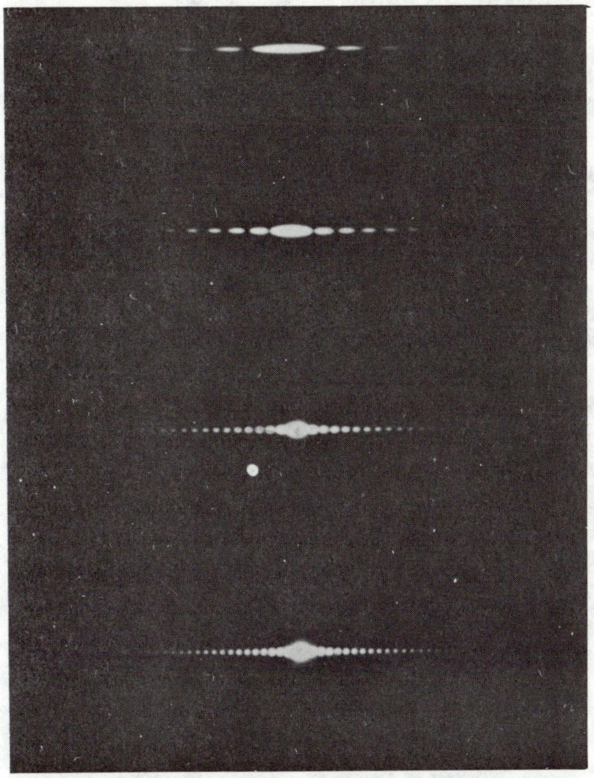

Fig. 36–17 Single-Slit Diffraction Patterns as seen on a screen. Slit widths: 0.088, 0.176, 0.35, and 0.70 mm.

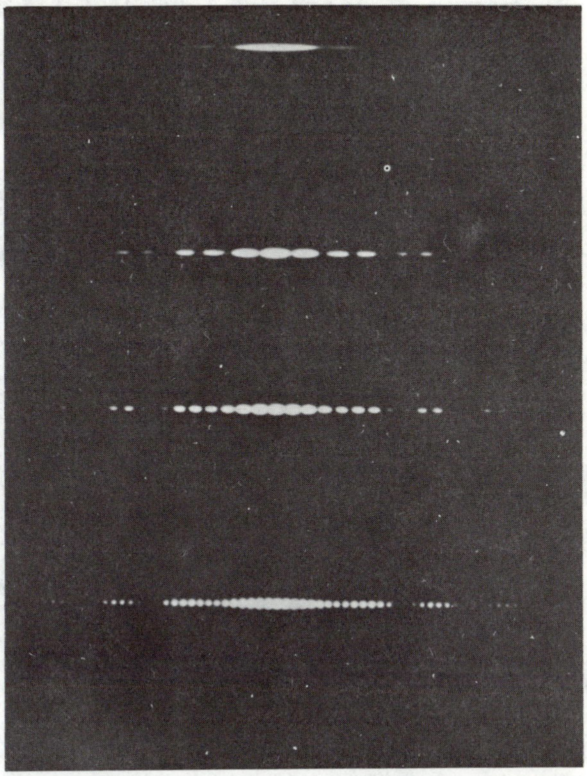

Fig. 36–18 Double-Slit Interference Patterns as seen on a screen. Slit width: 0.088 mm. Slit spacings: 0, 0.176, 0.35, and 0.70 mm.

behavior of a Fresnel zone plate can easily be illustrated by moving the screen to the primary focal point and then to the various secondary focal points. A zone plate may be made by taking a picture of one in an optics book and making a transparency of it.

7. *Double slit* (Fig. 36–18). This is the classic interference experiment, which can be performed with double slits ruled on smoked glass or old photographic plates. More convenient is the slide made by The National Press (Palo Alto, California) which has two rulings that start together as a single slit and gradually diverge. This gives a double slit of continuously variable spacing.

8. *Multiple slits* (Fig. 36–19). A series of slits with constant width and spacing but varying number is useful in showing how the maxima in the diffraction pattern become narrower and more intense as the number of slits increase approaching a diffraction grating in nature.

9. *Two-dimensional grating.* A handkerchief stretched tightly across the beam forms a useful two dimensional diffraction grating.

The photographs in Figs. 36–17, 36–18, and 36–19 have equal exposures which have been adjusted to show about as many fringes as are visible to an observer fifteen feet from the screen. The visibility of the brightest fringes increases markedly with the number of slits, although this may not be obvious from the photographs.

36–5.3 The Diffraction Grating

Another demonstration that has proved useful exhibits the paths of the diffracted rays from a grating. This has the advantage that the geometry of the demonstration is the same as the geometry usually used in the discussion and analysis: it shows ray paths in a plane, not just spots on a screen, representing the different orders of diffraction. Of course, only a single wavelength is available. The method is simply to place the grating in front

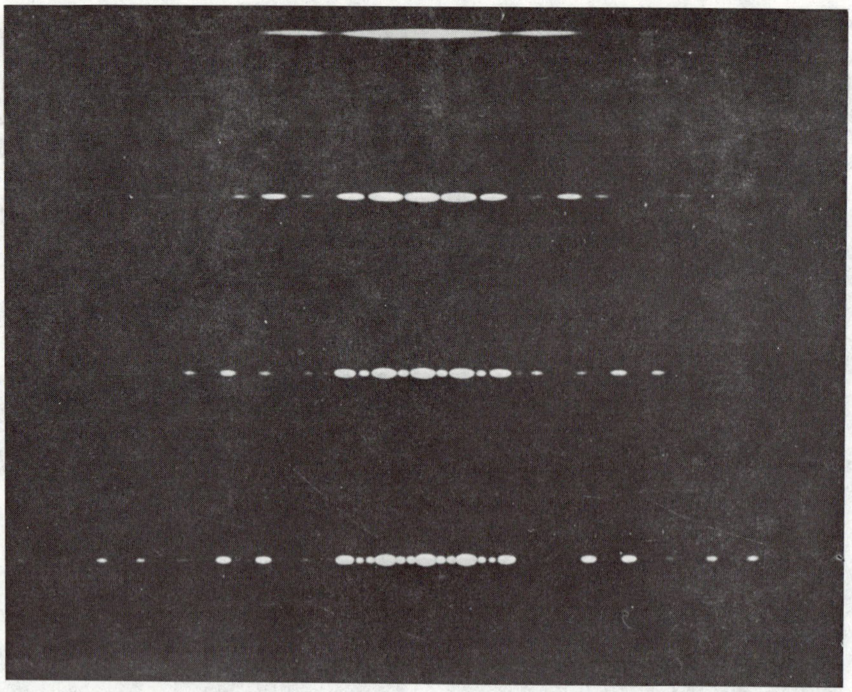

Fig. 36–19 Multiple-Slit Interference Patterns as seen on a screen. Slit width: 0.044 mm. Slit spacing: 0.132 mm. Number of slits: 1, 2, 3, and 4.

of the laser, with the beam aimed at a slight angle to the blackboard or other vertical viewing surface (Fig. 36–20). It is then possible to see from the side the various rays traveling out from the grating, and the diffraction angles of the various orders can be measured directly. It is quite difficult to find a viewing surface smooth enough to prevent shadows, but if these are objectionable, they can be removed by placing a cylindrical lens before the grating to make the beam diverge in a horizontal plane and allow the laser to be aimed at a greater angle to the surface.

36–6 Demonstrations of Polarization

36–6.1 Polarization of the Laser Beam

If the laser has external mirrors and Brewster's angle windows, then the output is linearly polarized, as may easily be demonstrated by projecting the beam on a wall and rotating a piece of Polaroid about the beam axis (shown in Fig. 36–21). If θ is the angle of rotation of the Polaroid measured from the direction of polarization of the light, then the intensity on the screen follows Malus' law

$$I(\theta) = I_0 \cos^2 \theta \tag{42}$$

which may be derived from elementary optics. If the laser mirrors are within the cavity, the output is not necessarily polarized, and this demonstration is not useful.

36–6.2 Brewster's Angle

It is well known that, for light incident from air on the boundary of a transparent medium and polarized in the plane of incidence, there exists an angle of incidence θ_i such that there is no reflected beam: all the light is transmitted. This angle is Brewster's angle,

$$\theta_B = \arctan(n), \tag{43}$$

where n is the index of refraction of the medium relative to air. θ_B is about 57 deg for ordinary glass.

This phenomenon can be easily demonstrated with a laser to a large class. The laser is oriented to provide horizontally polarized light, and a thick glass plate is placed on edge in the beam. As the plate is rotated about a vertical axis, the reflected spot can be seen moving along the wall of the lecture room. However, when Brewster's angle is reached, the intensity of the reflected beam clearly drops to zero. With the glass set at θ_B, the laser can be rotated about its own axis to show that zero reflection only occurs for polarization in the plane of incidence, while reflection is appreciable for polarization normal to the plane of incidence. In order to get a good null in the reflected beam, it is necessary to have reasonably good alignment of the system.

36–6.3 Scattering[22]

The use of a laser allows the demonstration of optical scattering to a large group. The laser is aimed across the front of the room, and a test tube containing the scattering material in solution is placed in the beam. The audience views the light scattered at right angles to the beam. As the (polarized) laser is rotated about its own axis, the dependence of scattering on the direction of observation relative to the direction of polarization is evident.

A weak colloidal suspension of $AgNO_3$ or $AuNO_3$ in water is a suitable scattering medium for observing Rayleigh scattering, which requires that the scattering object be small compared to an optical wavelength. This proceeds via an electric dipole mechanism, as may be verified by noting the lack of scattered light in a direction parallel to the polarization of the incident electric vector.

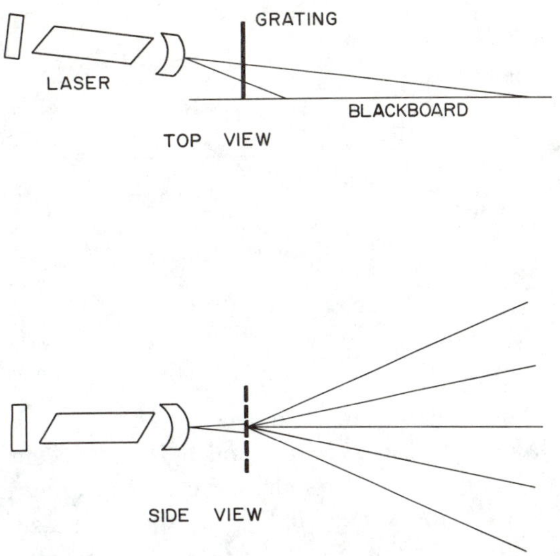

Fig. 36–20 Demonstration of Diffracted Rays from a Grating.

36–7 Demonstrations of Image Forming Properties of Lens System[23]

36–7.1 Diffraction at a Finite Aperture

The laser beam appears to come from a point source on the axis where it meets the plane mirror. A lens may be used to form an image of this point on a screen, but if the aperture is reasonably large, no diffraction rings are found around the image as would be the case with an ordinary source. The reason is the Gaussian intensity distribution across the laser wavefront. However, if a variable iris is placed between the laser and the lens and slowly closed down, the image will spread, and the Airy diffraction rings will appear. In this case we have nearly uniform illumination in the aperture corresponding to the usual conditions with an ordinary source.

36–7.2 Lens Aberrations

For these demonstrations the optical system is shown in Fig. 36–22. Lenses M_1 and M_2 are microscope objectives of focal length of 32 and 16 mm, respectively, while T_1 and T_2 are telescope objectives. The distance between the laser and M_1 is about 0.5 m, and that between M_2 and the screen (4) is 10 m. In order to eliminate extraneous aberrations, it is essential to use good quality optics. In particular the telescope objectives must be well corrected. Suitable air-spaced doublets of 356 mm focal length and 56 mm diam are available from the Edmund Scientific Corp. at reasonable cost. In

(22) This demonstration is a modification of Experiment Six in *Laboratory Manual for ULI Lasertron Instrument*, University Laboratories, Inc., Staff, Berkeley, California, Nov. 1966, and is used here by permission.

(23) The demonstrations in this section have been adapted, by permission, from the article by D. Dutton, M. P. Givens, and R. E. Hopkins, *Am. J. Phys.*, **32**, 355 (1964). Thanks are extended to them for allowing these demonstrations to be included here and for supplying the original figures.

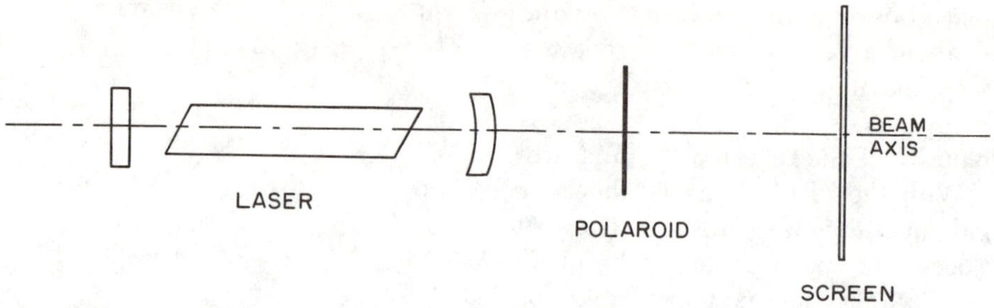

Fig. 36–21 Polarization Demonstration. The Polaroid can be rotated about the system axis.

addition the microscope objectives should be kept as free of dust as possible, and it is convenient to have the entire system mounted on a sturdy optical bench. In the system indicated, the spot size on the screen is 625 times the size at (1) and (3), or about 0.4 cm in diam, and if well-corrected lenses of adequate aperture are used, the spot has a smooth Gaussian intensity distribution.

The telescope objectives T_1 and T_2 are designed for an infinite object distance with the parallel light entering the positive elements. If either one is turned around, spherical aberration is introduced, and a number of diffraction rings are obtained instead of the Gaussian distribution, in the image, as shown in Fig. 36–23. If both T_1 and T_2 are turned around, the effect becomes worse, indicating that spherical aberration is additive.

If the lenses are returned to the correct position and then one of them is turned at a slight angle, coma is introduced in the image, as shown in Fig. 36–24.

If both T_1 and T_2 are turned by the same angle, astigmatism will appear in the system so that new (and different) positions of M_2 must be found for vertical focus, midfocus, and horizontal focus, with image configurations as shown in Fig. 36–25.

The lens system of Fig. 36–22 also has adequate magnification for projection of diffraction patterns due to macroscopic apertures placed in region (2). Such apertures may be formed by punching holes in an IBM card for example. Further information on this method may be found in reference 19. With slight modifications it may also be used to demonstrate the Abbe theory of the microscope and the technique of spatial filtering which are described in the same article.

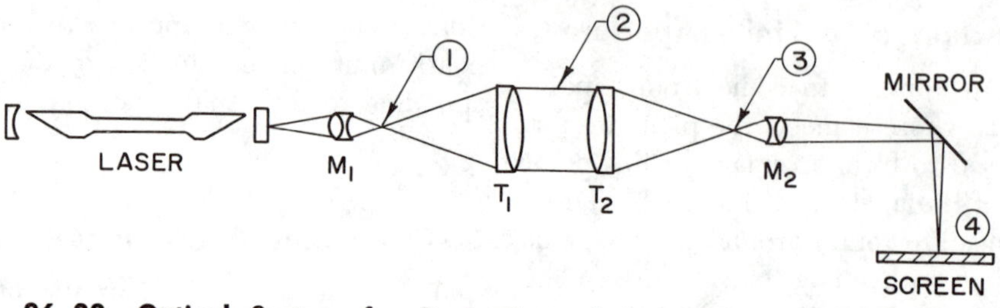

Fig. 36–22 Optical System for Projection of Lens Aberrations. (Courtesy of D. Dutton.)

36–8 Laser Safety[24]

No discussion of laser demonstrations would be complete without a few words on laser safety. Although the danger of burns from high-powered pulsed lasers is well known, it is not so well known that a 1 mW CW gas laser may also cause retinal burns under certain conditions.

Because of its small diameter, it is quite possible for the entire beam from a small gas laser to enter

[24] A valuable source of further data on the subject is the *Proceedings of the Seminar on Laser Safety,* May 1966, sponsored by the Martin-Orlando Division of Martin-Marietta Corp. and the U.S. Army Surgeon General's Office.

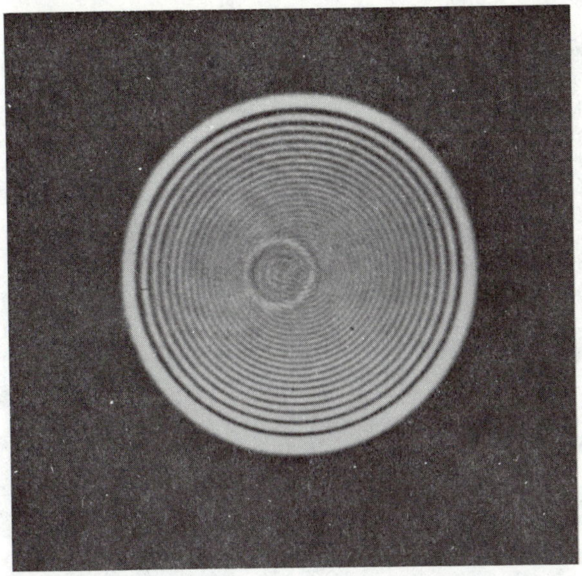

Fig. 36–23 Image of a Point Source Showing Spherical Aberration. (Courtesy of D. Dutton.)

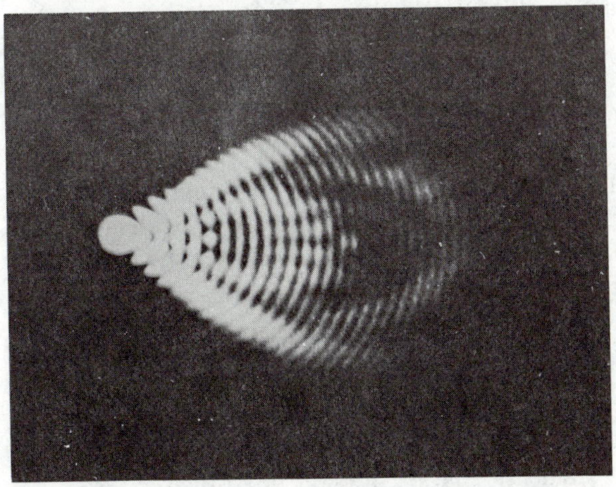

Fig. 36–24 Image of a Point Source, Showing Coma. (Courtesy of D. Dutton.)

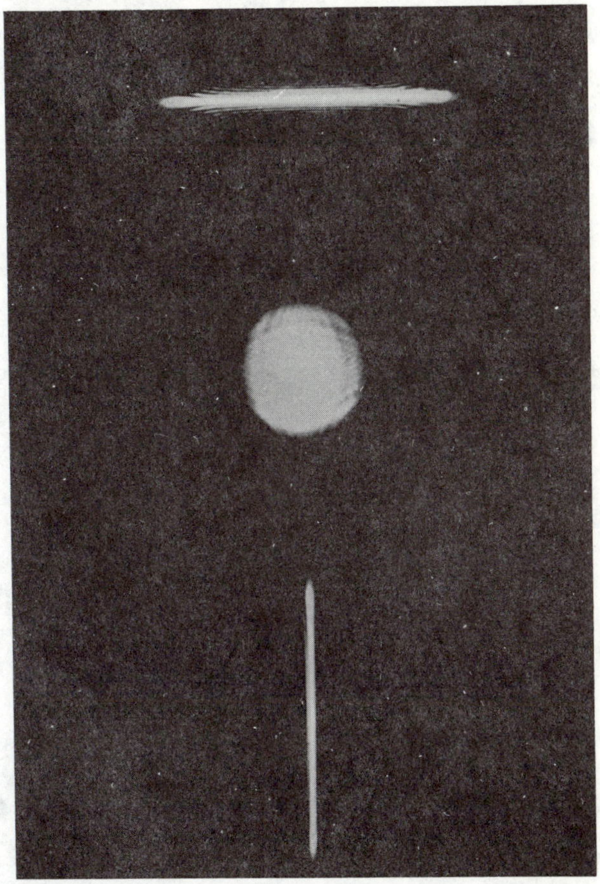

Fig. 36–25 Images of a Point Source, Showing Astigmatism. (Courtesy of D Dutton.)

the pupil of the eye, even when the eye is adapted to bright light so that the iris is closed. In the worst case the beam will be focused to the diffraction limit of the eye, about 10 μ in diam,[25] or an area of 10^{-6} cm^2. For a 1-mW laser this implies a possible power density of 10^3 W/cm^2 over the area of about 20 retinal receptors (rods or cones). Tests on rabbits indicate a retinal burn threshold[26] of 2.3 W/cm^2, so the danger of a burn is obvious.

Whether a burn will affect vision is another matter. In the peripheral region it might not, although several adjacent burns could cause a "blind spot." On the other hand, if the image of the laser were to fall on the optic disc, optic nerve fibers could be destroyed, and blindness over a large area of the retina would be a distinct possibility.

Therefore, extreme care must be taken when setting up laser demonstrations to keep the beam away from eye level and to physically prevent students (or anyone else) from looking into it. Remember too that the beam reflected from a polished surface is almost as dangerous as the direct beam itself.

(25) W. T. Ham, cited in "Laser Safety," *Scientific Research*, **1**, 26 (1966), McGraw-Hill Inc., New York.

(26) C. B. Smoyer, *The Hazards of Laser Radiation to the Human Eye*, IBM Watson Research Center Research Report RC-1036, Sept. 19, 1963.

36–9 List of Manufacturers

Listed below are several manufacturers of continuous gas lasers of classroom quality. There may be others unknown to the author, and in this rapidly growing field there surely will be others in the future.

As for auxiliary equipment, the usual optics supply houses, such as Edmund Scientific Corp. or Central Scientific Co., carry most of the necessary items.

The Bendix Corporation, Cincinnati Div.
Cincinnati, Ohio

Electro Optics Associates
Palo Alto, California

Maser Optics Inc.
Boston, Massachusetts

Optics Technology Inc.
Palo Alto, California

Perkin-Elmer, Electro-Optical Division
Norwalk, Connecticut

Semi-Elements Inc.
Saxonburg, Pennsylvania

Spectra Physics, Inc.
Mountain View, California

University Laboratories Inc.
Berkeley, California

36–10 Bibliography

Two books that provide a good, brief introduction to laser theory and techniques are

O. S. Heavens, *Optical Masers* (Methuen and Co. Ltd., London, 1964)

B. A. Lengyel, *Introduction to Laser Physics* (John Wiley & Sons, Inc., New York, 1966)

Useful papers by almost all the leaders in the field of laser research may be found in the reports of the first three Quantum Electronics Conferences:

C. H. Townes, ed., *Quantum Electronics* (Columbia University Press, New York, 1960)

F. Singer, ed., *Advances in Quantum Electronics* (Columbia University Press, New York, 1962)

P. Grivet and N. Bloembergen, eds., *Quantum Electronics III* (Columbia University Press, New York, 1964)

Two well-known collections of review articles appeared in

Applied Optics, Supplement on Optical Masers (1962)

Proceedings of the IEEE (January 1963)

There are two resource and reprint booklets published by the American Institute of Physics:

Quantum and Statistical Aspects of Light

Masers and Optical Pumping

At a more elementary level, but still enlightening, are two articles in *Scientific American:*

A. L. Schawlow, "Optical Masers," June 1961, p. 52

A. L. Schawlow, "Advances in Optical Masers," July 1963, p. 34.

Laser demonstrations have previously been published in the following articles as well as in the instruction books commonly supplied by laser manufacturers:

D. Dutton, M. P. Givens, and R. E. Hopkins, "Some Demonstration Experiments in Optics Using a Gas Laser," *Am. J. Phys.*, **32**, 355 (1964)

H. Whiteside, "Laser Optics Experiments and Demonstrations," *Am. J. Phys.*, **33**, 487 (1965)

A. L. Schawlow, "Measuring the Wavelength of Light with a Ruler," *Am. J. Phys.*, **33**, 922 (1965)

I. D. Abella et al., "Laser Experiments and Apparatus for Teaching Purposes," *Am. J. Phys.*, **34**, 98, 1966

Three standard works on optics that may be useful are

M. Born and E. Wolf, *Principles of Optics*, 2nd ed. (Pergamon, 1964)

R. W. Ditchburn, *Light*, 2nd ed. (John Wiley & Sons, Inc., New York, 1963)

F. A. Jenkins and H. E. White, *Fundamentals of Optics*, 3rd ed. (McGraw-Hill Book Co., New York, 1957)

37

HOLOGRAMS

By Tung H. Jeong
Lake Forest College

37–1 Introduction

The basic ideas of holography were originated by Prof. Dennis Gabor[1,2,3] in 1948. With the advent of the laser and an improved process by Leith and Upatnieks,[4,5] it caught the excitement and imagination of the scientific community and has now become an active field of applied research.

Briefly, holography is a process through which a three-dimensional image of an object can be completely recorded on a photographic film or plate without the use of any intermediate imaging devices. For this reason, it is also called lensless photography. The processed photoplate is called a hologram. Some major properties of holograms are the following:

1. The light arriving at the eyes of the observer from an illuminated hologram is precisely the same as that which would come from the original object. Therefore, this three-dimensional image can be photographed from various perspectives or scrutinized by any other ordinary optical means.

2. If a hologram is broken into many pieces, each piece contains a complete view of the entire object. One looks through the hologram as if it were a window behind which the object is situated. If this window is closed, one can peep through a knothole and still see the entire scene but with a more restricted perspective.

3. More than one independent scene can be recorded on the same photoplate. They can be viewed one at a time, without cross interference, by rotating or tilting the finished hologram with respect to the viewing light.

4. Although any hologram can be viewed by a monochromatic point source (not necessarily from a laser), a special kind, called the Lippmann-Bragg hologram,[6,7,8] can be viewed by unfiltered thermal sources such as the sun or tungsten filament lamps.

(1) D. Gabor, *Nature,* **161,** 777 (1948).

(2) D. Gabor, *Proc. Roy. Soc.* (London), **A197,** 454 (1949).

(3) D. Gabor, *Proc. Phys. Soc.* (London), **B64,** 449 (1951).

(4) E. N. Leith and J. Upatnieks, *Sci. Am.,* **212,** 24 (1965).

(5) E. N. Leith and J. Upatnieks, *J. Opt. Soc. Am.,* **52,** 1123 (1962), and **54,** 1295 (1964).

(6) Yu. N. Denisyak, *Soviet Phys.–Doklady,* **7,** 543 (1962).

(7) G. W. Stroke and A. Labeyrie, *Phys. Letters,* **20,** 368 (1966).

(8) L. H. Lin, K. S. Pennington, G. W. Stroke, and A. Labeyrie, *Bell System Tech. J.,* **45,** 659 (1966).

Fig. 37–1

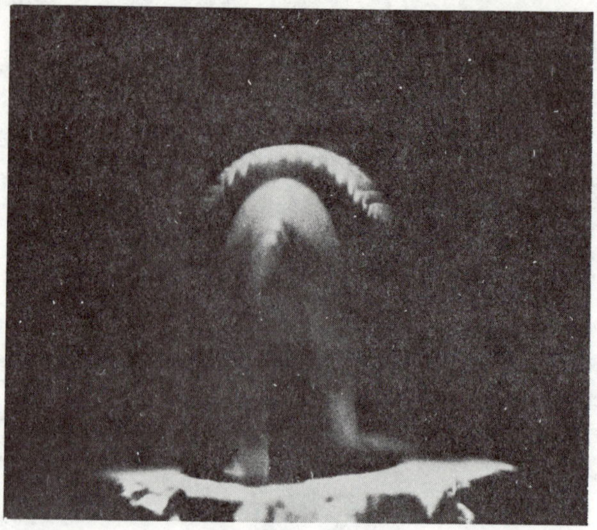

Fig. 37–2

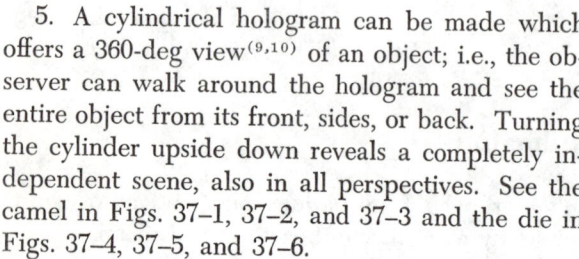

5. A cylindrical hologram can be made which offers a 360-deg view[9,10] of an object; i.e., the observer can walk around the hologram and see the entire object from its front, sides, or back. Turning the cylinder upside down reveals a completely independent scene, also in all perspectives. See the camel in Figs. 37–1, 37–2, and 37–3 and the die in Figs. 37–4, 37–5, and 37–6.

6. They can be encoded so that only the person in possession of a decoding device can see the true image.[11]

7. Three-dimensional multi-color scenes can be recorded and reconstructed holographically.[8,12]

Because of these and many other dramatic characteristics, various possible applications of holography in science and industry can be foreseen. However, the basic aim of this chapter is to present holography as an effective motivational device for students. Since it involves many basic principles in physics and mathematics, an instructor, having the enthusiasm of his students, can guide them to learn a great deal.

Fig. 37–3

A rigorous and comprehensive treatment[13] on this topic is out of the present scope. What specifically will be presented is a guide for teachers for using a particular approach to demonstrate the basic theory of holography and some practical hints for student projects.

[9] T. H. Jeong, P. Rudolf, and A. Luckett, *J. Opt. Soc. Am.*, **56**, 1263 (1966).

[10] T. H. Jeong, Paper presented at April, 1967, Meeting of American Optical Society at Columbus, Ohio.

[11] H. Kogelnik, *Bell System Tech. J.*, **44**, 2451 (1965).

[12] A. A. Friesem and R. J. Fedorowicz, *App. Opt.*, **6**, 529 (1967).

[13] G. W. Stroke, *An Introduction to Coherent Optics and Holography* (Academic Press, New York, 1966).

37–2 Mathematical Background

Initially, students will need to be reminded of the principle of superposition; i.e., in a linear system (a system which obeys Hooke's law), any complex periodic wave can be constructed by taking the sum of pure sine waves of definite frequencies, amplitudes, and phase relationships.

Consider the sine wave of Fig. 37–7a. Its frequency spectrum (also called a Fourier spectrum), as shown in Fig. 37–7b, consists of a zero frequency ("dc") component of amplitude $A(f_0) = 1$, and a single sine function of amplitude $A(f_1) = 1$. In other words,

$$f(x) = 1 + \sin(2\pi f_1 x).$$

Next, consider Figs. 37–8a and 37–8b. These represent a "beat note" which is obtained by adding two sine waves of different frequencies. If the curve $f(x)$ is not symmetrical with respect to the horizontal axis, it merely means that there is a "dc" component $A(f_0)$. Analytically,

$$f(x) = A(f_0) + A(f_1)\sin(2\pi f_1 x) + A(f_2)\sin(2\pi f_2 x).$$

It can be shown that the square wave of Fig. 37–9a can be obtained by summing all the sine waves designated in the Fourier spectrum (Fig. 37–9b). The instructor can actually draw a few of the sine waves on the chalkboard and add the amplitudes together point by point to prove this. The dotted line of Fig. 37–9a is the center of symmetry of the wave, and its amplitude is represented in Fig. 37–9b as the "dc" component. The f_0 component is the "fundamental" of $f(x)$, $3f_0$ is the third "harmonic" (or third order), etc. For a regular square wave, only odd harmonics are present. The dotted line on Fig. 37–9b is the envelope of the spectrum which depends on the ratio between the widths of the top and the bottom of a cycle.

The process of finding the frequency spectrum of a given complex wave is called Fourier analysis, and the corresponding process of finding the complex wave from a given set of sine and/or cosine waves is called Fourier synthesis. To attain a rigorous understanding of holography, detailed knowledge of this branch of mathematics is required. However, for beginning students, it suffices to understand what has been presented above.

For more advanced students, the spectrum of a single pulse can be discussed. This can be considered as a periodic wave with an infinite period. The spectrum consists of a continuous distribution of frequencies; $f(x)$ is a Fourier integral summing all frequencies of given amplitudes and phases, i.e., $f(x)$ is a Fourier transform of $A(f)$.

Fig. 37–4

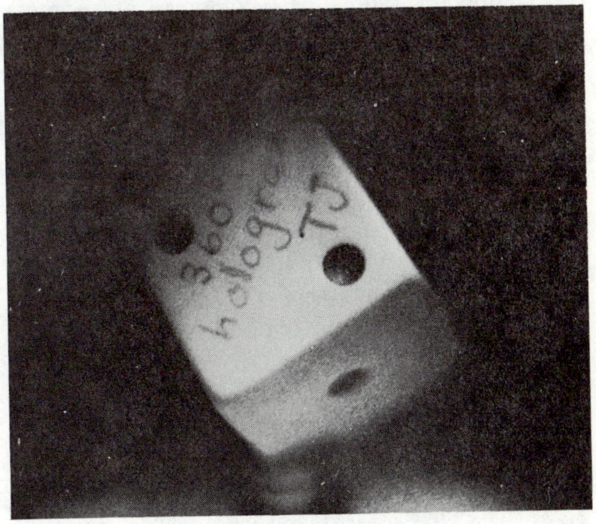

Fig. 37–5

37–3 Physical Demonstration

Having provided the above background, a physical demonstration can now begin. The material involved consists of a sine grating, a "beat" grating, an alternating bar replica grating, a Gabor zone plate, and a hologram of a three-dimensional scene.[14]

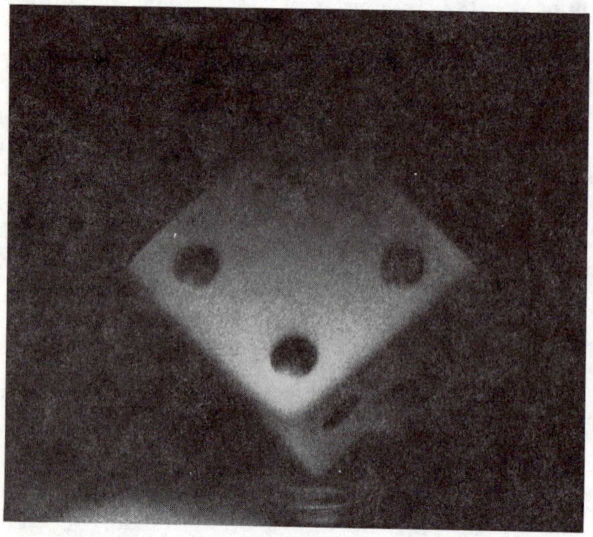

Fig. 37–6

37–3.1 The Sine Grating

The arrangement shown in Fig. 37–10a is used to make the sine grating. A laser beam is split into two components and then recombined at an angle θ to one another on a sheet of photographic film or plate. The intensity of the interference pattern across the plate has a sinusoidal distribution, essentially the same as that caused by a Lloyd mirror or a Fresnel biprism. The spatial frequency increases with the angle θ between the beams according to $f = \sin\theta/\lambda$, where λ is the wavelength of the light. If the laser beam were split into three components, and the third component recombined with the first two, as shown by the dotted line in Fig. 37–10a, the interference pattern would not be significantly changed. Since the film does not "know" whether this component is present or not, the diffraction pattern from it has a symmetrical order on each side. (If the emulsion is thick, the situation will be different. This point will be discussed later.) One is called the complex conjugate of the other. This sine grating can be said to be a hologram of a parallel beam of light, or of a point object located at infinity, since the reconstructed wavefronts are the same as those used to expose the plate.

The basic principle illustrated is that the Fraunhofer diffraction pattern represents the Fourier analysis of the diffracting aperture, in our case a grating or a hologram. If we plot the transmittance (fraction of light energy transmitted) versus distance across the sine grating, the curve would look like Fig. 37–7a, a pure sine wave having a certain number of cycles per millimeter (spatial frequency). As discussed above, such a wave has only one Fourier component, plus a "dc" term. When a beam of parallel and monochromatic light is diffracted by this grating, it can be seen that the diffraction pattern consists of an undeviated beam (the "dc" component) plus one order of diffraction on each side (Fig. 37–10b).

[14] This entire package, including a 360-deg hologram, is being distributed by the Welch Scientific Company.

37–3.2 The Beat Grating

The diffraction pattern from a beat grating further demonstrates this principle. The transmittance curve in this case is represented by Fig. 37–8a. As expected, the diffraction pattern of this grating consists of two beams on each side of the "dc" beam, represented by Fig. 37–8b. The beat grating is made in the same manner as the sine grating, with the addition of another object beam from a different angle. The pattern on the grating is just a superposition of two sine gratings of different frequencies. If more than two object beams are used, the beat pattern on the film gets more complex and the diffraction pattern from it merely reconstructs all the object beams. Furthermore, the object beams do not have to be in the same plane.

In the foregoing description, the film is performing a Fourier synthesis while being exposed, i.e., it adds together the individual sine waves caused by the interference between the reference and the object beam(s). The result is a complex periodic wave pattern. When monochromatic parallel light is incident on the processed film, Fourier analysis

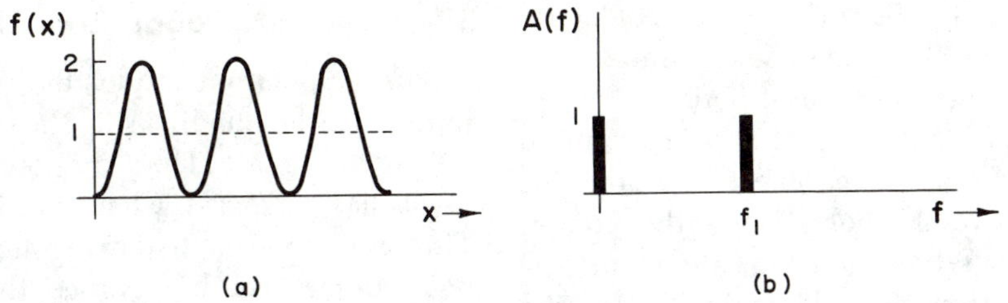

Fig. 37–7

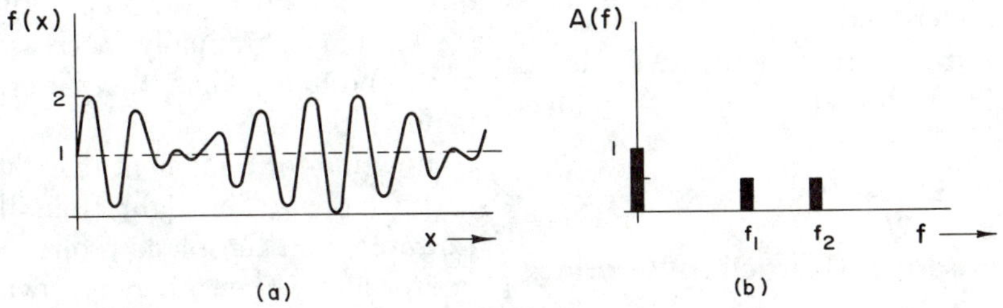

Fig. 37–8

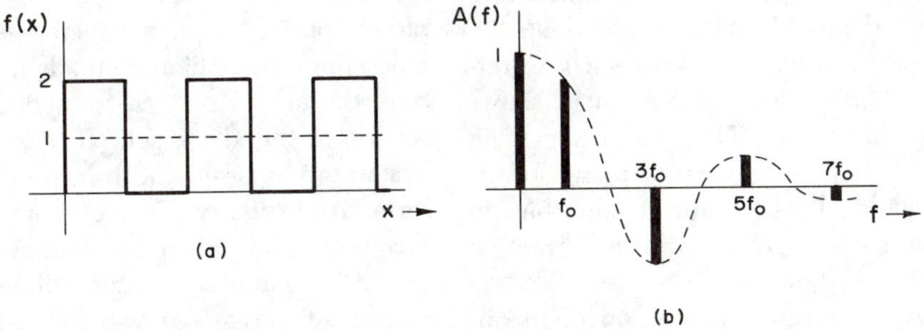

Fig. 37–9

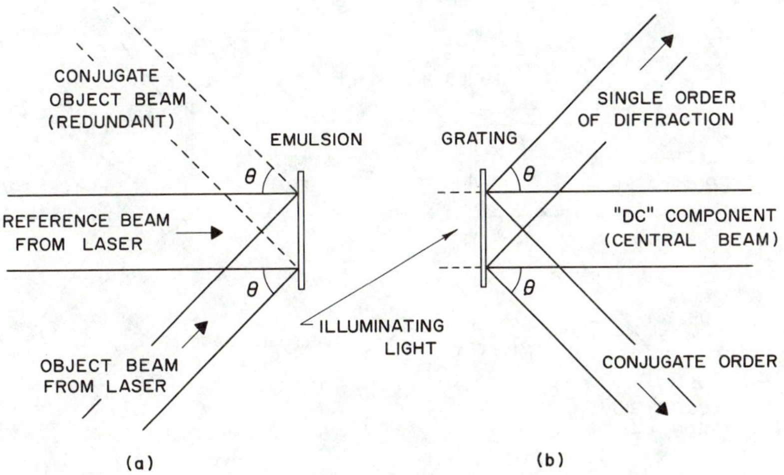

Fig. 37–10

takes place and the light is spread into a configuration similar to that used to make the exposure. This is the process of wavefront reconstruction.

Technically, one can say that the hologram is the Fourier transform of an object function; the diffraction pattern of the hologram is the Fourier transform of the grating function; therefore, the diffraction pattern from the hologram is the Fourier transform of the Fourier transform of the object function, i.e., the object. However, this is more easily said than understood and care should be exercised by the instructor so that the students do not merely substitute this statement for understanding.

37–3.3 The Standard Diffraction Grating

It follows that if a square wave intensity pattern (alternating opaque and transparent bars) is desired on the film, one *could* obtain it by bringing together many object beams of correct amplitudes from predetermined angles. This is not done because there are easier ways of making such a grating. The familiar diffraction pattern, then, shows the spectrum of this wave. Thus, such a grating can be said to be a hologram of many point objects at infinity. It should also be realized now that, in principle, we can make a grating whose diffraction patterns looks like anything we wish.

So far we have concentrated only on point objects at infinity (parallel beams). The same ideas hold when the objects are near the film (divergent beams).

37–3.4 The Gabor Zone Plate

For simplicity, consider the interference pattern formed on the film by using the configuration shown in Fig. 37–11a. Here, the object beam of Fig. 37–7a has traversed a lens and focused at a point. This is equivalent to having light coming from a point object nearby. Notice that the angle, and therefore the spatial frequency, is dependent on the location on the film. For example, the spatial frequency on top of the film (Fig. 37–11a) is $f_1 = \sin\theta_1/\lambda$ and gradually decreases to $f_2 = \sin\theta_2/\lambda$ at the bottom. Thus, the pattern subtends a finite bandwidth.

Imagine for a moment that the film is an infinite vertical plane, the light from the point object is isotropic, and the plane reference wave covers the entire film. Then there is cylindrical symmetry about a horizontal axis drawn through the point object (dotted line, Fig. 37–11a). The diffraction pattern on the film would be an infinite set of concentric rings, centered on the axis, and whose radial spatial frequency increases with the radius. Since our actual film subtends a small off-axis section, the actual pattern formed on it is an off-axis section of the above set. If these rings were an alternatingly opaque and transparent set, they would form an ordinary Fresnel zone plate. However, this set is sinusoidal in character, and it will be given the name Gabor zone plate.

The reconstructed wavefront from this plate is shown in Fig. 37–11b. One way to explain this pattern is as follows: Consider an area on the film small in dimension compared to the distance from

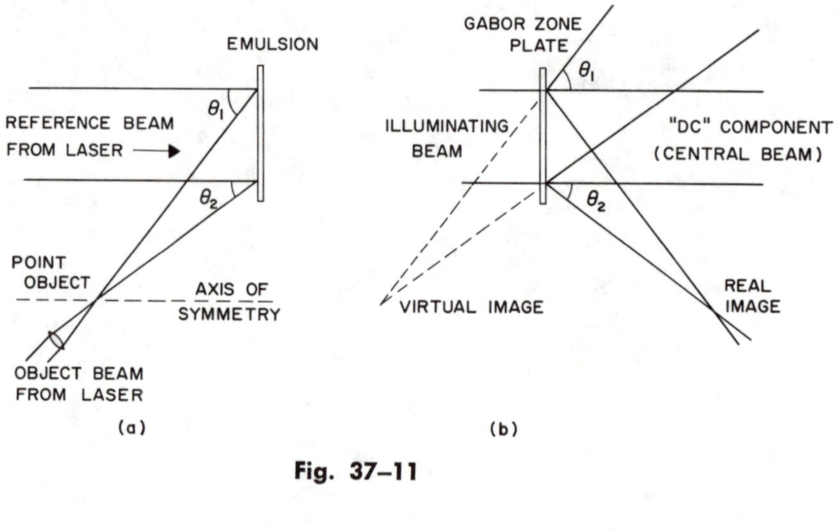

Fig. 37–11

it to the object point. During the exposure, both the reference beam and an element of the object beam arriving at this area can be considered to be parallel. Thus, *locally*, the interference pattern on the film is a pure sine wave. In the reconstruction, light impinging on any part of the processed film will have only one order of diffraction. However, the higher frequency regions have a larger dispersion, diffracting light to a larger angle. Therefore, light diffracted off the top of this grating diverges more than that diffracted from the bottom of the grating. By cylindrical symmetry, half of the diffracted light will converge to a point, forming a real image of the original point, while the other half diverges and forms a virtual image of the same point.

Photographically, an object can be considered as a set of point sources of light located at various distances from the film. If a three-dimensional figurine is substituted for the point object and illuminated by laser light, each point on it will reflect light onto the film and form a system of rings described above. The film would add together, or integrate, all the sets formed by each point on the object, i.e., the interference pattern formed is the superposition of all individual sets. In the reconstruction, each set of rings forms a real and virtual image of a point, thus creating in total a real and virtual image of the entire object.

The simplest possible arrangement to use in making a hologram of a three-dimensional object is shown in Fig. 37–12. Light from a laser is diverged by a lens; part of the light intersects the film directly and serves as the reference beam while the other parts illuminate the object. Light scattered from each point on the object will form with the reference beam a set of Gabor zones on the film. By facing the film directly toward the object rather than toward the reference beam, the virtual image is much easier to observe during the reconstruction.

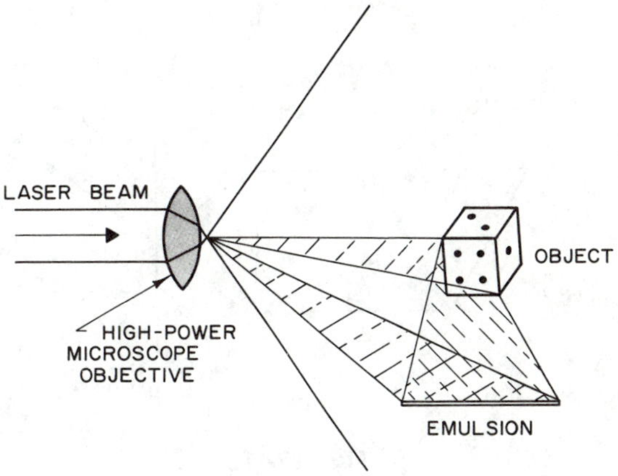

Fig. 37–12

By taking a figure of revolution of Fig. 37–12, using the original laser beam as the axis, a geometry exists with which a cylindrical hologram described in Section 37–1 (5) can be made.

37–3.5 The Hologram

At this point, an actual hologram of a three-dimensional scene can be shown.

37–4 Relationship Between Holography and Radio

For those who are familiar with radio theory, it should be realized at this time that it is strikingly similar to the theory of holography. Although radio is a time-dependent wave phenomenon and holography is space-dependent, both are described by the same communication theory. The almost trivial mathematical difference between the two is that one operates in the time domain (having t as a variable) while the other operates in the space domain (having $\mathbf{r}(x,y,z)$ as a variable) of the Fourier transform theory.

To illustrate this point, let us enumerate some phenomena exhibited in radio and relate them one by one to holography.

37–4.1 Bandwidth

As described previously, a hologram has in general a continuous range of spatial frequencies. The bandwidth of the arrangement shown in Fig. 37–11a is $\sin \theta_2/\lambda < f < \sin \theta_1/\lambda$. For a three-dimensional object, the bandwidth depends on the extreme angles between the references beam and the rays from various parts of the object as they intersect

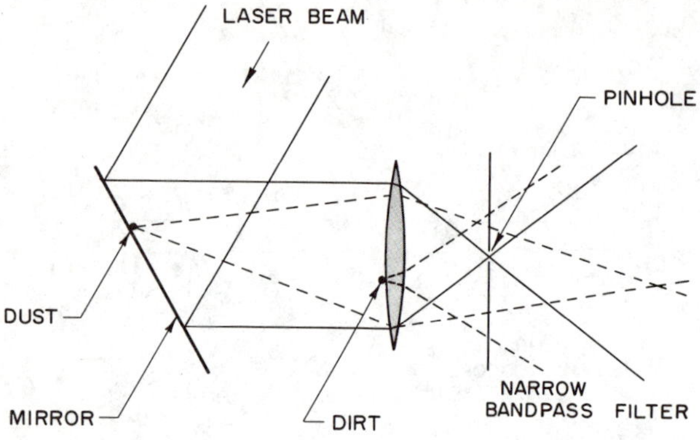

Fig. 37–13

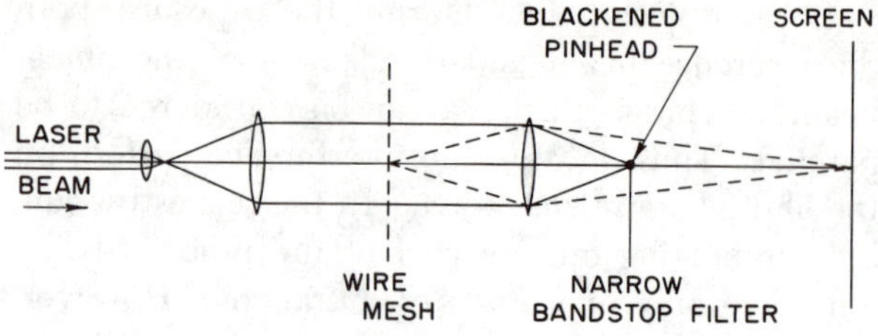

Fig. 37–14

on the film. Therefore, for a given reference beam direction, the bandwidth of the system increases with the physical dimensions and the proximity of the object and the film.

37–4.2 Noise

Static in radio is well known. The spatial equivalence is the smudgy appearance, the whirls and rings that can be seen on a hologram surface which have no relation to the pertinent information recorded. They are caused by the diffraction of dust particles and dirty spots on mirrors or lenses used in making the hologram.

37–4.3 Filtering

A narrow bandpass filter can be used in radio to eliminate the noise in the carrier. Similarly, this can be done in holography. Consider Fig. 37–13. A parallel laser beam is reflected by a dirty mirror and then focused through a dirty lens. The unscattered beam converges to a point of typically a few microns in diameter at the focal plane. The images of the dust particles, however, will occur at the conjugate foci, not coincident with the focal point. Therefore, a simple pinhole[15] located at the focal point will selectively pass the original parallel beam and stop the noise.

A meaningful demonstration on dark field illumination in microscopes can be done using the simple arrangement shown in Fig. 37–14. In this case, instead of the narrow bandpass filter described above, a narrow bandstop filter, i.e., a small blackened pinhead is used. The laser beam is first diverged by a microscope objective and rendered parallel again with a larger lens. The object (a fine wire mesh, for example) is placed in the larger beam. The next lens converges the undiffracted "dc" beam into a point, which is blocked out by the pinhead. The diffracted light, however, is focused at the screen, placed at the conjugate focal plane to the object. An image of the object will then be seen without the bright direct light. By carefully moving the pin along the optical axis of the system, phase reversal can be seen as various orders of diffraction are cut off.

37–4.4 Fidelity

The amplifier and speaker system in radio is considered "Hi-Fi" if it has a wide passband without distortion. This is also true in optics. A larger lens has better resolution (fidelity) than a small one because it gathers a larger number of orders of diffraction (more harmonics) from the object and recombines them at the focal point. The lower limit of resolution is realized when the aperture is so small that only the "dc" component, which carries no information, gets through. The lens here is performing a Fourier synthesis and a Fourier analysis all at once.

The size of the hologram, then, determines to a great extent the ultimate resolution of the image. Although every piece of a hologram gives a complete view of the object, the resolution decreases with the decrease in dimension of the piece—because the bandwidth is being narrowed. When a piece is small enough, only the "dc" component can come through. This point can be demonstrated by directly covering a hologram with a black card with various sizes of holes in it and observing the image through the individual holes.

[15] T. H. Jeong, *Am. J. Phys.*, 35, No. 5 (1967).

37–4.5 Encoding

A radio message can be "scrambled" by giving an auxiliary modulation to the carrier. The same can be done to a hologram by inserting a very irregular piece of glass in the path of the reference beam during exposure.[15a] The plane reference wavefront is now warped. In viewing the finished hologram with a plane wave, the image is "scrambled." However, if the same piece of glass is used in the reconstruction, the true image of the object is recovered.

37–4.6 Sideband Suppression and Multiplexing

A sinusoidal carrier has two sidebands. A radio channel can be multiplexed by modulating each sideband independently. How this can take place in holography will be explained in two steps.

1. Sideband Suppression. Figure 37–15 depicts a more realistic picture of a hologram construction because it shows the emulsion having a finite thickness. For example, Kodak 649-F plates have emulsion thickness of about 16 μ. For simplicity, consider two waves interacting on the emulsion as shown, where the dotted lines indicate the crest of the waves. On the surface a sinusoidal diffraction pattern occurs with antinodes located at lines (into the drawing) where the crests meet. On the plane immediately behind the surface, the same pattern occurs but is slightly shifted upwards. As the waves travel through the emulsion, nodal planes are formed as shown by the heavy diagonal lines. When the emulsion is exposed and processed, these darkened planes behave as venetian blinds and suppress one of the sidebands. The effectiveness of the suppression depends on the emulsion thickness and density after development. If the plane object beam is substituted by a three-dimensional object, and a hologram is made, the real image is suppressed, but not lost. By turning the hologram backward, which reverses the direction of the blinds, the real image can be projected onto a screen, and the virtual image is suppressed. In practice, this can be done easily by illuminating the backward hologram with a *narrow* laser beam. The small spot of the hologram used causes a sacrifice in resolution, but depth of field is gained in the projected image.

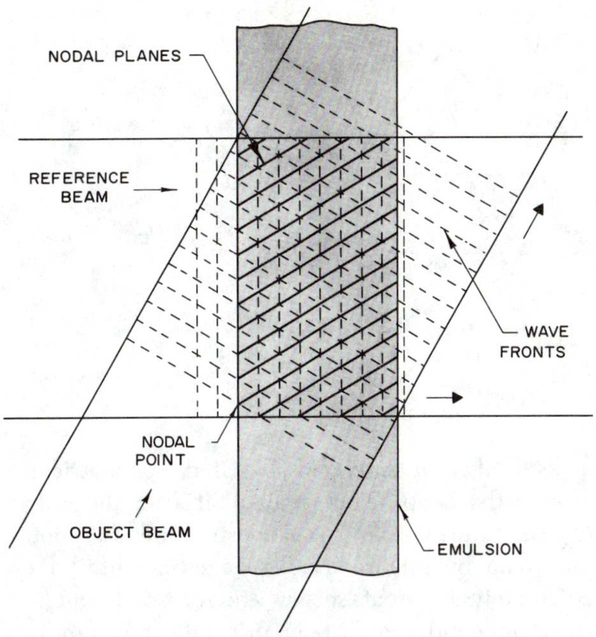

Fig. 37–15

2. Multiplexing. During construction, the once exposed photoplate can be turned upside down and a different object is used for another exposure. In this way, each sideband is modulated separately and the finished hologram will show two completely unrelated pictures, depending on its orientation with respect to the illumination. To avoid "cross-talk," the angle between the reference beam and the object beam should be sufficiently large, and the physical dimensions of the object should be sufficiently small. (In other words, the frequency domains of the two scenes must not overlap.)

At this point we can utilize the fact that the space domain has three dimensions, whereas the time domain has only one. When one looks out at night through a square mesh wire screen window at a street lamp, one sees a diffraction pattern resembling a cross—there are many orders of diffraction both in the vertical and the horizontal directions. This is so because the screen is a two-dimensional square wave, which has many Fourier components. If the transmittance of the screen had been sinusoidal in both directions, there would be

[15a] H. Kogelnik, *Bell System Tech. J.*, **44**, 2451 (1965).

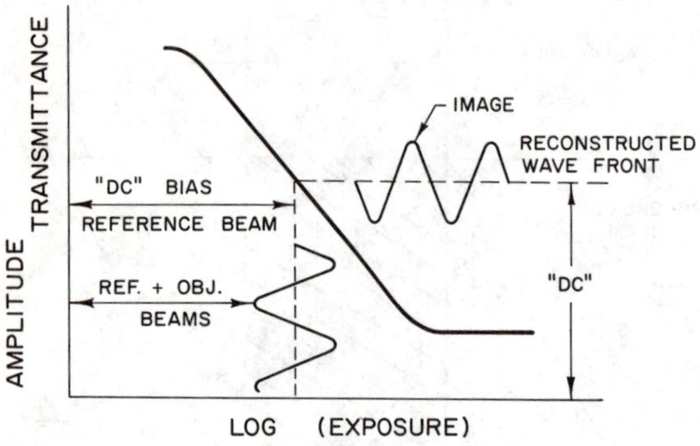

Fig. 37–16

a total of four sidebands, located symmetrically around the lamp. This means that after the initial exposure during a hologram construction, the photoplate can be rotated 90 deg at a time until four completely different scenes are recorded, one on each sideband. In fact, in principle, any number of independent scenes can be recorded by making a smaller rotation on the photoplate after each exposure, the limit being the over-exposure of the emulsion and the occurrence of cross-talk between adjacent scenes. Over-exposure, however, can be remedied to a degree by bleaching[16] the finished hologram in potassium ferricyanide or mercuric chloride, changing the "amplitude hologram" into a "phase hologram." The latter is similar to a transparent grating having on it a pattern of variations in thickness and/or index of refraction.

37–4.7 Tube Characteristic vs Film Characteristic

Everyone is, to a degree, familiar with the characteristic curve of vacuum tubes and how it affects the transmission of information in radio. A strikingly similar consideration occurs in holography.[17] Figure 37–16 shows the characteristic curve of a typical emulsion. We can consider the reference beam as a "dc bias" which exposes the emulsion uniformly to the center of a linear region of the curve (vertical dotted line) and the object beam as a modulation. For a given emulsion, the total exposure, as well as the ratio between the intensities of the reference and object beams, must be correct in order to transmit the strongest signal without distortion. For Kodak 649-F, the correct intensity ratio between the reference and the object beams is 7.5.[17] The exposure, of course, depends further on the total intensity at the plane of the photoplate. In practice, before exposure, the plate can be substituted by a white card. The brightness as judged by eye caused by the reference beam alone should be approximately twice that caused by the total scattered light from the object. Since the response of the eye is logarithmic, this gives a ratio of ten to one.

We have already mentioned that because of the additional dimensions available in the space domain, there are events in holography which cannot be related to radio directly. Examples so far given are the existence of more than two sidebands and the 360 deg hologram. There is still another class of holograms which utilizes the fact that the emulsion has a finite thickness. This is the white light reflection hologram, also called the Lippmann-Bragg hologram.

Referring back to Fig. 37–15, we note that the nodal planes formed become more parallel to the plane of the photoplate as the object beam is made more parallel to the reference beam. To the limit that both beams incident normally on the plate, the nodal planes would be parallel to the plate (standing waves). Because the emulsion is many microns thick, whereas the visible light has wavelengths around $\frac{1}{2}\ \mu$, many planes are formed *into* the emulsion (assuming that we are not limited by the grain size of the emulsion). Precisely the same effect occurs if the reference and object beams impinge on the photoplate from opposite sides, which is experimentally more easily arranged.[17a] This hologram then behaves precisely as a crystal, and produces Bragg diffraction when illuminated by white light from the sun or a tungsten filament lamp from the direction of the original reference beam. In effect, this hologram provides its own color filter because only one frequency in the visible region is coherently reflected, while all others are incoherently scattered, absorbed, or transmitted. Because of the large number of planes, the "Q" of the resonant system is quite high, having a reflected bandwidth of approximately 100 Å. The resonant frequency

(16) K. S. Pennington, *Microwave*, Oct., 1965.

(17) F. G. Kaspar and R. L. Lamberts, paper presented at San Francisco Meeting of American Optical Society, 1966.

(17a) G. W. Stroke and A. Labeyrie, *op. cit.*

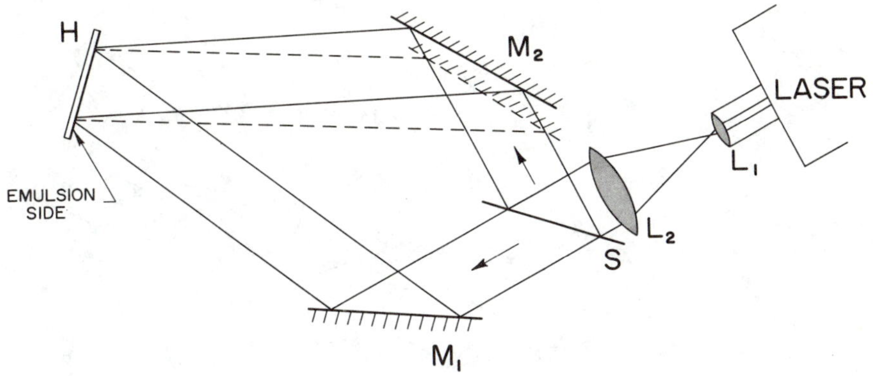

Fig. 37–17

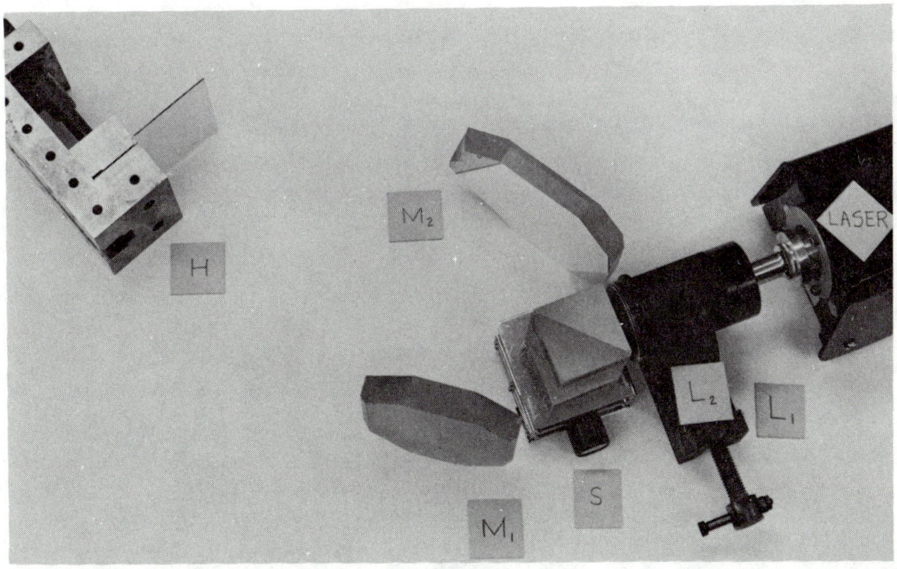

Fig. 37–18

can by varied slightly by moving the white light source, thus changing the angle of incidence, or by changing the temperature of the hologram. Usually, the emulsion shrinks slightly during the processing, and the color of the reconstruction is shifted to that of a higher frequency. This is fortunate, in a way, because the human eye is more sensitive to green than to red, assuming a He–Ne laser is used. If the shrinkage must be remedied, the finished hologram can be soaked in a solution of triethanolamine.[18]

If the light used in the construction of a hologram comes from more than one type of laser, say a He–Ne and an argon laser, multicolor holograms can be made.[18a] Here each frequency causes one set of three-dimensional diffraction patterns to be formed, and the finished hologram becomes a multiple resonant system.

37–5 Practical Hints for Constructing and Demonstrating Holograms

Few projects are as exciting to students as the construction of a good hologram. Depending on the quality and complexity desired, the equipment

(18) Private communication with G. W. Stroke.

(18a) L. H. Lin *et al., op. cit.;* A. A. Friesam and R. J. Fedorowicz, *op. cit.*

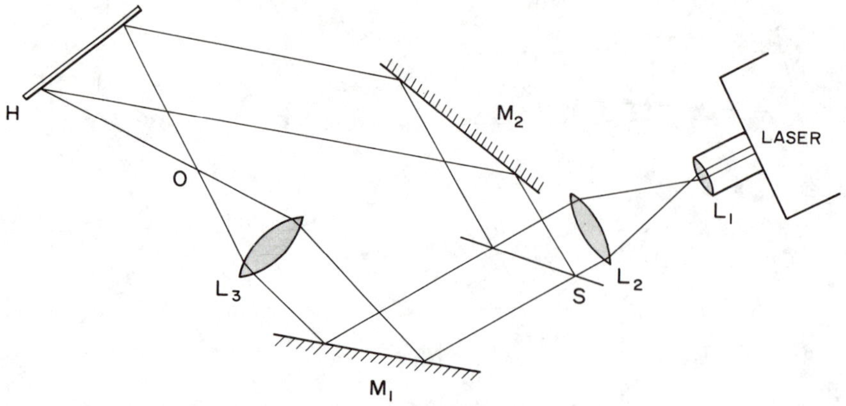

Fig. 37–19

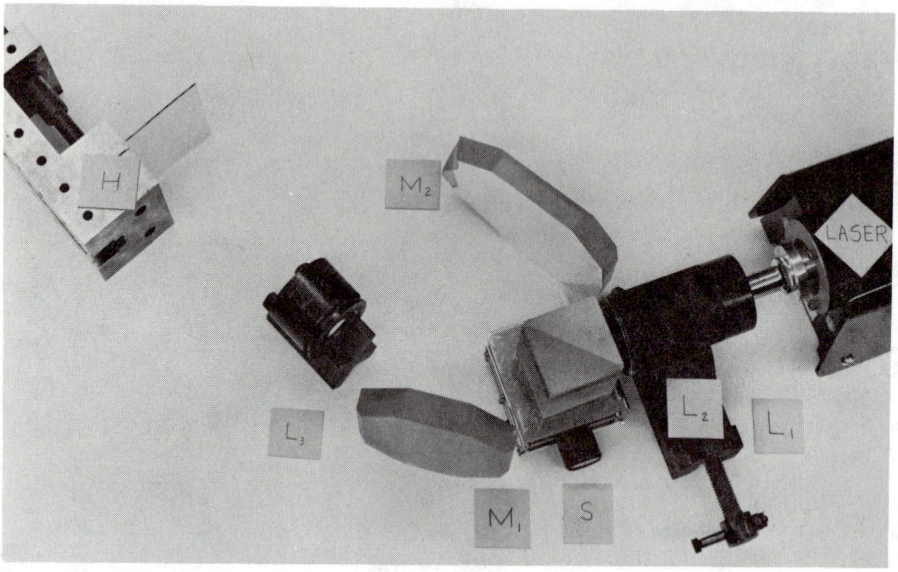

Fig. 37–20

requirements[19] vary from standard laboratory odds and ends to that costing tens of thousands of dollars. But always, a laser is a practical necessity.

We have made good holograms of varying complexity using a commercial 2 mW He–Ne laser,[20] as well as ones home-built according to the suggestions of Vander Sluis, *et al.*[21] The power requirement is unimportant if a mechanically stable system is available in a thermally stable environment. Following are some important points to remember, in addition to what has already been discussed.

(19) A medium price integrated set of apparatus capable of making high quality holograms of various kinds is sold by the Gaertner Scientific Corp., Chicago, Illinois.

(20) Model LAS-101 made by Electro Optical Assoc., Palo Alto, California.

(21) K. L. Vander Sluis, *et al., Am. J. Phys.*, **33**, 225 (1965).

37–5.1 Stability Requirement

This depends *entirely* on the angle between the reference and the object beams. In general, the smaller the angle, the lower the spatial frequency and the less stringent the stability requirement. Actually, holograms can be made with the film hand-held.[22,23] For angles exceeding 10 deg, support

(22) G. W. Stroke, D. Brumm, and A. Funkhouser, *J. Opt. Soc. Am.*, **55**, 1327 (1965).

(23) G. W. Stroke *et al., J. Opt. Soc. Am.*, **57**, 110 (1967).

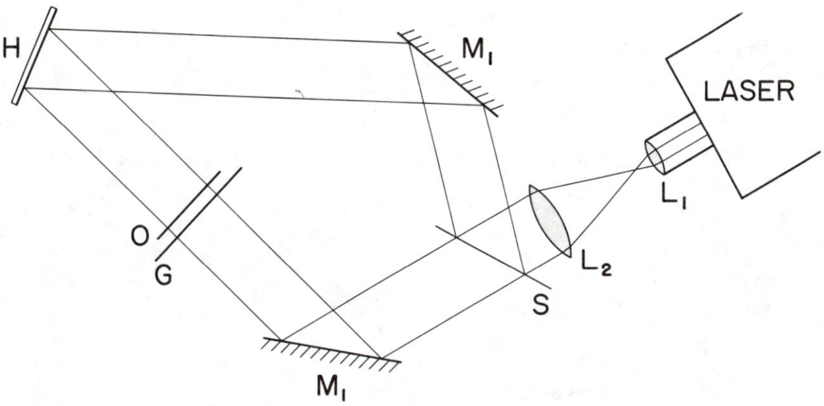

Fig. 37–21

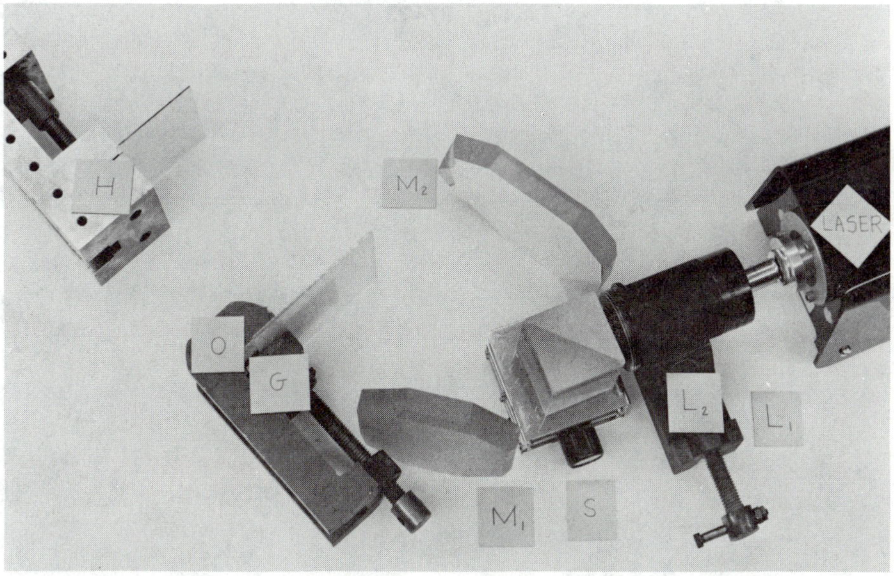

Fig. 37–22

the system on a heavy solid surface resting on a vibration-absorbing medium such as an inflated inner tube. If the photoplate has been handled, it should be allowed to reestablish thermal equilibrium with its environment before making the exposure. If a pulsed laser is used, the stability requirement becomes almost nonexistent. But other problems will come up.

37–5.2 Coherence Requirement

Ordinary lasers have coherent lengths not exceeding a few tens of centimeters. In general, the higher the operating TEM mode, the shorter the coherent length. For this reason, carefully arrange the geometry so that the optical paths for the reference and the object beams are approximately equal, each being measured from the beam splitter. If the paths must be different, they should differ by integral multiples of $2L$, L being the length of the laser cavity.

37–5.3 Emulsion Resolution

The spatial frequency on the prospective hologram dictates the resolution required of the emulsion. In general, use Kodak 649–F film or plate, and process as recommended. (It has a resolution of several thousand lines per millimeter.) On the

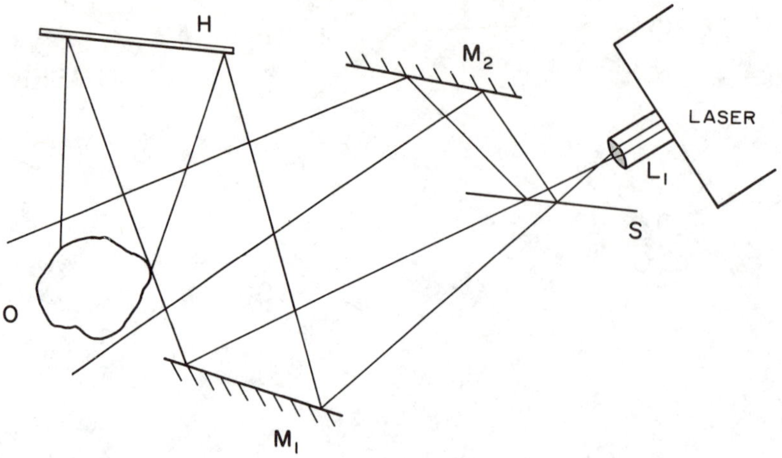

Fig. 37–23

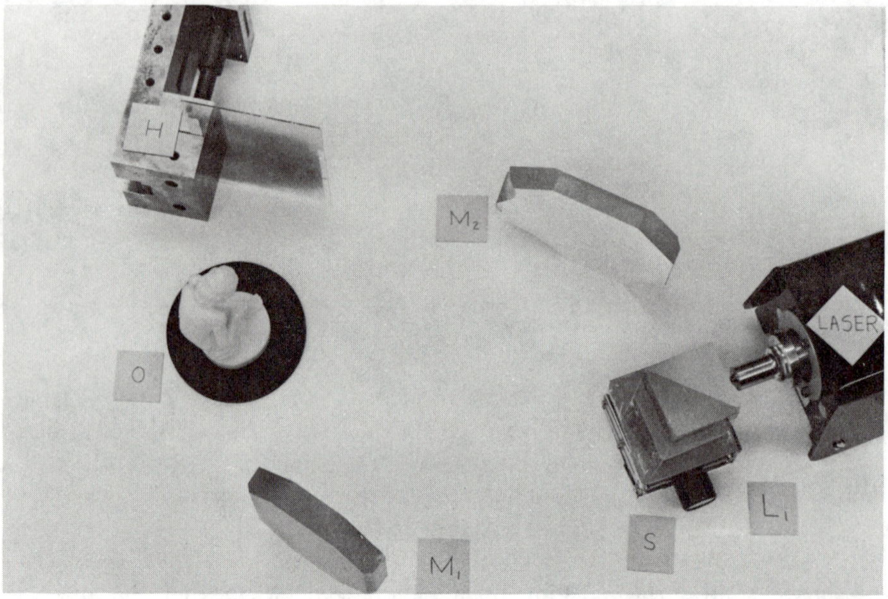

Fig. 37–24

other hand, if the spatial frequency is not to exceed 50 lines/mm, Polaroid P/N film can be used.[22]

37–5.4 Actual Optical Arrangements

Figures 37–17 (solid lines) and 37–18 show an actual setup for constructing a sine grating. In these and figures to follow, the symbols and their meanings are:

L_1–20× microscope objective

L_2 and L_3–large aperture, short focal length lenses

M_1, M_2, and M_3–first surface totally reflecting mirrors

S–beam splitter

O–object

H–emulsion for making grating or hologram

G–ground or opal glass

Note that the paths SM_1H and SM_2H are approximately equal. With a one-milliwatt (1-mW) laser and beam diameters of about 3 cm, the exposure time on Kodak 649-F emulsion is approximately 1 sec. Because of the thickness of this emulsion,

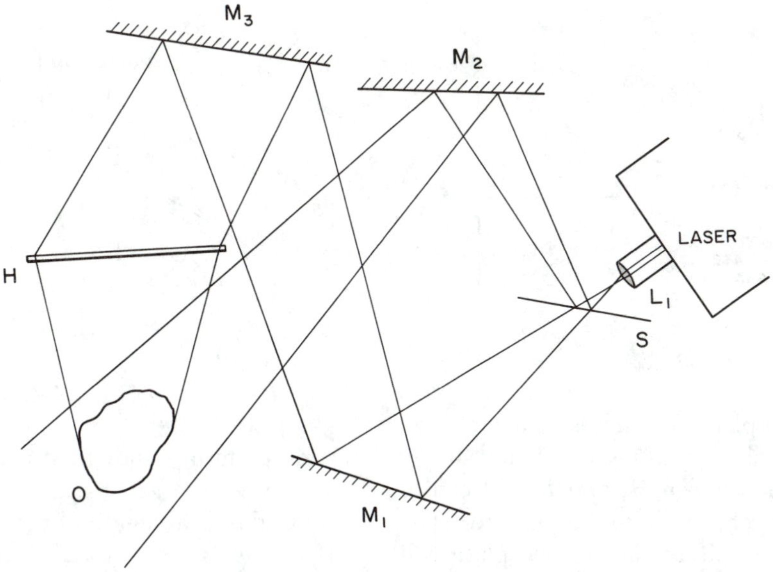

Fig. 37–25

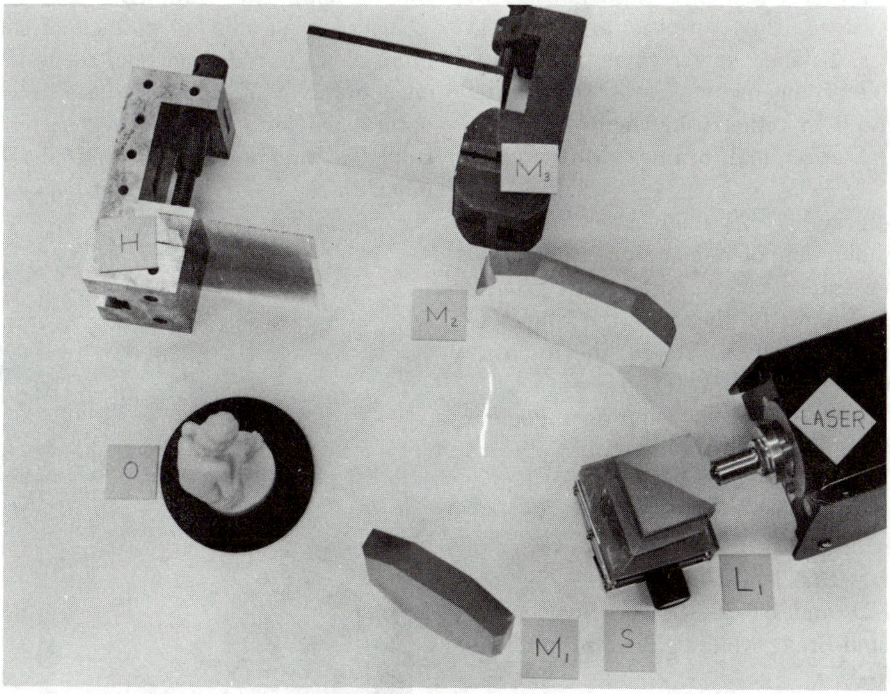

Fig. 37–26

the resultant grating will show sideband suppression (Sec. 37–4.6). A lower resolution emulsion can be used if θ is decreased. The exposure time, as well as the intensity ratio, has to be experimentally determined.

To make a beat grating, the same setup is used. After the initial exposure, M_2 is moved 2 or 3 cm, as shown by the dotted lines on Fig. 37–17, and exposed again after the mechanical vibration caused by the adjustment has subsided. The increase in the total exposure time will not significantly affect the quality of the grating.

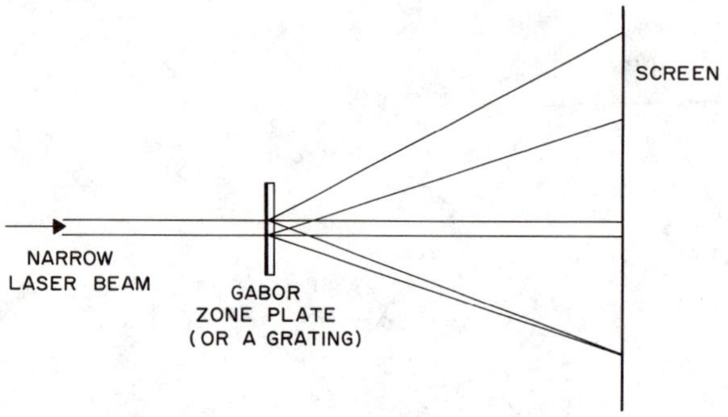

Fig. 37–27

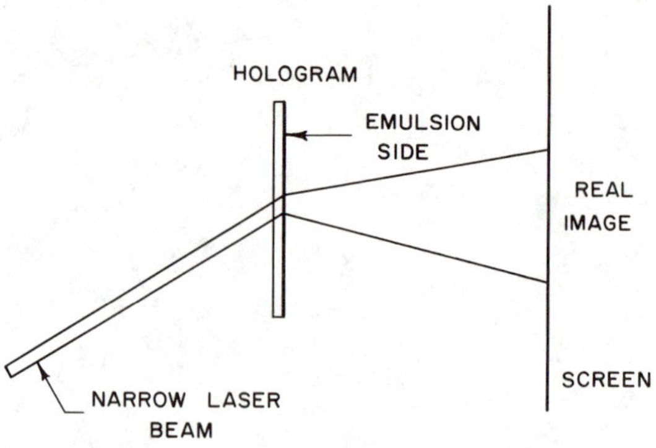

Fig. 37–28

The Gabor zone plate demands a minor change from Fig. 37–17, as shown in Figs. 37–19 and 37–20. A lens L_3 is placed between M_1 and H and causes an apparent point object O to be formed. The focal length of the resultant Gabor zone plate will be the distance between O and H. If a longer focal length is desired (which will be more suitable for demonstration to a large audience), a negative lens can be used instead, thus forming an apparent object point at a distance longer than the dimension of the whole arrangement. The intensity ratio should be adjusted by either using neutral density absorbers or by additional beam splitters (Sec. 37–4.7).

Figures 37–21 and 37–22 show a configuration for making a hologram of a complex transparent object—a 35-mm transparency. A piece of ground glass G is placed directly ahead of the object to provide diffused illumination. Again, the intensity ratio should be adjusted before exposure.

Figures 37–23 and 37–24 show an arrangement for making holograms of three-dimensional objects. Paths SM_2OH and SM_1H are approximately equal. As in all cases, the intensity ratio must be adjusted. Exposure time for Kodak 649–F is in the order of several minutes with a 1-mW laser.

Figures 37–25 and 37–26 show a method of making Lippmann-Bragg white light reflection holograms (Sec. 37–4.7).

37–5.5 Demonstrating

The gratings can be demonstrated to a large audience (of several hundred) using a laser having an output of 2 mW or more. The simple arrangement is shown in Fig. 37–27. The natural laser beam impinges directly on the gratings, and the diffraction pattern is shown on a screen. The separation between the grating and the screen should be equal to the focal length of the Gabor zone plate.

If a laser is not available, the demonstrator can pass the grating around the audience and have them look through it at a spectral discharge (a sodium lamp, for example).

There is no efficient method of showing a hologram of a three dimensional scene to a large audience unless a 50–100-mW laser is available. The natural narrow beam is directed on the hologram from the direction of the original reference beam, *but with the hologram turned backwards,* as shown

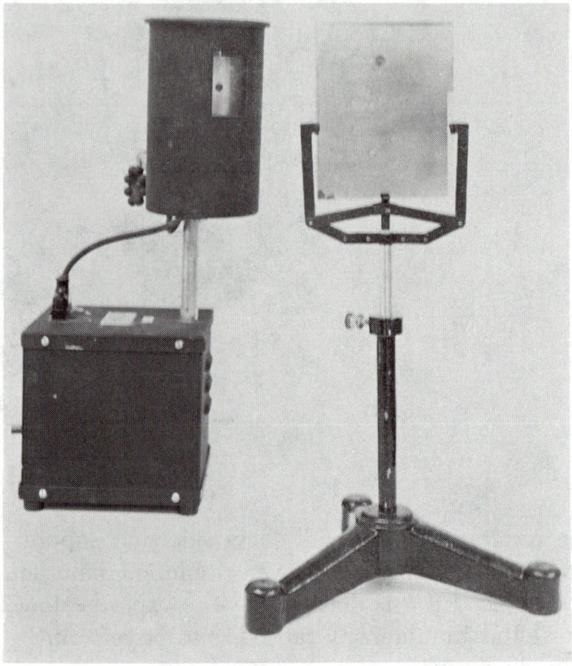

Fig. 37–29

in Fig. 37–28. A small screen (a white card) is positioned as shown where the real images are to be projected. In general, the depth of field is improved by making the illuminating beam narrow, with a corresponding sacrifice in resolution. By moving the beam about the hologram, changes in perspective can be observed. The reason for this procedure was explained in Sec. 37–4.6.

It should be noted that a laser is not necessary for viewing any hologram. The simple reason is that in the reconstruction the diffraction takes place locally and only a very short coherent length is required. This certainly is not the case during construction. A convenient corridor demonstration can be set up as shown in Fig. 37–29. A mercury discharge is used, and a small aperture is placed near the tube to approximate a point source. The entire system should be situated in a darkened area, or it should be placed in a dark cabinet. Without filters, each line in the mercury discharge will produce a separate set of image sizes directly proportional to the wavelength.

38

SPECIAL RELATIVITY AND QUANTUM EFFECTS

38–1 Special Relativity

38–1.1[1]

Often, when introducing relativistic kinematics, the teacher draws the analogy of ripples on a river to illustrate waves in a moving medium. With the flow ripple tank[2] in Fig. 38–1, he can also illustrate the analogy. In the apparatus, water flows in a smooth sheet across the tank, over a smooth lip, and down a screen to a trough, from which a pump returns the water to the ripple tank. This water is smoothed by the jersey tube through which it enters the tank and by a baffle. Wire-screen beaches further smooth the water by damping the reflections from the tank sides. Waves generated by the variable-phase generator[3] attached to the tank are projected by a light source[4] onto a screen beneath the tank.

For uniform flow, the tank must be level. It is designed so that the water depth remains close to 1 cm for all flow speeds in a level tank. At this depth, the wave propagation speed u in water is about 23 cm/sec and does not change appreciably for frequencies between 10 and 20 cps. At most flow speeds, the side reflection beaches alone will permit good wave patterns; however, for very slow speeds dampers may be needed at the inflow and outflow ends. It is more convenient to measure wavelengths on the screen than in the tank, so if flow speed is measured in the tank, it should be corrected to give screen flow speed, which can then be compared to the speed of propagation computed from screen wavelength.

When testing uniformity of flow speed by timing the motion of small bits of floating paper, agreement within ten percent over the central area of the tank is acceptable. Flow speeds v up to 4 or 5 cm/sec, which give a maximum $v/u \approx 0.2$, are fast enough. At higher flow speeds, some students will be confused by the small variations which are due to second order effects $(v/u)^2$.

Wave Propagation Speed Upstream and Down. The wave pattern of a single-point source in moving water appears elongated downstream and compressed upstream. Individual waves are circular nevertheless, but the circles are not concentric. (Students may have to see the pattern through a stroboscope from upstream to believe the waves are circular.) Since the stroboscope stops the motion of the entire wave pattern, the frequency is clearly the same both upstream and down. Therefore, in the observer's reference frame, the speed of propagation is greater downstream, since the wavelength in that direction is noticeably larger. Downstream

[1] Adapted, with permission, from *PSSC Advanced Topics Supplement* (D. C. Heath Publishing Co., Boston, 1964).

[2] MSC 4100, Macalaster Scientific Corp.

[3] MSC 3301, Macalaster Scientific Corp.

[4] MSC 1201, Macalaster Scientific Corp.

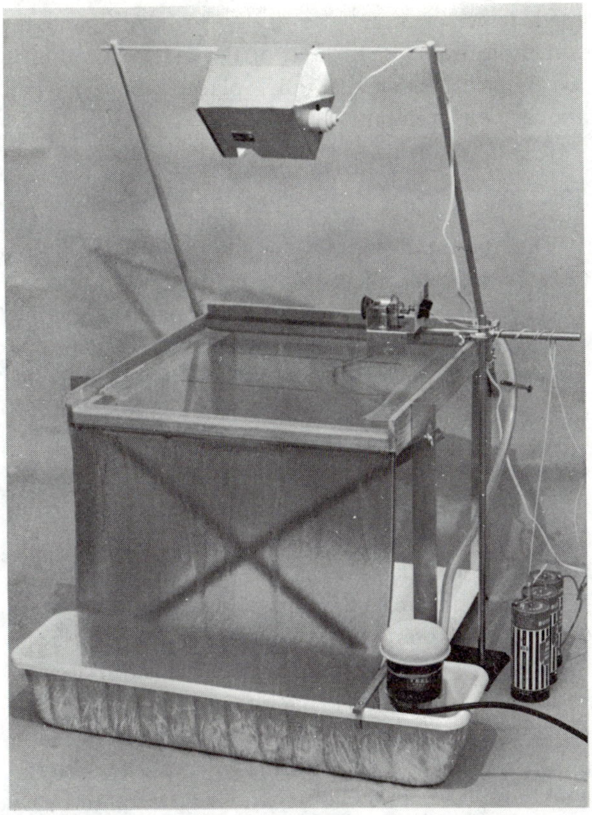

Fig. 38–1

the wave propagation speed u in water is in the same direction as the speed of the water v. Thus, the speed relative to the observer should be $u + v$, and, similarly, the upstream speed should be $u - v$. (It is, however, this additivity of velocities that will fail at high speed.) Since the frequency is the same for both directions, it is expected that $\lambda_u/\lambda_d = (1 - v/u)/(1 + v/u)$, where λ_u is the upstream wavelength and λ_d is the downstream wavelength.

38–1.2[5]

To demonstrate twin source interference in a moving medium, two point sources are used in the ripple tank discussed in Demonstration 38–1.1. The variable phase wave generator (see footnote 3) is so placed that the line between the sources is parallel to the direction of flow of the water. These two sources create an interference pattern that is symmetrical about the line between them.

(5) Adapted, with permission, from *PSSC Advanced Topics Supplement* (D. C. Heath Publishing Co., Boston, 1964).

Synchronized sources in still water give rise to a pattern with a straight nodal line which is the perpendicular bisector of the source line. If the sources are started in phase in still water and the water flow is increased, nodal lines arise which are curved but still symmetrical. The directions of the nodal lines are different from their rest directions, but the amount of change is the same both upstream and downstream. Thus, the motion of the medium relative to the sources and the observer is not detected by comparing the nodal lines. However, when there is a phase delay between the two sources in still water, the pattern closely resembles the same pattern as is obtained with synchronous sources in moving water. It is not possible, therefore, to determine, by looking at the interference pattern only, whether the medium is moving.

After starting with in-phase sources in still water and then distorting the interference pattern by letting the water flow, the flowing-water interference pattern can be restored to the still-water position by advancing the downstream source. This is done by turning the phase pointer on the wave generator through a positive angle (the battery polarity may have to be reversed). The ratio λ/d, wavelength to source separation, should be kept small, about 0.2. At this value with a water flow of 5 cm/sec a complete change of phase is needed to restore the interference pattern. Under these conditions, the shift in the central maximum, $\theta = p\lambda/d$, will be within 5 percent, where p = phase delay. In still water, end dampers as well as side dampers are needed to minimize reflections. When the flow speed v reaches about 2 cm/sec, reflections from the spillway and the upstream smoother are no longer troublesome, so the end dampers should be removed to permit faster flow. Uniformity of flow is better when a gap of 1–2 cm occurs between the damper baffle and the plastic-screen smoother.

Observations are made by noting either the position of the middle of the central maximum or the positions of the nodal lines on each side of it. In a process similar to tuning a radio, the phase angles are noted for both sides at which the pattern begins to depart from the original position; an average value is then calculated. This tuning process is necessary because the pattern position may look normal within a phase range of about 20 deg.

By plotting v/u against $p = \alpha°/360°$ and comparing the value of the slope with λ/d, one can show that a moving medium will cause a central-maximum shift of

$$\theta = v/u, \quad \text{for} \quad v << u; \quad \text{i.e.,} \ \frac{v}{u} = p\frac{\lambda}{d}.$$

The fact that the above relation can also be written,

$$p = \frac{vd}{f\lambda^2}$$

emphasizes the need for high accuracy in measuring λ, which is checked with the measured frequency f and the value of u. The flowing water shifts the interference pattern in the direction of flow, but the nodal line spacing appears unaltered.

From the given equation for p, it is possible to calculate the flow rate. If the two sources were in phase in still water, the flow rate can be calculated from the phase change necessary to make the central maximum perpendicular to the line between the sources:

$$v = u \cdot \frac{p\lambda}{d} = u \cdot \frac{\lambda}{d} \cdot \frac{\alpha}{360°}.$$

If the sources were in phase in still water, then one must change their phase difference in order to make them appear in phase in the moving medium. Moreover, even if the sources were out of phase in still water, it would be possible to make them appear in phase in a medium moving at a given rate simply by altering the phase delay. In any case simultaneity depends on the observer's motion relative to the sources and medium.

Although patterns can be shown during lecture and the concepts discussed, it is also helpful to use photographs. Ease and accuracy of analysis will depend on the quality of the photographs; therefore, care in photography is important. In preparation for photographing, the pattern of each source should be stabilized, reflections should be eliminated, and bubbles should be removed. A Polaroid camera with Polaroid 3000 film and a shutter speed of 1/100 sec give good results. With the screen beneath the tank, photographs taken obliquely can be analyzed if an angular scale is placed around the edge of the screen. With the light below the tank and a screen above, vertical photographs can be made and a protractor used on the print. Several kinds of screens can be used: ground glass or tracing paper held a few millimeters above the print; rag paper taped to a light wooden frame and dampened, which contracts on drying to form a smooth rear-projection screen; and opal glass (glass thinly coated with barium sulfate) placed about 25 cm above the ripple tank.

38–1.3[6]

The device described here is a visual aid intended primarily for use in a lecture on the Special Theory of Relativity, or more specifically, the Lorentz Transformation. It offers a visible representation of the space and time coordinates of two reference frames in uniform relative motion. Originally developed for use in an educational television program, this device has proved to be instructive, or at least entertaining, in talks given to a variety of audiences, including graduate physics classes. It could serve as the basis for a pseudo-experiment performed by individual students in an elementary laboratory. It might be suitable for a permanent operating exhibit in a science museum. It must be emphasized that the apparatus does not provide a real display of relativistic effects, but only mimics those effects in slow motion. As with a motion-picture, the demonstrator may run through the history of the physical process at any convenient rate, and can at any instant halt the development of the story to examine in detail the prevailing situation.

The essential features are shown in Fig. 38–2. As seen by the audience, the apparatus presents two panels, one above the other. The lower panel will represent a reference frame S at rest (relative to the audience), while the upper panel represents a frame S′ in uniform motion (relative to S) toward the right (of the audience). To operate the machine the demonstrator turns a crank on the back side of the machine, and thus drives from left to right the carriage that carries the upper panel.

To represent the measuring instruments that are the bases for the space and time coordinates, each panel is provided with: (1) markings (in relief) to represent a measuring rod with its graduations, placed parallel to the direction of relative motion, and (2) three dials with rotating pointers simulating clocks, located respectively at the ends and the center of the measuring rod. We will refer to the clocks on the lower panel (from left to right)

[6] J. F. Streib, *Am. J. Phys.*, **31**, 802 (1963).

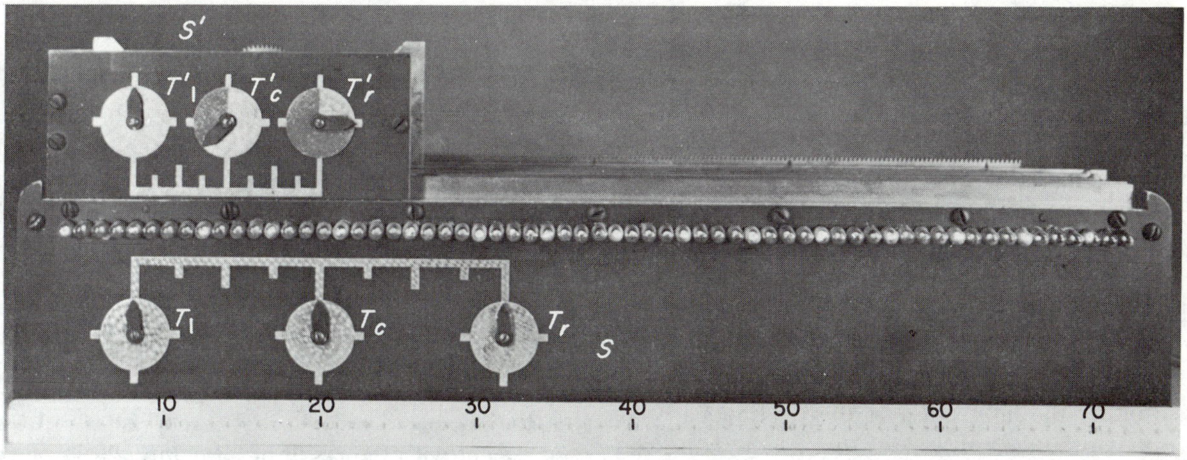

Fig. 38–2

by T_1, T_c, T_r, sometimes using these symbols simply to designate the point at which the clock is located. Similarly the clocks of the upper panel will be designated T'_1, T'_c, T'_r. A value of "time" will be written t or t' according as it is read on a clock of the lower panel or of the upper panel.

A horizontal row of closely spaced small incandescent lamps simulates a light signal. By means of a switch (S_S in Fig. A38–52) the demonstrator may select either of two patterns: (1) a single lamp is lit to represent a "light pulse"; or (2) every fourth lamp is lit to represent a continuous light wave. When the machine is operated, the pulse pattern propagates toward the right. But if the wave pattern is selected, the demonstrator may with a second switch (S_D in Fig. A38–52) determine whether it shall propagate toward the right or toward the left. See Appendix, page 1355, for construction details of the apparatus.

The lengths of the measuring rods and the operation of the clocks are correlated with the relative motion and the propagation of the light signal as required by the Lorentz Transformation for a relative velocity given by $\beta = \sqrt{3}/2$; these correlations will be described in detail below. It has been shown[7] that the relativistic contraction and the finite speed of light combine to give a moving object an apparent distortion and rotation. This device is not concerned with such appearances, however, but presents the actual configuration of S' relative to S.

[7] J. Terrell, *Phys. Rev.*, **116,** 1041 (1959); see also V. F. Weisskopf, *Phys. Today,* **13,** 24 (1960).

Even a brief exhibition of the apparatus with an explanation of its meaning will convey a substantial appreciation of the principles of the space-time transformation and its departure from the intuitive Galilean conception. It should be explained to the audience that: We must understand that the two sets of instruments—(1) the measuring rod and the three clocks on the lower panel, and (2) their counterparts on the upper panel—are constructed to identical specifications, and that the three clocks in each set are mutually synchronized; the observations that appear to belie these assertions are results of the relative motion. Specifically, the audience will note: S' moves with a velocity slightly less than the speed of light, and (1) the moving measuring rod has shrunk to one-half the length of its stationary counterpart, (2) the moving clocks run only half as fast as the stationary ones, and (3) the moving clocks are not synchronized.

But if more time is available, it is interesting to carry out a number of quantitative observations. In the following paragraphs is described a sequence which the writer has found convenient.

1. The "Speed of Light" (in S). To avoid the distractions from the clocks on the upper panel, which are not involved in these observations, it is sometimes desirable to cover that panel with a small screen. Then the machine is operated and the "speed of light" is determined by timing its propagation from T_1 to T_r, using those two (synchronized) clocks to indicate the starting time and the arrival time, respectively. To obtain numerical results, one could use the actual separation of T_1 and

T_r (24 cm), and announce that one revolution of the clock pointer signifies 9.2×10^{-10} sec. However, it is more satisfactory to ignore the standard units, and we will simply designate the length of the measuring rod (from T_1 to T_r) as one "unit of length," and let one revolution of the clock pointer measure one "unit of time." We observe that the light signal propagates over this unit distance in the interval from $t = 0$ to $t = 0.866$ $(= \sqrt{3}/2)$. Then, for the speed of light we obtain $c = 1.15$ $(= 2/\sqrt{3})$. In this first measurement, with the light signal propagating toward the right, we may use either the light pulse or the continuous wave. To make the demonstrations that follow more convincing, it is desirable to measure the speed for the reverse direction of propagation; then, as noted above, it is necessary to use the continuous wave pattern. To be more realistic, we may note that these two measurements really serve to verify the synchronization of T_1 and T_r. In practice the speed of light is measured in a "round trip" experiment; this may be demonstrated by using the switch to reverse the direction of propagation when the light signal reaches T_r, thus "reflecting" it back to T_1. This measurement, in a sense, serves to calibrate the space-time representation.

2. The Velocity of S′ (in S). In a similar manner we measure the velocity of T'_1 as it moves from T_1 to T_r, and find that the velocity of S' relative to S is $v = 1$. Thus we calculate $\beta = v/c = 1/1.15 = 0.866$ $(= \sqrt{3}/2)$. It must be emphasized to the audience that the machine always depicts conditions associated with this particular value of β, regardless of the actual speed of the carriage, or even if the carriage is motionless. In this respect the functioning of the machine is comparable with the presentation of a motion picture.

3. The Lorentz Transformation. We compute the contraction factor $\sqrt{(1-\beta^2)} = 0.500$. Now, when we uncover the upper panel, we find that the measuring instruments in S' exhibit the behavior required by the Lorentz Transformation: (1) The measuring rod has shrunk to one-half its proper length; (2) The clocks run at half speed; and (3) T'_c is three-eighths of a unit behind T'_1 and three-eighths of a unit ahead of T'_r. In this connection it must be explained that when the clock pointer lies within the sector of the dial that is colored red, a negative value of time is indicated.

The most entertaining part of the demonstration comes when we "verify" directly the two postulates of the Theory of Relativity by demonstrating in several respects the symmetry of the relations connecting the two reference frames and the "universal" value of the speed of light.

4. The Velocity of S (in S′). While the machine is operated we time the motion of T_c (for example) as it "moves" from T'_r to T'_1. This motion, over a unit distance, occurs during the interval from $t' = -\frac{3}{4}$ to $t' = +\frac{1}{4}$. The velocity, then, is $v' = 1$, in agreement with the result of Paragraph 2.

5. The Longitudinal Contraction (of S). We now measure the length of the lower measuring rod relative to the upper rod, or, more precisely, relative to the frame S'. To do this we must note the position on the upper rod of T_1 and T_r in two observations that are simultaneous (in S'). We note that T_1 lies at T'_1 at $t' = 0$ (as indicated by T'_1), and that T_r lies at T'_c at $t' = 0$ (as indicated by T'_c). It follows that, relative to S', the lower measuring rod has only one-half the length of the upper rod.

6. The Rate of Moving Clocks (of S). Taking the clocks of S' as standards, we determine the rate of T_c (for example). When T_c is adjacent to T'_r, we take the readings $t = 0$ and $t' = -\frac{3}{4}$. Then, when T_c is adjacent to T'_1, we take the readings $t = \frac{1}{2}$ and $t' = +\frac{1}{4}$. We now conclude that the clocks in S are operating at one-half the normal rate when referred to the clocks of S'.

7. The Discrepancy of Moving Clocks (of S). Again, taking the clocks of S' as standards, we compare T_1 and T_r. When T_1 is adjacent to T'_1, we read $t = 0$ and $t' = 0$. When T_r is adjacent to T'_c, we read $t = \frac{3}{4}$ and again $t' = 0$. From these two sets of data (taken simultaneously in S'), it follows that T_1 reads three-fourths of a unit behind T_r when referred to the clocks of S'.

8. The Speed of Light (in S). The propagation of the light signal from T'_1 to T'_r is timed by means of those two clocks, and we obtain again for the speed of light $c' = 1.15$. It is interesting to repeat this measurement with the light signal propagating toward the left; appearances are quite different, but the same value for c' is obtained. Again it is worthwhile to emphasize that the two separate measurements demonstrate the synchronization of the clocks (now T'_1 and T'_r), which is in this case not obvious. As before, a "round trip" measurement of c' may be carried out.

9. The Doppler Effect. The "experiments" described above contain the essentials of the Lorentz Transformation, but since the machine permits a simple representation of the Doppler effect, we will also describe this demonstration. The machine is operated with the continuous-wave pattern propagating toward the right, and the frequency (in S) is determined by counting the waves that pass T_1 during two units of time (indicated by T_1). We find the frequency $f = 6.25$. Now the procedure is repeated with reference to T'_1 instead of T_1, and we find the new frequency $f' = 1.67$, which agrees with the formula for the relativistic Doppler shift: $f' = f\sqrt{[(1-\beta)/(1+\beta)]}$. Similar measurements may be made with the light signal propagating toward the left. Then we find $f' = 23.3$, in agreement with the formula which now takes the form $f' = f\sqrt{[(1+\beta)/(1-\beta)]}$.

38–2 Photoelectric Effect

38–2.1

The photoelectric material is a zinc sheet, about 10 x 10 cm, to which a small lug is soldered. Rub the zinc with sandpaper (never emery paper) and wipe with paper toweling just before placing it in the socket at the top of the electroscope. Charge the system negatively by induction, and project the image of the leaf on a screen where a scale is drawn. Illuminate the zinc with a point-source direct current carbon arc with no glass between the arc and the zinc.

The falling leaf shows that electrons are emitted, the rate of fall being proportional to the current. Place the source at different distances from the zinc and measure the rate of discharge. The inverse square law is followed, demonstrating that the photoelectric current is proportional to the intensity of radiation.

When a thin piece of glass, say the cover glass of a lantern slide, is placed close to the arc, the discharge of the electroscope ceases immediately. The photoelectric current may then be started or stopped by removing or inserting the cover glass, thus demonstrating that the threshold lies in the ultraviolet spectrum. The electroscope will function properly whatever the weather if polystyrene insulation is used.

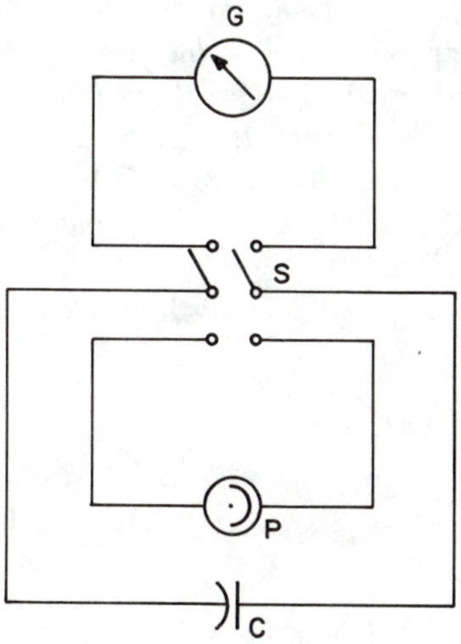

Fig. 38–3

38–2.2

A covered vacuum phototube[8] P and an oil filled capacitor[9] C are mounted on a nonconducting plastic sheet and connected through a double-pole, double-throw switch S so that the capacitor can be connected first to the photo cell and then to a galvanometer G. See Fig. 38–3. When the tube is exposed to a lamp's radiation, the capacitor is charged. The tube is then covered to prevent charge leakage, and the capacitor is discharged through a projection galvanometer to show the photoelectric effect. The demonstration is performed qualitatively with different light sources (mercury, xenon, sodium, and tungsten lamps). If a mercury arc or a xenon lamp is used, a large galvanometer deflection is observed. A sodium lamp produces a smaller deflection, and a tungsten source covered with ruby glass shows the least

[8] RCA 929 with an S–4 response; it requires a standard octal socket.

[9] 1 μF at 200 V.

effect. If the room lights are dimmed when a tungsten lamp is used, a smaller deflection occurs. A quantitative lecture experiment can also be performed with the apparatus.

38–2.3

This demonstration shows how the maximum energy of photoelectrons, ejected from a photosensitive cathode, increases as one goes from red light to blue light. It is not really quantitative, but it does exhibit the essential trend; its advantage is the use of a cheap, commercial phototube. The demonstration requires a bright continuous spectrum thrown on the wall of the lecture room. In a fair-sized lecture room this spectrum needs to be several feet long and several inches high, such as can be produced from a carbon arc and prisms at a distance of about 10 m. A 929 phototube is mounted inside a metal shield with a hole opposite the photocathode. A lead passes directly, inside the shield, from the photocathode to the insulated terminal of a Keithley electrometer.(10) The collector wire of the phototube is connected to a clip lead that is brought out through the metal shield. The shield is always connected to the grounded terminal of the electrometer. Meter readings are displayed on a projection meter(11) connected to the output of the electrometer.

Two types of measurements are made. First, the Keithley electrometer is set on a low-impedance range for measurement of current. A battery is placed between the phototube collector wire and ground. As the phototube is moved across the spectrum, the current passes through a maximum and falls away to very small values at the blue end. This represents a crude qualitative picture of the intensity distribution over the spectrum. Second, the Keithley electrometer is set on a high-impedance range for measurement of potential. The phototube collector wire is connected directly to ground. As the phototube is moved across the spectrum, the voltage rises steadily as one goes from red to blue. This gives a rough measure of the potential reached by the highly insulated photocathode when escape of any more photoelectrons is prevented.

This is *not* a clean experiment. The mixed character of the photocathode material is one factor. Still more important is the ill-defined character of the second situation, involving non-infinite impedance at the photocathode, and stray currents not due to the photoelectrons. Typically, the value of the voltage will be found to rise from about 0.6 V at the red end of the spectrum to about 1.2 V at the violet end—about half as big a change as one ought to find. The behavior of the system becomes especially dubious at the extreme red and violet ends of the spectrum, where the light intensities are very low and the photo-current may be small compared to stray current.

38–2.4

To show evidence for associating particle properties with photons, light of discrete frequency is allowed to fall on the cathode of the 1P 39 phototube (see Fig. 38–4). The photon may be completely absorbed by an electron in this surface, in which case it gives all its energy to this electron. If this energy is greater than the energy with which the electron is bound in the metal, it may escape. If the energy E of this escaped electron is measured for several different light frequencies, energy and frequency are related by

$$E = h(\nu - \nu_0),$$

where ν_0 is the frequency at which electrons no longer leave the surface, and h is Planck's constant.

Such a result is readily interpreted if the protons are regarded as carrying an energy which is related to their frequency by $E = h\nu$. For energies below some characteristic energy $E_0 = h\nu_0$, no electrons can be knocked from the surface of the metal. Thus E_0 represents the depth of the potential well in which the electron in the metal surface finds itself.

When light of sufficiently high frequency strikes

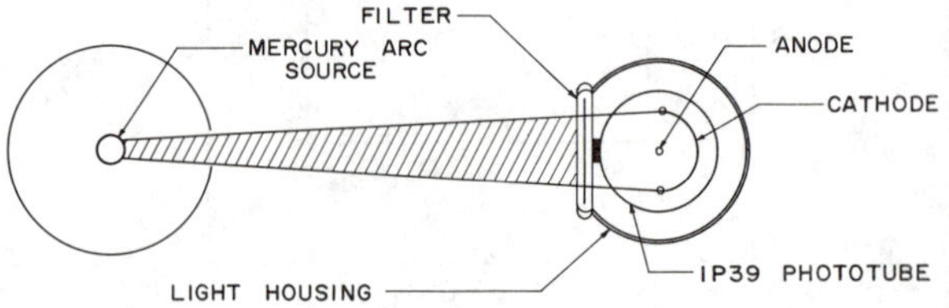

Fig. 38–4

(10) Model 200 or 200B, Keithley Instruments, Inc., Cleveland, Ohio.

(11) MB5, Keithley Instruments, Inc., with part of the case cut away and replaced with clear plastic.

the cathode, it causes the emission of photoelectrons. The central wire (the anode) of the photocell is shielded from illumination. If the anode and cathode are connected through an external circuit, an electrical current is observed to flow.

A battery and external circuit as shown in Fig. 38–5 are used to apply a variable electrical potential V to the electrodes of the photocell so as to make the anode negatively charged with respect to the cathode. Thus photoelectrons emitted from the cathode are repelled from the anode, and the product Ve represents the loss in kinetic energy by an electron in moving from the cathode to the anode. The potential difference V is read directly from the voltmeter. Electrons with kinetic energy less than Ve fail to reach the anode. As V is increased, the flow of electrons, and thus the flow of current in the external circuit, decreases and becomes zero when $Ve = V_0 e = E$. The stopping potential V_0 is just sufficient to reduce the current to zero.

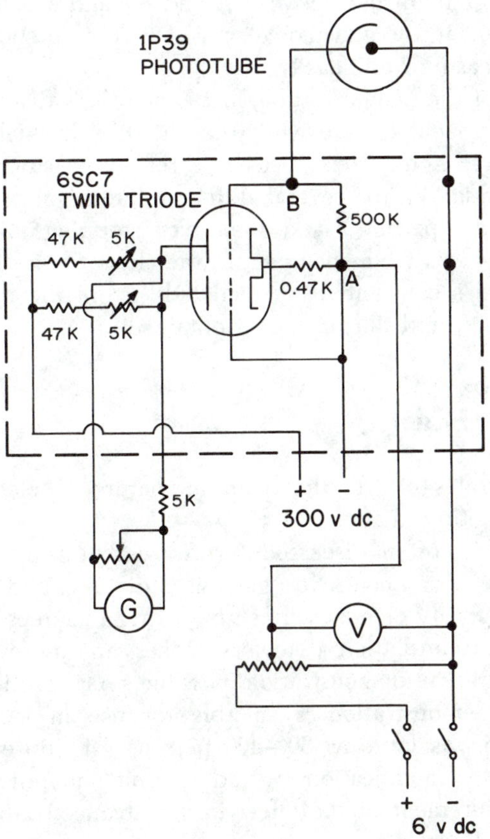

Fig. 38–5

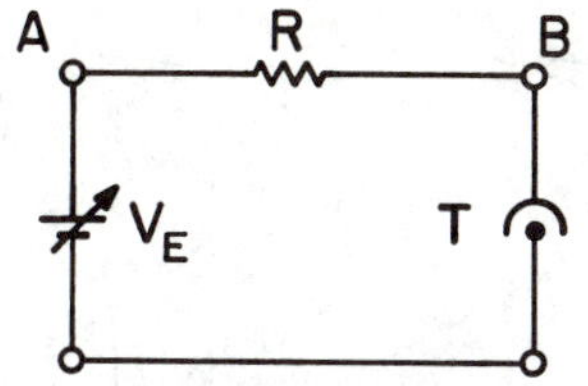

Fig. 38–6

Unfortunately, an ordinary ammeter is not sensitive enough to determine the point at which current ceases to flow. The circuit within the dotted lines of Fig. 38–5 is a device for amplifying the small currents which flow in the phototube circuit and therefore through the resistor between terminals A and B. Analysis of the circuit can be simplified by studying Fig. 38–6.

There are three cases to consider:

1. If $V_0 > V_E$, the anode (the dot in the diagram) collects electrons. The tube T acts like a battery of potential higher than E, and electrons flow around the circuit in a clockwise direction. This current results in a drop in potential of $I \times R$ across the resistance R. Thus A and B are at different potentials.
2. When $V_0 = V_E$, electrons are not quite energetic enough to reach the cathode. No current flows, and A and B are at the same potential.
3. When $V_0 < V_E$, one might think that the electrons would flow counterclockwise. This is not so, for there is no light striking the anode. Thus the tube T acts like an open circuit, and A and B are again at the same potential.

The mercury arc lamp is used as a source of light because different frequencies of monochromatic light can be obtained using different transmission filters, each with a pass band overlapping just one of the spectral lines of mercury. Using the relation between frequency and wavelength, one sees that the shortest wavelength corresponds to the highest frequency and thus the largest energy. It is this maximum energy which is determined. Various frequencies can be selected by passing the light through a sharp-cutoff filter, which selectively absorbs light below a certain wavelength. Figure 38–7 shows the percentage of light transmitted plotted on a log scale as a function of the wavelength for a series of sharp-cutoff filters. The wave-

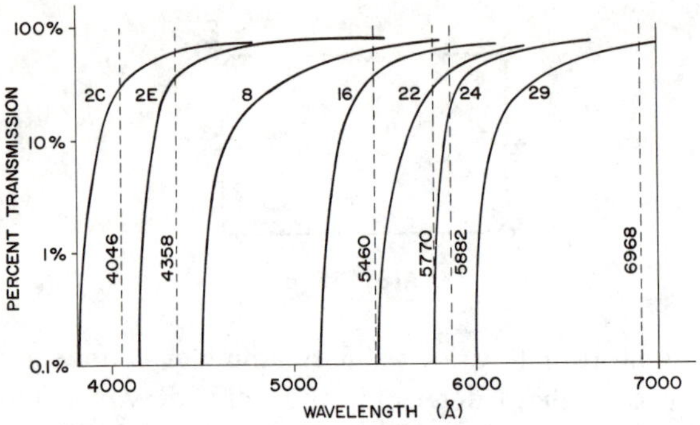

Fig. 38–7

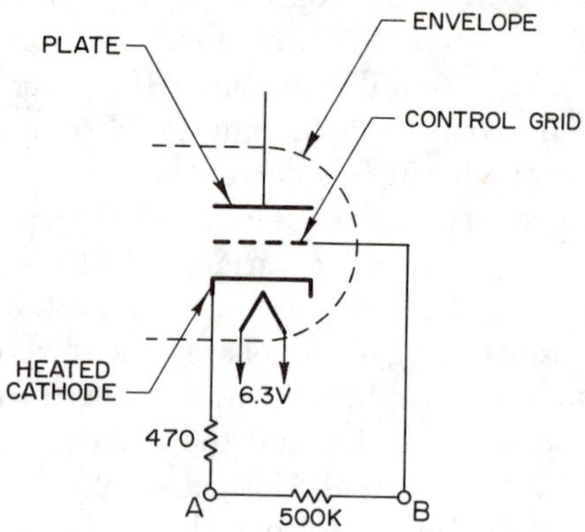

Fig. 38–8

lengths of the dominant lines of the mercury spectrum appear as vertical dotted lines. In using the filters, it is necessary to keep in mind that the frequency of the shortest wavelength line which passes through the filter corresponds to the most energetic photons in the beam striking the cathode of the photocell. This line, therefore, determines E_{max} of the photoelectrons, and V_0, the potential which is just able to stop the most energetic electrons.

The circuit enclosed by the dotted lines (Fig. 38–8) is an example of a grid-controlled vacuum tube amplifier. By cutting off all light to the photocell, A and B (and the two grids) can be brought to the same potential. This can also be done by short-circuiting A and B. Because of minor plate-circuit differences, this short-circuiting does not necessarily bring the two plates to the same potential, but the variable resistances (5k in Fig. 38–5) can be adjusted until the plates are at the same potential, as indicated by zero current in the galvanometer. If now the phototube is illuminated and if $V_0 > V_E$, electrons flow through the 500 kΩ resistor and on to the cathode of the phototube to take the place of the ones ejected by the light. The resulting difference of potential between A and B is transmitted to the two grids, and this transmission in turn results in a difference in plate potential (larger than that between A and B) and a flow of current in the galvanometer, which is large enough to measure quite easily.

For the demonstration, first adjust the galvanometer current to zero while the photocell is shielded from all light. Next, measure carefully the stopping potential V_0 for several different filters (as many as time permits) and then plot stopping potential versus frequency. From the slope of the line, Planck's constant h is calculated. From the intercept, V_0 and thus E can be obtained.

38–3 Compton Effect

38–3.1

When monochromatic electromagnetic radiation is scattered from a material, the scattered beam is found to consist of two lines, one line of the same wavelength as the incident radiation and the other of longer wavelength. Experimentally, it is found that the longer wavelength depends only on the incident wavelength and the scattering angle and is not dependent on the particular scattering material. The correct explanation of this phenomenon was first given by A. H. Compton[12] in 1923 and illustrates clearly the quantum nature of electromagnetic radiation.

Most often discussed for X-ray scattering, the effect also occurs for gamma radiation, and the availability of relatively strong gamma sources and good scintillation detectors makes an interesting qualitative demonstration possible. As described, the demonstration is suitable for use in lecture groups as large as 300–400 persons. If desired, a simple modification would permit quantitative measurement of the effect in an advanced laboratory.

Basically, the apparatus consists of the following:

(12) A. H. Compton, *Phys. Rev.*, **21**, 715 (1923).

Fig. 38–9

A NaI (Tl) scintillation crystal is mounted on a suitable photomultiplier. The unit, together with a Pb shield over the crystal as shown in Fig. 38–9 is mounted on a movable stand, which is arranged so that the detector assembly can rotate about an axis perpendicular to the lecture table. The scattering material is mounted on this rotation axis. The gamma source is put in a good Pb shield with a small opening facing the scatterer.

Output pulses from the photomultiplier cathode follower are amplified in a linear amplifier and applied to the y-input of a 5-in. oscilloscope. The most economical method of making the scope trace visible to a large class is to fit the bezel with a simple projection lens. An acceptable image large enough for viewing by 350–400 students can be obtained. An alternate method, somewhat more expensive, is to use a large-screen slave scope (17–23 in.) driven by a 5-in. scope, as shown in Fig. 38–10. For the projection method a Tektronix 541 works very well. Good results also are obtained from the Dumont 436 oscilloscope-slave system. See Appendix, page 1357, for further construction details and diagrams of the apparatus.

To use the apparatus for lectures to large groups of students, the detector assembly is mounted on the rotating frame unit. As mentioned above, the axis of rotation is perpendicular to the plane of the lecture table, and the scattering material is mounted on a rod which is coincident with this axis. It is convenient to mount the rotating frame member to

Fig. 38–10

a plywood sheet so that it can be clamped to the lecture table; this precaution prevents misalignment of the source and scatterer during the demonstration. The source, a solution containing several millicuries of Cs^{137}, is placed on Pb bricks at the proper height and surrounded by a Pb shield so that the detector, when viewing the scatterer at angles other than zero degrees, will see a minimum of primary radiation.

The pulses provided by the detector and displayed on the oscilloscope are a record of the fraction of the corresponding gamma photon energy absorbed in the scintillation crystal. Thus, if only the primary radiation from Cs^{137} is incident on the crystal, one sees on a properly adjusted oscilloscope an intense envelope representing the 662 keV photopeak and a fainter, more-or-less continuous distribution of lower energy pulses, representing photons whose entire energy was not absorbed by the crystal for one reason or another (including Compton effect in the crystal itself). Proper selection of triggering level and contrast adjustment can make the continuous distribution nearly invisible.

When the detector is moved off the source-scatterer axis to non-zero scattering angles, the oscilloscope pattern clearly changes: as the angle is increased from zero, a second intense envelope detaches itself from the envelope due to the primary radiation. The amplitude, and hence the corresponding photon energy, of this second envelope is always smaller than the original one and decreases steadily as the scattering angle is increased from 0–90 deg. This means, of course, that the second line occurring in the scattered beam is of longer wavelength than the original line. A typical oscilloscope display is diagramed in Fig. 38–11; *A* is the short wavelength envelope and *B* is the long wavelength envelope. The student thus sees for himself the qualitative features of the Compton effect.

For smaller classes, up to as many as 10 or 12 students, the use of the wall projection or the slave scope would not be essential. Conversion to an advanced laboratory experiment could be readily accomplished by providing a graduated, angular scale for the rotating frame unit and by replacing the oscilloscope with a single channel pulse height analyzer and a scaler. (These also can be used in demonstrating.) The student then could quantitatively verify the effect's independence of scattering material, the cosine dependence of the wavelength shift, and the broadening of the long wavelength line in the scattered beam.

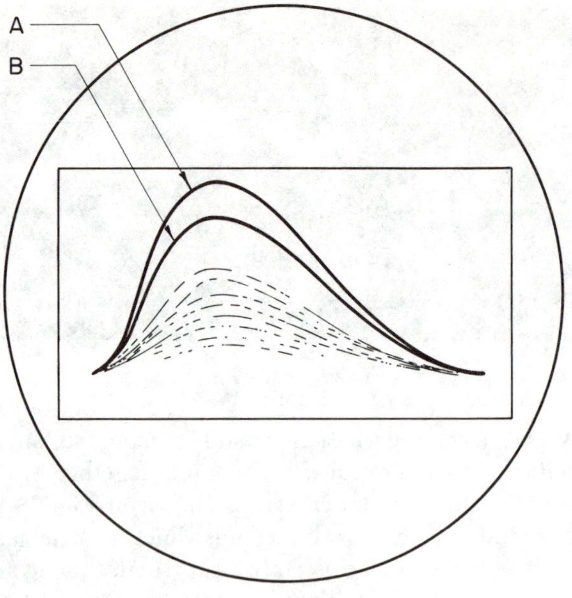

Fig. 38–11

38–3.2[13]

An apparatus which gives a qualitative demonstration of the Compton effect consists of a 1-cm-thick Al plate, placed at an angle of 45 deg to an X-ray beam produced in a tube at around 60,000 V. The plate scatters the beam into an ionization chamber. The change in wavelength upon scattering can be demonstrated by an absorption measurement. For this purpose an 0.7-mm-thick copper plate is inserted into the beam either at the tube before scattering takes place or in the scattered beam just before entering the ionization chamber.

If no change in wavelength occurred on scattering the two measurements should be equal. Actually the copper plate in the second position gives only half the reading shown by the plate when it is in the first position. It follows that the average absorption constant of the scattered light is greater than the corresponding constant for the direct beam. This means that the average wavelength has been increased by the scattering.

[13] Robert W. Pohl, *Optik und Atomphysik* (Springer Verlag, Berlin, 1967), 12th ed., pp. 287–289.

38–4 Pair Production and Annihilation

38–4.1

Since the positron was the first anti-particle discovered,[14] its properties are of considerable interest. It is stable in the absence of matter, but if a negatron (ordinary electron) is nearby, the particle-anti-particle pair will spontaneously annihilate with the appearance of two photons which share between them the momentum and the total energy (rest energy plus kinetic energy) of the annihilating pair. If the annihilation takes place at rest, then the photons have equal energy, and equal and opposite momentum.

Besides showing the conversion of mass into radiant energy, and the intrinsic interest in the annihilation phenomenon, the positron can serve as a probe in the study of matter. The lifetimes of positrons in condensed materials can yield information about electron densities[15] and about the chemistry of the positron as a positive ion and the positronium (Ps) atom and an "element."[16] The angular correlation of the annihilation radiation demonstrates the momentum distribution of the electrons which take part in the annihilation process; in metals these are mostly the electrons in the Fermi distribution.[17] The positronium atom formed by a positive and a negative electron in free space has a mean life against annihilation of about 10^{-7} sec in the $^3S_1$ state and about 10^{-10} sec in the $^1S_0$ state. It is the simplest quantum mechanical bound system in which the forces are completely known, and its properties can be predicted exactly by the Dirac theory; their detailed observation[18] constitutes a strong confirmation of the validity of the Dirac theory.

The demonstration uses a pair of scintillation counters[19] in coincidence to record the simultaneous appearance of two photons when positrons are stopped between the counters. It also demonstrates the angular correlation of the annihilation radiation, for the coincidences vanish if one moves the stopper to a position where it still intercepts the positron beam but it does not lie between the scintillation counters.

It is convenient to use radioactive substances as sources of high energy positrons and electrons. Phosphorus-32 (15 days) is a convenient source emitting only negative electrons. Sodium-22 (2.3 years) emits positrons, but each positron is accompanied by a coincident high energy gamma-ray which may be scattered into one of the scintillation counters or penetrate the lead shielding and give rise to a spurious coincidence signal. Copper-64 (12 hr) emits about equal numbers of positrons and negatrons and has no coincident gamma-rays, but of course it is less convenient because of its short half-life.

Figure 38–12 shows schematically two shielded scintillation counters *a* which face each other across the collimated electron beam. The introduction of a card *b* interrupts the electron beam and causes

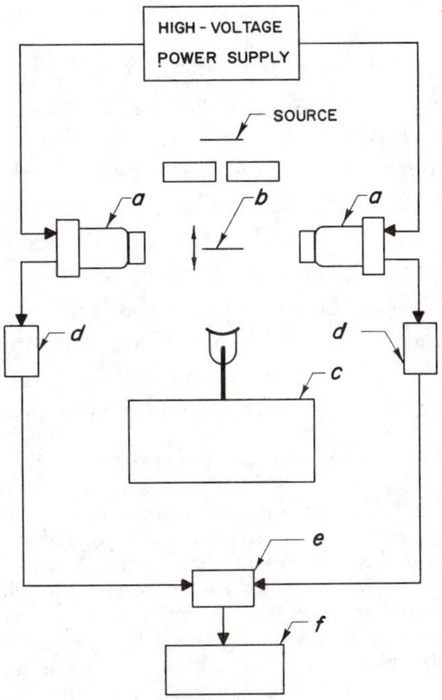

Fig. 38–12

[14] C. D. Anderson, *Science*, **76**, 238 (1932); *Phys. Rev.*, **43**, 491 (1933).

[15] See for example, R. E. Bell and M. H. Jorgensen, *Can. J. Phys.*, **38**, 652 (1960).

[16] J. D. Jackson and J. D. McGervey, *J. Chem. Phys.*, **38**, 300 (1963); and **39**, 487 (1963).

[17] J. D. McGervey, *Am. J. Phys.*, **31**, 713 (1963).

[18] Martin Deutsch, *Progress in Nuclear Physics*, **3**, 131 (1953).

[19] Harshaw Integral-line scintillation counter assemblies from Harshaw Chemical Co., Cleveland, Ohio.

the counting rate in the end-window Geiger counter *c* to diminish.[20] If positrons are stopped between the scintillation counters, the coincidence scaler starts recording coincidences. As the beam stopper is moved up and down the beam, the counting rate in the coincidence scaler varies, thus demonstrating the unique angular correlation of the annihilation radiation. Of course, no coincidences are observed if negatrons are stopped.

To de-emphasize the electron equipment used, and to make the entire apparatus more easily transportable, the amplifiers and pulse shapers *d* are transistorized, battery-operated, and mounted in the chassis containing the phototube sockets. The coincidence circuit *e* and its batteries are also contained in a small auxiliary box. The coincidence circuit has a three-position switch so that either input to it can be counted in the coincidence scaler *f*.

The demonstration also requires a conventional power supply and scaler or ratemeter for the end-window Geiger counter, a high-voltage power supply for the scintillation counters, and a scaler to count the coincidences. Construction details are given in the Appendix, page 1359.

38–5 Black Body Radiation

38–5.1

Three small tin cans with different outside coatings are filled with boiling water so that the heat radiating efficiency of each can be compared. The first can is painted black, the second is covered with asbestos, and the third is left shiny and uncoated. The cans are placed on an insulating mat, filled with boiling water, and each is covered with a wooden lid holding a thermometer. Periodic temperature readings are taken to determine which is the best radiator.

38–5.2

Very satisfactory experiments can be performed with two thermoscopes made by inserting lengths of glass tubing through corks in the necks of narrow-neck flasks. The lower ends of the tubes are immersed in beakers of colored water, as shown in Fig. 38–13. Partial evacuation is effected by means of a "T" closed by a short length of rubber tube and a pinchcock. One of the bulbs is painted white, the other flat black. When a lighted lamp bulb is brought near the thermoscopes, the behavior of the air thermometer is demonstrated as well as the relative absorption of the black and white surfaces.

38–5.3

One half of the side of an electric stove-heating element (110 V) is painted as black as possible, and the other half as white as possible. When a piece of white paper is held close to the hot element, the paper will char most rapidly in the proximity of the blackened half of the element. Other combinations of good radiators, absorbers, and reflectors can be obtained by using paper which has been half blackened on one side.

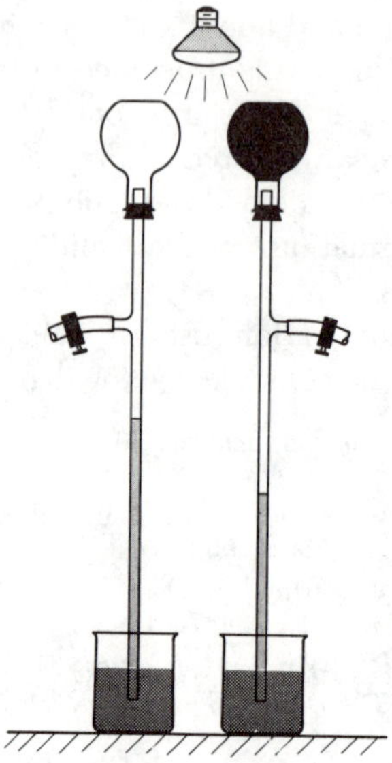

Fig. 38–13

[20] There is a large flux of gamma-rays from the annihilation of the positrons, and if sodium-22 is used, the scattering of the 1.28-MeV gammas also contributes to the background. The Geiger counter has a small but non-negligible efficiency for counting these gamma-rays and it also counts Compton electrons and photoelectrons ejected from the shielding, the table top, and other nearby matter. Therefore the change in the Geiger counter counting rate is less dramatic than one might otherwise expect. In addition, a large amount of stray gamma-radiation is scattered into the scintillation counters.

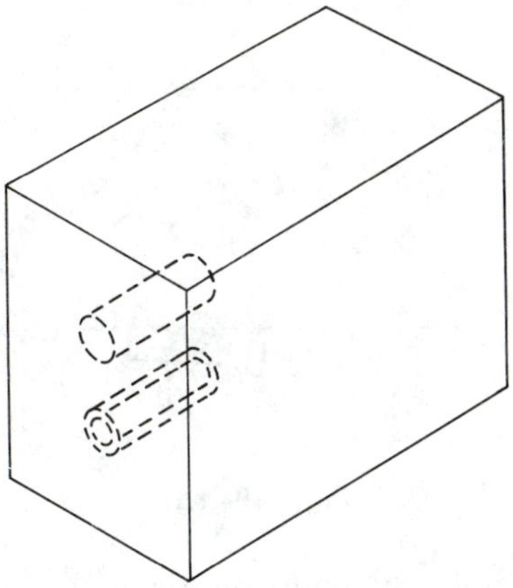

Fig. 38–14

38–5.4

Two 3/8-in.-diam holes are drilled in a 1¼ x 1¾ x 2-in. block of carbon as shown in Fig. 38–14. A white porcelain insulator is inserted into one of them. At room temperature one hole appears black, the other, white. A welder's torch is applied to the block until it is nearly white hot. Now the empty hole appears white, while the porcelain appears to be black.

38–5.5

A 3-in.-diam iron ball is heated until it is red hot. A 1/8-in.-diam hole to its center appears to glow much more brightly than the outside of the sphere because the hole closely approximates a black body.

38–5.6[21]

A helical platinum wire inside a quartz tube can be used to show that transparent objects radiate much less than other bodies at the same temperature. A platinum wire is heated inside of a quartz tube by means of a torch. The wire appears much brighter than the surrounding transparent quartz tube, which is certainly not at a lower temperature. If a glass tube is used, care must be taken that the glass does not flow too much during the experiment, and filters should be interposed to eliminate the bright light of the sodium flame. The wire should preferably have a characteristic shape—for example, a helix—so that it cannot be mistaken for the surrounding tube. An oxygen torch causes the wire to be bright enough for projection of its image on a screen.

[21] M. Iona Jr., *Am. J. Phys.*, **15**, 196 (1947).

38–5.7[22]

A carbon arc with a 10-cm-focal-length condenser has mounted in front of it a flat sided glass vessel containing water to act as an infrared absorber. The light which is essentially parallel falls on a concave mirror of short focal length. A small card is used to find the focus of the mirror which lies on the same side of the mirror as the arc lamp. A match is mounted at the focus. A filter for cutting out the visible light but passing the infrared radiation (MnO containing glass) is now mounted between the match and the arc. When the water filter is removed, the match head will ignite. It may be necessary to use a blackened match.

An interesting variation of this experiment can be performed as follows: The condenser is removed from the arc and replaced by a metal plate having a circular opening slightly smaller than the diameter of a 200-cc round flask. The flask filled with a solution of iodine and carbon disulfide (sufficient iodine is used so that the solution appears black) is then placed next to the opening so that it will act as a lens and focus an (invisible) infrared image of the arc light filament close to the flask. A small piece of Celluloid will burst into flame when placed at the image, the position of which can easily be found with the finger. (Be careful of burns!) CS_2 is inflammable and the flask should be tightly stoppered with a wood or glass stopper (not rubber). A comparison demonstration using water in place of the CS_2 solution is also of interest.

38–5.8

Simple but quantitative experiments on thermal radiation are easily carried out by using a radiation source such as Leslie's cube[23] filled with hot water or oil, and a thermal receiver such as a Moll Thermopile[24] connected to a suitably sensitive

[22] Robert W. Pohl, *Mechanik, Akustik und Wärmelehre* (Springer Verlag, Berlin, 1962), 15th ed.

[23] Leybold Cat. No. 38926.

[24] Leybold Cat. No. 55736.

galvanometer. (Be sure that the glass disk in the front of the thermopile is removed for all experiments using thermal radiation.) If a cavity radiator (black body) is desired, the Leslie's cube can be replaced by a small electric furnace.(25) In either case, screens must be inserted between the radiator and the thermopile in order to shield the latter from all radiation not emitted by the hot body. A screen is preferred to a cardboard tube. The screen has either a 10-cm-diam circular opening for the Leslie cube or a 2-cm-diam opening for the simple black body. The screen must be separated from the hot body in order to prevent the former heating up and to allow one to work behind it without affecting the thermopile. In general, a 50-cm-square screen is sufficient for distances up to 1 m. If a Leslie's cube is used, the hot oil or water must be constantly stirred during measurements. The temperature can be measured with a thermometer or with a calibrated thermocouple. A thermocouple is convenient for the simple black body.

Lambert's law is demonstrated by rotating Leslie's cube about a vertical axis while maintaining the cube at constant temperature and the thermopile at constant distance. The galvanometer scale reading A should be plotted against angle θ. The inverse square law is demonstrated by moving the thermopile away or towards Leslie's cube while maintaining the latter at constant temperature. Galvanometer scale readings should be plotted against $1/(d)^2$, or d should be plotted against $A^{-1/2}$.

To show the effect of total emissivity of the emitting body on the radiation flux, quick measurements are made of radiation (galvanometer scale values) from each of the four sides (black varnish, white varnish, mat metal, and polished metal) at a single temperature and several other temperatures as the cube cools down. Plots of scale readings A vs $(T°K)^4$ show the effect. If absolute values of total emissivity are to be found, one must divide the galvanometer scale readings obtained when the thermopile is exposed to a radiating surface by the readings obtained when the thermopile is exposed to the radiation of a black body at the same temperature. The fourth power law is demonstrated by measuring the radiation from an emitting body at several different temperatures and plotting galvanometer scale readings vs $(T°K)^4$. The range can be considerably extended if a cavity resonator is used instead of a Leslie's cube.

(25) Leybold Cat. No. 55581/82 with accessory No. 38943. All of the above items are available from: J. Klinger Scientific Apparatus and the LaPine Scientific Company.

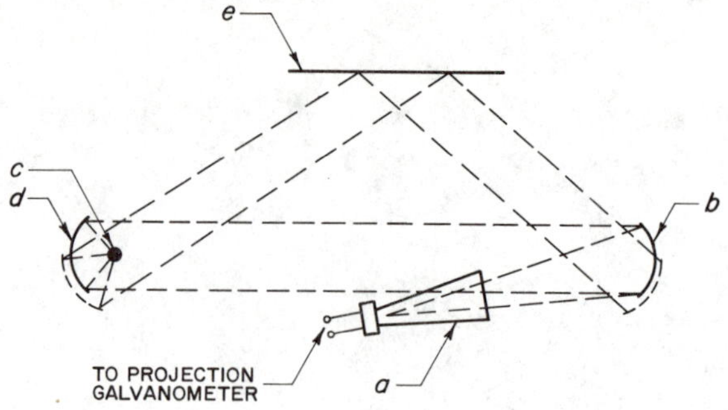

Fig. 38–15

38–5.9

In Fig. 38–15 a thermopile a is mounted at the focal point of a concave mirror b. The radiation from a heated metal ball c located at the focus of a parabolic mirror d is measured, first directly and then after being reflected from an aluminum sheet e. The thermopile and the mirror b are mounted on a common rod, which is free to rotate without changing the position of one of them with respect to the other. After the heat is measured directly (reflected only by the parabolic and the concave mirror), the apparatus is rotated so that the radiation is also reflected from a 48 x 4-in. sheet of aluminum before it is measured. The two values are compared.

38–5.10

Apparatus for the study of black-body radiation is shown in Fig. 38–16. A thermopile(26) a (3 mV/°C) is mounted at the focus of a parabolic mirror b on a tripod, and is connected to a PYE Galvanometer Amplifier(27) c (1 mV full scale). Beakers containing hot or cold water are placed on a second tripod about 20 ft away.

One beaker is covered with aluminum foil, another with soot from a turpentine flame, and a third is left untreated. Each beaker contains water

(26) CENCO Radiation Thermopile #81070, obtainable from Central Scientific Company.

(27) DC Microvoltmeter and Amplifier, Cat. No. 11341/11351 obtainable from W. G. Pye and Co., Ltd., "Granta Works," Cambridge, England.

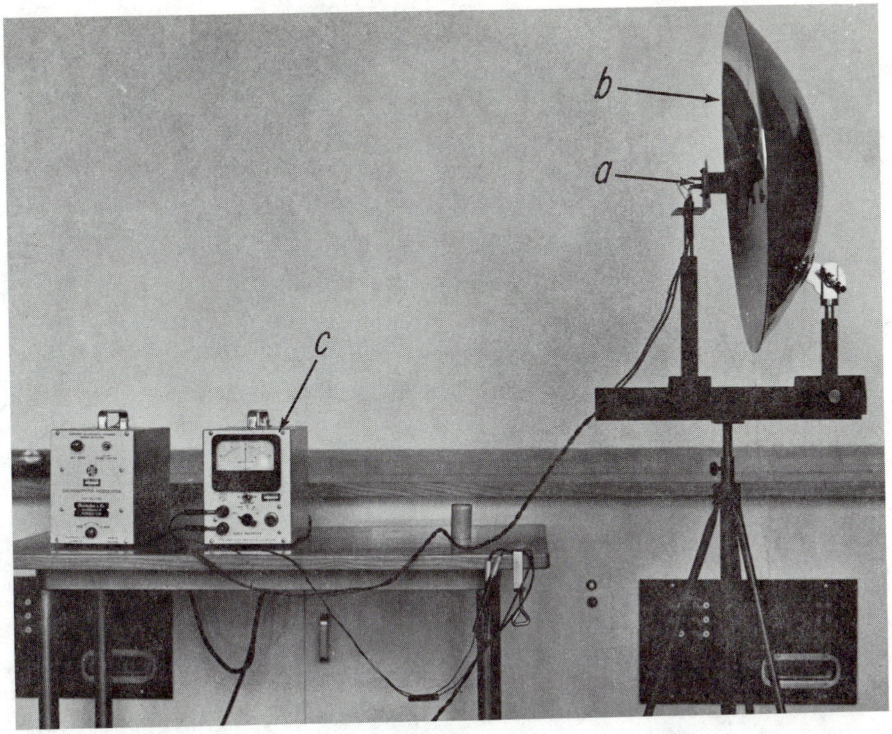

Fig. 38–16

at the same temperature. The galvanometer measures the corresponding heat radiation. This apparatus is so sensitive that a hand placed in the path of the radiation 20 ft away will cause a distinct deflection in the galvanometer. For this reason a sheet of Styrofoam is placed behind the beaker. Care must be taken that the table supporting the beakers does not heat up and start radiating itself.

38–5.11

In order to demonstrate thermal radiation intensity as a function of wavelength, it is necessary to set up both a monochromator, capable of delivering essentially monochromatic light between about .4 and 2 mμ, and an energy measuring device such as a Moll thermopile (Leybold 55736) or a photodiode (Leybold 55812), either of which should be connected to a sensitive galvanometer. The monochromator must be calibrated for wavelength. In addition, for accurate work a correction must be made for the fact that over the spectral region involved the spectral slit width $\Delta\lambda$ varies with λ. A suitable monochromator, put together from component parts, is described in the Appendix, page 1362.

The demonstration consists of measuring the galvanometer deflection $A(\lambda)$ due to a given wavelength as a function of wavelength λ. A deflection-wavelength plot gives only a very rough approximation to the true curve. A much better presentation is obtained by plotting $A(\lambda)/\Delta\lambda$ vs λ. A method for obtaining $\Delta\lambda$ is given in the Appendix. The difference between the two sets of data is shown in Fig. 38–17. A true curve drawn from values given by the Planck radiation formula can be compared.

The temperature of an incandescent light source can be estimated from the position of the measured maximum. However, it should be pointed out that an incandescent filament has an emissivity which depends on λ—the emissive power being considerably less in the infrared than in the visible. For this reason the measured maxima lie at somewhat shorter wavelengths than would the maxima of a true black body at the same temperature.

38–5.12

The Stefan-Boltzmann equation in its simplest form relates the radiant emittance W in W/m^2 of an emitting surface with its absolute temperature

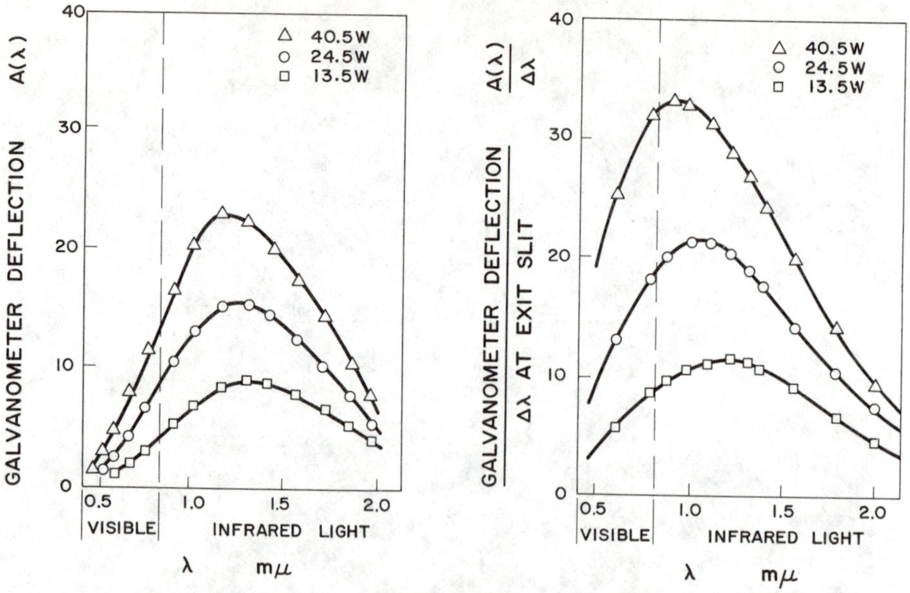

Fig. 38–17

T in °K. In particular, if T_0 is the ambient temperature of the space into which the emitting surface is radiating, then $W = e\sigma(T^4 - T_0^4)$, where e is the emissivity of the surface and σ is a universal constant. The emissivity is unity for a cavity emitter (black body), and it can be shown from Planck's radiation law that

$$\sigma = 5.67 \times 10^{-5}\ \text{W/m}^2\,{}^\circ\text{K}^4 = \frac{2\pi^5 k^4}{15c^2 h^3},$$

where k is Boltzmann's constant, c is the velocity of light, and h is Planck's constant.

It is difficult to measure σ absolutely, but it can be measured by a relative method using a standard radiator. The apparatus consists of a cavity radiator, a thermopile with a suitable galvanometer, and a standard amyl acetate (Hefner) lamp with its accessories.[28] This lamp has an 8-mm-diam wick in a German-silver sheath and is operated by burning pure amyl acetate with a flame height of exactly 40 mm. The flame height can be set with the lamp's sighting device. A diaphragm containing a 1.4 x 5.0-cm slit is set 10 cm in front of the flame, in order to shield the thermopile from the hot gases arising above the flame. Under these conditions, the irradiance, or the radiant flux, received on a 1-cm² area (by the thermopile), located 1 m from and normal to the flame, is 9.47×10^{-5} W.

In measuring Stefan's constant, it is necessary to have a clear understanding of several factors. One must differentiate between ϕ_σ, the total energy per second emitted by a plane radiator into the half space immediately in front of it, and ϕ_{th}, the total energy per second absorbed by the receiver, and find the relationship between the two. Since $E_{th} = \phi_{th}/S_{th}$ (where E_{th} is the irradiance on the thermopile in W/m² and S_{th} is the area of the thermopile receiver), it can easily be shown, by combining Lambert's law and the inverse square law, that

$$E_{th} = \frac{S_r W_r \cos\theta}{d^2},$$

where S_r is the area of the radiator source, W_r is the radiant emittance of the source, θ is the angle between the center line of the thermopile and the normal to the emitting surface, and d is the distance from the emitting surface to the thermopile, see Fig. 38–18. (In the Leybold thermopile, 35 mm is used as the effective distance between the inner recess of the front opening in the casing and the thermocouples.)

The energy ϕ_{th} falling on the thermopile per second is only a small portion of the total energy ϕ_σ emitted by the radiator. By a simple integration for a flat emitting surface, $\phi_\sigma = \pi S_r W_r$. Since

[28] Cavity radiator, Leybold Cat. No. 55581/82 with accessory No. 38943; thermopile, Leybold Cat. No. 55736; Hefner lamp, Leybold Cat. No. 46471; available from: J. Klinger Scientific Apparatus and the LaPine Scientific Company.

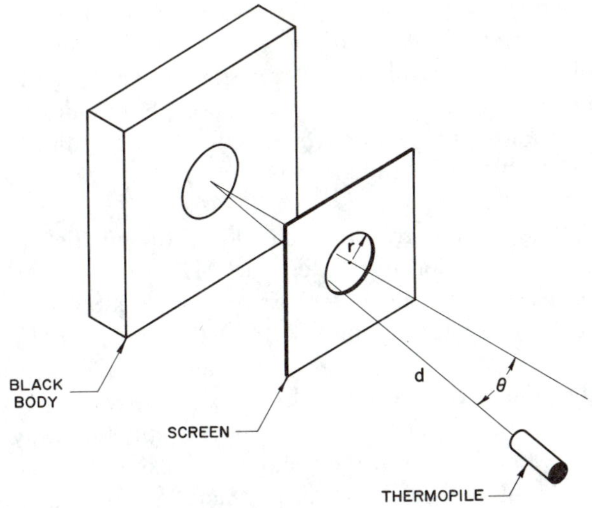

Fig. 38–18

in this demonstration the thermopile is located normal to the emitting surface, $\cos\theta = 1$ and

$$E_{th} = \frac{\phi\sigma}{\pi d^2}.$$

Further, if S_r be a circular opening of radius r, then from the Stefan-Boltzmann law

$$e\sigma = \frac{d^2 E_{th}}{r^2 (T^4 - T_0^{\,4})},$$

where $e = 1$ for a cavity radiator, and is less than 1 for any other body. In setting up the demonstration, the thermopile is first irradiated at a distance of 1 m by the Hefner lamp under the standard conditions. The irradiance of the thermopile under these conditions is E_s W/m²; this value is known for the Hefner lamp to be .947 W/m² at a distance of 1 m (the area refers to the active receiver area). The Hefner lamp is now replaced by the (black body) cavity radiator, again with $\cos\theta = 1$ and $d = 1$ m. The irradiance of the thermopile will now be E_{th}, which must be found. In both cases, the thermopile is connected to a sensitive galvanometer, which gives respectively readings D_s and D_{th}; clearly,

$$\frac{E_{th}}{E_s} = \frac{D_{th}}{D_s} \quad \text{and} \quad E_{th} = E_s\frac{D_{th}}{D_s} = .947\frac{D_{th}}{D_s}.$$

As a result, for the cavity radiator

$$\sigma = \frac{.947\, d^2 D_{th}}{(T^4 - T_0^{\,4})\, r^2 D_s}.$$

38–5.13[29]

The energy density radiated by a black body at absolute temperature T in the spectral range $d\nu$ centered on the frequency ν is

$$\rho\, d\nu = \frac{8\pi h}{c^3}\,\frac{\nu^3\, d\nu}{e^{h\nu/kt} - 1}\ \text{ergs/cm}^3.$$

The total energy u is obtained by integrating this expression over all frequencies

$$u = aT^4\ \text{ergs/cm}^3, \quad \text{where}\quad a = \frac{8\pi^5 k^4}{15c^3 h^3}.$$

Therefore, the energy radiated in one particular direction or upon a particular detector by a black body depends on the fourth power of the temperature. If a thermocouple is used to detect the radiation, the voltage produced will be over a small range proportional to the temperature rise of the thermocouple. A sensitive galvanometer will show a deflection proportional to that voltage.

But as the thermocouple is warmed by the radiation falling upon it, it loses energy more rapidly to its surroundings, which are at room temperature, and equilibrium will be reached when the amount lost by the thermocouple to its surroundings equals the amount radiated to it by the black body. If the final thermocouple temperature T_t is θ°C above room temperature T_r, the rate of energy loss is

$$W_1 = K_1\,[T_t^{\,4} - T_r^{\,4}] = K_1\,[(T_r + \theta)^4 - T_r^{\,4}]$$

or in the approximation

$$W_1 \approx 4K_1T_r^{\,3}\theta \approx K_2\theta \quad \text{ergs/sec}.$$

The rate at which energy is being received from a source at temperature T_s is

$$W_2 = K_3\,(T_s^{\,4} - T_t^{\,4}) \approx K_3\,(T_s^{\,4} - T_r^{\,4}).$$

But at equilibrium $W_2 = W_1 = K_2\theta = K_4 d$, where d is the final galvanometer deflection. Then by substitution we obtain a simple relation:

$$(T_s^{\,4} - T_r^{\,4})/d \approx \text{constant},$$

which should apply for a fairly large range of T_s. This form of the Stefan-Boltzmann law can be tested experimentally simply by observing the response of a thermopile detector to radiation from a black body whose temperature can be varied.

(29) C. Harvey Palmer, *Optics Experiments and Demonstrations* (The Johns Hopkins Press, Baltimore, Md., 1962), Experiment B 27, pp. 246–250.

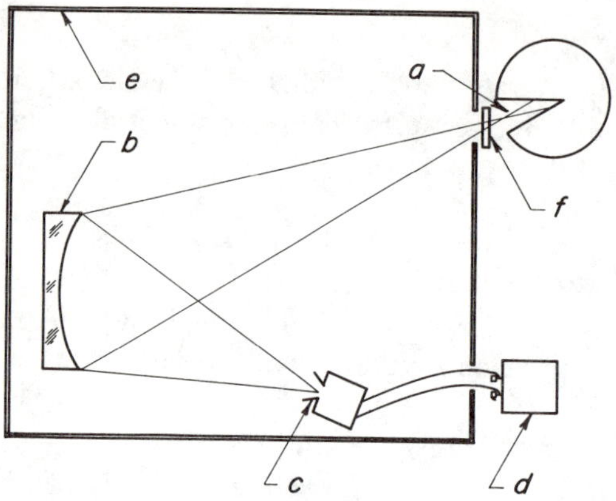

Fig. 38–19

Because no real surface is actually close to being "black," the detector is pointed at a cavity whose opaque walls are held at some known temperature. Such a cavity is a good approximation to a black body because any radiation which is not absorbed upon striking a wall inside is likely to be reflected onto another part of the wall, and so on. If a large fraction of the radiation is absorbed each time, there will be little unabsorbed radiation remaining after a few reflections. Consequently, the aperture leading into the cavity may be thought of as a "black" area, an area which absorbs all of the radiation falling upon it. Such a "black" area will also radiate perfectly.

The black body cavity *a*, concave mirror *b* (about 10 cm diam, 25 cm focal length), thermopile detector *c*, and a galvanometer *d* are set up as shown in Fig. 38–19. When a sensitive detector is used, a can-type body is chosen as the black body; for a less sensitive detector, a heater-coil black body is used. The thermopile detector consists of approximately 20 thermocouples connected in series, with all junctions of one kind blackened and exposed to the radiation, and all balancing junctions shielded from the radiation. This detector should produce at least 20 μV when exposed to a black body at 100°C. A mirror galvanometer with a sensitivity of about 0.2 μV/mm, perhaps with the help of a dc amplifier, is suitable for the can with a black body cavity. If a less sensitive galvanometer is available, the experiment may be performed over the greater temperature range, with correspondingly higher output voltages. With a heater-coil black body, a galvanometer that will indicate 10 or 20 μV/mm is adequate. Both mirror and detector are enclosed in cardboard or wooden box *e* to eliminate external heat and light. A removable reflecting shutter *f* is placed over a small opening made in the box near the cavity aperture. Both sides of the shutter are then covered with shiny aluminum foil to block the radiation without raising the temperature of the shutter.

The black body cavity should be tipped slightly so that a uniformly heated part of the source will be imaged on the detector. A small light source such as a flashlight bulb held at the center of the black body aperture is used to adjust the mirror so that the radiation will fall on the sensitive area of the detector. The light is removed and the stray heat and light from the room are blocked off. The enclosing box itself will also radiate, but that radiation is constant.

If the available detection system is sensitive enough to operate in the range from room temperature to 100°C, the can-type black body is set in position in front of the mirror and filled with boiling water. Then a thermometer is suspended in the liquid next to the cavity. With the shutter closed, set the galvanometer to zero. Remove the shutter, wait a half minute for the thermopile to reach its equilibrium temperature, and record the maximum galvanometer deflection. With the shutter closed, record the water temperature and the room temperature (near the thermopile). These measurements of water temperature are repeated at intervals of 10°C until the water has cooled to room temperature. Repeat with a mixture of Dry Ice and alcohol in place of the water, or with a mixture of ice and salt if Dry Ice is not available.

If the heating coil black body is used, connect the coil to the power line and wait until it has reached its highest temperature, as indicated on the thermopile. Read its temperature, and room temperature (near the thermopile), then open the shutter and record the maximum galvanometer deflection. With the shutter closed, reduce the voltage on the coil, wait for it to reach its new equilibrium temperature, and repeat the measurements. Readings should be made for temperature intervals of about 50°C.

For each galvanometer reading $(T_s^4 - T_r^4)/d$ should be calculated and plotted against T_s. The variations in the value $(T_s^4 - T_r^4)/d$ from a constant should be discussed, bearing in mind the various approximations made in the derivation, the reflective properties of the mirror's surface, the relative distances from the mirror to the cavity, and the type of detector used.

For demonstration purposes, the output of the galvanometer can be projected on any size screen. For a sensitive detector, cut a hole in the middle of the side of a tin can, and solder a small cone inside, with its large end toward the hole, to serve as a cavity. The cone is cut from a tin sheet rolled into shape and soft soldered along the overlap. The cavity should be an inch or more in diameter and blackened in the interior.

To obtain larger signals required for a less sensitive detector, a cone wrapped with an insulated heating coil is used; the radiation emitted from the large end of the cone is then measured. The interior of the cone is blackened with paint made of a sodium silicate solution (water glass) and lamp black. The temperature within the cone may be measured with a simple thermocouple thermometer of the type used with small ceramic kilns. A calibrated thermocouple is mounted in the cone touching a wall so that it can be read without disturbing the optical setup. The temperature inside the heating coil may be 700°C or more. If the black body is hot enough to glow, the color serves as a crude indicator of its temperature. The temperature can be reduced by using an autotransformer (of sufficient power capacity) to lower the voltage.

38–6 Wave Mechanics

38–6.1

With the inclusion of wave mechanics in undergraduate curricula and its use in describing the behavior of material particles, which the student is accustomed to thinking of as occupying definite points in space, it is imperative that the concept of the wave packet be clearly explained. Although particle wave packets result from the superposition of waves with a large number of frequencies, the principle involved can be illustrated by an analogy using two audio frequencies. In addition, discussion of the group and phase velocities can be aided with this demonstration.

The demonstration illustrates visually on an oscilloscope the resultant of adding two sine waves of different frequencies generated by a pair of audio oscillators. Both oscillator outputs are applied to the y-input of the scope and the resultant wave form is displayed. Both oscillators are adjusted to the same amplitude, and one oscillator is run at 785 cps and the other at 830 cps or 890 cps. At the points along the time axis where the two waves are in phase, the envelope of the resultant wave will have its maximum amplitude; i.e., it will have a bulge. (See Fig. 38–20.) At the points where the two waves are exactly out of phase, the resultant wave form will have zero amplitude.

The bulges in the resultant wave correspond to the wave packet. The train of bulges displayed on the oscilloscope can be used to discuss the positions of particles in an infinitely long wave-train. The demonstration also can be used to illustrate the probability amplitude for finding the particle at various points within an individual wave packet. Note, too, that the wavelengths of the component waves are much less than the width of the well-defined packet. A simple adjustment of the scope triggering causes the wave-train to move across the screen, and this can be used to discuss the meaning of the group and phase velocities.

Figure 38–21 shows the apparatus. The oscillators[30] used must be drift-free and of good quality. For best results the outputs are applied to the y-input of a Dumont model 401 A equipped with an external cathode follower adapter (type 343/395) which drives a large Dumont slave monitor (type 345).[31] As an audio aid, the outputs also can be amplified by a good audio amplifier, such as Bell model 2122-C[32] or equivalent, and applied to a suitable speaker, such as an Electro-voice model SP12B.

[30] Two Hewlett–Packard model 202-C or equivalent units.

[31] This equipment available from DuMont Laboratories, Clifton, New Jersey.

[32] Available from Bell Sound Systems, Inc., Cleveland, Ohio.

Fig. 38–20

38–6.2

The properties of a wave packet and group velocity can also be investigated with two audio oscillators and a Tektronix dual trace oscilloscope. This scope has two vertical inputs and a switch which enables: (1) a display of a single wave a or another wave b alone; (2) display of both a and b superimposed, but not added (the alternate position is more satisfactory than the "chop" position); and (3) the sum of waves a and b ("added" position). Both the single-wave forms and their sum are shown in Fig. 38–22. The scope sweep can be triggered by the signal from either a or b audio oscillators. If it is triggered by a, the a wave will stand still, i.e., $U_a = 0$. In general U_b is then not zero. If $U_a = 0$, then only a certain frequency of b will give clear sine waves. These frequencies are found by trial and error and depend on the sweep setting as well as the frequencies of a and b.

The experimental procedure is as follows:

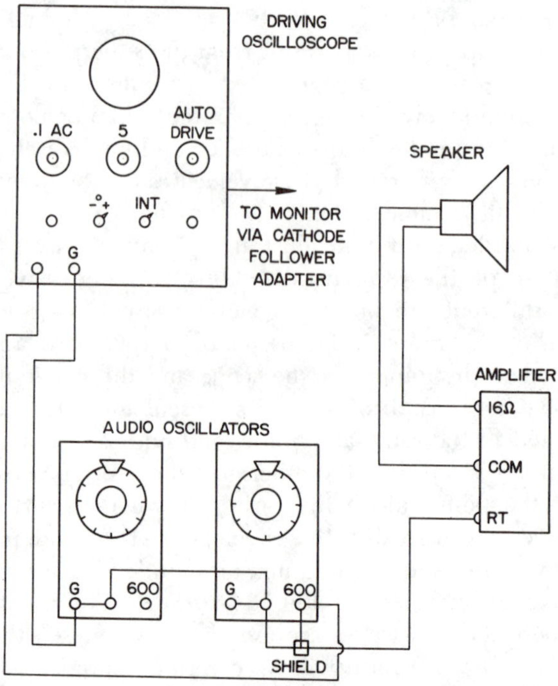

Fig. 38–21

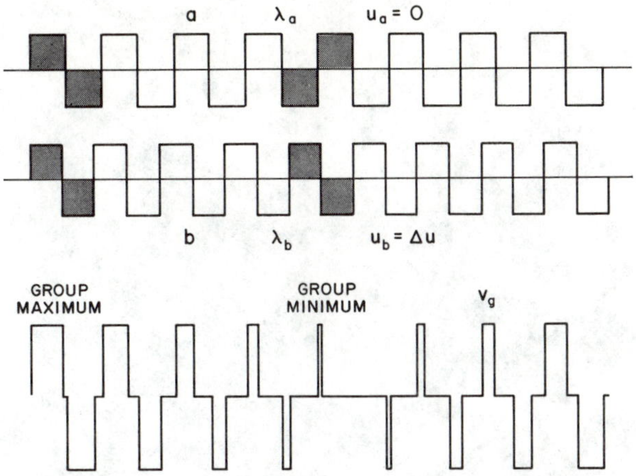

Fig. 38–22

1. The audio oscillators are connected to the a and b inputs of the oscilloscope. The oscillators should be allowed to operate at least one half hour before taking quantitative measurements.

2. The external sync is connected to oscillator a and the level is set near zero to obtain a sweep.

3. The frequency of oscillator a is set at 25 x 100 cps, and b at 32.8 x 100 cps. The sweep frequency is set at 0.2 msec/cm with the red dot on the fine control up. The a and b gains are set about 2 V/cm.

4. With the fine control on oscillator b and the scope on "b only," the frequency is adjusted until a good clear sine wave is obtained, traveling to the right with a velocity of about one screen width in 15–40 sec.

5. The scope is then set on "alternate," and the oscillator gain is adjusted to give an amplitude of about 2 divisions for each wave. The vertical position controls are used to superimpose both waves slightly below the center of the screen. It is then noted how long it takes for a b peak to move a distance λ_a.

6. Using the position "b only" the velocity of b in screen widths per second is measured; i.e., the time t_b is measured for a peak to cross the screen.

7. The scope is then set at "add," producing the bottom waveform of Fig. 38–22. The behavior of the waves is studied, and the group velocity is measured by determining the time t_g for a group maximum or minimum to cross the screen. The group wavelength λ_g is also measured.

8. The position "a only" is selected and the number N_a of waves across the screen is counted.

9. The sweep sync lead to audio oscillator b is changed, and with "b only" the number N_b of b waves across the screen is counted. An estimate is made for the nearest half wavelength.

10. Using the screen width as a unit of distance, then

$$\lambda_a = \frac{1}{N_a}, \quad \lambda_b = \frac{1}{N_b}; \qquad U_a = 0, \quad U_b = \frac{1}{t_b}$$

$$V_g = \frac{1}{t_g}, \quad \Delta\lambda = \lambda_b - \lambda_a, \quad \Delta U = U_b - U_a.$$

The measured value of V_g is then compared to the calculated value

$$V_g = -\lambda_a \frac{\Delta U}{\Delta \lambda}.$$

The time it takes b to go a distance λ_a is compared to the time it takes the group to go a distance λ_g.

A projection oscilloscope[33] can be used to perform this experiment in lecture.

38–6.3

The inclusion of wave mechanics in introductory and intermediate physics textbooks has made possible a more satisfactory explanation of many phenomena in modern physics to the beginning student. Unfortunately, at this relatively early stage of development, the student often has difficulty in grasping the significance of the results of wave mechanical calculations for even very simple systems, because he attempts to picture the situation in his mind and to compare the picture with others in the realm of his experience. The lack of classical analogs for comparison then leads to a lack of understanding of the wave mechanical problem.

This demonstration is designed to provide classical analogies illustrating:

1. The standing electron waves in Bohr orbits, and

2. The correlation between the positions of the radial nodes and the characteristic frequency and boundary conditions of the system. It provides the

(33) H. F. Meiners and G. Huse, *Am. J. Phys.*, **27**, 445 (A) (1959).

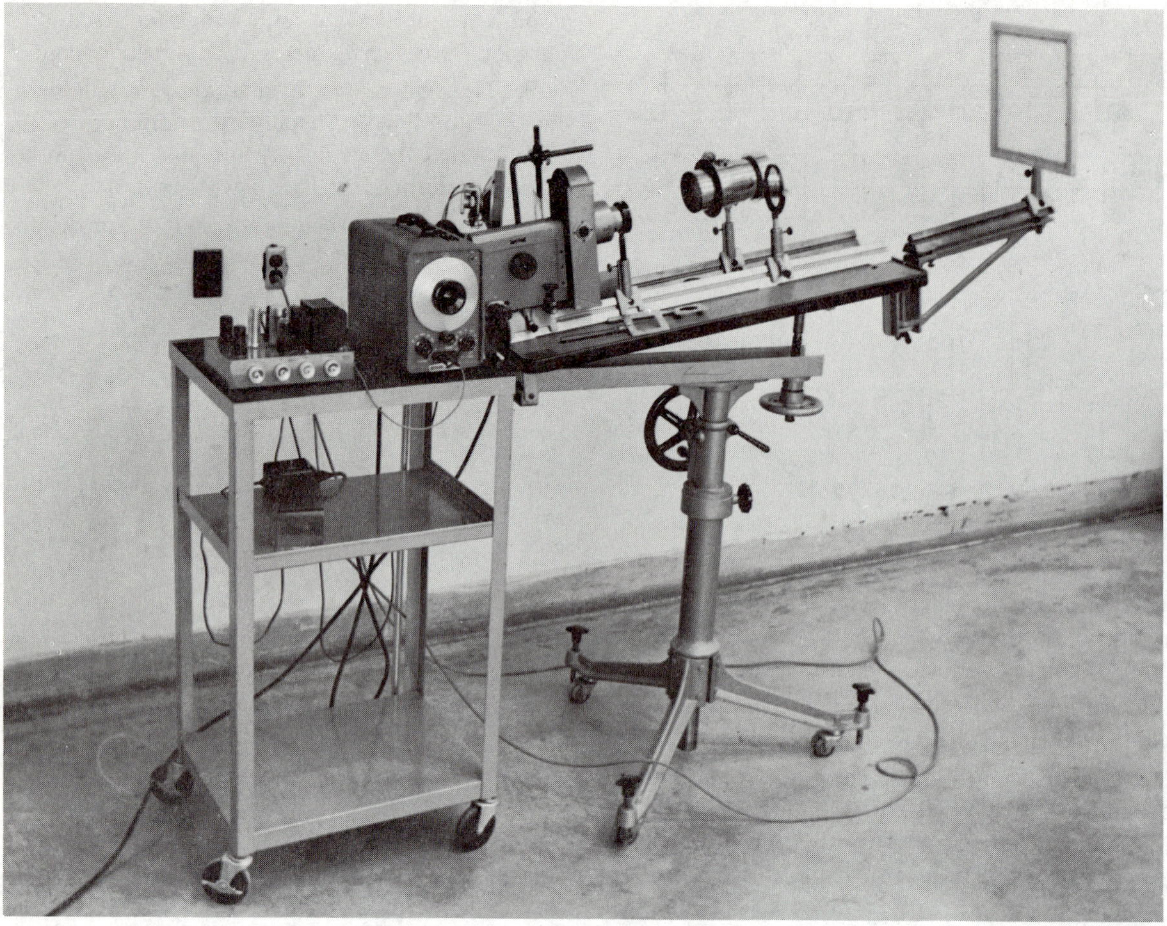

Fig. 38–23

student with a clear picture with which he can compare the results of the corresponding wave mechanical problems.

A photograph of the apparatus in operation is shown in Fig. 38–23. Basically, the demonstration consists of reflecting a beam of parallel light from the surface of a thin soap film, which is being made to vibrate by the sound waves produced by a small speaker mounted behind the film. The reflected beam is projected to a translucent screen a short distance away.

The frequency and amplitude of the sound wave driving the film are varied with the controls of a good quality sine wave generator, whose output after amplification drives the speaker. Soap films of different shapes are used to point up the effects of different boundary conditions.

The apparatus necessary for this demonstration is all readily available commercially with the exception of two simple film holders; these can be quickly fabricated. See Appendix, page 1363.

A transparency similar to the schematic setup diagram shown in Fig. 38–24 is projected on an overhead projector, and a careful description is given about precisely what the demonstration will show. It is well to review briefly how light reflected from fixed and vibrating surfaces will appear in projection.

A soap film then is prepared by dipping the film holder into the solution prepared as described in the Appendix. After placing it in position, the film should be allowed a few seconds to "drain." It is convenient to mount the apparatus on a rolling optical bench equipped with both tilt and elevation adjustments. This permits the two optical benches required to be set up and adjusted to the proper position in a preparation area away

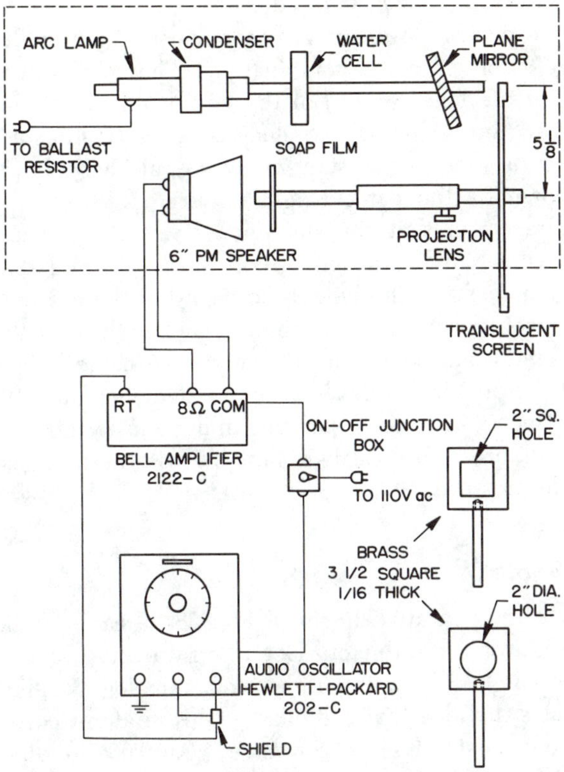

Fig. 38–24

from the lecture room without having to repeat tedious alignment.

The light beam is reflected from the surface of the square soap film to the screen. As the frequency of the driving wave is varied, the positions of the radial nodes are observed to shift in accordance with the requirements of the boundary conditions. They stand out sharply in projection as shown in Fig. 38–25, which is a typical pattern,

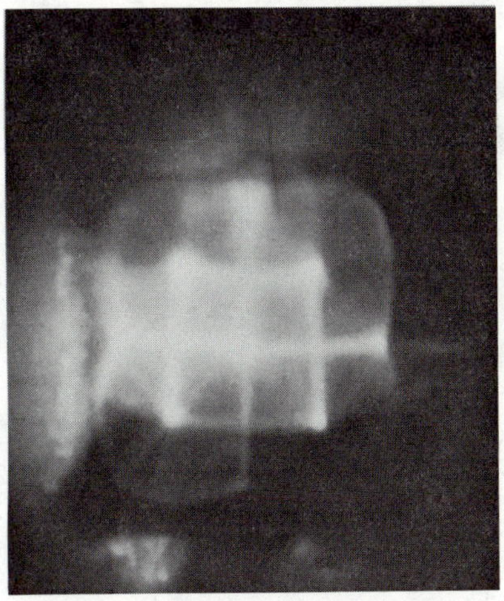

Fig. 38–25

photographed directly from the apparatus projection screen. The configuration of the nodes can be related to the eigenfunctions of the vibrating square membrane.

With the circular film in place, the radial node patterns (see Fig. 38–26 a and b), which are obtained at the characteristic frequencies, can be thought of as analogous to the standing electron waves in the Bohr orbits. It should be remembered in discussing the analogy that the electron wave is not in reality a standing wave along a line, but it extends through all space.

38–6.4

The concept of standing waves occurs in several subject areas of undergraduate curricula. In particular, it arises in discussions of the stationary

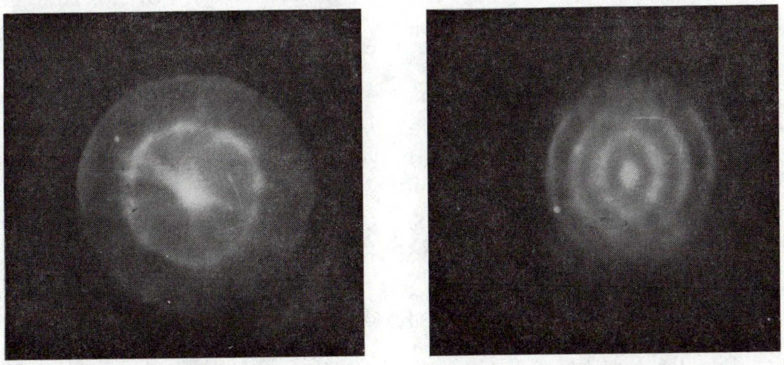

(a) (b)

Fig. 38–26

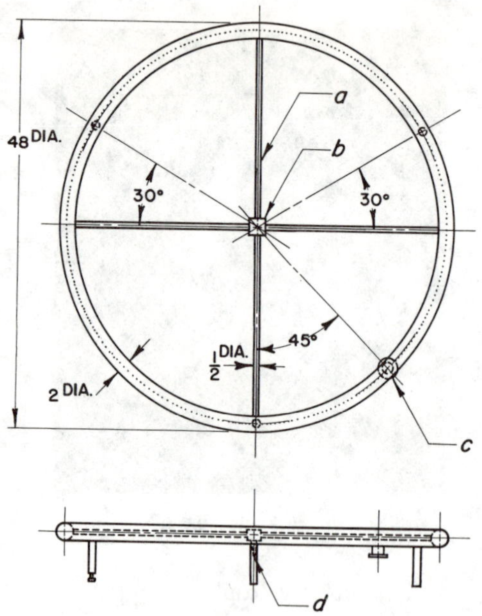

Fig. 38–27

orbits in the hydrogen atom. Often, the idea of a standing circular wave and the accompanying interdependence of the frequency and radius is difficult for the student to grasp. The apparatus described below and illustrated in Fig. 38–27 is designed to demonstrate a standing circular wave analogous to the orbital electron waves.

The principal part of the equipment is a 4-foot diam toroid,[34] which is formed from 2-in. i.d. copper tubing. Ordinary illuminating gas is supplied to the toroid at four points 90° apart by means of a simple brass distribution manifold and is forced by the line pressure through approximately 175 holes spaced uniformly around the top. The unit is supported 6 in. above the lecture table by three legs, one of which is adjustable for leveling the tube. See Appendix, page 1363, for construction notes.

The standing wave is set up in the tube by a 4-in. speaker driven at 10 to 15 watts and attached to a simple flange soldered to the underside of the toroid midway between two adjacent gas inlet points. The speaker can be driven by any convenient audio oscillator–amplifier combination. The gas issuing from the holes on top of the toroid is ignited, and when the oscillator frequency is properly adjusted, the presence of a standing wave in the tube is shown by the flame height pattern. It should be emphasized that the standing wave is a pressure wave set up in the gas inside the toroid and that the flames are simply a visual record of the pressure at each point. The maximum flame height is at the pressure antinodes and the minimum flame height at the pressure nodes. Maximum heights of an inch or more with corresponding minimum heights of ¼ in. or less are easily obtainable with normal gas-line pressure.

[34] Figure 38–28 is a photograph (Courtesy of M. P. S. de Peo) of apparatus developed independently by Dr. D. C. Sutton of the University of Illinois. This device was constructed of elliptical copper tubing with approximately the same diameter as that used at Purdue. A University PA-30 speaker was attached externally to the circular manifold. A gas inlet joint was brazed to the manifold approximately 20 deg from the speaker. The manifold contained 105 equally spaced holes, 0.055 in. in diameter.

38–6.5[35]

The apparatus shown in Fig. 38–29 can also be used in an introductory lecture on wave mechanics. Although only a two-dimensional analog, it illustrates the idea that stationary states in atoms correspond to standing matter waves. An 0.008-in.-diam wire *a* (violin E string) is bent to a 2.2-in.-diam circle and fastened by a screw to a ¼-in.-diam, ½-in.-long, Plexiglas rod *b* attached to a ⅛ x 3¼ x 4-in. Plexiglas plate *c*. The wire is excited by the magnetic field from a 650-Ω relay coil *d* with a 7/16-in.-diam steel rod core. To obtain the proper impedance for a 10-W amplifier *e*, the relay coil is connected to the primary C.T. of a 5-W 22S86 (Universal) Thordarson voice coil *f*. With fine tuning of the audio oscillator *g*, eigen-frequencies similar to the photos (a, b, c and d) in Fig. 38–30 are obtained; the higher eigen-frequencies are not integral multiples of the fundamental.

The apparatus was constructed at Rensselaer from diagrams provided by Dr. P. Huber of the Universität Basel, who obtained the details from the Universität Bern. It can be used on an overhead projector, or by direct projection.

38–6.6[36]

Frustrated total internal reflection, which is analogous to quantum mechanical tunneling, is demon-

[35] H. F. Meiners, *Am. J. Phys.*, **33**, No. 10, xiv (October 1965).

[36] W. J. Rhein, *Am. J. Phys.*, **31**, 808 (1963).

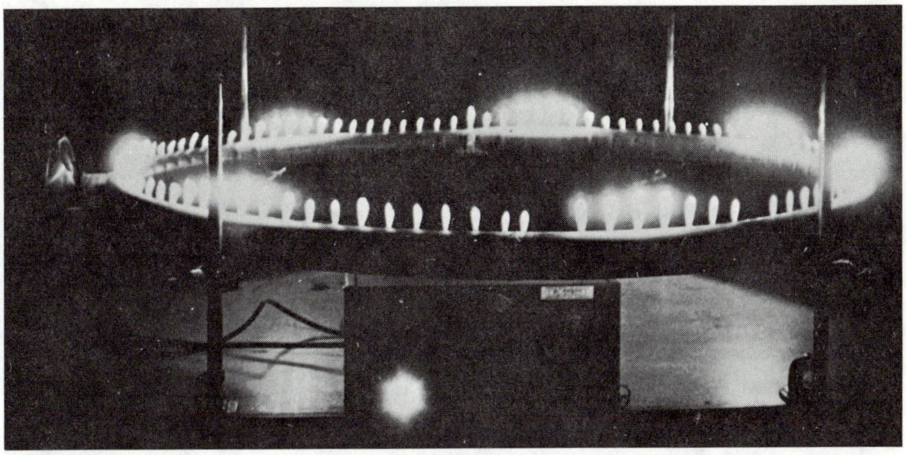

Fig. 38–28

strated with microwaves and two right-angle paraffin prisms. The prisms, whose small faces are about 20–25 cm squares, are made by molding paraffin in boxes with Masonite sides.

A beam of 3-cm microwaves is directed through a small face of a prism so that it strikes the large face at an angle of about 55 deg. Total internal reflection occurs as can be verified by placing a microwave receiver first in the reflected beam as shown in Fig. 38–31 and then in line with the incident beam. Since an attenuated wave exists in the air on the far side of the reflecting surface, the receiver should be placed several wavelengths away from that surface to avoid detecting this wave.

With the receiver in line with the incident beam, a second prism is now placed between the first prism and the receiver, as shown in Fig. 38–32, so that the large faces of each prism are parallel. As the second prism is moved closer to the first, detectable power begins to reach the receiver when the prisms are about 5 cm apart, measured along the line of the incident beam. Further reduction in the separation of the prisms produces a remarkable increase in the transmitted intensity. That the transmitted power is increased at the expense of the power in the reflected beam can be shown by placing the receiver in the reflected beam and moving the second prism.

The electromagnetic wave in the air behind the "totally reflecting" surface is propagated along the surface. The amplitude of this wave decreases exponentially with the perpendicular distance from the surface,[37] becoming negligible within a few wavelengths of the surface. In this demonstration the large surface of a second prism is placed in the region of attenuation and a wave is generated in the second prism which propagates away from the surface and reaches the receiver. The behavior of

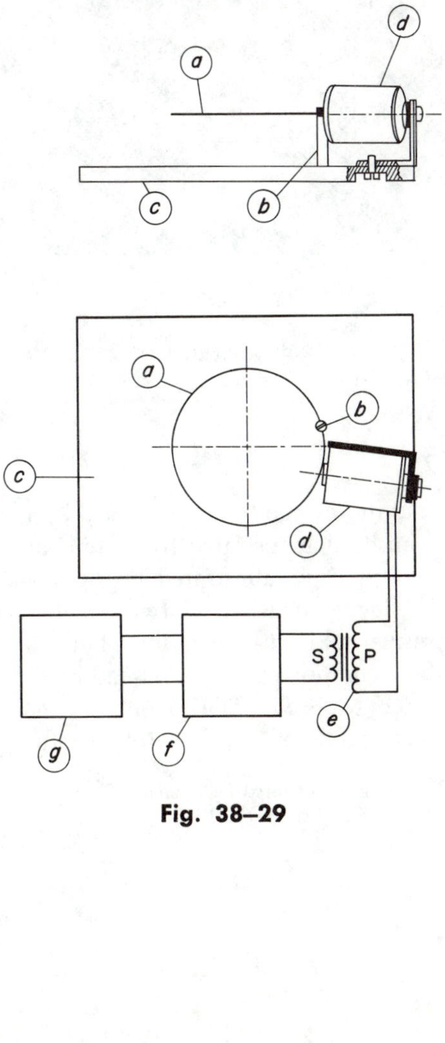

Fig. 38–29

[37] J. D. Jackson, *Classical Electrodynamics* (John Wiley & Sons, New York, 1962), p. 221.

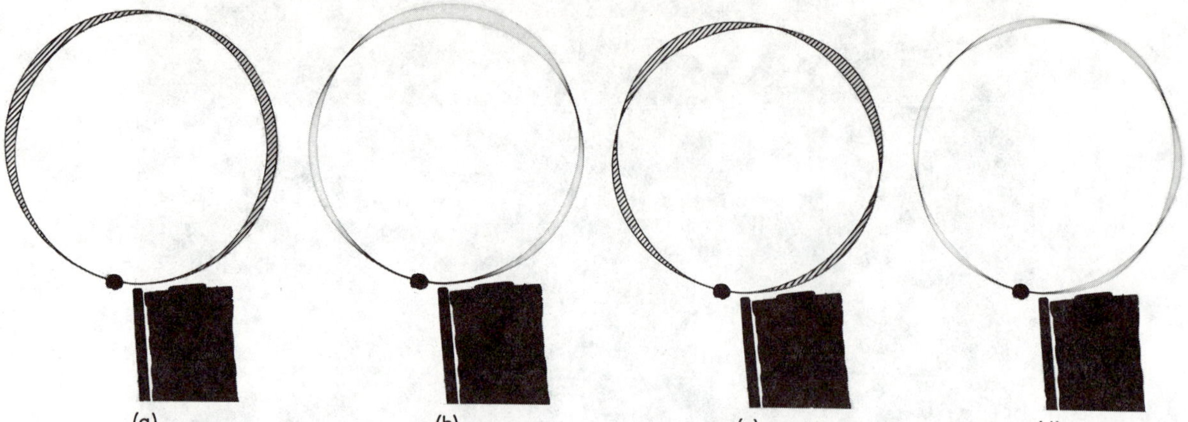

Fig. 38–30 Standing Circular Waves in a 2.2-in.-diam circle made from a violin E string. (Photos courtesy of Dr. P. Huber, Universität Basel, Switzerland. Crosshatching in photos (a) and (c) added by project artist to emphasize amplitude.)

Fig. 38–31

Fig. 38–32

the wave in the direction normal to the large surfaces of the prism is analogous to the behavior of matter waves striking a potential barrier with a total energy less than the potential energy within the barrier.[38]

Sommerfeld points out that a similar demonstration with microwaves was done at the Bose Institute in Calcutta before 1927 (perhaps as early as 1897) with 20-cm waves and asphalt prisms.[39] In an Appendix to Strong's optics book, Hull describes a quantitative experiment on microwaves of reduced intensity that emerge from a totally reflecting surface.[40]

38–6.7

The nature of quantum-mechanical barrier penetration can be demonstrated by an optical analogy, using the apparatus illustrated in Fig. 38–33. The "barrier" device consists of two identical 90-deg glass prisms held with their long faces in contact in a steel compressing jig. Assembly details are shown in Fig. 38–34. The prisms are mounted in a yoke by means of which a force normal to the

(38) R. M. Eisberg, *Fundamentals of Modern Physics* (John Wiley & Sons, New York, 1961), p. 235.

(39) A. Sommerfeld, *Optics* (Academic Press Paperback Edition, New York, 1964), p. 32.

(40) J. Strong, *Concepts of Classical Optics* (W. H. Freeman and Co., San Francisco, 1958), p. 517.

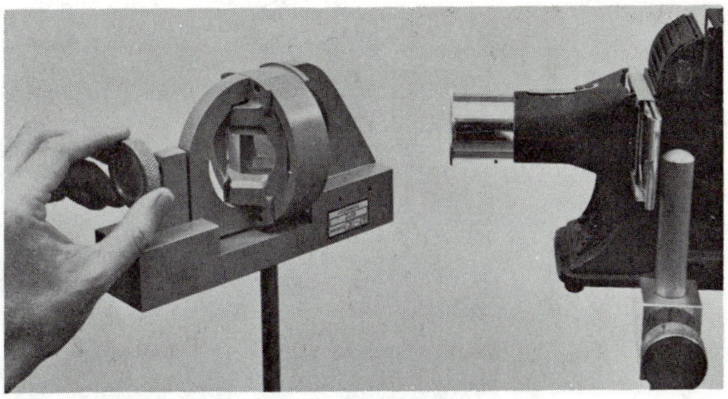

Fig. 38–33

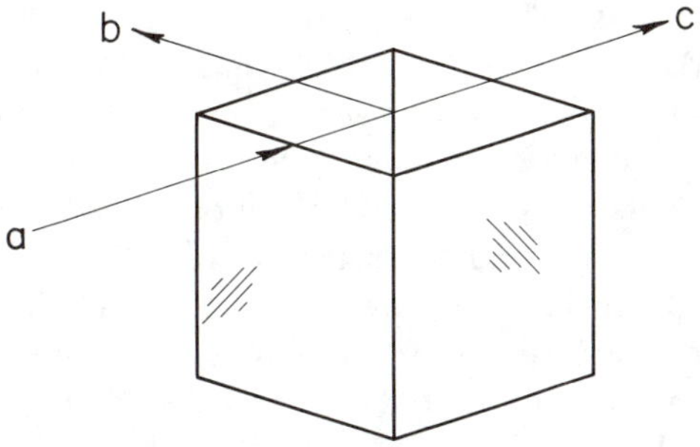

Fig. 38–35

interface can be applied with a compression screw. When a light beam a from a projector is incident normally on an external prism face as in Fig. 38–35 it meets the interface at an angle greater than the critical angle for glass and air. Hence, if the contact is light, the beam b will be totally reflected. When pressure is applied until the interstice between the prisms is sufficiently thin, partial transmission occurs, and some light c emerges from the second prism. This situation is analogous to the quantum-mechanical situation in which two regions that will support oscillatory propagation are separated by a region that does not. The transmission of a wave across a thin "barrier" is governed by a similar set of boundary conditions.

A simpler apparatus is shown in Fig. 38–36.[41] The prisms are held together with their long faces in contact by a C-clamp. Aluminum V-wedges w support the prisms so that a compression force normal to the interface can be applied by turning the adjustment screw on the clamp.

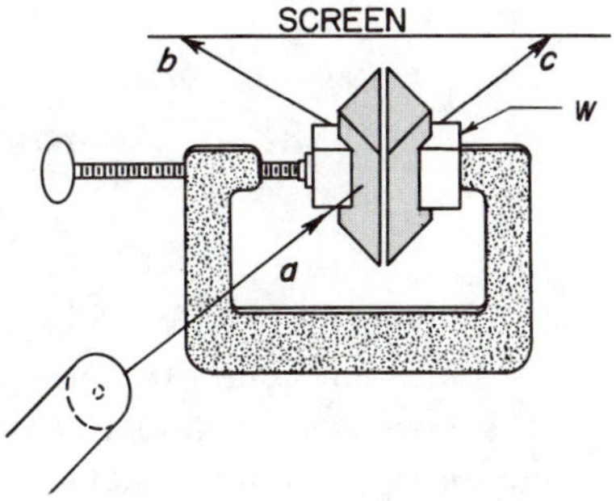

Fig. 38–36

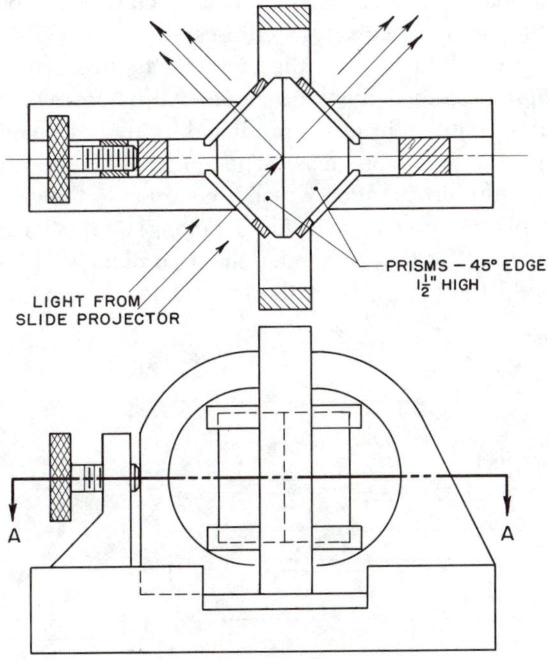

Fig. 38–34

38–6.8[42]

Two pieces of transparent material of index of refraction n_B are placed close enough together so as to form a thin film of index n_A and thickness d between them. Total reflection of a beam of light approaching the boundary at an angle ϕ to the normal will only take place if two conditions are simultaneously fulfilled:

1. $\phi \geqq \phi_T$ where $\sin \phi_T = n_A/n_B$

2. d is either of the order of magnitude of the light wave or is greater. If d is less than the wavelength, it is possible for the wave to pass through the film (although lessened in intensity) even though $\phi > \phi_T$.

(41) H. F. Meiners, *Am. J. Phys.*, 33, No. 5, p. xviii (1965).

(42) Robert W. Pohl, *Optik und Atomphysik* (Springer Verlag, Berlin, 1967), 12th ed., pp. 148–149.

It is possible to demonstrate this tunnel effect as follows: The crater of a carbon arc is focused with the aid of two rocksalt lenses onto a thermopile. The parallel rays between the lenses are divided into two separate beams by use of a small opaque stop b_1. A second vertical movable diaphragm b_2 can be moved up or down in such a way as to cut out one or the other of the two beams. The beams fall on one large and two small 90-deg prisms of rock salt arranged as in Fig. 38–37. The base planes of the two prisms are separated from that of the large prism by washers of thin metal foils. The upper prism is separated by a space of 15 μ thickness, the lower by a space 5 μ thick. In either case (i.e., whichever beam is passed by diaphragm b_2) the visible spectrum will be totally reflected and leave the prism in the direction of the arrows. The infrared spectrum (to 16 μ) will also be totally reflected as can be easily demonstrated since the thermopile M will show no deflection when the diaphragm b_2 allows the upper beam to pass. When the lower beam passes through b_2, it will also pass through the two prisms and the thermopile will show a sizable deflection. Thus a 5-μ-thick air-film will hinder the total reflection where a 15-μ-thick film allows reflection to take place totally.

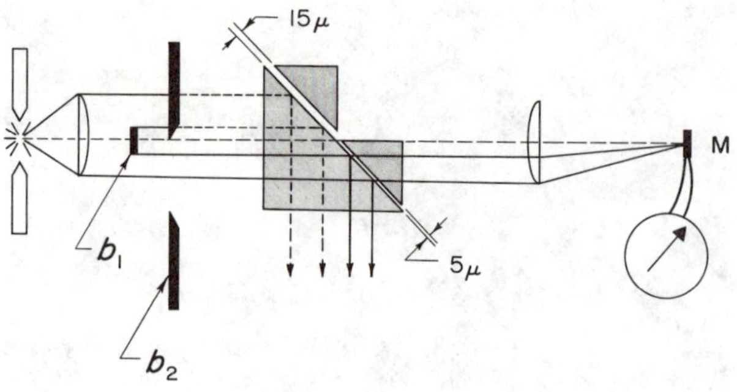

Fig. 38–37

38–7 X-ray and Electron Diffraction

38–7.1[43]

The apparatus shown in Fig. 38–38 is used to demonstrate how Laue diffraction patterns are obtained. It consists of a wood cylinder *b* which is 8–10 in. long and about ¾ in. in diameter. Twelve glass vanes *c* are glued into slits about ⅛ in. deep along the axis of the cylinder. Photoplates (5 x 7 in.) with the emulsion removed are used for vanes. One end of the cylinder acts as an axle and is supported by bearings *a* set in a support stand.

The apparatus is set up 20–28 in. away from a screen so that its axis will form a 60-deg angle with the screen. A beam of light is passed through a small round diaphragm whose image is projected onto the screen. When the beam of light falls on the apparatus, it is reflected from the plates, producing a series of spots on the screen. The spots lie on an ellipse which passes through a central spot, and as the cylinder with the plates is rotated, the spots will change places along the ellipse. Also, use an actual Laue diffraction pattern, and demonstrate that the spots corresponding to the same crystal zones are distributed in ellipses which pass through a central spot.

[43] Translated by K. V. Robinson and H. A. Robinson (Adelphi University) from *Lecture Demonstrations in Physics*, A. B. Mlodzeevskii, ed., M. V. Lomonosov State University, Moscow, U.S.S.R.

38–7.2

X-ray diffraction Laue patterns consist of spots which may be located on ellipses; these ellipses pass through the center of the picture (as given by the point produced by the incident X-ray beam). All spots on one ellipse are produced by planes belonging to a given zone axis. The zone axis is the crystallographic direction which is common to the group of planes which are said to belong to it. To get a feeling for this, imagine a line in a plane and then rotate the plane about the line. The line acts as a

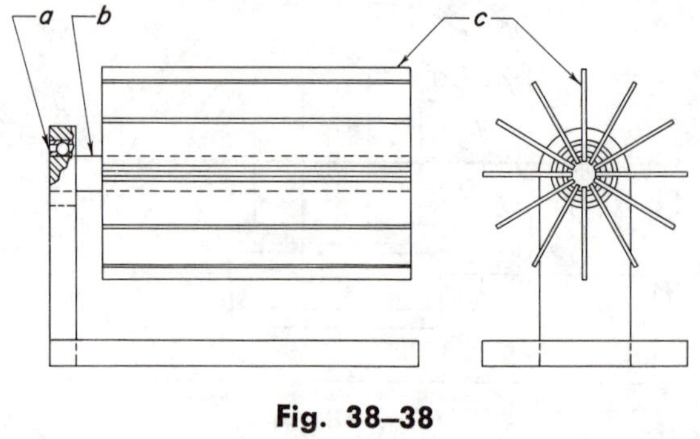

Fig. 38–38

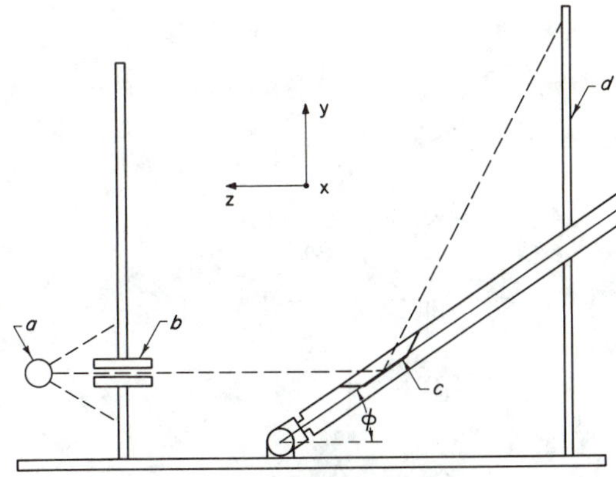

Fig. 38–39

zone axis and all planes belonging to it are produced in the rotation.

In the Laue pattern, the zone axis is not directly observable, nor is the continuous ellipse. A proof that Laue spots fall on ellipses is obtainable by vector methods. However, students have some difficulty in obtaining a feeling for the result of the mathematical derivation. For this reason, optical analog techniques have significance in instruction. Bragg[44] gives a method which utilized a single polished rod in a broad parallel light beam. The method described here is discussed in the text by Henry, Lipson, and Wooster.[45] Another general reference is Cullity.[46]

The apparatus in Fig. 38–39 consists of a light source *a*, a collimator *b*, a mirror *c* mounted with its reflecting surface on the rotation axis of a rod, a means for changing the direction of the rotation axis, and a screen *d* for receiving the reflected light beam.

The light source can be a flashlight, although greater intensity can be obtained with a focusing lens. The collimator need only be a rod with a hole drilled through it. A baffle around the collimator reduces background illumination. Two degrees of freedom for the rotation rod are suggested. The direction of the zone axis can be varied by allowing the rod to rotate about its lower end, the direction of rotation being the x-axis. (The x-axis in Fig. 38–39 is perpendicular to the page.) A lock nut should be provided to hold the rod in any position ϕ up to 90 deg. A swivel should also be provided just above the lock nut so that the rod with mirror can be rotated around the rod's axis. It is suggested that the mirror be polished metal so that the reflecting surface is at the axis. Although not apparent in Fig. 38–39, the white screen must contain a vertical slot to allow the rod to swing to a small ϕ. A second screen for "back reflection Laue" patterns may be provided at or near the collimator location.

The instructor should show the class the ellipses which result from various zone axes by fixing ϕ and rotating the rod about its own axis. He should, of course, explain that here continuous ellipses are obtained, whereas the Laue patterns show only spots on the ellipses. He should point out that the intersection of the zone axis with the "film" (screen) is given by the rod axis intersection. He should also mention that the central spot is obtained only when the reflecting surface is vertical.

The shape of the path of the moving spot of light on the screen depends on ϕ. The smaller ϕ is, the smaller are the size and eccentricity of the ellipse. When $\phi = 45$ deg, the curve becomes a parabola. When $\phi > 45$ deg, the curve becomes a hyperbola and back reflection patterns are obtained.

38–7.3

To demonstrate X-ray diffraction, a rock salt crystal *a* (about 2 x 2 x ½ in.) is mounted on a goniometer *b* (an instrument to measure angles) as shown in Fig. 38–40, with the front face of the crystal exactly on the turning axis. A collimator *c*, consisting of two finely adjustable lead plates 4 in. apart, is attached on the fixed arm of the goniometer. The other arm carries a Geiger-Müller counter *d* placed in a brass tube which has a ⅛-in.-thick wall and a 3/16 x 1½-in. slit pointing towards the crystal. The counter is connected to a scaler *e* and a loudspeaker. The two arms are coupled to the crystal carrier by a parallelogram guide *f*, so that the normal to the front face of the crystal bisects the angle between the arms.

A conventional X-ray tube *g* is mounted and screened, so that a small beam of collimated X rays just grazes the face of the crystal and is detected

(44) L. Bragg, *The Crystalline State, A General Survey* (G. Bell and Sons Ltd., London, 1949), pp. 24–25.

(45) N. F. M. Henry, H. Lipson, and W. A. Wooster, *The Interpretation of X-Ray Diffraction Photographs* (Macmillan and Co., London, 1953), p. 72.

(46) B. D. Cullity, *Elements of X-Ray Diffraction* (Addison–Wesley Publishing Co., Reading, Mass., 1956), pp. 90–91.

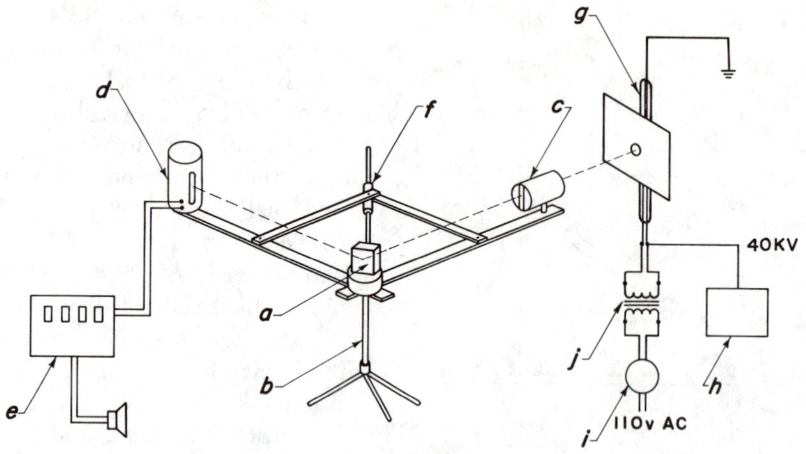

Fig. 38–40

by the counter upon reflection. When the arm with the GM-tube is turned slowly, between three and four sharp peaks can be observed at regular intervals.

The X-ray tube is supplied by 40 kV from an electrostatic high voltage generator *h*. A Variac *i*, connected to a high voltage insulated heating transformer *j*, is used to adjust the intensity of the beam. Care must be taken to keep hands out of the X-ray beam!

38–7.4[47]

The requirement for particle concepts, or for wave concepts, can be demonstrated by phenomena which are familiar or at least readily constructed. Experiments which force a union of these two kinds of concepts are much more difficult, but no less essential, to provide. In particular, the interference properties exhibited by a stream of particles (say electrons), whose ordinary kinetic behavior has been shown and accepted, ought to be displayed to students of physics. As a way of making this interference acceptable, the demonstration apparatus described below is most useful. Since it is constructed almost completely of commercially available materials (television components), the cost of the unit is quite reasonable.

The television tube used is a 16JP4, round, 16-in. black and white tube purchased from a local manufacturer of rebuilt tubes. With the cooperation of the manufacturer,[48] it was possible to obtain the electron gun, mount the diffracting film, and return the modified gun to be sealed in the tube in the ordinary manner. The gun is an electrostatic focusing electron gun of the type normally used in a 17AVP4 (17-in.), black and white tube.

The expected diffraction pattern for gold can be calculated from an expression given by Harnwell and Livingood[49]

$$r_{H,K,L} = \left(\frac{\lambda s}{a}\right)(H^2 + K^2 + L^2)^{1/2}, \qquad (1)$$

where $r_{H,K,L}$ is the radius of the ring corresponding to a particular set of integers *H,K,L*. For a face centered cubic lattice, such as gold, rings will occur for integer choices of *H,K,L* that are either all even or all odd; i.e., rings will occur for the sets (1,1,1), (0,0,2), (1,1,3), etc., λ is the deBroglie wavelength of the electron: $\lambda = 12.26/(V)^{1/2}$ Å. The distance from the gold film to the screen is *s*, and *a* is the lattice constant. For gold, $a = 4.07 \times 10^{-8}$ cm.

The rings observed on the television tube face are shown in Fig. 38–41. The two brightest rings at radii of 2 and 3 cm are the only ones easily

[47] J. M. Fowler, W. Warren, E. D. Lambe, *Am. J. Phys.*, **30**, 891 (1962). The authors wish to express their appreciation to Professor B. D. Pate of the Washington University Department of Chemistry who prepared the gold foil, to Eugene Dorrier who constructed the power supply and was of general assistance in preparing and operating the apparatus, and to the CRT Manufacturing Company of St. Louis, who cooperated willingly in the design and construction of this apparatus. We also wish to express our appreciation to Herbert Weitman, who is responsible for the photographs of the demonstration.

[48] CRT Manufacturing Company, St. Louis, Missouri.

[49] G. P. Harnwell and J. J. Livingood, *Experimental Atomic Physics* (McGraw-Hill Book Company, Inc., New York, 1933), p. 176.

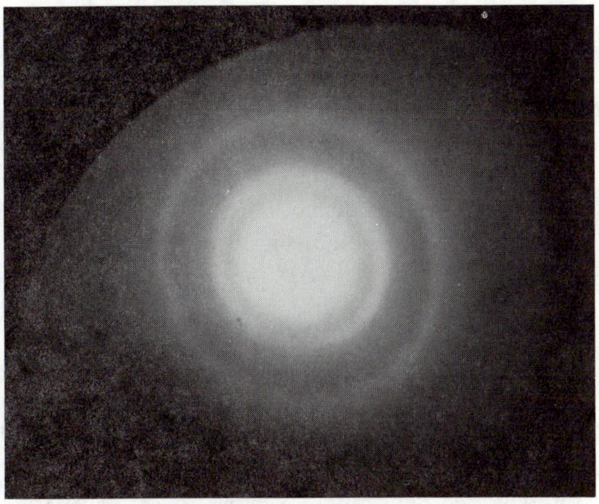

Fig. 38–41

visible to the class. The faint outer rings can only be seen by a dark-adjusted observer close to the tube.

The variation of ring diameter with voltage can be seen qualitatively by observers in a lecture room for a voltage variation in the 10–14-kV range. However, for quantitative results, it has been found necessary to make measurements of the radii of the rings at close range. A convenient way to do this is by taping a flexible, transparent, plastic ruler on the face plate of the tube in such a way that one edge lies along a diameter of the concentric rings. The radii of the various rings may then be measured and plotted against $(\text{voltage})^{-1/2}$. Such a plot is shown in Fig. 38–42. The solid lines are calculated from Eq. (1) for all allowed values of H,K,L up to (4,4,4). In cases where the rings lie very close together for two or more sets of integers, the average radius has been plotted. The indicated error is due primarily to the width of the diffraction rings and to a slight asymmetry of the pattern caused by imperfect alignment of the gun and foil.

The two inner rings, which are much more intense than the outer rings, can be identified with the sets (1,1,1), (0,0,2), and (1,1,3)–(2,2,2). This is consistent with the results of White,[50] who found that the brightest rings formed by 30-keV electrons were, in order of decreasing intensity, (1,1,1), (1,1,3)–(2,2,2), and (0,0,2). The pair of rings (1,1,1)–(0,0,2) cannot be resolved with the television tube apparatus and combine to form the bright inner ring. The appearance of the faint outer

[50] P. White, *Phil. Mag.*, **9**, 641 (1930).

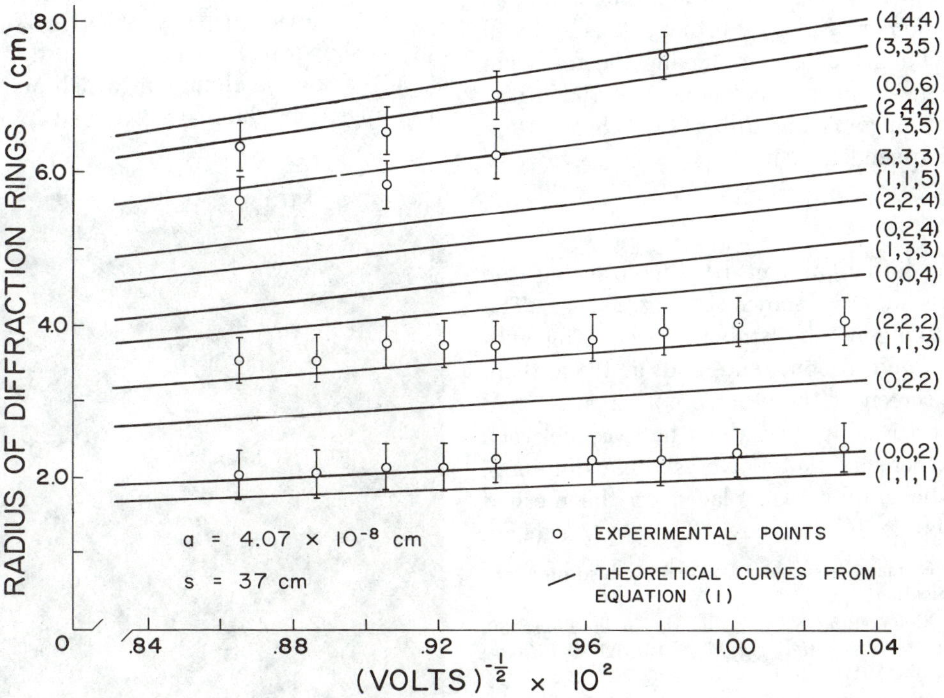

Fig. 38–42

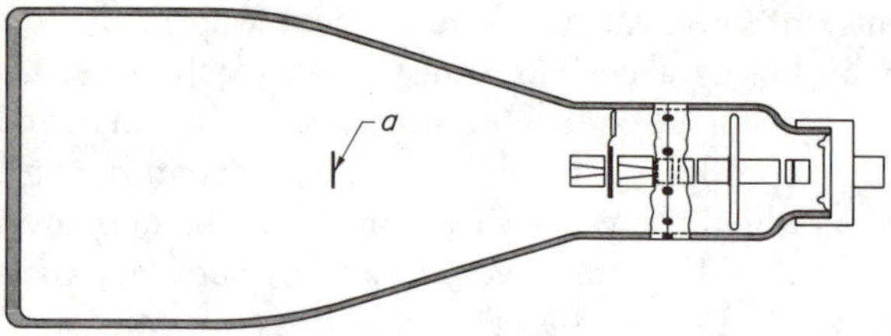

Fig. 38–43

rings and the absence of the (0,2,2) ring are somewhat puzzling since White's results indicate that (0,2,2) is the fourth most intense ring. Meiners[51] reports this ring missing also and suggests the possibility of preferred orientation in the gold foil.

A further demonstration is useful to help convince those students who doubt that the diffracted beam actually consists of electrons. A small magnet brought near the face of the tube will deflect and distort the diffraction pattern, demonstrating that the pattern is formed by charged particles rather than by a form of electromagnetic radiation.

The demonstration would be made much more convincing, at the cost of considerable mechanical complication, if it were arranged so that the gun and foil unit could be removed from the tube for display and examination. Alternatively, it is probably desirable to have a duplicate unit available for examination or display. Further refinement will aim at providing means of removing the foil from in front of the beam and changing material in the foil without opening the tube. For construction details see Appendix, p. 1364.

38–7.5[52]

Electron diffraction can be shown using the cathode ray tube[53] shown in Fig. 38–43. The electron diffraction tube is about 50 cm long with an electron gun at one end and a 16-cm-diam fluorescent screen at the other. A 2 x 2-cm etched mesh target *a* is supported about halfway between the gun and screen. Both polycrystalline aluminum and pyrolytic graphite are placed on the meshed target to double its usefulness. The electron beam is accelerated with potentials up to 10 kV and can be directed to probe any area of the target. Enough of the glass envelope of the tube is transparent to permit visual examination of all parts of the internal construction. The focusing and deflecting of the beam are electrostatically controlled. Connections to the deflection plates are made by external buttons on the neck of the tube.

The diffraction patterns produced on the fluorescent screen are clearly observable, sharply defined, and easily measured with sufficient accuracy using a millimeter scale. The intensity can be made high enough for class demonstration. The phosphor is particularly well suited for pickup on closed circuit television. The wavelength of the electrons or the lattice constant, if preferred, can be determined within 2 percent. Pattern dimensions can be varied by changing the accelerating voltage to give an adequate collection of data.

Two different target materials are provided. The electron beam can be positioned anywhere on the target, and the target is large enough so that if one point should be damaged by the beam, the beam may simply be moved to a new location, thereby insuring long useful target life. For example, about 300 small elements of polycrystalline aluminum are available.

A transmission pattern for polycrystalline aluminum is shown in Fig. 38–44. The thin film targets of polycrystalline aluminum which are several hundred angstroms thick are prepared by rapid evapo-

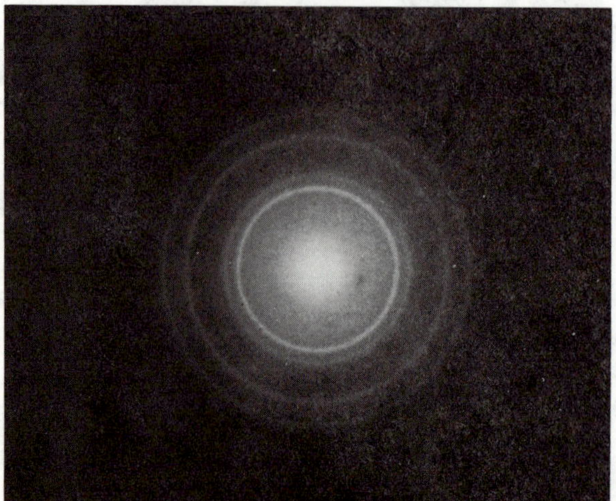

Fig. 38–44

[51] H. F. Meiners, Rensselaer Polytechnic Institute (private communication).

[52] H. F. Meiners and S. A. Williams, *Am. J. Phys.*, **30**, A549 (1962); *Acta Crystallographica,* **16**, A164 (1963); *Am. J. Phys.*, **32**, A81 (1964).

[53] Power supply and tube are distributed by the Welch Scientific Co., Skokie, Illinois.

Fig. 38–45

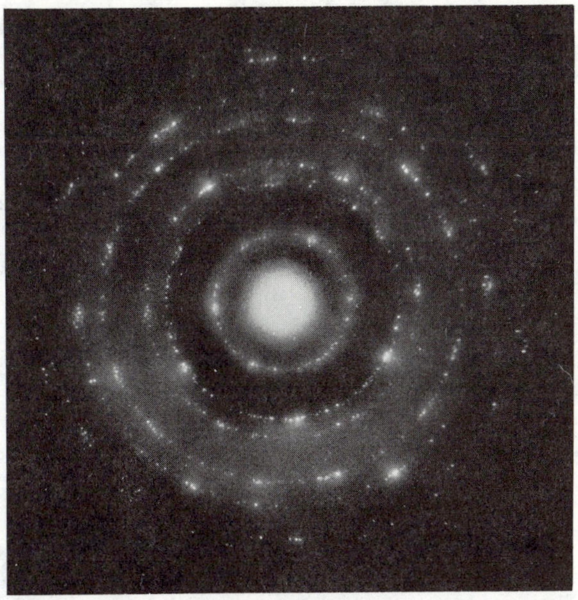

Fig. 38–46

ration onto a substrate of Celluloid. The substrate is dissolved away and the film of aluminum transferred to the mesh.

Because of the high specific heat of pyrolytic graphite which enables it to absorb heat and resist the thermal shock, *p*-graphite will withstand extensive exposure to the electron beam. As a result, a smaller number of elements of *p*-graphite are needed for long time use. A transmission pattern of two-dimensional pyrolytic graphite is shown in Fig. 38–45.

When held at high temperature during the manufacturing process, it "anneals" and the usual random planes adjust to form crystalline graphite. The planes consist of hexagonal networks of aligned carbon atoms in sheets 10 to 100 atoms thick. Since the thickness of the material varies, the spot diffraction pattern can be changed by slight adjustment of the centering controls to obtain sets of spots arranged on circles. Figure 38–46 shows one of the many patterns obtained from the polycrystalline hexagonal *p*-graphite.

The wavelength of electrons for a particular voltage can be calculated from Bragg's relationship when the lattice geometry and the radii of the rings are known. Alternatively, the lattice constant can be calculated by applying the de Broglie wavelength to Bragg's relationship.

It can be shown quickly in lecture that an increase in accelerating potential, which decreases the wavelength, will cause the radii of the ring diffraction patterns to decrease. Typical data for polycrystalline aluminum is given in Table 38–1. Table 38–2 contains data for *p*-graphite.

Table 38–1

FCC Aluminum

λ = wavelength computed from de Broglie's relationship
λ^* = wavelength computed for small angles from Bragg's relationship
a^* = 4.04 Å, the lattice constant from X-ray data

Accelerating Potential (V)	Reflection Plane	$(h^2+k^2+l^2)^{1/2}$	$\bar{r}$(cm)	λ(Å)	λ^*(Å)	Error
8000	111	1.732	1.06	0.137	0.134	2%
	200	2.000	1.24		0.136	1
	220	2.828	1.75		0.135	1
	311	3.316	2.05		0.135	1
6000	111	1.732	1.25	0.158	0.157	1
	200	2.000	1.42		0.155	2
	220	2.828	2.02		0.155	2
	311	3.316	2.42		0.158	0

38–7.6[54]

As Bragg has shown, X rays and γ rays can be diffracted by a crystal lattice, obeying the law

[54] Work performed under auspices of U.S. Atomic Energy Commission.

Table 38–2
Pyrolytic Graphite

$a^* = 2.46$ Å, the known lattice constant from X-ray data.
$a =$ the lattice constant as calculated from spot measurements by first using Bragg's relationship and then substituting for the wavelength λ calculated from de Broglie's equation. It is also necessary to consider the lattice geometry of hexagonal pyrolytic graphite.

Accelerating Potential (V)	$\bar{r}$(cm)	D(cm)	λ(Å)	d(Å)	a(Å)	Error
8000	1.18	18.47	0.137	2.14	2.47	.4%
7500	1.22	18.47	0.141	2.14	2.47	.4
7000	1.26	18.47	0.146	2.14	2.47	.4
6500	1.30	18.47	0.152	2.16	2.49	1.

NOTE: D (distance from target to screen) varies lightly for each tube and is marked on the tube.

Percentage variation between calculated a and a^* is shown. Conversely, if the known lattice constant a^* is used, the wavelength of the electrons can be determined from spot measurements and compared with λ calculated from de Broglie's equation.

$n\lambda = 2d \sin \theta$, where n, λ, d, and θ are, respectively, order, wavelength, lattice spacing, and grazing (Bragg) angle. Since the atomic planes are parallel to one another, the diffracted rays of a given wavelength are parallel and so can never converge to a focus. On the other hand, if atomic planes are normal to the crystal surface and this surface is bent to a cylindrical curvature, the planes, if prolonged, will intersect at the axis of the cylinder and all diffracted rays of a given wavelength will come to a focus, provided the incident rays are convergent. This is the principle of the bent crystal spectrometer first proposed by DuMond in 1927 and built by Y. Cauchois five years later.

The Cauchois spectrometer at Lawrence Radiation Laboratory measures the energies of γ-ray photons arising from the capture of thermal neutrons in various materials. These materials are placed in a 4-in. capsule and situated near the reactor core where there is a dense field of thermal neutrons. The gamma rays emitted in the general direction of the spectrometer pass into a Soller collimator. The function of the collimator is to absorb all rays except those forming a convergent beam. This beam next encounters a single crystal of quartz, germanium, or topaz, for instance, which is 2 mm thick and bent to a radius of 2 m. The crystal is usually 5 or 10 cm on a side, although it intercepts less than 1 cm width of beam. Due to the low efficiency of the crystal, only a small fraction ($\sim 10^{-7}$) of the beam is diffracted away to the detector, which in this case is a nuclear emulsion. The remainder of the beam is prevented from reaching the emulsion by a lead absorber in its path. The emulsion is fixed to a glass plate and bent to half the radius of curvature of the crystal for best focusing. The nuclear emulsion is 600 μ thick and mounted on a glass plate $0.030 \times 2 \times 10$ in. Although 10^{13} photons/sec of a given energy may be emitted from the capsule due to the (n, γ) process, an exposure time of a day or week is required for a sufficient number of diffracted photons to reach the emulsion and register as a perceptible line. Line widths are less than 200 μ, insuring high precision to the wavelength or energy determination.

Wavelengths or energies can be computed from Bragg's law, since the Bragg angle, θ, is determined by the displacement of a line from the plate center, or β-point, and this displacement can be measured to within a micron by setting the processed emulsion plate in a comparator. In practice, the Cauchois spectrometer is set to the approximate Bragg angle for the wavelength desired. Because incident rays are not precisely convergent, there will be at least some rays in the beam that exactly meet the Bragg criterion. Lines of the same wavelength are registered on the emulsion on opposite sides of the β-point by diffracting from both sides of the crystal planes. The distance between the two line centers, of course, gives an exact value for twice the Bragg angle.

A model (Fig. 38–47) of the bent crystal spectrometer of the transmission type described above can be easily built using light rays instead of X rays or γ rays, since the focusing property depends on the law of reflection and is identical for glass surfaces and atomic planes. In the model being described, a series of microscope slide cover glasses are arranged on an arc of radius R. The slides are equally spaced, normal to the arc, and if prolonged would intersect at the β-point (Fig. 38–48). A slide projector provides a convergent incident beam and the reflected rays focus upon a curved screen. The screen and block of cover glasses are attached to opposite ends of an arm which rotates about the vertical axis of the block. As the arm turns, the incident beam remains stationary but sweeps along

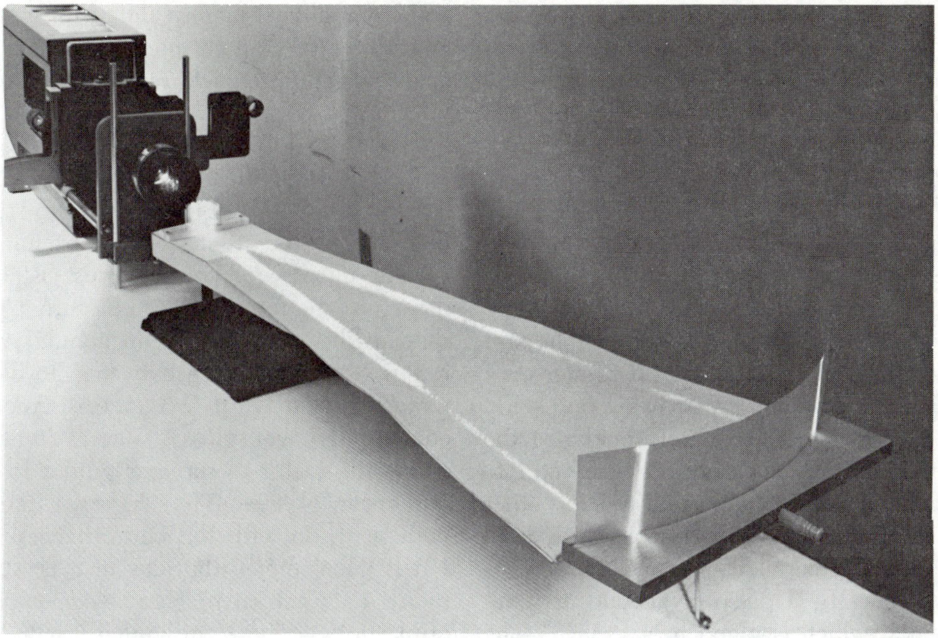

Fig. 38–47 Model of Cauchois Bent Crystal Spectrometer. (Courtesy of LRL Livermore Graphic Arts Department.)

the screen, toward or away from the β-point. Also the glancing angle θ changes so that the reflected beam sweeps along the screen on the other side of the β-point, the direct beam and its conjugate coming together or retreating. No matter how large the glancing angle becomes, the individual rays of the reflected beam always converge into a sharp line.

The model is small enough to be placed on a bench or table where it can be operated by the student or it can be used in a demonstration. By rotating the crystal arm, it is immediately evident why the incident beam and image beam are conjugate to each other. One can see the light rays as they reflect from the glass surfaces and converge at the screen, just as X rays or γ rays do from the crystal lattice. Also it becomes apparent that one can obtain line images on either side of the β-point by using either side of the crystal planes.

A caption in the vicinity of the glass surfaces is helpful if it reads somewhat as follows: "Model of bent crystal with planes perpendicular to curved surface. Instead of 9 planes/in. as in this model, there are 200 million planes/in. in the quartz crystal."

A shortcoming of this model is that it does not

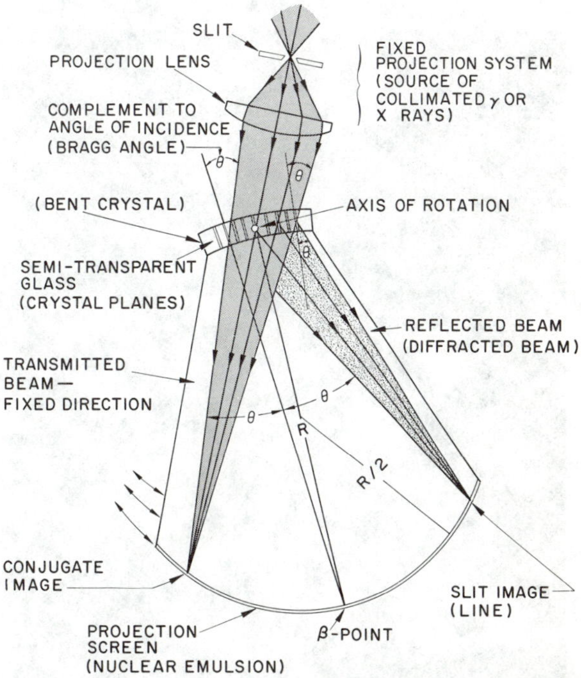

Fig. 38–48 Principle of the Spectrometer Model. As the screen is rotated through the angle θ, the slit image and conjugate image remain equidistant from the β-point. (Courtesy of LRL Livermore Graphic Arts Department.)

demonstrate selective reflection and show why the wavelength of the diffracted radiation changes with grazing angle. In other words, the light from the projector represents a continuous rather than a discrete spectrum of X or γ radiation.

The model shown in Fig. 38–47 is approximately 8 in. high, 2 ft wide, and 3 ft long. A plywood board or arm ties the block of cover glasses to the curved screen. It is covered with felt to show the paths taken by the various rays. A laboratory-type floor flange is attached to the underside of the arm directly below the block of glasses. It rotates on a ½-in. round rod, 2½ in. long, that has a ring stand for its base. A curtain rod support holds up the other end of the arm. One end of it is screwed into the board directly below the screen and the other end is bent at right angles and serves as the axle to a small ball bearing that rides on the table top.

The screen may be a piece of acetate or semi-opaque material which is glued to a groove. The groove is prepared by cutting a strip of wood on a bandsaw to a radius of $R/2$ and fastening the resulting two pieces, slightly separated, to the plywood arm. A standard lantern slide projector throws the image of a vertical slit onto the screen. The slit is made by mounting two razor blades over an aperture in a sheet of metal which then is slipped into the slide carrier.

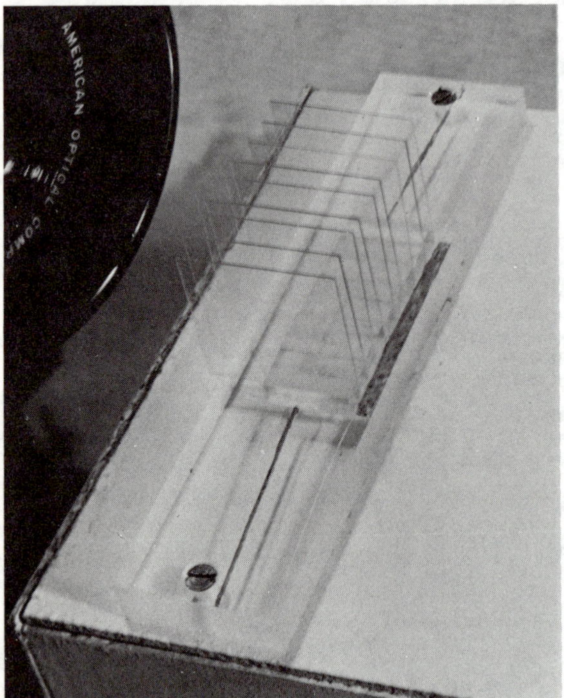

Fig. 38–49 Detail of Block Simulating Bent Crystal to show how cover glasses are mounted. (Courtesy of LRL Livermore Graphic Arts Department.)

The heart of the spectrometer model is a block of semireflectors (Fig. 38–49) to simulate the bent crystal and its lattice. The semireflectors are 1 x 1-in. microscope cover glasses. Eight or nine of them are mounted on a transparent plastic base shaped like the sector of an annulus with a mean radius R (the distance between the block and the screen) and a width ΔR (the distance along one edge of a cover glass). Radial lines are drawn across the sector to serve as guides for positioning the cover glasses. The shape of the sector has nothing to do with the performance of the model but it does convey the idea of a crystal bent to a radius R. Each cover glass is cemented to a piece of plastic ⅛ x ⅛ x ½ in. so that it will stand upright when placed on one of the radial lines. The cover glasses are set into final position with the aid of tweezers and slide projector. The projector is turned on and the slit is focused to some point on the screen. Cement is applied to the first cover glass support and the glass is adjusted until the midpoint between the direct and reflected beams falls at the center of the screen. The second glass is set next to the first in such a manner that the reflected image of the slit coincides with the first image. Successive cover glasses are treated in the same manner which results in a configuration of glass slides that simulates a bent crystal.

A small block of wood (not shown) may be placed close to the screen to show how a lead absorber intercepts the direct beam.

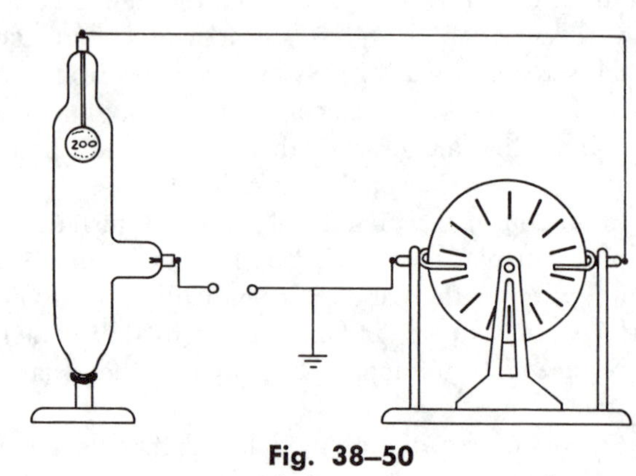

Fig. 38–50

38–7.7[55]

A very simple device for demonstrating principles of a field emission microscope is shown in Fig. 38–50. A metal coin impressed with a design is hung inside a highly evacuated glass tube and electrically connected to an outside electrode. A second electrode is also sealed into the tube. The two electrodes are connected in series with a spark-gap to a high voltage source such as a Wimhurst or Van de Graaff machine. When the voltage is high enough, electrons emitted from the coin will give rise to a fluorescent image of the coin on the glass envelope. The magnification is of course small.

(55) Robert W. Pohl, *Optik und Atomphysik* (Springer Verlag, Berlin, 1967), 12th ed., pp. 297–298.

Demonstrations	*Contributors*
38–1.1, 38–1.2	Physical Science Study Committee
38–1.3	J. F. Streib, University of Washington
38–2.1	G. B. Welch, Northeastern University
38–2.2	W. T. Ogier, Pomona College
38–2.3	R. Hulsizer and A. B. French, Massachusetts Institute of Technology
38–2.4	J. M. Fowler and E. D. Lambe, Washington University
38–3.1	R. A. Llewellyn, Purdue University
38–3.2	R. W. Pohl, Universität Göttingen, Göttingen, West Germany
38–4.1	B. L. Robinson and J. W. Weinberg, Case Western Reserve University
38–5.1	E. B. Nelson and W. E. Nickel, Iowa State University of Science and Technology
38–5.2	R. L. Weber, Pennsylvania State University
38–5.3, 38–5.4	University of Minnesota
38–5.5	E. M. Lyman, University of Illinois
38–5.6	M. Iona, Jr., University of Denver
38–5.7	R. W. Pohl, Universität Göttingen, Göttingen, West Germany
38–5.8	K. Hecht, E. Leybold's Nachfolger
38–5.9	G. Schwarz, Florida State University
38–5.10	Swiss Federal Institute (ETH), Zürich, Switzerland
38–5.11, 38–5.12	K. Hecht, E. Leybold's Nachfolger
38–5.13	W. T. Plummer, The Johns Hopkins University
38–6.1	R. A. Llewellyn, Purdue University
38–6.2	F. Oppenheimer, University of Colorado
38–6.3	R. A. Llewellyn, Purdue University
38–6.4	R. A. Llewellyn, Purdue University; D. C. Sutton, University of Illinois
38–6.5	P. Huber, Universität Basel, Switzerland
38–6.6	R. J. Rhein, Spring Hill College
38–6.7	A. Hudson, Occidental College
38–6.8	R. W. Pohl, Universität Göttingen, Göttingen, West Germany
38–7.1	H. A. Robinson, Adelphi University
38–7.2	L. Muldawer, Temple University
38–7.3	Swiss Federal Institute (ETH), Zürich, Switzerland
38–7.4	J. M. Fowler, W. Warren and E. D. Lambe, Washington University
38–7.5	H. F. Meiners, Rensselaer Polytechnic Institute
38–7.6	B. G. Saunders, University of California, Livermore
38–7.7	R. W. Pohl, Universität Göttingen, Göttingen, West Germany

39

ATOMIC PHYSICS

39–1 Spectra

39–1.1

The line spectra of five gases can be displayed using a large rectangular replica grating which forms the front wall of a lamp housing. Commercially available[1] spectrum tubes (Pluecker tubes) are modified and mounted vertically on the rear wall of the lamp housing.

To reduce the height of the apparatus, each tube is bent at opposite right angles to its longitudinal axis, leaving the capillary section (about 4 in. long) straight. The tubes are arranged symmetrically in line on the rear wall of the light box, as shown schematically in Fig. 39–1. To avoid distracting attention from the capillary sections, the bulbous portions of the Pluecker tubes and the housing interior is painted flat black (Fig. 39–2).

The tubes can be operated individually or all together from a switch panel, Fig. 39–3, on the display table. A parallel lamp circuit, with five momentary push-button switches, operates the unit.

The replica grating panel,[2] shown to the left of the switch panel in Fig. 39–3, is about 9½ in. wide

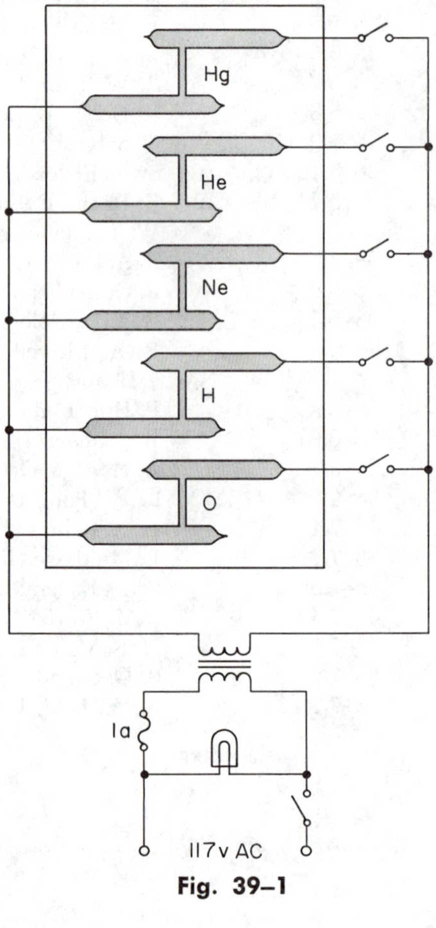

Fig. 39–1

(1) Hg, He, Ne, H_2, and O_2 tubes, and the power supply can be purchased from Cenco under the following catalog numbers: Hg–No. 87260; He–No. 87215; Ne–No. 87220; H_2–No. 87235; O_2–No. 87245; and Power Supply–No. 87208.

(2) Can be obtained in sheet form from the Edmund Scientific Co., Barrington, New Jersey.

Fig. 39–2

Fig. 39–3

and 32 in. high. For scratch protection, it is sandwiched between two thin sheets of Lucite.

The light box is made of $\frac{1}{8}$-in.-thick Masonite sheets and aluminum strips and angles. The overall dimensions of the apparatus are $10\frac{1}{8} \times 21\frac{3}{4} \times 32\frac{1}{4}$ in.

39–1.2

Many of the weaker spectral lines of a mercury light source can be made visible if the spectrum is projected on a white screen, the upper half of which has been painted with fluorescent paint.[3] The lower half of the screen is left unpainted for comparison. To obtain the spectrum, a mercury light source in a slit housing, a condenser lens, and a quartz prism are used.

39–1.3[4]

A photographic image of the ultraviolet cyanogen bands can be obtained in the lecture room by using ultraviolet-sensitive photographic paper dampened with developer. With the room lights low, light from a carbon arc is projected onto a slit by a condenser lens. A 30-cm-focal-length simple lens then forms an image of the slit on a plate glass screen covered with the photographic paper. The photographic paper is covered with an opaque screen while adjustments are being made. A small prism is placed in front of the lens to produce the spectrum. If the lens is not achromatic, the screen must be tilted.

The cyanogen bands will appear on the wetted paper shortly after the screen is removed. The blackening will spread through the action of the violet and blue rays into the visible range. The paper can be washed and fixed.

The same experimental arrangement can be used to observe the ultraviolet mercury lines if the carbon arc is replaced by a high-intensity mercury lamp.

39–1.4

Sodium vapor is created by heating large rock-salt crystals in the flames of multiple Meeker burners. Light from an arc lamp passes through the

(3) E.g., Sunset Gold, Krylon #3103.

(4) L. Grillet and P. Duffieux, *Bull. Union des Physiciens,* No. 298, 140 (1936).

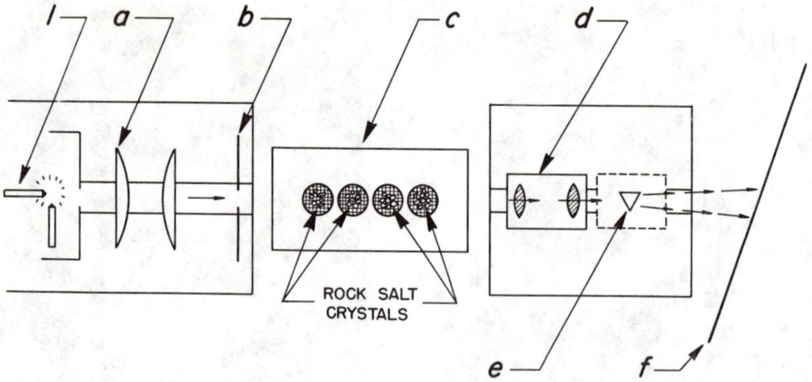

Fig. 39–4

flames, then through a projecting optical system and a housed prism to an angled screen. See Fig. 39–4. The screen *f* is angled to the optical axis for the purpose of spreading out the black absorption line of sodium. Two short optical benches are set up to support an arc lamp *l*, condensing lenses *a*, shielded adjustable slit *b*, projection lens system *d*, and a housed, direct vision prism *e*. Four Meeker burners *c* with screened mouths are mounted on a board which is arranged between the benches as shown in Figs. 39–4 and 39–5. The burners' bases are modified so that the mouth of each burner is touching the next. In operation the burner-board assembly is covered by a sheet metal housing to prevent the intense yellow flames from interfering with the spectrum on the screen.

Fig. 39–5

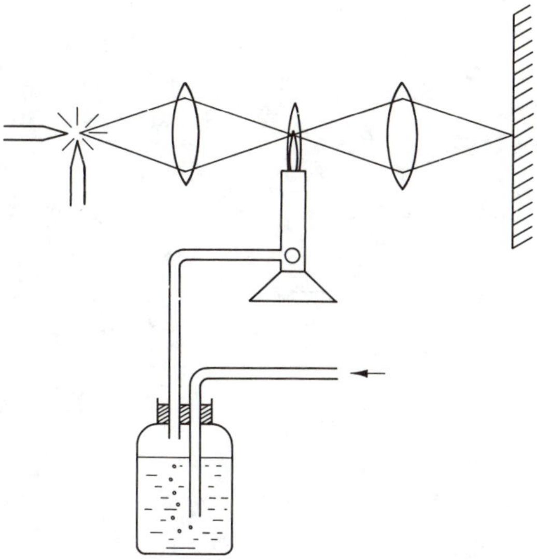

Fig. 39–6

Fig. 39–7

39–1.5

An ultraviolet lamp[5] is arranged so that it shines on a plywood screen coated with zinc sulfide.[6] When a small watch glass containing mercury is heated and placed before the screen, the mercury vapor casts a shadow due to absorption of the mercury 2537 Å line. Since mercury is a dangerous poison, use caution in handling the liquid and avoid inhaling the vapors.

39–1.6

The monochromator described in Demonstration 38–5.11 also can be used to show the presence and shape of absorption bands in the visible or infrared. The non-transmission of thin metal foils can be demonstrated, and the transparency of rock salt and glass to 2 μ can be shown. Water at about 1.45 mμ produces absorption bands that are easy to demonstrate. These bands should be shown in a glass cell of about 1-mm layer thickness which can be made from glass slides and sealing wax. It is sufficient in these demonstrations to measure and plot transmittance (galvanometer deflection) as a function of wavelength without correcting for slit width. Reasonably reliable absorption coefficients, as a function of wavelength, can be calculated from $A = A_0 e^{-kd}$, where A is the deflection with water, A_0 the deflection through the empty glass cell, d the absorbent thickness, and k the absorption coefficient.

39–1.7

An arc lamp is so arranged that its light is focused on a Bunsen burner flame at the focal point of a projection lens (Fig. 39–6). Since it absorbs the radiation of the lamp, the flame casts a shadow. The burner gas is fed through a jar filled with benzol, which results in a brighter flame and greater absorption.

39–1.8

The Hanle effect[7] provides a way of measuring the lifetimes of excited atomic states in a demonstration experiment. Mercury atoms are excited to the 6^3P_1 level by linearly polarized light of wavelength $\lambda = 2536.5$ Å, subsequently re-emitting linearly polarized radiation from this excited state provided all disturbing factors such as magnetic

[5] The lamp must produce light in which the 2537-Å line dominates. A mercury lamp with a filter to cut out visible light may be used or Mineral Light, Ultra-Violet Products Inc., San Gabriel, California, model 361.

[6] Shannon Luminous Materials Co., Inc., Los Angeles, California.

[7] See W. Hanle, *Naturwiss.*, **11**, 690 (1923), and *Zeitschrift für Physik,* **30**, 93 (1924); R. W. Wood and A. Ellet, *Proc. Roy. Soc.*, **103**, 396 (1923); G. Breit, *Phil. Mag.*, **47**, 832 (1924); v. Keussler, *Annalen der Physik,* **82**, 793 (1927), and *Zeitschrift für Physik,* **73**, 649 (1932).

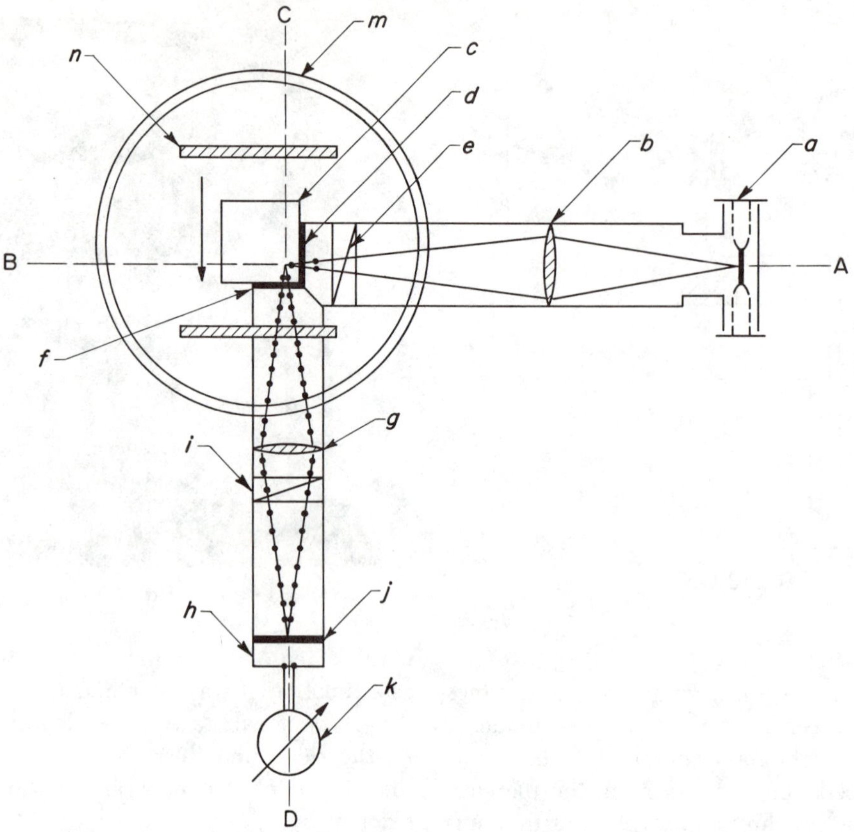

Fig. 39–8

fields, gas-kinetic collisions, and multi-fluorescence are eliminated. The resonance light is observed to be perpendicular to the direction of incidence. When the magnetic field is oriented in the direction of observation, the oscillator precesses uniformly around the axis of the magnetic field in the plane vertical to this axis with the Larmor frequency ω_0. The resonance light is therefore circularly polarized. The two linear components of this circularly polarized light can be measured separately, and the degree of polarization P is calculated from its definition

$$P = \frac{I_z - I_y}{I_z + I_y}, \qquad (1)$$

where I_y and I_z are the intensities of the respective components of the circularly polarized light.

If the degree of polarization P is sought as a function of the magnetic field H, the following relationship is readily deduced:

$$P = \frac{P_0}{1 + 4\omega_0^2\tau^2}, \qquad (2)$$

where P_0 is the proportion of the polarization without magnetic field; τ is the lifetime of the excited state, and $\omega_0 = eH/2m$ is the Larmor frequency.

The apparatus is shown in Figs. 39–7 and 39–8. A low pressure mercury vapor lamp *a*,[8] which radiates light of wavelength 2536.5 Å, serves as the light source directed along axis *A–B*. A quartz lens *b* fitted in front of the resonance box *c*, forms an image of the capillary tube of the mercury lamp on a slit screen *d*. A calcite polarizer *e* is mounted closely in front of the slit screen. The resonance box is a cube-shaped quartz vessel containing one drop of mercury. A second slit screen *f* is placed at the exit side of the resonance vessel, and a second quartz lens *g* forms an image of the exit slit on the photo cathode of an ultraviolet-sensitive photomultiplier *h*.[9] In front of the photomultiplier another calcite polarizer *i* is placed with a movable

[8] Type NK 4/4, Quarzlampengesellschaft, Heraeus, Hanau, Germany.

[9] RCA 1P28, maximum sensitivity at 3400 Å, with an additional maximum at 2500 Å.

slit j allowing for individual measurements of the intensity components. A sensitive galvanometer k,(10) connected to the photomultiplier, serves as the intensity indicator. A Helmholtz coil m provides compensation for the earth magnetic field, while a second Helmholtz coil n provides the variable applied magnetic field required for depolarization. The compensating Helmholtz coil has a radius of 32 cm and about 70 turns. It can produce fields up to 4 G and can be rotated around axis A–B. The applied magnetic field is produced by a Helmholtz coil of 12-cm radius and about 350 turns of 0.7-mm-diam copper wire No. 22 SWG, which can produce fields up to 30 G without overheating the coil.

The apparatus is first aligned with the exit optical axis C–D in the direction of the earth's field. The compensating Helmholtz coil is then turned on its axis and the current adjusted until the earth's magnetic field is neutralized at the resonance vessel. This is tested by bringing a declination and inclination needle to the position of the resonance vessel.

With the mercury lamp and photomultiplier turned on, an initial measurement of P_0 without magnetic field is made. The intensities of the components of polarization are measured individually by cutting out one of the two images with the movable slit j and then repeating this procedure for the other. P_0 can then be computed from Eq. (1). With the magnetic field turned on, the procedure is repeated in order to determine P. Then, since P, P_0, and H are known, the lifetime of the 6^3P_1 state may be determined from Eq. (2). For a temperature of −21°C, a typical value for τ is 1.19×10^{-7} sec.

Since the temperature of the resonance vapor affects the polarization through molecular collisions, a set of calibration curves (Fig. 39–9) is necessary in order to relate the observed polarization and the applied field at various temperatures.

The length of the capillary tube of the Hg-lamp is 10 mm and its diameter is 5 mm. Spurious emission amounts to less than 3 percent of the total lamp output.

The resonance source is a cube-shaped quartz vessel, 3 cm on a side with an attachable mount about 10 cm long. The cube surfaces are polished. By cooling this attachable mount with liquid oxy-

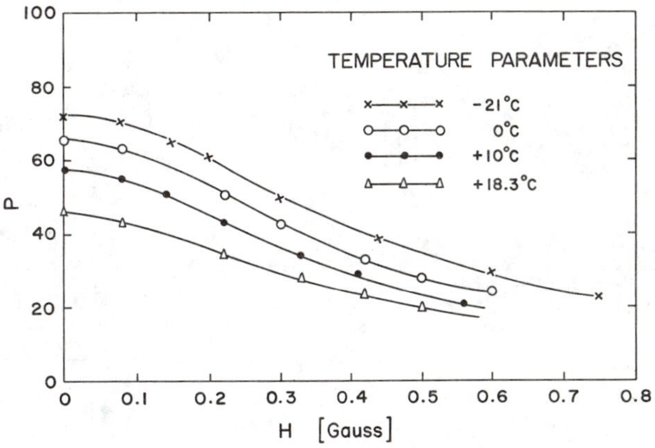

Fig. 39–9

gen, a Dry Ice acetone mixture, or with an ice–NaCl mixture, various Hg vapor pressures can be produced. The resonance vessel is cast under a vacuum of $3(10^{-4})$ mm Hg so that interference effects due to foreign gases are almost completely eliminated. The side walls of the resonance source are painted black. Two slits are left free (exit and entry). In order to eliminate multifluorescence, the distance between two slits should only be a few millimeters.

39–1.9(11)

A carbon arc a (Fig. 39–10) with a 7.5-cm condenser b is used to illuminate a vertical slit c. The

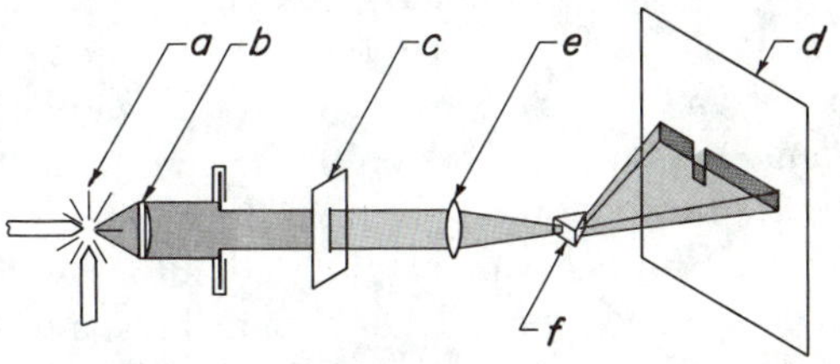

Fig. 39–10

slit is focused on a screen d with a 30-cm-focal-length lens e. A constant deviation prism or a 60-deg prism f, adjusted for minimum deviation, is inserted in the beam. The negative (vertical) carbon electrode has a 2-cm-deep, 2.5-mm-diam hole drilled in it. The hole is packed with anhydrous sodium carbonate. The arc is focused (with the condenser) so that only the end of the (upper) positive carbon

(10) Type A 75, Kipp and Zonen, Delft, The Netherlands. Keithley Instruments, Inc., Cleveland, Ohio, make a galvanometer suitable for use in large lectures.

(11) Hermann Daniel, *Demonstrations Versuche Aus der Optik* (Musterschmidt-Verlag, Göttingen, 1960), 2nd ed., p. 126.

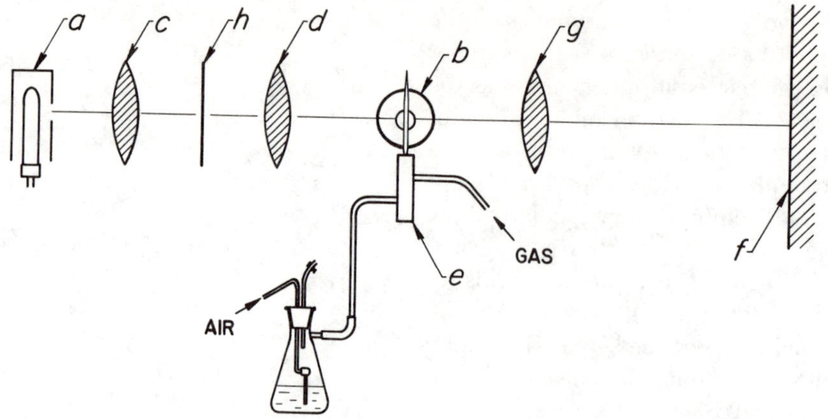

Fig. 39–11

is imaged at the top of the slit. A continuous spectrum will appear on the screen above which the D lines appear clearly. Their continuation into the lower (continuous) spectrum is seen to be dark, i.e., the absorption lines appear directly below the bright lines.

39–2 Zeeman Effect

39–2.1

Light from a sodium lamp[12] *a* (see Fig. 39–11) is focused on the air gap of an electromagnet *b* (65-mm-diam pole pieces, 5-mm gap, 15-kG field) by means of lenses *c* (f = 3 in.) and *d* (f = 8 in.). A torch[13] *e*, which is fed with city gas and NaCl saturated vapor, is adjusted so that it produces a thin, bright flame in the air gap. The image of the gap is projected on a 4–6-ft square screen *f* by means of an objective lens *g*. With the magnet off, the flame will appear dark on the screen, indicating that it is absorbing the light from the sodium lamp. A Polaroid filter *h* can be inserted into the light beam to enhance this effect. When the magnet is turned on, the resonance lines in the sodium vapor will shift, and the flame will appear bright. The magnet current switch should be screened by a black cloth so that the sparks will not disturb the experiment.

The apparatus which provides the NaCl saturated vapor is shown in Fig. 39–12. Based on the atomizer principle, it will provide a steady stream of air saturated with NaCl vapor. Air is forced into the apparatus by means of a compressed air source which is connected to an inlet tube *i* (4–5 mm in diam). The air is directed over one end of a small tube *j* (1.5–2.5 mm in diam); the other end of the tube *j* lies below the surface of the NaCl solution. A short length of hose *k* and a hose clamp

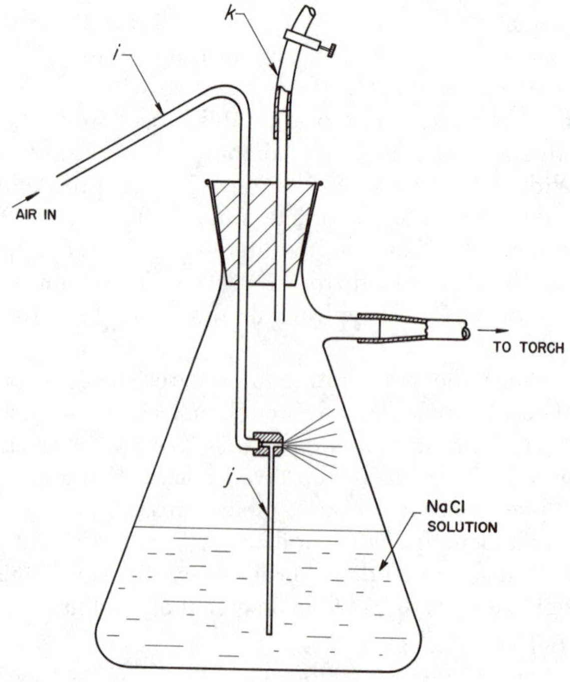

Fig. 39–12

[12] Philips type 61025 SO; 45-W or equivalent.
[13] Cenco #11298.

Fig. 39–13

are attached to the outlet tube in the stopper and are used to regulate the amount of vapor reaching the torch. The compressed air source is adjusted to produce a steady stream of atomized NaCl. The vaporizer should be cleaned after use so that it does not get clogged with crystals. Figure 39–13 shows the apparatus.

39–2.2

The light from a standard sodium light source[14] *a* is projected through the flame from a borax bead *d* which is positioned between the poles of an electromagnet *e* to demonstrate the Zeeman effect (Figs. 39–14 and 39–15). The bead is heated by a gas–air blowpipe *c*. A lantern slide projector *h*, holding a projection ammeter *i*, allows students to associate the effect of an increasing magnetic field with the shadows of the bead flame and the poles of the magnet on the same screen *g*. When no current flows through the magnet coils, the bead flame and poles appear black against a background of bright orange on the screen. With maximum field (approximately 6000 G) the center of the bead flame brightens while the remaining portion of the flame and the poles appear black against the same bright orange background. The blowpipe is held by a clamp and stand so that its mouth is about 1½ in. below the centerline between the poles (separated ½ in.). Just above it, but below the poles, the borax bead is fixed on a platinum wire support *b*. The shadows of the bead and magnet poles are focused on the screen *g* with a 9-in. plano-convex lens *f*. The electromagnet uses 117 V and 25–30 A dc to produce a *B* field of about 6000 G.

(14) George W. Gates Co., Franklin Square, New York.

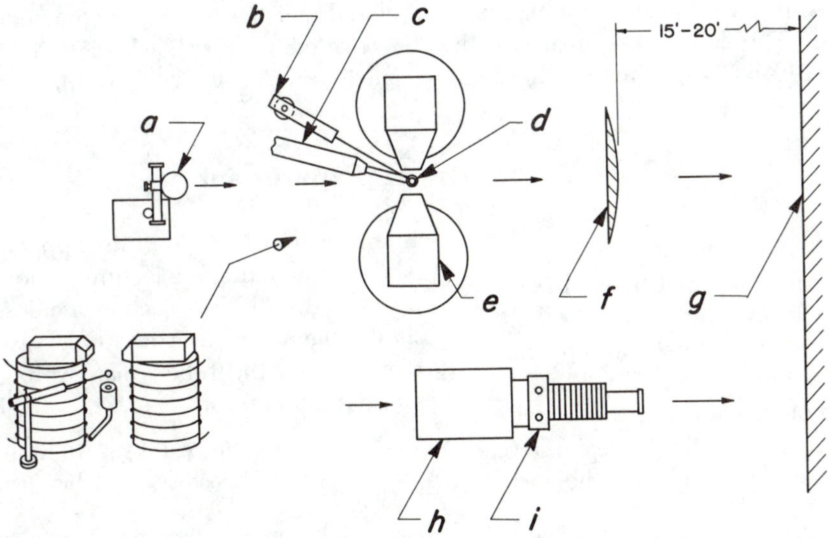

Fig. 39–14

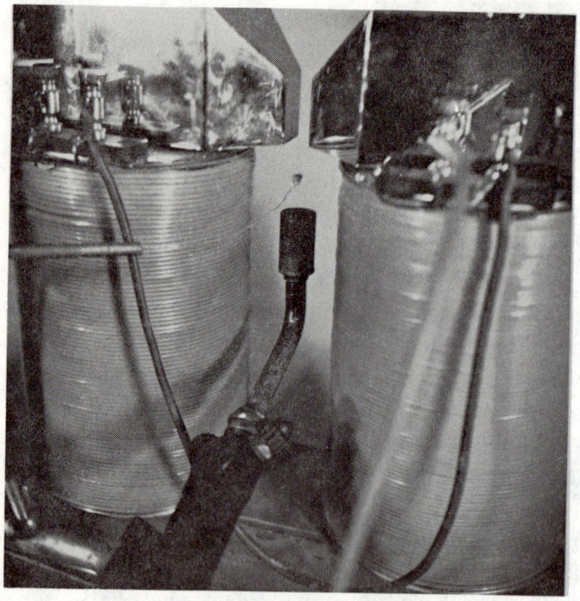

Fig. 39–15

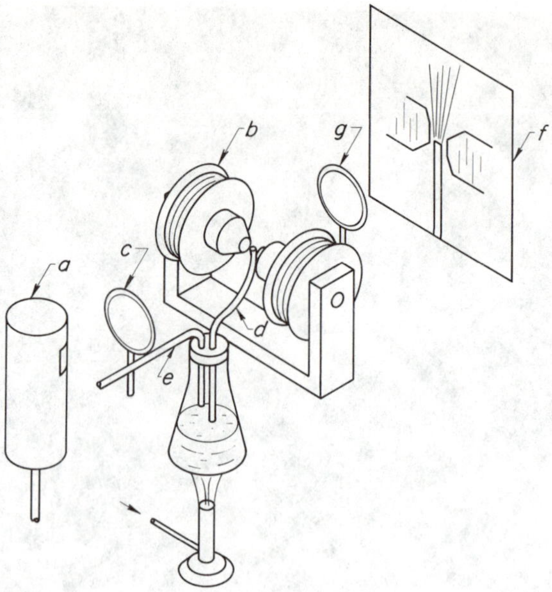

Fig. 39–16

39–2.3

The ultraviolet light from a low pressure mercury lamp[15] *a* (Fig. 39–16) is focused on the air gap of a large electromagnet *b* (¼-in. gap, 25 kG) by means of a quartz lens *c* (f = 6 in.). An air stream containing mercury vapor is blown up through the air gap through a stainless steel tube *d* connected to a flask containing liquid mercury. The input tube *e* for the flask is connected to a compressed air source which is adjusted to get the optimum amount of mercury vapor into the air stream moving through the air gap. The flask can be heated with a Bunsen burner to 40–50°C to improve the mercury vapor concentration. The air gap is focused on a fluorescent screen *f* about 5 ft away by means of a quartz lens *g* (f = 6 in.).

With the magnet off, the air stream emerging from the tube shows up on the screen as a black cloud, indicating that the mercury vapor in the air stream is absorbing the light from the lamp. When the magnet is turned on, the cloud disappears, indicating a shift in the resonance line within the vapor. Since mercury vapor is very toxic, this demonstration should be done under a hood or in a closed quartz system. The magnet switch should be covered with a black cloth to prevent sparks from interfering with the results.

39–3 Franck–Hertz Experiment

39–3.1

This demonstration shows that atoms have definite energy levels. The mercury atoms in the tube only change their energy by definite amounts, as shown by the oscilloscope signal in Fig. 39–17. The potential difference of the peaks in the signal (excitation potential) represents the energy difference between two such energy levels for the mercury atom.

The Franck–Hertz experiment is usually performed in the laboratory with commercially available apparatus.[16] If this apparatus is slightly modified, a qualitative lecture demonstration can be performed with a projection oscilloscope[17] which shows the characteristic Franck–Hertz curve obtained when the tube's plate voltage is swept by alternating current. See Figs. 39–18 and 39–19.

(15) Philips Type TUV, 6W, 57416 E140 or equivalent.

(16) J. Klinger Scientific Apparatus, Jamaica, New York, Catalog PH 32, p. 139.

(17) Tektronix Inc., Beaverton, Oregon, model 531A. See also H. F. Meiners and G. Huse, *Am. J. Phys.*, **27**, 445 (1959).

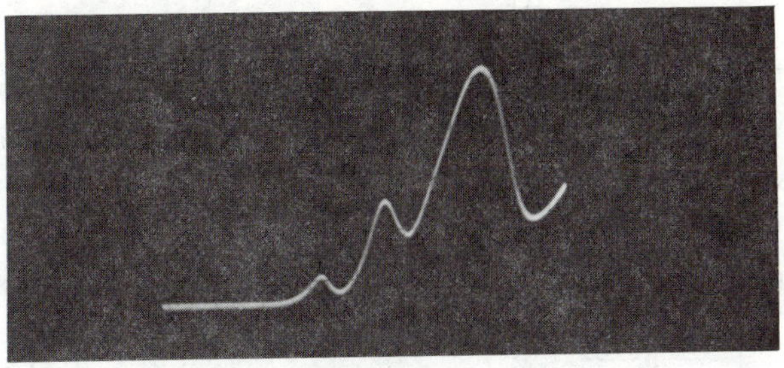

Fig. 39–17

Although the oven and temperature measuring system used are homemade, commercial equipment can be substituted. The tube circuitry is modified as shown in Fig. 39–20.

A cross section of the oven is shown in Fig. 39–21. The heating element is non-inductively wound. Its components are: a sheet metal envelope *a*, asbestos *b*, nichrome wire *c* connected to resistor and power line, glass tube *d*, inner grounded copper sleeve *e* to minimize electrostatic effects, and porcelain insulators *f*.

Fig. 39–18

After calibrating with a mercury thermometer, a projection microammeter can be modified to show the correlation between tube temperature and the oscilloscope pattern. The tube is heated to approximately 200°C to make certain that all mercury is vaporized. Mercury remaining on the electrodes shorts out the tube. A Fenwal thermoswitch[18] controls the temperature of the oven. Once this temperature has been reached, oven power is cut, and the tube is removed to cool slightly. Sharper

[18] Fenwal Electronics Inc., Framingham, Massachusetts.

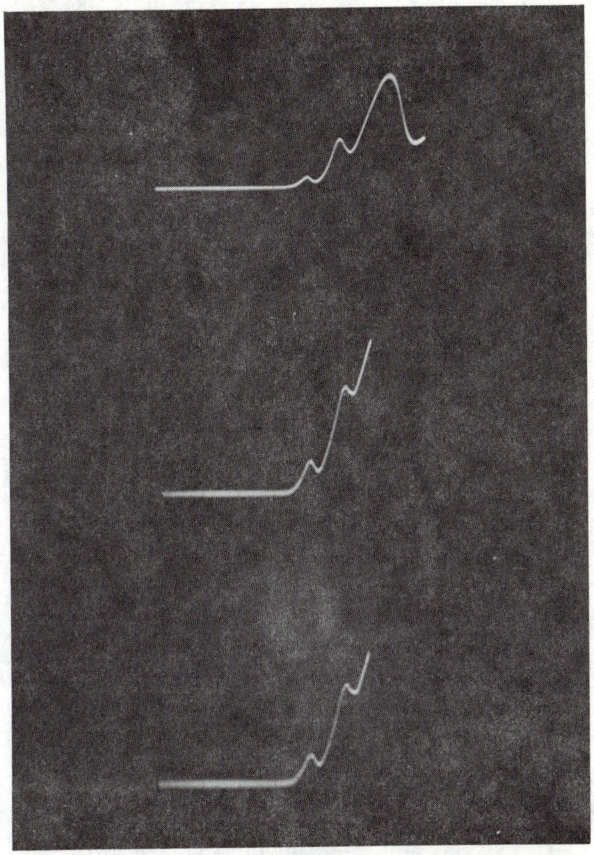

Fig. 39–19

minima are obtained at temperatures slightly below 200°C.

39–3.2

With an argon-filled GTIC thyratron, apparatus can be constructed which uses the Franck–Hertz Principle to demonstrate the ionization potential of argon. On a 2 x 3-ft piece of hardboard the circuit shown in Fig. 39–22 is layed out using white adhesive tape. The hardboard is painted black. The thyratron holder, battery clip, and potentiometer are attached to the front of the board, while the actual wiring is located at the back. Sockets are provided at the meter connections so that large lecture type meters or projection meters can be used.

The other half of the board is covered with a 2 x 3-ft sheet of tinned iron (Fig. 39–23). The iron is painted black, and attached to it by means of magnets is a diagrammatic layout of the thermionic tube. Strips of hardboard represent the electrodes, and colored disks represent electrons and argon

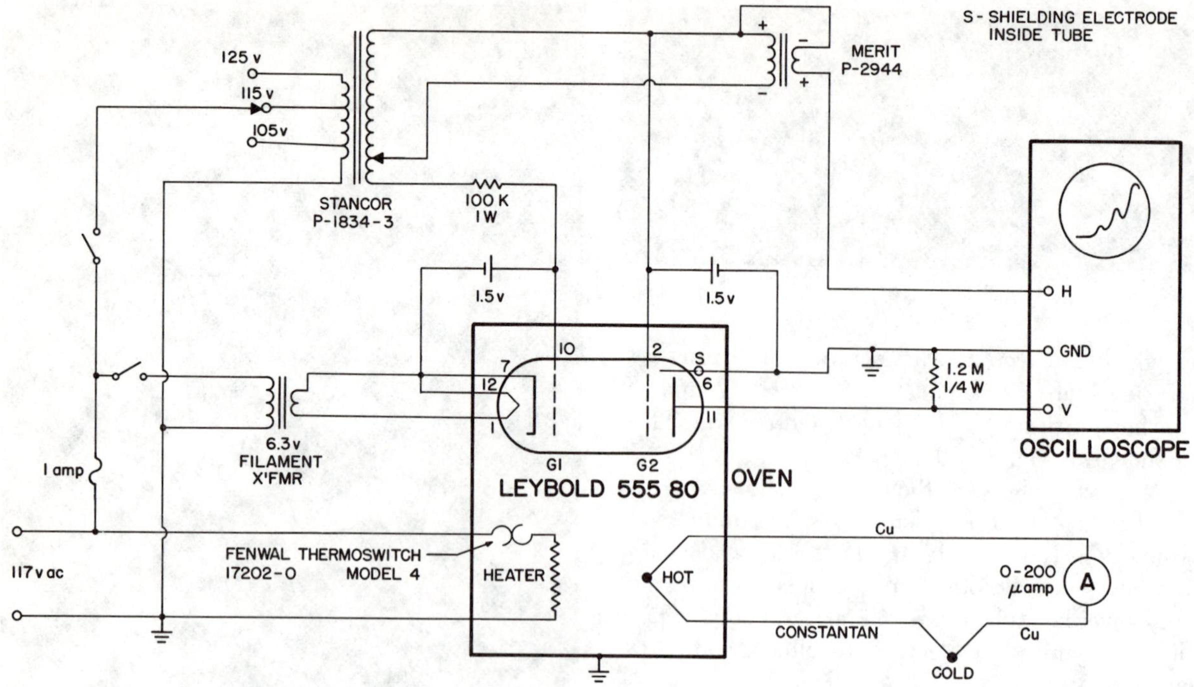

Fig. 39–20

atoms. This apparatus board, when completed, is clamped to a standard teaching chalkboard.

The circuit and tube functions can be explained by using the taped diagram and magnetic cut-outs. A 2-V battery maintains the anode negative with respect to the cathode. The grid potential is supplied through a potentiometer R from a 30-V battery and is measured by the voltmeter V. The arrival of positive ions at the anode is measured by the meter G.

Current is supplied to the cathode, and the tube is allowed to warm up. The potential difference

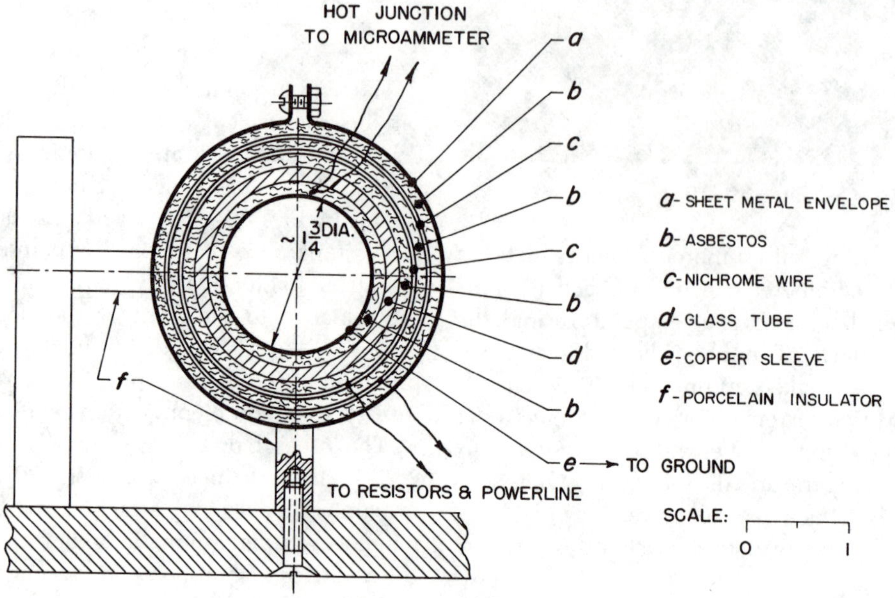

Fig. 39–21

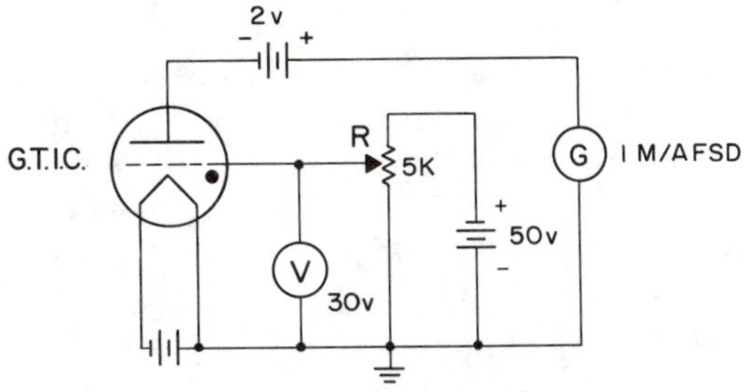

Fig. 39–22

between the grid and cathode is increased gradually until it is about 18 V. Ionization takes place, and there is a sudden increase in the current shown on meter G in the anode circuit. At the same time the voltage drops to 15.7 V, which is the ionization potential of argon. The characteristic glow discharge of argon can be seen.

A slightly modified circuit (Fig. 39–24) developed at Rensselaer can be used to measure the ionization potential of a 2D21 xenon-filled Thyratron.(19) A 3-V battery maintains the anode negative with respect to the cathode. The positive grid potential is provided by a 30-V battery through a 100-K potentiometer and measured by a voltmeter *a*. The arrival of positive ions at the anode is measured by a microammeter *b*.

Current is supplied to the filament, and the tube is allowed to warm up. The potential difference between the grid and the cathode is increased gradually until it is about 17 V. Ionization takes place, and there is a sudden increase in the current ($\sim$1–10 μA) in the anode circuit. At the same time, the voltage drops to about 12 V, which is approximately the ionization potential of xenon. The commercial-gas-filled 2D21 contains traces of other gases, but its principal constituent is xenon.

If 3-V and 30-V transistor batteries are used, the components can be assembled in a mini-box chassis or on a small square of plastic, so that the circuit elements can be viewed by a large group of students with the aid of an overhead projector. Projection meters can also be used and the experiment quickly performed in lecture.

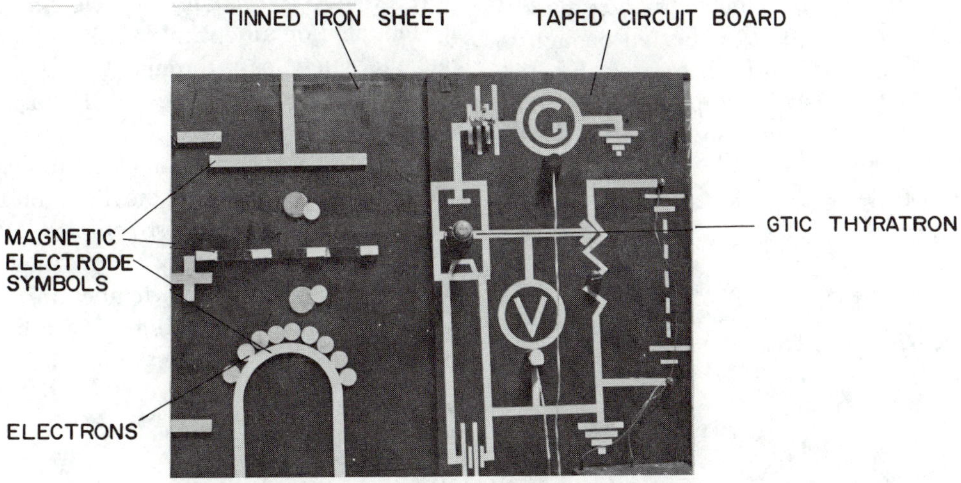

Fig. 39–23

39–4 Resonance Radiation

39–4.1

As shown in Figs. 39–25, 39–26, and 39–27, the resonance radiation of iodine vapor is strikingly visible to a large audience when a carbon arc beam is focused on a large evacuated Florence flask containing a few iodine crystals. Crossed Polaroids can be used with the apparatus to show various side effects.

(19) H. F. Meiners, *Am. J. Phys.*, 33, 5, pp. xvii–xviii.

39–4.2

To demonstrate the resonance radiation of potassium vapor, an evacuated flask containing a small piece of the metal is gently heated, and a con-

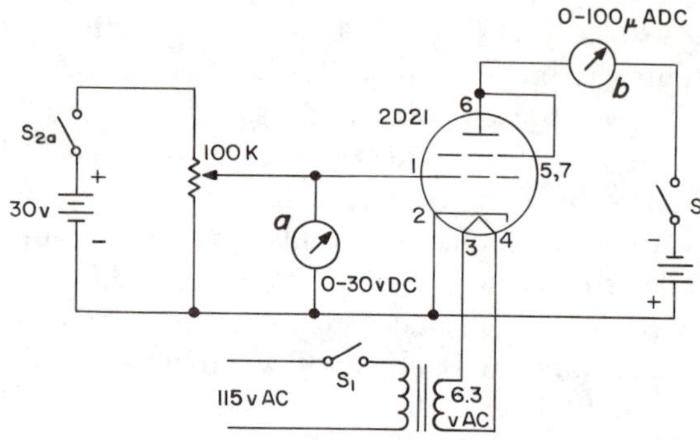

Fig. 39–24

densed arc lamp beam is passed through it. The cone of white light turns purple when the resonance point is reached. The apparatus components consist of an arc lamp, an iris diaphragm, a condensing lens system, a 1-liter, round Pyrex flask, a mechanical pump, and a pea-sized (approximately ½ cm³) portion of potassium metal. See Fig. 39–28. The flask is first thoroughly dried, and the potassium is introduced. The arc is started and the iris adjusted so that the light beam emerging from the condenser is 1 to 1½ in. in diameter. The flask is evacuated, and the potassium is heated enough to form vapor. The entire surface area of the flask must be kept hot so that the vapor does not condense. This apparatus can also be employed to demonstrate sodium resonance by simply replacing the potassium pellet with sodium.

39–4.3

A furnace is made with an angle iron frame, the base of which is about 28 x 28 cm and the height about 32 cm. The base, cover plate, and three sides are made from 5-mm-thick asbestos pads. At a height of 6 cm there is an asbestos platform that has a round 12-cm-diam hole, in which a glass bulb lies (see Fig. 39–29). The two opposite asbestos sides have 6-cm-diam holes, which are covered with glass to allow a light beam to pass through the glass bulb perpendicularly to the direction of observation. The front side is closed with glass strips (a large glass cover would break when the furnace is heated). The heating element consists of resistance wire masked by insulating beads. The masking is necessary so that observation is not disturbed by the glowing wire. Caution: when the heating element is switched on, it is necessary to put a guard in front of the glass side.

Before the bulb is placed in the furnace, the following procedure must be carried out: About 0.5 cm³ of sodium is distilled under vacuum over a small vessel into the 20-cm-diam spherical glass bulb (Jena glass). After distillation the bulb is sealed off.

A sodium vapor lamp and a mercury vapor lamp are mounted on an optical bench so that their light can alternately pass through the glass bulb from the lateral holes in the walls of the furnace. When the sodium vapor has reached a temperature of about 110°C, it can be stimulated to bright yellow, resonant fluorescence by the light of the sodium vapor lamp, but it is almost completely transparent to the light of the mercury vapor lamp. When the sodium vapor reaches a temperature of 300 to 350°C, a resonance spectrum of sodium vapor can be stimulated by light from an arc lamp.

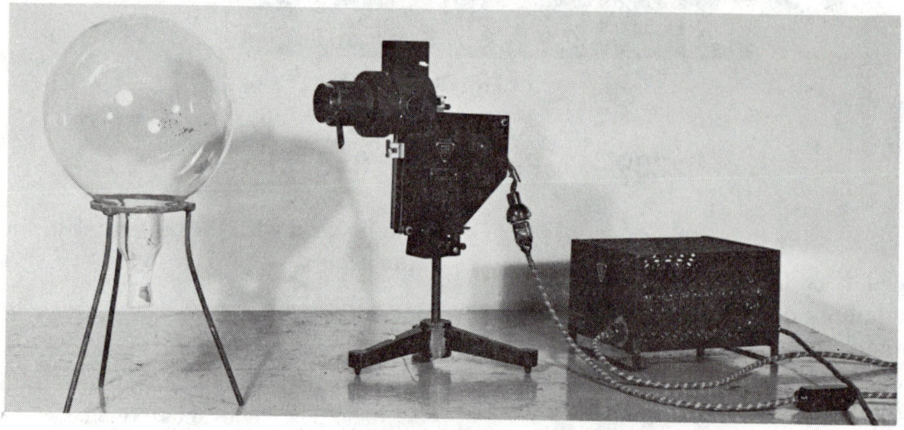

Fig. 39–25

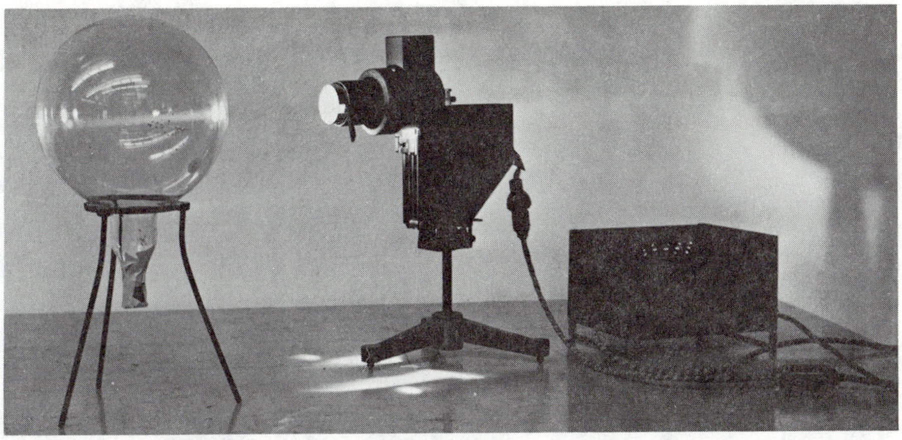

Fig. 39–26

39–4.4

As shown in Fig. 39–30, a small electric furnace heats a small quantity of sodium ($\sim$28 mm^3) in an evacuated bell jar to establish a collimated vertical beam of sodium atoms. When the encasing bell jar is illuminated by a mercury vapor lamp, the beam remains invisible. If a sodium lamp is used, a long thin "pencil" of sodium atoms, because of resonance reradiation, can be observed by a large group.

The furnace (Fig. 39–31) is stainless steel. It is $1\frac{1}{16}$ in. in diam and $1\frac{3}{8}$ in. long, coated with Lavite, a ceramic which has good high temperature properties. Heat to vaporize the sodium is provided by a tungsten element wound on the furnace. Vacuum levels are in the diffusion pump range. All pumping and electrical control gear is mounted on a wheeled cart. Construction details are given in the Appendix, page 1366.

39–4.5[20]

Dry, chemically pure zinc sulfide is used as a base for phosphorescent materials. The zinc sulfide is obtained by precipitation from an acid solution by means of hydrogen sulfide; chemically pure sodium chloride is used as a fusing agent, and a solution of a metal salt (for example, a nitrate) as an activator.

[20] Translated by K. V. Robinson and H. A. Robinson (Adelphi University) from *Lecture Demonstrations in Physics*, A. B. Mlodzeevskii, ed., M. V. Lomonosov State University, Moscow, U.S.S.R.

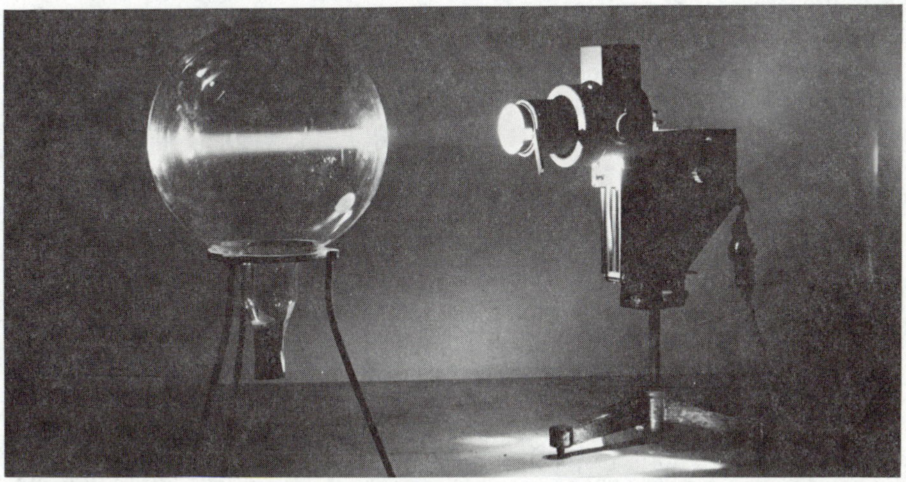

Fig. 39–27

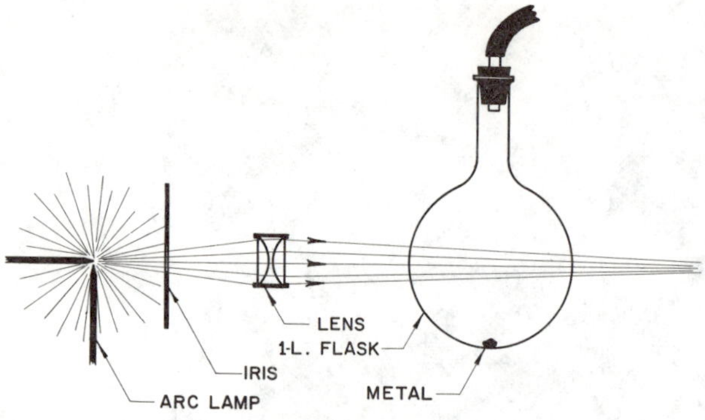

Fig. 39–28

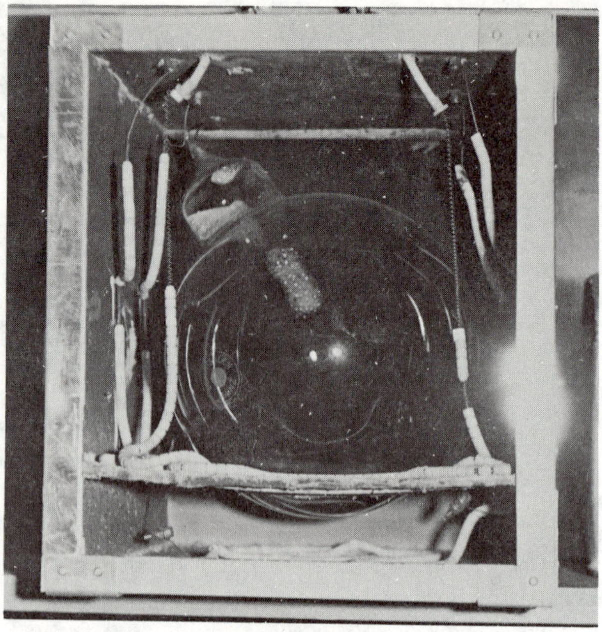

Fig. 39–29

1. Green luminescence with afterglow:

	ZnS	10g
Fusing agent	NaCl	0.5g
Activator	Cu	5×10^{-4}g
	(based on metallic copper)	

2. Orange luminescence and triboluminescence:

ZnS	10g
NaCl	0.5g
Mn	10^{-2}g
(based on metallic manganese)	

3. Light blue luminescence:

ZnS	10g
NaCl	0.5g
Ag	10^{-3}g
(based on metallic silver)	

The charge is prepared by careful mixing and slight drying. Quartz, corundum, or nonglazed porcelain crucibles are filled with the mixture. The crucibles are covered with lids, and the gaps filled with clay. The clay is dried (in a stove, muffle furnace, etc.) by heating to a temperature of 300°C and holding it at that temperature for about 30 min. Roasting is carried out at a temperature of 900°C (1200°C for longer afterglow and better triboluminescence) for 15–30 min by placing the crucibles in a preheated furnace. The crucibles are cooled in air after heating and are treated in ultraviolet light (see below).

The material, thus prepared, is kept in tightly sealed glass tubes either in the form of powder or fine broken-up pieces. Before the demonstration, light an arc lantern and turn off the light in the auditorium. The tubes containing the phosphorescent substances are held for several seconds close to the arc light; the lantern is then turned off and the luminescence demonstrated in the darkened auditorium. For triboluminescence the powder, ground in a mortar, is sprinkled between two pieces of glass (preferably watch glasses).

To show the quenching of phosphorescence by infrared light, expose a phosphorescent screen (ground powder glued on cardboard) first in the violet range of a spectrum and then in the red. Since intense illumination is necessary, obtain a spectrum as close to the lantern as possible.

Another demonstration consists of concentrating the light from a projection lantern through a condenser onto a previously light-struck screen in a darkened auditorium. A light filter is placed in the path of the light beam; the filter can be either a piece of ruby glass or a container with solution of iodine in carbon bisulfide. If the screen is moved about, one can obtain a dark design on the luminous background of the screen.

To show "flare-up," i.e., a bright flash of phosphorescence before phosphorescence is extinguished, a heated metal disk is placed behind a phosphorescing screen. A light circle appears on the screen and grows darker than the surrounding background.

Fig. 39–30

(The disk should have a handle so that the disk can be heated in a gas flame.)

A more intense flare can be shown by using a screen of calcium sulfide. The phosphorescent substance is prepared in the same way as described above; for 10 g of CaS, a mixture of 0.5 g NaCl, 0.3 g $Na_2B_4O_7$, and 0.2 CaF_2 is used as a fusing agent, and 0.0024 g Bi as an activator (based on metallic bismuth).

Phosphorescent screens can be made using the preparations of zinc sulfide listed above. A finely ground powder of the luminescent substance is mixed with glue and spread over a piece of cardboard. The screen is illuminated directly by an arc light as indicated above in the case of phosphorescent substances. To make the demonstration even more graphic, place some nontransparent object, for example, a hand, in front of the screen; then a dark outline of the hand will appear on the luminous screen.

39–5 Atomic Models

39–5.1[21]

In the Thomson model of the atom, negative electrons are balanced by an appropriate amount of positive electricity, and negative electricity is corpuscular while positive electricity is a continuous field. Thus, this model of the atom is a sphere of uniform positive charge in which embedded electrons arrange themselves so that they are in equilibrium under the attractive force of the sphere and their own mutual repulsions.

The forces of the Thomson model can be qualitatively demonstrated by the apparatus from St. Augustine High School, shown in Fig. 39–32, which consists of 2-in. magnetized-steel needles inserted into small (½ in. x ⅜ in. diam) corks. They are placed like vertical poles in a 6-in. diam x 3-in.

[21] See Alfred M. Mayer, *Phil. Mag.*, V, 397 (1878); G. Gamov and J. Cleveland, *Physics, Foundations and Frontiers* (Prentice-Hall, Englewood Cliffs, N.J., 1960), p. 365; Karl F. Herzfeld, *Wien Ber.*, CXXI, 2a, p. 593 (1912); R. W. Wood, *Physical Optics* (Macmillan, New York, 1934), p. 124; Robert K. Duncan, *The New Knowledge* (A. S. Barnes Co., 1912), pp. 154–158; J. J. Thomson, *The Corpuscular Theory of Matter* (Charles Scribner's Sons, New York, 1907), Chaps. VI and VII.

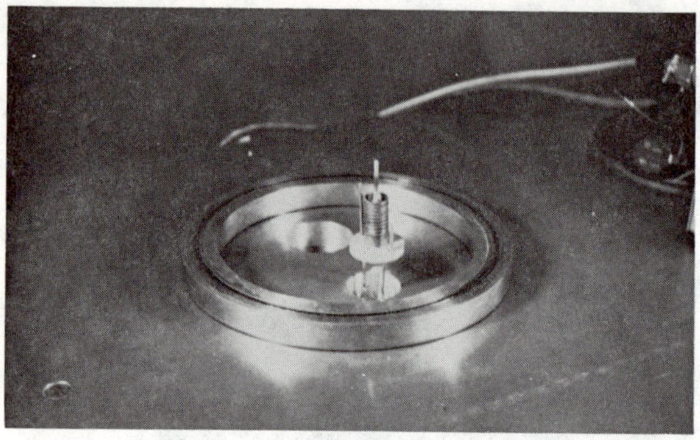

Fig. 39–31

crystallizing dish of water wrapped with 50 turns (~ 75 ft) of No. 22 or higher insulated Cu wire. Distilled water is used in the dish to a height of about 2 in. If tap water is used, it must be clear. A drop of liquid detergent aids in floating the corks. The needles are magnetized by stroking them in the same direction with the north pole of a strong magnet, and they are inserted into the corks with like poles in the same direction. The action of the magnetic poles can be projected with either an overhead projector, as was done at St. Augustine High School, or a large mirror which was employed at the University of Chattanooga. In place of the dish the apparatus developed by the University of Chattanooga uses a 9½-in.-diam, 4¾-in.-deep clear plastic tank resting in a box housing. The mirror is attached to the under side of the top of the box, which is hinged for correct angling.

This apparatus differs from the Thomson model

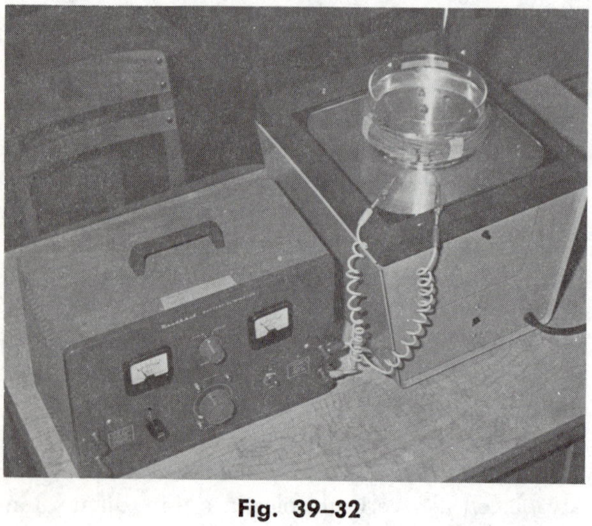

Fig. 39–32

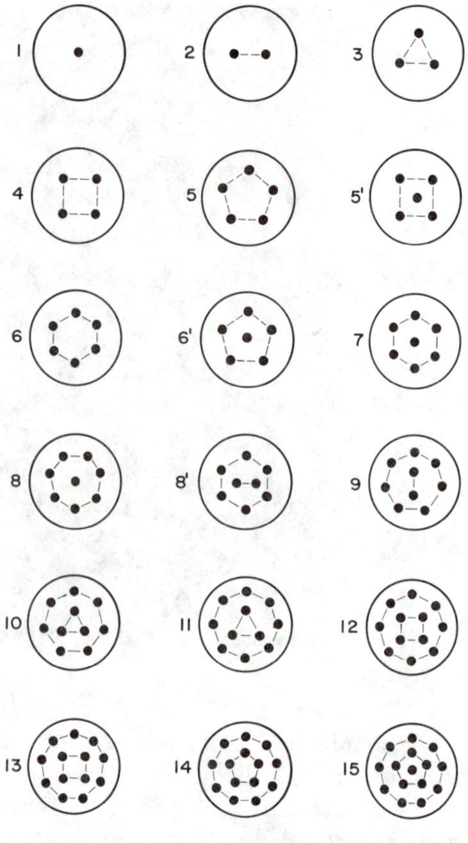

Fig. 39–33

since the electrons move in any direction in the positive sphere while the floating magnets are moving only in a plane cross section of the sphere. Insofar as both the magnetic field and the electrical field of the sphere are central and vanish at the center, the analog holds.

Even though the force laws involved are different, the corks assume equilibrium distributions completely analogous to those of the Thomson theory since the shapes of these distributions are determined essentially by the properties of the forces which are similar and by the symmetry. Figure 39–33 shows the configurations assumed by up to fifteen "electrons." When the number of electrons exceeds five, they break into two groups. The group containing the smaller number of electrons occupies the surface of an inner ring (or shell in three dimensions), and the others form an outer ring concentric with the inner one. As the number of electrons increases still further, equilibrium eventually cannot be achieved with two groups; the electrons, therefore, divide into three groups.

39–5.2

A simple model of the classical electron which falls into the nucleus because of emission of electromagnetic radiation is made from a glass funnel and a small steel ball. The steel ball does not stay in "orbit" because of frictional losses.

Demonstrations	Contributors
39–1.1	J. I. Swigart, University of Utah
39–1.2	R. Rollefson, University of Wisconsin
39–1.3	L. Grillet and M.-T. Grillet-Amy, University of Rennes, France
39–1.4	H. E. Anderson, Massachusetts Institute of Technology
39–1.5	E. R. Worley, University of Nevada
39–1.6	K. Hecht, E. Leybold's Nachfolger
39–1.7	Swiss Federal Institute (ETH), Zürich, Switzerland
39–1.8	K. Hirschberg and A. Scharmann, I. Physikalisches Institut der Universität, Giessen, West Germany
39–1.9	H. Daniel, Musterschmidt-Verlag, Göttingen, West Germany
39–2.1	Swiss Federal Institute (ETH), Zürich, Switzerland
39–2.2	H. A. Medicus and H. E. Anderson, Massachusetts Institute of Technology
39–2.3	Swiss Federal Institute (ETH), Zürich, Switzerland
39–3.1	J. A. Earl, E. R. Nye and R. L. Cain, University of Minnesota
39–3.2	The Royal Institution of Great Britain and Rensselaer Polytechnic Institute
39–4.1	University of Minnesota
39–4.2	Massachusetts Institute of Technology
39–4.3	H. Werheit, University of Köln, West Germany
39–4.4	W. Walters, Washington University
39–4.5	H. A. Robinson, Adelphi University
39–5.1	C. Francis and W. P. Darby, St. Augustine High School, Brooklyn, N.Y., and M. S. McCay, University of Chattanooga
39–5.2	Swiss Federal Institute (ETH), Zürich, Switzerland

40

SOLID STATE PHYSICS

40–1.1 Charge Motion Demonstrator

Some of the concepts associated with the mobility of charge carriers in conductors and semiconductors may be clarified by using marbles or small balls to simulate the carriers. A board (Fig. 40–1) with nails represents the conductor and scattering centers, and a component of gravity replaces the uniform electric field. Approximately 300 nails are driven in an almost random pattern into (but not through) a 1 x 6-ft board. Both the nailed and the smooth sides are edged. The nails must not be placed too close to one another or to the edge or they will form traps for the balls.

Marbles are rolled down the smooth side of the board first to simulate the motions of free particles in a uniform field. The marbles are then rolled down the nailed side to illustrate the concepts of mobility, mean free path, and current density. The mobility may be changed by using balls of the same diameter but more or less elastic.

Caution must be used in interpreting the demonstration. Each nail, for example, does not necessarily represent an individual atom. Also, in the actual case, the drift velocity may under certain conditions be proportional to the field so that the mobility will be constant. Ohm's law is obeyed in this case if the density and sign of the carriers is fixed. For the demonstration board, however, the drift velocity is proportional to the square root of the field so that only a modified "Ohm's" law applies. In this case, the apparatus simulates the motion of charge carriers in semiconductors under high fields.[1]

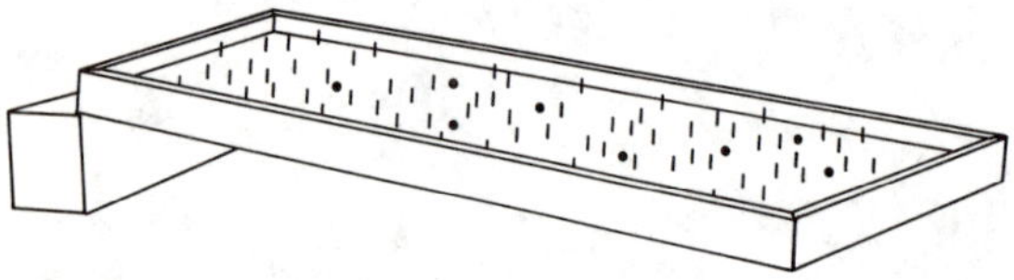

Fig. 40–1

40–1.2 Model of a Semiconductor

The interesting properties of crystals employed as semiconductors depend upon the presence of minute amounts of impurities in the crystal. For example, atoms of germanium each share four of their outer electrons with other atoms in the crystal; but if an atom of indium with only three outer electrons were to replace a germanium atom, there would be a deficiency of electrons—a "hole" results in this part of the crystal. If an electric field is applied, electrons from other parts of the crystal move towards the "hole."

A model section of a germanium crystal is con-

[1] E. J. Ryder, *Phys. Rev.*, **90**, 766 (1953); E. M. Conwell, *Phys. Rev.*, **90**, 769 (1953).

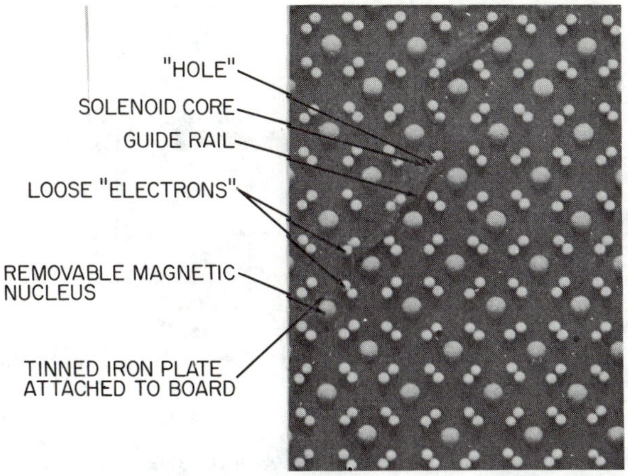

Fig. 40–2

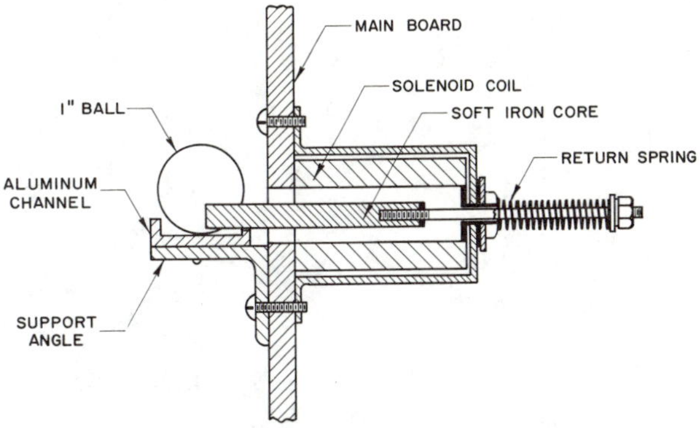

Fig. 40–3

structed from pegboard and wooden balls. (See Fig. 40–2.) The electrons are represented by 1-in. balls, and the nuclei are 3-in. wooden balls cut in half. A regular structure is laid out on a piece of pegboard, 4 x 3 ft. The electrons are mounted onto the board by dowel sticks, and the nuclei are glued directly to the board.

One nucleus near the middle left of the board is attached magnetically. This is achieved by fastening a thin tinned iron plate at the site of the nucleus and embedding a small permanent magnet into the back of the nucleus. Two of these magnetic nuclei are required, one representing germanium, the other, the donor atom, indium.

Along the path which the hole and electrons will travel, the fixed 1-in. balls (electrons) are removed, and small solenoids mounted—one for each ball removed. The coil of the solenoid is attached to the rear of the board, with the soft iron core protruding through a hole in the board in such a way that the 1-in. ball will rest upon the core. (See Fig. 40–3.) The modified solenoids are wired to the circuit shown in Fig. 40–4.

A guide rail along which the electrons will move is constructed from aluminum channel. Mount this rail, bent to conform to the path along which the electrons are required to move, on the board, ensuring that the balls, electrons, move freely along it. Any desired electron path can be built onto the model. (Curtain rail track can be used for guide paths.) Loose balls are used to fill the lattice. They will be held in position by the guide rail and by resting against the solenoid cores.

The board can be given a coat of flat black paint; fluorescent paint can be used to color the electrons yellow, the nuclei green, and the extra nuclei bright red.

The board, displayed vertically and arranged to show the regular structure of germanium, is caused to fluoresce under ultraviolet light (the room in darkness). The magnetic germanium nucleus is replaced by the red indium nucleus, and one electron removed (Fig. 40–2). The crystal now has an impurity, and an electron deficiency or hole. If the push-button (Fig. 40–4) is pressed with the selector switch at the first position, the first solenoid core is pulled in, and its electron released. The released electron moves along the guide rail, and occupies the position of the hole. By switching each solenoid in turn, and releasing the electrons, one can make the hole move through the crystal.

40–1.3[2] Test for Type of Conductivity

A simple demonstration of the two types of conductivity for semiconductor materials can be made with a small soldering iron and a dc microammeter with a 50-μA full-scale movement (preferably zero center). The resulting apparatus is called a hot-point probe. The experimental arrangement is shown in Fig. 40–5. Suitable samples are 0.1–1-Ω/cm *p*- and *n*-type germanium or silicon wafers with the approximate dimensions given in Demonstration 40–1.7.[3] To avoid ac pickup, the solder-

[2] From *Semiconductor Electronics*, p. 94, by James F. Gibbons. Copyright © 1966 by McGraw-Hill, Inc. Used by permission of McGraw-Hill Book Company and the author.

[3] Semi-Elements Inc., Saxonburg, Pennsylvania, sells an AAPT set consisting of a germanium *p*-type bar, a germanium *n*-type bar, and an *n*-type wafer. Cenco Semiconductor Demonstration Kit, No. 80394 can also be used.

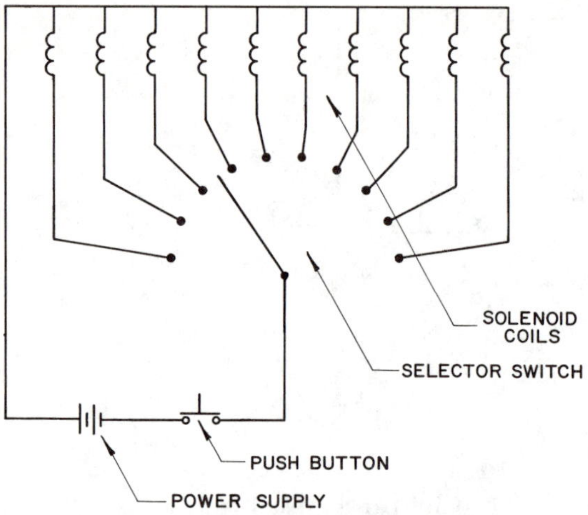

Fig. 40–4

ing iron should be turned off while the conductivity-type measurements are made.

40–1.4[4] Thermistors

A good kit of commercial thermistors for demonstration purposes is the Fenwal G200 Experimental Kit.[5] The differential negative resistance of a "fast" thermistor can be displayed with a Tektronix type 575 Transistor Curve Tracer or the simpler scheme shown in Fig. 40–6. A Wheatstone-bridge thermometer also provides an interesting demonstration.

40–1.5[6] Electron and Hole Injection in Alkali Halides

In connection with certain studies of color centers in alkali halides, an apparatus for the electrolytic coloration of large, transparent single crystals can be used. Demonstrations before solid-state physics colloquia and visitors have always been well received because of the visual nature of the phenomena. The particular feature which will be emphasized here is its educational value to students and specialists alike in depicting semiconducting processes. The cost of the equipment is quite small and the setup has natural built-in projection features

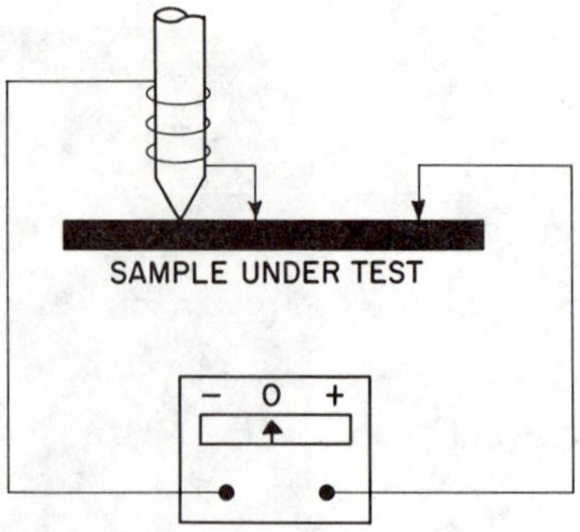

Fig. 40–5 (Courtesy James F. Gibbons and the McGraw-Hill Book Company.)

which make it demonstrable before a large group of spectators.

Electrons or holes are injected into an alkali halide crystal at high temperatures. This results in the formation of color centers.[7] Similar experiments have been performed many times before. Unfortunately they are not part of the usual elementary physics course in this country. The vivid colors which result from the injection in the transparent crystal are projected on a screen. The velocity of the carrier front as a function of electric field, temperature, time, and nature of electrode, can be demonstrated and studied.

A crystal is suitably clamped between two electrodes in a furnace enclosure and electrolyzed at high temperature as shown in Fig. 40–7. The clamp is made of stainless steel for low thermal conductivity and is spring-loaded outside the furnace in order to maintain a constant pressure of electrodes on the crystal. The fulcrum is made of Micalex in order to insulate the two arms which serve as electrical conductors. The electrodes are made of platinum due to its high work function.[8] One of these may be flat, while the other should be a point. Since platinum wire is too soft, the pointed electrode is made by melting old thermocouple wire into a ball and fashioning it into a right circular cylinder; this is then forced into a hollow ¼–20

[4] From *Semiconductor Electronics*, p. 143, by James F. Gibbons. Copyright © 1966 by McGraw-Hill, Inc. Used by permission of McGraw-Hill Book Company and the author.

[5] Can be obtained from a local supplier or from McKnight and Co., Menlo Park, California.

[6] H. N. Hersh, L. Bronstein, *Am. J. Phys.*, **25**, 306 (1957).

[7] Since this subject has a long development, mostly as a result of work done in Germany by Professor R. W. Pohl and his school, references are given here only to comparatively recent review articles in English, wherein references to the original literature may be found: (a) F. Seitz, *Rev. Mod. Phys.*, **18**, 384 (1946); **26**, 7 (1954). (b) N. Mott and R. Gurney, *Electronic Processes in Ionic Crystals* (Oxford University Press, New York, 1948). (c) R. Pohl, *Proc. Phys. Soc.* (London) **49** (extra part), 4 (1937).

[8] Y. Uchida and Y. Nakai, *J. Phys. Soc.* (Japan), **8**, 795 (1953).

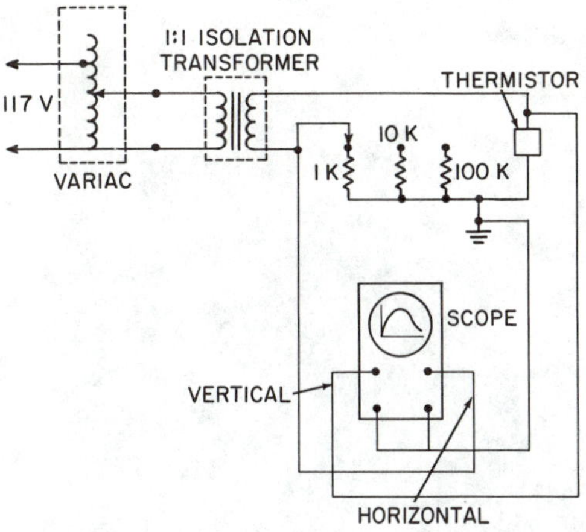

Fig. 40–6 (Courtesy James F. Gibbons and the McGraw-Hill Book Company.)

stainless steel screw and trimmed to an included angle of 45 to 90 deg. Dry cells in series, plus a reversing switch serve as a power supply of 90–720 V. The milliammeter (10 mA) is optional.

The horizontal tubular furnace has two oppositely positioned viewing holes drilled into the furnace insulation and alundum core at right angles to the core. The image of the electrodes and crystal is projected onto a screen by means of lenses and a projection lamp. One can use two convex lenses of focal lengths 33 and 15 cm, and a 6-V, 108-W projection lamp can be used to project the image on a white illustration board two feet away. The focal length of the lens near the lamp should be equal to the lens-to-crystal distance. From the crystal parallelepiped, 1 x 1 x 0.5 cm, an image about 15 x 15 cm is obtained. The furnace temperature is adjusted by means of an auto-transformer. Table 40–1 shows optimum temperatures for injection. Figure 40–8 is a photograph of the whole assembly mounted on a movable table. Figure 40–9 is a close-up view of the crystal holder and a colored crystal of CsI. The upper electrode has a platinum tip; a piece of foil is used between the crystal and the lower, flat electrode.

When the reversing switch is thrown in such a direction that the pointed electrode is negative, a coloration (blue for KI) emerges from the cathode and spreads out in an everwidening fan until it reaches the anode. Depending on the temperature and current, this process may take from 2 to 30 sec. At this point the crystal is almost completely colored due to the formation of *F* centers where electrons are trapped at halide vacancies. That this results from the migration of negative charges can be seen by subsequently reversing the polarity of the electric field, for then the "trapped-electron-cloud" is swept out; the "wave front" moves from the flat platinum foil electrode (now the cathode) back to the pointed electrode (now the anode). At any stage in the migration of the colored cloud its direction may be reversed by changing the polarity of the electrodes; the wave front moves toward the positive electrode. The cloud is fan-shaped because of the higher electric field near the point;

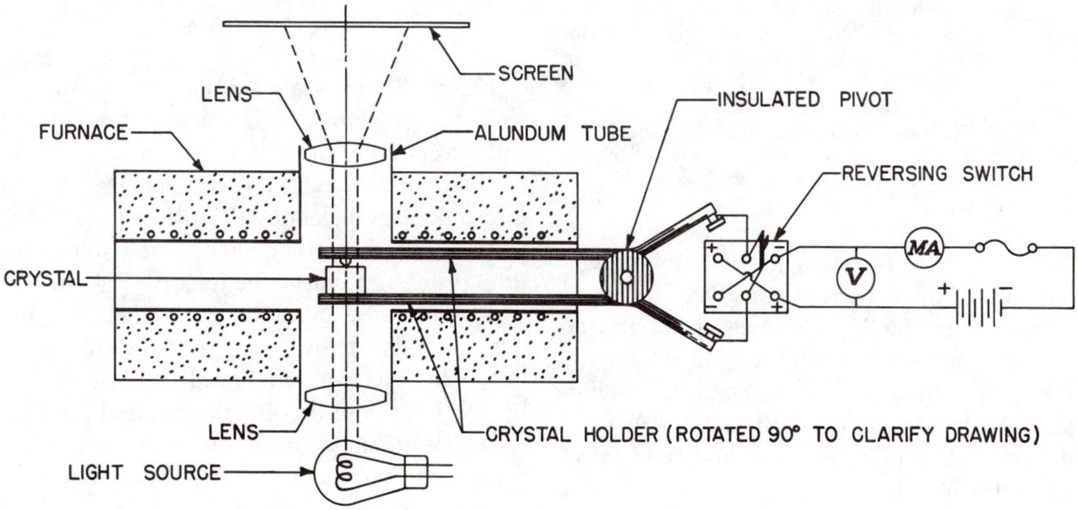

Fig. 40–7

Fig. 40–8

Table 40–1
Coloration Induced in Several Alkali Halides

	Coloration of Crystal at 600°C[a]		
Crystal	Electron Injection	Hole Injection	Optimum Temperature (°C[b])
KI	blue	yellow	600
KBr	blue	. . .	600
NaCl	red	. . .	640
CsI	blue	yellow	450
CsBr	blue	. . .	450
KCl	blue	. . .	625
KI(TlI)	black	yellow	600

[a] These colors describe the appearance of the crystal by transmitted white light at the coloration temperature. Upon cooling the crystal a color change may be noted; KCl, for example, goes from dark blue at 600°C to magenta at room temperature. The "blue" mentioned in the table has different qualities for the several alkali halides, but in order to avoid ambiguous terms like turquoise, aquamarine, royal blue, they are called "blue."

[b] These temperatures may vary somewhat, depending on the purity of the crystal.

there are more "lines of force" per unit solid angle and therefore a higher current density. The shape of the cloud mirrors the lines of force as can be seen by using two pointed electrodes, in which case the colored region of the crystal is shaped like an ellipsoid. Commonly available alkali halides[9] yield characteristic colors upon electron and hole injection (see Table 40–1).

[9] Crystals can be obtained from Harshaw Chemical Co., Cleveland, Ohio.

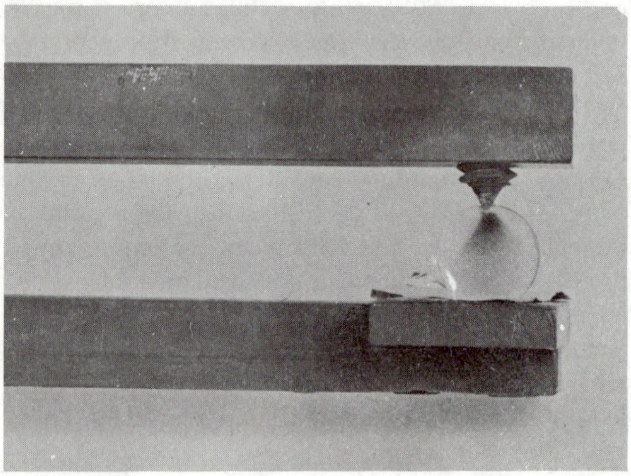

Fig. 40–9

Figure 40–10 (sequence a–h) shows the progress of the color cloud in KI during electron injection and reversal of field.

If the field intensity is decreased below the threshold for injection, the already existent color cloud leaves the cathode and merely moves through the crystal and out at the anode (Fig. 40–11).

Holes can be injected into KI and CsI by making the point positive.[10] A deficiency of *bound* (valence) electrons in the alkali halide is created by the high field around the pointed anode where an electron jumps from the valence band of the crystal into the metal. In the energy band descrip-

[10] Y. Uchida and Y. Nakai in 1954 were the first investigators to demonstrate that holes can be injected in KI.

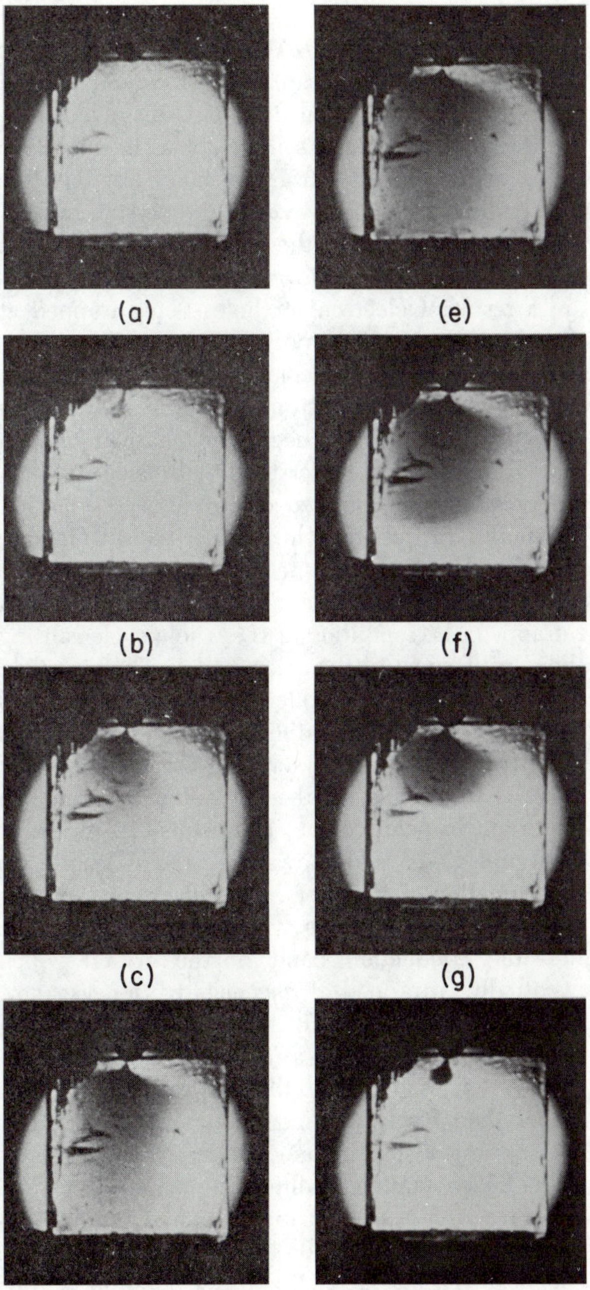

Fig. 40–10

tion this means that holes are created in the crystal which drift in the electric field and are trapped at imperfections. This gives rise to a brownish yellow cloud which emerges from the anode and advances toward the cathode. CsI is much more amenable to hole injection than KI. It is possible to inject

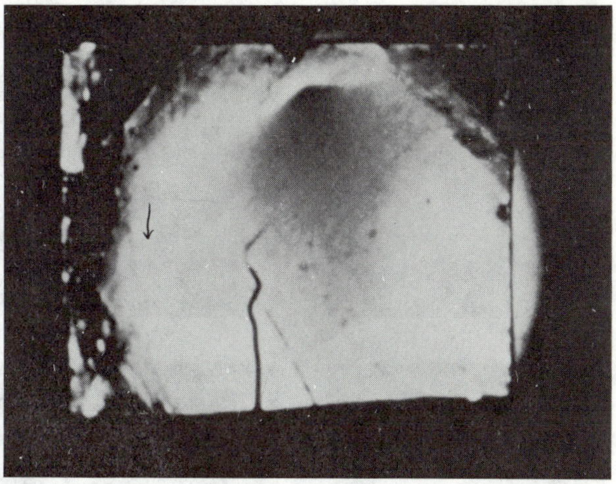

Fig. 40–11

both electrons and holes simultaneously using two pointed electrodes so as to observe recombination processes. A typical result is shown in Fig. 40–12.[11] At the moment of meeting of the blue and yellow cloud the field is removed and a recombination process instantly takes place. This is observed as a retrograde motion of the two color clouds with a resulting intermediate zone of no color, which is due to the annihilation of the two carriers by recombination.

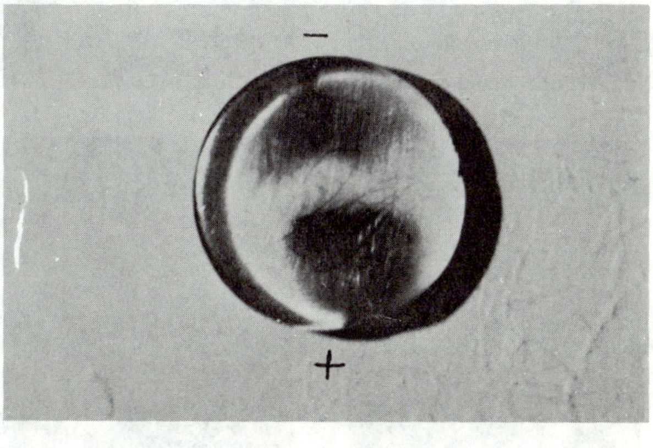

Fig. 40–12

The addition of a small quantity of thallium iodide to potassium iodide changes the color produced during electron injection markedly. In the pure KI, the color is blue and is due to the formation of *F* centers, whereas the color is jet black in

[11] Since yellow does not reproduce well in black-and-white, the color was deepened before the photograph was made.

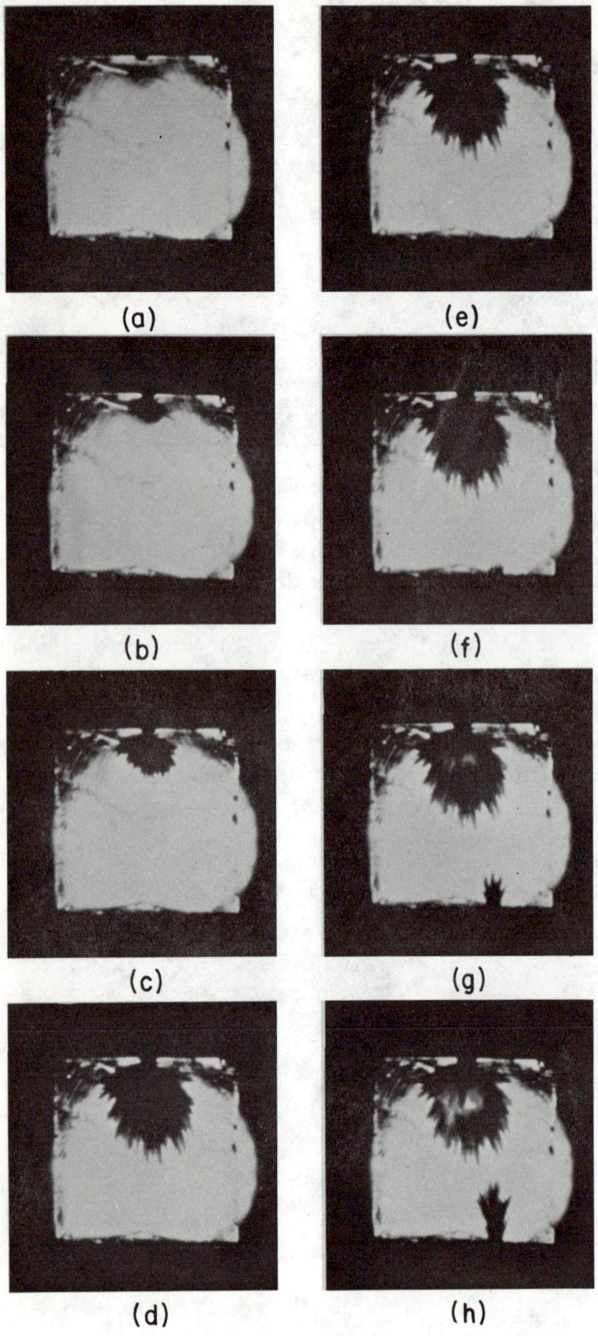

Fig. 40–13

the impure KI and emerges dendritically in dramatic fashion, as can be seen in the sequence in Fig. 40–13 (a–h).

Alkali halides are transparent over a large range of wavelengths, including the visible. This has made them the objects of many studies: for slight electrical or magnetic changes often show up as concomitant optical changes in the visible region, rendering them colored. When an alkali halide is subjected to ionizing radiation or when a small amount of one of its constituents is dissolved in it, the crystal acquires the same color as in the above experiments when electrons or holes are injected. The entity which forms within the crystal and absorbs light in a part of the visible spectrum is said to be a color center. It is believed that color centers are a result of electrons or holes being trapped at one or more negative or positive ion vacancies. The best understood of the color centers, the *F* center, is thought to be an electron trapped at a lattice site normally occupied by a negatively charged halogen ion. It may then be understood that the color which emerges from the pointed electrode represents not the mobile carriers but electrons or holes which are introduced and then trapped at imperfections; these are thermally released and drift in the field until retrapped, the repetition of the whole cycle causing the advancement of the color cloud. The high field required from the introduction of the carrier arises from the IR drop due to the constriction of the lines of current flow in the crystal near the contact. This effect is referred to as the spreading resistance so important in point-contact transistors.

Figure 40–14, a–d shows the process leading to the formation of *F* centers. The electrons source is the hot pointed cathode which injects an electron into the conduction band of the crystal which eventually is trapped at *one* negative ion vacancy, forming one *F* center. In the figure, the crystal at these high temperatures already contains the necessary vacancies. Most of the current is electrolytic rather than electronic and many of the vacancies are introduced by this mechanism.

In KI containing thallium impurities, evidently other electron traps contributed by the thallous ions have a larger cross section for electron capture than the negative-ion vacancies; this coloration is then not due to *F* centers as in the pure KI but is due to a chemical change represented by the reaction $Tl^{+} + e^{-1} \rightarrow Tl^{\circ}$. At the coloration temperature the *Tl* atoms coagulate and form metallic thallium.

40–1.6 Electrons in KCl Crystal

The injection of electrons into a transparent potassium chloride crystal at high temperatures results in the formation of color centers (see Demonstra-

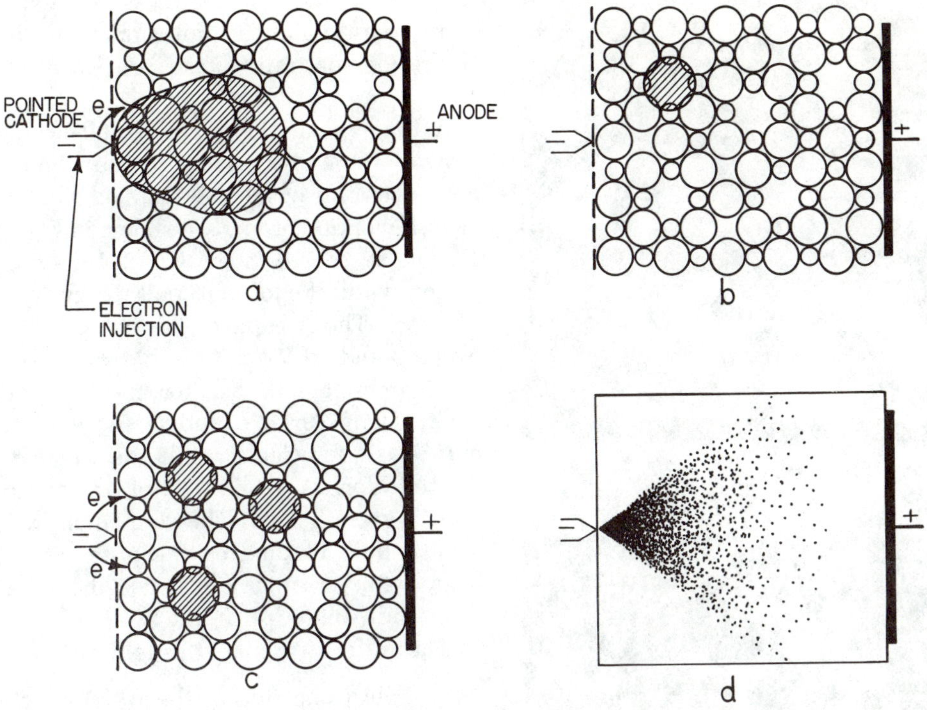

Fig. 40–14

tion 40–1.5). These color centers, when projected on a screen, demonstrate the velocity of the electron carrier as a function of electric field, temperature and time. (See Figs. 40–15 through 40–18.)

The crystal *a* (½ x ½ x ¼ in.) is held between two pieces of stainless steel *b* fastened to a ceramic support *c*, as shown in Fig. 40–19. A platinum sheet *d* is placed between one side of the crystal and the stainless steel, and is connected to one of the input jacks *e* with stainless steel wire. A ⅛-in.-diam screw *f* with a .04-in.-diam platinum tip is connected to the other input jack and touches the other side of the crystal. The ceramic support is set approximately half way into a tubular electric oven *g* open at both ends, whose temperature can be varied. To prevent heat loss, the oven should be closed with insulating plates which can be removed immediately before the demonstration. A

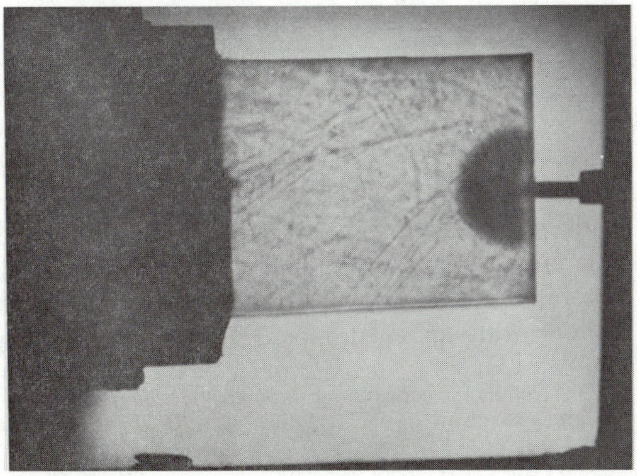

Fig. 40–15

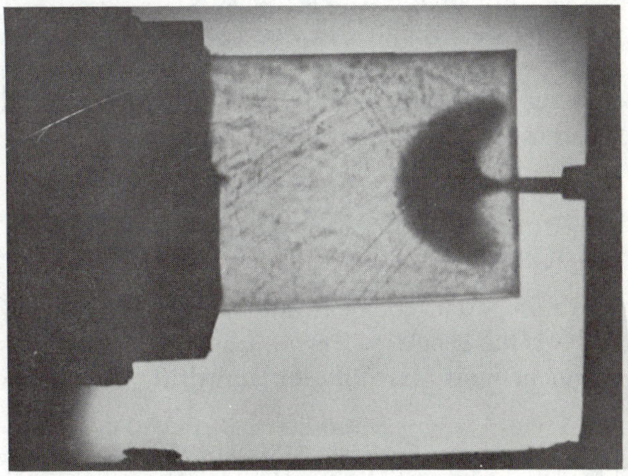

Fig. 40–16

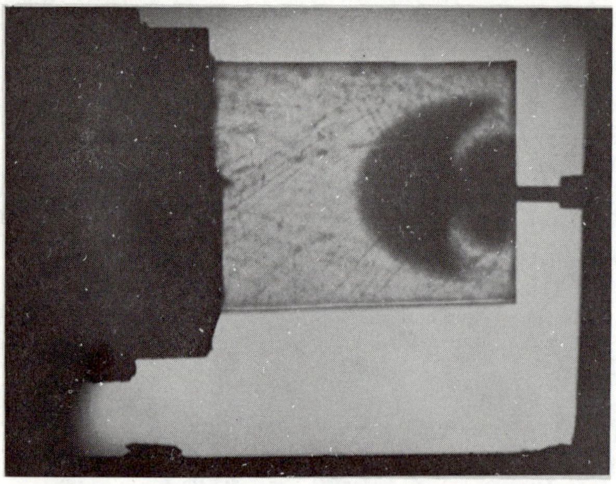

Fig. 40–17

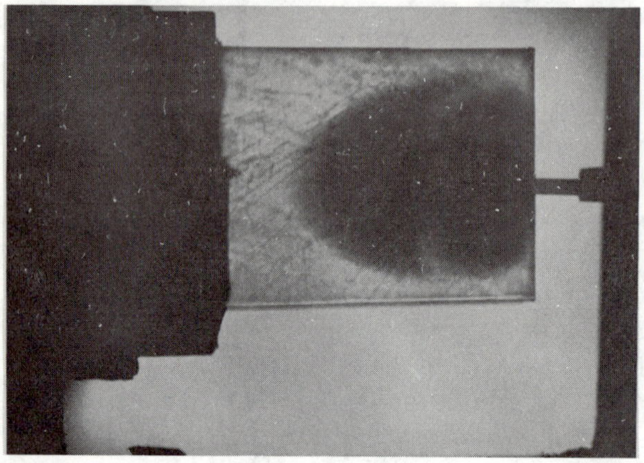

Fig. 40–18

strong, collimated light source is directed through the oven and projects the image of the crystal on a screen through a lens (Fig. 40–20).

When the crystal is heated to about 600°C and a 300-V potential is applied to the input jacks so that the platinum point is negative, a blue cloud emerges from the cathode and spreads out until it reaches the anode.

40–1.7[12] Shockley–Haynes Experiment

An important demonstration of physical principles is the Shockley–Haynes experiment. This experiment illustrates diffusion and drift phenomena very nicely and also forecasts the importance of narrow base width in transistors. The preparation of this demonstration takes time, but it is worth it.[13]

A diagram of the experimental arrangement is shown in Fig. 40–21, and a reproduction of the output waveforms which are obtained when the experiment is working properly is shown in Fig. 40–22 (a). The pulse source is an HP 212A, or equivalent, pulser, with the output pulse adjusted to about 0.5 μsec. The sweeping field should be variable between 2 and 10 V/cm.

The germanium bar for the experiment must have a long surface lifetime so that the injected pulse has appreciable amplitude when it arrives at the collector. A detailed set of preparation instructions is given below. The germanium wafer should have a bulk lifetime on the order of 100 μsec to begin with. A suitable wafer is about 1 to 2 cm in diam and 0.02 to 0.1 in. thick.[14]

The wafer is prepared as follows:

1. Polish one side of the wafer on a wet cloth or a wet glass plate with polishing alumina such as Linde A lapping powder (a fine emery paper is a possible substitute). After polishing, rinse the sample in distilled water and dry on a clean filter paper, always handling the sample with a clean pair of tweezers.

2. Coat the polished rod with polystyrene cement, except for about 0.5 cm on each end.

3. Roughen the uncoated ends of the rod with steel wool.

4. Make contacts on the end surfaces. This may be done in either of two ways. The most satisfactory way is to electroplate gold or rhodium onto the roughened ends. However, a solder contact can be applied. To do this, a coreless solder (for example, American Smelting and Refining Company NEATPAK) and an appropriate soldering flux should be used. The soldering flux can be made by mixing two parts zinc chloride with one part ammonium chloride and diluting with water. Several drops of the flux can be applied to the sample with an eyedropper. The solder is then

[12] From *Semiconductor Electronics,* pp. 143–146, by James F. Gibbons. Copyright © 1966 by McGraw-Hill, Inc. Used by permission of McGraw-Hill Book Company and the author.

[13] Details for constructing the Shockley-Haynes apparatus are also given in W. J. Leivo, *Am. J. Phys.,* 22, 622 (1954).

[14] Can be obtained from Monsanto or Cenco (Semiconductor Demonstration Kit Number 80394).

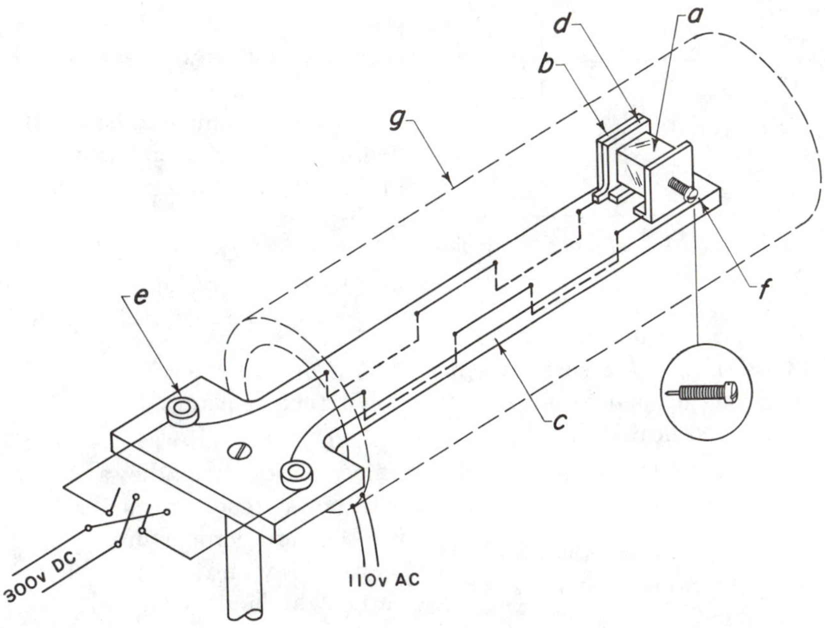

Fig. 40–19

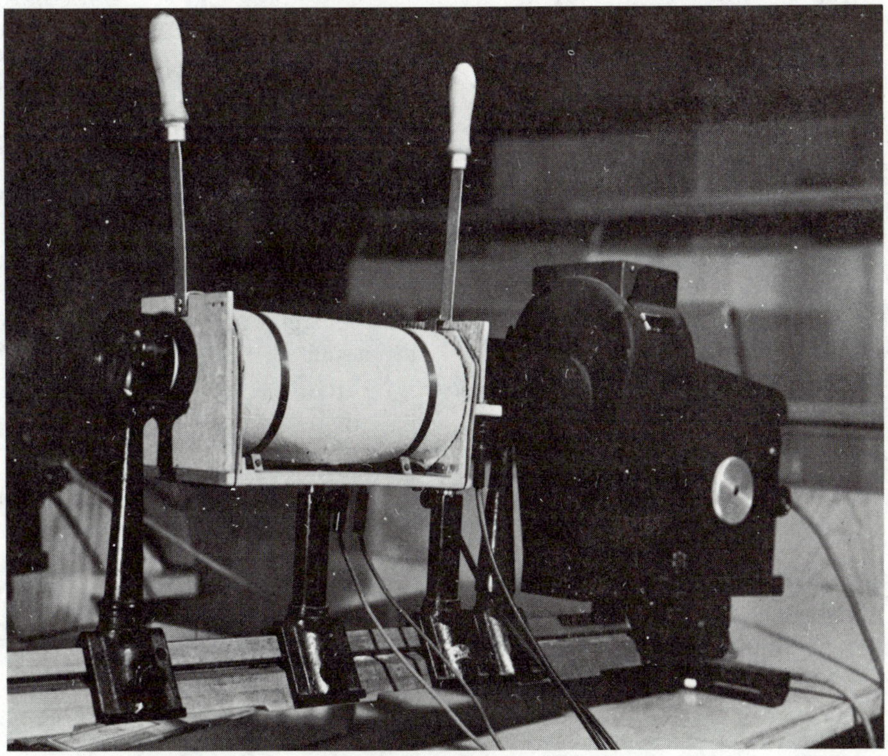

Fig. 40–20

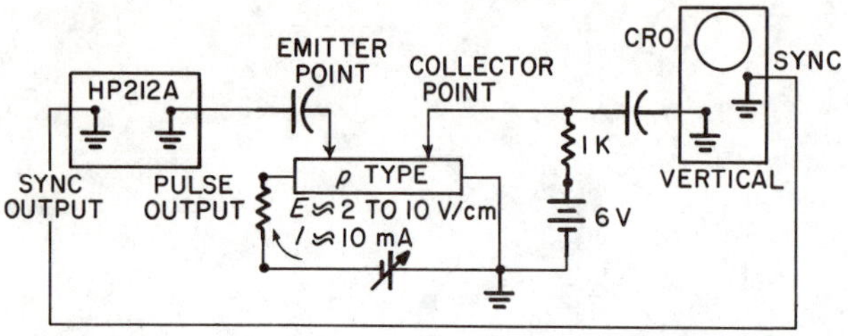

Fig. 40–21 (Courtesy James F. Gibbons and the McGraw-Hill Book Company.)

melted on the tip of a clean, well-tinned soldering iron and applied to the roughened surface. The iron should be applied long enough to insure that the germanium becomes well heated.

5. Remove polystyrene on the middle of the sample with toluene.

At this point there are two possible ways to proceed. A sample can be prepared with a minimum of further treatment; though its surface properties will not be stable and repreparation will be necessary, unless the experiment is done immediately. The minimum further treatment consists in etching the sample in boiling 3-percent hydrogen peroxide or C.P.4 (directions given below) for 2½ to 3 min. The wafer is then ready for contact application.

Or, a surface preparation which has excellent long-term stability (at least 5 yr on a well-prepared sample) can be made as follows:[15]

6. Cover the electrodes made in step 4 with polystyrene cement.

7. Etch the rod in C.P.4 for 2½ to 3 min. It is convenient to attach the sample to a platinum wire with polystyrene cement for this operation. C.P.4 is a standard germanium-etching solution, which is prepared by mixing five parts nitric acid, three parts hydrofluoric acid, and three parts acetic acid.

8. Rinse the sample in distilled water and dry on clean filter paper.

9. Hang platinum wire and the sample in toluene to loosen the sample. Remove the sample and rinse it thoroughly in toluene, making sure it is very clean.

10. Prepare a saturated antimony oxychloride suspension by placing a few crystals of antimony chloride in distilled water.

11. Place a negative electrode of tantalum or platinum in the suspension and connect it to the negative terminal of a 1½-V cell. Connect the positive terminal to a pair of tweezers, holding the sample by an electrode. Lower the sample into the suspension until all of the etched surface has been covered. Allow the sample to remain in the suspension for 5 min.

12. Remove the sample from the suspension, rinse thoroughly with distilled water, and dry on clean filter paper.

13. Place the sample in hot synthetic ceresin wax and allow the sample to come to the temperature of the molten wax.

14. Remove the sample from the wax, and while the sample is kept warm over a hotplate, remove the

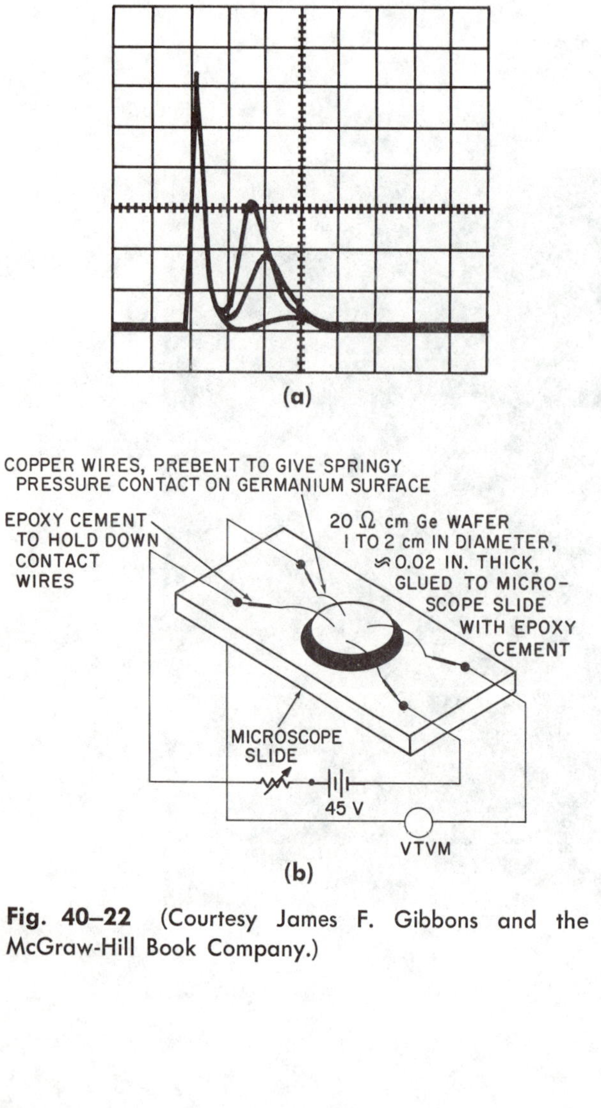

Fig. 40–22 (Courtesy James F. Gibbons and the McGraw-Hill Book Company.)

[15] A prepared sample of this type with ohmic contacts attached to its ends can also be obtained from the author for a nominal charge. Address inquiries to the author, Electrical Engineering Department, Stanford University, Stanford, California.

excess wax with a clean filter paper. When the sample cools, it is ready for experimentation.

Four contacts are required for the experiment. A mounting arrangement like that shown in Fig. 40–22 (b) is suitable. The two outer contacts (either copper or phosphor-bronze wires) are brought down on the prepared electrodes. The two inner contacts are the emitter and collector points. They must break through the wax and press solidly against the sample. These points should be a few millimeters apart and approximately in line with the end contacts.

It is necessary that the points make rectifying contacts to the semiconductor. This can be checked with an ohmmeter. A forward-to-backward resistance ratio of 100 or more for each point is desirable, though the experiment will work with less.

The diode with the highest resistance ratio is selected as the collector. It may work adequately without further adjustment, though if a noisy signal results the collector contact must be "formed." This is done by charging a 1-μF capacitor to 100 V and then discharging it through the diode, making sure that the direction of positive current flow in the capacitor discharge is the same as the direction of positive current flow when the ohmmeter reads low resistance.

40–1.8 Photoconduction

In a solid insulator, the electrons are bound to the atoms and, therefore, their energies are within the valence band. In the process of the photoelectric effect, an incoming photon can transmit its energy $E = h\nu$ to such an electron which can acquire enough energy to go into the conduction band, leading to conductivity. For this to occur, the photon energy must be greater than the energy gap between the valence and conduction bands.

The experiment illustrated in Fig. 40–23 shows qualitatively that conductivity occurs at a sharp value of the wavelength as photon energy is increased from infrared to violet. The photoconduc-

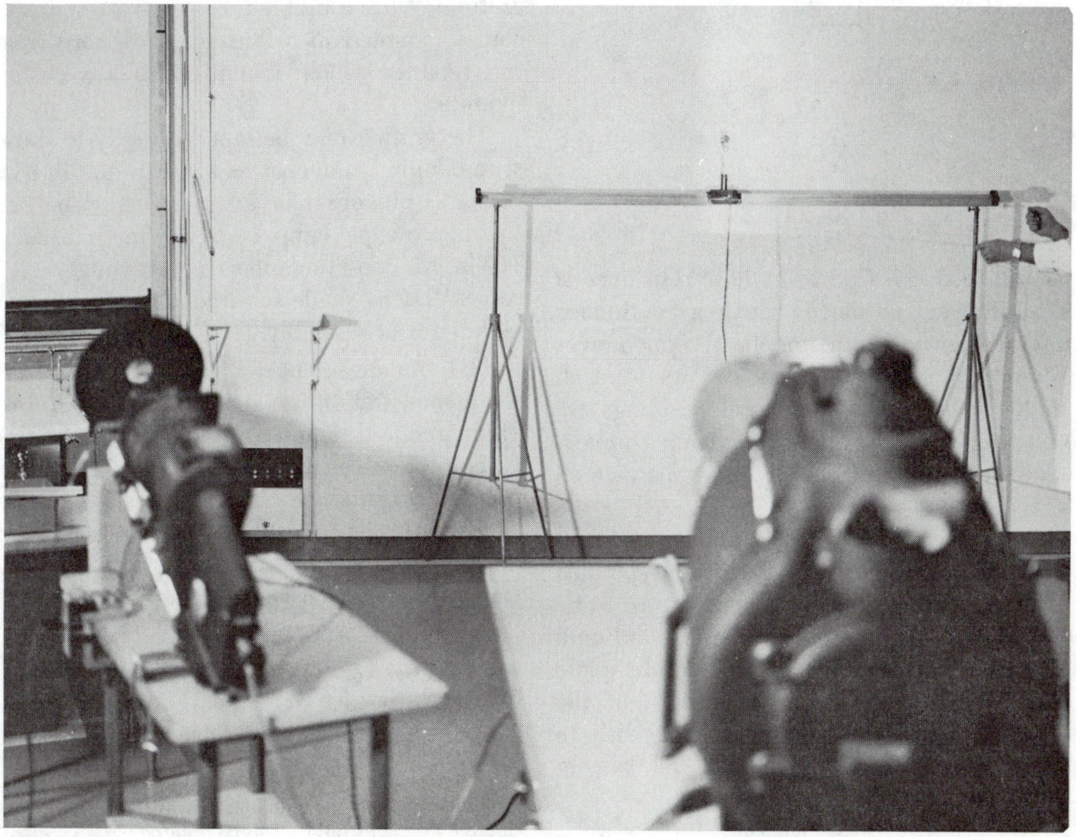

Fig. 40–23

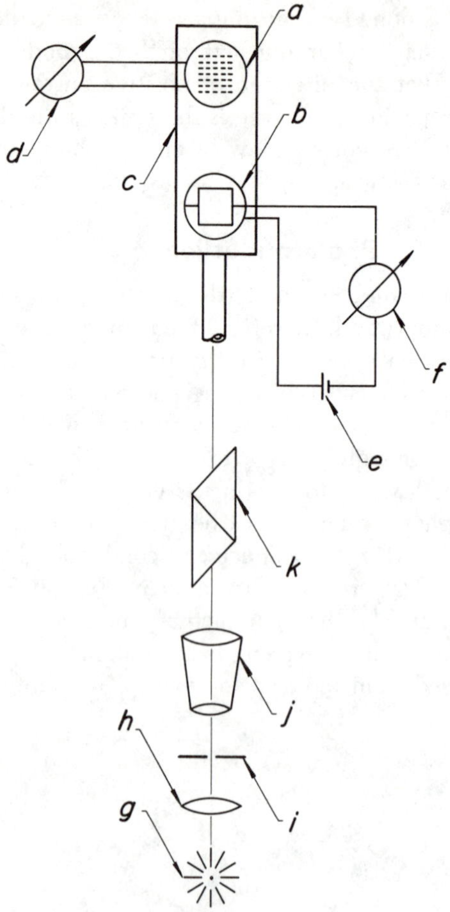

Fig. 40–24

tive material used is a CdS cell whose resistance is measured. The cell is moved along a continuous spectrum together with a thermopile[16] which gives the reference of spectral energy density at each wavelength. Whereas the thermopile readings are largest in the infrared region (arc lamp source), photoconduction sets in sharply at yellow-green light.

As shown in Fig. 40–24, the thermopile *a* and CdS cell *b* are mounted on a Plexiglas plate *c*, which in turn is mounted on an optical bench so as to be movable. A white light spectrum is projected on a screen and the bench is placed so that the plate can be moved through the entire length of the spectrum. The thermopile is connected directly to a mirror galvanometer *d*. The CdS cell is connected in series with a 2-V battery *e* and a 100-μA, 10,000-Ω, projection instrument *f*. The optics which provide the white light spectrum are also shown in Fig. 40–24. A 3000-W arc lamp *g* is used here; a smaller arc lamp would necessitate a smaller enlargement. The condenser *h* focuses the light onto the collimating slit *i*, and the objective *j* (f = 150 mm) renders the light parallel into the prism *k*, which produces the spectrum.

40–1.9[17] Photoelectric Threshold

The photoelectric threshold demonstrator shown in Fig. 40–25 consists of three main parts: (a) a light source and diffraction grating for producing a color spectrum, (b) a sample and sample holder, and (c) a translucent projection screen. These parts can be assembled as follows:

(a) A suitable spectrum for the demonstration may be obtained from an American Optical Company model 651 Microscope Light by fixing a transmission-type diffraction grating across its face. The grating[18] has a usable area of about 7/8 sq. in. and comes mounted in a 2-in.-sq. cardboard holder of the type in which 35-mm slides are customarily mounted.

The grating may be taped directly to the microscope lamp holder or cut down to fit a Kodak series-VI photographic adapter ring. The barrel of the microscope lamp is 1 17/64 in. in diam, so a 1 1/4-in. friction-fit adapter ring fits snugly. A Tiffen series-VI lens shade screwed into the adapter ring makes a neat assembly.

(b) A convenient sample and sample holder for the demonstration is a small wafer of gallium phosphide about 1/4 in. on a side and 0.035 in. thick.[19] It is preferable that the sample be lapped and polished to produce clear surfaces. Cadmium sulfide can also be used for the demonstration.[20]

[16] Cenco Radiation Thermopile No. 81070, available from Central Scientific Co., or Leybold No. 55736, LaPine Scientific Co.

[17] From *Semiconductor Electronics*, pp. 92–93, by James F. Gibbons. Copyright © 1966 by McGraw-Hill, Inc. Used by permission of McGraw-Hill Book Company and the author.

[18] Stock No. 40,272, Edmund Scientific Company, Barrington, New Jersey.

[19] Available from either Monsanto Chemical Co., St. Louis, Missouri; or Semi-Elements, Saxonburg, Pennsylvania.

[20] A polished CdS wafer with ohmic contacts which can be used for both the photoelectric threshold and photoconductivity demonstrations can be obtained at a nominal charge from Professor Gibbons, Electrical Engineering Department, Stanford University.

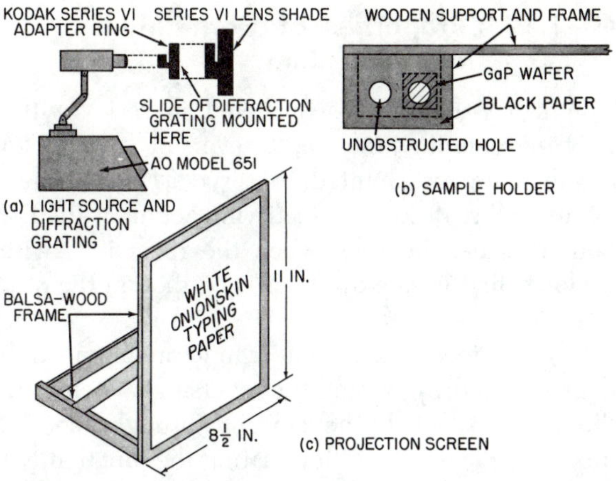

Fig. 40–25 (Courtesy James F. Gibbons and the McGraw-Hill Book Company.)

The gallium phosphide wafer can be mounted on a microscope slide or a piece of cardboard, using epoxy cement. As suggested in Fig. 40–25, the sample holder is mounted on black construction paper and has two 2¼-sq.-in. glass slides mounted glued to the front and back of the wooden supporting frame, to protect the sample.

(c) The projection screen (Fig. 40–25) is a piece of 8½ x 11-in. white onion skin typing paper glued to a balsa wood frame. If the light source is mounted about 2 to 3 ft from the screen, a usable spectrum can be displayed over almost the entire screen.

40–1.10[21] Photoconductivity

The light source and screen described in Demonstration 40–1.9 are useful also for demonstrating photoconductivity. The simplest demonstration is obtained from a commercial cadmium sulfide photoconductor (for example, GE A33 CdS photocell). The resistance of the photoconductor is measured with an ohmmeter as the sample is passed through the spectrum. Photoconductivity in germanium and silicon can be measured by pressing ohmmeter leads against a wafer of the material and turning the microscope light on and off.

Care must be taken to observe the photoconduc-

[21] From *Semiconductor Electronics*, pp. 93–94, by James F. Gibbons. Copyright © 1966 by McGraw-Hill, Inc. Used by permission of McGraw-Hill Book Company and the author.

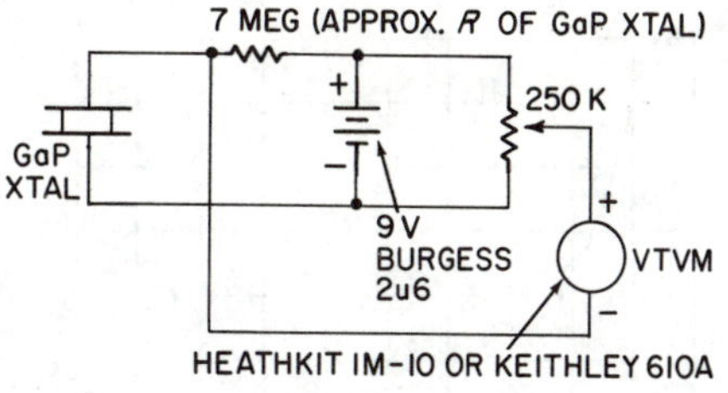

Fig. 40–26 (Courtesy James F. Gibbons and the McGraw-Hill Book Company.)

tivity of the gallium phosphide, since it is a very-high-resistance material. Satisfactory ohmic contacts can be made by the following procedure. Rinse the sample in trichlorethylene, acetone, and then distilled water. Place the clean specimen on a heating element in an inert atmosphere (nitrogen or argon) with two small globules of tin resting on top of it in the positions where contacts are to be made. A small amount of hydrazine dihydrochloride solution is used as a flux. The material is heated until the tin melts and then allowed to cool. Leads can be attached to the tin "dots" using a low melting point solder (such as an indium base solder). After attaching the leads, the contacts, and the area on the crystal around them should be coated with an opaque insulating material such as General Cement Anti-Corona Dope. A clear polystyrene cement applied to the exposed surface reduces surface leakage.

When leads have been attached, the photoconductivity can be measured as the gallium phosphide crystal is passed through the spectrum. However, the intensity of the projected spectrum is low, so a variation of about 200 kΩ (compared to a dark resistance of several MΩ) is all that can be obtained. To make this variation readily observable, a bridge circuit (Fig. 40–26) consisting of the GaP crystal, a

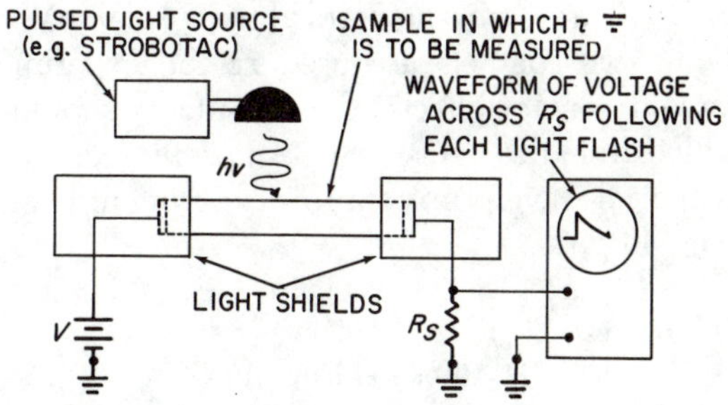

Fig. 40–27 (Courtesy James F. Gibbons and the McGraw-Hill Book Company.)

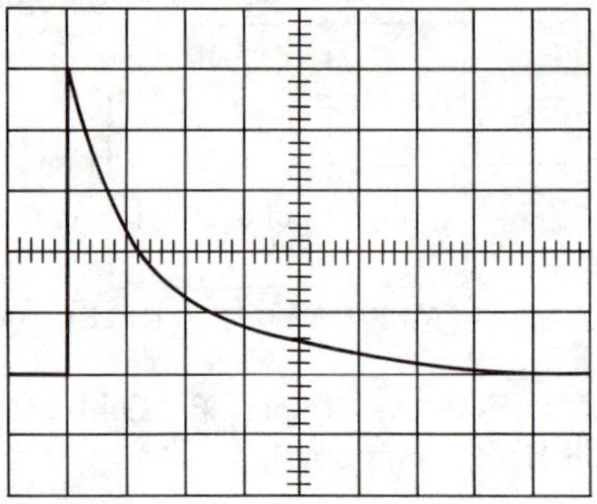

Fig. 40–28 (Courtesy James F. Gibbons and the McGraw-Hill Book Company.)

9-V battery,[22] a 7-MΩ resistor (the approximate resistance of the GaP crystal), and a VTVM,[23] can be used.

40–1.11[24] Carrier Recombination and Lifetime

Carrier recombination and lifetime can be demonstrated using a sample prepared as described in Demonstration 40–1.7. The experiment consists of exciting the sample with a burst of light and measuring the photoconductivity decay with time. The effect of the light burst is to create excess carriers in the bar, increasing its conductivity, and decreasing the voltage drop across it. The excess carrier densities decay according to an exp $(-t/\tau)$ law, resulting in an exponential decay of the voltage across the sample R_s.

The experimental setup is shown in Fig. 40–27. The battery is a 6-V or other convenient size. The resistor in Fig. 40–27 is made roughly equal to the resistance of the sample. A convenient light source is the General Radio Strobotac type 1513-A, which emits 100 light pulses per second. Each pulse lasts about 10^{-8} sec. The horizontal scale on the oscilloscope is 10 μsec/cm; the vertical scale is 50 mV/cm.

Only two contacts are used in the experiment. To avoid contact effects, these contacts should be shielded from the light burst. An oscilloscope trace of a typical waveform from this experiment is given in Fig. 40–28.

(22) Burgess.

(23) Heathkit 1M–10 or Keithley 610A.

(24) From *Semiconductor Electronics,* p. 147, by James F. Gibbons. Copyright © 1966 by McGraw-Hill, Inc. Used by permission of McGraw-Hill Book Company and the author.

40–1.12 Properties of Germanium and Selenium

A 4½-V battery is connected in series with a P–N germanium junction, a reversing switch, and a galvanometer shunted to serve as an ammeter (4 mA full scale). The rectifying action of the P–N boundary can be seen when the reversing switch is closed first in one direction, and then in the other direction.

The photoconductive effect in germanium can be demonstrated by closing the reversing switch in the direction in which the current through the P–N junction is least, then illuminating the junction with a lamp. There will be an increase in the current due to the photoconductive effect.

The same effect is noted when infrared light is focused on the P–N boundary by means of a selenium lens which is opaque to visible light. Therefore, the photoconductivity of the germanium must be due to the infrared light.

40–1.13 Lorentz Force Acting on Conduction Electrons

The action of the Lorentz force on the free electrons in a moving metal strip can be easily demonstrated by the apparatus pictured in Fig. 40–29. The circuit diagram is shown in Fig. 40–30. A metal strip *a* is drawn through a magnet. The Lorentz force causes a voltage between two sliding contacts. The induced voltage is found to be independent of the metal used.

Strips of brass and stainless steel with polished

Fig. 40–29

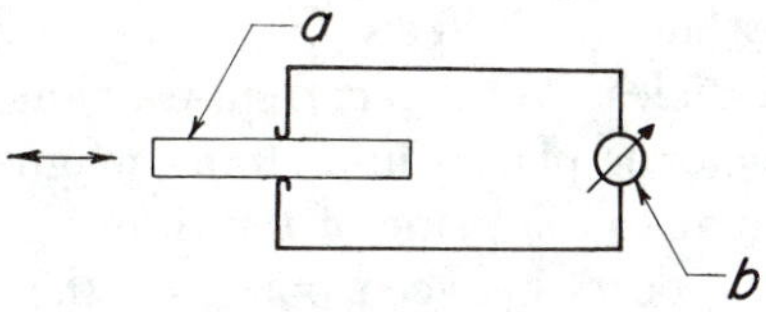

Fig. 40–30

sides (each 2 x 20 x 1/32 in.) are used. Two brass blocks (or carbon brushes) (Fig. 40–31), in which slits are cut, serve as sliding contacts. One of these, *c*, is mounted on a Plexiglas insulating holder *d*. The other one, *e*, is mounted on a resilient, hook-shaped brass strip *f*.

The magnet used was a surplus radar magnet with an air gap of 7/8 in., although an electromagnet can also be used. The magnet is mounted over the sliding contacts so that its field (about 1 kG) is perpendicular to the metal strip. A projection instrument *b* (sensitivity 5 mV, internal resistance 10 Ω) is connected directly to the sliding contacts. The instrument is placed on an optical bench and is made visible in shadow projection.

40–1.14 Resistance Change in Semiconductors

A resistance change of materials in the presence of a large magnetic field is found in almost all metals and is especially noticeable in semiconducting materials. Magnetoresistance, which is one of several galvanomagnetic phenomena including the Hall effect, is apparent both when the field is parallel and transverse to the conducting wire. It is a positive effect in most materials except for the noble metals and alloys, transitional metals and ferromagnetic materials above saturation. The change in resistance is usually proportional to the square of the field intensity. This is true only to a certain maximum value of the field where the change becomes directly proportional to the field strength. The effect is often very noticeably anisotropic in single crystals. The phenomena can be used to yield information about the band structure and current carrier mobilities in conductors and semiconductors.

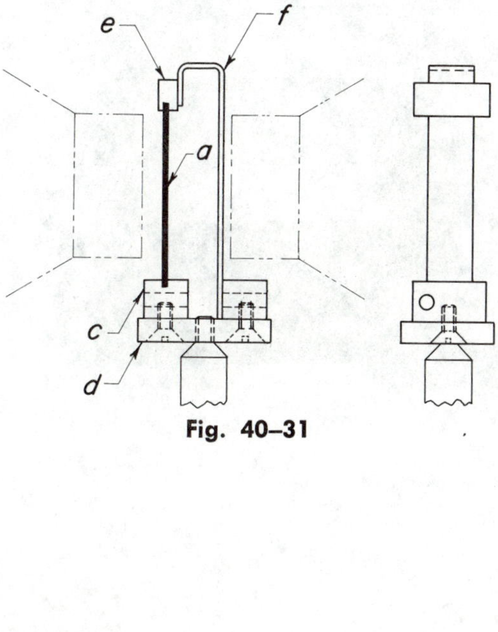

Fig. 40–31

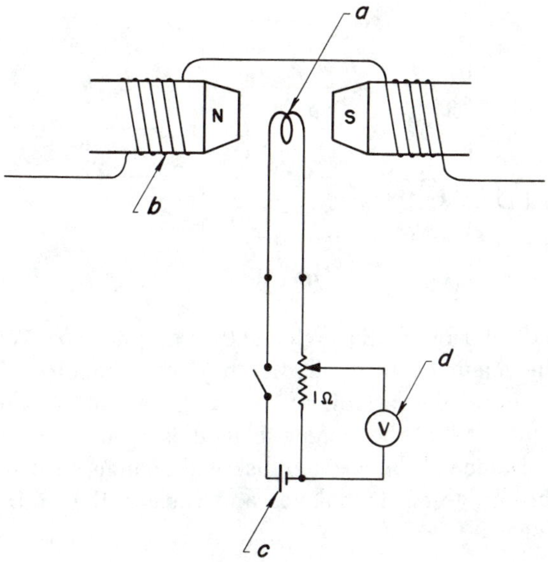

Fig. 40–32

In the demonstration, a spiral of bismuth *a* is placed in the gap of a large 25-kG electromagnet *b* as shown in Fig. 40–32. When the spiral is connected to a 2-V battery *c*, and through a 1-Ω shunt resistor potentiometer for a 300-mV measuring instrument *d*, the large magnetoresistance in Bi is qualitatively demonstrated very simply. The current in the spiral is about 0.1 A. The Bi spiral is commercially available.[25]

40–1.15 Resistance Change in a Corbino Disk

A corbino disk (InSb) forms one arm of a Wheatstone bridge and is placed between the poles of a large electromagnet, as shown in Fig. 40–33. The output of a Wheatstone bridge is connected to an

[25] This may be purchased from Hartman & Braun, Frankfurt a/M, West Germany.

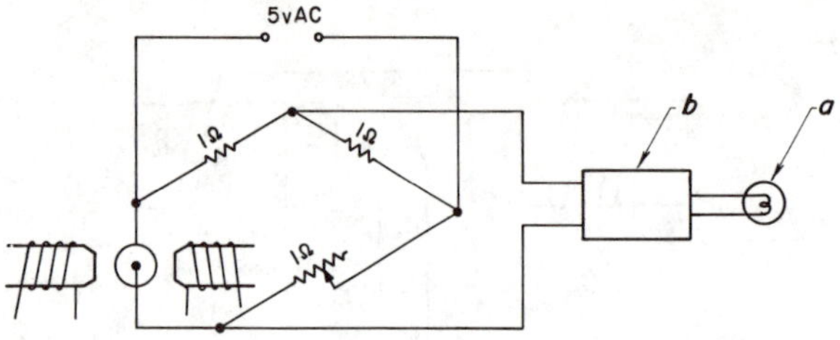

Fig. 40–33

indicator bulb *a* through a 100-W amplifier *b*. With the magnetic field off, the bridge is balanced by means of the variable resistor and the bulb does not light. When the magnetic field is turned on, the resistance of the corbino disk will change, throwing the bridge out of balance and causing the bulb to light.

40–1.16 The Hall Effect

The Hall effect, discovered in 1879,[26] can be used to demonstrate the effects of magnetic fields on charged particles. When a current-carrying, ribbon-like conductor is placed in a strong magnetic field perpendicular to the plane of the ribbon, a voltage difference appears between points on the opposite edges of the ribbon. This effect is small in metallic conductors, but considerably larger and easier to detect in semiconductors.

Apart from its many modern applications, the Hall effect can be used to determine the sign of the charge carriers and their number per unit volume in conductors and semiconductors.

The apparatus is used on the stage of an overhead projector for qualitative measurements only. It can easily be modified for quantitative measurements in the laboratory.

The components, shown in Fig. 40–34, are mounted on a clear acrylic plastic base placed on an overhead projector. The circuit consists of a miniature battery connected in series with a knife switch, a potentiometer used to control the current, a projection ammeter with its appropriate shunt, and a

[26] E. H. Hall, *Am. J. Math.*, 2, 287 (1879).

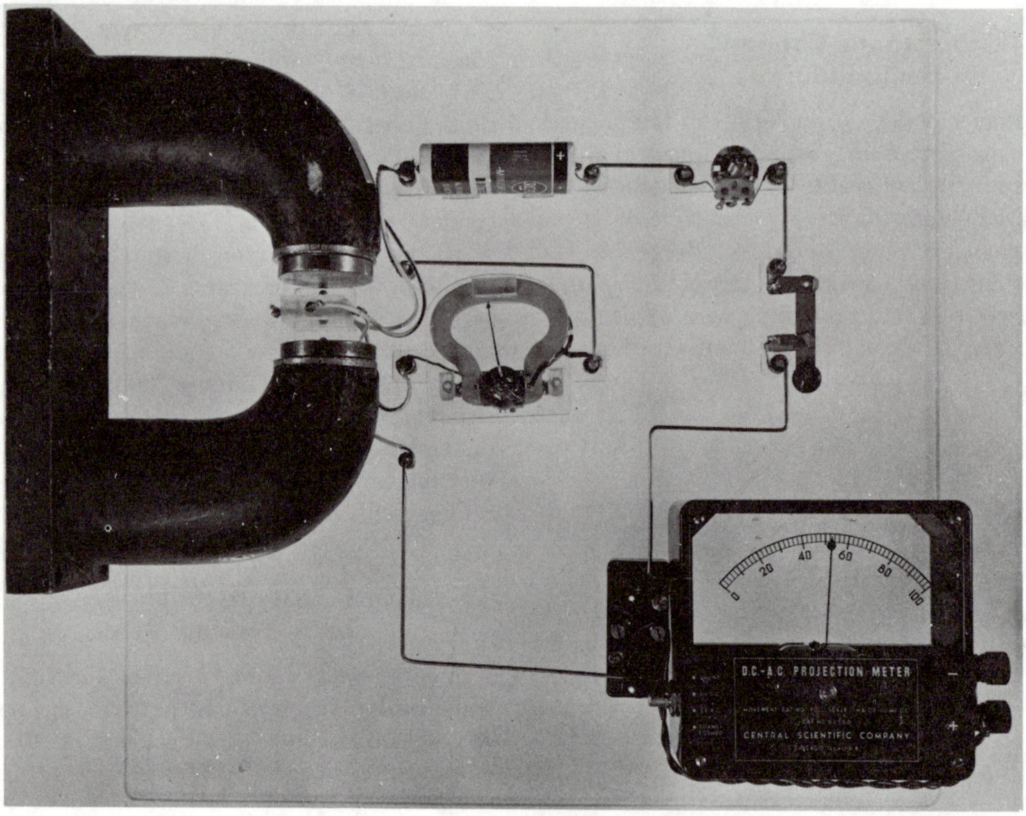

Fig. 40–34

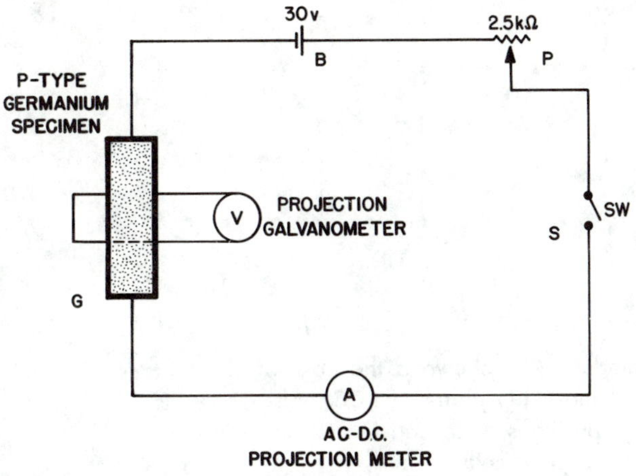

Fig. 40–35

germanium sample in a specially constructed mount.

A projection galvanometer is connected across the germanium sample and a magnetic field is applied in a direction perpendicular to the germanium strip by placing a magnetron magnet across the sample.

When lecturing to large groups of students, the demonstration is shadow-projected by means of an overhead projector. For this purpose it has been found best to first project and discuss a 10 x 10-in. transparency of the circuit diagram, shown in Fig. 40–35. The actual circuit is then superimposed and the projector refocused. The mount with the germanium sample is taken off the plastic base and shown in projection, making the probes visible to the students. After the mount has been replaced, the switch is closed and the potentiometer adjusted until the ammeter indicates a current of about 30 mA.

The magnet is now placed onto the projector so that the germanium mount is at the center of the gap. Several soft iron pole pieces can be used in the magnet gap to increase the magnetic field. The deflection of the galvanometer is noted—especially the direction. The polarity of the magnetic field is now reversed by rotating the magnet through 180 deg, and the deflection of the meter will be in the opposite direction.

If the directions of the magnetic field and of the control current are known, and the polarity of the galvanometer is indicated, the direction of the Hall voltage and the sign of the charge carriers can be determined. Because of the heavy drain on the miniature battery, the switch should not be left closed any longer than necessary. With the switch open, it can be shown that the effect of the magnetic field on the projection galvanometer is negligibly small.

When used for quantitative measurements in the laboratory,[27] the miniature battery is replaced by a 6- or 12-V power supply, and the galvanometer by a Leeds and Northrup student potentiometer. The student is thus able to prove that negative charge carriers are responsible for the current through a metal conductor, while either positive or negative charge carriers can be involved when a current flows through a semiconductor. In the laboratory the magnitude of the magnetic field can also be changed by inserting soft iron pole pieces into the gap of the magnet and calibrating by means of a gauss meter.

Construction details are given in the Appendix, page 1367.

40–1.17 Sphere Packing

Sphere packing models, in which "atoms" can be added or removed to reveal special features, can be constructed easily and are very useful. Each model consists of a board, rods mounted vertically on the board at predetermined locations, and a number of balls drilled through the center to fit the rods. For demonstrations, the balls should be 1½ to 2 in. in diam.

Crystal structure determines the locations of the rods. Body centered cubic crystals (BCC), Fig. 40–36 (a), have cubic unit cells with an added atom at the cell center. The unit cell size (lattice constant) is $a = (2\sqrt{3})D$, where D is the diameter of the balls. Hexagonal close packed crystals (HCP), Fig. 40–36 (b), have unit cells which are 120-deg equilateral parallelograms on the base and which have a third side of height $(\sqrt{8/3})a$ perpendicular to the base. For HCP, $a = D =$ side of parallelogram. There must be an inside atom which sits at one of two possible sites, as in (b). Thus, HCP can be obtained by piling close-packed planes

[27] See *The Taylor Manual of Advanced Undergraduate Experiments in Physics* (Addison-Wesley Publishing Co., Inc., Reading, Mass., 1959), pp. 453–457; R. Resnick and H. F. Meiners, *Demonstration and Laboratory Apparatus Report, 1960 Summer Visiting Professor Workshop,* Rensselaer Polytechnic Institute (1961), pp. 95–106; and R. G. Marcley, *Am. J. Phys.*, **29**, 29 (1961).

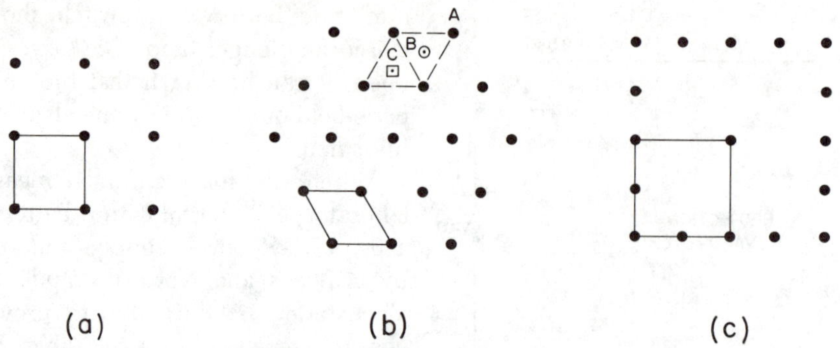

Fig. 40–36 (a) Rod positions for the BCC model. Also shown is the base of one unit cell. (b) Rod positions for the HCP model and the (111) plane of FCC. Also shown is the base of one unit cell and the close packing positions *A, B,* and *C.* (c) Rod positions for the FCC model. Also shown is the base of one unit cell.

(every atom is surrounded by six other atoms) in the order ABAB . . . or ACAC . . . or BCBC. . . .

Face centered cubic crystals (FCC), Fig. 40–36 (c), have unit cells which are cubic and have added atoms at the center of each face of the cube. The unit cell size a is $\sqrt{2}D$. The (111) planes are close-packed and have two sets of crystallographically different interstices, which give three positions (A, B, C) for the locations of these planes (see Fig. 40–36 (b)). It can be shown that the FCC can be obtained by piling (111) planes in the order ABCABC . . . or ACBACB. . . .

The crystal models should be demonstrated to the class by dropping the balls in place and building the crystal atom by atom. During this process, various planes can be built up, and body-centering and face-centering shown. In particular, the close-packed (111) planes of the FCC model should be noted. HCP should be built up in several ways, the tetrahedral and octahedral interstices pointed out and their sizes calculated. Finally, FCC can be constructed on the HCP model by using ABC stacking. The instructor should demonstrate that this is truly FCC by revealing a face-centered plane; however, this may be difficult and may take some practice.

40–1.18[28] Colored Overlays Showing Crystal Structure

While the single transparency is a substitute for slides or blackboards, the colored overlay uses techniques that could not be possible by any other means. A complicated picture or diagram may be built up in front of the students by using up to six or eight transparencies of different colors. With a basic transparency mounted in a frame, colored overlays can be flipped on or taken off from all four sides. In Fig. 40–37 a crystal model is shown.[29]

Fig. 40–37

(28) Walter Eppenstein, *The Overhead Projector in the Physics Lecture,* Department of Physics and Astronomy, Rensselaer Polytechnic Institute.

(29) Available from McGraw-Hill Text Films, New York, N.Y.

By means of overlays one atomic plane after another is projected. Up to four different sets of planes can be shown, flipped over from different sides. The lecturer can, of course, write on the transparency or indicate atomic distances. The planes appear in different colors for better visibility. Naturally, each overlay absorbs some light. For this reason, it is advisable not to exceed 6–8 overlays.

Demonstrations	***Contributors***
40–1.1	H. S. Story, State University of New York at Albany
40–1.2	The Royal Institution of Great Britain, London
40–1.3, 40–1.4	J. F. Gibbons, Stanford University
40–1.5	H. N. Hersh and L. Bronstein, Zenith Radio Corporation
40–1.6	Swiss Federal Institute (ETH), Zürich, Switzerland
40–1.7	J. F. Gibbons, Stanford University
40–1.8	Swiss Federal Institute (ETH), Zürich, Switzerland
40–1.9, 40–1.10, 40–1.11	J. F. Gibbons, Stanford University
40–1.12	R. W. Lefler, Purdue University
40–1.13, 40–1.14, 40–1.15	Swiss Federal Institute (ETH), Zürich, Switzerland
40–1.16	W. Eppenstein, Rensselaer Polytechnic Institute
40–1.17	L. Muldawer, Temple University
40–1.18	W. Eppenstein, Rensselaer Polytechnic Institute

41

NUCLEAR PHYSICS

41–1 Radioactivity

41–1.1

Radioactive contamination is demonstrated with a neutron source and a half dollar coin. First, a Geiger counter is used to test the half dollar for radioactivity (when the neutron source is shielded). The coin and neutron source are then placed together on a paraffin block for 30–60 sec. Then the neutron source is removed and shielded, and the coin is tested again and found to be radioactive. The following transformations are demonstrated:

$${}_{47}\mathrm{Ag}^{107} + {}_{0}n^{1} \longrightarrow {}_{47}\mathrm{Ag}^{108} \longrightarrow {}_{48}\mathrm{Cd}^{108} + {}_{-1}e^{0} + \nu$$

$${}_{47}\mathrm{Ag}^{109} + {}_{0}n^{1} \longrightarrow {}_{47}\mathrm{Ag}^{110} \longrightarrow {}_{48}\mathrm{Cd}^{110} + {}_{-1}e^{0} + \nu$$

41–1.2[1]

The height of water in a column discharging through a capillary filter can be used as a model for simple radioactive decay. Since many elementary physics students have no understanding of an exponential function, a simple physical case which quantitatively demonstrates the exponential behavior of radioactive decay is useful. The decreasing height h of the discharging water column as a function of time is given by $h = h_0 e^{-\gamma t}$. That this relationship holds can be shown as follows:

[1] K. Hecht, *School Sci. Rev.*, **41**, 439 (1960). *Note:* Demonstrations 41–1.2 through 41–1.6 and 41–2.11 were adapted from the contributions of Dr. K. Hecht, E. Leybold, Köin, West Germany.

The volume V of the water is

$$V(t) = Ah, \tag{1}$$

where A is the area. The rate of discharge i is

$$i = -\frac{dV}{dt} = -A\frac{dh}{dt}. \tag{2}$$

The minus sign arises because h decreases with time. From Poiseuilles equation,

$$i = \frac{\pi r^4}{8\mu l}\Delta P = K\rho g h, \tag{3}$$

where r is the radius of the tube, μ is the viscosity of the fluid, l is the length of tube, ΔP is the difference of pressure that equals ρgh, ρ is the density of the fluid, and K is $\pi r^4/8\mu l$. From Eqs. (2) and (3), we obtain

$$\frac{dh}{dt} = -\frac{K\rho g h}{A}, \tag{4}$$

and

$$h = h_0\, e^{-\gamma t}, \tag{5}$$

where γ is $r^2\rho g/8\mu l$, and h_0 is the initial height at $t = 0$. A plot of dh/dt vs h yields γ. With γ calculated from the plot, e can be experimentally determined.

The setup for this experiment is shown in Fig. 41–1, where a is the capillary filter tube,[2] b is a

[2] Leybold catalog No. 36170/74; also available from the LaPine Scientific Co.

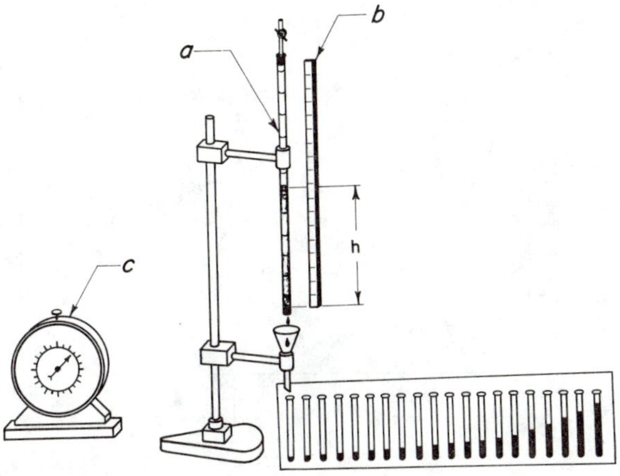

Fig. 41-1

meter stick, and c is a stop clock. Mount the capillary filter tube in such a way that the bottom surface of the filter coincides with the zero mark on the meter stick, and measure h from this surface. Close the upper end of the filter tube with a rubber stopper, into which a stopcock is inserted. The stopcock is opened at time = 0. The height h of the water column is recorded at equal time intervals. The water used for the experiment is carefully filtered and kept in a stoppered bottle before use.

A graduate placed under the filter tube to collect water represents the build-up of the daughter element. If h_G represents the height of the water in the graduate at time t, then

$$\frac{d_G{}^2}{d^2} h_G = h_0 - h = h_0 (1 - e^{-\gamma t}) \qquad (6)$$

where d_G and d are the diameter of the graduate and the efflux tube respectively. The experiment is particularly simple if $d_G = d$.

A somewhat more graphic demonstration can be made by successively catching the efflux in a series of 10–20 test tubes supported on a rack. The test tubes are moved into position, one after another, at equal time intervals. The time interval should be about 1/5 the half-value period (time required to reach $h_0/2$) for the filter tube selected. The water can be colored, and the test tubes shown against a white card.

A method used by P. Sprawls at Emory University to demonstrate radioactive decay is shown in Fig. 41-2. The apparatus can be used in either lecture or laboratory. Ordinary tap water drips from a standard Leybold No. 361-72 decay tube through two screens into a receptacle. The screen arrangement is analogous to a Geiger counter. Copper window screen is cemented over a 1½-in.-diam hole in a 4 x 4 x ⅛-in. clear plastic plate, which rests on another identical assembly. The two form a "sandwich" so that the water drips from the decay tube through the screens into the waste receptacle. The upper screen is at a potential of 500 V and the lower is grounded. These are connected to a Model No. 151 Nuclear-Chicago scaler.

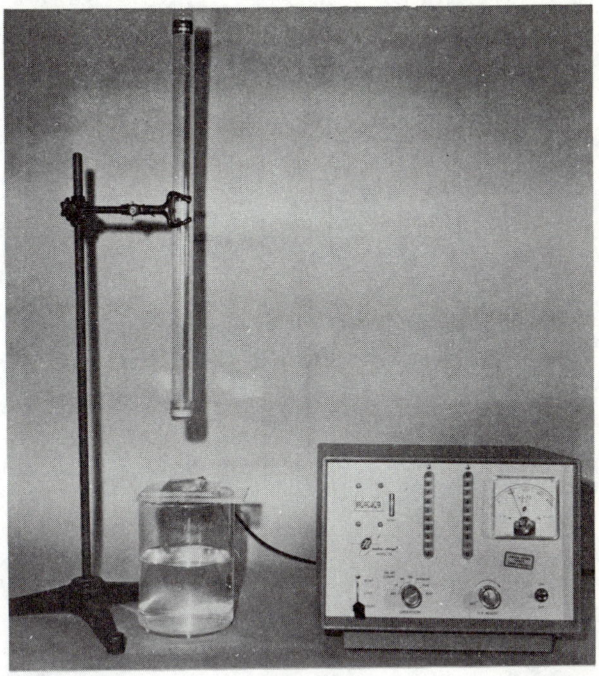

Fig. 41-2

The filter supplied with the decay tube is soaked with soapy water to increase the half-lifetime from 20 sec to about 2 hr for laboratory use, and from 20 sec to about 10 min for lecture. An X–Y plotter can be used to plot points in lecture if proper amplification is used. A current rate meter is attached to an amplifier such as Nuclear Corporation's Model No. 613A Classmaster and is adapted to the X–Y plotter.

41–1.3

A simple model for the simultaneous decay of a mixture of two radioactive isotopes can be constructed by a variation of the setup described in

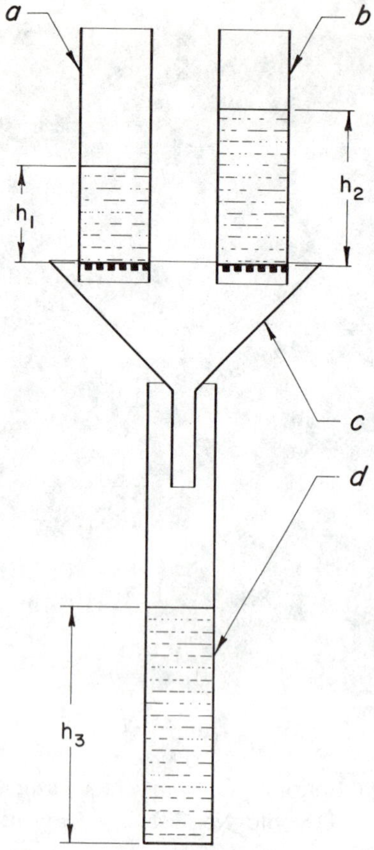

Fig. 41–3

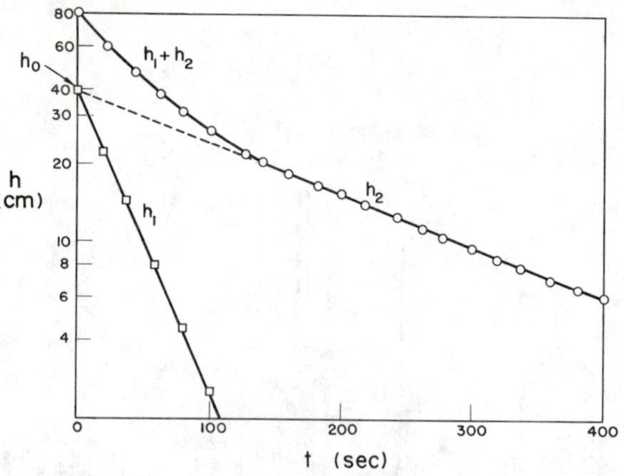

Fig. 41–4

Dem. 41–1.2. As shown in Fig. 41–3 two capillary filter tubes (Leybold 36174) a and b are mounted above a funnel c so that they discharge simultaneously into a single collecting tube d, identical in diameter to the two filter tubes. The collecting tube should be twice as long as the filter tubes. The filter tubes should have different efflux rates corresponding to two different half-lives. If h_1 and h_2 are the heights of water in the efflux tubes at some time t, h_0 is the initial water height in both tubes ($t = 0$), and h_3 is the height in the collector tube at time t. From Eq. (5) in Dem. 41–1.2,

$$h_1 = h_0 e^{-\gamma_1 t},$$

and

$$h_2 = h_0 e^{-\gamma_2 t}.$$

Therefore,

$$h_3 = h_0 - h_1 + h_0 - h_2 = 2h_0 - h_0 (e^{-\gamma_1 t} + e^{-\gamma_2 t}),$$

where $-h_1$, h_2, and h_3 must be read simultaneously. Note that $h_1 + h_2 = 2h_0 - h_3 = h_0 (e^{-\gamma_1 t} + e^{-\gamma_2 t})$.

If one plots $h_1 + h_2$ logarithmically against t as in Fig. 41–4, a single straight line is not formed, because of the two different values of γ. However, if the efflux tubes are so chosen that γ_2 is less than γ_1, the log curve eventually straightens out with a log slope equal to $-\gamma_2$, and at long times the behavior is dominated by the tube with the lower efflux rate (analog of long-lived isotope). The straight line position of the long-time curve now is extrapolated back to the intercept h_0 ($t = 0$). This curve (longtime straight position plus straight line extrapolation) represents the contribution to h_3 from the slower tube, which is $h_2 = f_2(t)$. The differences between the plotted curve for $h_1 + h_2$ and this curve gives $h_1 = f_1(t)$. γ_1 and γ_2 then can be evaluated from the two slopes, and the two half-lives can be computed for the tubes: $T_{1/2} = \ln 2/\gamma$.

41–1.4

As described in Demonstrations 41–1.2 and 41–1.3, it is possible to simulate various types of radioactive decay curves with models made from capillary filter tubes. A model for secular equilibrium can also be constructed. Water is supplied to a capillary filter tube (Leybold #36174) at a constant rate (as from a Mariotte Container). Because of the constant cross section of the tube, it is possible to express this rate of supply z in terms of the height h of the water column. If the filter tube is simultaneously discharging at rate v, then $dh/dt = z - v$. The experiment is begun with an empty filter tube; the water level arises at an almost constant rate since $v = \gamma h$ is small for small h. As h increases, v likewise increases until $z = v$. At this point equilibrium is reached and $h = h_\infty$. Since $dh/dt = z - \gamma h$,

$$h = \frac{z}{\gamma}(1 - e^{-\gamma t}) = h_\infty\ (1 - e^{-\gamma t}). \quad (1)$$

This model corresponds to secular equilibrium; the half-life of the daughter is much less than that of the parent.

It is best to use a filter tube of as short a half-life as possible; the limit is usually set by the maximum rate of constant flow z obtainable. Measure z, h_∞, and h at several intermediate t values and calculate γ from the relation $\gamma = z/h_\infty$.

In the model for transient equilibrium, the constant flow (Mariotte Container) is replaced by a capillary filter tube whose efflux is exponential. The rate of supply to the lower tube varies with functions of time and with $z = \gamma_1 h_1$. If h_1 and h_2 represent the water column heights in the upper and lower tubes respectively,

$$\frac{dh_2}{dt} = \gamma_1 h_1 - \gamma_2 h_2 = \gamma_1 h_0 e^{-\gamma_1 t} - \gamma_2 h_2. \quad (2)$$

From Eq. (2), we obtain

$$h_2(t) = h_0 \frac{\gamma_1}{\gamma_2 - \gamma_1}(e^{-\gamma_1 t} - e^{-\gamma_2 t}). \quad (3)$$

For the case of transient equilibrium $\gamma_2 > \gamma_1$; the daughter product is shorter lived than the parent. In carrying out the experiment, it is convenient to have measured γ_1 and γ_2 as in Demonstrations 41–1.2 and 41–1.3 and to measure h_1 and h_2 simultaneously as succeeding intervals of time. An important case of transient equilibrium is the decay of radium to RdTh.[3]

41–1.5

The model experiments on series disintegration that can be demonstrated with capillary filter tubes include the observation of the heights of the water columns in the various tubes, or their rates of discharge, when secular equilibrium exists between them. These values then correspond to the quantities and activities of the individual members of a radioactive series.

An example of this type of experiment can be demonstrated using the arrangement of Fig. 41–5. Three capillary filter tubes discharge successively into one another. To the uppermost one (tube 1) water is supplied at a constant rate from a constant level reservoir. This means that the parent of this series had a very long half-life period which may be considered infinitely long for practical purposes. Figure 41–6 shows a graphical representation of the results for an arrangement with $(t_{1/2})_1 = 22$ sec, $(t_{1/2})_2 = 124$ sec, and $(t_{1/2})_3 = 13$ sec. These show that the saturation height h_1 of the water column in tube 1 is attained after about 100 sec, $h_1(\infty) = z/\gamma_1$. The rate v_1 at which water is discharged from tube 1 then also assumes its saturation value $v_1(\infty)$. Obviously, this must equal the rate of supply, z:

$$v_1(\infty) = \gamma_1 h_1(\infty) = z.$$

The central tube 2 has a longer half-life of 124 sec. The curve representing the exponential increase of the height h_2 of the water column in tube 2 reaches saturation after the saturation height $h_1(\infty)$ in tube 1 has been attained. The saturation height $h_2(\infty)$ of the water column in tube 2 is attained after about 900 sec; it is much greater than that of the water columns in tubes 1 (and 3), for the saturation height $h_2(\infty) = z/\gamma_2 = z(t_{1/2})_2/\ln 2$ is directly pro-

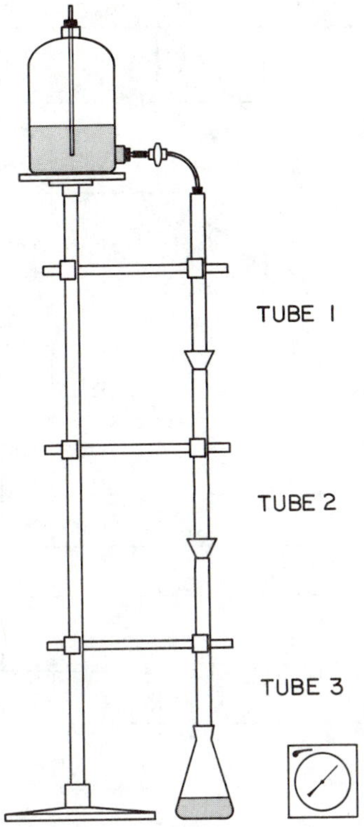

Fig. 41–5

[3] Robley D. Evans, *The Atomic Nucleus* (McGraw-Hill Book Co., Inc., New York, 1955), p. 482.

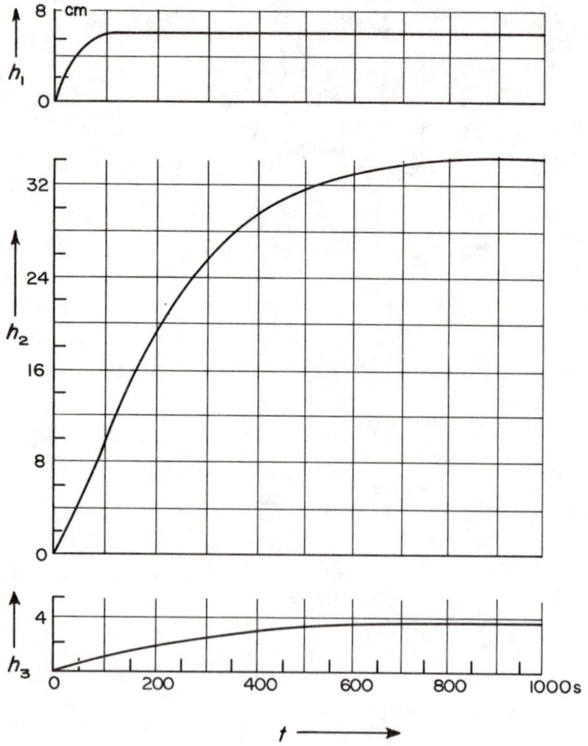

Fig. 41–6

portional to the half-life period. When the saturation height $h_2(\infty)$ has been attained, the rate of discharge $v_2(\infty)$ from tube 2 has also assumed its limiting value: $v_2(\infty) = \gamma_2 h_2(\infty) = z$. It must equal the rate of supply z, for once the saturation height of the water column has adjusted itself, that is, once a state of secular equilibrium exists between the heights of the water columns in the tubes, no water is lost or accumulated in any one of them.

The lowest tube 3 has a half-life of 13 sec. After as little as 60 sec, transient equilibrium is established between the height h_3 of its water column, and h_2. Notwithstanding its short half-life period, therefore, the saturation height $h_3(\infty)$ of the water column in tube 3 will not be attained earlier than $h_2(\infty)$, i.e., before about 900 sec. But because of the short half-life of tube 3, its value $h_3(\infty) = z/\gamma_3 = z(t_{1/2})_3/\ln 2$ is rather small. Once secular equilibrium exists between the heights of the water columns in the three tubes, the rate of discharge $v_3(\infty)$ from tube 3 will be the same as that from the other tubes, and equal to the rate of supply: $v_3(\infty) = \gamma_3 h_3(\infty) = z$.

Once, therefore, the longest-lived member of the series, which is tube 2 in this case, has attained its saturation height and its saturation rate of discharge, the water will pass all the tubes at a constant rate of flow z. In particular, we have as was explained above:

$$z = v_1(\infty) = v_2(\infty) = v_3(\infty).$$

Since, now,

$$v_1(\infty) = \gamma_1 h_1(\infty); \; v_2(\infty) = \gamma_2 h_2(\infty); \; v_3(\infty) = \gamma_3 h_3(\infty),$$

we have furthermore for this equilibrium state

$$\gamma_1 h_1(\infty) = \gamma_2 h_2(\infty) = \gamma_3 h_3(\infty)$$

or:

$$h_1(\infty) : h_2(\infty) : h_3(\infty) = \gamma_3 : \gamma_2 : \gamma_1 = (t_{1/2})_1 : (t_{1/2})_2 : (t_{1/2})_3.$$

This means: in the state of secular equilibrium the saturation heights of the water columns in the three tubes are directly proportional to the half-lives and inversely proportional to the decay constants of the tubes. The rates of discharge from all three tubes are equal at secular equilibrium.

These experiments illustrate the conditions prevailing in a radioactive decay chain where secular radioactive equilibrium exists between its members. If the height h_n of the water column in a capillary filter tube is taken to correspond to the available quantity of a nuclide, expressed say by the number N_n of its nuclei, then the rate of discharge $dh_n/dt = v_n$ can be compared to the activity of this nuclide, which is $dN_n/dt = A_n = \lambda_n N_n$. Once, therefore, the longest-lived daughter within the chain has built up its saturation ratio of nuclei and activity, the activity of all the nuclides present in the chain will be constant and equal to that of the parent A_0, assuming the latter to be infinitely long-lived. Therefore, we have

$$A_0 = A_1(\infty) = A_2(\infty) = A_3(\infty).$$

Since, on the other hand,

$$A_1(\infty) = \lambda_1 N_1(\infty); \quad A_2(\infty) = \lambda_2 N_2(\infty); \quad A_3(\infty) = \lambda_3 N_3(\infty),$$

we also have for this state of equilibrium

$$\lambda_1 N_1(\infty) = \lambda_2 N_2(\infty) = \lambda_3 N_3(\infty),$$

or

$$N_1(\infty) : N_2(\infty) : N_3(\infty) = (t_{1/2})_1 : (t_{1/2})_2 : (t_{1/2})_3 = \lambda_3 : \lambda_2 : \lambda_1$$

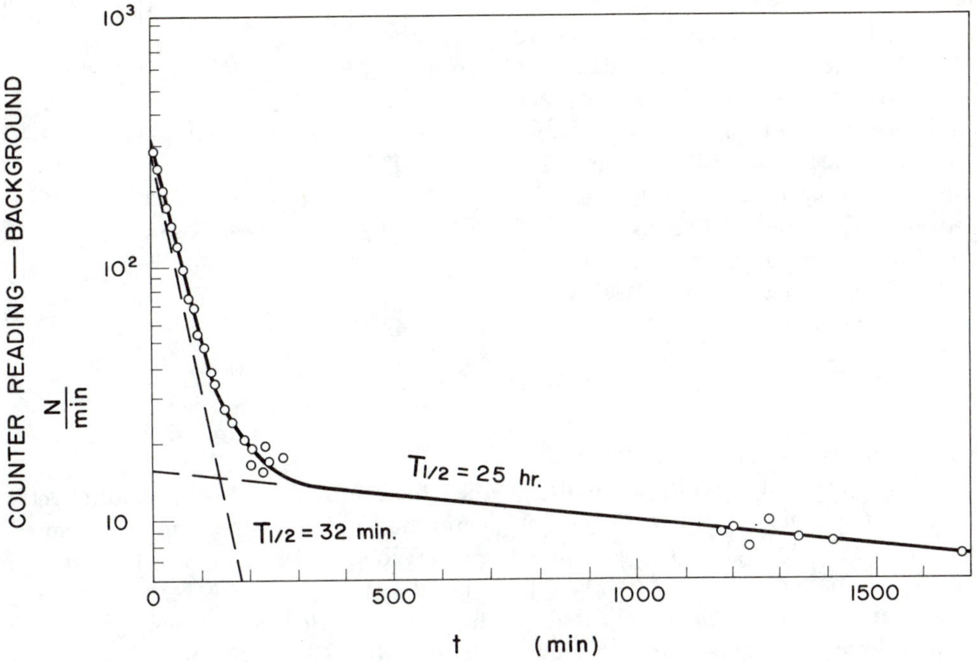

Fig. 41–7

This means: when (secular) radioactive equilibrium exists between the members of a radioactive decay chain, then the ratio of their numbers of atoms will be equal to the ratio of their half-lives and to the inverse ratio of their decay constants, and all the members of the chain will have the same activity.

In the case of a radium source, such as the Leybold plate with protected radioactive source, all the short-lived daughter products are in radioactive equilibrium with radium, and the activity of each of them equals that of Ra-226. But these daughter products are present in very different quantities. Thus a plate with 10 μg of Ra-226 contains about 6.5×10^{-5} μg of Em-222 (Rn), about 3×10^{-7} μg of Pb-214 (RaB) and about 3×10^{-14} μg of Po-214 (RaC′). But all these nuclides have 3.7×10^{-5} decays/sec each.

41–1.6[4]

Nearly half of the ions formed in air are due to the decay of ${}_{86}Rn^{222}$ (radon) and ${}_{86}Tn^{220}$ (thoron) and of their short-lived daughter products. These isotopes of a rare gas type emanation are "exhaled" from the soil into the atmosphere, and their daughter products are brought to the ground with rain or dust after having attached themselves to aerosols. The very small concentration of these substances in the atmosphere makes their detection difficult, and thus enrichment methods are necessary.

Enrichment is most easily carried out by attaching a large (12 x 12 cm) filter around the wire mesh part of a cylindrical intake tube, which can be attached to a vacuum pump. The paper, fastened with rubber bands, should cover the wire mesh completely, with no air gap. One can draw either fresh or building air through the pump. Depending on the building materials used, building air may contain more daughter products of thoron than fresh air. When a few cubic meters of air have been drawn through the pump, the filter can be removed from the mesh and wrapped around a thin glass-walled Geiger counter, with the contaminated surface next to the glass. The difference between the counter reading and the background reading should be plotted logarithmically against times in minutes. (See Fig. 41–7.)

In general, when carrying out the experiment, two active substances having half-lives of about 32 min and 10–20 h will be found. The log rate-time curves, therefore, will not be straight lines.

[4] K. Hecht, *Praxis,* **6**, 63 (1957), **8**, 64 (1959).

The activities can be correlated with the 26.8-min half-life of $_{82}Pb^{214}$, the daughter product of radon, and the 10.6-h half-life of $_{82}Pb^{212}$, the daughter product of thoron. Intermediate isotopes of very much shorter lives arising in both cases probably will not be detected. Moreover, the longer half-life measurement for thoron probably will not be accurate, because other radioactive isotopes of even longer half-lives also will be present and distort the results. Special means are necessary for their accurate measurement.

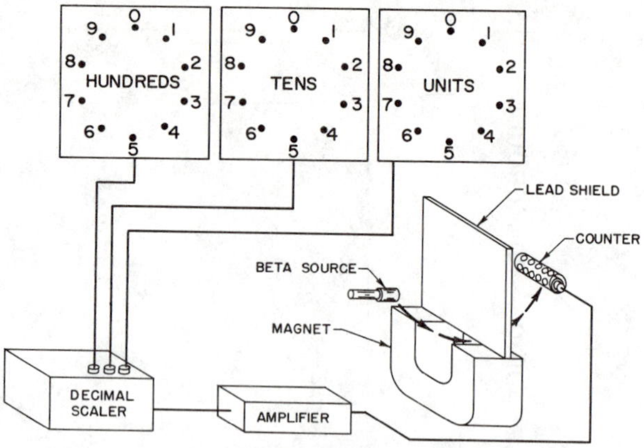

Fig. 41–8

41–1.7

This demonstration is a quantitative illustration of exponential absorption and provides an analogy to the attenuation of gamma rays in matter, or of a beam of atoms passing through gas at low pressure. A home-movie projector bulb (100 W) is mounted about 10 in. from a 929 phototube that is so housed as to exclude light from other directions. The phototube is connected in series with a battery and a projection meter (about 500 μA full scale). Small (1-in.) squares of neutral photographic filter (e.g., Wratten No. 96, N.D. 0.10) can be slipped into a simple frame between the light source and phototube. The class can be asked to record photocurrent as the number of filters is increased from zero to six or more. The data are plotted first on ordinary squared paper (or better, transparency with an overhead projector) and then in semi-log representation. The latter gives a good straight line, whereas the former gives an exponential curve.

41–1.8

The approximate straight line emission of electrons is shown at Princeton University with the arrangement in Fig. 41–8. A lead shield is interposed between a beta-ray source and a counter tube connected to an appropriate counting system. The counting equipment consists of the tube, an amplifier, a modified Berkeley decimal scaler, and a large display panel. The counting display panel consists of three ¼-in.-thick plywood boards 18 in. square, painted black and hinged together. Each unit of the panel has mounted on it 10 neon lamps, arranged at equal intervals on a circle 12 in. in diam and numbered consecutively 0 to 9. The flashing of the neon lights shows the counts in units, tens, and hundreds. The lead shield effectively intercepts the beta rays, resulting in a small counting rate. When a large Alnico horseshoe magnet is placed below the lead screen, as shown, the electron beam is deflected around the screen, passing between the poles of the horseshoe magnet. Hence, the counting rate is greatly increased. Turn the magnet around, north for south, and the counting rate drops to a small value. Change to a source of positrons, and again there is a large counting rate. The usual precautions must be taken for handling and storing the beta-ray source.

At the Swiss Federal Institute the arrangement shown in Fig. 41–9 is used. A 5-cm-thick lead block a is mounted between a particle counter b[5] and two radioactive sources c (a Na^{22} positron source and a radium electron source, each of about the same strength, no greater than 2 or 3 μCi) in such fashion that none of the particles emitted by either source will reach the counter by traveling in a straight line; i.e., they are absorbed in the lead. A strong permanent magnet d ($\sim$600 G) is placed over the lead block so electrons traveling through the gap will be deflected onto the counter and register. When the magnet is in a fixed position, counts are received from one source and not the other. If the direction of the B field is changed by 180 deg, counts are obtained from the source which previously did not produce counts.

41–1.9

Of the principal types of radioactive decay generally discussed in introductory courses in modern

(5) Philips Geiger Müller 4032.

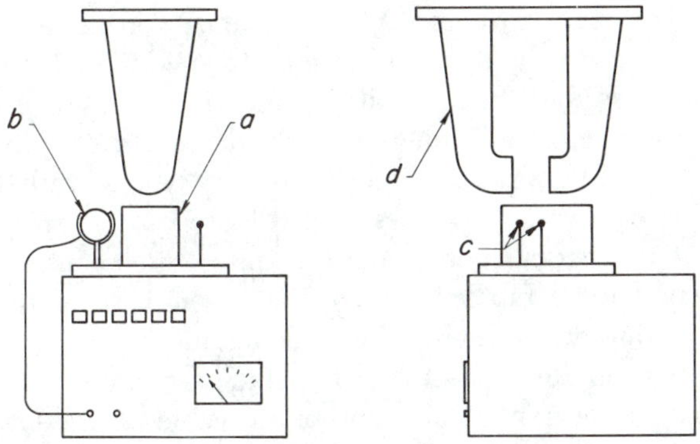

Fig. 41–9

physics, beta decay is unusual in several ways. The particles emitted do not have a sharp, well-defined energy as do those emitted in alpha and gamma decay, but rather a continuous spectrum of energies. Some beta emitters emit negative particles (electrons), some emit positive particles (positrons), some emit no charged particles at all (electron capture), and a few decay by two or even all three of these different modes. Some of these unusual features of beta decay can be demonstrated with the apparatus described in this article. In addition, the demonstration gives the student an introduction to the methods of detecting charged particles and to the important research tool of analyzing charged particle momentum by deflection in a magnetic field.

Experiments involving measurement of the beta spectra of the various radioactive isotopes have yielded a substantial part of what is currently known about the structure of the nucleus. Work is continually in progress in many research laboratories on problems concerning beta decay and its relation to current theories of nuclear structure and the nature of the nuclear force. For example, it was the continuous energy feature of beta spectra which eventually led to the postulating of the neutrino by W. Pauli in 1927.(6) The neutrino itself was not directly observed until 1959.(7)

The demonstration described is suitable for lecture groups as large as 300–400 students. Because of the "open-air" design of the beta ray spectrometer, the apparatus in Fig. 41–10 provides only qualitative measurements. This same reason makes it rather difficult to modify it for use as a quantitative laboratory experiment.

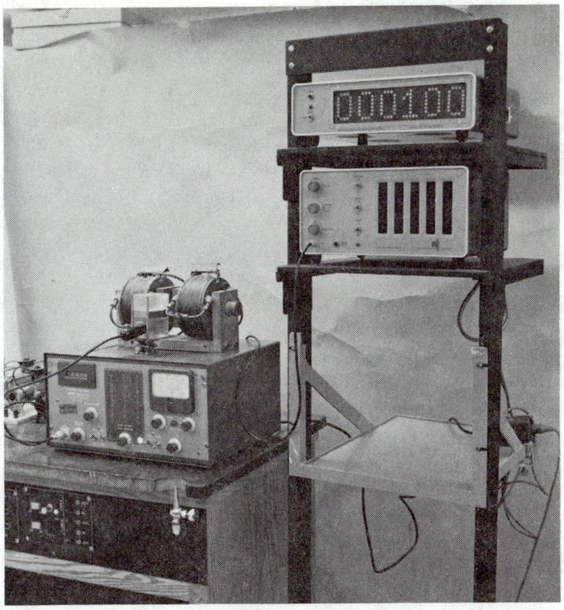

Fig. 41–10

Fig. 41–11

(6) The neutrino postulated by W. Pauli was included in E. Fermi's beta decay theory, which first appeared in Z. *Physik,* **88,** 161 (1934).

(7) F. Reines and C. L. Cowan, Jr., *Phys. Rev.,* **113,** 272 (1959).

Fig. 41–12

An important feature of this apparatus is that beta spectrum measurements are made without the usual necessity of enclosing the radioactive source and the beta particle detector in a vacuum system. The detector and source are mounted in a single clear plastic block between the flat pole faces of an electromagnet, as shown in Figs. 41–11, 41–12, and 41–13.

The magnet circuit (see Fig. 41–13) is arranged in the following way. Direct current is supplied to the center contacts of a DPDT knife switch, which is wired to permit reversal of the current direction through the circuit. In series with the coils are connected a center-zero projection ammeter, a quick-opening switch, and a group of four rheostats for controlling the coil currents. Specially designed switch mounts are used to permit expeditious circuit setup.

The beta detector circuit uses an end window Geiger-Müller tube, whose output pulses run a decade scaler with preset time mode of operation. The running total of accumulated counts stored in the scaler is displayed on an inline digital readout unit visible to the class.

After a preliminary discussion of radioactivity and a review of the effect of a uniform magnetic field on a motion of a charged particle, a transparent diagram of the apparatus arrangement (shown in Fig. 41–13) is projected with an overhead projector, and the general features and functions of the individual components are described.

The magnet is placed on the lecture table with the G–M tube window and the source facing the group. The switches are clamped to the edge of the lecture table, and each is identified with its counterpart in the projected diagram. The ammeter is projected with a projection stage-equipped arc lamp, so that the apparatus diagram is on the overhead projector throughout the demonstration for easy reference. The quick-opening switch, rather than the knife switch, should be used to interrupt the magnet current so that the knife switch contacts will not be burned by the arc accompanying the opening of the high inductance circuit.

To obtain a point on the beta spectrum, set the magnet current at an appropriate value on the low energy side and start the scaler (operating in preset time mode). As the count is being recorded, a transparent diagram of a set of axes for a plot of β^- (p) vs magnet current is placed on the overhead projector, and at the conclusion of the count the corresponding point is plotted. The magnet current is then increased and the process repeated for the second point; the sequence is continued until the counting rate decreases to about the background value. About six well spaced points are

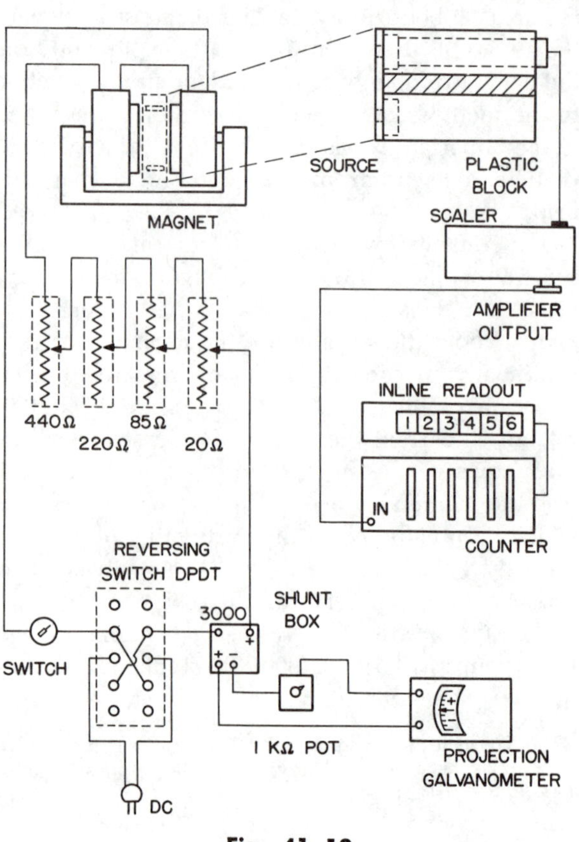

Fig. 41–13

Fig. 41–14

sufficient to obtain an adequate spectrum. The source should be strong enough so that counting times are not excessive; in practice, they should be kept in the range from 30 sec to 1 min. It is found that Sr^{90} is a good isotope to use because it decays with almost no accompanying gamma radiation. The daughter nucleus, Y^{90}, decays directly to the ground state of Zr^{90}, which is stable; therefore, no gamma background is introduced by this transition.

After completion of the beta spectrum, it is well to return to about the maximum of the curve, and to show, by reversing the direction of the magnet current, that the betas carry a negative charge. That the beta spectrum of a positron emitter is similar to that of a negatron emitter can be readily demonstrated by replacing the Sr^{90} source with Na^{22}. For construction details, see the Appendix, page 1370.

41–2 Nuclear Reactions

41–2.1[8]

Several properties of nuclear reactions can be shown with the overhead projection version of a $1/r$ surface shown in Fig. 41–14. A 12-in.-sq. black Formica frame surrounds an 8-in.-diam clear plastic $1/r$ surface. In the center of this surface is a 1¾-in.-diam well containing two loose-fitting brass disks, which provide for varying the depth of the well (or for completely filling it). Small steel balls (½ in. diam) are rolled down a Formica chute from different heights. The chute is mounted on the edge of the frame and pivots about a point located on a diameter of the surface.

For lecture demonstration, the balls are given low velocities and are reflected from the center. Next, the brass disks are removed from the center and the chute is directed toward the center. Balls are released fairly high on the chute to show capture of particles. With six or eight balls in a shallow well and the chute directed toward the center, a ball is released from the highest point on the chute. It is thus given sufficient energy to send some of the balls (particles) in the well flying. Rear projection on a vertical screen is used. Construction details are given in the Appendix, page 1372.

[8] See Demonstration 7–2.8.

41–2.2

The model (Fig. 41–15) employs a stream of steel balls, representing alpha particles, mechanically fed over an inclined chute *b* leading to a large glass tray *a* and impinging on an atom model (1) or (2). The particles experience little or no deflection when they strike a Thomson atom model (1) but exhibit Rutherford type scattering as they strike a nuclear model (2). An arc lamp beneath the tray projects the action onto a screen inclined to the ceiling or on the ceiling itself.

The Thomson model (1) is a Celluloid disk *d* shaped to form a mound. Electrons are simulated by about a dozen randomly oriented 3⁄16-in.-diam steel balls, which are held in place by wax. The nuclear atom model (2) consists of a large clear plastic board *e* and a random array of thin carpenter nails projecting vertically from the bottom which represents nuclei. The ball feed mechanism[9] *c*

[9] R. G. Marcley, *Am. J. Phys.*, **28**, pp. 666–9 (1960); R. G. Marcley, *Apparatus Drawings Project* (Plenum Press, New York, 1962), Project No. 9, pp. 69–75.

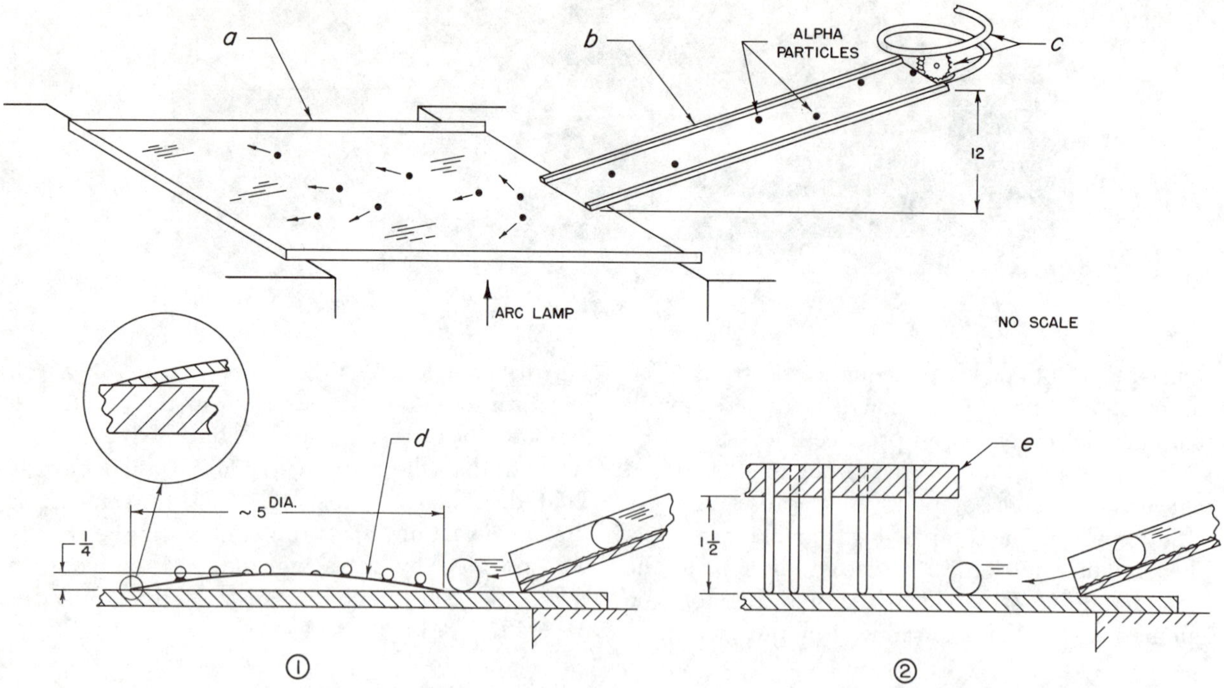

Fig. 41–15

consists of a motor driven toothed wheel arranged to feed ½-in.-diam steel balls from a helical tubular reservoir to the sloping chute. To randomize the paths of the balls, the chute is covered with crinkled aluminum foil.

The Celluloid disks for the Thomson model are formed by placing a thin flat blank over a circular opening and applying heat until the center sags about ¼ in., thus forming a watch glass shape. Beveling the edges allows a ½-in. ball to ride smoothly up and over the disks.

41–2.3

An electromagnet is suspended from the ceiling by a long aluminum pole which pivots at its upper end. The lower end of this swinging magnet represents an alpha particle. Below this, and stationary, is a much larger electromagnet that has a single loose pole piece painted yellow to represent the nucleus of a gold atom.

When the pendulum is swung at small amplitudes (low velocities) over the nucleus, the large scattering of slow speed alpha particles is illustrated. Large amplitudes (high velocities) of the electromagnet demonstrate that high speed alpha particles show small scattering except at very close approach. Professor Eric M. Rogers while at Cambridge saw Rutherford perform this demonstration in his undergraduate lectures (Private communication to the Editor).

41–2.4[10]

The so-called "chemical heart" is used to demonstrate a vibratory motion which is analogous to the oscillations an excited nucleus might undergo. It is made[11] by placing a pool of mercury about ¾ in. in diam in a clean 4-in. watch glass. The mercury is covered with 6 M H_2SO_4 to which is added 1 ml of 0.1 M $K_2Cr_2O_7$. The watch glass should be tilted slightly so that the mercury will not rest in the exact center. A clean iron nail or iron wire, which should be filed clean before each use, is placed radially in the watch glass in such a way that the mercury will just rest against the tip of the iron.

From 0.5 to 2 ml of 18 M H_2SO_4 is now added slowly, drop by drop, to the mercury. The mer-

(10) G. F. Powers and J. E. Smith, *Am. J. Phys.*, **28**, 561 (1960).

(11) J. A. Campbell, *J. Chem. Educ.*, **34**, 105 (1957).

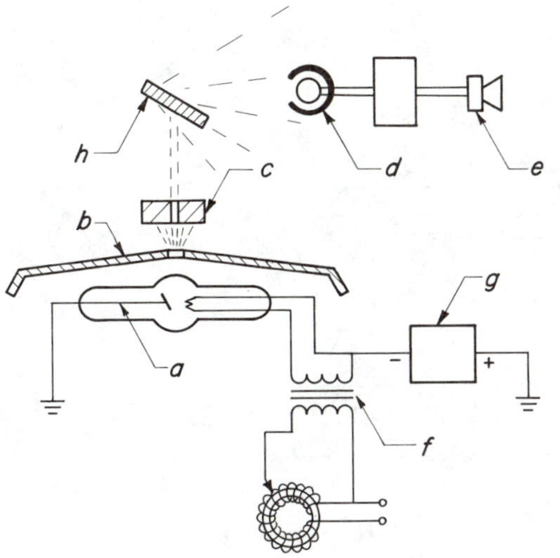

Fig. 41–16

cury will soon begin to quiver, and if the proper adjustment is made with respect to the amount of "tilt" of the glass, the mercury will begin to pulsate. When this occurs, no more acid is added.

As will be obvious when the system is adjusted for optimum conditions, the mercury, upon touching the iron, tends to contract into a tighter, more spherical shape. In so doing, contact is broken with the iron and the mercury then "relaxes" into a flatter shape, touching the iron again. This cycle is repeated. Although a triangle is the most common shape, a five leaved rosette vibrating mode can also be obtained. This demonstration can be performed on an overhead projector.

A similar experiment was contributed by W. Kundig, University of California. The Editor observed the experiment performed at Los Angeles. It works very well.

41–2.5

The apparatus in Figs. 41–16 and 41–17 demonstrates the scattering of 40-kV X rays by paraffin. An X-ray tube *a* is screened by a 2-mm lead shield *b*. The collimator *c* is 50-mm thick with a hole 12-mm in diam. (The rotating diaphragm in the picture is not used in this experiment.) A Geiger–Müller counter, shielded by a 4-mm brass tube with a 20-mm diam hole, is placed at *d*. The cathode of the X-ray tube is connected to a high voltage supply *g* and heated through an isolation transformer *f* until the current is 1 mA at 40 kV. Until the 4 x 4 x 1-in. paraffin block *h* is inserted, the counting rate heard on the loudspeaker *e* is very slow. Upon insertion of the paraffin, the rate increases by a factor of ten.

Fig. 41–17

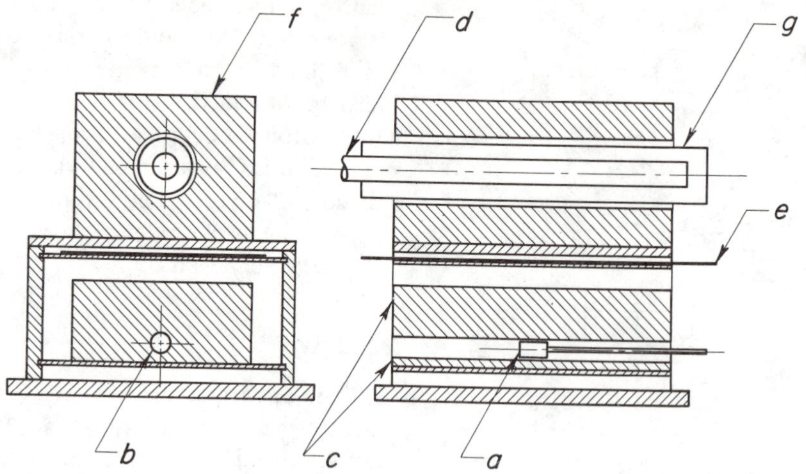

Fig. 41–18

41–2.6

The equipment in Fig. 41–18 used at the University of Basel to show moderation of fast neutrons in paraffin yields a mixture of thermal and fast neutrons. Fast neutrons show marked differences in their effects from slow neutrons, which only induce reactions with positive energy values (the operation of thermal uranium reactors depends on slow neutron action). As opposed to the slow neutron case, fast neutrons can produce reactions with negative energy values. Many nuclei possess capture cross sections sufficient for slow neutrons, but not for fast neutrons.

Suitable neutron sources for the demonstration are: Pu–Be, Ra–Be, D(d,n)^{3}He, and T(d,n)^{4}He. The source *a* is inserted in the hole *b* in the paraffin block *c* which slows down the neutrons. The boron counter *d* can be shielded from these slow neutrons by a cadmium sheet *e* (1 mm thick). The unmoderated fast neutrons that emerge from the cadmium are slowed by a second paraffin block *f*. A cadmium cylinder *g* with a wall thickness of 1 mm also can be used to shield the counter.

If the source is placed near the boron counter, it will show a few counts since the ^{10}B(n,α)^7 Li reaction cross section is too low for the fast neutrons. When the source is placed in the hole *b* in the paraffin block *c* and the counter placed near the block, it will record a large number of slow neutrons. The counting rate again sinks to a low value if the cadmium sheet is used to screen the counter from the slow neutrons emerging from block *c*.

When the counter is placed in the second paraffin block *f*, the counting rate increases sharply because the fast neutrons coming from the first block are slowed. The counter again returns to a low reading when it is placed in the cadmium cylinder *g*.

When D(d,n)^{3}He or T(d,n)^{4}He reactions are used to produce the neutrons, the paraffin block *c* can be replaced by the container of water. The counter then is placed below the target.

At the Swiss Federal Institute the apparatus shown in Fig. 41–19 is also used to demonstrate the difference between fast and slow neutrons for absorption and detection, as well as how neutrons can be thermalized. A Ra–Be fast neutron source

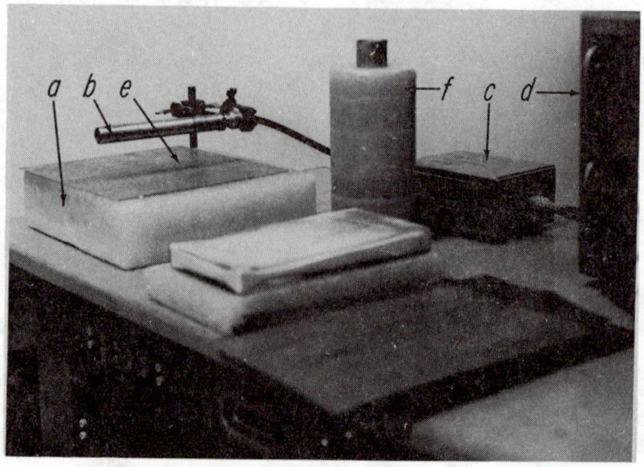

Fig. 41–19

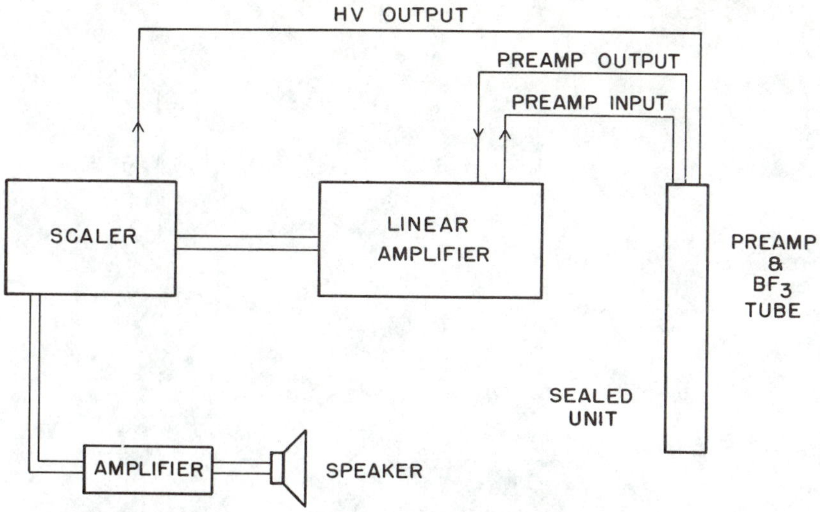

Fig. 41–20

(about 300 mg, 4.4×10^6 n/sec) is placed in a hole in the block of paraffin *a* that is about as deep as half the thickness of the paraffin block. The paraffin serves as a moderator for the fast source neutrons (energy of a few MeV). A BF_3 G-M counter *b* (enriched B^{10} Philips ZP 1000)[12] sensitive only to slow neutrons is mounted about two inches above the top of the block. In order to work the counter, a preamplifier *c* (Philips Type 4071) and a scaler *d* (Philips Type PW 4032) with a high voltage supply (Philips Type 4022, at least 2000 V) is needed. The pulses are made audible by the click box described in the Appendix to Demonstration 41–2.10 (Fission Chamber).

Different absorbers *e* such as lead, copper, or cadmium sheets or boron-plastic or borax plates; and different moderators such as water, paraffin sheets, a paraffin tube *f* or wood are used.

At Princeton University the same demonstration is shown using an audioamplifier or an oscilloscope. The source used is Ra–Be with a strength of 20–25 mCi or, much safer, a Pu–Be source. A block diagram of the counting circuit is shown in Fig. 41–20. A BF_3 counter tube with its preamplifier is mounted on rubber cushions in a 25-in.-long brass tube as in Fig. 41–21 and connected to a power supply, discriminator, scaler, and loudspeaker. The details of the BF_3 counter tube and the circuit for the preamplifier are shown in the Appendix, page 1372. Neutrons from the source are moderated successively by a paraffin block, a cadmium plate, and lead sheet, and a tank of water. If the demonstration is shown on an oscilloscope, a piece of black tape on the screen is used to blank out the glare from the time base trace. Note that the source must be handled by authorized personnel only and proper measures taken in handling and storing it.

(12) From Philips AG., Holland.

41–2.7[13]

The Mössbauer effect is a quantum mechanical phenomenon not explainable from classical principles. However, Frauenfelder has pointed out that two of the primary features of the effect can be effectively illustrated with a simple mechanical analogy. They are the following:

1. The sharpness of the Mössbauer line corresponding to the absence of Doppler broadening, and

2. The unshifted energy resulting from recoilless emission.

These can be demonstrated with a model "gamma ray emitting nucleus" consisting of specially suspended gun firing steel balls. Such a model, shown in the accompanying photographs and diagrams, is described below.

(13) W. Eppenstein and R. Resnick, *Report on the Workshop for the Development of Apparatus for College Physics at Rensselaer Polytechnic Institute, 1964–1965,* March 1966, pp. 61–66.

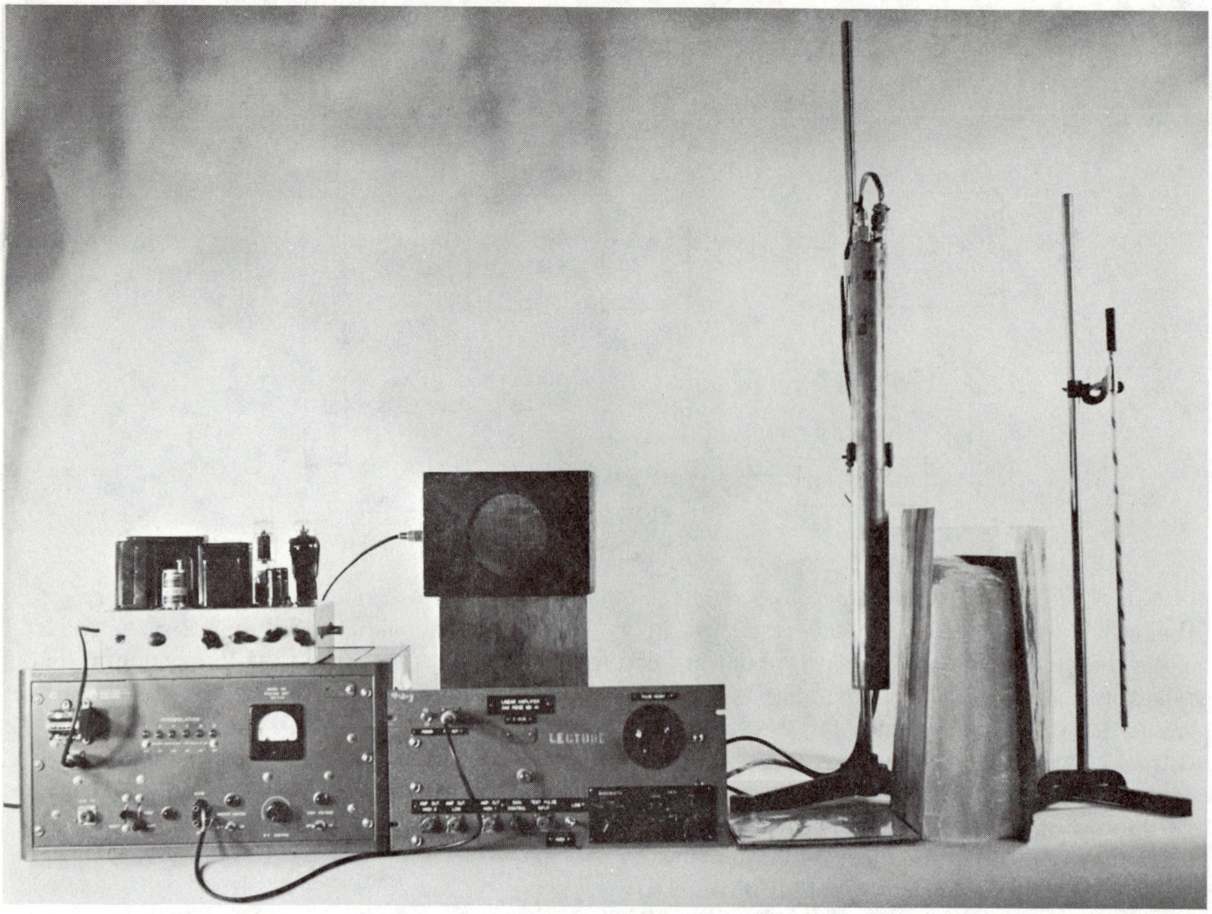

Fig. 41–21

Photographs of the apparatus in both its configurations, i.e., bound and unbound, are shown in Figs. 41–22 and 41–23. The apparatus is basically a light-weight spring gun with a carefully made spring and a trigger that can be activated with negligible effect on the gun. The gun has two mounting configurations: (1) suspended so as to recoil freely, and (2) clamped rigidly to the equipment base. The projectile energy is recorded with a strip of pressure sensitive spark tape in the catcher shown in Fig. 41–24.

The analogy is based on the fact that the nearly complete separation between the momentum transfer and energy transfer that occurs in the Mössbauer effect also occurs in several classical situations. In particular, a mechanical gun rigidly mounted in an assembly fastened to the earth corresponds to a nucleus bound in a solid, and a projectile fired by the gun plays the role of a gamma ray emitted by the bound nucleus. In the nuclear case the recoil momentum associated with the gamma emission is taken up by the solid as a whole. An exactly equivalent transfer accompanies the firing of a projectile from the gun, the recoil momentum being taken up by the gun mount and the earth.

The energy transfer in the Mössbauer effect is more complicated, the Mössbauer line being emitted when the lattice remains in its initial state after the transition. The mechanical analogy is not direct at this point; however, it is true that the "transition energy" of the bound gun system (i.e., the potential energy of the firing system) is transferred almost entirely to kinetic energy of the projectile. The kinetic energy of the projectile can be inferred from a measurement of its range. A negligible amount is transferred to the mount, a small amount is emitted as sound, and the remainder is lost due to friction in the firing system.

Fig. 41–22

If the gun is free to recoil, then the momentum transferred to the gun results in a significant recoil velocity and a significant recoil energy. Thus, the "transition energy" is divided between the projectile and the gun with small amounts again appearing as sound and frictional loss. Clearly then, if the emitting system is permitted to recoil, the energy of the emitted particle is shifted to a lower value than it would have had had it been emitted from a system whose state was unchanged by the transition. The range of the projectile is correspondingly shortened. This is illustrated by the following familiar calculation:

$$E_T = E_p + E_R,$$

where E_T is the transition energy, E_p is the kinetic energy of the projectile, and E_R is the recoil energy. From the conservation of momentum, the relationship

$$v_R^2 = \frac{v_p^2\, m_p^2}{M^2}$$

can be derived, where v_R is the recoil velocity, v_p is the projectile's velocity, m_p is the mass of the projectile, and M is the mass of the system without the projectile. Since $E_R = \frac{1}{2} M v_R^2$, the transition energy is given by

$$E_T = E_p\left(1 + \frac{m_p}{M}\right),$$

where the relationship $E_p = \frac{1}{2} m_p\, v_p^2$ has also been used. If the gun is clamped rigidly to the base, then M is very large and the projectile kinetic energy E_p is essentially equal to the transition energy E_T. If the gun is free to recoil, then M is

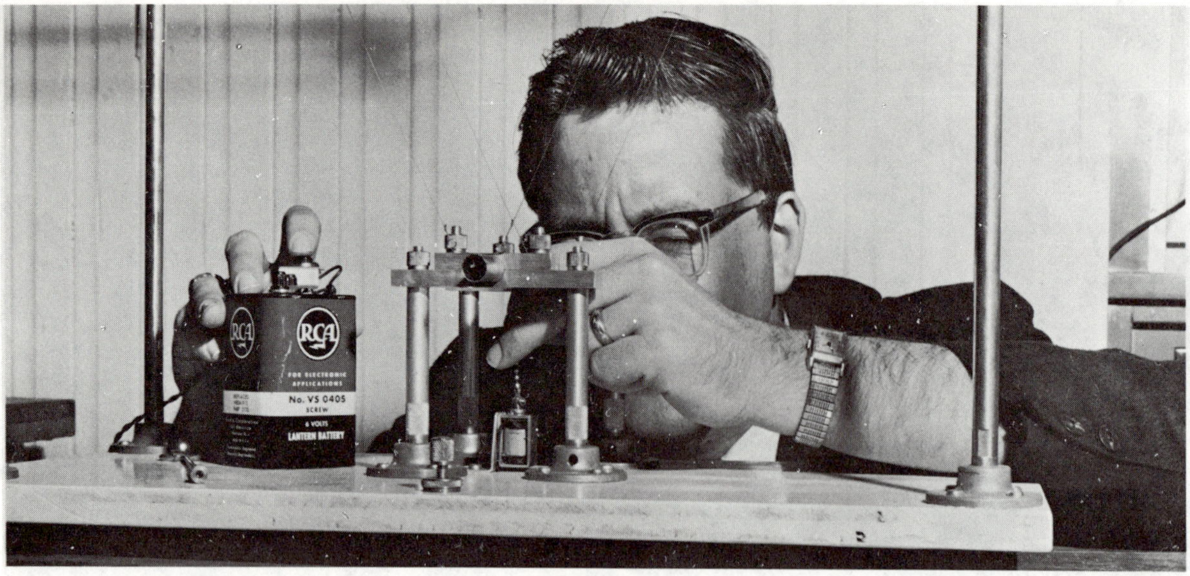

Fig. 41–23

the mass of the gun. The ratio $m_p/M = 0.2$ for the gun described in this article.

A gamma-ray line has a natural width determined by the life-time of the transition. The kinetic energy of the projectiles in the mechanical analog also displays a natural breadth due to slight differences in frictional effects. In addition, the Doppler broadening present in all but Mössbauer lines can be introduced in the analogy by imposing a small vertical vibration to the gun muzzle. Both of these effects cause a corresponding spread to the range values.

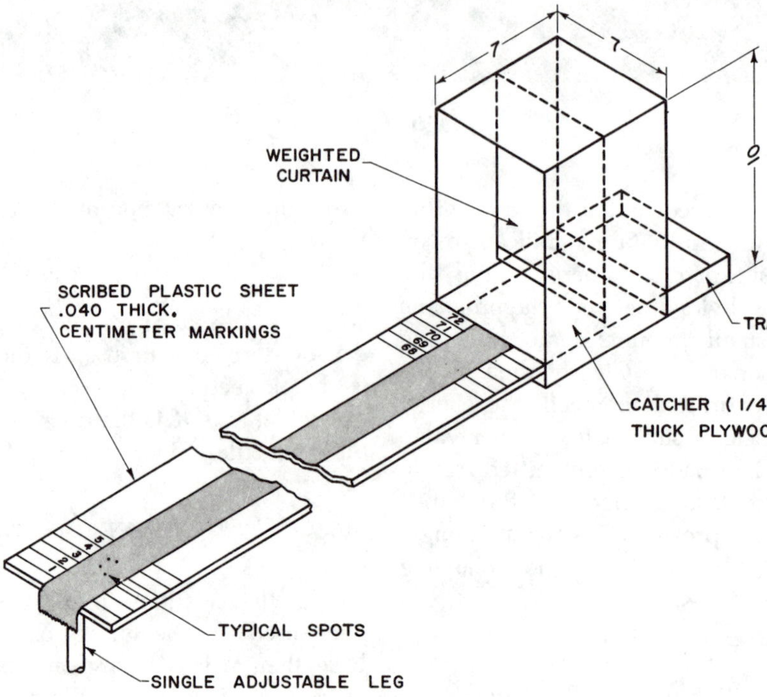

Fig. 41–24

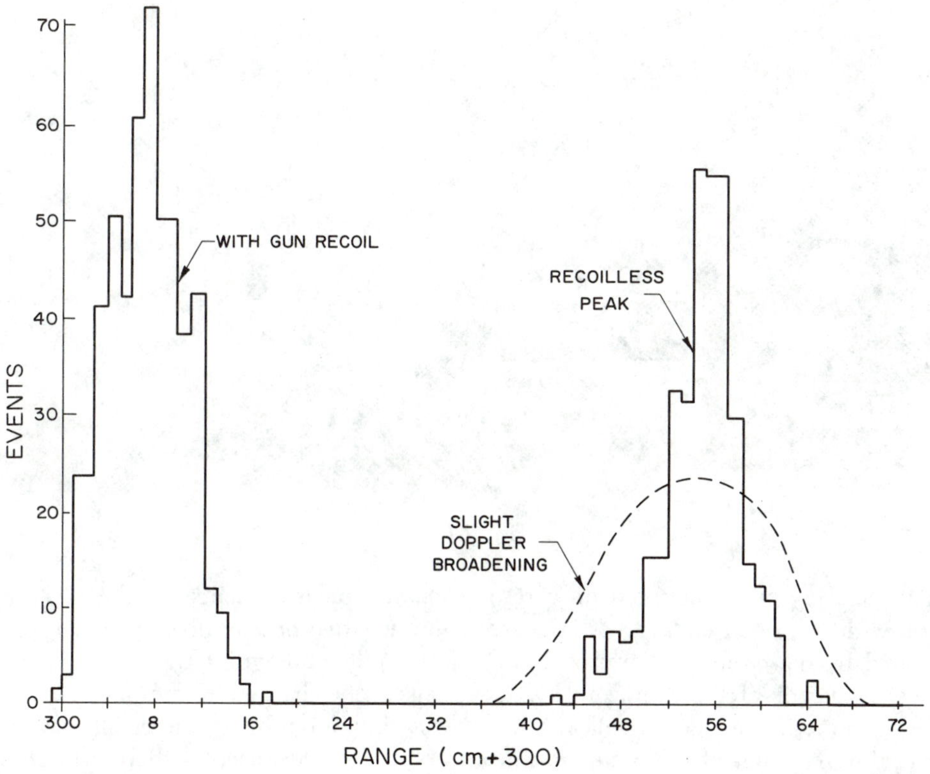

Fig. 41–25

Figure 41–25 shows the results of a typical series of runs. Events vs range number are plotted. The range number is in centimeters and indicates the projectile range in excess of 3 m. The left peak is that corresponding to gun recoil, the right peak to recoilless emission. Both are without Doppler broadening. There are approximately 600 counts in each peak of this particular data set. The observed peak separation, 48 cm, agrees with the prediction of classical mechanics assuming the projectiles in the right peak are emitted without recoil.

The broken line centering on the right peak shows the result of a slight Doppler broadening, i.e. a small vertical motion of the gun muzzle. A vertical amplitude of a few millimeters is sufficient to broaden the peak several centimeters.

The analogy to the Mössbauer effect is, of course, the principal usefulness. The demonstration could fill an important spot in introducing students to the effect. However, it does have the disadvantage that making full use of it would require considerable time, more than is generally available for lecture demonstration. For this reason plans are being made to film the analogy experiment together with a demonstration of the Mössbauer effect. The film would, of course, be generally available. A student laboratory group could readily perform the principal features of the mechanical analogy experiment and thus gain some feel for the phenomena of recoilless emission and Doppler broadening.

The gun used in the apparatus is itself a useful piece of gear since it is more carefully designed than commercially available spring guns, a fact leading to excellent data reproducibility. In particular, the conservation of momentum could be illustrated much better in the introductory laboratory than with current spring gun devices but probably not as conveniently as with air tracks. For construction details see the Appendix, page 1373.

41–2.8

The apparatus in Fig. 41–26 can be used to demonstrate sympathetic or induced vibrations. An audio oscillator *a* is connected to a 10-W amplifier *b* and an electromagnet *c*. The electromagnet is an ordinary headphone magnet from which the metal diaphragm has been removed. The tuning fork *d*,

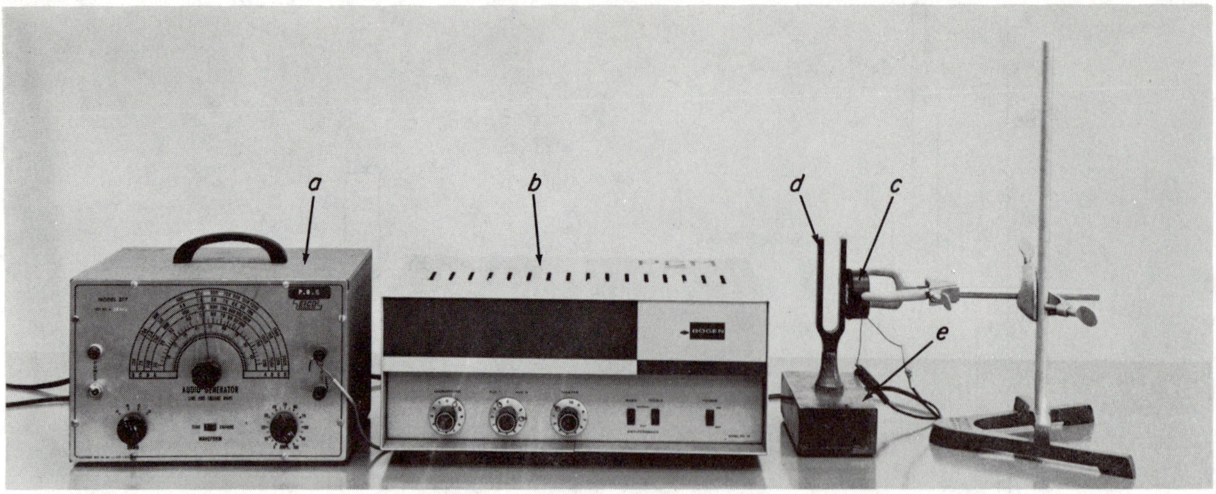

Fig. 41–26

which resonates at 512 cps, is mounted on a resonance cavity e with one open end. As ω (the frequency generated by the oscillator) approaches ω_0 (the resonance frequency of the tuning fork), the tuning fork begins to resonate. For significant resonance to occur, the frequency of the oscillator must be roughly in the range

$$\omega \approx \omega_0 \pm \frac{1}{\tau},$$

where τ is the time it takes the amplitude of the tuning fork to dampen to $1/e$ (roughly 1/3) of its initial value (in free vibration). By using tuning forks with different amounts of damping, the above formula for the sharpness of tuning can be verified.

This demonstration can be used to exemplify the phenomenon of resonance absorption of light by atoms and the resonance absorption of γ rays by nuclei. The extremely fine tuning of resonance absorption indicates why certain sharp γ rays which have been detuned slightly by the recoil of the emitting nucleus or by thermal Doppler broadening are not within the absorption range of similar nuclei. The Mössbauer effect which prevents recoils limits the detuning so that reabsorption can occur.

41–2.9

The method employed to determine the rest mass of a Λ° particle (hyperon) from bubble-chamber tracks can be demonstrated to a class by making measurements on the projected image of the negative of a bubble-chamber photograph. To obtain the photographs, a beam of very high energy, negatively charged K mesons is allowed to penetrate into a bubble chamber filled with a mixture of organic liquids, so that there are large numbers of the more ordinary constituents of matter (e.g. protons, neutrons, electrons) present in the chamber. These K^- mesons combine with the protons and neutrons to produce pions and uncharged Λ° hyperons in the following reactions:

$$K^- + p \rightarrow \Lambda^\circ + \pi^\circ, \qquad (1)$$

$$K^- + n \rightarrow \Lambda^\circ + \pi^-. \qquad (2)$$

Since the Λ° is uncharged, its track in the chamber will not be visible. In a very short time, however, the Λ° decays into a proton and a π^- meson according to the reaction

$$\Lambda^\circ \rightarrow p + \pi^-. \qquad (3)$$

Since both p and π^- are charged, their tracks will be visible in the chamber.

Figure 41–27 represents a typical series of events. Here the K^- tracks are easily identified since all the K^- mesons have the same velocity, and therefore their tracks all have the same curvature in the magnetic field which is applied to the chamber. At point A, a K^- combines with either a proton or a neutron producing a Λ° particle which travels to B where it decays. Since the decay products are oppositely charged, their tracks will curve in opposite directions.

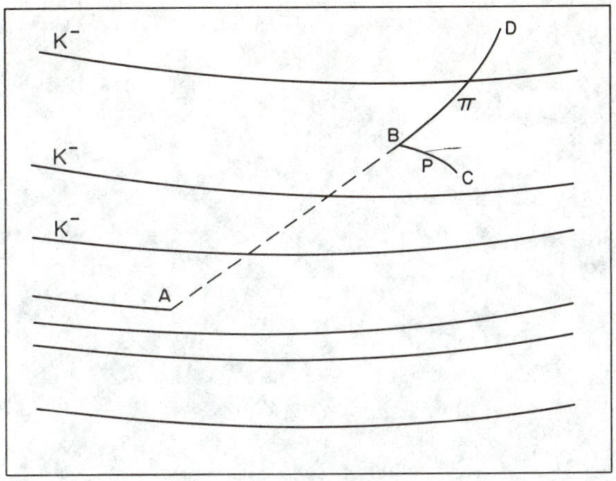

Fig. 41–27

The p and the π^- gradually lose their kinetic energies in collisions with the molecules of the liquid in the chamber, and will finally come to rest at points C and D respectively. The track length, or range, for any given particle moving in any particular medium is a well-defined function of the initial energy of the particle. By measuring the range, the energy can be obtained whenever previously determined range-vs-energy curves are available.

Figures 41–28, 41–29, and 41–30 are examples of carefully selected bubble chamber photographs in which the Λ^0 particles travel in the plane of the photo. In Figs. 41–28 and 41–29, the Λ^0 particle was produced when a K^- meson reacted with a proton, and in Fig. 41–30 the reaction was with a neutron.

In analyzing the photos (or projected images on the screen), the first step is to draw a straight line representing the path of the Λ^0. Then make the best estimate of the initial directions of the velocities of the pion and proton, and draw straight lines

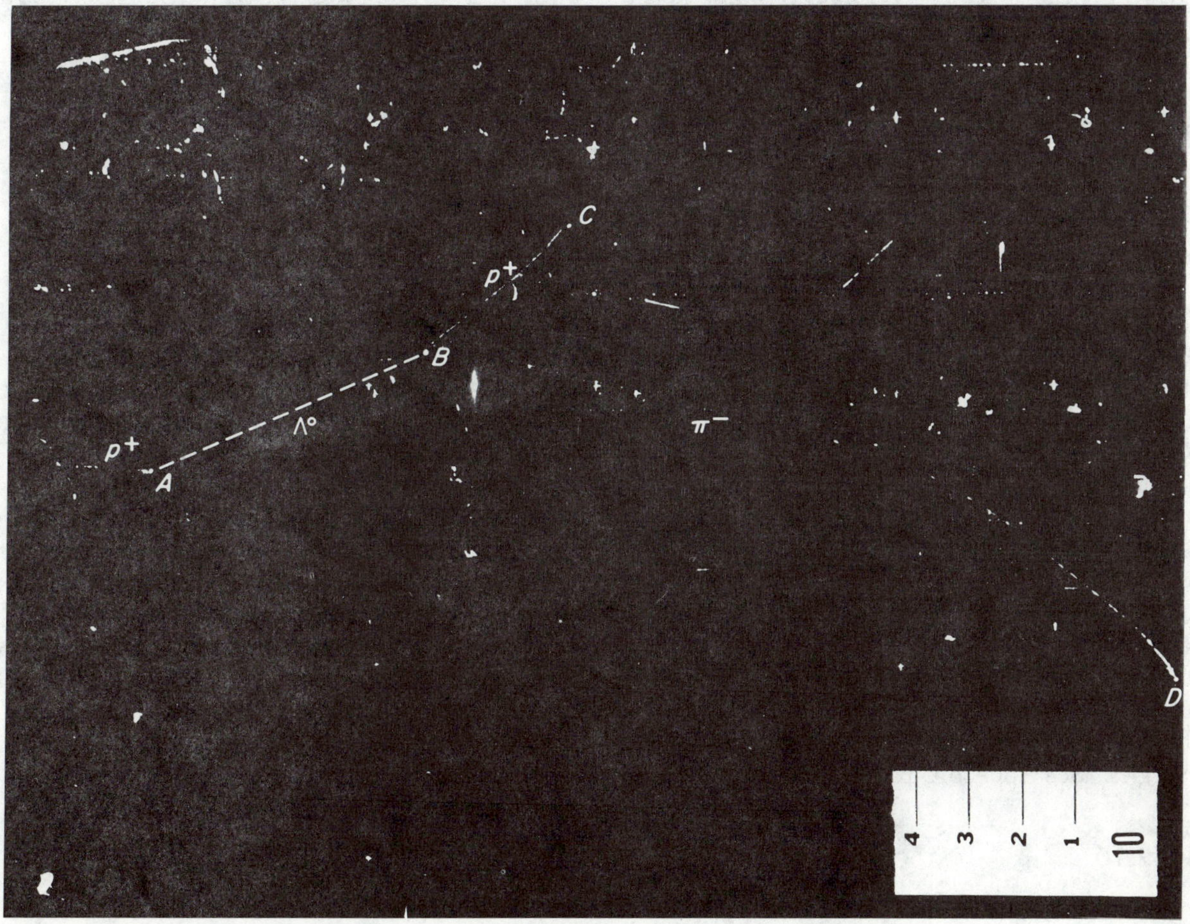

Fig. 41–28

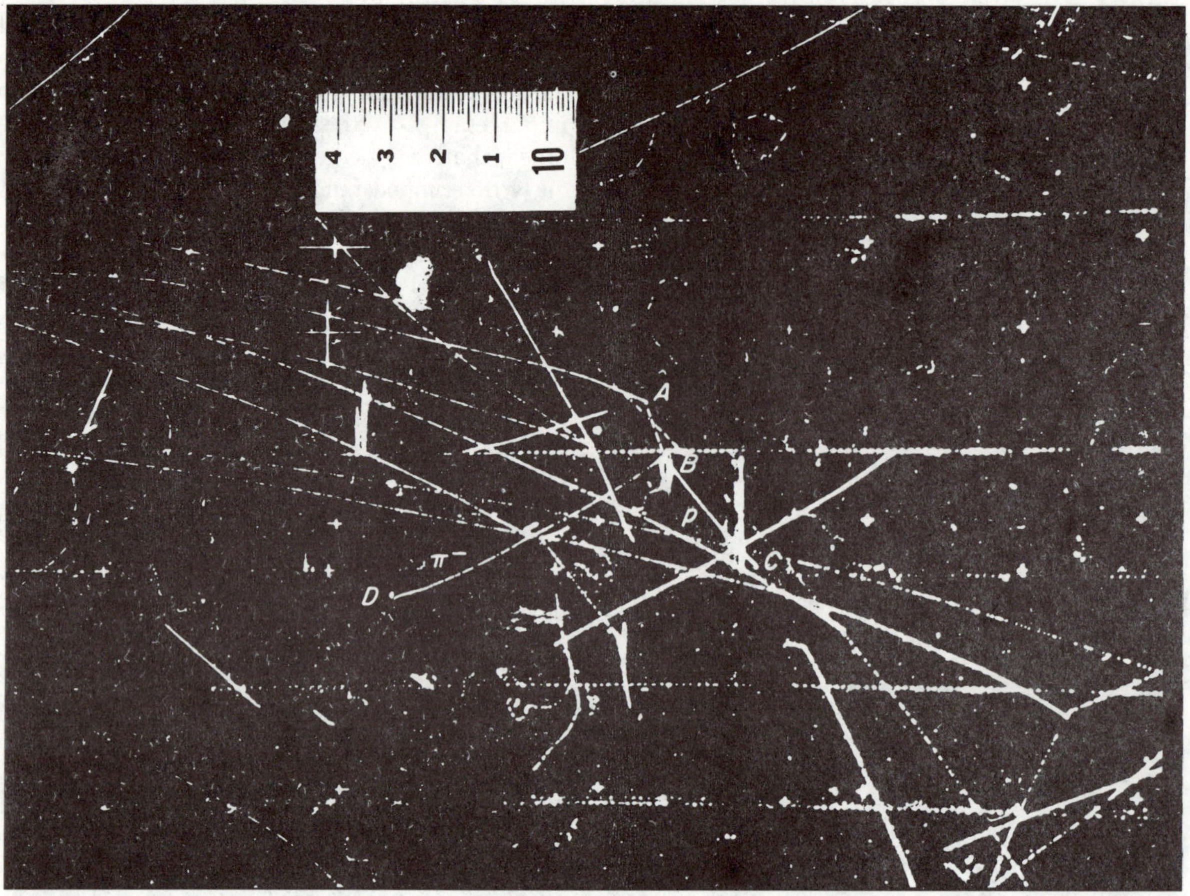

Fig. 41–29

representing these directions. The required angles are then measured with the protractor.

The ranges are found by laying a flexible plastic rule along the curved paths. The ranges measured in this way must be corrected slightly since the reproductions are not exactly life-size. A calibration scale is reproduced on the photo for this correction. Using range-momentum curves, the ranges are then converted to momentum using the proper range-momentum curve for the chamber. The rest mass of the Λ^0 hyperon is calculated from the measured angles, the momenta of the pion and the proton, and the relativistic relations for their energies. See Appendix, page 1373.

41–2.10

Apparatus for the detection of fission-produced neutrons is shown in Fig. 41–31. This setup was developed by the Swiss Institute for Reactor Research and has been modified to make it sensitive to the small neutron fluxes of the Ra–Be source (300 mg Ra). The source is embedded in a paraffin moderator *a* that makes the emitted neutrons thermal. The fission chamber *b* contains three plates electrolytically coated with UO_2 that has been enriched to 20 percent U^{235}. It also contains two electrodes (insulated and made gas-tight by Teflon sleeves) that are electrically connected to a discriminator–amplifier *c* that is mounted directly on the chamber to avoid the capacitance of a cable. Two coaxial cables lead to an oscilloscope and to a loudspeaker *d*. Details of the chamber and the electrical system are shown in the Appendix, page 1375.

Thermal neutrons from the source penetrate the chamber and strike the UO_2. The resulting fissions cause ionizations in the chamber that are detected by the discriminator–amplifier. The discriminator

Fig. 41–30

(potentiometer) eliminates the pulses formed by alpha particles, and the amplifier lengthens the short fission pulses to make them more visible on the oscilloscope. The oscilloscope traces are displayed by television, and the fission pulses are also displayed acoustically through the loudspeaker.

Before the chamber is used, it should be thoroughly cleaned and evacuated. It is then flushed and filled with argon to a pressure that allows long-range fission products to completely cross the chamber. During operation, the operator must guard against "pile up" of alpha particles that give large pulses which can be confused with the fission pulses. As a variation to the demonstration, the neutron source may be used without the paraffin moderator.

Note: Suitable shielding precautions should be observed at all times.

41–2.11

In Demonstrations 41–1.2 and 41–1.3 it was pointed out that the height of the water column in a capillary filter tube as a function of time could be used as the basis for models of simple radioactive decay. More complicated models of more complicated decay systems can be constructed by using such tubes in series or in parallel.[14] It is possible to construct such a model illustrating the behavior of a thermal neutron reactor suffering from xenon poisoning, as shown in Fig. 41–32. In this model a constant water flow apparatus *a* such as a Mariotte Container allows water to flow at a constant velocity into the top of a capillary filter tube *c* of suitable flow characteristics (Leybold #36174) through a stopcock *b*. The efflux from

[14] K. Hecht, *Nukleonik*, 2, 37 (1960).

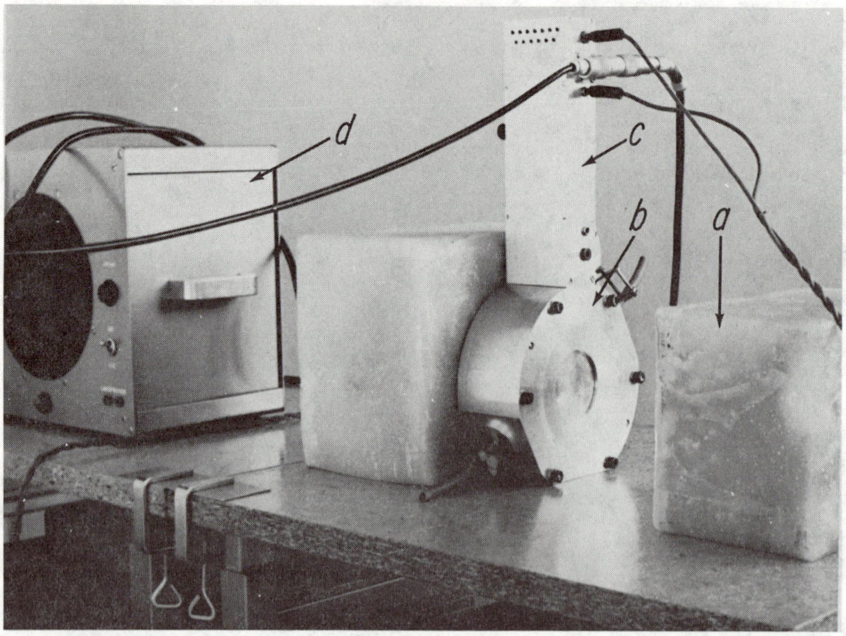

Fig. 41–31

filter tube c flows into a second capillary filter tube d whose flow characteristics are suitably different from those of tube c. Water from this filter tube can leave the tube by one or both of two paths: (1) through the stopcock e at a constant rate, or (2) through the capillary filter at a rate which is proportional to the height of the water column in that tube.

The analogs of the model are the following: (1) The constant flow of water from a is the analog of the constant flux of neutrons in a thermal reactor; closing stopcock b corresponds to shutting down the reactor. (2) The height of water h_1 (as a function of time) in the capillary filter tube c corresponds to the number of $_{53}I^{135}$ nuclei produced by the constant neutron flux. (3) The rate of discharge of water from tube c corresponds to the rate of radioactive breakdown of $_{53}I^{135}$. (4) The height of water h_2 in capillary filter tube d (as a function of time) is an analog of the number of $_{54}Xe^{135}$ nuclei arising from β-decay of $_{53}I^{135}$. (5) The flow out of tube d through the capillary filter corresponds to the amount of β-decay of $_{54}Xe^{135}$ into $_{55}Cs^{135}$. (6) The discharge through tube e corresponds to the decrease in the number of $_{54}Xe^{135}$ nuclei through capture of thermal neutrons ($_{54}Xe^{135}$ $(n,\gamma)_{54}Xe^{136}$) during the time the reactor is in operation.

In order for the model to be strictly accurate, the stopcock should be equipped with a small capillary filter corresponding to a very short half-life, since neutron capture is in fact an exponential function of time. However, since the time rate of growth of the $_{54}Xe^{135}$ nuclei is not essential for the experiment and since the number of nuclei at saturation (i.e., height h_2 in column d) can be regulated by stopcock e, the simpler arrangement is often preferred.

By using this model, one can simulate the ratio between the $_{54}Xe^{135}$ and the $_{53}I^{135}$ nuclei at initiation of the reaction, during reactor operation, and after shutting down the reactor. This ratio is clearly h_2/h_1. The behavior is shown in Fig. 41–33. In this figure, section (a) shows the heights h_1 and h_2 as a function of time after opening both stopcocks b and e; section (b) shows the same heights during operation; section (c) shows the behavior following shut down, i.e., closing both stopcocks b and e.

On starting up, stopcock b can be so arranged that the height h_1 at saturation is 36.5 cm, and stopcock e so arranged that height h_2 at saturation is only 3.5 cm. The half-life for discharging of filter tube c then is 31 sec and for tube d 23 sec. Note that the ratio $31/23 \approx 1.35$, and the ratio of the

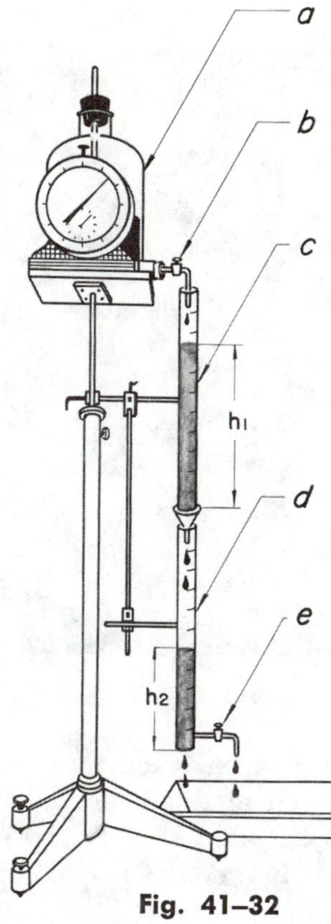

Fig. 41–32

half-lives of $_{54}Xe^{135}$ and $_{53}I^{135}$ is $9.2/6.7 \approx 1.37$. The model, therefore, operates on a much shorter time cycle; in Fig. 41–33 the figures in parentheses along the abscissa correspond to an actual reactor cycle in hours.

The model shows clearly the effect of shutting off the supply of thermal neutrons at shut down: the $_{53}Xe^{135}(n,\gamma)_{54}Xe^{135}$ reaction can no longer take place and the number of $_{53}Xe^{135}$ nuclei increases to a saturation value far above its value during operation. Following saturation, a slow decrease takes place because of the normal β-decay of $_{53}Xe^{135}$; the height h_2 correspondingly goes through a maximum and then decreases exponentially, in full agreement with actual reaction behavior.

41–2.12

The capture of a slow neutron by U^{238} leads to the formation of the β-emitting U^{239} which changes into Pu^{239} by means of two successive β emissions. This process is demonstrated by the model in Fig. 41–34.

The U^{238} nucleus is represented by a large wooden sphere (1). The 15-cm-diam sphere has two halves, which can be fastened together. One half of the sphere contains a mechanism (see Fig. 41–35) consisting of the clockwork of an alarm clock (3), an entrance opening for the iron sphere (4), and two exit openings (5) and (6) for the model electrons, which are symbolized by small

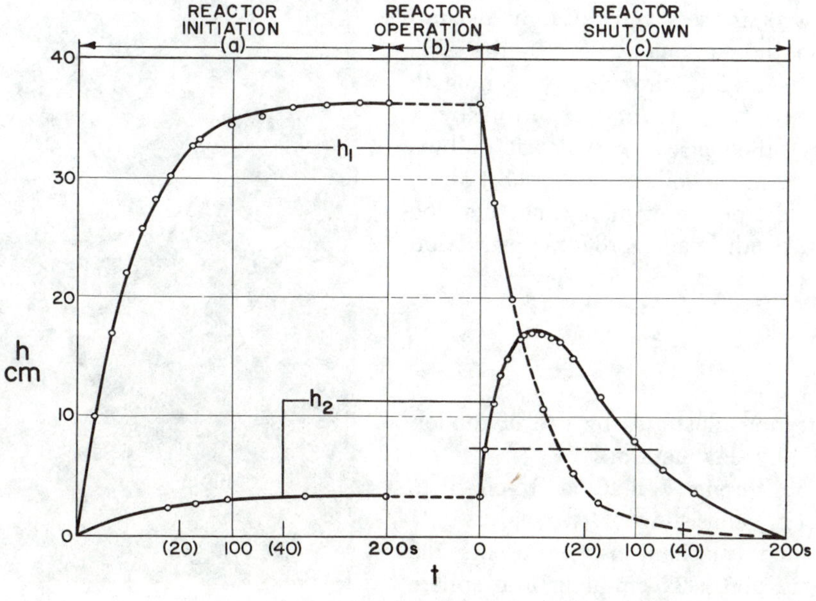

Fig. 41–33

Fig. 41–34

lighted lamp bulbs. In each exit opening there is a spiral spring which can be compressed and locked into position by means of an auxiliary rod.

The clockwork is wound up through opening (7), the springs are stressed, and the lighted lamp bulbs inserted in their openings. When the 2½-cm-diam iron ball (2) falls through its opening in the wooden sphere (capture of a neutron by a Uranium nucleus), the clockwork, which initially is not running, is set in motion. After a short time, one lamp is ejected from its opening, and then the other. See the Appendix, page 1378 for construction details.

41–2.13

This demonstration illustrates by use of models the fission of U^{235} by slow neutrons.

In Figs. 41–36 through 41–40, a 15-cm-diam wooden sphere (1), consisting of two halves (2) and (3) with inner mechanisms, represents the fissionable nucleus, and a 2.5-cm-diam iron sphere (4) represents the neutron. The wooden sphere

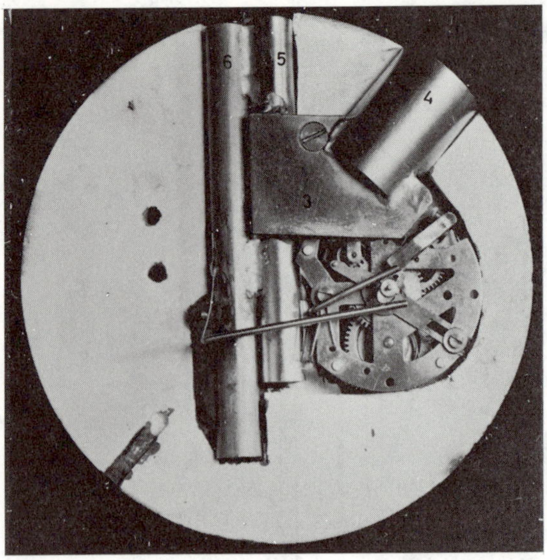

Fig. 41–35

has a 29-mm-diam opening on the top (5), into which the iron sphere can be introduced. Each hemisphere also has a 3.2-cm-diam opening (2.1) and (3.1) for the insertion of a spiral spring with

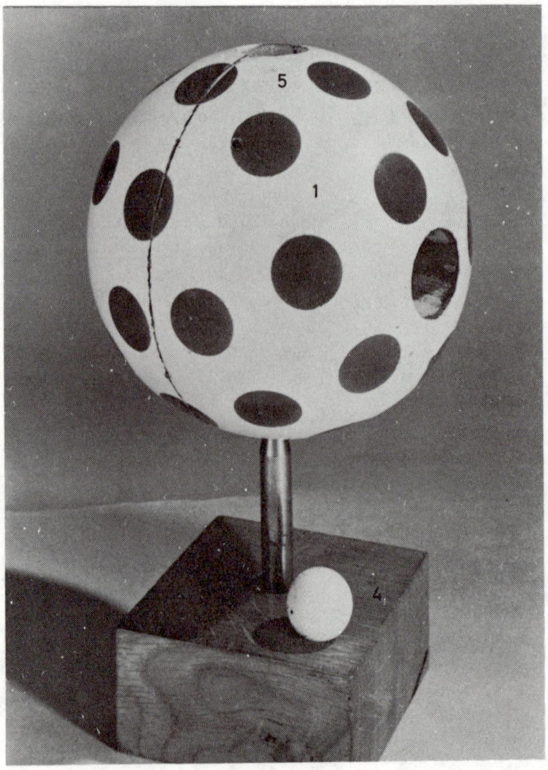

Fig. 41–36

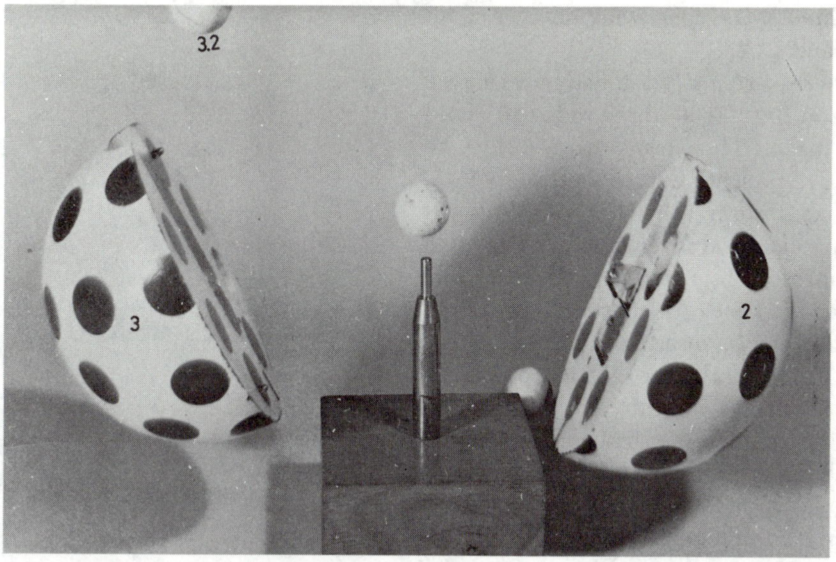

Fig. 41–37

a few turns of wire. A smaller wooden sphere (2.2 and 3.2) of the same diameter as the iron sphere is pressed into each opening and held by pins (2.3) and (3.3). In addition, each wooden hemisphere contains a second 3.2-cm-diam opening (2.4) and (3.4) in which a spiral spring (31-mm-diam spiral, 2.5-mm steel wire) can be placed to split the model. One hemisphere is furnished with a razor blade (2.5), over which an aluminum stirrup (2.6) for cutting the string can be lowered. A cross pin (3.5) is used in order to prevent accidental damage to the string by the razor. The stirrup is held in its normal position by a weak spring (see Fig. 41–38). Both hemispheres are placed together in such a way that both pins (3.6) fit into the holes (2.7).

To prepare the model for operation, an auxiliary apparatus is used (see Fig. 41–39). A spiral spring

Fig. 41–38

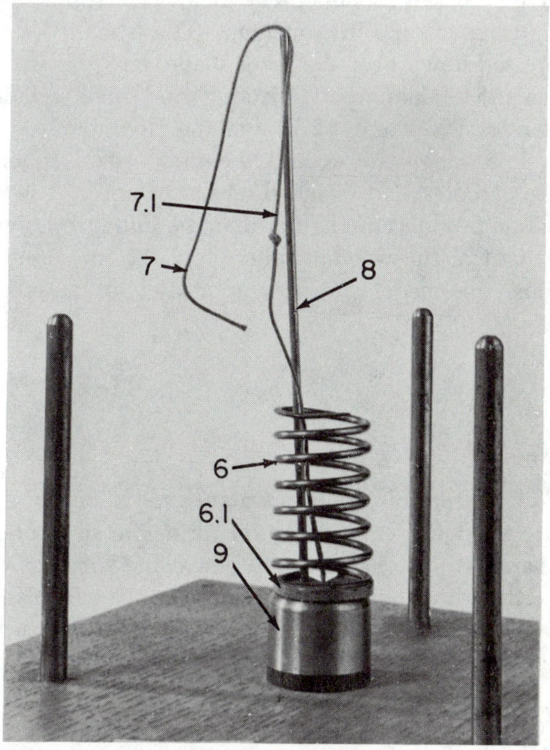

Fig. 41–39

(6) is soldered onto a slotted brass disk (6.1), and a string (7) is held in the slot by a knot (7.1). An end notched rod (8) holds the string in a vertical position to facilitate threading it into the model. Spring, string, and rod rest on a removable cylinder

(9), whose diameter is somewhat smaller than openings (2.4) and (3.4).

The hemispheres are placed together over the spring as shown in Fig. 41–40; three rods (10) hold the sphere in place. The string, carried by the auxiliary rod (8), projects out of opening (2.4). Into this opening a spring similar to (6) also is inserted. The auxiliary rod is removed, and the two springs are compressed by using a pressure handle (11) so that a second knot (7.1) in the string, 11 cm from the first knot, can be inserted in the slot of the upper spring. (Two marks are scratched on the auxiliary rod in order to indicate the correct distance between the knots in the string.) The model is cocked and the pins (2.3) and (3.3) removed.

After the model is cocked and the large wooden sphere and small iron sphere are tared on a balance, the model is placed as shown in Fig. 41–36. (All breakable objects must be removed from the vicinity of the cocked model.) The small iron sphere (4) is dropped into its opening. The stirrup (2.6), pressed down over the razor blade, cuts the string; the model flies apart, ejecting the small wooden spheres (2.2) and (2.3) and the iron sphere.

Both hemispheres and the three small spheres are gathered and weighed on the balance. The reaction products are lighter than the initial combined weight of the wooden sphere and the iron sphere; thus, the model provides an analog of mass defect. If both springs and the cut string now are laid on the balance, the initial weight is restored.

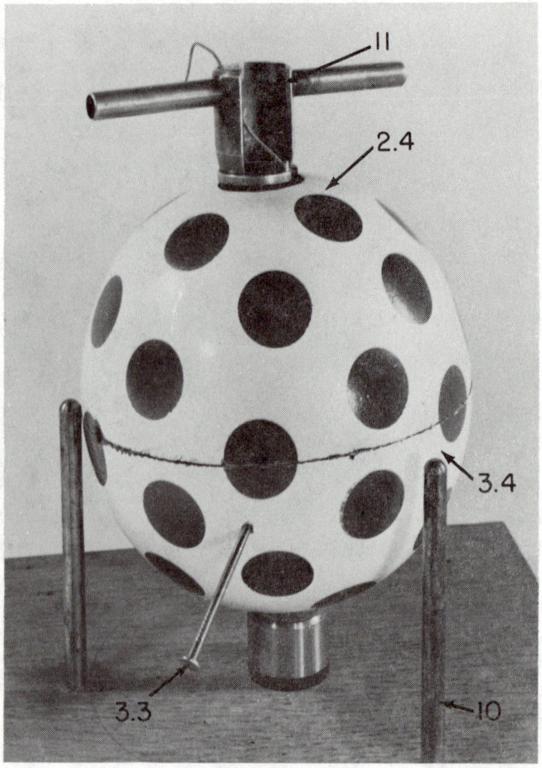

Fig. 41–40

See the Appendix, page 1380, for a list of materials.

41–3 Particle Detectors

41–3.1

A kettle of boiling water and a spark gap are used to demonstrate the principle of a cloud chamber. The vapor cloud above the spout of the kettle is shadow projected and a spark gap placed in the vapor stream. Sparks will greatly intensify the density of the cloud because of the ionization.

41–3.2

The apparatus described here provides an effective continuous cloud chamber. It is suitable for use in a high school physics laboratory. Alpha tracks appear due to the formation of ion pairs which serve as condensation nuclei in supersaturated vapor.

The chamber itself is made from a transparent plastic refrigerator jar. A satisfactory one is commercially available under the name "Freeze-r-Jar" from the Tri-State Plastic Molding Company of Henderson, Kentucky (Cat. No. C–40). This particular jar has a rubber gasket in a threaded cover which insures a tight seal and eliminates convection currents. The inside bottom of the jar is painted flat black. A piece of blotter paper about 1 in. wide is fitted around the inside bottom so that its ends overlap. A paper clip makes a convenient fastener. The source is a very small piece of the radioactive material from the hands or numerals of a luminous dial clock. The best physical size for the source material is determined by the content of radioactive material in the numerals. This size should be chosen such that no more alpha tracks are yielded than can be counted by the student. If this is not done, many properties will be obscured simply by sheer numbers of tracks present. The source is glued to a rectangular piece of file card painted flat black. It is bent at right angles for the purpose of holding it vertical. This also facilitates handling and permits observation of alpha particles traversing the file card with no deflection.

To operate the device, the source is placed in the center of the chamber. The blotter is saturated with isopropyl alcohol, the top screwed tight, and the chamber is placed on a small cake of Dry Ice. The chamber should be level to prevent the tracks from drifting. The room is darkened and the chamber is illuminated from the side with either a flashlight or a clear 150-W light bulb (painted black excepting for a small spot on the filament level). In less than five minutes the alpha tracks appear as straight, heavy trails. Since the material used for luminous dial clocks is a mixed radioactive source, beta tracks can also be seen as thin, broken, circuitous trails. However, these are usually difficult to see because of the prominence of the alphas. Gamma tracks are also present, but positive identification of these is impossible for the same reason.

After several minutes of operation there may be too many ions present for clear viewing. To clear the chamber of these it is only necessary to stroke the cover with silk, creating a sufficient potential difference to sweep the unwanted ions from the chamber bottom. Several hours of operation are possible before resaturation of the blotter is necessary.

41–3.3

The tracks of α-particles and β-particles can be readily observed with the apparatus of Fig. 41–41. A cylindrical (or rectangular) glass walled container *a* is 15–30 cm in diam (or width) and 3 to 6 cm in height. It is closed at the top and bottom by metal plates. The bottom plate *b* is cooled with streaming water (12°C), while the cover *c* is a brass plate heated to a temperature of about 100°C by a heating coil *d*. (Steam or hot water can be substituted, or the heating element of a flatiron can be used.) Air at atmospheric pressure is the enclosed gas. The source of the vapor is 1–2 glycol contained in sheets of filter paper *e* held against the top plate by small magnets *f*. Iron lugs g soldered to the brass cover plate facilitate holding the paper. The tracks are produced by a *very weak* Po α-source *h* and must be observed under a strong source of light. Clearing is produced by a dc potential of about 50 V applied between the top and bottom plates. In practice two or three sheets of filter paper are soaked in glycol and attached to the cover. Moistening the glass walls with glycol will keep them clear. The tracks will appear about 15 min after the proper temperatures have been reached.

41–3.4

The cloud chamber shown in Fig. 41–42 is similar in construction to most other cloud chambers. Its main features are its long operation time (4–5 h, continuously) and its ability to be viewed on closed circuit TV.

The apparatus is made up of two major assemblies: the Dry Ice box and the chamber and light guide assembly. The box *a* (see Fig. 41–43) is made of wood lined with Styrofoam *b* and contains a brass spring loaded sliding Styrofoam plate *c* which presses a cake of Dry Ice against the bottom plate *e* of the chamber. The wall of the chamber

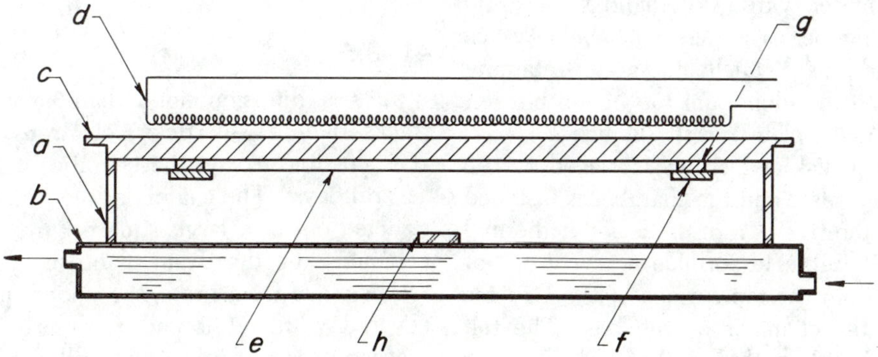

Fig. 41–41

Fig. 41–42

is a glass cylinder d ($5\frac{1}{2}$ in. in diam, 3 in. high) which fits in a milled groove in the aluminum bottom plate e. The cover is made of two aluminum plates. A felt ring f is glued to the bottom of the lower plate g. A recess milled into the lower surface of the upper plate h accommodates the viewing glass i, which is sealed with rubber gaskets j. The alpha source k (Po) is set in the bottom plate of the chamber. The Lucite light guide m is fastened to the cylinder with a steel band n (see Fig. 41–42). The chamber is attached to the Dry Ice box by means of pins p which slide under retaining blocks located on the aluminum top of the box. A resin-paper-laminate plate r rests on box a.

In operation, a disk of Dry Ice is placed on the sliding Styrofoam plate and the chamber is fastened onto the top of the box. To insure a better thermal contact, it is advisable to sprinkle a few drops of methyl alcohol onto the top surface of the Dry Ice before placing the chamber on the box. The felt ring is saturated with methyl alcohol and the cover is placed on the chamber. A clearing field of 3000 V is put across the chamber and light from a lamp is directed into the chamber through a slit, lens, and the light guide. Leveling screws are provided on the bottom of the box to insure that the chamber is horizontal. The trails (see Fig. 41–44) can be shown through closed-circuit TV by mounting the camera vertically over the viewing glass or by using a prism.

41–3.5

A large diffusion cloud chamber is used to show tracks due to α-particles, electrons, and protons. It is possible to run the chamber continuously for several days. The chamber (see Fig. 41–45) is constructed out of a large, square Pyrex container (1) cemented together from plane sheets. The top lid (2) is made from brass, the under side of which is covered with a felt pad (supported by threads) whose edges rest in a trough filled with alcohol, in

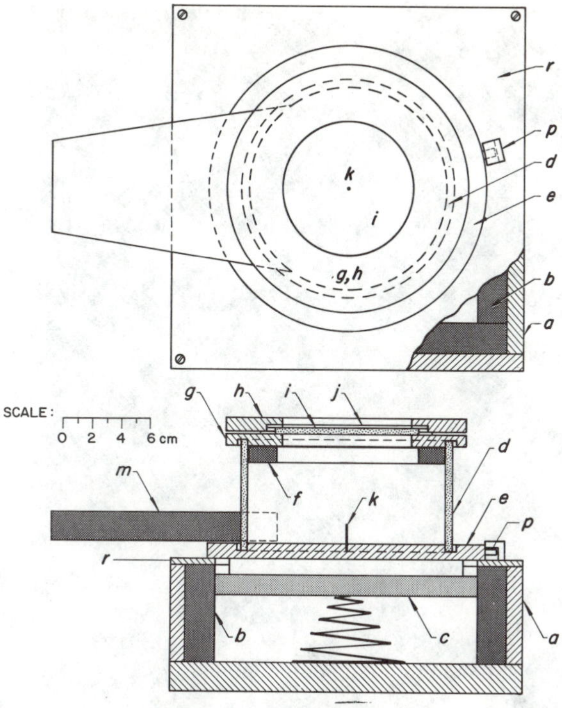

Fig. 41–43

order to keep it uniformly damp. Opening (2.4) is used to fill the trough with alcohol. The Pyrex container rests on a thermal insulating frame, the floor of which is filled to a depth of about 8 mm with alcohol. The floor of the frame (supported by four legs made of german silver) carries the cooling coils. The coils and a circulating pump keep the floor at a temperature between −45°C and −50°C. Alternatively, Dry Ice can be used for cooling. The heat exchanger (see Appendix, page 1382, for construction details) is contained in a Dewar flask. Alcohol is used as the refrigerant. The bottom of the chamber is packed with thermal insulating material (Styrofoam). The lower chamber is enclosed by metal walls (6). In order to keep the Pyrex container free from ice, a heating wire is built into a groove in the thermal insulating frame. It gives just enough heat to prevent ice formation. Illumination of the chamber is carried out by two 12-V, 45-W, automobile fog lights (10) which are cooled by fans.

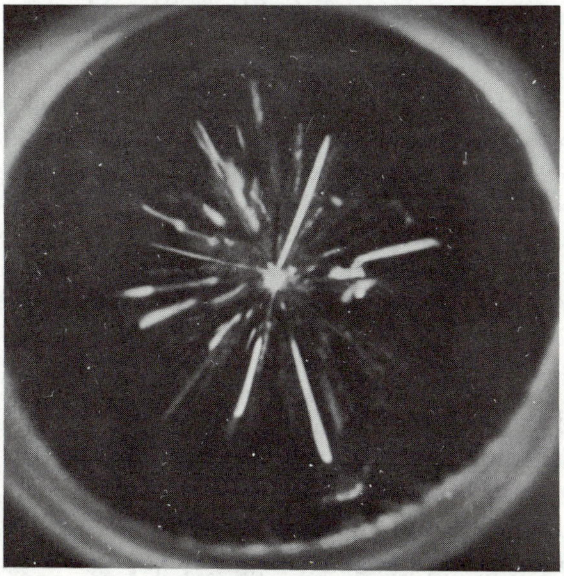

Fig. 41–44

The operation of the chamber is quite simple. The cooling apparatus is started (or the Dewar flask is filled with Dry Ice) and the circulating pump is turned on. The floor of the chamber and the trough attached to the roof of the chamber are filled with alcohol. The chamber is ready for operation after 20–30 min; traces of cosmic ray tracks will appear. If the chamber is run for several days the supply of alcohol in the trough will gradually be diminished, since alcohol condenses on the floor of the chamber. This alcohol must be replaced and a similar amount extracted from the bottom.

For demonstrating α-particle tracks, the source (Po, for example) is fastened to the outer edge of a source holder (8.1). Figure 41–46 shows sharp α-particle tracks (15.1) and diffuse α-particle tracks (15.2). A motor (9) rotates the source in a circle so that spatial saturation of the chamber cannot occur. In order to show recoil protons arising from neutron bombardment, a piece of hydrogen-containing material (3.5) (polyethylene) is mounted in the critical zone on the wall. When neutrons, produced for example from a $T(d,n)^4He$ reaction, impinge on the polyethylene, recoil protons arise, whose tracks traverse the entire chamber. Figure 41–47 shows tracks of neutrons (15.3) and electrons (15.4).

41–3.6

Lecture groups can observe alpha particle tracks in an automatically cycling Wilson cloud chamber.[15] The construction details for the apparatus

[15] F. E. Christensen, *Am. J. Phys.*, 18, 149–50 (1950).

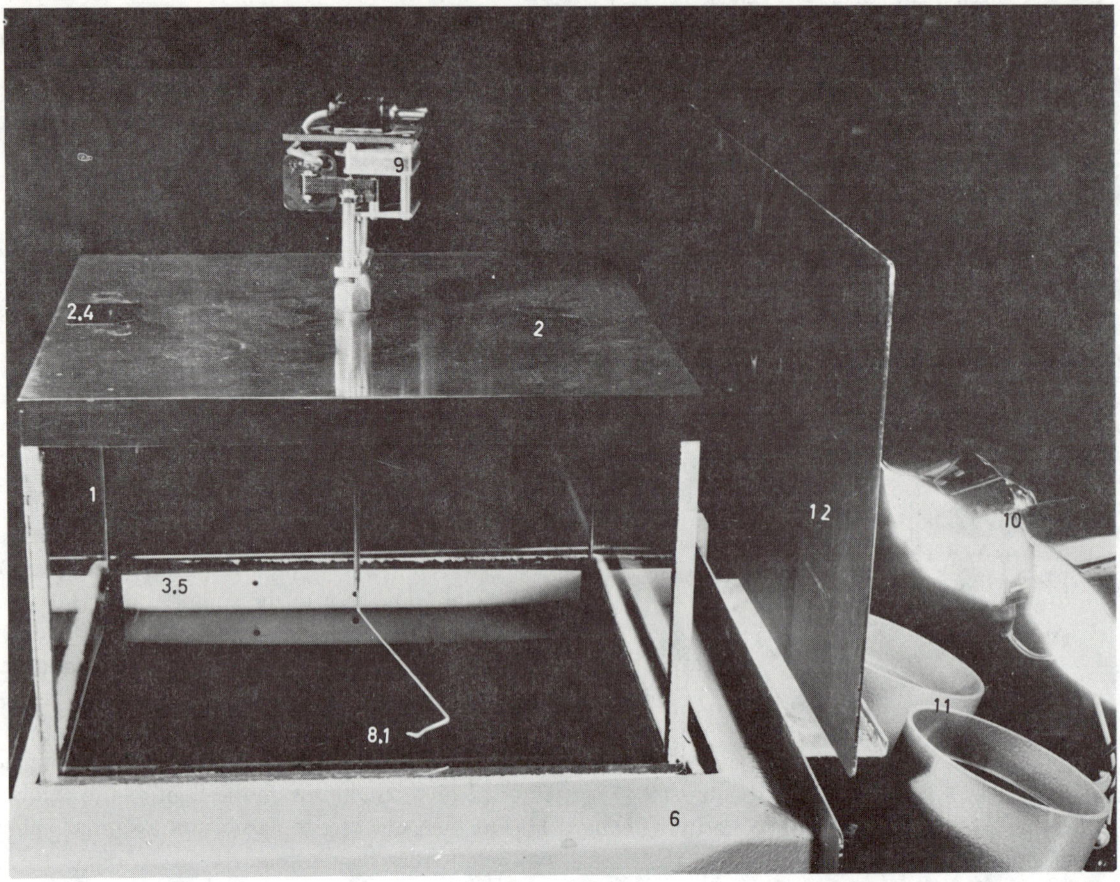

Fig. 41–45

shown in Figs. 41–48, 41–49, and 41–50 are found in the Appendix, page 1382.

41–3.7[16]

A Lucite plank with rows of steel balls on its surface (Fig. 41–51) simulates ionization by collision when a sufficient electric field is applied to a gas. When an ion is made, both it and the detached electron are driven by the field so much that they gain enough energy to make more ions at the next collision with gas molecules. Thus one ion makes others, and they in turn make others, in a growing avalanche or chain reaction.

In the model the gravitational field along the sloping plank simulates an electric field. The Lucite chute, $5\frac{1}{16} \times 35 \times \frac{1}{4}$ in. with side rails, has thin transverse wires at 5-in. intervals (one "mean free path" apart) on its surface. They serve as retainers for rows of ½-in.-diam steel balls, which represent molecules. When the chute is tilted by raising one end, a single ball, or "ion," rolled down from the top dislodges another ball, while the rolled ball is itself held against the wire. The process is repeated until a slope is reached where the single ball can create an avalanche of balls, or "spark." A simpler form that works well can be made with a wooden plank. The wires are replaced by rubber bands around the plank and glass marbles may be used instead of steel balls.

[16] Eric M. Rogers, *Physics for the Inquiring Mind* (Princeton University Press, Princeton, N.J., 1960), p. 462.

41–3.8

The alpha-particle counter illustrated in Fig. 41–52 (Cenco photo), consists of a grid of very fine wires and a plate mounted on a Bakelite disk. A steady voltage (from rf supply), just below the sparking potential is applied between the grid and

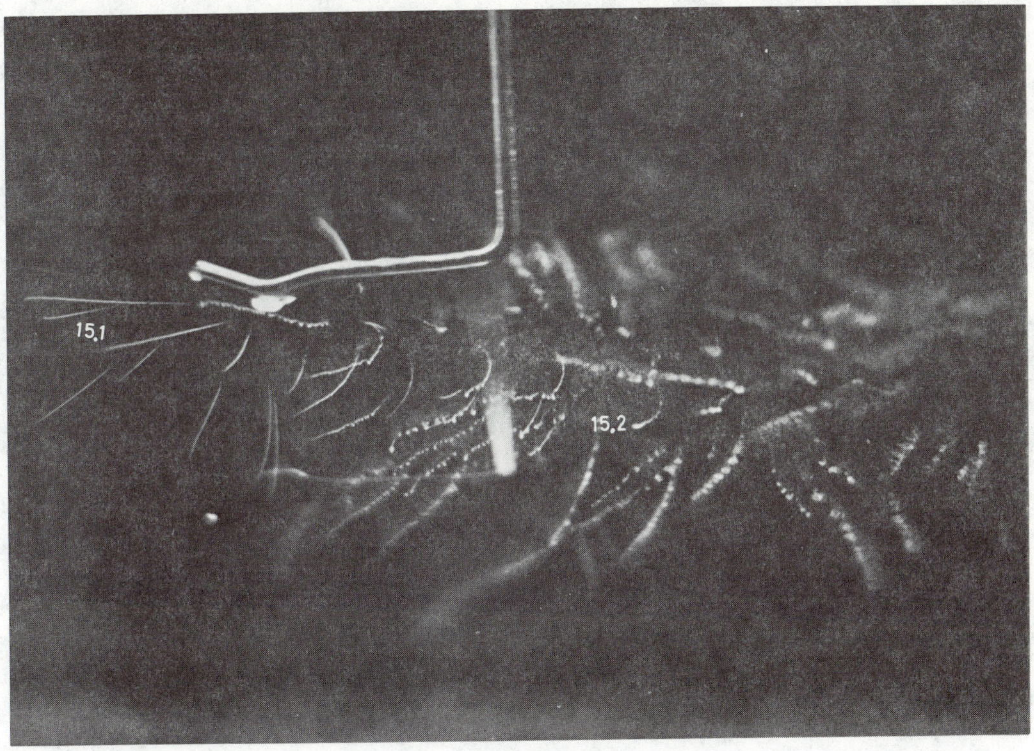

Fig. 41–46

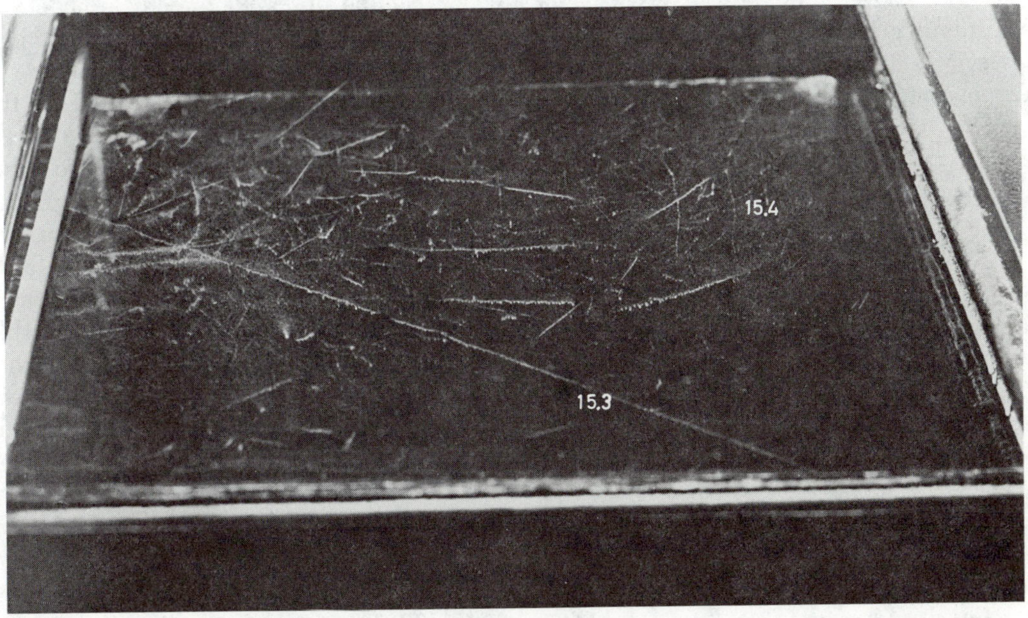

Fig. 41–47

Fig. 41–48

Fig. 41–49

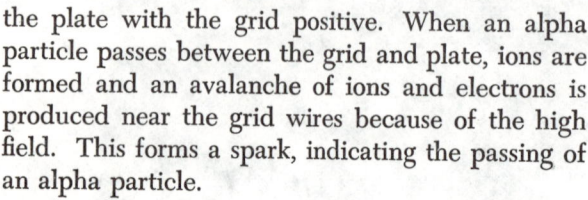

the plate with the grid positive. When an alpha particle passes between the grid and plate, ions are formed and an avalanche of ions and electrons is produced near the grid wires because of the high field. This forms a spark, indicating the passing of an alpha particle.

The alpha counter[17,18] has a small alpha source, which can be brought near the grid. A clear Lucite dome protects the wire grid from damage and the operator from shock, and permits a good view of the sparks. This dome, as well as the alpha source support, may be removed thus allowing the detector to be used in any experimental arrangement. The counter is simple to operate and requires only a 5000-V dc power supply.

41–3.9[19]

Recently spark chambers have come into common use as detectors of high-energy particles produced by large accelerating machines.[20] The spark chamber described here is a reliable and flexible device which has been successfully used as a lecture demonstration to show the tracks of sea-level cosmic rays.

In principle, the operation of spark chambers is very simple. If a high voltage is suddenly applied to the gap between a pair of parallel plates shortly after an ionizing particle has passed through the rare gas between them, a spark jumps along the path of the particle. If a number of gaps are stacked vertically and operated simultaneously, the track of the particle is defined by a line of sparks which can be observed directly or photographed.

To achieve successful operation in practice, the main problems that must be overcome are: (1) the high-voltage pulse must be applied to the gaps within about 0.5 μsec after the passage of an ionizing particle, and (2) the mechanical design of the gaps must be simple and convenient, and yet maintain a high degree of parallelism between the plates. Although the circuit described here is fairly complex (a consequence of the first requirement), its components are standard, readily available electronic parts. The design of the gaps involves the use of a number of basic modules, each of which comprises two gaps. This type of construction is fairly simple, and has the further advantages that (1) each gap is electrically isolated from all the others, and (2) the geometrical configuration of the modules is very flexible.

Figure 41–53 shows the front and rear views of the apparatus. The spark chamber proper is a stack of 9 two-gap modules. Trays of Geiger–

(17) Available from Central Scientific Co., Chicago, Illinois; Cat. No. 71248.

(18) W. Y. Chang and S. Rosenblum, "A Simple Counting System for Alpha-Ray Spectra and the Energy Distribution of Po Alpha Particles," *Phys. Rev.*, **67**, Nos. 7 and 8, pp. 222–227 (April 1–15, 1945).

(19) This work supported by the Office of Naval Research under Contract No. 710 (19).

(20) Spark Chamber Symposium, *Rev. Sci. Instr.*, **32**, 489 (1960).

Fig. 41–50

Müller counters—one above the chamber and one below—are operated as a cosmic-ray telescope and indicate that a particle has passed through the apparatus. The spark-chamber driver chassis contains the circuitry needed (1) to form a coincidence between pulses from the two trays, and (2) to use this coincidence signal as a trigger for the high-voltage pulse (6 to 8 kV) required by the spark chamber. All power supplies for the circuits also are included on this chassis. A tank of helium or argon supplies gas for flushing and filling the gaps.

The spark chamber's operation is reliable and setup is nominal. Meson tracks, which occur at a

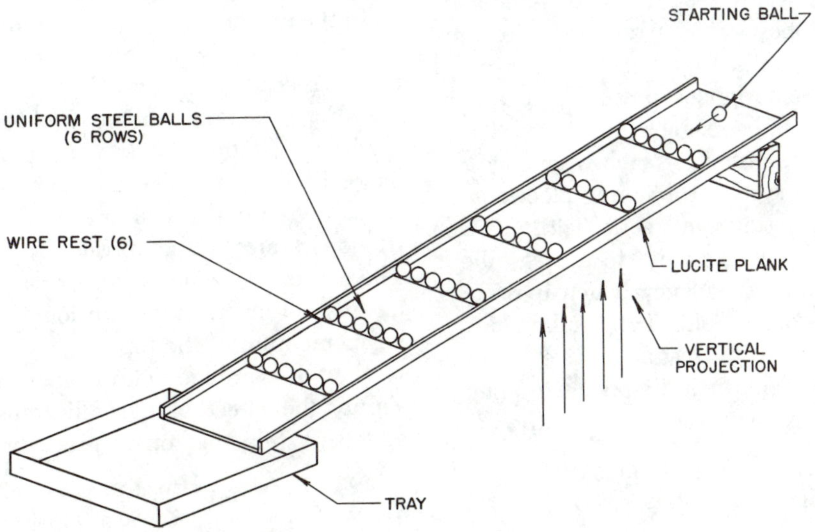

Fig. 41–51

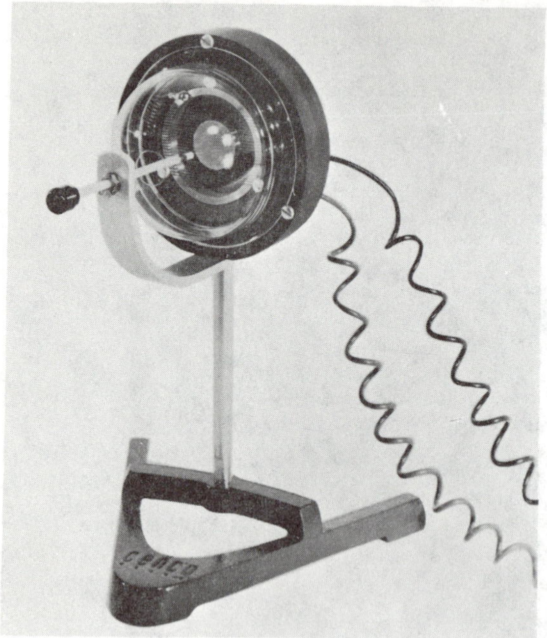

Fig. 41–52

rate of about 10/min, are readily viewed by students in a large lecture hall through three sides of the chamber. It is usually necessary, however, to tilt the chamber, so that the plates are parallel to the line of view. This lining up is not very critical, since sparks seen as reflections from the surfaces of the plates give the same impression of tracks as those seen by direct vision. This chamber can also be used for laboratory exercises.[21]

A very attractive feature of the modular arrangement is the ease with which gaps can be shifted into various configurations. For example, lead plates of various thicknesses can easily be introduced between gaps to show the great penetrating power of sea level mesons. Or, as another example, gaps can easily be placed both below and above the trays of Geiger counters.

The high voltage involved in the operation of the spark chamber is *dangerous*, and *safety precautions must be taken*. See the Appendix, page 1390, for construction details.

41–4 Nuclear Magnetic Resonance

41–4.1

Nuclear magnetic resonance and electron paramagnetic resonance have become important tools for the physicist and chemist in the investigation of molecular and solid structure. This demonstration[22] consists of a model intended to clarify some of the mechanical features of these resonance experiments. Descriptions of several more complex devices have been published.[23]

Electrons, and some nuclei, possess both a spin angular momentum and a magnetic moment. These two properties are essential for such resonance experiments. When a system containing electrons or nuclei experiences a static magnetic field, the net magnetization of the system tends to precess about the field direction at a frequency which is proportional to the magnetic field. If, simultaneously, there is a small oscillating magnetic field perpendicular to the static field, then the precessing nuclei may exchange energy with the source of the oscillating field. The rate of energy exchange will be a maximum when the frequency of the oscillating field matches the precession frequency.[24,25]

The apparatus, shown in Fig. 41–54 (a), consists of a gyroscope or top supported by a needle bearing on a stand, a wire loop, rubber bands, and a frame.

In the analogy, the top can represent an electron or nucleus (if we are content to treat such a particle classically), but it is better to let it represent the total magnetic moment of the system. The earth's gravitational field corresponds to the static magnetic field. When the top is placed on the stand and spun, it will precess freely (neglecting friction), due to the torque by the earth's field. The rate of precession is directly proportional to the torque and inversely proportional to the spin angular momentum of the top.

Now the wire loop and rubber bands, which will supply the effect of a small transverse field, are added. While the top is precessing, the frame is

[21] J. A. Earl, *Am. J. Phys.*, **31**, 571 (1963).

[22] R. A. Fowler and H. S. Story, *Am. J. Phys.*, **29**, 709 (1961).

[23] R. W. Christy, *Am. J. Phys.*, **24**, 523 (1956).

[24] G. E. Pake, *Sci. Am.*, **199**, 2 (1958).

[25] G. E. Pake, "Solid State Physics," *Advances in Research and Applications* (Academic Press, Inc., New York, 1956), Vol. 2, pp. 2–12.

Fig. 41–53

moved back and forth with a very small amplitude. If the frequency of this motion is above or below the precession frequency, little effect will be noted. But if the oscillation frequency matches the precession frequency, the axis of the top will rapidly change its angle with respect to the vertical. This motion may be such as to increase or decrease the angle of tilt depending on the relative phase of the applied oscillations and the precession. Such a motion involves an energy exchange between the top and the oscillator, and this constitutes the resonance effect.

If the lecturer wishes to pursue the matter further, some interesting points might be made about nuclear induction, and the relaxation times associated with the longitudinal and transverse spin components of a collection of similar tops.

The apparatus originally used was built around a Maxwell top made by the Welch Scientific Co., but this item is not now listed in their catalog. Figure 41–54 (b) shows details of construction of a similar top. The most important point is that the center of gravity of the top must be below the point of support, so that the top is inherently stable. Almost any metal can be used as long as it is readily machinable. None of the dimensions is critical, except that the top should be free to tilt at least 45 deg. The steel shaft *a* is ¼ in. in diam and

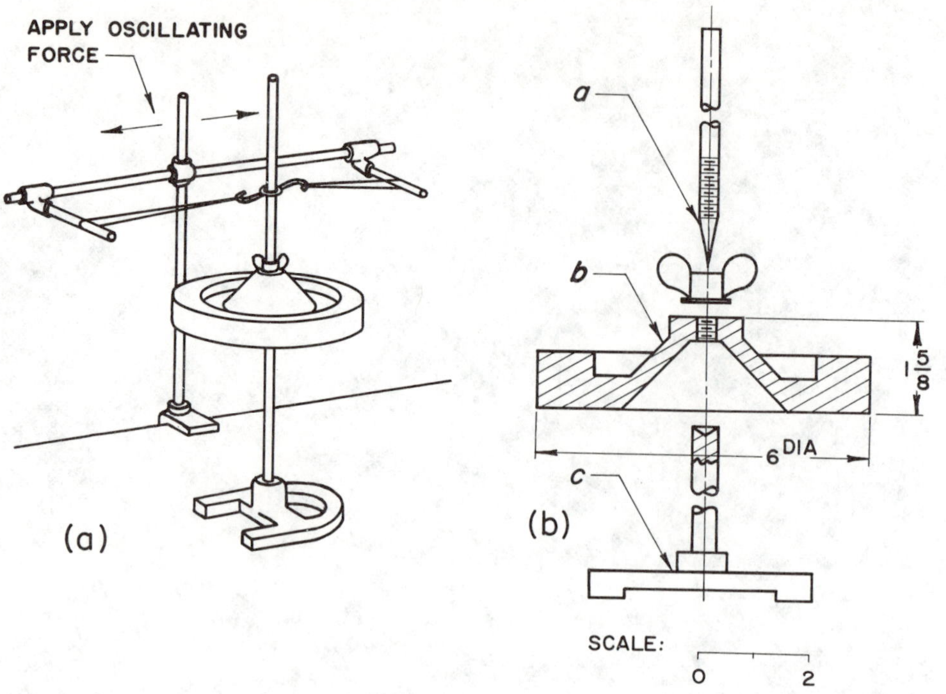

Fig. 41–54

6 in. long. One end is pointed 30 deg and is threaded (1/4–20 UNC) for 1 in. from the base of the point. This rod screws into rotor b where it rotates on a rod[26] attached to the base c.[27]

41–4.2

When a gyroscope a is equipped with a permanent magnet b, as illustrated in Fig. 41–55, and placed upon like poles of a powerful electromagnet c, the gyroscope precesses about its pivot post d. The electromagnet used is the same as the one used for Demonstration 39–2.2. It is operated on 110 V dc and provides a magnetic field of 6000 to 8000 G. The gyroscope consists of a gyro wheel on the extremity of a pivoted shaft carrying a counterweight e. A bar magnet[28] (180 mm long x 15 mm diam), taped to the horizontal shaft, gives the magnetic field interaction that produces the Larmor precession.

41–4.3

The model in Fig. 41–56 is designed to show the magnetic transitions that are produced in an atom when the frequency of the H_1 field (f_{H_1}) and the Larmor frequency (f_L) are identical. The model further demonstrates that the correct circularly polarized H_1 field is necessary for magnetic transitions. The simplest method of presentation would be to use shadow projection or closed circuit TV.

The model uses a gyroscope (1) driven by air jets (1.1), and riding on an air cushion. Gravity

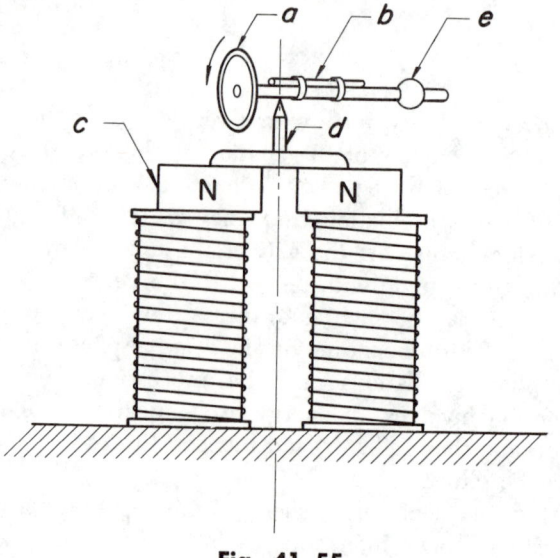

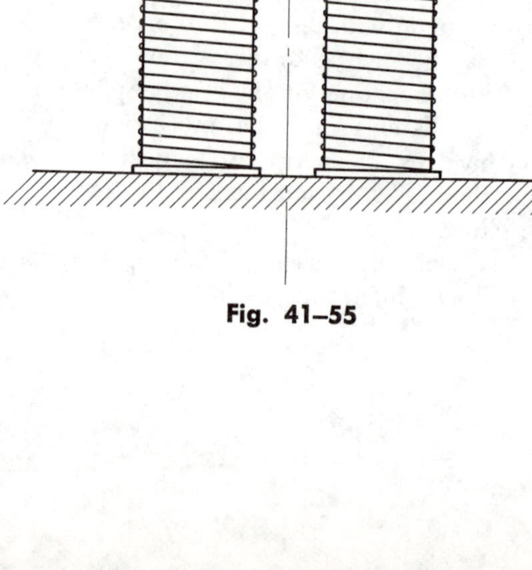

Fig. 41–55

(26) Cenco No. 72175.

(27) Cenco No. 72002.

(28) Cenco, Catalog No. 78291 Alnico.

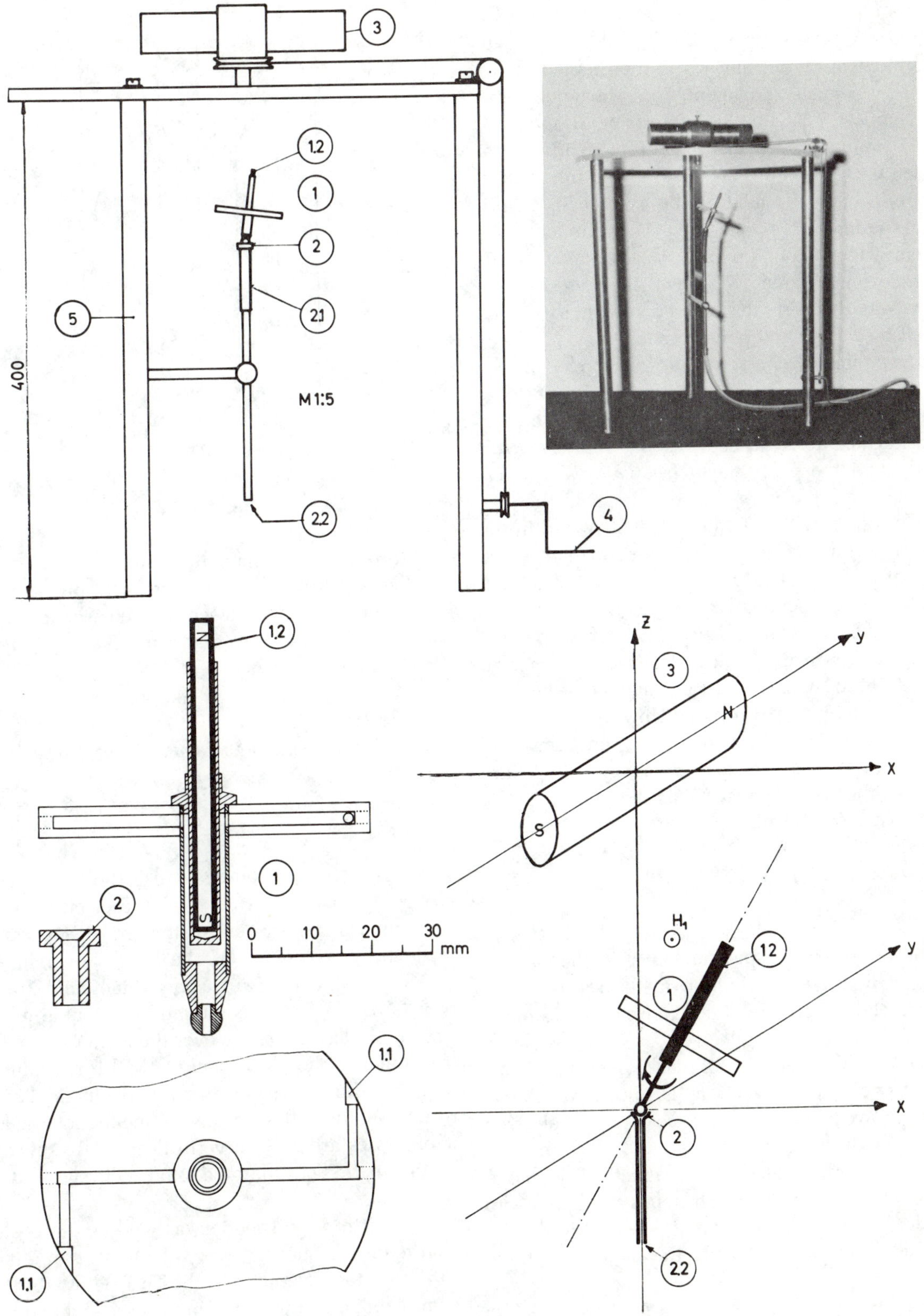

Fig. 41–56

induces a precession, which serves as an analog for the Larmor precession in the H_1 field. The magnetic moment of the nucleus is represented in the model by a small permanent Alnico magnet (1.2). The support (2) for the gyroscope is made of rubber tubing (2.1), which allows for steady operation of the gyroscope, and has an air inlet (2.2). To produce the H_1 field at the gyroscope, a horizontal permanent magnet (3) 17.0 cm long and 3.4 cm wide, made up of 7 smaller magnets, is mounted on a frame (5) above the gyroscope in such a way that it rotates about a vertical axis when turned by a cable and pulley system.

If H_1 remains constant, the gyroscope, on the average, experiences no disturbing force. If H_1 is made to rotate by rotating the crank (4), there is no disturbing force as long as the rotational frequency and the precessional frequency are different. Resonance occurs, however, when the bar magnet rotates with the same frequency and in the same directional sense as the gyroscope precesses, at either a lead or lag of 90 deg. At resonance the field H_1 is perpendicular to the vertical plane through the gyroscope's axis. The gyroscope lifts (absorption) when the field H_1 is directed as in Fig. 41–56 and falls (induced emission) when the field is oppositely directed. However, even at identical frequencies ($f_L = f_{H_1}$), if the sense of rotation of the magnet and of the gyroscope are opposed, there is no effect. This shows that the H_1 field must show the correct circular polarization for resonance.

41–4.4

The fact that the spectrum lines in the Balmer series of hydrogen, and those in the series of the alkali metals, are doublets, is interpreted to mean that each of the energy levels of these atoms is double.[29] The doubling of energy levels in atoms having one valence electron, is due to the interaction between the magnetic field of the electron orbit, and the magnetic field of the electron spin. A bar magnetic located in a magnetic field has a torque exerted on it which tends to line it up parallel to the field. A spinning electron in a magnetic field behaves in exactly the same way; there

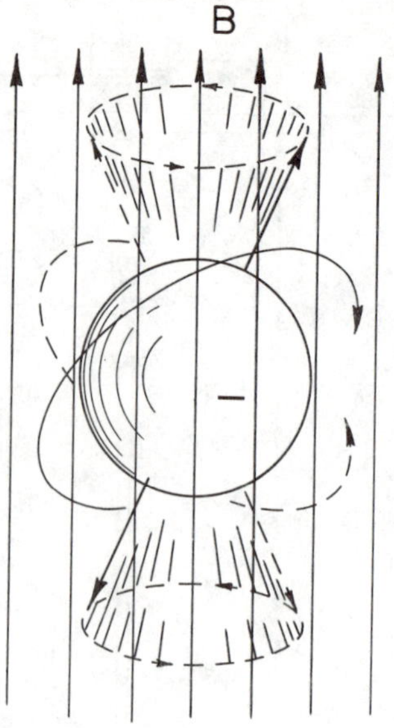

Fig. 41–57

is a torque acting upon it, trying to turn its axis parallel to B. Due to the mechanical properties of a revolving mass, the electron precesses around B in much the same way that a mechanical top precesses in a gravitational field.

A schematic diagram of the precession of a spinning electron is given in Fig. 41–57. Such a motion is called a Larmor precession, and its frequency f is given by

$$f = \frac{e}{4\pi m} B,$$

where B is the magnetic field, e is the charge of an electron, and m is the mass of the electron.

A good demonstration of this precession can be made by a gyroscope of the kind shown in Fig. 41–58. A nonconducting sphere, mounted free to turn in ball-bearings, is mounted in double gimbal rings and placed directly over the center of an electromagnet as shown. When the ball is set spinning in the position shown, and the magnetic field is turned on, the ball will retain its inclination angle as it precesses around the vertical axis. By reversing the magnetic field the precession will reverse direction.

Owing to orbital motion, as well as the positive charge on the nucleus, every electron in an atom is subjected to a magnetic field. To see how this field comes about, consider the simple case of a hydrogen atom with its one electron in an orbit around a posi-

[29] Harvey E. White, *Modern College Physics* (D. Van Nostrand Co., Inc., Princeton, N.J., 1962), 4th ed., pp. 560–1.

Fig. 41–58

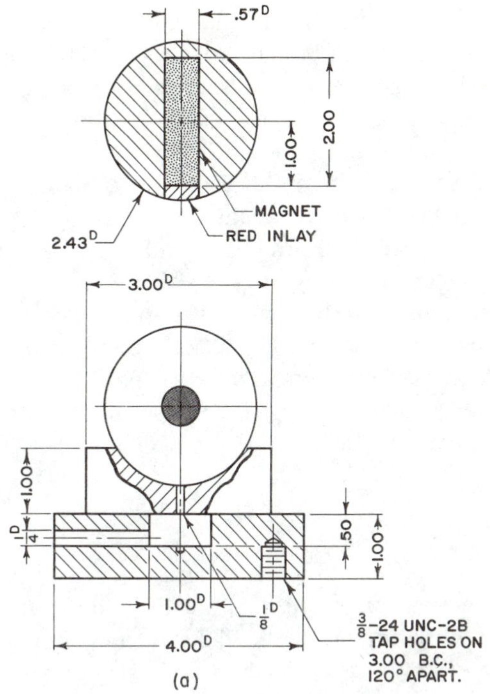

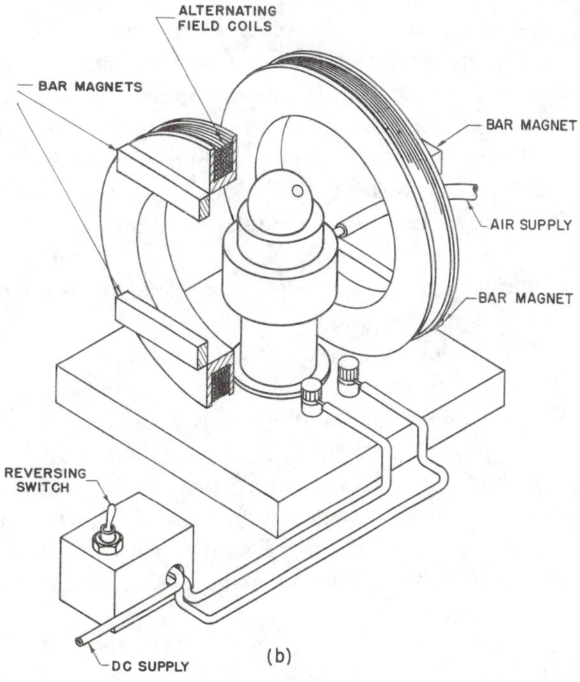

Fig. 41–59

tive charge. If we imagine ourselves riding around with the electron, looking in at the nucleus, we see this positive charge as though it were moving in an orbit around us. This moving charge gives rise to a magnetic field at the electron. In this field the electron carries out a Larmor precession. For construction details, see the Appendix, page 1392.

41–4.5[30]

A demonstration of the principles of magnetic resonance can be given with a modification of the air bearing gyro described in Dem. 11–2.2. A billiard ball is used as the rotor, Fig. 41–59 (a). A hole is drilled partway through the ball along a diameter to accommodate a cylindrical Alnico magnet. The magnet is epoxied in the hole and the hole may be capped with a lid machined from another billiard ball. There should be no gravitational torque on the final configuration about the center of the ball. A static magnetic field is supplied by four 6-in. Alnico bar magnets placed as in Fig. 41–59 (b), with their magnetic moments parallel. An arrangement of the coils of an electromagnet[31] (from Cenco e/m experiment) connected through a reversing switch to a dc supply allows the application of

[30] H. A. Daw, *Am. J. Phys.*, **33**, 322 (1965).

[31] Cenco, Cat. J200, Item #71264.

a slowly alternating magnetic field perpendicular to the static field. The ball in the air bearing is centered between the coils. It can be set spinning around its magnetic axis by slightly tilting the bearing arrangement to one side.

If the current in the coils is alternated at the proper frequency, the magnetic axis of the ball will start to precess around the direction of the static field at an increasing angle. The apparatus permits a qualitative study of the basic principles of magnetic resonance, including absorption of energy from the alternating field. The demonstration can be made visible to a large group by means of closed circuit television.

41–4.6[32]

The concept of a spinning magnetic dipole in an external magnetic field and its interaction with an alternating magnetic field is commonplace in atomic and nuclear physics. It is basic to the understanding of such phenomena as Russell–Saunders coupling, Zeeman splitting of spectral lines, molecular-beam magnetic resonance, nuclear magnetic absorption, and nuclear magnetic induction. The dynamics involved in these phenomena are capable of being demonstrated on a macroscopic scale. An air-driven spinning magnetic dipole placed in a magnetic field exhibits a Larmor precession whose dependence on magnetic and gravitational parameters can be demonstrated. An ac magnetic field of variable frequency permits a visual demonstration of magnetic resonance.

The model is a magnetic top, supported and spun by compressed air. It contains a vertically oriented Alnico 5 rod-magnet and operates within a dc horizontal and an ac vertical Helmholtz coil combination. See Fig. 41–60. This demonstration can be made visible to a large group through the use of closed circuit television. The construction details are found in the Appendix, page 1393.

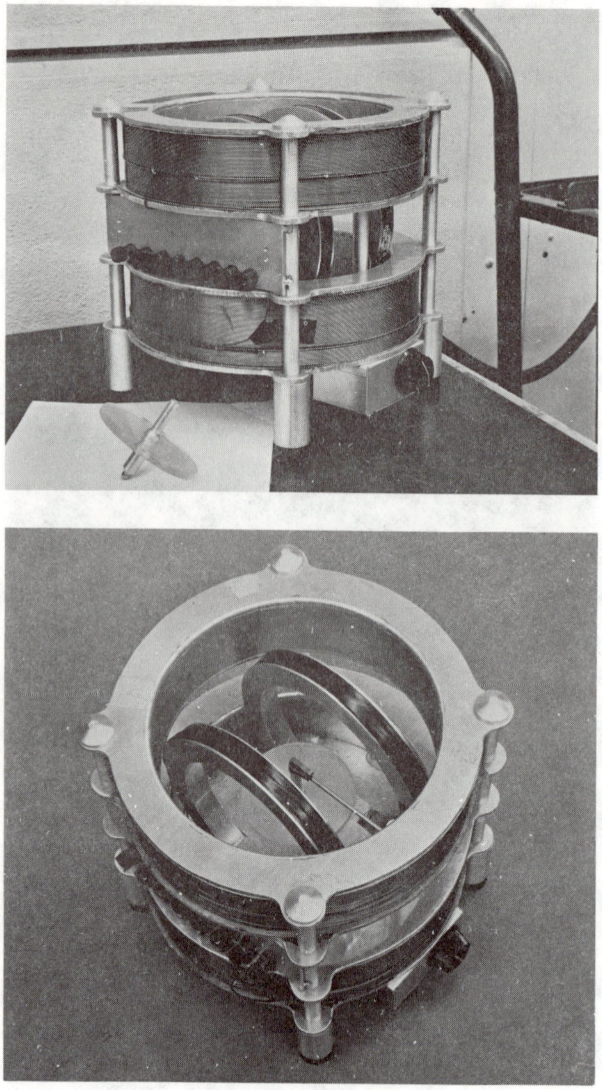

Fig. 41–60

41–4.7

Although it is relatively easy to demonstrate the resonance principle with a marginal oscillator, that method affords only limited insight into the practical measurements possible with the powerful tool of nuclear magnetic resonance. A simplified spin-echo[33] instrument can be used to determine the gyromagnetic ratio and both spin-spin and spin-lattice relaxation times. The theory of nuclear magnetic resonance and the spin-echo method are discussed in the references, particularly the first one.

An Arenberg PG-650-C pulsed oscillator is the heart of the apparatus. Other oscillators may be used, but the Arenberg is chosen since it can be found in most laboratories in which research in nuclear magnetic resonance or ultrasonics is done. Further, the PG-650-C has the advantage of producing two variable length pulses from a single

(32) F. Verbrugge, *Am. J. Phys.*, **21**, 603 (1953).

(33) E. L. Hahn, *Phys. Today*, **6**, 4–9 (1953).

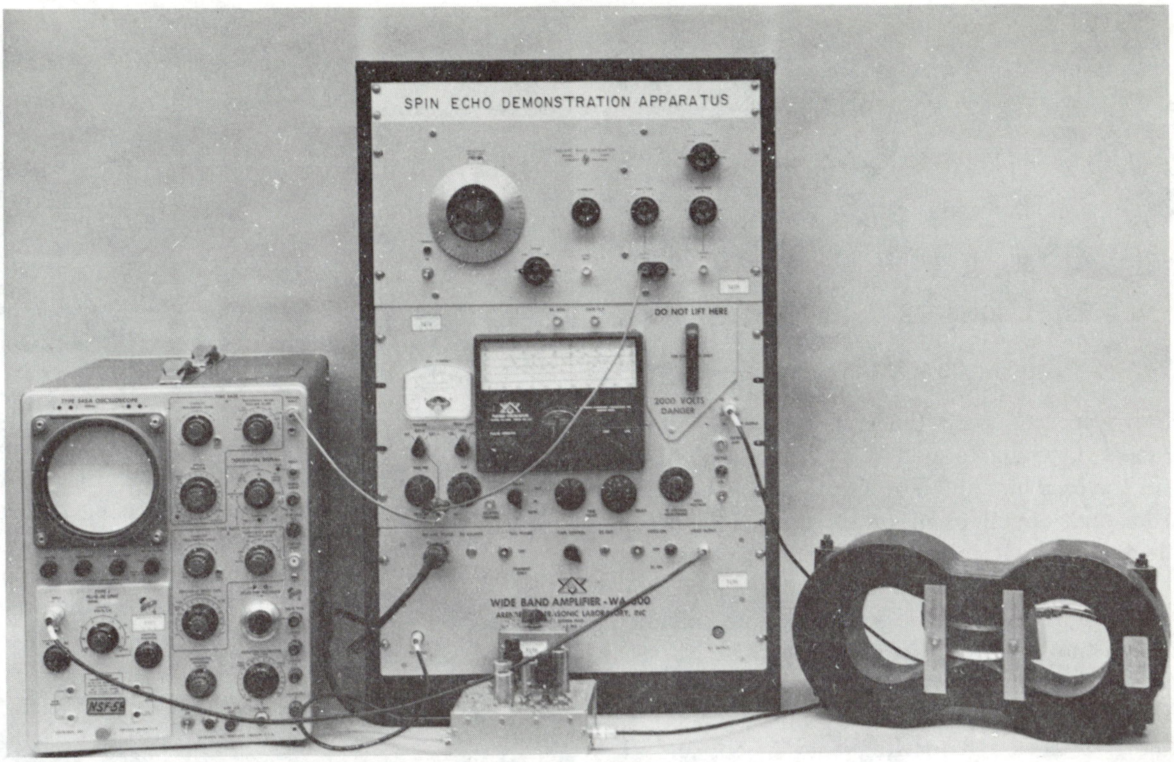

Fig. 41–61

trigger, with a variable delay between them. To provide slower repetition rates, an external square-wave generator is used. Figure 41–61 shows the complete apparatus consisting of the square-wave generator, the PG-650-C, and an Arenberg WA-600 broad-band amplifier. The small chassis in the foreground is an Arenberg preamplifier. The sample coil is mounted to a small aluminum box containing a variable capacitor, and the box is mounted to a permanent magnet. If a high voltage oscilloscope is used, the trace of the echo can be projected.

For the several experiments described below, the apparatus is connected as shown in Fig. 41–62, using the dotted line. The square wave generator is adjusted to produce 5-V–peak-to-peak square waves at 20 cps. An alternate scheme, requiring a pulsed oscillator producing only single pulses rather than the sophisticated PG-650-C, is described in Sec. IV, below.

I. Determination of the Gyromagnetic Ratio. Place the sample to be measured in the sample coil and adjust the pulsed oscillator as follows: set the pulse amplitude to produce rf pulses of about 400 V peak-to-peak ("60" on the PG-650-C); set the pulse delay switch to produce one rf pulse per trigger pulse ("out" on the PG-650-C); adjust the pulse length control to produce 5-μsec pulses, and set the oscilloscope control at 2 msec/cm with the vertical gain at 10 mV/cm. Adjust the gain of the broad-band amplifier until the background noise on the oscilloscope is 10 mV; reduce the vertical gain on the oscilloscope to 0.5 V/cm. The pulsed oscillator

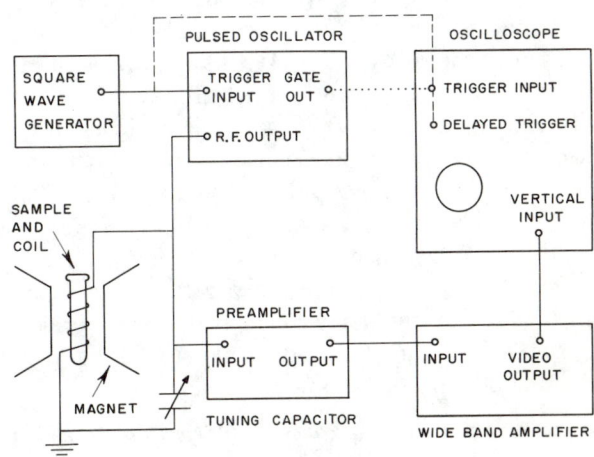

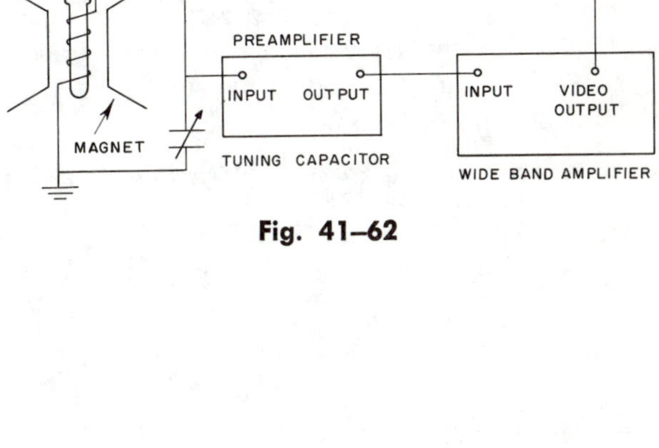

Fig. 41–62

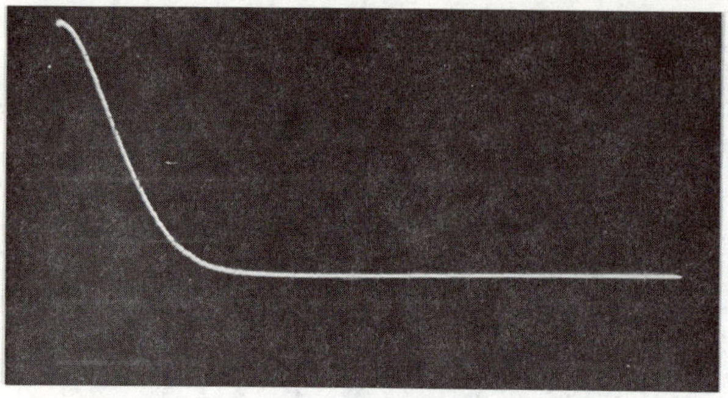

Fig. 41–63

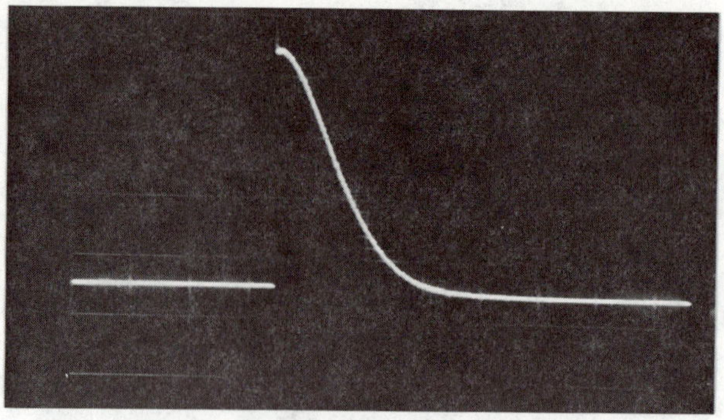

Fig. 41–64

is slowly turned until the rf pulse broadens into an induction tail,[34] such as shown in Fig. 41–63. Tune the oscillator and adjust the variable capacitor in the tank circuit to maximize the height of the induction tail. It may be necessary to peak the tuning of the preamplifier. The gyromagnetic ratio, γ, of the nuclei contained in the sample coil then can be computed from the relationship

$$2\pi f = \gamma H_0,$$

where f is the oscillator frequency in cps and H_0 the magnitude of the static magnetic field in gauss.

II. Determination of the Spin–Spin Relaxation Time.[35] Produce the induction tail as described in I, and advance the pulse-length control until the induction tail reaches a maximum height. This produces a 90-deg pulse of about 8 μsec duration. With the pulse delay switch still set to "out," set the delay-range switch to the 11k μsec position and the pulse delay vernier to 100. This produces a pulse occurring about 1.1 msec after the trigger pulse. Set the delayed pulse length control to a minimum and slowly increase it until an induction tail is clearly visible (Fig. 41–64). As this control is advanced still further, the tail height passes through a maximum (90-deg pulse) and begins to decrease, finally disappearing entirely. This is the condition for a 180-deg pulse, about 16 μsec long in this case. Now, set the pulse delay switch to the "both" position which produces a 90-deg pulse followed by a 180-deg pulse, separated by a delay determined by the delay range switch and vernier. A spin-echo should now be visible, appearing after the 180-deg pulse by the same amount that the 180-deg pulse occurs after the 90-deg pulse. (See Figs. 41–65 and 41–66.) Increasing the pulse delay diminishes echo amplitude, and decreasing the delay increases it. To insure that the amplifier is not saturated, reduce the gain of the amplifier until the height of the induction tail is slightly less than that of the 90-deg pulse. The spin-echo now has less

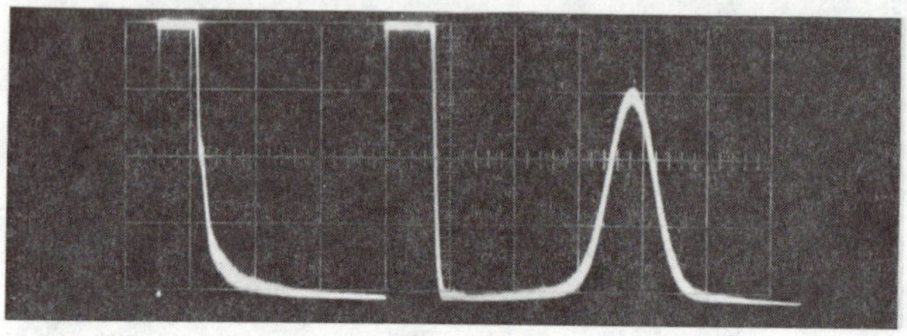

Fig. 41–65

amplitude than the induction tail. As shown in Fig. 41–66, the spin–spin relaxation time is the time constant of the decay of the spin-echo amplitude and is given by[36]

$$T_2 = \frac{2T}{\ln(h_2/h_1)},$$

where h_1 is the amplitude of the induction tail, h_2 the height of the spin-echo, and T the pulse delay. As in Fig. 41–65, a spin-echo of approximately the same width as the rf pulses, is produced by moving the sample into a very homogeneous portion of the field near the center of the pole faces (see Appendix, page 1395) and reducing the pulse delay to about 50 μsec. In this manner, the oscilloscope sweep can be expanded to show the relation between the 90- and 100-deg pulses.

III. Determination of the Spin-Lattice Relaxation Time. Set the delay switch to the "in" posi-

[34] A. Abragam, *The Principles of Nuclear Magnetism* (Oxford Press, London, 1962), p. 63.

[35] A. Abragam, *op. cit.*, pp. 58–59.

[36] A. Abragam, *op. cit.*, p. 59.

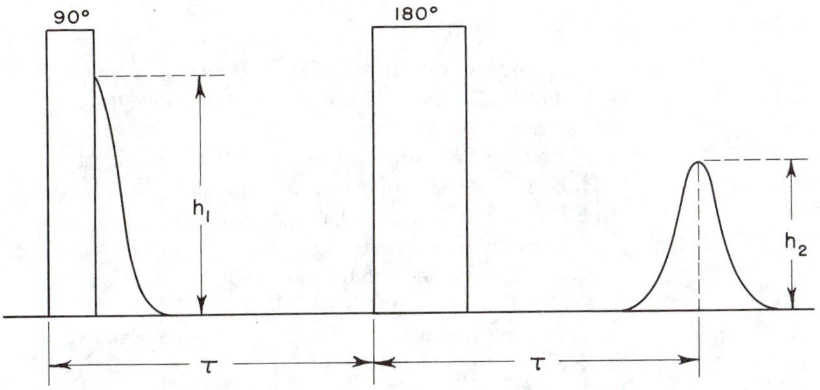

Fig. 41–66

tion to produce only the delayed pulse and adjust its length to 90 deg, using the methods of I and II. Then, set the delay switch to "out," and tune the first pulse to produce a 180-deg pulse. Finally, set the switch to the "both" position, producing a 180-deg pulse and a 90-deg pulse with an induction tail. As the pulse delay is increased, the amplitude of the induction tail diminishes to zero and begins to increase again. The delay, t, between the 180- and 90-deg pulses for zero amplitude provides a very simple determination of the spin-lattice relaxation time, T_1,(37) which can be calculated from

$$T_1 = \frac{t}{\ln 2}.$$

IV. Spin-Echo Demonstration Using a Pulsed Oscillator Producing Only Single rf Pulses. If a pulsed oscillator producing two variable-length pulses is not available, the spin-echo technique still can be demonstrated by using a pulsed oscillator producing single pulses and an oscilloscope with a built-in sweep delay. The apparatus arrangement is set up as in Fig. 41–62, using the broken- rather than dotted-line connection. Both the pulsed oscillator and the oscilloscope are triggered initially by the square-wave generator, and then the pulsed oscillator is triggered by the delayed trigger of the oscilloscope to produce two pulses of equal length, followed by an echo as shown in Fig. 41–67. This technique is somewhat more difficult to analyze(38) than the 90-deg, 180-deg pulse sequence, but it does make the demonstration possible with somewhat less complicated apparatus.

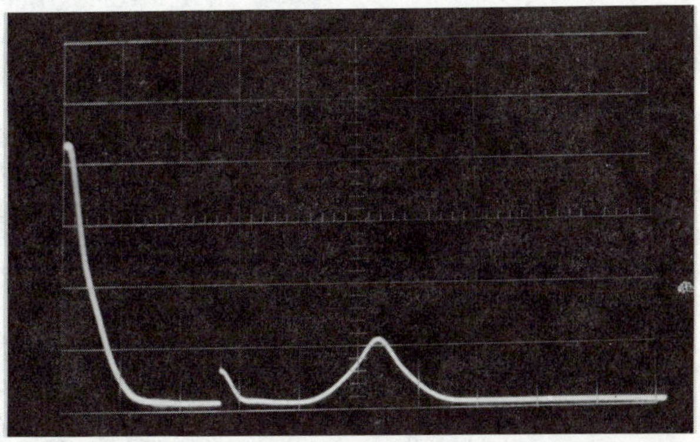

Fig. 41–67

(37) A. Abragam, *op. cit.*, p. 64.

(38) E. L. Hahn, *Phys. Rev.*, **80**, 580 (1960).

Demonstrations	***Contributors***
41–1.1	University of Illinois
41–1.2	K. Hecht, E. Leybold's Nachfolger; P. Sprawls, Emory University
41–1.3 to 41–1.6	K. Hecht, E. Leybold's Nachfolger
41–1.7	A. P. French, Massachusetts Institute of Technology
41–1.8	Princeton University, Swiss Federal Institute (ETH), Zürich, Switzerland
41–1.9	R. A. Llewellyn, Purdue University
41–2.1	V. E. Eaton, Wesleyan University
41–2.2	Eric M. Rogers and H. M. Waage, Princeton University

Demonstrations	***Contributors***
41–2.3	Eric M. Rogers and H. M. Waage, Princeton University
41–2.4	G. F. Powers, Northeast Louisiana State College and J. E. Smith, Arkansas State Teachers College
41–2.5	Swiss Federal Institute (ETH), Zürich, Switzerland
41–2.6	P. Huber, University of Basel, Switzerland; Swiss Federal Institute (ETH), Zürich, Switzerland; M. G. White, Princeton University
41–2.7	R. A. Llewellyn, Purdue University
41–2.8	E. Brown, Rensselaer Polytechnic Institute
41–2.9	N. Ashby, University of Colorado
41–2.10	Swiss Federal Institute (ETH), Zürich, Switzerland
41–2.11	K. Hecht, E. Leybold's Nachfolger
41–2.12	P. Huber, University of Basel, Switzerland
41–2.13	P. Huber, University of Basel, Switzerland
41–3.1	H. A. Medicus, Rensselaer Polytechnic Institute
41–3.2	Elmer L. Galley, Oak Ridge Institute of Nuclear Studies
41–3.3	P. Teunissen and S. Wynobel, Technical University, Delft, Holland
41–3.4	Swiss Federal Institute (ETH), Zürich, Switzerland
41–3.5	P. Huber, University of Basel, Switzerland
41–3.6	F. E. Christensen, E. P. Ney, F. Oppenheimer, E. Lefgren, University of Minnesota
41–3.7	E. M. Rogers, H. M. Waage, Princeton University
41–3.8	H. M. Waage, Princeton University
41–3.9	J. A. Earl, University of Minnesota
41–4.1	H. S. Story, State University of New York at Albany
41–4.2	F. Bitter, Massachusetts Institute of Technology
41–4.3	P. Huber, University of Basel, Switzerland
41–4.4	H. E. White, University of California, Berkeley
41–4.5	H. A. Daw, New Mexico State University
41–4.6	F. Verbrugge, University of Minnesota
41–4.7	C. E. Tarr, C. V. Briscoe, and E. Merzbacher, University of North Carolina

APPENDICES TO DEMONSTRATIONS

Construction Details and Materials Lists

Chapter 26 Heat and Thermodynamics

A26–3.10

Figures A26–45 and A26–46 show the heat conduction apparatus with an aluminum base a ($1\frac{1}{2}$ x $\frac{3}{16}$ x $12\frac{5}{8}$ in.). The forward $1\frac{1}{8}$ in. is reduced to a thickness of $\frac{5}{32}$ in. Holes $\frac{1}{8}$ in. in diameter are drilled at b. The base plate is fastened to the frame c by a $\frac{3}{16}$-in. screw d so that the apparatus can be turned around this axis. A piece of rubber hose e is placed between the frame and base plate to provide friction to allow the plate to be held in any position.

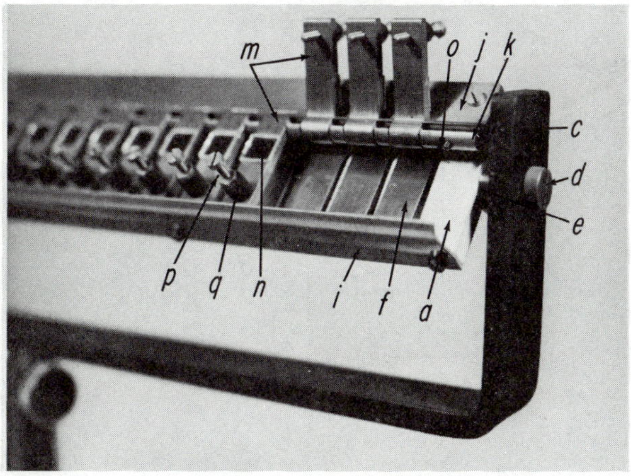

Fig. A26–45

Twenty-three bimetallic plates f ($\frac{3}{8}$ x 0.02 x $1\frac{1}{2}$ in.) are attached with $\frac{1}{8}$ in. between plates to the base plate by a clamping device g made from an aluminum sheet ($\frac{3}{8}$ x $\frac{1}{16}$ x $12\frac{5}{8}$ in.). An aluminum washer h ($\frac{3}{8}$ x $\frac{3}{8}$ x $\frac{3}{32}$ in.) provides clearance for the screw heads. An aluminum sheet i is attached to the front end for protection from the flame.

The dimensions of the mirror support bearing j are shown in Fig. A26–46. In the round part, a slit $\frac{3}{32}$ in. wide and $\frac{1}{8}$ in. deep is milled to hold the axis k, which is made of $\frac{3}{32}$-in. steel wire, $12\frac{5}{8}$ in. long. On both ends, $\frac{9}{16}$ in. is left so that a slit $\frac{5}{16}$ in. wide and $\frac{9}{32}$ in. deep can be milled. Everything is then mounted on the base plate by the screws at each end of the mirror support bearing.

The mirror support m is $\frac{5}{16}$ in. wide and has a $\frac{3}{32}$-in.-diam hole drilled for the axis. A small mirror n ($\frac{5}{16}$ x $\frac{5}{16}$ in. x $\frac{1}{32}$ in.) is glued to the upper part. A screw o attaches the axis to the support. The image on the screen is aligned by an adjusting screw p ($\frac{3}{32}$ x $1\frac{1}{2}$ in.), which has a nut and a rubber sleeve q to retain it. Since the mirrors cannot be glued with sufficient precision, they must be selected in order to provide a good alignment of the images.

When the horizontal slit f in Fig. 26–17 (page 765) and the mirrors have been adjusted so that a straight line of light is imaged onto the screen, a pointed gas flame is directed from behind base plate a (Fig. A26–46) at its center. Immediately, a curve similar to h in Fig. 26–18 is obtained. After the flame is shut off, the curve decreases in height according to the heat conduction equation. After a long time, homogeneous temperature is obtained, but at a higher temperature level. A reference line should be placed at the height of the row images before the experiment is started.

A26–6.2 Ratchet and Pawl

Construction details for the ratchet and pawl can be seen in Figs. A26–47 and A26–48. The ratchet wheel is made from a 16-in.-diam disk or $\frac{1}{4}$-in. hardwood. Twenty-four teeth are cut around the

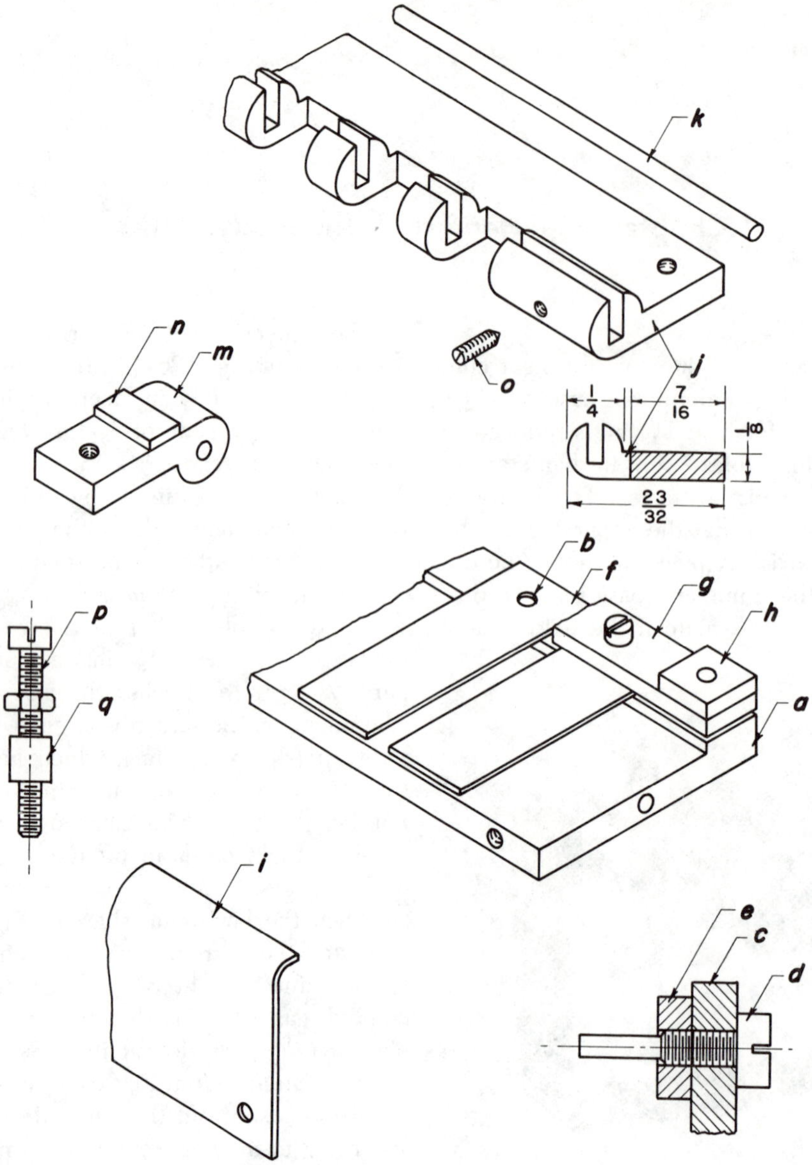

Fig. A26–46

perimeter of this circle. Each tooth has a rise of ½ in. and a run of 1⅞ in. On the underside of the ratchet wheel is a concentric 4½-in. plate. Extending from the center of this plate is a ¼-in.-long by 1-in.-diam brass cylinder. This cylinder fits in the center of the roller bearing, 2 in. o.d., 1 in. i.d., ⅛ in. thick, on which the brass plate revolves. This bearing is held by a ⅛-in.-thick brass ring screwed to the 16-in. by 21-in. platform as shown in the photographs.

The pawl measures 4 x ¾ x ⅜ in. It has a single roller bearing fastened at one end. It is mounted so it is free to swivel on a 3½-in. by 5-in. plate, and it is held in tension by a spring as illustrated. The pivot of the pawl is set back ½ in. from one edge and 1¾ in. from the other.

Figure A26–49 is a full-size tracing of the four-lobed cam that drives the pawl. It is mounted as shown and is manually turned by means of a crank handle. Turning this handle causes the pawl to

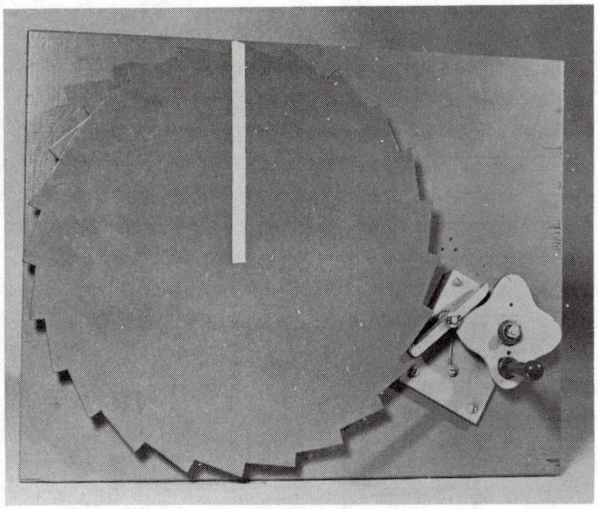

Fig. A26–47

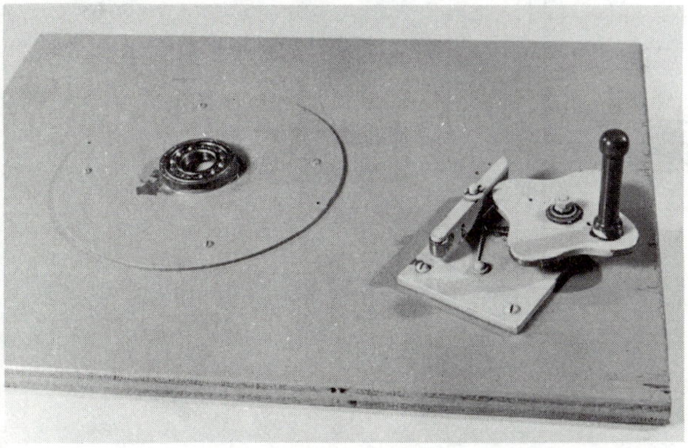

Fig. A26–48

move in a manner analogous to the way a smaller pawl would move under thermal agitation causing the ratchet to revolve in a direction opposite to its normal mode of operation as described in the reference.

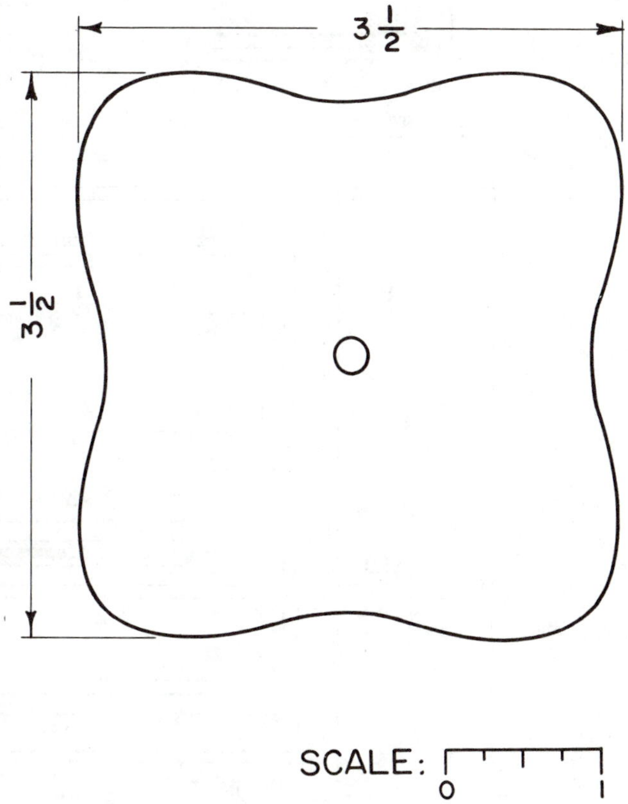

Fig. A26–49

Chapter 27 Gases and Kinetic Theory

A27–2.5

Details of the water cell are shown in Fig. A27–82. The cell measures 6 in. square by 1½ in. thick. Its square sides are glass plates (4^{15}⁄$_{16}$ x 4^{15}⁄$_{16}$ x ⅛ in.), which are sealed to the aluminum frame with rectangularly arranged ⅛-in. O-rings *m*. The ⅛-in.-thick aluminum glass retainers *n* are held to the 1½ x ⅞-in. aluminum bar by 10-24 NC flat head brass screws. All pipe fittings should be centered in the sides. The scale is a section of a transparent centimeter rule mounted on a ¾ x 1⁄16-in. piece of aluminum which is jam fitted against the walls.

The heat exchanger uses a 9¾-in-long coil of ¼-in.-o.d. copper tubing immersed in a 3¾-in.-o.d. brass tube, as shown in Fig. A27–83. A flat rubber gasket is used to seal the top of the tube. Rubber tubing is used to carry the water between the cell, exchanger, and pump. The pump is set upon a rubber pad in a sink and covered with sound-insulating material. It can be primed by squeezing the rubber tubing.

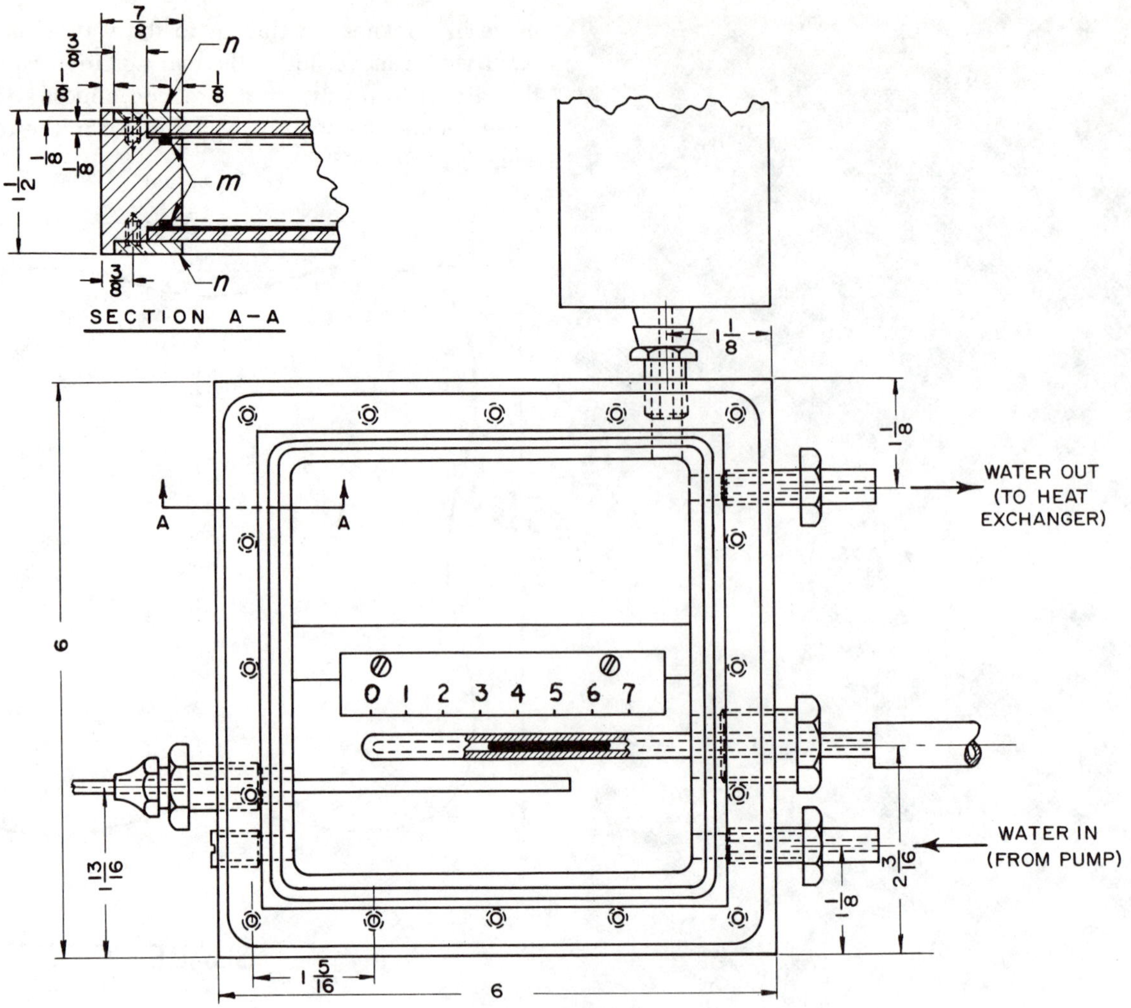

Fig. A27–82

Insertion of Dry Air and Mercury into the Capillary Tube

Figure A27–84 shows the arrangement to be used to insert mercury into the capillary tube *p* (1 mm i.d., 6 mm o.d., 8½ in. long). Copper tubing *q* (¼ in. i.d., 2 in. long) and thick-wall rubber tubing *r* are used as connectors. Desiccants are held in the 1½ x 17-in. glass tube *s* with glass wool, and the ends of the glass tube are sealed with pierced No. 8 rubber corks *t*. The system is evacuated for about 10 min, and air is then permitted to enter the system by means of the three-way valve *u*. This is done five or six times to flush out the capillary tube. Then the capillary tube is heated and disconnected from the system under mercury. Heating is continued to drive some of the air from the tube. Be sure to keep the open end of the capillary under the mercury. Then the heat source is removed and the tube allowed to cool. A column of mercury will be drawn into the capillary. After 2 in. of mercury has been drawn into the tube, it is withdrawn from the mercury bath and held vertical with the open end up. A 10-in. glass pipette is then used to withdraw some of the trapped air until 2⅜ in. of air remains in the tube.

A27–4.2

Two brass end plates *a* (1/16 x 4 in. diam) and a brass cylinder *b* (7 11/16 x 4 in. o.d., 3¾ in. i.d.) form the structure of the apparatus (Fig. A27–85). Sections of the cylinder sides are cut away for accessibility so that three equally spaced ¾ x 6 9/16-in.

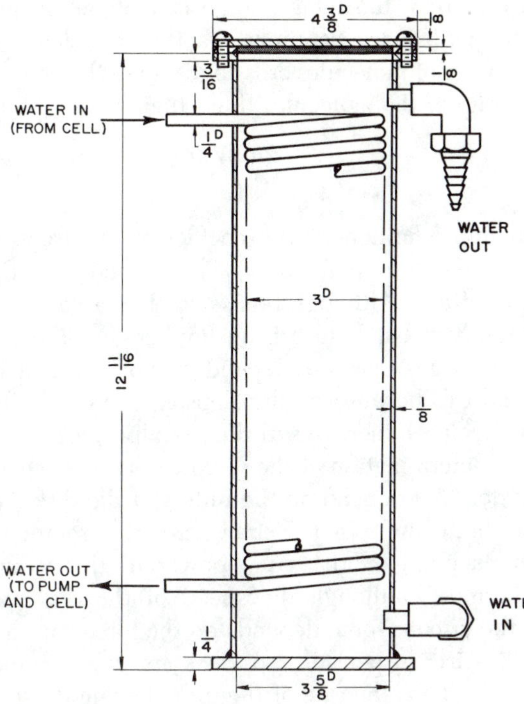

Fig. A27–83

pillars support two end caps. A $1\frac{7}{8}$-in. vertical slit wide enough for the $\frac{3}{16}$-in.-diam rods *c* to slide freely are centered in the pillars. Brass stops *d* ($\frac{1}{8}$ x $\frac{5}{16}$ x $\frac{3}{4}$ in.) are screwed across the slots with their centers $3\frac{13}{32}$ in. above the bottom of the apparatus; the screw-holes in these stops are $\frac{3}{32}$ in. long to allow for adjustment. The apparatus rests on rubber feet.

In the center of the apparatus is the oscillating aluminum disk *e* (0.035 x $1\frac{3}{4}$ in. diam). This disk is epoxied to a short piece of ceramic tubing *f*, $\frac{1}{16}$ in. o.d., which is mounted axially in the disk and which has *two* holes through its length for the .003–.004-in. spring steel suspension wire *g*.

The two fixed plates *h* mounted parallel to and on either side of the disk are machined from $\frac{3}{8}$ x $1\frac{11}{16}$-in.-diam pieces of brass. After machining, the plates have a squat, U-shaped cross section; the face is $\frac{1}{8}$ in. thick with a $\frac{1}{4}$ in.-diam hole in its center, and the sides are $\frac{7}{32}$ in. thick with three 10-24 NC threaded holes placed 120 deg around the circumference and centered $\frac{1}{8}$ in. from the open edge. The plates open away from the disk.

Into the threaded holes are screwed six (three holes in each of two plates) brass rods ($1\frac{3}{4}$ x $\frac{3}{16}$ in. diam with $\frac{1}{4}$ in. of 10-24 NC thread). Between each rod and a screw, a spring *i* made from 0.02-mm-diam wire (39 turns, 7 mm o.d., 20 mm long) is stretched to maintain maximum spacing between the plates. The maximum spacing is controlled by the use of brass locking blocks *j*, which allow the springs to move the plates only 0.06 in. apart. These blocks are $\frac{3}{8}$ x $\frac{3}{8}$ x 1 in., with two $\frac{1}{4}$-in.-deep round-end slots ($\frac{9}{32}$ x $\frac{7}{32}$ in.), which are spaced $\frac{5}{32}$ in. apart along one side of each block and which fit over the ends of the rods. When these blocks are removed, the springs pull the plates apart until the rods stop at the ends of the slots.

One end of the suspension wire is wrapped around a small bolt between two nuts *k*. The other end is let down through a spring, a sleeve, and one hole of the tube to a plug *m* at the bottom. Then it is led back up to the small bolt along the same path, this time through the other tube hole. The center of the disk is $3\frac{13}{32}$ in. from the bottom of the apparatus. The sleeve at the top has one spline into which a setscrew (not shown in Fig. A27–85) fits to allow the sleeve to be adjusted up and down, but not turned. Also this sleeve is threaded at the top to accept an open cylinder *n*. Turning the cylinder moves the sleeve up or down, decreasing or increasing the tension on the suspension wire.

Theory

When two surfaces move past one another with a layer of air between, the air exerts a retarding force due to its viscosity. This effect is very differ-

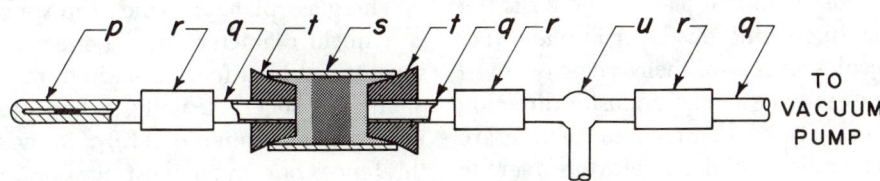

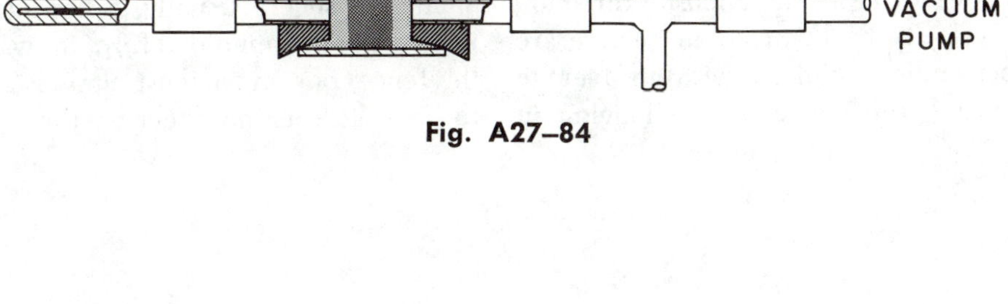

Fig. A27–84

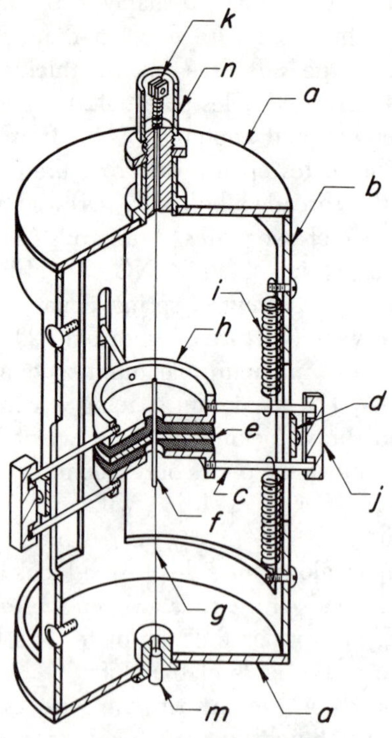

Fig. A27–85

ent from the viscosity of a liquid in which the viscosity is caused by the intermolecular forces among the molecules of the liquid. It is a kinetic effect caused by the molecules of air, which hit the moving plate increasing their momentum in the direction of motion and, therefore, exerting a force. The retarding force per square centimeter is

$$\frac{F}{\text{cm}^2} = \eta \frac{V}{s},$$

where V is the relative velocity of the plates, s is the separation of the plates, and η is the coefficient of viscosity (in poises). A coefficient of viscosity of one poise would exert a force of 1 dyne/cm² with a velocity gradient of 1 cm/sec².

To calculate the coefficient of viscosity in terms of the properties of the gas, we can think of a layer of gas thickness l between two plates, where l is the mean free path. In crossing this layer in one direction, the molecules increase their velocity by l (V/s). The ones moving in the opposite direction decrease their velocity by this amount. There are n molecules per cubic centimeter with an average velocity $\bar{v}$. Since $\frac{1}{6}$ $n\bar{v}$ molecules are moving in each direction, the total change in momentum per second per square centimeter across this layer is the number of molecules crossing it times the change in velocity of the molecules times their mass. Thus,

$$\frac{F}{\text{cm}^2} = \frac{2n\bar{v}lVm}{6s}.$$

From the definition of the coefficient of viscosity, we see that $\eta = nm\bar{v}l/3 = \rho\bar{v}l/3$, where ρ is the density of air. Although this seems like a very natural result, it has a surprising implication: the viscosity of a gas does not depend on how thick it is. This is true because as the density decreases, the mean free path increases in the same proportion.

The determination of the viscosity by this experiment does not depend on the radius of the disk, because both the viscous drag and the rotational inertia depend on the same power of the radius. Furthermore, although the speed of the disk and thus the viscous force depends on the force constant of the wire suspension, so does its angular momentum. Thus the rate of the disk slowing down is independent of the suspension. The viscosity is given by

$$\eta = ds\sigma(1/t_1 - 1/t_2),$$

where t_1 is the time required for the amplitude to decrease by $1/e$, when the plates are close together; t_2 is the time when the plates are far apart (i.e., when there is no viscous damping); d is the density of the disk; and σ its thickness. In this apparatus, $d = 2.7$ g/cm² and $\sigma = 0.06$ cm. The distance s between the aluminum and brass disks is set at about 0.04 cm. The experiment can be extended to obtain approximately the diameter of an air molecule. In this case the density of the air, Avogadro's number, the velocity of sound, and the value of the viscosity must be used.

A27–7.1 A Dynamic Hard Sphere Model

Model Design

The glass plate on which the spheres are placed is 9 in. in diameter and ⅛ in. in thickness. It is surrounded by a fence 2 in. high made from $\frac{1}{16}$-in. aluminum sheet. Usually, a corrugated aluminum liner with an amplitude of $\frac{3}{16}$ in. was placed inside the fence, but under most circumstances it seemed to have little or no effect on the operation of the

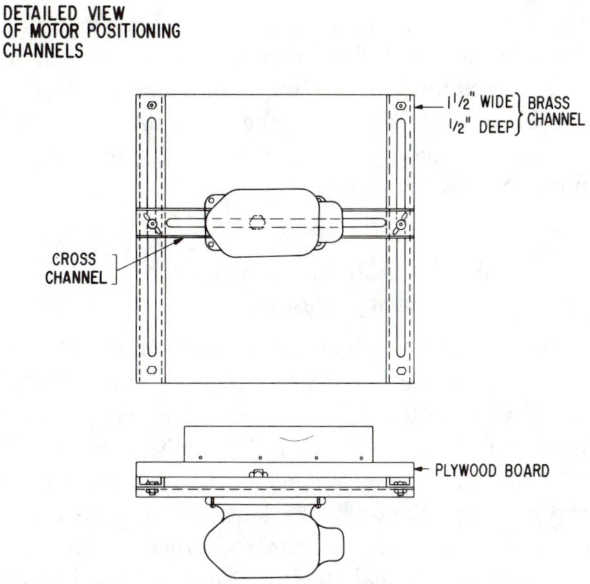

Fig. A27–86

model. The glass plate is positioned in the center of a plywood board, 15 in. square by $\frac{3}{4}$ in. thick, which is suspended by springs attached to each of its corners from an angle-iron framework. Leveling is accomplished by turnbuckle screws placed between the framework and each spring. Coil springs with a constant of 3.18×10^5 dyn/cm are used.

The eccentric weight which vibrates the board is an iron half-cylinder about 1 in. in radius and 1 in. in length. It is located under the center of the board and rotated by a universal ac-dc $\frac{1}{30}$-hp variable speed motor. The drive shaft is $\frac{1}{2}$ in. long and its maximum rate of rotation is about 1700 rpm. Figure A27–86 shows how the motor is mounted on the underside of the board. The motor is fastened in the slot of a brass crossbar and can be positioned anywhere along the bar. In turn the crossbar is attached to two slotted brass side bars located along opposite edges of the board. The crossbar can be put in any desired position parallel to one edge by sliding it along the side bars.

A vibration tester[(1)] can be used to impart vibrational energy to the model. To do this, the model is bolted securely to the vibration tester. This arrangement works very satisfactorily. The vibrational frequency and amplitude can be controlled more precisely and varied over a wider range than with the simpler arrangement described above.

The model can also be activated by a loudspeaker driven by the amplified output of an oscillator.[(2)] The loudspeaker is mounted over the plywood board which rests on squares of sponge rubber 0.25 in. thick. The horizontal plate on which the spheres are placed is coupled to the speaker through a cork ring about 2.5 in. diam. Both the frequency and energy of the vibration can be varied with the settings of the oscillator and amplifier. No "hot spots" are apparent in the model when it is activated in this way. Except where noted otherwise, the observations were made using the rotating eccentric weight.

Operation of Model

The motor speed is controlled by a variable auto transformer. In operating the model, a speed is selected which causes the board to vibrate near its resonant frequency. The amplitude of vibration and hence the energy of the spheres can be varied considerably by small changes in the motor speed.

These spheres have been used in the model: glass (3, 4, 5, or 6 mm diam); iron (4 mm diam), aluminum (4 mm diam), or polyethylene (4, or 5 mm diam). All have worked satisfactorily. The aluminum spheres acquire an electric charge from the glass base during their motion and repel each other. When the humidity is very low, the glass spheres behave similarly. Most observations were made using 4-mm-diam glass spheres supplied by the Fisher Scientific Company. Thus, there could be as many as 2400 spheres in the base layer of the model.

During the operation of the model, the spheres appear to move randomly and marked spheres appear to become distributed at random after a time. Usually the lateral velocity of the spheres averages around 1 cm/sec, corresponding to a kinetic theory temperature of the order of 10^{16} °K. Very long paths, such as are seen at a low sphere density, appear to be curved slightly. The average energy of the spheres seems to be uniform over most of the surface, but there usually remains a small area over which the average energy is noticeably higher than

(1) Model No. T-51-MC, type G, MB Electronics, New Haven, Connecticut. This method was suggested by F. D. Bundy, General Electric Research Laboratory, Schenectady, New York.

(2) Beat frequency oscillator, type 913A, General Radio Co., Cambridge, Massachusetts. This method was suggested by B. Egan, Education Development Center, Newton, Massachusetts.

elsewhere. These "warm spots" could not be eliminated consistently in the simple apparatus. However, no warm spots are apparent when the vibrations were imparted by the MB vibrational tester.

The most critical requirements for satisfactory operation of the model are that the plane of the board be kept horizontal and the eccentric weight be positioned under the center of the board.

In most of the work mutually unattracting spheres have been used. Nevertheless, even without attraction, the model illustrates a remarkable number of atomic interactions believed to occur in actual matter. The reason for this is that on the inside of dense matter the atomic attractions, to a large extent, cancel each other out.

Two methods have been found whereby the spheres are made to attract each other. One method is to magnetize steel spheres. In this case, the attractive force falls off relatively slowly with distance, but unfortunately the force law between spheres changes when they associate. A second method is to coat the spheres of any of the materials with a thin liquid film. This results in a very short range attraction due to capillary forces. Nucleation can be produced by dropping three drops of ethyl alcohol into the operating model containing aluminum spheres moving on an aluminum plate.

A27–7.3 Two-Dimensional Kinetic Theory Model

Figure A27–87 shows the details of the kinetic theory apparatus. A sheet of foam rubber *c* ($3\frac{1}{2}$ x 4 x $\frac{3}{4}$ in.) is placed between motor *b* and plywood base *a* (8 x 21 x $\frac{3}{8}$ in.). A Lucite push bar *g* ($\frac{3}{4}$ x $3\frac{5}{8}$ x $\frac{1}{4}$ in.) is connected to a Lucite push rod *f* ($\frac{1}{4}$ in. diam $4\frac{9}{16}$ in. long). The push rod is connected to another Lucite rod *e* made from $\frac{1}{4}$ x $\frac{3}{8}$-in. cross-sectional stock which is attached to an eccentric aluminum hub *d* (1 in. diam, $\frac{11}{16}$ in. long) that is connected to the shaft of the motor by setscrews.

The aluminum guide block *i* is 1 x 1 x $1\frac{1}{2}$ in. and

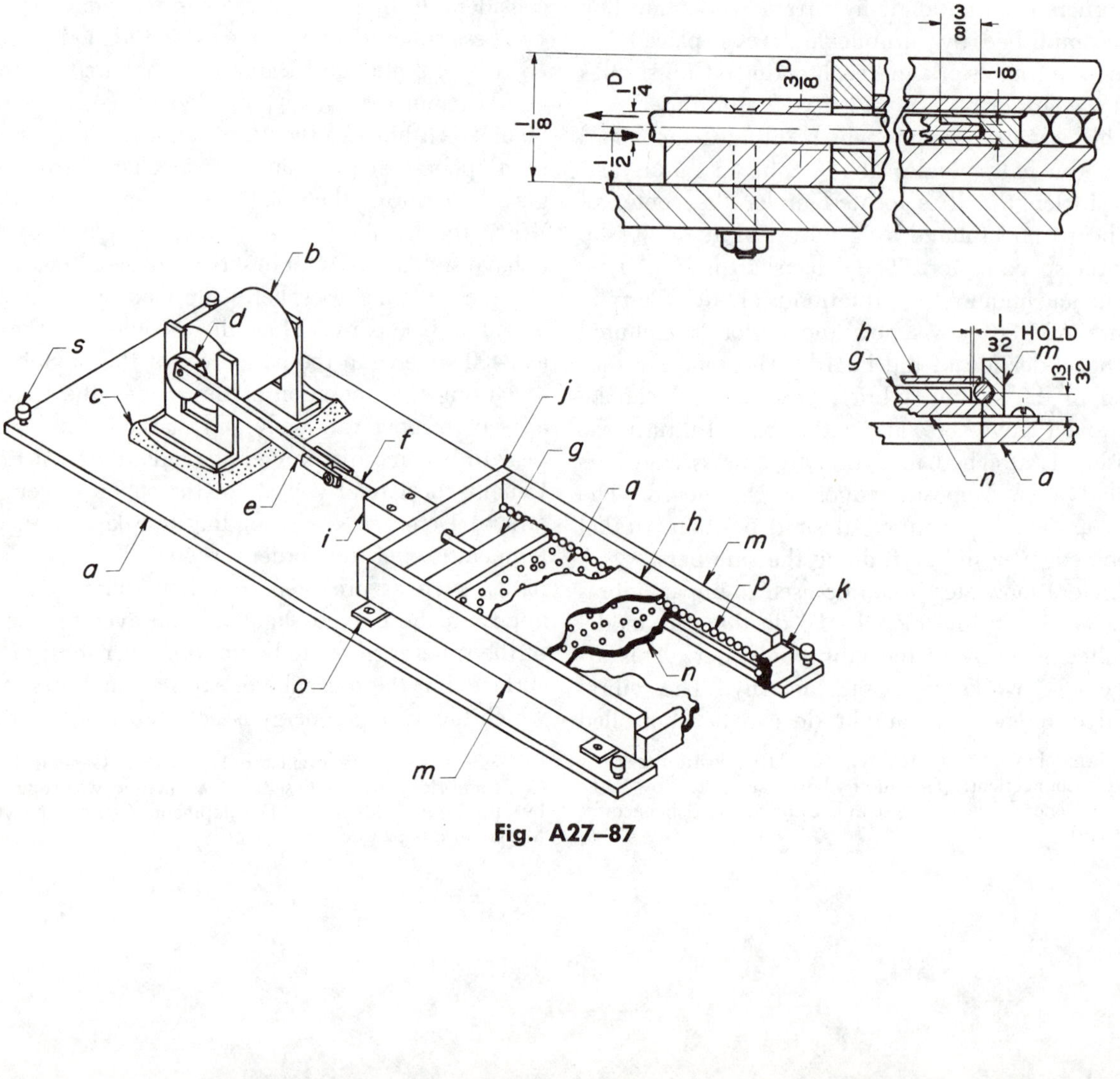

Fig. A27–87

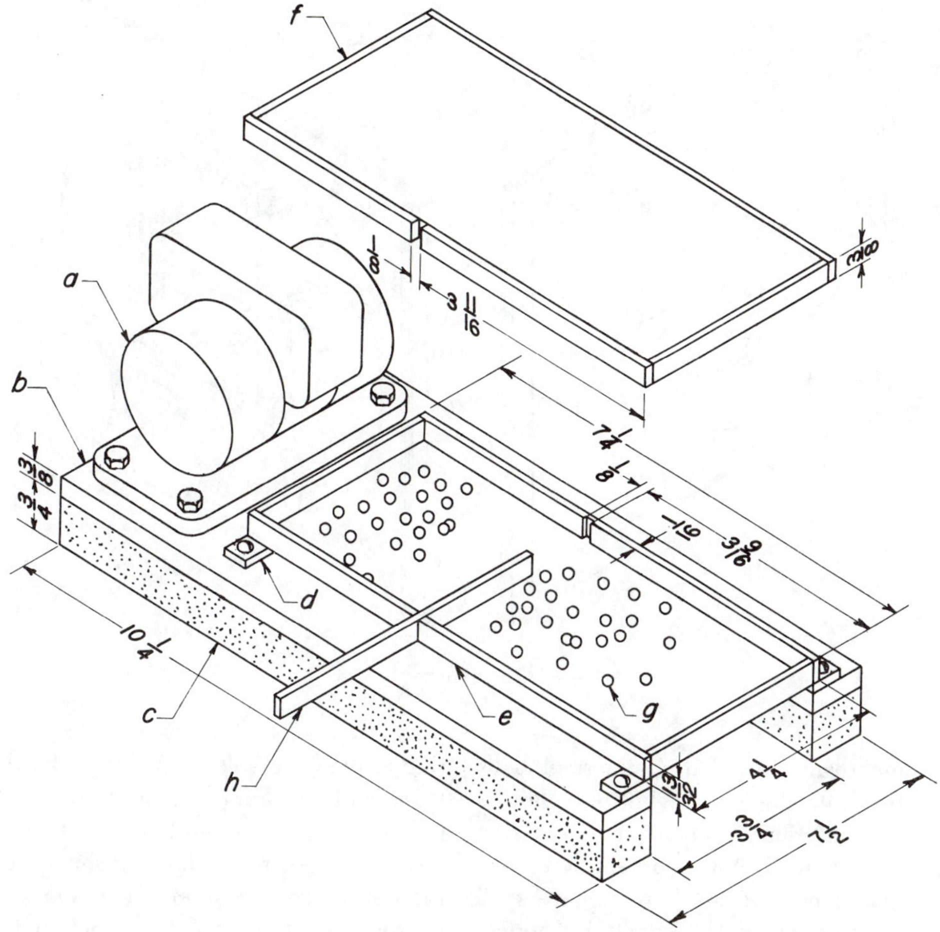

Fig. A27–88

has a hole drilled through its center for a sliding fit with the push rod. A Lucite cover plate *h* (3^{15}⁄$_{16}$ x 7¼ x ⅛ in.) fits between the two 1 x 7¼ x ⅜-in. sides *m* and the front piece *j* (1 in. high) and the rear piece *k* (⅝ in. high), both made from 4¾-in.-long, ⅜-in-thick Lucite. The small balls *q* roll on the glass plate *n* (4 x 7¼ x ¼ in.) between the larger bearings *p*. Four Lucite pieces *o* (½ x 1 x ⅛ in.) attach the frame to the base. The glass plate is kept level by four screws *s*.

A27–7.4 Gaseous Diffusion and Brownian Motion

Construction details for the kinetic theory model are shown in Fig. A27–88. The vibrator *a*[3] is mounted on a plywood base *b* which sits on a foam rubber cushion *c*. Four attachment ears *d* (½ x 1 x ⅛ in., Lucite) are cemented to the Lucite tray *e*. The Lucite cover assembly *f* fits over the tray. The ⅛-in. plastic spheres *g* should be of two different colors. A separator bar *h* (5⁄32 x 5 x ⅛ in.) is used to separate the "molecules" of the two different "gases." Note that the plywood and foam rubber are cut away directly below the tray, so that the motion can be projected from below onto an overhead screen.

[3] "Vitasurge" vibrator: SPN74207, 115 V, 45 W, 60 cps; Howard Industries Inc., Racine, Wisconsin.

A27–7.5 Kinetic Theory Demonstration

As shown in Fig. A27–89, aluminum angles *a* (¾ x ¾ x 1⁄16 in., 7 in. long) hold the top glass (9½ x 6^{13}⁄$_{16}$ x ¼ in.) to aluminum bars *b* (^{11}⁄$_{16}$ x ¾ x 5¾ in. and ^{11}⁄$_{16}$ x ¾ x 9^{29}⁄$_{64}$ in.) which have

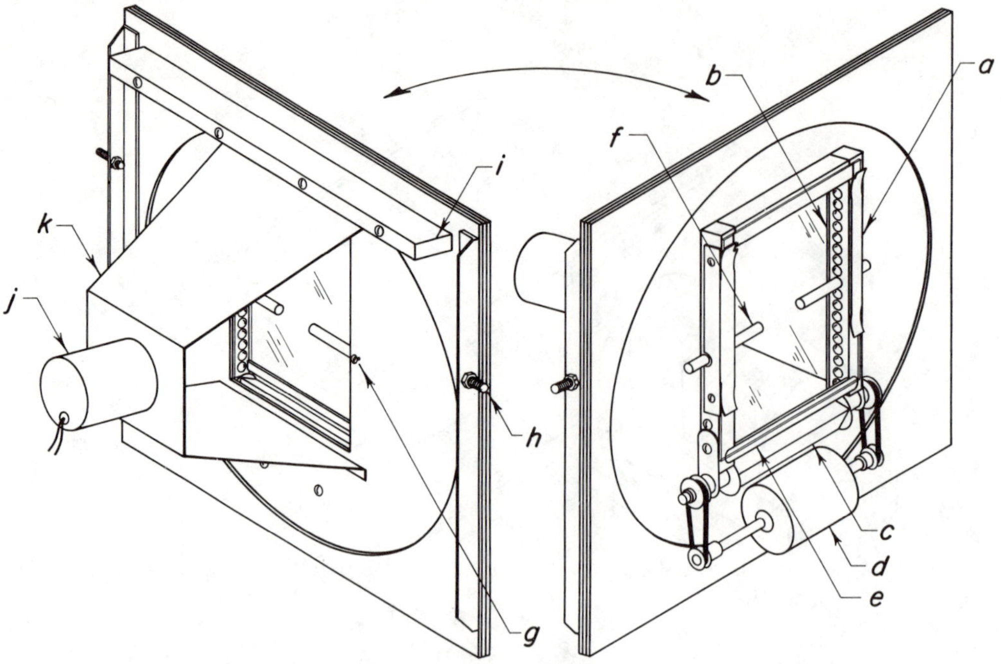

Fig. A27–89

been drilled and reamed so that 3/8-in. steel balls can be free-fitted into them. The bottom glass is 10 x 5½ x ¼ in. Motion is imparted to the free balls by a motor-driven[4] rotor *c* (5½-in. long, 8½ in. with shaft ends) made from ¾-in. square steel rod with its edges rounded to 1/16-in. radius (optical bench rod can be used). The motor *d* drives the rotor through rubber O-rings (2½-in. i.d. x ⅛ in.) and brass pulleys. To deflect any balls accelerated out of the plane of the apparatus by the rotor, a steel bar *e* (½ in. wide) is placed at the end of the top glass near the rotor. The steel diffusion pins *f* (3/8 in. diam x 4 in. long) have rounded ends and are held in place by thumbscrews *g*.

Three Masonite rectangles (18¾ x 20 x ⅛ in.) are needed for mounting the apparatus. Disks, 16 in. diam, are cut from the centers of two of the boards and discarded, and a 17-in.-diam disk is cut from the center of the other board and retained. A 5½ x 8-in. rectangle is cut in this disk so that the center of one of the short sides is 2 in. from the edge of the disk. The boards are stacked, and bolt holes are bored 2 in. apart in a 17¾-in.-diam ring around the hole. For assembly, the boards are stacked in this order: the board with the 17-in.-diam hole is placed on one of the small-hole boards; the disk is placed in the hole; washers are placed over the bolt holes; the other small-hole board is placed on top; and the boards are bolted together. The washers allow the disk to be rotated in the frame. Bolts *h* for mounting this frame are provided on its sides, and a steel bar *i* (16½ x 1½ x 1 in.) is attached to counterweight the motor.

Finally the apparatus is mounted on the disk, and a light source is attached to the frame. This light source consists of a bulb,[5] a 7½-V transformer, a cylinder *j* (2 in. diam x 4 in. long) and a light funnel *k* made from an aluminum strip (22½ x 7½ x 1/16 in.). The insides of the cylinder and funnel are painted flat black. For the Brownian-motion smoke particle, a "cap" is made by hollowing out one face of a brass disk. The finished cap measures 3/8 in. high x 13/8 in. o.d. and has 1/16-in.-thick side walls and a 1/32-in.-thick top. This cap rides freely on 73/8-in. ball bearings so that it can be pushed in any lateral direction by the 50 to 60¼-in. steel ball bearings vibrating inside the frame.

(4) AC-DC series motor, model 2M066, 115 V, 1/15 hp, 5000 rpm with a twin end shaft; Dayton Electric Mfg. Co., Chicago, Illinois.

(5) 50 cp, 6–8 V automobile bulb; Tung-Sol No. 1184.

A27–8.1

Materials List

Part and Description	Remarks
Brownian motion cell	Cenco catalog No. 71270, Central Scientific Co., Chicago, Ill.
Microscope; objective lens to have focal length of ⅔ in.	Make and model not critical
Carbon arc lamp, 4.5 A with condensing lens and cooling cell (filled with ⅖% solution $CuSO_4$ and distilled water)	Bausch and Lomb No. 31–33–73 or equivalent
Secondary condensing lens, focal length ~8¾ in., double convex	Exact focal length not critical. Adjust setup to suit lens used
Television camera	Model 70-B or equivalent, Dage-Bell Corp., Michigan City, Ind.
Monitor	Model K-59, Miratel Electronics, Inc., St. Paul, Minn.
Camera tripod	Model ATV, with pan head MH-2 or equivalent, Davis and Sanford Co., New Rochelle, N.Y.
Microscopic jack	Any commercially available model will suffice
Prism, 90 deg, opaque on hypotenuse	Commercially available

A27–10.3

Heat Exchanger. The heat exchanger is constructed from brass tubing,[6] 1.6 mm o.d. and 1.3 mm i.d. The windings of the exchanger are separated by twine wrapped around the tubing, one turn per centimeter of tubing (Fig. A27–90). The tubing is wound inside a form made of two 20-cm tubes and a wood or plastic disk (Fig. A27–91). One tube *a*, which has nearly the same inside diameter (3 to 4 cm) as the Dewar flask *b*, determines the outside diameter of the exchanger, and the other tube *c* (1 cm diam) determines the inside diameter; a 1-cm-thick disk *d* fits tightly into the larger tube. The smaller tube is inserted into the center of the disk and the disk is inserted into the larger tube. A 2.0 x 2.0-mm short spiral groove on the circumference of the disk holds the end of the brass tubing while the coil is formed (Fig. A27–92). The coil is wound as shown in Fig. A27–93; the tubing must not be bent sharply. Finished windings are periodically compressed into the form (Fig. A27–94) until 60 layers have been formed. When the coil is removed from the form, it must not uncoil or expand. Twine is used to tie it together (Fig. A27–95). A standard fitting is hard-soldered to the upper end of the coil to connect to the intermediate coil.

[6] Available from Wieland-Werke AG., Metallwerke, Olgastrasse 147, 79 Ulm/Donau, West Germany.

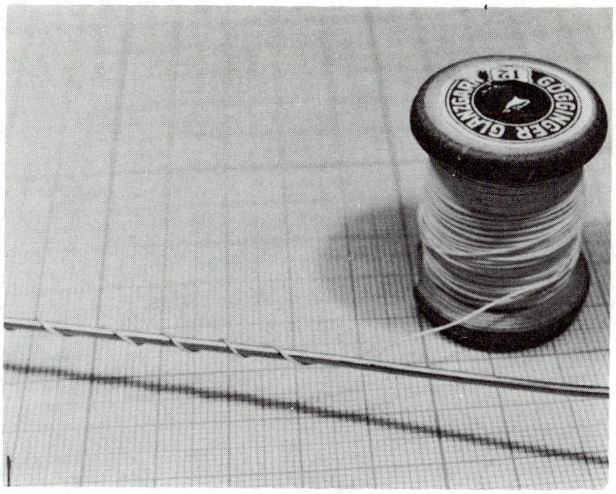

Fig. A27–90

The intermediate coil is a few turns of tubing, such as the tubing used for the heat exchanger, with standard fittings hard-soldered to each end for connecting to the exchanger and to the oxygen tank (Fig. A27–96).

Needle Valve. Construction of the needle valve is shown in Figs. A27–97, A27–98, and A27–99. A

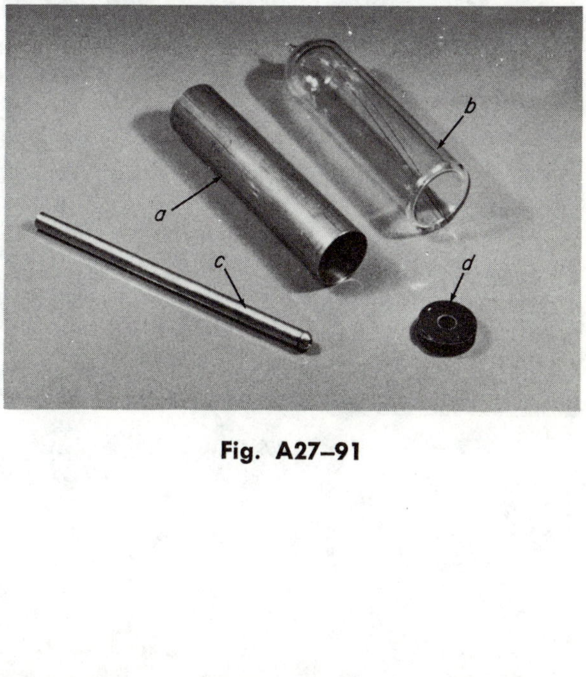

Fig. A27–91

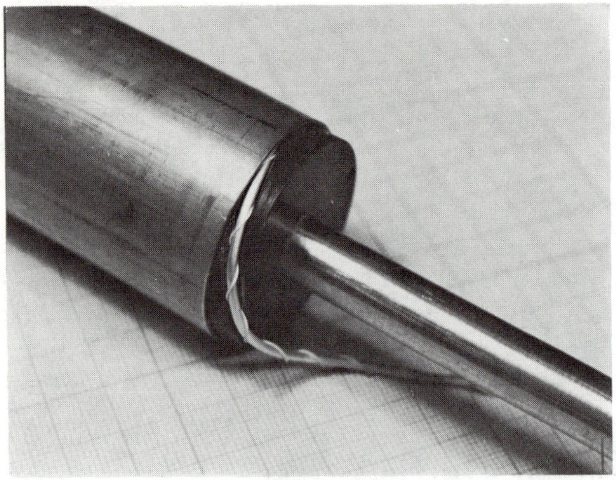

Fig. A27–92

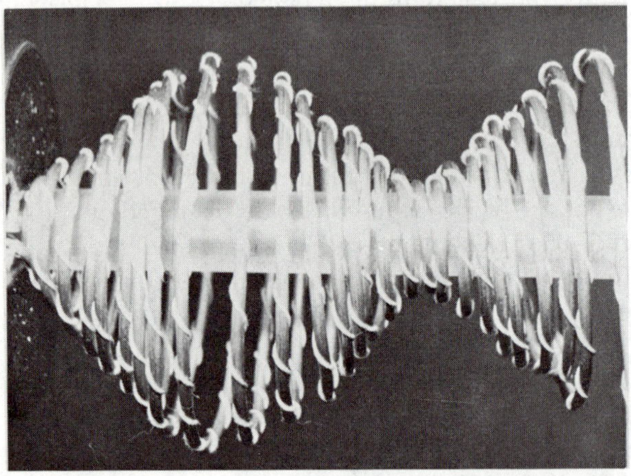

Fig. A27–93

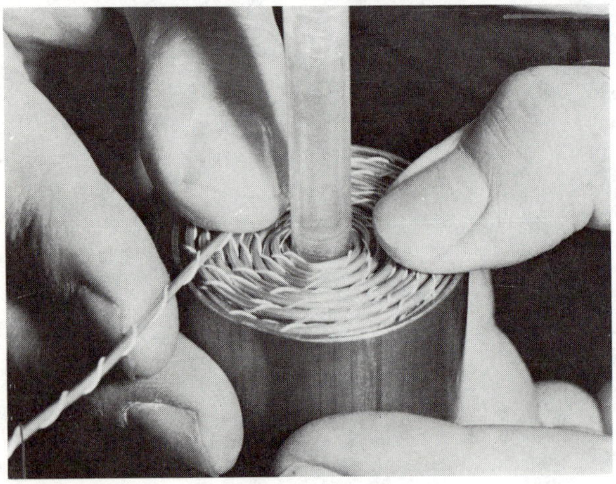

Fig. A27–94

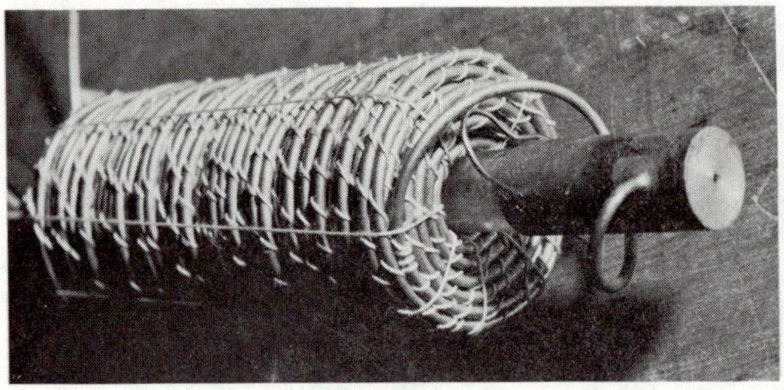

Fig. A27–95

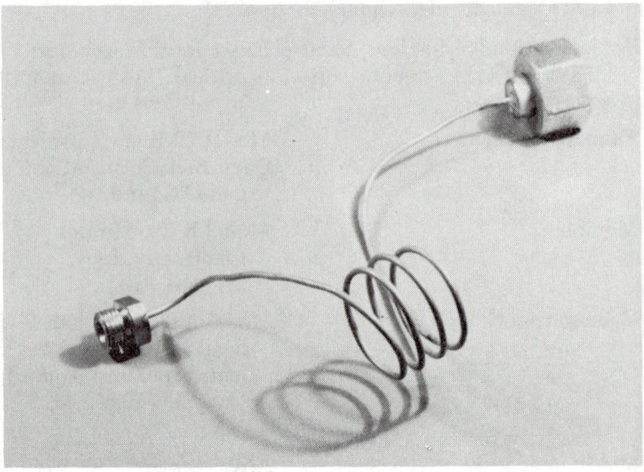

Fig. A27–96

phonograph needle *e* is hard-soldered into a hole bored in the end of threaded (1-cm) brass rod *f*, which screws into valve body *g*; the needle must sit exactly axial in its bore hole. Since no sealing compound can be used, the threads must be fine to insure a tight fit. Both the threaded rod and the valve body are made from brass to prevent loosening or binding (same thermal expansion). The lower end of the exchanger *h* is hard-soldered to the valve body *g*. The valve is operated by a German silver tube *i* hard-soldered to rod *f*. This tube (4 mm o.d.) passes through a Trolitul[7] tube *j* with outside diameter equal to the inside diameter of the exchanger and inside diameter equal to the outside diameter of the German silver tube. German silver and Trolitul are used to keep heat from the valve. A cotter pin *k* prevents the needle valve

(7) A kind of "polyvinyl-benzol" similar to polystyrene, available from Dynamit Nobel AG., 521 Troisdorf Bez., Köln, West Germany; Wieland-Werke AG., Metallwerke, Olgastrasse 147, 79 Ulm/Donau, West Germany; and Robert Bosch GmbH., Postfach 50, 7 Stuttgart 1, West Germany.

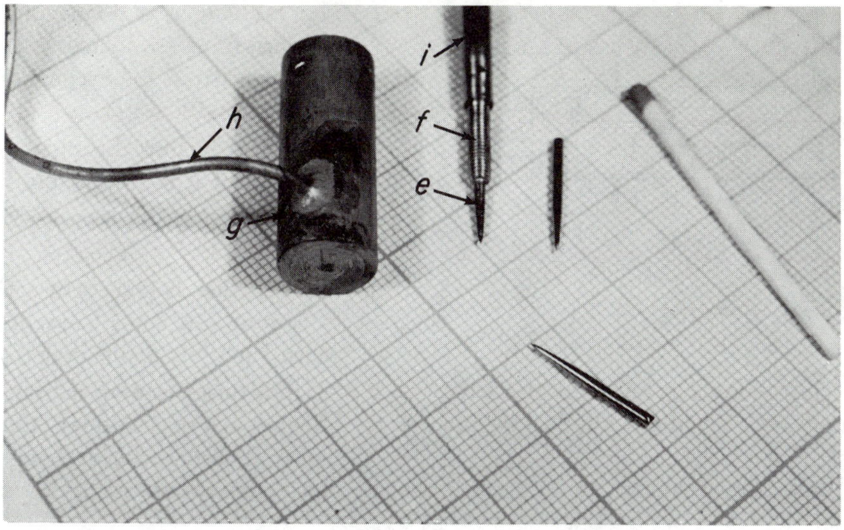

Fig. A27–97

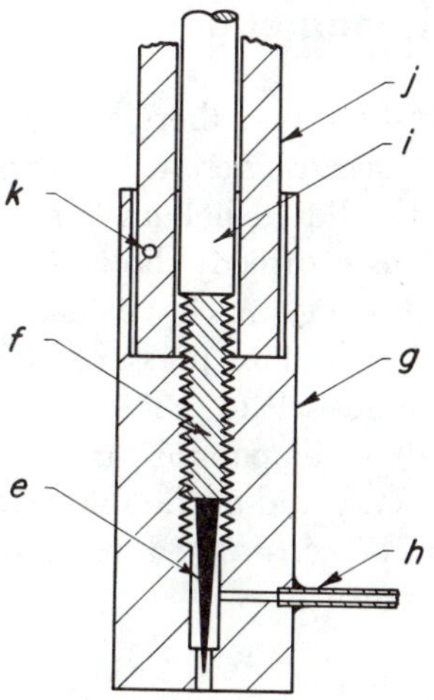

Fig. A27–98

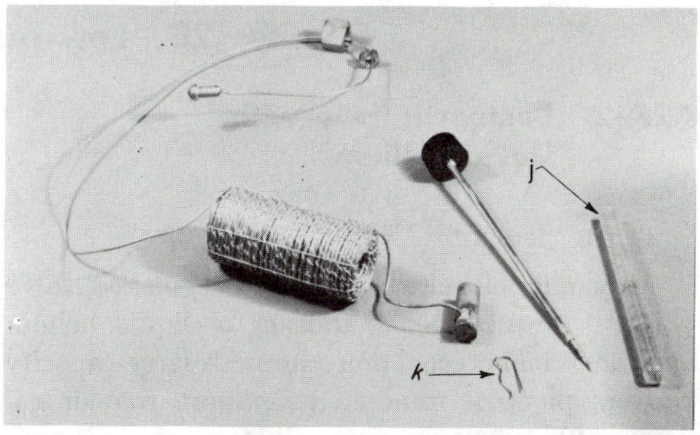

Fig. A27–99

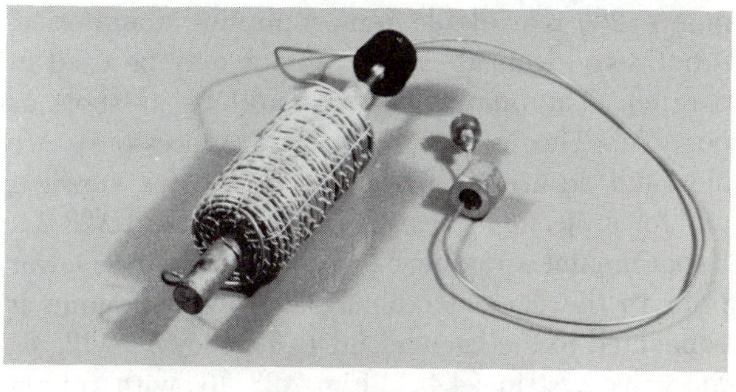

Fig. A27–100

from turning with respect to the exchanger when the German silver tube is turned. Figure A27–100 shows the assembled exchanger and valve.

Testing. The completed apparatus is tested at 320 atm, using a pump made for testing the injection nozzles of diesel engines.(8) The oil used as the pressure fluid in the pump must not enter the liquefier, because even tiny quantities prevent the liquefier from working; therefore, it is necessary to make a pressure-transfer device so that water can be used for the pressure fluid in the exchanger.

(8) Type: Düsenprüfvorrichtung EFEP 60; available from Robert Bosch (footnote (7)) for about $15.00.

This device (Fig. A27–101) is made from a 12-cm brass tube *m* (2 cm o.d., 1 cm i.d.) with outside threads at its ends for strong screw caps *n*. These caps press brass flanges *p*, into which the connecting pipes *q* are hard-soldered, against the ends of the

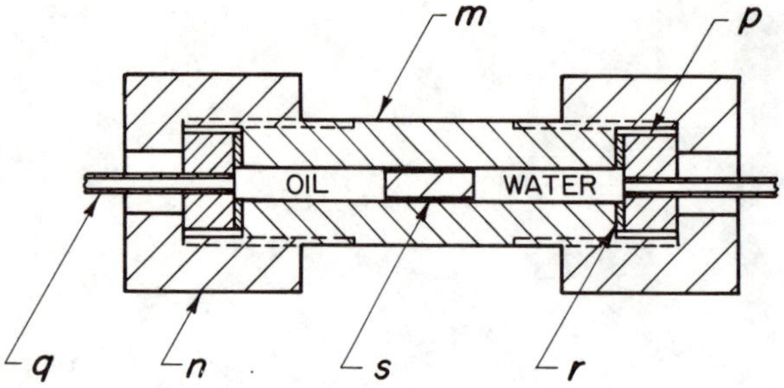

Fig. A27–101

brass tube with ring gaskets *r* between the caps and the tube ends. A Trolitul cylinder is the piston *s*.

The test procedure is as follows:

1. The exchanger, a connecting pipe, and the pressure-transfer device with piston at the end away from the exchanger are connected and filled with water.
2. Oil is forced into the pipe coming from the pump until it is completely filled.
3. The pipe from the pump is connected to the pressure-transfer device.
4. The liquefier is placed in a protective vessel to contain jets from leaks in the tubing, because such jets could penetrate the skin.
5. All moisture is removed from the liquefier by blowing oxygen through the liquefier. Any moisture left in the exchanger or valve will choke the apparatus by freezing.

Chapter 28 Low-Temperature Phenomena

A28–2 Demonstrations with Liquid Helium

General Notes

A cylinder of helium gas and a regulating valve (8B)[1] are used in the transfer of liquid helium and for certain precool procedures. A large-capacity vacuum pump is required (minimum free-air capacity 140 liters/min), especially for lecture demonstrations where there is a heat leak from projection light, as well as a time limitation. Usually such a pump (22) is available in most physics laboratories; if not, two identical smaller pumps may be used in parallel. The pumping line should be as short as possible. The Dewar flask and its accessories are mounted on a cart for easy servicing, for viewing, and for projection, Figs. 28–2, page 827, and A28–11. Experimental equipment is placed inside the Dewar prior to the demonstration. The vacuum pump is connected to the vacuum lid (16), Fig. A28–12, via the valve system (4,4a), Fig. A28–13, with rubber hose (7) to decrease vibration. The valve system can be replaced by a homemade heavy-duty pinchcock,[2] Fig. A28–14.

[1] Numbers in parenthesis refer to items listed in the Materials List for this section.

[2] As designed by H. Forstat, Michigan State University.

In demonstration work there is no great need for a thermometer, because here it is necessary to know only whether the liquid helium is above or below the λ-point. This is directly shown by the presence or absence of boiling. A convenient optional item is 0–50 mm Hg absolute pressure gauge (9). Its temperature range is 0 to 2.3°K. For quantitative laboratory studies, temperature is accurately measured by a mercury-and-oil double-manometer system, which can be constructed in the laboratory.

Construction Details

The Dewar flask (1) (Fig. 28–2) should have single-unit construction for rigidity. Its inner vacuum jacket must have a pumping port, because helium gas diffuses through glass at room temperature. The Dewar is cleaned with standard chemical methods. Iron is to be avoided in all parts of the cryogenic station area because of its interference in potential magnetic studies. The Dewar cart, therefore, is fabricated from nonmagnetic materials (Figs. 28–2 and A28–11). All points of contact between the Dewar and the cart's wooden top members are protected by rubber. The manufacturer provides the Dewar with a "hip" or "shoulder," by which it is supported.

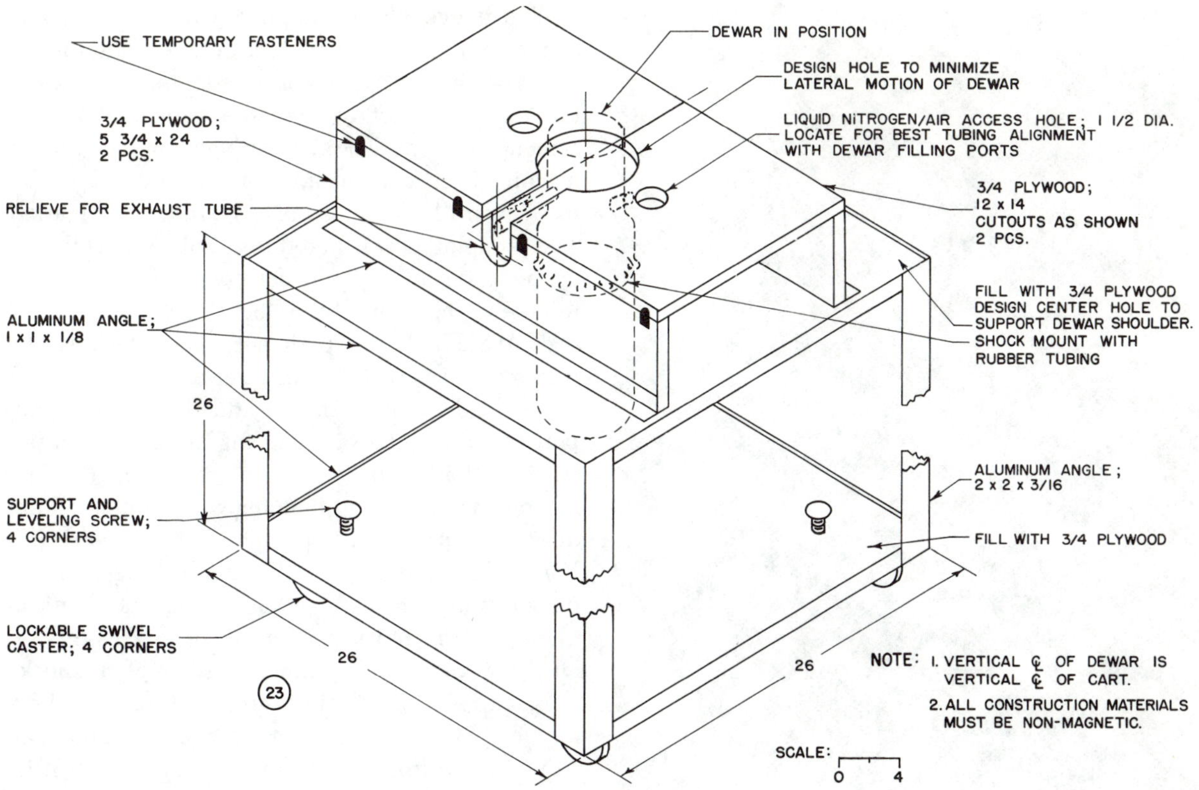

Fig. A28–11

Transfer tubes (3) are manufactured by the same companies which make helium storage Dewars (Fig. A28–15). They are semistandard, semicustom items. A ½-in.-o.d. tube which jackets a ⅛-in.-diam stainless steel tube with a movable helium-storage Dewar seal is recommended. The vacuum jacket (½ in. o.d.) has a pumping connection. Should a leak develop or should there be thermal contact between the inner and outer tube, it can then be serviced and reevacuated without returning it to the factory. The lengths of the three sections of the transfer tube assembly should be specified with the following considerations in mind: (a) The depths of both the storage and demonstration Dewars. The transfer tube should extend into each as far as possible. (b) The distance between the storage and demonstration Dewars. It should be minimized.

The tubing (5) used to suspend demonstration accessories from the vacuum lid (16) is cut into convenient lengths with a jeweler's saw. As a safety measure, a cotter pin *p* (Fig. A28–16) is installed in the lower end of each to prevent complete accidental removal from the Dewar. This is dangerous in operations below atmospheric pressure. A plain suspension tube is sealed by soldering a short brass bolt *q* (e.g., ¼-20 NC, ½ in. long) to the end. This type of tube can also be used as a "blank" to stop up the movable seal, or else a short rod *n* can be made for this purpose. To suspend accessories requiring electrical connections, Kovar glass feedthroughs or modified coaxial cable receptacles, such as *l*, can be used. To use the latter, a brass adapter is machined to fit closely over the tube and to take the threaded end of the receptacle. This is soldered to the tube. An O-ring between the shoulder of the receptacle and the adapter will provide an adequate seal. A cheaper way to seal wires emerging from the tube is to slip a short length of vacuum hose over the end with the wires separated and with vacuum grease (23) in the hose. A strong screw-type pinchcock *m* will seal it. The total number of tubes required depends upon the accessories used. During operation below 4.2°K, these tubes tend to be drawn inward. Small C-clamps o (17) can be made to prevent this. The

Fig. A28–12 Vacuum Dewar Lid. Access and transfer port to right. Pumping port and T with side nipple and gauge to left. Center shows the three movable vacuum seals *j* for ¼-in. nickel-silver tubing. Rear seal shows a brass plug *k*, which is mounted when a seal is not in use with tubing.

only critical aspects of such a clamp are the bearing length and the hole size. The length should be at least ¼ in., and the hole should be reamed to exactly fit the outside diameter of the tube.

The gauge (6) is supplied with a pipe-thread connection by the manufacturer. A close nipple (16c) with a threaded brass adapter is added to the pumping port to take the gauge. Pipefitting cement is used. The Dewar vacuum lid (16) is machined on its underside for a vacuum seal with the Dewar (1) using (2) (i.e., no radial tool marks). The pumping port is bored for a solder fit with 1-in. copper tube, and the transfer port is similarly bored to take ¾-in. copper tube. See Figs. 28–2, A28–12, and A28–17. For experiments below 4.2°K, the transfer port is sealed with a rubber stopper (16f). A small side nipple (16e) is added to the pumping port tee to facilitate introducing helium gas in precooling, for attaching a thermometer (9) (or manometers), and for pumping with a smaller pump during the precooling operation. It can be seen in Fig. A28–12. Three (this number depends on demonstration needs) sliding vacuum seal assemblies (18) are fabricated as shown in Figs. A28–12 and A28–18. One assembly consists of a knurled brass cap (18a), a ¼-in O-ring (18c), and a brass nipple (18d) threaded on both ends. These are Edward seals made to fit the ¼-in.-o.d. suspension tubes. The O-ring and tubing are *lightly* greased with silicone grease (23) for sealing and movement.

The superleak accessory (12) is a modified Büchner funnel with an ultrafine fritted disk. All funnel glass below the disk is removed and three equally spaced eyelets are added. Plastic monofilament thread is used to connect the eyelets to a suspension tube. One of these can be seen in Fig. A28–16. An alternative procedure is to have a glassblower fuse a fritted disk to a short length of glass tubing and then to add eyelets as before.

The static fountain (13) is fabricated by having

Fig. A28–13 Valve System.

Fig. A28–14 Heavy-Duty Pinchcock.

a glassblower attach a bulb to the end of a filter tube and add three suspension eyelets at the other end (Fig. A28–19). The bulb is packed tightly with jeweler's rouge. For the dynamic fountain (14), a short length of glass tubing (14c) is bent to the dimensions shown in Fig. A28–20, and a 1¼-in. piece of ½-mm-bore capillary tube (14a) is fused to it, thus forming a J. The curved portion is packed with rouge and mounted in a clear plastic holder (14b) (Fig. 28–7). A very small hole is drilled in the leg of the holder, which is then slit to provide a spring-grip on the fountain. The holder is connected to a suspension tube with a $\frac{1}{16}$-in. cotter pin.

The persistent-current coil (11) is wound on a wooden coil-form with about 12 layers of plastic-clad 10-mil niobium wire (resistance per meter, 2 Ω). The core is ¼-in.-diam x 1⅛-in.-long wood with ⅝-in.-o.d. wooden enddisks. A 4¾-in.-long dowel is imbedded in the core and the parts are glued together. The long dowel is wrapped with tape about 2 in. from the coil to facilitate a push fit in the end of the suspension tube. After the coil is wound, it too is wrapped with tape (Figs. A28–4 and A28–21). The coil leads are shunted by a short piece of 20-mil tantalum wire at a point 1 in. from the coil. Connections are spot-welded. A drop of alcohol at the contact just before welding avoids oxidation and breakage of the wires. External leads can be enameled copper wires connected above the shunt (by ordinary soldering), and these issue from the Dewar through a suspension tube sealed, if desired, as in Fig. A28–16*m*.

The Rollin film accessory (15) is a ¾-in.-diam dish made from the bottom of a No. 6 Pyrex test tube. Three suspension eyelets are added as in (13) and (14). The surface is slightly etched with hydrofluoric acid to increase the rate of bulk flow (Fig. 28–10).

The arrangement of the projection system is illustrated in Fig. A28–22. A carbon arc (21f) provides an adequate high-intensity light source. A water cell (21d) protects a blank of infrared-absorbing filter glass (21c) against cracking. Together, their effect is to substantially decrease the amount of heat transmitted to the liquid helium by the arc. Lens apertures must be large because of the Dewar's dimensions, and also because of the astigmatism produced by the Dewar (especially strong when it is filled with liquid nitrogen or air). A large inversion prism (21a) is used to produce an upright image. Two smaller prisms can be combined side by side for this purpose.

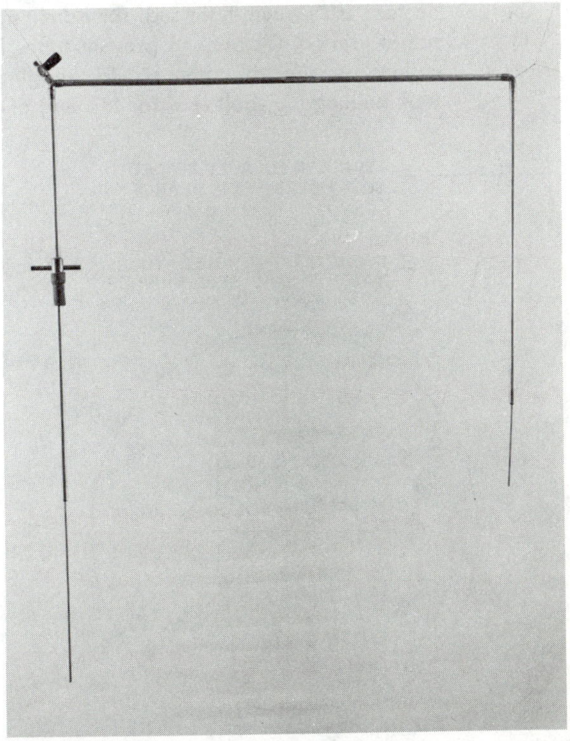

Fig. A28–15 A Typical Liquid Helium Transfer Tube. Valve on upper left allows access to the vacuum space. Rubber hose fits the spout of the liquid helium storage tank, and the nipple above it allows exerting pressure with helium gas during transfer.

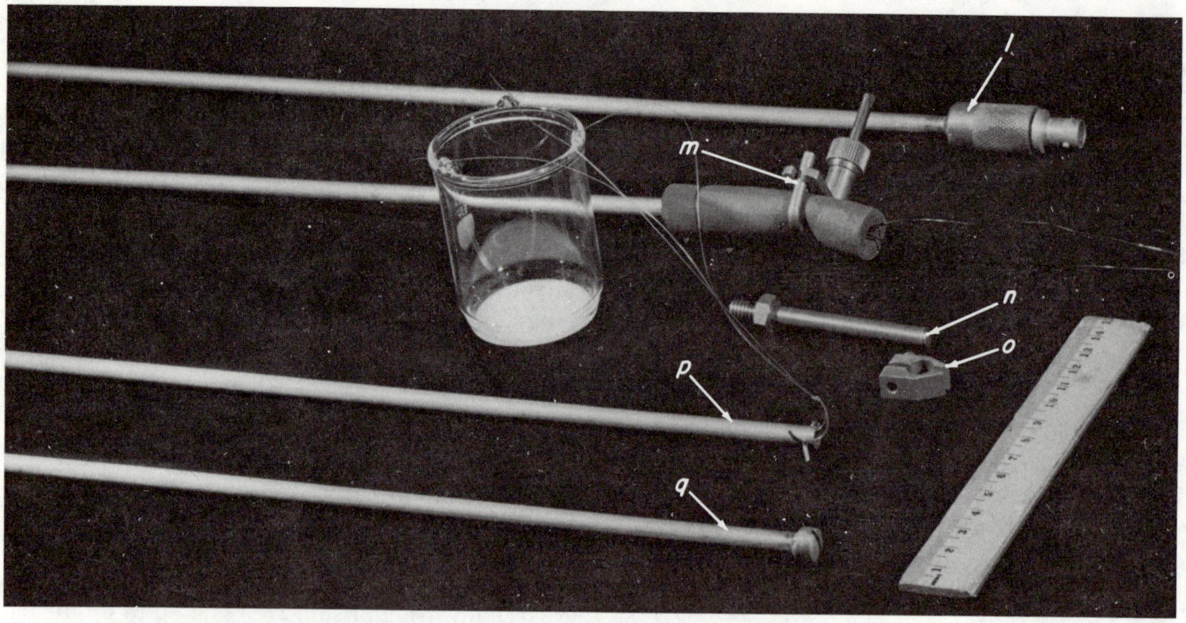

Fig. A28–16 Illustration of the Uses of Nickel-Silver Tubing which fits through the movable vacuum seals and permits raising or lowering of apparatus in the Dewar at low pressure. *Top to bottom: l*, vacuum-tight seal for top end of tube with BNC O-ring connector. *m*, vacuum-tight seal for admitting wire leads. *n*, brass plug for movable seals not in use. o, C-clamp to prevent tubes from being sucked inward when the Dewar is at low pressure. *p*, cotter pin for hanging apparatus from bottom of tube (near superleak beaker). *q*, bolt seal for top end of tube.

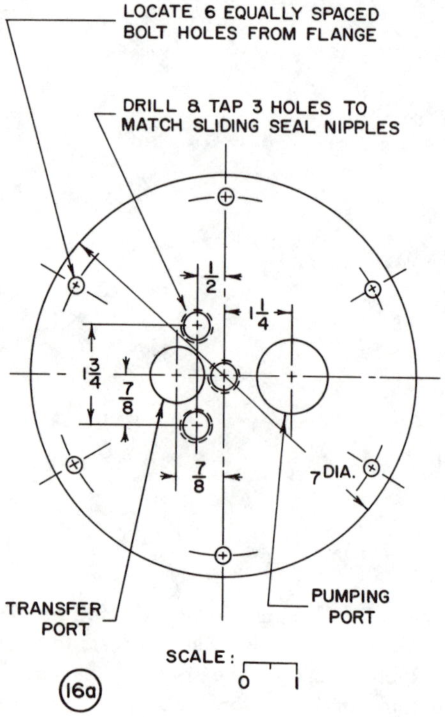

Fig. A28–17

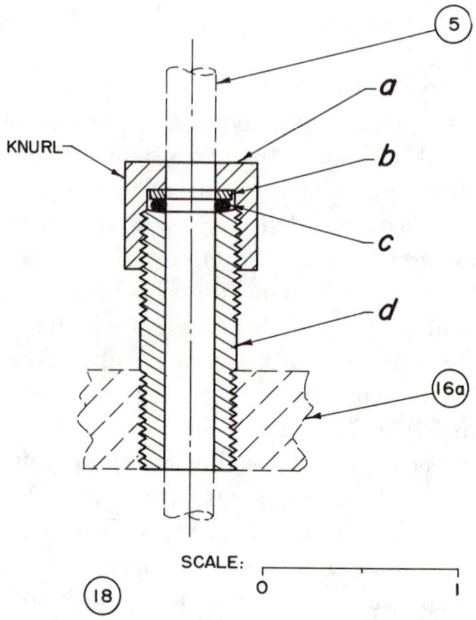

Fig. A28–18

Materials List

Detail	Quantity	Part and Description	Remarks
1	1	Double Dewar, glass, with pipe joint to allow vacuum-tight attachments	Modification of model M2-3940: no "finger," no "foot," no pitch, unsilvered, H. S. Martin Co., Evanston, Ill.
2	1 assembly	Aluminum-glass pipe flange, cover, bolts, and rubber gasket for details 1 and 16	H. S. Martin Co.
3	1 assembly	Helium transfer tube	Superior Air Products, Inc., Newark, N.J.
4	1	Bellows valve, 1-in. size	Kinney OBTM, commercially available
4a	1	Needle valve, ⅜-in. size	Ideal-Aerosmith 535, commercially available
5	10 ft	Suspension tube, nickel-silver; ¼ in. o.d. (0.010-in. wall) x 24 in. long. (Note: Number required depends on demonstration needs.)	Uniform Tubes, Inc., Collegeville, Penna. (Note that minimum order may exceed requirements.)
6	1	Compound Bourdon gauge; 0–30-in. vacuum; 0–30 lb overpressure; 6-in. dial	U.S. Gauge Co., Figure 1800, commercially available
7	10 ft	Vacuum hose, extra heavy wall, 13/16-in. bore	Cenco Catalog No. 18204 (No. 11), Central Scientific Co., Chicago, Ill.
8		Liquid helium	1. Gardner Cryogenics Corp., Hightstown, N.J. 2. Union Carbide, Linde Div. Note that storage tank size and make should be specified. Ask for depth gauge
8a		Liquid nitrogen (or air)	Welding supply house
8b		Helium gas with regulating valve	Welding supply house
9	1	Thermometer (optional); absolute pressure indicator FA 160110; 0–50 mm Hg. Factory-calibrated to NBS He tables	Wallace and Tiernan, East Orange, N.J.
10	1 assembly	Floating magnet accessory	
10a	4	Bar magnet, 1/16 x 1/16 x ½ in.; 22S5A sintered Alnica 2	Permag Corp., Jamaica, N.Y.
10b	1	Pyrex pipe, 75 mm o.d. (standard wall) x 1½ in. long	Cenco Catalog No. 14101. See Remarks, detail 7
10c	1	Lead plate, 3 1/16 in. diam x ⅛ in. thick, slightly concave	
10d	1	Styrofoam hemisphere, 3 in. diam	
10e		Stirrup, copper wire	Make to suit
10f		Fine chain (silver), 24 in. long	
11	1 assembly	Persistent-current coil accessory	Fabricate
11a	1	Suspension tube (detail 5)	

Materials List (*Cont.*)

Detail	Quantity	Part and Description	Remarks
11b	1	Wood dowel; ⅛ in. diam x 4¾ in. long	
11c	1	Core, wood dowel, ¼ in. diam x 1⅛ in. long	
11d	2	End disk, nonmagnetic, ¼ in. i.d. x ⅝ in. o.d. x 1⁄16 in. thick	
11e		Coil wire, niobium, 0.010 in. diam, plastic clad; resistance: 2 ohms per meter; 100 turns per layer; 12 layers	Fansteel Metallurgical Corp., North Chicago, Ill.
11f		Shunt wire, tantalum, 0.020 in. diam, 1 in. long	See remarks, detail 11e
11g		Lead wires, enameled copper, 0.020 in. diam, 6 ft long	
11h		Plastic tape	
12	1	Superleak accessory; modified Cenco Pyrex Büchner funnel with ultrafine porosity fritted disk; 60-ml capacity	Catalog No. 15184-3. See Remarks, detail 7
13	1	Static fountain effect accessory; modified Cenco No. 4 Pyrex filter tube	Catalog No. 15170. See Remarks, detail 7
		Absorbent cotton	
		Red jeweler's rouge	Catalog No. 606; stock No. 40,015. Edmund Scientific Co., Barrington, N.J.
14	1 assembly	Dynamic fountain effect accessory	Fabricate
14a	1	Capillary tube, Pyrex, 0.5-mm bore, 1¼ in. long	Cenco Catalog No. 14106 (No. 1). See detail 7. (Note that minimum order may exceed requirements.)
14b	1	Holder; clear acrylic plastic, 1 1⁄16 x 3¾ x 7⁄16 in.	Material commercially available under trade names Lucite and Plexiglas
		Red jeweler's rouge	See detail 13
14c	1	Pyrex tube, 8 mm o.d., standard wall	Cenco Catalog No. 14095. See remarks, detail 7
	1	Cotter pin, steel, 1⁄16 in. diam x ¾ in. long	
15	1 assembly	Rollin creeping film accessory, modified Pyrex test tube, 18 mm o.d	Cenco No. 6. See Remarks, detail 7
		Fine plastic monofilament fishline for suspending details 12, 13, and 15	
16	1 assembly	Vacuum lid assembly	Fabricate
16a	1	Brass lid, 7 in. diam x ½ in. thick	
16b	1	Transfer port tube, copper, ¾ in. diam x 3½ in. long	
16c	3	Copper tube, 1½ in. long	

Materials List (Cont.)

Detail	Quantity	Part and Description	Remarks
16d	1	Copper tee, 1 in.	
16e	1	Copper tube, 3/8 in. diam x 2 in. long	
16f	1	Rubber stopper for 3/4-in.-diam copper tube (transfer port)	
17		Miniature C-clamp, brass	See Appendix 2
18	3 assemblies	Sliding vacuum seal	Fabricate
18a	3	Vacuum cap, brass 3/4 in. diam x 1/2 in. long	
18b	3	Light washer, modified standard	
18c	3	O-ring; neoprene or rubber, 1/4-in.	Locally available
18d	3	Close nipple, brass, 1/2 in. diam x 1 1/4 in. long	
19	1 assembly	Feedthrough seal	
19a	1	Bulkhead type BNC coax receptor	Commercially available
19b	1	O-ring to fit detail 19a; neoprene or rubber	
19c	1	Receptor coupling, brass, 1/2 in. diam x 1/2 in. long	
	1	Suspension tube	See detail 5
20	1 assembly	Tube for liquid nitrogen transfer	Fabricate
20a	1	Copper tube, 3/8 in.	Length determined by depth of storage Dewar
20b	1	Brass cap, 1 1/2 in. long	Diameter determined by storage Dewar spout
20c	1	Pressure connection nipple, brass	Fabricate
20d	1	Rubber hose to fit detail 20b	
21	1 assembly	Projection system	
21a	1	Inversion prism	
21b	1	Projection lens	
21c	1	Infrared-absorbing filter glass	Corning 4600. Corning Glass Works, Optical Development Dept., Corning, N.Y.
21d	1	Water cell	
21e	1 unit	Condenser lens system	
21f	1 unit	Arc lamp	
22	1 unit	Vacuum pump, large capacity, minimum 140 liters/min free air	Cenco Hypervac 23 or Welch Duo-Seal 1402B. See Remarks, detail 7. Sargent-Welch Scientific Co., Skokie, Ill. Two smaller pumps in parallel may be used instead
23	1 8-oz jar	Silicone high-vacuum grease	Dow-Corning, Cenco catalog, No. 15517-2. See Remarks, detail 7

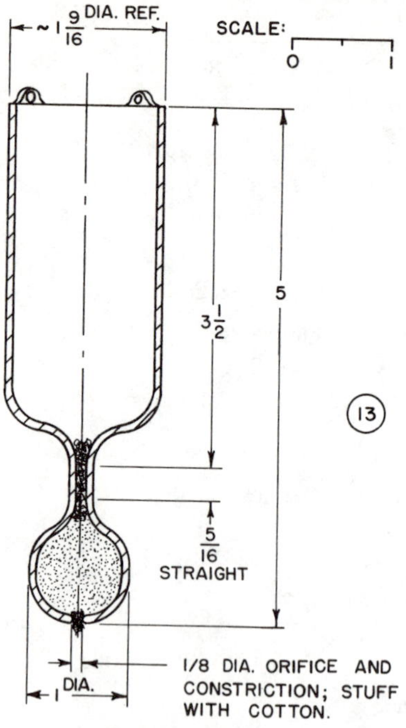

Fig. A28–19

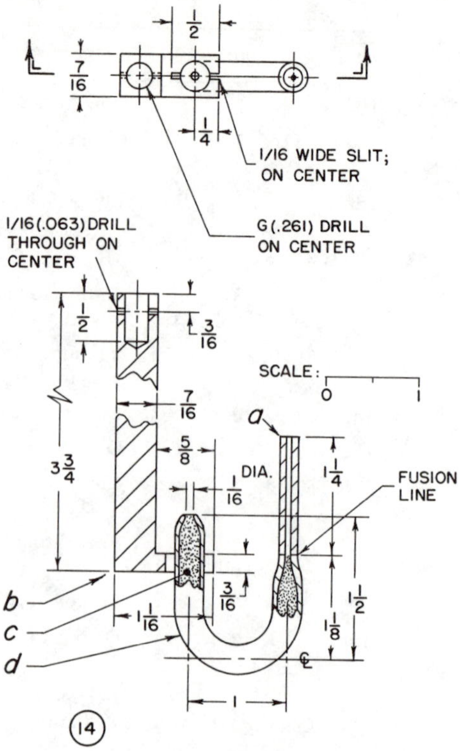

Fig. A28–20

Operational Procedures and Liquid Helium Techniques

Precooling the Dewar

(a) Since helium gas will diffuse through glass at room temperature, the inner vacuum jacket will be contaminated with it from a previous experiment. It should be evacuated with a small mechanical pump.

(b) All traces of water and water vapor are removed from both the liquid helium space as well as the liquid nitrogen space with a heat lamp and/or an electric hair dryer.

(c) The Dewar lid (16) is bolted to the flange (2), and the rubber gasket and vacuum grease (23) is used to seal it.

(d) All hose connections are made, and the helium space is sealed to prevent entry of moisture.

(e) Liquid nitrogen (or air) is introduced into the outer Dewar with a cardboard funnel and a small Thermos. The Dewar is slowly filled and kept full for two hours. Unshielded Thermos bottles must be wrapped with masking tape for safety.

(f) Shortly, the air sealed into the inner Dewar by operation (d) will be at low pressure. This chamber is then evacuated and filled with helium gas from cylinder 8b to just over one atmosphere, as indicated by the gauge (6). Several refills are required as the gas cools. This method keeps the inner Dewar free of solid air and moisture (icing) until the time of transfer, when liquid helium is introduced. This necessitates opening the transfer port (16b) to room pressure (discussed further in step (b) under "Transfer").

(g) Just before transfer, any helium gas which may have diffused into the inner vacuum chamber (during operation (f)) is pumped out at the gas pumping port of (1).

Transfer

(a) The helium gas hose from cylinder (8b) is now attached to that port on the transfer tube (3) which connects to the helium space in the storage Dewar. No pressure is applied at this point.

(b) The storage Dewar and the transfer port to the inner demonstration Dewar are opened. Since the latter contains helium gas at slight overpressure (due to operation (f) of precool), helium gas will rush out rather than moist air in. This prevents icing at this point and subsequent solid air formation.

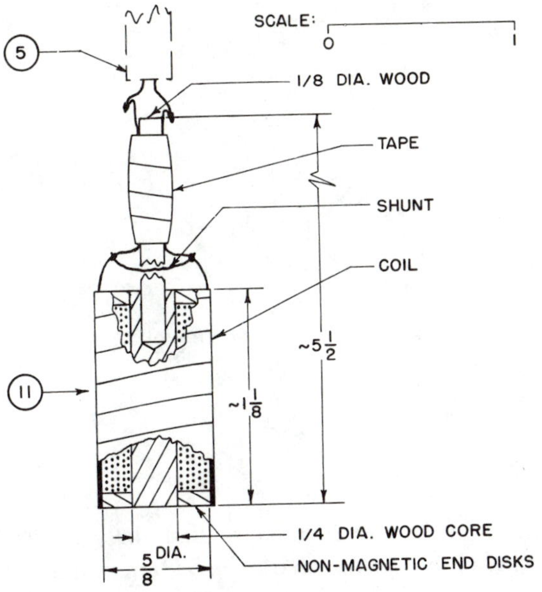

Fig. A28–21

(c) The transfer tube is inserted into both Dewars simultaneously with a slightly vibrating motion to prevent the tube from "freezing" to the storage Dewar with solid air. Gloves are required for this operation. The open end of the transfer tube in the demonstration Dewar should extend as far down into it as possible.

(d) The seal on the transfer tube should sit on the neck of the storage Dewar. Vapor pressure will rise in the storage Dewar and push liquid helium through the transfer tube. For the first two or three minutes, increasingly colder gas will rush through as the transfer tube and demonstration Dewar cool. If the outside of the transfer tube frosts over, then either its vacuum jacket leaks, or the inner tube touches the outer tube, causing a heat leak. It must be immediately withdrawn. If the tube has a pumping port and valve (as purchased), it can be repaired in the shop and pumped out again.

(e) The pressure (not to exceed about 1 lb/in.^{2}) can be increased by adjusting the regulating valve on the helium gas cylinder. Liquid helium flow is speeded up in this manner. The valve should be closed before the transfer tube is withdrawn.

(f) When the demonstration Dewar is full, the tube is withdrawn, and both Dewars are capped (not too tightly).

(g) An alternative method of transfer is to seal the pressure port on the transfer tube which leads to the helium space in the storage Dewar. Rubber tubing and a screw type pinchcock are used. The procedure is the same as before, but without external helium gas pressure. This transfer method is time consuming, particularly when the storage Dewar is nearly empty. For each liter of liquid helium transferred, about two liters are lost in transfer.

(h) The Dewar (1) should be filled to about the level of the liquid nitrogen, but not above it.

(i) The supply of liquid nitrogen (or air) is constantly replenished during an experiment. In lecture (especially when projecting demonstrations), about one hour's use can be expected from an initial filling before the liquid helium level becomes impractically low. Several hours usage can be expected from one filling in laboratory usage.

Warmup

After a run at 4.2°K, the pressure will be 1 atm. No special precautions are needed at this point, and the supply of liquid nitrogen (or air) is merely cut off. If the experimental run ends at low pressure, the *pump should be left running*, but the valves

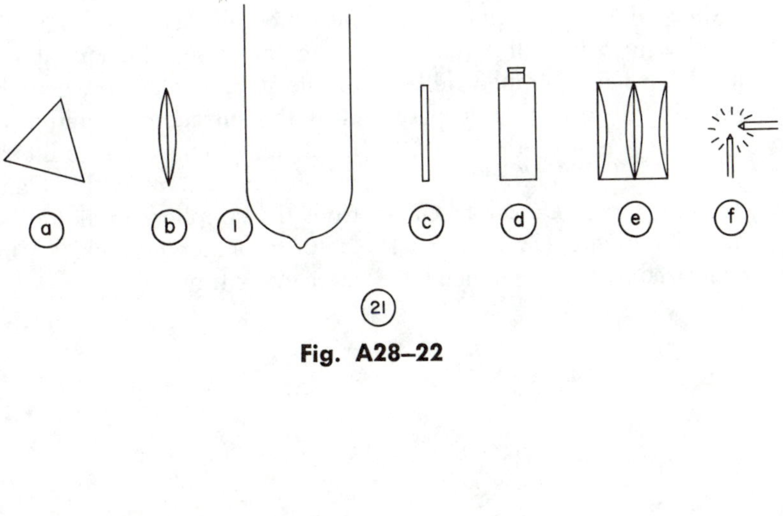

Fig. A28–22

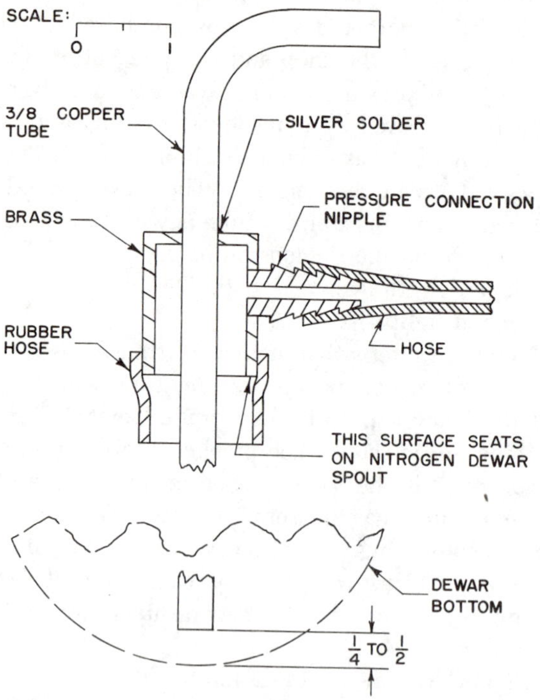

Fig. A28–23

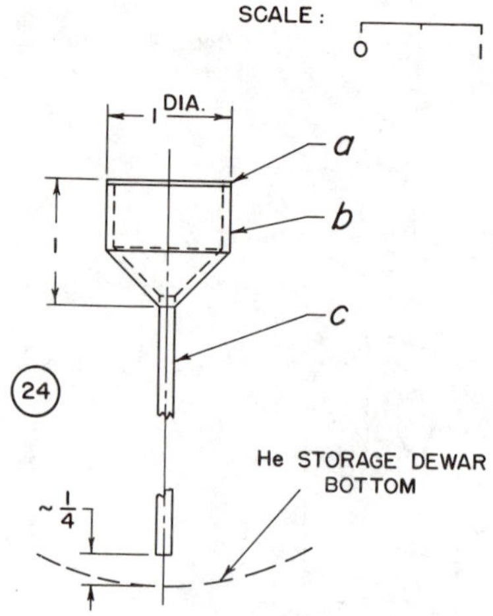

Fig. A28–24

should be closed. When the pressure just exceeds 1 atm (indicated by gauge 6), the pump is turned off and the valves opened. The pressure rise can be accelerated by judicious use of a heat lamp.

Liquid Nitrogen Storage Tanks

Either open-spout or pressure storage Dewars are used. To pressurize the outflow from the common storage Dewar, the transfer tube arrangement shown in Fig. A28–23 can be used. The diameter of the brass caps should match that of the storage Dewar spout, and when the cap is seated, the copper tube should reach to about ⅜ in. from the bottom. The cap has an access port for providing external pressure to the vapor space above the liquid nitrogen. An overpressure of 2 to 3 lb will force the liquid out quickly enough. A hand-pumped rubber bulb or dry compressed air will serve this purpose.

Care of the Liquid Helium Storage Dewar

When the liquid helium storage Dewar first arrives, its supply of liquid nitrogen (or air) must be replenished. A pressurized flow is recommended. The outer Dewar is then refilled daily until the liquid helium is expended. The helium output spout must be kept closed with a one-way gas escape valve (supplied with the storage Dewar) at all times other than during transfer or when taking depth measurements.

If a depth-sounding gauge is not supplied by the dealer, one can be fabricated as illustrated in Fig. A28–24. The top surface of the gauge *a* is hard-soldered to the body *b* to form a diaphragm. The length of the tube *c* is determined by the depth of the storage Dewar and should exceed it by a few inches. It is soldered to the body. When this device is lowered slowly into the helium space of the storage Dewar, an oscillation can be heard, or felt with the finger. Its frequency changes from high to low at the instant when the bottom end of the tube passes through the liquid surface. A reference mark is scratched or painted on this tube to mark the point which is just at the top, or neck, of the storage Dewar when the lower end of the tube touches the bottom of the storage Dewar. One now measures the distance from this mark to the point along the tube which is at the level of the neck of the storage Dewar when the tube's lower end is at the surface of liquid helium. This measurement can be converted to liters of liquid helium remaining in the tank with the aid of graphs available from the suppliers of liquid helium or the manufacturers of storage tanks. A measurement is made after every transfer.

Chapter 29 Electrostatics

A29–1.19

The construction details of the electrostatic balance are shown in Figs. A29–87 (front view) and A29–88 (top view). All parts are made of aluminum except for the Lucite sphere arm *a*, $\frac{3}{8}$ in. diam, and the damper arm *b*, a $\frac{5}{16}$-in.-diam Lucite rod. There are four 1 x 2-in. mirrors *m* cemented to the $\frac{1}{4}$-in.-diam aluminum shaft as shown. Two of these mirrors are placed back to back, and located directly above these is another set of two mirrors oriented at right angles to the first pair. The assembly as a whole is suspended from a large support by a piano wire ($\sim$0.64 mm) which is fastened with pairs of setscrews.

A29–1.23

The principal parts and dimensions are shown in Fig. A29–89. Experience indicates that collar dimensions, distance from collar to nozzle, and inside diameter of nozzle opening are critical and should be held to fairly close tolerances. Other dimensions can have considerable latitude without affecting performance. The structure needed to support the tank, nozzles, and so on, is not shown in detail here. The experimenter can supply the details to suit himself. Although much of the structure can be made of wood, some parts must be easily dried and must not allow surface leakage of charge. Plexiglas or Lucite works quite well. The platform, which is made of this material, is fastened to a sheet of Lamacoid, or some other strong, waterproof insulating panel, which is fastened to the aluminum angles. These angles in turn are screwed to the upright wood panels of the structure. The other Plexiglas part is the bar which supports the collar rods.

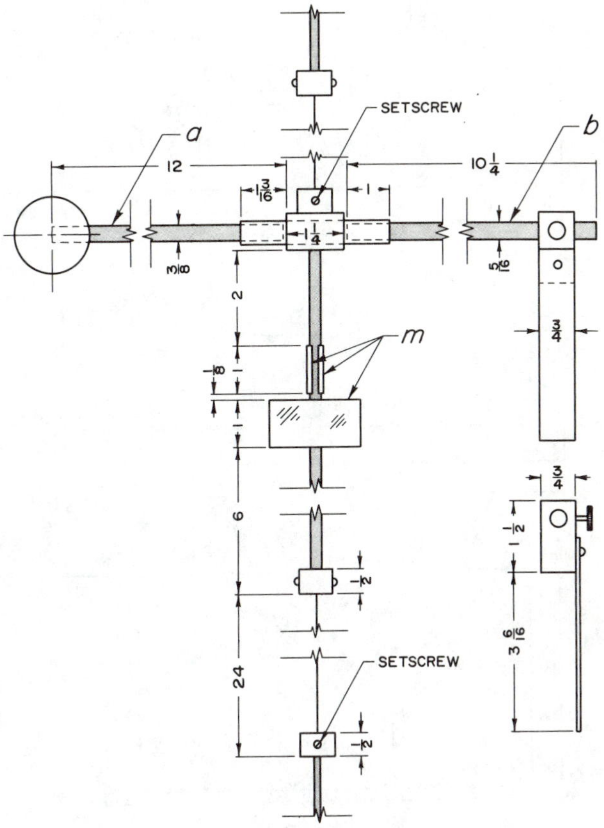

Fig. A29–87

Although the specific size, shape, and material of the water tank *a* are not important, the depth should be great enough (about $\sim$6 in.) to insure sufficient water pressure. The rubber water tubes *b* ($\frac{1}{4}$ in. o.d., $\frac{1}{8}$ in. i.d.) end in nozzles which have been drawn from glass tubing *c* ($\frac{3}{16}$ in. o.d., $\frac{3}{32}$ in. i.d.). At their openings these nozzles should have $\frac{1}{16}$ in. i.d. As a means of controlling water flow, two valves are made from rods *d* and washers: the inside ends of the rods fit loosely in holes in

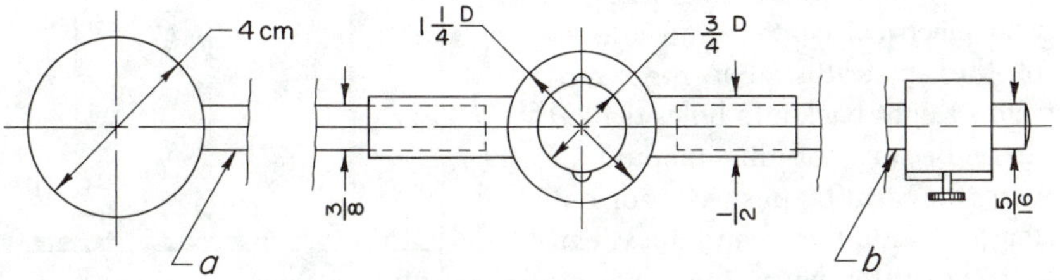

Fig. A29–88

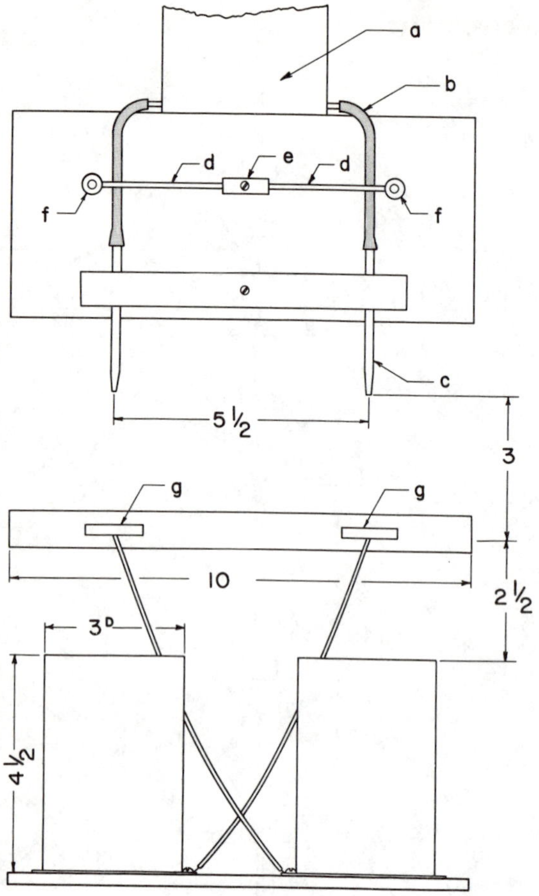

Fig. A29–89

the sides of a brass block *e*, and the outside ends are soldered to washers *f* which fit over bolts protruding through the wood faceplate. Tightening the knurled nuts on the bolts squeezes the rubber tubes, decreasing the flow rate. With these valves, the jets can be controlled individually.

To make the collars *g*, remove the paint from wooden curtain rings, saturate them with waterproof varnish, then wrap them with Nichrome ribbon. Collars can also be made of wire screen, bent to form tubes an inch in diameter and about ½ in. high. The waterproof collars made in this way always provide good electrical contact and add to the appearance of the apparatus. The rods supporting the collars go straight back into holes drilled in the Plexiglas bar. Because of this method of support of the collars, they can be pushed in or out to center the water jets with them, and they can be rotated to direct the water when the spraying limit is reached. Rubber-covered wire connects these rods, and therefore the collars also, to the aluminum plates (3 x 9 in.) on which the cans stand. The cans are about 3 in. in diameter and 4½ in. high. They are coated with aluminum paint to inhibit rusting. Finally, an electric light is attached to the frame under the tank (Fig. A29–90). This extra illumination is especially advantageous for showing the spray.

A more dramatic neon light flasher can be made from several neon bulbs mounted on a Plexiglas plate as shown in the foreground of Fig. 29–26. There is a small gap at each end terminal. When the bulbs are connected to the generator, they will all flash at once.

A29–1.25

Five different dirods were constructed. Their characteristics are given in the Materials List. Dirod Junior is more compact than Radial Dirod Junior, although the latter is simpler and produces a slightly higher voltage. Dirod Junior can be "stretched," with rods 8 in. long instead of 4, and the other parts lengthened axially to correspond. This will double the current output at the same speed and voltage.

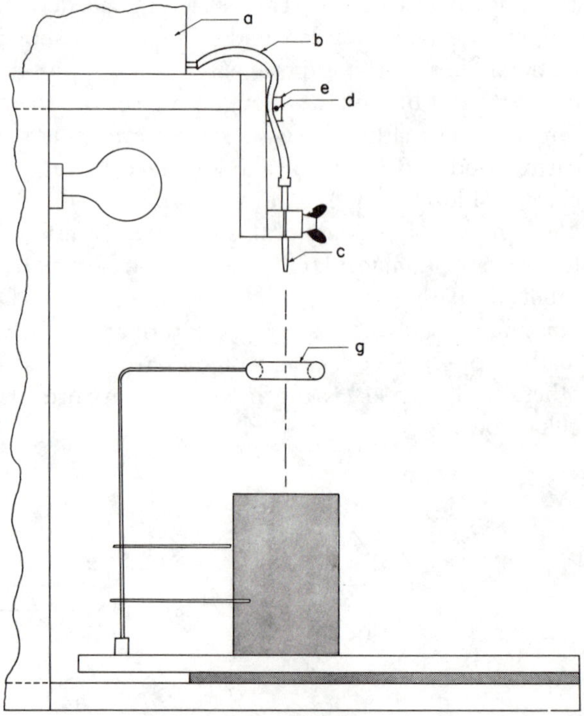

Fig. A29–90

Dirod I and Dirod II can also be stretched. These are heavy rotors, however, and they should then be between bearings instead of overhanging. The minor end plate should be moved to the front, and the longer rods should extend equally each way from the disk as in Dirod Junior. The neutral wire and brushes could also be brought to the front.

The big Radial Dirod, the largest and most spectacular dirod, has the highest output.

Complete construction details for Dirod Junior and Radial Dirod Junior follow. From this information and that given in the Materials List, the larger dirods can also be constructed.

A subbase of plywood supports both the generator and the drive motor of Dirod Junior. A 1-in.-thick wood base *b*, Fig. A29–91(a), is screwed to the subbase. Attached to the base are two Plexiglas end plates *e* and *d*, Fig. A29–91(b)—and minor end plate in front, the main end plate behind. The minor end plate supports the neutral brushes, and the main end plate supports the round collector plates C which are attached by screws. Each collector supports its own inductor. The corona shields *s* are supported by the main end plate, each attached by a single screw. The rotor end play and rotor position are controlled by round Plexiglas pieces on the shaft.

Base *b* is 1 x $4\frac{1}{4}$ x $5\frac{3}{8}$ in. The main end plate *d* is $\frac{1}{2}$ x $4\frac{1}{4}$ x 8 in., and the minor end plate *e* is $\frac{1}{2}$ x $1\frac{1}{2}$ x $6\frac{3}{4}$ in. The $\frac{3}{8}$-in.-diam shaft *f* is centered 2 in. down from the top of the minor end plate. Plexiglas disk *a* is 4 in. in diameter and $\frac{1}{2}$ in. thick (Fig. A29–92). Brass rods *g* are $\frac{1}{8}$ x 4 in., extending equally each way from the disk. The disk center plane is $2\frac{1}{4}$ in. from the main end plate, bringing the rod ends $\frac{1}{4}$ in. from that plate.

Collector plates C are 6 in. in diameter and are made from 0.042-in.-thick aluminum. A line joining the centers of the collectors passes through the disk center. Attachment screws for the collector plate are $1\frac{1}{2}$ in. apart. Ears (terminals) are made from $\frac{1}{4}$-in. copper tubing bent to an inside radius of $\frac{1}{2}$ in. or more. The shaft is 9 in. long and the pulley gives a speed reduction of about 4:1. The corona shields S (Fig. A29–93) are pieces of $\frac{1}{8}$-in.-thick window glass or Plexiglas, 4 x $5\frac{1}{4}$ in. They are

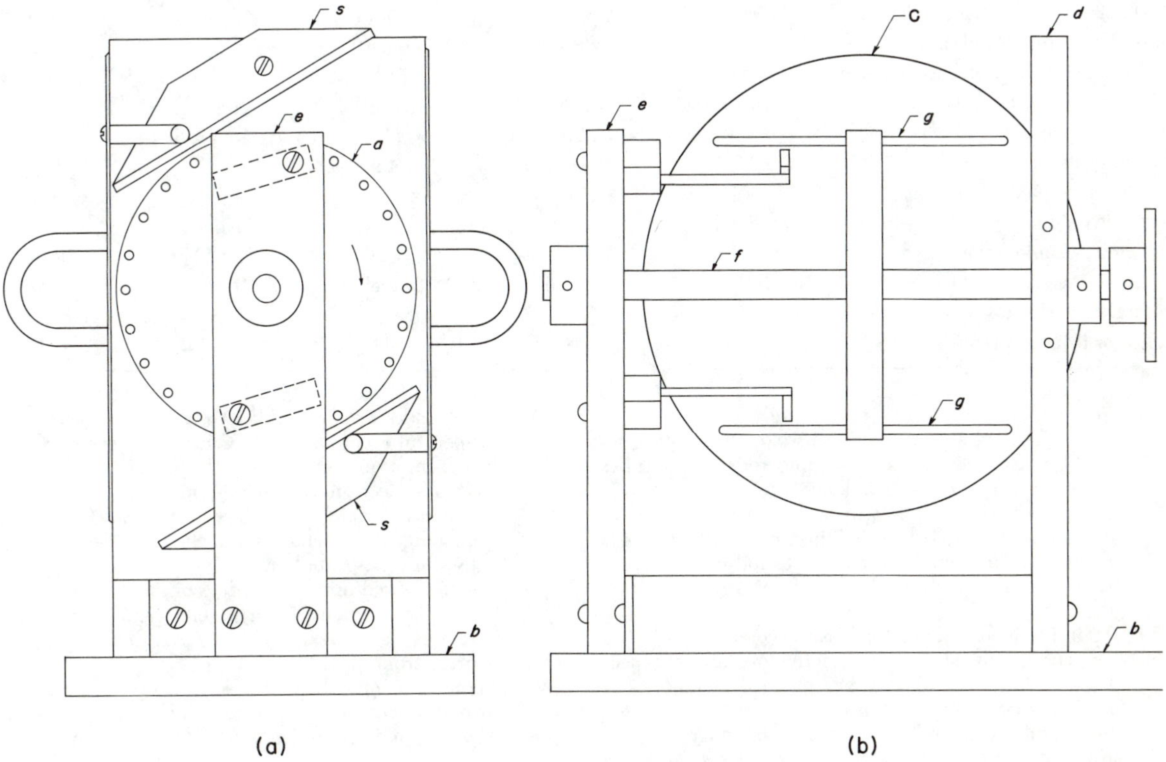

Fig. A29–91

Materials List

(Footnotes referred to by letters in parentheses.)

Specification or Part	Dirod Junior	Dirod I	Dirod II	Radial-Dirod Junior	Radial Dirod
Suggested top speed, rpm	1200	1200	1200	1200	1200
Suggested usual speed (or less)	600	600	600	600	600
Topmost kV measured	60	85+	85+	70–75	90–95
Steady spark rate, kV	50	75	75–80	60–65	80–85
Approx. μA, 600 rpm, 20 kV	2	3.2	2.3	2	8
Shaft size, in.	3/8	1/2	1/2	1/4	1/2
Bearings	Plexiglas	Brass	(A)	Plexiglas	(A)
Disk diameter and thickness, in.	4 x 1/2	6 x 3/8	6 x 1/2	5 x 1/8	6 x 1/2
Clockwise rotation	Yes	Yes	Yes	Yes	Yes
Rods	24 brass	36 brass	24 aluminum	48 brass	30 aluminum
Rod length and diameter, in.	4 x 1/8	4 x 1/8	4 x 1/4	2 x 1/8	4 x 1/4
Rods glued in or on with epoxy	No	No	No	On disk	Into holes
Rod ends covered against corona	No	Both ends (B)	Front ends (C)	No	No
Rods extend back of, onto, into disk, in.	(D)	1/2	3/4	1/2	1/2
Rod outer surface to disk edge, in.	0.1	1/8	1/8	–	–
Rear ends of rods to main end plate, in.	1/4	1/2	3/8	–	–
Corona shield	Plexiglas	Glass	Plexiglas	Plexiglas (E)	Plexiglas (F)
Corona shield is sloped 30 degrees	Yes	Yes	Yes	–	–
Corona shield support pieces	Plexiglas	Fiber angle	Fiber angle	–	–
Shield size (axial dimension last), in.	1/8 x 4 x 5 1/4	1/8 x 7 x 6	1/8 x 7 x 6	–	(G)
Main end plate, Plexiglas, in.	1/2 x 4 1/4 x 8	3/8 x 8 x 9 1/4	1/2 x 9 x 10 1/2	3/16 x 12 x 12	–
Shaft center to top of end plate, in.	3 1/2	4	5 1/4	5	–
Minor end plate, Plexiglas	In front	In rear	In rear	In rear	–
Minor end plate, in.	1/2 x 1 1/2 x 6 3/4	3/8 x 2 x 6 1/2	1/2 x 3 x 6	1/4 x 3 x 8	–
Collector plates, aluminum, in.	0.042 x 6	1/16 x 4 x 6	1/16 x 4 x 5 1/2	(H)	(I)
Plate support pieces, Plexiglas, in.	(J)	1/2 x 3/4 x 4	1/2 x 1 1/8 x 5	(K)	–
Corona trim, in.	None	None	1/4	None	1/4
Inductors, aluminum, in.	1/4 x 3	0.2 x 3/4 x 2 1/2	3/8 x 2 1/4	5/8 x 2 1/4	1/2 x 3
Neutral connector	In front	In rear	In rear	In rear	In front
Neutral brush radius position, in.	–	–	–	3	3 1/4
Collector brushes placed	(L)	(M)	(M)	(N)	(O)

(A) Sintered bronze bearings, permanently oiled.
(B) Tygon tubing, 1/8 in. i.d., 1/4 in. o.d. Ends plugged for 1/8 in. with epoxy, slipped onto rods. Rear tubes 3/8 in. long; front tubes 1 in. long.
(C) Front ends covered for 1/4 in. with two heavy coats of corona dope (applied by holding rotor with rod ends *down*, bringing small dip cup *up* to rod ends.)
(D) Rods extend equally each way, front and rear, from disk.
(E) Main Plexiglas end plate iself is the corona shield.
(F) No corona shields; instead, inductors are encapsulated in Plexiglas tubes; tubes plugged at inner ends for 1/2 in. with RTV 102 Silicone Rubber.
(G) Shield tubes 1/2 in. i.d., 1 in. o.d., 6 1/4 in. long.
(H) Front plates 1/16 x 1 1/2 x 2 in.; rear plates 1/16 x 1 1/2 x 2 1/2 in.
(I) Front and rear plates alike, 1/16 x 1 x 4 in., with 1/4 in. copper tube corona trim. Outer ends trimmed with 1/2 x 2-in. aluminum pieces. Plate pair rigidly joined by brass piece of rod 1/2 x 3/4 in. long. Front plates attached to Plexiglas blocks 1/2 in. thick, 1 1/2 in. axially, 2 in. long. These blocks attached to the 1/2 x 1 1/2 x 11-in. Plexiglas slant bar.
(J) Plates screwed directly to edges of main end plate.
(K) Front plates against front of end plate. Rear plates spaced back by Plexiglas block 3/8 in. thick.
(L) 1/2 in. in front of disk.
(M) About 1/4 in. or so behind disk.
(N) At rear, about 1/2 in. out from disk edge.
(O) About 5/8 in. from rod ends.

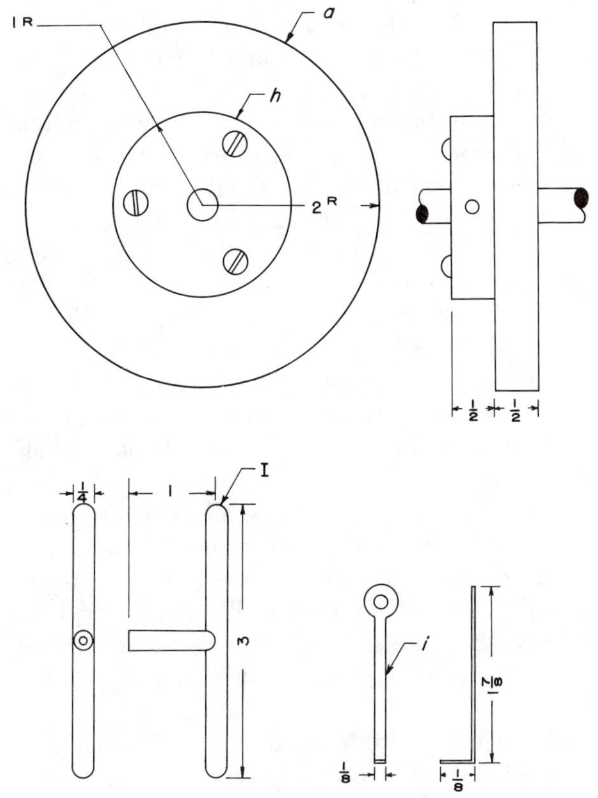

Fig. A29–92

filled halfway through with a nib ($\frac{1}{8}$ x $\frac{1}{4}$ x $\frac{1}{32}$ in.) soldered to it. Conducting rubber brushes, $\frac{1}{8}$ x $\frac{3}{8}$ in., are tied to the nib. The screw holding the block is seen in front view in Fig. A29–91(a). It is $\frac{3}{8}$ in. down from the top of the minor end plate and the same distance in from the side. Each soft aluminum collector brush rig *i* (Fig. A29–92) is about 2 in. long and $\frac{1}{8}$ in. wide with its foot bent at right angles for $\frac{1}{8}$ in. The conducting rubber brush is tied to each foot by thread. A $\frac{1}{4}$-in. hole in the collectors lets the brush stick through to touch the rods at their closest approach. The piece is bent in or out for brush adjustment. (The Radial Dirod Junior has brushes rigged in a still simpler fashion.) The right-hand collector has the head of the brush rig mounted above center; the other has it below center. The neutral connector is a piece of pull chain looped around the neutral brush holder rods.

Both end plates can be screwed directly to the epoxy-glued to the $\frac{1}{2}$-in.-thick Plexiglas support, which was $\frac{3}{4}$ x 4 in. before the corners were leveled off.

The attachment screws are on a vertical centerline, $\frac{3}{8}$ in. down from the top of the end plate. The shields are sloped at an angle of 30 deg. Inductors I are brass or aluminum, $\frac{1}{4}$ in. in diameter. They are 3 in. long and are attached to the 1-in.-long support rod made from the same material. The end of the support is hollowed out $\frac{1}{8}$ in. to fit to the inductor. The joint is epoxy bridged with silver paint. The center of the top inductor is about $1\frac{1}{4}$ in. down from the top end plate.

Disk *a* is screwed to a Plexiglas hub *h*, $\frac{1}{2}$ x 2 in. in diameter. The hub, if needed, is either forced onto the shaft, held by an Allen screw, or epoxied on. Pieces of card between the disk and the hub are used for adjustment to make the rotor run true. It may run true enough without a hub. The neutral brush assembly has a Plexiglas block, $\frac{1}{2}$ x $\frac{3}{8}$ x $1\frac{3}{8}$ in., holding a brass rod $\frac{1}{8}$ x $2\frac{1}{4}$ in. A threaded screw hole at the other end of the block is $\frac{7}{8}$ in. away from the rod hole. The far end of the rod is

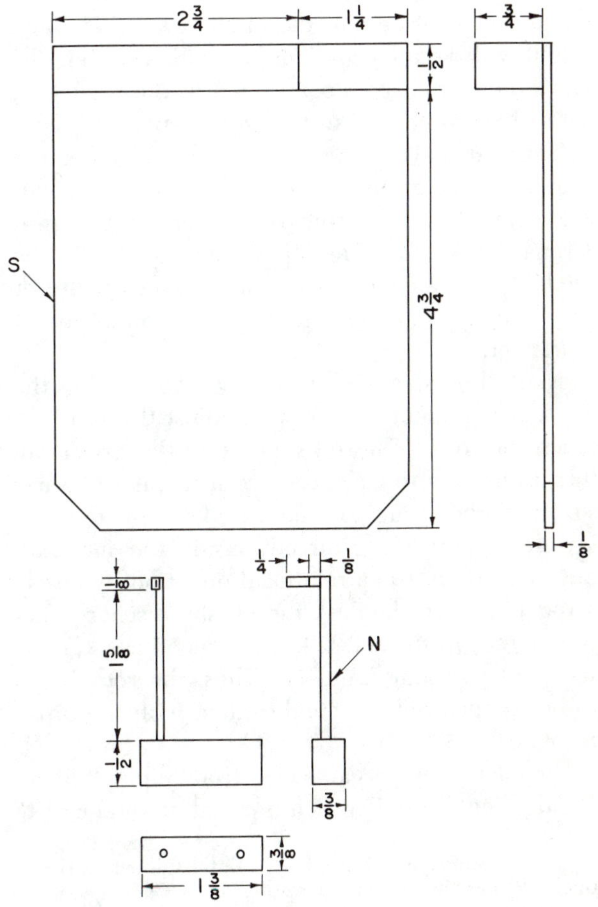

Fig. A29–93

box. Screw holes in the minor (front) end plate must then be larger than the screws, permitting some rocking to bring the two end plates and shaft holes into alignment. A refinement is to add the *metal* plate, Fig. A29–91(a), with it screwed to the base and the minor end plate permanently attached to it by machine screws. This allows the end plate to be removed any number of times, with accurate replacement.

Steel shafts normally have metal bearings—bronze, sintered bearings, ball bearings, etc. Dirod Junior has the shaft running in holes in the Plexiglas. With a drop or two of oil once in a while, these bearings last for a long time.

The collector plates are aluminum blanks, 0.042 in. thick, 6 in. diam. The edges are carefully rounded by file strokes, and then further smoothed and polished. The inductors are aluminum or brass with their ends rounded and polished. One end of the support rod is drilled and threaded for screw attachment to the collector. The other end is hollowed out to fit around the inductor. A V is filed onto it with a flat file and finished with a rat-tail file. The support rods are epoxy-glued to the inductors, with silver paint used to bridge the epoxy.

The rear ends of the disk rods come under the rear end of the corona shields. Therefore, the Plexiglas supports for the shields are part of the anticorona protection. The epoxy glue holding the shield to the support piece must be complete, or gap free, since a hole through it would permit sparkover.

Neutral brush assembly N (Fig. A29–93) lets the block swing about the screw to adjust the brush to touch the rods. The rod supporting the brush can turn in its hole to make the brush radial, or nearly so. In Dirods I and II, and Dirod Junior, the best brush position is not directly inside the inductor, but away from it in a rotational direction by nearly a rod pitch. In the dark the brushy discharge between the brush and disk rods sometimes can be seen to be as long as ¼ in. Thus, the rods begin to be charged *before* actual contact with the brush is made.

Excellent brushes can be cut from Slipknot Semi-Conducting Tape,(1) which is "used as shielding to reduce stresses around irregular points in high-voltage power cable splices and terminations." In the absence of this tape, wire brushes can be used. To make wire brushes, bend about six wires into a V and tie them together with thread about ½ in. from the bottom of the V. Tie the bundle to the nib of the brush. The tuft of wire should be about ¾ in. long; no shorter wires should be used, shorter wires soon fatigue and break off.

A V-groove can be cut in the Plexiglas pulley to hold a V-belt, which can be obtained from any sewing machine agency. O-rings also make good dirod belts. Since they eventually break, it is necessary to keep a spare. The pulley should have a half round groove.

Except for the wood screws, all the screws used are 8-32 machine screws. They are all roundhead screws except for Allen screws. The end-stop pieces, the pulley, and the hub all have Allen screws. Ordinary machine screws, with the heads sticking out, can be used instead, but they may slip on the shaft.

Although the dirods listed in the Materials List run clockwise, they can be easily modified to run counterclockwise. Dirod Junior has been run up to 1500 rpm. For safety, a new machine should initially be run at twice the speed at which it will normally be run, with proper precautions being taken while doing this.

Although all the dirods can be hand cranked if a large pulley and handle are provided, it is far better to have a speed-controlled motor. The best motor is the "universal," which runs on both ac and dc. Such motors are used on sewing machines and malted milk machines. For best speed control use a Powerstat or Variac rated at 1 A, or adapt the resistor control of a sewing machine. Another method of speed control is to connect several lamp sockets in parallel, with the entire bank in series with the motor. Experiment with bulbs of different wattages to get a suitable range of speeds.

Many construction details for Radial Dirod Junior have been given above and in the Materials List. There are no rod holes to drill in Radial Dirod Junior; the rods are epoxy-glued to the face of the disk. A drawing, with radial lines for each rod position, is laid on a flat surface and the disk is laid on the drawing. Rods 2 in. long extend in onto the face of the disk by ½ in. Short pieces of rod are laid around to support the outer ends of the rods to

(1) No. 5360 Cured Unsupported Butyl Rubber, 0.015 in. thick, Plymouth Rubber Company, Inc., Canton, Massachusetts. A few inches of tape can probably be obtained from the manager's offices of any electrical power utility.

make them level. About 1/2 in. of one end of each rod is dipped in epoxy and then laid onto the disk, with that part of the disk roughened for better adhesion.

The main end plate is structural in function and also serves as a corona shield. It is made from a sheet of Plexiglas between 1/8 and 3/16 in. thick. The thinner sheet would bring the inductors closer to the rods, which should increase the rod charge induction and raise the current output.

The end plate is screwed to the base. This alone is structurally insufficient, and stiffeners are needed, one at each side. These stiffeners are 1/8 by 1-in. Plexiglas braces, whose ends are heated and bent at 45 deg. Screw holes are made large enough for adjusting the end plate to be parallel to the disk.

The inductors are aluminum rods (2 1/4 in. long, 5/8-in. diam) and are epoxy-glued to the end plate. The inner ends are at a 2 1/4-in. radius. Rods 1/4 in. in diameter can also be used, but this limits the top

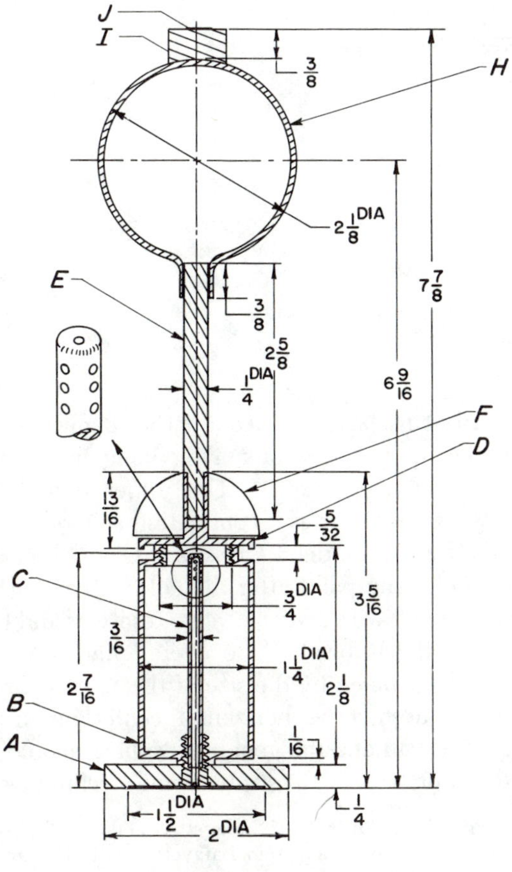

Fig. A29–94

voltage to 45 kV and reliable sparking voltage to 35 kV.

The inductors are connected to the collectors and terminals with flexible insulated wire. The connection is made by drilling a short hole big enough to get the wire insulation inside, going further with a smaller drill for the bared wire end, and then using an Allen screw to hold the wire. The screw can be eliminated by jamming the wire into its small hole and using epoxy.

The brush arms are Plexiglas strips, 1/8 x 1 x 3 1/2 in., held back from the end plate by 3/8-in. Plexiglas blocks. They swing up or down for brush adjustment and swing out of the way of the rotor when it must be shifted back for cleaning. These are for the neutral brushes.

The dimensions of the collectors are given in the Materials List. The rear plates are held behind the end plate by 3/8-in. blocks. A screw through the end plate connects the front and rear collector.

All four brushes are at a 3-in. radius. They all stick through from the rear, through 1/8-in. holes in the neutral brush arms and in the rear collector plates. These are double-strip brushes, each made of two strips of 1/8-in.-wide conducting rubber tied together. The tie comes through the hole by about 1/8 in. On the outside, or rear, surface, the two strips are bent apart flat against the surface and taped to the surface. If the inner, or contact, ends are made too long, they must be trimmed to rub properly. These brushes wear so well that one adjustment lasts for many operations.

Every dirod needs a neutral connector. A machine screw is placed on the brush arm so that a nut on it screws down on a brush strip. The neutral wire, which can be *bare*, attaches to the two screws. All the machine screws used are 6-32.

Plexiglas bearings last surprisingly well. Serious wear can be handled later by new Plexiglas bearing blocks glued to the end plate.

A safety guard is recommended for the demonstrator working with the dirod to lessen the possibility of an accident. The guard can be made of Plexiglas, Celluloid, or similar material as a 2-in.-wide band up both sides and across the top of the main end plate, reaching back 2 in.

There are two Plexiglas end stops on the shaft held by Allen screws, one in front and one in the rear, the rear one also being the pulley. They are set to give little or no end play and to give about

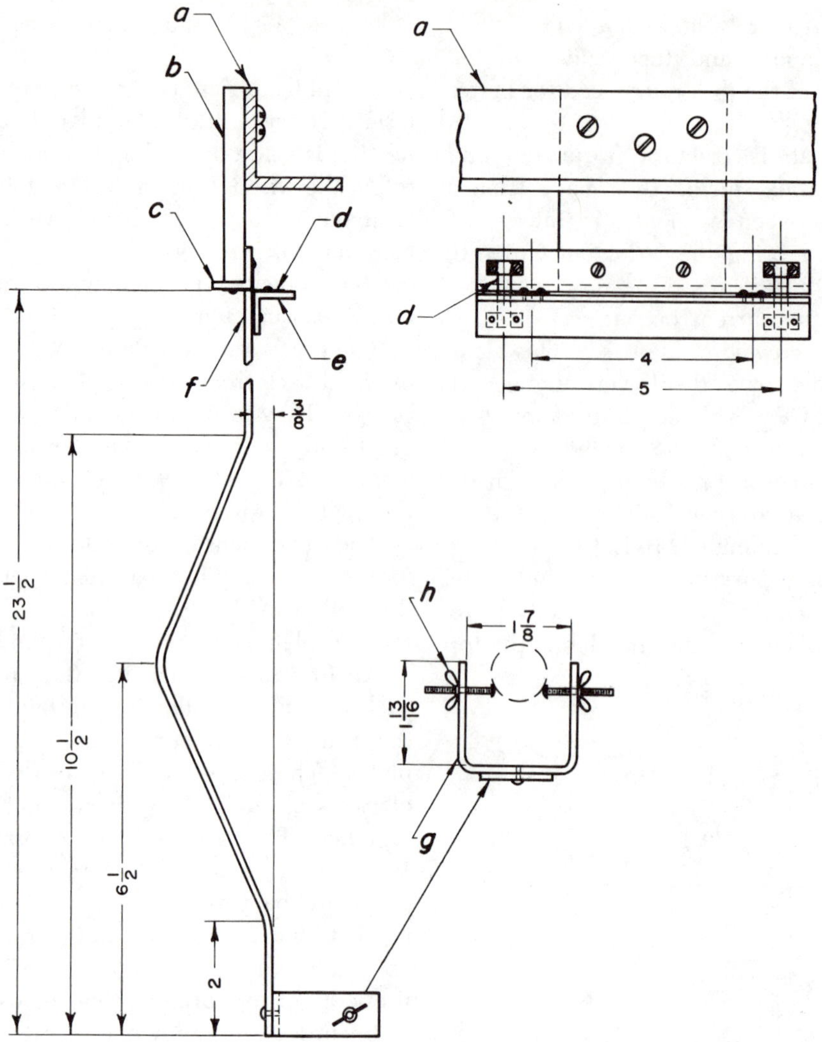

Fig. A29–95

1/32-in. clearance between the rods and the main end plate.

A29–1.28

The construction of the puck is shown in Fig. A29–94. The base *A* is a clear acrylic plastic disk 2 in. in diameter and 1/4 in. thick. A cylindrical Dry Ice container *B* is 2 in. high and 1 1/4 in. in diameter and is made of thin brass. It is attached to the base by a metal-to-plastic seal. The CO_2 gas flows to the 1/8-in. outlet in the base through a brass tube *C* with orifices near the top. The Dry Ice reservoir is closed at the top by a threaded cover *D*. A 1/4-in. polystyrene rod *E* fits into the cover and is surrounded at its base by a conducting hemisphere *F* consisting of one half of a Ping Pong ball coated with conducting paint.(2) The upper end of the polystyrene rod carries a conducting sphere *H* consisting of a 2 1/8-in.-diam Christmas tree ball coated with conducting paint after the colored coating has been removed with acetone. The reservoir and base are painted flat black. The overall height of the puck is determined by the size of the Van de Graaff generator used. The horizontal centerline of the Ping Pong ball and the generator sphere should coincide. For use with stroboscopic techniques, a

(2) Print-Kote Silver Paint, Lafayette Catalogue No. P-313 GC-21-1, can be obtained from Lafayette Radio Electronics, Syosset, New York.

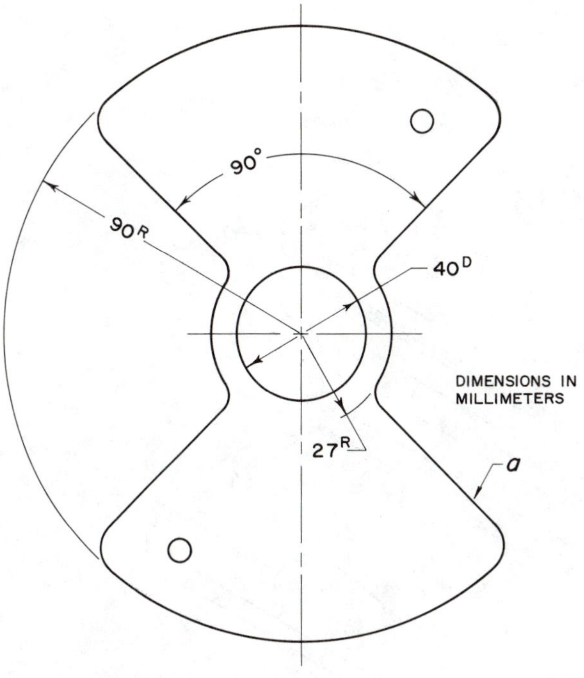

Fig. A29–96

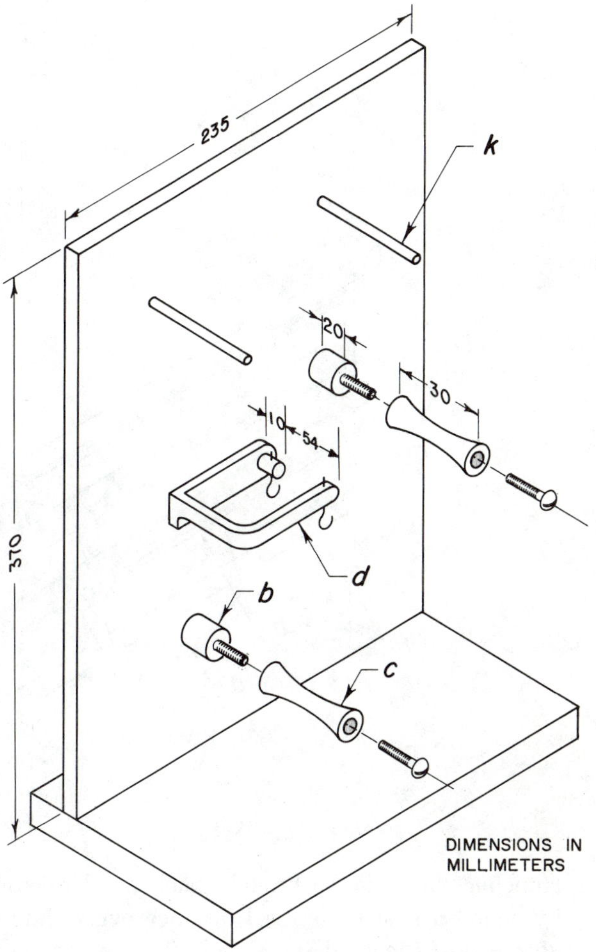

Fig. A29–98

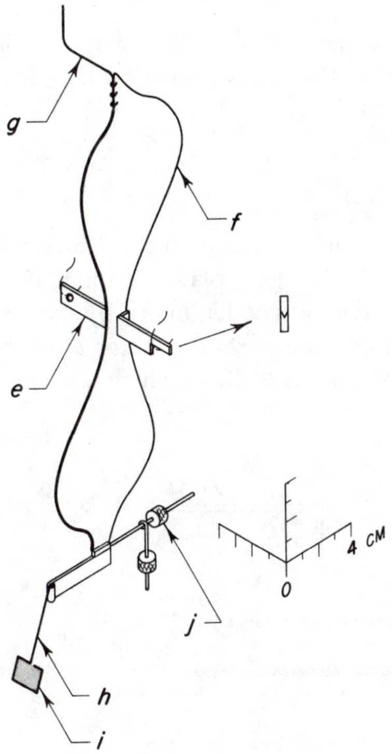

Fig. A29–97

$\frac{5}{8}$-in.-diam, $\frac{3}{8}$-in.-long piece of wax *I* with a $\frac{1}{4}$-in.-diam spot of electrical tape *J* is stuck to the ball by heating.

The puck launcher mechanism shown in Fig. A29–95 is simple, assures stability of direction, and provides reproducibility of paths and initial puck velocities. It consists of a 25-in.-long strip of $\frac{3}{16}$ x $\frac{1}{2}$-in. aluminum *f*, attached to a hanger block *b* by two $\frac{3}{4}$ x $\frac{3}{4}$ x $\frac{1}{8}$ x 6-in.-long aluminum angles *c* and *e*. These angles are connected by four pieces of blued spring steel *d* to an upper support *a*. The U-shaped lower piece *g* is made from a $\frac{3}{4}$ x $\frac{3}{16}$-in. aluminum strip, with two light screws with wing nuts *h*.

The glass riding surface measures 36 x 48 x $\frac{1}{4}$ in. and is supported and leveled with jack screws. Directly below and taped to the glass from beneath is a piece of graph paper, marked off so that the

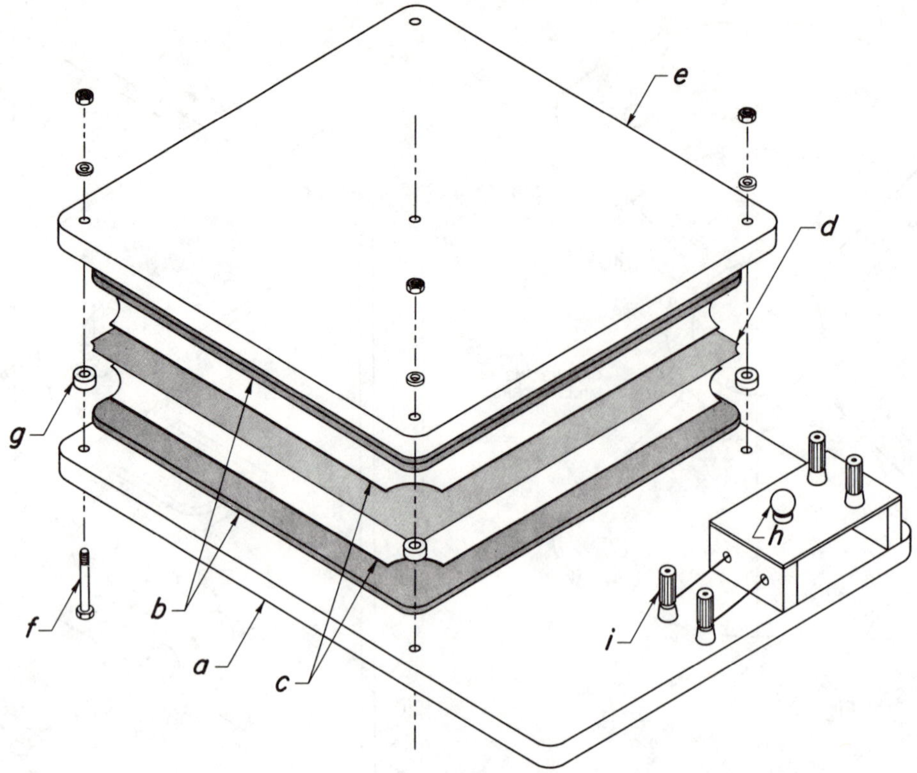

Fig. A29–99

launcher claw can be brought back (and released by hand) a nearly identical distance every time to give almost identical initial velocities to the puck. A temporary steel angle framework attached to a supporting table can be employed for hanging the launcher, the strobe light, and a Polaroid camera. Black cloth is used as a canopy to prevent unwanted reflections from ceiling lights when quantitative work is performed with the apparatus.

The silver conductivity paint should be applied with a camel's-hair brush. All the brass parts should be machined so that they have as little mass as possible. Pea-sized grains of Dry Ice are used in the puck.

A29–3.3

The construction of the voltmeter is straightforward. A Plexiglas plate (9 in. x 15 in. x 3/8 in.), mounted on a wooden board by means of an aluminum angle, supports the stator *a* and rotor *f* (Figs. A29–96 and A29–97), which are drawn to scale.

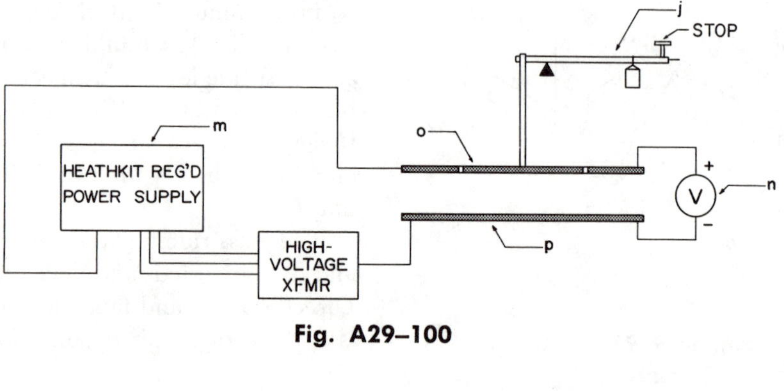

Fig. A29–100

Thus, after appropriate scaling, these measurements can be transferred to a piece of sheet metal directly and then cut out. The two 2-mm-thick brass plates of the stator are screwed onto the Plexiglas plate by means of columns (*b* and *c*, Fig. A29–98). A U-shaped support *d*, made of brass sheet metal, is equipped with hooks of steel wire; the lowest point of the support will be the center for the stator plates (point of touch for the shaft *e* in Fig. A29–97).

The rotor is made of hard aluminum sheet metal ($\frac{1}{2}$ mm thick). The shaft *e* is put through the center hole of the rotor, which is held by a small plate put on the rotor and soldered onto the shaft *e*. The shaft has a knife edge on one end and a hole in the other. Pointer g and the silencer *h* are made of aluminum wire. They are fastened to the rotor by means of a little string and some glue (resin). Damper *i* is made of a bent frame covered with cellophane tape; it is put into a vessel filled with water. This prevents oscillations of the instrument. A small, movable weight *j* serves as a zero point adjustment. Allowance must be made for the buoyancy of the water. Two end stops *k* limit the deflections of the rotor.

The instrument can be calibrated, and, if desired, the measuring range can be extended by putting different weights on the silencer. The scale is made by gluing pieces of paper to the Plexiglas plate. The columns (supporting the stator) and the U-shaped aluminum hanger for the rotor serve as electrical contacts. Whenever possible, the latter should be grounded because the water vessel is poorly insulated.

A29–4.3

The capacitor is built in a sandwich arrangement as shown in Fig. A29–99. The base *a* is plywood, 20 x 28 x 1 in. A sheet of sponge rubber *b*, 17 x 22 x $\frac{3}{8}$ in., is placed on top of the base, and then, in turn, a 17 x 22-in. sheet of aluminum foil *c*, a 17 x 22-in. dielectric *d* (Mylar-thin plastic sheet), another sheet of aluminum foil identical to the first, another piece of sponge rubber, also identical to the first, and finally a top of plywood *e*, 20 x 24 x 1 in. Carriage bolts *f*, $2\frac{1}{2}$ x $\frac{1}{4}$ x $\frac{1}{4}$ in., are used to bolt the capacitor together. Lucite spacers *g* are used on the bolts. They are $\frac{9}{32}$ in. i.d. x $\frac{3}{4}$ in. o.d. x $\frac{5}{16}$ in. long. The small neon bulb *h* is mounted on a sheet of aluminum $2\frac{1}{2}$ x 6 x $\frac{1}{16}$ in., which is

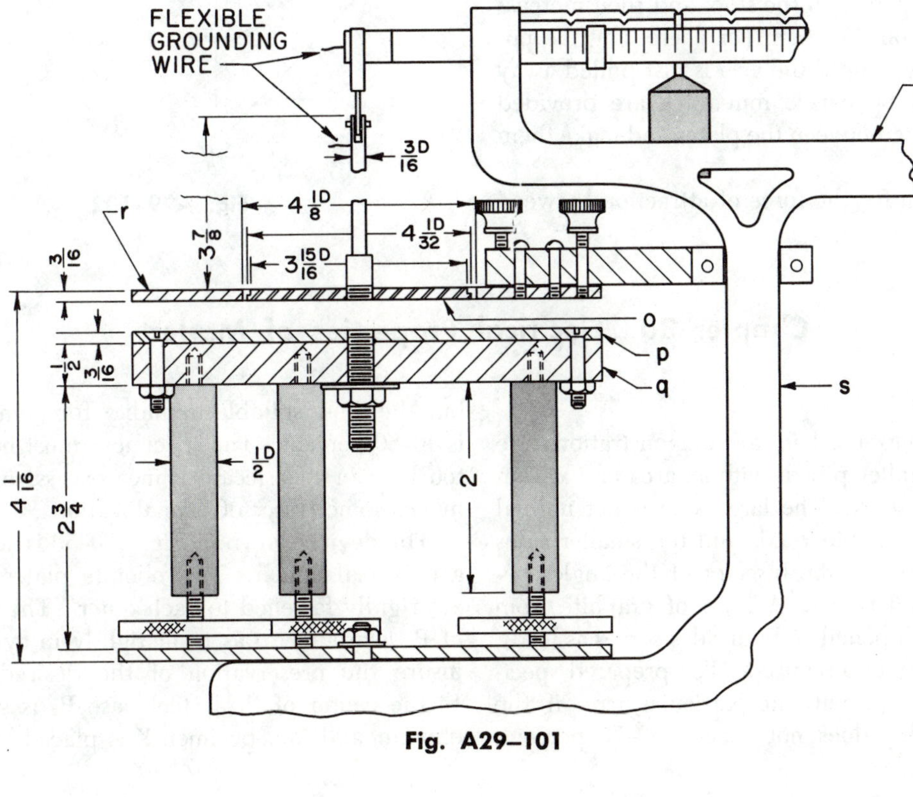

Fig. A29–101

mounted, in turn, about 1½ in. above the base. Connections are made using standard binding posts *i*.

A29–4.13

The Colorado apparatus, shown in Fig. A29–100, is composed of a No. 4030 Welch triple beam balance *j*, a parallel plate arrangement mounted on it, a 4.5-kV model 4574 Spellman step-up transformer, a Heathkit regulated power supply *m*, and a Simpson No. 260 voltmeter *n*. An aluminum plate *o* is hung from the balance arm and takes the place of a stock weight pan. Below this a 6-in.-square aluminum plate *p* is screwed to a thick Bakelite plate *q*, which can be raised, lowered, and leveled (Figs. A29–101 and A29–102). In order to eliminate edge effects, the upper plate is surrounded by a 6-in-square guard *r* in the same plane and at the same potential as the disc.

Attach the guard, which has provisions for leveling, to balance post *s*. Both the disc and guard are grounded. The lower plate is insulated and can be raised to a potential of 2 to 4 kV. The balance has a stop which does not let the disk go below the plane of the guard. Measurements are made by lifting the arm gently to the stop, and then increasing the weight on the beam (or decreasing the potential difference) until the arm is just pulled away from the stop. Spacers 3 mm thick are provided to set the spacing between the plates and make them parallel.

The equation for the force of attraction between charged parallel plates is (if no dielectric is used, permittivity is ϵ_0).

$$F = \frac{\epsilon V^2 A}{2S^2},$$

where F is the attractive force, V is the potential difference, A is the area of the movable plate, S is the distance between the two plates, and ϵ is the permittivity constant.

The dielectric constant K can be obtained from

$$K = \frac{F'/F_0}{2 - F'/F_0},$$

where F_0 is the force of attraction without the dielectric and F' is the force with the dielectric in place, the thickness of the dielectric being half the plate separation.

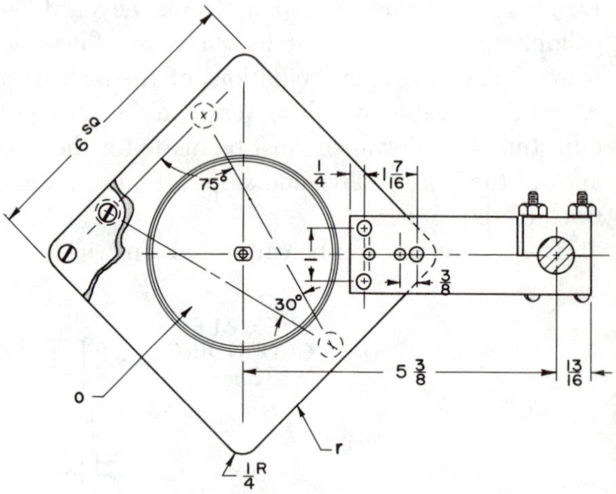

Fig. A29–102

Chapter 30 Electrical Properties of Matter

A30–6.3

The specimens used in the demonstrations are rectangular parallelepipeds with an area of 2 x 2 cm and 3 x 5 mm thick. The large side is cut normal to the crystallographic *x* axis and the smaller sides are perpendicular to the bisector of the angles between the *y* and *z* axes. A layer of graphite from rubbing of a soft pencil on both sides serves as leads, or gold leaf may also be used. The prepared specimens can be kept without particular care if the relative humidity does not exceed 60–70 percent, but the most suitable humidity for demonstrations is 40–50 percent. The specimens must not be dried too long in a desiccator, since excessive drying removes some water of crystallization.

The device shown in Fig. A30–59 is suggested as a universal holder. Two ebonite plates P_1 and P_2 are rigidly fastened to each other. The upper part of P_2 is ground so as to fit tightly in cylinder A to insure the preservation of the desired humidity. In the center of P_1 a steel base P_3 is sunk in the ebonite, and the specimen K is placed upon it. To

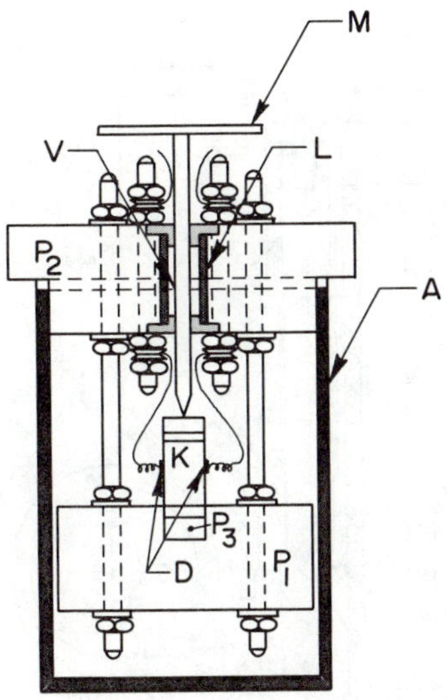

Fig. A30–59

avoid shorting, a thin ebonite plate is fastened to the press on the side facing the crystal. Lead plates measuring 4 x 4 x 1 mm are fastened to the springs D to ensure better contact between the springs and the electrodes. Rod V passes through bushing L set into the ebonite plate P_2. To obtain the desired mechanical stresses, iron cylinders of various weights are loaded upon M.

To achieve the proper humidity, the specimen is brought in contact with moist air, and then a glass vessel containing crushed calcium chloride is placed near the sample on the lower plate within the cylinder. The sample is connected to a battery of 200–300 V and a shunted galvanometer. When the surface conductivity of the specimen decreases to 10^{-2}–10^{-3} μA, the drying is considered complete. This takes 10–15 min with a few tenths of a gram of calcium chloride.

Chapter 31 Magnetism

A31–1.14

The trough-shaped gap magnet shown in Fig. A31–98 is 34 in. long and constructed from 1 x 4 x ⅛-in. plates *f* of low-carbon steel. It is held together with five 3/16-in. brass rods *c* and brass end plates *e*, and it is wound with 200 turns of No. 16 magent wire *d*. For demonstration purposes, a ⅛-in.-thick Plexiglas disk *b* is fastened on top of the magnet.

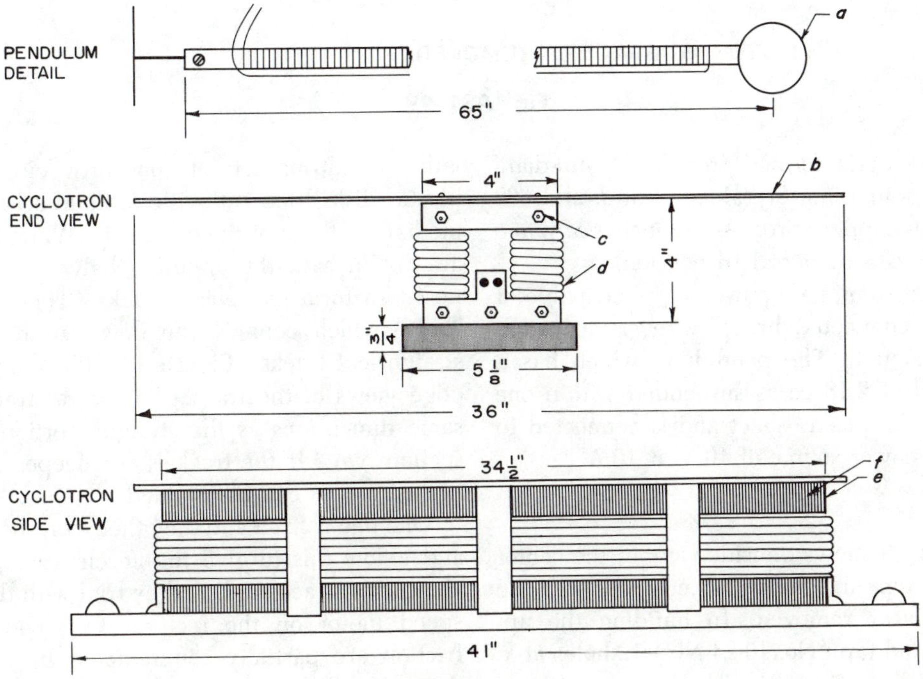

Fig. A31–98

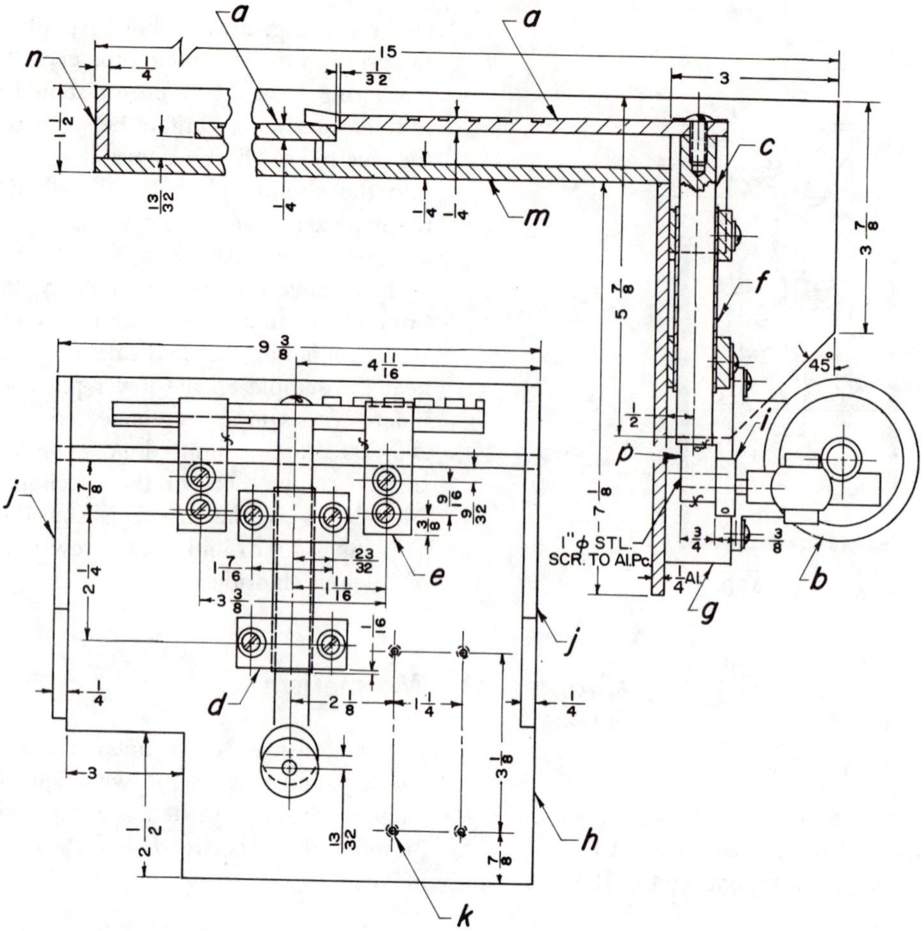

(MOTOR & SPACER REMOVED)

Fig. A31–99

The pendulum is formed from a $\frac{7}{16}$-in.-diam steel rod, a $1\frac{1}{2}$-in.-diam Styrofoam ball *a*, and 662 turns of No. 16 copper wire. A few inches of piano wire are used as a universal suspension.

To use the system, a dc power supply capable of 30 V at 10 A is connected through a reversing switch to the gap magnet. The pendulum, which has a natural period of 2.18 cc, is suspended within one centimeter of the plastic sheet and is connected to a constant dc power source of 10 V at 10 A.

A31–1.15

Figure A31–99 shows the side view of the center section at the top, and the right end view with the motor and spacer removed. In building the apparatus, drill and tap (No. 10-24 NC) 12 holes at *k*. The spiral track *a*, Fig. A31–100, is produced on a lathe by cutting sets of concentric circles on two plastic disks, one set with radii alternately intermediate to those of the other set. When these disks are cut in half, the opposing halves can be juxtaposed to form two spiral tracks. The accelerating "hills" which connect the dees are formed from straight-cut tracks. Figure A31–100 shows top and edge views of the track. The spiral track has the same dimensions as the straight portion shown in section A-A. If the track is cut deeper and wider, higher speeds can be attained.

One dee is fixed and the other reciprocates above and below this to give the acceleration. The synchronizing is accomplished by trial with the variable speed motor on the reciprocator. The effects of friction are partially counteracted by sloping the dees very slightly in opposite directions.

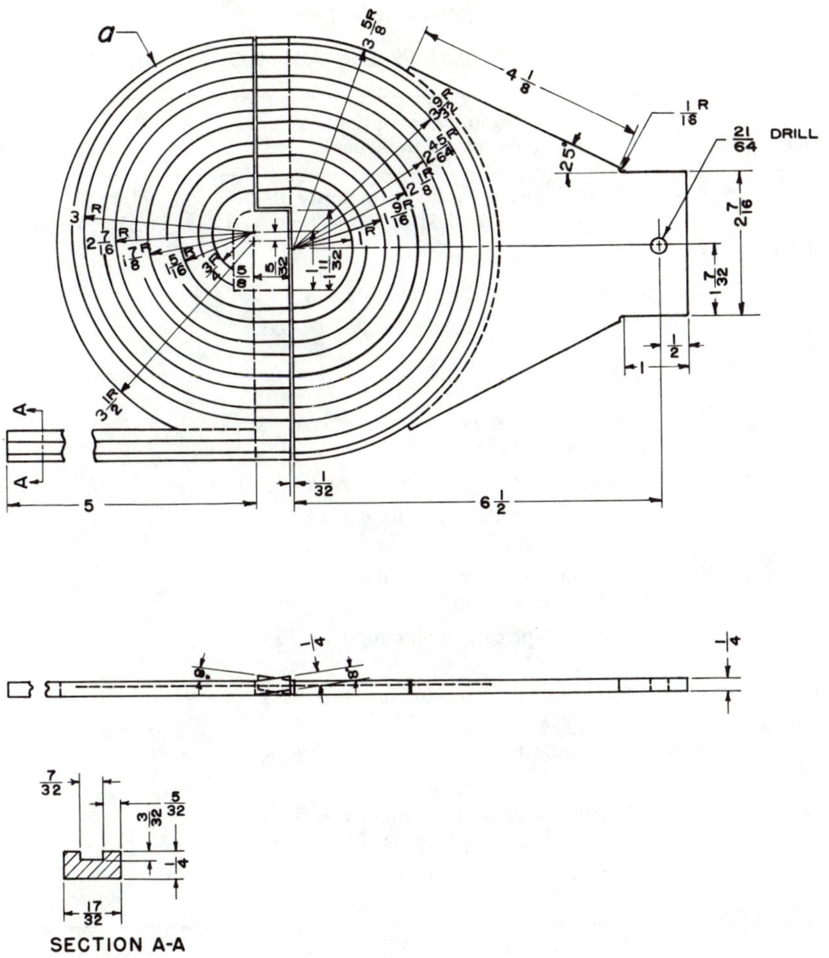

Fig. A31–100

A31–1.16

Figure A31–101 shows the construction details of the inverted pendulum. A brass flywheel *a* (4 in. diam, $1\frac{3}{16}$ in. thick) is connected to the shaft of an electric motor *b*, $\frac{1}{30}$ hp, 115 dc, shunt-wound, 1725 rpm. A tapered steel arm *c* with bearings at each end ($6\frac{1}{4}$ in. from center to center) is attached to the flywheel at the bottom and to a 1 x 0.325-in.-diam steel shaft *d* at the top. This shaft moves up and down in a 5 x $\frac{1}{2}$-in. slot cut in a steel or aluminum support plate *e* (4 x $16\frac{3}{4}$ x 0.375 in.), which is fastened to the steel base plate *f* (8 x 6 x $\frac{1}{2}$ in.) and supported by two pieces of aluminum g ($4\frac{5}{8}$ x $18\frac{7}{8}$ x 0.650 in.).

The steel drive pipe *h* (0.625 in. o.d., $16\frac{3}{4}$ in. long) moves through two 1-in.-long brass tubes *i*, which are fastened to the support plate. The shaft *d* fits through a hole in the pipe and is connected to the pendulum roller bearing (0.875 in. o.d., 0.300 in. i.d.). The pendulum consists of an aluminum tube, $26\frac{1}{2}$ in. long, $\frac{1}{2}$ in. diam, with a $2\frac{3}{4}$-in.-diam disk made from $\frac{1}{16}$-in. aluminum fastened to it.

The apparatus must be clamped firmly to the lecture bench and must run smoothly. To prevent the pivot of the pendulum from coming loose, it should be fastened with lock nuts. The motor should be mounted on a larger metal plate, preferably the same size as the baseboard of the apparatus.

A31–2.2

The primary and secondary coils consist respectively of 285 turns of No. 24 and 600 turns of No. 38 coated wire (Formvar). The primary coil is wound

Materials List

Figs. A31–99 and A31–100

Detail	Quantity	Part and Description (dimensions in inches)	Remarks
a	2	Track, Lucite; ¼ thick; quantity dependent on method of fabrication	Model 2T60, G. K. Heller & Co., Las Vegas, Nev.
b	2	Controller and motor	
c	1	Follower, brass; ⅝ diam x 5⅜ long	
d	2	Follower guide block, Lucite; 1⅛ x 2 x ⅞	
e	2	Track guide block, Lucite; 1 x 2$\frac{5}{16}$ x ⅞	
f	1	Follower bushing, Lucite; ¾ o.d. x ⅝ i.d. x 3¼ long	
g	1	Motor spacer, wood; 2½ x 3¾ x 1$\frac{3}{16}$	
h	1	Motor and guide mounting plate, aluminum; 7⅛ x 9⅜ x ¼ thick	
i	1	Cam mounting wheel, aluminum; 1 diam, ⅜ thick	
j	2	Tray side pieces, aluminum; 5⅞ x 15 x ¼ thick	
m	1	Tray bottom plate, Lucite; 9⅜ x 12 x ¼ thick	
n	1	Tray end piece, aluminum; 1¼ x 8⅞ x ¼ thick	
p	1	Follower cam, steel; 1 diam x ¾ thick Suitable hardware for assembly	
	2	Ball bearing, Steel; 1 pc. ⅜ diam; 1 pc. ½ diam	

on a cylindrical form, Fig. A31–102, upper, made of Bakelite tubing. Two small holes *A* which angle to the outer surface are drilled in one end of the tubing and a groove B is machined along the outside from one of them toward the other end, where two more holes are drilled at such an angle as to meet without breaking through the inner surface. The wire is threaded through the holes C, along the groove B, and through one of the holes *A* in the end of the cylinder, and then is wound in a single layer from the end at *C* to *A*, where it is threaded through the outer hole. The whole coil is then covered (encapsulated) with clear epoxy resin. This construction shields all the wire from accidental breakage. The primary is joined to the base by a No. 8-32 screw, ⅞ in. long, inserted into a ¼-in. plug, as shown in Fig. A31–102, lower.

The secondary coil is wound on another piece of Bakelite tubing (a coil winder is useful here), and a piece of clear plastic containing holes ($\frac{19}{32}$ in. diam, $\frac{5}{16}$ in. deep) for banana jacks *a* and provisions for support of the secondary is cemented to the coil form. The ends of the wire are soldered to the jacks and are covered, along with the coil, with clear epoxy resin.

The secondary assembly slides by means of steel ball *b*, a No. 10-32, $\frac{5}{32}$-in.-long screw, and spring *c* (20 turns/in., ¼-in.-long compression type) on a vertical post of cold rolled, "bright" plated steel. The post is scored, using a lathe, at 1-cm intervals beginning with the zero mark, at such a position as to indicate centering of the secondary coil on the primary. Both ends of the post have a No. 10-32 thread and are 0.190 in. diam. The upper portion is ⅜ in. long and threaded to fit a knob with No. 10-32 thread.

The secondary coil mount consists of two parts, shown in Fig. A31–103, joined by a No. 8-32 screw

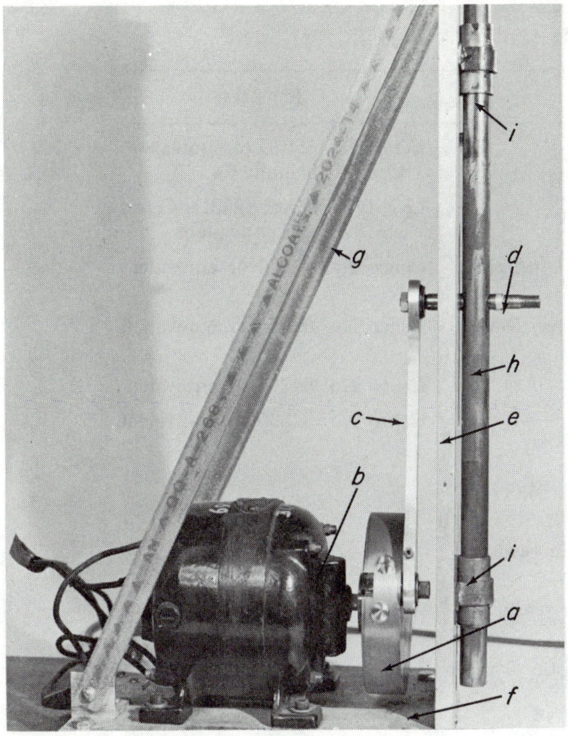

Fig. A31–101

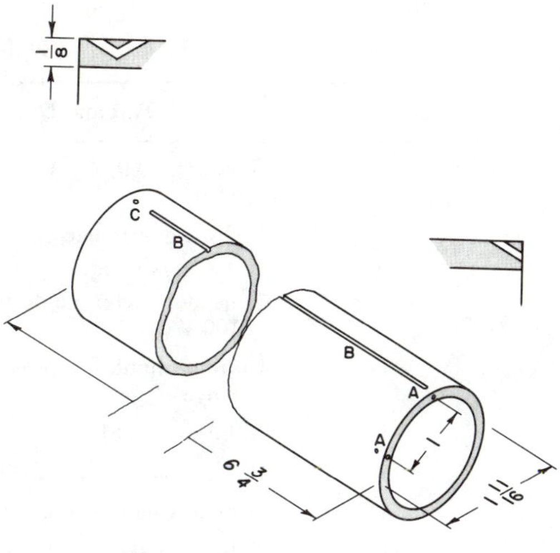

Fig. A31–102(a)

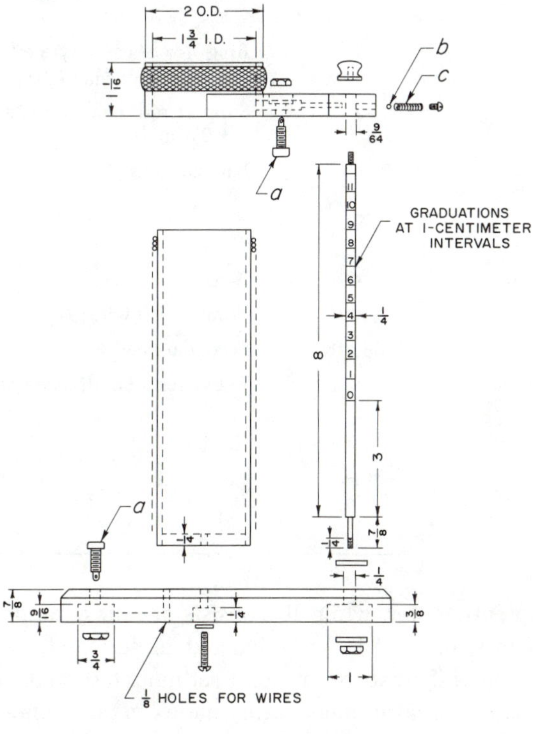

Fig. A31–102(b)

with a compression spring inserted into the central shaft. The central shaft is a No. 18 hole with the wider portion 1/4 in. diam and 3/8 in. deep, as in part A. This part is made from clear plastic. Made from "bright" plated brass, part B is drilled to receive the graduated post and spring and ball assembly described above.

A hardwood base is constructed as shown in Fig. A31–104. Hole D is 1/4 in. diam to accommodate the vertical post. Hole E is for attachment of the primary coil. Wires are passed through 1/8-in.-diam holes *F*. Banana jacks are inserted into holes *G*, 19/32 in. diam.

A31–2.3

The solenoid, shown in Fig. A31–105, is wound with 16-gauge enameled copper wire on a clear acrylic plastic tube, 6 in. o.d. x 5½ in. i.d. There are three layers wired in parallel with 837 turns in each layer. A 3/4-in.-thick plywood board with a 1 3/4-in.-diam hole in the center is cut to fit in the back of the tube. An appropriate wooden framework supports the tube.

The various loops which can be inserted in the solenoid are shown in Fig. A31–106. Plywood, 3/4 in. thick, is used for the forms, and the loops are made with No. 12 enameled copper wire, which is held in place with insulated clips.

Resistor R_2 (Fig. 31–55) is a 6.5-A, 0–10 Ω

Materials List

Quantity	Part and Description	Remarks
1	Rheostat, 90 Ω, 2.2 A	Cenco No. 82910 or equivalent, Central Scientific Co.
1	Ballistic galvanometer	L&N type P, No. 2239D; Cenco No. 82075 or equivalent
1	Projection meter, ac-dc full scale 1 mA or 100 V dc	Cenco No. 82550 or equivalent
1	Current shunt for projection meter, 5-A range	Cenco No. 82554 or equivalent
1	Knife switch, SPST	Cenco No. 84315 or equivalent
1	Pinch-type switch, DPDT	Cenco No. 84335 or equivalent
1	Base, any hard wood, 6 x 4 x 7/8	
1	Tubing, Bakelite, 1 11/16 in. o.d., 1/8-in. wall, 6 3/4 in. long	
1	Tubing, Bakelite, 2 in. o.d., 1/8-in. wall, 1 1/16 in. long	
1	Mounting plug for primary, clear plastic, 1 7/16 in. diam x 1/4 in. thick	
1	Secondary coil support, clear plastic, 2 x 2 x 7/16 in.	
1	Ring for secondary coil support, brass, 1 in. diam x 7/16 in. thick	
1	Support rod, cold rolled steel, 1/4 in. diam x 9 1/4 in. long	
4	Banana jacks	E. F. Johnson Co., No. 108–903 or equivalent, available from Allied Radio Corp., Chicago, Ill.
4	Banana plugs	E. F. Johnson Co., No. 108–303
140 ft	Primary coil wire	No. 24 Formvar
340 ft	Secondary coil wire	No. 38 Formvar
1 can	Epoxy resin bonding agent	R-313, Carl H. Biggs Co., Santa Monica, Calif.
1 bottle	Hardener –B	
1	Coil winder	Morris "Coilmaster" or equivalent; Allied Radio Corp., No. 61 G 053

potentiometer, while R_1 consists of four ceramic resistors, 0.3 Ω, 25 W, connected in series. S_2 is an automobile distributor point set mounted on a small board. Measurements are made on an ammeter, 0–5 A dc, a voltmeter, 0–50 V dc, and a Tektronix oscilloscope.

A31–2.11

Details of the Helmholtz coils are shown in Fig. A31–107. Each coil is wound with 30 turns of No. 20 enameled wire on a piece of 1/8-in. aluminum channeling (type 606-3TS) which has been rolled into a 30-in.-diam hoop.[1] The wire is covered with black electrical tape. The type of suspension shown in Fig. A31–107 consists of three aluminum rods, 3/4 in. diam, 16 in. long, used as braces for the coils, and a 26-in.-long rod across the middle of the apparatus is used for the attachment of the suspending wire. The suspension shown in Fig.

(1) The coil rings for the original apparatus were rolled by the Chicago Metal Manufacturing Co., Chicago, Illinois.

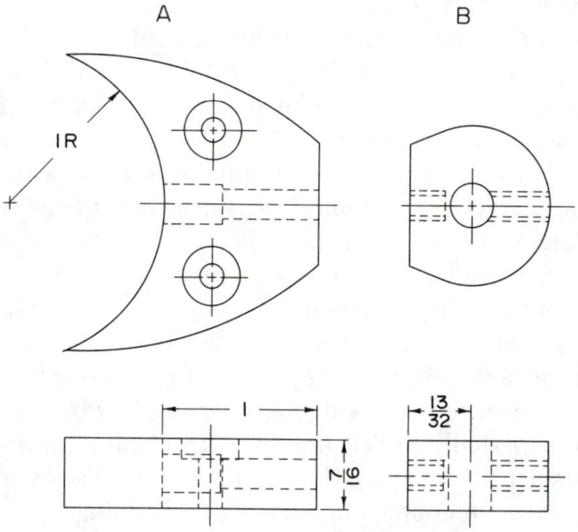

Fig. A31–103

31–64 replaces the single 26-in. rod by two rods joined to a suspending rod. The apparatus is entirely of screw construction. The bottom bracing rod can be left off, if necessary, to provide space for the various devices to be placed in the field. The coils are excited by a six-volt automobile battery.

A31–2.20

The stator *a* is fashioned from six pieces of soft iron, 2 mm thick, cut as shown in Fig. A31–108, with an outside diameter of 4 in. and an inside diameter of 1½ in. They are mounted, using spacers, on a

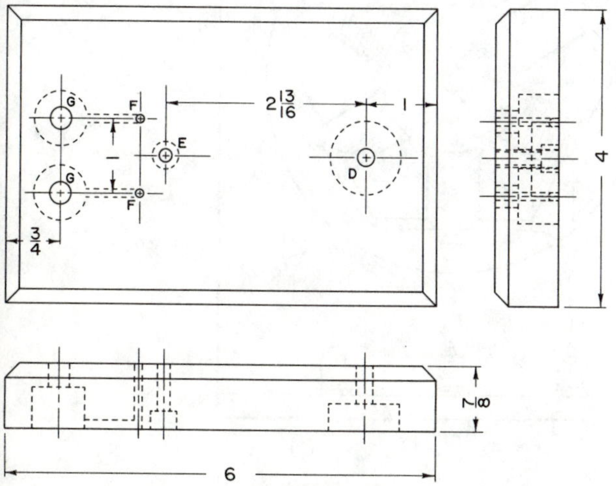

Fig. A31–104

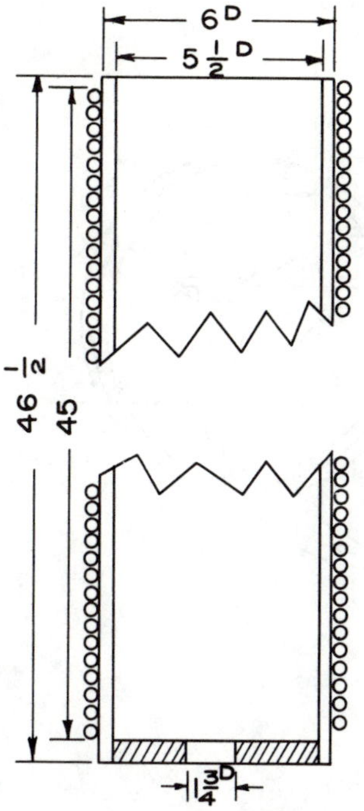

Fig. A31–105

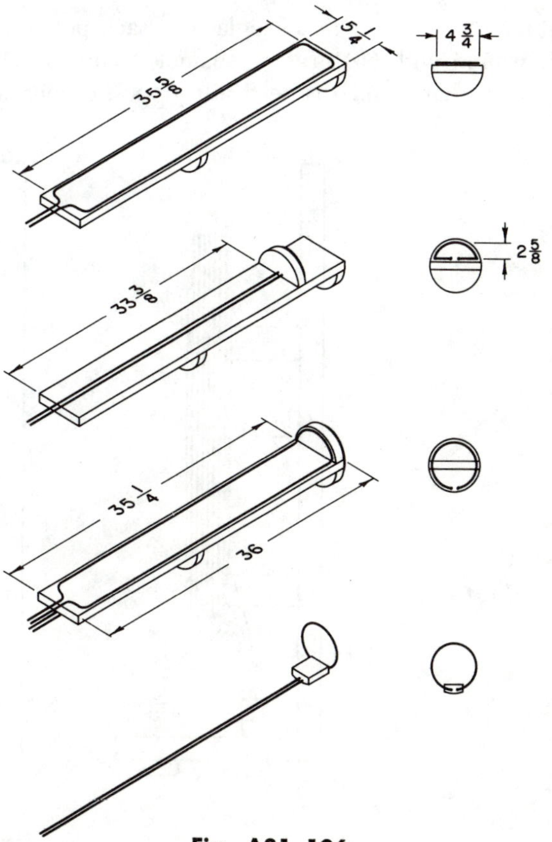

Fig. A31–106

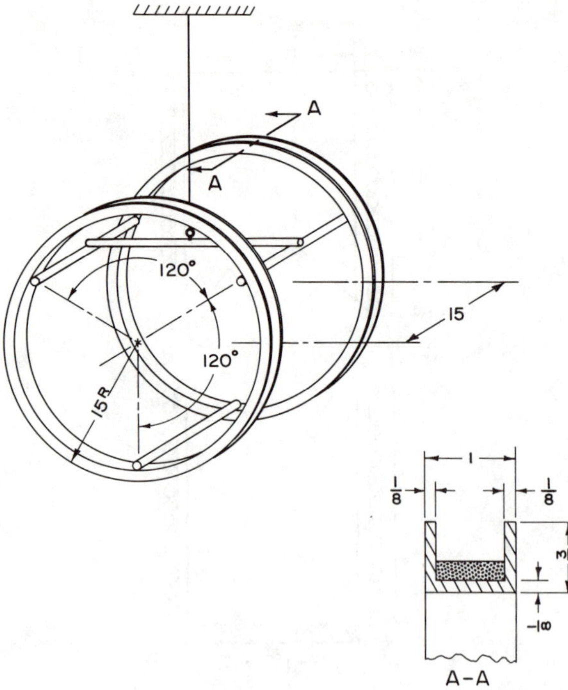

Fig. A31–107

piece of 5 x 6 x ¼-in. Plexiglas *b*. Each pole piece is wound with 60 turns of enameled copper wire, $\frac{1}{32}$ in. diam, and diametrically opposite coils are connected in series to produce opposite poles toward the center when a voltage is applied to them. A piece of 2 x 5 x ¼-in. Plexiglas in which six banana plug receptacles *d* have been mounted is glued to the stator mounting plate. A bearing *e*, made from a phonograph needle inserted in a machine screw, is mounted on the stator mounting plate, centered within the pole pieces. The 1-mm-thick rotor *f* (Fig. A31–109) is used to show synchronous motor operation. It is made from blue steel, which is permanently magnetized after it is fabricated. The bearing *g* for this rotor consists of a ¼-in. length of ⅛-in.-diam brass rod which has been partially drilled through. The bearing is soldered to the rotor. The rotor *h*, which shows induction motor operation, is made from a single piece of aluminum. The bearing is made with a drill point in the underside of the rotor. The drilled holes in the top are put in so that rotation of the rotor can be seen when the apparatus is projected.

The three-phase generator which creates the varying voltages 120 deg out of phase with each other is constructed as shown in Fig. A31–110. A 360-deg continuously wound potentiometer *i* (0.5 mm Constantan wire on a ceramic 2-in.-diam ring) is used with three rotating pickups *j* mounted 120 deg apart on a short section of Plexiglas rod *k*. The

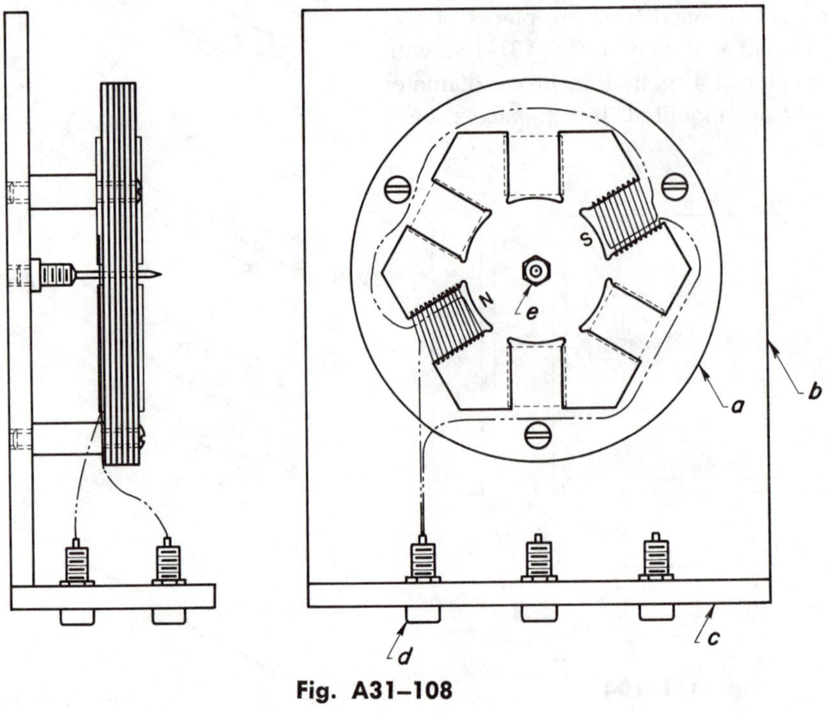

Fig. A31–108

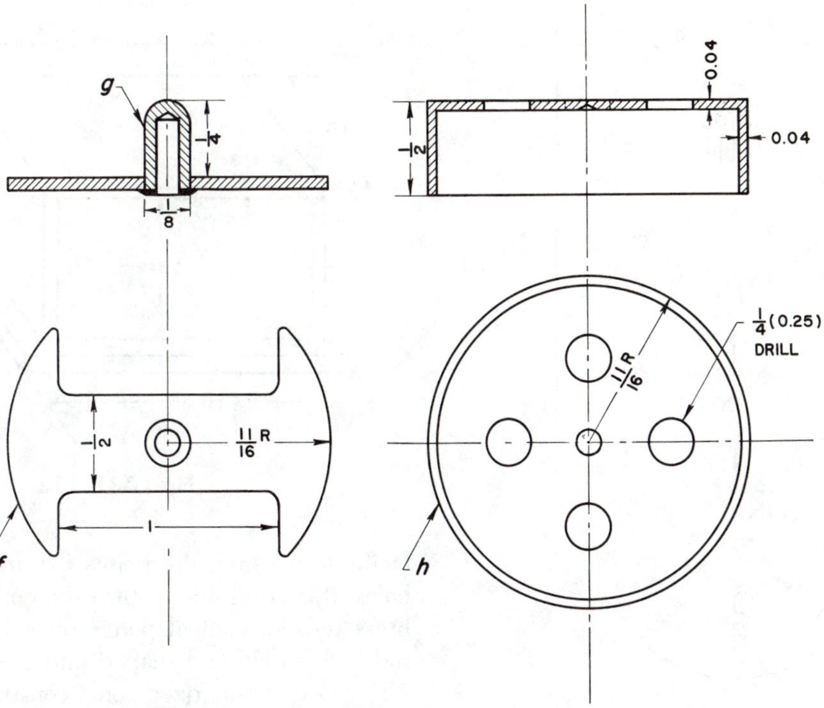

Fig. A31–109

input leads to the potentiometer are connected to a pair of banana jacks, which in turn are connected to a 12-V battery. Grooves are milled in the rod for contact wires which connect the pickups with the commutators *m*. The commutators are made from brass pipe having an inside diameter equal to the outside diameter of the Plexiglas rod. The Plexiglas rod is mounted on a ¼-in.-diam steel shaft *n*, which is ball bearing mounted in the Plexiglas frame. A pulley wheel *o*, fastened to the end of the shaft, is connected by a rubber band to a pulley wheel on the shaft of a variable speed motor. Three spring-steel pickup brushes *p* are mounted on the bottom of the frame and are connected to banana jacks set into the back of the frame. The output from these jacks is fed to the stator coils through the upper three jacks mounted on the stator frame. The lower three jacks are interconnected, as shown in Fig. A31–111.

The pulley wheel can be turned by hand to

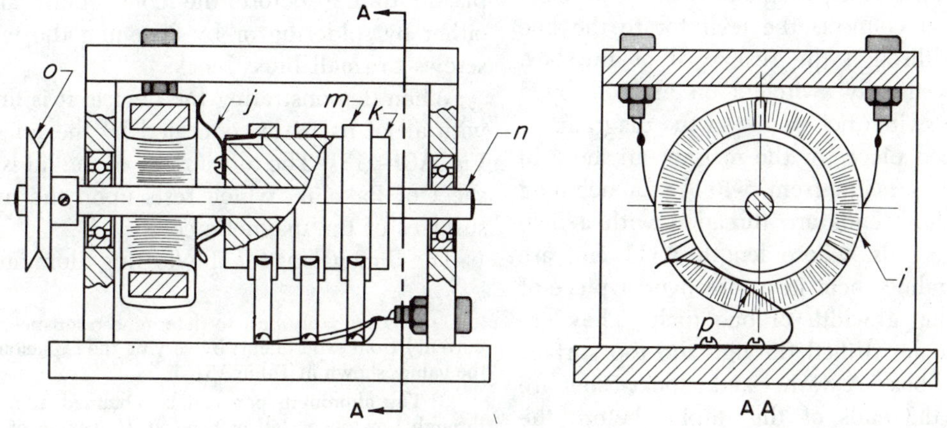

Fig. A31–110

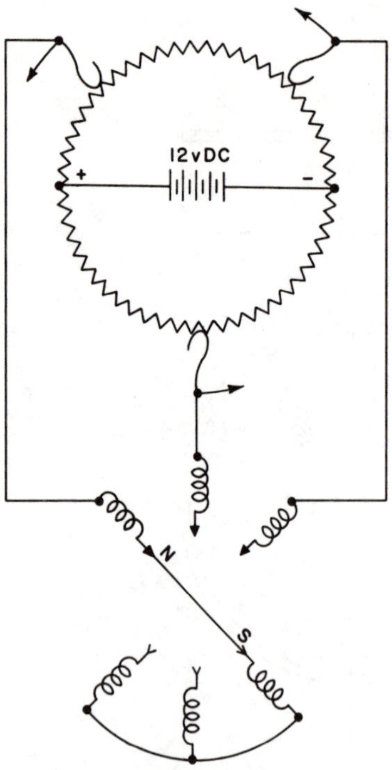

Fig. A31–111

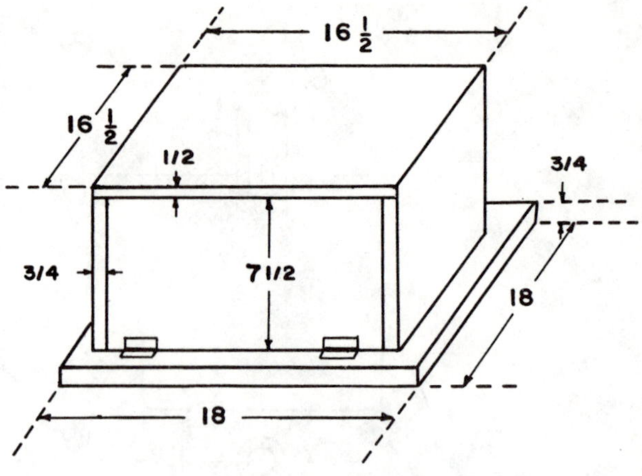

Fig. A31–112

show synchronous motor operation, but a pulley speed of at least 60 rpm is necessary to show induction motor operation, and thus it must be connected to the motor.

A31–2.22

Figure A31–112 is a sketch of the plywood box needed for assembling the levitator. The space housing the bank of capacitors is accessible for wiring and for inspection through a hinged door held closed with a small magnetic lock. The electrical cable that connects the levitator to the line is brought in through the rear wall of the box. Power is conveniently switched on and off by a surface toggle switch not visible in the diagram.

The L-shaped pieces of the magnet are held in place by staples made from 5-40 free machining brass stock whose ends are threaded with a 5-40 die before the rods, whose length is 11 cm, are shaped by bending them in a vise over a piece of flat steel having a width of one inch. They are inserted into holes drilled through the top surface of the plywood box. A washer and a lockwasher are slipped over the ends of the staples before the staples are made tight with 5-40 nuts. In order to facilitate marking the points for drilling the staple holes, it is advisable to turn the ends of one of the brass rods to conical points on a lathe before the rod is threaded and shaped into a staple. In Table A31–1 are summarized some construction data.[2]

The coils are taped and varnished. In order to avoid the possibility of reducing the inside diameter of coil 1 beyond the point of the coil's fitting over the center of the magnetic core, one must use 1-in. surgical bandage. The bandage is strong enough to be wrapped tightly around the coil and makes a very satisfactory and tough covering after having been heavily coated with Glyptal varnish.

As the coils are put in place over the core, the 12-in. leads are passed through small Pyrex feed-through insulators into the space above the capacitors, in which the wires are eventually connected. It is advisable to identify the wires with colored plastic tubing before the connections are made, either by soldering or by fastening the wires with screws to small brass blocks.

When demonstrating the device, it is first shown with all of its parts concealed by the lid shown in Fig. A31–113. The top surface of the lid is a 1/16-in. sheet of Bakelite, which rests upon the uppermost surfaces of the magnetic core.

The dimensions for the floating aluminum pan[3]

(2) It is probably best to determine resonance (minimum current) from experiments by varying the capacitance around the values shown in Table A31–1.

(3) The aluminum pan can be obtained at nominal cost through Lawrence Hall of Science, University of California, Berkeley, California.

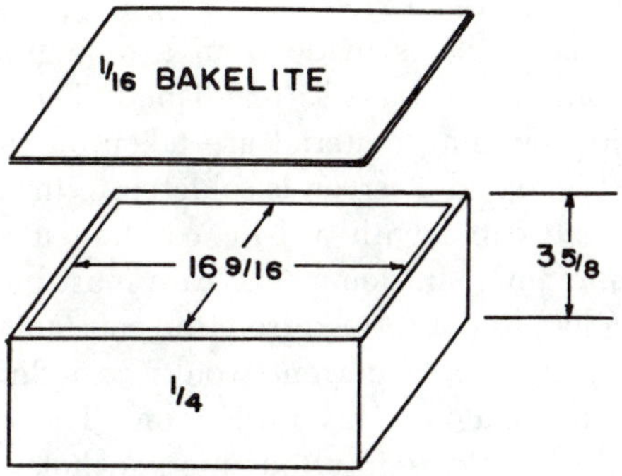

Fig. A31–113

shown in Fig. A31–114 should be followed rather closely. If the pan is too flat near its periphery, there results a rather stable equilibrium position in the magnetic field and the visual effect of the pan's floating in space is entirely lost. If the diameter of the pan is too small, or if its sides are too steep, the pan finds its equilibrium position somewhere near the periphery of the magnet with both the lift and the visual impression of floating nearly destroyed. Similarly, a reduction in thickness of the material of which the pan is made, quickly reduces the necesary eddy currents to the point where the lift becomes insignificant.

It is hardly necessary to add that if a little water is added to the dish and the latter is pushed down against the magnet, the power consumed by the levitator will increase and the water soon begins to boil.

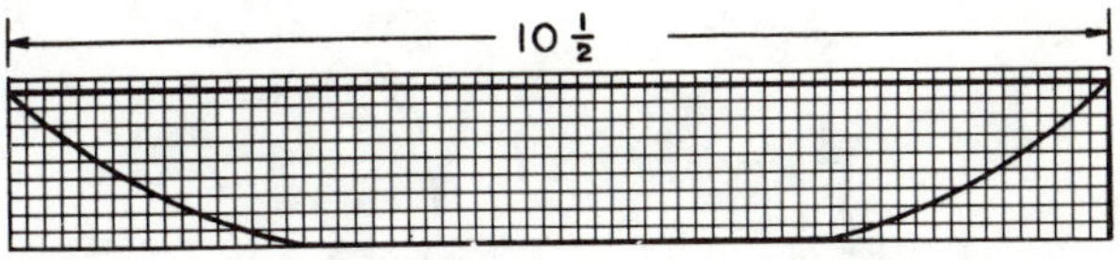

Fig. A31–114

Table A31–1

	Inside Diameter (in.)	Outside Diameter (in.)	Length (in.)	Coil Turns	Turns per Layer
Coil 1	2¾	9	1¾	303	17
Coil 2				297	17
Coil 3	10½	13¼	1¾	135	17
Coil 4				150	17

Occasionally a city power plant can supply a burned out pole transformer from which core laminations can be obtained and cut.

Wire size: No. 12 B & S double cotton-covered copper magnet wire.

Capacitors: 46, surplus General Electric Pyranol, 4 mF, 330 V.

Covering: 1-in. surgical bandage coated with Glyptal.

Chapter 32 Magnetic Properties of Matter

A32–3.18

Materials List

Quantity	Part and Description	Remarks
1	Oscilloscope, 5-in.	Dumont model 401A, Dumont Laboratories, Clifton, N.J.
1	Cathode follower adapter	Dumont model
1	Large-screen monitor	Dumont model 345
1	1000-turn winding transformer coil	Leybold apparatus available from J. Klinger Scientific Apparatus
1	23,000-turn winding transformer coil	Leybold, from J. Klinger
6 ft	#20 Nichrome wire	Commercially available
1	Rheostat, 40 Ω, 5 A	Welch Scientific
1	Potentiometer, 1000 Ω	Allied Radio
1	Resistor, 4700 Ω, 2 W	Allied Radio
1	Resistor, 220 kΩ, 2 W	Allied Radio
2	Capacitors, 0.2 μF and 0.5 μF, 600 V dc	Allied Radio
	(For hysteresis only; in addition to oscilloscopes and 1000-turn winding coil)	
1	Rheostat, 10.7 Ω	Welch
2	Rheostat, 24 Ω	Welch
1	Potentiometer, 5 kΩ	Allied Radio
1	Capacitor, 16 μF, 400 V dc	Allied Radio
1	250-turn transformer coil	Leybold, from J. Klinger
1	Iron yoke	Leybold, from J. Klinger

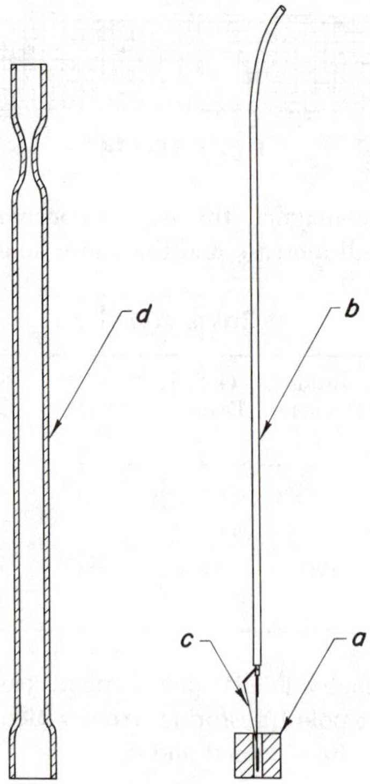

Fig. A32–72

A32–5.1

A 0.04-in.-diam hole is drilled through one of the longer sides of a block of gadolinium (Gd) metal *a* (Fig. A32–72), which is 3/8 x 3/8 x 3/16 in. The thermocouple is made from a thin grounded microphone cable *b* (~1/8 in. diam). The insulation and grounding material are taken off the end (about 1 in.), and a screen is soldered to the center wire of the cable with a 1-in. constantan wire *c* (0.01 in. diam). In doing this, care must be taken so that a loop is not formed from the constantan and copper wire, since a current would be induced as soon as the magnet was turned on. The end of the thermocouple is inserted into the hole in the Gd block, and this is then filled with solder, which provides good thermal conduction.

The Plexiglas container *d* is made from a pipe with a 1/4-in. diameter. A brass block with the same shape (cross section) as the Gd block is sharpened at one end. The end of the Plexiglas tube and the brass block are then heated in an oil bath at 120°C until the Plexiglas is soft. The brass mold is then pushed into the end of the tube to a depth of 3/8 in. to form a space for the Gd block. They are cooled under a stream of water, and when the brass form is removed, the Plexiglas keeps its new shape. The cable is then inserted through the tube and the Gd piece glued into place. In order to prevent damage to the thermocouple, the tube is pinched with a heated pair of pliers (~150°C) so that the wire can not be torn away. Banana plugs are then attached to the ends of the cable.

Chapter 33 Electromagnetic Oscillations

A33–1.1

The general layout of the LCR circuit mounted on a plywood board is shown in Fig. A33–83, which corresponds to the Materials List. The construction is straightforward except for the interrupted assembly, illustrated in Figs. A33–84 and A33–85. This device is made of epoxy board, parts *c, d* and *e*, supporting a 6½-in.-long phenolic drive shaft *f*, which rotates on two ball bearing assemblies. Mounted on the drive shaft is a counterbalanced Phenolic pulley *g*, which is driven by an O-ring drive belt connected to another 2½-in.-diam phenolic pulley mounted on the induction motor. The three commutator brush assemblies *h* are mounted on a 3/8-in.-thick epoxy board *a*, which is connected by screws to part *d* as shown. Part *b* is brass except for the insert, which is epoxy board. A tapered pin is used to fasten *b* to the drive shaft *f*. The carbon commutator brushes may be purchased complete with a spring, insulated lead, and spade lug. The brush casings are made from thin-walled brass tubing. Each is 1 in. long and is formed approximately square by inserting a piece of 1/4-in.-square steel

Fig. A33–83

Materials List

Fig. A33–83

Detail	Part and Description	Remarks
a	Interrupter assembly	
b	Rheostat, 0–35 Ω, 1.19 A max	Ohmite model
c	Induction motor, 1725 rpm	
d	Ferrite cores, ½ in. diam	Lafayette Radio Co.
e	Capacitors (2), 0.012 μF, 1000 V, type 72 D Dubilier, 4 A; 0.12 μF, 1000 V, Sprague	
f	PSSC inductor; wind 5 layers of No. 16 wire	Macalaster Bicknell Co., Cat. No. 100
g	Inductor; wind 54 turns over length of 1 21/32 in., on core 1 13/16 in. i.d. x 2 in. o.d. x 3 5/16 in. long	
h	Small choke coil	
i	B Battery (22.5 V)	Everready No. 762-S
j	Terminal block with 6 phone jacks	
	Plywood board, 24 x 24 x ⅝ in.	
	Oscilloscope	Tektronix type 561, with type 63 differential amplifier and type 67 time base

stock and hammering it to shape. A thin piece of brass with a small slit in it is mounted on part *a* and holds the spring and brush against the interrupter disk *b*, while allowing the insulated lead to pass through.

A33–2.1

The soap bubble generator consists of a reservoir can on the front of the table, a compressed-air line

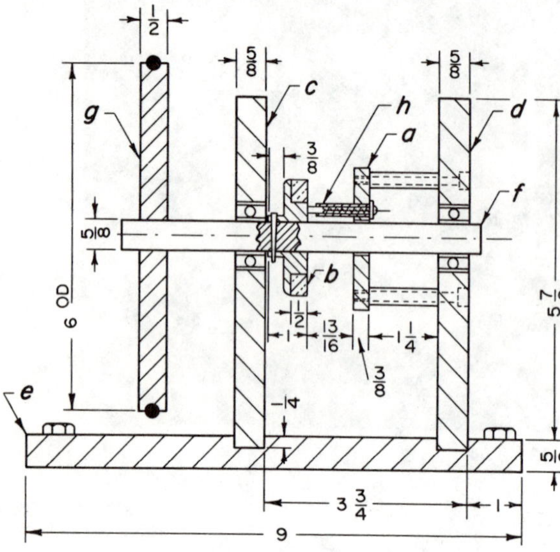

Fig. A33–84

with a regulating valve in the rear, light copper tubing running beneath the table to link these two components, and a vertical orifice tube rising through the center of the table from a T-joint in the copper linking tube (Fig. A33–86). The soap solution (glycerine and detergent) is fed into the line by gravity. Its flow is controlled by an adjustable pinchcock on a short length of rubber hose which connects the reservoir to the copper line ($\frac{1}{8}$-in.-o.d. tube). Very little air pressure is required to create the bubbles. The orifice pipe is brass, $\frac{1}{8}$ x $\frac{3}{16}$ x 6 in. long. Its mouth is chamfered externally to correctly form a stream of bubbles of uniform size. A light brass well set on short legs encircles the pipe to catch any overflow of liquid.

The design of the reservoir is arbitrary. The reservoir shown is 4 in. in diameter x 1$\frac{5}{8}$ in. deep. An outlet pipe near its bottom leads to the rubber hose, and a provision is made for height adjustment. The 24 x 24 x $\frac{3}{4}$-in. plywood table, is covered with sheet metal tacked to its top surface. Four 3$\frac{1}{2}$-in.-long wooden legs support it. Two $\frac{3}{4}$-in.-diam phenolic rods, spaced 15$\frac{1}{4}$ in. apart (symmetrically about the orifice) support the horizontal plate 21 in. above the table. All sheet metal plates are No. 18 gauge. The overhead plate is 17 in. o.d. and 5$\frac{1}{2}$ in. i.d. Each of the parallel vertical plates is 6 in. in diameter; they are separated by 14$\frac{1}{4}$ in. Charge leakage is minimized by splitting $\frac{1}{8}$-in. copper tubing longitudinally and forming it over all sharp plate edges. Details of the overflow well are shown in Fig. A33–87. Part *a*, which forms the wall of the well, is brass tubing, 2$\frac{3}{16}$ in. o.d., 2 in. i.d. The orifice tube *b*, $\frac{3}{16}$ in. o.d., $\frac{1}{8}$ in. i.d., has a chamfered top end. The well bottom *c* is $\frac{1}{16}$-in.-thick brass and is sized to push fit into part *a*. Three equally spaced holes in the bottom of *c* provide a means of attaching the three $\frac{1}{4}$-in.-long brass legs *d*.

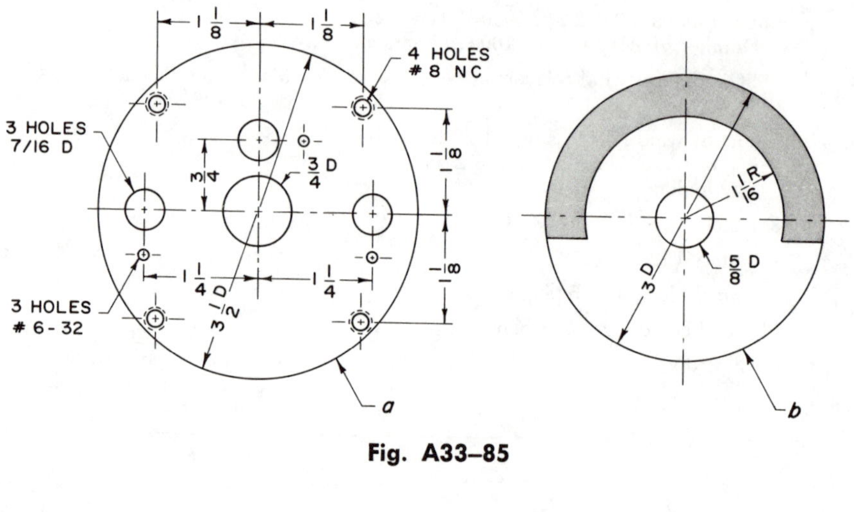

Fig. A33–85

A33–2.2

Materials List

Fig. 33–7 (p. 993)

Detail	Component and Description	Remarks
V_1	Vacuum tube diode	117Z3
D_1	Selenium rectifier, 110 V, 65–75 mA	IT&T No. 1234AH
D_2	Silicon rectifier	Starkes Tarzian No. M150
I_1	Neon bulb	GE No. NE51
I_2	Incandescent night lamp bulb 7 W, 110 V, candelabra base	
R_1	Resistor, 8.2 kΩ, 2 W, ±20%	
R_2	Resistor, 100 Ω, 2 W, ±20%	
R_3	Resistor, 5.1 kΩ, 2 W, ±20%	
S_1	Wafer switch, break-before-make type	
	Small component mounting strip, clear acrylic plastic, ½ x 2 x ⅛ in.	
	Tube mounting strip, clear acrylic plastic, 2¼ x 2¼ x ⅛ in.	
	Switch mounting strip, clear acrylic plastic, 4¼ x 5 x ⅛ in.	
	Tube socket, 7-pin miniature	
	Small knob for switch	
	Sockets (2) to fit bulbs I_1 and I_2	
	Phone tip jacks (38), modified	Herman H. Smith Inc., Brooklyn, N.Y., No. 107
	Solder type phone tips (105), modified	Herman H. Smith Inc., No. 108
	Connecting wire, brazing rod, 1⁄16 in. diam, ~5 ft	
	Outlet plugs (2), ac, grounding type	
	Line cord, ac, ~4 ft	

A33–2.10

The electronic details of the 12-in. oscilloscope are given in Figs. A33–88 through A33–94.

A33–3.7

Materials List

Component	Quantity	Description
		Resistors
R_1	1	Potentiometer, 10 kΩ, 2 W
R_2	1	18 kΩ, ½ W
R_3	9	Potentiometer, 5 kΩ, 2 W
R_4	9	5.6 kΩ, ½ W
R_5, R_{12}	2	2.2 MΩ, ½ W
R_6	1	56 kΩ, ½ W
R_7	1	56 kΩ, 1 W
R_8, R_9	2	27 kΩ, 1 W
R_{10}, R_{13}	2	1 kΩ, 1 W
R_{11}	1	22 kΩ, ½ W
R_{14}	1	33 kΩ, 1 W
R_{15}	1	Potentiometer, 100 kΩ, 2 W, linear taper
R_{16}	1	5 kΩ, 5 W
		Capacitors
C_1	1	1 μF, 600 V
C_2	1	0.25 μF, 600 V
C_3	1	0.11 μF, 600 V

Component	Quantity	Description
C_4	1	0.0625 μF, 600 V
C_5	1	0.04 μF, 600 V
C_6	1	0.025 μF, 600 V
C_7	1	0.02 μF, 600 V
C_8	1	0.0156 μF, 600 V
C_9	1	0.0125 μF, 600 V
C_{10}	1	0.01 μF, 600 V
C_{11}	1	16 μF, 450 V
C_{12}, C_{13}, C_{14}	3	0.5 μF, 450 V
C_{15}	1	8 μF, 450 V
C_{16}	1	40 μF, 450 V
		Tubes
V_1	1	6SN7
V_2	1	6J5
V_3	1	0D3
V_4	1	0C3
V_5	1	5Y3
		Miscellaneous
T_1	1	Transformer; Stancor, PC8403
L_1	1	Choke; Stancor, C1002
L_2	1	Toroid, inductance 0.5 H; UTC MQA-10
S_1	1	Switch, 2-pole, 12-position; Centralab, PA2005
	4	Binding post, 5-way
	1	Switch, SPST
	1	Fuse holder
	5	Octal socket
	1	Chassis, 8 x 12 x 3 in.; BUD AC-424

A33–7.4

In performing demonstrations with microwaves, one can pick the wavelength to suit the demonstration. For demonstrations of Fresnel diffraction, 12-cm wavelengths are preferred. For Fraunhoffer and Bragg diffraction, shorter wavelengths are more convenient. With the source of 4-cm wavelength described here, qualitative microwave interferometers may be made small enough to demonstrate on an overhead projector. This same simple source may be combined with a high-gain horn or parabolic reflector for use in large-scale, quantitative demonstrations.

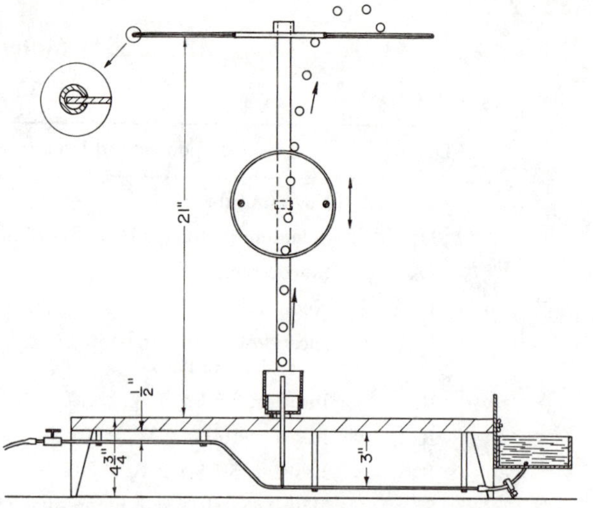

Fig. A33–86

The source employs the 7486 triode tube which was found to operate at the highest frequency available from commercial triode tubes at the time, 6.5 gigacycles per second (6.5 x 10^9 cps). The corresponding wavelength is 4.6 cm. The oscillator is effectively an amplifier with feedback from output to input resonant cavities. The coaxial resonant cavity at the output end of the tube is formed by fitting a brass thimble over the plate and grid disks of the disk-seal triode tube. Figure A33–95(a) is a

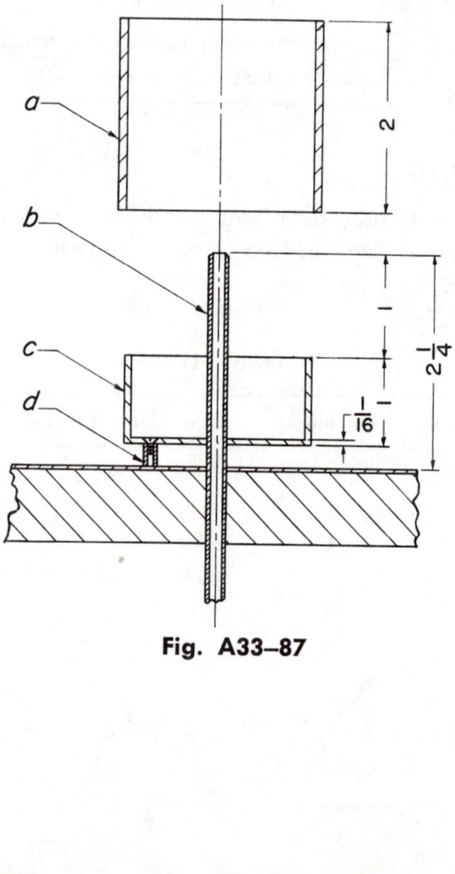

Fig. A33–87

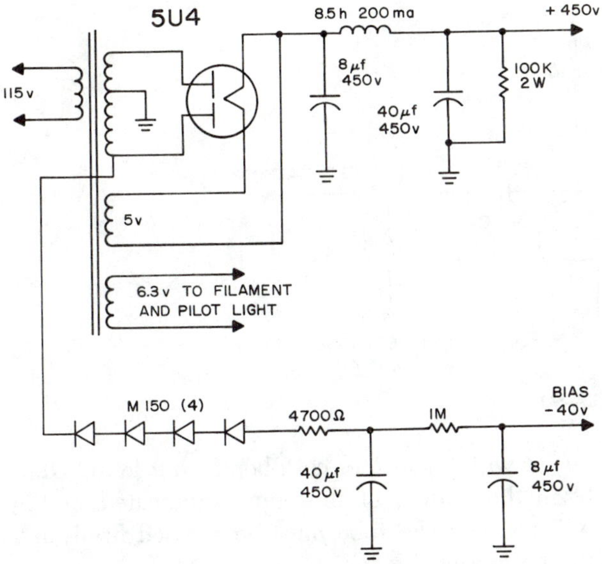

LV AND BIAS POWER SUPPLY

Fig. A33–88

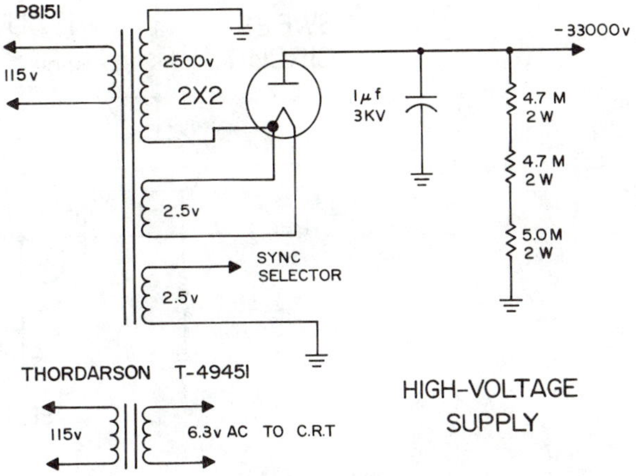

Fig. A33–89

drawing of the outer cross section of the 7486 triode tube, and Fig. A33–95(b) shows the tight-fitting thimble with a mica washer to insulate the grid from the anode. The enclosed quarter-wave coaxial cavity determines the frequency of the oscillator. The grid-cathode resonant cavity, known as the input cavity in amplifier circuits, consists of a piece of rectangular copper wave guide 1 x ½ in. in outer cross section and 1 in. long with one end closed. To tune this rectangular waveguide cavity to the frequency of the grid-anode cavity, the length of the rectangular cavity had been adjusted to yield maximum power from the oscillator.

Figure A33–96 is a cross section drawing showing the tube mounted in the rectangular waveguide. The brass screw extending from the left-hand wall of the waveguide to the grid-anode cap grounds the grid. The cathode and grid are both at ground potential in the dc circuit. The screw extending through the insulating washers in the bottom of the waveguide to the anode is the anode lead. The *amplitude* of the feed-back wave that leaks from the grid-anode cavity to the grid-cathode cavity is determined by the thickness of the mica washer. The screw which extends from the right-hand wall of the waveguide toward the grid-anode cap distorts the electric and magnetic fields. The adjustment of the gap at the end of the screw determines the *phase* of the wave which is fed back from the output to input cavity.

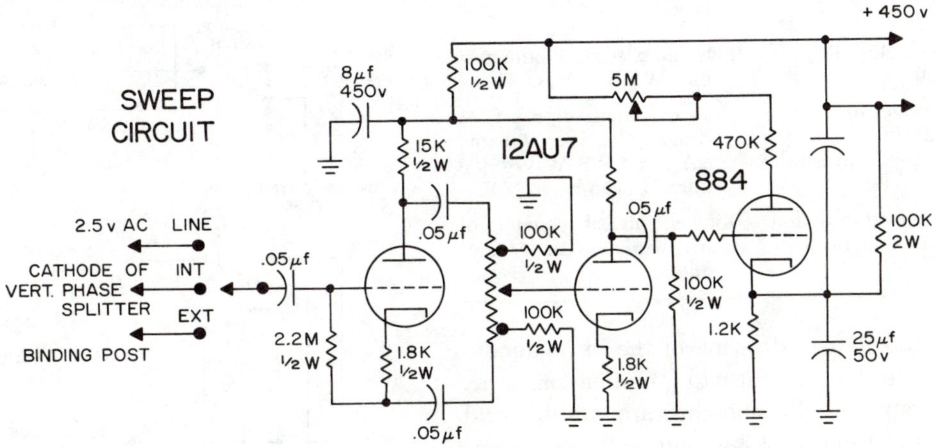

Fig. A33–90

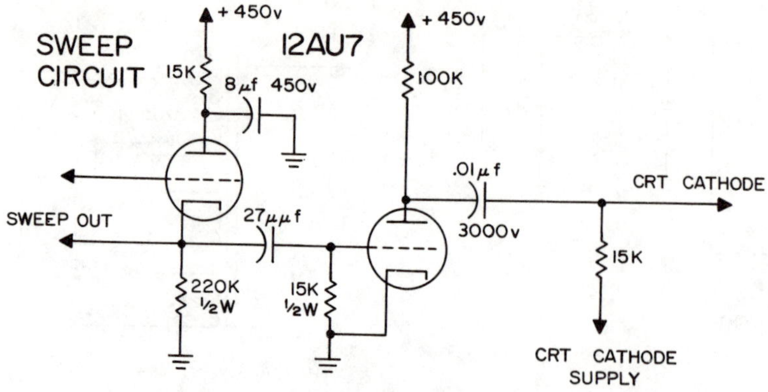

Fig. A33–91

Materials List

Component	Remarks
1. Microwave triode tube, 7486	General Electric Co., Receiving Tube Dept., Owensboro, Ky.
2. Rectangular copper waveguide for x-band	Sold only in 12-ft lengths. Utility Brass and Copper Corp., Brooklyn, N. Y.
3. Extruded washers, FW384	Waldon Electronics, Chicago, Ill., or any local radio supply dealer
4. Teflon sheet, 0.005 in. thick	Comes in rolls 6 in. wide and sold in lots as low as 2 lb
5. Brass rod, ⅝ in. diam	To be used in machining anode-grid caps
6. Copper sheet, 0.020 in. thick	From which to make washers
7. Plexiglas or Lucite sheet, ¼ in. thick, 1 ft square	
8. Power transformer, P-8181 plate supply 150-V ac, filament supply 6.3 V	Stancor Electronics Inc., Chicago, Ill.
9. Semiconductor diodes, D3530	Sylvania Electric Products, Inc., Woburn, Mass.
10. Cover flanges of brass, RG521U for DBG-1000 waveguide	Demornay Bonardi, c/o J. A. Reagan Co., Inc., Albany, N.Y., or CWS Waveguide Corp., Lindenhurst, N.Y.
11. Funnel, tinned sheet iron, No. 30 (6 in. diam)	Kitchen utensil counter of any hardware or general store

Figure A33–97 is a drawing of the brass anode-grid cap. The slots are cut with a milling machine. Since the cap must fit smoothly around the grid washer of the tube, hacksaw cuts will not suffice. A Teflon washer 0.005 in. thick and $\frac{3}{16}$ in. i.d. may be cut with scissors, or cork borer, to 0.48 in. o.d. to fit in the bottom of the cup as indicated in Fig. A33–95(b). The tube must be pressed firmly into the bottom of the cup.

As a cathode connector, a washer of copper, brass or aluminum is cut from a 0.020-in.-thick sheet of $\frac{9}{32}$ in. i.d. and of ⅝ in. i.d., cut in two in the middle and pressed firmly against the cathode washer of the tube as indicated in Fig. A33–98.

Figure A33–99 is a drawing of the rectangular copper waveguide with machined holes. The brass or copper end plate is machined to fit conveniently for soldering. It is of such precision as to be a press fit; the plate need not be soldered in.

The anode lead consisting of a ½-in. #4-40 screw with three nuts, a 0.010-in.-thick metal washer, a 0.010-in.-thick Teflon or mica washer, an extruded fiber washer, and a soldering lug may extend into the cavity about $\frac{5}{32}$ in., ready to press against the

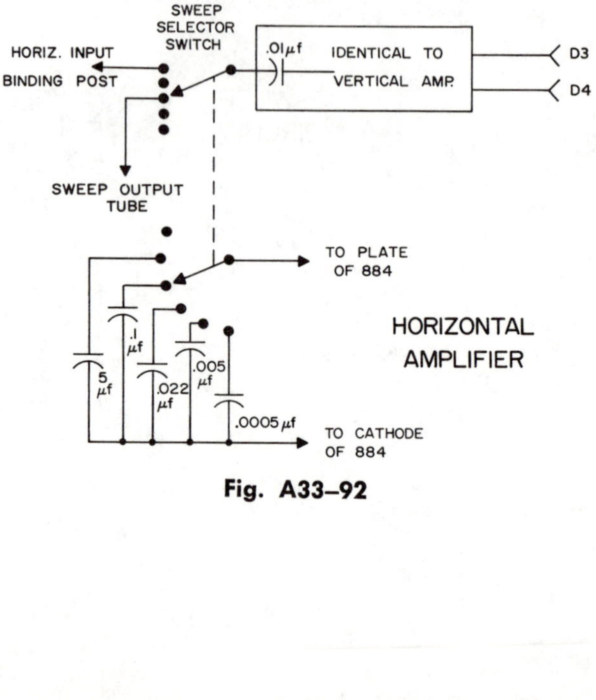

Fig. A33–92

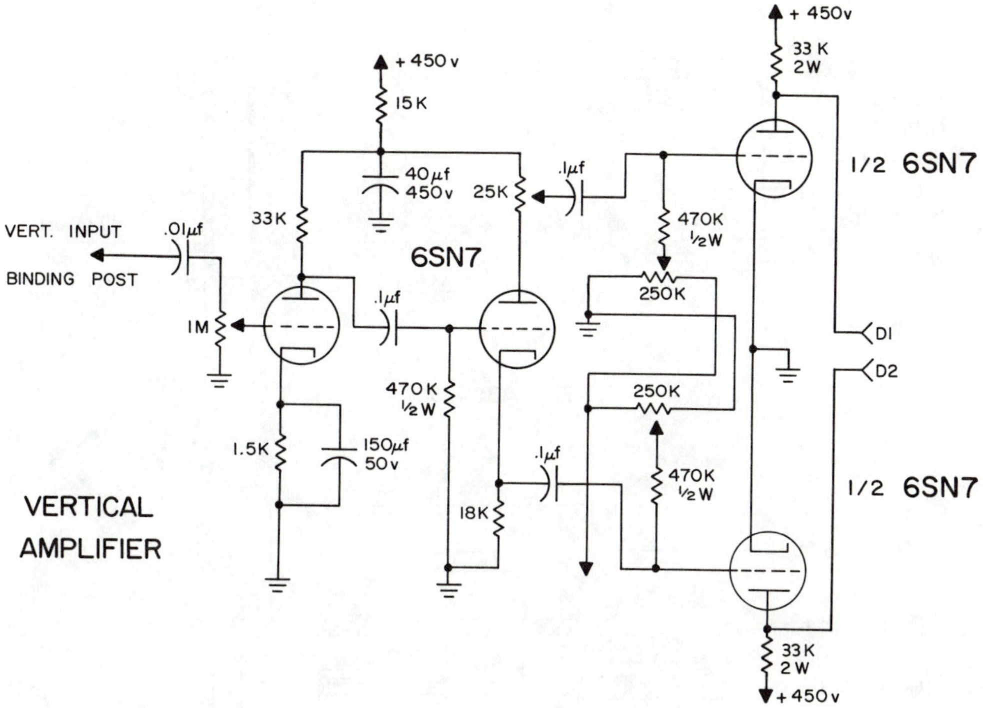

Fig. A33–93

anode disk of the tube. The Plexiglas block of Fig. A33–100 serves the dual role of holding the tube in place and holding two 0.045-in. phosphor bronze wires as heater leads. If a small soldering tip is employed so as to avoid heat shock, the heater leads may be soldered directly to the pins of the tube. The tube with its grid-anode cap and split cathode washer are inserted in the waveguide and held in center by the Plexiglas block with four No. 2-56 roundhead brass screws 5/16 in. long, as indicated in Fig. A33–101. The anode screw and grid screw can be turned to make good contact with the tube and locked in place by nuts. The phase adjusting screw may be set so that the gap is about 1/32 in. Its final adjustment will be made to cause the system to oscillate and maximize the output power when it is in the circuit.

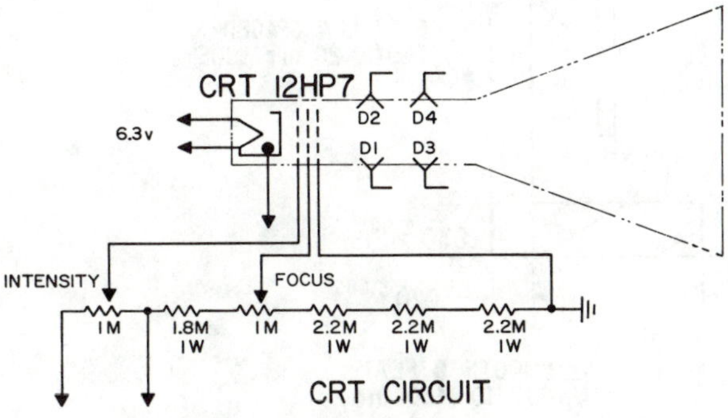

Fig. A33–94

The power supply may be either dc or ac. If a 60-cps ac source is used, the generated microwaves will be modulated so that the output of a crystal detector will have an ac as well as a dc component. The recommended Stancor P-8181 transformer gives 150-V plate potential and 6.3-V filament potential. If a dc plate supply is used, some of the tubes may require as much as 140 V for oscillation.

Since 1959, crystal detectors with wire leads have been available which are sensitive in the microwave region. To make an intensity meter for use near the source, connect the D-3530 crystal detector between the terminals of a microammeter. Figure A33–102 indicates a crystal detector used as a half-wave dipole probe. The capacitance between the wires of the twisted leads of fine insulated wire is sufficient to confine the microwave frequency currents to the small loop at the detector. The length of the antenna should be cut to a half wavelength, about 2.3 cm, to maximize its sensitivity. The recti-

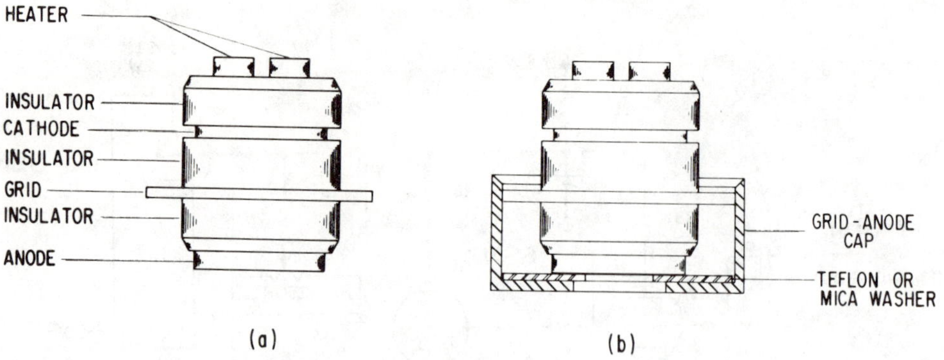

Fig. A33–95

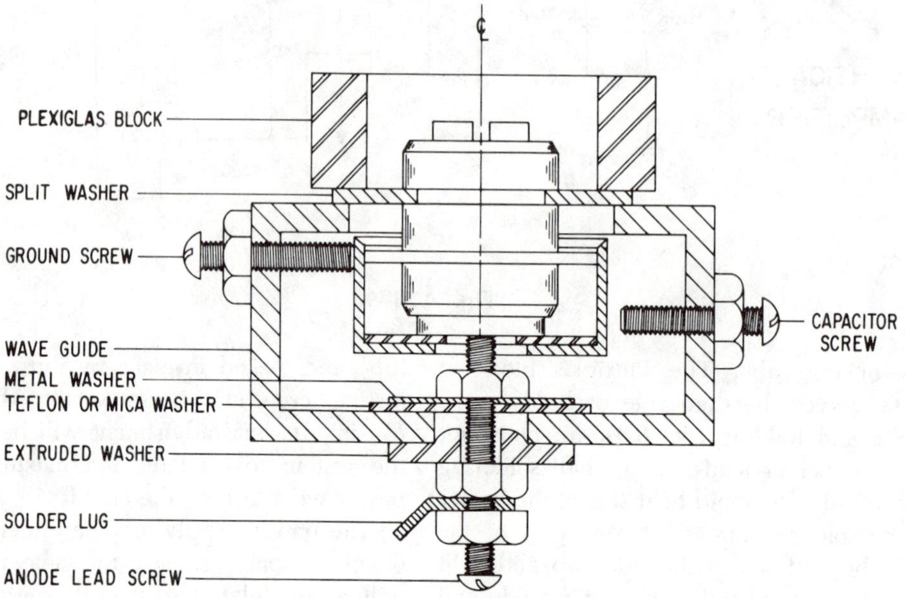

Fig. A33–96

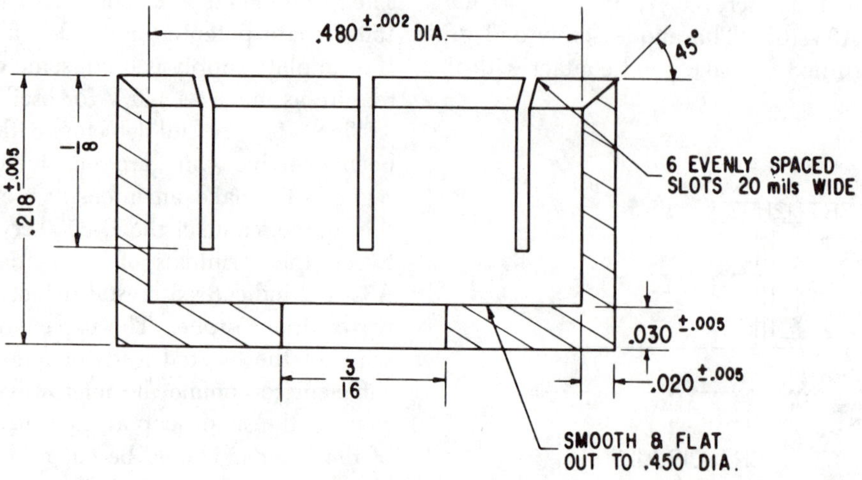

Fig. A33–97

fied current with its 60 cps, as well as dc components, may be indicated on a microammeter, a galvanometer, a cathode ray oscilloscope, earphones, or an audio amplifier and speaker. Figure A33–103 is an oscilloscope trace of rectified current against time, indicating that oscillations are taking place during only about one quarter of the ac cycle when the potential of the plate of the microwave oscillator tube is sufficiently positive.

With the detector antenna and meter placed an inch or two in front of the open end of the oscillator and with the antenna parallel to the axis of the tube, the phase screw may be adjusted until the source begins to oscillate. During the adjustment the lock nut should be held with pliers so that the turning of the screw is somewhat firm. For most tube the gap between the screw and grid-anode cylinder will be less than $\frac{1}{32}$ in. when the output power is a maximum.

Before turning to audio amplifiers to amplify the output power of the crystal, one should become acquainted with the optical gain of parabolic re-

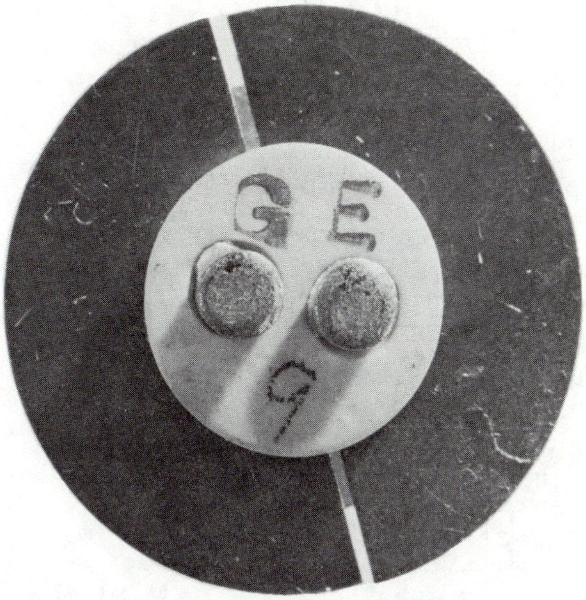

Fig. A33–98

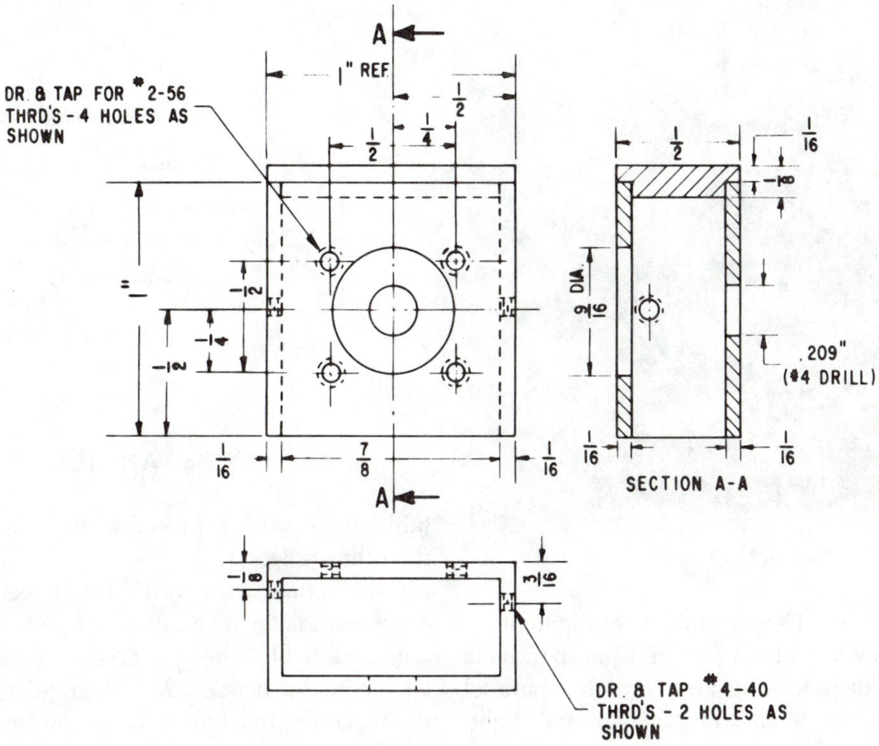

BRASS END PLATE TO BE SOLDERED INTO END OF WAVE GUIDE

Fig. A33–99

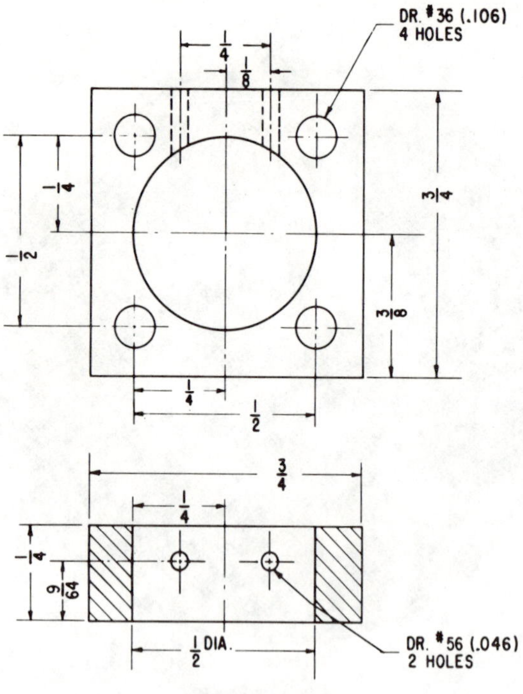

Fig. A33–100

Fig. A33–101

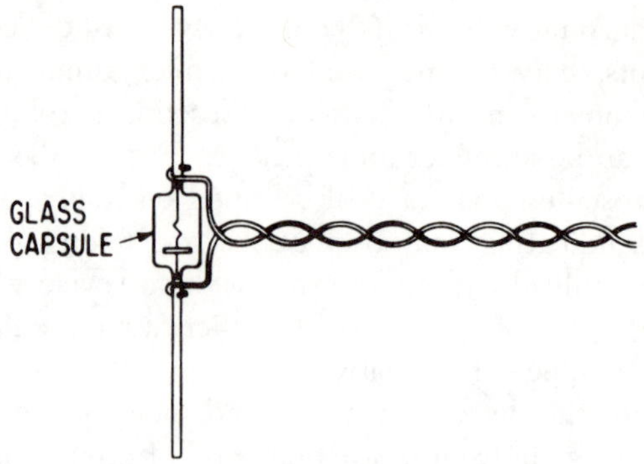

Fig. A33–102

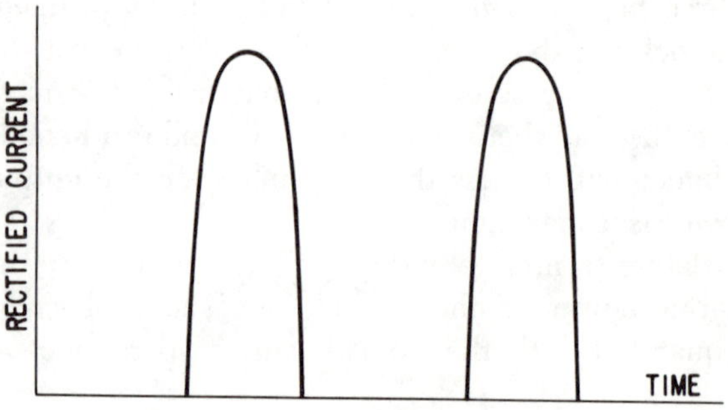

Fig. A33–103

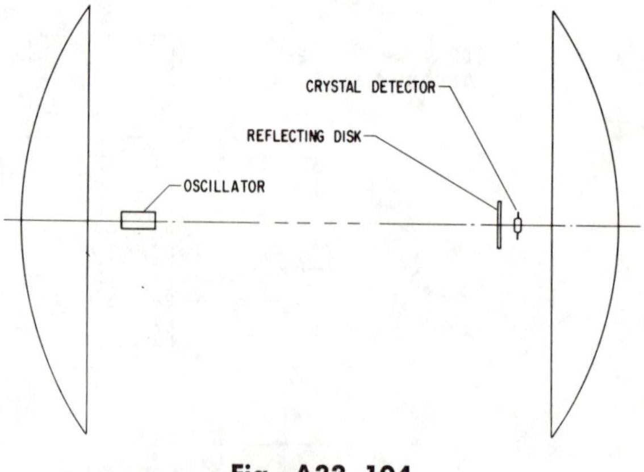

Fig. A33–104

flectors and horns. Demonstration equipment in physics generally includes a pair of 12-in. parabolic reflectors used in producing and receiving parallel beams in a series of demonstrations of sound, light, and infrared waves. The open end of the source is faced toward one parabolic reflector at the principal focus. At the other end of the demonstration table, the crystal is placed at the principal focus of the other reflector.

The students can watch the detected current on a microammeter or oscilloscope as four adjustments are made: (1) the position of the receiving reflector in the beam, (2) the angular directivity of the receiving reflector, (3) the position of the crystal detector, and (4) the position of a metal disk which is about 20 percent more than a half wavelength in diameter at a distance of about a quarter wave from the crystal detector on the opposite side of the crystal from the receiving parabolic reflector, as indicated in Fig. A33–104. The disk may be held from the side on the end of a wooden dowel. The presence of this disk will multiply the current reading by as much as a factor of two. A wave reradiated by the antenna will move one quarter wave to the reflecting disk, undergo 180-deg phase change upon reflection, and travel a quarter wave back to the antenna to be in phase with the wave incident upon the antenna from the receiving parabolic reflector.

Although horns may be soldered directly to the waveguide oscillator, the variety of uses of the oscillator can be increased by soldering a standard waveguide flange to the open end of the waveguide. The flange itself gives a gain of two in intensity at the center of the beam in the far field. Figure A33–105 indicates the oscillator with the cylindrical portion of the flange machined off so that the flange is 3/32 in. thick. As shown in the figure, a No. 30 household funnel with its tube removed has been soldered to the cylindrical end of a second waveguide flange to provide a 6-in.-diam horn.

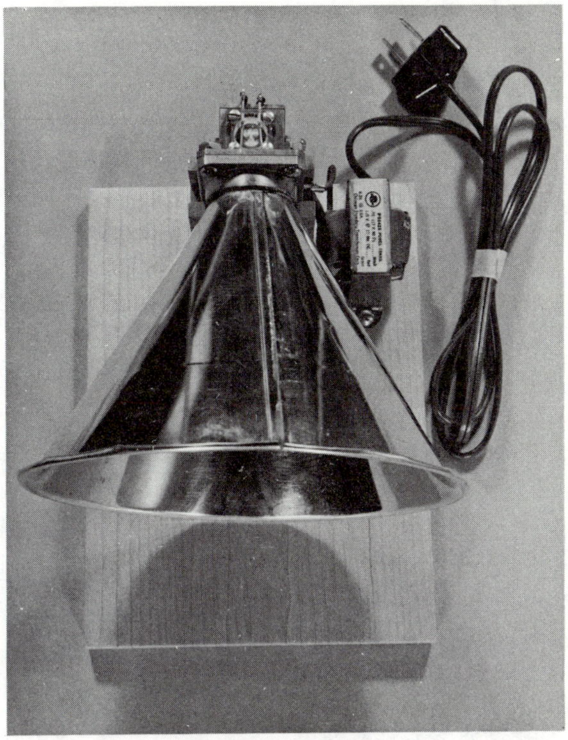

Fig. A33–105

Chapter 34 Geometrical Optics

A34–1.11

The camera consists of a 2 x 4½ x 5-in. sheet metal box with one of the sides replaced by a clear piece of 3/8-in.-thick Lucite. A 1-in. hole is drilled opposite the Lucite face, and a 1-mil-thick piece of shim stock with a 7-mil-diam pinhole is fixed over the opening. Care should be taken that the rim of the pinhole is sharp, to limit glancing reflections. A piece of Lucite is fixed over the pinhole. (The 1-in. hole also serves as a filling hole.) The metal box is mounted on a Graflex press camera back[(1)] so that its Lucite side presses firmly against the Polaroid film. This allows the box to be easily replaced by one which is air-filled. Alternatively, the camera could be constructed with a removable block of Lucite in place of the water. The apparatus also might be adapted for use with a roll-type Polaroid film holder for which transparency film is available.

(1) Polaroid Film Holder No. 500 for 4 x 5-in. Polaroid Film Packets, Polaroid Corp., Cambridge, Massachusetts.

A34–1.21

The device for demonstrating reflection and refraction at a plane surface is shown in Fig. A34–52. Ten elastic fluorescent strings[(1)] are butt-cemented to ten strands of black button thread with Dekophane cement.[(2)] The threads, connected to a lead weight *a*, pass over a Lucite pulley *b* (¾ in. diam x 4 in. long) and pass through five holes in the Lucite box (two per hole). Then the strings enter the surface *c* through holes which make a 45-deg angle with the surface. One of the two strings goes through the hole, while the other string (the "reflected" one) passes through another hole which is perpendicular to the original hole. The "reflected" threads pass over another Lucite pulley *d* and then join the other threads at pulley *e*. All the threads are then attached to a variable speed motor drive *f*. The box is made of five pieces of ¼-in.-thick Lucite,

(1) Available in red, orange, lemon yellow, cerise, green and blue from Cooper Bros. Co., Port Washington, New York.

(2) Crystal Essence Corp., Plainfield, New Jersey.

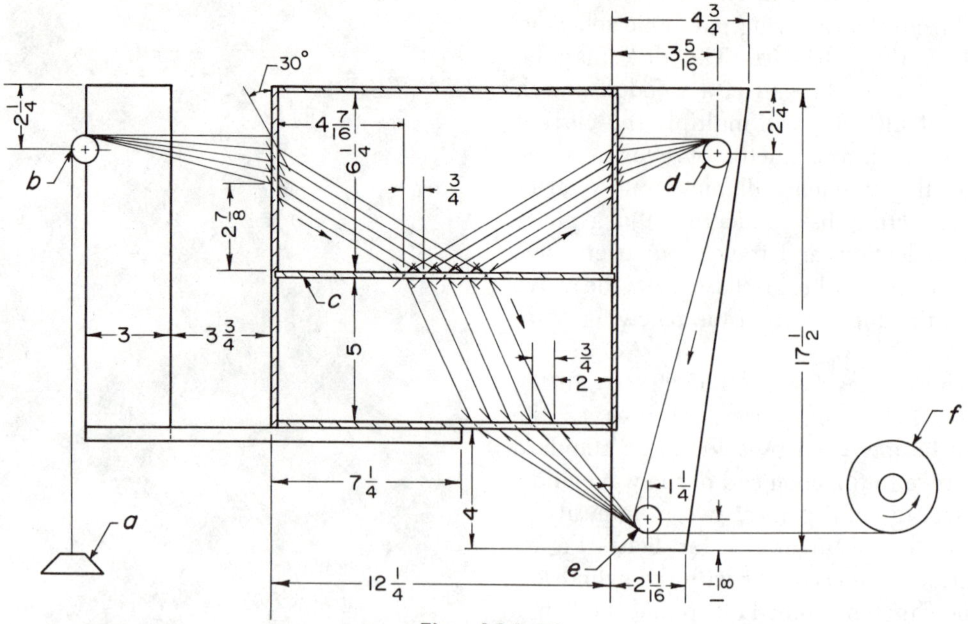

Fig. A34–52

4 x 12 in. The rear of the bottom half of the box is covered with 1/8-in.-thick frosted Lucite. An ultraviolet lamp[3] illuminates the fluorescent strings in a dark room. Pulleys *d* and *e* are mounted on a 3/16-in.-thick piece of Masonite.

The plastic lenses are made from 1-in.-thick Plexiglas. They were turned on a lathe, rough-cut to a precut template of proper curvature by hand control of cross and longitudinal lathe motions. Finishing was done first with a wood-turning gauge, then a file, emery cloth (of increasingly finer grades), household cleanser with water and cloth, and finally toothpaste.

Bases for the models are made of 1/2-in. plywood painted flat black. The end guides are made from 1/4-in. Plexiglas to allow precision drilling and smooth holes for free string movement. For photography in ultraviolet light, these end guides are covered with flat black paper. The elastic fluorescent strings are laced back and forth in a continuous fashion for the still models.

Figure A34–53 shows the simple lens apparatus, and Fig. A34–54, the astigmatism arrangement. Dots in the lens detail indicate where 1/16-in.-diam holes are to be drilled. Dispersion by a prism is illustrated with the apparatus in Fig. A34–55, and the components necessary to show chromatic aberration are shown in Fig. A34–56. To show movement, the fluorescent threads are attached to black threads. For example, in Fig. A34–55, the threads are wound on a Lucite spool *g*, and pass through holes in end-pieces located on each side of the prism. They are then connected to a variable speed motor on the left side of the prism (not shown). The spool is made from 1/2-in. diam Lucite rod, threaded at one end. Disks on spools are Lucite, 1 in. in diameter and 1/2 in. thick, with one end threaded. The spool support is made from 1/2-in.-thick plywood.

The lenses in the models for chromatic acceleration (Fig. A34–56), spherical aberration, and coma (Fig. A34–57) are each arranged as shown in Fig. A34–54 (astigmatism). On each side of the lens are shown the *end* guides for the fluorescent threads. When motion is desired, the fluorescent threads are attached to black threads and then pulled slowly through the lens by a variable speed motor, as mentioned above.

A34–1.25

A very practical way to set up the demonstration is to secure the various optical components to metal or wood blocks of suitable size using a smoking hot mixture of beeswax and rosin applied with a medi-

(3) Strobolite 250 W U.S. Reflector Lamp, Strobolite Company, New York, N.Y.

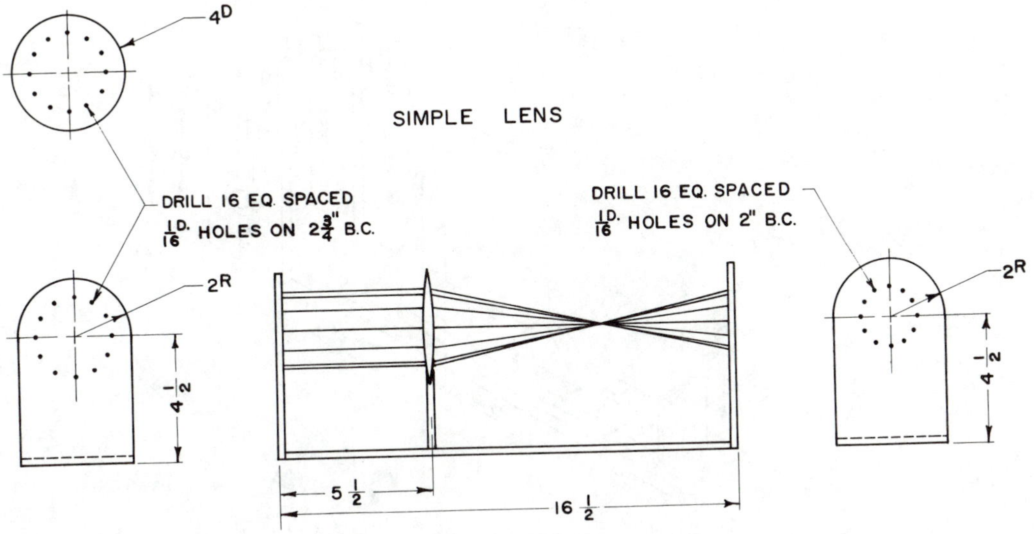

Fig. A34–53

cine dropper. After the components are located on the base plate, they are secured either temporarily with Tackiwax or more permanently with the hot beeswax and rosin mixture. Errors are easily corrected by using a razor blade to remove the waxed-on components—the hard wax can be melted up again.

The hot mixture is made in about equal proportions of beeswax and rosin so that, when cool, it is just possible to make an impression in the wax with one's fingernail. The beeswax tempers the rosin so that the mixture is not brittle. Too much beeswax will make the mixture too soft. The wax should be applied smoking hot so that it will adhere well, but when very hot, the beeswax tends to distill off so that the mixture becomes brittle. The addition of more beeswax retempers the mixture. The mixture burns, and it is well to have a cover to put out fires

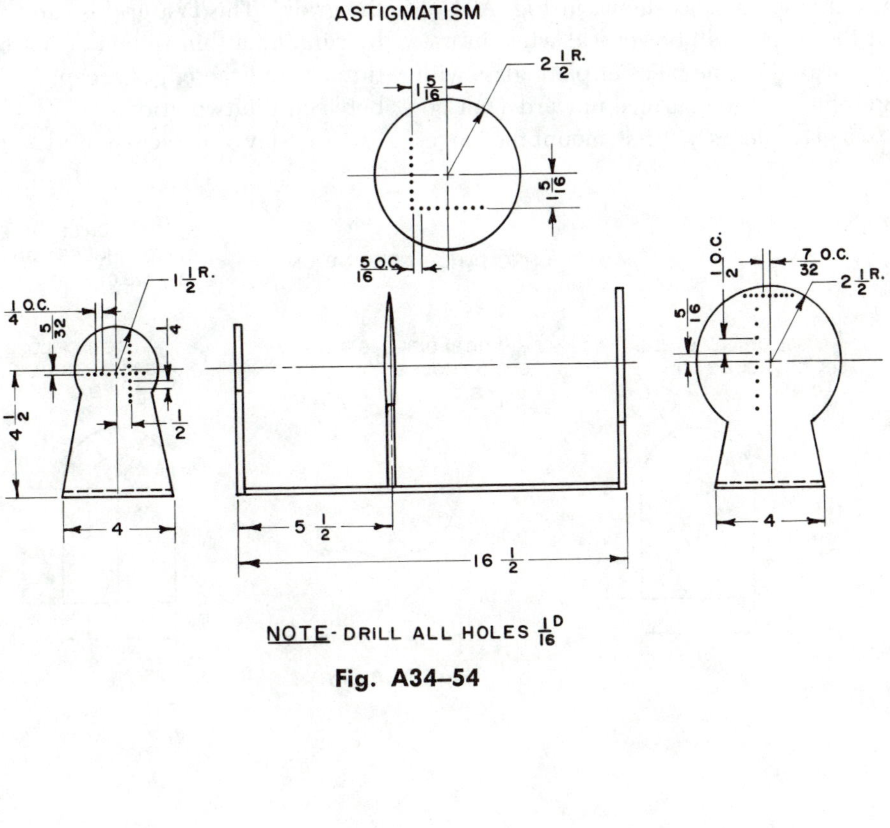

Fig. A34–54

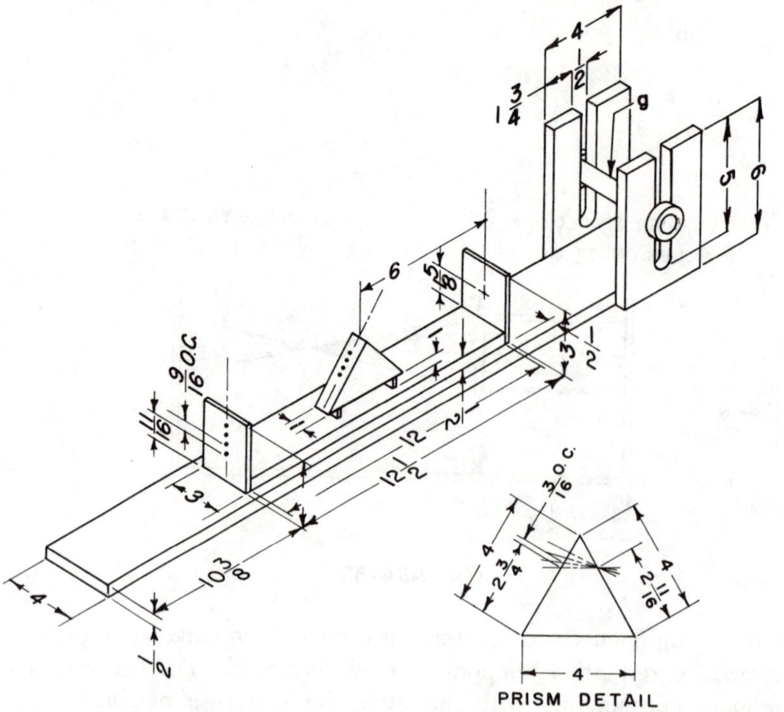

Fig. A34–55

should they occur. About 20 cc of the mixture should be prepared, preferably melted on a hotplate whose temperature can be controlled by a Variac. (A Bunsen burner will also do.)

The grids are mounted as shown in Fig. A34–58 (A) so that the stripes will be vertical when in use. The grids are waxed to one piece of plate glass with the photographic or ruled surface upward (not between the two glass pieces). First, mount the longer piece of grid *a* (1 in Fig. 34–42) on the glass base plate *b*. Carefully position it so that there is adequate room for the other grid and so that the base plate can be mounted on a block without injury to the grids. This first grid is carefully waxed in place by running a thin ribbon of hot beeswax-rosin mixture around the edges except for the common edge between the two grids.

Next, lay the second grid *c* (2 in Fig. 34–42)

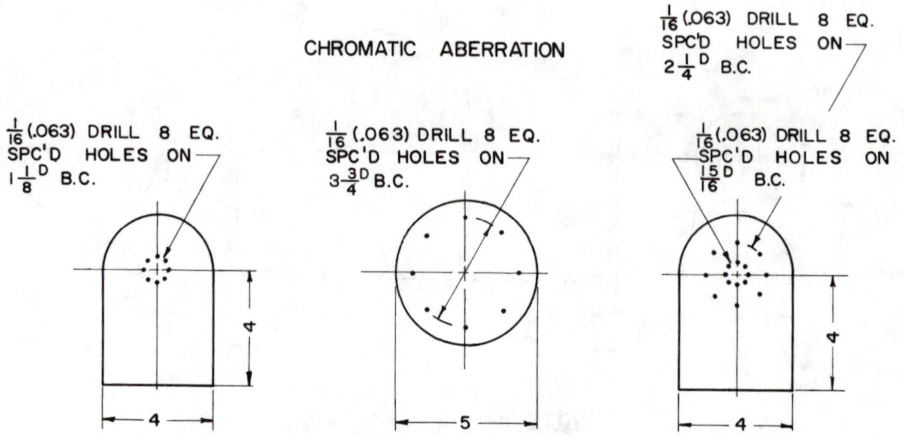

Fig. A34–56

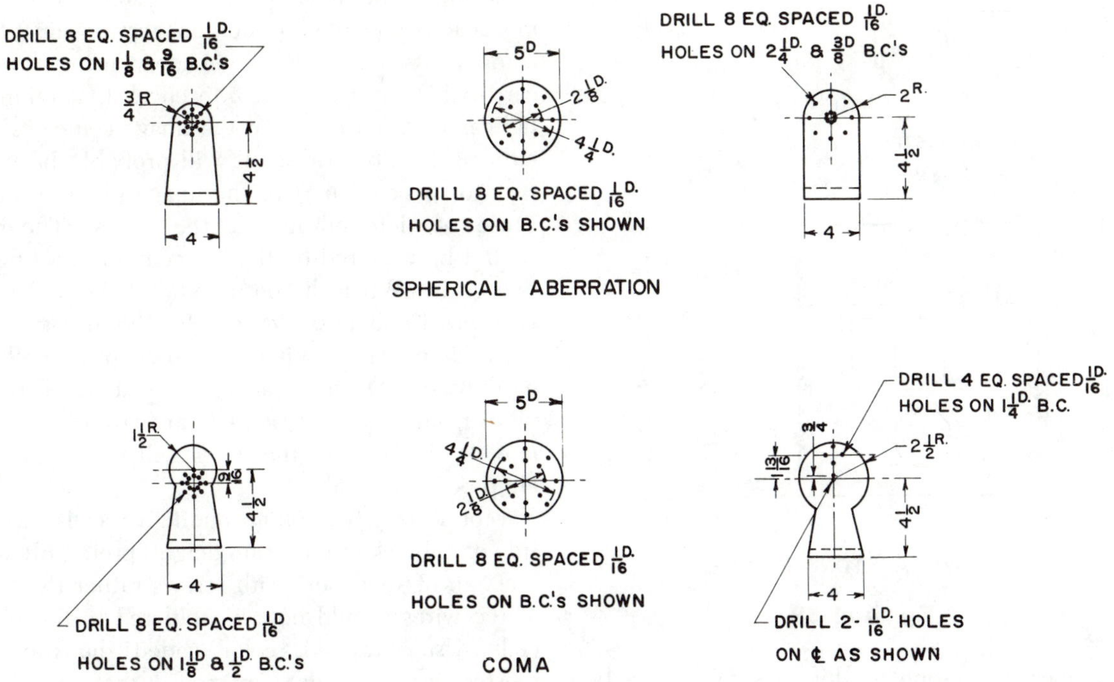

Fig. A34–57

carefully in place as close to the first grid as possible, but out of phase. The exact position of the second grid is determined by the use of the auxiliary piece of test grid *d* laid over the other two, photographic emulsion face down for more intimate contact. By use of the moiré fringe pattern, the correct phase relation is established as indicated in Fig. A34–58 (B) even though it may be difficult to see the individual lines of the grids. The moiré fringes should be straight over the whole surface, shifting from black to white at the junction of the grids. If there is a bend in the fringes at the junction, the stripes of the two grids (1 and 2) are not parallel. When the second grid is accurately positioned, wax it in place along three edges.

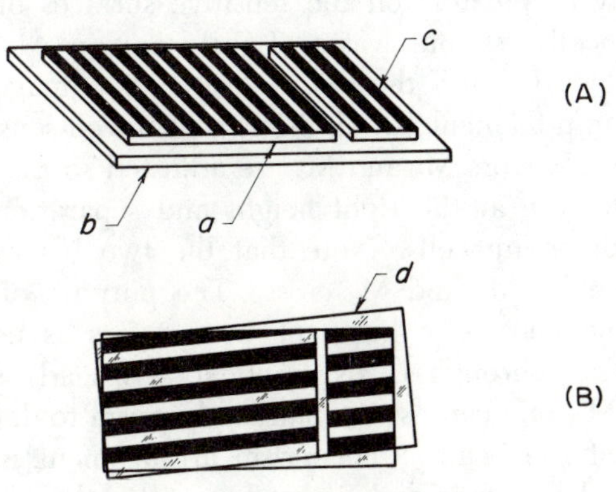

Fig. A34–58

The grid unit is now waxed at the proper height to a block (which will later be fixed to the base plate). (The axis of the optical path should be arranged so that it is parallel to the surface of the base plate and about 2 in. above it.) When waxing the unit to the block, keep in mind that the grid should be close to lens L_1 and mirror M_2. The emulsion surface should face lens L_2 (see Fig. A34–59 and Fig. 34–42, page 1056).

The demonstration should be set up on a sturdy table which will not shake when someone walks across the floor. Locate the heavy metal base plate so that the optical parts are readily accessible and the electrical components will also be within reach. The galvanometer G should be placed where its scale is easily observed. Place the metal base plate on three corks, one midway along the left edge, and the other two in the corners at the right edge.

Wax the achromatic doublet L_2 to a support block so that its center will be at the chosen height of the optical beam axis. Keep in mind that mirror M_1 should be about ½ in. behind L_2, so that the support block should be mostly to the right of the lens. Note

(B)

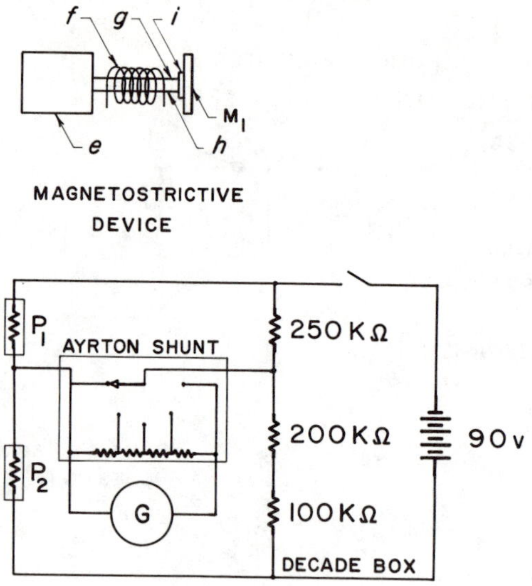

MAGNETOSTRICTIVE DEVICE

(C)

WHEATSTONE BRIDGE

Fig. A34–59

the proper orientation of this lens. Wax the two plano-convex lenses of longer focal length to a mount to make L_1. (The focal length of the combination should be about 2–3 in.) The two lenses may be waxed to a ring spacer for better alignment. Next, mount the auto lamp S at approximately the right height on a block with its filament in a vertical plane. The lamp should be secured to its mount by means other than wax, for when it gets hot the wax will obviously soften and the lamp will move. With soft wax, mount a small mirror (¾ x ¾ in.) on a block to serve as M_1 for preliminary adjustments.

Wax L_2 on the left side of the base plate about 4 in. from the edge, keeping the lens axis along the desired line (say parallel to the front edge of the base plate). The lens should be carefully positioned to avoid astigmatism in the imaging produced. Put M_1 about ½ in. behind L_2 and locate the approximate focal plane of the lens, using the lamp source and a white card to find the reflected image. Put the grid unit in approximately the proper place with the emulsion side toward L_2. Locate L_1 about ½ in. behind the grid and about ¾ in. off the axis of lens L_2. Find the proper location of the lamp S so that an image of the filament is formed on M_1. Tilt M_1 so that a reflected image of the grid falls symmetrically on the grid, both at the right height and so that the center of the illuminated circle of the image falls on the junction of grids 1 and 2. It is important that the optical system be as symmetrical as possible for good imagery. Adjust these components for optimum imagery using a white card behind grids 1 and 2 to check that minimum light is transmitted when the image coincides with either grid 1 or grid 2. It will probably be necessary to mask off part of the image of the grid because of deterioration at the edges. The mask should be centered on the boundary separating the two grids. When the positioning of these elements is as precise as possible, wax L_2, the grid structure L_1, and the lamp unit to the base plate with the hard wax. (Do not wax M_1 in place). Tape the wires from S to the base plate at two places so that moving the ends of the wires will not disturb the lamp position at all.

Mount the two photoconductive cells side by side on a block at the beam height, preferably using a clamp of some sort with screws rather than wax. Three wires should now be soldered to the photocells. Use stranded, color coded, hookup wire twisted into a cable (or special three conductor shielded cable if available). The cable should be long enough to reach from the cell position to the bridge circuit with perhaps a foot to spare. One stranded wire is soldered to the common junction of one lead from each photocell, and one to each of the other leads of the two cells.

Mount M_2 and M_3 on a block with soft wax so that M_2 may be as close as possible to the grid and so that the block will not get in the way of the mounts for L_1 or L_3. Mount two remaining plano-convex lenses on another block (hard wax) to make L_3. The focal length of L_3 should be less than that of L_1 so that the final lamp images will be smaller than the lamp filament and will fall as nearly as possible on the sensitive surfaces of the photocells without waste. Locate mirrors M_2 and M_3, lens L_3, and the photocells so that images of the lamp filament fall on the two photocell-sensitive areas. Mirrors M_2 and M_3 are adjusted so that the images are at the right height and separation for the detecting cells. Note that the two beams reflected by M_2 and M_3 cross. The purpose of this arrangement is to have the beams pass as nearly centrally through L_3 as practical. (Instead of M_2 and M_3, one can use a small angle prism to deviate one of the beams. The prism arrangement is described by Jones and Richards.) After the images falling on the photocells are optimum, wax M_2 and

M_3 in place with hard wax (do not remove the soft wax), and wax the mirror, lens, and photocell mounts to the base plate. Tape the photocell wires to the base plate at two different places so no motions of the wires disturb the cell positions.

Blackened cardboard or wood blocks (flat black Krylon spray or poster paint, etc.) should be placed so as to shield the light path both from stray light and drafts. Leave enough area around M_1 so that the magnetostrictive device, Fig. A34–59 (B), can be put in. Also leave a gap about 4 in. wide near C_2 so that this plate (see below) can be adjusted. A large sheet of heavy, blackened cardboard should be used for the top. Be sure that light from the lamp cannot directly reach the photocells and is not scattered to them. It is best to make at least part of the shielding removable for minor adjustments.

Set up the Wheatstone bridge circuit as indicated in Fig. A34–59 (C). The Ayrton shunt (see Stout) is used to reduce the bridge sensitivity while keeping the galvanometer properly damped. A 90-V B battery is used for the bridge source, though 180 V would be suitable across the two photocells.

Since the image of the grid formed by L_2 has an arbitrary phase with respect to the remainder of grid 1 and grid 2, a pair of compensating plates C_1 and C_2 are used to restore the phase. These plates are simply located in the beam near the grid unit where the incoming and return beams do not overlap. Compensator C_1 may be waxed to the base plate normal to the beam direction through it and C_2 left free for adjustment to displace the beam slightly in either direction.

Set the Ayrton shunt so that the sensitivity of the galvanometer is zero. Check the bridge circuit carefully to make certain that the wiring is correct and that all connections are tight. Connect the battery and turn on lamp S. Close the bridge circuit switch and set the decade resistance box at 50 kΩ or the approximate value required to make the two resistance arms of the bridge about equal to each other. Set the Ayrton shunt for minimum bridge sensitivity, and balance the bridge reasonably well by adjusting the rotation of the compensator plate C_2 which displaces the image slightly either way. When the bridge is balanced (or nearly so), increase the sensitivity with the Ayrton shunt and obtain a better balance. At same galvanometer sensitivity it will be found that the required adjustment of the compensator is too delicate to make, and at this point, make the bridge balance adjustments with the decade resistance box. It may well be impossible to balance the bridge at full sensitivity, for the exact balance point will drift slowly. But full sensitivity is not needed. Now turn the shunt to zero sensitivity, disconnect the bridge battery, and turn off the galvanometer light (if any).

Determine the smallest useful mirror size for M_1. The size should be about ¾ in. high x ⅜ in. wide (though a smaller one could be used with little loss of sensitivity).

Use a solenoid coil or wind a many-turn coil for the production of the magnetic field. The unit shown schematically in Fig. A34–59 (B) is mounted on a single block *e*. Coil *f* may be waxed in place or otherwise secured so that the wires can be mounted horizontally at about the right height along the axis of the coil. Wires *g* and *h*, which support the mirror, are about 2½ to 3 in. long. One wire is copper and one soft iron. They are soldered at one end to a small block of copper *i* about ¼ in. square and 1/16 to 1/32 in. thick. The wires should be about a millimeter apart at the copper block. The other ends, passed through the coil, are fastened to the mounting block either with a clamp or with wax. The small mirror M_1 is waxed to the copper endpiece with its long direction vertical. Two stranded wires are soldered to the coil terminals so that they may be connected to the magnetostriction drive battery later.

Remove the temporary mirror unit M_1 and replace it with the magnetostrictive unit. Position the magnetostrictive unit so that the mirror reflects the grid image back to the proper place, and wax the unit in position.

Connect lamp S to its battery source, activate the bridge, and balance it as before, using the compensator as much as possible. With the bridge sensitivity as high as practical (the drift of the galvanometer should be less than 1 cm per 10 sec), energize the magnetostrictive coil with one or more dry cells. The galvanometer deflection, which is almost instantaneous, should be at least half scale, probably off scale. With the author's unit, the deflection was perhaps twice full scale when 2 V dc was applied to the 100-Ω coil. (The mirror deflection observed directly with a telescope and scale at a distance of one meter was not more than 0.1 mm at most.)

Materials List

1. Beeswax and rosin mixture, stove to heat it, and Variac to control the heat
2. Medicine dropper to apply wax
3. Soft Tackiwax
4. Heavy aluminum (Dural) or steel base plate, ½ x 12 x 20 in. or larger
5. Set of blocks for mounting optical components, either metal or wood, in the shape of rectangular prisms
6. 6-V auto lamp, 21 cp, with small filament (for S)
7. 2 plano-convex lenses, 1¼ in. diam, 3-in. focus (for L_1)
8. 2 plano-convex lenses, 1¼ in. diam, 2-in focus (for L_3)
9. Achromatic doublet, 1 in. diam, 10-in. focus (for L_2)
10. First surface mirrors (may be cut from one larger piece), 1/16 in. thick
    (a) ¾ x ⅜ in. (for M_1)
    (b) ¾ x ¾ in. (for preliminary M_1)
    (c) Two 1½ x 2 in. (for M_2 and M_3)
11. Ronchi ruling, 50 to 150 lines per inch for grid (a 4 x 4-in. piece ruled diagonally can be cut up). The minimum sizes required are approximately
    (a) 1½ x 2 in. (for grid No. 1)
    (b) 1½ x ¾ in. (for grid No. 2)
    (c) 1¼ x 3 in. (for testing)

    The ruling in each case must be parallel or perpendicular to the edges, not diagonal
12. 3 pieces of plate glass, ⅛–3/16 in. thick
    (a) Two compensators, 2 x 3 in. (for C_1 and C_2)
    (b) One mount for grids, 4 x 2 in.
13. Two Clairex CL-3 photoconductive cells or the equivalent (for P_1 and P_2)
14. Sensitive galvanometer: period 1–3 sec, resistance 1–10 kΩ, sensitivity about 0.001 μA/mm
15. Ayrton shunt for the galvanometer
16. Two fixed resistors, ½ W or 1 W, 200 kΩ and 250 kΩ (carbon—other values are possible)
17. Decade resistance box with steps of 100 kΩ, 10 kΩ, 1 kΩ, 100 Ω, 10 Ω, and possibly 1 Ω
18. Batteries:
    (a) 7½ V dry cell or 6-V storage battery for auto lamp. Note: It is not satisfactory to use a dc power line supply for the lamp. Fluctuations in line voltage upset the balance of the bridge.
    (b) 90-V B battery for bridge circuit
    (c) 1½ to 6 V for magnetostrictive unit
19. Copper and iron wire of roughly the same size—18–24 gauge
20. Coil of 100- to 500Ω resistance from a solenoid or other device (for magnetostrictive unit)

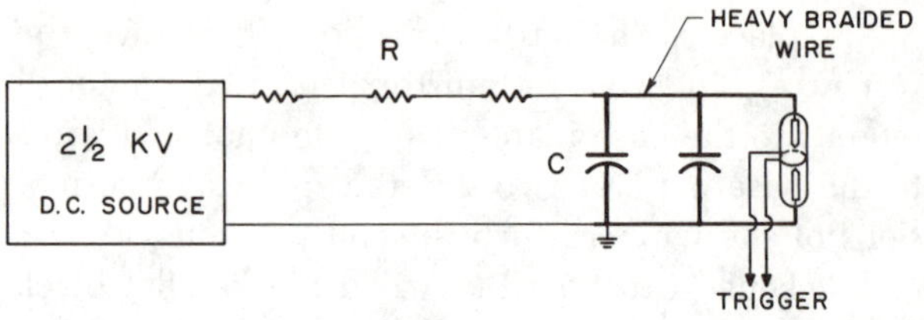

Fig. A34–60

A34–1.27

If the patterns are to be photographed, one must use some sort of high-intensity flash tube. The tube will, of course, require an electronic power supply. A circuit suitable for operation of the GE type FT 230 flash tube is shown in Fig. A34–60. Peak currents of the order of several thousand amperes are involved in the discharge of the capacitors through the flash tube, and consequently any appreciable resistance in this part of the circuit will be detrimental to the energy of the flash. It is, therefore, important to connect the flash lamp and the capacitors with very heavy wire—heavy braided ground strap wire is satisfactory. Even so, the length of these wires should be kept to a minimum. The remainder of the circuit may be put together with ordinary wire. Any dc source which supplies the requisite high voltage (2½ kV for the tube mentioned) is satisfactory. Detailed instructions on firing the tube come with it. A wire ring wound around the waist of the tube and triggered with the output of an old Ford coil can be used to trigger the tube.

Requirements of the circuit for the GE type FT 230 flash lamp are the following:

R: ~2 MΩ. Use a series of 1-W carbon resistors to avoid voltage breakdown.

C: ten 100 μF, 3000-V dc capacitors intended for flash tube purposes.

RC: time constant, ~1 min.

Trigger: use Ford coil spark generator to trigger flash gap (one turn of hook-up wire around lamp).

Caution: *High Voltage. Danger.*

Chapter 35 Physical Optics

A35–1.1

Materials List

Fig. 35–1 (p. 1065)

Detail	Component and Description
1.	Arc lamp with copper plated high-intensity carbons
2.	Lens, focal length 7 cm
3.	Lens, focal length 18 cm
4.	Iris diaphragm, 15 mm diam
5.	Lens, focal length 35 cm
6.	Source slit
7.	Partially reflecting mirror, middle section (about 5 mm wide) clear in order to let through the direct rays (1.2)
8.	Rotating mirror with a reversible motor (Leybold). The reversing mechanism was rebuilt by the University of Basel for this demonstration.
9.	Lens, focal length 45 cm
10.	Iris diaphragm
11.	Condensing lens, focal length 7 cm
12.	Lamp, 12 V
13.	Lens, focal length 4 m
14.	Surface mirror, plane front
15.	Screw, pitch 1.5 mm
16.	Entrance slit for photomultiplier
17.	Mounting slide and track for photomultiplier
18.	Pointer
19.	Millimeter scale, immovable
20.	Multiplier impedance matcher (cathode follower) and scaler
21.	6-V lamp for projecting pointer and scale
22.	Condensing lens, focal length 7 cm
23.	Lens, focal length 8.5 cm
24.	Prism

A35–1.4

A schematic diagram of the light source is shown in Fig. A35–94. Two 0.5-mm-diam platinum wires with the ends melted to form spheres are connected to the shortened lead wires of the resistors to form a spark gap with a separation of 0.5 mm. The capacitor is connected across this gap and the high voltage applied. The increase in light intensity

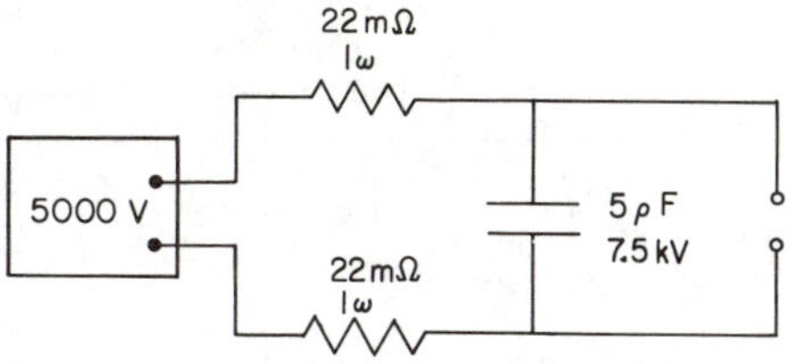

Fig. A35–94

takes place during less than 4 nsec, and the exponential decay has about a 10 nsec decay time.

A35–1.5

The following items used in the apparatus are interchangeable without further changes:

The described four stage 99.2 Mc/sec amplifier may be replaced by any amplifier with the same electrical properties. It is even better to choose one with a broader bandwidth, e.g. the tunable amplifier type ASV BN1372 (Rohde u. Schwartz, München, West Germany), which is variable between 27 and 300 Mc/sec.

The germanium diodes type OA 81 which are used in the phase detector may be replaced by any type of general purpose diode, e.g. HG 1005 (Hughes), IN47, IN58, IN61, IN63, IN68, or IN70.

The germanium PNP transistors used in the 13-V power supply have the essential characteristics given in Table A35–1 (page 1354) and may be replaced by equivalent types.

A35–2.8

The interferometer for the demonstration is conveniently built of plywood. The base is 1 in. thick and about a foot square. The mounts for the optical parts are made of $\frac{3}{4}$-in. and $\frac{1}{4}$-in. plywood. The mirrors *f* are simply waxed to the adjustable plywood mirror-mounts (Fig. A35–95). Each of the two mounts required is equipped with an adjusting screw *g* (such as 6-40 threaded through the wood) and a phosphor bronze spring *h*. The base support has three supporting feet *i* which are waxed to allow easy rotation about a screw pivot *j* which holds the mount in place. The axis of rotation is approximately in the plane of the mirrors.

Table A35–1

Type	Collector Current (dc amperes)	Power Dissipation (at 45°C case temperature)	Collector Emitter Breakdown (voltage BVcex)	Equivalent Types
AD 103	15	22.5	40	2N 443
GFT 3108	3	8.0	40	2N176, 2N351, 2N1183
TF 78	0.6	2.0	32	2N2139 (Motorola)
GFT 29	0.2	0.3	8	2N270
GFT 22	0.03	0.07	15	2N109, 2N591, 2N1309

Z 10 is the Zener diode, Zener voltage 8.8–11 V, temperature coefficient $+5 \times 10^{-4}/°C$, 250 mW power dissipation at 25°C ambient air temperature.

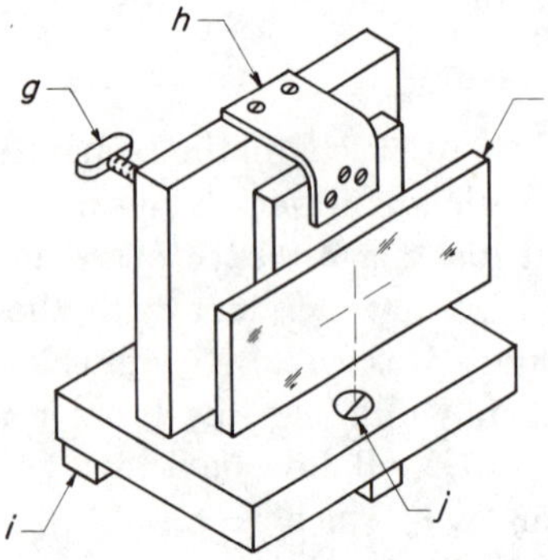

Fig. A35–95

The support for the beam splitter (Fig. A35–96) need not be adjustable, but should be provided with three feet *k* for semikinematic design. The beam splitter mount should be arranged to hold either one of two plates so that both symmetrical and unsymmetrical beam splitters can be tried. The beam splitter *m* is held in place by a clamp screw and bracket assembly *n*, and the mount is held to the base by a screw *o*. The angle of the beam splitter can be fixed at about 45 deg to the direction of the incident beam. It is perhaps best to lay out the parts symmetrically, making the exit beam emerge at right angles to the incident beam.

A35–5.2

The absorption cell is constructed of thin-walled steel or brass tubing 1/16 in. thick, 1½ to 2 in. diam, and about 20 in. long. Figure A35–97 shows two views of the ends of the cell. Two holes are drilled in the side of the main tube *a* about 2 in. from either end to accommodate the two side tubes *b* for flushing and evacuating the cell. The holes should provide a fairly snug fit for two 6-in. lengths of ¼-in.-o.d. copper tubing to be silver-soldered in place. The ends of the cell tube should be squared on a lathe or milling machine and the edges smoothly rounded with emery paper. The end plates *c* and flanges *d*

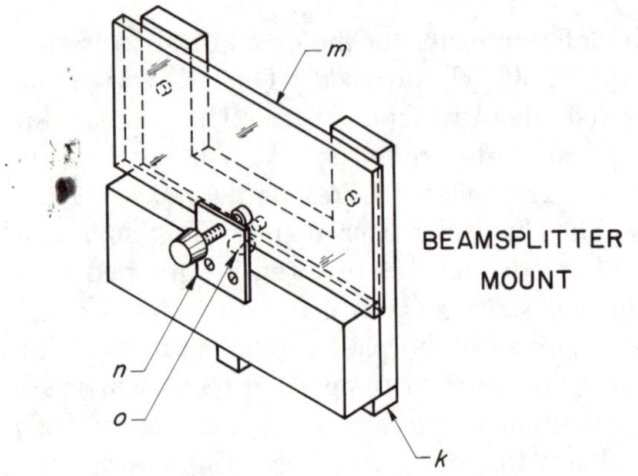

Fig. A35–96

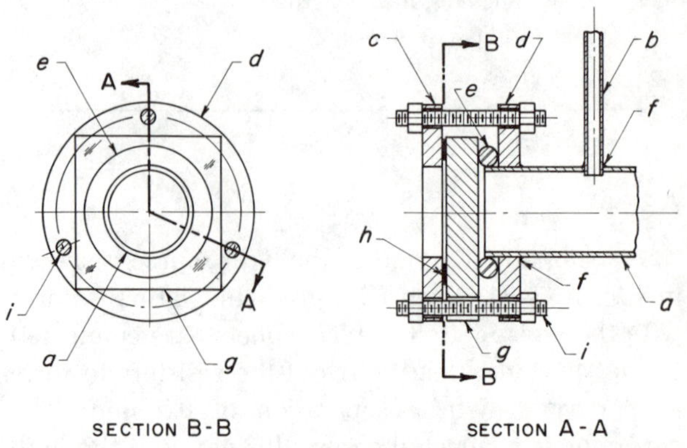

Fig. A35–97

are made from four disks $\frac{3}{16}$ to $\frac{1}{4}$ in. thick, 4 in. diam. The flanges are to be slid over the end of the tube and, accordingly, have holes machined to give a smooth snug fit over the cell tube. The end plates, holding the windows in place, require center holes which have the same diameter as the flange center holes. The flanges and end plates are then clamped together and three $\frac{17}{64}$-in. holes drilled around the circumference as in the figure. One side of the flanges and end plates should be machined or sanded flat and smoothed with emery paper.

It is assumed that neoprene O-rings *e* of a size to just slide over the cell tube are available. If they are not, grooves are machined in the flanges to accommodate such rings as are available. In this case, the tubing should not project through the flange plates as it does in Fig. 35–97.

The cell should now be assembled for silver soldering. The two 6-in. side tubes *b* are fitted into the main tube *a*—they may project about $\frac{1}{8}$ in. inside. The flanges are clamped in place so that the tube projects through them about two thirds of the thickness of the O-rings to be used. The flanges should be carefully squared with the tube. Make sure that the silver solder runs freely over the indicated joints *f* to ensure that they will be vacuum tight. When the cell is cool, clean it thoroughly and remove the soldering flux.

Windows *g* for the cell are cut from plate glass $\frac{3}{8}$ in. or more thick. One can use 3-in.-diam disks of plate glass (mirror-making tools) from Edmund Scientific Co. Four windows (two sets) are recommended so that while one set is being cleaned the other set covers the ends of the cell. Either round or square windows are satisfactory.

The absorption tube should first be tested to see that it is vacuum tight. Assemble the cell using neoprene O-rings, which may be lightly greased with silicone high-vacuum grease. A heavy paper or thin rubber spacer *h* should be used between the end plates and the windows so the latter will not crack when clamped. Use three 2-in., $\frac{1}{4}$-20 bolts *i*, and tighten the flanges until the O-rings are about $\frac{1}{3}$ compressed. The windows should almost, but not quite, touch the ends of the tube.

After the cell is assembled, it should be evacuated and the pressure checked with the manometer. When the pressure is reduced below 1 mm, close the connection to the pump and observe how quickly pressure builds up. The vacuum system is very small and the pressure will very likely rise to a few centimeters within an hour or so. If it is no worse, the system is probably sufficiently tight. Any more serious leaks must be located and sealed. (Glyptal varnish, a General Electric Co. product, is useful for sealing very small leaks).

Chapter 38 Quantum Effects

A38–1.3

The mechanical construction is shown in Fig. A38–51. The machine consists of two parts: (a) the stationary unit, which carries the lower panel, and (b) the carriage, which carries the upper panel.

The carriage rides on three wheels. Two of the wheels are located immediately behind the upper panel. They turn freely on the shafts that serve as their axles, and their rounded rims run in a V-track fastened to the top of the lower panel. Their functions are simply to carry the carriage and to keep it aligned. The third wheel is at the center of the rear of the carriage; it is a large gear that rolls on a long rack mounted on the stationary unit, and it provides the timing for all the mechanism in the carriage. This gear runs freely on its shaft, but it is coupled to the two outershafts by sprocket-chain systems which reduce the angular velocity by the factor one-half. This reduced angular velocity is transmitted unchanged from each outer shaft to the center shaft by sprocket-chain systems. The sprocket-chain system is closed, and by properly setting the sprockets on their shafts it is possible to eliminate virtually all backlash. The three shafts carry the clock pointers.

On the stationary unit a sprocket chain forms a loop extending almost from one end to the other. At one end of the unit the chain runs on a sprocket which is driven by means of a crank on the back side when the demonstrator operates the machine. At the other end the chain runs over an idler. The bottom of the carriage is fastened to a special link in the upper run of this chain by means of a single

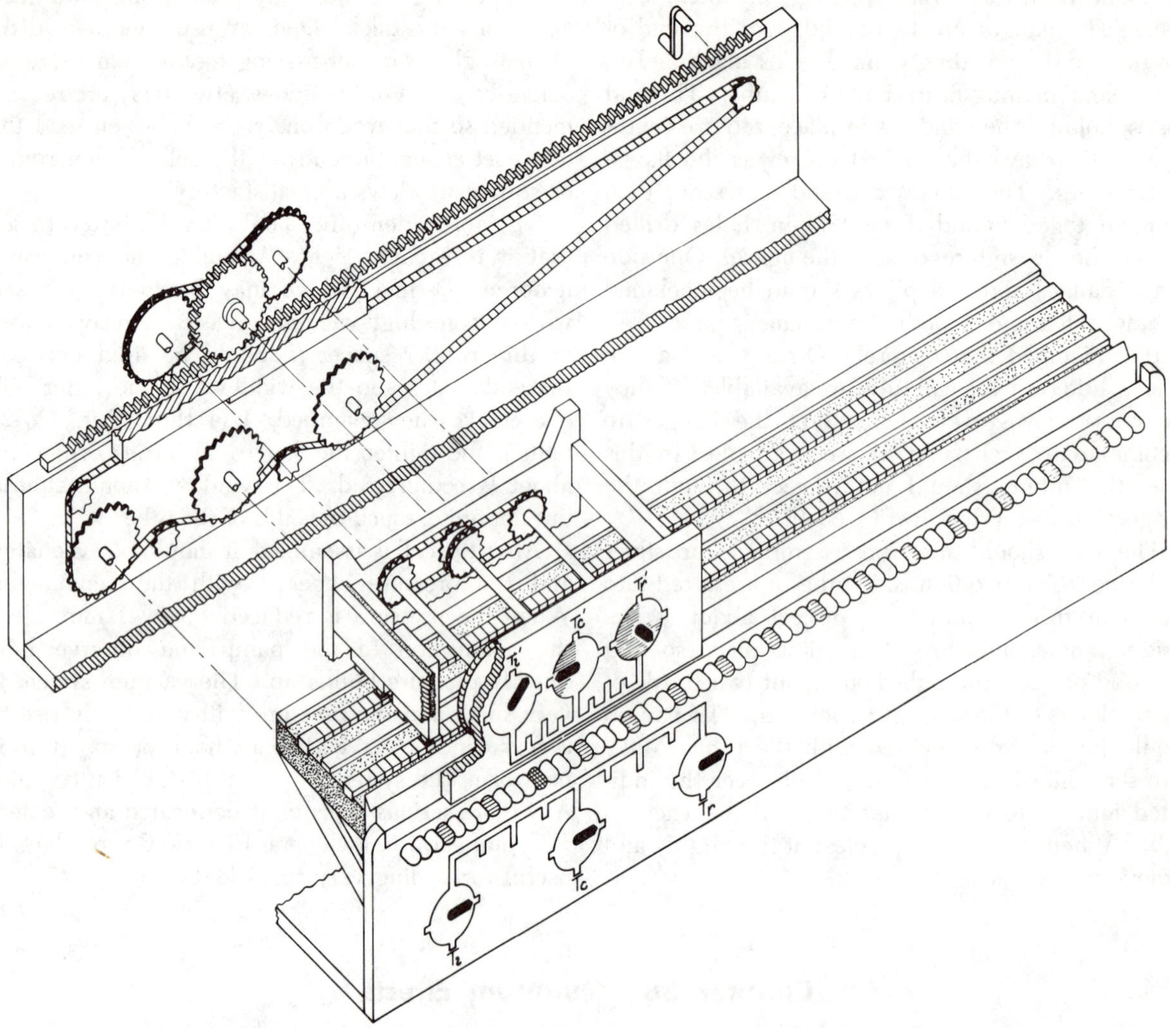

Fig. A38–51 An Exploded Isometric View. The two panels seen by the audience are shown in front. The panels are black; the measuring rod markings and dial faces are bright aluminum; parts of the upper dials are red, as indicated by shading; the pointers are black. When the carriage is at its extreme left-hand position, T_1 *and* T_1' are in line, and the lamp which is in line with them is lit. At this time T_1, T_c, and T_r all read $t = 0$, while T_1', T_c', and T_r' read $t' = 0$, $t' = -3/8$, and $t' = -3/4$, respectively. As the carriage moves to its extreme right-hand position, the pointer on each lower clock executes two revolutions, and the pointer on each upper clock executes one revolution. The shading indicates those lamps that are lit to represent the continuous wave when the carriage is at the position shown. If the pulse pattern is selected, only the single lamp shown with double shading is lit. The switches are on a panel on the right end of the stationary unit. The dimensions are: 30 in. (length) x 9.5 in. (total height) x 5.75 in. (depth). Following are some critical specifications: The spacing between T_1 and T_r is 9.42 in.; the gear has a pitch diameter of 3 in.; the chain has 65 links/ft; the three special sprockets on the lower shafts have 51 teeth each.

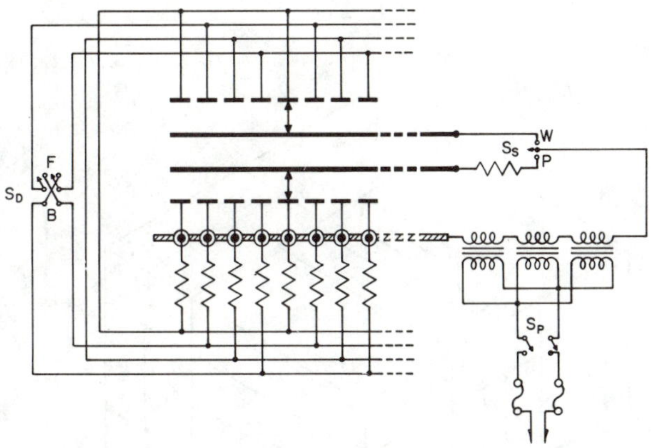

Fig. A38–52 Electrical Circuit. There are 63 incandescent lamps, of which eight, represented by the larger circles, are shown. The four slip-bars are represented by the heavy black lines, and the large arrows represent the brushes of the sliding contacts. The pattern of connections to the lamps and to the slip-bar segments is repeated in each sequence of four. The lamps are set into holes in the panel. The shell of the base is thus "grounded," the "ground" being indicated by the shaded area. The other connection to each lamp is made of a lead soldered to the center terminal of the base. Ratings are as follows: lamps, 28 V, 0.17 A; resistors, 82 Ω, 2 W; each of three transformers, 12.6 V, 2 A.

screw. In disassembling for transportation, this connection is easily broken and the carriage can be lifted off the stationary unit. The lower run of the chain, guided by some idlers, engages the sprockets which turn the three shafts carrying the pointers of the lower clocks.

The electrical circuit to the lamps is shown in Fig. A38–52. The lamps are connected in four groups, each group consisting of every fourth lamp. The power is distributed to the lamps by sliding contacts. The brushes are mounted on the carriage, and the slip-bars run the length of the stationary unit. To achieve the versatility of the light signal, it is necessary to employ four sliding contacts, two with continuous bars and two with segmented bars. With the switch S_P closed, and switches S_S and S_D in positions W and F, respectively, all the lamps of one group are lit. As the machine is operated, the sliding contact energizes the four groups successively, and the pattern of lit lamps propagates toward the right. If switch S_D is thrown to the B position, the second and third groups of lamps are interchanged in their connections to the sliding contact, and as a result, the pattern of lit lamps propagates toward the left. If switch S_S is thrown to the P position, the resistors are brought into play and serve to isolate the lamps in each group. As a result, only one lamp is lit, and as the machine is operated, this simple pattern propagates toward the right. In order that the speed of the carriage be less than the speed of the light signal by the factor $\beta = 0.866$, the spacing of the segments of the sliding contacts is correspondingly less than the spacing of the lamps.

A38–3.1

Because the apparatus used in this demonstration includes several electronic units, its total cost represents more than many institutions could invest for a single experiment. However, many of the items can be used in other demonstrations in modern physics so that multiple use can justify purchase. Either the oscilloscope-slave system or the projection oscilloscope system has a wide range of applicability throughout all areas of physics.

A detector assembly can be obtained commercially from several companies; however, a completely satisfactory unit can be constructed simply and more economically in your own shop as follows: The photomultiplier envelope first is made light-tight up to the flat surface over the photocathode by completely wrapping the tube with black plastic electrical tape. A commercially canned NaI(T1) scintillation crystal can be either bonded to the face of the photomultiplier with a transparent epoxy resin[(1)] or set with a transparent vacuum grease. Once the crystal is in place, it is taped to the tube with the black plastic tape.

Figure A38–53 shows the details of the detector assembly *a* and lead shield *b*. The detector is adjustable vertically, and horizontally along its brass base *c* (3 in. wide and ⅜ in. thick). It also revolves about the aluminum scatterer *d*, which is 3 in. in diameter and 3¼ in. long. The scatterer is vertically adjustable; it is connected to a ¼-in.-diam steel rod which moves in a brass cylinder ¾ in. diam, 3⅛ in. long. The lead shield is constructed from 2 x 4 x 8-in. lead bricks as desired; it rests on

(1) E.g., bonding agent R-313 manufactured by Carl H. Biggs Company, Santa Monica, California.

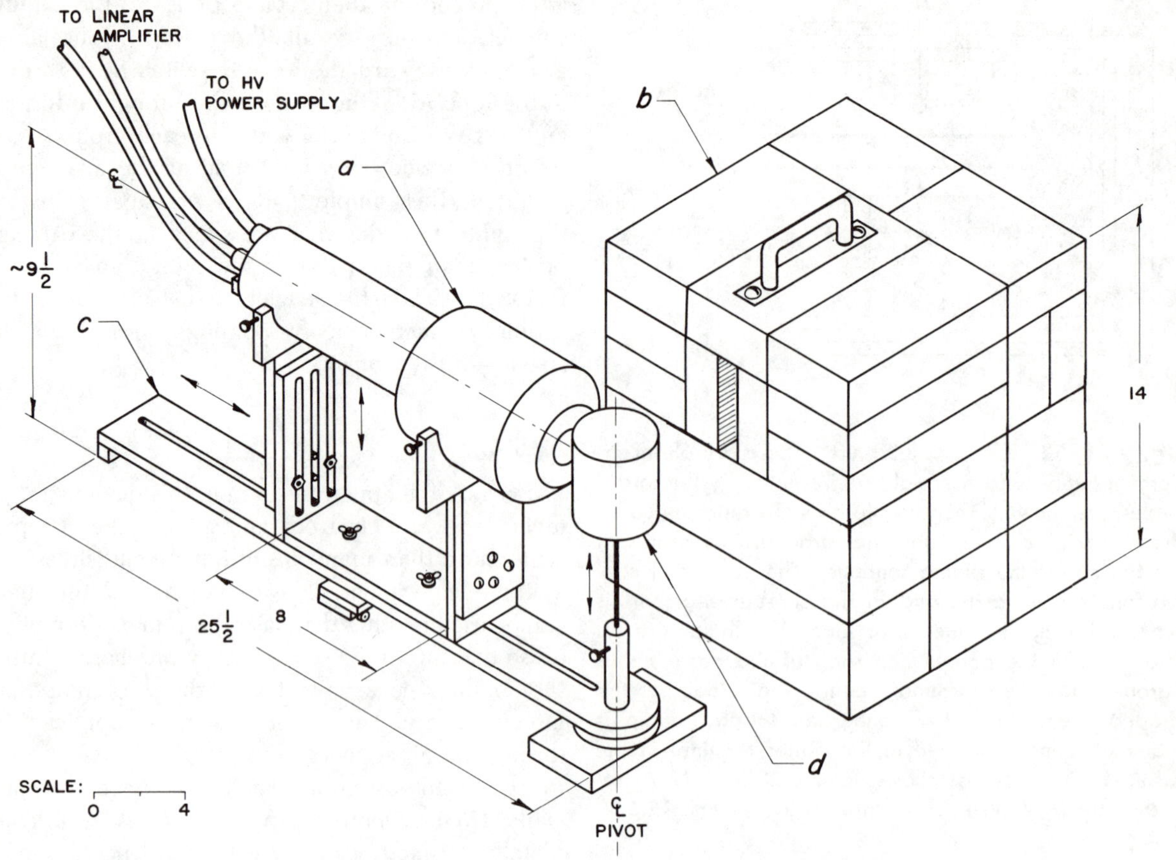

Fig. A38–53

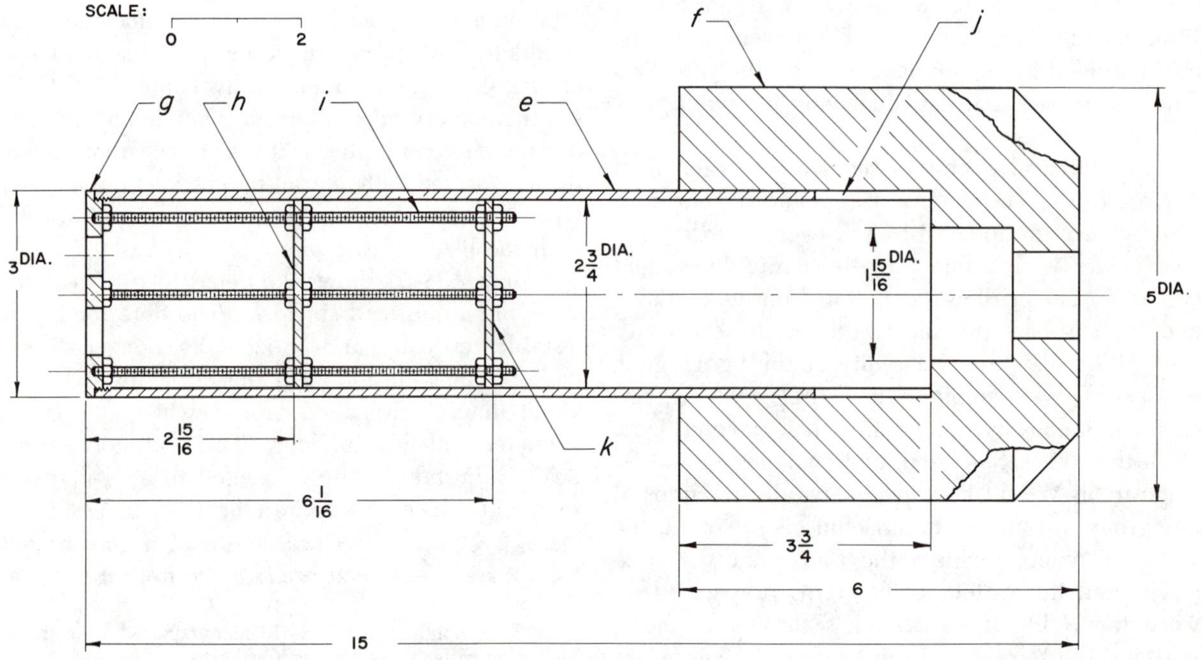

Fig. A38–54

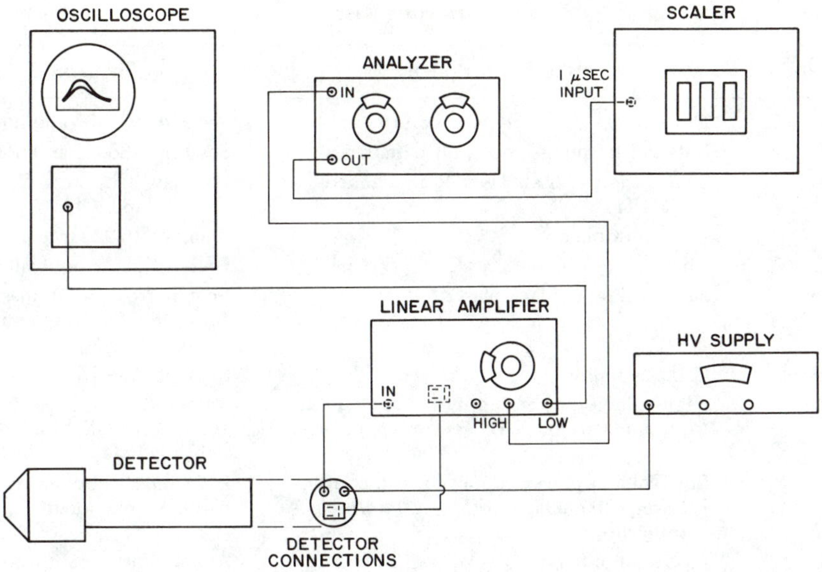

Fig. A38–55

a base made of two rows of 3½ x 3½ x 8½-in. wooden blocks. The entire block is approximately 12 in. square and 14 in. high. The opening for the source is approximately 1 x 4 in.

Figure A38–54 is a detailed diagram of the detector mount, crystal housing, and cathode follower frame. The brass cylinder *e*, 3 in. o.d., makes a tight sliding fit with the lead shield *f*, 5 in. o.d. The 1 15/16-in. hole contains the crystal. The ¼-in. brass end plate *g* contains two ⅝-in.-diam holes for the coax connectors and a 11/16 x 15/16-in. hole for mounting a six-prong Jones plug. The two brass plates *h* and *k* are 2¾ in. diam and ⅛ in. thick. They contain holes for mounting the cathode follower circuit; the holes are located and drilled at assembly. The four brass rodes *i* are No. 8-threaded and approximately 6½ in. long. In the brass cylinder are two cutaways *j*, 1 in. wide, 1¾ in. long, and 180 deg apart.

A diagram for the complete apparatus is given in Fig. A38–55, and the cathode follower is diagramed in Fig. A38–56. The cathode follower is simple to construct and has the advantage that it can easily be matched to any coaxial cable by replacing the 90-Ω resistor to output (BNC) with one whose value is equal to the cable impedance. Since the output impedance of the cathode follower up to the resistor is approximately 2 Ω, a 90-Ω resistor will match to a 92-Ω cable. This cathode follower requires 250 V dc and 15 mA electronically regulated, which can be supplied either from a separate power supply or from the linear amplifier power supply, since many commercial linear pulse amplifiers provide regulated 250 V dc at external outputs. The cathode follower provides a positive output pulse.

All other units except the rotating frame unit already described are commercially available and are described completely in the Materials List, p. 1360.

A38–4.1

The 9 x 5 x 2-in. chassis containing the phototube socket also contains a transistorized amplifier[2] with a gain of about 50 and a tunnel diode discriminator and pulse shaper. Figure A38–57 is a schematic diagram of the phototube base. The circuit for the preamplifier, tunnel diode discriminator, and squaring amplifier is shown in Fig. A38–58. The coincidence circuit shown in Fig. A38–59 is mounted on a 6 x 4 x 2-in. chassis.

Separate batteries are used for each chassis to avoid coupling between circuits through the internal impedance of common power supplies and to avoid the use of long cables from a common battery box. The batteries selected for these circuits have

(2) K. J. Casper (private communication, 1964).

Materials List

Quantity	Part and Description	Remarks
1	Detector assembly, all parts brass	See Fig. A38–54 and notes
1	Rotating frame assembly, all parts brass	See Fig. A38–53 and notes
1	Scintillation crystal, 1½ x ½ in. canned NaI(Tl)	Harshaw Chemical Co., Cleveland, Ohio
1	Photomultiplier	Dumont 6292 or equivalent
1 tube	Transparent vacuum grease	Cell-o-Seal or equivalent
1 role	Electrical tape, black plastic	Scotch No. 33, Minnesota Mining & Manufacturing Co.
1 assembly	Cathode follower	See Fig. A38–57
1	High-voltage power supply for photomultiplier, 500–3000 V dc @ 1 mA regulated	Model 312A or equivalent, Baird-Atomic, Inc., Cambridge, Mass.
1	Low-voltage power supply for cathode follower, 200–325 V dc @ 20 mA electronically regulated	Lambda Electronics, model 89, or equivalent
1	Linear amplifier	Atomic Instrument, model 204-B, no longer commercially available; however, equivalent units are commercially available from several sources.
1 set	5-in. oscilloscope and 17-in. monitor with connecting cables	Model 436 or equivalent, DuMont Laboratories, Clifton, N.J.
1	Scatterer, aluminum, 3 in. diam x 3¼ in. long	
20	Lead bricks for gamma source shield, standard 2 x 4 x 8 in.	
	2 millicuries Cs 137	Union Carbide Corp., Nuclear Div., Tuxedo, N.Y. An AEC license is required.
2	Coaxial connectors, BNC type UG-260/u for photomultiplier high voltage	
15 ft	Coaxial cable, type RG59/u for photomultiplier high voltage	
1	Multiwire connector, plug, for cathode follower power, 4-wire	
1	Multiwire connector, socket, for cathode follower power, 4-wire	
15 ft	Rotor cable, for cathode follower power, 4-wire	
2	Coaxial connectors, UHF type 83-1SP for pulse cable	
15 ft	Coaxial cable, type RG114/u or similar for pulse cable	
1	Oscilloscope, Tektronix 541	Tektronix, Inc., Beaverton, Ore.
6	Banana plugs, nylon insulators	Allied Radio Corp., Chicago, Ill.
20 ft	Flexible wire, No. 16, with plastic or rubber insulation	
1	Single-channel analyzer	Model 510, Baird-Atomic, Inc., Cambridge, Mass.
1	Scaler, 1 μsec	

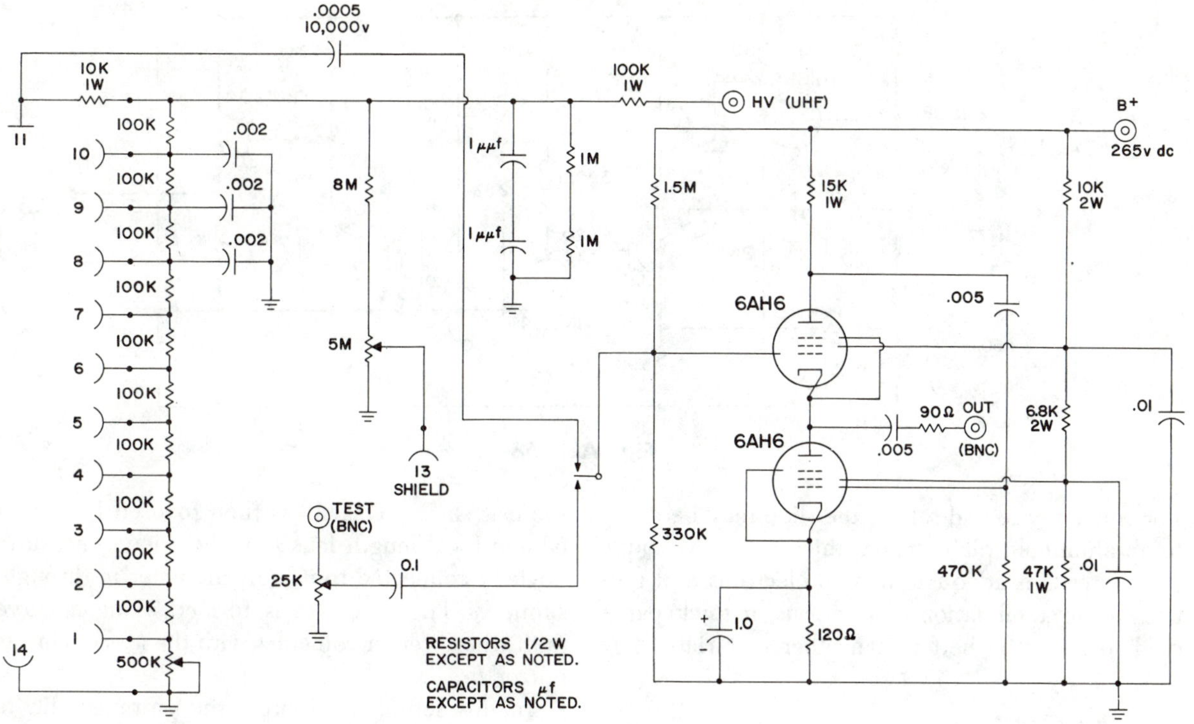

Fig. A38–56

adequate capacity for about 100 hr of operation. The scintillators used in this demonstration are 2 x 2-in. sodium iodide crystals mounted on 2-in. phototubes. Smaller detectors can of course be used at some sacrifice in efficiency.

The system should be adjusted so that the same high voltage power supply can be used for the two phototubes. This entails an adjustment of the over-all gain of the system. This is accomplished by controlling the voltage of the focusing grid of the phototubes, but the gain of the amplifier or the sensitivity of the tunnel diode trigger circuit might as well be used for this purpose. For best operation, the adjustment should be made using a fast oscilloscope triggered on the discriminator output displaying the amplifier output. The discriminator should accept the full-energy peak of the annihilation radiation.

A sodium-22 source[3] of less than 0.1 mCi will suffice. The source should be evaporated to dryness on a glass slide or microscope cover glass and covered with one layer of Scotch Magic Mending Tape.

(3) Available from Nuclear Science and Engineering Corp., Pittsburgh, Pennsylvania.

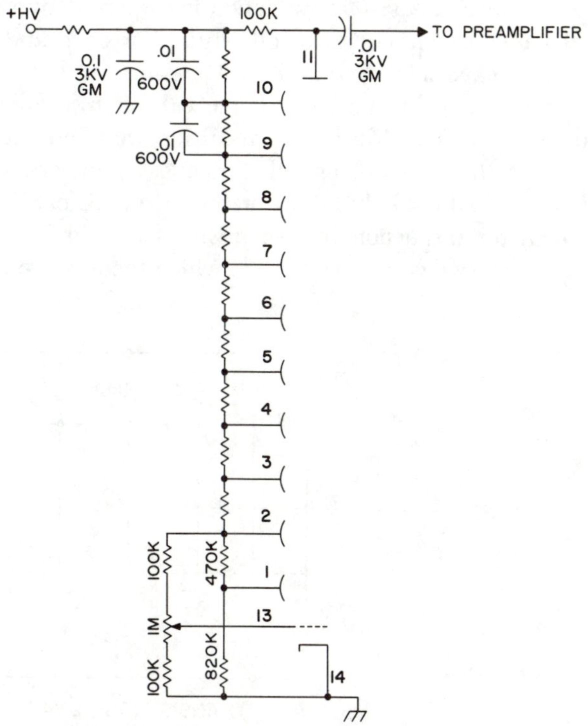

Fig. A38–57

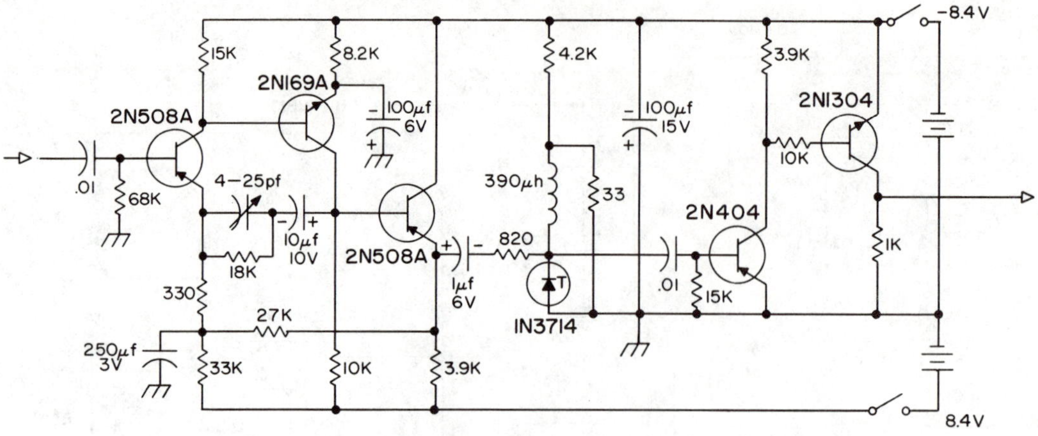

Fig. A38–58

For convenience and safety, the slide may be glued to an aluminum plate or to a thick card. Although any material is adequate to stop electrons and give rise to the annihilation of positrons, a thick paper card seems to be better than other materials.

A38–5.11

A simple monochromator can be constructed as shown in Fig. A38–60. Radiation is dispersed by a 50 x 60-mm equilateral or other glass prism *a* and two concave mirrors *b* and *c*. The entrance slit *d* is illuminated by the light source *i* through a condenser lens *h* with a 50-mm focal length. The slit is set at the focus of one of the concave mirrors *c* in order that the light falling on the prism *a* is parallel. After dispersion by the prism, the light falls on the second concave mirror *b*, which focuses it on the exit slit *e*. This slit is then focused by another 50-mm focal length lens *j* on the energy-measuring device *f* connected to its galvanometer *m* through a shunt *k*. The spectrum is formed along a curved path whose center coincides with the axis of concave mirror *b*.

In this setup, the source, the entrance slit, the prism, and both concave mirrors remain stationary at all times. The exit slit *e*, condensing lens *j*, and the energy-measuring device *f* are mounted on a second optical bench connected by a bar to the extended vertical mount of mirror *b* in such a way that the whole bench can rotate about the bar. (The support for the optical bench should be placed on a smooth surface such as glass or plastic for easy movement.) With this setup, the exit slit can pass through various parts of the spectrum, one after the other. In order to be able to measure the displace-

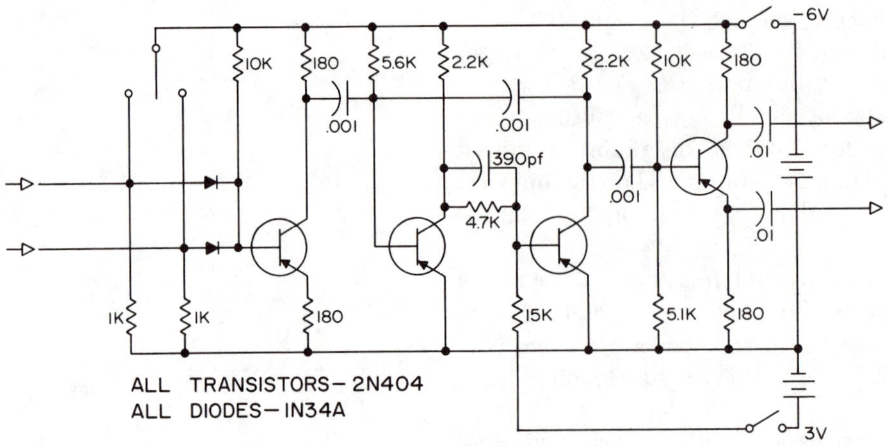

Fig. A38–59

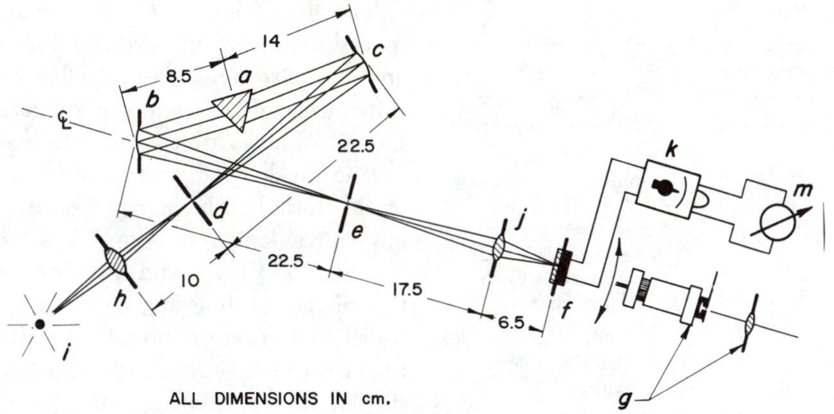

Fig. A38–60

ment of the exit slit, a projection device, consisting of a lamp and lens *g*, is mounted on the same bench as the exit slit and receiver. This device, directed along the bench, focuses the image of a filament on a scale several meters away. The position of this image thus is a measure of the displacement of the exit slit.

In assembling the monochromator care must be exercised in leveling and adjusting the several heights. It is also best to measure the focal lengths of the several lenses and mirrors beforehand in order to be sure that the entrance and exit slits are at the correct foci and that the mirror and prism are filled with light. The prism should be adjusted for minimum deviation at some intermediate wavelength such as the yellow lines of mercury. A mercury lamp is convenient for calibrating wavelength against the scale position of the projected filament. These calibration data also are necessary to determine the spectral slit width, which is needed for reducing the spectrum to standard slit width. For the calibration as described, the entrance and exit slits should be set at about 0.2 mm.

In order to measure the energy distribution from an incandescent solid, the mercury lamp should be replaced either by a high-current-density carbon arc or by an incandescent lamp. A thermopile makes a better receiver than the more sensitive photodiode because its sensitivity is less dependent on wavelength. For a constant width slit the admitted spectral region $\Delta\lambda$ is much smaller in the short wavelengths region than in the long wavelength region. To obtain comparable results, all measurements must be referred to equal values of $\Delta\lambda$ along the spectrum; for this, a plot of $\Delta\lambda$ against λ (or scale position) is necessary. This plot can be evaluated graphically or by fitting an interpolation formula to the wavelength calibration data (λ vs scale reading) obtained from the Hg lamp and then differentiating.

A38–6.3

The construction diagram for the two film holders is shown in Fig. 38–24, p. 1187. These are made of $\frac{1}{16}$-in. brass stock, $3\frac{1}{2}$ in. square. The remainder of the apparatus is readily available from commercial suppliers.

While not essential for the operation of the apparatus, it was found convenient to place light shields between the various components of the system. These tubes cut out a considerable amount of background light, making the projected images more visible, particularly when viewed by the class from the back of the translucent screen.

After considerable experimentation with both commercial and locally concocted soap solutions, the best solution was found to be a simple one made of equal parts of distilled water, pure glycerine, and Trend (a commercial household preparation). See the Materials List on p. 1364 for additional details.

A38–6.4

The toroid is the most difficult part to construct since care must be taken to ensure that the tube does not buckle during forming and that the joint

Materials List

Quantity	Part and Description	Remarks
2	Film holders, brass 1/16 in.	Model 03001 or equivalent
1	Arc lamp	Spindler-Hoyer model 030001 or equivalent. Available from Klinger Scientific Apparatus Corp., Jamaica, N.Y.
1	Ballast resistor for arc lamp	Spindler-Hoyer model 030053 or equivalent
1	Condenser lens	Spindler-Hoyer model 035000 or equivalent
1	Plane mirror	Spindler-Hoyer model 014100 or equivalent
1	Water cell	
1	Projection lens	
1	Iris diaphragm	Leybold model 460-26
2	Optical bench, Zeiss type	
6	Apparatus mounts for optical benches	
1	PM speaker, 6-in. diam, 8 Ω	
1	Audio oscillator	Hewlett-Packard model 202-G or equivalent
1	Audio oscillator	Bell model 2122-C or equivalent
1	Translucent projection screen, bench mount	
1	Soap film solution holder	

is not under excessive stress. The buckling can be prevented by filling the tube with sand during forming, and the stress on the joint can be minimized by initially bending the tube with a radius somewhat smaller than that desired for the final toroid. The toroid should be finished before beginning work on any of the accessories.

The gas distribution system in Fig. 38–29, p. 1189, is easy to construct: the manifold *b* is a drilled 1½-in. brass cube to which are soldered the four brass distribution tubes *a*. The gas inlet fitting *d* is a standard gas valve, threaded on one end to screw into the manifold and provided with a hose fitting on the other end.

The gas holes on top of the toroid should be spaced uniformly and drilled 0.015 to 0.025 in. The optimum size depends somewhat on the individual system; therefore, some experimenting is necessary. 150 to 200 holes are adequate for a 4-ft-diam toroid.

The speaker flange *c* is soldered into the bottom of the toroid. The flange holes are drilled to fit any convenient 3- or 4-in. PM speaker. The three legs are solid brass and 6 in. long, with the exception of the leveling leg, which is ½ in. shorter and is drilled and tapped to take a ¼-24 machine screw to provide the leveling adjustment. These are shaped to the tube contour and hard-soldered in place.

A38–7.4

The gold film was vacuum evaporated onto a polyvinyl sheet.[(4)] The thickness of the gold film is roughly 500 Å, which is equivalent to a surface density of about 100 $\mu g/cm^2$. This weight has worked satisfactorily, although a film with a surface density of about 50 $\mu g/cm^2$ should improve the intensity and resolution of the diffraction rings, without causing the central maximum of the pattern to be so bright that the rings cannot be seen. Since supporting polyvinyl sheet has a surface density of about 5 $\mu g/cm^2$; its effect on the diffraction pattern may be ignored. The gold plated polyvinyl was placed in a brass "sandwich" with a hole in the center to expose the gold to the beam. One piece of this "sandwich" is shown in Fig. A38–61(a). Two 3/32-in. holes were drilled in the end of the electron gun, and the "sandwich" was mounted using two 3/16-in. cylindrical spacers as shown in Fig. A38–61(b). The gun, with the mounted foil, was placed in the tube and the tube was evacuated and sealed by the manufacturer.

The schematic diagram of the power supply and cathode ray tube is shown in Fig. A38–62. The power supply provides up to 14 kV to the high voltage input on the envelope of the tube and 6 V to the cathode heater. All other terminals of the tube are grounded. The applied high voltage is measured by a 100-μA ammeter in series with a 140-MΩ

(4) Polyvinyl chloride–acetate copolymer–is available commercially in powdered form under the name VYNS. A discussion of this resin and the procedure for making suitable backing films is to be found in a paper by B. D. Pate and L. Yaffee, *Canadian Journal of Chemistry,* 33, 15 (1955).

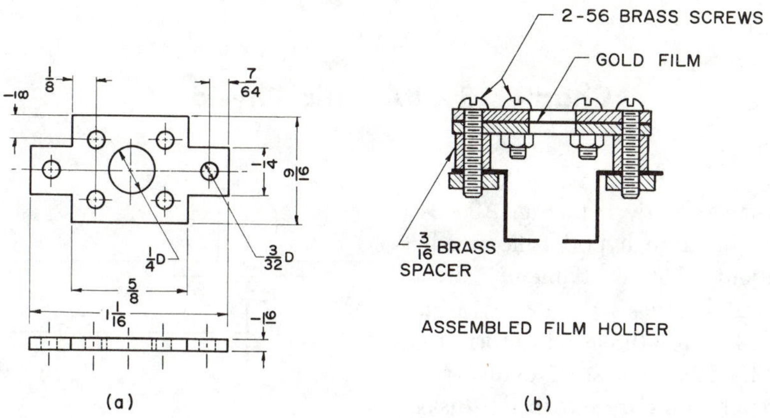

Fig. A38–61

resistor. The meter is calibrated for 14 kV at maximum deflection.

Two external magnets, both of which are common components of commercial television receivers, are used to improve the electron beam. The first is a single "ion trap"—a small permanent magnet which clamps around the thin neck of the tube near the cathode to suppress the flow of ions released from the cathode. The second external magnet is a beam-position magnet. This magnet is a yoke which fits around the neck of the tube and is used here in lieu of the electrostatic focusing electrode of the gun to focus the beam. It has been found that the sharpest pattern appears on the screen when the plane of the focusing yoke makes an angle of about 60° with the axis of the tube.

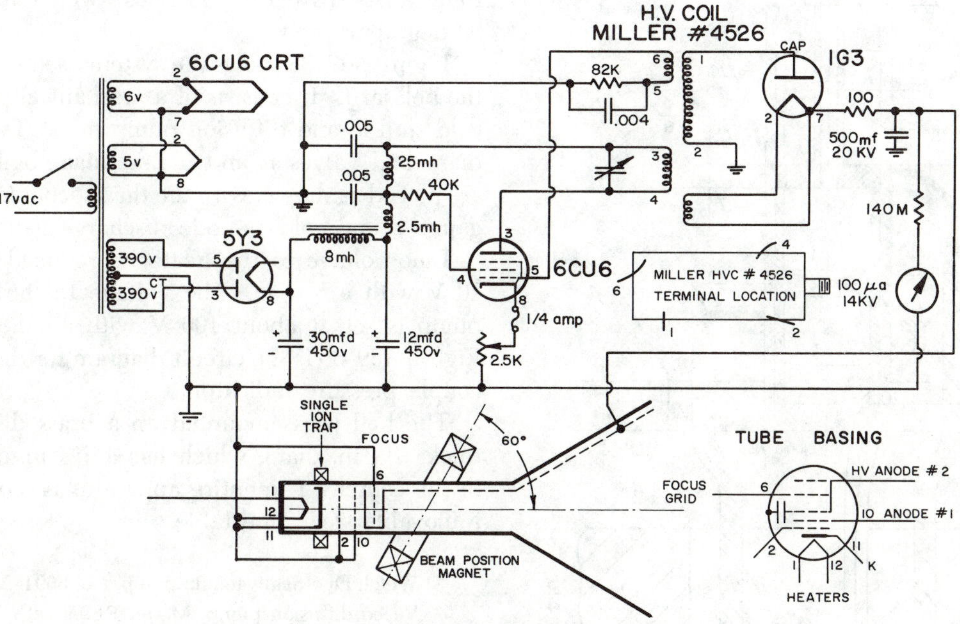

Fig. A38–62

Chapter 39 Atomic Physics

A39–4.4

Details of the furnace are given in Fig. A39–34. Tube *a* is made from a large hypodermic needle with a center hole about 0.1 in. in diameter. Brass tubing, 1/8 in. o.d., can also be used. Two 7/16-in.-diam stainless steel disks *b* are threaded to form the top and bottom of the stainless steel cylinder *c*. Suitable spanner wrench holes are cut in the disks. Stainless steel is used because of its resistance to corrosion. The inner cylinder is machined to the shape in the diagram, with a 0.0995-in.-diam (No. 39 drill) collimating hole cut in the center.

Grooves are cut in the Lavite ceramic cylinder *d* so that between 8 and 10 turns of 0.012-in.-diam tungsten wire *e* can be wound around it. The furnace fits into the ceramic base *f*. The tungsten wires are connected to 1/8-in.-diam copper wires *g*.

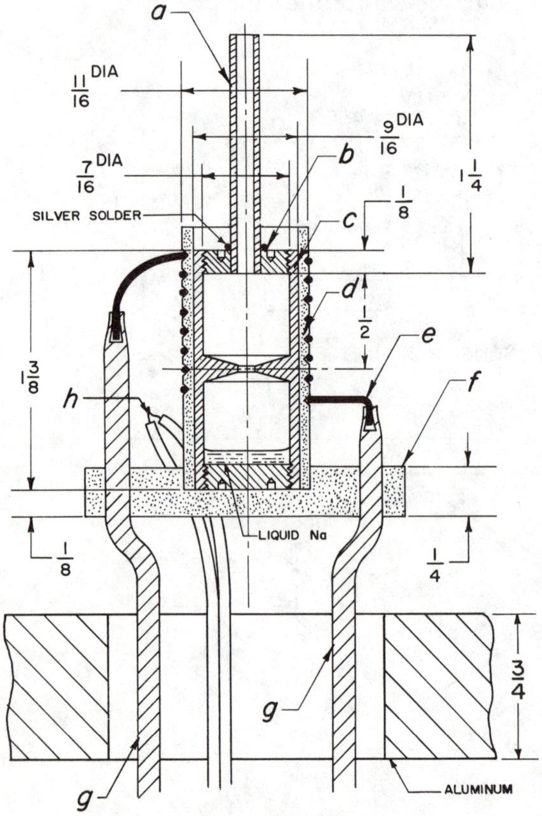

Fig. A39–34

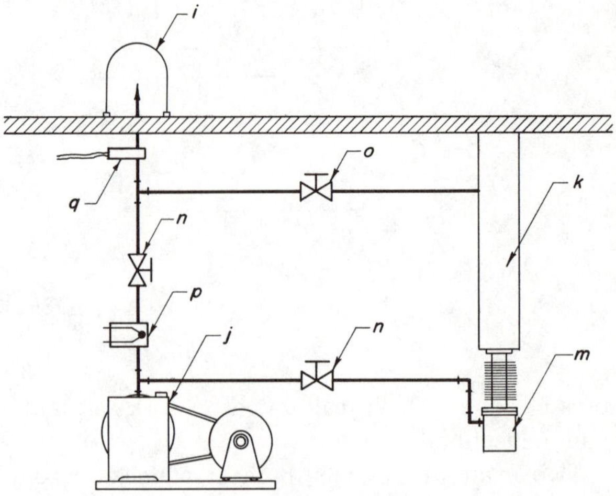

Fig. A39–35

This is done by splitting the copper and then squeezing it around the tungsten wire. The two copper wires and the two terminals of the chromel-alumel thermocouple *h* leave the vacuum system through four glass seals (Stubikoff seals) which are located 90 deg apart.

Figure A39–35 shows the system used to evacuate the bell jar *i*. It consists of a mechanical pump *j*[(1)], cold trap *k*, and diffusion pump *m*[(2)]. Two 3/4-in.-diam ball valves *n* and a 2-in.-diam ball valve *o* are placed as shown, with the thermocouple pressure gauge *p* and cold cathode discharge gauge *q*.[(3)]

Line voltage to the heater is reduced to about 10 V with a Variac. The voltage to the diffusion pump is set at about 100 V with another Variac. Figure A39–36 is the circuit diagram for the thermocouple pressure indicator.

The bell jar is mounted on a brass disk, 3/4 in. thick, 9 1/4 in. diam, which has a 7 1/4-in.-diam hole in the center. The entire apparatus is mounted on a movable steel pipe table.

(1) Welch Duo-Seal vacuum pump, No. 6691–2.

(2) Veeco diffusion pump, Model EP2A with blower and cold trap, Vacuum Electronics Mfg. Co., New Hyde Park, New York.

(3) Veeco cold cathode discharge gauge DG2; pressure gauge, part No. 543, Ser. No. DG2-1677, Vacuum Electronics Corp., Plainview, New York.

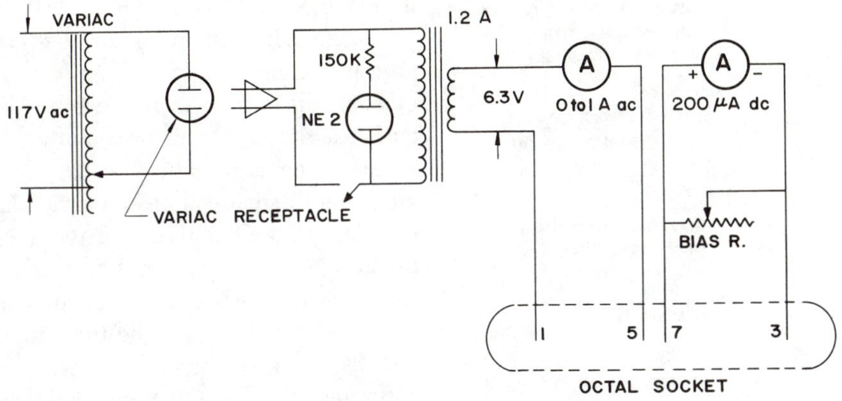

Fig. A39–36

Chapter 40 Solid State Physics

A40–1.16

Because of the enlarged image obtained when shadow-projecting the circuit, it is necessary to make all components small, especially those parts of the circuit not playing an important role. Components useful for other demonstrations are not mounted directly on the plastic base, but are put on

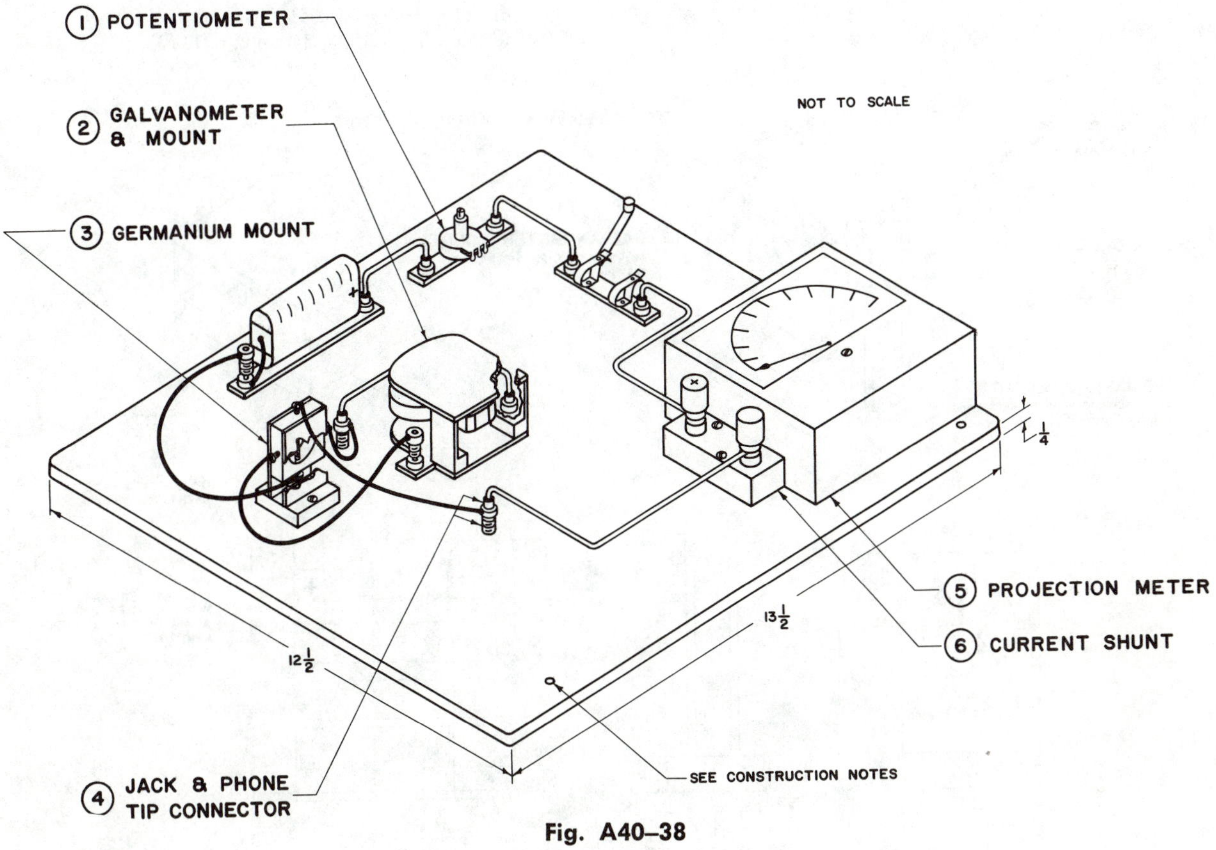

Fig. A40–38

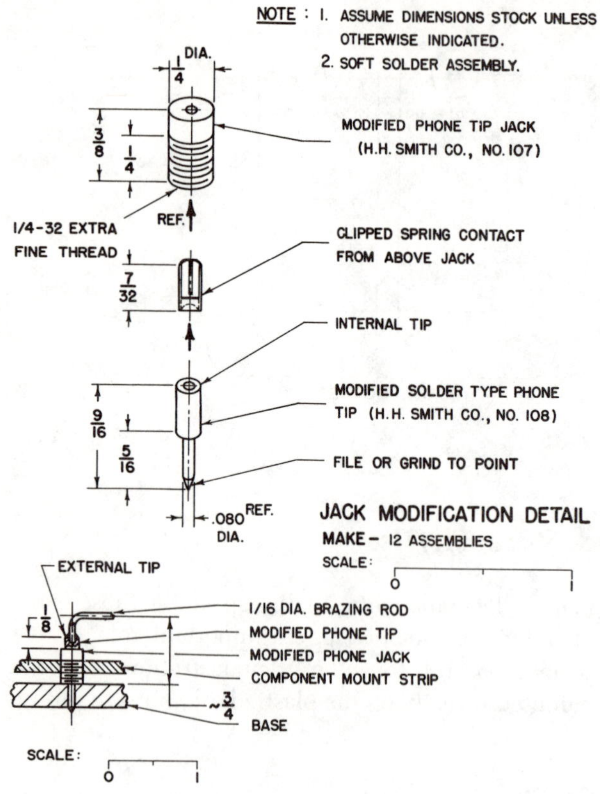

Fig. A40–39

interchangeable strips (Fig. A40–38). The components, such as the miniature battery, knife switch, plug-in connectors, potentiometer, and galvanometer, are also used in many other breadboard circuits for the overhead projector.[(1)]

Component mounting strips vary in size. They are generally made of clear acrylic plastic, 1/8 x 1/2 in. in cross section, and they have a 1/4-32 extra-fine threaded hole in each end to accept the connector jacks. Strips with attached components are held on the base by jack tips. The tips are an easy push-fit into holes drilled into the base to allow insertion and removal. The recommended drill size is No. 45 (.082 in.).

Commercially available jacks were found to be too large for projection purposes and were modified. For the plug-in electrical connectors used in this and many other demonstrations on the overhead projector, phone-tip jack No. 107 and solder-type phone tip No. 108, manufactured by the H. H. Smith Company,[(2)] were obtained (Fig. A40–39).

(1) W. Eppenstein, *The Overhead Projector in the Physics Lecture,* Rensselaer Polytechnic Institute, Troy, New York, 1961.

(2) *Radio Electronic Master,* United Publishers, Inc., Hempstead, New York, 1960, 24th ed., p. 1179.

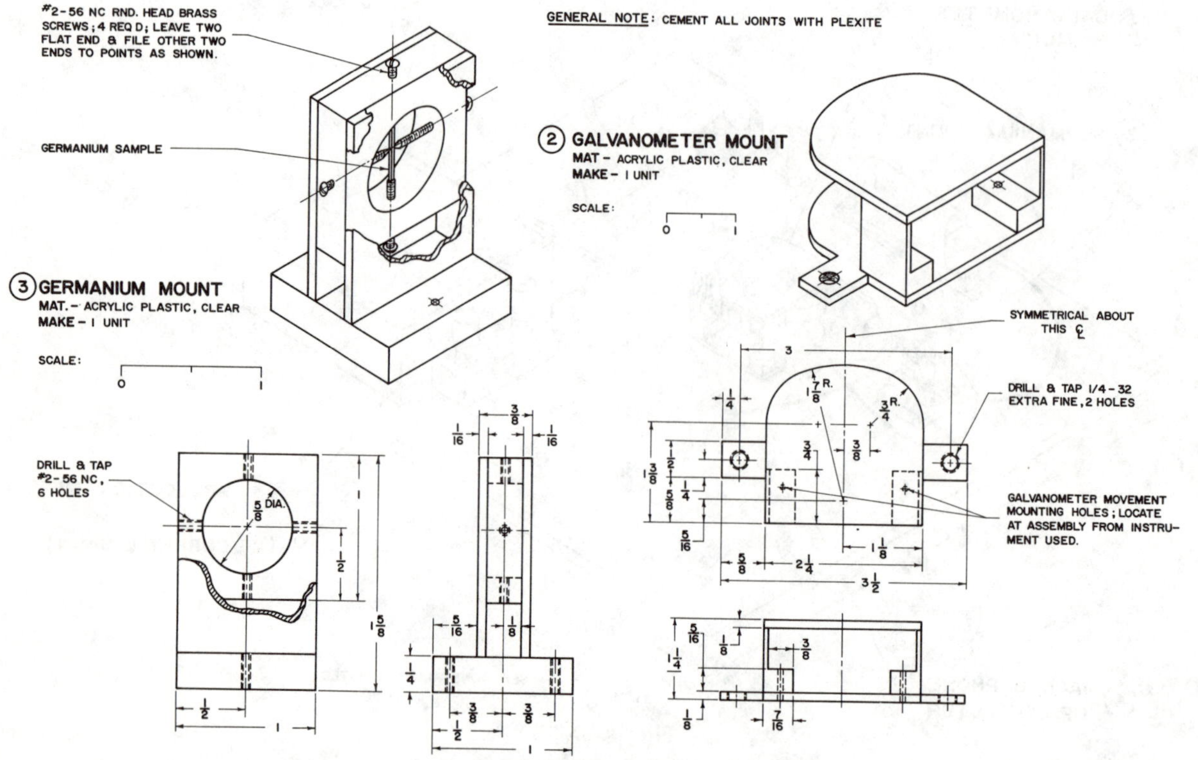

Fig. A40–40

Materials List

Fig. A40–38

Detail	Quantity	Part and Description (dimensions in inches)	Remarks
1	1	Miniature potentiometer; 2500 Ω, 1 W	Commercially available
2		Galvanometer mount assembly, all parts clear acrylic plastic:	Material commercially available under trade names Lucite and Plexiglas
	1	2³⁄₁₆ x 2¼ x ⅛	
	1	2³⁄₁₆ x 3½ x ⅛	
	2	¾ x 1 x ⁷⁄₁₆	
	1	Galvanometer movement, permanent magnet type, sensitivity 2–5 μA per division	Weston model No. 440, No. 699, or equivalent
3		Germanium mount assembly, all parts clear acrylic plastic:	See Remarks, detail 2
	2	1 x 1 x ¼	
	2	1 x 1⅜ x ¹⁄₁₆	
	1	Germanium sample, ¹⁄₃₂ x ⁷⁄₁₆, P-type	Available on a sample card holding two pieces N-type and one piece P-type from Semimetals Inc., Brooklyn, N.Y.
4	12	Modified phone tip jack	H. H. Smith Co., No. 107
	20	Modified solder type phone tip	H. H. Smith Co., No. 108
5	1	Projection meter, ac-dc; full scale 1 mA or 100 V dc	Cenco No. 82550 or equivalent, Central Scientific Co.
6	1	Current shunt for meter above, 100 mA range	Cenco No. 82554 or equivalent
	1	Base, clear acrylic plastic, 12½ x 13½ x ¼	See Remarks, detail 2
		Component mounting strips, clear acrylic plastic	See Remarks, detail 2
	2	For knife switch and potentiometer, ½ x 2 x ⅛	
	1	For battery, ½ x 3½ x ⅛	
	1	Modified knife switch, SPST	Birnbach No. 6100 or equivalent
	1	Battery, 30 V miniature B-type	RCA No. VS085 or equivalent
	1	Magnetron magnet, nominal minimum field strength, 1750 G	Cinandagraph No. 20-1825, type 15A185 or equivalent, Cinandagraph Corp., Stamford, Conn. Commonly available as a surplus item
	4	Pole pieces, soft iron, 1½ in. diam, ¼ in. thick	
	12 in.	Connecting wire, brazing rod, ¹⁄₁₆ in. diam	Commercially available

The hexagonal head on the jack is lathe-turned down to a ¼ in. diam and its body cut into two parts with a parting tool on a lathe. The spring contact is then removed, and its solder-end is clipped and sanded smooth. The phone tip is cut to size and the parts are solder-assembled. The heavy connecting wire used is 1/16-in.-diam brazing rod, bent to shape and soldered to the phone tips.

Base specifications are given under Detail 6 in the Materials List. When the base is used on a Charles Beseler No. 6600 Master Vu-graph, two clearance holes for ⅛-in.-diam pins on the top stage of the overhead projector are drilled 10⅝ in. apart (Fig. A40–38), in line, and 9/16 in. from the side of the base.

Construction details for the germanium and projection galvanometer mounts are given in Fig. A40–40 and the Materials List. The galvanometer is a commercial movement removed from its case and assembled within a clear plastic mount. Other components in Fig. A40–38 are commercial items with sources given in the Materials List. The knife switch was modified to open and close in a plane parallel to the base.

Chapter 41 Nuclear Physics

A41–1.9

The holder for the G-M tube and source is designed to allow easy construction, a simple demonstration setup, and quick source changes. As shown in Fig. A41–68, the holder consists of two clear acrylic plastic blocks, one bored through to receive the G-M tube and one bored to accept 1-in.-diam source rings. The blocks, separated by a Pb block that shields the G-M tube from direct radiation, are fastened together by a brass plate over the end containing the source and the end-window of the tube. The entrance and exit slits are cut in this brass plate, and a plastic defining baffle, which defines the trajectories of the beta particles that are accepted by the detector, is mounted on its outer side. Two brass strips screwed to the opposite end complete the assembly.

The knife switches indicated in Fig. 41–13 are mounted on clear acrylic plastic blocks. The switch is cross-wired to provide for polarity reversing in applications such as the magnet current switching.

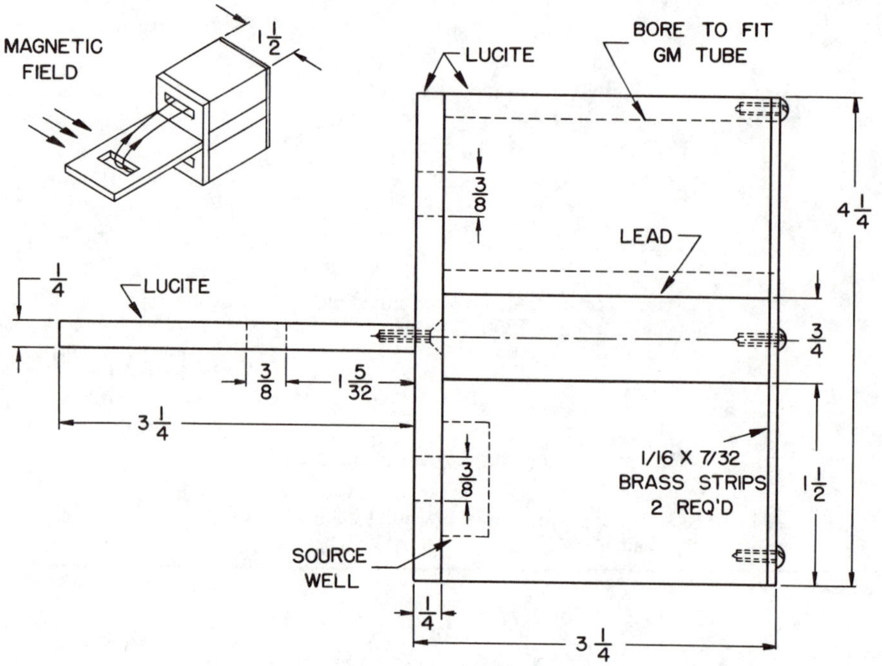

Fig. A41–68

Materials List

Figs. A41–68; 41–13

Quantity	Part and Description (dimensions in inches)	Remarks
1 assembly	Mount for G-M tube and source	
1	Block, clear acrylic plastic, 3 x 2⅛ x 2⅛	Material commercially available under trade names Lucite and Plexiglas
1	Block, clear acrylic plastic, 3 x 2⅛ x 1½	
1	Lead block, 3 x 2⅛ x 11/16	
1	Brass plate, 4 3/16 x 2⅛ x ¼	
2	Brass strip, 4 3/16 x ¼ x 1/16	
1	Clear acrylic plastic, 1 x 3¼ x ¼	
1	Knife switch, DPDT, on special base	
As needed	Source rings, aluminum, 1 in. diam x ¼ in., ⅛ in. wall	
As needed	Mylar, ¼ mil	
1 assembly	Magnet	Atomic Laboratories, Inc., model 79641, or equivalent
1	Geiger-Müller tube, end window	Tracerlab TGG-2 or equivalent, Tracerlab, Waltham, Mass.
1	Switch, quick-opening	Leybold type 504-45 or equivalent, available from Klinger Scientific Apparatus Corp., Jamaica, N.Y.
1	Galvanometer, projection type	New England Scientific Instrument Corp., Cambridge, Mass.
1	Shunt box for galvanometer above	New England Scientific Instrument Corp.
1	Potentiometer, 1000 Ω, 1 watt	Commercially available
1	Rheostat, 440 Ω	Welch Scientific Co., Skokie, Ill.
1	Rheostat, 220 Ω	Welch Scientific Co.
1	Rheostat, 85 Ω	Welch Scientific Co.
1	Rheostat, 20 Ω	Welch Scientific Co.
1	Scaler with self-contained GM tube high-voltage supply	Baird-Atomic "Ten hundred" series or equivalent, Baird-Atomic, Inc., Cambridge, Mass.
1	Digital in-line readout unit, six digit	Computer Measurements model 200B frequency counter plus model 401A in-line readout, Computer Measurements Co. Inc., San Fernando, Calif.

The magnet used for the demonstration can be any iron-core electromagnet of suitable configuration and capable of producing an approximately uniform field of the order of 2000 gauss throughout the region of the electron trajectories. One can be constructed or obtained commercially. For example, an Atomic Laboratories, Inc., model 79641 would be suitable.(1)

The source is prepared by depositing the source

(1) This unit comes with 3 x 5-in flat pole faces; however, it is rated at 50 V dc. If 50-V dc is not normally available, an accessory power supply is needed.

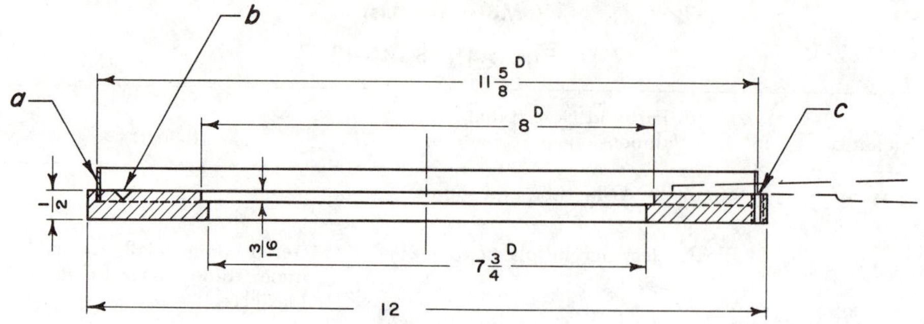

Fig. A41–69

material on Mylar[2] film bonded to an aluminum ring 1 in. in diameter. The Mylar is bonded to the ring in the following way: A small sheet of Mylar is stretched on a piece of scrap plywood and held with thumbtacks. A thin ribbon of Duco cement is laid on the surface of the ring, which is then inverted on the Mylar film and weighted down. After drying for two to three hours, the source ring is cut free and is then ready to receive the source material.

A41–2.1

Figures A41–69 and A41–70 are diagrams of the frame and the $1/r$ surface respectively. A guard *a*, made from a $\frac{1}{32}$-in.-thick brass sheet, $\frac{1}{2}$ x $35\frac{1}{2}$ in., surrounds the frame except where the chute enters. It is held tightly in a $\frac{1}{16}$-in.-wide x $\frac{1}{8}$-in.-deep groove by four shims. A $\frac{5}{16}$-in.-wide, 90-deg V-groove *b* goes 30 deg each side of the center line. A hole *c* is tapped for a No. 6 screw at a radius of $5\frac{13}{16}$ in. from the center to hold the chute.

The bottom of the well of the $\frac{1}{4}$-in. surface is covered with aluminum foil. Above this is a $\frac{1}{16}$-in.-thick brass disk *d* and another brass disk *e*, $\frac{1}{4}$ in. thick.

(2) Manufactured by E. I. duPont de Nemours & Co.; available commercially.

Details of the chute are shown in Fig. A41–71.

A41–2.6

Figure A41–72 shows the detail for the container for the BF_3 counter tube and the preamplifier. The container consists of a tubular brass casing *b*, $2\frac{1}{4}$ in. o.d. x $2\frac{1}{8}$ in. i.d. x $25\frac{1}{4}$ in. long, closed at the bottom end by a brass disk *g*, $\frac{5}{16}$ in. thick machined with a $\frac{1}{16}$-in. shoulder and screwed into the tube. The joint is soldered to make a watertight seal. The cover plate *a*, shown in plan and section, is machined like the bottom plate and fitted with three holes for connectors as shown in the plan dimensions. A Lucite disk *c*, $\frac{1}{4}$ in. thick and 2 in. o.d., fits snugly inside the tube as shown. Four brass support strips *d* fit tightly over the purchased counter tube *f*.[3] They are attached by flathead screws to the Lucite disk *c* and to the brass ring *e*, 2 in. o.d. x $1\frac{3}{4}$ in. i.d. x $\frac{1}{2}$ in. long. The preamplifier is contained in the space above the Lucite disk *c*. Two holes in the cover receive Amphenol 1-pole coaxial connectors *h*. The third hole is fitted with a 5-pole coaxial connector *j* (female) which mates with No. AN-3106-16S-8S (male). The preamplifier circuit is shown in Fig. A41–73.

(3) Radiation Counter Laboratories, Inc., Skokie, Illinois; Mark 1, Serial No. 136.

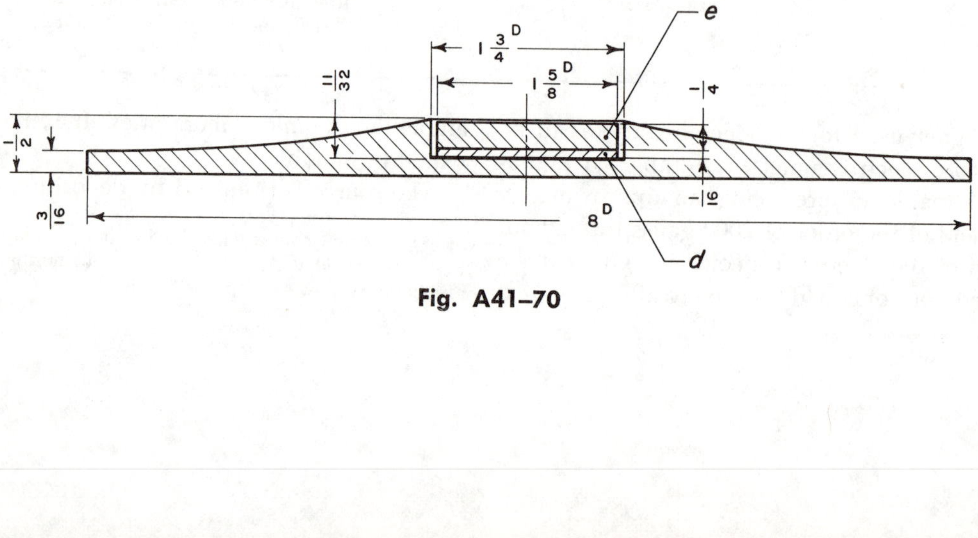

Fig. A41–70

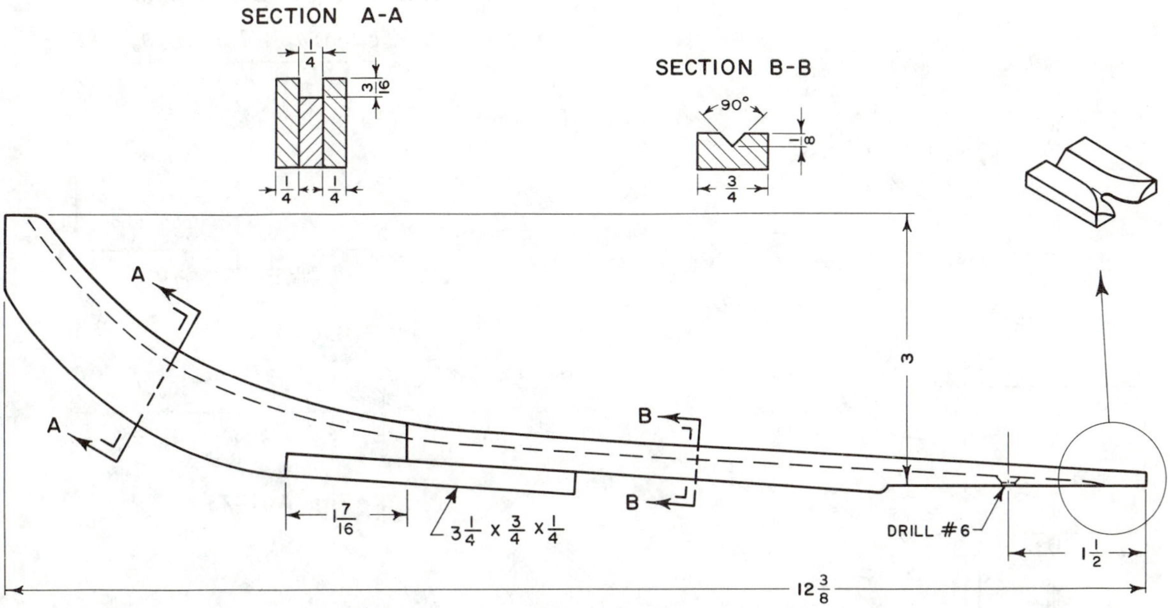

Fig. A41–71

A41–2.7

The gun is constructed from a $\frac{9}{16}$-in.-o.d. brass tube with 0.050-in. walls measuring $7\frac{1}{2}$ in. long. The firing mechanism, Fig. A41–74(a) and (b), consists of a spring inserted in the barrel and soldered in place at one end. The other end is fitted with a lightweight plunger shaped to contact the $\frac{1}{2}$-in. steel projectile at two points and grooved to receive the cocking lever. A small stud is mounted on the rear of the plunger to accept a rod used to compress the spring in cocking.

The barrel is suspended from an overhead frame by four threads 20 in. long, the threads being fastened to wire loops soldered to the barrel, two at each end, as can be seen in the photographs. These threads, which constitute the entire support in the unbound configuration, make angles of about 30 deg with the vertical. In the recoilless (i.e., bound) configuration the barrel is rigidly clamped near both the front and rear to a base, which is in turn fastened to a heavy table.

Firing is accomplished by pulling the cocking lever with a solenoid. The solenoid plunger is connected to the cocking lever with a lightweight metal chain of the sort used on lamp pulls. A wire loop from the chain to the lever falls away on firing so as to present not the slightest interference to recoil.

Range measurements are automatically recorded on a strip of spark tape mounted under a plastic sheet (.040 in. thick) scribed with centimeter range markings as shown in Fig. 41–24. This paper tape is pressure sensitive, and the impact of a $\frac{1}{2}$-in. steel projectile leaves a sharp spot about 1 mm in diameter. Behind the range recorder is a catcher fitted with a weighted curtain to slow down the projectiles and prevent rebound onto the recorder. The balls then roll into a tray behind the catcher.

A41–2.9

A typical range-momentum graph for a 45% CF_3Br - 55% C_3H_8 by volume bubble chamber is shown in Fig. A41–75. The units of momentum are given as MeV/c.

The dotted line in Fig. A41–76 represents the path of the $\Lambda°$. With an assumed velocity v_Λ, the energy of $\Lambda°$ is

$$E_\Lambda = \frac{(m_0)_\Lambda c^2}{\sqrt{1 - v_\Lambda{}^2/c^2}}. \qquad (1)$$

If the energies of the proton and pion are E_p and E_π respectively, then by the conservation of energy,

$$E_\Lambda = E_p + E_\pi. \qquad (2)$$

If θ_π and θ_p are the initial angles at which the pion

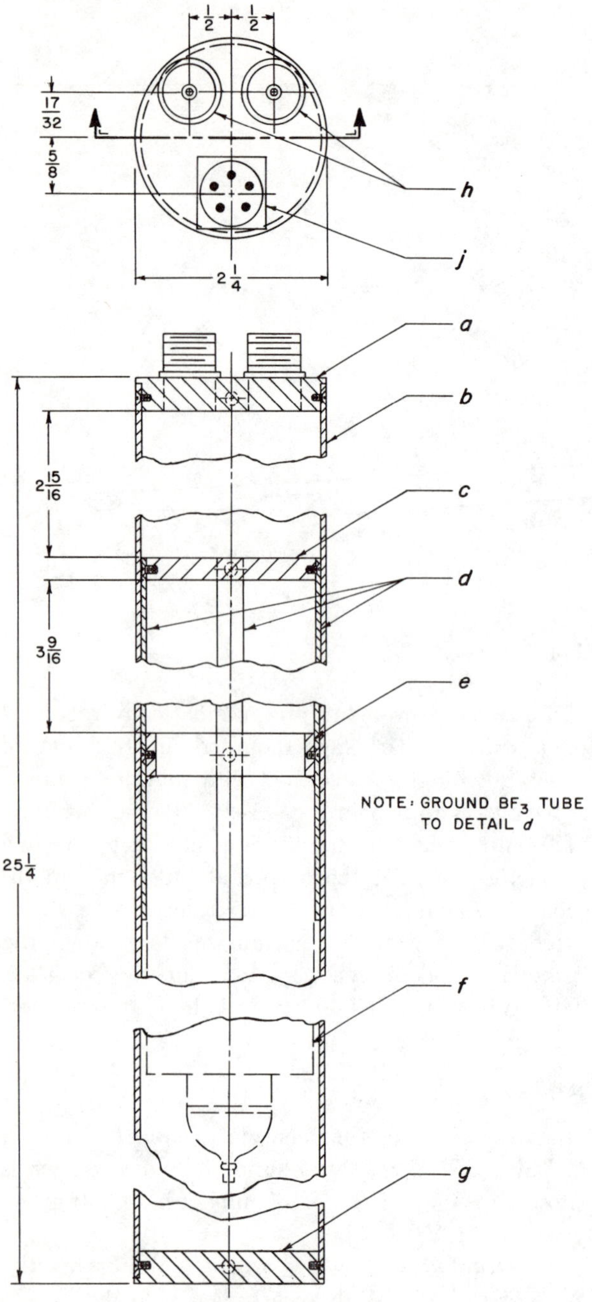

Fig. A41–72

and proton are projected, then the conservation of momentum (using $p = mv = Ev/c^2$)

$$\frac{E_\Lambda v_\Lambda}{c^2} = \frac{E_\pi v_\pi}{c^2}\cos\theta_\pi + \frac{E_p v_p}{c^2}\cos\theta_p, \qquad (3)$$

and

$$0_\Lambda = \frac{E_\pi v_\pi}{c^2}\sin\theta_\pi - \frac{E_p v_p \sin\theta_p}{c^2}. \qquad (4)$$

When (3) is divided by (2), we obtain

$$v_\Lambda = \frac{E_\pi v_\pi \cos\theta_\pi + E_p v_p \cos\theta_p}{(E_p + E_\pi)}, \qquad (5)$$

By combining (1) and (2),

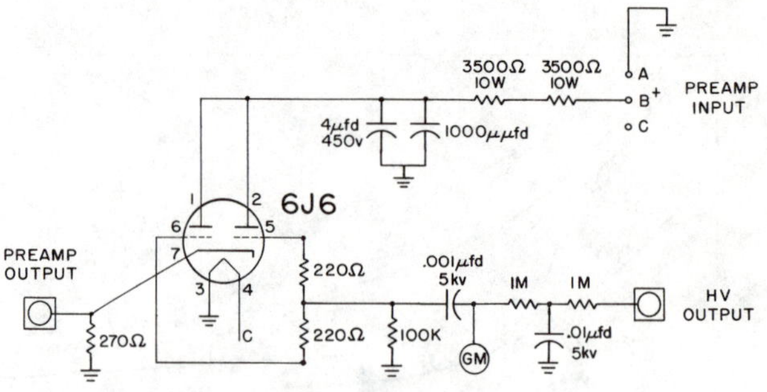

Fig. A41–73

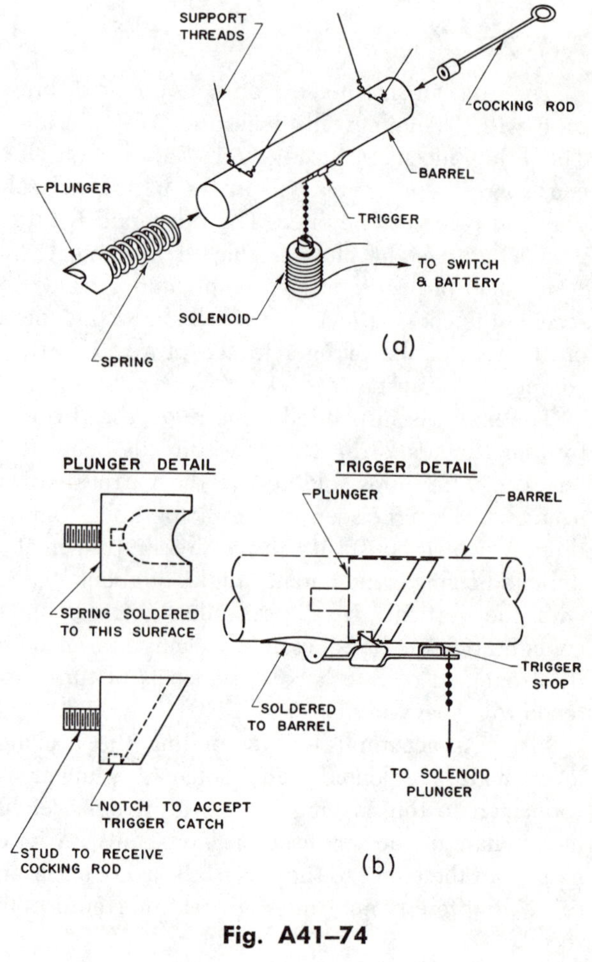

Fig. A41–74

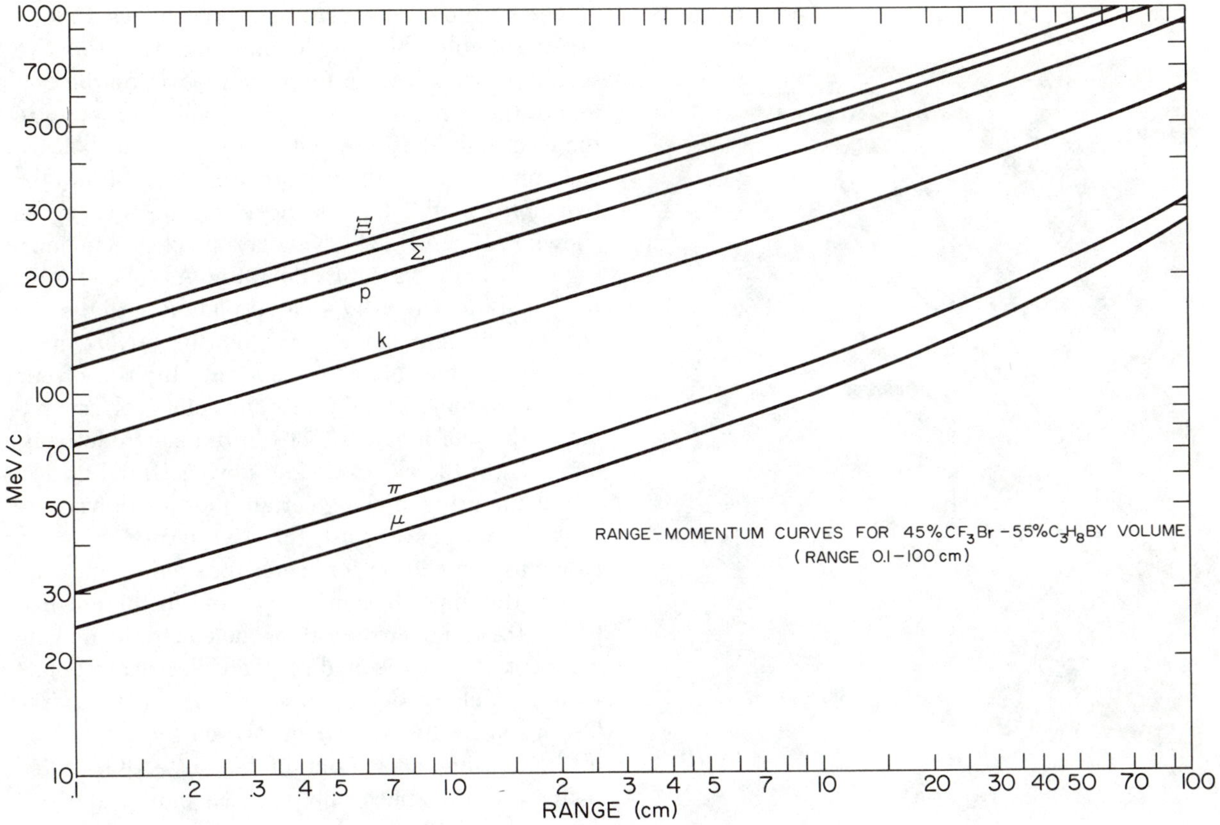

Fig. A41–75

$$(m_0)_\Lambda c^2 = (E_\pi + E_p)\sqrt{1 - v_\Lambda{}^2/c^2}. \qquad (6)$$

All the quantities on the right side of Eq. (5) can be obtained from the photographs. The energies of the proton and pion are calculated from

$$E = \sqrt{p^2c^2 + m_0{}^2c^4}, \qquad (7)$$

and their velocities from

$$E = \frac{m_0c^2}{\sqrt{1 - v^2/c^2}}. \qquad (8)$$

It is now possible to calculate v from Eq. (5), then, determined from Eq. (6), the rest mass of the Λ° hyperon.

A41–2.10

Before constructing or operating the fission chamber, it is advisable to refer to the description of fission chambers given by Rossi and Staub.[4]

[4] B. B. Rossi and H. H. Staub, *Ionization Chambers and Counters* (McGraw-Hill Book Company, Inc., New York).

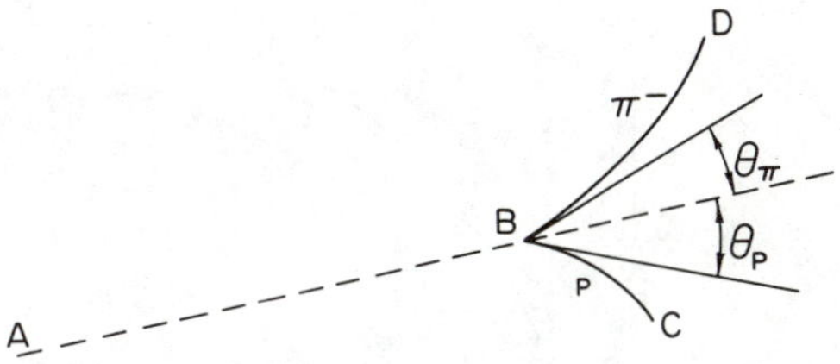

Fig. A41–76

Figure A41–77 shows a closeup of the fission chamber and the discriminator-amplifier. Figures A41–78, A41–79, A41–80, and A41–81 provide construction details of the chamber's various components. Necessary dimensions not given in the drawings may be found in the Materials List. All dimensions in the drawings are in millimeters. Figures A41–82 and A41–83 give the wiring diagrams for the discriminator-amplifier and the loudspeaker-amplifier respectively. All internal surfaces of the chamber should be polished before the chamber is assembled. The UO_2-coated plates are coated on the sides shown in Fig. A41–84.

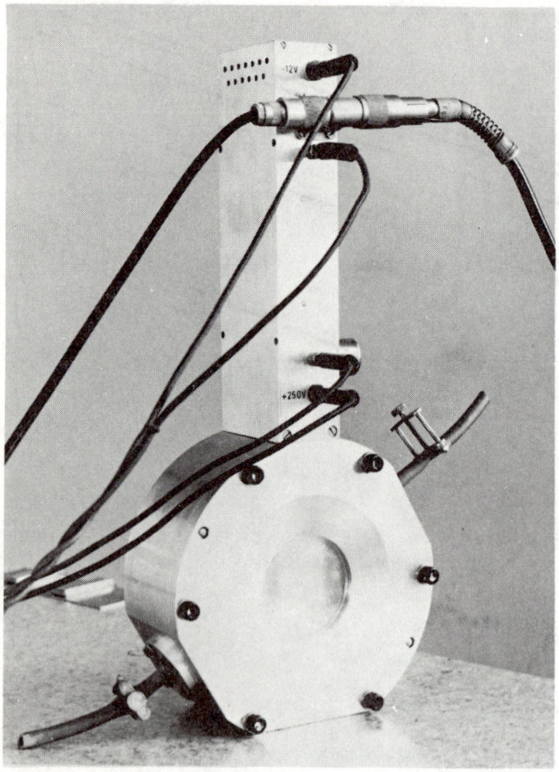

Fig. A41–77

To assemble the chamber, proceed as follows under suitable shielding conditions. Lay the two small O-rings in the appropriate grooves on the bottom of the two gas nipples *j*, and bolt the nipples to the electrode housing *f* with four hex-head bolts in each nipple. Push the two Teflon sleeves *n* into the two holes in the top of the electrode housing as shown in Fig. A41–84. Now lay the electrode housing into what is its right side in Fig. A41–84. Lower the small-diameter UO_2-coated plate *h* onto the rim cut for it in the center of the housing. Next, push three of the Plexiglas dowels *k* onto the rim of one of the electrodes *g* at equidistant points corresponding to the half-moon notches in the electrode housing. Lower the electrode and dowels straight down into the housing, making certain that the dowels are pushed in tight against the UO_2-coated plate *h*. Now push one of the connector rods *p* through what is now the upper Teflon sleeve, and fit the notched end of the rod over the rim of the electrode, making certain that the rod makes good electrical contact with the electrode. Next, lay one of the large O-rings *m* in the circular groove on the upper side of the housing. Place one of the large UO_2-coated plates *i* in the space cut for it in one of the end

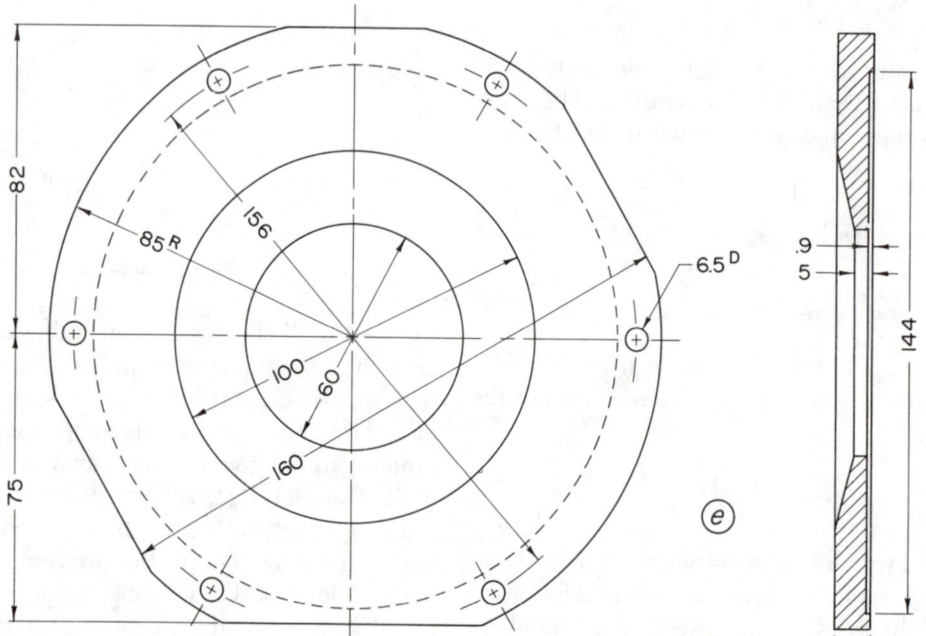

Fig. A41–78

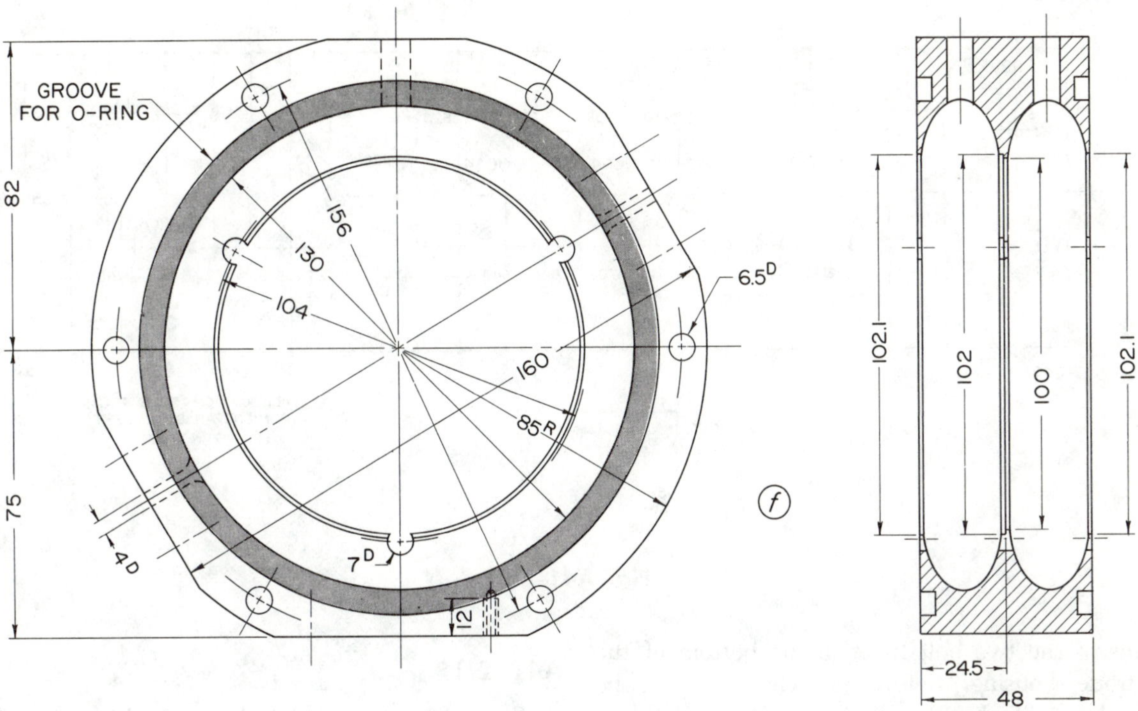

Fig. A41–79

plates *e*. Lower the end plate and the coated plate onto the housing, lining up the bolt holes, and insert six allen-head bolts into the bolt holes. Now turn the partial assembly upside down. Push the remaining dowels *k* onto the rim of the second electrode *g*, and lower the electrode and dowels into the housing, again making certain that the dowels are in as far as they will go. Push the second connector rod *p* through the other Teflon sleeve and fit the notched end of the rod over the rim of the electrode, again making certain that there is good electrical contact. Lay the remaining O-ring *m* in the appropriate groove in the face of the housing and place the last UO_2-coated plate *i* in the other end plate *e*. Lower the end plate and the coated plate onto the housing and screw six nuts onto the six bolts. Tighten all bolts with appropriate wrenches. Now connect lead wires from the discriminator-amplifier to the connector rods and fasten the discriminator-amplifier to the top of the chamber.

To prevent the assembly from tipping over, the chamber may be fastened to the work surface by

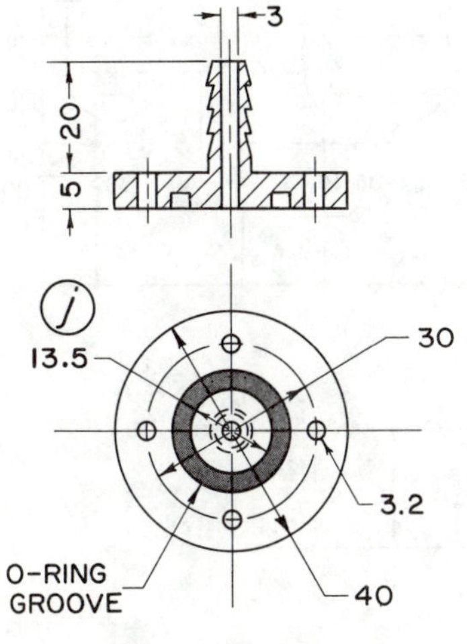

Fig. A41–80

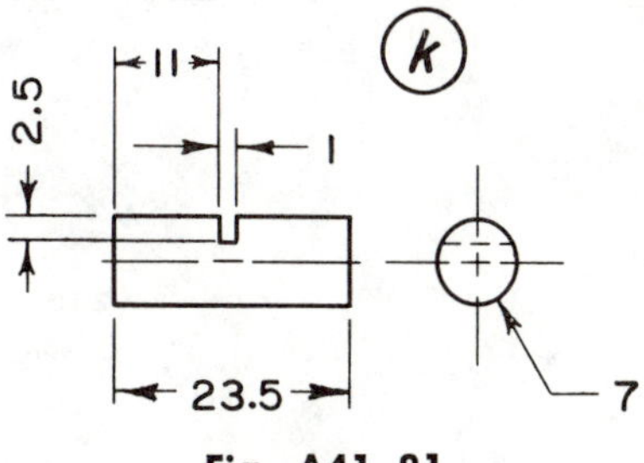

Fig. A41–81

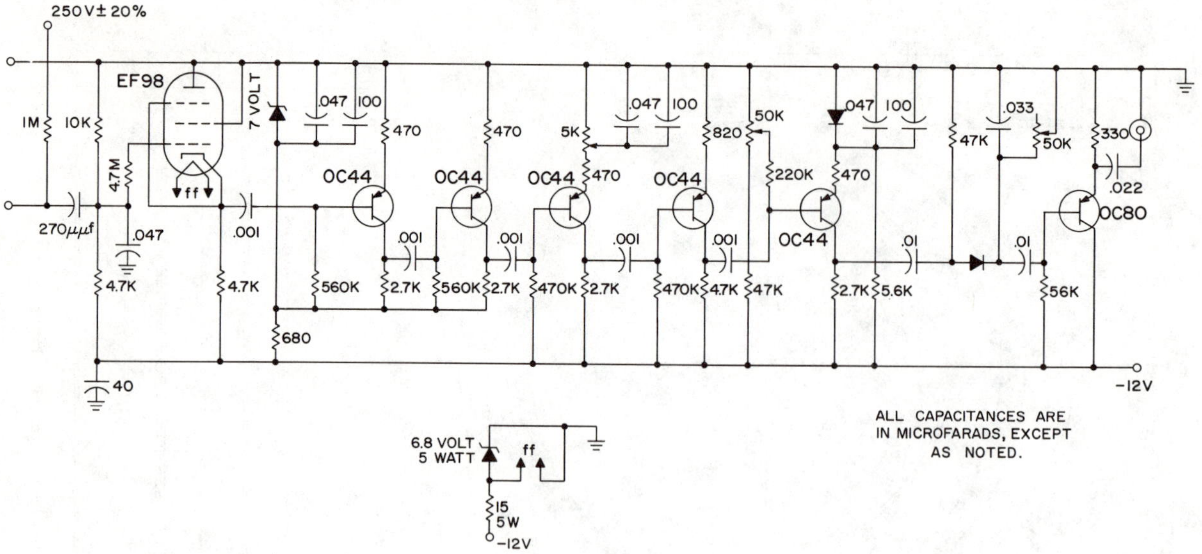

Fig. A41–82

means of the two bolt holes in the bottom of the electrode housing. Before the chamber can be flushed and filled with argon, two short lengths of rubber tubing must be fitted over the gas nipples and a clamp must be placed on each piece of tubing to regulate the flow of argon. The chamber is now ready for operation.

A41–2.12

To operate the model shown in Figs. A41–85, A41–86, and A41–87, the spiral spring is compressed from the outside by means of an auxiliary rod, and then located in position with the catch. When the iron sphere falls through the opening in the wooden

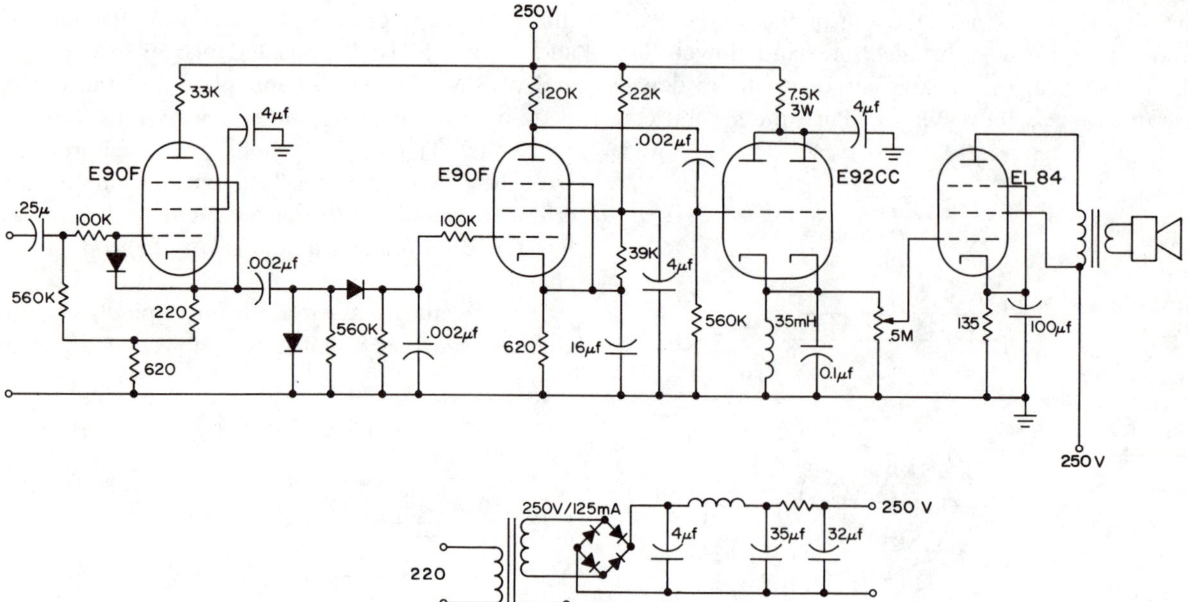

Fig. A41–83

Materials List

Figs. A41–78 to A41–81, A41–84

Detail	Quantity	Description
e	2	End plate, aluminum
f	1	Electrode housing, aluminum
g	2	Electrode, aluminum, 1 mm thick x 102 mm diam
h	1	Plate, UO_2-coated aluminum, 1 mm thick x 102 mm diam
i	2	Plate, UO_2-coated aluminum, 1 mm thick x 143.5 mm diam
j	2	Gas nipple, aluminum
k	6	Dowel, Plexiglas
m	2	O-ring, rubber, 130 mm i.d.
	2	O-ring, rubber, 13.5 mm i.d
n	2	Sleeve, Teflon
p	2	Connector rod, aluminum, 3 mm diam
	16	Assorted hex-head and Allen-head bolts, stainless steel
	6	Assorted nuts, stainless steel

Fig. A41–85

sphere, the clockwork is set in motion by the lever (4.12), which moves from its initial position I in Fig. A41–87 to position III. The cam on the arresting mechanism (3.5) is raised by means of the pin (4.11); the stopping pins (3.6) and (3.7) of the clockwork are loosened, and the clockwork set in motion. The cam (3.1) turns and releases by means of the levers (3.2) and (3.3) and the catches (5.2) and (6.2) on the compressed springs (5.1) and (6.1) one after the other, so that the lighted lamps are ejected from the openings, one following the other. After one revolution of the cam (3.1), the clockwork again is stopped by the pins (3.6) and (3.7). If the iron sphere (2) is removed from its opening (4), spring (4.2) brings back the lever (4.1) to its initial position; simultaneously pin (4.11) slides over cam (3.51) to position II in Fig. A41–87, and level (4.12) slides back over cam (3.4).

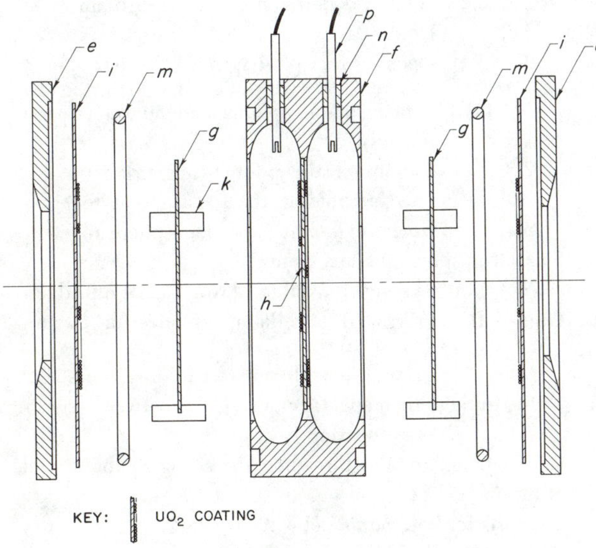

Fig. A41–84

Materials List

Figs. 41–34, 41–35 (p. 1264)

Detail	Component
1	Wooden sphere, 15 cm diam
2	Iron sphere, 2.5 cm diam
3	Clockwork from alarm clock, 5.5 cm diam
3.1	Cam for releasing stop on spring

Fig. A41–86

Detail	Component
3.2	Lever for releasing stop 5.2
3.3	Lever for releasing stop 6.2
3.4	Cam for driving clockwork
3.5	Spring-control for stopping the clockwork
3.51	Cam for releasing the stop
3.52	Twin leaf spring
3.6, 3.7	Stopping pin
3.8	Balance wheel, the teeth (3.82) all removed except for last two
3.81	Vane for decelerating the clockwork run
3.9	Hour wheel of the clockwork with disk soldered on for stopping
3.91	Soldered-on disk with slit
4	Entrance opening for iron sphere
4.1	Starting mechanism for clockwork
4.11	Spring-driven pin for releasing the stop
4.12	Lever (leafspring)
5.6	Exit openings for lamps
5.1, 6.1	Spiral springs
5.2, 6.2	Stops
7	Opening for winding the clockwork
8	Small lamps housing
8.1	Lamps
8.2	Battery (1.5 V) with socket soldered on

A41–2.13

Materials List

Figs. 41–36 to 41–40 (p. 1264)

Detail		Component
1		Wooden sphere made from two hemispheres, 15 cm diam
2, 3		Hemispheres with inner mechanisms
	2.1, 3.1	Opening with spiral spring, 3.2 cm diam
	2.2, 3.2	Small wooden spheres, 2.5 cm diam
	2.3, 3.3	Holding nails
	2.4, 3.4	Opening for spiral spring (No. 6), 3.2 cm diam. To avoid damage to string by razor blade, openings contain cross pins.
	2.5	Razor blade
	2.6	Aluminum stirrup for cutting string
	3.6	Pins for holding the hemispheres together
	2.7	Holes for holding the hemisphere together
4		Small iron sphere, 2.5 cm diam
5		Hole for introducing iron sphere (No. 4), 29 mm diam
6		One of the springs, 31 mm diam, for releasing the reaction. 2.5 mm steel wire
	6.1	Brass disk with slot for holding the string
7		String for holding model together (11 cm between knots)
	7.1	Knot in string
8		End notched rod for threading the string in the model
9		Mounting cylinder
10		Supports for holding model while cocking
11		Pressure handle for compressing springs (No. 6)

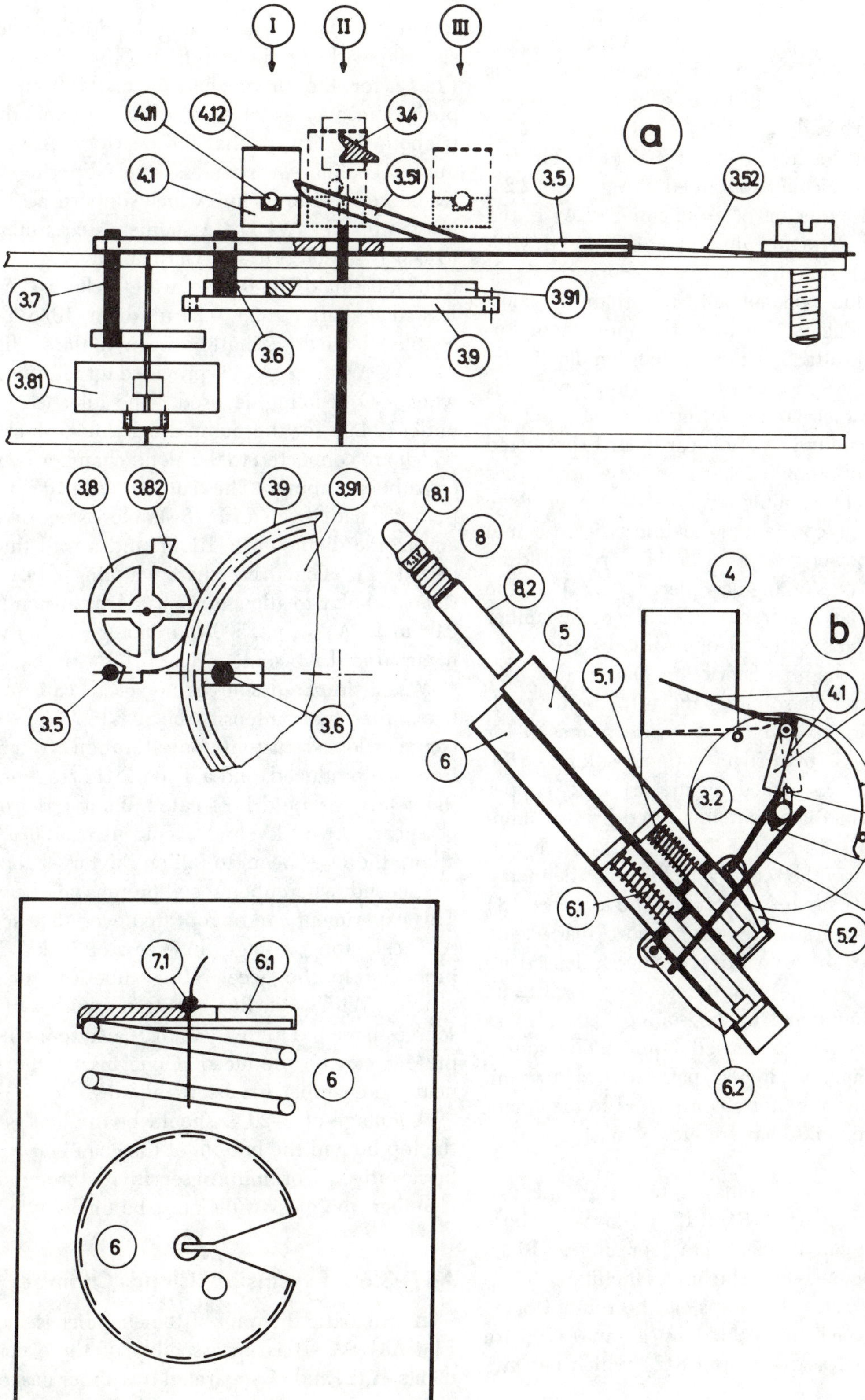

I
II
III
4.11
4.12
3.4
a
4.1
3.51
3.5
3.52
3.7
3.91
3.6
3.9
3.81
3.8
3.82
3.9
3.91
8.1
8
8.2
4
5
b
5.1
4.1
4.12
3.5
3.6
6
4.2
3.2
3.4
3.1
6.1
5.2
3.3
6.2
7.1
6.1
6
6

Fig. A41–87

A41–3.5

The wall (1) of the chamber (Fig. A41–88, is made from plane sheets of Pyrex cemented together with epoxy. Threads (2.1) hold a felt pad (2.3), Fig. A41–89, against the chamber roof (2). The felt is dampened by alcohol contained in a gutter (2.2), which is filled by means of an opening (2.4) in the chamber roof. The opening may be closed with screws or Durexband. Another opening (2.5) is used to introduce alcohol onto the chamber floor; the same opening is also used to mount a motor. A frame (3.1), attached to the chamber floor (3), supports the Pyrex wall (1). The frame, which is made from a thermal insulating material such as Formica (resin-impregnated cardboard protected with an alcohol-proof paint can also be used), has a groove (3.11) running around its upper edge which guides a constantan heating wire (1 mm diam). The power generated in the wire should be of about 7.5 W. Triangular pieces (3.12) at the corners strenthen the frame. The copper chamber floor (3.2) contains the alcohol. Plates (3.3) are fastened to the chamber floor by means of pins (3.4) and soft solder. These plates make for better cooling in the corners and also provide a means for fastening four German silver support columns (7).

The chamber is cooled by means of 12 copper coils (4) (0.4 in. diam) which are fed by two stainless steel tubes (4.1) (18/8 alloy, 12 in. long). These tubes carry the refrigerant through the metal-covered (6) Styrofoam base (5). The α-source (8) is carried by a variable length support, made from 2-mm Ni wire. In order to prevent condensed alcohol from draining onto the source, a 0.012-in. wire (8.1) is attached to the source carrier so as to direct away the condensed liquid. The source is carried around a circular path with a 100-mm radius by means of motor (9) attached to the chamber roof. The rotational frequency of the source is about 4 rpm.

The chamber is illuminated by means of two automobile fog lamps (10) (12 V, 45 W) whose supports (10.1) are connected to door hinges (10.2), as shown in Fig. A41–90. The hinges facilitate changing the bulbs, since the lamps can be rotated back. The lamps, which are cooled by a fan (11), are directed through a diaphragm (12) which narrows the beam.

The main part of the pump (14), Fig. A41–91, which circulates the cooling liquid (alcohol), is situated within a Dewar flask (13). The bearing (14.1) for the drive shaft is made from Delrin pressed in a brass case. The play around the axis is about 0.003 in. so that the bearing will not bind during cooling. A stainless steel slip ring (14.2) is soldered to a brass hub which supports a 12-vaned brass propeller (14.3). A stainless steel stuffing box (14.4) is soldered to the German silver drive shaft (14.5), 0.3 in. diam, 0.02 in. wall thickness. A stainless steel shaft (14.6), 0.28 in. diam, 18/8 alloy, is coupled to the drive shaft by means of the stuffing box (14.4). A sieve (14.7) is provided for filtering liquid when CO_2 cooling is used. The inlet (14.8) and outlet (14.9) for the pump are stainless steel tubes, which are connected to the cloud chamber by means of rubber tubing. The support (14.10) for the pump is made from an 18/8 stainless steel tube with an outside diameter of 1.1 in. and a wall thickness of 0.04 in. The drive shaft coupling (14.11) has some end play to allow for thermal expansion (about 0.08 in.). A ¼-hp, 2800-rpm motor (14.12) is used to turn the drive shaft.

When prepared sources are used, care must be taken that their intensity be not too high, as the chamber loses sensitivity on saturation. When neutrons are produced, and a $T(d,n)^4He$ reaction used, the T target should be located about 1 ft from the chamber. At 30 kV and a current of 100 μA, one allows the D_2+ beam to fall on the target for about one second, whereupon recoil protons will be visible. This experiment can be repeated every five seconds. At accelerating voltages in excess of 50 kV, X rays produced in the accelerating tube can be made visible. With (Pu-Be) neutron sources (Ra-Be sources have too high a gamma background), recoil protons can be produced if one inserts the source near the chamber for a few minutes.

A voltage of 5–20 C should be applied between the top lid and the bottom of the chamber, with top lid negative. For uniform sensitivity throughout the chamber, the glass walls must be uncharged.

A41–3.6 Expansion Cloud Chamber[5]

A schematic diagram of the chamber is shown in Fig. A41–92. It is an assembly of three compartments, A, B, and C, separated by rubber gaskets and

(5) F. E. Christensen, *Am. J. Phys,* **18,** 149 (1950).

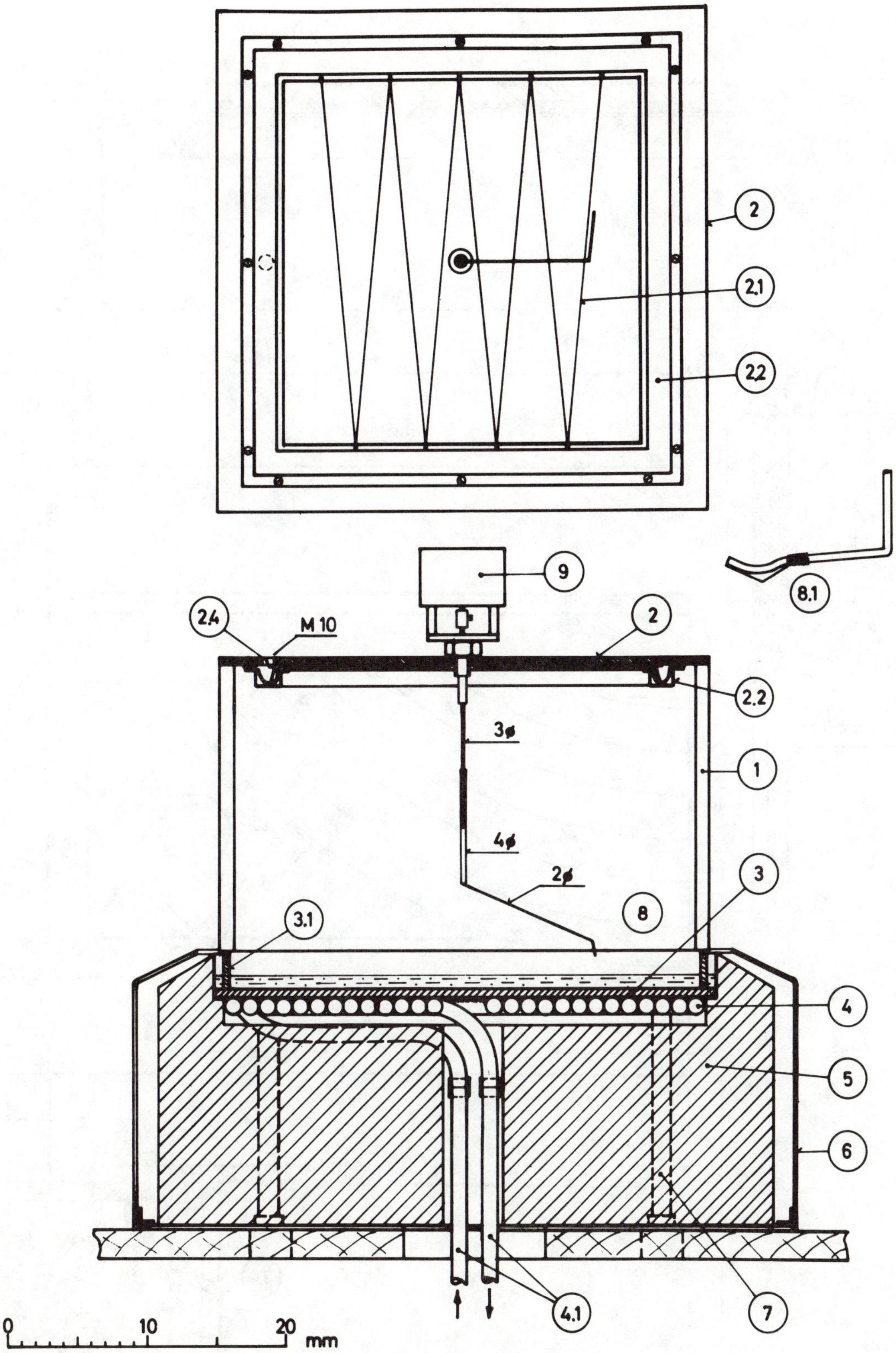

Fig. A41–88

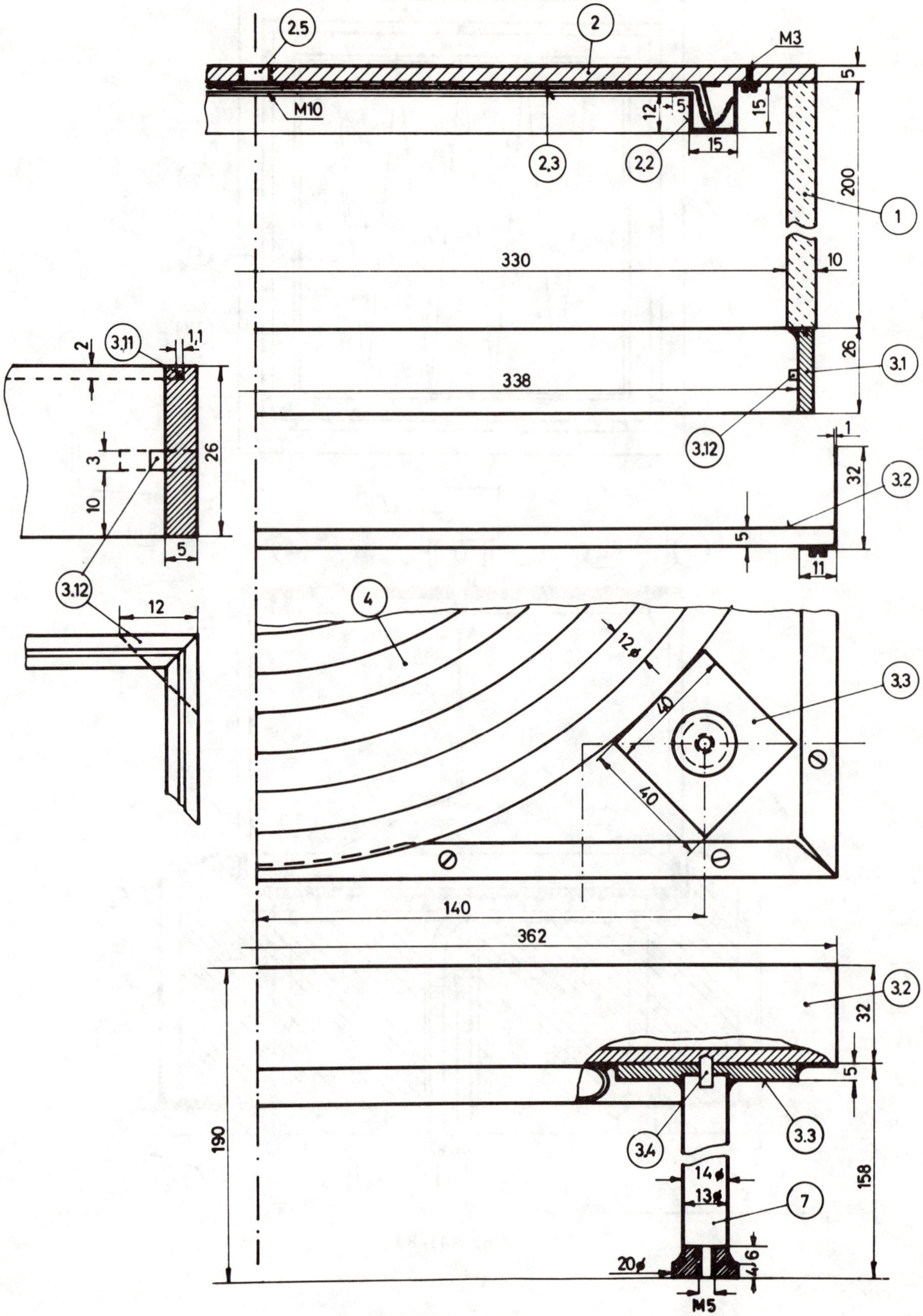
2.5
2
M3
5
M10
12
5
15
15
2.3
2.2
200
1
330
10
3.11
1.1
2
26
338
3.1
3
26
10
5
3.12
1
32
3.2
5
11
3.12
12
4
12ø
3.3
40
40
140
362
3.2
32
5
190
3.4
3.3
158
14ø
13ø
7
4.6
20ø
M5

Fig. A41–89

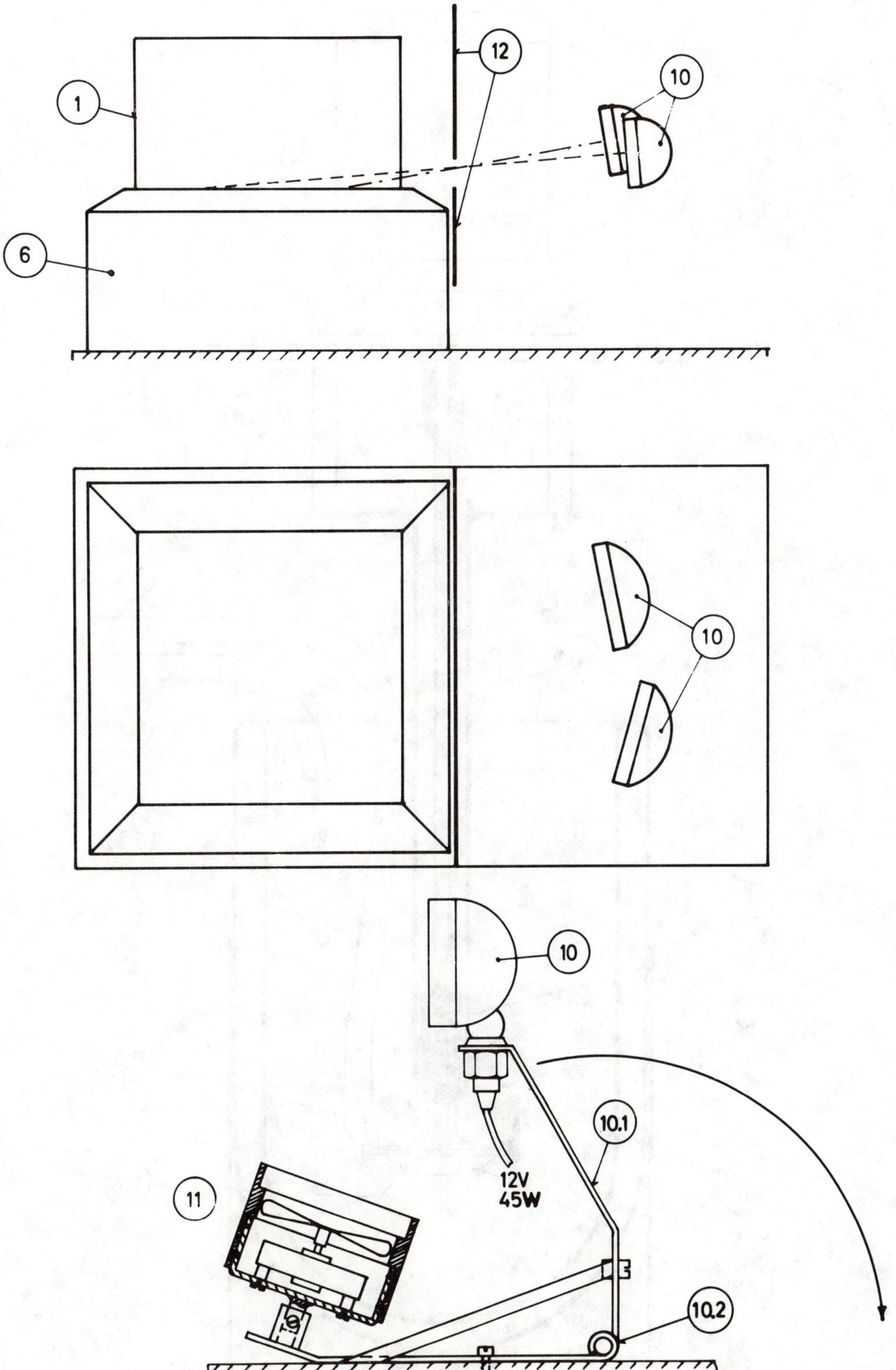

Fig. A41–90

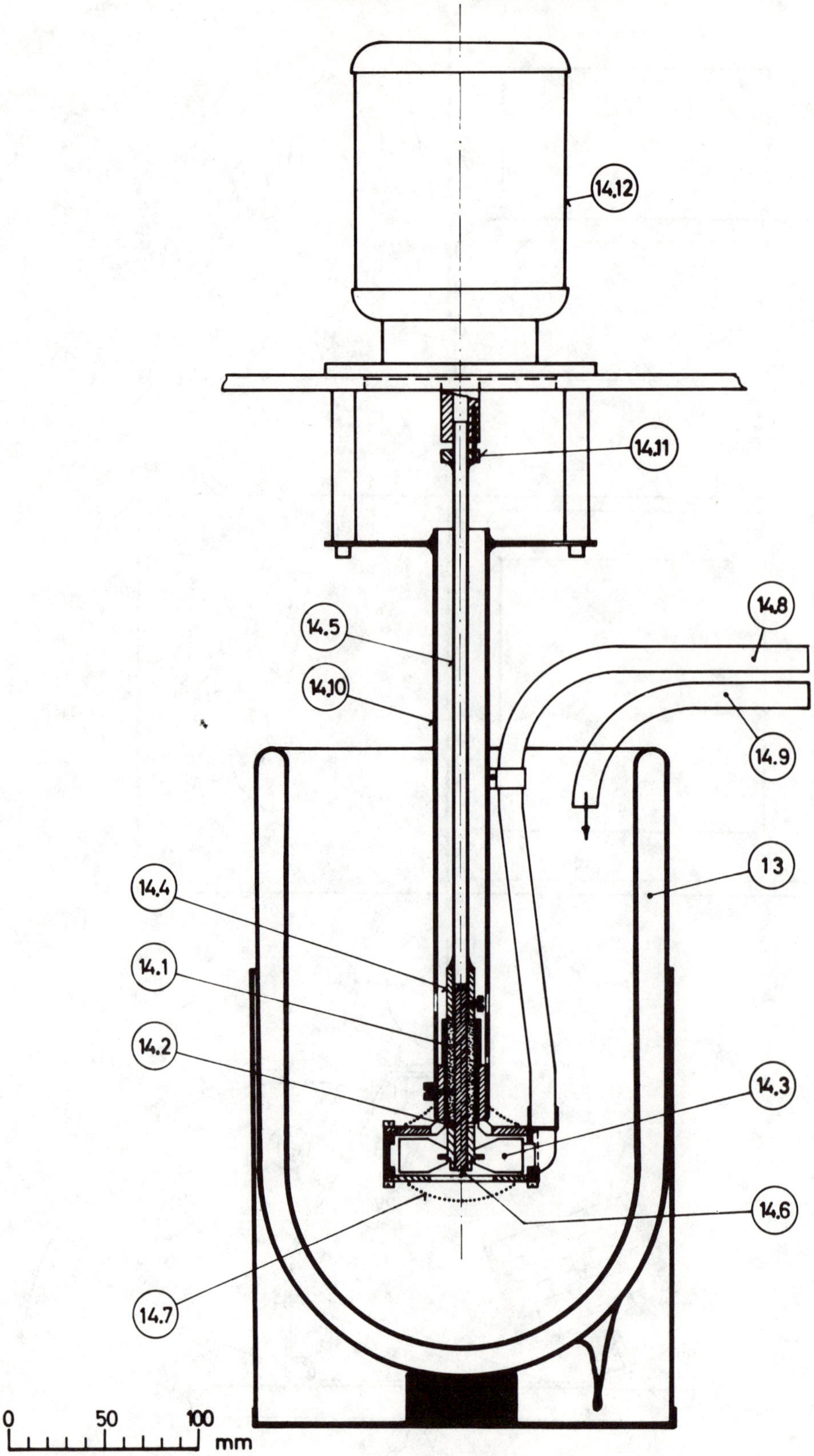

Fig. A41–91

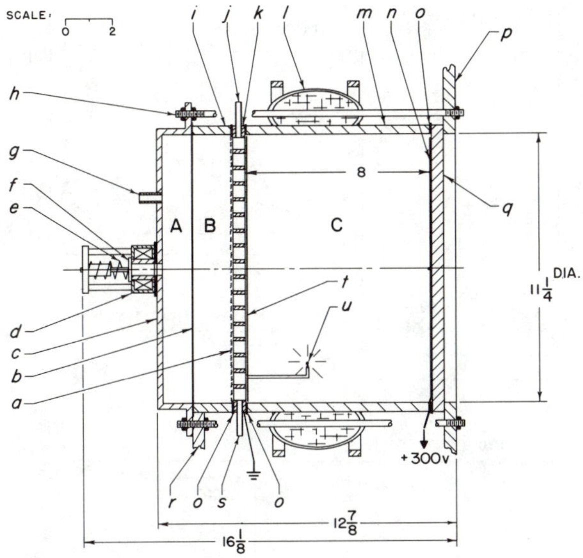

Fig. A41–92 Front View of Spark Chamber and Geiger Tubes (at left) and Driver Chassis (at right).

held together by 18 brass tension rods *h*. Front compartment C is a 12 in. diam x ⅜ in. wall x 8 in. long glass[(6)] cylinder. A brass plate *k*, 12 in. diam x ½ in. thick, perforated with a large number of ½-in.-diam holes, separates compartment C from B.

The front end of C is a 12-in.-diam Pyrex glass[(6)] window *q* through which the tracks are observed. The "C" side of plate *k* is covered with black velvet to provide a background for the white tracks. A wire screen *a* (18 mesh stainless steel) is tied to the other side of the plate. Its purpose is to keep a rubber compression diaphragm *b* from being forced into the holes in plate *k* by gas pressure from compartment A. Compartment B, *i*, is a brass cylinder, 12 in. o.d. x 11¼ in. i.d. x 1$^{13}/_{16}$ in. long.

Gas under pressure is admitted to A through inlet *g*, the diaphragm expands toward C to compress helium in B and C. (Plate *k* has an inlet valve *j* and an outlet valve *s* for outside connections.) There are several ways of controlling the compression. A reducing valve operating off a compressed air line is a simple control device. It has been found, however, that helium is more advantageous in that it has a shorter escape time than air. Further, a helium tank with a flow valve attached is a common piece of demonstration equipment. An electromagnetic valve *f* allows the compressed gas in A to escape quickly during the expansion.[(7)] The electromagnet *d* is energized by four 1½-V dry cells in series. A spring *e* returns the valve to its normally closed position.

A radioactive source *u* is located in the front compartment C. It is an L-shaped stainless steel rod with a fragment of an old radon bulb cemented to its tip.

A clearing field is set up between the grounded plate *k* and an aluminum foil ring[(8)] *n*, which is cemented to the inside of the observation window *q*. A No. 36 tungsten wire is cemented to the ring diametrically horizontal. Both the wire and a tab on the ring are connected to the positive terminal of a 300-V B battery. The negative terminal is connected to some convenient point on plate *k*.

An operating pressure of 12–14 cm of mercury above one atmosphere is fairly well fixed. Deviations from this pressure may be attributed to temperature changes. An open tube manometer for indicating pressure changes in B and C is mounted on a wooden board *p* framing the window.

Four sealed-beam auto headlights *l* provide the necessary illumination for the chamber. They are arranged inside the four corners of a square framework of light brass strips which is concentric with the chamber proper.

The electrical control system operating the chamber consists of three separate circuits. These are shown schematically in Fig. A41–93. A synchronous clock motor running at 1 rpm turns a shaft on which three hard-fiber disk cams are fastened. Each cam opens and/or closes a leaf-type switch in proper sequence to repeat a cycle continuously. One cycle constitutes closing the lighting circuit contacts a few seconds before an expansion, opening the electromagnetic valve circuit, and opening the clearing field circuit. (When the clearing field circuit is opened, the field is shorted through a 220-kΩ resistor to remove any residual field.) The (normally open) lighting switch cam is designed for an 11-sec dwell. A 7-sec-dwell cam operates the magnetic

(6) The contributor suggests using clear acrylic plastic in lieu of glass.

(7) Commercially available solenoid-operated valves can perform this function as well.

(8) The original model employed a clearing field arrangement as described. The model presently in use at the University of Minnesota eliminates the aluminum foil ring and tungsten wire on the window by applying the 300-V potential to a long wire probe which enters compartment C at an angle through a Kovar seal in the rim of plate *k*. The source probe arrangement is also slightly different.

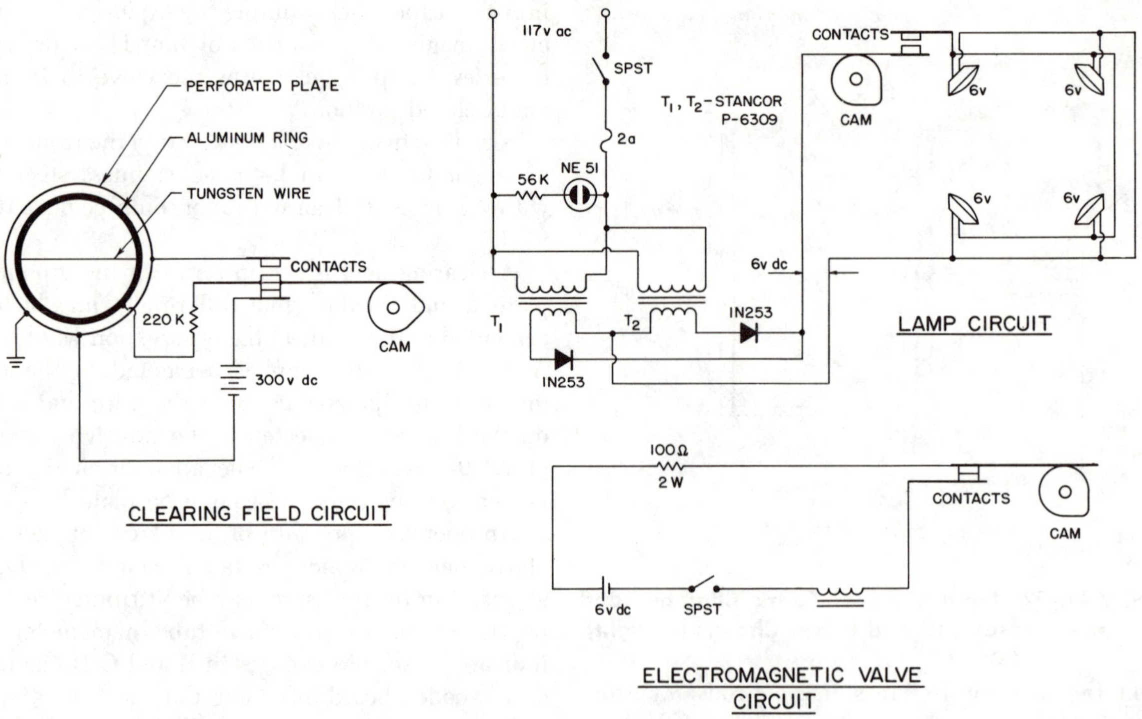

Fig. A41–93 Rear View of Spark Chamber, showing how electrical and flushing gas connections are made.

valve switch (normally closed). A SPDT switch coupled with an 11-sec-dwell cam controls the clearing field. Once the gas pressure in compartment A is satisfactorily regulated, the cams (lockable on the shaft) are adjusted with relation to one another to give one expansion per minute.

The chamber and all electrical accessories are mounted on a wedge-shaped base to make a single unit and also to make observation easier. Its bottom measures 29 in. wide x 16½ in. deep. It is 4 in. high in front and 1½ in. high in the rear. The front is sheet metal (for mounting two toggle switches, a pilot lamp, and a fuse) and the remainder is plywood. A large sheet of ½-in.-thick plywood, framing the window, blocks off extraneous light. It is painted flat black for contrast with the white tracks (5 to 6 in. long).

A definite procedure is followed to prepare the chamber for use. All internal parts are meticulously washed with a solution of Aerosol and warm tap water. Several alcohol-and-water (pure methyl or ethyl alcohol will do) rinsings follow. Without wiping or blowing dry, the chamber is assembled. Compartments B and C are flushed with helium gas (argon can be used, but much shorter tracks are produced) and left at a pressure of 1 atm, as indicated by the manometer. A mixture of 5 cc methyl alcohol and 5 cc water is next injected into these compartments through valve *j*. If additional refillings are found to be necessary, the mixture used is 6 cc alcohol, 4 cc water. As previously mentioned, the operating pressure is 12 to 14 cm Hg above atmosphere. Trial and error determines an optimum value. For each pressure setting tried, the chamber should cycle three or four times. Any condensation formed on the inside of the window can be removed with an electric hair dryer.

A radioactive source is made by crushing an old radon bulb and cementing a fragment to the end of the L-shaped rod. A Geiger counter is used to choose a fragment with an average emission of four to six alpha particles per expansion.

The electromagnetic valve and coil mounting are made of Armco ingot magnetic metal. About 2000 turns of No. 30 enameled copper wire are used for the coil. Valve *f* is a thin disk of Armco metal,

Materials List

Fig. A41–92

Detail	Amount	Part and Description	Remarks
a	1	Wire screen, stainless steel, 18 mesh, 11⅛ in. diam	Commercially available
b	1	Diaphragm, flexible rubber, 12 in. diam x 1⁄16 in. thick	
c	1	Compartment A, brass, 11¼ in. i.d.	Fabricate to suit; see text
d	1 assembly, ~1500 ft	Electromagnet, No. 30 enameled copper wire, Armco ingot magnetic metal	Fabricate or purchase
e	1	Valve return spring	
f	1	Valve, steel	
g	1	Compression gas inlet	Use commercially available hose connector
h	18	Tension rod, brass, 3⁄16 in. diam x 12½ in. long	
i	1	Compartment B, brass or stainless steel, 12 in. o.d. x 11¼ in. i.d. x 1 13⁄16 in. long	If brass is used, chrome plate
j, s	2	Air valve	Hoke No. 480 or equivalent
k	1	Perforated plate, brass, 12 in. diam x ½ in. thick	
l	4	Headlight, auto sealed-beam, 6 V	Commercially available
	~13 ft	Supporting strips for lamps, brass, ⅛ x ⅝ in.	
m	1	Compartment C, glass tubing, 12 in. o.d. x 11¼ in. i.d. x 8 in. long	Clear acrylic plastic can be substituted
n	1 assembly	Positive electrodes	
	1	Aluminum foil; 12 in. o.d. x 10¼ in. i.d.	
	11¼ in.	Tungsten wire; No. 36	
o	3	Gasket, pure gum arabic rubber, 12 in. o.d. x 11¼ i.d. x 1⁄16 in. thick	
p	–	Face frame, plywood, ½ in. thick	
q	1	Window, Pyrex glass, 12 in. diam x ½ in. thick	See Remarks, detail *m*
r	1	Rear chamber support, plywood, ½ in. thick	Fabricate to suit
t	1	Black velvet, 11⅛ in. diam	
u	1 assembly	Radioactive source; radon bulb fragment	
		Support rod; ⅛ in. diam x 4 in. long	
	1 assembly	Base, plywood, ½ in. thick	Fabricate to suit
		Miscellaneous electrical components	See text and Fig. 41–49
		Helium gas	Commercially available
		Pressured regulator and flow valve for helium supply tank	Commercially available
		Manometer with 55-cm scale	Fabricate or purchase
		Suitable hardware for assembly	

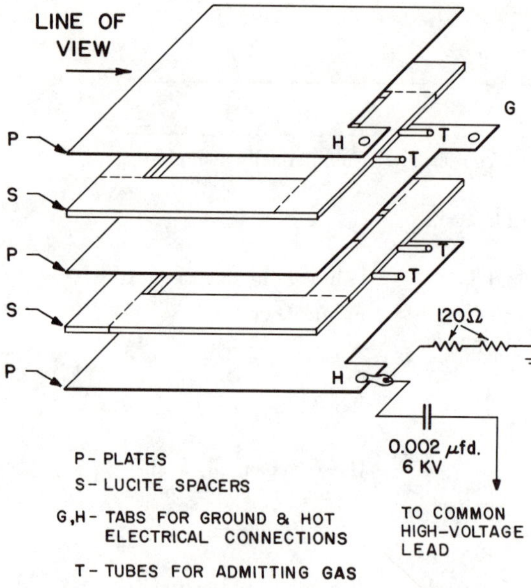

Fig. A41–94 Exploded View of the Basic Two-Gap Module.

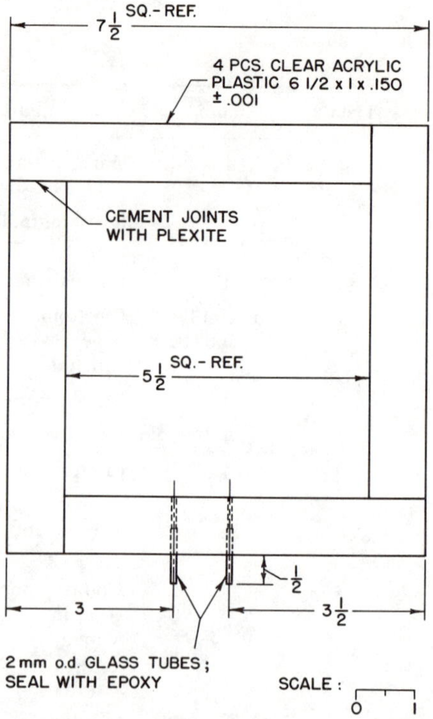

Fig. A41–95 Assembly Drawing of Lucite Frames. Two are required for each module.

ground on one side to make a metal-to-metal vacuum seal with the hollow magnet core and mounted on a spring-loaded guide rod.

Cycling time is indicated on a clock face screwed to the frontal board. A clock hand is mounted on the end of the camshaft (Fig. 41–48).

The size of compartment A is not critical. It can be fabricated from brass to form a short cylinder with a flange as shown in Fig. 41–49. Details *i* and *k* are chrome plated after machining to minimize internal contamination. Stainless steel could be used to eliminate the plating operation.

A41–3.9

The features of the basic module are shown in Fig. A41–94. Each unit is a sandwich consisting of three 0.05-in.-thick aluminum plates and two square 0.15-in.-thick plastic spacer frames. The high-voltage pulse is applied to the two outer plates, which are separated from the common grounded center plate by the spacers. The units are sealed together with double-sided Scotch tape (¾ in. wide) under hydraulic pressure.

Figure A41–95 shows a spacer frame in more detail. The four component strips of the frame are sawed from a 1-in.-thick slab of clear Lucite. The ends of the strips are milled square and cemented together. This procedure eliminates optical polishing, since the edges of the frame were the already polished top and bottom surfaces of the stock slab. The sawed surfaces that form the upper and lower frame surfaces are milled flat to a standardized thickness of 0.15 in. (±0.001). A homemade vacuum chuck simplifies this phase of construction. All interior edges and holes for filling tubes are carefully deburred. A final soap and water wash removes dirt and fingerprints from the polished windows (frame edges). This also ensures that the Scotch tape will adhere properly to the plastic surfaces.

The aluminum plates are made of the hard aluminum alloy 7075-T6. They are cut on a brake and degreased. No other surface preparation is necessary if the large sheets from the plates are cut are reasonably free from scratches. It is important that the plates be free of small dents or bends, but a gentle overall curvature of the plates is unimportant because of the excellent edge support provided by this method of construction.

Two glass tubes, 2 mm o.d., are sealed into the rear wall of each gap frame with epoxy. They ad-

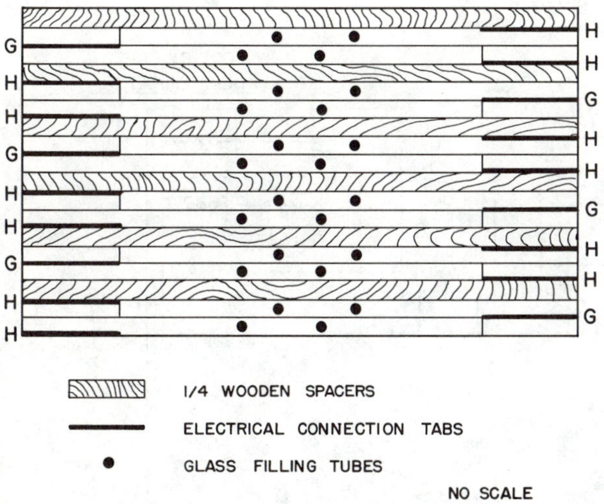

Fig. A41–96 Rear View of Stacked Modules. Note how electrical and gas connections are staggered to avoid interference between components of different modules.

mit and exhaust the gas filling. The chambers are connected in parallel to the gas supply manifold (made of copper tubing soldered together) with small-bore rubber tubing. Good results have been obtained with only an input manifold; the gas is allowed to discharge directly to the open air. Either helium or argon flowing through the chamber at a rate of about 0.5 cu ft/min gives good results, but the minimum operating voltage is higher with argon than with helium.

Since the chambers are continuously flushed with gas during operation, they do not need to be perfectly gas-tight. Sealing with Scotch tape seems to be a satisfactory method, even though the modules appear quite leaky when tested under water. Precautions must be taken, however, to ensure that the pressure inside the chambers never gets high enough to blow them apart.

The spark chamber in its final form consists of a stack of several modules separated by insulating spacers made of ¼-in.-thick plywood. The number and size of the modules is not critical. Electrical connections are made to tabs cut in the plates, Fig. A41–94, and the resistor and condenser associated with each gap are mounted near the module as shown in Figs. 41–54 and A41–94. It is necessary to provide common high voltage and ground leads to all the gaps. Figure A41–96 shows a schematic rear view of the assembled spark chamber. The filling tubes and electrical connections are staggered so as to leave adequate space between components.

A schematic diagram of the spark-chamber driver circuit is shown in Fig. A41–97. Standard circuits discussed in the literature are used throughout.[9,10] Pulses from the two trays of Geiger counters are amplified by a factor of 20 (to reduce delay time) and fed to the inputs of a conventional diode coincidence circuit. The coincidence discriminator (a conventional Schmitt circuit) triggers a blocking oscillator when a large coincidence pulse comes from the diode circuit. The blocking oscillator[11] gives a fast, high-power pulse, that triggers a small thyratron (2050), which in turn triggers a hydrogen thyratron (Type 3C45). The hydrogen thyratron applies an 8000-V pulse to the spark chamber gaps.

The components are standard except for the filament transformer T_2, which is an ordinary 6.3-V filament transformer whose secondary windings were replaced with well insulated windings for the Type 1X2 high-voltage rectifier tubes. (The 1X2 filaments can be supplied by batteries if some loss in convenience is tolerable.) The high-voltage transformer T_1 can be any transformer which has a secondary rated to deliver 10 mA at 4000 to 6000 V. The variac controlling the output voltage of the high-voltage supply is essential, because it allows the voltage applied to the chamber to be set at the value for optimum performance. Since the spark chamber operating voltage is higher than the maximum voltage rating of the 3C45 hydrogen thyratron, it was necessary to try several of these tubes in order to find one that withstood the voltage used. A thyratron rated for higher voltage such as a 4C35 or 5C22 can be substituted with no change in performance.

The high-voltage supply[12] for the Geiger tubes is small and battery-operated. Its voltage must be set at the operating value for whatever Geiger tubes

[9] W. C. Elmore and M. L. Sands, *Electronics: Experimental Techniques* (McGraw-Hill Book Company, Inc., New York, 1949).

[10] B. Chance, A. C. Hughes, E. F. MacNichol, N. A. Sayre, and F. C. Williams, *Waveforms,* MIT Radiation Laboratory Series (McGraw-Hill Book Company, Inc., New York, 1950), Vol. 19.

[11] Suitable blocking oscillator transformers are available from Lectronic Research Laboratories Inc., Philadelphia, Pennsylvania.

[12] Universal Transistor Products Corporation, Westbury, New York, Model UAC-1150/6/50.

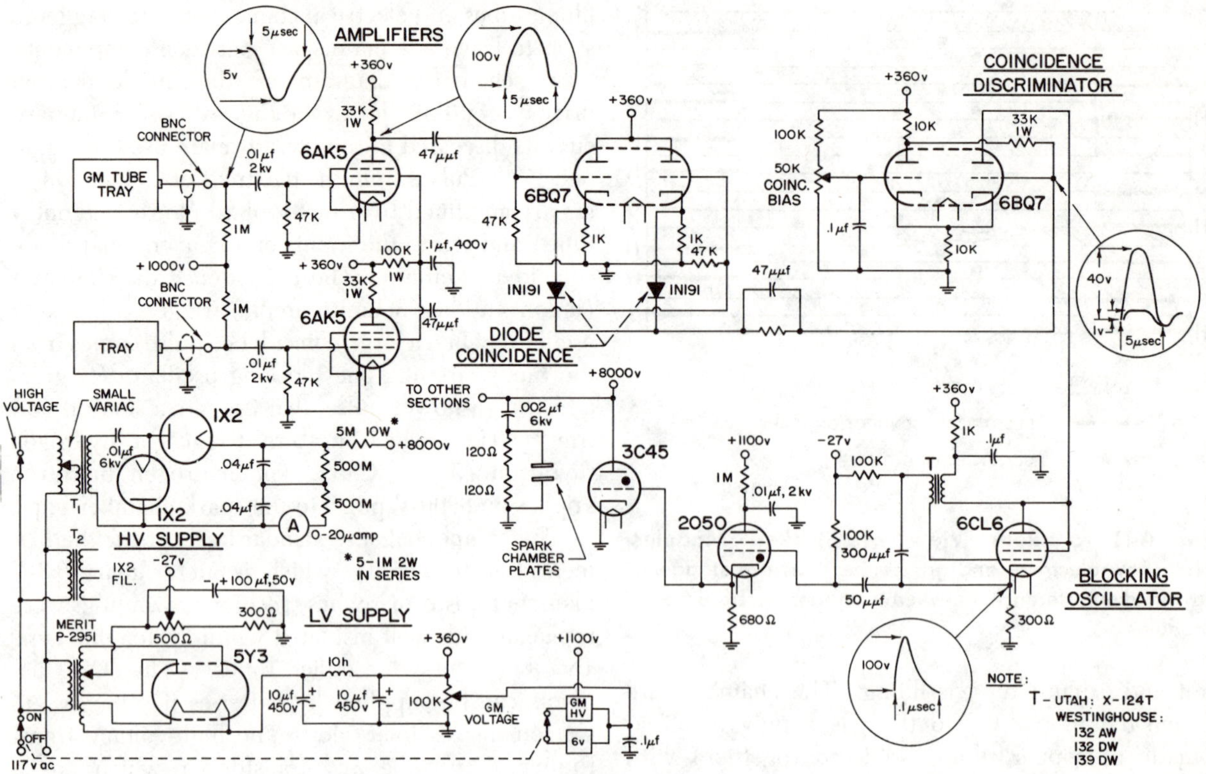

Fig. A41–97 Schematic Diagram of Spark Chamber Driver and Power Supply Circuits.

are used. This can be done by a series rheostat in the primary circuit of the supply. It is more convenient to operate the supply at a fixed voltage to which is added a fraction of the 360-V output by means of a voltage divider potentiometer.

One high-voltage capacitor and two 120-Ω, 1-W carbon resistors are mounted on the spark chamber unit for each gap. The capacitors are inexpensive ceramic units rated at 0.002 μF at 6000 V (Centralab DD60-202). Operating the capacitors well above their rated voltage caused no trouble, except that a few of them failed on the first application of higher than rated voltage. Such failures were easily detected by feeling for warm capacitors with the high voltage turned off.

For the initial tune-up of the driver circuit, it is convenient to use an oscilloscope to compare waveforms at various points with those given in Fig. A41–97. This should be done with the high-voltage switch turned off, since pickup from sparks (if any) seriously distorts the waveforms. Once proper operation of the circuits is verified, the high voltage should be set at about 8000 V and the flow of gas should be started. Tracks should become apparent as soon as most of the air is flushed out of the spark chamber.

A41–4.4

In Figs. A41–98 and A41–99 are shown the details of the gyroscope and sphere. The $4\frac{3}{16}$ x $4\frac{3}{16}$-in. brass frame which supports the double gimbal is mounted on a 5 x 5-in. brass base. Both the outer ring *a* ($3\frac{5}{8}$ o.d. x $3\frac{5}{16}$ i.d.) and inner ring *b* ($3\frac{1}{6}$ o.d. x $2\frac{3}{4}$ i.d.) of the double gimbal are aluminum. The $2\frac{9}{16}$-in.-diam sphere is made from laminated Teflon. A nylon sleeve *c* through its bearing axis contains a Alnico permanent magnet *d* ($\frac{5}{8}$ in. diam x $1\frac{1}{2}$ in. long) and a Teflon spacer *e*. The laminations are aligned with two clear acrylic plastic dowels *f* ($\frac{3}{16}$ in. diam) and are held together by a nylon spanner nut on the sleeve. No cement is used. A rubber disk, $\frac{3}{4}$ in. x $2\frac{1}{2}$ in. diam, driven by a 2000-rpm motor, is used to start the sphere rotating. The finished ball must be very carefully balanced. The magnet coil is wound with seven layers (507

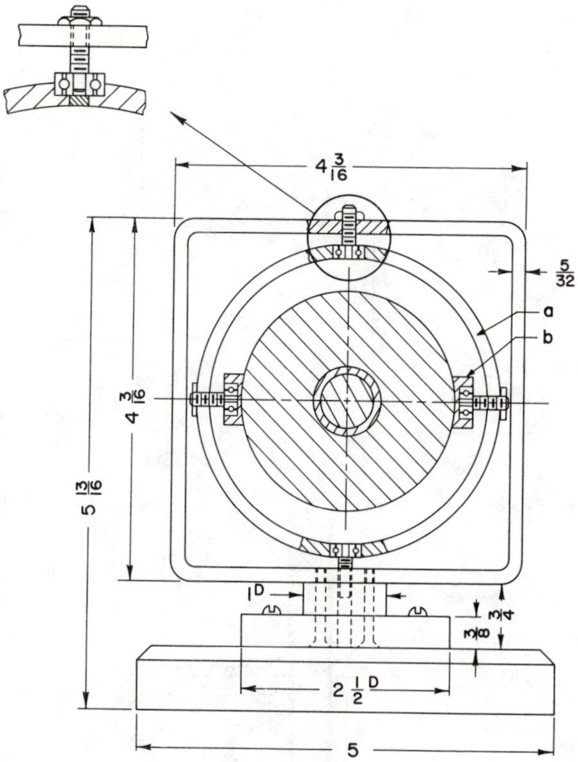

Fig. A41–98

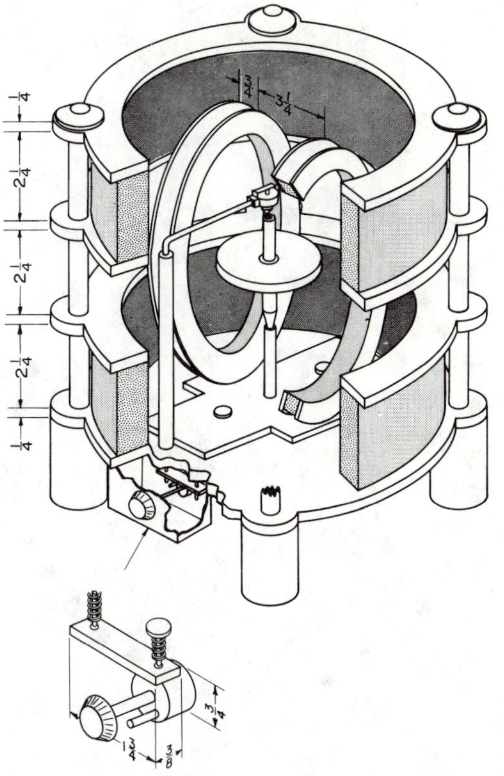

Fig. A41–100

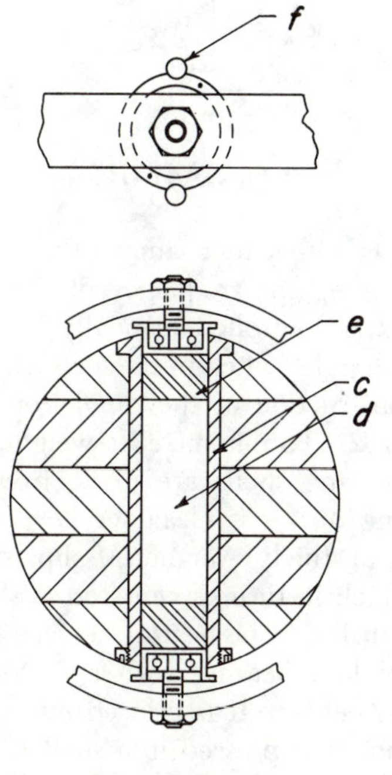

Fig. A41–99

turns per layer) of No. 10 double cotton covered wire on a steel core 2¾ in. diam x 8 in. long.

A41–4.6

For a maximum magnetic field strength and for arbitrary horizontal coil diameters, the size of copper wire to be used (Fig. A41–100) is determined uniquely by the voltage of the available source of dc power (assuming no limit to the current drawn through the source). This follows from the fact that H is proportional to the ampere-turns product NI, and the resistance of the coils is proportional to N. This information, combined with the volume of the windings, specifies the proper wire size for a given battery voltage. A unit producing a field strength of about 200 Oe and using a 120-V battery for two coils in series is adequate. Each of the two Helmholtz field coils has 1485 turns of No. 19 Formvar-covered copper wire on nonconducting formers. The inside and outside diameters of each coil are 8 in. and 10 in. respectively.

The ac coils are at right angles to the main coils.

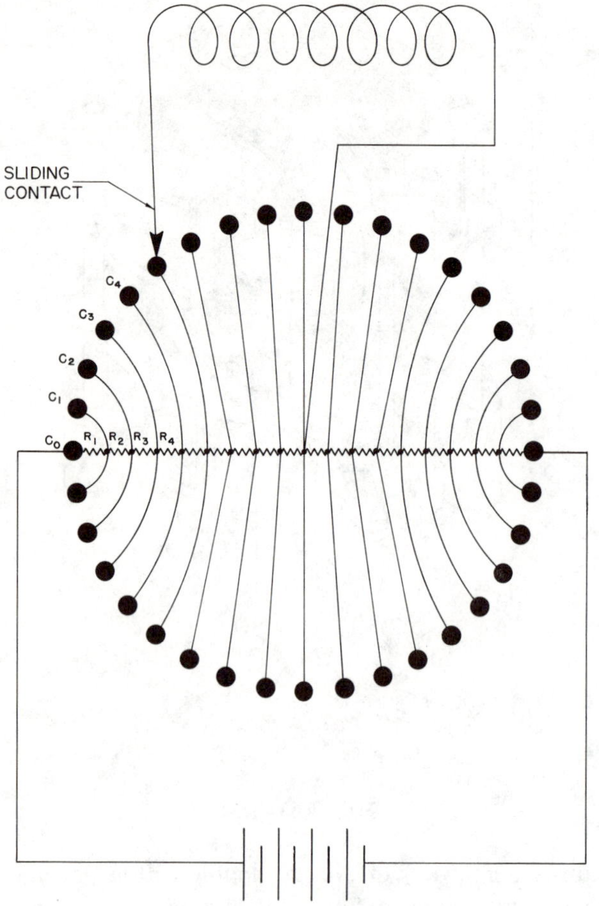

Fig. A41–101

Each coil has 4500 turns of No. 32 Formvar-covered copper wire and an average diameter of 6¼ in. Since the period of precession of the dipole is of the order of seconds, the applied field must have a frequency of less than 1 cps. To obtain the low frequencies needed, an ac generator-converter, motor-driven sinusoidal potentiometer is used (Fig. A41–101). The rotating contact is driven at constant speed, and the resistors are so chosen that the resulting potential difference across the coils is a step function which approximates a sine curve. For example, the contact points C_1, C_2, C_3, etc. are spaced at 10-deg intervals. Resistance R_1 is proportional to sin 10 deg, and resistance $R_1 + R_2$ is proportional to sin 20 deg, etc. The frequency of the ac voltage is varied by changing the motor speed.

Tuning the oscillating field to the precessing dipole is accomplished by varying the external magnetic field H, either by changing the spin frequency or by changing the frequency of the field. Phase adjustment is controlled manually by momentarily opening a toggle switch in the motor power line.

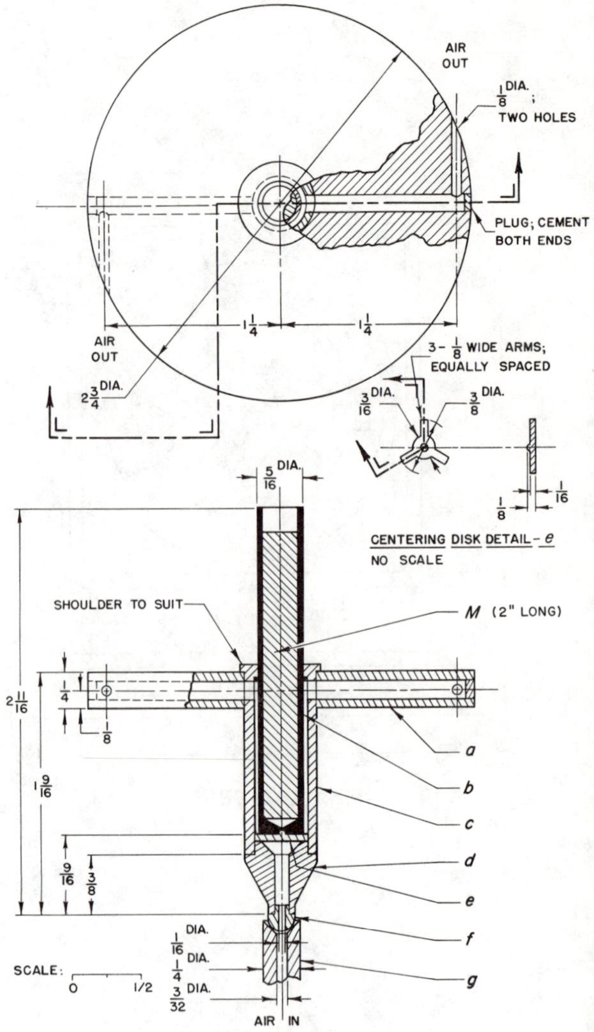

Fig. A41–102

The construction of the dipole-top is shown in Fig. A41–102. To minimize its weight, the vertical magnet housing shells are made from aluminum. The driving disk *a* is clear acrylic plastic, 2¾ in. diam x ¼ in. thick. Magnet *M* slips smoothly into shell *c*, which in turn, is centered in shell *b* by the centering disk *e*. The bearing surface *f* is a 3/16-in.-diam steel ball bearing with a 1/16-in.-diam hole through its center. It is pressed into body connector *d*. Cone *d* is pressed into shell *b*. The plastic disk is a snug-fit over *b* to make a tight assembly which is dynamically balanced. Air passageways ⅛ in. in diameter are drilled radially in toward the center of the disk from opposite directions and plugged as shown. Two $\frac{1}{16}$-in.-diam holes are drilled to meet the passageways perpendicularly at equal distances from the center (1¼ in.) to permit air to leave tangentially and drive the top.

A compressed air tube *g* is supported in the Helmholtz-coil frame by a rubber stopper (because rigid attachment to the frame causes air turbulence, preventing smooth precession). The bearing seat is conical as indicated in Fig. A41–102.

To attain a fixed angular velocity about a fixed axis, the top is supported at its upper end. Precession about this axis commences upon removal of the support.

A41–4.7

A wide variety of pulsed oscillators and receivers can be used, and an electromagnet can be substituted for the permanent magnet. A wide variety of samples can be used; glycerine with a small amount of a paramagnetic salt has been found to be satisfactory. The sample holder is a small (¼ x 3½ in.) test tube with a cork. The sample coil is wound on the test tube and fastened with banana jacks and plugs to the aluminum box, which contains a variable capacitor and two diodes. The diodes are back to back and are inserted between the sample coil and the oscillator to present a large back impedance to the small nuclear signals. To hold the sample coil in the middle of the field, secure the aluminum box to the magnet with brass plates and bolts. Two BNC connectors are provided on the box for attaching cables to the pulsed oscillator and receiver. The capacitor is chosen to resonate the coil over the desired frequency range. For the 5.1 kG magnet shown, the sample coil was six turns of No. 14 wire spaced $\frac{1}{16}$ in. and resonated with a 450 $\mu\mu$F variable capacitor in the range 21–22 Mc.

General requirements for apparatus needed for this demonstration are: a pulsed oscillator with an rf output of 200–500 V peak to peak and variable length pulses of 5–50 Ms, and an rf amplifier with a voltage gain of 90–100 and a 0.5–2 Mc bandwidth. High homogeneity is not required for the magnetic field because an absence of homogeneity leads only to a narrowing of the induction tail and echo, and is not significant for magnets ordinarily used.

ADDITIONAL REFERENCE SOURCES FOR PHYSICS DEMONSTRATIONS

Advanced Undergraduate Laboratory Experiments in Physics, Lloyd W. Taylor, Manual (Addison-Wesley Publishing Company, Inc., Reading, Mass., 1959).

Advice to a Lecturer, The Royal Institution of Great Britain, London, U.K. An anthology from *The Life and Letters of Faraday* by Bence Jones (1869).

Andrews, C. L., *Optics of the Electromagnetic Spectrum* (Prentice-Hall, Inc., Englewood Cliffs, New Jersey, 1960).

Daniel, H., *Demonstrations Versuche Aus der Optic* (Musterschmidt-Verlag, Göttingen, 1960), 2nd ed.

Experiments in Optics (Electro Optics Association, Inc., Palo Alto, 1968).

Experiments in Physical Optics Using Continuous Laser Light (Optics Technology, Inc., Palo Alto, 1964).

Feynman, R. P., R. B. Leighton, and M. Sands, *The Feynman Lectures in Physics* (Addison-Wesley Publishing Company, Inc., Reading, Mass., 1964), Volumes I, II, and III.

Freier, G. D., *University Physics, Experiment and Theory* (Appleton-Century-Crofts, New York, 1965).

Friedrich, A., *Handbuch der Experimentellen Schulphysik* (Aulis Verlag Deubner and Co., K G Koln, West Germany, 1963), 5 Vols.

Gathercole, L. J., *O Level Experiments* (Hutchinson & Co. [Publishers], Ltd., London, 1966).

Huggins, E. R., *Physics 1* (W. A. Benjamin, Inc., New York, 1968).

Ingard, U., and W. L. Kraushaar, *Introduction to Mechanics, Matter and Waves* (Addison-Wesley Publishing Company, Inc., Reading, Mass., 1960).

IPS Group, Education Development Center, *College Introductory Physical Science* (Prentice-Hall, Inc., Englewood Cliffs, N.J., 1969).

Leybold Physics Leaflets, E. Leybold's Nachfolger, Koln, West Germany.

Marcley, R. G., *Apparatus Drawings Project* (Plenum Press, Inc., New York, 1962).

Mark, H., and N. T. Olson, *Experiments in Modern Physics* (McGraw-Hill Book Co., Inc., New York, 1966).

Marousek, G., and T. W. Williams, II, *Experiments on an Air Table* (The Ealing Corporation, Cambridge, Mass., 1969).

Meiners, H. F., W. Eppenstein, and K. H. Moore, *Laboratory Physics* (John Wiley & Sons, Inc., New York, 1969).

Melissinus, A. C., *Experiments in Modern Physics* (Academic Press, Inc., New York, 1966).

Mlodzeevskii, A. B., ed., *Lecture Demonstrations in Physics,* M. V. Lomonosov State University, Moscow, U.S.S.R. Available in a condensation by H. A. Robinson (April, 1963) on *Microfilm* from the American Institute of Physics, New York, N.Y.

Novel Experiments in Physics, American Association of Physics Teachers (American Institute of Physics, New York, 1964).

Nuffield Foundation, *Physics Teachers Guides* (3) Vols); *Physics Guide to Experiments* (2 Vols); *Physics Questions Book* (4 Vols); (Longmans, Green and Co., Ltd., London, Penguin Books, Ltd., Harmondsworth, Middlesex, U.K., c. 1966).

Palmer, C. H., *Optics Experiments and Demonstrations* (The Johns Hopkins Press, Baltimore, Md., 1962).

Physical Science Study Committee, *Physics Laboratory Guide, Advanced Topics Supplement* (D. C. Heath & Company, Boston, 1966).

Physics Experiments and Demonstrations, An Annotated Subject Index, (American Institute of Physics, New York, 1965).

Physics Handbook (Bureau of Secondary Curriculum Development, The New York State Education Department, Albany, New York, 1959).

Pohl, R. W., *Elektrizitätslehre* (Springer-Verlag, Berlin, Heidelberg, New York, 1967), 20th ed.

Pohl, R. W., *Mechanik, Akustik, und Warmelehre* (Springer-Verlag, Berlin, Heidelberg, New York, 1964), 16th ed.

Pohl, R. W., *Optik und Atomphysik* (Springer-Verlag, Berlin, Heidelberg, New York, 1967), 12th ed.

PSNS Project Staff, *An Approach to Physical Science, Physical Science for Nonscience Students* (John Wiley & Sons, Inc., New York, 1969).

Resource Letters. A series of articles in selected subject areas intended as guides for physicists. Each article includes *references* to literature, apparatus, laboratory experiments, demonstrations and visual aids (from 1964). For list of resource letters, write to The American Institute of Physics, New York, N.Y.

Rogers, E. M., *Physics for the Inquiring Mind* (Princeton University Press, Princeton, N.J., 1960).

Science Masters Series, Part 1 of Series I (1931), Part 1 of Series II (1936), Part 1 of Series IV (1961), (John Murray Publishers, Ltd., London).

Scientific Instruments and Apparatus: Sources for the Fuller Documentation of the History of Physics, Pub. R-152, History of Physics Project, American Institute of Physics, New York, N.Y.

Shamus, M. H., *Great Experiments in Physics* (Holt, Rinehart & Winston, Inc., New York, 1959).

Sourcebook for Science Teaching (UNESCO, Paris, France, 1962).

Sutton, R. M., *Demonstration Experiments in Physics* (McGraw-Hill Book Co., Inc., New York, 1938).

Swezey, K. M., *After-Dinner Science* (McGraw-Hill Book Co., Inc., New York, 1961), rev. ed.

Teacher's Guide, Units 1 through 6, *Project Physics* (Holt, Rinehart & Winston, Inc., New York, 1968), interim version.

White, H. E., *Modern College Physics* (Van Nostrand Rheinhold Company, New York, 1966), 5th ed.

LIST OF CONTRIBUTORS

AINSLIE, D. S., University of Toronto, Canada, 29–1.12
ALBERT, D. E., Adelphi University, 29–1.2
ALLEN, W. K., Western Canada High School, Calgary, Alberta, Canada, 33–3.8
ANDERSON, H. E., Massachusetts Institute of Technology, 13–5.3, 16–3.2, 29–2.5, 39–1.4, 39–2.2
ANDREWS, C. L., General Electric Co., Schenectady, N.Y., 33–7.4
ARMSTRONG, H. L., Queens University, Canada, 13–5.12, 18–8.2, 35–3.10
ARONS, A. B., Amherst College, 18–8.6
ASHBY, NEIL, University of Colorado, 41–2.9

BANKS, M., University of Florida, 19–3.1
BARNETT, J. D., Brigham Young University, 29–1.19, 29–4.3
BARTLETT, A. A., University of Colorado, 19–5.1, 31–2.6
BARTUNEK, P. F., Colorado School of Mines, 12–3.10, 15–5.2, 15–8.1, 15–10.10
BRAULT, J., Princeton University, 30–4.3
BERKEY, D. K., Colgate University, 6–4.3, 31–3.1, 32–3.17
BICHSEL, HANS, University of Southern California, 13–7.7, 35–1.4
BITTER, FRANCIS, Massachusetts Institute of Technology, 41–4.2
BLACK, J. G., Eastern Kentucky University, 6–3.2, 8–2.3, 8–4.8, 12–4.1, 14–2.4, 16–2.2, 16–4.8, 17–2.10, 30–3.5
BOLSEY RESEARCH AND DEVELOPMENT, Scarsdale, N.Y., 34–1.19
BRISCOE, C. V., University of North Carolina, 41–4.7
BRONSTEIN, L., Zenith Radio Corporation, 40–1.5
BROOKS, J. T., Washington University, 7–1.18, 8–5.9, 9–5.13, 10–2.2, 10–2.14, 10–2.18, 13–7.13, 15–1.10, 18–6.2, 25–2.7, 27–7.5, 29–1.26
BROWN, EDMOND, Rensselaer Polytechnic Institute, 12–6.7, 41–2.8
BRYAN, A. B., Rice University, 8–4.9
BRYANT, H. C., University of New Mexico, 34–1.11
BUTLER, A. B., Washington State University, 25–2.3

CAIN, R. L., University of Minnesota, 9–2.4, 39–3.1
CALIFORNIA INSTITUTE OF TECHNOLOGY, 8–4.3, 8–4.4, 9–1.3, 9–2.1, 12–3.11, 12–4.4, 12–4.5, 15–1.17, 15–6.1, 15–9.3, 15–10.2, 17–9.1, 17–9.2, 17–9.3, 18–1.2, 18–3.5, 18–7.5, 19–4.2, 19–4.7, 26–4.1, 26–6.2, 27–3.1, 29–1.24, 29–3.2, 30–2.8, 31–1.7, 31–1.16, 31–2.16, 31–3.4, 33–2.7, 33–6.1, 33–7.1, 35–6.2
CAPECELATRO, ACHILLE, Newark College of Engineering, 8–1.4, 14–3.5, 18–3.3, 18–7.6, 25–2.1, 29–1.29, 34–1.10, 34–1.13, 34–1.15
CARDWELL, A. B., Kansas State University, 26–3.1
CARR, J., University of Minnesota, 27–8.1
CARVER, L., Princeton University, 34–2.1
CHRISTENSEN, F. E., St. Olaf College, 18–3.1; University of Minnesota, 41–3.6
CHRISTIE, J., Oak Ridge Institute of Nuclear Studies, 35–6.5
CLEVELAND, E. L., New Mexico State University, 35–2.1
COBURN, H. H., New Mexico State University, 13–8.1, 15–9.2
COOK, E. O., Frostburg State College, Md., 18–6.6, 27–7.3, 27–7.4, 30–6.5, 30–6.7
CORMIA, R. L., General Electric Co., Research Laboratory, Schenectady, N.Y., 27–7.1
CORRELL, MALCOLM, University of Colorado, 18–8.14
COTTS, R. M., Cornell University, 10–2.16

DANIEL, H., Musterschmidt Verlag, Göttingen, West Germany, 35–3.1, 35–4.1, 35–4.2, 35–4.3, 35–5.5, 39–1.9
DARBY, W. P., St. Augustine High School, Brooklyn, N.Y., 16–5.10, 18–6.5, 39–5.1
DAW, H. A., New Mexico State University, 9–5.14, 10–3.1 to 10–3.3, 11–2.2, 41–4.5

Delaup, P. S., University of Southwestern Louisiana, 34–1.9
Delsasso, L. P., University of California, Los Angeles, 9–5.6, 12–6.5, 26–5.12, 33–2.1
DeVany, A. S., Williams College, 35–2.12
Dillon, J. F., Bell Telephone Laboratories, 32–3.8
Donnally, B. L., Lake Forest College, 7–1.10, 27–6.6
Doran, R. L., University of Utah, 12–5.3
Dosso, H. W., Victoria College, Canada, 29–1.3, 31–1.1
Duke University, 27–6.3
Dursch, H. R., Skagit Valley College, Wash., 27–3.6
Duursema, C. H., Newark College of Engineering, 16–4.3

Eaton, V. E., Wesleyan University, Conn., 6–4.2, 7–1.12, 8–4.7, 12–4.9, 15–9.5, 15–10.1, 18–3.2, 18–6.1, 30–3.4, 31–2.17, 34–1.17, 34–1.18, 35–7.4, 41–2.1
Earl, H. E., Bell Telephone Laboratories, 32–3.8
Earl, J. A., University of Minnesota, 39–3.1, 41–3.9
Eastmond, E. J., Brigham Young University, 7–1.15, 31–1.15, 34–1.21
Edward, A., St. Augustine High School, Brooklyn, N.Y., 7–2.3
Ebbighausen, E. G., University of Oregon, 27–2.3
Eppenstein, Walter, Rensselaer Polytechnic Institute, 6–1.1, 6–4.4, 7–1.9, 7–2.8, 8–8.4, 13–7.6, 15–1.11, 18–8.4, 18–8.5, 31–2.14, 33–2.2, 40–1.16, 40–1.18
Evans, D. C., Oklahoma State University, 8–2.2, 16–4.5, 26–4.5, 30–3.6

Ferguson, W. F. C., New York University, 29–1.1
Ferlen, R., Carleton College, 15–10.9
Ferman, J. W., General Motors Corp., Warren, Michigan, 8–4.12
Fostoria Corporation, Fostoria, Ohio, 34–1.19
Fowler, J. M., Washington University, 7–1.18, 9–5.13, 10–2.2, 10–2.14, 10–2.18, 15–1.10, 18–6.2, 25–2.7, 29–1.28, 38–2.4, 38–7.4
Fox, F. E., Catholic University of America, 19–6.5
Francis, C., St. Augustine High School, Brooklyn, N.Y., 7–2.3, 16–5.10, 18–6.5, 29–1.20, 39–5.1
Freeman, I. M., Rutgers, The State University, 7–2.13
Freier, G. D., University of Minnesota, 7–1.16, 7–1.20, 7–2.4, 7–2.5, 8–2.4, 9–2.4, 9–4.10, 9–5.1, 14–3.8, 16–6.1, 26–3.3 to 26–3.5, 27–8.7, 29–4.10
French, A. P., Massachusetts Institute of Technology, 9–1.5, 9–1.6, 15–1.8, 15–2.1, 15–9.1, 18–3.4, 18–7.1, 18–7.2, 38–2.3, 41–1.7

Galley, Elmer L., Oak Ridge Institute of Nuclear Studies, 41–3.2
Gibbons, J. F., Stanford University, 40–1.3, 40–1.4, 40–1.7, 40–1.9, 40–1.10, 40–1.11
Glassman, D. M., Rensselaer Polytechnic Institute, 18–1.6, 34–2.7
Goldsmith, N. W., State University of New York, College at Oswego, 27–3.8
Gordon, G. D., Radio Corporation of America, 13–7.1
Gray, Robert L., University of Massachusetts, 29–4.9
Grillet, L., University of Rennes, France, 8–2.7, 15–3.4, 17–6.1, 18–1.5, 18–4.2, 18–4.3, 19–2.2, 26–5.14, 27–3.10, 29–4.6, 30–1.3, 30–2.4, 30–2.6, 30–2.7, 30–2.9, 32–3.15, 33–2.4, 34–1.3, 34–1.7, 34–1.8, 34–1.22, 35–2.6, 39–1.3
Grillet-Amy, M.-T., University of Rennes, France, 8–2.7, 15–3.4, 17–6.1, 18–1.5, 18–4.2, 18–4.3, 19–2.2, 26–5.14, 27–3.10, 29–4.6, 30–1.3, 30–2.4, 30–2.6, 30–2.7, 30–2.9, 32–3.15, 33–2.4, 34–1.3, 34–1.7, 34–1.8, 34–1.22, 35–2.6, 39–1.3
Günneberg, F., University of Köln, West Germany, 27–10.3, 31–1.12, 32–1.2, 32–1.3, 32–3.6, 32–4.4, 32–4.5

Hansen, A. W., The Citadel, 19–4.9
Hart, J. B., Xavier University, Ohio, 9–5.12, 13–3.1, 31–2.9
Harvard Project Physics, 6–4.1, 7–1.1, 7–1.4, 7–1.14, 7–2.1, 7–2.18, 7–3.1, 8–1.6, 8–2.5, 8–2.6, 8–2.8, 8–4.1, 8–4.2, 8–8.1, 8–8.6, 9–1.9, 9–4.5, 9–4.15, 15–1.2, 15–1.3, 19–4.12, 25–2.5, 25–2.8, 26–5.3, 27–2.1, 27–2.2, 31–2.18
Hatfield, T. N., University of Houston, 9–4.13, 9–5.11
Hecht, K. (E. Leybold's Nachfolger), 12–3.9, 12–4.6, 12–4.8, 13–2.1, 13–7.4, 15–7.1, 19–7.1, 38–5.8, 38–5.11, 38–5.12, 39–1.6, 41–1.2 to 41–1.6, 41–2.11
Henry, R. L., Wabash College, 15–10.9
Hersh, H. N., Zenith Radio Corporation, 40–1.5
Hilsch, R., I. Physikalisches Institut der Universität Göttingen, West Germany, 10–2.17
Hilton, W. A., William Jewell College, 8–5.3
Hirsch, Lester, California State College at Los Angeles, 7–1.13, 7–2.11, 13–7.11, 18–5.5, 18–8.3, 30–4.5, 30–5.3, 31–1.17, 31–1.20, 33–1.3
Hirschberg, K., I. Physikalisches Institut der Universität, Giessen, West Germany, 39–1.8
Hoag, J. B., University of Florida, 18–6.3, 18–8.11, 29–3.1
Howey, J. H., Georgia Institute of Technology, 11–1.16, 18–8.12, 19–5.3, 29–1.27, 32–3.22, 33–7.2
Huber, Paul K., University of Basel, Switzerland, 27–7.2, 35–1.1, 35–7.7, 38–6.5, 41–2.6, 41–2.12, 41–2.13, 41–3.5, 41–4.3
Hudson, A. M., Occidental College, 8–1.10, 18–4.1, 18–5.7, 34–2.2, 38–6.7
Hulsizer, Robert I., Jr., Massachusetts Institute of Technology, 38–2.3
Huse, Guy, Rensselaer Polytechnic Institute, 7–2.17, 8–8.9
Hwang, C., University of Minnesota, 9–4.10

Iona, Mario, University of Denver, 12–3.7, 12–3.8, 30–2.1, 33–1.4, 38–5.6

Jaumann, J., University of Köln, West Germany, 31–1.12, 32–1.2, 32–1.3, 32–3.6, 32–4.4, 32–4.5
Jensen, A. S., U.S. Naval Academy, 19–6.4
Jensen, H. C., Lake Forest College, 6–1.2, 13–7.5, 15–1.7, 15–1.16, 15–3.2, 15–3.3, 15–9.6, 16–4.7, 19–4.1,

26–5.10, 27–6.6, 30–1.1, 31–1.9, 31–1.13, 31–2.13, 31–2.19, 33–2.10, 33–3.7, 35–6.4
Jeong, T. H., Lake Forest College, 7–1.10
Johnson, R. C., Westminster College, Pa., 12–5.4

Kansas State University, 16–5.5, 18–8.1
Karlsches, A., I. Physikalisches Institut der Universität, Giessen, West Germany, 35–1.5
Katz, Robert, Kansas State University, 27–3.5
Kenworthy, R. W., University of Washington, 29–1.13
King, J. G., Massachusetts Institute of Technology, 19–4.16, 27–8.4
Kirkpatrick, Paul, Stanford University, 8–3.4
Koehl, G. M., George Washington University, 16–2.5
Kruse, O. E., Texas A & I University, 19–5.5

Lambe, E. D., Washington University, 7–1.18, 9–5.13, 10–2.2, 10–2.14, 10–2.18, 18–6.2, 25–2.7, 29–1.28, 38–2.4, 38–7.4
Landsman, Leon, Newark College of Engineering, 16–4.3
Lavatelli, L. S., University of Illinois, 9–5.15, 10–2.7, 10–2.8, 10–2.12
Lefgren, E., University of Minnesota, 41–3.6
Lefler, R. W., Purdue University, 40–1.12
Leifson, S. W., University of Nevada, 16–4.9
Leitner, Alfred, Michigan State University, 8–6.1, 8–7.1, 13–6.2, 17–5.1, 17–7.5, 18–5.9, 19–3.6
Leonard, R. W., University California, Los Angeles, 18–1.1
Likely, J. G., University of Minnesota, 26–5.11, 29–1.16
Lind, David A., University of Colorado, 12–4.7
Little, E. M., U.S. Navy Electronics Laboratory, San Diego, 31–3.3
Llewellyn, R. A., Purdue University, 9–4.4, 19–7.2, 32–3.18, 33–6.2, 38–3.1, 38–6.1, 38–6.3, 38–6.4, 41–1.9, 41–2.7
Louisiana State University, 14–2.3, 26–5.13, 31–1.3
Lyman, E. M., University of Illinois, 26–5.1, 27–8.6, 31–2.21, 38–5.5

Mackay, R. S., University of California, Berkeley, 13–4.4
Mainardi, Marcus N., Newark College of Engineering, 8–1.4, 8–4.9, 14–3.5, 16–4.3, 18–3.3, 18–7.6, 25–2.1, 30–1.7, 34–1.10, 34–1.13, 34–1.15
Mason, P. R., Rose Polytechnic Institute, 31–2.2
Massachusetts Institute of Technology, 9–2.2, 13–5.1, 27–8.5, 30–3.2, 32–3.7, 33–2.8, 39–4.2
Maxwell, H. N., Ohio Wesleyan University, 35–1.2
McCay, M. S., University of Chattanooga, 39–5.1
McGovern, J. D., Brother Rice High School, Birmingham, Mich., 19–2.1, 30–2.5
Medicus, H. A., Massachusetts Institute of Technology, 39–2.2; Rensselaer Polytechnic Institute, 41–3.1
Meiners, H. F., Rensselaer Polytechnic Institute, 7–1.8, 7–1.15, 7–2.6, 7–2.16, 7–2.17, 7–3.2, 8–3.15, 8–4.11, 8–8.7, 8–8.9, 9–3.8, 9–5.8, 9–5.16, 10–2.1, 10–2.4, 10–2.5, 10–2.10, 10–2.13, 12–5.6, 12–5.7, 12–6.6, 18–1.6, 19–3.4, 19–6.7, 19–6.8, 27–6.5, 27–7.7, 31–1.11, 33–7.5 to 33–7.16, 34–2.7, 38–7.5
Meissner, Hans W., Stevens Institute of Technology, 33–4.1
Mellen, W. R., Lowell Technological Institute, 29–1.8
Merzbacher, Eugen, University of North Carolina, 41–4.7
Meyerhof, W. E., Stanford University, 27–7.6
Michigan State University, 17–2.11, 17–3.3, 31–1.25
Miller, J. S., El Camino College, Calif., 8–5.2, 13–7.12, 16–2.3, 16–5.2
Minnegerode, G. v., I. Physikalisches Institut der Universität Göttingen, West Germany, 10–2.17
Mitchell, J. P., New Mexico State University, 9–5.14
Moore, A. D., University of Michigan, 17–8.1, 29–1.23, 29–1.25, 31–1.30
Morris, L. W., Louisiana State University, 9–5.5, 15–10.15, 17–4.3, 19–3.2, 29–1.18, 34–2.3, 34–2.5, 35–2.7
Mueller, Hans, Massachusetts Institute of Technology, 35–2.9
Muldawer, Leonard, Temple University, 18–8.8, 19–6.8, 38–7.2, 40–1.17

Nahshol, D., Tel Aviv University, Israel, 8–4.10, 32–1.1
Nickel, W. E., Iowa State University of Science and Technology, 8–2.4, 9–4.20, 13–7.8, 38–5.1
Nelson, E. B., Iowa State University of Science and Technology, 6–4.12, 8–2.4, 9–5.4, 16–4.4, 29–1.9, 30–2.3, 38–5.1
Nelson, R., Occidental College, 11–1.15, 15–10.7
Ney, E. P., University of Minnesota, 41–3.6
Nicodemus, D. B., Oregon State University, 9–4.11
Nolle, A. W., University of Texas, 9–1.7, 15–10.7
Nye, E. R., University of Minnesota, 9–2.4, 39–3.1

Oak Ridge Institute of Nuclear Studies, 33–1.2
Occidental College, 13–5.15
Ogier, W. T., Pomona College, 38–2.2
Ogura, T., Central Laboratory, Tokyo Shibaura Electric Co., Ltd., 19–7.4
Oppenheimer, F., University of Colorado, 8–5.8, 9–1.8, 9–4.14, 9–4.18, 15–10.6, 15–10.8, 27–4.2, 29–1.21, 29–4.13, 31–1.26, 31–2.3, 33–1.1, 34–1.1, 34–1.6, 35–3.2, 35–6.1, 38–6.2; University of Minnesota, 41–3.6
Orear, Jay, Cornell University, 10–2.16
Overbeck, C. J., Northwestern University, 15–1.9, 18–5.11

Palmer, C. H., Johns Hopkins University, 34–1.2, 34–1.23, 34–1.25, 34–1.27, 35–2.8, 35–5.1, 35–5.2, 35–7.8
Palmer, Frederic, Jr., Haverford College, 17–2.12
Pany, F. G., Dickson Electronics Corp., Scottsdale, Arizona, 9–1.1
Parmley, T. J., University of Utah, 12–4.3
Pearlstein, E. A., University of Nebraska, 16–2.10
Physical Science Study Committee, 38–1.1, 38–1.2

Plummer, W. T., Johns Hopkins University, 38–5.13
Pohl, R. W., University of Göttingen, West Germany, 6–2.11, 34–1.4, 34–1.20, 34–1.26, 35–2.11, 35–3.4, 38–3.2, 38–5.7, 38–6.8, 38–7.7
Pomona College, 6–2.5, 8–5.1, 17–2.9
Powers, G. F., Northeast Louisiana State College, 41–2.4
Princeton University, 8–4.9, 15–3.1, 41–1.8
Pryor, M. J., State University of New York at Albany, 27–2.6, 31–2.5, 34–1.14, 35–2.13

Randall, R. H., City College of The City University of New York, 19–6.6
Rensselaer Polytechnic Institute, 7–1.2, 9–4.22, 13–7.10, 19–3.5, 39–3.2
Resnick, Robert, Rensselaer Polytechnic Institute, 19–6.7
Rhein, R. J., Spring Hill College, 38–6.6
Rice, Michael W., Smith College, 7–2.20
Rice University, 15–1.5
Richards, H. F., Florida State University, 31–2.10
Robinson, B. L., Case Western Reserve University, 38–4.1
Robinson, G. F., Rensselaer Polytechnic Institute, 7–1.9
Robinson, H. A., Adelphi University, 8–1.3, 8–2.1, 8–3.11, 9–4.7, 9–4.8, 12–2.1, 12–2.2, 12–6.4, 13–5.6, 15–1.12, 15–1.15, 15–5.1, 15–10.12, 16–2.1, 16–2.11, 16–2.12, 16–3.3, 16–4.10 to 16–4.12, 17–2.2, 17–2.7, 17–2.8, 17–2.13, 17–8.3 to 17–8.5, 18–1.4, 18–5.3, 18–5.4, 18–8.10, 19–4.5, 19–9.1, 19–9.2, 19–9.3, 19–10.1, 26–4.7, 26–5.4 to 26–5.7, 26–5.16, 26–5.17, 26–5.18, 27–3.11, 27–5.2, 29–1.2, 29–2.6, 29–4.4, 29–4.5, 29–4.14, 30–3.3, 30–6.8, 30–6.9, 31–1.22, 31–1.29, 32–2.2, 32–3.1, 32–3.2, 32–3.3, 32–3.11, 32–3.13, 32–3.26, 32–4.1, 32–4.2, 32–4.3, 33–3.2, 34–1.5, 35–2.4, 35–2.5, 35–2.10, 35–3.9, 35–5.3, 35–6.7, 35–6.8, 35–6.10, 35–6.12 to 35–6.17, 38–7.1, 39–4.5
Rollefson, Ragnar, University of Wisconsin, 39–1.2
Rogers, Eric M., Princeton University, 6–4.11, 7–1.5, 7–2.7, 8–3.8, 8–3.12, 8–3.13, 10–2.9, 13–5.11, 16–5.1, 16–5.6, 16–5.10, 26–4.2, 27–7.8, 27–10.2, 27–9.1, 30–4.3, 31–1.2, 31–2.4, 32–3.5, 32–3.14, 33–3.1, 34–2.4, 35–3.5, 41–2.2, 41–2.3, 41–3.7
Rood, Paul, Western Michigan University, 29–1.10
Royal Institution of Great Britain, London, 15–10.3, 17–10.1, 18–8.15, 30–1.2, 30–2.10, 32–3.20, 32–3.21, 39–3.2, 40–1.2

Satterly, John, University of Toronto, Canada, 27–3.9
Saunders, B. G., University of Indonesia, 34–1.16; Lawrence Radiation Laboratory, University of California, Livermore, 38–7.6
Sherr, Rubby, Princeton University, 8–3.10
Scharmann, Arthur, I. Physikalisches Institut der Universität, Giessen, West Germany, 35–1.5, 39–1.8
Schwarz, Guenter, Florida State University, 7–1.3, 7–2.9, 11–2.1, 11–2.3, 15–2.2, 17–2.1, 18–6.4, 19–4.3, 19–5.4, 26–5.14, 32–3.10, 35–7.2, 38–5.9
Scott, G. D., University of Toronto, Canada, 13–6.1
Sears, F. W., Dartmouth College, 7–2.19, 13–4.2, 13–5.5, 15–1.6, 29–1.11
Sells, R. L., State University of New York, College at Geneseo, 7–2.13
Shive, J. N., Bell Telephone Laboratories, 18–2.1
Shoemaker, F. C., Princeton University, 10–2.6, 10–2.11, 10–2.15
Smith, J. E., State College of Arkansas, 41–2.4
Smyth, H. D., Princeton University, 7–1.5, 27–9.1
Soules, J. A., New Mexico State University, 33–2.9
Sprawls, Perry, Jr., Emory University, 9–5.2, 16–4.1, 18–5.1, 29–1.4, 31–1.18, 32–3.25, 41–1.2
Stanford University, 9–4.17, 13–5.15, 27–2.5, 30–1.9
Staub, H. H., University of Zürich, Switzerland, 8–1.8, 9–4.12
Stiteler, F., Rensselaer Polytechnic Institute, 27–6.5
Story, H. S., State University of New York at Albany, 40–1.1, 41–4.1
Straley, J. W., University of North Carolina, 7–1.17, 7–1.19, 9–5.7, 9–5.9, 18–4.4, 18–8.7, 26–5.9, 29–4.1, 30–1.8, 31–2.11, 32–3.16, 35–3.3
Streib, J. F., University of Washington, 38–1.3
Strickler, A., The Allen Strickler Company, Fullerton, Calif., 26–3.6
Strickler, T. D., Berea College, 31–1.6
Stull, J. L., Alfred University, 11–1.17
Sutton, D. C., University of Illinois, 38–6.4
Sutton, R. M., California Institute of Technology, 6–1.3 to 6–1.6, 6–2.2, 6–2.9, 6–2.10, 6–4.9, 7–1.6, 7–2.12, 7–2.14, 7–2.15, 8–1.1, 8–1.2, 8–2.4, 8–3.3, 8–3.9, 8–5.4, 9–1.2, 9–3.2, 9–3.3, 9–5.4, 12–5.1, 12–5.2, 12–5.5, 13–7.3, 13–7.9, 14–2.1, 14–3.1, 14–3.2, 15–8.2, 16–2.6 to 16–2.9, 16–4.6, 16–5.3, 16–5.8, 17–6.3, 19–2.4, 26–4.6, 27–3.2, 27–3.3, 27–3.4, 29–1.5, 31–3.2, 32–3.4, 35–5.4
Suzuoki, Z., Department of General Education, Nagoya City University, Japan, 19–7.4
Swigart, J. I., University of Utah, 14–3.3, 14–3.4, 14–3.6, 15–1.4, 15–6.3, 33–2.3, 39–1.1
Swiss Federal Institute (ETH), Zürich, Switzerland, 6–2.1, 6–2.4, 6–2.8, 6–4.5, 7–1.11, 7–2.2, 8–1.5, 8–1.7, 8–1.9, 8–3.1, 8–3.6, 8–3.14, 8–5.5, 8–5.6, 8–5.7, 9–3.4 to 9–3.7, 9–4.1, 9–4.2, 9–4.3, 9–4.6, 9–4.9, 9–4.16, 9–4.21, 9–5.10, 10–2.3, 10–2.10, 12–2.3, 12–3.1, 12–6.1, 12–6.3, 12–6.8, 13–4.1, 13–4.3, 13–5.7, 13–5.8, 13–5.9, 13–5.13, 13–7.2, 15–1.13, 15–1.14, 15–2.3, 15–2.4, 15–2.5, 15–10.5, 15–10.13, 15–10.14, 16–2.4, 16–5.7, 16–5.9, 17–2.4, 17–2.5, 17–2.6, 17–3.1, 17–3.2, 17–4.1, 17–4.2, 17–6.2, 17–7.2, 17–7.3, 17–7.4, 17–8.6, 17–8.7, 17–9.4, 17–9.5, 17–11.1, 18–1.3, 18–1.7, 18–3.6, 18–5.8, 18–5.10, 18–7.4, 19–2.3, 19–4.8, 19–4.11, 19–5.2, 19–6.1, 19–6.2, 19–7.3, 19–8.1, 25–2.2, 25–2.4, 26–2.1, 26–3.7 to 26–3.10, 26–4.3, 26–4.4, 26–5.2, 26–5.8, 26–6.1, 27–2.4, 27–2.9, 27–3.7, 27–5.1, 27–6.2, 27–6.4, 27–8.2, 27–9.2, 27–10.1, 29–1.14, 19–1.15, 29–1.17, 29–1.22, 29–1.30, 29–2.2, 29–2.3, 29–2.8, 29–3.3, 29–4.2, 29–4.7, 29–4.8, 29–4.11, 29–4.12, 30–1.4, 30–1.5, 30–1.6, 30–3.1, 30–4.1, 30–4.4, 30–4.6 to 30–4.8, 30–5.1, 30–5.2, 30–6.1, 30–6.2, 30–6.4, 30–6.6, 31–

1.4, 31–1.5, 31–1.10, 31–1.19, 31–1.21, 31–1.23, 31–1.24, 31–2.1, 31–2.12, 31–2.15, 31–2.20, 31–3.5, 31–3.6, 31–3.7, 32–2.1, 32–3.9, 32–3.12, 32–3.19, 32–3.23, 32–3.24, 32–4.6, 32–5.1, 33–2.5, 33–2.6, 33–3.3 to 33–3.6, 33–5.1, 33–5.2, 33–6.3, 33–7.3, 33–7.17, 34–1.12, 34–2.6, 35–1.3, 35–1.6, 35–2.3, 35–3.6, 35–3.7, 35–3.9, 35–6.6, 35–6.9, 35–6.18, 38–5.10, 38–7.3, 39–1.7, 39–2.1, 39–2.3, 39–5.2, 40–1.6, 40–1.8, 40–1.13, 40–1.14, 40–1.15, 41–1.8, 41–2.5, 41–2.6, 41–2.10, 41–3.4.

SYED, BASHIR A., University of Saskatchewan, Canada, 30–2.2

TARR, C. E., University of North Carolina, 41–4.7

TAYLOR, J. R., Materials Laboratory, General Electric Co., St. Petersburg, Florida, 9–2.5, 15–10.11, 17–2.3, 19–4.6, 30–6.3

TEA, P. L., JR., City College of The City University of New York, 15–4.1

TEEGARDEN, J. L., Rose Polytechnic Institute, 31–2.2

TELFAIR, DAVID, Washington University, 13–7.13

TENNISSEN, P., Technical University Delft, Holland, 41–3.3

TEXAS WESTERN COLLEGE, 19–4.13

TOTANI, SHINICHIRO, Nagoya City University, Japan, 18–2.2, 18–8.13

TRATHEN, R. H., Rensselaer Polytechnic Institute, 18–1.6

TUITE, D. F., Rensselaer Polytechnic Institute, 15–9.7

TURNBULL, DAVID, Harvard University, 27–7.1

UNIVERSITY OF CALIFORNIA, Berkeley, 6–4.7, 12–6.2, 13–5.15

UNIVERSITY OF COLOGNE, West Germany, 30–4.2

UNIVERSITY OF COLORADO, 15–7.2, 15–10.4, 25–2.6

UNIVERSITY OF ILLINOIS, 6–4.10, 8–4.6, 13–5.14, 15–6.2, 27–2.8, 31–1.8, 31–2.7, 41–1.1

UNIVERSITY OF MINNESOTA, 6–2.6, 6–4.8, 8–3.7, 8–4.5, 8–8.2, 9–2.3, 9–3.1, 12–3.3 to 12–3.6, 12–4.2, 13–5.4, 14–2.2, 15–1.1, 15–7.3, 16–3.1, 16–4.2, 16–5.4, 27–2.7, 27–4.1, 27–6.1, 38–5.3, 38–5.4, 39–4.1

UNIVERSITY OF NEVADA, 29–1.7

UNIVERSITY OF WASHINGTON, 9–5.3, 13–5.2, 19–4.4, 31–1.27, 31–2.8, 35–2.2, 15–9.4

VAN DEN AKKER, J. A., The Institute of Paper Chemistry, Appleton, Wisc., 8–3.2, 13–5.10, 35–7.6

VERBRUGGE, FRANK, University of Minnesota, 41–4.6

VIDAL, R. H., Victoria College, Canada, 29–1.3, 31–1.1

WAAGE, H. M., Princeton University, 7–1.5, 7–2.10, 8–3.8, 8–3.10, 8–3.12, 8–3.13, 8–8.5, 10–2.9, 17–8.2, 18–4.1, 26–4.2, 27–10.2, 29–1.6, 29–2.1, 29–2.4, 34–2.1, 35–7.5, 41–2.2, 41–3.7, 41–3.8

WADE, T. E., JR., University of Nebraska, 9–4.19, 29–2.7, 31–1.14

WALL, C. N., University of Minnesota, 8–3.5

WALTERS, W., Washington University, 39–4.4

WARD, G. H., Pennsylvania State University, 8–8.8, 9–1.4

WARREN, W., Washington University, 38–7.4

WATTERS, R. L., Research Laboratories, General Electric Co., Schenectady, N.Y., 19–2.1

WEBB, H. H., Pennsylvania State University, 10–3.4

WEBER, L. R., Colorado State University, 12–3.2, 14–3.7, 18–5.2, 19–3.3, 19–4.14, 19–4.15, 26–3.2, 35–7.1

WEBER, R. L., Pennsylvania State University, 6–2.3, 38–5.2

WEHR, M. R., Drexel Institute of Technology, 18–7.3, 18–8.9, 19–4.10, 27–8.3, 29–5.1, 35–6.11

WEIDNER, R. T., Rutgers University, 7–2.13

WEINBERG, J. W., Case Western Reserve University, 38–4.1

WELCH, G. B., Northeastern University, 38–2.1

WELTIN, HANS, University of California, Berkeley, 31–2.22

WERHEIT, H., University of Köln, West Germany, 27–10.3, 31–1.12, 32–1.2, 32–1.3, 32–3.6, 32–4.4, 32–4.5, 39–4.3

WEYMOUTH, J. W., University of Nebraska, 6–3.1, 6–4.6

WHITE, H. E., University of California, Berkeley, 6–2.7, 8–8.3, 18–5.6, 31–2.22, 35–6.3, 35–7.3, 41–4.4

WHITE, M. G., Princeton University, 41–2.6

WHITING, P., University of Minnesota, 9–4.10, 27–8.1

WILLIAMS, C. E., University of New Mexico, 34–1.11

WOOD, E. A., Bell Telephone Laboratories, 34–1.24

WORLEY, R. E., University of Nevada, 17–7.1, 39–1.5

WYNOBEL, S., Technical University Delft, Holland, 41–3.3

YOUNGER, PHILIP G., St. Cloud State College, Minn., 31–1.28

ZELENY, J., Yale University, 19–6.3

LIST OF MANUFACTURERS

Ace Plastic Company
Jamaica, New York 11418

Airmec Laboratories, Ltd.
Buckinghamshire, England

Aluminum Company of America
Los Angeles, California 90017

Allegri-Tech, Inc., Science Department
Nutley, New Jersey 07110

Allen-Bradley Company
Milwaukee, Wisconsin 53204

Allied Radio Corporation
Chicago, Illinois 60680

Amatom Electronic Hardware Company
New Rochelle, New York 10801

American Electrostatic Company
Tulsa, Oklahoma 74152

American Institute of Physics
New York, New York 10017

American Optical Company
Instrument Division
Buffalo, New York 14215

J. Ammann, Kunstverlag
Spiel-und Papeteriewaren en gros
Kartonagen, Uberlandstrasse 178
8057 Zurich, Switzerland

Applied Science, Inc.
Dallas, Texas 75219

Argonne National Laboratories, Film Center
Argonne, Illinois 60440

Atlas Sound Corporation
Brooklyn, New York 11218

Audio-Visual Center
Michigan State University
East Lansing, Michigan 48823

Audio-Visual Center
Purdue University
Lafayette, Indiana 47907

Awyco AG.
Otten, Switzerland

Baird-Atomic, Inc.
Cambridge, Massachusetts 02138

Balzers AG
Balzers, Liechtenstein

Bausch & Lomb, Inc.
Rochester, New York 14602

Beckman Instruments, Inc.
Fullerton, California 92632

Film Library, Bell Telephone Laboratories
Murray Hill, New Jersey 07971

The Bendix Corporation
Cincinnati Division
Cincinnati, Ohio 45241

Charles Beseler Company
East Orange, New Jersey 07018

Robert Bosch Gmbt.
7 Stuttgart 1, West Germany

Better Bubbles, Inc.
North Hollywood, California 91605

Carl H. Biggs Company
Santa Monica, California 90404

Bodine Electric Company
Chicago, Illinois 60618

The British Thomson Houston Company, Ltd.
Rugby, England

Charles Bruning Company
Mount Prospect, Illinois 60056

Bud Radio, Inc.
Willoughby, Ohio 44094

Buhl Optical Company
Pittsburgh, Pennsylvania 15233

Cadillac Plastic and Chemical Company
Detroit, Michigan 48203

Cambosco Scientific Company
Boston, Massachusetts 02135

Dr. A. Capecelatro
c/o Madison Associates, Inc.
Madison, New Jersey 07940

Canadian Laboratory Supplies, Ltd.
Montreal, Canada

Canadian Laboratory Supplies, Ltd.
Toronto 18, Canada

Central Scientific Company
Chicago, Illinois 60613

C & H Sales Company
Pasadena, California 91107

Chicago Metal Manufacturing Company
Chicago, Illinois 60632

Cohu Electronics
San Diego, California 92112

Communication Papers, Inc.
Scranton, Pennsylvania 18501

Computer Measurements Company
San Fernando, California 91342

Cornell–Dubilier Electronics
Newark, New Jersey 07101

Corning Glass Works
Corning, New York 14830

Convair Motion Picture Section
Department 98–80
San Diego, California 92112

Cooper Brothers Company
Port Washington, New York 11050

CRT Manufacturing Company
St. Louis, Missouri 63130

Curtis Automotive Devices, Inc.
Bedford, Indiana 47421

Dage-Bell Corporation
Michigan City, Indiana 46360

Dayton Electric Manufacturing Company
Chicago, Illinois 60612

The De Vilbis Company
Toledo, Ohio 43612

Dow Chemical Company
Midland, Michigan 48640

Dumaco, G. Manta, Ing.
Biel, Switzerland

DuMont Laboratories
Division of Fairchild Camera
Clifton, New Jersey 07012

E. I. DuPont de Nemours & Company
Polychemical Department
Wilmington, Delaware 19801

Dynamit Nobel AG
Köln, West Germany

Ealing Corporation
Cambridge, Massachusetts 02140

Eastern Industries
Hamden, Connecticut 06514

Eastern Metals Wholesale, Inc.
Albany, New York 12201

Eastman Kodak Company
Rochester, New York 14650

Eaton Manufacturing Company
Axle Division
Cleveland, Ohio 44108

Eaton Yale & Towne, Inc.
Cleveland, Ohio 44114

Eberbach Scientific Corporation
Ann Arbor, Michigan 48106

Edgecomb Steel of New England
Nashua, New Hampshire 03060

Edmund Scientific Company
Barrington, New Jersey 08007

Educational Television Office
State University of New York
New York, New York 10017

Electro Optics Associates
Palo Alto, California 94306

EMA AG
Meilen/Zurich, Switzerland

Electronic Associates, Inc.
West Long Branch, New Jersey 07764

Empire Magnets, Inc.
Maumee, Ohio 43537

Encyclopaedia Britannica
Educational Corporation
Chicago, Illinois 60611

Fansteel Metallurgical Corporation
North Chicago, Illinois 60064

Fenwal Electronics, Inc.
Framingham, Massachusetts 01701

Film Librarian
Educational Development Center
Newton, Massachusetts 02160

Fisher Scientific Company
Pittsburgh, Pennsylvania 19406

Forest Products, Inc.
Cambridge, Massachusetts 02139

Fostoria Corporation
Horsham, Pennsylvania 19004

Gaertner Scientific Corporation
Chicago, Illinois 60614

LIST OF MANUFACTURERS

Gardner Cryogenics Corporation
Hightstown, New Jersey 08520

George W. Gates Company
Franklin Square, New York 11010

General Aniline and Film Corporation
New York, New York 10020

General Electric Company
Metallurgy and Ceramics Research Department
Research Laboratory
Schenectady, New York 12305

General Electric Company
Receiving Tube Department
Owensboro, Kentucky 42301

General Electric Company
Apparatus Sales
Schenectady, New York 12305

General Precision, Inc., GPL Div.
Aerospace Group
Pleasantville, New York 10570

General Radio Company
West Concord, Massachusetts 01781

Glidden Glass Corporation
Cleveland, Ohio 44106

Gomar Manufacturing, Inc.
Linden, New Jersey 07036

B. F. Goodrich Company
Sponge Rubber Division
Shelton, Connecticut 06484

Guardian Electric Manufacturing Company
Chicago, Illinois 60607

Hans Hamberger AG
Zurich, Switzerland

Hardware Products Company
Boston, Massachusetts 02113

Harshaw Chemical Company
Cleveland, Ohio 44106

Hartford Steel Ball Company
Rocky Hill, Connecticut 06067

Hartmann and Braun
Frankfurt am Main, West Germany

Hastings Aluminized Films
Philadelphia, Pennsylvania 19103

The Heath Company
Benton Harbor, Michigan 49022

H. Heeb
Abt. Vorlesung, Physics Institute
Swiss Federal Institute of Technology (ETH)
8006 Zurich, Switzerland

Gerald K. Heller Corporation
Las Vegas, Nevada 89106

Hewlett-Packard Company
Palo Alto, California 93040

Hoke, Inc.
Cresskill, New Jersey 07626

Holt, Rinehart and Winston, Inc.
New York, New York 10017

Hoover Ball Division
Erwin, Tennessee 37650

Hoover Ball and Bearing Company
Ann Arbor, Michigan 48107

Houston Instrument Corporation
Bellaire, Texas 77401

Hunter Manufacturing Company, Inc.
Iowa City, Iowa 52240

Indiana General Corporation
Magnet Division
Valparaiso, Indiana 46383

Indiana University
Audio-Visual Center
Bloomington, Indiana 47401

Industrial Research
Annual Buyer's Guide
Beverly Shores, Indiana 46301

Institut für den Wissenschaftlichen Film
Göttingen, West Germany

International Film Bureau, Inc.
Chicago, Illinois 60604

Instructional Media Center
Michigan State University
East Lansing, Michigan 48823

A. Jaegers Optical Corporation
Lynbrook, New York 11563

E. F. Johnson Company
Waseca, Minnesota 56093

Welton V. Johnson Engineering Company
Summit, New Jersey 07901

Kerotest Mfg. Company
Pittsburgh, Pennsylvania 15222

Keithley Instruments, Inc.
Cleveland, Ohio 44106

Keuffel and Esser Company
Hoboken, New Jersey 07030

Kimberly-Clark Corporation
Neenah, Wisconsin 54956

Kimble Glass Company
Boston, Massachusetts

Kipp and Zonen
Delft, The Netherlands

Klinger Scientific Apparatus Corporation
Jamaica, New York 11432

Kojima Megane-Koki Company
Tokyo, Japan

Krupp Widia Fabrik
Essen, West Germany

Lab T.V.
New York, New York 10019

LIST OF MANUFACTURERS

Laboratory Management
New York, New York 10016

LaFayette Radio Electronics Corporation
Syosset, New York 11791

LaPine Scientific Company
Chicago, Illinois 60629

Lectronic Research Laboratories, Inc.
Philadelphia, Pennsylvania 19106

Leeds and Northrup Company
Philadelphia, Pennsylvania 19144

E. Leybold
5 Koln-Bayental, West Germany

Los Alamos Scientific Laboratory
Report Librarian
Los Alamos, New Mexico 87544

Macalaster Scientific Company
Nashua, New Hampshire 03060

A. D. Mackay, Inc.
New York, New York 10038

Madison Associates, Inc.
Madison, New Jersey 07940

Marsh Instrument Company
Skokie, Illinois 60076

H. S. Martin Company
Evanston, Illinois 60202

Maser Optics, Inc.
Boston, Massachusetts 02134

Matheson, Coleman and Bell
East Rutherford, New Jersey 07073

M. B. Electronics
New Haven, Connecticut 06515

McGraw-Hill Book Company
Text-Film Division (sales),
Contemporary Films (rental)
New York, New York 10036

McKnight and Company
Menlo Park, California 94025

Mettler Instrument Corporation
Hightstown, New Jersey 07901

Mettler, Pelikanstrasse 19
Zurich, Switzerland

Minnesota Mining and Manufacturing Company
3M Visual Products
St. Paul, Minnesota 55119

Miratel Electronics, Inc.
St. Paul, Minnesota 55119

Modern Learning Aids
New York, New York 10036

C. V. Moffoletto
Baltimore, Maryland 21234

Monsanto Company, Inorganic Chemicals
St. Louis, Missouri 63166

Motion Picture Division
The Ohio State University
Columbus, Ohio 43210

*Movie*Mite Company
Detroit, Michigan

Nathan Manufacturing Division
Wegner Machinery Corporation
Long Island City, New York 11106

National Audio-Visual Association, Inc.
Fairfax, Virginia 22030

National Film Board of Canada
New York, New York 10019

National Instructional Films
Nanuet, New York 10954

National Research Council of Canada
Ottawa, Ontario, Canada

National Research Corporation
Equipment Division
Newton Highlands, Massachusetts 02161

Norton Company
(Formerly Behr-Manning Corporation)
Watervliet, New York 12189

New England Scientific Instrument
Natick, Massachusetts 01760

Nuclear Science and Engineering Corporation
Pittsburgh, Pennsylvania 15236

Office of Institutional Research
Rensselaer Polytechnic Institute
Troy, New York 12181

Optics Technology, Inc.
Palo Alto, California 94304

John Oster Manufacturing Company
Milwaukee, Wisconsin 53217

Patterson's Source Guide for Educational
Materials and Equipment
Educational Directories, Inc.
Mount Prospect, Illinois 60056

PEK Laboratories
Sunnyvale, California 94086

Park Chemical Company
Detroit, Michigan 48204

Perkin-Elmer Corporation
Electro-Optical Division
Norwalk, Connecticut 06852

Permag Corporation
Jamaica, New York 11418

Phywe AG.
Göttingen, West Germany

Philips AG.
The Netherlands

Plasteel Corporation
Inkster, Michigan 48141

Plymouth Rubber Company, Inc.
Canton, Massachusetts 02021

LIST OF MANUFACTURERS

Polaroid Corporation
Cambridge, Massachusetts 02139

Popular Science Publishing Company, Inc.
Audio Visual Division
New York, New York 10017

Potter and Brumfield
Division of American Machine and Foundry Company
Princeton, Indiana 47570

Precision Film Laboratories, Inc.
New York, New York 10036

Projection Optics Company, Inc.
East Orange, New Jersey 07018

W. G. Pye and Company, Ltd.
Granta Works
Cambridge, England

Pyrotronics
Union, New Jersey 07083

Quartzlampengesellschaft Heraeus
Hanau, Germany

Radiation Counter Laboratories, Inc.
Skokie, Illinois 60076

Ransom and Randolph Company
Toledo, Ohio 43604

Peter M. Robeck and Company, Inc.
New York, New York 10017

Rohm and Haas Company
Philadelphia, Pennsylvania 19105

Rosco Laboratories
Brooklyn, New York 11228

Sager-Spuck Supply Company, Inc.
Albany, New York 12207

Sargent-Welch Scientific Company
Skokie, Illinois 60076

Science, Annual Guide to Scientific Instruments
Washington, D.C. 20005

Science Electronics
Nashua, New Hampshire 03060

Seal-A-Metic Corporation
Patterson, New Jersey 07522

The Search Corporation
Tallahassee, Florida 32302

Semi-Elements, Inc.
Saxonburg, Pennsylvania 16056

Semimetals, Inc.
Brooklyn, New York 11209

Shannon Luminous Materials Company, Inc.
Los Angeles, California 90046

Shell Oil Company
Shell Films Library
Indianapolis, Indiana 46204

Shure Brothers, Inc.
Evanston, Illinois 60202

Simplex Products Company
Cleveland, Ohio 44120

Sinclair-Koppers Company
Pittsburgh, Pennsylvania 15219

Herman H. Smith, Inc.
Brooklyn, New York 11207

SMS Instrument Company
Castleton-on-Hudson, New York 12033

Spectra Physics, Inc.
Mountain View, California 94040

Sprague Electric Company
North Adams, Massachusetts 01247

Stancor Electronics, Inc.
Chicago, Illinois 60618

Standard Electric Time Company,
Standard Scientific Company
Springfield, Massachusetts 01102

Stark Electronic Instrument, Ltd.
Ajax, Ontario, Canada

L. S. Starrett Company
Athol, Massachusetts 01331

Allen Strickler Company
Fullerton, California 92631

Stocker and Yale, Inc.
Marblehead, Massachusetts 01945

Strobolite Company
New York, New York 10036

Superior Air Products, Inc.
Newark, New Jersey 07105

Sylvania Electric Products, Inc.
Woburn, Massachusetts 01801

Systron-Donner Corporation
Concord, Massachusetts 94520

Tecnifax Corporation
Holyoke, Massachusetts 01042

Tektronix, Inc.
Beaverton, Oregon 97005

Tracerlab
Richmond, California 94804

Technical Animations, Inc.
Port Washington, New York 11050

Arthur H. Thomas Company
Philadelphia, Pennsylvania 19105

Thordarson-Meissner Manufacturing Company
Mt. Carmel, Illinois 62863

Trans-Sonics, Inc.
Lexington, Massachusetts 02173

Trüb, Täuber and Company
Zurich, Switzerland

Ultra-Violet Products, Inc.
San Gabriel, California 91776

Uniform Tubes, Inc.
Collegeville, Pennsylvania 19426

LIST OF MANUFACTURERS

Union Carbide Corporation
Nuclear Division
Oak Ridge, Tennessee 37831

Union Carbide, Linde Division
Indianapolis, Indiana 46224

United Transparencies, Inc.
Binghamton, New York 13902

Universal Education and Visual Arts
New York, New York 10003

Universal Relay Corporation
New York, New York 10013

Universal Transistor Products
Copiague, New York 11726

University Laboratories, Inc.
Berkeley, California 94703

Utility Brass and Copper Corporation
Brooklyn, New York 11231

Vacuum Electronics Corporation
Plainview, New York

Vacuum Electronics Manufacturing Company
New Hyde Park, New York 11040

Vacuum Research Mfg. Company
San Ramon, California 94583

Van der Heem, N.V.
Hague, The, Netherlands

Visual Systems, Inc.
Milwaukee, Wisconsin 53201

Waldom Electronics
Chicago, Illinois 60632

Wallace and Tiernan, Inc.
East Orange, New Jersey 07018

Wang Laboratories, Inc.
Tewsbury, Massachusetts 01876

CWS Waveguide Corporation
Lindenhurst, New York 11757

Western Metals Wholesale, Inc.
Albany, New York 12201

Western Union Telegraph Company
Development and Research Dept.
New York, New York 10013

Wham-O Corporation
La Jolla, California 92037

Wieland-Werke AG., Metallwerke
79 Ulm/Donau, West Germany

Henry Wiggin and Company
Birmingham, England

Yellow Springs Instrument Company
Yellow Springs, Ohio 45387

Zero Max Company
Minneapolis, Minnesota 55408

INDEX

Asterisks (*) indicate demonstration experiments.

***Boldface numbers** indicate material in the chapter appendices at the end of each volume.*

Abbe refractometer, 1038
Abbe theory of microscope, reference for, 1144
aberration in a lens, 1049–50, ***1345–46***
  astigmatism, 1050, 1144
  chromatic, 1050
  coma, 1049–50, 1144
  spherical, 1049, 1144
    lack of at aplanatic points, 1051–52
absorption coefficient, 506–7
  of water, calculation of, 1205
absorption of color, selective, 1113–15
absorption of gamma rays
  model, 1246
  resonance, model, 1257–58
absorption of light, 1096–97, 1205
  resonance, by atoms, model of, 1257–58
absorption of neutrons, 1240
absorption of radiant energy, 759–63
  connection of with emission, 763
  dem. of using a thermicon card, 759–62
  dependence of on wavelength, 760–62
  infrared, 1177
  relative, of black and white surfaces, 1176
  resonance, in infrared region, 762
absorption spectra, 1203–5, 1207–8
acceleration
  by air jet: the rocket principle, 259–60
  angular
    calculation of by graphical differentiation, 281–82
    * *as a function of applied torque, 268*
    measurement of for different mass distributions, 275
    in Newton's Second Law, 284
    of a torsional pendulum, 355–56

acceleration (*cont.*)
  apparent change of weight due to, 147–48, 150–51
  of a body in a fluid, 143–44, 146–47
  of a body undergoing rotational and translational motion, 288–89
  centripetal
    body in a fluid, 145–46
    * *candle on a revolving air puck, 229–30*
    meas. of using an accelerometer, 160
  and change of momentum, 201–3
  in circular motion, 337
  correlation with displacement and velocity, 106
  deformation produced by, 136–37
  due to gravity
    * *using an electromagnetic timer, 116–17*
    * *using an inclined air track, 256*
    * *oscillation in a potential well, 256–57*
    * *using a phonograph turntable, 114*
    * *using a photoelectric timer, 118–19*
  on an inclined plane, 105
    using air pucks, 224–26
  * *linear and angular, 230–31*
  meas. of, 102–5
  of points on a hinged stick, 272
  production of, 135
  and reaction forces, 199
  recording of, 338
  relationship between force and, 137–38
  and relative velocity, 131–33
  of a rotating body with viscous resistance, 407–8
  in simple harmonic motion, 337
  uniform, 101–2, 114–15

acceleration carts and track demonstration, 102–5
accelerometer, principle of, 160
* *acoustic baffles, resonators, and radiators, 502–3*
acoustic Doppler shift experiments, 516–19, ***649–52***
  theory of, ***649–50***
acoustic resonance apparatus, 501–2
acoustical impedance, 901–2
activity, 1244
  at secular equilibrium, 1244–45
addition, vector
  simultaneous velocities, 97
  uniform motion, 97
  of velocities, 92–3, 98
adiabatic compression
  * *changes in thermal energy of air undergoing, 800–1*
  * *liberation of heat in, 800*
  * *temperature change in a gas undergoing, 800*
adiabatic demagnetization, 988–89, ***1334***
  discussion of, 988–89
adiabatic expansion, temperature change in a gas undergoing, 801
adiabatic processes, 800–4
after-image, 88
air
  conduction of electrons in, due to visible carriers, 39
  electrical conductivity due to heat, 38, 39, 895
  index of refraction of, 1039
    observation of small changes in, 1056–59, ***1352***
  pressure, 383
    effect of on boiling point of water, 792, 793
    force of, 383–84

air (*cont.*)
thermal conductivity of, qualitative dem., 798–800
* *weight of, 382*
* *air-bearing gyro, 266–67,* ***587–88***
air-bearing gyroscope, 307–8, 309–11, ***594,*** 1277–78
air-bearing pulley, 255, 263–64
U. of Southern California, 264
U. of Florida, 264–65
air-bearing table, 267
air bearings, rotary, 263–65
air-stream generator, 398
air-supported pucks, 222–24
captive, 222–23
compressed air, 223
dry ice, 223
electrically charged, ***1318–19***
free, 222, 223–24
launching ram for, 224, 1319–20
liquid air, 224
magnetic, 227
air table
New Mexico State U., 242–43, ***580–81***
Pennsylvania State U., 243–44, 248, ***581–82***
air tracks
deformation of, 256–57
magnetic release for glider, 252, 254
radius of curvature of, 257
Stull–Ealing linear, 251–52
air troughs
Cal. Inst. Tech., 250–51, ***582–87***
jet construction of, ***584–85***
jet pattern of, ***585–86***
Search linear, 251
Airy disk, 1131
alpha-particle counter, 1270–71
alpha particles
observation of
in an alpha-particle counter, 1270–71
in a cloud chamber, 1266–70
scattering of, 49
model for, 1249–50
alternating current
circuits, 992–96
effect of on a capacitor, 869–70
effects on impedance of high-frequency, 1007–9
* *of a capacitor and an inductor, 1007*
* *in four concentric conductors, 1008–9*
fluctuating nature of, 880–81
generation of, 908
Ampere's law, 924–27
Ampere's motor apparatus, 909
* *amplification of sound, 524–26*
using a piezoelectric crystal, 901–2, 903, 905–7
amplification of travelling waves, 449–50
theory of, ***623–25***
amplitude of forced oscillations, ***604–9***
equations for, simple harmonic motion, ***604–7***
equations for, torsion pendulum, ***608–9***
analog computer
analysis of coupled systems, 361–63
analysis of particle motion, 127, ***546–47***
basic principle of, 887–88
angular acceleration (*see* acceleration, angular)
angular correlation, of pair annihilation radiation, 1175–76
angular impulse, 297
angular momentum (*see* momentum, angular)
angular velocity (*see* velocity, angular)
anharmonic behavior in a lattice, 263
annealing of a nickel rod, 985
antistatic polish, 845
aplanatic points of a sphere, 1051–52
arc, electric, importance of a hot cathode in sustaining, 882
Archimedes, 143
Archimedes' Principle, 143ff, 378–80
* *block immersed in water, 378*
* *boat in a lake, model, 378*
* *buoyant effect of air, 379*
* *disappearance of buoyant force, 379–80*
* *iron floating on water, 378–79*
* *tube in water and kerosene, 379*
* *variation of buoyant force with temperature, 380*
Arkad'ev capacitor-bank transformer, principle of, 871
astigmatism, 1050, ***1346–47,*** 1144
atmospheric pressure
meas. of, 384
role of in the operation of a siphon, 386–88
atomic collisions, 56
atomic mass, 55
atomic models, 1217–19
* *instability of orbital motion of a classical radiating electron, 1219*
* *Thomson model, forces of, 1217–18*
atomic physics, 1203ff
need for dem. in, 55–57
suggestions for dem. in, 54–55
atomic spectra, 1202–8
absorption, 1203–5, 1207–8
* *absorption bands of water, 1205*
* *absorption and emission spectrum of sodium, 1207–8*
* *absorption of radiation by a Bunsen flame, 1205*
* *absorption spectrum of mercury, 1205*
* *absorption spectrum of sodium, 1203–4*
emission, 1202–3, 1207–8
* *lifetime of 6^3P_1 mercury state by Hanle effect, 1205–7*
atomic spectra (*cont.*)
* *line spectra of gases, 1202–3*
* *ultraviolet cyanogen bands, 1203*
* *weak lines of mercury, 1203*
atomic states
excited, lifetime of 6^3P_1 mercury, 1205–7
stationary, as standing matter waves, analogy, 1188
atoms, 55
trajectories of, at low pressure, 821
* *Atwood machine experiment, variation of, 137*
Atwood's machine using air bearings, 265–66
automation of teaching, 63–66, 73–75, 700–2, 717–18
Avogadro's number
observations for a computation of, 819
rough estimate of, from diffusion of bromium, 821–23
axes of rotation, 272–77
through the center of mass: Steiner's theorem, 276–77
instantaneous, 272–73
of a gyroscope, 310
for nonparallel rotations, 272–73
for parallel rotations, 272
principle, stability of rotations about, 274

"baby bulldozer," 131
baffle
distinction between resonator and, 502
function of, 502
balance, gravitational torsion, 168–71
impulse, 198–99
Harvard trip, 196, ***557–59***
ball feed mechanism, ***532–34, 543–44***
ballistic galvanometer, 869
ballistic meter, gravity pendulum as, 31–32
Barkhausen effect, 967–68, 968–69
Barlow's wheel, 911–12
barometer
demonstration, 376, 384–85
plastic, 384, ***610***
barrier penetration, optical analogy, 1190–91
Barus, Carl, 707
Beyer, Robert T., 707–8
beats, 455–59, 510–14
to demonstrate Doppler effect, 511–14
to demonstrate interference of light, 1136–37
difference tone, 510
model for dem. of, 477
period of, 458–59
produced by Doppler effect, 511
production of, 360, 455–56, 510, 995–96
recording of, electrical method, 456–58

beats (*cont.*)
  recording of, optical method, 458–59
  recording of, simultaneous visual and audio, 510
Bell, Alexander Graham, 40
bent crystal spectrometer, 1197–1201
Bernoulli effect
  paradox, 46
  principle, 399–405
  * *air stream over a model house, 403*
  * *apparatus for throwing curve balls, 406*
  * *ball suspended in an air stream, 399*
  * *curving of a spinning ball, 401–2*
  * *flow around a Rayleigh disk, 406*
  * *hydrodynamic attraction, 405*
  * *Magnus effect, 402–3*
  * *stability of an object in a stream, 405*
  * *two balls on a curved track, 405*
Bernoulli's Theorem, 425
beta decay, 1246–49, ***1370–72***
  energy spectrum of, 1246–49
  straight-line emission of electrons, 1246
  in uranium-239, model, 1263–64, ***1378–81***
beta particles, 56
  observation of in cloud chambers, 1267–69
bifilar pendulum
  * *oscillations of a, 357–58*
  period of a, 357–58
bimetallic thermometer, 750
binary counter, 86, ***534–37***
birefringence
  with colloidal solution, 420–22
  in polymorphic transformations, 779
bistable thermochromism, 763
black body
  construction of, 1182
  relative brightness of, 1177
black body radiation, 1176–83
  * *infrared absorption and heat, 1177*
  * *meas. of Stefan's constant, 1179–81*
  * *outside coating affecting radiation properties, 1176*
  * *properties of using Leslie's cube, 1177–78*
  * *reflection of radiation from aluminum, 1178*
  * *relative absorption of black and white surfaces, 1176*
  * *relative brightness of a black body, 1177*
  * *relative emission of black and white surfaces, 1176*
  * *relative emission with a sensitive galvanometer, 1178–79*
  * *relative emission of transparent and opaque surfaces, 1177*
  * *Stefan–Boltzmann law, 1181–83*
black body radiation (*cont.*)
  * *thermal radiation intensity as a function of wavelength, 1179,* ***1362–63***
black boxes
  measurements of, 92
  use of in dem., 4, 47, 60
blackboards
  compass for, 87
  disadvantages of for physics lecture-demonstrations, 720–21
  protractor and slope indicator for, 86
  reference systems for, 85
  templates for drawing sine curves, 87
Blake, Clarence D., 40
blinky, 92, 99, 131
Bohr orbits, standing electron waves in, analogy, 1185–88, ***1363–64***
boundary conditions, correlation between radial node position and, 1185–87, ***1363–64***
Boyle's Law, 785–90
  * *using air and mercury in a capillary tube, 787–88,* ***1289–90***
  * *using air in a tube, 785*
  * *using air and water in a closed container, 786–87*
  * *dem. of for a large class, 785–86*
  * *using a glass stocking-shaped tube, 789–90*
* *brachistochrone, 120–21*
Bragg diffraction, 1193–1201
  in a bent crystal spectrometer, 1197–1200
  with electrons, 1194–97
  in a halographic emulsion, 1156
  with microwaves, 1026–29
  model, using light, 1198–1200
  with water waves, 472, 473, ***634–35***
Bragg's law, 1198
Braun electroscope, 835
Brewster's angle, 1024–25, 1102–4, 1142–43
Brewster's law, 1025, 1102, 1142–43
bridge circuit, 886
brightness
  of image, using laser light, 1130–31
  relative, of a black body, 1177
British Association for the Advancement of Science, 20
Bronk, Detlev W., 21
Brownian motion, 58, 816–20
  analog, 813–14
  apparatus, 814
  * *and Avogadro's number: corridor dem., 819*
  model of, 811, ***1295***
  * *in smoke particles, 816,* ***1297***
  * *in a suspension of polystyrene latex, 817–19*
  * *in a water suspension of rutile, 816–17*
bubble chamber tracks, analysis of, 1258–60, ***1373–75***
Buechner funnel, 829
bullet
  flight of using stroboscopic light passing through a soap bubble, 733–34
    shock waves produced by, 734
    splitting a playing card, 733
  speed of, 99–102, 105, 258
* *buoyancy in liquids, a qualitative model experiment, 33*
buoyant effect, 379
buoyant force, 145–46, 379–80
  disappearance of in a liquid, 379–80
  hydrostatic, 145–46
  variation with temperature, 380
Burkhardt, J. L., 711–12

calipers
  micrometer, dem., 85
  vernier, dem., 85
Calthrop resonance pendulum, 369
camera tube, closed-circuit television
  image orthicon, 657–58
  plumbicon, 657
  vidicon, 657
capacitance, 868–74
  behavior at constant charge
    * *dependence of on plate area, 869*
    * *dependence of potential difference on dielectric medium, 869*
    * *dependence of potential difference on plate spacing, 869*
  * *effect of physical dimensions on, 871*
  functional relationship of in a circuit, 994
  * *of an isolated sphere, 869*
  meas. of
    * *using a ballistic galvanometer, 869*
    * *using a gas-filled bulb, 870–71,* ***1321–22***
  in an RC series circuit, 890–91
  * *result of charge growth on, 869*
capacitive reactance, 873, 890–91
capacitor, 868–75
  effect of AC on a, 869–70
  * *effect of the dielectric constant on the impedance of a, 872*
  * *effect of dielectric polarization on a, 871–72*
  and electrical energy, 869–79, 873
  * *energy storage in a, 869–70*
  force between plates of a, 873–75, ***1322***
  equations for, ***1322***
  large scale, for dem., 868–69
  Leyden jar, 871–72
  phase of current in, 993
  * *use of to produce short-lived high voltages, 871*
capillarity, 44, 389–90
capillary waves, 429
Carnot cycle, model, 757

Carnot engine, hot air motor as a, 779–80
* *carrier recombination and lifetime, 1234*
cartesian diver, 376
cathode
glow, 895
temperature of
in a carbon arc, 882
effect on mode of gas discharge, 894–95
cavity resonator, 503–5
celestial sphere, model of, 171–73
center of gravity
* *dem. of, 328*
effect on the period of a physical pendulum, 353–54
rotation about, 328
of a stool, 328
center of mass
acceleration of, 288–89
* *dem. of, 328*
* *determination of, with an air table, 246–47*
motion of
* *in the absence of an external force, 185*
in a coupled system, 254
* *under an eccentric impulse, 232–33*
* *using the "falling board" apparatus, 185–86*
* *in a many-particle system, qualitative, 184*
* *in a pendulum-cart system, 185*
* *of a rod, 282*
* *in a two-body collision, 184–85*
rotations about, 286
Steiner's theorem, 276–77
system, motion in, 215–16, 235
* *translation and oscillation, 253–54*
* *two colliding masses under equal forces, 210,* ***565–66***
center of percussion
* *of a baseball bat, 355*
* *dem. of the concept of, 355*
* *determination of, 355*
central force
conservation of momentum under a, 323–24
inverse square law
electrical, 52
gravitational, 53
quantitative dem. of orbits under a linear, 163–65
quantitative dem. of orbits under a quadratic, 165
stability of orbits under a, model, 323–24
centrifugal force, 145–46, 157–62, 293
* *in emptying a jug, 157*
* *geotropism in plants, 159*
* *in a rotating liquid, 157*
* *surface of a rotating liquid, 158–59*
centripetal force, 145, 157–62, 229–30
* *principle of the accelerometer, 160*
centripetal force (*cont.*)
* *properties of a conical pendulum, 162–63*
Cerberus smoke detector, 897–98
chain reactions, 44
characteristic triangle, 350
charge (*see* electric charge)
charge generator, water drop, 847–49, 1311–12
* *charge motion demonstrator, 1220*
* *Charles' law, 790*
Chew, V. K., 702–4
Chladni figures, 496
Chladni's plates, 491
Christmas lectures
Rockefeller Institute, 21
Royal Institution, 16
chromatic aberration, 1050, ***1346, 1348***
circle of least confusion, 1049
circuits, electrical (*see* electrical circuits)
circuits, magnetic (*see* magnetic circuits)
circular motion
acceleration during, 337
* *on an overhead projector, 120*
uniform, relation to simple harmonic motion, 334
circular orbits
quantitative dem. of stability of, under a linear restoring force, 163–65
quantitative dem. of stability of, under a quadratic restoring force, 165
circulation, in a velocity field, 425
closed-circuit television (*see* television, closed-circuit)
cloud chamber, 55–57, 1266–73, ***1382–90***
expansion, 1269–70, 1272–73, ***1382, 1387–89***
using a glass-walled container, 1267
large diffusion, 1268–69, ***1382–86***
long-operating, 1267–68
using a plastic refrigerator jar, 1266–67
principle of, 1266
CLR circuit, 990–92, ***1334–36,*** 998–1000
coaxial cable, velocity of propagation in, 1010–12
* *coefficient of cubical expansion, 750*
coefficient of friction, 152, 153
apparatus, construction of, 153
dynamic, 153
kinetic, meas. of, 155–56, ***550***
static, 153
statistical nature of, 155
coefficient of restitution, 209, 210, 248–49
coefficient of viscosity, of a gas, ***1292***
coherence
function, partial, 1128–29
image brightness, 1130–31
length, 1129, 1136
coherence (*cont.*)
longitudinal, 1138
requirements of, in holography, 1159
spatial, 1129–30, 1138
time, 1129
transverse, 1138
coherent radiation (*see also* lasers)
comparison of laser light to conventional, 1131–32
directional response of, 519–20
image brightness, 1130–31
from opposite sides of a source, 1137–38
properties of, 1127–32
cohesive forces, secondary role of in siphon operation, 386–88
cold engine, hot air motor as a, 780
collisions
atomic, 56
between two air pucks, 228–29
and center of mass, 210
conservation of angular momentum in, 217–18
conservation of mechanical energy in, 183–84
conservation of momentum in, 208–12, 217
elastic, 212, 215–16
impulse and momentum during, 194–95
inelastic, 212–13, 216, 218
conversion of kinetic energy to heat, 751–52
low-friction surface for, 151–52
on an overhead projector, 123
producing ionization, model, 1270
* *quantitative study of, using a ball and pendulum, 212*
random, of a system of particles, 244–46
spring gun for use in dem. of, 128–30
inelastic collisions, 131
nearly elastic collisions, 130–31
* *a study of using two swinging balls, 215–16*
* *transfer of momentum and energy in a many-body system, 214–15*
of two suspended balls, for stroboscopic light, 739
viewed by stroboscopic light
one-dimensional, 743–44
two-dimensional, 744–45
colloidal solutions, dem. of birefringence with, 420–22
color
additive process, 115–16
complementary, 1115–18, 1120–21
* *appearance of in an "inverted" spectrum, 1120–21*
* *effect of an overabundance of, 1122*
under crossed polarizers, 1109
composition of spectral, 1109
dependence of, on illumination, 1114, 1115
fatigue, 88

color (*cont.*)
ICI monochromatic method of specifying, 1120–21
* *and illumination, 1114*
luminosity, 1120–21
* *mixing of primary, 1113*
* *mixing, with small glass wedges, 1116–18*
* *phenomena, 1114–16*
and polarization, 1108–10, 1116
primary, additive, 1115, 1120
production of in crystals, by electron and hole injection, 1222–28
psychological, 87–88
purity of, 1120–21
relation between spectral distribution and, 1120
* *selective absorption, 1113–14*
selective absorption, 1113–14, 1115
subjectivity of, 88–89, 1115, 1121, 1122
subtractive process, 1116
synthesis, 1120
* *synthesizer, 1118–21*
color centers, 1222–28
discussion of, 1226
F center, 1223, 1225, 1226
use of to dem. velocity of charge carriers, 1226–28
coma, 1049–50, ***1346, 1349,*** 1144
communication in lectures (*see also* demonstrations, communication)
informed, 79–80
subliminal, 79–80
unplanned, 80
compass, blackboard, 87
compressibility of water
* *dem. of, 380,* ***610***
* *near-compressibility of water, 380–81*
* *piezometer, 381–82*
compression, adiabatic, 800, 801
compression heating of a gas, 33–34
Compton effect, 123, 1172–74
* *qualitative dem. by absorption measurements, 1174*
* *scattering from a scintillation crystal, 1172–74,* ***1357–60***
condensation, model, 810
condenser (*see* capacitor)
conducting paint, 1318
conducting paper, 222
conduction
of electricity
in circuits, 885–91
in a crystal due to photoelectric effect, 1231–32, 1233–34
in a gas, 894–98
* *in gases, using visible carriers, 39*
electrolytic, 891–94
Faraday's first law of electrolysis, 891–92
of heat, 758–66
analogy to, transmission line pump, 1012
conduction (*cont.*)
of heat (*cont.*)
dem. of using a thermicon card, 759
thermal and electrical, differences between, 765
conductivity
electrical
of air, due to heat, 38, 39, 895
of gases using visible carriers, 39
n- and *p*-type, in semiconductors, 1221–22
thermal
comparison of in different materials, qualitative, 758, 764
of a gas, 798–800
relative, using temperature-sensitive paint, 750, 751, 758
conductor, electrical
charge carriers in
mobility of, 1220
sign of: Hall effect, 1236–37, ***1367–70***
electric field of a, 864–65
force on a current-carrying, 908–11
magnetic field of a current-carrying, 920–23
shielding action of a closed, 841
shielding of from magnetic forces, 916–18
conical pendulum
construction of, 157–58
properties of, 162–63
conservation
of angular momentum, 217–18, 232–34, 247–48, 267–69, 318–24, 739–40
of energy, 105, 128, 179, 181–84, 203–6, 234–35, 258, 269, 359, 360
of linear momentum, 48, 105, 195–98, 200–1, 203–11, 217–19, 227–29, 247–48, 739
in fission, 105
of mass, 143
convection, using a thermicon card, 759
cooling
curves, fusion, 770–71
* *curve of a melted material, 770*
by evaporation, 772
by the formation of a solution, 772
by sublimation, 772
coordinate systems
for blackboards, 85
for overhead projector, 85
for slide projector, 85
slides for projecting, 87
core magnets for moving coil instruments, 956
Coriolis force, 290–94, 300–2
demonstrator, CENCO, 271
Cornell plates, 1087
corridor demonstrations, 10–11, 57, 698–718
computer monitoring of, 717–18
corridor demonstrations (*cont.*)
difficulties with passive, 716
at educational institutions, 708–15
examples of
anomalous dispersion, 1097–98
Brewster's angle, 1103–4
Brownian motion and Avogadro's number, 819
comparison of two frequencies: epicycle patterns, 710–12
field emission microscope, 708–9
forced vibrations and resonance, 703–4
geotropism in plants, 159
hologram, 1163
inverted pendulum, 704
magnetic flux meas., 710
meas. the wavelength of light with a ruler, 712, 1138–39
mechanical system for producing Lissajous figures, 347–48
Peltier effect in semiconductors, 714
radiation pressure of light, 709–10
superposition of waves, 704–5
thin lens formula, 708
velocity of light, 711–12
"World's smallest watch," 90
maintenance of, 57
as "more effective" than lecture demonstrations: debate, 702
requirements for effective, 716–18
Rüchhardt's method for determining gamma, as an ideal corridor dem., 716–17
at Science museums, 699–708
Cranbrook Inst. of Science, Bloomfield Hills, 699–700
Deutsches Museum, Munich, 705–7
George Eastman House, Rochester, 708
Hall of Science and Industry, Corning Glass Center, 707–8
Lawrence Hall of Science, Berkeley, 701
Palais de la Découverte, Paris, 700–1
Science Museum in S. Kensington, London, 702–5
types of
automated, 699, 700
open laboratory, 699–700
self-instruction demonstration exhibits, 700–2, 710
static, 699, 700
use of luminous graphs, 707, 708
uses of
dynamic outline of steps in a meas., 715
opportunity to put together experiments, 711
stimulate interest, 700–1, 707
to supplement the lab, 702, 710

corridor demonstrations (*cont.*)
variability under the control of the operator, 702, 716
Cottrell precipitator, 896
Coulomb, Charles A., 28
Coulomb scattering, 853–54, ***1318–20***
Coulomb's law, 844–47
dem. of, for an overhead projector, 845
electrostatic balance, 844, ***1311***
counterpole elements, 898–99
counters
binary, 86, ***534–37***
Geiger, model of, 49–50
spark, 57
time intervals, 99–101
couple, effect of, 231–32
coupled systems, 358–63
analysis of with an analog computer, 360–63
conservation of energy in, 358–59
weak, 359
* *coupled systems, 254*
creeping film, of liquid helium, 831, ***1300–10***
critical point, gas, 790–91
critical temperature, 790–91, 810
cryophorus, for use on an overhead projector, 773–74
crystal detectors, for microwaves, 1014–17, ***1341–43***
crystal growth
by evaporating, 775–76
hard sphere model of, 809–10
by heating, 775
by melting and recooling, 776
in supercooled liquids, 776–77
crystal-like state, hard sphere model
defects in, 807–8
transition to, 807
crystallization, 770–77
heat release during, 770-72, 776–77
hard sphere model of, 809–10
of iodine, 772–73
of lead carbonate, 819–20
of organic compounds, 775–76
of a supercooled liquid, 771–72, 776–77
of tartaric acid and benzoic acid, 776
of a vapor, 772, 773
of water, 775
crystals
amplification of sound with, 901–3, 905–7
color centers, formation in, 1222–26
use of to dem. velocity of charge carriers, 1226–28
cubic, interplanar spacing of, 1028
diffraction in
electron, 1194–97
gamma ray, 1197–1201
x-ray, 1192–94
double refraction in a uniaxial, 1107–8
crystals (*cont.*)
ferroelectric, 900–1
hysteresis in, 903, 905
properties of, 902–3, ***1322–23***
resonance in, 905
ice, formation of at the triple point, 773, 774
interference colors in, 1108–10
polarization in
domains of, 905
pyroelectric, 899–901
stress-induced, 901–6
polycrystalline sheets of piezoelectric, creation of plate textures of, 905
pyroelectric
discussion of, 900–1
effect of temperature change on polarization of, 899–901
semiconducting, impurities in, model, 1220–21
sign of, 1111–12
structure of
body-centered cubic, 1237
* *colored overlays showing, 1238–39*
face-centered cubic, 1238
hexagonal close-packed, 1237–38
sphere-packing model, 1237–38
zone axis of, 1192–93
Curie point temperature
disappearance of ferromagnetic susceptibility at, 974–76
for electric polarization, 905
for magnetic materials, 967, 973–74, 976, ***1333***
phase change in iron at, 974–76
upper, in a ferroelectric crystal, 903
Curie–Weiss law, 988
curl of a vector, film for dem. of, 671
current, air, display of, due to wind and heat by means of an interferometer, 1084–85
current, electrical
alternating
in circuits, 992–96
effect on impedance of high-frequency, 1007–9
* *fluctuating nature of, 880–81*
generation of, 908
characteristics of electronic components, 992–93
in a circuit, 865–91
direct
* *continuous nature of, 880–81*
* *displacement current, 1005–7*
distribution, in an LC parallel circuit, 998
eddy, 934–36, 942–48
effect of on a hysteresis curve, 973
motor, 943–44
production of a rotating magnetic field, 943, 944
in an electrolyte, 891–94
induced, 908, 910, 932–51
* *vaporization of a coil by, 882*
current, electrical (*cont.*)
of ions in a gas, 894–98
factors affecting, 897–98
large, for magnetic field dem., 927
as a movement of charge, 855–56
phase relation between voltage and, in a parallel resonant circuit, 1000
produced by crystals, 901
rise in an inductor, 949–50
currents, electrical
* *control of, 39–41*
persistent, 828–29
curving of a spinning ball, 401–2, 406
effect of a rough surface, 402
cycloidal pendulum, 352–53, ***603–5***
isochronous property of, 352–53
cyclotron
gravity model of, 919, ***1324***
magnetic pendulum model of, 918–19, ***1323–24***

D'Alembert force, 149
Dalton's law, 794–96
damped oscillations, 346–47, 356, 366–69, 371–73, 801–3, 904–7
damping
aperiodic, 346
coefficient, in a Calthrop resonance pendulum, 369
constant, effect on resonance of, 35
critical, 346
due to domain boundary hysteresis, 986–87
force proportional to velocity, 138–39
due to hysteresis in a magnetic substance, 984–86
role of, in motion at resonance, 606
sub-critical, 346
in a system of forced oscillations, 366–68, 371–73
vane support apparatus, ***550***
* *of vibrational mechanical energy in solids. Snoek effect, 156–57,* ***550–51***
viscous, in air, 796, 797, ***1290–92***
theory of, ***1291–92***
de Broglie, Louis, 28
de Broglie wavelength, 1194
decay, radioactive, 1240–49
beta, 1246–49, ***1370–72***
energy spectrum of, 1246–49
straight line emission of electrons, 1246
in uranium-239, model, 1263–64, ***1378–81***
decibel, size of, 507
defects in crystals, behavior of, hard sphere model, 807–9
deformation, elastic, 441
force of, meas. of, 441–42
* *deformation produced by acceleration,* 136–37
demonstration barometers, 376

demonstration calipers
  micrometer, 85
  vernier, 85
demonstration capacitor, 868–69
demonstration thermometers, 749–50
demonstrations
  aims of
    for delight, 46, 73
    to enliven lectures, 44, 73
    show crucial tests, 48
    show the experimental basis of physics, 48
    show experiments too difficult for laboratory, 46
    show important phenomena, 47–48
    show the value of models, 48–49
    simplicity, 50–51
    teach factual knowledge, 45
    widening the field of vision, 79–80
  analysis, relation between dem. and, 6
  atomic physics, 54–57
  closed-circuit television, 10, 18, 655–69 (*see also* television, closed-circuit)
  conveying a lasting impression, 11, 16
  corridor, 10–11, 57, 698–718 (*see also* corridor demonstrations)
  delays and failures, 59
  dramatic, role of, 4, 8, 10, 11, 74
  equipment
    availability of, 9
    black boxes, 4, 47, 60
    impact on dem. of new, 5
    specialized vs unsophisticated, 4, 7, 14, 44
    use of authentic research equipment, 69
  a European tradition, 11
  "experiments of principle," 59
  films (*see also* films)
    as a lecture aid, 670–97
    use of short, 7–8, 10
  first-hand experience provided, 6–7
  future of, 21
  illustration of well-defined concepts, 5
  inherent limitations, 67
  laboratory, 57–58, 673, 702, 710
  laboratory work, correlation with, 7
  models
    relation between phenomena and, 50
    use of, 60, 62, 70
    value of, shown by dem., 48–50
  pedagogical benefits, 6, 8–9
  philosophy of
    attitudes, importance of conveying, 7, 8, 41, 77–78
    communication, 4, 58, 60, 62–63, 65, 77–78, 79–80
    dissociation of phenomena from natural context, 62, 68, 69
demonstrations (*cont.*)
  philosophy of (*cont.*)
    exercising self-discipline, 47, 54–55, 58, 59–60, 62, 67
    pedagogy vs cost, 6, 8, 10, 44
    qualitative vs quantitative, 5, 7
    role in total program, 4, 5, 6, 11
    showing "first-order reality," 71–72
  preparation, importance of, 4, 14
  problems with
    in complex experiments, 9
    confusion of students, 4, 9, 47, 70
    performance and presentation, 4, 8, 14, 15, 62
    precision, 4, 11, 47, 59
    vision, 11, 29, 59–60
  reaction of young teachers to, 8–9
  rolling lecture table, 10
  shadow projection, advantages of, 18
  showmanship in, 58
  stimulation of thought, 5, 6, 10, 44, 73
  technician teams, use of in England, 14
  use of to make science "alive," 5, 6, 10
dendrites, 779
densimeter, 485
density
  of air, effect of on viscosity, 796–97, ***1290–92***
    theory of, ***1291–92***
  of liquids, comparison, 383
  uniform, fluid of, 143–44
"Design for ETV," 663
de-spin device, 318–19, ***596–99***
Dewar flask, for low temperature dem., 826–27, ***1300, 1305, 1310***
dew-point
  apparatus, 783–84
  * *and relative humidity, 796*
Diaghilev, 73
diamagnetism, 959–60
  in a weak paramagnetic substance, 960
diatomic lattice
  as a band elimination filter, 262
  dispersion in, 262–63
  energy gap in, model of, 241–42
  frequency gap in, 262–63
  modes of vibration, 262–63
  velocity of an elastic wave in, 263
diatomic molecule, model of, using an air puck, 235, ***571–73***
diazo process, 724–25
diazo-sensitized films, 724–25
dichoic compounds, 1103
didactic reality, 68
dielectric
  effect of on the force between plates of a capacitor, 874–75, ***1322***
  effect of on the potential difference in a capacitor, 869
dielectric (*cont.*)
  force on, due to a homogeneous electric field, 875
  index of refraction of, using Brewster's law, 1023–25
  polarity of charges on the surfaces of a, 871–72
dielectric constant
  calculation of, 870–71, 874–75, ***1221–22, 1322***
  effect of, on the impedance of a capacitor, 872
  equations for calculation of, from force measurements, ***1322***
dielectric polarization, in a Leyden jar, 871–72
dielectric susceptibility, 875
differentiating circuit, 889–90
diffraction
  Bragg, 472, 473, ***634–35,*** 1026–29, 1156, 1193–1200, ***1364–65***
  dem. of "white spot," 46
  of electrons, 56, 1194–97, ***1364–65***
  Fraunhofer, 522–24, 1020–22, 1087–88
  Fresnel, 522–24, 1026
  with a laser
    dem. techniques for large classes, 1140
    diffraction grating, 1141–42
    random, 1139–40
    suggested objects, 1140–41
  of light, 1087–95
    * *around small particles, 1090*
    * *from a circular disk, 1090*
    * *in a coarse grating, 1090–93*
    * *Cornell plates, 1087–88*
    * *due to the edge of a stop, 1088–90*
    at a lens aperture, 1143
    * *from periodic density variations, 1094–95*
    * *using random arrays of single and double slits, 1088*
    * *and resolving power, 1092*
    * *resolving power of a microscope, 1092–94*
    * *single slit, 1087*
  of microwaves
    Bragg, 1026–29
    double slit, 1020–22
    through a circular aperture, 1026
  observation of with closed circuit television, difficulties with, 665–66
  phasor model of, for overhead projector, 729–39
  of sound waves, 519–24
    * *diffraction pattern of a piston, 520–22*
    * *directional response of coherent and incoherent radiation, 519–20*
    grating for, 522
    * *ultrasound camera, 522–24*

diffraction (*cont.*)
of water waves, 470
Bragg, 472, 473, ***634–35***
of x-rays, 1192–94
Laue patterns, 1192
diffraction grating
formed by holography, 1150–53
beat grating, 1150–51
Gabor zone plate, 1152–53
sine grating, 1150
standard diffraction grating, 1152
for light, 1088, 1141–42
coarse, 1093
two-dimensional, 1141
for sound waves, 522
diffusion, 821–23
of gases
apparatus for the dem. of, 764
bromium, 821–23
to estimate molecular size, 55
model, 811, ***1295***
in liquids
using potassium permanganate, 410
pressure difference caused by, 410
of a ring vortex, 422–23
using slowly diffusing salts, 410
neutron, model of, using magnetic pucks, 227–28, ***570***
pump, cooling arrangement for, ***610–13***
Shockley–Haynes experiment, 1228–31
surface, hard sphere model, 809
dilution experiments, need to avoid, 55
diode rectifier, 994
dipole moment, electric, formation of by piezoelectric effect, 904
dipole moment, magnetic, of a bar magnet, 955
dipoles, dielectric, scattering of microwaves by, 1030
dipoles, magnetic
force between, 956–57
force on, in an inhomogeneous field, equations for, 958
spinning, in an external magnetic field, 1280
direct current, continuous nature of, 880–81
dirod, 850–52, ***1312–18***
discharge, electrical
gas, modes of, 894–95
effect of cathode temperature on, 894–95
gas, Paschen's law, 894
to a pointed object, 838
of an RC circuit, 886
* *discharging stream, effects of, 195, 196*
dispersion, in a lattice, 239–42, 261–63
theory of, ***575–77***
dispersion curves, 261–63
effect of defects on, 263
dispersion of light
anomalous dispersion
dispersion of light (*cont.*)
anomalous dispersion (*cont.*)
* *with a fuchsin dye prism, 1097–98*
* *in a sodium absorption cell, 1099–1100,* ***1354–55***
* *in sodium vapor, 1100–1*
* *curve of a prism, 1101*
discussion of, 1096
* *multiple prism dem., 1101–2*
in a prism, 1050, ***1346, 1348***
displacement
angular
meas. of, 282–83
proportionality to torque, 284–85
of a ball in a tube, 96
correlation with velocity and acceleration, 106
of a falling body, 115–16, 117
of a lighted object, 120–21
of vectors, 95
displacement current, 1005–7
distribution field phenomena, isolated sandbed analogy for, 418
domains, magnetic
deformation of, 986–87
effect of external field on, model, 964
effect of temperature on, 967
structure of in a thin ferromagnetic garnet, 964–66
Weiss, 966–67, 974
Doppler broadening, absence of in Mossbauer effect, 1253, 1256, 1257
Doppler effect, 510–19
* *acoustic experiments, 516–19,* ***649–52***
* *dem. of using beats, 511–14*
difficulties with dem. of, 511–12, 513, 514, 516
* *moving reflector, 511*
relativistic, visual representation of, 1169, ***1355–57***
* *rotating pulsed speaker, 514–15*
* *rotating whistle, 511*
theory of, ***649–50***
use of in satellite tracking, 515–16
* *whistling rocket, 510–11*
double refraction, 1104, 1107–8
drift velocity, 38, 39
driving mechanism for a pendulum, use of a resonant LCR circuit as, 302–5, 369
* *drop formation, model, 810*
Dubos, René, 21
Dulong–Petit rule, 757
failure of at low temperatures, 757–58
Duroc dental stone, use for fluid mappers, ***614***
dylite, 141
dyna-jet, 191
* *dynamic hard sphere model, kinetic theory, 805–10,* ***1292–94***
dynamometer, 155
eddy current, 934–36, 942–48
effect of on hysteresis curve, 973
motor, 943–44
production of a rotating magnetic field, 943–44
Edgerton, H. E., 731, 734, 745
Ehrenhaft, F., 38
Einstein coefficients, 1133
elastic collisions, 56, 130–31, 212, 215–16
elastic deformation
bending, 441
meas. of force of, 441–42
in a one-dimensional potential problem, 180–81
shear, 441
stretching, 441
torsion, 441
elastic waves
diatomic lattice, 262–63
monatomic lattice, 261–62
in a solid, 439–40
velocity of, 261–63, 440, 448–54
elasticity
* *calculation of Young's modulus for a bending mode, 441*
* *dem. of Poisson contraction, 442*
* *elastic deformation dem., 441*
* *fundamental frequency of a continuous body, 440*
* *lateral contraction of a body, 440–41*
* *meas. of force of deformation, 441–42*
* *production of standing waves in a cylindrical bar, 439–40*
* *electric arc, importance of a hot cathode in sustaining, 882*
electric charge
carriers
mobility of in conductors and semiconductors, 1220
recombination and lifetime, 1234
sign of: Hall effect, 1236–37, ***1367–70***
velocity of, in a crystal, 1226–28
concentration of at points, 838–39
decay of, in an RC circuit, 886
effect of, on the surface tension of a liquid, 843
electroscopes for dem. of, 835–37
* *improvement of a toy Van de Graaff generator, 852–53*
interaction of like and unlike, 837–39
meas. of e/m for an electron, 914–16
* *Millikan experiment, 862–64*
* *model of, 860–62*
theory of, 862–63
* *motor operated by, 853*
* *movement of, as current, 856*
polarity of, on the surfaces of a dielectric, 871–72
* *rotating electrostatic generator–dirod, 850–52,* ***1312–18***

electric charge (*cont.*)
* *separation of, 847*
techniques for producing, in humid weather, 840–41
* *transfer of, in a Faraday cage, 855–56*
* *water drop charge generator, 847–49,* ***1311–12***
electric dipole moment, formation of by piezoelectric effect, 904
electric field, 853–65
* *air bubbles in an inhomogeneous, 857–58*
analog model of, 876–78
* *of a conductor, 864–65*
configurations of, for TEM modes, 1134
effect of, on the velocity of a charge carrier in a crystal, 1226–28
of an electrostatic voltmeter, 868, ***1320–21***
force on a dielectric in a homogeneous, 875
ion drift in an, 896, 897
lines, drawing of, 865–67
* *meas. of, around a sphere, using brass plates, 858*
motion of a charged particle in an, 858–60, 914–16, 918
* *motion of a charged puck in an, 853–54,* ***1318–20***
* *motion of a stream of charged particles in an, 858–60*
electric furnace, 1178
electric lines of force, 865–67
electric motor
principle of, 908
shielding of conductors from magnetic forces in, 916–18
electric polarization
in a ferroelectric crystal
domains of, 905
influence of a constant displacing field on, 903
of a pyroelectric crystal, 899–901
discussion of, 900–1
stress induced, 901–6
electric potential, 865–68
* *electrostatic voltmeter, 868,* ***1320–21***
* *gradient, meas. of, 867–68*
mapping of equipotential lines, 865–67
electric vector in an electromagnetic field, model, 93–94
electrical circuits, 885–91
ac, 992–96
* *addition of waveforms of different frequency, 995–96*
* *analysis of, using Thevenin's theorem, 886–87*
* *back EMF of a motor under no load, 891*
* *basic principle of an analog computer, 887–88*
* *bridge circuit, 886*
electrical circuits (*cont.*)
control, for closed-circuit television, key aspects of, 658–60
* *current-voltage characteristics of electronic components, 992–93,* ***1337***
dem. for overhead projector, 728–29
extended, ambiguity of flux ideas in, 911–12
* *functional relation between R, L, and C, 994*
* *half-wave and full wave rectification, 994*
LC, parallel, currents in, 998
* *meas. of high resistances using a discharging RC circuit, 886*
* *operation of an LC filter, 994*
parallel resonant, response of, 1000
RC
differentiating, 889–90
discharging, 886
integrating, 889–90
as a relaxation oscillator, 991–92
series ac, 890–91
* *RC integrating and differentiating circuits, 889–90*
* *RC phase-shift oscillator, 994–95*
* *reciprocity theorem, 889*
RLC, 990–92, 998–1000, ***1334–36***
* *phase relation between currents, 993*
* *study of an RC series circuit, 890–91*
* *superposition theorem, 888–89*
Thevenin equivalent, 887
electrical energy
conversion of, to heat, 759, 768–69
conversion of, to mechanical energy, 873
* *lifting of a weight, 873*
* *running of a motor, 873*
storage of, in a capacitor, 869–70
Electr cal Engineers, Institution of, 19
electrical equivalent of heat, 768–69
electricity
conduction of
in circuits, 885–91
in a gas, 894–98
in gases using visible carriers, 39
conversion of heat to, 901
debate on one- or two-fluid model, 28
Faraday lecture on, 19
electrolysis, 44, 880
Faraday's first law of, 891–92
of a sodium sulfate solution, 893–94
electrolytic conduction, 891–94
* *electrolysis of sodium sulfate solution, 893–94*
* *Faraday's first law of electrolysis, 891–92*
* *migration of ions on moist filter paper, 893*
* *migration of permanganate ions, 892–93*
* *in organic matter, 894*
electromagnetic pump, 912–14
electromagnetic oscillations, 990–92, 997–1005
modes of, TEM, 1134
electromagnetic radiation
coherent, 1127–31
light (*see* light)
microwaves, polarization of, 1023
modes of oscillation, 1134
electromotive force, 885–91
back, developed by an electric motor under no load, 891
back, in an inductor, 949–50
induced, 727–28, 932–51, 973
factors affecting, 932–34, ***1325–28***
in an inductor, 948–51
motional, 937–39
electron
beam, deflection of by magnetic fields, for overhead projector, 729
charge of
Millikan experiment, 56, 67, 860–64
meas. of e/m, 914–16
classical radiating, instability of orbit, model, 1219
flow of, in a vacuum tube, model, 992, ***1376, 1378***
mass of, in meas. of e/m, 914–16
straight line emission of, in beta decay, 1246
* *electron and hole injection in alkali halides, 1222–26*
electron diffraction, 56, 1194–97
* *in aluminum and graphite, 1196–97*
* *in gold foil, 1194–96,* ***1364–65***
electron paramagnetic resonance, 1274–83 (*see also* nuclear magnetic resonance)
electronic satellite planetarium, 173–76, ***552–56***
* *electrons on KCl crystal, 1226–28*
electrophorus, 28, 837, 840–41
electroscopes, 835–37
aluminum foil attached to a rectifier diode, 837
aluminum foil as a substitute for pith balls, 837
Braun, 835
large scale, using two aluminized tennis balls, 835
large scale, using Mylar tape, 837
with nonconducting supports, 835
using two sheets of aluminum foil, 837
use of in dem. radioactivity, 55
for use with an overhead projector, 836–37
electrostatic balance, 844, ***1311***
electrostatic force, 38–39, 52, 837–56, 873–74
* *between a charged ball and a grounded plate, 842–43*

electrostatic force (*cont.*)
 * *between plates of a capacitor, 874–75,* ***1322***
 * *Coulomb's law, 845–47*
 * *Coulomb's law dem. for overhead projector, 845*
 * *on a dielectric in a homogeneous electric field, 875*
 * *discharging effects of points, 838–39*
 * *dispersion of a water drop due to electrostatic pressure, 843*
 * *electrostatic balance, 844,* ***1311***
 * *electrostatic breakdown of surface tension, 843*
 * *floating bubble experiment, 862*
 * *image force between a charged ball and a conduction plate, 839*
 mechanics of an electrically charged particle, 853–54, ***1318–20***
 Millikan experiment, 862–64
  model of, 860–62
  theory of, 862–63
 * *oscillating ping pong ball, 841*
 * *repulsion of charged strands of nylon, 843–44*
 * *repulsion of unlike charges, 837–38*
 * *shielding action of a closed conductor, 841*
electrostatic generator
 rotating electrostatic generator—di-rod, 850–52, ***1312–18***
 Van de Graaff generator, toy, improvement of, 852–53
 water drop charge generator, 847–49, ***1311–12***
electrostatic pressure, 843
electrostatic projection voltmeter, 864
electrostatic voltmeter, principle of, 868, ***1320–21***
electrostatics, 835ff
 Laplace's equation analog, 876–78
ellipsoid of inertia, 268
emission of gamma rays, Mossbauer effect, model, 1253–57, ***1373***
emission of radiation, 1132–33
 resonance, 1213–17
 spontaneous, 1133
 stimulated, 1133
 thermal
  relative, of black and white surfaces, 1176, 1178–79
  relative, of transparent and opaque bodies, 1177
emission spectra, 1212–13, 1217–18
emissivity, 1178
endless screw, 91
energy
 conservation of
  in coupled systems, 359–60
  * *lead-in dem. using cart, 179*
  * *mechanical energy using an air trick, 258*
  mechanical energy of an oscillating puck, 234–35, ***570–71***
  * *mechanical energy in a pendulum, 181–82,* ***557***

energy (*cont.*)
 conservation of (*cont.*)
  * *mechanical energy in a spring system, 182*
  * *mechanical energy in a two-body collision, 183–84*
  * *using a rotating disk, 269*
  in a variable mass cart, 204–6
 * *conversion of elastic potential to kinetic rotational, 178–79*
 electrical
  conversion of to heat, 759, 768–69
  conversion of to mechanical energy, 873
  storage of, in a capacitor, 869–70, 873
 equipartition of, model, 816
 exchange, in a ferromagnetic substance, 973
 gap in a diatomic lattice, model, 241–42
  theory of, ***575–77***
 * *heavy pendulum dem., 179*
 increase and decrease in forced oscillations, 34–36, 364
 kinetic, 105, 215–16
  * *dependence of on the square of velocity, 179–80*
  * *loss of in inelastic collisions, 213–14*
  loss of in inelastic collisions, 213–14, 228–29, 248–49, 751–52
  rotational, 178–79, 203
 magnetic, 949–50
  absorption of in nuclear magnetic resonance, 1276–80
 mechanical
  conservation of, 181–84, 234–35, 258, 259
  conversion of electrical energy to, 873
  conversion of to heat, 751–52, 763, 766–68
  vibrational, damping of in solids, 156–57, 181–84, 234–35, 258, 259
 * *one-dimensional potential problem, 180–81*
 of a photon, relationship of to frequency, 1170–72
 potential, 105, 128
  elastic, 173–74, 259
  gravitational, 259
  for small oscillations, 254–55
 principle for stable equilibrium, 354, 392–93
 radiant
  absorption of, 759–62, 763, 1176
  conversion of mass to, 1175–76
  emission of, 763
  resonant absorption of, 762
  spectral distribution of, 761–62, 1120
 thermal, 800–1

energy (*cont.*)
 transfer in coupled systems, 358–60, 366, 373
 transfer, in a many-body system, 214–15
  equations for, 214–15
 transfer in a recoilless system, 1253–56
 translational, 203
 transport of by waves, 443
 in a zero-momentum system, 215–16
 of a zero-momentum system, 215–16
energy level diagram, for neon in a gas laser, 1132–33
energy levels, existence of, 1210–12
engine
 cold, hot air motor as a, 780
 heat, hot air motor as a, 780
 heat, model, 757
 pulse jet, 191
 rubber band, 766
entrance pupil, 1048–49
epicycle motion, 168
epicycle patterns to compare two frequencies: corridor dem., 710–12
equations of state, 785–91
 Boyle's law, 785–91
 Charles' law, 790
 van der Waals' equation, 790–91
equilibrium
 * *concurrent forces dem., 330–31*
 equipment for the dem. of, 327–28
 neutral, 327, 353, 354
 * *objects balanced on a glass, 332*
 in radioactive decay
  secular, 1242–43
  transient, 1243
 rotational, 330, 331–32
 * *second degree lever, 329*
 stable, 327, 354
 * *static force apparatus, 329*
 translational, 329, 330, 331–32
 * *translational and rotational, 331–32*
 * *translational and static rotational forces apparatus, 330*
 unstable, 327, 354
 * *in walking and standing, 232*
* *equipartition of energy, model, 816*
equipment
 blackboxes, in dem., 4
 in demonstrations (*see* demonstrations, equipment)
 need for improved, 63, 73
 television, closed-circuit (*see* television, closed-circuit)
equiphase cell, 750
equipotential lines
 electric, mapping of, 865–67
 fluid mapper analogy for, 418
escape velocity, 166
Euler's angles, model, 324–26
evaporation
 cooling by, 772
 crystal growth by, 775–76

evaporation (*cont.*)
  rapid, freezing by, 774–75
  of single molecules, model, 810
exchange energy in a ferromagnetic substance, 973
excitation, parametric, of a simple pendulum, 344
excited states, atomic, lifetime of 6^3P_1 of mercury, 1205–7
exhaust velocity
  calculation of, 187–88
  direction of, 188–89
exit pupil, 1048–49
expansion, adiabatic, 800–1
expansion, thermal
  apparatus for dem. of, 755–56
  coefficient of cubical, 750
"experiment of principle," 59
experimental reality, 68
exponential curves
  absorption of gamma rays, 1246
  height of a discharging water column, 1240–41
    equations for, 1240
extraordinary wave, 1107–8, 1110
eye, human, optics of, apparatus for dem. of, 1059–60
eye as a radiation detector, 90

F center, 1223, 1225, 1226
  discussion of, 1226
Fabray–Perot cavity, 1133–34
falling ball apparatus, 117
falling bodies (*see* free fall)
Faraday cage, 854–56
Faraday effect, 1112–13
Faraday Lecture, 19
Faraday's first law of electrolysis, 891–92
Faraday's law, 882, 908, 910–12, 932–48, 973
  * *with a coil in a solenoid, 933–34, **1327–28***
  compass needle apparatus for dem. of, 948
  * *construction of a levitator, 945–48, **1332–33***
  * *eddy-current motor, 943–44*
  * *forces produced by eddy currents, 935*
  * *generator for an overhead projector, 940*
  * *induced EMF by a moving magnet, 932*
  * *induction and synchronous motors, 944, **1329–32***
  * *inductive effects in a secondary coil, 932–33, **1325–27***
  * *Lenz's law, common misuse of, 936–37*
  * *Lenz's law using rotating iron cores, 934–35*
  * *Lorentz force and motional EMF, 938–39*
  * *motion in a magnetic field, 937*
  * *operation of a generator, 939–40*
Faraday's law (*cont.*)
  * *rotating magnetic field, 943*
  * *shielding against a changing magnetic field, 935–36*
  * *spinning ring in a toroid, 944–45*
  * *study of a generator, 940–42*
  * *torque on a cylinder, due to eddy currents, 935*
  * *with two Helmholtz coils, 934*
  * *two-phase induction motor, model, 943*
  * *two revolving disks, 942–43*
Faraday's number, 891
fatigue
  color, 88
  retina, 88
feedback, principle of for generating oscillations, 371
ferroelectric crystals, 901–5
  amplification of sound, 903, 905
  hysteresis in, 903, 904
  piezoelectric effect in
    dynamic, 903
    static, 903
  polarization of
    domains of, 905
    influence of a constant displacing field, 903
  upper Curie point, 903, 905
ferromagnetic susceptibility, disappearance of at the Curie point temperature, 974–76
ferromagnetism, 960–81
  * *Barkhausen effect, 967–68*
  * *Barkhausen effect in a stretched wire, 968–69*
  * *breakdown of the hysteresis loop at the Curie temperature, 973–74, **1333***
  discussion of, 973–74
  * *display of hysteresis on an oscilloscope, 972–73*
  * *display of hysteresis on a television tube, 970–71*
  * *domain structure in a thin ferromagnetic garnet, 964–66*
  * *effect of an external field on magnetic domains, 964*
  * *effect of heat on magnetic behavior, 976–77*
  * *effect of orientation on magnetic behavior, 969*
  * *effect of varying the air gap between two magnets, 978*
  * *Gilbert's experiment, 962*
  hysteresis, 968–69
  * *hysteresis with a dc generator, 971*
  * *loss of magnetization at the Curie point, 976*
  magnetic circuits, 978–81
  * *magnetization curves, 969–70*
  * *magnetoscope, 960–61*
  * *meas. of magnetic flux, 971–72*
  * *model of a magnet, 962*
  * *phenomena in iron at the Curie point temperature, 974–76*
ferromagnetism (*cont.*)
  * *poleless magnet, 962*
  * *thermomagnetic motor, 978*
  * *in a tube of iron filings, 961*
  * *Weiss domains, 966–67*
field emission microscope, 1201
  as a corridor dem., 708–9
fields of flow, 414–26
  analogy to electric fields, 414–19
  analogy to heat flow, 414–19
  * *fluid mappers, 414–19, **614–20***
  obeying Laplace's equation, 414–19
  obeying Poisson's equation, 414–19
  patterns, inverse square law, 419–20
  * *potential vortex, 425–26*
films
  film distributors, list of, 693–94
  film list
    8 mm. film loops, 674, 683–93
    16 mm., 674–83
  film vs live demonstrations, 670, 695–96
  film loop in a cartridge, 671, 673–74
    advantages of, 673
    list of available, 672, 683–93
    need for in 16 mm., 673–74
    problems with projectors, 673
    use of in lab, 673
  made by physicists, recent advances in, 671
  moving picture, use of
    to extend laboratory experience, 695
    introductory, to stimulate interest, 694–95
    short demonstrations, 695–96
    for specialized topics, 696
    as a supplement, 696–97
  not made by physicists, pros and cons, 671–72
  16 mm. with sound
    demonstration lecture, 672
    documentaries on physics and physicists, 672–73
    lists of available, 674–83
    well-known physicists, 672
  use of for lecture demonstrations, 7–8, 10, 673, 695–96
  use of in modern physics, 54
filter
  band elimination, 262
  "bandpass," for holography, 1154
  interference, 1080
  LC, 994
  LC circuit, analog, parallel resonance in, 997
  light, half-spectrum, 1075
  low pass, 261–62
fission chamber, for detection of neutrons, ***1375–78***
fission in U-235 by slow neutrons, model, 1264–66, ***1380***
flame, sensitive, 490
flash tube, circuit for, ***1352***

flask, Dewar, for low-temperature dem., 826–27, ***1300, 1305, 1310***
flask, Franklin's, 783
Fletcher's trolley system, 321
flexural mode, 473–74
  relation of resonance frequency of to material parameters, 439–40
flow
  * *advantages of streamlining, 403–4*
  * *around a body, with a Krebs apparatus, 404–5*
  * *birefringence, 420–22*
  * *change in character of, with velocity, 412*
  * *critical Reynolds number, 412–13,* ***614***
  laminar, 410–14
  potential, with a singularity, 425–26
  * *ring vortices in a liquid, 422–23*
  * *semicircular vortex in water, 423–24*
  * *signficance of Reynolds number, 410–12*
  steady-state, Laplace's equation analog, 876–78
  * *streamline, 399–401*
  * *streamline, between glass plates, 403*
  * *streamline, and field pattern dem., 419–20,* ***620–21***
  * *supersonic, of an air stream, 430–31*
  * *transition from smooth to turbulent, 413–14*
  turbulent, 410–14
  viscous, Newton's law of, 407
flow rate, effect of viscosity on, 796
fluid
  accelerating, 143–44
  buoyant force in an accelerating, 146–47
  flow (*see* flow)
fluid catalytic process, 431–32
fluid dynamics, 397ff
  equipment for dem. in, 397–98
fluid friction, 398
fluid mappers, analogies to heat flow and electric fields, 414–19, ***614–20***
  references on, ***620***
fluid statics, 375–95
  Archimedes principle, 378–80
  equipment for dem. in, 375–77
  liquid for fluid density measurements, 379
  Pascal's law, 377–78
fluidization, 431–32
flux, magnetic, 925, 926
  change of, in a magnetic circuit, 951
  and Faraday's law, 932–34, ***1325–28***
  in a magnetic circuit, 979–81
  meas. of, corridor dem., 710
  meas. of, in iron, 971–72
flux lines, fluid mapper analog for, 418
flux linkages, mistaken ideas about, 911–12
focal length, 1049
  calculation of by electrical analog, 887–88
  of a plano-convex lens, 1050
* *fog, formation of on ions, 796*
force
  of air pressure, 383–84
  apparatus for resolution of, 96
  arising in connection with a barometer, 384–85
  Bernoulli, 406
  buoyant, 145–46, 379–80
    hydrostatic, 145–46
  central
    conservation of angular momentum under a, 323–24
    quantitative dem. of orbits under a linear, 163–65
    quantitative dem. of orbits under a quadratic, 165
    stability of orbits under a, model, 323–24
  centrifugal, 157–62
    body in a rotating fluid, 145–46
  centripetal, 157–62, 229–30
    body in a rotating fluid, 145
  due to a change of momentum, 201–3, ***562–63***
    * *water stream on an impulse balance, 198–99*
  cohesive, secondary role of in siphon operation, 386–88
  concurrent, 330–31
  constant, motion due to a, 255
  Coriolis, 290, 291, 293, 294, 300–2
    demonstrator, CENCO, 271
  D'Alembert, 149
  damping, proportional to velocity, 138–39, ***549–50***
  dynamic, 330–31
  of elastic deformation, meas. of, 441–42
  electrostatic, 38–39, 390–91, 837–47, 853–54, 860–64, 873–75, 918
    breakdown of the surface tension of a liquid, 843
    shielding of, by a closed conductor, 841
  external, and conservation of momentum, 197–98, ***556–60***
  fictitious, 293
  focusing and defocusing to produce a net focusing force, 919–20, ***1325***
  frictional, 143, 151–57, 322, 336
    meas. of, 155, 281
  gravitational, 53
    potential well for, construction of, 166
  impulsive, and inertia, 136, 140–41
  inertial, 143–51
  inverse square law
    dem. of, 236–37
    electrical, 52
force (*cont.*)
  inverse square law (*cont.*)
    gravitational, 53
    motion under an, 235–39
  lines of, electric, 865–67
  Lorentz, 938–39, 1234–35
  magnetic, 908–31
    on a charged particle, 914–16, 918–20
    on a current-carrying wire, 908–11
    involving magnetic matter, 956–58
    between two current-carrying wires, 927–30
    use of in a sustained Foucault pendulum, 302–5
  magnetomotive, 980–81
  Magnus, 402–3
  measurement of, 105, 198–200
  perpendicular to motion, to reduce kinetic friction, 32–33
  produced by eddy currents, 935
  reaction, 140
  * *reaction, of a ball falling in a liquid, 199–200*
  relation between acceleration and, 137–38
  restoring
    linear, 163–65
    quadratic, 165
  static, 329, 330–31
  static rotational, 330
  Stokes', 200
  surface, in a liquid, 388–90
  in the Thomson model of the atom, 1217–18
  of vectors, 97–98
force constant for Hooke's law, calculation of, 259
forced oscillations, 34–36, 257, 364–73
  electromagnetic, 990ff
* *forced vibrations, 257*
forced vibrations and resonance, corridor dem., 703–4
* *forced vibrations of a torsion pendulum, 269*
Foucault, Leon, 302
Foucault pendulum, 299–305, ***592–94***
  * *behavior of at selected longitudes and latitudes, 299–300,* ***592–93***
  * *dem. of with an ink pendulum, 299*
  drive mechanism for, 302–5
  * *meas. of the earth's rotation, 300–2,* ***593–94***
  model of, 296
  sustaining motion of, 302–5
fountain effect, with liquid helium, 830–31, ***1300–10***
Fourier analysis, in holography, 1149, 1150
Fourier computer, 174, ***552–56***
Franck–Hertz experiment, 56, 1210–13
  * *Franck–Hertz experiment with mercury, 1210–11*

Franck–Hertz experiment (*cont.*)
* *ionization potential of argon, 1211–13*
Franklin, Benjamin, 28
Franklin's flask, 783
Fraunhofer diffraction, 522–24, 1020–22, 1087–88
free fall, 113–18
acceleration
meas. of, 114, 116–17, 118–19
* *uniformity of, 114–15*
* *falling ball apparatus, 117*
gravitational pressure gradient in, 146
inertial force in, 147
inertial force in a fluid undergoing, 145
* *motion of body in, 115–16, **543–44**, 735–36*
Pohl apparatus, 105–6
* *velocity, meas. of, 116*
* *using wooden blocks, 113*
* *freezing by rapid evaporation, 774–75*
frequency
acoustic, in a diatomic lattice, 262–63
addition of waveforms having different, 995–96
beat, 511
Doppler shift of, 511–12
change of, due to Doppler effect, 510–19
comparison of two, epicycle patterns, a corridor dem., 710–12
cutoff, 240–41, 262
expressions for, in forced oscillations, ***604–7***
gap, in diatomic lattice, 262–63
Larmor, 1206
modulation, 46
natural, of a vibrating system, effect of mass and stiffness on, 260
optic, in a diatomic lattice, 262–63
relation to energy, for a photon, 1170–72
relationship between pitch and, 497
relationship of to wave number
derivation of, ***575–80***
in a diatomic lattice, 262–63
in a linear mass chain, 239–42, ***575–80***
in a monatomic lattice, 261–62
resonant, 34, 366, 474–75
of a coupled system, 366
in a cubical cavity, 503–5
derivation of equations for, ***604–7***
in a driven harmonic oscillator, 367–68
fundamental longitudinal mode, 440
of a mass attached to a vibrating arm, 366–67, 604–7
of a quartz plate, 528
relationship to material parameters of a solid, 439–40
frequency (*cont.*)
* *spectrum of a cubical cavity, 503–5*
threshold, for photoelectric effect, 1169, 1170–72
velocity of a wave, dependence on frequency, 453–54
Fresnel diffraction, 522–24, 1026
Fresnel mirrors, 1084
Fresnel zones, 1026
friction
coefficient of, 152, 153, 155–56, ***550***
apparatus, construction of, 153
and frictional forces, 152
independence of weight of, 152
statistical character of, 155
distinction between rolling and sliding, 152
dynamic
coefficient of, 153
properties of, 155
external, 32
kinetic
coefficient of, 155–56, ***550***
effects of on a moving vehicle, 33
* *reduction of, by an additional perpendicular force, 32–33*
static
coefficient of, 153
properties of, 155
friction-minimizing devices, 221
air-supported pucks, 222–42
air tables, 242–49
air tracks, 250–63
air troughs, 250–63
rotary air bearings, 263–69
friction surface, low, glass beads as, 151–52
frictional forces, 143, 151–57, 322, 336
* *automobiles and braking, 153*
* *coefficient of friction for a chain, 152*
* *coefficient of friction and frictional forces, 152*
* *construction of coefficient of friction apparatus, 153*
* *dem. of using a movable surface, 153–54*
* *distinction between rolling and sliding friction, 152*
* *glass beads as a low-friction surface, 151–52*
* *independence of weight of coefficient of friction, 152*
* *low-friction surface for corners, 151*
meas. of, 155, 281
* *meas. of coefficient of kinetic friction, 155–56, **550***
* *properties of dynamic and static friction, 155*
fringe visibility and partial coherence function, 1128–29
furnace, electric, 1178

G, calculation of, 168–71
Gabor zone plate, 1152–53
construction of, 1162
Galilean experiment, 50–51
Galilean relativity
circular motion, 131–32
linear motion, 131
Galileo, 45
Galton's whistle, 37, 490
galvanometer, ballistic, 869
gamma (ratio of C_p to C_v), meas. of by Rüchhardt's method, 801–3
as an ideal corridor demonstration, 716–17
gamma rays
absorption of, model, 1246
emission of, Mossbauer effect, 1253–57
gas
* *apparent surface of a, 823*
* *compression heating of a, 33–34*
conduction of electricity in
* *Cerberus smoke detector, 897–98*
* *Cottrell precipitator, 896*
* *effect of cathode temperature on discharge modes, 894–95*
factors affecting, 897–98
* *ion drift in an electric field, 896*
* *ionization currents, 895–96*
* *Paschen's law, 894*
* *recombination of ions, 896–97*
* *thermally induced ionization, 896*
using visible carriers, 39
difference in behavior under compression of a vapor and, 823–24
diffusion of
bromium, 821–23
to estimate molecular size, 55
model, 811, ***1295***
equipment for demonstrating properties of, 783–84
general behavior of, model, 812–13, ***1295–96***
kinetic theory of, 48–49
kinetic theory, model of, 814–16
ideal, 814–15
modified van der Waals, 815–16
laser (*see* lasers)
laws, ideal
Boyle's law, 785–90
Charles' law, 790
study of, 814–16
liquefaction of, 823–25
behavior of, close to, 790–91
using the Joule–Thompson effect, 824–25, ***1297–1300***
* *as a liquid without a surface, 34*
low pressure, source of, 785
molecular model of, 33–34
obeying van der Waals equation, 790–91
* *particle analog, 263*
spectra of, 1202–8
thermal conductivity of, 798–800
thermometer, 751
constant-pressure, 754

gas (*cont.*)
thermometer (*cont.*)
constant-volume, comparison of, 752–53
viscosity of
effect of density on, 796–97
effect of temperature on, 796
theory of, ***1291–92***
* *gas lens, 524*
gas-like behavior, hard sphere model, 806
gaussmeter, 978
Gay–Lussac's law, 789–90
Geiger counter, model of, 49–50
generator
homopolar, 938–39
operation of, 939–40
for an overhead projector, 940
principle of, 908
study of, 940–42
unipolar, 911
geometric optics, 1035ff
Gilbert's experiment: iron bar in the earth's magnetic field, 962
glider, for air troughs, 250, 251, 252
magnetic release for, 252, 254
grating, diffraction
formed by holography, 1150–53
beat grating, 1150–51
Gabor zone plate, 1152–53
sine grating, 1150
standard diffraction grating, 1152
for light, 1088, 1141–42
coarse, 1093
two-dimensional, 1141
replica, 1202
for sound waves, 522
gravitation, 166–76
* *calculation of gravitational constant, 168–71*
* *celestial sphere, model of, 171–73*
* *electronic satellite planetarium, 173–76,* ***552–56***
* *epicycle motion, 168*
* *Kepler's second law, 166–67*
* *model of Earth–Jupiter motion about the sun, 167,* ***551–52***
* *satellite motion using model potential, 166*
gravitational constant, calculation of, 168–71
gravitational field and an accelerated reference frame, 288
gravitational force
inverse square nature, 53
potential well for, construction of, 166
gravitational mass and inertial mass, 142, 252–53
gravitational potential energy, 259
conversion of, to rotational kinetic energy, 286
gravitational pressure, 382
gradient, 146
gravitational torsion balance, 168–71
gravity
acceleration due to, 114–17, 118–19, 256, 257
center of, 328
effect on period of a physical pendulum, 353–54
rotation about, 328
* *gravity pendulum on a long string as a static balance and ballistic meter, 31–32*
group velocity, 241, 478–79, 481–82, 485–88, **575–77**
dem. of on an overhead projector, 479–80, 635–36
in a dispersive medium, 241, **575–77**
equations for, **575–77**
near the cutoff frequency, 241
and phase velocity, 241, 479–81, 1183
of a wave packet, 1184–85
Guericke, Otto von, 38
* *Guericke's levitation experiment, 38–39*
gun, spring, for projectile motion and collision dem., 128–30
gun for vector presentation, 93–94
* *gyration, radius of, 271*
gyro, air-bearing, 266–67, ***587–88***
gyrocompass
correction of, for true north, 316–17
north-seeking, 315–16
gyromagnetic ratio, determination of, 1281–82, ***1395***
gyroscope
air-bearing, 307–8, 309–11, ***594***
for demonstrations, 296–97
Poinsot's construction for angular momentum, 309–11
stabilization by, 308
gyroscopic motion
* *ball on a rotating turntable, 314–15*
* *change of angular momentum due to an impulse, 313–14*
* *of a corrected gyrocompass, 316–17*
* *effective change of weight due to precession, 311–12*
* *fundamentals of, using a bicycle wheel, 306–7*
* *of a gyroscope suspended from a spring balance, 308–9*
* *of a gyroscope on a turntable, 309*
as a model of nuclear magnetic resonance, 1274–78, 1279–80
as a model of spin-orbit interaction, 1278–79, ***1392–93***
* *north-seeking gyrocompass, 315–16*
* *nutation, 313*
* *precession of the equinoxes, model, 312–13*
* *precession and nutation, model, 313*
relation to linear dynamics, 305–6
* *relative movement of associated vectors in, 309–11,* ***594***
* *of a shell fired from a gun, 312,* ***595–96***
* *stabilization of a system, 308*
gyroscopic motion (*cont.*)
* *three-dimensional, 306*
* *two coupled gyroscopes, 312*

half-life, 1240–46
half-spectrum filter, 1075
half-wave plate, 1110, 1111
Hall effect, 1236–37, ***1367–70***
Hanle effect, 1205–7
harmonic analyzer, 1000–1, ***1337***
harmonic motion
damped, 257, 346–47, 801–3
* *differences in, 345*
simple, 234, 257, 334–45, 356
harmonic oscillator
analysis of motion of, 339–40
driven, 366–67, 371–73
normal modes of two coupled, 365–66
pictorial representation of solution of the equations of, 340–42
harmonic series of notes, 45
harmonics, 474
Hartl optical disk, for large audiences, 1046–47
Harvard trip balance, 196, ***557–59***
Hatt, Robert T., 699–700
heat
absorption of
connection of with emission, 763
dependence of on wavelength, 760–62
infrared radiation, 1177
relative, of black and white surfaces, 1176
resonant, in infrared region, 762
using a thermicon card, 759–62
and changes of state, 763
freezing by rapid evaporation, 774–75
conduction
* *comparison of in iron and aluminum, 758*
* *comparison of in iron and copper, 764*
* *comparison of in a metal and nonmetal, 758*
* *comparison of using temperature-sensitive paint, 758*
* *differences between electrical and thermal, 765*
* *in stainless steel, 758*
using a thermicon card, 759
* *in water, 758–59*
conductometers, 756
convection, using a thermicon card, 759
conversion of electrical energy to, 759, 768–69
conversion of, to electricity, 901
conversion of mechanical energy to, 763
by the forging of lead, 769–70
in an inelastic collision, 751–52
* *J-value demonstration, 766*

heat (*cont.*)
conversion of mechanical energy to (*cont.*)
* *mechanical and thermal energy, 767–68*
currents, display of with an interferometer, 1084–85
* *dem. in, using a thermicon card, 759–63*
* *detection by temperature-sensitive paint, 750–51*
effect of on magnetic behavior, 976–78
* *electrical equivalent of, 768–69*
engine
hot air motor as a, 780
model, 757
equipment for dem. of fundamental concepts of, 755–57
* *of evaporation, 772*
* *of fusion, latent, two experiments, 770–71*
* *of fusion of water, 770*
* *mechanical equivalent of, 769–70*
mechanical equivalent of
apparatus for dem. of, 755
Joule measurements, 767
* *of melting, 771–72*
pump, hot air motor as a, 780
release during crystallization, 770–72, 776–77
release by crystallization of a vapor, 772
release by the formation of a solution, 772
* *of solution, 772*
specific
apparatus, Tyndall's, 755
at constant pressure and volume, ratio of, 716–17, 801–3
at low temperature, 757–58
molar, of solids, 757
transport of by radiation, detection of with a thermicon card, 759–62
waves, propagation of
analogy to, of signal propagated in a tranmission line, 1012
* *in copper, 763–64*
* *in a long thin rod, 765–66,* ***1287***
and work
* *dust explosion, 769*
* *rubber band motor, 766*
* *work from heat, 769*
heat capacity (*see* specific heat)
heat exchanger, construction of, ***1289, 1297–1300***
heater, stage, for a microprojector, 778
heats of transformation, 770–78
Hefner lamp, 1180
helium, liquid, dem. with, 827–32, ***1300–10***
* *floating magnet experiment, 827–28*
* *fountain effect, 830–31*
* *lambda-point transition, 829*
* *the persistent current, 829–30*

helium, liquid, dem. with (*cont.*)
* *resistance vs temperature, 831–32*
* *Rollin creeping film, 831*
* *superleak, 829*
* *Helmholtz resonators, 499*
Helmholtz vortex theorems, 424
Hero's engine, 323
Hero's fountain, 397
high-frequency oscillations, 1007–9
hinged stick and falling ball apparatus, 272
holes
injection of in alkali halides, 1222–26
movement of in a crystal, model, 1220–21
holograms
cylindrical, 1148
formation of, physical dem. of theory, 1150–53
beat grating, 1150–51, 1161
diffraction grating, 1152
Gabor zone plate, 1152–53, 1162
sine grating, 1150, 1160
Lippmann–Bragg, 1147, 1156–57, 1162
practical hints for dem. and construction of, 1157–63
actual optical arrangements, 1160–63
coherence requirement, 1159
demonstrating, 1162–63
emulsion resolution, 1159–60
stability requirement, 1158–59
properties of, 1147–48
holography, 1147–63
discussion, 1147–48
mathematical background, 1149
physical dem. of the theory of, 1150–53
relation between radio and, 1153–57
bandwidth, 1153–54
encoding, 1155
fidelity, 1154
filtering, 1154
multiplexing, 1155–56
noise, 1154
sideband suppression, 1155
tube characteristics vs film characteristics, 1156–57
homopolar generator, 938–39
Hooke's law, 259
calculation of the force constant
conservation of mechanical energy, 259
dynamic method, 259
elastic and gravitational potential energy, 259
static method, 259
hot air motor
as a cold engine, 780
as a heat engine, 780
as a heat pump, 780
as a refrigerator, 779
hourglass system, 197–98, ***557–60***
Howey, J. H., 710

Huggins, Elisha R., 738
Hughes, D. E., 40
Huygen's construction, 1107–8
hydrodynamic attraction, 405
hydrogen atom, Bohr orbits in, classical analogies to, 1185–88, ***1363–64***
hydrogen molecule model, for overhead projector, 729
hydrostatic paradox, 384–86
hydrostatic pressure, 386
hydrostatics, basic equation of, 145
hysteresis
damping by magnetic, 984–86
in domain boundary deformations, 986–87
damping by, 986–87
in a ferroelectrical crystal, 903, 904
in ferromagnetic substances, 969–73
effect of strain of, 968–69
loop, for ferromagnetic substances
breakdown of, at the Curie temperature, 973–74, ***1333***
effect of eddy currents on, 973
in a magnetic circuit, 979–80
in a thermicon card, 763

ideal gas
laws, 785–90
model of, 814–15
image
aberrations, 1049–50, ***1345–46***
brightness, discussion, 1130–31
location of by parallax
real, 1042–43
virtual, 1043–44
pinhole, 1040
real, formation of, 44
with a convex lens, 1049
in a mirror, 1035–37
with a pinhole camera, 1040
with an ultrasound camera, 522–24
retina fatigue, 88
size of, in a concave mirror, 1036–37
three-dimensional (*see* holograms)
image orthicon tube, 657–58
impact time
constancy of, 193–94
measurement of, 193–94, 194–95
impedance
acoustical, 901–2
of a capacitor
calculation of, in an RC series circuit, 890–91
relation between dielectric constant and, 873
effect on of high-frequency ac, 1007–9
impedance matching, on a wave machine, 448
impedometer, low frequency, principle of, 890–91
impulse
angular, 297

impulse (*cont.*)
* *constancy of impact time, 192–94*
meas. of, 31–32, 258
* *meas. of time of contact in a collision, 194–95*
and momentum, 192–208, 258
impulse balance, 198–99
inclined plane
combined translational and rotational motion, 288–89
"dilution of gravity," conversion of potential energy to rotational kinetic, 286
measurement of acceleration on, 105
* *motion up an, using air pucks, 224–26*
incoherent radiation, 1127–32
directional response of, 519–20
index of refraction
of air, using a Raleigh refractometer, 1039
bending of light through a medium of varying, 1040–42
of glass, using total internal reflection, 1038
of a liquid, using an Abbe refractometer, 1038
meas. of, using microwaves
of a dielectric, 1023–25
with an interferometer, 1018–19
of a liquid, 1020
of a paraffin prism, 1022
observation of small changes in, 1056–59, ***1352***
periodic variation of, to diffract light, 1094–95
of a transparent substance, with a telescope, 1038–39
induced current, 882, 908, 910, 932–51
eddy, 934–36, 942–48, 973
induced EMF, 932–51, 973
dem. for overhead projector, 727–28
factors affecting, 932–34, ***1325–28***
in an inductor, 948–51
motional, 937–39
inductance, 948–51
* *delay circuit, 948*
functional relationship of, in a circuit, 994
mutual inductance
geometry, effect of, 949
iron core, effect of, 949
large, difficulty in finding, 948
* *meas. of, 948*
* *qualitative dem. of on an oscilloscope, 948–49*
* *properties of an inductor, 949–50*
current rise in, 949
energy storage in, 950
* *self-inductance, 950*
induction motor
principles of operation, 944, **1329–32**
two-phase, model, 943
inductors, 949–50
current rise in, 949–50
energy storage in, 950
phase of current in, ac circuit, 993
inelastic collisions, 131, 212–14, 216
conversion of kinetic energy to heat in, 751–52
inertia
demonstrations of effects of, 141
ellipsoid of, 268
* *and impulsive forces, 136, 140, 141*
moment of (*see* inertia, rotational)
rotational, 274–81
demonstrator, CENCO, 272
dependence of period of physical pendulum on, 352
determination of, 268, 279–80
effect of mass distribution on, 271, 275, 278–79, 284
and impulsive forces, 297
* *of a steel triangle, 269*
Steiner's theorem, 276–77, 280–81
inertia pump, 432–34
inertial balance, clamped rule, 142–43
inertial balance, low friction puck, 141–42
inertial force, 143–51
* *apparent change of weight due to acceleration, 147–48, 150–51*
* *bodies in a falling container, 147*
* *body in an accelerating fluid, 143–44*
* *body in a falling fluid, 145*
* *body in a rotating fluid, 145, 145–46*
* *candle in a falling container, 146*
* *falling yo-yo, 149*
* *liquid in a container, 146*
* *paper between two weights, 143*
* *two balloons in an accelerating system, 147*
* *two bodies in an accelerating fluid, 146–47*
inertial mass, and gravitational mass, 142, 252–53
inertial reference frames
Lorentz transformation and its consequences, visual representation of, 1166–69, ***1355–57***
* *and non-inertial frames, 159–60*
* *properties of, 132–33*
infrared radiation
absorption of, 761–62, 1177
quenching of phosphorescence by, 1216
infrared transparency, meas. of, 762
instantaneous speed, 118–19
integrating circuit, 889–90
intensity, light, 1075
image, discussion of, 1130–31
intensity, sound
decay of, 506–7
unit of, feeling for the size of, 507
intensity of thermal radiation, as a function of wavelength, 1179
interference
* *beats, electrical method of recording, 456–58*
* *beats, optical method of recording, 458–59*
colors of a crystalline plate, 1108–10
colors, from thin film reflection
turpentine on water, 1079
oxide on heated metal, 1079–80
filter, 1080
fringes, colored
from an interferometer, 1080–85
from thin film reflection, 1077–78
of light
* *double-slit, 1076*
* *using a Fresnel mirror, 1084*
* *with a home-made Jamin interferometer, 1082–83*
* *interference filter, 1080*
* *with a Jamin apparatus, 1083–84*
* *meas. of angular separation of two artificial stars, 1085–87*
* *projection of fringes with an interferometer, 1080*
* *reflection from a thin mica sheet, 1076–78*
* *showing wind and heat currents, 1084–85*
* *temper colors, 1079–80*
* *thin film transmission, 1076*
* *with a triangular interferometer, 1080–82,* ***1353–54***
* *turpentine on water, 1079*
* *light spots from moving pendula, 455–56,* ***625–26***
of microwaves, using interferometers, 1014–17, ***1338–45***
model for dem. of, 478
moiré patterns, 485–86, 1052–56, ***1346–52***
observation by closed circuit television, difficulty with, 665–66
phasor model for overhead projector, 729–30
rings from thin film reflection, 1077–78
equations for, 1077
of sound waves, 508–10, 511–14
* *synthesis of a signal, 459,* ***626***
thin film
using a laser, 1138
using light, 1076–80
using microwaves, 1025–26
of travelling waves, 445–47
tubes, Quincke, 490
washing out of, 507–8
of water waves, 470–72
interferometers
with laser light
double ended, 1137–38
Michelson long-path, 1136–37
light
Jamin, 1082–84
to show effects of wind and heat currents, 1084–85

interferometers (*cont.*)
light (*cont.*)
triangular, 1080–82, ***1353–54***
microwave
grid-detection, 1014–17, ***1338–45***
Lloyd's mirror, 1014–17, ***1338–45***
Michelson's, 1014–17, ***1338–45,*** 1017–19
to project interference fringes, 1080
inverse square law
for electrical forces, 52
field patterns for, 419–20, ***620–21***
force dem., 236–37
for gravitation, 53
for infrared radiation, 52
for light, 51
orbital motion under, 235–39, ***571–75***
for thermal radiation, 1178
ionization
chamber, model, 1270
currents, 894–98
factors affecting, 897–98
thermally induced, 38, 39, 395, 396–97
ionization potential
of argon, 1211–13
of xenon, 1213

jack screw, model, 91
Jamin interferometer, 1082–83, 1083–84
superiority of to the Fresnel biprism, 1083–84
* *Joule-Kelvin coefficients, 801*
Joule measurements, 47–48, 766–67
Joule–Thomson effect, 824–25, ***1297–1300***
Jupiter, orbital motion of, 167

Kelvin's water drop charge generator, 847–49, ***1311–12***
* *Kepler ellipse, 237–39, **574–75***
Kepler's laws, 236–39
first law, 237–39
* *and Rutherford scattering, 236–37*
second law, 166, 237–39
third law, 237–39
kinetic energy, 105, 215–16
dependence of on the square of the velocity, 170–80
interchange with potential energy
in a pendulum, 181–82
in a two-body collision, 183–84
loss of in inelastic collisions, 213–14, 228–29, 248–49
meas. of using a thermister, 751–52
rotational, 178–79, 269, 286
kinetic friction
coefficient of, 155–56, ***550***
effects of on a moving vehicle, 33
* *reduction of by an additional force, 32–33*
* *kinetic theory apparatus, 814–16*
* *kinetic theory demonstration, 812–13, **1295–96***
kinetic theory of gases, 48–49
kinetic theory models, 805–16
* *Brownian movement analog, 813–14*
for drop formation, 810
a dynamic hard-sphere model, 805–10, ***1292–94***
defects in crystal-like state, 807–9
liquid-like behavior, 807
nucleation and growth of crystals, 809–10
surface diffusion, 809
transition from gas-like to liquid-like to solid-like behavior, 806–7
equipartition of energy model, 816
* *gaseous diffusion and Brownian motion, 811, **1295***
general behavior of gaseous systems, 812–13, ***1295–96***
study of the gas laws, 814–16
* *two-dimensional, 810–11, **1294–95***
Kirchhoff's theorems, applied to fluid fields, 424
Kleist, E. G. von, 27
Klockounter gate unit, 202, 205, ***562***
Koenig's manometric flame vibrator, 490
Kollergang, 311–12
Krebs apparatus, 404–5
Kundt, August, 28
Kundt's apparatus, for overhead projector, 493

"lab-gas" units, 785
laboratory demonstrations
advantages of, 57–58, 710
as an economical means of performing experiments, 702
examples of, 58
film loops as, 673
laboratory visits, value of, 68
lambda particle, rest mass of, from bubble-chamber track analysis, 1258–60, ***1273–75***
lambda point, of helium, 829
Lambert's law, 1178
laminar flow, 410–14
* *Laplace's equation, analog of, 876–78*
Laplace's equation, fields satisfying, 418–19
Larmor frequency, 1206, 1278
Larmor precession, 1274–78
lasers
coherence, discussion of, 1127–31
coherence, properties of
* *double-ended interferometer, 1137–38*
* *long-path interferometer, 1136*
* *long-path interferometer, second version, 1136–37*
* *meas. of the wavelength of light with a ruler, 712, 1138–39*
* *sparkling dots and random diffraction, 1139–40*
lasers (*cont.*)
coherence, properties of (*cont.*)
* *thick reflecting plate, 1138*
* *transverse coherence, 1138*
comparison of, with conventional light, 1131–32
diffraction grating, 1141–42
image forming properties of lens systems, 1143–44
polarization demonstrations
* *Brewster's angle, 1142–43*
* *polarization of the laser beam, 1142*
* *scattering, 1143*
safety, 1144–45
suggested diffracting objects, 1140–41
techniques for diffraction dem. with a large class, 1140
theory of operation of
emission of radiation, 1132–33
Fabry–Perot cavity, 1133–34
laser action, 1133
linewidth, 1135
polarization, 1134–35
stimulated and spontaneous emission, 1133
latent heat of fusion, 770–71
lattice
anharmonic behavior, 263
constant, calculation of by electron diffraction, 1196–97
defects, effect of, 263
diatomic
as a band elimination filter, 262
dispersion in, 262–63
energy gap in, model, 241–42, ***575–80***
frequency gap in, 262–63
modes of vibration, 262–63
velocity of an elastic wave in, 263
distant neighbor behavior, 263
monatomic
dispersion in, 261–62
as a low-pass filter, 261–62
modes of vibration, 262
velocity of an elastic wave in, 262
* *lattice dynamics, analog for, 261–63*
Laue diffraction patterns, 1192
launching ram, for air-supported pucks, 224, ***1319–20***
laws of motion, Newton's
first law, 44
dem. of using carts, 137
second law, 48
for angular motion, 284, ***588–89***
application of, with damping force, 138–39, ***549–50***
dem. of, 137–38, 255, 740–41
third law, 102
* *dem. of, using buoyant force, 139*
* *dem. of, using cart and pulley, 140*
* *dem. of, using magnets, 139–40*
* *dem. of, using a roller conveyor, 140*

* *"laws of nature," acceptance on faith of, 143*
LC circuit
filter, analog, 997
parallel, current in, 998
LCR circuit, 990–92, ***1334–36***, 998–1000
lecture demonstrations
conversation, need for, 8
depiction of reality, 69–71
development of in Germany, 27–29
difficulties with
complexity of modern phenomena, 63, 67
impossibility of isolating phenomena, 67, 68
lack of experienced teachers, 63
as an effective teaching method, 8–9
effectiveness of, compared to corridor dem.: debate, 702
in England, 13ff
experimental and didactic reality, 68–69
faults of
displacement of attention, 62
dissociation of phenomena from natural context, 62, 68
distortion of phenomena, 62, 72–73
recognizing levels of reality, 72
films, as an aid (*see* films), 670–97
films, short, use of for, 7–8, 10, 673, 695–96
human vs automation, 64–66, 73–75
laboratory work, correlation with, 5, 7
levels of reality, 68–71
purposes, 3ff, 64–65, 67, 70, 73, 79–80
as a ritual, 81
at the Science Museum, London, 20–21
student participation in, 66
techniques
change of pace, 10, 73
diagrams, use of, 17–18
"fooling" the student, 9, 143
making the audience fellow-experimenters, 15
unconscious learning, value of, 76–77, 79–80
uses, 3ff, 20
lecture-equipment industry, 63, 73
lecture rooms
construction of, for physics, 29
at Göttingen, Germany, 28–29
at the Royal Institution, 14
lecturer
audience and subject as a coupled system, 75–78
classical theory of, 65
function of, 74–75, 80–81
personal involvement of, 74, 80
styles of
augmentors, 75
exegetic, 78
lecturer (*cont.*)
styles of (*cont.*)
reducers, 75
sinaistic, 78
lectures
balance and composition, 16
classical theory of, 65
delivery of
diagrams, use of, 17–18
importance of proper background of knowledge, 15–16
maintaining interest, 16
personal involvement, 80
reading vs extemporaneous, 15
showmanship, 58
styles, 59, 75, 78
Faraday, 19
modification of classical theory of, 65
popular, 15–16, 19
as ritual, 81
Royal Institution, 14, 16–17
sample lecture on magnetism, 22–26
slides, avoidance of, 17, 29
sponsored by Museums, 20
timing of, 15
unidirectional aspect of traditional, 66
the unplanned happenings, 79–80
varieties of, 30
length contraction, in relativity, visual representation of, 1168, ***1355–57***
length standard, calibration of using microwaves, 1018–19
lens
astigmatic and biconvex, 46
gas, 524
microwave, 1013
plastic
advantages of over glass, 1047–48
construction of, ***1346***
types of, 1048
types of for closed-circuit television, 658
zoom, advantages of, 658
lens systems
* *aplanatic properties of a sphere, 1051–52*
* *diffraction at a finite aperture, 1143*
* *entrance and exit pupil, 1048–49*
* *focal length of a plano-convex lens, 1050*
* *lens aberrations, 1143–44*
* *locations of images and objects by parallax, 1042–45*
* *Moiré patterns, 1052–53*
* *principles of image formation, 1049–50, **1345–46***
thin lens formula, corridor dem., 708
* *use of Moiré patterns in delicate measurements, 1053–56, **1346–52***
Lenz's law
common misuse of, 936–37
Lenz's law (*cont.*)
in a levitator, 945–48
in a rotating iron core, 934–35
Leslie's cube, 1177–78
lever, second degree, 329
leverarm, continuously changing, 30
Levey screens, 1054
levitator, 945–48, ***1332–33***
Leyden jar, 27, 871–72
resonant apparatus, 997
Lichtenberg, G. Charles, 27–28
light
absorption of, 1096–1102
* *absorption bands, 1096–97*
and anomalous dispersion, 1096–1102
* *blue light, by colloidal sulfur, 1096*
* *blue light by gum mastic, 1096*
coherence, 1127–32
color, 1113–22
diffraction of, 1087–95
circular disk, 1090
from the edge of a stop, 1088–90
grating, 1088, 1090–93
multiple slit, 1087–88
from periodic density variations, 1094–95
single slit, 1087, 1088, 1092–94
from small particles, 1090
dispersion of, anomalous
discussion of, 1096
in sodium, 1099–1100, ***1354–55***, 1100–1
dispersion of, in a prism, 1050, 1101–2, ***1346, 1348***
* *intensity of, 1075*
interference of, 1076–87
double slit, 1076
with an interferometer, 1080–85
to meas. angular separation of artificial stars, 1085–87
thin film, 1076–80
inverse square law for, 51
polarization of, 1102–13
Brewster's law, 1102, 1103
circular, 1104–5
discussion of, 1102
interference colors of a crystalline plate, 1108–10
model, 93–94
plane, construction of, 1103
plane, production of, 1102–3
rotation of the plane of, 1104, 1112–13
due to scattering, 1105–6
polarized, use of to observe magnetic domains, 964–66
radiation pressure of, 37
sources of
movie theater arc lamp, 1061
point, 1060
straight line, 1061
speed of, 1064–75
corridor dem., 1064
discussion of meas. attempts, 1067

light (*cont.*)
speed of (*cont.*)
* *Leybold apparatus, techniques for finding motor speed of, 1065–66*
in a moving frame, visual representation of, 1168, ***1355–57***
* *by phase differences, 1068–75, **1353***
* *rotating mirror apparatus, 1064–65*
* *by time difference measurements, 1066–67, 1067–68, **1353***
lighting, room and camera for closed circuit television, 664
linear momentum (*see* momentum, linear)
* *linear oscillator, 234–35, **570–71***
linear restoring forces, 163–65
lines of force, electrical, 865–67
"lines of no work," 417
linewidth
Gaussian, 1135
of a laser, 1134–35
liquefaction of a gas, 823–25
* *behavior of a gas and vapor under compression, 823–24*
* *condensation of oxygen, 823*
* *using the Joule–Thomson effect, 834–35, **1297–1300***
liquid-gas surface, properties of, 388–89
liquid-like behavior, hard sphere model, 806, 807
liquid surface
perpendicularity to forces, 146
shape of under rotation, 158–59
liquids
buoyancy in, 33, 378–80
density of, comparison, 383
diffusion in, 410
helium, dem. with, 826–32, ***1300–10***
index of refraction of, meas. of, 1020
surface forces in, 388–90
surface tension, 388–94
electrostatic breakdown of, 843
Lissajous figures, 347–51
* *construction of a Lissajous ellipse, 350–51*
for a coupled mechanical system, 362–63
mechanical system for producing, 347–48
* *projection of space curves to form, 348*
properties of a Lissajous ellipse, 350–51
superposition of two sinusoidal functions on an *x-y* plotter, 348–49
Lloyd's mirror interferometer, 1014–17, ***1338–45***
longitudinal modes
relationship of resonant frequency of to material parameters, 439–40
longitudinal modes (*cont.*)
velocity of, 491–93
Young's modulus for, 440
longitudinal waves, 242, 260–61, 482–83
sound, 490ff
standing, 493–96, 503–5
velocity of, 440, 491–93
loop-the-loop apparatus, 271, 280
Lorentz force, 938–39
* *acting on conduction electrons, 1234–35*
Lorentz transformations and their consequences, visual representation of, 1166–69, ***1355–57***
low-frequency impedometer, principle of, 890–91
low-friction surfaces (*see also* friction minimizing devices)
for corners, 151
glass beads as, 151–52
low-temperature phenomena, 826–32, ***1300–10***
apparatus, 826–27, ***1300–10***
behavior of liquid helium, 829–31
fountain effect, 830–31
lambda-point transition, 829
Rollin creeping film, 831
superleak, 829
superconductivity, 828, 829, 831–32

Mach angle, 429
Mach cone, 429
Mach disk, 90
machines, simple, 91–92
Magdeburg hemispheres, 376
magnet
effect of agitation of, 961, 962
effect of varying air gap between, 978
model of, 962
without a pole, 962
magnet experiment, floating, 827–28, ***1300–10***
magnetic circuit, 951
* *change of reluctance in, 978–79*
* *hysteresis loop of a, 979–80*
* *reluctance and magnetic flux, 980–81*
magnetic dipole moment, of a bar magnet, 955
magnetic dipoles
force between, 956–57
force on in an inhomogeneous field, equation for, 958
spinning, in an external magnetic field, 1280
magnetic domains, 964–67
deformation of, 986–87
effect of external fields on, model, 964
effect of temperature on, 967
structure of, in a thin ferromagnetic garnet, 964–66
Weiss, 966–67, 974
magnetic energy, 949–50
in nuclear magnetic resonance, 1276–83
magnetic field, 908–31
* *Ampere's law, dem. of, 925–26*
of a current-carrying conductor, 920–23
* *on the axis of a solenoid, 922–23*
* *exterior to a coil, 922*
* *using metallic, liquid, and gas conductors, 921–22*
* *straight wire, using compasses, 919–20*
* *straight wire, using iron filings, 919–20*
current-carrying conductors in a, 908–11
energy storage in, 949–50
focusing by, 919–20
pendulum analog, ***1325***
large currents for dem. of, 927
lines, mapping of, 919–20, 954
meas. of
using a magnetometer, 923–25
between two magnets, 978
motion of a charged particle in a, 914–16, 918, 919–20
for overhead projector, 729
production of, in a volume visible to a lecture audience, 937–38, ***1328–29***
* *properties of, for overhead projector, 953–56*
* *Rogowski coil and magnetic potential difference, 926–27*
rotating, 923
in synchronous and induction motors, 944, ***1329–32***
rotation of the plane of polarization by a, 1112–13
shielding from, 954, 962–63
of conductors in motors, 916–18
time-varying, 935–36
time-varying and Faraday's law, 727–28, 932–48
magnetic field intensity and magnetic induction vectors, 925–26
in a hysteresis curve, 972–73
magnetic flux, 925, 926
* *change of in a magnetic circuit, 951*
and Faraday's law, 932–34, ***1325–28***
linkages, mistaken ideas about, 911–12
in a magnetic circuit, 979–81
meas. of
corridor dem., 710
in iron, 971–72
magnetic force, 908–20, 927–31, 956–58
on a charged particle, 914–16
between two current-carrying wires, 927–30
* *using aluminum foil strips, 928–29*
* *using six vertical wires, 928*

magnetic force (*cont.*)
between two current-carrying wires (*cont.*)
* *using torque meas. of a suspended wire loop, 928–29*
* *using two coils, 929–30*
cyclotron
* *gravity model of, 929,* ***1324***
* *magnetic pendulum model of, 918–19,* ***1323–24***
* *on a dc arc, 912*
on a dipole in an inhomogeneous field, equation for, 958
* *electromagnetic pump, 912–14*
and flux linkages, mistaken ideas about, 911–12
* *between a magnet and soft iron, 957–58*
between a magnetic field and a current-carrying element, 908–11
* *Barlow's wheel as a unipolar generator, 911*
* *moving U-shaped wire, 908*
* *rotation of a wheel, 910*
* *variation of, with current, 908–9*
* *wheel and axle system rolling on rails, 909–10*
* *wire rolling on rails, 909*
* *particle beam focusing using, 919–20*
pendulum analog, 919–20, ***1325***
shielding from, 916–18, 954, 962–63
* *shielding of conductors from, in motors, 914–16*
shielding from a time-varying, 935–36
* *standing waves of small magnetized balls, 930–31*
* *between two dipoles, 956–57*
use of, in a sustained Foucault pendulum, 302–5
magnetic permeability, 926–27, 930, 956
ratio of, of iron and air, 990
magnetic potential difference
independence of path of, conditions for, 926
meas. of, 925–27
magnetic resonance, nuclear, 1274–83 (*see also* nuclear magnetic resonance)
magnetic vector, in an electromagnetic field, 93–94
magnetism (*see also* ferromagnetism)
sample lecture outline on, 22–26
magnetization
change in due to local heating, 978
curves, 970, 980
discussion of, 973–74
effect on polarized light, 964–66
effect of stress on, 984–86
meas. of, 969–70, 972
saturation, 970, 972
and Young's modulus, 984–86
magnetocaloric effect, 988–89, ***1334***
magnetometer, 922–23, 954–55
weak-field
construction of, 923–24
principle of, 924–25
magnetomotive force, 980–81
magnetoresistance, 988, 1235, 1236
magnetoscope, 960–61
magnetostriction, 981–87
* *dem. of using Newton's rings, 982–83*
* *inverse effect, 986–87*
* *magnetic oscillations of a ferromagnetic rod, 981–82*
* *of a nickel wire in a constant magnetic field, 983–84*
* *Young's modulus and magnetization, 984–86*
magnets, core, for moving coil instruments, 956
Magnus effect, 402–3
Malus' law, 1023
manometers, 398
Mariotte container, 377
as a constant velocity reservoir, 377
mass
atomic, 55
center of, 210, 328
acceleration of, 288–89
determination of, 246–47
motion of, 184–86, 231–33, 246–47, 253–54, 282
rotation about, 286
Steiner's theorem, 276–77
conservation of, 143
conversion of, to energy, 1175–76
distribution, effect on rotational inertia, 271, 275, 284
effect on natural frequency of a vibrating system, 360
of an electron meas. of e/m, 914–16
gravitational, 142, 252–53
inertial, 142, 252–53
calibration curve, 141–42
* *inertial and gravitational, 252–53*
inertial properties of, 140–41
reduced, 234
rest, of a lambda particle, meas. of, 1258–60, ***1373–75***
variable, effect on momentum conservation, 204–6, ***563–64***
"massless spring," 803–4
Maxwell-Boltzmann distribution, of velocities, 244–46, 810–11
Maxwell's inertia disk and cycloid-trochoid demonstrator, 272
* *mean free path, 820–21*
mean free path, 246
* *trajectories of atoms at low pressure, 821*
in a two-dimensional kinetic theory model, 810–11
measurement
aids for teachers, 85–86
in lectures, 47–48
"no-change," dangers of, 52
subjective effects, 87–90
measuring instruments, 85
mechanical advantage, 91
mechanical energy (*see* energy, mechanical)
mechanical equivalent of heat
apparatus for dem. of, 755
dem. of by the forging of lead, 769–70
Joule-measurement dem., 766–67
mesons, observation of in a spark chamber, 1272–74, ***1390–92***
meters, projection techniques for, 882–85
using a commercial slide projector, 883–85
using a lantern slide projector, 883
using a mirror-and-lens system, 882–83
Michelson's interferometer, 1014–17, 1017–19, 1136, ***1338–45***
micrometer calipers, demonstration, 85
microphone, graphite, 40
microphysical world, 54
microprojector, construction of, 777–78
microscope
Abbe theory of, reference for, 1144
field emission, 1201
corridor dem., 708–9
microscope slides, sealant for, 819
microwave absorbent material, 1028
microwaves, 1012–30
accessories for dem. of optical properties of 3-cm., 1013
* *Bragg diffraction, 1026–29*
calibration of a platinum-iridium bar with, 1019
* *double-slit diffraction, 1020–22*
* *Fresnel diffraction, 1026*
* *index of refraction using Brewster's law, 1023–25*
* *index of refraction of a liquid, 1020*
* *index of refraction of a prism, 1022*
* *interference of, using interferometers, 1014–17,* ***1338–45***
* *interference from thin films, 1025–26*
introduction to experiments in, 1017
* *Michelson's interferometer, 1017–19*
* *opaqueness of, to water, 1013*
* *polarization of, 1023*
* *resonance, using a beer can, 1012–13*
* *rotation of plane of polarization due to sugars, 1029–30*
* *scattering of, by a dielectric dipole, 1030*
wavelength, meas. of
using an interferometer, 1017–19
* *using standing waves, 1019–20*
Millikan, R. A., 28, 38
Millikan experiment, 56, 67, 862–64
model of, 860–62
theory of, 862–63
mirrors
concave, 1036–37

mirrors (*cont.*)
Fresnel, 1084
plane, 1035–36
mobility of charge carriers in conductors and semiconductors, 1220
models
relationship between phenomena and, 50, 72
use of in dem., 60, 62, 70
value of shown by dem., 48–50
moderator, neutron, 56
neutron diffusion in, model of, using magnetic pucks, 227–28, ***570***
modern physics
need for dem. in, 54–55
suggestions for dem. in, 55–57
modes
normal
lower, of a many-particle system, 473
of a rubber tube, 473
of two coupled, driven oscillators, 365–66
of two coupled oscillators, 358, 359
of oscillation, electromagnetic waves, TEM, 1134
modes of vibration
* *dem. of using a cello bow on aluminum plates, 497*
* *dem. by Chladni figures, 496*
in a diatomic lattice, 262
flexural, 439, 473–74
stress produced by, 474
Young's modulus for, 440, 441
in a free rubber tube, 463
longitudinal, 439, 440, 981–82
velocity of, 491–93
Young's modulus for, 440
in a monatomic lattice, 262
sonic, in a cubical cavity, 503–5
of a stretched circular membrane, 505–6
in a stretched wire, 465–66
torsional, 439
modulation
frequency, 46
* *of sound waves, 527–28*
Moiré patterns, 485–86, 1052–53
use of in making delicate measurements, 1053–56, ***1346–52***
use of in observing arrays of atoms, 1053
molecular diameter, determination of, 393–94
molecules
average speed of gas, qualitative, 821–23
diatomic, model of using air pucks, 235, ***571–73***
properties of, shown by dem. experiments
average speed of air molecule, 55
length, 55
size, 55
Moll thermopile, 1177
moment of inertia (*see* rotational inertia)
momentum, angular, 293ff
* *of a body moving in a straight line, 268*
change of in a gyroscope due to an impulse, 313–14
conservation of
* *air-jet Hero's engine, 323*
* *using an air table, 247–48*
* *colliding pucks, 233–34*
* *de-spin device on a flying craft, 318–19,* ***596–99***
* *de-spin device on a model satellite, 319–20*
equations for, 319–20
* *execution of ski turns, 322*
* *falling weight dem., 320*
* *Fletcher's trolley system, 321*
* *isolated body, 232*
* *for masses subject to a central force, 323–24,* ***598–602***
* *on an overhead projector, 322*
* *using photographs of a two-body collision, 217–18*
* *using a pocket watch, 322*
* *rubber ball on a strip of paper, 322–23*
* *simple Hero's engine, 323*
* *using strobe light photography, 739–40*
* *two rotating disks, 321,* ***599***
* *using two rotating disks, 267–68*
due to an eccentric impulse, 232–33
Poinsot's construction for, in a gyroscope, 309–11, ***594***
role of, in ski turns, 322
rotating stool for demonstrations of, 322
momentum, linear
and angular momentum, 232–33
* *change of, due to a force, 201–3,* ***562–63***
conservation of
* *using two air pucks, 228–29*
* *using an air table, 247–48*
* *using two carts with no external forces, 218–19*
* *in a discharging stream, 195–96*
* *using duck pin balls, 211,* ***566***
* *dumbbell on an inclined plane, 203*
* *and external forces, 197–98,* ***557–60***
in fission, 105
* *using a gun on an air puck, 227*
* *using the Penn. St. air table, 248*
* *using photographs of a two-body collision, 217*
* *recoil of a gun, 206–7*
* *recoil of a gun-carriage assembly, 207–8*
* *using a rotating mass system, 208*
* *using a row of balls, 210–11*
momentum, linear (*cont.*)
conservation of (*cont.*)
* *using a spring-loaded gun and golf ball, 204*
* *of two suspended balls using strobe light, 739*
* *using two swinging balls, 203*
* *using two swinging balls, for an overhead projector, 209*
* *using eleven swinging billiard balls, 209*
* *in a thrust-producing stream, 195*
* *using a toy cannon on a cart, 206,* ***564–65***
* *using a 22-caliber bullet, 218*
* *in a two-body collision, quantitative, 211–12*
* *in a variable mass cart, 204–6,* ***563–64***
force due to a change of, 198–99
and impulse, 258
* *principle, using falling sand, 200–1,* ***560–61***
systems of zero-momentum, 232–33
transfer, in a many-body system, 214–15
equations for, 214–15
transfer, in a recoilless system, 1254–56
transmission of, in elastic and inelastic collisions, 212
monatomic lattice
dispersion in, 261–62
as a low pass filter, 261–62
modes of vibration, 262
velocity of an elastic wave, 262
monel metal tape, 977
monitors for closed circuit television, 660–61, 664–65
distribution system, 664–65
lighting for, 664
number of required, criteria for, 664
"monkey and bananas" problem, 287–88
"monkey and hunter" apparatus, 125
"monkey and hunter" demonstration, 130
monochromator, ***1362–63***
Mossbauer effect, classical analogs, 1253–57, ***1373***
absence of line broadening, 1256–57
unshifted energy from recoilless emission, 1253–56
Mosteller, Frederick, 63
motion
accelerated, 105, 230–31, 337–38
of a ball bouncing on a rubber band, 740–41
of a ball in a vertical circle apparatus, 289
Brownian, 58, 811, 813–14, 816–20
about the center of gravity, 328
of the center of mass, 184–86, 231–33, 253–54, 282
circular
on an overhead projector, 120

motion (*cont.*)
circular (*cont.*)
uniform, relation to simple harmonic motion, 334
* *due to a constant force, 255*
of crystal defects, model of, 807–9
of a diatomic molecule, model, 235, ***571–73***
in an electric field
of a charged air puck, 853–54, ***1318–20***
of a charged particle, 858–60, 914–16, 918
epicycle, 168
equations of, for a rocket, 189
of exhaust gases of a rocket, 188–89
of falling bodies
one-dimensional, 115–16, ***543–44***
two-dimensional, 122–31
gyroscopic, 305–15
as a model for nuclear magnetic resonance phenomena, 1274–80
of a harmonic oscillator, analysis of, 339–40
* *of a lighted object, 120–21*
in a magnetic field, charged particle in, 914–16, 918
Newton's laws of, 44, 48, 102, 135–40, 255, 284
in one dimension, 99–118, 106–8
orbital, 166, 167, 168
equations of, for an artificial satellite, 175–76
under an inverse square law, 235–39
simulated, of an artificial satellite, 173–76
of a particle in a rotating frame, 293
of a particle on the surface of a water wave, model, 482–83, 488, ***636–39, 644–45, 648***
planetary, 167, 237–39, ***551–52, 574–75***
projectile, 122–31
on an overhead projector, 122–23, ***544–45***
random, of gas particles, model of, 263, 806, 810–11, 812–13
relative, 131–33, 188–89, 290–91, 1164–69
of a rolling ball on a rotating turntable, 314–15
equations of, 314–15
rotational
using an air puck, 230–31
combined translational and, 286–89, 322–23
and rectilinear, 272
simple harmonic, 234, 257, 334–45, 356
damped, 257, 346–47, 356, 801–3
thermal
of gas molecules, 810
illustration of buoyancy in liquids, 33

motion (*cont.*)
turbulent, 401, 410–14
in two dimenisons, 119–31
between two points, 121–22
uniform, 99
circular, 334
and vector addition, 97
as viewed from a rotating frame, 291–93
up an inclined plane, using air pucks, 224–26
wave, 449–53, 483
apparatus for the dem. of, 476–77
longitudinal, 242, 260–61
one-dimensional, 239–42
study of, using a ripple tank, 467–73, 1164–66
of wave packets, 486–88, ***639–47,*** 1183
* *motion-study stroboscope, 121*
electric
principle of, 908
shielding of conductors from magnetic forces in, 916–18
induction
model, 943
principles of operation, 944, ***1329–32***
operated by electrostatic charge, 853
synchronous, 944, ***1329–32***
thermomagnetic, 978
unipolar, 911
motor, Ampere's apparatus, 909
multiplexing, holography, 1155–56
* *musical notes, 498–99*
* *musical sounds, 497*
mutual inductance, 948, 949
geometry, effect of, 949
iron core, effect of, 949
large, difficulty in finding, 948
meas. of, 948
Mylar metallized tape, 837

neutron absorption, 1240
neutron diffusion, model of, using magnetic pucks, 227–28, ***570***
neutron moderator, 56
neutron diffusion in, model, using magnetic pucks, 227–28, ***570***
neutrons, 56
capture of slow, model, 1263–64, ***1380***
fast, 1252–53
fission, induced by slow neutrons, model, 1264–66, ***1380***
fission-produced, apparatus for detection of, 1260–61, ***1375–78***
moderation of, 1252–53, ***1372, 1375–78***
observation of, in a cloud chamber, 1268–69
thermal, 1252–53
in a reactor suffering from xenon poisoning, model, 1261–63

Newton's laws of motion
first law, 44
dem. of using carts, 137
second law, 48
for angular motion, 284, ***588–89***
application of, with damping force, 138–39, ***549–50***
demonstration of, 137–38, 255, 740–41
third law, 102
using a buoyant force, 139
using a cart and pulley, 140
using magnets, 139–40
using a roller conveyor, 140
Newton's law of viscous flow, 407
Newton's rings, 982–83
next-nearest neighbor interaction, dependence on of Poisson contraction, 442
nodes, 445–47
position of, 473–74
* *noise generators, 507–8*
Nollet, Abbé, 28
normal modes
of two coupled driven oscillators, 365–66
of two coupled oscillators, 358, 359
* *lower, of a many-particle system, 473*
* *of a rubber tube, 473*
nuclear magnetic resonance, 1274–83
discussion of, 1274
* *dynamics of, model, 1280,* ***1293–95***
* *gyroscopic model of, 1274–76*
* *Larmor precession of a gyroscope, 1274–76*
* *magnetic transitions in, model, 1276–78*
* *model of, using an air-bearing gyro, 1279–80*
* *spin-echo method, 1280–83,* ***1395***
determination of the gyromagnetic ratio, 1281–82
determination of spin-lattice relaxation time, 1282–83
determination of spin-spin relaxation time, 1282
spin-echo dem., 1283
nuclear reactions
* *apparatus for detection of fission-produced neutrons, 1260–61,* ***1375–78***
* *moderation of fast neutrons, 1252–53,* ***1372, 1375–78***
* *Mossbauer effect, aspects of, classical analog, 1253–57,* ***1373***
* *neutron-induced beta decay in U-239, 1263–64,* ***1378–81***
neutron-induced fission in U-235, model, 1264–66, ***1380***
oscillations of an excited nucleus, simulation, 1250–51
* *resonance absorption of photons, simulation, 1257–58*
* *rest mass of the lambda particle, 1258–60,* ***1373–75***

nuclear reactions (*cont.*)
1/*R* surface for the dem. of, 1249, **1372–73**
* *thermal neutron reactor suffering from xenon poisoning, model, 1261–63*
nucleation, hard sphere model of, 809–10
nucleus
* *oscillations of, simulation using a chemical heart, 1250–51*
recoilless emission of gamma rays, simulation, 1253–57, **1373**
resonance absorption of gamma rays, simulation, 1257–58
Nuffield Foundation, "Science Teaching Project" of the, 21
nutation, 311, 313

object
real, difference between virtual and, 1045
virtual
difference between real and, 1045
location of by parallax, 1044–45
* *one-dimensional wave motion, 239–42, **575–80***
O'Neill, John J., 1005
optic axis, 1107
optical gain, of a parabolic reflector, **1343–45**
optical illusions, 88
optics
apparatus for dem. in, 1059–62
geometrical, 1035ff
physical, 1065ff
Schlieren, 430–31
to observe small changes in the index of refraction, 1056–59, **1352**
orbital motion, 166, 167, 168
of an artificial satellite, simulated, 173–76, **552–56**
equations of, for an artificial satellite, 175–76
instability of, for a classical radiating electron, model, 1219
under an inverse square law, 235–39, **573–74**
* *orbital motion under an inverse square law, 235–36, **573–74***
orbits
Bohr, standing electron waves in, analog, 1185–87, 1187–88, **1363–64**
quantitative study of
* *under a linear restoring force, 163–65*
* *under a quadratic restoring force, 165*
* *in a spherical cavity, 344–45*
stability of, model, 322–23, **598–602**
ordinary wave, 1107–8, 1110
oscillation
* *center of, in a skewed oscillator potential, 181*

oscillation (*cont.*)
fundamental mode of, 474
modes of, electromagnetic waves, TEM, 1134
period of, 181, 257, 259
of a ball in a spherical cavity, 344–45
of a bifilar pendulum, 357–58
of a cycloidal pendulum, 352–53
as a function of mass, 803–4
nonisochronism in pendula, 343–44
of a physical pendulum, 277–78, 352–54
in a relaxation oscillator, 991
of a torsion pendulum, 269, 275, 278–79
of a trifilar pendulum, 279–80
oscillations
about the center of mass, in a coupled system, 254
change of longitudinal into simple, 342
coupled, 358–63, 873
damped
* *aperiodic damping of a closed tube system, 346*
* *critical and sub-critical damping, 346*
due to domain boundary hysteresis, 987
with a driving force, 366–69
due to magnetic hysteresis, 986
* *of a metal strip, 346–47*
* *oscillations of a strip of spring steel, 347*
Rüchhardt's method for determining gamma, 716–17, 801–3
* *vibrations of a tuning fork, 346*
electromagnetic, 990ff
* *two coupled LCR circuits, 999–1000*
* *current distribution in an LC parallel circuit, 998*
high frequency, 1007–9
* *induced by a high-frequency generator, 998–99*
* *in a Leyden jar circuit, 997*
* *parallel resonance analog, 997*
* *properties of an RLC circuit, 990–91, **1334–36***
* *relaxation oscillator, 991*
* *relaxation oscillator, model of, 992*
* *relaxation siren oscillator, 991–92*
* *response of a parallel resonant circuit, 1000*
* *Tesla high-frequency transformer, 1001–5*
of an excited nucleus, simulation using a chemical heart, 1250–51
forced, 257, 364–73, 997–1005 (*see also* oscillations, electromagnetic)
* *with a block and tape, 364*
* *calthrop resonance pendulum, 369*

oscillations (*cont.*)
forced (*cont.*)
* *of a cart-spring system, 371–73*
* *in a driven harmonic oscillator, 367–68*
* *eccentric mass on a shaft, 365*
* *energy build-up in an oscillating system, 364*
* *energy increase and decrease in, 34–36*
of a ferromagnetic rod, 981–82
* *in the mountings of a motor, 369–70*
* *normal modes of a coupled system, 364–65*
* *resonance using a mass attached to a vibrating arc, 366–67, **604–7***
* *resonance using steel strips, 365*
* *resonance in a tuned system, 364–65*
theory of, **604–7**
* *in a torsional oscillator, 368–69, **607–9***
* *vibrational modes of a braid cable, 474*
* *of a water system emptying through a siphon, 370–71*
* *of a weakly-coupled system, 373*
* *generated by feedback, 371*
of an inverted pendulum, 919–20, **1325**
potential energy for small, 354
rotary, 357–58
stability of, in a physical pendulum, 354
three types of, in a bifilar pendulum, 357–58
use of a hair clipper to generate, 476
of a water drop, 391–92
oscillators, coupled
* *analysis of coupled mechanical systems with an analog computer, 361–63*
* *two charged pith balls, 873*
* *demonstrations with two coupled pendulums, 358*
* *double pendulum, 360–61*
driven, 365–66, 373
energy transfer between, 358–60, 365–66
* *energy transfer among seven coupled balls, 359–60*
* *energy transfer in a coupled pendulum system, 359*
normal modes of, 358, 359, 365–66
* *properties of, 358*
* *properties of vibrating systems, 360*
weak coupling, 359, 373
oscillators, linear
conservation of mechanical energy in, 234–35, **570–71**
driving frequency of
group velocity as a function of, 241, **575–77**

oscillators, linear (*cont.*)
driving frequency of (*cont.*)
phase velocity as a function of, 240–41, ***575–77***
wavelength as a function of, 240–41, ***575–77***
oscillators, RC phase shift, 994–95
oscillators, relaxation, 991–92
model of, 992
oscillators, torsional, forced vibrations of, 368–69, ***607–9***
equations for, ***608–9***
oscilloscope, for large lecture halls, 996, ***1337, 1339–41***
Oudin coil, 1001–5
overhead projector, 719–30
accessories for
demonstration calipers, 85
motors, 726
polarizing spinners, 724
potential well
$1/R$ surface, 166, 1249, ***1372–73***
$1/R^2$ surface, 166
probability board, 86, ***532–34***
reference systems, 85
slides of coordinate systems, 87
transparencies, availability of, 725–26
transparencies, design of, 724
transparencies, mounting of, 725
transparencies, reproduction of, 724–25
transparencies, technamated, 723–24
transparencies, use of, 721–24
x-y plotter, adaptation of Heath servo-recorders, 111
x-y plotter, adaptation of Houston *x-y* recorders, 112–13
x-y plotter, construction of, 110–11, ***537–43***
experiments for use with
Brownian movement analog, 813–14
circuit analysis using Thevenin's theorem, 886–87
circular motion, 120
collisions, 123
conservation of angular momentum, 322
conservation of momentum, 209
Coulomb's law, 845
cryophorus, 773, 774
cyclotron, gravity model of, 919, ***1324***
deflection of an electron beam by a magnetic field, 729
determination of molecular diameter, 393–94
electrical circuits, 728–29
electroscope, 836–37
generator, electric, 940
Hall effect, 1236–37, ***1367–70***
hydrogen molecule model, 729
overhead projector (*cont.*)
experiments for use with (*cont.*)
illustration of group velocity: densimeter, 485–86
illustration of phase and group velocities, 479–80, ***635–36***
Kundt's apparatus, 493
Lissajous figures, 348
magnetic field, properties of, 953–56
magnetic field of a current-carrying wire, 920–21, 921, 922
magnetic fields and induced EMF's, 727–28
microwave interferometers, 1014–17, ***1338–45***
model for drop formation, 810
motion of charged particles in an electric and magnetic field, 918
nuclear reactions, 1249, ***1372–73***
polarization dem., 726–27
projectile motion, 122–23, ***544–45***
ripple tank dem., 472–73
rotating vector model for interference effects, 729–30
scattering, 123
semicircular vortex in water, 423–24
significance of Reynolds' number, 410–12
simple harmonic motion, 340–42, ***602–3***
standing waves, 460
synchronous and induction motors, 944, ***1329–32***
water pressure demonstrator, 382
wave motion dem., 477
sources of equipment and materials, 730
* *oxidation of ferrous iron, 894*

paint, conducting, ***1318***
pair annihilation, 1175–76, ***1359, 1361–62***
description of, 1175
radiation, angular correlation of, 1175–76
paper, conducting, 222
parabolic reflector, optical gain of, ***1343–45***
parabolic trajectory
of a ball, using carbon paper, 126
of a bomb, 125–26
of a projectile launched from an air puck, 226–27
of a water-drop stream, 124–25
of a water stream, 123–24
parallax, location of images and objects by, 1042–45
parallel axis theorem, 276–77, 280–81
application of, 277–78
paramagnetic resonance, electron, 1274–83
paramagnetic susceptibility, Curie–Weiss law, 988
paramagnetism, 959–60
behavior of a weak paramagnetic substance, 960
effect of orientation on magnetic behavior, 969
particle detectors, 1266–74
alpha-particle counter, 1270–71
cloud chambers, 1266–70, ***1382–90***
ionization chamber, model, 1270
spark chamber, 1272–74, ***1390–92***
particle dynamics, 106ff, 135ff
* *particle scattering: magnetic pucks, 227–28,* ***570***
Pascal's law
* *blowing up a paint can, 377–78*
* *gas in a heated light bulb, 379*
* *water in a bottle, 378*
Pascal's vase apparatus, 375
Paschen curve, 894
Paschen's law of gas discharge, 894
path length, average between collisions, 246, 810–11, 820–21
Peltier effect, 898–99
in semiconductors, corridor dem., 714
pendulum
bifilar, 357–58
calthrop resonance, 369
conical
construction of, 157–58
properties of, 162–63
cycloidal, 352–53, ***603–5***
double, 360–61
Foucault, 299–305
model of, 296
gravity
as a ballistic meter, 31–32
as a static balance, 31–32
* *nonisochronism of, 343–44*
oscillation of an inverted, 919–20, ***1325***
corridor dem., 704
* *parametric excitation of a simple, 344*
physical
period of, 277–78, 352–53, 354
potential energy for small oscillations, 354
stability of oscillations, 354
spring, 342–43
torsion, 269, 275, 278–79, 355–56
forced vibrations of a, 269
trifilar, 279–80
percussion, center of, 355
periodic impact, timing by, 256
permeability, magnetic, 926–27, 930, 956
ratio of, of iron and air, 990
Perrine, J. O., 512
persistence of vision, 88, 89–90
persistent currents, 828, 829
Petit–Dulong rule, 757
failure of, at low temperatures, 757–58
phase angle in a series RC circuit, 890–91

phase detector, 1073–74, ***1353***
phase relations
  between current and voltage in a parallel resonant circuit, 1000
  between currents in an LCR circuit, 993
  of a driven torsion pendulum, 269, 368–69
    equations for, ***608–9***
  in forced oscillations, 257, 366, 367–68
  and interference of waves, 470–72
  Lissajous figures, 347–51
  in longitudinal wave motion, 483
  of quantities in simple harmonic motion, 334
  at resonance, 35–36, 368
phase shifter, mechanical, 36
phase velocity, 36–37, 240, 478–79
  dem. of on an overhead projector, 479–80, ***635–36***
  and group velocity, 241, 479–81, 1183
  theory, ***575–77***
phasors, model for overhead projector, 729–30
phonautograph, 40
phosphorescence, 1215–17
  preparation of materials, 1215
  quenching of, by infrared light, 1216
* *photoconduction, 1231–32*
* *photoconductivity, 1233–34*
photoelectric effect, 56, 1169–72
  calculation of Planck's constant, 1172
  conductivity in an insulator due to, 1231–32, 1233, 1234
  * *with a phototube and a capacitor, 1169–70*
  * *relationship of photoelectron energy to frequency, 1170–72*
  stopping potential, 1171
  threshold frequency for, 1169, 1170
  * *in a zinc sheet connected to an electroscope, 1169*
* *photoelectric threshold, 1232–33*
photoelectric timing, 103, 117–18, 255
photons
  energy of, relation to frequency, 1170–72
  particle properties of, 1170–71
physical optics, 1065ff
physical pendulum
  period of, 276–77, 354
    dependence of on rotational inertia, 352
    effect of position of center of gravity on, 352–53
  potential energy for small oscillations, 354
  stability of oscillations, 354
piezoelectric crystals, 901–7
  polycrystalline plate textures of, 905–6
* *piezoelectric effect, 192*
piezoelectric effects, 901–7
* *amplification of sound with a piezoelectric crystal, 901–2*
* *domains of polarization, 905*
* *ferroelectric crystals, 902–3,* ***1322–23***
* *hysteresis in a ferroelectric crystal, 904*
* *inverse, with a crystal of Rochelle salt, 906–7*
* *in polycrystalline sheets, 905–6*
* *principle of, model, 904*
* *resonance in ferroelectric crystals, 905*
* *stress-induced surface charge on a crystal, 901*
piezometer, 381–82
pinhole camera, 1040
  water-filled, 1040
pinhole image, 1040
pitch
  change in due to temperature change, 493
  relationship between frequency and, 497
Pitot tube, 398
Planck, Max, 62
Planck's constant, calculation of, 1170–72
Planck's radiation law, 1181
planetarium, electronic satellite, 173–76, ***552–56***
planetary motion, dem. of
  using an air puck, 237–39, ***574–75***
  model of Earth–Jupiter motion about the sun, 167, ***551–52***
plastic lenses, 1047–48
  advantages of over glass, 1047–48
  construction of, ***1346***
  types of, 1048
plotter, *x-y, for an overhead projector,* 110–13
  adaptation of Heath servo-recorders, 111
  adaptation of Houston *x-y* recorders, 112–13
  construction of, 110–11, ***537–43***
Pluecker tubes, 1202
plumbicon tube, 657
Pohl free fall apparatus, 105–6
Poinsot's construction, 309–11
Poiseuille's law, 410
Poisson, S. D., 28
Poisson contraction, 442
Poisson's equation, fields satisfying, 419
Poisson's ratio, 440
polarization, electric, 93–94, 899–906
  dielectric, 871–72
  of a ferroelectric crystal
    domains of, 905
    influence of a constant displacing field, 903
  of a pyroelectric crystal
    discussion of, 901
polarization, electric (*cont.*)
  of a pyroelectric crystal (*cont.*)
    effect of temperature change on, 899–901
  stress-induced, 901–6
polarization, magnetic
  in a diamagnetic substance, 959–60
  in a ferromagnetic substance, 964–67
    effect of temperature on, 967
    Weiss domains, 966–67, 974
  in a paramagnetic substance, 959–60
polarization of electromagnetic radiation
  Brewster's law, 1024–25, 1102, 1103, 1142–43
  * Brewster's law corridor dem., 1103–4
  circular, 1029, 1206, 1278
  * *circular effect in a sugar solution, 1104–5*
  * *color change in a double-refracting material, 1104*
  degree of, 1206
  dem. on an overhead projector, 726–27
  * *depolarization due to diffuse scattering, 1106–7*
  discussion of, 1102
  * *double-refraction in a uniaxial crystal, 1107–8*
  * *effects of a half- and a quarter-wave plate, 1111*
  * *function of a half-wave plate, 1110–11*
  * *interference colors of a crystalline wedge, 1108–9*
  * *interference colors in various crystalline materials, 1109–10*
  in a laser, 1134–35, 1142–43
    scattering with polarized light, 1143
  of microwaves, 1023
  for photoelastic stress analysis, 1116
  plane, 1029
    construction of, 1103
    production of, 1102–3
  * *rotation of the plane of, 1104*
  * *rotation of the plane of polarization by a magnetic field, 1112–13*
  * *rotation of the plane of, due to sugars, microwaves, 1029–30*
  * *scattering of light in a cloudy medium, 1105*
  * *scattering from a rosin solution, 1105–6*
  * *scattering from soap and water, 1106*
  * *sign of a crystal, 1111–12*
  * *spectral analysis of the colors of a quartz wedge, 1109*
  use of, to observe magnetic domains, 964–66
polarizing spinners, for an overhead projector, 724

* *polymorphism*
in ammonium nitrate, 779
in mercury iodide, 778–79
population inversion, 1133
positronium, 1175
positrons, 56, 1175
potential
electric, 865–68 (*see also* electric potential)
ionization
of argon, 1211–13
of xenon, 1213
one-dimensional, 180–81
stopping, 1171
potential difference
across a capacitor
dependence of on dielectric medium, 869
dependence of on plate spacing, 869
magnetic, meas. of, 925–27
potential energy, 105
elastic, 178–79
gravitational, 259
interchange with kinetic energy
in a pendulum, 181–82
in a two-body collision, 183–84
for small oscillations, 354–55
potential well
for the earth and moon, construction of, 166
representation of gravitational, 166
$1/R$ surface for overhead projector, 166, 1249, ***1372–73***
$1/R^2$ surface for overhead projector, 166
power developed by a rocket, 189–90
Prandtl rotatable disk, 296
Prandtl tube, 398
precession
of equinoxes, 45, 312–13
of a gyroscope, 308–14
Larmor, 1274, 1275, 1276, 1278
and nutation, nodel, 313
relation of to linear dynamics, 305–6
pressure
* *air, 383*
* *air, force of, 383–84*
Bernoulli's principle, 399–406
Boyle's law, 785–90
calculation of, for a ram pump, 433–34
Charles' law, 790
* *comprehensive demonstration barometer, 384–85*
critical point, 790–91
difference, caused by diffusion, 410
effect of, on the boiling of water, 792–93
effect of, on discharge in a gas tube, 894
electrostatic, 843
gradient, gravitational, 146
* *gravitational, 382*
* *hydrostatic, of a liquid on the side walls of a container, 386*
pressure (*cont.*)
* *hydrostatic paradox, 385–86*
independence of thermal conductivity on, 800
inside a soap bubble, 389, 392
* *liquid density comparisons, 383*
meas. of, 382–88
Pitot tube, 298
* *using a plastic barometer, 384,* ***610***
Prandtl tube, 398
* *operation of a siphon, 386–88*
of radiation
light, 37, 709–10
sound, 37–38
relationship to temperature at constant volume, 753–54
trajectories of atoms at low, 821
* *transfer of water under an evacuated bell jar, 388*
vapor, 791–96
apparatus for measurement of, 782
Dalton's law dem., 794–96
effects of temperature on, 792, 793
variation of with depth, in a liquid, 382
* *water, demonstrator, 382*
waves, standing, 494, 495–96
pressure detector, photoelectric, 803
pressure syringe, 375
pressure-temperature curves, 753–54, 792
pressure-volume-temperature curves, three-dimensional
for carbon dioxide, 783
for water, 783
principle axes, stability of rotations about, 274
prism
angle of, 1022
angle of minimum deviation, 1022
dispersion in, 1050, 1101, 1102, ***1346–48***
microwave, 1013
index of refraction of, 1022
paraffin, 1022
probability board, for overhead projector, 86, ***532–34***
projected coordinate systems, 85
slides for, 87
projectile motion, 122–31
apparatus for use with multiple-flash photography, 738–39
* *using a billiard ball cannon, 127,* ***545–46***
* *independence of velocity components, 126*
* *"monkey and hunter" apparatus, 125*
on an overhead projector, 122–23, ***544–45***
* *parabolic descent of a bomb, 125–26*
projectile motion (*cont.*)
* *parabolic path, using carbon paper, 126*
* *parabolic path of a water-drop stream, 124–25*
* *parabolic path of a water stream, 123–24*
* *parabolic templates, 126*
* *prediction of point of impact, 127–28*
spring-gun, for use in dem. of, 128–30, 226, ***548***
independence of horizontal and vertical components of motion, 130
"monkey-and-hunter" dem., 130
* *prediction of trajectories and projectile range, 130*
vertical puck gun, 226–27
use of an analog computer for, 127, ***546–47***
projection, shadow (*see* shadow projection)
projection screen, translucent, 1061
projection systems, 1061–62
projection techniques for meters, 882–85
using a commercial slide projector, 883–85
using a lantern slide projector, 883
using a mirror and lens system, 882–83
projector, overhead (*see* overhead projector)
projector, slide (*see* slide projector)
prony brake, constant torque type, 281
propulsion, rocket (*see* rocket propulsion)
protractor and slope indicator for blackboards, 86
Pryor, Marvin J., 708
psychological colors, 87–88, 90
psychological effects, 85–90
psychological error, 89
* *puck molecule demonstration, 235,* ***571–73***
pucks, air-supported, 222–24
captive, 222, 223
compressed air, 223
dry ice, 223
electrically charged, ***1318–19***
free, 222, 223–24
launching ram for, 224, ***1319–20***
liquid air, 224
magnetic, 227, ***570***
pulley, demonstration set, 91
pulse jet engine, 191
pulses, 442, 452–53
speed of, in a spinning lariat, 452
pump
diffusion, cooling arrangement for, ***610–13***
electromagnetic, 912–14
inertia, 432–34
lift and force, 397
ram, 432–34, ***621–22***

pyr-a-larm, 897–98
pyroelectric coefficient, 900
pyroelectric crystals
  discussion of, 900–1
  effect of temperature change on the polarization of, 899–901
Pythagoras, 45

Q, definition of for an oscillating system, ***606***
Q value, at resonance, 1003
quadratic restoring forces, 165
quarter-wave plate, 1111
* *quasi-static expansion and contraction of a coil spring, 30–31*
Quincke interference tube, 490

radiant energy
  absorption of
    connection of, with emission, 763
    dependence of, on wavelength, 760–62
    infrared, 762, 1177
    light, 1096–97, 1205
    relative, of black and white surfaces, 1176
    resonant, in infrared region, 762
    using a thermicon card, 759–62
  emission of
    relative of black and white surfaces, 1176, 1178–79
    relative, of transparent and opaque bodies, 1177
  spectral distribution of, 761–62, 1120, 1179, 1181, ***1362–63***
radiation
  absorption of
    gamma rays, 1246, 1257–58
    infrared, 1177
    light, 1096–97, 1205, 1257–58
    thermal, 759–63, 1176
  black body, 1176–83
  coherent, 519–20, 1127–32
    coherence length, 1129
    coherence time, 1129
    partial coherence function, 1128
    spatial coherence, 1129–30
  comparison of conventional and laser, 1131–32
  emission of, 1132–33
    gamma rays, Mossbauer effect, model, 1253–57, ***1373***
    spontaneous, 1133
    stimulated, 1133
    thermal, 1176–79
  flux, effect of total emissivity on, 1178
  incoherent, 519–20, 1127–32
  infrared, 1177
    quenching of phosphorescence by, 1216
    resonant absorption of, 761–62
  Planck's law, 1181
  from radioactive nuclei (*see* radioactive decay and radioactivity)
radiation (*cont.*)
  resonance, 1213–17
  spectral distribution of, 761–62, 1120, 1179, 1181, ***1362–63***
  thermal
    absorption of, 759–63
    emission of, 1176–79
    inverse square law for, 1178
    properties of, 760–62
radiation detector, the eye as, 90
radiation pressure
  of light, 37
    corridor dem., 709–10
  * *of sound, 37–38*
radio, relation between holography and, 1153–57
radioactive decay, 1240–49
  beta decay, 1246–49, ***1370–72***
    energy spectrum of, 1246–49
    straight line emission of electrons, 1246
    in U-239, model, 1263–64, ***1378–81***
  use of a capillary filter tube system as a model for, 1240–45, 1261–63
  equilibrium
    secular, 1242–43
    transient, 1243
  model of, 1240–41
  series disintegration, 1243–45
    activity of nuclei undergoing, 1244
radioactivity, 55, 56, 1240–49
  * *beta decay, 1246–49,* ***1370–72***
  * *daughter products of radon and thorium in the air, 1245–46*
  * *radioactive contamination, 1240*
  * *radioactive decay, model, 1240–41*
  * *secular and transient equilibrium, model, 1242–43*
  * *series disintegration, model, 1243–45*
  * *simultaneous decay of two isotopes, model, 1241–42*
  * *straight line emission of electrons in beta decay, 1246*
radiometer, sound, 38
radius of gyration, determination of, 271
rainbow, formation of
  mechanical model, 1046
  string model, 1045–46
Raleigh refractometer, 1039
ram pump
  construction of, ***621–22***
  operation of, 432–34
random walk, 246
range, prediction of, 127–28
* *ratchet and pawl, 780–81,* ***1288–89***
ray, principal, 1049
Rayleigh disk, 406
Rayleigh scattering, 1105, 1143
RC circuits
  differentiating, 889–90
  discharging, decay rate of, 886
RC circuits (*cont.*)
  integrating, 889–90
  as a relaxation oscillator, 991–92
  series, ac, 890–91
    vector diagrams of, 890
reactance, capacitive, 873, 890–91
reaction time, meas. of, 89
reactions, nuclear (*see* nuclear reactions)
reactor, thermal neutron, suffering from xenon poisoning, model, 1261–63
real image, formation of, 44
  with a convex lens, 1049
  location of by parallax, 1042–43
  in a mirror, 1035–37
  with a pinhole camera, 1040
  with an ultrasound camera, 522–24
real object, difference between virtual and, 1044
reality, levels of, 68–73
  depiction, 69–70
  didactic, 68–69
  experimental, 68–69
  models, 70
  others, 70–71
  pedagogic consequences of recognizing, 71–73
reciprocity theorem, for electrical circuits, 889
recoil of a gun, to show momentum conservation, 204, 206–8
recoil of a gun, rigidly and freely mounted, to show momentum and energy transfer, 1254–56
recorder, *x-y*, for an overhead projector, 110–13
rectification
  full-wave, 994
  half-wave, 994
rectilinear and rotational motion, addition of, 272
reduced mass, 235
reference frames
  center of mass, motion in, 215–16, 235
  Euler's angles, 324–26
  gravitational field and accelerated, 288
  inertial
    Lorentz transformations, visual representation of, 1166–69, ***1355–57***
    and non-inertial, 159–60, 293–94
    properties of, 132–33
  rotating, 289–94
    * *bending of a rotating water tube, 291*
    * *Coriolis force dem., 290,* ***590–91***
    * *dart gun fired from a swivel chair, 289*
    * *effect of the Coriolis force, 291*
    * *fictitious forces, 293,* ***591–92***
    and the Foucault pendulum, 300–2

reference frames (*cont.*)
rotating (*cont.*)
* *relative motion, 290–91*
* *transformation of vectors, 293–94*
* *uniform motion viewed from, 291–93*
reference systems
for blackboards, 85
for overhead projector, 85
for slide projector, 85
slides for projecting, 87
reflection, 1035–38
Bragg, 473, ***634–35,*** 1026–29, 1193–1201
* *from a frosted surface, 1037*
* *from a glass rod, 1037*
* *image from a concave mirror, 1036–37*
interference from a thin film, 1077–78, 1138
* *multiple, from two mirrors, 1035*
from a plane surface, 1035–36, 1049, **1345–46**
prevention of unwanted, in a ripple tank, 470, ***633***
* *properties of a plane surface: signal mirror, 1035–36*
of radiant energy, 1178
* *Snell's law, 1037–38*
total internal, 1038
frustrated, 1188–90, 1191–92
meas. of index of refraction using, 1038
wavelength dependence of, 1038
of waves in a wave machine, 445, 448
reflectors
parabolic, optical gain of, ***1343–45***
Scotchlite screen, 735
refraction, 1038–50
* *Abbe refractometer, 1038*
double, 1104, 1107–8
* *fish-eye camera, 1040,* ***1345***
fluid mapper analogy for, 419
* *through heated air, 1045*
index of
* *of air, 1039*
of a dielectric, 1023–25
of glass, using total internal reflection, 1038
of a liquid, 1020, 1038
meas. of with a microwave interferometer, 1018–19
* *meas. of by telescope focus, 1038–39*
periodic variation of, to diffract light, 1094–95
of a prism, 1022
index of, observation of small changes in, 1056–59
* *direct shadow method, 1057–58*
* *Schlieren method, 1058–59,* ***1352***
* *Toeppler Schlieren apparatus, 1056–57*
in lenses, 1042–45, 1049–50

refraction (*cont.*)
* *of light through a medium of varying index of refraction, 1040–42*
from a plane surface, 1049, ***1345–46***
* *rainbow, mechanical model of, 1046*
* *rainbow, string model of, 1045–46*
* *Raleigh refractometer, 1039*
of sound waves, 524
* *"spherical lens" effect, 1042*
of water waves, 470
refractometer
Abbe, 1038
Raleigh, 1039
refrigerator, hot air motor as a, 779
Reis, Philip, 39–40
relative humidity, 796
relative motion, 131–33, 188–89, 290–91
reference frames in, relativistic phenomena, 1166–69, ***1355–57***
waves in a moving medium, 1164–66
relative velocity, 131–33, 188–89
relativity
of black and white, 88
Galilean
circular motion, 131–32
linear motion, 131
of simultaneity, visual representation of, 1168, ***1355–57***
special, 1164–69 (*see also* special relativity)
relaxation oscillators, 991–92
model of, 992
relaxation time
spin-lattice, 1282–83, ***1395***
spin-spin, 1282, ***1395***
reluctance, 951, 978–81
replica gratings, 1202
resistance, electrical
change of, due to an applied magnetic field, 988
* *change in a Corbino disk, 1235–36*
* *change in semiconductors due to magnetic field, 1235*
effect of low temperature on, 831–32, ***1300–10***
equivalent, of resistors in parallel, 886
functional relationship of, in a circuit, 994
internal, of a voltage source, 885
meas. of, using a discharging RC circuit, 886
molybdenum-copper sulfate liquid, 871
resistance, viscous, 407–8
resistivity, 881
resistors
decade box for use with large classes, 881
in parallel, voltage drop across, fluid analog, 885
phase of current in, ac circuit, 993

resistors (*cont.*)
in series, voltage drop across, fluid analog, 885
resolution
of forces apparatus, 96
of a television image
horizontal, 658
vertical, 658
of vectors, 94
resolving power, 1092–94
resonance
apparatus, acoustic, 501–2
in a cubical cavity, 503–5
curves, 35
of a cubical cavity, 503–5
of a driven harmonic oscillator, 366–67
effect of damping on, 366–68, 371–73
of a stretched string, 475–76
of a torsion pendulum, 269, 368–69
electron paramagnetic, 1274–83
energy increase and decrease at, 34–36
in a ferroelectric crystal, 905
and forced vibrations: corridor dem., 703–4
function, behavior, ***604–7***
in a laser cavity, 1134
in an LCR circuit, 998–1000
in magnetically induced oscillations of a rod, 981–82
in a mechanical system, 34–36, 364, 365, 367, 371–73
microwave, 1012–13
in the mountings of a motor, 369–70
nuclear magnetic, 1274–83
parallel, in an LC filter, analog, 997
phase relations at, 35–36, 368
as a possible cause of damage, 369–70
sound, using Helmholtz resonators, 499–500
spin, 35
tuning fork, 500–1
resonance absorption of radiation
gamma rays and light, model for, 1257–58
infrared, 762
resonance radiation, 1213–17
* *of a collimated beam of sodium atoms, 1215,* ***1366***
* *of iodine vapor, 1213*
* *phosphorescence, 1215–17*
* *of potassium vapor, 1213–14*
* *of sodium vapor, 1214*
resonant frequency, 34–36
of a cubical cavity, 503–5
derivation of equations for, ***604–7***
fundamental longitudinal mode, 440
fundamental transverse oscillation, 474
* *quantitative study of, in a string, 475–76*
of a quartz plate, 528

resonant frequency (*cont.*)
  relationship of, to material parameters of a solid, 439–40
  of an RLC circuit, 990
resonators
  cavity, 503–5
  conical, 499–500
    specifications for, 500
  distinction between baffle and, 502
  function of, 502
  Helmholtz, 499
restitution, coefficient of
  * *of two metal balls, 209*
  * *of pucks on an air table, 248–49*
  * *of a steel ball, 210*
reverberation time
  concept of, 506–7
  expression for, 507
  meas. of, 506–7
reversible changes of state, 30
Reynolds' number, 410–14
  critical, 412–13, 414
  equations for, 411
  meas. of, 411
  significance of, 410–11
rigid body
  combined translational and rotational motion of, 286–89
  equilibrium of, 327–32
  rotational dynamics of, 281–86
ripple tank, 466–73, ***626–35,*** 1164–66
  flow, 1164–66
  operation of, 466–67, ***626–29***
  for an overhead projector, 472–73
ripple tray, 469, ***633–34***
RLC circuit, 990–92, 998–1000, ***1334–36***
Rockefeller Institute, Christmas lectures at, 21
rocket, mechanical construction of, 186
rocket, whistling, 510
rocket principle, 227, 259–60
rocket propulsion, 186–92
  * *using carbon dioxide, 186*
  * *dependence of power on velocity, 189–90*
  * *gas-propelled glider: rocket thrust, 260*
  * *independence of thrust of surrounding medium, 187*
  * *meas. of thrust, 187–88*
  * *principles of, using a toy rocket, 186*
  * *relative motion of exhaust gases, 188–89*
  * *using rubber bands, 186*
  * *thrust in the pulse jet engine, 191*
Rogowski coil, 924–27
Ronchi rulings, 1054
rotary air bearings, 263–65
rotary oscillations, 357–58
rotating apparatus, motor-driven, 163
rotating electrostatic generator—dirod, 850–52, ***1312–18***
rotating magnetic field, 943
  in synchronous and induction motors, 944, ***1329–32***
rotating reference frames, 289–94
  and the Foucault pendulum, 300–2
rotating vectors, 94–95, **536,** 729–30
rotational dynamics, 231, 232, 271ff
  * *apparatus, 267–69*
  * *effect of a couple, 231–32*
  * *monkey and bananas problem, 287–88*
  of a rigid body, 281–86
    * *calculation of torsional modulus, 284–85*
    * *length of a moment arm and torque, 287*
    * *meas. of angular displacement, 282–84*
    * *meas. of a torque arm, 286*
    * *moment arm of a bicycle, 281–82*
    * *motion of the center of mass of a rod, 282*
    * *Newton's second law, 284*
    * *principles of, 282*
  * *similarity of an accelerated frame and a gravitational field, 288*
  * *vertical circle apparatus, 289,* ***589–90***
rotational equilibrium, 330, 331–32
rotational inertia
  and angular impulse, 297
  demonstrator, CENCO, 272
  dependence of period of physical pendulum on, 352
  * *determination of, 268*
  * *of a disk using a trifilar pendulum, 279–80*
  effect of mass distribution on
    * *using two equal mass tubes, 275*
    * *using a hoop and disk, 271*
    * *using a light rod with an adjustable mass, 284,* ***588–89***
    * *using two masses on a rotating shaft, 275*
    * *effect of, on the period of a torsion pendulum, 275*
  * *parallel axis theorem, qualitative, 280–81*
  * *period of a physical pendulum, 277–78*
  rotation about the axis of maximum
    * *using a bar, 275*
    * *using a suspended cigar box, 274*
  * *Steiner's theorem, 276–77*
  * *of a torsion pendulum, as a function of its period, 278–79*
rotational kinematics, 272–73
  * *addition of angular velocities, 272–73*
  * *axes of rotation, 272*
  study of, using a vertical circle apparatus, 289
rotational kinetic energy, 178–79, 269, 286
rotational motion
  using an air puck, 230–31
  combined translational and, 322–23
rotational motion (*cont.*)
  combined translational and (*cont.*)
    * *conversion of potential energy into rotational kinetic energy, "dilution of gravity," 286*
    * *quantitative study of, 288–89*
  and rectilinear, 272
rotations
  addition of parallel, 272
  about the axis of maximum rotational inertia, 274, 275
  about the center of gravity, 328
  about the center of mass, 286
    Steiner's theorem, 276–77
  of the earth, meas. of, 300–2, ***593–94***
  about two nonparallel axes, 272–73
  * *about principal axes, stability of, 274*
  small, meas. of using Moiré fringes, 1055–56, ***1346–52***
Royal Institution
  Christmas lectures, 16, 21
  Christmas lectures published as books, 21–22
  founding, 13
  Friday evening discourses, 14
  lecture room of, 14
  schools lectures, 16–17
    sample outlines of, on magnetism, 22–26
  traditions, 14ff
* *Rüchhardt's method for determining gamma, 801–3*
  as an ideal corridor dem., 716–17
Rumford, Count, 13
Rutherford scattering, 123
  * *demonstration of, 236–37*
  * *using electromagnets, 1250*
  * *from a model of a Rutherford atom, 1249–50*

sagittal focus, 1050
sagittal plane, 1050
sandbed, isolated, analogy for distribution field phenomena, 418, ***617–18***
  references, ***620***
satellite, artificial, 173–76, ***552–56***
  display of motion of, 173–75, ***552–56***
  equations of motion for, 175–76
  motion of, in orbit, 175–76
  simulator, 173–74
  speed of, 173
  * *tracking, an audio-visual dem. of, using the Doppler effect, 514–15*
scale, high-impedance spring, ***560–61***
scattering
  of alpha particles, 49
  Compton, 1172–74
  Coulomb, 853–54, ***1318–20***
  depolarization by diffuse, 1106–7
  of two equal masses: 90-degree law, 226

scattering (*cont.*)
low-friction surface for dem. of, 151–52
of microwaves by a dielectric dipole, 1030
neutron diffusion in a moderator model using magnetic pucks, 227–28, ***570***
on an overhead projector, 123
polarization of light caused by, 1105–6, 1142
Rayleigh, 1105, 1142
Rutherford, 123, 236–37, 249–50
through small angles: identity of angular distribution from positive and negative forces at large distances, 228
from a Thomson atom, 1249–50
of *x*-rays by paraffin, 1250
Schilling's principle, 59–60
Schlieren apparatus, Toeppler, 1056–57
Schlieren optics, 430–31
to observe small changes in the index of refraction, 1056–57, 1058–59, ***1352***
Schmidt, Johann A., 27
Schools Lectures
at the Royal Institution, 16–17
sample outline of, on magnetism, 22–26
Science Museum, London, lectures arranged by, 20
Scotchlite screen, 735
screen, translucent projection, 1061
screw apparatus, 91
screw model, 91
sealant for microscope slides, 819
second law of thermodynamics, 30
* *hot air motor, 779–80*
"violation" of, with a ratchet and pawl, 780–81, ***1288–89***
Seeback voltage, 899
Seignette salt crystal, 901
self-inductance, 950
semi-conductors
magnetoresistance in, 1235, 1236
mobility of charge carriers in, 1220
* *model of, 1220–21*
Peltier effect in, corridor dem., 714
* *properties of germanium and selenium, 1234*
photoconductivity, 1234
rectifying action, 1234
sign of charge carriers in: Hall effect, 1236–37, ***1367–70***
* *test for type of conductivity, 1221–22*
shadow projection
as a dem. technique, 18, 29
experiments which can advantageously employ, 30–40
as a test for equipment arrangement, 29
shear modulus, 439, 440
shielding action of a closed conductor, 841
shielding from magnetic fields, 954
of conductors, 916–18
* *produced by magnets and current-carrying conductors, 962–63*
time-varying, 935–36
shock waves
* *dem. of using a water-filled champagne bottle, 428*
* *effect of, in argon, 427–28*
produced by a bullet, stroboscopic photograph of, 734
* *from a source moving at supersonic velocity, 428–30*
* *supersonic flow of an air-stream, 430–31*
* *in a water trough, 426–27*
* *Shockley–Haynes experiment, 1228–31*
sidebands, suppression of, in holography, 1155
signal generator, low-frequency, mechanical, 459, ***626***
use of with Fourier analysis, 459
simple harmonic motion, 334–45
* *acceleration during, 337*
* *using an air puck, 234*
ball in a spherical cavity, 344–45
* *change of longitudinal oscillations into simple oscillations, 342*
* *characteristic appearance of, 334*
characteristics of, torsion pendulum, 356
* *correspondence with uniform circular motion, 334*
* *free and damped, 257*
* *harmonic oscillator I, 339–40*
* *harmonic oscillator II, 340–42,* ***602–3***
* *phase relationships of position, velocity and acceleration, 334*
* *plate on two rotating drums, 336–37*
* *recording of, on newsprint, 337–38*
recording of on an overhead projector, 340–42, ***602–3***
* *recording of on a paper towel roll, 338–39*
* *recording of on a roll of paper, 339*
* *synchronization of periods, 335–36*
simple machines, 91–92
simultaneity, relativity of, visual representation of, 1168, **1355–57**
simultaneous velocities, 96
sine and cosine curve templates, 87
sinusoidal curve
as a picture of a travelling wave, 36–37
as a trace of simple harmonic motion, 337–39, 340–42
siphon, operation of, 386–88
size, judgment of, 88
slide projector
demonstration calipers for, 85
reference systems for, 85
slides
avoidance of in lectures, 17, 29
for projecting coordinate systems, 87
sling psychrometer, 784
Slipknot semiconducting tape, ***1316***
Smith, Elbridge, 62
smoke detector, Cerberus, 897–98
Snell's law, 1037–38
Snoek effect, 156–57, ***550–51***
solenoid, magnetic field of, 956
on its axis, 922–23
sound
amplification
dem. of with a tuning fork, 526–27
with piezoelectric crystals, 901–3, 905, 906–7
water stream and a membrane, 523–25
* *Fourier spectrum of, 501*
* *frequency spectrum, 503–5*
* *intensity of, feeling for the size of the unit of, 507*
production of musical notes, 498–99
radiometer, 38
relationship between frequency and pitch, 497
resonance effects
in a cubical cavity, 503–5
using Helmholtz resonators, 499–500
using a tuning fork, 500–1
* *reverberation time, 506–7*
sources of, 496–508
noise generator, 507–8
velocity of, 491–93
* *in air, 491*
* *dependence of on temperature, 493*
* *for the fundamental mode in a bar, 491–92*
* *in a rod, 492–93*
sound waves, 490ff
diffraction of, 519–24
Frauenhofer, 523–24
Fresnel, 523–24
gratings for, 522
equipment for dem. of, 490–91
interference of, 508–10
* *beats, 510*
* *dem. of for large classes, 508*
* *using two loudspeakers, 508–9*
* *using two trombone slide assemblies, 509*
* *visual and audio recording of beats, 510*
mechanical effects of, 37–38
modulation of, 527–28
radiation pressure of, 37–38
refraction of, 524
spark chamber, 1272–74, ***1390–92***
principle of, 1290
spark counters, 56–57
spark recording and timing, 228–29, 254–55
spatial coherence, 1129–30

special relativity, 1164–69, ***1355–57***
Doppler effect, visual aid for, 1169
Lorentz transformations and their consequences, visual representation of, 1166–69
twin source interference in a moving medium, 1165–66
wave propagation speed, upstream and downstream, 1164–65
specific heat
apparatus, Tyndall's, 755
* *at low temperatures, 757–58*
molar, of a gas
at constant volume, 803
at constant pressure, 803
molar, of solids, 757
spectra (*see* atomic spectra)
spectral distribution of radiation, 761, 1179, ***1362–63***
dependence of absorption on, 761–62
Planck's radiation law, 1181
relationship between color and, 1120
speed
average
of air molecules, 55
of gas molecules, qualitative, 821–23
of a system of Maxwell particles, 245–46
of a bullet, 99–101, 100–2, 105, 258
of light, 1064–75
corridor dem., 711–12
discussion of measurement attempts, 1066, 1068–69
Leybold apparatus, techniques for finding motor speed, 1065–66
in a moving frame, visual representation of, 1168, ***1355–57***
rotating mirror apparatus, 1064–65, ***1353***
time-distance measurements, 1066–75, ***1353***
* *at a point, 119–20*
wave, 450–54
in water, 468–69, 1164–65
* *speed of molecule dem. and a rough estimate of Avogadro's number, 821–23*
* *sphere packing, 1237–38*
spherical aberration, 1049, 1144, ***1346, 1349***
elimination of at aplanatic points, 1051–52
spin-echo dem. using a single-pulsed RF oscillator, 1283, ***1395***
spin-echo instrument, for nuclear magnetic resonance, 1280–83, ***1395***
spin-lattice relaxation time, 1282–83, ***1395***
spin-orbit interaction, model, 1278–79, ***1392–93***
spin-spin relaxation time, 1282–83, ***1395***
spontaneous emission of radiation, 1133
spring, "massless," 803–4
spring coil, expansion and contraction, 30–31
spring gun
for gyroscopic motion dem., 312, ***595–96***
for projectile motion and collision dem., 128–30, 226, ***548***
for rotating reference frame and Coriolis force dem., 290, ***590–91***
* *spring pendulum, 342–43*
spring scale, high impedance, ***560–61***
stability
* *of a flame, 413*
of flow of a fluid, 410–14
of an object in a fluid stream, 405
of orbits
under a central force, model, 323–24, ***598–602***
quantitative dem. of, under a linear restoring force, 163–65
quantitative dem. of, under a quadratic restoring force, 165
of oscillating systems, test for, using an analog computer, 362
of a system by gyroscopic action, 308
stage heater for a microprojector, 778
standing waves, 446–48, 460–66
* *due to a changing magnetic field, 462–63*
* *in a clamped rubber tube, 460*
compressional, 460
* *dem. of for an overhead projector, 460*
electron in Bohr orbits, analog, 1185–87, 1187–88, ***1363–64***
* *in a free rubber tube, 463*
* *in a heated nichrome wire, 460–61*
* *in an iron wire, 461*
longitudinal, 493–96
* *in a closed pipe of natural gas, 496*
in a cubical cavity, 503–5
* *in a grooved aluminum rod, 496*
* *on a nichrome wire, 494–95*
* *pressure nodes and antinodes, 494*
to measure the wavelength of microwaves, 1019–20
in a quartz plate, 528
* *in the rim of a large bowl, 461*
* *shape of, on a stretched wire, 464–65*
of small magnetized balls on water, 930–31
* *in a soap bubble, 461–62*
in a solid, 439–40, 473–74
* *in a spring, 460*
* *in a stretched circular membrane, 505–6*
* *stroboscopic projection of an illuminated wire, 465–66*
in water, 470
state, change of
of bromine, 774–75
and heat, 763
state, equations of, 785–91
Boyle's law, 785–90
Charles' law, 790
van der Waals' equation, 790–91
states
atomic, excited, lifetime of 6^3P_1 mercury, 1205–7
stationary, as standing matter waves, analogy, 1188
static balance, gravity pendulum as, 31–32
static force apparatus, 329
stationary states, as standing matter waves, analogy, 1188
Stefan–Boltzmann law, 1178, 1179–83
Stefan's constant, 1180–81
Steiner's theorem, 276–77, 280–81
application of, 277–78
stimulated emission, 1133
classical analog, 35
Stokes' force, 200
Stokes' law, 408–9
apparatus for dem. of, 409
* *four balls of varying specific gravity, 409*
* *two balls of different diameters, 408–9*
Stoney, G. J., 28
stopping potential, in the photoelectric effect, 1171
streamline flow, 399–401, 403
and field pattern demonstration, 419–20, ***620–21***
streamlining, advantages of, 403–4
stress, compression, producing electric polarization, 901–6
stress, mechanical
change in magnetic field due to, 984–86
photoelastic analysis of, 1116
piezoelectric effect, 901–7
stroboscope, for motion-study, 121
stroboscope, properties of
type 1531-A Strobatac
exposure data, 736
light characteristics of, 732–36
light output data for, 732, 733
operation without the reflector, 732–33
triggering, 737–38
type 1532-D Strobolume
exposure data, 736
light characteristics of, 734–36
light output data for, 735
techniques for using, 736
stroboscopic effects, 89–90, 731–45
of a bullet in flight
passing through a soap bubble, 733–34
shock waves produced by, 734
splitting of a playing card, 733
list of experiments involving, 731
motion of falling bodies, 735–36

stroboscopic effects (*cont.*)
multiple flash photography, 738–45
collision of two suspended balls, 739
conservation of angular momentum, 739–40
motion of a ball bouncing on a rubber band, 740–41
one-dimensional collisions, 743–44
projectile motion apparatus, 738–39
two-dimensional collisions, 744–45
techniques for triggering, 737–38
contact sensors, 737
microphone and audio amplifier, 738
photoelectric cell and light beam, 737–38
signal through a wire, 737
subjective effects, 87–90
subjective measurements, 87–90
subjectivity of colors, 88–89
* *sublimation of ammonium chloride, 772*
superconductivity, 828, 829, 831–32
* *supercooling of liquids, 776–77*
* *superheating of liquids, 777*
superposition, principle of, 445, 1148
superposition of waves
as a corridor dem., 704–5
to form a wave packet, 1183
models for dem. of, 477, 478–79, 486–88
principle of, 445, 1148
in a signal generator, 459
in a stringed instrument, 466
superposition theorem in electrical circuits, 888–89
surface diffusion, hard sphere model, 809
surface of a gas, apparent, 823
surface tension, 388–94
* *determination of molecular diameter, 393–94*
discussion of, 388–90
* *effect of water-proofing on a fabric, 391*
electrostatic breakdown of, 843
equipment for meas. and dem. of, 376
* *lowering of by electrostatic forces, 390–91*
* *in mercury and water, comparison, 390*
* *minimum-surface properties of a free soap film, 392–93*
* *oscillations of a water drop, 391–92*
* *pressure inside a soap bubble, 392*
* *in a soap film, 390*
* *support of a piece of aluminum by, 391*
* *volume of water picked up by a sponge, 390*
susceptibility, dielectric, 875
susceptibility, ferromagnetic, disappearance of at the Curie point, 974–76
susceptibility, paramagnetic, and the Curie–Weiss Law, 988
Symner, Robert, 28
sympathetic vibration, 365, 457–58, 500, 1257–58
synchronization of clocks, relativistic, visual representation of, 1168, ***1355–57***
synchronizing generator, for closed-circuit television, 663
synchronous motor, principles of operation, 944, ***1329–32***

* *tangential velocity, 120*
tape, for electrostatics dem.
Mylar metallized, 837
Slipknot semiconducting, ***1316***
tape, Monel metal, for magnetization experiments, 977
teaching, automation of, 63–66, 73–75, 702
technamated transparencies, 723–24
"technological classroom," 63–65
telephone, invention of, 40
television
broadcast, lectures by, 18–19
tubeless, 90
television, closed-circuit, 655–69
accessories, 661–63
multiplexer, 662
recording equipment, 663
remote control apparatus, 662
synchronizing generator, 663
tripod, 661
vertical mount, 662
video switcher, 662–63
wide screen, 662
control circuitry, key aspects of, 658–60
control of target voltage, manual or automatic, 658
horizontal resolution, 659
interlace system, 658
lines per field, 658
vertical resolution, 658
disadvantages of lectures by, 10, 18
equipment, basic
image orthicon tube, 657–58
lenses, 658
monitor, 660
plumbicon tube, 657
recent modifications in, 661
videcon tube, 657
equipment, cost vs quality, 656
equipment, sources of, 656
systems planning
distribution system, 664–65
information on, sources of, 663
lighting, room and camera, 664
monitors, number required, 664
use of, in physics dem., 665–69
diffraction and interference effects, difficulty with, 665–66
television, closed-circuit (*cont.*)
use of, in physics dem. (*cont.*)
graphs and diagrams, 667
microscopy, 668
quality of system as determining factor, 668
reading of meters and scales, 665
reading of an oscilloscope face, 665
as a supplement to shadow projection, 667
temperature
change, demonstrations involving, 750–54
change of, meas. of in a gas undergoing adiabatic compression, 800, 801
critical, 790–91, 810
Curie point
electric polarization, 905
for ferromagnetic materials, 967, 973–74, ***1333***
effect of, 974–76
in an electric arc, 882
on magnetic domains, 966–67
on polarization of a pyroelectric crystal, 899–901
on resistance, at low temperatures, 831–32, ***1300–10***
on resistivity, 881
on vapor pressure, 792, 793
on the velocity of a charge carrier in a crytsal, 1226–28
on viscosity, 408
equipment for the dem. of the idea of, 749–50
lambda point of helium, 829, ***1300–10***
of a long rod as a function of distance and time, equation for, 766
low, phenomena, 826–32, ***1300–10***
* *P-T by immersion, 753–54*
relationship to pressure at constant volume, 753–54
Stefan–Boltzmann law, 1179–80
triple point, production of as a reference, 750
temperature-sensitive paint, 750–51
templates for drawing sine curves, 87
tension
effect of, on internodal distance, 460
effect of, on wave velocity, 451, 453
meas. of, in a falling yo-yo, 149
in a string, 149
surface (*see* surface tension)
terminal velocity, 138–39, 408, 409
Tesla high-frequency coil, 1001–5
thermal agitation of gaseous molecules, apparatus for the dem. of, 784
thermal conductivity
comparison of in different materials, qualitative, 758, 764
of a gas, 798–800

thermal conductivity (*cont.*)
of a gas (*cont.*)
* *of air and hydrogen, qualitative, 798*
* *comparison of for different gases, 798–800*
pressure, independence of, 800
relative, using temperature-sensitive paint, 750–51, 758
thermal energy, change of in air undergoing adiabatic compression and expansion, 800–1
thermal expansion
apparatus for dem. of, 755–56
coefficient of cubical, 750
thermal motion
buoyancy in liquids, illustration of, 33
of gas molecules, model, 810
thermal radiation
absorption of, 759–63, 1176
black body, 1176–83
emission of, 1176–79
inverse square law for, 1178
properties of, 760–62
spectral energy distribution of, 761–62, 1120, 1179, 1181, ***1362–63***
thermally-induced ionization, 38, 39, 295–97
thermicon card, 759
dem. in heat using, 759–63
thermisters, 1222
use of as a thermometer, 751–52
thermochrome temperature-sensitive crayons, 749
thermochromism, bistable, 763
Thermodynamics, Second Law of, 30
* *hot air motor, 779–80*
"violation" of, with a ratchet and pawl, 780–81, ***1288–89***
thermoelectric effects, 898–901
* *conversion of heat to electricity, 901*
* *effect of temperature change on the polarization of a pyroelectric crystal, 899–901*
* *Peltier effect, 898–99*
thermomagnetic motor, 978
thermometers
bimetallic, 750
demonstration, 749–50
gas, 751
constant-pressure, 754
constant-volume, comparison of, 752–53
projection, 767, 768
* *supersensitive, 751–52*
thermopile, 1178
thermopile, Moll, 1177
Thevenin's theorem, 886–87
thin-film interference, 1025–26, 1076–80, 1138
thin lens formula, corridor dem., 708
Thomson model of the atom, forces of, 1217–18
Thomson model of the atom, scattering from, 1249–50
threshold frequency, for the photoelectric effect, 1169, 1170–72
throttling process, to liquefy a gas, 824–25, ***1297–1300***
thrust
independent of surrounding medium, 187
meas. of, 186, 187–88, 191, 260
* *resistance and reversal effects, 194*
static, 191
thyratron, 507–8
time dilation, visual representation of, 1168, ***1355–57***
time of impact, meas. of, 193–94, 194–95
time of reverberation, 506–7
timer
photocell, 103, 105
spark, 181
switch, 99–101
timing
by periodic impact, 256
photoelectric, 103, 117–18, 255
spark, 228–29, 254–55
stroboscopic (*see* stroboscopic effects)
Toepler Schlieren apparatus, 1056–57
* *top, tippe, direction of spin of, 297–98*
torque
angular acceleration as a function of applied, 268
arm, measure of, 286
and length of the moment arm, 287
proportionality to torsional deflection, 284–85
* *torque rod, 329*
torsion balance, gravitational, 168–71
torsion constant, meas. of, 444
torsion pendulum, 269, 275, 278–79
* *calculation of angular velocity and acceleration in a, 355–56*
forced vibrations of, 269
* *oscillating chain, 356*
* *stroboscopic photographs of an os--cillating body, 356*
torsional mode, relation of resonant frequency of to material parameters, 439–40
torsional modulus, 284–85
torsional oscillator, forced vibrations of, 368–69, ***607–9***
equations for, ***608–9***
track, for use in one-dimensional motion dem., 107–8
transformer, Arkad'ev capacitor bank, principle of, 871
translation
distance of, meas. of with Moiré patterns, 1054–55, ***1346–52***
of vectors, 95
translational equilibrium, 329, 330, 331–32
translational motion, combined rotational and, 286–89, 322–23
transmission lines, 1009–12
* *analogy to heat conduction, 1012*
transmission lines (*cont.*)
* *behavior of, using a series of inductors and capacitors, 1009–10*
* *signal propagation velocity in a coaxial cable, 1010–12*
transparencies, 721–26
availability of, 725–26
design of, 724
mounting of, 725
relation between physical dem. and model, 722–23
reproduction of, 724–25
diazo process, 724–25
electrostatographic process, 725
photographic process, 724
thermal process, 725
technamated, 723–24
use of, 721–24
avoiding the presentation of too much material, 722
colored overlays, 721–22
transverse waves, 36, 239–42, 482–83
behavior of at an interface, 449–50
velocity of, 450–54
travelling waves, 36–37, 443–48
amplification of, 450
theory of, ***623–25***
cut-off frequency for, 240–41, **575–77**
motion of, model, 483
* *and phase velocity, 36–37*
properties of, 443–48, **622–23**
tribometer, direct reading, construction of, 153
trifilar pendulum, 279–80
triggering techniques for a stroboscope, 737–38
triple-point
cell, 750
* formation of ice at the, 773, 774
temperature, production of as a reference, 750
tubeless television, 90
tuning fork drive, 704
* *tuning fork resonance, 500–1*
tunneling, analogy, 1189–90
turbulent flow, 401, 410–14
Tyndall's friction cylinder, 755
Tyndall's specific heat apparatus, 755

ultrasonic vibrations producing a small liquid fountain, 528–29
* *ultrasonic vibrations of quartz, 528–29*
ultrasonic waves, 523–24, 1094–95
ultrasound camera, 522–24
ultraviolet cyanogen bands, 1203
uniaxial crystal, double refraction in, 1107–8
uniform acceleration, 101–2
due to gravity, 114
uniform circular motion
acceleration dem., 337
relation to simple harmonic motion, 334
uniform motion, 99
and vector addition, 97

unipolar generator, 911
unipolar motor, 911

vacuum system, movable, dem. using, 395–96, ***610–13***
vacuum tube
  characteristics vs holographic film characteristics, 1156–57
  flow of electrons in, model, 992, ***1376, 1378***
Van de Graaff generator, toy, improvement of, 852–53
van der Waals' equation, 790
van der Waals' gas, 790–91
vapor, difference in behavior under compression, of a gas and a, 823–24
vapor pressure, 791–96
  * *using air and ether, 792*
  apparatus for meas. of, 782
  * *boiling under reduced pressure, 792–93*
  * *in a bottle of soda pop, 791–92*
  * *curve for water, 793–94*
  * *Dalton's law, 794–96*
  * *dependence of, on temperature, 793*
  * *of two dissimilar liquids, 792*
  * *effect of pressure on the boiling point of water, 792*
  effect of temperature on, 792, 793
  * *water fountain due to reduced, 792*
vector
  addition
    simultaneous velocities and, 97
    uniform motion and, 97
    of velocities, 92–93, 98
  curl of a, film for dem. of, 671
  diagrams, of a series RC circuit, 890–91
  displacement, 95
  electric, 93–94
  force, 97–98
  magnets, 93–94
  models, 92–98
  presentation, gun for, 93–94
vectors, 92–98
  forces of, 97–98
  resolution of, 94
  resultant of, 94–95
  rotating, 94–95, ***536***, 729–30
  transformation of, to a rotating frame, 293–94
  translation of, 95
velocity
  addition of, dem., 92–93, 98
  angular
    addition of, 272–73
    damping force proportional to, 407–8
    and impulse, 297
    meas. of, 282–83
    of a torsional pendulum, 355–56
  * *of a bullet, using an air trough, 258*
  of a charge carrier in a crystal, 1226–28
velocity (*cont.*)
  correlation with displacement and acceleration, 106
  damping force proportional to, 138–39
  drift, in an electric field, 38, 39
  of an elastic wave, 448–49
    in a diatomic lattice, 263
    longitudinal, 440
    in a monatomic lattice, 262
    transverse, 450–54, 475
  escape, dem. of, 166
  exhaust, calculation of, 187–88
    direction of, 188–89
  field, for a fluid, 425–26
  flow, effect on flow characteristics, 401, 410–14
  free fall, meas. of, 116, 117
  group, 241, 478–82, 485–88, 1183–85
    theory of, ***575–77***
  independence of horizontal and vertical components in projectile motion, 126, 127, 130
  instantaneous, 119–20
  of light
    corridor dem., 711–12
    discussion of meas. attempts, 1066, 1068–69
    Leybold apparatus, techniques for finding motor speed, 1065–66
    in a moving frame, visual representation of, 1168, ***1353–57***
    rotating mirror apparatus, 1064–65, ***1353***
    time-distance measurements, 1066–75, ***1353***
  Maxwell–Boltzmann distribution, 244–46, 810–11
  * *meas. of on an air track, 254–55*
  meas. of by ballistic meter, 31–32
  most probable, 811
  phase, 37, 240–41, 478–81, 1183
    theory, ***575–77***
  of propagation for a signal in a coaxial cable, 1010–12
  relative, 131–33, 188–89
  of a rocket, equation for, 189
  of sound, 491
    temperature-dependence of, 493
  tangential, 120
  terminal, 138–39, 408, 409
  of a vortex ring, factors influencing, 422
  of water waves, 468–69, 1164–65
    equations for, 468
* *velocity distribution and path length, 244–46*
Venturi tube, 398
vernier
  caliper, for demonstrations, 85
  large scale model, 85
  model for overhead projector, 85
  model for slide projector, 85
vertical circle apparatus, 289, 589–90
* *vertical puck gun, 226–27*
vibration tester, ***1293***
vibrations
  acoustic, 262
  flexural, 439, 473–74
  forced, 257
    and resonance, corridor dem., 703–4
    of a torsion pendulum, 269
  longitudinal, 439, 440, 492–93, 981–82
  modes of
    in a diatomic lattice, 262
    in a monatomic lattice, 262
    in a solid, 439–40
  optic, 262
  * *stress produced by transverse vibrations, 474*
  sympathetic, 365, 456–57, 500, 1257–58
  torsional, 439
  ultrasonic, of quartz, 528–29
vibrator, hair clipper as a, 476
videcon tube, 657
virtual image, location of by parallax, 1043–44
virtual object
  distinction between a real object and, 1045
  location of by parallax, 1044–45
viscosity, 407–9
  of a gas
    coefficient of, ***1292***
    * *effect of density of, 796–97,* ***1290–92***
    * *effect of temperature on, 796*
    theory of, ***1291–92***
  * *of a liquid, effect of temperature on, 408*
viscous flow, Newton's law of, 407
viscous resistance, 407–8
visual aids in the physics lecture, 719–20
  as an aid to, not the center of, a lecture, 721
  relation between physical dem. and a model, 722–23
Volta, Alessandro, 28
voltage, high, production of short-lived, by electric switching of capacitors, 871
voltage drop
  across a gas tube, Paschen's law, 894
  due to the Hall effect, 1236, 1237
  meas. of, in a series RC circuit, 890–91
  across resistor combinations, fluid analog, 885
voltmeter, electrostatic
  principle of, 868, ***1320–21***
  projection, 864
volume, Boyle's law for pressure and, 785–90
vortex
  formation of, by flow past a body, 404

vortex (*cont.*)
ring, characteristics of, 424
ring, in a liquid, 422–23
semicircular, in water, 423–24
shape of water surface around a, 424–25
* *smoke rings, 424–25*
theorems of Helmholtz, 424

water
boiling point of, effect of pressure on, 792
compressibility of, 380–82
heat of fusion of, 770
near-incompressibility of, 380
velocity of waves in, 468–69, 1164–65
water drop charge generator, 847–49, ***1311–12***
Watson, Fletcher, 63
Watson, William, 28
wave generation
using an air pulse generator, 467–68, ***630***
with a ripple tank, 466–73, ***626–29***
with a wave machine, 443–48, ***622–23***
wave machine, 443, ***622–23***
speed of waves in, 443–44
wave mechanics, 1183–92
* *barrier penetration, optical analogy, 1190, 1191*
* *classical analogs to radical wave functions, 1185–87,* ***1363–64***
* *construction of a wave packet, 1183*
frustrated total internal reflection
* *with light, 1191–92*
* *with microwaves, 1188–90*
* *properties of a wave packet, 1184–85*
standing circular waves
* *in a gas-filled totoid, 1187–88,* ***1363–64***
* *in a wire, 1188*
wave models, 476–88
* *addition of waves, 478–79,* **635**
for dem. of interference and beats, 478
* *difference between phase and group velocity, 480–81*
* *for group velocity, 481–82*
* *for group velocity: densimeter, 485–86*
* *to illustrate basic properties of waves, 486–88,* ***639–47***
* *illustration of phase and group velocities, 479–80,* ***635–36***
* *for motion of travelling waves, 483*
* *for a particle on a sea wave, 488,* ***644–45, 648***
* *simulation of periodic properties of waves, 476*
slinky, for wave dem., 476
steel chain, for wave dem., 476
* *superposition of sine waves, 477*

wave models (*cont.*)
* *for theory of wave motion, 482–83,* ***636–39***
wave motion
apparatus for the dem. of, 476–77
on an overhead projector, 477
longitudinal, 242, 260–61, 483
one-dimensional, 239–42, **575–80**
of packets, 486–88, ***639–47,*** 1183, 1184–85
* *slow pulse in a bead chain, 452–53*
* *study of using a ripple tank, 467–73,* ***629–35***
* *Bragg reflection, 473,* ***634–35***
* *interference and diffraction, 470–72*
* *observation of wave motion, 467–68,* ***629–33***
* *qualitative study of wave properties, 468–70,* ***633–34***
* *twin-source interference in a moving medium, 1165–66*
* *wave propagation speed, upstream and downstream, 1164–65*
theory of, apparatus for dem., 482–83, ***636–39***
transverse, 482
of travelling waves, 483
* *velocity of propagation and amplification of travelling waves, 449–50,* ***623–25***
* *velocity of a transverse wave, 450–52*
* *wave motion apparatus, air trough, 260–61*
wave number, relation of to frequency
in a diatomic lattice, 262–63
in a linear mass chain, 239–42, **575–80**
in a monatomic lattice, 261–62
theory, linear mass chain, **575–77**
wave packets, motion of, 486–88, ***639–47,*** 1183
wave speed
dependence of on frequency, 453–54
* *of a disturbance in a rope, 454–55*
* *of a pulse in a long spring, 453*
* *of a pulse in a long wire, 453–54*
* *of a pulse in a spinning lariat, 452*
in water, upstream and downstream, 1164–65
of a water wave, 469–70
waveforms
addition of, of different frequencies, 995–96
* *harmonic composition of complex, 1000–1,* ***1337***
* *production of, of any shape, 996*
wavelength, 37, 240
Compton shift, 1174
de Broglie, of electrons, 1194
calculation of, 1196–97
dominant, 1120–21
of elastic waves
in a diatomic lattice, 262

wavelength (*cont.*)
of elastic waves (*cont.*)
in a monatomic lattice, 261
of light, meas. of with a ruler, 1138–39
of microwaves, meas. of, 1018, 1019–20
thermal radiation intensity, as a function of, 1179
waves
addition of, 477, 478–79, 486–88, ***635,*** 996
capillary, 429
continuous, 442
elastic
in a diatomic lattice, 262–63
in a monatomic lattice, 261–62
in a solid, 439–40
velocity of, 261–63, 440, 448–54
extraordinary, 1107–8
group velocity, 241, 478–82, 485–88, 1183–85
heat, propagation of
analogy to, of signal propagating in a transmission line, 1012
in copper, 763–64
in a long thin rod, 765–66, ***1287***
interference, 445–47, 455–59
light, 1076–87
microwaves, 1014–17, 1025–26
sound, 508–14
water, 470–72
light
diffraction of, 1087–95
interference of, 1076–87
radiation pressure of, 37–38
speed of, 1064–75
longitudinal, 242, 260–61, 441, 482–83
standing, 493–96, 503–5
velocity of, 491–93
microwaves
diffraction of, 1020–22, 1026–29
interference of, 1014–17, 1025–26
refraction of, 1020–25
ordinary, 1107–8
phase velocity, 36–37, 240, 478–79, 479–80, 1183
reflection of, 444–45, 448, 470, 1035–38
refraction of, 470, 524, 1040–50, 1056–59
shock, 426–30, 734
formation of by an object moving at a supersonic velocity, 428–30
sound, 491ff
amplification and modulation of, 524–28
diffraction of, 519–20
interference of, 508–14
radiation pressure of, 37–38
velocity of, 491–93
standing, 446–48, 460–66
compressional, 440

waves (*cont.*)
standing (*cont.*)
electron, analog, 1185–87, ***1363–64***
longitudinal, 494–96, 503–5
microwaves, 1019–20
of small magnetized balls on water, 930–31
in a solid, 439–40
in a stretched circular membrane, 505–6
in water, 470
superposition of, 445, 459, 486–88
in a stringed instrument, 465–66
transmission of, at a discontinuity, 476–77
in a transmission line, 1009–10
transport of energy by, 443
transverse, 36, 239–42, 450–52, 475–76, 482–83
behavior of at an interface, 449–50
travelling, 36–37
amplification of, 450, ***623–25***
cut off frequency for, 240–41, ***575–77***
properties of, 443–48, ***622–23***
ultrasonic, 523–24, 1094–95

waves (*cont.*)
water
diffraction of, 470
interference of, 470–72
motion of a particle in a, 482–83, ***636–39***, 488, ***644–45***, ***648***
reflection of, 470
refraction of, 470
velocity of, 468–69
Weber, William, 28
wedge apparatus, 91
weight
apparent change of, due to acceleration, 147–48, 149
and mass, 140–43
Weiss domains, 966–67, 974
"whistling rocket," 510
wind tunnel, 398
work, unorganized, 769
work and heat, 766–70
"world's smallest watch," a corridor dem., 90

x-rays, 56
diffraction of
* *Laue patterns, 1192*
* *x-ray diffraction, 1193–94*
* *zone axis and the Laue pattern ellipse, 1192–93*

x-rays (*cont.*)
* *scattering by paraffin, 1250*
x-y projector plotter, 110–13
adaptation of the Heath servo-recorder, 111
adaptation of the Houston *x-y* recorder, 112–13
construction of, 110–11, ***537–43***

Young, Thomas, 15
Young's modulus, 439–40
for bending modes, 440, 441
for stretching modes, 440
variation with the intensity of magnetization, 984–86

Zeeman effect, 1208–10
* *shift of resonance lines in mercury, 1210*
shift of resonance lines in sodium
* *through the flame from a borax bead, 1209*
* *through a sodium chloride saturated vapor, 1208–9*
zero-momentum systems, 215–16, 235
zone axis of a crystal, 1192–93
zoom lens, advantages of, 658